METHODS IN ENZYMOLOGY
VOLUME VI

METHODS IN ENZYMOLOGY

Edited by

SIDNEY P. COLOWICK   and   **NATHAN O. KAPLAN**
Vanderbilt University School of Medicine   *Brandeis University*
Nashville, Tennessee   *Waltham, Massachusetts*

Volume VI

1963

ACADEMIC PRESS   New York and London

ACADEMIC PRESS INC.
111 Fifth Avenue, New York, New York 10003

United Kingdom Edition published by
ACADEMIC PRESS INC. (LONDON) LTD.
Berkeley Square House, London W.1

LIBRARY OF CONGRESS CATALOG CARD NUMBER: *54-9110*

First Printing, 1963
Second Printing, 1966

PRINTED IN THE UNITED STATES OF AMERICA

Preface

The present volume completes the planned two-volume supplement to the original four-volume treatise on Methods in Enzymology. A cumulative index to all six volumes is in preparation and will be published shortly as a separate volume.

The Editors regret the long delay in publication of Volume VI, which was earlier planned for publication in the summer of 1962. In order to avoid similar delay in future volumes, we plan tentatively that future supplements will be smaller volumes dealing with specific areas of research. This will not only insure less obsolescence prior to publication, but will also simplify the organization of future volumes.

Again it is our pleasure to thank the authors who have contributed so generously to this treatise. We are especially indebted to those who were at least willing, if not happy, to make drastic revisions in proof to avoid obsolescence of their articles.

August, 1963

SIDNEY P. COLOWICK
NATHAN O. KAPLAN

Contributors to Volume VI

Article numbers are shown in parentheses following the names of contributors. Affiliations listed are current.

ROBERT A. ALBERTY (124), *University of Wisconsin, Madison, Wisconsin*

DANIEL I. ARNON (37), *University of California, Berkeley, California*

CLINTON E. BALLOU (70), *University of California, Berkeley, California*

ROBERT G. BARTSCH (55), *University of California, San Diego, La Jolla, California*

CARLOS BASILIO (1, 98), *New York University School of Medicine, New York, New York*

HELMUT BEINERT (59), *Institute for Enzyme Research, University of Wisconsin, Madison, Wisconsin*

A. A. BENSON (122), *University of California, Los Angeles, California*

MAURICE J. BESSMAN (19, 20, 99), *Mc-Collum-Pratt Institute, Johns Hopkins University, Baltimore, Maryland*

SIMON BLACK (91), *Public Health Service, National Institutes of Health, Bethesda, Maryland*

ARNOLD F. BRODIE (35, 36), *Harvard Medical School, Boston, Massachusetts*

THOMAS C. BRUICE (88), *Cornell University, Ithaca, New York*

PAOLO BRUNETTI (68), *University of Michigan Hospital, Ann Arbor, Michigan*

CLARK BUBLITZ (42), *University of Colorado School of Medicine, Denver, Colorado*

ENRICO CABIB (107), *Instituto de Investigaciones Bioquímicas, Fundacion Campomar, Buenos Aires, Argentina*

E. S. CANELLAKIS (4), *Yale University School of Medicine, New Haven, Connecticut*

CHARLES E. CARTER (110), *Yale University School of Medicine, New Haven, Connecticut*

SEYMOUR S. COHEN (16), *The School of Medicine, University of Pennsylvania, Philadelphia, Pennsylvania*

MINOR J. COON (80), *University of Michigan Medical School, Ann Arbor, Michigan*

L. A. COSTELLO (127), *Brandeis University, Waltham, Massachusetts*

HENK DE KLERK (55), *University of California, San Diego, La Jolla, California*

CHESTER DE LUCA (43), *Sinai Hospital, Baltimore, Maryland*

LUTHER E. ERICKSON (124), *Grinnell College, Grinnell, Iowa*

GERALD D. FASMAN (126), *Brandeis University, Waltham, Massachusetts*

D. S. FEINGOLD (108), *University of Pittsburgh, Pittsburgh, Pennsylvania*

I. H. FINE (127), *Brandeis University, Waltham, Massachusetts*

JOEL G. FLAKS (9, 17, 18, 69, 96), *School of Medicine, University of Pennsylvania, Philadelphia, Pennsylvania*

MARTIN FLAVIN (77), *Department of Health, Education, and Welfare, National Institutes of Health, Bethesda, Maryland*

MORRIS FRIEDKIN (15), *Tufts University School of Medicine, Boston, Massachusetts*

HERBERT C. FRIEDMANN (22), *The University of Chicago, Chicago, Illinois*

SIDNEY FUTTERMAN (112), *Massachusetts Eye and Ear Infirmary, Boston, Massachusetts*

PAUL M. GALLOP (93), *Albert Einstein College of Medicine, Yeshiva University, New York, New York*

DAVID E. GREEN (58), *Institute for Enzyme Research, University of Wisconsin, Madison, Wisconsin*

D. M. GREENBERG (53, 54), *University of California School of Medicine, San Francisco, California*

SANTIAGO GRISOLIA (71), *University of Kansas Medical Center, Kansas City, Kansas*

PHILIP HANDLER (44), *Duke University School of Medicine, Durham, North Carolina*

ISAAC HARARY (40), *University of California Medical Center, Los Angeles, California*

E. HASLAM (74), *University of California, Davis, California*

W. Z. HASSID (108), *University of California, Berkeley, California*

R. D. HAWORTH (74), *University of California, Davis, California*

OSAMU HAYAISHI (47), *Kyoto University Faculty of Medicine, Kyoto, Japan*

J. W. HEALY (8), *Brandeis University, Waltham, Massachusetts*

LEON A. HEPPEL (14), *National Institute of Arthritis and Metabolic Diseases, National Institutes of Health, Bethesda, Maryland*

EDWARD HERBERT (4), *Massachusetts Institute of Technology, Cambridge, Massachusetts*

P. P. K. HO (114), *Scripps Clinic and Research Foundation, La Jolla, California*

TAKEKAZU HORIO (55), *Institute for Protein Research, Osaka, University, Osaka, Japan*

F. M. HUENNEKENS (48, 49, 50, 113, 114), *Scripps Clinic and Research Foundation, La Jolla, California*

JERARD HURWITZ (3a), *New York University College of Medicine, New York, New York*

JOHN IMSANDE (44), *Western Reserve University, Cleveland, Ohio*

VERNON M. INGRAM (118), *Massachusetts Institute of Technology, Cambridge, Massachusetts*

ANDRÉ T. JAGENDORF (60), *McCollum-Pratt Institute, Johns Hopkins University, Baltimore, Maryland*

WILLIAM P. JENCKS (104, 125), *Brandeis University, Waltham, Massachusetts*

MARY ELLEN JONES (81), *Brandeis University, Waltham, Massachusetts*

JOHN JOSSE (101), *McCollum-Pratt Institute, Johns Hopkins University, Baltimore, Maryland*

ELLIOT JUNI (72), *Emory University, Atlanta, Georgia*

MARTIN D. KAMEN (38, 55), *University of California, San Diego, La Jolla, California*

DONALD L. KEISTER (61), *McCollum-Pratt Institute, Johns Hopkins University, Baltimore, Maryland*

EUGENE P. KENNEDY (109), *Harvard Medical School, Boston, Massachusetts*

H. GOBIND KHORANA (94), *Institute for Enzyme Research, University of Wisconsin, Madison, Wisconsin*

W. WAYNE KIELLEY (33), *National Heart Institute, Bethesda, Maryland*

PAUL F. KNOWLES (74), *University of California, Davis, California*

S. R. KORNBERG (31), *Stanford University School of Medicine, Palo Alto, California*

JOSEPH S. KRAKOW (1, 2), *Yale University, New Haven, Connecticut*

EDWIN G. KREBS (45), *University of Washington, School of Medicine, Seattle, Washington*

STEPHEN A. KUBY (27), *Institute for Enzyme Research, University of Wisconsin, Madison, Wisconsin*

R. A. LANDOWNE (76), *Yale University School of Medicine, New Haven, Connecticut*

J. LARNER (46), *Western Reserve University School of Medicine, Cleveland, Ohio*

WILLIAM LEE (59), *Institute for Enzyme Research, University of Wisconsin, Madison, Wisconsin*

YA PIN LEE (12), *Washington University, School of Medicine, St. Louis, Missouri*

I. R. LEHMAN (5, 6, 7), *Stanford University School of Medicine, Stanford, California*

ALBERT L. LEHNINGER (32, 42), *Johns Hopkins University, School of Medicine, Baltimore, Maryland*

LUIS F. LELOIR (107), *Instituto de Investigaciones Bioquimicas, Fundacion Campomar, Buenos Aires, Argentina*

LAWRENCE LEVINE (8, 119), *Brandeis University, Waltham, Massachusetts*

IRVING LIEBERMAN (11), *University of Pittsburgh School of Medicine, Pittsburgh, Pennsylvania*

S. R. LIPSKY (76), *Yale University School of Medicine, New Haven, Connecticut*

JOHN M. LOWENSTEIN (95, 121), *Brandeis University, Waltham, Massachusetts*

OLIVER H. LOWRY (111), *Washington University School of Medicine, St. Louis, Missouri*

LEWIS N. LUKENS (9, 96), *Yale University, New Haven, Connecticut*

WILLIAM D. MCELROY (23, 63, 106), *McCollum-Pratt Institute, Johns Hopkins University, Baltimore, Maryland*

BORIS MAGASANIK (13), *Massachusetts Institute of Technology, Cambridge, Massachusetts*

J. MARMUR (100), *Albert Einstein College of Medicine, Yeshiva University, Bronx, New York*

C. K. MATHEWS (48, 113), *Scripps Clinic and Research Foundation, La Jolla, California*

ALTON MEISTER (102), *Tufts University School of Medicine, Boston, Massachusetts*

KIVIE MOLDAVE (103), *Tufts University School of Medicine, Boston, Massachusetts*

STANFORD MOORE (117), *The Rockefeller Institute, New York, New York*

HARRIS S. MOYED (84), *Harvard Medical School, Boston, Massachusetts*

ALVIN NASON (57), *McCollum-Pratt Institute, Johns Hopkins University, Baltimore, Maryland*

E. F. NEUFELD (108), *University of California, Berkeley, California*

MARSHALL W. NIRENBERG (3), *National Institute of Arthritis and Metabolic Diseases, Bethesda, Maryland*

LAFAYETTE NODA (27), *Dartmouth College, Hanover, New Hampshire*

SEVERO OCHOA (1, 2, 98), *New York University School of Medicine, New York, New York*

JANET V. PASSONNEAU (111), *Washington University, St. Louis, Missouri*

HARVEY S. PENEFSKY (34), *The Public Health Institute of the City of New York, New York, New York*

LEWIS L. PIZER (16), *University of Pennsylvania, Philadelphia, Pennsylvania*

G. W. E. PLAUT (28, 39), *College of Medicine, University of Utah, Salt Lake City, Utah*

BURTON M. POGELL (41), *Vanderbilt University School of Medicine, Nashville, Tennessee*

JACK PREISS (44), *National Institute of Arthritis and Metabolic Diseases, National Institutes of Health, Bethesda, Maryland*

W. E. PRICER, JR. (51), *National Institute of Arthritis and Metabolic Diseases, Institutes of Health, Bethesda, Maryland*

MAYNARD E. PULLMAN (34), *The Public Health Institute of the City of New York, New York, New York*

JESSE C. RABINOWITZ (10, 51, 52, 97, 115, 116), *University of California, Berkeley, California*

MURRAY RABINOWITZ (26), *The University of Chicago, Chicago, Illinois*

EFRAIM RACKER (64), *The Public Health Research Institute, Inc., New York, New York*

W. E. RAZZELL (29), *Syntex Institute for Molecular Biology, Palo Alto, California*

PETER REICHARD (21), *Kemiska Institutionem, Karolinska Institutet, Stockholm, Sweden*

HELEN R. REVEL (24), *The Public Health Research Institute, Inc., New York, New York*

PHILLIPS W. ROBBINS (105), *Massachusetts Institute of Technology, Cambridge, Massachusetts*

EUGENE ROBERTS (89), *City of Hope Medical Center, Medical Research Institute, Duarte, California*

WILLIAM G. ROBINSON (80), *University of Michigan Medical School, Ann Arbor, Michigan*

SAUL ROSEMAN (68), *University of Michigan Hospital, Ann Arbor, Michigan*

SANFORD M. ROSENTHAL (90), *National Institutes of Health, Bethesda, Maryland*

JOEL A. ROTHSCHILD (73), *The Rockefeller Institute, New York, New York*

ANTHONY SAN PIETRO (61, 62), *The Charles F. Kettering Research Laboratory, Yellow Springs, Ohio*

MICHAEL SCHRAMM (25), *The Hebrew University Hadassah Medical School, Jerusalem, Israel*

K. G. SCRIMGEOUR (48, 49, 50, 113, 114), *Scripps Clinic and Research Foundation, La Jolla, California*

SAM SEIFTER (93), *Albert Einstein College*

of Medicine, Yeshiva University, Bronx, New York

ARNOLD M. SELIGMAN (123), Sinai Hospital, Baltimore, Maryland

OLA SKÖLD (21), Karolinska Institutet, Stockholm, Sweden

MICHAEL SMITH (94), Institute for Enzyme Research, University of Wisconsin, Madison, Wisconsin

OLIVER H. SMITH (86), Stanford University, Palo Alto, California

LEONARD SPECTOR (81), The Rockefeller Institute, New York, New York

M. SPRECHER (73), Columbia University College of Physicians and Surgeons, New York, New York

D. B. SPRINSON (73), Columbia University College of Physicians and Surgeons, New York, New York

WILLIAM H. STEIN (117), The Rockefeller Institute, New York, New York

JAKOB A. STEKOL (83), Institute for Cancer Research, Philadelphia, Pennsylvania

DAVID STOLLAR (8, 119), Biological Sciences Division, AIR FORCE OFFICE OF SCIENTIFIC RESEARCH, United States Air Force, Washington, D. C.

LARS SVENNERHOLM (65, 66), Göteborgs Universitet, Medicinsk-Kemiska Institutionen, Göteborg, Sweden

ANN SWANSON (68), Bowman Gray School of Medicine, Winston-Salem, North Carolina

MORTON SWARTZ (101), Massachusetts General Hospital, Boston, Massachusetts

CELIA WHITE TABOR (90), Public Health Service, National Institutes of Health, Bethesda, Maryland

HERBERT TABOR (85), Public Health Service, National Institutes of Health, Bethesda, Maryland

HARRIS H. TALLAN (82), Geigy Research Laboratories, Ardsley, New York

T. T. TCHEN (75), Wayne State University, Detroit, Michigan

SIDNEY UDENFRIEND (87), National Heart Institute, National Institutes of Health, Bethesda, Maryland

KOSAKU UYEDA (52), University of California, Berkeley, California

P. ROY VAGELOS (79), National Institutes of Health, Bethesda, Maryland

FRANK D. VASINGTON (57), Johns Hopkins University School of Medicine, Baltimore, Maryland

BIRGIT VENNESLAND (56), University of Chicago, Chicago, Illinois

C. VILLAR-PALASI (46), Western Reserve University School of Medicine, Cleveland, Ohio

JEROME VINOGRAD (120), California Institute of Technology, Pasadena, California

CHARLES L. WADKINS (32), Johns Hopkins School of Medicine, Baltimore, Maryland

SALIH WAKIL (78), Duke University Medical Center, Durham, North Carolina

LEONARD WARREN (67), National Institutes of Health, Bethesda, Maryland

HERBERT WEISSBACH (87), National Heart Institute, National Institutes of Health, Bethesda, Maryland

F. R. WHATLEY (37), University of California, Berkeley, California

ELIZABETH WORK (92), Twyford Laboratories, London, England

CHARLES YANOFSKY (86), Stanford University, Stanford, California

D. M. ZIEGLER (58), University of Texas, Austin, Texas

STEVEN B. ZIMMERMAN (30), Stanford University Medical Center Palo Alto, California

Outline of Supplementary Volumes

VOLUME V

PREPARATION AND ASSAY OF ENZYMES

VOLUME VI

PREPARATION AND ASSAY OF ENZYMES (*Continued*)

PREPARATION AND ASSAY OF SUBSTRATES

SPECIAL TECHNIQUES

VOLUME VI

Table of Contents

PREPARATION AND ASSAY OF ENZYMES (*Continued*)

Section I. Enzymes of Nucleic Acid Metabolism

Section II. Enzymes of Phosphate Metabolism

Section III. Enzymes of Coenzyme and Vitamin Metabolism

Section IV. Respiratory Enzymes

PREPARATION AND ASSAY OF SUBSTRATES

Section I. Carbohydrates

Section II. Lipids and Steroids

Section III. Proteins and Derivatives

Section IV. Nucleic Acids, Coenzymes, and Derivatives

SPECIAL TECHNIQUES

Outline of Volumes I, II, III, IV, and V

VOLUME I
PREPARATION AND ASSAY OF ENZYMES

VOLUME II
PREPARATION AND ASSAY OF ENZYMES

VOLUME III

PREPARATION AND ASSAY OF SUBSTRATES

Section I. Carbohydrates

A. Polysaccharides. **B.** Monosaccharides. **C.** Sugar Phosphates and Related Compounds. **D.** Unphosphorylated Intermediates and Products of Fermentation and Respiration.

Section II. Lipids and Steroids

A. Isolation and Determination of Lipids and Higher Fatty Acids. **B.** Preparation and Analysis of Phospholipids and Derivatives. **C.** Fractionation Procedures for Higher and Lower Fatty Acids. **D.** Preparation and Assay of Cholesterol and Ergosterol.

Section III. Citric Acid Cycle Components

A. Chromatographic Analyses of Organic Acids. **B.** Specific Procedures for Individual Compounds.

Section IV. Proteins and Derivatives

A. General Procedures for Determination of Proteins and Amino Acids. **B.** General Procedures for Preparation of Peptides and Amino Acids. **C.** Specific Procedures for Isolation and Determination of Individual Amino Acids.

Section V. Nucleic Acids and Derivatives

A. Determination, Isolation, and Characterization of Nucleic Acids. **B.** Determination, Isolation, Characterization, and Synthesis of Nucleotides and Nucleosides.

Section VI. Coenzymes and Related Phosphate Compounds

A. General Procedures for Isolation, Determination, and Characterization of Phosphorus Compounds. **B.** Specific Procedures for N-Phosphates and Individual Coenzymes.

Section VII. Determination of Inorganic Compounds

VOLUME IV

SPECIAL TECHNIQUES FOR THE ENZYMOLOGIST

Section I. Techniques for Characterization of Proteins (Procedures and Interpretations)

A. Electrophoresis; Macro and Micro. **B.** Ultracentrifugation and Related Techniques (Diffusion, Viscosity) for Molecular Size and Shape. **C.** Infra-red Spectrophotometry. **D.** X-ray Diffraction. **E.** Light Scattering Measurements. **F.** Flow Birefringence. **G.** Fluorescence Polarization and Other Fluorescence Techniques. **H.** The Solubility Method for Protein Purity. **I.** Determination of Amino Acid Sequence in Proteins. **J.** Determination of Essential Groups for Enzyme Activity.

Section II. Techniques for Metabolic Studies

A. Measurement of Rapid Reaction Rates; Techniques and Applications, Including Determination of Spectra of Cytochromes and Other Electron Carriers in Respiring Cells. **B.** Use of Artificial Electron Acceptors in the Study of Dehydrogenases. **C.** Use of Percolation Technique for the Study of the Metabolism of Soil Microorganisms. **D.** Methods for Study of the Hill Reaction. **E.** Methods for Measurement of Nitrogen Fixation. **F.** Cytochemistry.

Section III. Techniques for Isotope Studies

A. The Measurement of Isotopes. **B.** The Synthesis and Degradation of Labeled Compounds (Including Application to Metabolic Studies): Monosaccharides and Polysaccharides; Citric Acid Cycle Intermediates; Glycolic, Glyoxylic and Oxalic Acids; Purines and Pyrimidines; Porphyrins; Amino Acids and Proteins; Steroids; Methylated Compounds and Derivatives; Sulfur Compounds; Fatty Acids; Phospholipids; Coenzymes; Iodinated Compounds; Intermediates of Photosynthesis; O^{18}-Labeled Phosphorus Compounds.

VOLUME V

PREPARATION AND ASSAY OF ENZYMES

Section I. General Preparative Procedures

A. Column Chromatography of Proteins. **B.** Preparative Electrophoresis. **C.** Preparation and Solubilization of Particles (Bacterial, Mammalian, and Higher plant). **D.** Mammalian Cell Culture. **E.** Protoplasts.

Section II. Enzymes of Carbohydrate Metabolism

A. Polysaccharide Cleavage and Synthesis. **B.** Disaccharide, Hexoside and Glucuronide Metabolism. **C.** Metabolism of Hexoses, Pentoses and 3-Carbon Compounds. **D.** Hexosamine and Sialic Acid Metabolism. **E.** Aromatic Ring Synthesis.

Section III. Enzymes of Lipid Metabolism

A. Fatty Acid Synthesis and Breakdown. **B.** Acid Activating Enzymes. **C.** Phospholipid Synthesis and Breakdown. **D.** Steroid Metabolism.

Section IV. Enzymes of Citric Acid Cycle

A. Krebs Cycle. **B.** Krebs-Kornberg Cycle. **C.** Related Enzymes.

Section V. Enzymes of Protein Metabolism

A. Proteolytic Enzymes. **B.** Amino Acid Dehydrogenases and Transaminases. **C.** Amino Acid Activating Enzymes. **D.** Other Enzymes of Amino Acid Breakdown and Synthesis. **E.** Enzymes of Sulfur Metabolism.

PREPARATION AND ASSAY OF ENZYMES
(*Continued*)

Section I

Enzymes of Nucleic Acid Metabolism

[1] Polynucleotide Phosphorylase from *Azotobacter vinelandii*

$$n\text{XDP} \rightleftarrows (\text{XMP})_n + n\text{P}_i$$

By Severo Ochoa, J. S. Krakow, and Carlos Basilio

Assay Methods

Principle. Two methods are used for the assay of polynucleotide phosphorylase. In one, the rate of exchange of P_i^{32} with ADP is measured;[1,2] in the other, the rate of formation of ADP resulting from the phosphorolysis of polyadenylic acid is determined spectrophotometrically by coupling with the reactions catalyzed by pyruvic kinase and lactic dehydrogenase.[2]

P_i^{32} Exchange Assay

Procedure. The sample contains: K_2HPO_4, 3.5 micromoles; ADP, 2.5 micromoles, enzyme (not over 0.35 unit), and 0.3 ml. of a mixture of Tris–HCl buffer, pH 8.1, 100 micromoles; $MgCl_2$, 2 micromoles; EDTA, 2 micromoles; and $KH_2P^{32}O_4$ with 1 to 2×10^5 c.p.m. in a volume of 1.0 ml. Enzyme dilutions are made in $0.1\,M$ Tris–HCl buffer, pH 7.4. A blank, lacking either ADP or enzyme, is always run with the assay samples. After 15 minutes of incubation at 30° the reaction is stopped by addition of 0.1 ml. of 40% trichloroacetic acid, and the mixture is centrifuged to remove the precipitated protein. Then 1.0 ml. of the supernatant is brought to 3.0 ml. with water, and 0.3 ml. of $10.0\,M$ H_2SO_4, 1.5 ml. of 5% ammonium molybdate, and 5.0 ml. of isobutanol are added. A slow stream of air is bubbled through the mixture to obtain good mixing. After the liquid phases are allowed to separate by standing, the upper isobutanol layer, containing the radioactive orthophosphate as ammonium phosphomolybdate, is removed by aspiration and discarded. If the aqueous phase is not clear, a second extraction with isobutanol is carried out as before. The aqueous phase is then washed with 3 ml. of ether; the ether is removed by aspiration and discarded. Then 1.0 ml. of the aqueous phase is pipetted into a stainless-steel planchet, the material dried, and the radioactivity determined. The radioactivity in the blank is subtracted from that of the experimental values. The final value for

[1] M. Grunberg-Manago, P. J. Ortiz, and S. Ochoa, *Biochim. et Biophys. Acta* **20**, 269 (1956).
[2] S. Ochoa and S. Mii, *J. Biol. Chem.* **236**, 3303 (1961).

the radioactivity is used to calculate the amount of orthophosphate incorporated from the expression:[1]

Micromoles phosphate incorporated =
$$\frac{\text{c.p.m. incorporated (micromoles } P_i + \text{micromoles ADP)}}{\text{c.p.m. } P_i}$$

Definition of Unit and Specific Activity. One unit is defined as the amount of enzyme catalyzing the exchange of 1.0 micromole of $P_i{}^{32}$ in 15 minutes at 30° and pH 8.1. Specific activity is expressed as units per milligram of protein.

Optical Assay

Procedure. The reaction mixture contains in 1.0 ml. the following components (in micromoles per milliliter): potassium phosphate, pH 7.4, 10; glycylglycine buffer, pH 7.4, 5; $MgCl_2$, 5; EDTA, 1; crystalline egg albumin, 2 mg.; phosphoenolpyruvate, 1.6; an excess of crystalline pyruvic kinase and lactic dehydrogenase (each in an amount sufficient to give a rate of DPNH oxidation of 0.3 micromole/ml./min. at 30° under the conditions of their respective optical assays[3,4]; DPNH,[5] about 0.13 (initial absorbancy at 340 mμ, about 0.8); poly A (aqueous solution), 0.18 (calculated as AMP); and up to 0.3 unit of enzyme. The mixture, lacking poly A, is equilibrated at 30° for 1 to 2 minutes, after which the absorbancy at 340 mμ is determined. (This is done by means of a Beckman spectrophotometer having a cell compartment fitted with a jacket, through which water is circulated at a temperature of 30°.) If the absorbancy remains constant and no turbidity develops, the enzyme is added, and readings are taken at 30-second intervals for 3 to 5 minutes, during which time the oxidation of DPNH proceeds at an approximately constant rate. This is proportional to the concentration of enzyme up to about 0.3 unit/ml.

Definition of Unit and Specific Activity. One mole of ADP is produced per mole of DPNH oxidized, and, for correlation with the exchange assay unit, one optical unit is taken as the amount of enzyme catalyzing the formation of 1 micromole of ADP in 15 minutes at 30° and pH 7.4. As determined experimentally, 1 optical unit is equivalent to approximately 1.2 $P_i{}^{32}$ exchange units. Specific activity is expressed as units per milligram of protein.

---

[3] A. Tietz and S. Ochoa, *Arch. Biochem. Biophys.* **78**, 477 (1958).
[4] See Vol. I [67].
[5] Fresh DPNH solutions are prepared every few days, adjusted to pH 10 with NaOH, and kept frozen when not in use.

Purification Procedure

Through stage 4 of purification (see the table) protein is determined by the biuret method[6] or by the method of Lowry et al.[7] because of the high nucleic acid content of the enzyme fractions. Thereafter, it is determined spectrophotometrically with use of the table given by Layne[8] to correct for the nucleic acid content. For correspondence of the values obtained by the colorimetric and spectrophotometric methods, the protein concentration of crystalline egg albumin solutions used as standard for the former is determined spectrophotometrically.

Growth of Cells. Azotobacter vinelandii (strain 0)[9] is carried on agar slants with frequent transfers (approximately once a month). Inoculations are made into 150-ml. portions of Burk's medium[10] in 1-l. conical flasks, and the cells are grown at 32° for 18 to 20 hours on a rotary shaker[11] set at maximal speed. The absorbancy of the cell suspension at 500 mμ should not be lower than 2.5. Fresh 150-ml. portions of medium are inoculated with 5 ml. of the culture thus obtained, the cells are grown as above, and the new culture is used to inoculate fresh medium again. This operation is repeated several times. Approximately 1 l. of this culture is used to inoculate 180 l. of medium in a vat fermenter. Cells are grown at 28° with vigorous agitation for 18 to 20 hours and harvested at about 0°, with an industrial-type refrigerated Sharples centrifuge within 1 hour. The yield of wet cells is about 1.5 kg. After harvesting, the cells are frozen until used. Under these conditions, polynucleotide phosphorylase activity remains unchanged on storage for several months.

Step 1. Extraction. This is carried out in the cold room (4°) in a stainless-steel Waring blendor essentially as described by La Manna and Mallette[12] for the preparation of *Escherichia coli* extracts. Frozen cells (1 kg.) are mixed in a 1-gallon blendor with 450 to 500 ml. of ice-cold 0.01 M potassium phosphate buffer, pH 7.4, and 3 kg. of glass beads,[13] and extracted by stirring first slowly for a few minutes until a thick slurry is produced, then for 15 minutes at about two-thirds maximal speed. Then 2 l. of buffer are added, and stirring is continued as before

[6] A. G. Gornall, C. S. Bardawill, and M. M. David, *J. Biol. Chem.* **177**, 751 (1949).
[7] O. H. Lowry, N. J. Rosebrough, A. L. Farr, and R. J. Randall, *J. Biol. Chem.* **193**, 265 (1951).
[8] See Vol III [73].
[9] Obtained from the American Type Culture Collection, Washington, D. C.; listed as No. 9104 in the sixth edition (1958) of the ATCC Catalogue.
[10] J. W. Newton, P. W. Wilson, and R. M. Burris, *J. Biol. Chem.* **204**, 445 (1953).
[11] New Brunswick Scientific Company, Model V 855370.
[12] C. Lamanna and M. F. Mallette, *J. Bacteriol.* **67**, 503 (1954).
[13] Superbrite No. 100 (average diameter 200 μ), obtained from Minnesota Mining and Manufacturing Company, St. Paul, Minnesota.

for 10 minutes more. After the beads have settled, the extract is poured off. The residue is re-extracted in the same way with 2 l. of buffer for 1 minute. The combined supernatants yield approximately 4.5 l. of greenish-brown turbid extract with about 20 mg. of protein per milliliter. The enzyme is somewhat unstable at this stage, and the next step is therefore started at once.

Step 2. Ammonium Sulfate Fractionation. After cooling to 0°, the extract (4.6 l.) is diluted with ice-cold 0.01 M potassium phosphate buffer, pH 7.4, to a protein concentration of 10 mg./ml. with addition of enough 0.5 M EDTA to give a final concentration of 0.001 M; volume, 10 l. To the diluted extract are added 2320 g. of solid, finely powdered ammonium sulfate (to give approximately 0.33 saturation) over a period of about 60 minutes with mechanical stirring, the temperature being maintained at 0° and the pH kept at 7.4 (glass electrode) by occasional dropwise addition of from 50 to 60 ml. of 6.0 N potassium hydroxide. Stirring is continued for another hour, and, after adjustment of the pH to 7.4 if necessary, the mixture is allowed to stand at 0° overnight. The precipitate, containing over 60% of the protein and 30 to 40% of the units (specific activity, 0.6 to 0.7), is removed by centrifugation for 1 hour at 0° and maximal speed of the large rotors of the refrigerated Servall or Lourdes angle centrifuge and discarded. Solid ammonium sulfate (910 g.) is added as above to the ice-cold supernatant fluid (to give approximately 0.46 saturation) over a period of 30 minutes, the temperature being maintained at 0°, and the pH at 7.4. This requires about 12 ml. of 6.0 N potassium hydroxide. After the mixture is stirred for a further 60 minutes, it is centrifuged as before, and the supernatant fluid is discarded. The precipitate is dissolved in 170 ml. of ice-cold 0.01 M potassium phosphate buffer, pH 7.4, and dialyzed at 0° with stirring overnight against 6 l. of 0.033 M succinate buffer, pH 6.3, containing 0.5 $\times$ 10^{-3} M cysteine. The clear, reddish-brown dialyzed solution (470 ml.) contains 66 mg. of protein per milliliter. Usually, several batches of cells are worked up through step 2, and the dialyzed solutions are stored in the frozen state ($-18°$) before the next step is performed. However, storage for periods longer than 3 months should be avoided. It has been found that, after storage for about a year, the enzyme retains its original activity but can no longer be successfully fractionated with ethanol.

Step 3. Low-Temperature Ethanol–Zinc Acetate Fractionation. The solution from the previous step is diluted with 0.033 M succinate buffer, pH 6.3, containing 0.5 $\times$ 10^{-3} M cysteine, to a protein concentration of 10 mg./ml., and the diluted solution (3.1 l.) is cooled to 0°. To this solution are added, in small alternating fractions, 570 ml. of absolute ethanol (chilled to $-15°$) and 156 ml. of $0.1 M$ zinc acetate (cooled to $0°$) with vigorous mechanical stirring over a period of 40 minutes, the temperature being gradually lowered to $-6°$. The concentration of ethanol is approximately 15% by volume, and that of zinc acetate, $0.004 M$. After being stirred for a further 30 minutes, the mixture is centrifuged for 2 hours at $-10°$ and maximal speed of the large rotors of the refrigerated Servall or Lourdes angle centrifuge. The precipitate, containing 60 to 80% of the protein and 30 to 40% of the units (specific activity about 0.5) of the ammonium sulfate fraction, is discarded. To the supernatant fluid are added, as above, 420 ml. of ethanol and 520 ml. of $0.1 M$ zinc acetate over a period of 30 to 45 minutes, the temperature being lowered gradually to $-8°$. The concentration of ethanol is approximately 20% by volume, and that of zinc acetate is $0.014 M$. The mixture is stirred for a further 30 minutes and centrifuged for 2 hours at $-15°$ as before; the clear, pale-pink supernatant is discarded. The precipitate is immediately dissolved in about 160 ml. of ice-cold $0.1 M$ potassium phosphate buffer, pH 7.4, containing $0.01 M$ cysteine and $0.03 M$ EDTA, and dialyzed with stirring at once against 6 l. of $0.01 M$ potassium phosphate buffer, pH 7.4, containing $0.001 M$ EDTA, at $0°$ overnight. The clear, reddish-brown dialyzed solution (235 ml.), containing 23 mg. of protein per milliliter, can be stored in the frozen state with little loss of activity.

Step 4. Adsorption and Elution from Calcium Phosphate Gel. The solution from the previous step (235 ml.) is brought to 540 ml. with $0.01 M$ potassium phosphate buffer, pH 7.4, giving a protein concentration of 10 mg./ml., cooled to $0°$, and brought to pH 5.4 (glass electrode) by the dropwise addition of $1.0 N$ acetic acid (about 6 ml.) with mechanical stirring. Calcium phosphate gel (130 mg. of $Ca_3(PO_4)_2$ per milliliter), 25 ml., is then added; after the gel has been stirred for 15 minutes, it is centrifuged off at $0°$ and discarded. To the supernatant (pH 5.6) are added 50 ml. of the calcium phosphate gel, and the gel is discarded after stirring and centrifugation as above. The supernatant (pH 5.7) is brought to pH 5.5, with about 0.4 ml. of $1.0 N$ acetic acid, and a further 50 ml. of gel are added. After 15 minutes of stirring, the gel is collected by centrifugation, and the supernatant is discarded.[14] This gel is eluted four times at $0°$, each with 30 ml. of $0.1 M$ potassium phosphate buffer, pH 6.0, and the eluates are combined to give a clear, pale-yellow solution with 3.0 to 3.5 mg. of protein per milliliter. It is dialyzed over-

[14] As followed by optical assay, which at this stage could be used as a rough guide of the adsorption of the enzyme, only the third addition of gel removed most of the enzyme from the supernatant. Since different batches of gel often give different results, the supernatant should always be assayed after each addition of gel.

night at 0° against 3 l. of 0.02 M potassium phosphate buffer, pH 6.8, containing 0.001 M EDTA. The dialyzed eluate (115 ml.) can be stored in the frozen state for periods up to 2 or 3 months with little loss of activity.

Step 5. Protamine and Ammonium Sulfate Fractionation. A freshly prepared 2% solution of protamine sulfate is added dropwise to 41 ml. of the dialyzed eluate at 0° with mechanical stirring. Just enough protamine sulfate solution, in this case 2.5 ml., is used to precipitate most of the enzyme as ascertained by optical assay of the supernatant. After a further 20 minutes of stirring, the bulky precipitate is collected by centrifugation at 0° and 15,000 × g, and the faintly yellow supernatant, containing 4% of the units and about half of the protein of the eluate, is discarded. The yellow precipitate, which becomes gummy on centrifugation, is washed with 20 ml. of 0.02 M phosphate buffer, pH 6.8, containing 0.001 M EDTA, and dissolved in 6.5 ml. of a solution containing 0.1 M glycine and 20% saturated ammonium sulfate, the pH of which had been adjusted to 6.3 with potassium hydroxide. A small amount of insoluble residue is removed by centrifugation. To the clear yellow supernatant (7.4 ml.), cooled to 0°, are added 3.0 ml. of saturated ammonium sulfate with magnetic stirring, bringing the ammonium sulfate concentration to approximately 0.43 saturation. After being stirred for 10 minutes, the precipitate, which has little activity, is removed by centrifugation for 15 minutes at 0° and 15,000 × g and discarded. A further 2.0 ml. of saturated ammonium sulfate are added to the supernatant as above, bringing the ammonium sulfate concentration to approximately 0.52 saturation. The precipitate is collected by centrifugation, washed twice, each time with 5 ml. of 60% saturated ammonium sulfate containing 0.001 M EDTA, and dissolved in 1.5 ml. of 0.02 M phosphate buffer, pH 6.8, containing 0.001 M EDTA. A light precipitate forms whenever the dark-yellow solution is brought to 0°. From previous trials, this precipitate is known to carry down much of the enzyme, as the activity of the supernatant diminishes considerably, and enzyme can be recovered from the precipitate by extraction with the 0.1 M glycine–20% saturated ammonium sulfate solution. Since it was found that the material responsible for precipitate formation could be removed with charcoal, a pinch of acid-washed Norit A is added to the solution of the ammonium sulfate precipitate at about 10°, at which temperature the solution is quite clear. Addition of too much Norit leads to substantial losses of enzyme and should be avoided. After a few minutes of stirring, the Norit is removed by centrifugation at 0° and washed with 0.5 ml. of the 0.02 M phosphate–0.001 M EDTA buffer, pH 6.8. The combined supernatants are dialyzed overnight at 0° against the same buffer, yielding 2.0 ml. of dark-yellow solution containing 9.5 mg. of protein per milliliter.

Step 6. Chromatography on Hydroxylapatite. Hydroxylapatite, prepared by the method of Tiselius *et al.*,[15] is packed by gravity into a 1×40-cm. column and washed overnight in the cold room (3° to 4°) with $0.02\,M$ phosphate buffer, pH 6.8, containing $0.001\,M$ EDTA. The flow rate is adjusted to roughly 20 ml./hr. by applying slight pressure (20 to 50 mm. Hg).

The enzyme solution from the previous step is passed through the column, whereby all the protein is retained by the gel. Elution is carried out stepwise with each 30 ml. of $0.02\,M$, $0.04\,M$, and $0.06\,M$ and each 60 ml. of $0.11\,M$ and $0.2\,M$ sodium phosphate buffer, pH 6.8, containing $0.001\,M$ EDTA. The volume of the individual fractions collected in a fraction collector at 10-minute intervals in the cold room varies between 2.5 and 3.5 ml. The elution of protein is followed spectrophotometrically by determining the absorption of light at 280 mμ and that of enzyme by optical assay. Small amounts of inactive protein are released by phosphate buffer up to $0.06\,M$ and the first half of the $0.11\,M$ buffer. The enzyme is eluted as a sharp band with $0.11\,M$ buffer; these eluates are colorless. A nonfluorescent yellow protein, responsible for the color of the protamine fraction, is eluted with $0.2\,M$ buffer. Fractions with specific activities of 300 or higher are combined to give 20 ml. of solution with 0.4 mg. of protein per milliliter. This solution is dialyzed overnight at 0° against 3 l. of $0.5 \times 10^{-3}\,M$ EDTA, adjusted to pH 7.4, and concentrated by lyophilization to about 1.0 ml. The concentrated enzyme is dialyzed for 5 hours at 0° against 1 l. of $0.02\,M$ sodium phosphate buffer, pH 6.8, containing $0.001\,M$ EDTA. The dialyzed enzyme is rechromatographed as above. Fractions of specific activity above 400 are combined (9.5 ml.), and the enzyme is precipitated at 0° by addition of 3.2 g. of solid ammonium sulfate. The precipitate is washed with 5 ml. of 50% saturated ammonium sulfate containing $0.001\,M$ EDTA, dissolved in 1.5 ml. of $0.01\,M$ glycylglycine buffer, pH 6.8, containing $0.001\,M$ EDTA, and dialyzed at 0° against 1 l. of the same buffer overnight. Two milliliters of enzyme solution containing 5 mg. of protein are obtained. This solution is stored at $-20°$. The enzyme is relatively stable under these conditions. A summary of the purification procedure is given in the table on page 10.

The above procedure was carried out several times with similar results during 1958 and 1959, and a total of 10 kg. of cells was worked up in this manner. More recently, the reproducibility of steps 3 and 4, as regards both purification and yield, has been poor. The following

[15] A. Tiselius, S. Hjertén, and Ö. Levin, *Arch. Biochem. Biophys.* **65**, 132 (1956).

PURIFICATION OF POLYNUCLEOTIDE PHOSPHORYLASE OF *A. vinelandii*

Step	Volume, ml.	Units[a]	Protein, g.	Specific activity, units/mg. protein	$\dfrac{280^b}{260}$	Yield, %
1. Extract[c]	4600	101,360	101.4	1	0.60	100
2. (NH₄)₂SO₄ fractionation	470	58,262	31.1	2	0.60	58
3. Ethanol fractionation	235	42,287	5.4	8	0.60	42
4. Ca₃(PO₄)₂ gel eluate	115	17,850	0.4 mg.	45	0.65	18
4. Ca₃(PO₄)₂ gel eluate	41	6,300	142	45	0.65	100
5. Protamine and (NH₄)₂SO₄ fractionation	2	4,275	19	226	0.91	68
6. Hydroxylapatite chromatography	20	2,812	8	350	0.92	45
6b. Chromatography repeated	2	2,447	5	495	0.93	38

[a] P_i^{32} exchange assay used through step 4; optical assay used thereafter. Specific activity expressed in terms of optical assay throughout (1.0 unit $\backsimeq$ 1.2 P_i^{32} exchange units).

[b] Ratio of light absorption at 280 mμ to 260 mμ.

[c] Prepared from 1-kg. cells.

average specific activities and percentages of over-all yields (given in parentheses) were obtained in several small-scale runs between June and December, 1960; step 2, 2.0 (60%); step 3, 4.5 (25%); step 4, 17.0 (5%); step 5, 65.0 (2.5%). These values are to be compared with the corresponding values in the table. The cells were grown from agar slants that had been kept in the laboratory with occasional transfers for over a year, and bacterial variation may have occurred. It should be pointed out that, in any case, polynucleotide phosphorylase preparations of specific activity 40 to 60 are quite suitable for the preparation of polyribonucleotides.

Properties

Purity. The enzyme has been purified 500-fold and appears to be essentially devoid of nuclease activity. However, when assayed viscosimetrically with polyuridylic acid as substrate, 7 μg. of enzyme brought about a decrease in viscosity at about the same rate as 0.01 μg. of crystalline pancreatic ribonuclease.

The purified enzyme contains small amounts (1.7% of the enzyme protein) of a firmly bound oligoribonucleotide. Attempts to remove this material with activated charcoal, Dowex 1 resin, DEAE-cellulose chromatography, or by incubation with ribonuclease followed by dialysis have proved unsuccessful. Base analysis of the polyribonucleotide gives the following molar ratios: AMP, 1.0; GMP, 1.2; UMP, 0.85; CMP, 0.89. End-group analysis indicates an average chain length of about 12.

Although preparations of *Azotobacter* polynucleotide phosphorylase react with individual ribonucleoside 5'-diphosphates to give the corresponding homopolymers or with mixtures of ribonucleoside diphosphates to form different copolymers, a single enzyme seems to be involved. This is indicated by the fact that, when preparations at different stages of purification are assayed by P_i^{32} exchange with individual ribonucleoside diphosphates, the degree of purification at each step is approximately the same for five different substrates, namely ADP, GDP, UDP, CDP, and IDP.

Specificity. The enzyme specifically requires the ribonucleoside 5'-diphosphates; no reaction is observed with the deoxyribonucleoside 5'-diphosphates or with ribonucleoside mono- or triphosphates. For a more detailed discussion of specificity with regard to precursor as well as primer, refer to the article by Basilio and Ochoa.[16]

[16] See Vol. VI [98].

[2] RNA Polymerase from *Azotobacter vinelandii*[1-3]

$$\begin{matrix} n\text{ATP} \\ n\text{UTP} \\ n\text{GTP} \\ n\text{CTP} \end{matrix} \xrightarrow[\text{DNA}]{} \begin{bmatrix} \text{AMP} \\ \text{UMP} \\ \text{GMP} \\ \text{CMP} \end{bmatrix}_n + 4n\text{PP}$$

By Joseph S. Krakow *and* Severo Ochoa

Assay Methods

Principle. The method for the assay of RNA polymerase measures the incorporation of an acid-soluble radioactive nucleotide into an acid-insoluble polyribonucleotide. To exclude incorporation due to polynucleotide phosphorylase present in the crude enzyme (up to step 5), the reaction is carried out in the presence of a suitable concentration of inorganic phosphate.[1]

[1] S. Ochoa, D. P. Burma, H. Kröger, and J. D. Weill, *Proc. Natl. Acad. Sci. U.S.* **47,** 670 (1961).

[2] D. P. Burma, H. Kröger, S. Ochoa, R. C. Warner, and J. D. Weill, *Proc. Natl. Acad. Sci. U.S.* **47,** 749 (1961).

[3] J. S. Krakow and S. Ochoa, *J. Biol. Chem.* in preparation.

Assay A: DNA-Primed

Procedure. Each incubation mixture (0.25 ml.) contains (in micromoles): Tris buffer, pH 8.1, 20; potassium phosphate buffer, pH 8.0, 2; $MgSO_4$, 5; mercaptoethylamine, 8; putrescine, 10; calf thymus DNA, 60 μg.; CTP, 0.5; GTP, 0.5; UTP, 0.5; ATP-8-C^{14} [4] (specific activity about 2×10^5 c.p.m./micromole), 0.5; and enzyme. The mixture is incubated for 5 minutes at 37°, and the reaction is terminated by the addition of 5 ml. of ice-cold 0.4 N perchloric acid. The tubes are placed in ice, and serum albumin is added as carrier to bring the protein in each tube to approximately 700 μg. The precipitate is collected by centrifugation at 2°, and the supernatant solution is discarded. The acid-insoluble precipitate is dispersed, 5 ml. of cold 0.4 N perchloric acid is added, and the centrifugation is repeated. The supernatant solution is discarded, and the washing procedure is repeated once more. The acid-washed pellet is dispersed, and 0.5 ml. of 5 N NH_4OH is added; the resulting solution is pipetted into a stainless-steel planchet. The tube is rinsed with another 0.5 ml. of 5 N NH_4OH, and this is added to the planchet. The material is dried, and the radioactivity is measured.

Assay B: Polyribonucleotide-Primed

Procedure. Each incubation mixture (0.2 ml.) contains (in micromoles): Tris buffer, pH 7.1, 20; $MgCl_2$, 2; $MnSO_4$, 0.5; mercaptoethylamine, 8; poly U, 10 μg.; ATP-8-C^{14} (specific activity about 2×10^5 c.p.m./micromole), 0.2; and enzyme. The mixture is incubated for 10 minutes at 37°, the reaction is terminated by the addition of 5 ml. of ice-cold 0.4 N perchloric acid, and the procedure outlined in assay A is followed to determine the amount of incorporation of radioactive AMP into acid-insoluble form.

Definition of Unit and Specific Activity. One unit of enzyme is defined as the amount of protein catalyzing the incorporation of 1.0 millimicromole of C^{14}-AMP into acid-insoluble form per minute under the assay conditions described. The specific activity is expressed as units per milligram of protein. Protein is determined in steps 1 and 2 by the biuret method;[5] in subsequent steps in the purification the protein concentration is determined spectrophotometrically with the following formula;[5]

$$\text{Protein concentration (mg./ml.)} = 1.55D_{280} - 0.76D_{260}$$

Purification Procedure

Unless otherwise noted, all operations are carried out at ice-bath temperature.

---

[4] Obtained from Schwarz BioResearch, Inc., Orangeburg, New York.
[5] See Vol. III [73].

The growth of the *Azotobacter* cells used for the isolation of RNA polymerase, step 1 (extraction), and step 2 (ammonium sulfate fractionation) are identical with those described in the article on the purification of polynucleotide phosphorylase.[6]

Step 3. Precipitation with Protamine[7] and Ammonium Sulfate Fractionation. The protein concentration of the solution from step 2 is adjusted to 25 mg./ml. by dilution with 0.03 M succinate buffer, pH 6.3, containing 0.002 M mercaptoethylamine. To 310 ml. of the diluted solution is added dropwise, with stirring, 62 ml. of a freshly prepared 1% solution of protamine sulfate in the buffer and at the pH indicated above (the 1% protamine solution is kept at room temperature). After 10 minutes of stirring, the solution is centrifuged for 20 minutes at 15,000 $\times g$, and the precipitate is discarded. The enzyme is precipitated by dropwise addition of 62 ml. of the 1% protamine solution used earlier. After 10 minutes of stirring, the bulky precipitate is collected by centrifugation for 20 minutes at 15,000 $\times g$, and the supernatant is discarded. The precipitate is washed with 150 ml. of 0.03 M succinate buffer, pH 6.3, containing 0.002 M mercaptoethylamine. The washing and the subsequent elution are accomplished by dispersing the precipitate with the aid of a Serval Omnimix operated at 50 volts for 2 minutes. The suspension is centrifuged at 15,000 $\times g$ for 20 minutes, and the supernatant is discarded. To elute the enzyme, the protamine pellet is dispersed in 100 ml. of 0.2 M potassium phosphate buffer, pH 6.8, containing 0.002 M mercaptoethylamine. The solution is centrifuged at 15,000 $\times g$ for 20 minutes, and the supernatant is collected. The protamine pellet is re-extracted as before, and the resultant supernatant is added to the first. To the pooled extract is added ammonium sulfate to 0.3 saturation. After 10 minutes of stirring, the precipitate is removed by centrifugation for 20 minutes at 15,000 $\times g$. The supernatant is brought to 0.55 saturation by addition of solid, finely powdered ammonium sulfate. After 10 minutes of stirring, the precipitate is collected by centrifugation at 15,000 $\times g$ for 20 minutes. The supernatant is discarded, and the precipitate is dissolved in 15 ml. of 0.02 M potassium phosphate buffer, pH 6.8, containing 0.001 M EDTA, and is dialyzed against 2 l. of 0.02 potassium phosphate buffer, pH 6.8, containing 0.001 M EDTA, and 0.001 M cysteine. (The dialysis tubing is soaked in 0.2 M EDTA, pH 8.0, for several hours and rinsed with distilled water before using.)

Step 4. Adsorption and Elution from Calcium Phosphate Gel. The di-

---

[6] See Vol. VI [1].

[7] The amount of protamine required in step 3 varies with each preparation, aliquots of dialyzed step 2 solution should be titrated with protamine, centrifuged, and polymerase assayed in the supernatant (assay **A**) to determine the proper amounts of protamine.

alyzed solution is diluted with buffer to 115 ml. (protein, 5 mg./ml.), 800 mg. of calcium phosphate·gel is added, and the suspension is stirred for 10 minutes. The suspension is centrifuged for 10 minutes at $15,000 \times g$, and the supernatant is discarded. The gel is eluted with 100 ml. of $0.2 M$ potassium phosphate buffer, pH 6.8, containing $0.001 M$ EDTA, by first dispersing the gel for 1 minute in the buffer solution with the Omnimix operated at 50 volts and then stirring the suspension for an additional 20 minutes. The calcium phosphate gel is removed by centrifugation at $15,000 \times g$ for 15 minutes. Ammonium sulfate is added to the supernatant to 0.55 saturation, and, after 10 minutes of stirring, the precipitate is collected by centrifugation at $15,000 \times g$ for 20 minutes. The precipitate is dissolved in 5 ml. of 0.02 potassium phosphate buffer, pH 6.8, containing $0.001 M$ EDTA, and is dialyzed as in step 3.

Step 5. DEAE-Cellulose Fractionation. The dialyzed enzyme is passed through a column of DEAE-cellulose[8] (1.5×30 cm.) which has been previously equilibrated with the above buffer. The column is eluted with $0.11 M$ potassium phosphate buffer, pH 6.8, containing $0.001 M$ EDTA, until the absorbancy of 280 mμ of the effluent falls to approximately 0.1. To elute the enzyme, $0.1 M$ potassium phosphate buffer, pH 6.8, containing $0.2 M$ KCl and $0.001 M$ EDTA, is run through the column, and 5-ml. fractions are collected. The enzyme appears after approximately 20 ml. of effluent has been collected and is eluted with passage of an additional 30 ml. of the buffer solution. The fractions are pooled after being assayed for polymerase activity, and the enzyme is concentrated by precipitation with ammonium sulfate as indicated previously. The precipitate is dissolved in 1 ml. of $0.02 M$ potassium phosphate buffer, pH 6.8, containing $0.001 M$ EDTA, and dialyzed as in step 3.

Step 6. Chromatography on Hydroxylapatite. Hydroxylapatite[9] is packed by gravity into a 1×30-cm. column and is washed overnight with $0.02 M$ potassium phosphate buffer, pH 6.8, containing $0.001 M$ EDTA. The dialyzed enzyme (2 ml.) is passed through the hydroxylapatite column; all the protein is retained by the gel. Elution of inactive protein is accomplished by passing $0.11 M$ potassium phosphate buffer, pH 6.8, containing $0.001 M$ EDTA, through the column at a rate of 4 ml./hr. After 60 ml. is collected, the enzyme is eluted with $0.2 M$ potassium phosphate buffer, pH 6.8, containing $0.001 M$ EDTA. Fractions of 2 ml. are collected by means of a fraction collector, and the elution of protein is followed spectrophotometrically at 280 mμ and the enzyme by assaying the fractions as described (assay A). Fractions with

[8] Whatman DE-50, capacity 1.0 meq./g.
[9] A. Tiselius, S. Hjertén, and Ö. Levin, *Arch. Biochem. Biophys.* **65,** 132 (1956).

specific activities of **140** or higher are pooled. The enzyme is concentrated by precipitation with ammonium sulfate and is dissolved in 1 ml. of **0.02 M** potassium phosphate buffer, pH 6.8, containing 0.001 M EDTA, and stored at −10°.

A summary of the purification procedure is presented in Table I.[3]

TABLE I

PURIFICATION OF RNA POLYMERASE OF *A. vinelandii*

Step	Volume, ml.	Units[a]	Protein, mg.	Specific activity, units/mg protein	$\frac{280^b}{260}$	Yield, %
1. Extract[c]	1040	7750	25,800	0.3	0.60	100
2. (NH₄)₂SO₄ fractionation	154	5460	7,700	0.7	0.60	71
3. Protamine and (NH₄)SO₄ fractionation	30	5250	583	9	0.65	68
4. Ca₃(PO₄)₂ gel eluate	10	4000	91	44	0.80	52
5. DEAE-cellulose column	30	1960	27	73	1.45	25
6. Hydroxylapatite column	6	1380	8.6	160	1.58	18

[a] Assay A.

[b] Ratio of light absorption at 280 mμ to 260 mμ.

[c] Prepared from 250-g. cells.

Properties

Stability. The purified enzyme dissolved in 0.02 M potassium phosphate buffer, pH 6.8, containing 0.001 M EDTA, retains more than 85% of its initial activity after 2 months at −10°. The enzyme shows no marked loss of activity on repeated freezing and thawing.

Purity. During ultracentrifugation, the purified enzyme exhibits a major component ($S_{20} = 10.4\,S$) comprising approximately 90% of the total protein. In addition there are two minor components (6.4 S and 2.3 S). The enzyme is free of polynucleotide phosphorylase but contains some ribonuclease activity.

Specificity. The enzyme requires ATP, CTP, GTP, and UTP as well as native DNA for maximal activity.[2] With C^{14}-ATP alone, incorporation of C^{14}-AMP occurs to the extent of 25% of that observed with a complete complement of ribonucleoside triphosphates. For the concentrations used in the assay, heat-denatured DNA is only one-fourth as effective in priming the reaction as is native DNA. The enzyme can also use synthetic polyribonucleotides as primers[10]; with poly U as primer and C^{14}-ATP incubated under the conditions optimal for the polyribonucleo-

[10] J. S. Krakow and S. Ochoa, *Proc. Natl. Acad. Sci. U.S.* **49**, 88 (1963).

tide-primed reaction, the incorporation of C^{14}-AMP occurs to the extent of about 12% of that observed for RNA synthesis with DNA as primer in the standard assay (Table II). The activity of the enzyme in response

TABLE II
EFFICIENCY OF VARIOUS POLYNUCLEOTIDES AS PRIMERS

Polynucleotide	Assay	Nucleotide(s) incorporated	Nucleotide incorporated in 10 minutes at 37°, mμmoles/mg. protein[a]
Calf thymus DNA, native	A	AMP, CMP, UMP, GMP	5600[b]
Calf thymus DNA, denatured	A	AMP, CMP, UMP, GMP	1400[b]
Tobacco mosaic virus RNA	B	AMP, CMP, UMP, GMP	44[b]
Poly U	B	AMP	650
Poly A	B	UMP	420
Poly C	B	GMP	410
Poly I	B	CMP	29
Poly FU[c]	B	AMP	22

[a] Hydroxylapatite enzyme.
[b] Based on total nucleotide incorporated, calculated from base ratio of primer.
[c] Polyfluorouridylic acid.

to a variety of RNA preparations tested is poor, approximately one-thirtieth of that elicited by poly U. Both the DNA- and polyribonucleo-tide-primed activity appear to be a property of the 10-S component.

Kinetic Properties. Maximal enzyme activity is obtained with 0.002 M for each of the ribonucleoside triphosphates and 0.0003 M DNA (as constituent mononucleotides). The pH optimum for the DNA-primed reaction is pH 8.1, and that for the polyribonucleotide-primed reaction is approximately pH 7.1. The reaction rate is constant for approximately 20 minutes at 37°, and then the reaction reaches a plateau.

Activators and Inhibitors. The presence of mercaptoethylamine or 2-mercaptoethanol is required for maximal activity; glutathione and cysteine are less active in this respect. On the inclusion of $4 \times 10^{-6}\ M$ PCMB in the reaction mixture (assay A), an 80% inhibition is observed.

With native DNA as primer, 0.04 M putrescine gives a twofold in-crease in the amount of RNA synthesized during a 40-minute incubation. Putrescine is without marked effect when the reaction is primed with heat-denatured DNA and inhibits to the extent of 80% when the reaction is primed with polyribonucleotides. With 10 μg. of poly U as primer, the addition of 5 μg. of polyfluorouridylic acid results in an inhibition of AMP incorporation to the extent of 90%. With 60 μg. of native DNA as primer, the addition of 1 μg. of polyfluorouridylic acid results in 35% inhibition of RNA synthesis. With the same concentrations of DNA and polyfluorouridylic acid, but with putrescine omitted, the reaction is inhibited to the extent of 90% as compared to a reaction carried out in the absence of both putrescine and polyfluorouridylic acid. Polyinosinic acid behaves similarly to poly FU in that both inhibit the synthesis of polyribonucleotides in the presence of suitable primers.

[3] Cell-Free Protein Synthesis Directed by Messenger RNA

By Marshall W. Nirenberg

Assay Method

Principle. Cell-free extracts of *Escherichia coli* incorporate C^{14}-amino acids into protein. Under certain conditions this amino acid incorporation is dependent on the addition of synthetic[1-3] or natural[1,4] messenger RNA, thus providing a sensitive assay for such RNA fractions. In the absence of added messenger RNA, amino acid incorporation presumably is directed by preformed and newly synthesized endogenous messenger RNA. Preincubation of extracts with DNase prevents new messenger RNA synthesis and depletes endogenous messenger RNA. C^{14}-Amino acid incorporation by these extracts is therefore dependent on added messenger RNA and over a certain range is proportional to the amount added.

E. coli extracts ("preincubated" S-30 fractions) are incubated with an ATP-generating system, a C^{14}-amino acid, nineteen C^{12}-amino acids, and messenger RNA. After incubation, proteins are precipitated and treated to remove C^{14}-amino acyl transfer RNA and free C^{14}-amino acid. The C^{14} in the washed protein precipitates then is determined.

Reagents

1. Mix I contains: 10.0 ml. of $2M$ Tris buffer, pH 7.8; 2.0 ml. of 1.4 M magnesium acetate; 5.0 ml. of $2M$ KCl; 3.0 ml. of 6.66 ×

[1] M. W. Nirenberg and J. H. Matthaei, *Proc. Natl. Acad. Sci. U.S.* **47**, 1588 (1961).
[2] J. H. Matthaei, O. W. Jones, R. G. Martin, and M. W. Nirenberg, *Proc. Natl. Acad. Sci. U.S.* **48**, 666 (1962).
[3] J. F. Speyer, P. Lengyel, C. Basilio, and S. Ochoa, *Proc. Natl. Acad. Sci. U.S.* **48**, 63 (1962).
[4] A. Tsugita, H. Fraenkel-Conrat, M. W. Nirenberg, and J. H. Matthaei, *Proc. Natl. Acad. Sci. U.S.* **48**, 846 (1962).

$10^{-2}\ M$ ATP, Na salt, and $2.0 \times 10^{-3}\ M$ GTP, Na salt. Store in 5-ml. aliquots at $-20°$.

2. Phosphoenolpyruvate kinase, crystalline, 10 mg./ml., obtained as a suspension in $(NH_4)_2SO_4$ solution (California Biochemical Corp.).

3. $7.5 \times 10^{-2}\ M$ phosphoenolpyruvate, K or Na salt, crystalline (California Biochemical Corp.) Store in small aliquots at $-20°$. The tricyclohexylammonium salt of phosphoenolpyruvate inhibits C^{14}-amino acid incorporation and should be converted to the potassium salt as follows: 5 g. of phosphoenolpyruvate, tricyclohexylammonium salt, are dissolved in minimal quantities of H_2O at $3°$, and the solution is added to a washed Dowex 50 (H^+) column of approximately 2.5×15 cm. Phosphoenolpyruvic acid is eluted with H_2O, and the pH of the eluate is determined rapidly with pH paper. When the pH of the eluate drops to 2, collection of the solution is begun, after the pH rises to 4, collection is discontinued. The phosphoenolpyruvic acid solution is converted to the potassium salt by the addition of M KOH with vigorous stirring. The pH of the final solution should be 5.5 to 6.0.

4. 2-mercaptoethanol.

5. C^{14}-L-Amino acid. $10^{-2}\ M$, 4 mc./millimole.

6. Mixture of nineteen C^{12}-L-amino acids, minus the appropriate C^{14}-amino acid, $2 \times 10^{-3}\ M$ (each amino acid).

7. Poly U (Miles Chemical Co., Clifton, New Jersey), 30 O.D. units/ml. at 260 mμ, or other messenger RNA preparation. Poly U solutions should be stored at $-20°$ in small aliquots. It is important to note that poly U undergoes a conformational change at $6°$, so that solutions which have been frozen and thawed may have lower absorbancies at 260 mμ.

8. "Preincubated" S-30 or S-100 and W-RIB enzyme fractions, described below.

Procedure. (*a*) For ten reaction mixtures, prepare mix II each day as follows: 250 μl. of mix I; 1 μl. of 2-mercaptoethanol; 250 μl. of phosphoenolpyruvate, 4 μl. of phosphoenolpyruvate kinase suspension; 250 μl. of C^{12}-amino acid mixture, and 50 μl. of C^{14}-amino acid solution.

(*b*) Add components of each fraction mixture in the following sequence: 80 μl. of mix II; appropriate amount of H_2O to make final volume 0.25 ml.; 5 to 20 μl. of poly U solution (15 to 20 μg. of poly U or 100 μg. of natural messenger RNA usually saturates a reaction mixture); and *E. coli* extracts (0.5 to 2 mg. of protein). Reaction mixtures are incubated at $37°$ for 15 minutes. C^{14}-Amino acid incorporation should be dependent on the addition of messenger RNA, and, when limiting amounts of messenger RNA are added (see Remarks), incorporation should be proportional to the amount of messenger RNA added.

(c) Protein is precipitated at the end of incubation by the addition of 3.0 ml. of 10% trichloroacetic acid at 3°, and precipitates are washed by the method of Siekevitz.[5]

For exploratory work and preparation of extracts, the following fast procedure[6] for washing protein precipitates may be used for all C^{14}-amino acids incorporations *except C^{14}-tryptophan*. Reaction mixtures deproteinized by the addition of 10% trichloroacetic acid are placed in a water bath at 90° to 95° for 20 minutes to hydrolyze amino acyl transfer RNA. Tubes then are chilled in ice for 30 minutes. Protein precipitates are dispersed by stirring or by vigorous agitation, and each suspension is filtered under suction through a Millipore filter (HA Millipore filter, 25 mm. in diameter, 0.45-μ pore size, held in an appropriate Millipore filter funnel, Millipore Co., Bedford, Massachusetts). Each precipitate then is washed rapidly with five 5-ml. aliquots of cold 5% trichloroacetic acid. If the radioactivity of the washed protein precipitate is to be counted in a thin-window gas-flow B-scaler, the Millipore filter is glued with rubber cement to a disposable planchette, dried for 5 to 10 minutes under an infrared lamp, and counted. Curled Millipore filter edges should be reglued, for they will tear windows of gas-flow counters.

Growth of *E. coli*

Escherichia coli W-3100 (other strains may be used) are grown at temperatures between 25° and 37° (30° is advantageous) in a medium containing 8 g. of nutrient broth (Difco Co., Detroit, Michigan) and 5 g. of glucose per liter. A 60% solution of glucose is autoclaved and is added to autoclaved nutrient broth solution after both have cooled. A typical preparation consists of five 19-l. carboys, each containing 15 l. of nutrient broth–glucose solution and equipped for aeration. About 1 g. of Antifoam A (Dow-Corning Corp.) is suspended with stirring in approximately 50 ml. of H$_2$O, and the suspension is centrifuged at 3000 $\times$ g for 10 minutes to remove undissolved lumps. The supernatant solution of antifoam is autoclaved, and approximately 1 to 2 ml. are added to each carboy before inoculation with bacteria. Additional antifoam may be needed to prevent foaming during growth of bacteria; however, minimal amounts should be used. One liter of nutrient broth–glucose solution in a 6-l. Ehrlenmeyer flask is inoculated with *E. coli* grown on the surface

[5] P. Siekevitz, *J. Biol. Chem.* **195, 549** (1952).
[6] M. W. Nirenberg, unpublished data.

of an agar slant, and the flask is aerated by shaking at 25° for 8 to 12 hours. While the culture is still in the logarithmic growth phase, 200-ml. aliquots are added to each carboy. The carboys are vigorously aerated for 1.5 to 6 hours, depending on the temperature of the medium and the degree of aeration. At 30°, 2.5 to 3.0 hours of growth will suffice. *It is important to follow cell growth carefully so that aeration may be stopped at the proper time.* Aliquots of cultures should be removed at intervals so that turbidities may be determined spectrophotometrically. Aeration should be stopped when a reading of 25 to 35 Klett spectrophotometer units at 520 mμ is obtained, compared to a sterile nutrient broth–glucose solution set at zero. The carboys then are chilled in ice water, and the cells are harvested by means of a refrigerated Sharpless centrifuge. A yield of 0.5 to 0.75 g. of packed cells, wet weight, per liter of medium is optimum. *When more cells are obtained, the extracts often are less active.* The packed cells are washed by rapid suspension in 3 vol. (w/v) of 0.01 M Tris, pH 7.8, 0.014 M magnesium acetate, and 0.06 M potassium chloride at 3°, and are centrifuged at $15,000 \times g$ in preweighed tubes for 15 minutes at 3°. The supernatant solutions are decanted, and the pellets are drained. Centrifuge tubes again are weighed to obtain wet weights of packed cells. Washed, packed cells may be frozen quickly and stored for several months at —20° if desired.

Preparation of Extracts

Step 1. All operations, unless specified, are carried out in a cold room or at 3°. Cells may be disrupted either by grinding with alumina in a mortar, or with a French press (American Instrument Co.). If the alumina grinding method is to be used, about 30 g. of fresh or thawed packed cells are transferred to a large prechilled, unglazed porcelain mortar (20 cm. in diameter), and 30 g. of alumina A-301 (Aluminum Corporation of America) are added. The cells are ground vigorously with a pestle into a thick paste; cracking, popping noises should be produced during grinding. As the cells break, the paste becomes more fluid, and an additional 30 g. of alumina are slowly added to keep the paste thick and to maintain the popping noise. After about 15 minutes of grinding, 60 ml. of "standard buffer" containing 0.01 M Tris, pH 7.8, 0.014 M magnesium acetate, 0.06 M potassium chloride, and 0.006 M 2-mercaptoethanol freshly prepared are added to the mortar, and the paste is suspended evenly by gently stirring. Alumina, intact cells, and debris are removed by centrifugation at $20,000 \times g$ for 20 minutes. The supernatant fluid is decanted, and the pellet is discarded. Cells to be broken in a French press should be evenly suspended in 1.5 to 2.0 vol. of "standard buffer" (w/v) and disrupted with 18,000 p.s.i. in a prechilled cylinder. Additional 2-mercaptoethanol (6 micromoles/ml. of extract) is added after the cells have been broken.

Step 2. Extracts disrupted in either manner are treated as follows: Two micrograms of pancreatic DNase (twice crystallized, Worthington, Biochemical Co.) are added with gentle mixing to each milliliter of *E. coli* extract. (*Caution:* Some commercial DNase preparations are contaminated with pancreatic RNase, which strongly inhibits C^{14}-amino acid incorporation into protein.) Before the DNase is added, the extract obtained by alumina grinding should be noticeably viscous and should contain some gelatinous clots. Although the extract is maintained at 3°, 5 minutes after addition of the DNase the viscosity and clots disappear. The extract is centrifuged at $30,000 \times g$ for 30 minutes at 3°. The supernatant solution is removed by aspiration to within 1 cm. of the pellet and again is centrifuged at $30,000 \times g$ for 30 minutes. The upper four-fifths of the supernatant solution is removed by aspiration, and henceforth will be referred to as the *S-30 fraction*.

At this stage the activities of 25 to 150-μl. aliquots of the S-30 fraction should be assayed rapidly to determine how much C^{14}-phenylalanine can be incorporated into protein in the presence and in the absence of poly U. One milligram of S-30 protein added to a 0.25-ml. reaction mixture containing 15 to 20 μg. of poly U should incorporate 1 to 20 millimicromoles of C^{14}-phenylalanine into protein in 15 minutes at 37°. If, *in the absence of poly U*, 1 mg. of S-30 protein directs more than 0.1 millimicromole of C^{14}-phenylalanine into protein, the "preincubation" procedure described in step 3 should be followed. (Step 3 usually is necessary.) If less than 0.1 millimicromole of C^{14}-phenylalanine is incorporated in the absence of poly U, the S-30 fraction may be dialyzed for 8 hours as described in step 3.

For routine messenger RNA assays and many amino acid incorporation studies, ribosomes need not be separated from the S-30 fraction. If the S-30 preparation at this stage can be used for assays, the dialyzed extract should be divided into 1- to 3-ml. aliquots (one tube of extract per experiment), frozen quickly, and stored until needed.

Step 3. "Preincubation." S-30 extracts carried through this step are almost completely dependent on added messenger RNA.

The components of a reaction mixture appropriate for "preincubation" of 100 ml. of S-30 are: 10 ml. of M Tris, pH 7.8; 2.0 ml. of $0.14 M$ magnesium acetate; 4.0 ml. of $0.02 M$ ATP (neutralized); 12 ml. of $7.5 \times 10^{-2} M$ phosphoenolpyruvate, K or Na salt; 1 mg. of pyruvate kinase, crystalline; 0.04 ml. of 2-mercaptoethanol; and 1 micromole each of twenty L-amino acids. After addition of 100 ml. of S-30 fraction, the reaction mixture is incubated at 37° for 80 minutes. The reaction mixture then is cooled to 3° and dialyzed against 120 vol. of "standard buffer" at 3° for 8 hours. The dialyzing medium should be changed once. In most cases the "preincubated" S-30 fraction can be used after dialysis without further preparation to assay messenger RNA. The extract is divided into 1- to 3-ml. aliquots and frozen quickly for storage.

One milligram of dialyzed "preincubated" S-30 protein added to a 0.25-ml. reaction mixture containing 15 to 20 μg. of poly U (see Remarks) should incorporate 1 to 40 millimicromoles of C^{14}-phenylalanine into protein during 15 minutes of incubation at 37°.

Step 4. If washed ribosomes and supernatant solution are required, the dialyzed S-30 extract is centrifuged at 105,000 × g for 120 minutes in a Spinco Model L preparative ultracentrifuge at 3°. The upper four-fifths of the supernatant solution is aspirated (S-100 fraction). The lower fifth of the supernatant solution is decanted and discarded. Ribosomal pellets are suspended in the initial volume of "standard buffer" by *gentle* homogenization (four to five passes) in a Potter-Elvejehm homogenizer. The ribosomal suspension is centrifuged again at 105,000 × g for 2 hours, and the supernatant solution is decanted and discarded. The ribosomal pellets are suspended, as described previously, in one-fourth the original volume of "standard buffer," and the suspension is centrifuged at 10,000 × g for 5 minutes to remove aggregates. The washed ribosomal fraction will be designated W-RIB. W-RIB and S-100 fractions are divided into small aliquots, frozen quickly, and stored until needed.

Remarks

Stability of E. coli Extracts. The extracts appear to retain more activity if they are frozen rapidly in either dry ice–acetone mixture or in liquid N_2 than if they are frozen slowly at −20°. Little loss in activity of S-30 fractions can be detected after storage in liquid N_2 refrigerators (Linde Corp.) for 6 months and longer. S-30 fractions stored at −20° lose less than 5% activity per week. S-100 and W-RIB preparations also retain more activity when stored under liquid N_2 than when stored at −20°.

Some amino acid-activating enzymes may be more labile than others; thus, S-30 and S-100 fractions are not frozen and thawed more than once. However, W-RIB fractions may be frozen and thawed several times without undue loss of activity.

Template Activities of Poly U Preparations. Optimal amounts of poly U or natural messenger RNA required for reaction mixtures may depend on the nuclease content of *E. coli* extracts.[2] Another variable factor related to the template activity of poly U is the average molecular weight of each poly U preparation. Poly U molecules composed of less than one hundred uridylic acid residues are almost completely inactive.[2] Therefore, poor C^{14}-phenylalanine incorporations need not be ascribed always to inactive *E. coli* extracts.

Most poly U preparations contain traces of nucleases; to prevent degradation, solutions may be stored in small aliquots at $-20°$ and kept at $3°$ when thawed.

[3a] RNA Polymerase

$$n(\text{ATP} + \text{GTP} + \text{CTP} + \text{UTP}) \overset{\text{DNA}}{\rightleftharpoons}$$
$$(\text{AMP} - \text{GMP} - \text{CMP} - \text{UMP})n + 4n\text{PP}_i$$

By JERARD HURWITZ

The enzyme RNA polymerase was first detected in rat liver nuclei by Weiss and Gladstone[1] and later shown to involve DNA as a primer.[2] It was found to be widely distributed and has been studied in plant,[3] bacterial,[4-8] and animal tissues.[9]

Assay Method

Principle. The polymerization of ribonucleotide to RNA can be measured by the incorporation of acid-soluble ribonucleoside triphosphates, labeled with C^{14} or P^{32} in the α-phosphate group, into an acid-insoluble form. With any of the above ribonucleoside triphosphates labeled, the rate of incorporation of ribonucleotides into RNA is directly proportional to the amount of RNA polymerase added. The assay requires the availability of labeled ribonucleoside triphosphates. The four common natu-

---

[1] S. B. Weiss and L. Gladstone, *J. Am. Chem. Soc.* 81, 4118 (1959).

[2] J. Hurwitz, A. Bresler, and R. Diringer, *Biochem. Biophys. Research Communs.* 3, 15 (1960).

[3] R. C. Huang, N. Maheshwari, and J. Bonner, *Biochem. Biophys. Research Communs.* 3, 689 (1960).

[4] A. Stevens, *Biochem. Biophys. Research Communs.* 3, 92 (1960).

[5] S. B. Weiss and T. Nakamoto, *J. Biol. Chem.* 236, PC18 (1961).

[6] S. Ochoa, D. P. Burma, H. Kröger, and J. D. Weill, *Proc. Natl. Acad. Sci. U.S.* 47, 670 (1961).

[7] M. Chamberlin and P. Berg, *Proc. Natl. Acad. Sci. U.S.* 48, 81 (1962).

[8] J. J. Furth, J. Hurwitz, and M. Anders, *J. Biol. Chem.* 237, 2611 (1962).

[9] S. B. Weiss, *Proc. Natl. Acad. Sci. U.S.* 46, 1020 (1960).

rally occurring ribonucleoside triphosphates labeled with C^{14} are commercially available from Schwarz Biochemicals, Mt. Vernon, New York. Although commercial C^{14}-ATP normally does not require further purification, the other labeled ribonucleoside triphosphates should be further chromatographed on Dowex 1–Cl⁻ to remove impurities by the technique of Lehman et al.[10] The commercially available tritiated ribonucleoside triphosphates are contaminated with nucleoside monophosphates which do not react in the kinase assay. These materials can be removed by the above chromatographic procedure. The general method for preparing α-P^{32}-labeled ribonucleoside triphosphates has been described in detail elsewhere.[8, 11]

Reagents

0.5 M potassium maleate buffer, pH 7.5.
0.2 M 2-mercaptoethanol.
0.1 M $MnCl_2$.
0.001 M C^{14}-ATP containing about 1×10^6 c.p.m./micromole.
0.004 M GTP.
0.004 M UTP.
0.004 M CTP.
1 mg. of calf thymus DNA[12] per milliliter.
Escherichia coli W.[8]

Procedure. The reaction mixture (0.5 ml.) contains 0.05 ml. of maleate buffer, 0.01 ml. of 2-mercaptoethanol, 0.02 ml. of $MnCl_2$, 0.08 ml. of C^{14}-ATP, 0.02 ml. of GTP, UTP, and CTP, 0.04 ml. of calf thymus DNA, and RNA polymerase. Incubation is for 20 minutes at 38°, after which the reaction is stopped by the addition of 2 ml. of ice-cold 5% TCA. The entire solution is filtered through a millipore filter (Schleicher and Schuell Co., Keene, New Hampshire), and the reaction tube is rinsed twice with 2-ml. aliquots of 1% TCA which are also filtered through the millipore. The millipores are dried with gentle heat and placed in an aluminum planchet, and radioactivity is measured in a windowless Geiger-Mueller counter. Control tubes lacking enzyme or including either RNase or DNase are included in each series of assays. A unit of enzyme is defined as that amount of enzyme which catalyzes

[10] I. R. Lehman, M. J. Bessman, E. S. Simms, and A. Kornberg, *J. Biol. Chem.* 233, 163 (1958).
[11] G. M. Tener, *J. Am. Chem. Soc.* 83, 159 (1959).
[12] E. R. M. Kay, N. S. Simmons, and A. L. Dounce, *J. Am. Chem. Soc.* 74, 1724 (1952).

the incorporation of 1 millimicromole of labeled ribonucleotide into RNA in 20 minutes at 38°.

Purification Procedure

Crude Extract. Four hundred grams of *E. coli* W were mixed with 260 ml. of a solution containing $0.01 M$ Tris buffer, pH 7.5, $0.01 M$ $MgCl_2$, and $0.001 M$ mercaptoethanol in a large Waring blendor with 1000 g. of No. 100 Superbrite glass beads (Minnesota Mining and Manufacturing Company) for 15 minutes at −10°. The glass beads were precooled to −10°, and during the grinding the temperature of the mixture did not rise above 8°. The mixture was then treated with 500 ml. of the solution containing Tris buffer, $MgCl_2$, and mercaptoethanol, and the mixing was continued for 3 minutes. The glass beads were permitted to settle, and the supernatant fluid was decanted. The residue was then re-extracted with 400 ml. more of the buffer solution by a 1-minute mixing in the Waring blendor and allowed to settle; and the supernatant fluid was combined with the first extract. The extract was then centrifuged for 90 minutes at $78,000 \times g$ (average) in the Spinco Model L ultracentrifuge in plastic tubes containing glass wool. The supernatant fluid was decanted, and the precipitate discarded.

Protamine Eluate. The crude extract (1000 ml.) was treated with 250 ml. of a 0.5% solution of protamine sulfate. After 5 minutes at 0°, the stringy precipitate was collected by centrifugation at $6000 \times g$ in the Servall centrifuge. The pellet was then washed two times by suspension in 500 ml. of a solution containing $0.1 M$ β,β-dimethylglutarate buffer, pH 7.0, and $10^{-3} M$ mercaptoethanol and recentrifugation. The washings contained negligible amounts of enzyme and were discarded. RNA polymerase was then eluted from the protamine precipitate by washing with 250 ml. of a solution containing $0.5 M$ sodium succinate buffer, pH 6.0, and $10^{-3} M$ mercaptoethanol to yield the protamine eluate.

Ammonium Sulfate. The protamine eluate (246 ml.) was adjusted to 20% saturation by the addition of 26.3 g. of solid ammonium sulfate. After 15 minutes at 0°, the precipitate was removed by centrifugation. The supernatant solution (254 ml.) was then adjusted to 35% saturation with 20.4 g. of solid ammonium sulfate. This precipitate, containing most of the RNA polymerase, was dissolved with approximately 7 ml. of $0.1 M$ Tris buffer, pH 8.4, containing $10^{-3} M$ mercaptoethanol (AS-IB fraction).

DEAE-Cellulose Eluate. A column of DEAE-cellulose (20×3 cm.) was washed with 80 ml. of a solution containing $0.1 M$ Tris buffer, pH 8.4, and $10^{-3} M$ mercaptoethanol. The enzyme solution (AS-IB, 6 ml.) was diluted to 60 ml. with $0.1 M$ Tris buffer, pH 8.4, containing $10^{-3} M$

mercaptoethanol, and applied to the column. The column was then washed with 250 ml. of a $0.5 M$ Tris buffer, pH 8.4, containing $10^{-3} M$ mercaptoethanol. This removed approximately 35% of the activity, with little purification. The remaining enzyme was eluted with 125 ml. of $0.5 M$ Tris buffer, pH 7.5, containing $10^{-3} M$ mercaptoethanol. Approximately ten 12-ml. fractions were collected; fraction 7 (DEAE-cellulose eluate) contained 50% of the initial activity placed on the column.

Ammonium Sulfate II. The DEAE-cellulose eluate (11.4 ml.) was treated with 4.1 g. of solid ammonium sulfate. The precipitate obtained after centrifugation was extracted with 5 ml. of a solution containing 60% ammonium sulfate, adjusted to pH 7.8 with NH_4OH, and $10^{-3} M$ mercaptoethanol. The residue remaining after centrifugation was re-extracted successively with 50%, 40%, 30%, and 20% ammonium sulfate solutions as described above. All the activity was associated with the 40% and 30% ammonium sulfate fractions. These enzyme fractions had approximately the same specific activity and were combined. This fraction is AS-IIB.

The crude extract and ammonium sulfate IB fractions could be stored overnight at 2°, whereas the protamine eluate and DEAE-cellulose eluate fractions lost up to 50% of the initial activity after storage for 16 hours at 2°. The AS-IIB fractions lost no detectable activity after 6 weeks at 2°, although freezing and thawing led to marked inactivation.

PURIFICATION OF RNA POLYMERASE

Enzyme fraction	UMP incorporation, total units	Specific activity, units/mg. protein
Crude extract	60,000	6–10
Protamine eluate	40,000	170
Ammonium sulfate IB (AS-IB)	28,000	440
DEAE-cellulose eluate	15,000	2200
Ammonium sulfate IIB (AS-IIB)	12,000	2800

The purification procedure is outlined in the table. The RNA polymerase activity has been purified approximately 300-fold with a 20% over-all yield.

Properties

Specificity. RNA polymerase appears to be relatively specific for the ribonucleoside triphosphate. The addition of labeled GDP in place of labeled GTP reduced the incorporation approximately 20-fold. A variety of ribonucleoside triphosphates are active as substrates, provided they possess the same hydrogen-bonding potentialities as the naturally occurring ribonucleoside triphosphate they replace; i.e., ITP will replace GTP, rTTP will replace UTP, etc. A study of analogs and their ability to replace naturally occurring ribonucleotides has been published.[13]

The enzyme is not specific for DNA as a primer, since RNA will replace DNA.[14] The significance of the role of RNA as a primer in this reaction remains to be elucidated.

Stability. The RNA polymerase preparation shows marked instability at stages prior to the final step. Purified enzyme fractions, over a 6-month period, gradually lose activity until only 10 to 15% of the activity remains. Freezing and thawing accelerated the inactivation, and for this reason enzyme preparations are stored at 2° and not frozen. The enzyme activity is completely inhibited by $10^{-4} M$ p-hydroxymercuribenzoate and is protected by SH compounds. The enzyme activity is stimulated by the addition of K^+, and for this reason potassium maleate buffer is used. The enzyme activity can be further increased approximately 20% by the addition of $5 \times 10^{-4} M$ spermine. Enzyme fractions, measured in the presence of K^+ and spermine, have yielded preparations with specific activities approaching 4000.

pH. The pH optimum for this enzyme is relatively broad. There is very little difference in activity between pH 7 and pH 7.8. At pH 6.5 and 8.5 the reaction is about 40% of that seen at pH 7.5.

Effect of Substrate Concentration. The reaction velocity reaches a maximum at a concentration of ribonucleoside triphosphates of about $1 \times 10^{-4} M$, and the K_s for each of these substrates is about $1 \times 10^{-5} M$. The binding of DNA to the enzyme varies and depends on the DNA primer. Heat-denatured DNA or ϕX 174 DNA saturates the enzyme at concentrations about one-tenth of native DNA. However, although affinity of the enzyme for denatured DNA is increased, the maximum velocity is reduced 2-fold.

Inhibitors of the Reaction. The RNA polymerase reaction is inhibited by a variety of agents. All reagents which bind SH groups inactivate the enzyme. RNA added before the reaction or RNA generated during the course of the reaction markedly inhibits RNA polymerase activity. Actinomycin D, a naturally occurring antibiotic, inhibits the enzyme only by first combining with DNA.[15,16] DNA primers lacking dGMP do not combine with actinomycin D, and RNA polymerase reaction mixtures primed by such DNA preparations are unaffected by the antibiotic.

[13] F. Kahan and J. Hurwitz, *J. Biol. Chem.* **237**, 3778 (1962).
[14] T. Nakamoto and S. B. Weiss, *Proc. Natl. Acad. Sci. U.S.* **48**, 880 (1962).
[15] I. H. Goldberg and M. Rabinowitz, *Science* **136**, 315 (1962).
[16] J. Hurwitz, J. J. Furth, M. Malamy, and M. Alexander, *Proc. Natl. Acad. Sci. U.S.* **48**, 1222 (1962).

[4] Purification and Assay of Soluble Ribonucleic Acid-Enzyme Complexes Isolated from Rat Liver

By Edward Herbert and E. S. Canellakis

Introduction

$$XTP + S\text{-}RNA \rightarrow XMP\text{-}RNA + P\text{-}P \qquad (1)$$

Enzyme systems occur in the soluble fraction of rat liver which incorporate ribonucleotide units (XMP) derived from ribonucleoside triphosphates (XTP) into terminal positions of the soluble ribonucleic acid (S-RNA) as shown in Eq. (1).[1-5] During purification these enzyme systems are fractionated along with S-RNA and can be separated into three distinct ribonucleoprotein components all of which exhibit enzymatic activity.[7] The method of purification of these components and the method of assay are presented.

Assay Method

Principle. After incubation of the S-RNA–enzyme complex with a radioactive ribonucleoside triphosphate $(C^{14}\text{-}XTP)$, the reaction is stopped by the addition of prechloric acid. To ensure complete precipitation of all the acid-insoluble components, a small amount of partially purified fraction (from step 3) is added as carrier[8] to the perchloric acid-treated incubation mixture. After repeated washing with perchloric acid, RNA is extracted from the acid-insoluble residue with NaCl and further treated so as to remove adsorbed acid-soluble compounds. The amount of radioactivity incorporated into RNA isolated in this manner is used as a measure of the activity of the fraction.

[1] E. S. Canellakis, *Biochim. et Biophys. Acta* 25, 217 (1957).
[2] M. Edmonds and R. Abrams, *Biochim. et Biophys. Acta* 26, 226 (1957).
[3] L. I. Hecht, P. C. Zamecnik, M. L. Stephenson, and J. F. Scott, *J. Biol. Chem.* 233, 954 (1958).
[4] E. Herbert, *J. Biol. Chem.* 231, 975 (1958).
[5] It should be pointed out that the reactions catalyzed by these systems are not limited to the terminal addition of ribonucleotide units to S-RNA. Incubation of these systems with combinations of ribonucleoside triphosphates results in sequential addition of ribonucleotide units to S-RNA.[3,6]
[6] E. Herbert, *N. Y. Acad. Sci.* 81, 679 (1959).
[7] E. S. Canellakis and E. Herbert, *Proc. Natl. Acad. Sci. U.S.* 46, 170 (1960).
[8] The carrier is not required in the assay of the material isolated up to step 3.

Reagents

0.08 M potassium phosphate buffer, pH 7.2.
0.10 M $MgCl_2$.
0.002 M radioactive ribonucleoside triphosphate (C^{14}-XTP) (0.7 to 1.2×10^6 c.p.m./micromole).[9]
0.2 N perchloric acid.
1.5 N perchloric acid.
Carrier protein–RNA.[8]
10% NaCl.

Enzyme. The enzyme should be dialyzed against 100 vol. of the potassium phosphate buffer for 3 hours with two changes of the buffer solution. For assay of the original crude extracts (material isolated up to step 3), about 10 to 20 mg. of protein (based on extinction at 280 mμ) should be used per assay tube. The ribonucleoprotein components should be diluted to give an extinction of 20 at 260 mμ.

Procedure. To 0.8 ml. of the buffer are added 0.2 ml. of the enzyme solution, 0.02 ml. of $MgCl_2$, and 0.02 ml. of C^{14}-XTP. Incubation is carried out for 10 minutes at 37°. The tubes are cooled in an ice bath, and the reaction is stopped by the addition of 0.5 ml. of 1.5 N perchloric acid. Then 0.05 ml. of carrier is added,[8] the contents of the tubes are mixed, and the tubes are allowed to stand in an ice bath for 5 minutes. After centrifugation, the supernatant liquid is decanted and the precipitate is washed three times in 8.0 ml. of 0.2 N perchloric acid each time. The residue is then washed once in 10 ml. of 95% EtOH, and the tubes are carefully drained. Next 2.0 ml. of 10% NaCl are added to the residue, and the mixture is carefully neutralized[10] with phenol red as an internal indicator. The neutralized mixture is heated at 100° for 30 minutes and chilled in an ice bath. After centrifugation, the residue is washed once with 0.5 ml. of the NaCl solution, and the wash and extract are combined and chilled. Then 2.5 vol. of chilled EtOH are added to the combined extract and wash, and the mixture is allowed to remain at −20° for 2 hours. After centrifugation, the residue is washed once in 10 ml. of 95% EtOH, and the tubes are carefully drained by allowing them to stand inverted over paper towels for a few minutes. Then 0.5 ml. of water or of 1 N NH_4OH is added, and the material is plated on aluminum planchets. Another 0.5 ml. of liquid is added to the tubes as wash and plated as before. The planchets are dried in air or under infrared lamps and the

[9] E. S. Canellakis, M. E. Gottesman, and H. O. Kammen, *Biochim. et Biophys. Acta* 39, 82 (1960).
[10] E. Herbert, V. R. Potter, and L. I. Hecht, *J. Biol. Chem.* 225, 659 (1957).

radioactivity determined. The results reported here were obtained with a Nuclear-Chicago thin-window counter with an efficiency of 18%.

Application of Assay Method to Crude Tissue Preparations. In order to demonstrate full activity of the system in crude liver preparations, it is necessary to add larger amounts of C^{14}-XTP to minimize the dilution of the radioactive precursor by endogenous substrate.[6] The C^{14}-XTP may be maintained in the triphosphate form either by adding pyruvate kinase and phosphoenolpyruvate or by adding creatine kinase and creatine phosphate. (If the radioactive substrate is a nucleotide other than ATP, a small amount of nonradioactive ATP will be necessary as phosphate donor.)

Purification Procedure

The purification procedure described below has been carried out successfully with rat liver. Attempts to apply this procedure to chicken, calf, and pigeon liver have yielded less satisfactory results with considerable loss of activity occurring in the early stages of purification and on storage. All the procedures described below should be carried out at temperatures between 0° and 4°.

Reagents

Sucrose–salt solution (0.35 M sucrose–0.004 M MgCl$_2$–0.025 M KCl).
Ammonium sulfate (fine granular form).
0.08 M potassium phosphate, pH 7.2.
0.01 M potassium phosphate, pH 7.2.
2 N acetic acid.
Hydroxylapatite column.[11]

Step 1. Preparation of the Soluble Cytoplasmic Fraction (or Soluble Enzyme Fraction). Livers from 25 male albino rats (weighing 450 to 500 g. each) are excised, minced with scissors, and homogenized in a blendor for 30 seconds in 2.5 vol. of the sucrose–salt solution. The homogenate is centrifuged at 85,000 × g for 1 hour in a Spinco Model L preparative ultracentrifuge (No. 30 centrifuge head), to sediment the particulate fractions. The supernatant liquid is carefully decanted to

[11] The method used for the preparation of the hydroxylapatite is described in Vol. V [94a]. The column is prepared by adding sufficient hydroxylapatite in the form of a slurry in 0.001 M potassium phosphate buffer, pH 7.2, to form a column 20 × 2.5 cm. after packing under pressure.

avoid contamination with microsomes and diluted with 1 vol. of distilled water.[12]

Step 2. Ammonium Sulfate Fractionation. Solid ammonium sulfate is added slowly over a period of about 10 minutes to 55% saturation; during this operation the pH of the suspension drops to 6.0. The solution is centrifuged at 15,000 × g for 10 minutes in a Lourdes Model SL centrifuge. The precipitate is resuspended in 100 ml. of 0.05 M potassium phosphate buffer (pH 7.2) and dialyzed for 16 hours against 100 vol. of the 0.01 M buffer with two changes of the buffer. Solid ammonium sulfate is added to the supernatant liquid until it is 85% saturated. The mixture is centrifuged as above, and the precipitate is resuspended in 250 ml. of buffer and dialyzed as described above. About 80 to 90% of the activity of the initial soluble cytoplasmic fraction is recovered in the 55 to 85% saturated fraction.

Step 3. Precipitation at pH 5.2. The 55 to 85% ammonium sulfate fraction is acidified to pH 5.2 by dropwise addition of 2 N acetic acid with constant stirring at 0°. The approach of the end point of the titration is easily observed because the solution becomes cloudy at pH 5.3. The mixture is centrifuged at 15,000 × g for 10 minutes. About 80% of the activity of the ammonium sulfate fraction is in the precipitated material. The precipitate is dissolved in 20 ml. of 0.5 M buffer and dialyzed for 16 hours against 100 vol. of 0.01 M buffer with two changes of the buffer solution.[13] The solution contains 20% of the protein and 60 to 80% of the RNA present in the 55 to 85% ammonium sulfate fraction. The E_{260}/E_{280} ratio is approximately 1.3, indicating that this material contains protein and RNA in a ratio of approximately 10:1 by weight.

Step 4. Fractionation on an Hydroxylapatite Column. To 10 ml. of solution of the pH 5.2 precipitate is added 90 ml. of distilled water. The extinction of this solution (at 260 mμ) is now 15. The sample is added to the column without pressure (a rate of 2 to 4 drops per minute is possible providing the particle size of the hydroxylapatite has been properly graded). The elution may be carried out either by a stepwise procedure or by a linear gradient procedure. The stepwise procedure shown in Fig. 1 yields three distinct ribonucleoprotein fractions designated ribo-

---

[12] A clean separation of sediment from supernatant liquid by centrifugation of the 0.55 to 0.85 ammonium sulfate fraction is difficult to achieve without prior dilution of the soluble cytoplasmic fraction.

[13] The dialysis time determines in part the pattern of elution of the three ribonucleoproteins from the hydroxylapatite column. If the dialysis time is shortened, the β and γ peaks tend to come off earlier and merge with the α peak.

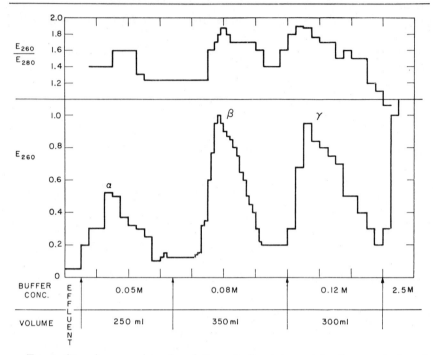

Fig. 1. Stepwise procedure for elution of ribonucleoproteins from a hydroxyl-apatite column. The width of the steps in the chromatogram represents the volume of eluate collected in each fraction. These volumes are as follows, in the order of buffer concentrations listed under the chromatogram (0.05 M, 0.08 M, and 0.12 M); 20 ml., 10 ml., and 25 ml.

nucleoproteins α, β, and γ.[14] The contents of the peak tubes in each of these fractions have E_{260}/E_{280} ratios of 1.6 to 1.9, indicating that about one-half of the ultraviolet-absorbing material is RNA. Over 90% of the total activity recovered from the column is associated with these three fractions (see the table). The linear gradient procedure (shown in Fig. 2) is recommended for routine preparations because it is rapid and gives more reproducible results than the stepwise procedure. The stepwise method has the advantage, however, of giving a more complete separation of ribonucleoproteins α, β, and γ.

The fractions from each peak which have E_{260}/E_{280} ratios of 1.6 to 1.9

[14] The relative amounts of orcinol-reacting material in ribonucleoproteins β and γ vary somewhat from preparation to preparation (see the table), but the total amount of this material in the two peaks is constant. Rechromatography of the three peaks individually on hydroxylapatite columns shows that the α peak is free of the β and γ peaks but that the β and γ peaks are either contaminated with each other or interconverted to a small extent in the process of preparing them for rechromatography.

SUMMARY OF PURIFICATION PROCEDURE

Fraction	Specific activity[a]	Yield, % of soluble cytoplasm	RNA,[b] mg.	Mg. protein[c] Mg. RNA
Soluble cytoplasm	5–10	"100"	135	100
Ammonium sulfate (0.55–0.85)	30–50	80	70	60
pH 5 precipitate	100–150	60	35	15
Hydroxylapatite				
Ribonucleoprotein α	80–100	4	3	1
Ribonucleoprotein β	750–1000	15–20	15–20	1
Ribonucleoprotein γ	400–600	10–15	10–15	1

[a] The specific activity unit is expressed as the micromicromoles of AMP-C^{14} incorporated into RNA in the presence of 1 mg. of protein in 6 minutes at 25°.

[b] RNA is determined by an orcinol method.[4]

[c] Protein is determined by the Lowry method.[15]

are pooled separately and lyophilized. The remaining fractions may be lyophilized separately and refractionated on a hydroxylapatite column.[14] Prior to assay, the lyophilized powders may be suspended in a minimum amount of water and dialyzed against the 0.08 M buffer.

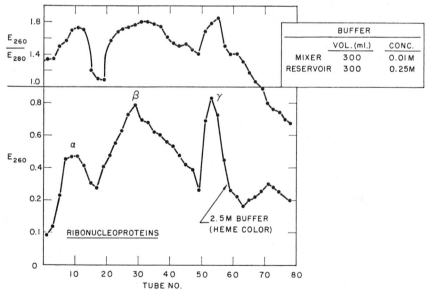

FIG. 2. Linear gradient procedure for elution of ribonucleoproteins from an hydroxylapatite column. The volume of eluate collected per tube is 10 ml. The 2.5 M buffer is added directly to the column and collected in 10-ml. portions.

[15] O. H. Lowry, N. J. Rosebrough, A. L. Farr, and R. J. Randall, *J. Biol. Chem.* **193**, 265 (1951).

Properties

Specificity. Under the conditions described here, the enzyme system attaches a specific sequence and number of nucleotides to RNA. It is characteristic of this reaction that, once AMP has been added to the RNA, the reaction ceases, whether other nucleoside triphosphates are present in the medium or not. The purified systems appear to exhibit small but significant pyrophosphatase and phosphatase activities. Assay of fractions by the pyrophosphate–ATP exchange technique in the presence and absence of a mixture of amino acids demonstrates that ribonucleoproteins α, β, and γ are free from amino acid-activating enzymes known to be present through step 3.[6]

Activators and Inhibitors. Incorporation of AMP and CMP does not occur in the absence of Mg^{++}. The optimum level of Mg^{++} is 0.002 M. Incorporation of AMP is strongly inhibited by levels of pyrophosphate above 0.001 M. Incorporation of CMP, on the other hand, is stimulated three- to sixfold by 0.0005 M pyrophosphate and strongly inhibited by pyrophosphate concentrations above 0.003 M. The activity of ribonucleoproteins β and γ is sensitive to heat (85°; 5 minutes causes complete inactivation), ribonuclease, and trypsin (concentrations above 1 μg./ml.).[16]

[16] Separation of the S-RNA component from the enzyme component has now been achieved with ribonucleoprotein β and ribonucleoprotein γ. Recombination of the two components in the proportions in which they exist in the unfractionated ribonucleoprotein form restores full activity. [E. S. Canellakis and E. Herbert, *Biochim. et Biophys. Acta* 45, 133 (1960).]

[5] DNA Synthesis (Bacterial)

$$\begin{bmatrix} dTPPP \\ dGPPP \\ dCPPP \\ dAPPP \end{bmatrix} + DNA \overset{Mg^{++}}{\rightleftarrows} DNA - \begin{bmatrix} dTP \\ dGP \\ dCP \\ dAP \end{bmatrix} + 4(n)PP$$

By I. R. LEHMAN

Assay Method

Principle. The enzyme from *Escherichia coli* responsible for DNA synthesis will be referred to as "polymerase."

The assay of polymerase measures the conversion of acid-soluble P^{32}-labeled deoxynucleoside triphosphates into an acid-insoluble product.[1]

[1] I. R. Lehman, M. J. Bessman, E. S. Simms, and A. Kornberg, *J. Biol. Chem.* 233, 163 (1958).

Reagents

dTP³²PP, prepared as described by Lehman *et al.*,[1] 0.5 micromole/ ml., 1.5×10^6 c.p.m./micromole.

dATP, dGTP, dCTP, synthesized by the method of Smith and Khorana,[2] each 0.5 micromole/ml.

$MgCl_2$, 0.1 *M*.

Calf thymus DNA, isolated by the method of Kay *et al.*,[3] 0.5 mg./ml.

Glycine buffer, pH 9.2, 1 *M*.

2-Mercaptoethanol, 0.01 *M*.

"Carrier" DNA: calf thymus DNA, 2.5 mg./ml.

Perchloric acid, 1 *N*.

NaOH, 0.1 *N*, and NaOH, 0.2 *N*.

Enzyme. Dilute the enzyme in 0.02 *M* Tris buffer, pH 7.5, containing 0.1 mg. of thymus DNA per milliliter.

Procedure. To tubes containing 0.02 ml. each of glycine buffer and $MgCl_2$, 0.03 ml. of 2-mercaptoethanol, 0.01 ml. each of dATP and dCTP, 0.02 ml. of dGTP, 0.01 ml. of dTP³²PP, and 0.02 ml. of thymus DNA are added 0.005 to 0.05 unit of enzyme and sufficient water to give a final volume of 0.30 ml. This is incubated for 30 minutes at 37°. The reaction is terminated by placing the tubes in ice and adding 0.2 ml. of cold "carrier" DNA solution and 0.5 ml. of cold perchloric acid. After 2 to 3 minutes, the precipitate is thoroughly broken up with a snug-fitting glass pestle, 2 ml. of cold distilled water are added, and the precipitate is thoroughly dispersed. After centrifuging for 3 minutes at $10,000 \times g$, the supernatant fluid is discarded. The precipitate is dissolved in 0.3 ml. of 0.2 *N* NaOH, the DNA is reprecipitated by the addition of 0.40 ml. of cold perchloric acid, 2.0 ml. of cold water are added, and the precipitate is again thoroughly dispersed. After centrifugation, this precipitate is again dissolved, reprecipitated, and recentrifuged. Finally, the precipitate is dissolved by the addition of 0.2 ml. of 0.1 *N* NaOH, the entire solution transferred to a planchet and dried, and the radioactivity measured.

Controls for the crude enzyme fractions (I to III in the table) are incubation mixtures to which the enzyme fraction is added after completion of the incubation period, but just before the perchloric acid is added. With more-purified fractions (fractions IV to VII), an incubation mixture lacking Mg^{++} or one of the deoxynucleoside triphosphates

[2] M. Smith and H. G. Khorana, *J. Am. Chem. Soc.* **80**, 1141 (1958).
[3] E. R. M. Kay, N. S. Simmons, and A. L. Dounce, *J. Am. Chem. Soc.* **74**, 1724 (1952).

serves as well. Precipitates obtained from control incubation mixtures contain less than 0.10% of the total radioactivity added, and in most assays they contain radioactivity of the order of 2% of the experimental values.

PURIFICATION OF "POLYMERASE"

Fraction	Step	Units		Protein, mg./ml.	Specific activity, units/mg. protein
		Per ml.	Total		
I	Sonic extraction	2.0	16,800	20.0	0.1
II	Streptomycin	13.0	19,500	3.0	4.3
III	DNase, dialysis	12.1	18,100	1.80	6.7
IV	Alumina gel	15.4	12,300	0.78	19.8
V	Concentration of gel eluate	110.0	9,900	4.90	22.4
VI	Ammonium sulfate	670.0	6,030	8.40	80.0
VII	Diethylaminoethylcellulose[a]	120.0	3,600	0.60	200.0[b]

[a] This step was actually carried out many times on a smaller scale (see text), and these values are calculated for the large-scale procedure.

[b] In some runs, values as high as 400 have been obtained.

Definition of Unit and Specific Activity. A unit of enzyme is defined as the amount causing the incorporation of 10 millimicromoles of the labeled deoxynucleotide into the acid-insoluble product during the period of incubation. The specific activity is expressed as units per milligram of protein. Protein is determined by the method of Lowry *et al.*[4]

Application of Assay Method to Crude Tissue Preparations. Assays of crude extracts, because of their high levels of deoxyribonucleases and phosphates, yield only an approximate value, and the levels of activity assayed should not exceed 0.02 unit. With the exception of fraction I (crude extract), proportionality of enzyme addition with the amount of labeled substrate incorporated into the product is obtained. With fraction IV, for example, the addition of 0.5, 1.0, and 1.5 μg. of the enzyme preparation yielded specific activities of 21.0, 18.7, and 20.5, respectively.

Purification Procedure

Step 1. Growth and Harvest of Bacteria. E. coli strain B or ML30 is grown in a medium containing 1.1% K_2HPO_4, 0.85% KH_2PO_4, 0.6% Difco yeast extract, and 1% glucose. Cultures, usually 60 l., are grown at 37° with vigorous aeration in a large growth tank[5] and are harvested

[4] O. H. Lowry, N. J. Rosebrough, A. L. Farr, and R. J. Randall, *J. Biol. Chem.* **193**, 265 (1951).

[5] The Biogen, available from American Sterilizer Company, Erie, Pennsylvania.

about 2 hours after the end of exponential growth. The cultures are chilled by the addition of ice. The cells are collected in a Sharples supercentrifuge, washed once in a Waring blendor by suspension in 0.5% NaCl–0.5% KCl (3 ml./g. of packed cells), centrifuged, and then stored at −12°. The yield of packed wet cells is approximately 8 g./l. of culture, and cells stored as long as 1 month have been used. All subsequent operations are carried out at 0° to 3° unless otherwise specified.

Step 2. Preparation of Extract. The cells are suspended in 0.05 M glycylglycine buffer, at pH 7.0 (4 ml./g. of packed cells), and are disrupted by treatment for 15 minutes in a Raytheon 10-kc. sonic oscillator. The suspension is centrifuged for 15 minutes at 12,000 × g, and the slightly turbid supernatant liquid is collected. The protein content is determined and adjusted to a concentration of 20 mg./ml. by addition of the same glycylglycine buffer (fraction I).

Step 3. Streptomycin Precipitation. Fraction I (8400 ml. obtained from 2 kg. of packed cells) is treated in the following manner. To a 525-ml. batch are added 525 ml. of Tris buffer (0.05 M, at pH 7.5); then slowly, with stirring, 81 ml. of 5% streptomycin sulfate (Merck and Co.) are added. After 10 minutes, the precipitate is collected by centrifugation for 10 minutes at 10,000 × g. This precipitation with streptomycin is carried out four times on this scale, and the precipitate is collected in the same centrifuge cups. The precipitates, with potassium phosphate buffer (0.05 M, at pH 7.4) added to a total volume of 430 ml., are homogenized in a Waring blendor for 30 minutes at low speed. This suspension is centrifuged for 2 hours at 78,000 × g in a Spinco Model L centrifuge, and the supernatant fluid is collected (fraction II).

Step 4. Deoxyribonuclease Digestion. To 1500 ml. of fraction II (derived from 8400 ml. of fraction I) are added 15 ml. of 0.3 M MgCl$_2$ and 150 μg. of pancreatic deoxyribonuclease (once recrystallized, Worthington and Co.). This mixture is incubated at 37° for about 5 hours, until 85 to 90% of the ultraviolet-absorbing material is rendered acid-soluble.[6] A considerable amount of protein settles out during the digestion but is not removed at this time. The digest is dialyzed for 16 hours against 24 l. of Tris buffer (0.01 M at pH 7.5), then centrifuged for 5 minutes at 10,000 × g, and the supernatant fluid is collected (fraction III).

Step 5. Alumina C_γ Gel Adsorption and Elution. Enough aged alumina gel[7] (195 ml. containing 15 mg. dry weight per milliliter) is

---

[6] The optical density at 260 mμ is determined before and after precipitation with an equal volume of cold 1 N perchloric acid.

[7] R. Willstätter and H. Kraut, *Ber. deut. chem. Ges.* **56**, 1117 (1923).

added to 1500 ml. of fraction III in order to adsorb 90 to 95% of the enzyme. The mixture is stirred for 5 minutes and then centrifuged. The supernatant fluid is discarded, and the gel is washed with 400 ml. of potassium phosphate buffer (0.02 M at pH 7.2). The gel is then eluted twice with 400-ml. portions of potassium phosphate buffer (0.1 M at pH 7.4) in order to remove the enzyme, and the eluates are combined (fraction IV).

Step 6. Concentration of Alumina C_γ Eluate. To 800 ml. of fraction IV are added 16 ml. of 5 N acetic acid and then 480 g. of ammonium sulfate. After 10 minutes at 0°, the resulting precipitate is collected by centrifugation (30 minutes, 30,000 $\times$ g) and dissolved in 90 ml. of potassium phosphate buffer (0.02 M, at pH 7.2) (fraction V).

Step 7. Ammonium Sulfate Fractionation. To 90 ml. of fraction V are added 9 ml. of potassium phosphate buffer (1 M at pH 6.5) and 0.90 ml. of a 0.10 M solution of 2-mercaptoethanol. Then 24.7 g. of ammonium sulfate are added, and after 10 minutes at 0° the precipitate is removed by centrifugation at 12,000 $\times$ g for 10 minutes. To the supernatant fluid an additional 9.6 g. of ammonium sulfate are added, and after 10 minutes at 0° the precipitate which forms is collected by centrifugation at 12,000 $\times$ g for 10 minutes. This precipitate is dissolved in 9 ml. of potassium phosphate buffer (0.02 M at pH 7.2) (fraction VI).

Step 8. Diethylaminoethylcellulose Fractionation. A column (11 $\times$ 1 cm.) is prepared from diethylaminoethylcellulose[8] which has previously been equilibrated with 0.02 M K_2HPO_4. Next 1.2 ml. of fraction VI are diluted to 8.0 ml. with 0.02 M K_2HPO_4 and passed through the column at a rate of 12 ml./hr. The column is washed with 3.0 ml. of the same buffer and then eluted (flow rate of 9 ml./hr.) with pH 6.5 potassium phosphate buffers as follows: 8 ml. of 0.05 M, 10 ml. of 0.10 M, 3 ml. of 0.20 M, and finally 4 ml. of 0.20 M. Approximately 60% of the enzyme applied to the adsorbent is obtained in the last elution with 0.20 M buffer (fraction VII).

Samples of fraction VII have been further purified by another treatment with diethylaminoethylcellulose or by treatment with a phosphocellulose adsorbent.[8] Specific activities of 250 to 350 have thus been obtained.

With the exception of the alumina gel eluate (fraction IV) which in some instances lost activity on storage, each of the enzyme fractions has been stored for at least several weeks at −12° without any significant loss of activity. Some preparations of ammonium sulfate fraction (fraction VI) have been stored for as long as 3 years without loss in activity.

[8] E. A. Peterson and H. A. Sober, *J. Am. Chem. Soc.* **78,** 751 (1951).

Properties

Specificity. The purified enzyme requires, for maximal incorporation of deoxyribonucleotide into DNA, polymerized DNA primer, Mg^{++}, and the deoxynucleoside triphosphates of thymine, cytosine, guanine, and adenine. Thymidine diphosphate is inert in replacing the triphosphate, and the diphosphates of deoxycytidine and deoxyguanosine are inactive in place of the respective triphosphates. Adenosine triphosphate fails to substitute for deoxyadenosine triphosphate, and when labeled with C^{14} it is not incorporated to a detectable extent. Cytidine triphosphate labeled with P^{32} (cytidine $P^{32}PP$) is not incorporated.[9]

The triphosphates of several purine and pyrimidine analogs (uracil, hypoxanthine, 5-bromouracil, 5-bromocytosine, and 5-methylcytosine) can substitute specifically for the nucleoside triphosphate of the naturally occurring base it closely resembles with respect to the hydrogen-bonding properties required in the DNA structure as proposed by Watson and Crick.[10, 11]

Inhibitors. High concentrations of salt interfere with the assay; for example, NaCl at a final concentration of $0.2\ M$ produces a 97% inhibition. Fluoride at a concentration of $0.05\ M$ produces a 90% inhibition.[1]

pH Optimum. Activity is maximal at about pH 8.7; at pH 6.5, 8.0, and 10.0, the respective rates are 15%, 70%, and 15% of that observed at pH 8.7.[1]

Distribution. Enzymatic incorporation of deoxynucleoside triphosphates into a DNA fraction under conditions used in assaying the *E. coli* polymerase has been observed with sonic extracts of other bacterial species (*Hemophilus influenzae* and *Aerobacter aerogenes*).[12]

This activity has also been observed with extracts of several types of animal cell including HeLa cell cultures, lymph glands, leukemic blood cells,[13] regenerating rat liver,[14, 15] and Erlich ascites tumor cells.[16] An enzyme similar in properties to the *E. coli* polymerase has been partially purified from extracts of calf thymus tissue.[17]

[9] M. J. Bessman, I. R. Lehman, E. S. Simms, and A. Kornberg, *J. Biol. Chem.* **233**, 171 (1958).

[10] J. D. Watson and F. H. C. Crick, *Cold Spring Harbor Symposia Quant. Biol.* **18**, 123 (1953).

[11] M. J. Bessman, I. R. Lehman, J. Adler, S. B. Zimmerman, E. S. Simms, and A. Kornberg, *Proc. Natl. Acad. Sci. U.S.* **44**, 633 (1958).

[12] A. Kornberg, *Harvey Lectures,* **Ser. L** 111, 83 (1959).

[13] C. Harford and A. Kornberg, *Federation Proc.* **17**, 515 (1958).

[14] F. J. Bollum and V. R. Potter, *J. Am. Chem. Soc.* **79**, 3603 (1957).

[15] R. Mantsavinos and E. S. Canellakis, *J. Biol. Chem.* **234**, 628 (1959).

[16] R. M. S. Smellie, E. D. Gray, H. M. Keir, J. Richards, D. Bell, and J. N. Davidson, *Biochim. et Biophys. Acta* **37**, 243 (1960).

[17] F. J. Bollum, *J. Biol. Chem.* **235**, 2399 (1960).

[6] Nucleases of *E. coli*

I. A DNA-Specific Phosphodiesterase

By I. R. LEHMAN

Assay Method

Principle. Assay of the *E. coli* phosphodiesterase measures the conversion of P^{32}-labeled, heat-denatured DNA to acid-soluble products.[1]

Reagents

P^{32}-labeled *E. coli* DNA, 1 micromole of P per milliliter, 2 to 3 μc./micromole, isolated as described by Lehman,[1] heated at 100° for 10 minutes, then rapidly cooled. (See footnote 2 for method of making P^{32}-DNA.)

---

[1] I. R. Lehman, *J. Biol. Chem.* **235**, 1479 (1960).

[2] P^{32}-Labeled DNA was isolated from *E. coli* grown to a limit on P^{32}-orthophosphate. Glycerol–lactate medium (200 ml.), containing 0.6 micromole of orthophosphate per milliliter (specific activity, 50 μc./micromole), was inoculated with 0.2 ml. of a 7-hour nutrient broth culture of *E. coli*, strain B. After 18 hours of growth, the cells were harvested and washed twice with 0.9% KCl. The packed cells (0.7 g.) were suspended in 4.5 ml. of 0.14 M NaCl containing 0.01 M sodium citrate, and then 320 mg. of recrystallized sodium dodecyl sulfate were added. The suspension was stirred for 1 hour at room temperature during which time it became clear and extremely viscous. Two volumes of 95% ethanol were added, and the heavy fibrous precipitate which formed was transferred to a polyethylene tube and homogenized for several minutes by means of a glass pestle, with 5 ml. of 1.4 M NaCl–0.01 M sodium citrate. The suspension was centrifuged at 12,000 × g for 10 minutes, and the slightly opalescent, viscous supernatant fluid was drawn off. This extraction process was repeated twice on the residual precipitate, and the supernatant fluids were combined. On addition of 2 vol. of cold 95% ethanol to the supernatant fluids, a fibrous precipitate was formed which was dissolved in 5 ml. of 0.14 M NaCl–0.01 M sodium citrate. Sodium dodecyl sulfate (22.5 mg.) was added, and the solution was again stirred at room temperature for 1 hour. Solid NaCl was added to a final concentration of 1 M. After the resulting precipitate was removed by centrifugation for 20 minutes at 12,000 × g, the supernatant fluid was again treated with 2 vol. of ethanol. The precipitated DNA was dissolved in 3 ml. of 0.14 M NaCl–0.01 M sodium citrate. Pancreatic RNase (40 μg.) was added, and the solution was incubated at room temperature for 15 minutes. The solution was mixed with 1.5 ml. of a Norit suspension (20% packed volume) and stirred at 0° for 5 minutes; the Norit was removed by centrifugation. This process was repeated, and the Norit was washed with 2 ml. of 0.14 M NaCl–0.01 M sodium citrate. The supernatant solution obtained after Norit treatment and the wash were combined and treated with 2 vol. of ethanol. The DNA fibers which formed were dissolved in 4 ml. of 0.02 M NaCl and the solution was centrifuged for 30 minutes

Glycine buffer, pH 9.2, 1 M.

$MgCl_2$, 0.1 M.

2-Mercaptoethanol, 0.01 M.

"Carrier" DNA. Calf thymus DNA, 2.5 mg./ml., isolated by the method of Kay *et al.*[3]

Perchloric acid, 0.5 N.

KOH, 1 N.

Enzyme. Dilute the enzyme in a solution composed of 0.05 M Tris buffer, pH 8.0, 0.25 M ammonium sulfate, and crystalline bovine plasma albumin (Armour Laboratories), 1 mg./ml.

Procedure. To tubes containing 0.02 ml. each of glycine buffer and $MgCl_2$, 0.03 ml. of 2-mercaptoethanol, and 0.05 ml. of heat-denatured DNA, are added 0.05 to 0.25 unit of enzyme and sufficient water to give a final volume of 0.3 ml. The mixture is incubated at 37° for 30 minutes; then 0.2 ml. of "carrier" DNA and 0.5 ml. of cold perchloric acid are added. After 5 minutes at 0°, the precipitate is removed by centrifugation at 10,000 × g for 3 minutes, and 0.2 ml. of the supernatant fluid is pipetted into a planchet. After addition of a drop of KOH, the solution is taken to dryness, and the radioactivity is determined.

The supernatant fluids obtained from control incubations (enzyme omitted) usually contain 0.05 to 0.1% of the added radioactivity.

Definition of Unit and Specific Activity. A unit of enzyme is defined as the amount causing the production of 10 millimicromoles of acid-soluble P^{32} in 30 minutes. The specific activity is expressed as units per milligram of protein. Protein is determined by the method of Lowry *et al.*[4]

Application of Assay Method to Crude Tissue Preparations. In crude extracts of *E. coli*, the radioactivity made acid-soluble is proportional to the enzyme concentrations at levels of 0.05 to 0.25 unit. Thus, with the addition of 0.005, 0.01, 0.02, and 0.04 ml. of a 1:50 dilution of crude *E. coli* extract, 360, 367, 379, and 334 units of enzyme per milliliter of extract, respectively, are obtained.

---

at 105,000 × g. Two volumes of cold ethanol were added to the supernatant fluid, and the resulting DNA fibers were dissolved in 4 ml. of 0.02 M NaCl. The DNA prepared in this way had a molar extinction coefficient at 260 mμ of 6.9 × 10³ based on deoxypentose, and a reduced viscosity of 52 (g./per 100 ml.)$^{-1}$. Protein contamination was of the order of 2% or less.

[3] E. R. M. Kay, N. S. Simmons, and A. L. Dounce, *J. Am. Chem. Soc.* **74,** 1724 (1952).

[4] O. H. Lowry, N. F. Rosebrough, A. L. Farr, and R. J. Randall, as cited in Vol. III [73].

The RNA-inhibited endonuclease is quantitatively inhibited in crude extracts of *E. coli* (see Vol. VI [7]) and does not interfere with assay of the phosphodiesterase under the conditions described.

Purification Procedure

Growth and harvest of bacteria and preparation of the extract are carried out as described in the purification of the DNA polymerase from *E. coli* (see Vol. VI [5].) All operations are carried out at 0° to 4°.

Protamine Precipitation and Elution. To 800 ml. of extract are added, first, 800 ml. of glyclyglycine buffer (0.05 M, pH 7.0) and then slowly, with stirring, 160 ml. of a 2% protamine sulfate solution (Eli Lilly). After 10 minutes, the suspension is centrifuged at 10,000 × g for 15 minutes. The gummy precipitate is transferred to a Waring blendor and homogenized with 1600 ml. of potassium phosphate buffer (0.07 M, pH 8.0) for 5 minutes at low speed. The suspension is centrifuged for 3 hours at 44,000 × g in a Spinco Model L centrifuge, and the supernatant fluid is collected.

Concentration of Protamine Eluate. To 1600 ml. of protamine eluate are added, with stirring, 904 g. of solid ammonium sulfate. After 10 minutes at 0°, the precipitate is collected by centrifugation for 30 minutes at 10,000 × g and dissolved in 300 ml. of potassium phosphate buffer (0.05 M, pH 6.8).

Ammonium Sulfate Fractionation. To 300 ml. of concentrated protamine eluate are added, with stirring, 36 g. of ammonium sulfate. After 10 minutes at 0°, the resulting precipitate is collected by centrifugation for 10 minutes at 12,000 × g and dissolved in 36 ml. of potassium phosphate buffer (0.02 M, pH 7.5).

DEAE-Cellulose Chromatography. A column of DEAE-cellulose[5] (10 × 2.2 cm.) is prepared and equilibrated with potassium phosphate buffer (0.02 M, pH 7.5). Ammonium sulfate fraction (40 ml.), which has been previously dialyzed against potassium phophate buffer (0.02 M, pH 7.5), is added to the column, and the adsorbent is washed with 10 ml. of the same buffer. A linear gradient is applied with 0.1 M and 0.5 M potassium phosphate at pH 6.5 as limiting concentrations; 150 ml. of each buffer are used (flow rate 70 ml./hr.). Fractions are collected at 5-minute intervals. Over 90% of the activity is eluted in a discrete peak between 3.6 and 5.0 resin-bed volumes of effluent. The peak fractions which contain enzyme of specific activity of 1200 or greater (approximately 50% of the total) are pooled and then concentrated in the following way. To 36 ml. are added 10 g. of ammonium sulfate. After 10

[5] DEAE-cellulose—diethylaminoethylcellulose; E. A. Peterson and H. A. Sober, *J. Am. Chem. Soc.* **78**, 751 (1951).

minutes at 0°, the precipitate is collected by centrifugation for 20 minutes at $12,000 \times g$ and dissolved in 6.0 ml. of Tris buffer (0.05 M, pH 7.5). Residual orthophosphate is removed by dialysis of the preparation against the Tris buffer.

The concentrated DEAE fraction showed no significant loss of activity when stored at —20° for 2 months. A loss of about 60% in activity was observed after storage at —20° for 6 months.

A summary of the purification procedure is given in the table.

PURIFICATION OF *E. coli* PHOSPHODIESTERASE

Fraction	Units/ml.	Total units	Protein, mg./ml.	Specific activity, units/mg. protein
Extract	208	166,400	19.7	10.6
Protamine eluate	70	112,000	1.6	43.8
Concentrated protamine eluate	354	100,000	7.5	47.3
Ammonium sulfate	2000	68,000	12.6	159
DEAE-cellulose	1000	36,000	0.7	1430
Concentrated DEAE-cellulose	5850	36,000	4.0	1460

Properties

Specificity. The *E. coli* phosphodiesterase hydrolyzes native, double-stranded DNA at less than 1% the rate at which it attacks heat-denatured DNA. Hydrolysis proceeds in an exonucleolytic or stepwise manner beginning at the 3'-hydroxyl end of the chain. The enzyme is, however, unable to cleave free dinucleotides or the 5'-terminal dinucleotide portion of a polydeoxynucleotide chain.

In contrast to the phosphodiesterase from venom,[6] the *E. coli* phosphodiesterase is able to degrade bacteriophage DNA's bearing glucosylated hydroxymethylcytosine quantitatively to their constituent mononucleotides.

The enzyme is highly specific for polydeoxynucleotides and will not attack polyribonucleotides. The low level of ribonuclease activity in the DEAE fraction reported previously[1] can be attributed to a small amount of DNA contaminating the RNA used as substrate.

The purified enzyme is free of any detectable acid or alkaline nucleotidase activity.

Magnesium Requirement. Under the conditions of the assay, no significant difference in activity can be detected at Mg++ concentrations ranging from 1.5×10^{-3} to $3 \times 10^{-2} M$. In the absence of added Mg++,

[6] R. L. Sinsheimer and J. F. Koerner, *J. Biol. Chem.* **198**, 293 (1952).

one-third to one-fifth maximal activity is observed. This residual activity is completely eliminated by the addition to the reduction mixture of Versene $(0.02\,M)$, which suggests that the enzyme has an absolute requirement for some cation and that the residual activity is the result of traces of the metal ion present in the enzyme preparation. Mg^{++} cannot be effectively replaced by a number of other divalent cations. Thus Mn^{++} and Ca^{++} each, at concentrations of 3×10^{-4} or $6 \times 10^{-3}\,M$, completely abolish enzymatic activity. Zn^{++} gives 64% of maximal activity at a concentration of $3 \times 10^{-4}\,M$, and less than 10% at $6 \times 10^{-3}\,M$.

pH Optimum. The pH optimum for the purified enzyme is between pH 9.2 and 9.8 (0.07 M glycine buffer). At pH 10.7 (0.07 M glycine) about 30% of optimal activity is observed, and at pH 7.5 (0.07 M Tris buffer), about 20% of optimal activity. Enzymatic activity is enhanced by phosphate ion. Thus at pH 8.5 (0.07 M glycine), where the activity is 60% of optimal, the addition of 0.07 M K_2HPO_4 (pH 8.5) increases the activity to the optimal level.

[7] Nucleases of *E. coli*

II. An RNA-Inhibitable Endonuclease

By I. R. LEHMAN

Assay Method

Principle. Assay of the *E. coli* endonuclease measures the conversion of P^{32}-labeled, native DNA to acid-soluble products.[1] An alternative assay for the more-purified enzyme fractions (CM-cellulose[2] or Amberlite XE-64) measures the increase in optical density at 260 mμ accompanying hydrolysis of the DNA.

Reagents

P^{32}-labeled *E. coli* DNA, 1 micromole of P per milliliter, 2 to 3 μc./micromole.[3]
Tris buffer, pH 7.5, 1 M.
$MgCl_2$, 0.1 M.
Crystalline pancreatic RNase (Armour Laboratories), 5 mg./ml.

[1] I. R. Lehman, G. G. Roussos, and E. A. Pratt, *J. Biol. Chem.* **237**, 819 (1962).
[2] CM-cellulose-carboxymethylcellulose. DEAE-cellulose—diethylaminoethylcellulose. E. A. Peterson and H. A. Sober, *J. Am. Chem. Soc.* **78**, 751 (1951).
[3] I. R. Lehman, *J. Biol. Chem.* **235**, 1479 (1960); see Vol. VI [6].

Crystalline cytochrome c (Sigma Chemical Company), 1 mg./ml.
"Carrier" DNA. Calf thymus DNA, 2.5 mg./ml., isolated by the
    method of Kay *et al.*[4]
Perchloric acid, 0.5 *N*.
KOH, 1 *N*.
Enzyme. Dilute the enzyme in a solution composed of 0.05 *M* Tris
    buffer, pH 7.5, 0.25 *M* ammonium sulfate, and crystalline bovine
    plasma albumin (Armour Laboratories), 1 mg./ml.

Procedure. To tubes containing 0.02 ml. each of Tris buffer, $MgCl_2$,
and DNA is added 0.01 to 0.07 unit of enzyme. In assays of the crude
extract 0.01 ml. of RNase is added to destroy the inhibitory RNA
present in this fraction (see below). All subsequent fractions are free of
this inhibitor; however, 0.01 ml. of RNase or 0.04 ml. of cytochrome c is
added to stabilize the enzyme during incubation. The reaction mixture
(brought up to 0.3 ml. with water) is incubated for 30 minutes at 37°;
then 0.2 ml. of "carrier" DNA and 0.5 ml. of cold perchloric acid are
added. After 5 minutes at 0° the resulting precipitate is removed by
centrifugation, and 0.2 ml. of the supernatant fluid is pipetted into a
planchet. A drop of KOH is added directly to the aliquot on the planchet;
the solution is taken to dryness, and the radioactivity is determined.

The supernatant fluids obtained from control incubations, with enzyme omitted, usually contain 0.05 to 0.1% of the added radioactivity.

The spectrophotometric assay is performed with a Beckman Model
DU or a Zeiss Model PMQ II spectrophotometer, equipped with a
thermospacer through which water at 37° is circulated from a constant-
temperature bath. The components of the reaction mixture are the same
as those used in the isotopic assay with the exception that 0.03 ml. of
DNA (unlabeled) and 0.1 to 0.5 unit of enzyme are added. A blank
cuvette of high absorbancy is used to permit direct measurements of
optical density at the relatively high concentration of DNA used. Readings of optical density at 260 mμ are taken at 1-minute intervals. The
increase in optical density is proportional to the amount of enzyme
added. Thus with 0.01, 0.02, and 0.04 ml. of a 1:20 dilution of the
CM-cellulose fraction, the rates of increase in optical density are 0.0034,
0.0066, and 0.0138 per minute, respectively.

Definition of Isotope Assay Unit and Specific Activity. A unit of
enzyme is defined as the amount causing the production of 0.10 micromole of acid-soluble P^{32} in 30 minutes. The specific activity is expressed

[4] E. R. M. Kay, N. S. Simmons, and A. L. Dounce, *J. Am. Chem. Soc.* **74**, 1724
(1952).

as units per milligram of protein. Protein is determined by the method of Lowry et al.[5]

Application of the Isotope Assay Method to Crude Tissue Preparations. In crude extracts of *E. coli* the radioactivity made acid-soluble is proportional to the enzyme concentration at levels of 0.005 to 0.07 unit of enzyme. Thus, with the addition of 0.005, 0.01, 0.02, and 0.04 ml. of a 1:200 dilution of crude *E. coli* extract, values of 208, 229, 220, and 223 units of enzyme per milliliter of extract, respectively, are obtained.

Purification Procedure

Unless otherwise indicated, all operations are carried out at 0° to 4°. All centrifugations are at $15,000 \times g$ for 10 minutes.

Preparation of Extracts. E. coli strain B is grown as described in Vol. VI [5]. The cells (450 g. wet weight) are mixed with 300 ml. of potassium phosphate buffer, $0.15 M$, pH 7.0, in a large Waring blendor equipped with a cooling jacket and connected to a Variac. Stirring is begun at approximately one-third of maximal speed. To the suspension are gradually added 1350 g. of acid-washed glass beads (Superbrite, average diameter 200 μ, obtained from the Minnesota Mining and Manufacturing Company). When the mixture appears homogeneous, an additional 1500 ml. of buffer are added, and the homogenization is continued for 20 minutes at maximal speed. During this period, the temperature of the mixture should not rise above 12°. After the homogenization approximately 10 minutes are required for the beads to settle out. The broken cell suspension is then centrifuged, and the supernatant fluid (approximately 2000 ml.) is collected (fraction I).

Protamine Precipitation and Elution. To 800 ml. of extract are added, with stirring, 400 ml. of a 1% protamine sulfate solution. After standing for 10 minutes, the suspension is centrifuged. The precipitate is transferred to a Waring blendor regulated by means of a Variac and homogenized with 800 ml. of potassium phosphate buffer, $0.2 M$, pH 7.4, for 5 minutes at one-fourth maximal speed. The suspension is centrifuged, and the resulting precipitate is homogenized with 800 ml. of potassium phosphate buffer, $0.5 M$, pH 7.4, for 10 minutes, then centrifuged as before. To the supernatant fluid are added an equal volume of distilled water and 5 ml. of $0.5 M$ $MgCl_2$. The diluted solution is incubated at 37° for 2 hours. A heavy precipitate of inactive protein which forms during incubation is removed by centrifugation (fraction II).

Ammonium Sulfate Fractionation. To 1600 ml. of protamine eluate are added, with stirring, 565 g. of solid ammonium sulfate. The suspen-

[5] O. H. Lowry, N. J. Rosebrough, A. L. Farr, and R. J. Randall, as cited in Vol. III [73].

sion is allowed to stand for 10 minutes after the ammonium sulfate has dissolved and then is centrifuged. To the supernatant solution are added, with stirring, an additional 512 g. of ammonium sulfate. The suspension is allowed to stand and then is centrifuged as before. The resulting precipitate is dissolved in 80 ml. of potassium phosphate buffer, $0.02 M$, pH 7.4. This solution is dialyzed for 12 hours against 75 vol. of potassium phosphate buffer, $0.01 M$, pH 7.4. An inactive precipitate which forms on dialysis is removed by centrifugation (fraction III).

DEAE-cellulose Fractionation. A column of DEAE-cellulose[2] (16×3.3 cm.) is prepared and equilibrated with 2 l. of potassium phosphate buffer, $0.01 M$, pH 7.4. The dialyzed ammonium sulfate fraction is applied to the column at the rate of 80 ml./hr. The adsorbent is then washed with 130 ml. of potassium phosphate buffer, $0.01 M$, pH 7.4. Approximately 80% of the activity applied to the column is recovered in the wash (fraction IV).

CM-cellulose Chromatography. A column of CM-cellulose (16×2.2 cm.) is equilibrated with 1 l. of potassium phosphate buffer, $0.01 M$, pH 6.5. DEAE-cellulose fraction (126 ml.), to which 1.26 ml. of $1 M$ potassium dihydrogen phosphate has been added, is applied to the column at the rate of 40 ml./hr., and the adsorbent is washed with 25 ml. of potassium phosphate buffer, $0.01 M$, pH 6.5. A linear gradient of elution is applied with $0.04 M$ and $0.2 M$ potassium phosphate at pH 7.4 as limiting concentrations. The total volume of the gradient is 600 ml. The flow rate is 48 ml./hr., and fractions are collected at 10-minute intervals. Approximately 90% of the activity is eluted in a sharp peak between 4.4 and 6.4 resin-bed volumes of effluent. The peak fractions which contain enzyme of specific activity greater than 3500 (approximately 60% of the total) are pooled (fraction V). This fraction is purified 600-fold over the starting extract and contains 14% of the activity initially present. It can be concentrated by lyophilization. The pooled CM-cellulose fraction (64 ml.) is first dialyzed for 5 hours against two 4-l. changes of Tris buffer, $0.01 M$, pH 7.5, then lyophilized, and finally taken up in 4 ml. of distilled water. Approximately 80% of the activity in the CM-cellulose fraction is recovered after dialysis and lyophilization. Further purification of the enzyme can be achieved by chromatography on Amberlite XE-64 (IRC-50).[6] A column of Amberlite XE-64 (20.5×0.7 cm.) is washed with 1 l. of potassium phosphate buffer, $0.1 M$, pH 6.48. CM-cellulose fraction (34 ml.), which has been dialyzed for a period of 5 hours against the equilibrating buffer, is added at the rate of 10 ml./hr. and washed into the column with 5 ml. of this same buffer. A linear gradient from $0.3 M$ KCl to $1 M$ KCl is applied, both limiting

[6] C. H. W. Hirs, S. Moore, and W. H. Stein, *J. Biol. Chem.* **200**, 493 (1953).

solutions containing potassium phosphate buffer, $0.1\,M$, pH 7.4. The total gradient volume is 100 ml. The flow rate is maintained at 12 ml./hr., and 2-ml. fractions are collected. Approximately 80% of the activity initially applied to the column appears in a sharp peak midway through the gradient. This procedure removes most of the contaminating ribonuclease activity present in the CM-cellulose fraction. Such preparations are not significantly different in their physical properties or in their action on DNA from the CM-cellulose fraction.

The crude extract has been stored at $-20°$ for as long as 1 year with no loss in activity. Fractions II, III, and IV have shown no significant decrease in activity on storage at $-20°$ for 6 months. Fraction VI loses activity rapidly on storage at $-20°$; however, no loss in activity could be detected when this fraction was kept at $0°$ for a 4-month period. The lyophilized CM-cellulose fraction showed no decrease in activity when stored at $-20°$ for 2 months.

An apparent loss in activity of the CM-cellulose fraction has been observed on prolonged dialysis at $0°$ for 24 hours or more. This loss is probably due to passage of enzyme protein through the cellophane dialysis bag. After dialysis of the CM-cellulose fraction (1 ml. containing 50 μg. of protein) against 50 vol. of Tris buffer, $0.02\,M$, pH 7.5, containing 0.1 mg. of cytochrome c per milliliter, for a 72-hour period, approximately 5% of the activity initially present in the dialysis bag was detected in the surrounding fluid.

A summary of the purification procedure is given in the table.

PURIFICATION OF *E. coli* ENDONUCLEASE

Fraction	Step	Units/ml.	Total units, $\times 10^{-3}$	Protein, mg./ml.	Specific activity, units/mg. protein
I	Extract	177	141	17.4	10.2
II	Protamine eluate	42	67	0.30	140
III	Ammonium sulfate	444	42	1.70	260
IV	DEAE-cellulose	264	33	0.50	528
V	CM-cellulose	318	20	0.05	6360

Properties

Specificity. The *E. coli* endonuclease degrades native DNA at a sevenfold greater rate than thermally denatured DNA. It carries out an endonucleolytic attack on DNA; that is, it produces scissions at many points along the DNA chain, yielding a limit digest whose oligonucleo-

tides have an average chain length of approximately **7**; these are terminated by a 5'-phosphoryl group. The isolated oligonucleotides are neither
cleaved by high concentrations of the enzyme, nor can they, when added
in equal concentrations, inhibit the hydrolysis of DNA.

The enzyme is highly specific for DNA. A ribonuclease present in the
CM-cellulose fraction can be removed by chromatography on Amberlite
XE-64 (IRC-50) (see above). The purified enzyme (CM-cellulose frac-
tion) is free of any detectable acid or alkaline nucleotidase activity.

The enzyme exists in the crude extract bound to an inhibitory RNA
which is removed during purification. The purified enzyme can be inhibited by a variety of RNA's from *E. coli*, guinea pig liver, and tobacco
mosaic virus. Of the RNA's tested, the amino acid-acceptor RNA from
E. coli is the most active. The kinetics of the enzyme inhibited by the
amino acid-acceptor RNA obey the equations for a competitive inhibition, and the calculated K_i is of the order of $1 \times 10^{-8} M$ (RNA-
nucleotide).[7]

Requirement for Divalent Cations. The purified enzyme (CM-cellulose fraction) requires added magnesium ion for maximal activity. No
activity is detectable in its absence. Under the conditions of the assay,
maximal activity is observed at a Mg⁺⁺ concentration of $7 \times 10^{-3} M$. At
$1 \times 10^{-3} M$ and $3.4 \times 10^{-2} M$, 66% and 50%, respectively, of maximal
activity is observed. Magnesium ion can, to some extent, be replaced by
manganous ion. Thus, at $7 \times 10^{-3} M$ Mn⁺⁺, the activity is 57% of that
observed with an equal concentration of Mg⁺⁺. Ca⁺⁺ and Zn⁺⁺ are completely ineffective in replacing Mg⁺⁺. No activity is observed when these
ions are added at $3 \times 10^{-4} M$, $1.5 \times 10^{-3} M$, or $7 \times 10^{-3} M$.

pH Optimum. The pH optimum for the purified enzyme with Tris
buffer extends from 7.5 to 8.5.

[7] I. R. Lehman, G. G. Roussos, and E. A. Pratt, *J. Biol. Chem.* **237**, 829 (1962).

[8] Preparation of Lamb Brain Phosphodiesterase

By J. W. HEALY, D. STOLLAR, and L. LEVINE

Assay Method

Principle. The immunological assay method is based on the determination of denatured DNA serologically before and after treatment
with the phosphodiesterase. Antibodies produced in rabbits to T-even

coliphage DNA[1] or to purinoyl bovine serum albumin[2] and antibodies present in the sera of some lupus erythematosus patients[3] react *in vitro* more effectively with thermally denatured DNA than with native DNA. The serological reaction is sensitive to a decrease in the molecular weight of the thermally denatured DNA. Fragmentation of the DNA results in loss of serological activity.[4] Endonucleolytic cleavage of the DNA is more effective than exonucleolytic cleavage in decreasing the size of DNA as measured serologically. The immunological assay is more than one hundred times as sensitive as the measurement of nuclease activity by formation of PCA-soluble products. However, the latter enzymatic assay can be used.

Immunochemical Assay of Deoxyribonuclease Activity

Enzyme is incubated with 32 μg. of either heat-denatured or native DNA at 37°, in the presence of $5 \times 10^{-4} M$ $MgSO_4$, 0.01 M 2-mercapto-ethanol, and 0.01 M phosphate buffer, pH 7.4, in a total volume of 1.0 ml. Samples of this reaction mixture, withdrawn immediately after addition of the enzyme (zero time) and after various periods of incubation, are diluted 400-fold in ice-chilled Veronal buffer to terminate the reaction and to bring the DNA to a concentration suitable for the immunologic assay. If native DNA is the substrate, it is boiled after this dilution. Varying concentrations of the diluted samples are reacted with antibodies to denatured DNA, and quantitative C' fixation curves are obtained.[5] From these curves, the maximal amount of C' fixation is found for each sample, and the loss in maximal C' fixation is measured as a function of deoxyribonuclease activity. One unit of enzyme is arbitrarily defined as that amount which causes a 50% loss of maximal C' fixation in 20 minutes under the conditions described above.

Purification Procedure

All steps in the purification are carried out at 2° to 4°, with the exception of acetone fractionation, which is done at —5°.

Four frozen whole lamb brains are thawed and homogenized in 300 ml. of 0.15 M NaCl for 10 minutes in a Waring blendor. The homogenized brain (550 ml.) is centrifuged in a Spinco Model L centrifuge at 105,000

[1] L. Levine, W. T. Murakami, H. Van Vunakis, and L. Grossman, *Proc. Natl. Acad. Sci. U.S.* **46**, 1038 (1960).
[2] V. Butler, S. Beiser, B. Erlanger, S. Tannenbaum, S. Cohen, and A. Bendich, *Proc. Natl. Acad. Sci. U.S.* **48**, 1597 (1962).
[3] D. Stollar and L. Levine, *J. Immunol.* **87**, 477 (1961).
[4] W. T. Murakami, H. Van Vunakis, L. Grossman, and L. Levine, *Virology* **14**, 190 (1961).
[5] E. Wasserman and L. Levine, *J. Immunol.* **87**, 290 (1960).

$\times g$ for 180 minutes. The supernatant fluid (370 ml.) is saved, and acetone is added to it to give a final acetone concentration of 20%. After 10 minutes, the precipitate which forms is removed, and acetone is added to the supernatant fluid to give a final acetone concentration of 40%. The precipitate which forms is saved after centrifugation and is suspended in 100 ml. of 0.001 M phosphate buffer, pH 7.4, with gentle stirring. This suspension is centrifuged, and the supernatant fluid is saved and dialyzed for 18 hours against 0.01 M 2-mercaptoethanol in 0.001 M phosphate buffer, pH 7.4. To concentrate the enzyme, an equal volume of neutralized saturated $(NH_4)_2SO_4$ is added to the dialyzate; the resulting precipitate is saved and dissolved in 20 ml. of 0.15 M Tris buffer, pH 8.0. The dialyzate is added to a 40-ml. slurry of DEAE which has been equilibrated with Tris buffer (0.01 M, pH 8.0, and 2-mercaptoethanol 0.01 M). The mixture is stirred with a glass rod for 30 minutes; the supernatant fluid is collected after centrifugation.

A summary of the purification procedure is given in the table.

PURIFICATION OF LAMB BRAIN PHOSPHODIESTERASE

Enzyme fraction	Total volume, ml.	Activity, units/ml.	Protein, mg./ml.	Specific activity, units/mg. protein	Total units
Supernatant of homogenate	370	540	11	49	199,800
Acetone, 20–40% fraction	91	1250	6.8	191	113,750
Ammonium sulfate, 0–50% fraction	20	2750	12.1	227	55,000
DEAE-cellulose eluate	34	440	0.05	8780	14,920

Properties

The enzyme attacks thermally denatured DNA more effectively than native DNA. The products of an extensively digested calf thymus DNA preparation were hexonucleotides and larger.

Inhibitors. The enzyme is inhibited by p-chloromercuribenzoate and disodium ethylenediaminetetraacetate.

Effect of pH. The enzyme activity varies little between pH 7.3 and 9.0 but is reduced by half at pH 6.0.

Divalent Cation Requirement. The enzymatic activity required Mg^{++}, optimally $5 \times 10^{-4} M$ in concentration.

Stability. The purified enzyme is labile at pH 5.0 and is destroyed during lyophilization or freezing. It is stable at 2° to 4° in the presence of 0.01 M mercaptoethanol.

NaCl Concentration. The enzyme activity is optimal in 0.03 M NaCl and is reduced by half at 0.07 M and 0.005 M.

[9] The Enzymes of Purine Nucleotide Synthesis *de Novo*

By Joel G. Flaks and Lewis N. Lukens

Since the previous article on this topic by Goldthwait and Greenberg,[1] a considerable amount of information has been added to the knowledge concerning purine synthesis *de novo*. The present evidence suggests a reaction sequence consisting of nine intermediates between PRPP and IMP,[2] with the reactions catalyzed by ten distinct enzymes.[3] A summary of the reactions and intermediates is presented in Fig. 1. At the present writing, assay procedures have been developed for each of these enzymes, and the majority of them have been extensively purified. These procedures are detailed in the present article. Reference should be made to a companion article, Vol. VI [96], which details the procedures for the preparation and characterization of the intermediates, many of which are employed in the assay procedures described here.[4]

I. General Procedures

A. Preparation of Ethanol Fractions of Avian Liver Extracts

Most of the enzymatic work reported to date on purine biosynthesis has employed as source material avian livers, in which the requisite enzymes occur in the soluble cytoplasmic fraction of the cell. Avian livers have the highest known concentration of purine synthetic enzymes of any animal tissue studied, since the pathway is utilized by birds as a means of nitrogen elimination. As an initial step, crude preparations of considerable stability and high activity can be obtained from these livers

---

[1] D. A. Goldthwait and G. R. Greenberg, Vol. II [78].

[2] The following abbreviations have been used: AICAR, 5-amino-4-imidazolecarboxamide ribotide; AIR, 5-aminoimidazole ribotide; CAIR, 5-amino-4-imidazolecarboxylic acid ribotide; EDTA, ethylenediaminetetraacetate; formyl-AICAR, 5-formamido-4-imidazolecarboxamide ribotide; FGAM, formylglycinamidine ribotide; FGAR, formylglycinamide ribotide; GAR, glycinamide ribotide; PRA, 5-phosphoribosylamine; succino-AICAR, 5-amino-4-imidazole-*N*-succinocarboxamide ribotide; THFA, 5,6,7,8-tetrahydrofolic acid. For the systematic naming of these intermediates, the reader is referred to a recent review.[3]

[3] J. M. Buchanan and S. C. Hartman, *Advances in Enzymol.* **21**, 199 (1959).

[4] The authors wish to express their indebtedness to Drs. J. M. Buchanan, T. C. French, S. C. Hartman, R. L. Herrmann, and R. W. Miller for the availability of a good deal of information prior to its publication. They have also provided the details of recent modifications of a number of published procedures.

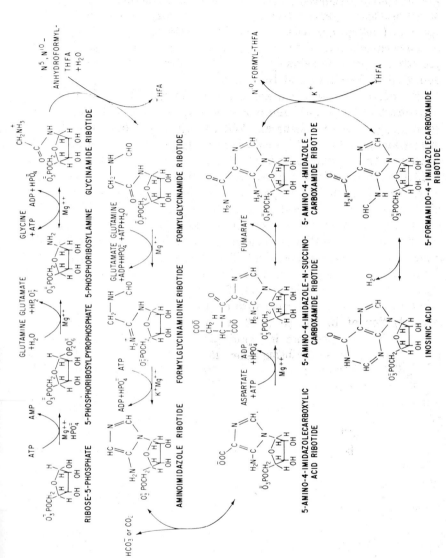

Fig. 1. Summary scheme of inosinic acid biosynthetic pathway.

by fractionation of particulate-free extracts with ethanol, between 15 and 30 vol. %, followed by lyophilization. Such fractions have been used as the starting material for a number of the enzyme purification procedures and assay methods detailed in this section, as well as for the preparation of some of the intermediates detailed in Vol. VI [96]. The procedure described below is modified somewhat from that originally described by Williams and Buchanan.[5]

Procedure. Pigeon or chicken livers are obtained as soon as possible after death of the animal and placed on cracked ice. The livers are freed from gross connective tissue and minced with a pair of scissors directly into a known volume of ice-cold homogenizing medium in a graduated cylinder. The medium consists of 0.035 M Na phosphate buffer, pH 7.4, 0.13 M KCl, and 0.01 M MgCl$_2$ (the KHCO$_3$ used in the original procedure may be omitted). One part of tissue is homogenized with 1.5 parts of the medium in an all-glass or glass–Teflon tissue homogenizer. The homogenate is first centrifuged for 10 minutes at 2000 $\times$ g, and the supernatant fluid is then centrifuged at 100,000 $\times$ g for 30 minutes to remove any remaining particles.[6] The temperature of the particle-free fluid is lowered to $-5°$ with the aid of a dry ice–acetone bath and continuous stirring (magnetic stirrer). Then 90% ethanol ($-15°$) is slowly added dropwise to the solution with continued stirring, to a final ethanol concentration of 15 vol. % (0.20 ml. of 90% ethanol per milliliter of extract). The temperature is gradually lowered from $-5°$ to $-10°$ during the addition, care being taken to avoid the formation of ice crystals. After 5 minutes of stirring at $-10°$, the mixture is centrifuged at $-10°$ for 10 minutes at 25,000 $\times$ g. The supernatant fluid is removed and cooled to just above the freezing point ($-12°$ to $-15°$), and 90% ethanol is added, as above, to a final ethanol concentration of 30 vol. % (0.30 ml. of 90% ethanol per milliliter of extract). The temperature should be held between $-15°$ and $-20°$ during this step. After 10 minutes of stirring, the mixture is centrifuged at $-15°$ to $-20°$ for 15 minutes at 25,000 $\times$ g, and the supernatant fluid is discarded. The precipitate is dissolved in a minimum volume of cold distilled water, shell-frozen, and lyophilized, with the lyophilizer maintained at 4° or lower. Lyophilized powders thus obtained retain their activity for a number of months when stored at $-15°$ with a suitable desiccant.

[5] W. J. Williams and J. M. Buchanan, *J. Biol. Chem.* **203**, 583 (1953).
[6] Spinco Model L ultracentrifuge. For larger-scale preparations, particularly with chicken livers, the homogenization was carried out for 1 minute in a Waring blendor, and the high-speed centrifugation for 1 hour in the Spinco No. 30 rotor at 30,000 r.p.m. Use of the Waring blendor somewhat reduces the activity of these preparations when compared to the other homogenization procedure.

B. Assay for Diazotizable Amines

Half of the intermediates in the synthetic pathway are derivatives of aminoimidazole. Since the amino groups of this class of compound may be diazotized, the aminoimidazoles may be estimated by the very sensitive colorimetric procedure devised by Bratton and Marshall.[7] In other cases the same assay procedure may be employed by coupling the reaction under study to those leading to the formation of a diazotizable amine. A general procedure for this assay is described here,[8] with 5-amino-4-imidazolecarboxamide as a standard. For the assay of the other diazotizable amines slight modifications of this procedure are necessitated and are indicated in the text where applicable.

Reagents

H_2SO_4, 2 N.
5-Amino-4-imidazolecarboxamide standard,[9] 10 μg./ml. in 0.01 N HCl.
Na nitrite, 0.1%.
Ammonium sulfamate, 0.5%.
N^1-naphthylethylenediamine dihydrochloride, 0.1% in 0.01 N HCl.

The latter three reagents are hereafter referred to as the diazotizable amine assay reagents. They are stable for several months if stored in the refrigerator, preferably in amber bottles.

Procedure. To a sample containing 0.3 to 5.0 μg. of 5-amino-4-imidazolecarboxamide in a final volume of 0.75 ml., 0.1 ml. of 2 N H_2SO_4 is added, followed by 0.05 ml. of 0.1% Na nitrite. The contents are mixed and allowed to sit at room temperature for 5 minutes. Then 0.05 ml. of 0.5% ammonium sulfamate is added with mixing, and after 3 minutes 0.05 ml. of 0.1% N^1-naphthylethylenediamine dihydrochloride is added with mixing. After 30 minutes the absorbancy at 540 mμ is determined against a reagent blank. A calibration curve should be constructed, since the absorbancy deviates from linearity above a level of 2 μg. of the 5-amino-4-imidazolecarboxamide. The molecular extinction coefficient for the colored product is 26,400. If necessary, the purity of the sample of 5-amino-4-imidazolecarboxamide may be estimated from its ultraviolet absorption spectra, with an ϵ_{267} of 12,700 at pH 7.0 and an ϵ_{266} of 11,200 at pH 1.0.

[7] A. C. Bratton and E. K. Marshall, Jr., *J. Biol. Chem.* **128**, 537 (1939).
[8] J. M. Ravel, R. E. Eakin, and W. Shive, *J. Biol. Chem.* **172**, 67 (1948).
[9] E. Shaw and D. W. Woolley, *J. Biol. Chem.* **181**, 89 (1949).

C. Purine Biosynthetic Enzymes from Microorganisms

All the enzymes described in this section have been found in extracts of *Neurospora crassa,* and many have been found in extracts of *Salmonella typhimurium, Escherichia coli, Aerobacter aerogenes,* and *Saccharomyces ceriviseae.* In many cases, however, a preliminary fractionation step may be necessary to remove interfering substances before the enzyme activity can be demonstrated. Thus, in assaying for the two initial reactions leading to the synthesis of GAR (i.e., PRPP amidotransferase and GAR kinosynthase), negligible activity was found in an extract of *N. crassa* (0.05 *M* Tris–chloride buffer, pH 8.0), whereas after fractionation of the same extract with ammonium sulfate (0.4 to 0.5 saturation) and dialysis, a highly active preparation was obtained, the activity of which was proportional to protein concentration.

With small-scale preparations the general methods of cellular disruption involving sonic vibration, grinding with alumina, or disruption in a Hughes press of cells frozen in liquid nitrogen or dry ice are all suitable. The disrupted cells are then extracted with the media used in Section I.A for homogenization of avian liver, and the extract is subjected to ultracentrifugation at 100,000 × *g* for 1 to 3 hours to remove the ribosomal fraction. The extract obtained may then be assayed or subjected to further fractionation with ammonium sulfate, ethanol, or streptomycin.

The preparation of acetone powders from bacterial cells is most convenient as an initial step in the purification of enzymes on a large scale. From 400 to 600 g. of washed cells are frozen in liquid nitrogen, and the frozen mass is pulverized in a chilled, gallon-size Waring blendor for 2 minutes. Two liters of cold acetone (—15°) are then added through an inlet in the cover during blending, and blending is continued for 30 seconds after the addition of acetone is completed. The suspension is filtered on a large Büchner funnel at 3°, and the insoluble residue is reblended with 2 l. of cold acetone for 1.5 minutes and collected by filtration as above. The filter cake is washed on the funnel with 2 l. of cold acetone and a liter of cold anhydrous ether and then sucked dry on the funnel. The filtered solids are partially dried by being spread out at room temperature, and the remaining ether is removed *in vacuo* at 3°. The powder is finally dried over P_2O_5 and stored at —20°.

II. 5-Phosphoribosylpyrophosphate Amidotransferase

PRPP + L-glutamine + H_2O → PRA + L-glutamic acid + PP

Assay Method

Principle. The PP formed in the reaction is isolated as the Mn^{++} salt according to the procedure of Kornberg,[10] hydrolyzed with acid, and estimated as P_i. Alternative procedures involve estimation of the L-glutamate formed with glutamic dehydrogenase and acetylpyridine-DPN,[11] or the enzymatic conversion of the PRA formed to GAR, the latter being estimated colorimetrically after its use as a substrate in a transformylation reaction with IMP.[12] None of these assay procedures is ideal, but the PP assay is preferred, since it does not depend on the presence of other enzymes and is applicable to crude tissue preparations. The chief error with this method occurs at very low levels of enzyme activity, where the small amounts of PP formed cannot be accurately estimated.

Reagents

L-Glutamine, 0.2 M.
PRPP,[13] Na salt, 0.04 M.
Tris–chloride buffer, pH 8.0, 1.0 M.
Na fluoride, 0.5 M.
$MgCl_2$, 0.1 M.
Acetate buffer, pH 5.0, 1.0 M.
$MnCl_2$, 0.1 M.
$MnCl_2$, 0.01 M.
Acetone, 10% aqueous solution.

Procedure. Each assay vessel contains, in a final volume of 0.6 ml.: 0.05 ml. of L-glutamine (10 micromoles), 0.05 ml. of PRPP (2 micromoles), 0.02 ml. of Tris–chloride buffer (20 micromoles), 0.05 ml. of $MgCl_2$ (5 micromoles), 0.02 ml. of Na fluoride (10 micromoles), and enzyme. The vessels are incubated for 15 minutes at 37°. Then 0.2 ml. of acetate buffer, pH 5.0, and 0.2 ml. of 0.1 M $MnCl_2$ are added to each vessel, the contents mixed, and the vessels allowed to sit at 0° for 15 minutes. The flocculent precipitate is centrifuged and washed with 1.0 ml. of 0.01 M $MnCl_2$ and 0.2 ml. of 10% acetone. After centrifugation, the precipitate is suspended in 2.0 ml. of 1 N H_2SO_4 and held at 100° for 15 minutes to hydrolyze the PP to P_i. Aliquots are then assayed for P_i.[14] It is important to run blanks here in which L-glutamine is omitted to correct for PP produced by the hydrolytic cleavage of PRPP. The assay is linear over the range of 0.1 to 1.5 micromoles of PP produced. Hartman recommends the addition of 0.3 micromole of carrier PP where

[10] A. Kornberg, *J. Biol. Chem.* **182**, 779 (1950).
[11] J. B. Wyngaarden and D. M. Ashton, *J. Biol. Chem.* **234**, 1492 (1959).
[12] S. C. Hartman and J. M. Buchanan, *J. Biol. Chem.* **233**, 451 (1958).
[13] J. G. Flaks, M. J. Erwin, and J. M. Buchanan, *J. Biol. Chem.* **228**, 201 (1957); see Vol. VI [69].
[14] C. H. Fiske and Y. SubbaRow, *J. Biol. Chem.* **81**, 629 (1929); see Vol. III [115].

low activity is encountered. Although this increases the blank, it allows a more complete recovery of the PP as the Mn++ salt.

Definition of Unit and Specific Activity. One unit of enzyme activity is defined as that amount catalyzing the formation of 1.0 micromole of PP under the assay conditions. Specific activity is the units of enzyme activity per milligram of protein. Protein is determined by absorbancy measurements at 280 mμ and 260 mμ.[15]

Application of Assay Method to Crude Tissue Preparations. The only trouble encountered with the PP assay procedure is that due to inorganic pyrophosphatase. Na fluoride has been added to the assay components to inhibit this activity, but it can be dispensed with in studies with the purified enzyme. Even though fluoride slightly inhibits the synthetic reaction, the assay procedure, with fluoride present, is sufficiently accurate with various crude liver preparations. It is not certain that the above conditions described for liver enzymes are satisfactory for studies with other preparations, particularly yeast extracts, which are notable for their high pyrophosphatase content. It may be necessary with such preparations to use an alternative assay procedure.[12] With extracts of bacteria and *N. crassa* a preliminary fractionation with ammonium sulfate has been found to be necessary, as noted in Section I.C.

Purification Procedure

The procedure described below represents recent unpublished modifications by S. C. Hartman of a previous procedure.[12] All steps are carried out at 3° unless otherwise indicated.

Step 1. Preparation of Crude Extract. Two kilograms of fresh chicken livers are treated in convenient portions in a Waring blendor for 1 minute with a total of 4 l. of 0.01 M dibasic ammonium citrate ($(NH_4)_2$ $HC_6H_5O_7$), pH 5.3, containing 0.003 M EDTA. It is important to maintain the pH of the homogenate at 5.3 for complete extraction of the enzyme. This mixture is centrifuged at 5000 × g for 0.5 to 1 hour (or long enough to obtain a clear supernatant fraction).

Step 2. First Heat Step. The pH of the supernatant solution is adjusted to 7.0 by the addition of 1.0 M K_3PO_4. The solution is divided into two parts and placed in 2-l. flasks (either a flat-bottomed culture flask or an Erlenmeyer flask), and the flasks are placed in a water bath at 65°. The temperature of the enzyme solution is allowed to rise to 60°, and the flasks are then rapidly cooled in an ice bath. The mixture is then centrifuged for 20 to 30 minutes at 5000 × g, and the precipitate is discarded.

Step 3. Ammonium Sulfate Fractionation. In an ice bath, with stir-

[15] O. Warburg and W. Christian, *Biochem. Z.* **310**, 384 (1942); see also Vol. III [73].

ring, 27.7 g. of solid ammonium sulfate (0.45 saturation) are slowly added for each 100 ml. of the supernatant solution from step 2. After centrifugation for 30 minutes at 5000 × *g*, the precipitate is taken up in 100 ml. of cold water.

Step 4. Dialysis Step. The solution from step 3 is adjusted at pH 5.3 with 10% acetic acid and dialyzed against 4 to 8 l. of cold distilled water for 4 hours. The pH of the enzyme is readjusted to 5.3, and the precipitate is collected by centrifugation and suspended in 50 ml. of 0.02 *M* K_2HPO_4 buffer, pH 6.9.

Step 5. Second Heat Step. The suspension of step 4 is brought to 60° and held at this temperature for 5 minutes. After chilling in ice, the inactive precipitate is removed by centrifugation.

The preparation at this point is about 91-fold purified over the initial extract. At this stage of purification the enzyme can be stored frozen indefinitely without loss in activity. For further purification, chromatography on columns of hydroxylapatite appears to be promising, but the conditions for this procedure have not as yet been accurately established.

A summary of the purification procedure is given in Table 1.

TABLE I

SUMMARY OF PURIFICATION OF PRPP AMIDOTRANSFERASE

Fraction	Total activity, units	Specific activity, units/mg. protein
Crude extract	13,600	0.10
First heat step	18,000	0.37
Ammonium sulfate step	19,400	0.61
Dialysis step	14,400	3.31
Second heat step	9,200	9.1

Properties

Specificity. The purified enzyme is free from GAR kinosynthase and inorganic pyrophosphatase activities but does, however, catalyze the decomposition of the product of the reaction, PRA. PRA has never been isolated from an enzymatic incubation, owing to its marked lability, but less direct evidence strongly favors its occurrence.[16] PRPP amidotransferase from avian liver and *S. typhimurium* has an absolute requirement for L-glutamine.[17] However, Nierlich and Magasanik have recently found that a similar enzyme purified from extracts of *A. aerogenes*, until free from glutaminase activity, cannot catalyze the reaction with L-gluta-

[16] D. A. Goldthwait, *J. Biol. Chem.* **222**, 1051 (1956).
[17] S. C. Hartman, unpublished results.

mine.[18] The enzyme isolated from this source has instead a specific requirement for ammonium ions.[18]

Activators and Inhibitors. Mg^{++} is required for the reaction,[12] the rate being 15% of optimal in its absence.[11] A K_m value of 3×10^{-4} M has been determined for Mg^{++}.[17] Zn^{++} can partially replace the Mg^{++} requirement as can Mn^{++}, the latter with low activity. There may be some inhibitory effect of inorganic phosphate, since the rate of the reaction carried out in phosphate buffer is approximately 60% that in Tris buffer at the pH optimum.[11] A sulfhydryl group is implicated in the action of the enzyme, since complete and irreversible inhibition is obtained in the presence of *p*-chloromercuribenzoate.[17]

Wyngaarden and Ashton[11] have studied the influence of a large number of purine and pyrimidine derivatives on the reaction. No inhibition was found with any pyrimidine nucleotide, any free purine or purine riboside, 2'- or 3'-AMP or GMP, dAMP, IDP, ITP, or GTP. Competitive inhibition was found with AICAR, IMP, AMP, GMP, ADP, GDP, and ATP. The dissociation constants of these inhibitors are listed in Table II. On the basis of this specificity of inhibition, these authors propose that purine biosynthesis may be regulated by means of a competitive feedback inhibition operating on this enzyme.

TABLE II

DISSOCIATION CONSTANTS OF PRPP AMIDOTRANSFERASE INHIBITORS[a]

Inhibitor	K_i
ATP	3.7×10^{-5} M
ADP	3.9×10^{-5} M
GMP	8.6×10^{-5} M
AMP	9.0×10^{-5} M
IMP	1.8×10^{-4} M
GDP	3.8×10^{-4} M

[a] J. B. Wyngaarden and D. M. Ashton, *J. Biol. Chem.* **234**, 1492 (1959).

L-Azaserine (*O*-diazoacetyl-L-serine) and 6-diazo-5-oxo-L-norleucine are antagonists of L-glutamine in the reaction, but the kinetics have not been studied with the purified enzyme.[19, 20] It has been observed, however, that the reaction of 6-diazo-5-oxo-L-norleucine with the enzyme can be increased 100-fold by the presence of PRPP and Mg^{++}.[17]

pH Optimum. The optimum pH is between 8.0 and 9.0. The rate is

[18] D. P. Nierlich and B. Magasanik, personal communication.
[19] B. Levenberg, I. Melnick, and J. M. Buchanan, *J. Biol. Chem.* **225**, 163 (1957).
[20] J. B. Wyngaarden, H. R. Silberman, and J. H. Sadler, *Ann. N. Y. Acad. Sci.* **75**, 45 (1958).

50% of the maximal value at pH 7.5.[17] Hartman[17] does not find the reaction rate falling above pH 8.0, as reported by Wyngaarden and Ashton;[11] in this latter case the influence may be on the glutamic dehydrogenase, which is a component of the assay system.

Equilibrium. The reaction appears to be irreversible. Attempts to demonstrate exchange with the reactants with either isotopic L-glutamate or PP have been unsuccessful.[12, 16]

Kinetic Properties. Evidence has been obtained for separate binding sites for PRPP and L-glutamine,[11] in confirmation of a proposed mechanism for the reaction.[12] The K_m for PRPP is $2.3 \times 10^{-4} M$, and the K_m for L-glutamine is $1.1 \times 10^{-3} M$.

III. Glycinamide Ribotide Kinosynthase

$$\text{Glycine} + \text{ATP} + \text{PRA} \rightleftarrows \text{GAR} + \text{ADP} + \text{P}_i$$

Assay Method

Principle. GAR formed in the reaction becomes the acceptor of a formyl group in an enzymatic transformylation with IMP as the formyl donor. The result is the stoichiometric formation of AICAR, which can be estimated colorimetrically as diazotizable amine. The transformylation reaction, $\text{GAR} + \text{IMP} \rightarrow \text{FGAR} + \text{AICAR}$, is irreversible and is catalyzed by a relatively crude enzyme system, the 15 to 30% ethanol fraction of pigeon liver. Other synthetic reactions are minimized by the presence of EDTA in the transformylation assay. Experience indicates this as the most sensitive and accurate means of determining GAR.[21, 22] Alternative methods involve the incorporation of isotopic glycine into nucleotide form[23] and a ninhydrin assay procedure.[24]

Reagents

Glycine, 0.25 M.
ATP, Na salt, pH 7.0, 0.05 M.
MgCl$_2$, 0.05 M.
Tris–chloride buffer, pH 8.0, 1.0 M.
PRA[16], 0.1 M (0.1 M with respect to total pentose[25]).

---

[21] S. C. Hartman and J. M. Buchanan, *J. Biol. Chem.* **233**, 456 (1958).

[22] L. Warren and J. M. Buchanan, *J. Biol. Chem.* **229**, 613 (1957).

[23] S. C. Hartman, B. Levenberg, and J. M. Buchanan, *J. Biol. Chem.* **221**, 1057 (1956).

[24] D. A. Goldthwait, R. A. Peabody, and G. R. Greenberg, *J. Biol. Chem.* **221**, 555 (1957); see also Vol. II [78].

[25] PRA is prepared according to Goldthwait's procedure[16] (see also Vol. VI [96]). On the basis of total pentose this preparation contains 10 to 30% PRA. Solutions of PRA are markedly unstable at room temperature and neutral pH, and as such

EDTA, 0.15 M, pH adjusted to 7.0 with KOH.

IMP, Na salt,[26] 0.05 M.

Trichloroacetic acid, 30%.

H_2SO_4, 1.0 N.

Acetic anhydride, reagent grade.

Diazotizable amine assay reagents as listed in Section I.B.

15 to 30% ethanol fraction of pigeon liver; the lyophilized powder is prepared as described in Section I.B. As needed, the powder is dissolved in 0.1 M Tris–chloride buffer, pH 7.4, to give a solution of 10 mg./ml. (by weight, not protein content).

Procedure. Each assay vessel contains, in a final volume of 0.3 ml.: 0.02 ml. of glycine (5 micromoles), 0.02 ml. of ATP (1 micromole), 0.02 ml. of $MgCl_2$ (1 micromole), 0.015 ml. of Tris–chloride buffer (15 micromoles), 0.05 ml. of PRA (5 micromoles of total pentose), and enzyme. The contents are mixed and incubated at 37° for 30 minutes. The reaction is terminated by the addition of 0.1 ml. of EDTA (K salt, 15 micromoles). For assay of the GAR formed, 0.05 ml. of IMP (2.5 micromoles) and 0.1 ml. of the 15 to 30% ethanol fraction of pigeon liver (1 mg.) are added. The contents are again mixed and incubated at 37° for 30 minutes. The second reaction is terminated by the addition of 0.1 ml. of 30% trichloroacetic acid, and the insoluble protein is sedimented by centrifugation. Then 0.15 ml. of 1.0 N H_2SO_4 and 0.05 ml. of acetic anhydride are added with mixing directly to the incubation vessels. After 10 minutes at room temperature, 0.05-ml. quantities of each of the diazotizable amine assay reagents listed in Section I.B are added after the proper time intervals. The absorbancy at 540 mμ is determined after 30 minutes. The quantity of GAR is equivalent to the AICAR produced and is calculated from the absorbancy values. As noted by Warren and Buchanan,[22] for unexplained reasons the transformylation from IMP to GAR, although irreversible, does not proceed to completion but abruptly ceases after 80% of the GAR is transformylated. Consequently an absorbancy at 540 mμ of 0.204 is equivalent to a corrected value of GAR of 0.01 micromole in this assay. The ϵ_{540} of the colored product formed from AICAR in the amine assay is 26,400.

Definition of Unit and Specific Activity. Hartman and Buchanan[12, 21] have noted that the synthesis of GAR is not proportional to the amount

---

the precautions noted in Vol. VI [96] regarding storage should be closely followed. For use in the above assay PRA should be pipetted into the assay vessels just prior to the incubation.

[26] See Vol. III [123].

of protein present. Therefore the activity of the 15 to 30% ethanol fraction of chicken livers measured at different protein concentrations is used as a standard curve of activity to which all subsequent fractions are referred. Specific activity is thus relative to the starting fraction in terms of protein content. Protein is estimated by absorbancy measurement at 280 mμ with the assumption that a concentration of 1.0 mg./ml. has an absorbance of 1.6 in a light path of 1 cm.[27]

Application of Assay Method to Crude Tissue Preparations. The assay is applicable to various animal tissue extracts that have been tested. With extracts of microorganisms a preliminary fractionation with ammonium sulfate or ethanol may be necessary, as noted in Section I.C.

Purification Procedure

All steps are carried out at 3° unless otherwise indicated.

Step 1. First Ammonium Sulfate Fractionation. Twenty grams of the lyophilized 15 to 30% ethanol fraction of chicken liver are dissolved in 1 l. of 0.1 M ammonium citrate buffer, pH 5.3. With continuous stirring and cooling in an ice bath, 277 g. of ammonium sulfate (0.45 saturation) are slowly added, and the resulting precipitate is removed by centrifugation and discarded. Then 166 g. of ammonium sulfate (0.65 saturation) are added as before, and the precipitate obtained by centrifugation is dissolved in 250 ml. of cold distilled water. The solution is dialyzed against continuously changing water for 6 hours. Any precipitate forming is removed by centrifugation and discarded.

Step 2. Second Ammonium Sulfate Fractionation. Five milliliters of 1.0 M Tris–chloride buffer, pH 8.0, are added to each 100 ml. of enzyme solution from step 1, followed by 37.4 g. of solid ammonium sulfate (0.58 saturation). The precipitate is removed by centrifugation and discarded, and 15.5 g. of ammonium sulfate (0.80 saturation) are added for each 100 ml. of supernatant solution. The precipitate is collected, dissolved in 60 ml. of cold distilled water, and dialyzed against 0.02 M Tris–chloride buffer, pH 8.0, for 8 hours.

Step 3. Acetone Fractionation. The dialyzed solution from step 2 is carefully adjusted to pH 5.3 with 1.0 N acetic acid. Acetone (−18°) is added in dropwise fashion to the continuously stirred solution in a room at −18° until the concentration reaches 30%. When the temperature of the mixture reaches −10°, the precipitate is removed by centrifugation at −10° for 15 minutes at 25,000 × g. To the supernatant solution, acetone is added as described above, to a final concentration of 50%, and the precipitate is again collected at −10°. This precipitate is dissolved in 50 ml. of 0.02 M Tris–chloride buffer, pH 8.0, and dialyzed

[27] J. M. Buchanan and C. B. Anfinsen, *J. Biol. Chem.* **180**, 47 (1949).

against 4 l. of the same buffer for 8 hours. The small amounts of residual acetone must be removed by this dialysis, since prolonged contact with the enzyme markedly reduces the activity.

Step 4. Alumina C_γ Gel Step. The dialyzed fraction from step 3 is diluted with distilled water to a protein concentration of 3 mg./ml.; and the pH is adjusted to 6.1 with 1 M Na acetate buffer, pH 5.0. Alumina C_γ gel suspension[28] is added at a gel-to-protein ratio of 0.25, conditions under which 35% of the protein is adsorbed. After centrifugation, the supernatant fraction is discarded, and the enzyme is eluted with a volume of 0.05 M K phosphate buffer, pH 7.6, equal to the volume of the fraction used from step 3.

The enzyme at this point has a purity sixty times as great as that of the initial 15 to 30% ethanol fraction. Both this fraction and the

TABLE III

SUMMARY OF PURIFICATION OF GAR KINOSYNTHASE

Fraction	Relative specific activity	Yield, %
15 to 30% ethanol fraction	1.0	100
First ammonium sulfate step	2.2	61
Second ammonium sulfate step	8.4	32
Acetone step	22.0	15
Alumina C_γ gel eluate	60.0	9

dialyzed acetone fraction from step 3 are stable for about a month at 2°.

A summary of the purification is given in Table III.

Properties

Specificity. The enzyme at steps 3 and 4 is free from PRPP amido-transferase activity and enzymes which will catalyze the exchange of the terminal phosphate of ATP with P_i. Small amounts of ATPase and appreciable adenylate kinase activities are present.

At higher levels (1 M), hydroxylamine can replace PRA with the formation of glycine hydroxamate, ATP still being required. The reverse reaction proceeds at equal rates with either arsenate or phosphate. In either case, however, there is an absolute requirement for ADP.

Activators and Inhibitors. Mg^{++} is required for the reaction, although the optimal level has not been reported. The presence of EDTA inhibits the enzyme activity, presumably by chelation with the Mg^{++}.

Equilibrium. The equilibrium favors the phosphorolysis of GAR; the

[28] See Vol. I [11].

constant for this reaction has not been reported, since the continued breakdown of PRA during the course of the reaction makes this measurement highly inaccurate.

Mechanism. The reaction bears many similarities to the synthesis of glutamine from glutamate and ammonia,[29] and the synthesis of glutathione.[30] A mechanism for reactions of this type has been proposed.[3,21]

IV. Glycinamide Ribotide Transformylase

$$\text{GAR} + N^{5,10}\text{-methenyl-THFA} + H_2O \rightarrow \text{FGAR} + \text{THFA} + H^+$$

Assay Method

Principle. Transformylation of GAR with $N^{5,10}$-methenyl-THFA results in the stoichiometric formation of FGAR and THFA.[22] The THFA formed is unstable in acid solution and cleaves spontaneously, and quantitatively, to form an uncharacterized pteridine derivative and *p*-aminobenzoylglutamic acid. The latter compound is determined colorimetrically as diazotizable amine.[7] Alternative procedures involve transformylation of GAR with IMP in the presence of an excess of purified AICAR transformylase and inosinicase[22,31] and the fixation of C^{14}-formate either into nucleotide form (FGAR)[23,24] or acid-stable form.[24]

Reagents

GAR,[32] 0.001 M (with respect to enzymatically active GAR).

$N^{5,10}$-methenyl-THFA,[33] 0.001 M; an aqueous solution of the chloride salt (isoleucovorin chloride) is prepared just prior to use. The compound is not stable in neutral or alkaline solution.

Tris-chloride buffer, pH 7.8, 0.2 M.

Trichloroacetic acid, 30%.

H_2SO_4, 0.8 N.

Diazotizable amine assay reagents as listed in Section I.B.

[29] L. Levintow, A. Meister, G. H. Hogeboom, and E. L. Kuff, *J. Am. Chem. Soc.* **77**, 5304 (1955).

[30] J. E. Snoke and K. Bloch, *J. Biol. Chem.* **213**, 825 (1955).

[31] J. G. Flaks, M. J. Erwin, and J. M. Buchanan, *J. Biol. Chem.* **229**, 603 (1957).

[32] See Vol. VI [96]; GAR prepared from FGAR by acid hydrolysis may be used. It is 50% active (on the basis of total GAR) in the above assay, and the inactive isomer (the α-ribotide configuration) has no influence on the reaction.

[33] D. B. Cosulich, B. Roth, J. M. Smith, Jr., M. E. Hultquist, and R. P. Parker, *J. Am. Chem. Soc.* **74**, 3253 (1952); see also Vol. VI [116]. The derivative used here has been variously referred to as N^5,N^{10}-anhydroformyl-THFA, anhydroleucovorin A and B, and isoleucovorin. The latter three represent different crystalline and hydrated salts originally isolated by Cosulich *et al.* In the above assay procedure the derivative designated as isoleucovorin chloride has been used since it is the most soluble of the various forms.

Procedure.[22] The assay vessel contains, in a final volume of 0.55 ml.: 0.06 ml. of $N^{5,10}$-methenyl-THFA (0.06 micromole), 0.04 ml. of GAR (0.04 micromole), 0.05 ml. of Tris–chloride buffer, pH 7.8 (10 micromoles), and enzyme. The vessels are incubated at 37° for 30 minutes, and the reaction is terminated by the addition of 0.1 ml. of 30% trichloroacetic acid. After centrifugation of the denatured protein, an aliquot of 0.4 ml. of the supernatant fluid is transferred to another tube, and 0.1 ml. of 0.8 N H_2SO_4 is added, followed by 0.05-ml. volumes of the three diazotizable amine assay reagents listed in Section I.B, added after the proper time intervals. After 30 minutes the absorbancy at 540 mμ is determined. The GAR transformylated is equivalent to the *p*-aminobenzoylglutamate formed; the amount of the latter is calculated from the molecular extinction of the colored product of the arylamine reaction (40,500 at 540 mμ).

Definition of Unit and Specific Activity. One unit of GAR transformylase is defined as the amount of enzyme required to convert 0.1 micromole of GAR to FGAR under the conditions of the assay. Specific activity is expressed as units per milligram of protein. The protein content is estimated by absorbancy measurement at 280 mμ as described in Section III.

Application of Assay Procedure to Crude Tissue Preparations. The present assay procedure is suitable for most animal tissue preparations. Difficulties might be encountered in enzyme preparations containing large endogenous amounts of formyl acceptors, or in those tissues possessing enzymes capable of removing $N^{5,10}$-methenyl-THFA by a hydrolytic reaction. The extent of these interferences can be estimated with suitable controls.

It is also important to point out possible difficulties encountered regarding the nature of the formyl donor. Although all the animal enzyme systems studied to date utilize formylated derivatives of THFA, a number of microbial systems have been reported which show greater activity with polyglutamyl derivatives of THFA. In such cases substitution of the proper formyl donor or use of the other assay procedures may be necessary.

Purification Procedure[22]

Step 1. Acetone Powder Extract. All operations are carried out at 3° unless otherwise noted. Chicken livers are treated in 1.5 vol. of distilled water for a few minutes in a Waring blendor. The mixture is rapidly added to 2 vol. of acetone (−15°) and filtered at room temperature. The filter cake is washed with cold acetone until the filtrate is almost colorless and then air-dried. Two hundred grams of chicken liver acetone powder are extracted with 2000 ml. of 0.01 M Tris–chloride buffer, pH 7.4, for 30 minutes at 4°. The suspension is centrifuged at 9000 $\times g$ for 15 minutes, yielding a clear supernatant fraction (1390 ml.) containing 30 mg. of protein per milliliter.

Step 2. Calcium Phosphate Gel Fractionation. Calcium phosphate gel[34] containing 4.17 g. of solids is centrifuged, and the supernatant fluid is discarded. The enzyme solution from step 1 (1390 ml.) is adjusted to pH 6.1 with 1 N HCl, added to the packed gel, and stirred mechanically until the suspension is uniform. The pH of the suspension is now readjusted to 6.1. After 1 hour of stirring, the suspension is centrifuged, and the supernatant fluid is discarded. The packed gel is washed by stirring for 10 minutes with 2 l. of cold distilled water followed by centrifugation. Further washing of the gel is similarly carried out with two 1600-ml. portions of 0.01 M K phosphate buffer, pH 7.4. The enzyme is then eluted from the gel by gentle stirring for 1 hour with 700 ml. of 0.1 M K phosphate buffer, pH 7.4, followed by centrifugation. The elution is repeated with 600 ml. of the same buffer. The combined gel eluates (1300 ml.) contain 3.9 mg. of protein per milliliter.

Step 3. Ammonium Sulfate Fractionation. To 1300 ml. of the enzyme solution from step 2 are slowly added, with stirring, 278 g. of ammonium sulfate (0.4 saturation). The precipitated protein is removed by centrifugation, and 127 g. of ammonium sulfate (0.5 saturation) are slowly added to the supernatant solution. The resulting precipitate is collected by centrifugation and dissolved in 130 ml. of 0.05 M Tris–chloride buffer, pH 7.4.

TABLE IV

SUMMARY OF PURIFICATION OF GAR TRANSFORMYLASE

Fraction	Volume, ml.	Protein concentration, mg./ml.	Specific activity, units/mg. protein	Total activity, units	Fold purification	Yield, %
Acetone powder extract	1390	30.0	0.92	38,200		100
Ca phosphate gel eluate	1300	3.9	6.6	33,100	7.2	87
Ammonium sulfate	130	5.0	26.5	17,250	28.8	45
Dialysis	130	1.8	53.6	12,600	58.5	33

Step 4. Dialysis Step. The enzyme solution from step 3 is dialyzed in 10-ml. quantities against 1 l. of distilled water which is changed twice

---

[34] D. Keilin and E. F. Hartree, *Proc. Roy. Soc.* **B124**, 397 (1937–1938); see also Vol. I [11].

during the dialysis period. After approximately 4 hours a small precipitate is detectable in the dialysis sac. This fraction is collected by centrifugation before the bulk of the protein precipitates and is dissolved in 10 ml. of 0.05 M Tris–chloride buffer, pH 7.8.

The preparation at this point is roughly 58-fold purified over the initial extract. There is no appreciable loss in activity when the enzyme is stored at $-10°$ for 8 months.

A summary of the purification procedure is given in Table IV.

Properties

Specificity. The purified enzyme is free from AICAR transformylase and inosinicase activities.[22] Most preparations contain cyclohydrolase activity (catalyzing $N^{5,10}$-methenyl-THFA $\rightleftharpoons N^{10}$-formyl-THFA), but storage at $2°$ for 3 to 4 months results in a complete loss of this activity with some loss in the GAR transformylase. Cyclohydrolase-free enzyme preparations will not catalyze the reaction with N^{10}-formyl-THFA as the formyl donor but have a specific requirement for $N^{5,10}$-methenyl-THFA.[35] In coupling the GAR transformylase and AICAR transformylase reactions, in which THFA is present in catalytic amounts, cyclohydrolase must be present to effect the interconversion of the THFA derivatives, since AICAR transformylase has a specific requirement for N^{10}-formyl-THFA.[35, 36]

The requirement for GAR is specific; no activity was found when this substrate was replaced with glycinamide or its riboside.[22]

Activators and Inhibitors. The nature of the buffer in which the reaction is carried out has a marked effect on the activity. In Tris–chloride and phosphate buffers the enzyme has greater activity and stability than in maleate buffers. However, the latter buffer may be preferred in certain studies, since the nonenzymatic interconversion of formylated-THFA derivatives proceeds at a slower rate in maleate.[35]

Certain preparations of the enzyme are markedly stimulated by a variety of compounds such as tryptophan, EDTA, and cyanide.[22, 35] These substances may effect a reversal of enzyme inhibition caused by heavy metal ions, since 95% of the enzyme activity is lost in the presence of $1 \times 10^{-5} M$ Cu[++].[22]

A second factor which must be taken into account in the coupling of GAR transformylase with the AICAR transformylase is the stimulatory effect noted with several reducing agents.[36] Of a number of reducing agents which show an activating effect, at levels of $6 \times 10^{-4} M$, DL-homocysteine, Na sulfide, and Na borohydride are the best. These effects are

[35] S. C. Hartman and J. M. Buchanan, *J. Biol. Chem.* **234**, 1812 (1959).
[36] L. Warren, J. G. Flaks, and J. M. Buchanan, *J. Biol. Chem.* **229**, 627 (1957).

noted only with some purified preparations of both enzymes. With impure enzyme preparations similar stimulations have been observed on the addition of TPN (at $2 \times 10^{-5} M$) and L-malate or DPNH (at $1 \times 10^{-4} M$). These effects are most probably related to a maintenance of the catalytic quantities of formylated-THFA coenzymes in a reduced state; since the reducing agents have no effects on the reactions catalyzed by either of the enzymes alone when substrate quantities of THFA derivatives are present.

pH Optimum. Maximal activity occurs at pH 7.8 in Tris–chloride buffer.[22]

Equilibrium. The reaction is essentially irreversible, and all attempts to date to demonstrate reversibility, including exchange reactions, have been unsuccessful.[22]

Kinetics. The K_m for GAR is $5.2 \times 10^{-5} M$; that for $N^{5,10}$-methenyl-THFA is $5.8 \times 10^{-5} M$.[22] This latter value is reported with the reservation that losses of $N^{5,10}$-methenyl-THFA occur by cyclohydrolase action, and nonenzymatically, during the determination.

V. Formylglycinamide Ribotide Amidotransferase

FGAR + L-glutamine + ATP + H₂O →
$$\text{FGAM} + \text{L-glutamic acid} + \text{ADP} + \text{P}_i$$

Assay Method

Principle. FGAM formed by amide transfer from L-glutamine to FGAR is enzymatically cyclized to AIR in the presence of ATP and a partially purified FGAM kinocyclodehydrase preparation. The AIR formed is then estimated as diazotizable amine.

Reagents

ATP, K salt, pH 8.0, 0.15 M.
L-Glutamine, 0.015 M.
MgCl₂, 0.15 M.
Tris–chloride buffer, pH 8.0, 0.3 M.
KCl, 0.6 M.
FGAR,[37] ammonium salt, 0.006 M.
FGAM kinocyclodehydrase, 25 mg. of protein per milliliter in 0.1 M K phosphate buffer, pH 7.4 (step 2 of the preparation described in Section VI).
K phosphate, pH 1.4, 1.33 M, containing trichloroacetic acid, 20%

---

[37] T. C. French, S. C. Hartman, and J. M. Buchanan, unpublished results; see also Vol. VI [96].

(18.3 ml. of 10.9 M phosphoric acid are adjusted to pH 1.4 with KOH, 30 g. of trichloroacetic acid are added, and the volume is made up to 150 ml.).

Diazotizable amine assay reagents as listed in Section I.B.

Procedure. It has recently been found that this enzyme is extremely sensitive to inactivation by heavy metal ions.[38] For reproducibility in the assay described below it is therefore recommended that the vessels be soaked in, and rinsed with, a solution of EDTA, $1 \times 10^{-4} M$, and then dried prior to use in the assay. For the same reason, the reagents listed above should be prepared in ion-free distilled water.

Each assay vessel contains, in a final volume of 0.3 ml.: 0.02 ml. of ATP (3 micromoles), 0.02 ml. of L-glutamine (0.3 micromole), 0.02 ml. of $MgCl_2$ (3 micromoles), 0.02 ml. of Tris–chloride buffer, pH 8.0 (6 micromoles), 0.02 ml. of KCl (12 micromoles), 0.02 ml. of FGAR (0.12 micromole), 0.03 ml. of FGAM kinocyclodehydrase, and enzyme. Dilutions of the enzyme should be made in K phosphate buffer, pH 7.4, 0.001 M, containing EDTA, pH 8.0, $10^{-4} M$. The vessels are incubated at 38° for 15 minutes and then chilled in ice, and the reaction is terminated by the addition of 0.1 ml. of the K phosphate–trichloroacetic acid solution. The denatured protein is sedimented by centrifugation, and 0.05-ml. volumes of each of the diazotizable amine assay reagents are added as indicated in Section I.B, with the modification that the ammonium sulfamate is added 3 minutes after the Na nitrite. Ten minutes after the addition of the final reagent the absorbancy at 500 mμ is determined against a reagent blank.

The modifications in the diazotizable amine determination are necessitated by the fact that the colored product obtained with AIR is less stable than most others, and both its wavelength maxima and extinction coefficient are highly pH-dependent.[39]

Definition of Unit and Specific Activity. A unit of activity is defined as the amount of enzyme necessary to give an absorbancy of 1.0 at 500 mμ under the assay conditions. The quantity of AIR formed may be calculated from the molecular absorbancy of the colored product of the assay reaction (24,600 at 500 mμ). Because of the nonlinearity of the assay, particularly with crude extracts, the units of enzyme activity are calculated from the assay absorbancy measurements made within some convenient range. The range of 0.4 to 0.6 optical density unit was chosen for the purification of the enzyme from *S. typhimurium* reported below.

---

[38] R. L. Herrmann and T. C. French, unpublished experiments.

[39] L. N. Lukens and J. M. Buchanan, *J. Biol. Chem.* **234**, 1799 (1959).

Specific activity is the units of activity per milligram of protein. Protein is estimated by the biuret procedure.[39a]

Application of Assay Method to Crude Tissue Preparations. Crude enzyme preparations, particularly those of chicken liver, contain inhibitors that interfere with the enzyme assay. Plots of enzyme activity versus protein concentration deviate from linearity with such preparations, but at later stages of purification linearity is approached. Crude extracts from microbial sources have proved less troublesome in this regard.

Purification Procedure

The enzyme has been extensively purified from both chicken liver[40] and *S. typhimurium*.[41] At present, a purer enzyme preparation is obtained from the latter source, and the procedure itself is less laborious. This preparation is recommended for studies on the mechanism of the reaction or on the enzyme itself. For the preparation of FGAM (Vol. VI [96]), and for the enzymatic assay of FGAM kinocyclodehydrase (see Section VI), an amidotransferase of lesser purity suffices, and this is most conveniently obtained after the first few steps of the preparation from chicken liver. Therefore both preparations are described below. All operations are carried out at 3° unless otherwise noted.

1. Preparation of the Enzyme from Chicken Liver

Fresh chicken livers (4.5 kg.) are ground in a chilled meat grinder, and 600-g. portions of the minced tissue are then blended with 3 l. of cold acetone (−17°) in a 1-gallon Waring blendor for 30 seconds at top speed. The combined suspension (approximately 35 l.) is filtered on four large Büchner funnels, and the precipitate on each funnel is washed with 5 to 10 l. of cold acetone until the filtrates are almost colorless. This operation is followed by washing with 1 pound of anhydrous ether for each funnel. When almost dry, the filter cake is crumbled and spread out in a hood to remove traces of the solvents. If necessary the powder is stored in an evacuated desiccator over P_2O_5 at −17°. Two 600-g. portions of the acetone powder are each extracted with 6.5 l. of $0.1 M$ K phosphate, pH 7.0, containing $0.001 M$ L-glutamine and $0.002 M$ EDTA, for 1 hour with mechanical stirring. The suspension is centrifuged at $5000 \times g$ for 1.5 hours in a Stock refrigerated centrifuge (Wilhelm Stock Co., Marburg, Germany), and the supernatant fluid obtained is strained through several layers of cheesecloth. To each liter of the

---

[39a] L. C. Mokrasch and R. W. McGilvery, *J. Biol. Chem.* **221**, 909 (1956).
[40] R. L. Herrmann and J. M. Buchanan, unpublished results.
[41] T. C. French and J. M. Buchanan, unpublished results.

acetone powder extract 242 g. of solid ammonium sulfate (0.4 saturation) are slowly added with stirring. The suspension is centrifuged for 30 minutes in the Stock centrifuge at $5000 \times g$, and the supernatant fluid is discarded. The precipitate is stored overnight at $-17°$. The precipitate is dissolved in 1 l. of $0.02 M$ K phosphate, pH 7.2, containing $0.001 M$ L-glutamine and $0.002 M$ EDTA, and the solution is dialyzed against 7.5 l. of the same buffer for 1.5 hours. The protein solution is then diluted to 2 l. and added to 4 l. of calcium phosphate gel[34] (30 mg. of solids per milliliter) with stirring. After 30 minutes of stirring, the suspension is centrifuged at $5000 \times g$ for 30 minutes in the Stock centrifuge, and the sediment is discarded. To each liter of the supernatant solution, 263 g. of solid ammonium sulfate (0.43 saturation) are added over the course of 1 hour with stirring, and the resulting precipitate is collected by centrifugation at $5000 \times g$ for 30 minutes. The precipitate may be stored at $-15°$, or taken up in a minimal quantity of $0.02 M$ K phosphate, pH 6.5, containing $0.2 M$ KCl, $0.0005 M$ L-glutamine, and $0.0005 M$ EDTA, pH 8.0.

The above procedure has also been successfully carried out on one-tenth the described scale. The enzyme at this point is approximately 90-fold purified over the acetone powder extract. It is free from FGAM kinocyclodehydrase activity and substances which interfere with the amidotransferase assay and is stable for several months when stored at $-15°$. Further purification, resulting in a preparation of approximately 25% purity, has been obtained by ethanol and ammonium sulfate fractionations coupled with chromatography on columns of hydroxylapatite and DEAE-cellulose.

2. Purification of the Enzyme from S. typhimurium

Step 1. Acetone Powder Extract. The medium for the growth of S. typhimurium contains 10 g. of $(NH_4)_2SO_4$, 5 g. of Na citrate·$2H_2O$, 1 g. of $MgSO_4·7H_2O$, 30 g. of KH_2PO_4, 70 g. of K_2HPO_4, 30 g. of casamino acids (Difco, technical grade), 5 g. of nutrient broth (Difco), 0.15 g. of uracil, 0.15 g. of thymine, 0.05 mg. of thiamine·HCl, 2 mg. of Ca pantothenate, 2 mg. of nicotinamide, and 4 mg. of p-aminobenzoic acid. The components are dissolved in 9.75 l. of distilled water in a 20-l. carboy and sterilized in the autoclave. The solution is brought to 37°, and 250 ml. of 40% glucose (sterilized separately) are added, followed by the addition of a small amount of sterile Antifoam B (Dow Corning). The medium is inoculated with the cells washed from a large nutrient agar slant of S. typhimurium (strain LT-2) with 0.85% NaCl. Incubation of the culture is carried out at 37° with forced aeration (40 l./

min.),[42] until the cell suspension reaches an optical density of 1.2 at 650 mμ. The cells are then harvested in an enclosed, refrigerated centrifuge yielding a cell paste of 110 to 120 g. The cells are washed by suspension in 200 ml. of 0.05 M K phosphate, pH 6.5, containing 0.0005 M L-glutamine and 0.01 M MgCl$_2$, followed by centrifugation. An acetone powder is prepared from the washed cells as described in Section I.C. Twenty grams of acetone powder are obtained from each 100 g. of washed cells, and the powder is stable indefinitely when stored under desiccation at −20°.

Twenty grams of the acetone powder are extracted for 20 minutes with 250 ml. of 0.02 M K phosphate, pH 6.5, containing 0.0005 M L-glutamine and 0.01 M MgCl$_2$. Then 0.1 mg. of deoxyribonuclease (Worthington) is added to lower the viscosity of the suspension. The extract is centrifuged at 25,000 × g for 15 minutes, and the supernatant fluid is cautiously decanted and retained. The extraction is repeated with 150 ml. of the same buffer, and the supernatant fluids are combined and made 0.02 M with respect to EDTA by the addition of 0.5 M EDTA, pH 8.0.

Step 2. First Ammonium Sulfate Fractionation. Per liter of the extract from step 1, 330 mg. of L-glutamine are added, followed by 350 g. of solid ammonium sulfate (0.55 saturation), the latter over a period of 5 minutes. The mixture is stirred for 5 minutes and then centrifuged for 10 minutes at 15,000 × g. A suspension of the precipitate is immediately made with 10 ml. of 0.02 M K phosphate, pH 6.5, containing 0.005 M L-glutamine and 0.0005 M EDTA, pH 8.0, and the suspension is dialyzed for 4 hours against a continuous flow, totaling 5 l., of 0.01 M K phosphate, pH 6.5, which is 2.5 × 10^{-4} M with respect to L-glutamine and EDTA, pH 8.0. The enzyme is unstable during the ammonium sulfate fractionation of the crude extract, and this step should be carried out rapidly. The dialyzed enzyme is stable at 3° or when frozen.

Step 3. RNase Treatment and Second Ammonium Sulfate Fractionation. Forty milliliters of 0.5 M K phosphate, pH 6.5, containing 0.0125 M L-glutamine and 0.025 M EDTA, pH 8.0, are added per liter of the dialyzed enzyme from step 2. Then 50 mg. of ribonuclease (Armour) are added for each liter of the solution, and the mixture is incubated at 38° for 10 minutes. After chilling in an ice bath, the mixture is centrifuged

[42] Strain LT-2 has been grown in kilogram quantities without any apparent health hazard. The following precautions should be employed. All connections to the aerating apparatus and the carboy are clamped securely with wire, and the air exhaust is flushed through a solution containing antifoam and a suitable disinfectant. The entire apparatus, centrifuge cups, and supernatant fluid remaining after the cells are harvested are similarly disinfected.

at 20,000 × g for 15 minutes, the supernatant fluid retained, and the precipitate washed by suspension in 15 ml. of the buffer used for the dialysis in step 2. After centrifugation, the supernatant solutions are combined and 0.2 of a volume of citrate–phosphate buffer, pH 5.2 (23.3 ml. of 1 M citric acid and 53.4 ml. of 1 M K_2HPO_4 per 100 ml.), is added dropwise over a 5-minute period. Then 350 mg. of L-glutamine are added per liter of solution, followed over the course of 5 minutes by 140 g. of ammonium sulfate (0.20 saturation). After 5 minutes of stirring, the suspension is centrifuged for 10 minutes at 15,000 × g, and the precipitate is discarded. Next 130 g. of ammonium sulfate (0.35 saturation) are added to the supernatant fluid in 5 minutes, and the mixture is stirred and centrifuged as before. The resulting precipitate is dissolved in 20 ml. of the same buffer used to dissolve the ammonium sulfate precipitate in step 2. The solution is then dialyzed for 2 hours, as in step 2, against 1600 ml. of the same buffer.

Step 4. First Chromatography on DEAE-Cellulose. The DEAE-cellulose (1 meq./g.) used in this and subsequent steps is treated as follows: For each gram of the cellulose powder, 20 ml. of 0.1 N KOH containing 0.001 M EDTA are added, and, after being stirred briefly, the suspension is filtered on a sintered-glass funnel. The procedure is repeated with a stirring period of 30 minutes. The residue is washed on the filter with 2 vol. of deionized water followed by 2 vol. of 10% ethanol. A slurry of the cellulose in water is made up into columns which are washed successively with 20 vol. of 10% ethanol containing 0.001 M 8-quinolinol, 10 vol. of deionized water, 20 vol. of 0.1 N HCl, 10 vol. of deionized water, and 0.1 M K phosphate, pH 7.0, until the pH of the effluent is 7.0. The columns are stored in the latter buffer at 3°. Columns that have been used are regenerated by washing with 0.1 M K_2HPO_4 containing 0.001 M EDTA. Just prior to use, the columns are washed with 4 bed vol. of the dialysis buffer used in step 2.

The enzyme solution from step 3 is diluted to 50 ml. with water and centrifuged at 20,000 × g for 15 minutes. The supernatant fluid is allowed to flow through a DEAE-cellulose column (26 cm. × 2.2-cm. diameter) under gravity, and the column is then subjected to a concave gradient elution. The cylindrical reservoir, 41 mm. I.D., contains 350 ml. of 0.32 M KCl containing 0.02 M K phosphate, pH 6.5, 0.0005 M L-glutamine, and 0.0001 M EDTA, pH 8.0. The cylindrical mixer, 50 mm. I.D., contains 0.01 M K phosphate, pH 7.0, and the same concentrations of L-glutamine and EDTA as the reservoir. Fractions of 20 ml. are collected with a flow rate of 2 ml./min. The amidotransferase activity appears after the elution of a yellow contaminant and has its peak at the 0.65 effluent volume fraction. Those fractions containing at least 25 units of enzyme activity per milliliter are pooled, yielding about 100 to 140 ml.

Step 5. Second Chromatography on DEAE-Cellulose. The pooled enzyme fractions from step 4 are diluted with 3 vol. of 0.0005 M L-glutamine containing 0.0001 M EDTA, pH 8.0, and the solution is applied to a DEAE-cellulose column (10 cm. × 2.2-cm. diameter) at a flow rate of 4 to 6 ml./min. The column is eluted by the concave gradient method, with 150 ml. of 0.18 M KH$_2$PO$_4$ containing 0.0005 M L-glutamine and 0.0001 M EDTA in the reservoir flowing into the mixer containing 0.03 M K phosphate, pH 7.0, with L-glutamine and EDTA in the same concentration as in the reservoir. Fractions of 5 to 7 ml. are collected every 10 minutes. The peak of enzyme activity appears at the 0.70 effluent volume fraction. Those fractions containing enzyme activity of at least 60 units/ml. at the beginning of the peak and at least 120 units/ml. at the end of the peak are combined. The pooled fractions contain 50 to 70 ml.

Step 6. Concentration on DEAE-Cellulose. One-tenth volume of 0.05 M L-glutamine and 0.005 vol. of 0.5 M EDTA, pH 8.0, are added to the pooled enzyme fractions from step 5. The solution is dialyzed for 2 hours as efficiently as possible (rocking dialysis) against four changes of 400 ml. of the dialysis buffer used in step 2. The dialyzed enzyme is applied to a DEAE-cellulose column (5 cm. × 1.0-cm. diameter) at a flow rate of 2 ml./min. The column is then eluted with 0.20 M KCl containing 0.02 M K phosphate, pH 6.5, 0.0005 M L-glutamine, and 0.0005 M EDTA, pH 8.0. The enzyme solution obtained is stable for many weeks when stored at 3°, or when frozen.

TABLE V
SUMMARY OF PURIFICATION OF FGAR AMIDOTRANSFERASE

Fraction	Total protein, mg.	Total activity, units × 10^{-2}	Specific activity, units/mg. protein	Recovery, %
Acetone powder extract	6010	332	5.52	100
First ammonium sulfate step	4740	251	5.30	76
RNase and second ammonium sulfate step	883	187	21.2	56
First DEAE-cellulose step	98	153	156	46
Second DEAE-cellulose step	21.3	132	620	40
Concentration step	20.5	125	610	38

The amidotransferase is purified about 110-fold over the acetone powder extract and is 50 to 70% pure at this stage. It is contaminated with a yellow substance which can be removed by further chromatography on DEAE-cellulose.

A summary of the purification procedure is given in Table V.

Properties

Specificity. This enzyme has been isolated and highly purified from avian liver and *S. typhimurium.* In either case the nitrogen donor has been found to be L-glutamine. The enzyme has not as yet been studied in other organisms such as *A. aerogenes* where ammonia is the nitrogen donor in the PRPP amidotransferase reaction.[18]

Activators and Inhibitors. Mg^{++} is required for the reaction. ADP at $1 \times 10^{-3} M$ inhibits the bacterial enzyme approximately 50%. As noted above, both enzymes are considerably stabilized in the presence of L-glutamine (0.001 M). Low concentrations of heavy metal ions markedly, and irreversibly, inhibit the activity; $1.4 \times 10^{-5} M$ Cu^{++} inhibits the *S. typhimurium* enzyme 50% in the assay procedure. Protection is afforded in this case by the presence of 0.001 M EDTA.

The most striking inhibitions are those obtained with the glutamine antagonists, L-azaserine (O-diazoacetyl-L-serine) and 6-diazo-5-oxo-L-norleucine,[19] which also inhibit, but much less effectively, the PRPP amidotransferase. In the presence of L-glutamine, and with low levels of either inhibitor, the early phases of the inhibition are competitive in nature. However, as the incubation progresses, enzymatic activity is lost. Incubation of the enzyme with the inhibitor in the absence of L-glutamine leads to a noncompetitive and irreversible inhibition. Only the L-forms of the inhibitors are active, and on a molar basis the norleucine derivative is about forty times as active as L-azaserine. With the avian liver enzyme the Ki for L-azaserine is $3.4 \times 10^{-5} M$; that for 6-diazo-5-oxo-L-norleucine is $1.1 \times 10^{-6} M$.[19] Similar effects have been observed with the enzyme from *S. typhimurium.*[41]

Equilibrium. The equilibrium is very far in the direction of FGAM synthesis, and no evidence has been obtained for reversibility. Similarly, no exchange has been found for Pi^{32} with ATP or for L-glutamate-C^{14} with L-glutamine.

Kinetics. K_m values have been determined with the avian liver enzyme.[19] The K_m for FGAR is $6.4 \times 10^{-5} M$; that for L-glutamine is $6.2 \times 10^{-4} M$.

VI. Formylglycinamidine Ribotide Kinocyclodehydrase

$$FGAM + ATP \rightarrow AIR + ADP + P_i$$

Assay Method

Principle. FGAM is cyclized to AIR in the presence of ATP and the enzyme. The AIR formed is estimated colorimetrically as diazotizable amine. Alternatively, when FGAM is not readily available, the assay procedure described in Section V may be used with the modifications noted below.

Reagents

FGAM,[43] ammonium salt, 0.001 M.
ATP, 0.06 M.
$MgCl_2$, 0.12 M.
Tris–chloride buffer, pH 7.4, 0.3 M.
KCl, 1.2 M.
K phosphate, pH 1.4, 1.33 M, containing trichloroacetic acid, 20% (prepared as described in Section V).
Diazotizable amine assay reagents as described in Section I.B.

Procedure. Each assay vessel contains, in a final volume of 0.3 ml.: 0.025 ml. of FGAM (0.025 micromole), 0.02 ml. of ATP (1.2 micromoles), 0.02 ml. of $MgCl_2$ (2.4 micromoles), 0.02 ml. of Tris–chloride buffer, pH 7.4 (6 micromoles), 0.025 ml. of KCl (30 micromoles), and enzyme. The vessels are incubated at 38° for 20 minutes, chilled in an ice bath, and 0.1 ml. of the K phosphate–trichloroacetic acid solution is added. After sedimentation of the denatured protein, the supernatant fluid is assayed for diazotizable amine exactly as described in Section V.

Alternative Procedure. Where FGAM is not available, a modified form of the FGAR amidotransferase assay of Section V is employed. The kinocyclodehydrase is made limiting, and the amidotransferase is present in excess. For the latter, the enzyme from Section V (second ammonium sulfate fraction purified 90-fold) is suitable.[44]

Definition of Unit and Specific Activity. One unit of activity is defined as the amount of enzyme giving an absorbancy of 0.1 at 500 mμ

---

[43] T. C. French and S. C. Hartman, unpublished results; see also Vol. VI [96].

[44] It should be noted that the assay procedures for FGAR amidotransferase and FGAM kinocyclodehydrase are dependent on one another in the sense that each requires either the other enzyme in excess, or the FGAM kinocyclodehydrase assay requires as a substrate FGAM, which is prepared with the amidotransferase. Delineation of the two reactions was facilitated by the initial observation [B. Levenberg and J. M. Buchanan, *J. Biol. Chem.* 224, 1019 (1957)] that a clean separation of these enzymes from each other can be obtained by a single ammonium sulfate fractionation step. Either an acetone powder extract of avian liver in 0.01 M K phosphate buffer, pH 7.4, or the lyophilized 15 to 30% ethanol fraction described in Section I, dissolved in the same buffer, is fractionated with solid ammonium sulfate. The FGAR amidotransferase precipitates between 0 and 0.35 saturation, and the FGAM kinocyclodehydrase between 0.45 and 0.60 saturation. The respective enzymes are then handled with the precautions noted in Section V and in this section. This fractionation should be the starting point for assays for both enzymes where FGAR is the only available material.

under the assay conditions. The actual amount of AIR produced may be calculated as described in Section V. Specific activity is units per milligram of protein. Protein is estimated by absorbancy measurement at 280 mμ as described in Section III.

Application of the Assay Method to Crude Tissue Preparations. No difficulty has been encountered in applying the assay with FGAM as substrate to a variety of extracts. The assay procedure with FGAR as the substrate is subject to the limitations described in Section V.

Purification Procedure

The kinocyclodehydrase has not been extensively purified owing for the most part to its instability. The procedure described here represents a recent unpublished method.[17, 41]

Step 1. Acetone Powder Extract. Pigeon liver acetone powder is prepared by treating the livers directly in a Waring blendor for 1 to 2 minutes with 20 vol. of cold acetone ($-15°$). The suspension is filtered with suction, washed with cold acetone until the filtrate is almost colorless, and finally washed with anhydrous ether. The powder is dried briefly in air and then *in vacuo* over P_2O_5. All subsequent steps are at 3°. Ten grams of acetone powder are extracted by stirring with 100 ml. of 0.01 M K phosphate buffer, pH 7.4, for 30 minutes. The resulting suspension is centrifuged for 15 minutes at 10,000 $\times$ g, and the precipitate is discarded.

Step 2. Ammonium Sulfate Fractionation. To each 100 ml. of the extract from step 1, 27.7 g. of solid ammonium sulfate (0.45 saturation) are slowly added with stirring, and the precipitate is removed by centrifugation. To the supernatant solution, 11.3 g. of ammonium sulfate (0.6 saturation) are added as above. The resulting precipitate is collected by centrifugation, dissolved in 10 ml. of 0.01 M K phosphate, pH 7.4, containing 0.1 M KCl, and dialyzed for 2 hours against the same solution.

Step 3. Hydroxylapatite Chromatography. The dialyzed enzyme solution from step 2 is applied to a hydroxylapatite column[45] at a flow rate of 1 ml./min., and the column is subjected to elution with a linear gradient of K phosphate, pH 7.4, from 0.01 to 0.25 M, all solutions containing 0.1 M KCl. The enzyme is eluted from the column in a narrow range. The combined fractions with activity are purified approximately 10-fold over the initial acetone powder extract.

The enzyme is rather unstable, with almost complete loss in activity occurring after 1 week at 2°. Somewhat better stability is obtained on freezing, particularly in the presence of 0.1 M KCl.

[45] A. Tiselius, S. Hjertén, and Ö. Levin, *Arch. Biochem. Biophys.* **65**, 132 (1956).

More recently it has been found that the kinocyclodehydrase from extracts of *N. crassa* mycelia is more stable than the avian liver enzyme.[41] Extracts of mycelia, obtained by disruption with a Hughes press, can be directly fractionated on a DEAE-cellulose column. The details of this procedure have not as yet been sufficiently established for publication.

Properties

Specificity. The partially purified enzyme is free from FGAR amidotransferase activity and has specific requirements for FGAM and ATP.

Activators and Inhibitors. The enzyme has a requirement for relatively high concentrations $(0.1\,M)$ of certain monovalent cations such as K^+, Rb^+, and NH_4^+. The cations Na^+, Li^+, and cesium are inactive. Aside from the requirement for activity, these cations also markedly increase the stability of the enzyme. Mg^{++} is also a requirement for activity.

ADP strongly inhibits the reaction at levels of $4 \times 10^{-3}\,M$.

Equilibrium. No evidence for reversibility of the reaction has been obtained, nor has exchange of $P_i{}^{32}$ with ATP been found to occur.

VII. 5-Aminoimidazole Ribotide Carboxylase

$$\text{AIR} + CO_2 \rightleftharpoons \text{CAIR}$$

Assay Method

Principle. In alkaline solution AIR possesses only end absorption in the ultraviolet region, whereas CAIR has an absorption peak at 249 mμ. The assay is therefore based on measurement of the increase in absorbance on formation of CAIR, with 270 mμ chosen as the most suitable wavelength. The alkaline pH serves to terminate the reaction and to stabilize the CAIR formed.

Reagents

> AIR,[46] K salt, $0.002\,M$.
> $KHCO_3$, $1.07\,M$, freshly prepared each day.
> Tris–chloride buffer, pH 8.0, $0.8\,M$.
> KOH, $10\,N$.

Procedure.[39] Each assay vessel contains, in a final volume of 0.4 ml.: 0.05 ml. of AIR (0.1 micromole), 0.15 ml. of $KHCO_3$ (160 micromoles), 0.05 ml. of Tris–chloride buffer, pH 8.0 (40 micromoles), and enzyme. The vessels are incubated at 38° for 10 minutes, and the reaction is

[46] B. Levenberg and J. M. Buchanan, *J. Biol. Chem.* **224**, 1005 (1957); see also Vol. VI [96].

terminated by the addition of 0.04 ml. of $10 N$ KOH. The absorbance at 270 mμ is determined as shortly thereafter as is feasible in a Beckman spectrophotometer equipped with a photomultiplier and an attachment for microcuvettes. The vessels are read against an incubated blank containing the complete assay system with the enzyme omitted. To correct for increases in absorbance due to the enzyme itself, an identical blank is run, and the enzyme is added after the addition of the KOH.

Definition of Unit and Specific Activity. A unit of activity is defined as the amount of enzyme giving an increase in absorbance of 0.100 at 270 mμ under the assay conditions. Specific activity is the units per milligram of protein. Protein is estimated by absorbancy measurement at 280 mμ as indicated in Section III.

Application of Assay Method to Crude Tissue Preparations. The assay procedure has not been tested with a sufficient variety of tissues to permit generalizations. With avian liver preparations, proportionality with time and enzyme concentration is obtained provided that the total increase in absorbance does not exceed about 0.3. It is conceivable that difficulties may be encountered with extracts of microorganisms where the nucleic acid content is relatively higher and shows increases in absorbance on depolymerization and degradation.

Purification Procedure[39]

Step 1. Chicken Liver Extract. All operations are performed at 3°. Fresh chicken livers (4.54 kg.) are minced in a meat grinder, and the resulting tissue is treated with 1.5 vol. of $0.05 M$ Na acetate buffer, pH 5.4, for 1 minute in a Waring blendor. The resulting suspension is centrifuged in a refrigerated Stock centrifuge (Wilhelm Stock Co., Marburg, Germany) at $5300 \times g$ for 1.5 hours. The supernatant fluid (4.5 l.) is poured through a funnel containing glass wool.

Step 2. First Ammonium Sulfate Step. To each liter of the supernatant solution from step 1, 176 g. of solid ammonium sulfate (0.3 saturation) are slowly added with stirring. After centrifugation for 1.5 hours in the Stock centrifuge, the precipitate is discarded, and 175 g. of ammonium sulfate (0.55 saturation) are added, as above, for each liter of the original extract from step 1 fractionated. The precipitate is collected, as above, and dissolved in 2.2 l. of $0.005 M$ K acetate, pH 6.0.

Step 3. Second Ammonium Sulfate Step. The fractionation is carried out in identical fashion to that in step 2 with collection of the fraction precipitating between 0.25 and 0.45 saturation of ammonium sulfate. The amounts of solid ammonium sulfate added per liter of the step 2 enzyme solution are 144 g. to 0.25 saturation and 133 g. to 0.45 saturation. The precipitate obtained is dissolved in 645 ml. of $0.005 M$ K acetate, pH 6.0.

Step 4. Dialysis Step. The enzyme solution from step 3 is dialyzed against 14 l. of 0.002 *M* K acetate, pH 6.0, with one change of the dialysis fluid. The dialysis is carried out for the length of time necessary to precipitate roughly half the protein. This time has been found to vary and will have to be judged by experience. The suspension is then centrifuged, and the supernatant fluid is retained.

Step 5. Calcium Phosphate Gel Step. The enzyme solution from step 4 is treated with sufficient packed calcium phosphate gel[34] to absorb 90% or more of the protein. The correct quantity of gel should be judged by a pilot run. The proper gel-to-protein ratio is of the order of 0.7:1. The enzyme solution and the packed gel are gently stirred for 30 minutes and then centrifuged, and the supernatant fluid is discarded. The gel is stirred for 30 minutes with a volume of 0.05 *M* K phosphate, pH 7.0, equal to the volume of the supernatant solution from step 4. After centrifugation, the packed gel is again eluted as above with 0.05 *M* K phosphate, pH 7.0. Both eluates are discarded. The enzyme is then eluted, twice, with similar volumes of 0.15 *M* K phosphate, pH 7.0, to that used above.

Step 6. Third Ammonium Sulfate Step. The fractionation is identical to that carried out in step 2 with the fraction precipitating between 0.35 and 0.50 saturation collected. The amounts of solid ammonium sulfate added per liter of the enzyme solution from step 5 are 209 g. to 0.35 saturation and 104 g. to 0.50 saturation. The resulting precipitate is dissolved in 80 ml. of 0.005 *M* Tris–chloride buffer, pH 7.2.

The enzyme at this state is purified 17-fold over the activity in the initial extract and is fairly stable on storage at −15°. Further purification to about 50-fold has recently been obtained by chromatography on DEAE-cellulose of the enzyme from step 5. This latter procedure has been carried only once, so that the results are only preliminary. DEAE-cellulose is washed as described in Section V, prepared into a column

TABLE VI
SUMMARY OF PURIFICATION OF AIR CARBOXYLASE

Fraction	Volume, ml.	Units, $\times 10^{-6}$	Specific activity, units/mg. protein, $\times 10^{-2}$	Yield, %
Crude extract	4500	43.6	1.97	100
First ammonium sulfate step	2200	31.4	3.17	72
Second ammonium sulfate step	645	12.9	3.85	30
Dialysis step	850	12.75	5.77	29
Calcium phosphate gel eluate	1430	11.8	28.4	27
Third ammonium sulfate step	79	7.6	36.4	17

18 × 1 cm., and the pH adjusted to 7.0 by percolation with $1\,M$ K phosphate, pH 7.0. After washing with water, 5 ml. of the enzyme solution from step 5, previously dialyzed overnight against 3 l. of $0.005\,M$ K phosphate, pH 7.0, are applied to the column. The column is then eluted with K phosphate solutions, pH 7.0, varying from 0.005 to $0.1\,M$. Elution of the enzyme begins at a buffer concentration of $0.035\,M$.

A summary of the purification procedure is given in Table VI.

Properties[39, 47]

Specificity. The purified enzyme is contaminated with small amounts of succino-AICAR kinosynthase.[48] The substrate requirements for the reaction are absolute; no activity has been found at the nucleoside level, nor can oxaloacetic acid or carbamyl phosphate substitute for the bicarbonate requirement.

Activators and Inhibitors. The enzyme differs from most other carboxylases in that there are no known activators or cofactors required for the reaction.

Equilibrium. The equilibrium of the reaction lies in the direction of decarboxylation, but the formation of CAIR can be readily demonstrated. The actual value of the equilibrium constant is not known, but in the presence of high levels of bicarbonate $(0.3\,M)$ approximately 50% conversion of AIR $(1 \times 10^{-3}\,M)$ to CAIR was obtained.

VIII. 5-Amino-4-imidazole-*N*-succinocarboxamide Ribotide Kinosynthase

$$\text{CAIR} + \text{L-aspartic acid} + \text{ATP} \rightleftharpoons \text{Succino-AICAR} + \text{ADP} + \text{P}_i$$

Assay Method[48]

Principle. Arsenolysis of succino-AICAR in the presence of ADP yields CAIR. The latter compound decarboxylates in the presence of excess AIR carboxylase to AIR, which is estimated colorimetrically as diazotizable amine. Under conditions of the diazotizable amine assay for AIR, neither succino-AICAR nor CAIR interferes with the colorimetric determination.

Reagents

Na₂HAsO₄, pH 7.4, $0.58\,M$.
ADP, Na salt, $0.03\,M$.
MgCl₂, $0.076\,M$.

[47] L. N. Lukens and J. M. Buchanan, *J. Biol. Chem.* **234**, 1791 (1959).
[48] R. W. Miller and J. M. Buchanan, *J. Biol. Chem.* **237**, 485 (1962).

Tris–chloride buffer, pH 7.4, 0.58 M.

Succino-AICAR,[49] Na salt, 1.34 $\times$ 10^{-3} M.

K phosphate, pH 1.4, 1.33 M, containing trichloroacetic acid, 20% (prepared as described in Section V).

Diazotizable amine assay reagents as described in Section I.B.

Procedure. The assay system contains, in a final volume of 0.6 ml.: 0.1 ml. of Na_2HAsO_4 (58 micromoles), 0.1 ml. of ADP (3 micromoles), 0.1 ml. of $MgCl_2$ (7.6 micromole), 0.1 ml. of Tris–chloride buffer, pH 7.4 (58 micromoles), 0.05 ml. of succino-AICAR (0.067 micromole), and up to 0.35 unit of enzyme (the acetone powder extract is diluted 10-fold, and 0.04 ml. is used). The vessels are incubated at 38° for 20 minutes, and the reaction is terminated by the addition of 0.1 ml. of the K phosphate–trichloroacetic acid solution. After removal of the denatured protein by centrifugation, an aliquot of 0.4 ml. is removed and assayed for diazotizable amine exactly as described in Section V. With the present assay the absorbance at 500 mμ is linear up to 0.35.

Definition of Unit and Specific Activity. A unit of activity is defined as the amount of enzyme giving an absorbance of 1.0 at 500 mμ under the assay conditions. Specific activity is expressed as units per milligram of protein. Protein is determined by absorbancy measurement at 280 mμ as described in Section III.

Application of Assay Method to Crude Tissue Preparations. No difficulties have been encountered with the present assay procedure, provided excess CAIR decarboxylase is present. This has been found to be the case with the present enzyme purified as far as 25-fold over the initial acetone powder extract.

Purification Procedure[48]

Details of an extensive purification of this enzyme are not available at present, but the following short procedure is sufficient for most purposes.

Chicken liver acetone powder is prepared in similar fashion as that described for pigeon liver in Section VI. Ten grams of chicken liver acetone powder are extracted with 100 ml. of 0.05 M Tris–chloride buffer, pH 7.5, for 30 minutes at 2°. The insoluble material is removed by centrifugation at 14,000 r.p.m. (Servall SS-1 centrifuge) for 20 minutes at 2°, yielding a clear supernatant fluid (83 ml.). Crystalline ribonuclease is added to give a final concentration of 0.2 mg./ml., and the mixture is incubated at 38° for 30 minutes. After the mixture is cooled to 2°, 14.6 g. of solid ammonium sulfate (0.3 saturation) are

[49] R. W. Miller and J. M. Buchanan, unpublished method; see Vol. VI [96].

slowly added, and the precipitate is removed by centrifugation and discarded. To the supernatant fluid, 11.4 g. of solid ammonium sulfate (0.5 saturation) are slowly added, and the resulting precipitate is collected by centrifugation and dissolved in 45 ml. of 0.05 M Tris–chloride buffer, pH 7.5.

The enzyme at this point is purified approximately 5-fold over the initial extract with an almost quantitative yield. It is stable when stored frozen. After the removal of salts by dialysis, further purification to about 25-fold has been attained by chromatography on DEAE-cellulose columns.

The enzyme has also been demonstrated in extracts of *N. crassa*, *E. coli*, and yeast.

Properties

Specificity. The enzyme (purified 25-fold) is free from the succino-AICAR cleavage enzyme but is contaminated with CAIR carboxylase activity.[48] When studied in the forward direction, the enzyme has an absolute requirement for CAIR and L-aspartic acid. With the nucleoside triphosphates, both UTP and CTP possessed about 33% the activity of ATP, but GTP and ITP were inactive.

When studied in the reverse direction, there are specific requirements for succino-AICAR and either P_i or arsenate. With arsenate present, the requirement for ADP is satisfied by catalytic quantities. CDP and UDP as replacements for ADP show slight activity for the reverse reaction. CDP is the more active of the two, but concentrations four times that of ADP are necessary, in the presence of either P_i or arsenate, to demonstrate a reaction rate about 20% that with ADP.

Activators and Inhibitors. Mg^{++} ($10^{-2} M$) is required for activity when the reaction is studied in either direction. Mn^{++} substitutes for the Mg^{++} in the forward direction but precipitates with P_i or arsenate in the reverse reaction.

pH Optimum. The influence of pH has been studied on the formation of succino-AICAR from AIR in the presence of excess CAIR carboxylase.[47] No difference in the rate of succino-AICAR formation was observed over the pH range 7.3 to 8.8. However, the influence of CAIR carboxylase has not been assessed.

Equilibrium. The equilibrium of the reaction is in the direction of succino-AICAR formation, but an exact determination of the equilibrium constant will necessitate studies with a preparation free from CAIR carboxylase.

Mechanism. The reaction bears some similarities to the GAR kinosynthase reaction and to the reaction leading to the synthesis of adenyl-succinic acid from IMP and aspartate.[50] A mechanism for reactions of this type has recently been proposed.[3]

IX. 5-Amino-4-imidazole-*N*-succinocarboxamide Ribotide Cleavage Enzyme

$$\text{Succino-AICAR} \rightleftarrows \text{AICAR} + \text{fumaric acid}$$

Assay Method

Principle. Cleavage of succino-AICAR yields fumaric acid and AICAR. The latter is estimated colorimetrically as diazotizable amine under conditions in which succino-AICAR does not respond to the diazotizable amine test. As an alternative procedure, the removal of AICAR may be estimated in the presence of a large excess of fumarate.

Reagents

Succino-AICAR,[49] Na salt, $1.22 \times 10^{-3}\, M$.
K phosphate, pH 7.2, 0.11 M.
Trichloroacetic acid, 30%.
H_2SO_4, 2 N.
Diazotizable amine assay reagents as described in Section I.B.

Procedure.[51] The assay system contains, in a final volume of 0.65 ml.: 0.05 ml. of succino-AICAR (0.061 micromole), 0.1 ml. of K phosphate, pH 7.2 (11 micromoles), and the enzyme. The reaction mixture is incubated at 38° for 6 minutes, and the reaction is terminated by the addition of 0.15 ml. of 30% trichloroacetic acid. After removal of denatured protein by centrifugation, an aliquot of 0.4 ml. is removed, 0.05 ml. of 2 N H_2SO_4 is added to it, and the assay for diazotizable amine is carried out as described in Section I.B. The final volume in the colorimetric determination is 0.6 ml., necessitating the use of a spectrophotometer equipped for microcuvettes for the absorbancy measurement. The absorbancy is measured at 540 mμ, and the quantity of AICAR may be calculated as indicated in Section I.B.

Definition of Unit and Specific Activity. A unit of activity is defined as the amount of enzyme causing an absorbance of 1.0 per minute at 540 mμ. Specific activity is expressed as units per milligram of protein. Protein is estimated by the procedure of Lowry *et al.*[52]

[50] I. Lieberman, *J. Biol. Chem.* **223**, 327 (1956).
[51] R. W. Miller, L. N. Lukens, and J. M. Buchanan, *J. Biol. Chem.* **234**, 1806 (1959).
[52] O. H. Lowry, N. J. Rosebrough, A. L. Farr, and R. J. Randall, *J. Biol. Chem.* **193**, 265 (1951); see also Vol. III [73].

Application of Assay Method to Crude Tissue Preparations. No difficulties have been encountered in applying the assay procedure to enzyme preparations from a variety of sources.

Purification Procedure[51]

The following purification procedure for the enzyme utilizes the first four steps of the yeast adenylosuccinase preparation of Carter and Cohen.[53] The yield and specific activity of the enzyme in the autolyzate and after the heat step are indicated in Table VII as steps 1 and 2.

Step 3. Chromatography on DEAE-cellulose. The supernatant fluid (175 ml.) from step 2 is thoroughly dialyzed near 0° against continuously changing $0.004\,M$ K phosphate, pH 7.8. This step should not be prolonged beyond 5 hours, since the enzyme is unstable at low ionic strengths. Tests should be run to ascertain when the removal of ammonium sulfate is complete. DEAE-cellulose is washed with $1\,N$ NaOH, water, $0.1\,M$ K phosphate, pH 7.8, and finally water. The cellulose is made up into a column 25×2 cm. and percolated with $0.005\,M$ K phosphate, pH 7.8. The dialyzed enzyme (about 200 ml.) is passed through the column, followed by 100 ml. of $0.005\,M$ K phosphate, pH 7.8. A gradient elution is carried out with use of a mixer of 500 ml. of $0.005\,M$ K phosphate, pH 7.8, and a reservoir containing 1 l. of $0.14\,M$ K phosphate, pH 6.8. The enzyme begins to appear in the eluate in a total volume of 300 ml. after approximately 600 ml. have passed through the column. Fractions having activity are pooled and adjusted to pH 7.8 with $0.1\,N$ KOH. It should be noted that the pH conditions for the chromatography step have been modified from that originally reported,[51] resulting in better recovery and greater stability.

Concentration of the enzyme is achieved by rechromatographing on a column of DEAE-cellulose (30×1.2 cm.). The enzyme is diluted with 3 vol. of cold distilled water and passed through the column which has been previously washed with $0.005\,M$ K phosphate, pH 7.8. Elution of the enzyme occurs rapidly, in a volume of about 50 ml., on passage of $0.2\,M$ K phosphate, pH 6.8, through the column. The preparation at this point is 190-fold purified over the initial autolyzate.

The enzyme is fairly stable on storage at $-15°$ and withstands repeated freezing and thawing. A preparation of lower activity, but similarly free from fumarase activity, has been purified 15-fold from chicken liver.[51] Activity has also been found in cell-free extracts of *N. crassa*[51] and *E. coli*.[54]

[53] C. E. Carter and L. H. Cohen, *J. Biol. Chem.* **222**, 17 (1956); see also Vol. VI [11].

[54] J. S. Gots and E. G. Gollub, *Proc. Natl. Acad. Sci. U.S.* **43**, 826 (1957).

TABLE VII

SUMMARY OF PURIFICATION OF SUCCINO-AICAR CLEAVAGE ENZYME (ADENYLOSUCCINASE)

Step	Fraction	Adenylosuccinase[a]			Succino-AICAR cleavage enzyme			Ratio, $\frac{a}{b}$
		Activity, units	Yield, %	Specific activity (a), units/mg. protein	Activity, units	Yield, %	Specific activity (b), units/mg. protein	
1	Yeast autolyzate supernatant	2300	100	0.055	9600	100	0.20	0.27
2	Supernatant fluid after ammonium sulfate and heat steps	1500	57	0.66	5900	61	2.8	0.24
3	DEAE-cellulose chromatography	1100	48	9.90	4400	46	38.	0.26

[a] Adenylosuccinase was assayed by the method of Carter and Cohen.[53]

Properties

Specificity. The purified enzyme is free from fumarase and arginosuccinase but contains appreciable adenylosuccinase activity. Several lines of evidence now strongly suggest that both adenylosuccinase and the succino-AICAR cleavage enzyme represent a single protein entity. As can be noted in Table VII, there is no change in the ratio of the two activities over the course of purification of the enzyme from yeast autolyzates 190-fold. When either reaction is studied in the reverse direction, there is an absolute requirement for fumarate.[51, 53] The succino-AICAR cleavage enzyme is more resistant to heat denaturation in the presence of adenylosuccinate than in its absence.

There is also strong genetic evidence for the identity of the two activities. Several single-step adenine-requiring "F" mutants of *N. crassa* have been shown to lack both adenylosuccinase and succino-AICAR cleavage activity.[51, 55] A similar finding has been made with certain purine-requiring mutants of *E. coli.*[54]

Activators and Inhibitors. Both enzymatic activities are strongly influenced as regards their stability and enzymatic activity by ionic strength. With monovalent salts optimal stability and enzymatic activity occur over the concentration range of 0.05 to 0.1 M.

The 6-mercapto analog of adenylosuccinate (6-(L-1,2-dicarboxyethylmercapto)-9-β-D-ribofuranosylpurine 5'-phosphate) inhibits both enzymatic activities.[51, 56] Adenylosuccinate itself will competitively inhibit the cleavage of succino-AICAR under conditions in which much larger amounts of argininosuccinate, AMP, succinate, or D,L-asparate are inactive.[51]

pH Optimum. The pH optimum for succino-AICAR cleavage is 7.3 in 0.01 M phosphate buffers. No activity is found at pH 4 or 9. In 0.01 M Tris buffers the optimum is at pH 7.8.[51] Adenylosuccinase has a similar pH optimum.[53]

Equilibrium. The equilibrium lies in the direction of cleavage with an equilibrium constant of 2.3×10^{-3} moles/l. in 0.01 M K phosphate, pH 7.2, at 37°. This value is quite similar to the equilibrium constant for the adenylosuccinase reaction.[53]

Kinetics. The K_m for succino-AICAR is $1.1 \times 10^{-4} M$ over the concentration range from 8×10^{-5} to $9 \times 10^{-4} M$ and in the presence of 0.04 M K phosphate, pH 7.2.

[55] N. H. Giles, C. W. H. Partridge, and N. J. Nelson, *Proc. Natl. Acad. Sci. U.S.* 43, 305 (1957).

[56] A. Hampton, M. H. Maguire, and J. M. Griffiths, *International Abstracts of Biological Sciences 40, 4th International Biochemical Congress*, Supplement, Pergamon Press, London, 1958.

X. 5-Amino-4-imidazolecarboxamide Ribotide Transformylase

$$\text{AICAR} + N^{10}\text{-formyl-THFA} \rightleftarrows \text{Formyl-AICAR} + \text{THFA}$$

Assay Method[31]

Principle. The transformylation of AICAR is estimated by the removal of nonacetylatable diazotizable amine. Under the assay conditions inosinicase is present and converts the formyl-AICAR formed to IMP. Alternatively, in the presence of excess GAR transformylase (see Section IV), the reaction may be studied in reverse, with measurement of the appearance of AICAR from either formyl-AICAR or IMP.[22]

Reagents

AICAR,[57] 0.002 M.

N^{10}-formyl-THFA,[33] 0.004 M; the reagent is prepared by dissolving isoleucovorin chloride in water, adjusting the pH to 10 with 1 N NH$_4$OH, and holding the solution anaerobically at 37° for 15 minutes.[35] N^{10}-Formyl-THFA is unstable, and aqueous solutions should be prepared shortly before use.

KCl, 0.2 M.

Tris–chloride buffer, pH 7.4, 0.6 M.

Trichloroacetic acid, 10%.

Acetic anhydride.

H$_2$SO$_4$, 0.2 N.

Diazotizable amine assay reagents as listed in Section I.B.

Procedure. The assay system contains, in a final volume of 0.5 ml.: 0.05 ml. of AICAR (0.1 micromole), 0.05 ml. of N^{10}-formyl-THFA (0.2 micromole), 0.05 ml. of KCl (10 micromoles), 0.05 ml. of Tris–chloride buffer, pH 7.4 (30 micromoles), and enzyme. The vessels are incubated at 38° for 20 minutes, and the reaction is terminated by the addition of 0.4 ml. of 10% trichloroacetic acid. The denatured protein is sedimented by centrifugation, and 0.1 ml. of acetic anhydride is added to each vessel with mixing. After 20 minutes, 4 ml. of 0.2 N H$_2$SO$_4$ are added, followed by 0.5-ml. quantities of each of the diazotizable amine assay reagents at the appropriate time intervals as described in Section I.B. After 30 minutes the absorbancy at 540 mμ is determined against a blank, with AICAR omitted, that has been carried through the incubation and amine assay. The amount of AICAR removed is estimated from the decrease in

[57] J. G. Flaks, M. J. Erwin, and J. M. Buchanan, *J. Biol. Chem.* **228**, 201 (1957); see also Vol. VI [96].

absorbancy at 540 mμ and may be calculated from the molecular absorbancy of 26,400 at this wavelength.

Definition of Unit and Specific Activity. A unit of activity is defined as the amount of enzyme catalyzing the removal of 0.1 micromole of AICAR under the assay conditions. Specific activity is expressed as units per milligram of protein. Protein is estimated by absorbancy measurement at 280 mμ as indicated in Section III.

Application of Assay Procedure to Crude Tissue Preparations. The rate of the reaction depends not only on the removal of AICAR but also on the removal of the product, formyl-AICAR, as catalyzed by inosinicase. The transformylase appears to be the limiting reaction in this sequence, since formyl-AICAR does not accumulate to any observable extent. In all tissue preparations thus far studied, and with the purified enzyme described below, the amount of inosinicase present has been found to be in excess. It has thus far been impossible to separate the two activities. Under the assay conditions and with excess inosinicase present, the rate of the reaction is proportional to both time and enzyme concentration, and no difficulties have been encountered in assaying the reaction in a wide variety of tissue preparations. However, in the event that inosinicase is absent the assay procedure may have to be revised, or an alternative procedure used. It is therefore suggested that assays for inosinicase (cf. Section XI) should be carried out along with the present procedure.

The statement regarding formyl donors made in Section IV applies to this enzyme as well.

Purification Procedure[31]

Step 1. Acetone Powder Extract. Acetone powders of fresh chicken livers are prepared as described in Section IV. All operations below are carried out at 2° unless otherwise noted. Fifty grams of chicken liver acetone powder are extracted for 30 minutes with 500 ml. of 0.01 M Tris–chloride buffer, pH 7.4. The insoluble residue is removed by centrifugation at 12,800 $\times$ g for 10 minutes.

Step 2. First Ammonium Sulfate Step. Thirty-six milliliters of 1 M Tris–chloride buffer, pH 8.0, are added to the acetone powder extract (356 ml.), followed by the slow addition of 95 g. of ammonium sulfate with stirring. After 30 minutes the precipitate is removed by centrifugation, and 67 g. of ammonium sulfate are slowly added to the supernatant solution. After 30 minutes the resulting precipitate is collected by centrifugation, dissolved in 70 ml. of 0.01 M Tris–chloride buffer, pH 7.4, and dialyzed overnight against 4 l. of the same buffer.

Step 3. Zinc-Ethanol Fractionation. Fifty milliliters of 0.01 M Tris–

chloride buffer, pH 7.4, are added to the dialyzed ammonium sulfate fraction (105 ml.), followed by 1.08 ml. of saturated ammonium sulfate. Slowly, 1.54 ml. of 1 M zinc acetate are added dropwise with stirring. After 10 minutes the precipitate is removed by centrifugation. 3 N NH$_4$ OH is added to the supernatant solution until the pH reaches 9.0, and the resulting precipitate is removed by centrifugation. The supernatant solution (140 ml.) is cooled to —5°, and 70 ml. of 90% ethanol are slowly added with stirring while the temperature is gradually lowered to —15°. After 15 minutes, the precipitate is collected by centrifugation at —15° and dissolved in 50 ml. of 0.01 M Tris–chloride buffer, pH 7.4, containing 0.01 M EDTA, pH 7.4.

Step 4. Second Ammonium Sulfate Step. Five milliliters of 1.0 M Tris–chloride buffer, pH 8.0, are added to the zinc ethanol fraction (50 ml.), followed by 16.1 g. of ammonium sulfate. After 30 minutes of stirring, the precipitate is removed by centrifugation, and 4.4 g. of ammonium sulfate are slowly added to the supernatant solution. The resulting precipitate is collected by centrifugation, dissolved in 10 ml. of 0.01 M Tris–chloride buffer, pH 7.4, and dialyzed overnight against 4 l. of the same buffer.

Step 5. Alumina C$_\gamma$ Supernatant. Forty milliliters of 0.01 M Tris–chloride buffer, pH 7.4, are added to the dialyzed ammonium sulfate fraction (17 ml.), followed by 5.7 ml. of 1.0 M Tris–chloride buffer, pH 8.0. This solution is stirred with 290 mg. of alumina C$_\gamma$ gel[28] for 10 minutes, after which time the gel is removed by centrifugation and discarded.

Step 6. Ethanol Fractionation. The alumina C$_\gamma$ supernatant solution (63 ml.) is cooled to —5°, and 46 ml. of 90% ethanol are slowly added in dropwise fashion with constant stirring. The temperature is gradually lowered to —15° during the ethanol addition. After 20 minutes the precipitate is removed by centrifugation at —15° and dissolved in 7 ml. of 0.01 M Tris–chloride buffer, pH 7.4.

Step 7. Alumina C$_\gamma$ Eluates. The enzyme solution from step 6 (7 ml.) is stirred with 18 mg. of alumina C$_\gamma$[28] gel for 10 minutes, the mixture is centrifuged, and the supernatant fluid is discarded. The enzyme is eluted from the gel by stirring it with 4.5 ml. of 0.02 M K phosphate, pH 8.0, followed by centrifugation. A second elution is similarly carried out with 4.5 ml. of 0.03 M K phosphate, pH 8.0. The first and second eluates are designated in Table VIII as E$_1$ and E$_2$, respectively.

The enzyme at this point is over 40-fold purified, is stable for a number of months on storage at —15°, and withstands repeated freezing and thawing. Two other enzymes involved in the formation of serine when coupled to the AICAR transformylase reaction are removed at

TABLE VIII

SUMMARY OF PURIFICATION OF AICAR TRANSFORMYLASE AND INOSINICASE

Fraction	Volume, ml.	Protein, mg./ml.	Inosinicase		AICAR transformylase		Ratio, $\frac{a}{b}$
			Activity, units $\times 10^{-3}$	Specific activity (a) units/mg. protein	Activity units, $\times 10^{-3}$	Specific activity (b) units/mg. protein	
Acetone powder extract	356	26.2	180	19.3	81.8	8.7	2.2
First ammonium sulfate step	105	30.4	126	39.2	60.0	18.7	2.1
Zinc–ethanol step	50	13.5	88.2	130	32.9	48.1	2.7
Second ammonium sulfate step	17	16.8	57.6	200	18.0	61.9	3.2
Alumina $C\gamma$ step	63	0.55	25.2	728	3.65	106	6.9
Ethanol step	7	1.70	13.5	1130	1.93	162	7.0
Alumina $C\gamma$ eluate E_1	4.5	1.06	8.5	1780	1.23	260	6.9
Alumina $C\gamma$ eluate E_2	4.5	0.47	5.0	2340	0.77	360	6.5

step 3. GAR transformylase is removed at step 5, as is the cyclohydrolase activity.[35] For most studies, purification through step 5 is sufficient, but the stability of the enzyme is somewhat better at steps 4 or 6.

Properties

Specificity. The enzyme is specific for the nucleotide AICAR; no activity was found with the corresponding nucleoside or free base.[58] The purified enzyme, free from cyclohydrolase, requires N^{10}-formyl-THFA as the formyl donor and is inactive with $N^{5,10}$-methenyl-THFA, N^5-formyl-THFA, or N^{10}-formyl folic acid.[35] In coupling this reaction to that catalyzed by GAR transformylase, where THFA is present in catalytic amounts, cyclohydrolase must be present to effect the interconversion of the formylated-THFA derivatives (N^{10}-formyl-THFA $\rightleftarrows$ $N^{5,10}$-methenyl-THFA), since GAR transformylase has a specific requirement for $N^{5,10}$-methenyl-THFA.[35, 36]

Formyl-AICAR has never been isolated from incubations of the AICAR transformylase system, owing to the presence of inosinicase in all enzyme preparations thus far studied. However, formyl-AICAR has been shown to be a formyl donor when the AICAR and GAR transformylases are coupled to each other. In this system specificity was shown for formyl-AICAR; the corresponding nucleoside and free base were inactive.[36]

The data of Table VIII suggest little resolution of AICAR transformylase from inosinicase. The slight resolution that does occur is achieved at step 5 in the purification, and this may be due to a selective destruction of the transformylase due to dilution (the transformylase is less stable relative to inosinicase at step 5 in the purification scheme than at other stages). A similar inability to resolve the two activities has been found in several strains of *E. coli.*[59] This has led to the view that both enzymatic activities may be the function of a single protein molecule. Genetic evidence supports this view. Purine-requiring mutants of *E. coli* which lack AICAR transformylase also lack inosinicase.[60] A similar situation has been found in *S. typhimurium*[61] and *N. crassa.*[62] With the *N. crassa* mutants, complementation tests have indicated that the genes for both activities are very closely linked.

Activators and Inhibitors. K^+ is required for the AICAR trans-

[58] J. G. Flaks, L. Warren and J. M. Buchanan, *J. Biol. Chem.* **228,** 215 (1957).
[59] T. Yokota and M. G. Sevag, personal communication.
[60] A. Ottolenghi and J. S. Gots, personal communication.
[61] H. Ozeki and M. Demerec, personal communication.
[62] N. H. Giles and J. M. Buchanan, unpublished results.

formylase reaction, with optimal activity being shown over the range 3×10^{-3} to $3 \times 10^{-2}\,M$.

pH Optimum. The pH optimum has been found to be 7.4, with the reaction rate falling off rapidly below 7.0 and above 7.8. These values are identical to those for inosinicase.

Equilibrium. Although the transformylase reaction is freely reversible, the equilibrium constant has not been determined owing to the presence of inosinicase and the difficulties involved with the side reactions of THFA and its derivatives. Inosinicase drives the transformylase equilibrium to the right, but it can be driven to the left by coupling either to the GAR transformylase or to the system leading to the synthesis of L-serine.[36, 58]

XI. Inosinicase

$$\text{Formyl-AICAR} \rightleftarrows \text{IMP} + \text{H}_2\text{O}$$

Assay Method[31]

Principle. The cyclization of formyl-AICAR to IMP results in the removal of diazotizable amine which is produced on mild acid hydrolysis of formyl-AICAR.

Reagents

> Formyl-AICAR,[31] $6 \times 10^{-4}\,M$; see also Vol. VI [96] for the preparation from AICAR.
> Tris–chloride buffer, pH 7.4, 0.6 M.
> H_2SO_4, 4 N.
> Diazotizable amine assay reagents as indicated in Section I.B.

Procedure. The assay system contains, in a final volume of 0.3 ml.: 0.05 ml. of formyl-AICAR (0.03 micromole), 0.05 of Tris–chloride buffer, pH 7.4 (30 micromoles), and enzyme. The vessels are incubated at 38° for 10 minutes, and the incubation is terminated by the addition of 0.2 ml. of 4 N H_2SO_4. The tubes are capped, placed in a boiling-water bath for 5 minutes to hydrolyze the residual formyl-AICAR, and cooled. At the appropriate time intervals, 0.1-ml. quantities of the amine assay reagents are added as indicated in Section I.B. After 20 minutes the absorbancy at 540 mμ is determined on a spectrophotometer equipped with microcuvettes, with readings taken against a blank with the formyl-AICAR omitted. The amount of formyl-AICAR removed is proportional to the decrease in absorbancy at 540 mμ and may be calculated from the molecular absorbancy of AICAR of 26,400 at this wavelength.

Definition of Unit and Specific Activity. A unit of activity is defined as the amount of enzyme catalyzing the removal of 0.1 micromole of formyl-AICAR under the assay conditions. Specific activity is expressed as units per milligram of protein. Protein is estimated by absorbancy measurement at 280 mμ as indicated in Section III.

Application of Assay Method to Crude Tissue Preparations. No difficulties have been encountered in applying the assay procedure to a wide variety of crude tissue preparations. The assay procedure is linear with respect to both time and enzyme concentration in all cases tested.

Purification Procedure

As indicated in Section X, inosinicase has not yet been resolved from AICAR transformylase, and present evidence suggests that both activities are the function of a single protein entity. As such, the purification procedure described in Section X is applicable for inosinicase, and the activities at various stages of purification are indicated in Table VIII.

The purified enzyme (from steps 4 to 7) is stable for several months at −15° and withstands repeated freezing and thawing.

Properties[31]

Reference should be made to Section X for some properties of the enzyme.

Specificity. The enzyme is specific for formyl-AICAR and will not catalyze the cyclization of the corresponding riboside or free base.

Activators and Inhibitors. There appears to be no other requirements for the reaction. K$^+$, which is required for AICAR transformylase activity, has no influence on the cyclization reaction.

pH Optimum. The optimum pH is 7.4 with the reaction rate falling off rapidly below 7.0 and above 7.8.

Equilibrium.[36] The equilibrium greatly favors cyclization with a constant of the order of 1.6×10^4 Formyl-AICAR has not been isolated as a product of the ring opening from IMP.

[10] Formiminoglycine Formimino Transferase[1]

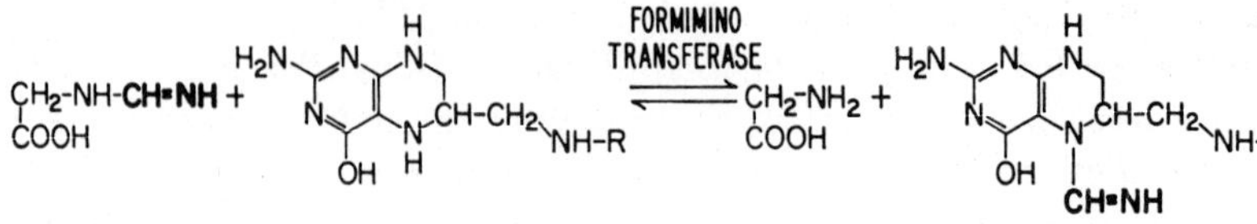

FORMIMINOGLYCINE   THF   GLYCINE   5-FORMIMINO-THF

By JESSE C. RABINOWITZ

Assay Method

Principle. 5-Formiminotetrahydrofolic acid formed in the enzymatic reaction shown by Eq. (1) is converted to 5,10-methenyltetrahydrofolic acid in the presence of an excess of formiminotetrahydrofolic acid cyclodeaminase, as shown in Eq. (2). The 5,10-methenyltetrahydrofolic acid

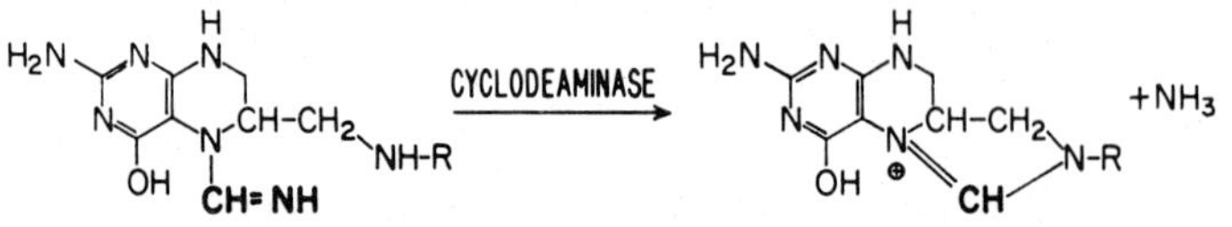

5-FORMIMINO-THF   5,10-METHENYL-THF

formed is determined by its absorption at 356 mμ, and the rate of its formation is a function of the concentration of formiminoglycine formimino transferase.

Reagents

0.1 M formiminoglycine.[2]

0.1 M 2-mercaptoethanol.

0.5 M potassium maleate buffer, pH 7.0.

0.5 M sodium versenate, pH 7.0.

0.01 M dl-tetrahydrofolic acid. Twenty-eight milligrams of tetrahydrofolic acid[3] is dissolved in 5 ml. of 1.0 M 2-mercaptoethanol and neutralized with about 4 drops of 1 N KOH.

Cyclodeaminase. Either the supernatant solution obtained in step 2 of this preparation or the purified enzyme described in Vol. VI [52] may be used.

---

[1] J. C. Rabinowitz and W. E. Pricer, Jr., *J. Am. Chem. Soc.* **78**, 5702 (1956).

[2] (a) F. Micheel and W. Flitsch, *Ann. Chem.* **577**, 234 (1952); (b) H. Tabor and J. C. Rabinowitz, *Biochem. Preparations* **VI**, 100 (1957); (c) see Vol. VI [97].

[3] J. C. Rabinowitz and W. E. Pricer, Jr., *J. Biol. Chem.* **229**, 321 (1957).

Enzyme. The enzyme is diluted in $0.05\,M$ maleate buffer at pH 7.0 for assay.

Procedure. An assay mixture is prepared by mixing 4 ml. of $0.1\,M$ formiminoglycine, 2 ml. of $0.1\,M$ 2-mercaptoethanol, 5 ml. of maleate buffer, 0.2 ml. of $0.5\,M$ sodium versenate, and water to bring the volume to 50 ml. To 2.9 ml. of this mixture, contained in a quartz cell with a 1-cm. light path, are added 50 μl. of the *dl*-tetrahydrofolic acid solution and 100 units of cyclodeaminase.[4] Then up to 50 μl. of a suitably diluted enzyme solution is added, and the increase in absorbance at 356 mμ is determined.

Definition of Unit and Specific Activity. One unit of enzyme is defined as that amount which causes an initial rate of increase in absorbance of 1.0 per minute under the above conditions. Specific activity is expressed as units per milligrams of protein. Protein is determined by using the Folin-Ciocalteu reagent with crystalline serum albumin containing 14.6% nitrogen as a standard.[5]

Application of Assay Method to Crude Tissue Preparations. The presence of cyclohydrolase interferes with the present assay, since it hydrolyzes the product to 10-formyltetrahydrofolic acid, which has decreased absorbance at 356 mμ. The amount of interference encountered in natural products, however, will depend on the relative amounts of the formimino transferase and the cyclohydrolase. In cases so far investigated, the cyclohydrolase is relatively inactive and does not seriously interfere with the assay of formimino transferase by this method.[6]

Purification Procedure

Step 1. Cell Autolyzate. Four grams of lyophilized cells of *C. cylindrosporum* (Vol. VI [97]) is suspended in 80 ml. of $0.05\,M$ potassium maleate buffer at pH 7.0, $0.05\,M$ with respect to 2-mercaptoethanol. The mixture is incubated for 30 minutes at 37°. The autolyzate is centrifuged for 10 minutes at $105,000 \times g$, and the residue is discarded.

Step 2. Isoelectric Precipitate I. Fifty milliliters of the autolyzate is

---

[4] This is usually equivalent to about 0.1 ml. of the pH supernatant fraction obtained in step 2 of this purification.

[5] Several alternative assays for the transferase have been used. One of these, described in Vol. V [105], depends on the conversion of formiminotetrahydrofolic acid to 5,10-methenyltetrahydrofolic acid by acid treatment. Another assay is based on the increase in absorbance at 280 mμ resulting from the conversion of tetrahydrofolic acid to 5-formiminotetrahydrofolic acid.[10]

[6] The assay based on the conversion of formiminotetrahydrofolic acid to 5,10-methenyltetrahydrofolic acid by acid described in Vol. V [105] is not affected by the occurrence of cyclohydrolase and may, therefore, be used in preference to the assay method described here when large amounts of cyclohydrolase are present.

added to 200 ml. of 0.1 M potassium acetate at pH 5.0 and allowed to incubate at 0° for 10 minutes. The mixture is centrifuged at 30,000 × g, and the precipitate is discarded. To the supernatant solution is added 2 N acetic acid to bring the pH to 4.1 (about 12.5 ml.). The mixture is centrifuged at 30,000 × g.[7] The pellet is washed with 40 ml. per tube of 0.01 M potassium acetate buffer at pH 4.1. The tubes are drained carefully, and the interiors wiped out with tissue. The wash is repeated. The precipitate in each centrifuge tube is dissolved in 20 ml. of water, and 1 drop of 1 N KOH is added to adjust the solution to pH 7.0. Solutions are combined and diluted to 125 ml. Any insoluble material is sedimented by centrifugation and discarded.

Step 3. Calcium Phosphate Treatment. To 125 ml. of the combined isoelectric precipitate obtained in step 2 is added 21 ml. of calcium phosphate gel (9 mg./ml.).[8] The suspension is stirred for 5 minutes with a magnetic stirrer. The precipitate is removed by centrifugation at 30,000 × g for 5 minutes.

Step 4. Alumina C_γ. To 140 ml. of the supernatant solution obtained in the previous step is added 50 ml. of alumina C_γ gel[9] (7.5 mg./ml.) diluted 1 to 5. The mixture is stirred for 5 minutes at room temperature with a magnetic stirrer.

Step 5. Isoelectric Precipitate II. To the supernatant solution obtained in the previous step is added 0.01 vol. (1.9 ml.) of 1 M potassium

SUMMARY OF PURIFICATION PROCEDURE

Step	Volume, ml.	Units	Protein, mg./ml.	Specific activity, units/mg. protein	Yield, %
1. Autolyzate	60	9600	13.2	12.1	100
2. Isoelectric precipitate I	145	4700	1.0	31.5	49
3. Calcium phosphate	169	5200	0.71	43.3	54
4. Alumina C_γ	190	4100	0.24	90.0	43
5. Isoelectric precipitate II	50	3750	0.69	109	39

acetate buffer at pH 5.0. The pH should be 5.0. The mixture is incubated in an ice bath for 10 minutes, and the precipitate is removed by centrifugation. The pellet is dissolved in 15 ml. of 0.05 M maleate buffer

[7] The supernatant solution is retained for the preparation of cyclodeaminase and cyclohydrolase and is stored at 0° after the solution has been adjusted to pH 6.8 with 6.9 ml. of 5 N KOH. This fraction may be used as a source of cyclodeaminase in the assay of formimino transferase.

[8] D. Keilin and E. F. Hartree, *Proc. Roy. Soc.* **B124**, 397 (1938).

[9] R. Willstätter and H. Kraut, *Ber.* **56**, 1117 (1923).

at pH 7, 0.01 M with respect to Versene. The insoluble material is removed by centrifugation. The enzyme may be stored in the frozen state in the presence of 90 mg. of dialyzed plasma albumin per 50 ml. of enzyme solution.

A summary of the purification procedure is given in the table.

Properties

Physical Properties. The molecular weight based on sedimentation measurements is approximately 160,000.

Specificity. The enzyme is entirely specific for formiminoglycine. The following compounds were tested and found to be completely inactive: formiminoalanine,[10] formiminoglycinamide,[10] formiminoglycine methyl ester,[10] formiminoglutamic acid,[2b] formiminoaspartic acid,[2b] and formylglycine.

Activators and Inhibitors. Under the conditions of the assay described here, the optimal activity is obtained at pH 7 in the phosphate or maleate buffers.[11]

Equilibrium. The enzymatic reaction is readily reversible, and the equilibrium constant for the reaction shown by Eq. (1) is about 0.2 in maleate buffer at pH 7.[1]

Kinetic Properties. The following kinetic constants were obtained with the most highly purified enzyme preparation and were found to be independent of the type assay used: K_m (formiminoglycine) $= 1 \times 10^{-4}$ M; K_m (*l*-tetrahydrofolic acid) $= 2.8 \times 10^{-6}$ M.

[10] K. Uyeda, Ph.D. Thesis. University of California, Berkeley, California, 1962.

[11] When the activity is determined by the direct assay with measurements of increase in absorbance at 280 mμ,[10] optimal activity is observed at pH 7.6 in Tris buffer or in diethanolamine buffer. With the assay in which 5-formiminotetrahydrofolic acid is converted to 5,10-methenyltetrahydrofolic acid with acid, optimal and comparable activity was noted in ethanolamine buffer at pH 9.0.

[11] Amination of IMP to AMP: Adenylosuccinate Synthase (*Escherichia coli*)

IMP   Adenylosuccinate

By Irving Lieberman

Assay Method

Adenylosuccinate synthesis is estimated most simply with a Beckman DU spectrophotometer by the increase in optical density at 280 mμ. To perform the assay properly, great care in pipetting is required, particularly of the ultraviolet-absorbing reagents. Greater sensitivity may be obtained with the use of C^{14}-aspartate. In this procedure, adenylosuccinate is measured as the radioactivity remaining on an anion-exchange column after elution of the aspartic acid.[1] Only the spectrophotometric assay will be described.

Reagents and Procedure. The test mixtures (0.7 ml.) contain 0.1 ml. of glycine buffer (1 M, pH 8.0), 0.04 ml. of MgCl$_2$ (0.1 M), 0.04 ml. of ATP (0.001 M), 0.04 ml. of phosphoenolpyruvate (0.01 M), 30 units of pyruvate phosphokinase, 0.05 ml. of L-aspartate (0.01 M), 0.01 ml. of GTP (0.001 M), 0.06 ml. of IMP (0.005 M), and the enzyme preparation. After 30 minutes at 37°, 0.5 ml. of perchloric acid (7%) is added, and insoluble material is removed by centrifugation. The increase in optical density (280 mμ) is determined by comparison with a control mixture to which one omitted component (IMP, GTP, aspartate, or enzyme) is added after the perchloric acid.

Crude Preparations. Enzyme levels of crude preparations cannot be accurately estimated by the procedures indicated. The major difficulty arises from the enzymatic hydrolysis of the reaction product to AMP and fumarate.[2]

[1] I. Lieberman, *J. Biol. Chem.* **223**, 327 (1956).
[2] C. E. Carter and L. H. Cohen, *J. Am. Chem. Soc.* **77**, 499 (1955).

Purification Procedure

The purification procedure is simple and reproducible. Recoveries of activity greater than 100% result from the increased efficiency of the assay with the gradual removal of adenylosuccinase. The final preparation appears to be free of this enzyme. Manipulations are carried out at 0° to 3° unless otherwise indicated.

Step 1. Preparation of Cell-Free Extract. Escherichia coli, strain B, is grown on a glucose and inorganic salts medium[3] at 37° with vigorous shaking. When the optical density, measured with a Coleman junior spectrophotometer at 540 mμ, is approximately 0.7, the cultures are cooled, and the cells are collected by centrifugation at 3°. Cell-free extracts are prepared by grinding with alumina[4] (Alcoa A-301, 2.5 g./g. of packed, wet cells) and extracting with potassium phosphate buffer (0.001 M, pH 7.2, 5.5 ml./g. of wet cells). Insoluble material is removed by centrifugation (approximately 10,000 $\times$ g).

Step 2. Treatment with Streptomycin. To 50 ml. of cell-free extract, 13 ml. of a 5% solution of streptomycin sulfate is added with stirring. After 5 minutes, the stringy precipitate is discarded by centrifugation.

Step 3. Low pH Treatment. I. The pH of the streptomycin fraction (56 ml.) is adjusted to 5.4 by the addition of 0.1 M acetic acid (6 to 7.5 ml.) with stirring. The solution is kept in a water bath at 37° for 10

PURIFICATION OF ADENYLOSUCCINATE SYNTHSEA

Enzyme fraction	Volume of solution, ml.	Total units	Total protein, mg.	Specific activity, units/mg. protein
Cell-free extract	50	2200	590	3.7
Streptomycin	56	3920	353	11.1
Low pH I	60	7100	226	31.4
Ammonium sulfate	40	4240	53	80.0
Low pH II	20	2060	9.0	229

to 15 minutes with constant stirring. It is then cooled, and the supernatant solution, obtained by centrifugation, is neutralized with 1 M KOH.

Step 4. Precipitation with Ammonium Sulfate. To the low pH fraction I (60 ml.), 30 ml. of glycine buffer (1 M, pH 9.6) followed by 30.9 g.

[3] One and one-half grams of KH_2PO_4, 13.5 g. of Na_2HPO_4, 0.2 g. of $MgSO_4 \cdot 7H_2O$, 2.0 g. of NH_4Cl, 10 mg. of $CaCl_2$, and 0.5 mg. of $FeSO_4 \cdot 7H_2O$ are dissolved in distilled water to 900 ml. After sterilization by autoclaving, 100 ml. of a sterile 4% solution of glucose is added.

[4] H. McIlwain, *J. Gen. Microbiol.* **2,** 288 (1948).

of ammonium sulfate are added with stirring. After 5 minutes, the precipitate is discarded by centrifugation, and an additional 8.4 g. of ammonium sulfate is added to the supernatant fluid. The precipitate, collected after 5 minutes, is dissolved in water to a volume of 40 ml.

Step 5. Low pH Treatment II. Acetic acid (2.5 to 3.5 ml., 0.1 M) is slowly added to the ammonium sulfate fraction with stirring to lower the pH to 4.3. After 10 minutes, the precipitate is obtained by centrifugation and dissolved in glycine buffer (0.05 M, pH 8.5) to a volume of 20 ml.

Properties

Specificity. The bacterial enzyme appears to be specific for the three substrates, IMP, GTP, and L-aspartate.

pH and Other Factors. With glycine buffers, the optimal pH range is 7.3 to 7.8. At pH 6.6 and 8.4, the reaction rates are 50 to 60% of that in the optimal range. In the absence of added Mg^{++}, no adenylosuccinate synthesis occurs. Concentrations below 0.003 M do not afford a maximum reaction rate. The Mg^{++} requirement is not satisfied by Co^{++} or Zn^{++}, but with Mn^{++} and Ca^{++} the rate of adenylosuccinate synthesis is about one-half of that with Mg^{++}.

Inhibitors. Approximately 50% inhibition of adenylosuccinate synthesis occurs with $10^{-4} M$ GDP, 0.01 M glycylglycine buffer, 0.08 M Tris, and 0.03 M NaF. The inhibition by glycylglycine can be almost completely reversed by increasing the Mg^{++} concentration.

[12] Crystalline Adenylic Acid Deaminase from Rabbit Skeletal Muscle

$$AMP \rightarrow IMP + NH_3$$

By Y. P. LEE

Assay Method

Principle. The activity of the enzyme can be measured either spectrophotometrically or colorimetrically. The spectrophotometric method is based on the decrease of optical density at 265 mμ or the increase at 240 mμ which was introduced by Kalckar.[1] Experiments at high substrate concentration (above 0.063 micromole/ml. of reaction mixture) can be carried out in test tubes in which the reaction will be stopped by the addition of 0.1 ml. of 60% perchloric acid. After proper dilution

[1] H. M. Kalckar, *J. Biol. Chem.* **167**, 429, 461 (1947).

with water, the optical density can be determined with a spectrophotometer at 265 mμ or 240 mμ. The colorimetric method is based on the formation of ammonia formed by the deamination with conventional Nessler's reagent.

Reagents

$1.35 \times 10^{-3}\ M$ 5'-AMP in H_2O.

$0.15\ M$ succinate buffer pH 6.4.

Enzyme solution. Dilute to approximately 10 units (5 to 20 units) per 0.1 ml. with $0.5\ M$ KCl solution—in the case of the muscle extract about 0.5 to 2 mg. per 0.1 ml.; in the case of crystalline preparation 0.5 to 2 μg. per 0.1 ml.

Procedure. One-tenth milliliter of diluted enzyme solution is added to 2.9 ml. of prewarmed (30°) reaction mixture which contains 0.1 ml. of $1.35 \times 10^{-3}\ M$ AMP, 2.0 ml. of $0.15\ M$ succinate buffer, compounds to be tested if any, and water to 2.9 ml.

Definition of Unit and Specific Activity. Under assay conditions, the reaction is first-order with respect to AMP, and the initial velocity is directly proportional to the enzyme concentration. Ten units is defined as that amount of enzyme which catalyzes an optical density change from 0.55 to 0.40 in 1 minute at 265 mμ. The specific activity is expressed as units per milligram of protein. Protein may be determined by the biuret reaction.

Purification Procedure[2]

Extraction of Muscle. The back and leg muscles of rabbits are excised immediately after decapitation of the rabbits and are chilled in ice. The muscle is passed through a chilled meat chopper and then homogenized for 1 minute in a Waring blendor with 3.5 vol. of a solution containing $0.3\ M$ KCl, $0.09\ M$ KH_2PO_4, and $0.06\ M$ K_2HPO_4 at pH 6.5. The deaminase is extracted with efficient stirring at 3° for 1 hour and is centrifuged at $1500 \times g$ for 30 minutes. The residue is re-extracted with 2 vol. of the same buffer for 1 hour at 3°. The combined supernatant liquid is passed through two layers of cheesecloth to remove the lipid layer.

Low Salt Fractionation. The combined extract is diluted with 9 vol. of chilled water with stirring over 15 minutes at 3° and then is stirred for 10 more minutes. The suspension is centrifuged (Sharples) or is allowed to stand in the cold room overnight, and the clear supernatant fluid is aspirated and then the lower layer centrifuged. The resulting

[2] Y. P. Lee, *J. Biol. Chem.* **227**, 987 (1957).

precipitate is dissolved in 0.5 M KCl (usually about 200 ml. per 100 g. of original ground muscle), and the protein concentration is adjusted to 1.5% by addition of 0.5 M KCl.

Heat Fractionation. The resulting viscous cloudy solution is brought to 0.02 M MgCl$_2$ by the addition of 1 M MgCl$_2$ solution, and the pH of the solution is adjusted to 6.8 with 1 M K$_2$HPO$_4$ solution. One-liter portions of this solution, efficiently stirred, are heated in a 2-l. stainless-steel beaker to 50 ± 1° in warm running water (57 ± 2°). This usually requires about 3 minutes. The suspension is maintained at 50 ± 1° for 2 minutes and then chilled in a −10° alcohol bath until the temperature drops to 3°. The denatured protein is centrifuged at 1500 × g for 30 minutes. The precipitate is suspended in 0.5 M KCl solution (100 ml. per 100 g. of original ground muscle) and again heated to 45°. The coagulated elastic protein is quickly filtered through one layer of cheesecloth. The supernatant fluid and the filtrate are combined.

Ethanol Fractionation. The pH of the combined cloudy solution is adjusted to 6.5 with 0.5 N acetic acid and chilled to −2° in a −10° bath. Ninety-five per cent ethanol is slowly added to 7% ethanol (v/v) at such a rate that a temperature of −2° is maintained. The suspension is centrifuged at 1500 × g at −2° for 30 minutes. The resulting cloudy supernatant liquid is filtered through a thin layer of Celite on a Büchner funnel under mild suction in order to get a clear solution, and the fluffy precipitate is again centrifuged at higher speed (19,000 × g) for 10 minutes. The slightly opalescent filtrate and the supernatant liquid are combined and brought to 23% (v/v) with 95% ethanol at −5° at such a rate that the temperature stays at −5° in a −10° bath. After the temperature is allowed to drop to −10°, the suspension is centrifuged at 1500 × g at −10° for 30 minutes. The precipitate is dissolved in 0.5 M KCl (about 20 ml. per 100 g. of original ground muscle) to give a protein concentration of about 0.5%.

Ammonium Sulfate Fractionation. Most of the deaminase activity is obtained between 1.26 and 2.26 M ammonium sulfate by the addition of solid ammonium sulfate at pH 6.5 at 3°. The pH is maintained by addition of ammonia solution during the addition of solid ammonium sulfate. The precipitate, collected by centrifugation, is dissolved in 0.1 M KCl (about 10 ml. per 100 g. of original ground muscle) to give a protein concentration of about 0.5%.

Low Salt Fractionation. The ammonium sulfate fraction is adjusted to pH 6.5 and dialyzed against 10 vol. of 0.02 M KCl solution with stirring at 3° for 8 hours. The precipitate is dissolved in 0.5 M KCl (about 5 ml. per 100 g. of original ground muscle) to give a protein concentration of 0.5%.

Calcium Phosphate Gel Fractionation. Calcium phosphate gel is prepared according to the method described by Keilin and Hartree.[3] Two milliliters of the gel (20 mg. of dry weight per milliliter) is added to each 10 ml. of the low salt fraction. The suspension is adjusted to pH 6.5 with 0.5 N acetic acid. After being stirred for 30 minutes at 3°, the precipitate is collected by centrifugation. If all the enzyme has not been absorbed, successive small amounts of Ca phosphate gel (0.3 ml. of gel suspension per 10 ml. of initial solution) are added and collected by centrifugation. Each of the gel fractions is washed individually with 0.3 M KCl solution (equal to the volume of the gel suspension added) and then eluted twice with 0.08 M K_2HPO_4, pH 8.5 (half the volume of the gel suspension which is added), at 3° for 2 hours in each elution. The combined gel residues are eluted with 0.1 M K_2HPO_4, pH 8.5, solution overnight, and the eluate is saved.

Crystallization. The eluates of high specific activity (more than 4000 units/mg.) are collected, and the pH of this clear solution is adjusted to 8.0. The solution is chilled to −3° in a −10° bath, and 0.15 vol. of 95% ethanol is added slowly with mild stirring. After the temperature is allowed to drop to −8° the suspension is centrifuged at 19,000 × g at −8° for 10 minutes. A small amount of 0.5 M KCl of

PURIFICATION OF 5′-ADENYLIC ACID DEAMINASE
(3 kg. of rabbit skeletal muscle)

Fraction		Total protein, mg.	Total units	Units per mg.	Purification	Yield, %
1	Muscle extract	340,000	3,730,000	11[a]	1	(100)
2	Low salt precipitate	168,000	3,360,000	20	1.8	90
3	Heat-treated fraction	49,500	2,470,000	50	4.5	66
4	Ethanol fraction	3,960	1,270,000	320	29	34
5	Ammonium sulfate fraction	1,320	960,000	720	65.5	25.5
6	Low salt fraction	850	930,000	1,100	100	24
7	Ca phosphate gel eluate	120	540,000	4,500	410	14.5
8	Crystals (first crop only)	18	183,000	10,200	920	5

[a] This value is not corrected for the inhibition caused by phosphate [see Y. P. Lee, *J. Biol. Chem.* **227**, 999 (1957)].

solution is added to the resulting precipitate to make a saturated protein solution at room temperature (22 ± 2°). This clear viscous solution is cooled very slowly with mild stirring. The crystals come out during the drop of the temperature. After standing at 0° overnight, crystals are collected by centrifugation and then washed with 2 vol. of 0.1 M KCl

[3] D. Keilin and E. F. Hartree, *Proc. Roy. Soc.* **B124**, 397 (1938).

solution. The packed crystals are stored at 0° after the centrifugation. More crystals can be obtained from the mother liquid, the 0.15 vol. ethanol supernatant fraction, and the Ca phosphate gel eluates of lower specific activity by repeating the fractionation with ethanol described in this step.

The results of a typical fractionation are summarized in the table. The over-all purification is about 800-fold after correction for inhibition by phosphate present in the initial KCl–phosphate muscle extract.

Properties

Molecular Weight.[4] A molecular weight of 3.2×10^5 was calculated, based on sedimentation coefficient 12.29×10^{-13} sec. and diffusion coefficient 3.76×10^{-7} cm.^{2} sec.$^{-2}$ The f/f_0 was calculated to be 1.22.

Isoelectric Point. The isoelectric point is approximately at pH 5.6.

Specificity.[5] The protein has been found to be free from the following enzymes: ATP-AMP transphosphorylase, nucleoside triphosphatase, ATP-creatine transphosphorylase, nucleotide pyrophosphatase, nucleotidase; it does not remove the amino group from adenine, 2,6-diaminopurine, adenosine, 2'-AMP, 3'-AMP, ADP, ATP, DPN, TPN, cytosine, and guanosine. 5'-Deoxyadenylic acid is slowly deaminated by crystalline enzyme, and the initial velocity (first 30 seconds) is about 1% of that with 5'-AMP as substrate.

The enzyme exhibits a sharp peak at pH 6.4 in 0.1 M succinate buffer and at pH 6.1 when citrate is used as a buffer.

The K_m is calculated to be 1.43×10^{-3} M at pH 6.4 and 30°.

For every 1.0 mole of AMP reacting, 1.0 mole of IMP and 1.0 mole of NH_3 are formed, but the reaction is not reversible.

[4] Y. P. Lee, *J. Biol. Chem.* **227**, 993 (1957).
[5] Y. P. Lee, *J. Biol. Chem.* **227**, 999 (1957).

[13] Synthesis and Reduction of GMP

By BORIS MAGASANIK

In Enterobacteriaceae IMP and GMP are linked by a cyclic series of irreversible reactions.

$$\begin{array}{ccc}
 & \text{DPN}^+ & \\
\text{IMP} & \xrightarrow{\hspace{2cm}} & \text{Xanthosine-5'-P} \\
\text{TPNH} \diagdown \; 3 \quad 1 \quad 2 \diagup & & \diagup \begin{array}{l}\text{NH}_3, \\ \text{ATP}\end{array} \\
 & \text{GMP} &
\end{array}$$

Reactions 1 and 2, catalyzed by the enzymes IMP dehydrogenase and xanthosine-5′-P aminase, respectively, are essential steps in the biosynthesis of GMP.[1,2] Loss of either enzyme by mutation results in a requirement for guanine. Mutants blocked in reaction 1 can use xanthine in place of guanine, whereas mutants blocked in reaction 2 require guanine specifically.

Reaction 3, catalyzed by GMP reductase, enables the organisms to use exogenously supplied xanthine or guanine for the biosynthesis of adenine nucleotides.[3] The loss of GMP reductase by mutation does not result in an altered growth requirement; however, such a mutant can incorporate exogenously supplied xanthine or guanine only into guanine nucleotides and not, like the wild strain, into both guanine and adenine nucleotides.[4]

Similar enzymes have been found in animal tissues. IMP dehydrogenase and xanthosine-5′-P aminase have been found in pigeon liver[5,6] and in rabbit bone marrow and thymus.[7] The bacterial and animal xanthosine-5′-P aminases differ in the nature of the preferred amino donor. The bacterial enzyme utilizes only NH_3 and not glutamine,[2] whereas the animal enzymes utilize glutamine well and NH_3 very poorly.[6,7] A recent report suggests that rat liver contains a GMP reductase similar to the bacterial enzyme.[8]

The three enzymes acting in concert would catalyze the conversion of IMP to IMP and obtain the energy for the operation of this useless cycle from the hydrolysis of ATP. However, it has been shown that IMP dehydrogenase is inhibited by GMP and that GMP reductase is inhibited by ATP.[3] These inhibitions regulate the flow of IMP to GMP and of GMP to IMP according to the physiological needs of the cell.[4]

I. IMP Dehydrogenase from *Aerobacter aerogenes*

$$IMP + DPN^+ + H_2O \rightarrow Xanthosine-5'-P + DPNH + H^+$$

Assay Method

Principle. The progress of the reaction is followed by measuring the increase in absorbancy at 340 mμ due to the formation of DPNH. It

[1] B. Magasanik, H. S. Moyed, and L. B. Gehring, *J. Biol. Chem.* **226**, 339 (1957).
[2] H. S. Moyed and B. Magasanik, *J. Biol. Chem.* **226**, 351 (1957).
[3] J. Mager and B. Magasanik, *J. Biol. Chem.* **235**, 1474 (1960).
[4] B. Magasanik and D. Karibian, *J. Biol. Chem.* **235**, 2672 (1960).
[5] U. Lagerkvist, *J. Biol. Chem.* **233**, 138 (1958).
[6] U. Lagerkvist, *J. Biol. Chem.* **233**, 143 (1958).
[7] R. Abrams and M. Bentley, *Arch. Biochem. Biophys.* **79**, 91 (1959).
[8] A. J. Guarino and G. Yuregir, *Biochim. et Biophys. Acta,* **36**, 157 (1959).

is also possible to follow the oxidation of IMP by measuring the increase in absorbancy at 290 mμ due to the formation of xanthosine-5′-P.

Reagents

> M Tris buffer, pH 7.55.
> 0.1 M reduced glutathione.
> M KCl.
> 0.03 M disodium IMP.
> 0.0125 M DPN$^+$.
> Enzyme.

Procedure. Into a quartz cell with a 1-cm. light path are pipetted 0.10 ml. of Tris buffer, 0.05 ml. of glutathione, 0.10 ml. of KCl, 0.10 ml. of IMP, enzyme, and water to 2.8 ml. The reaction is started by adding 0.2 ml. of DPN$^+$. A similar reaction mixture from which IMP has been omitted serves as control. The increase in absorbancy at 340 mμ is measured at room temperature at 1-minute intervals over a period of 10 minutes.

Definition of Unit and Specific Activity. One unit of enzyme is defined as that amount which causes the formation of 1 micromole of DPNH per minute under the above conditions. Specific activity is expressed as units per milligram of protein.

Purification Procedure

Growth of Cells. The guanine-less mutant of *A. aerogenes*, strain P-14, which lacks xanthosine-5′-P aminase (reaction 2), produces about forty times as much IMP dehydrogenase as the wild strain when grown on limiting guanine. It is therefore advantageous to use this organism as the source of the enzyme. The growth medium contains, per liter, 13.6 g. of Na$_2$HPO$_4$·12H$_2$O, 12.6 g. of KH$_2$PO$_4$, 0.4 g. of MgSO$_4$·7H$_2$O, 0.02 g. of CaCl$_2$, 2.0 g. of vitamin-free Casamino acids (Difco), 5 mg. of guanine, and 20 g. of glycerol, sterilized at 120° for 15 minutes. The inoculum, containing about 10^9 organisms per milliliter, is prepared by suspending the cells from a yeast extract–tryptone–agar slant culture grown at 37° for 24 hours in sterile distilled water. One-half milliliter of this suspension is added to 1 l. of the medium contained in a 2-l. Erlenmeyer flask. The flasks are shaken on a rotary shaker at 37° for 16 hours.

Step 1. Extraction. The cells from 24 l. of culture are collected by centrifugation in a steam-driven supercentrifuge (Sharples) and washed three times with 200-ml. portions of 0.03 M potassium phosphate buffer, pH 7.40. In a typical experiment 40 g. of cell paste were obtained. The cells are suspended to a volume of 400 ml. with 0.03 M potassium phosphate buffer, pH 7.40. All subsequent operations are carried out at a temperature between 0° and 4°; for the centrifugations a high-speed angle centrifuge (Servall) at 14,500 r.p.m. is used. The cells are disrupted by subjecting 25-ml. batches of the suspension to sonic vibration in a 10-kc. magnetostrictive oscillator (Raytheon) for 6 minutes. Intact cells and larger subcellular particles are removed by centrifugation for 15 minutes.

Step 2. Removal of Nucleic Acid and Inactive Protein. Twenty-three milliliters of 2% protamine sulfate are added slowly to 385 ml. of extract. The precipitate is removed by centrifugation for 10 minutes and discarded.

Step 3. First Ammonium Sulfate Precipitation. Ammonium sulfate solution (303 ml.) saturated at 0° is added to 388 ml. of the supernatant solution from step 2 with stirring. The precipitate is collected after 1 hour by centrifugation for 15 minutes. It is dissolved in distilled water to give a volume of 29.3 ml.

Step 4. Adsorption of Inactive Material with Calcium Phosphate Gel. The enzyme solution (29.3 ml.) is mixed with 15 ml. of calcium phosphate gel containing 19 mg. of solids per milliliter. The mixture is stirred for 30 minutes, and the gel is removed by centrifugation and discarded. This treatment does not result in a significant increase in specific activity but was found to be necessary for the success of the subsequent steps.

TABLE I

PURIFICATION OF IMP DEHYDROGENASE

Step	Volume, ml.	Activity, units	Protein, mg.	Specific activity units/mg. protein
1. Extraction	385	315	5790	0.054
2. Protamine	388	333	2900	0.12
3. First $(NH_4)_2SO_4$ precipitate	29	420[a]	922	0.46
4. Gel adsorption	41	420	763	0.55
5. Second $(NH_4)_2SO_4$ precipitate	8	192	137	1.40
6. Gel adsorption and elution	4	45	13	3.41

[a] The apparent increase in total activity in step 3 is probably due to the removal of a DPNH oxidase.

Step 5. Second Ammonium Sulfate Precipitation. Ammonium sulfate solution (16.5 ml.) saturated at 0° is added to 41.3 ml. of the enzyme solution from step 4. At the end of 1 hour, the precipitate is sedimented by centrifugation for 15 minutes and discarded. Then 4.8 ml. of the saturated ammonium sulfate solution are added to the supernatant fluid, and the mixture is held at 4° for 16 hours. The precipitate which has formed during this time is collected by centrifugation, dissolved in a small volume of distilled water, and dialyzed for 16 hours against 1 l. of distilled water.

Step 6. Gel Adsorption and Elution. Eight milliliters of the dialyzed enzyme solution are mixed with 0.72 ml. of 0.50 M potassium phosphate buffer, pH 6.52, and 1.74 ml. of calcium phosphate gel. The mixture is stirred for 30 minutes, after which the gel is collected by centrifugation and washed twice with 10-ml. portions of distilled water. The adsorbed enzyme is eluted by stirring the gel in 4.0 ml. of 0.005 M potassium phosphate buffer, pH 7.40, for 30 minutes. A summary of the purification procedure is given in Table I.

Properties

Stability. The purified enzyme can be stored at $-17°$ for several months without loss of activity.

Effect of pH. The enzyme is most active at pH 8.1. A change in either direction sharply diminishes the rate of the reaction.

Substrate Specificity and Affinity. The enzyme acts on IMP but not on hypoxanthine or inosine. It reduces DPN^+ but not TPN^+. K^+ or NH_4^+ is required for activity; Na^+ is not inhibitory at ten times the concentration of K^+. Glutathione usually stimulates the activity 3-fold, but some preparations are completely inactive without it. Cysteine can replace glutathione with the same molar efficiency but will not support a linear rate of IMP oxidation beyond the first few minutes, presumably because of its instability at alkaline pH. The substrate constants of the enzyme are summarized in Table II and show its relatively poor affinity for DPN^+. The affinity of the enzyme for IMP and DPN^+ does not depend on the level of glutathione.

TABLE II
COMPONENTS OF IMP DEHYDROGENASE SYSTEM

Component	Substrate constant
IMP	$1.39 \times 10^{-5} M$
DPN^+	$3.80 \times 10^{-3} M$
K^+	$1.58 \times 10^{-2} M$
Glutathione	$1.13 \times 10^{-3} M$

Apparent Irreversibility. IMP is completely oxidized to xanthosine-5'-P even at a low pH, unfavorable for the rate of the reaction. Addition of DPNH in 10-fold excess over DPN⁺ does not affect the rate or the extent of the reaction. After incubation of xanthosine-5′-P and DPNH with the enzyme, not a trace of IMP and DPN⁺ can be detected. The equilibrium of the enzymatic reaction appears to be overwhelmingly in favor of xanthosine-5′-P formation.

Inhibitors. The enzyme is inhibited by GMP, but not by adenine nucleotides. The inhibition can be overcome by an increase in the concentration of IMP.[3]

II. Xanthosine-5′-P Aminase from Aerobacter aerogenes

$$\text{Xanthosine-5′-P} + \text{ATP} + \text{NH}_3 \rightarrow \text{GMP} + \text{AMP} + \text{PP}$$

Assay Method

Principle. The amount of GMP produced in 30 minutes is estimated by measuring the increase in absorbancy at 290 mμ in the reaction mixture deproteinized with perchloric acid. Under these conditions the molar extinction of GMP at 290 mμ is 6.0×10^3, whereas that of xanthosine-5′-P, ATP, and AMP, the other ultraviolet components of the reaction mixture, is negligible. When purified enzyme preparations are used it is possible to follow xanthosine-5′-P disappearance at pH 8.6 by measuring the decrease in absorbancy at 290 mμ.

Reagents

M Tris buffer, pH 8.52.
0.05 M ATP, pH 7.0.
0.40 M MgCl$_2$.
0.10 M xanthosine-5′-P.
2 M (NH$_4$)SO$_4$.
3.5% perchloric acid.

Procedure. A mixture consisting of 0.04 ml. of Tris buffer, 0.02 ml. of ATP, 0.01 ml. of MgCl$_2$, 0.06 ml. of xanthosine-5′-P, enzyme, and water to 0.21 ml. is prepared. The reaction is started by the addition of 0.04 ml. of (NH$_4$)$_2$SO$_4$. A similar mixture from which xanthosine-5′-P has been omitted serves as control. The mixtures are incubated at room temperature for 30 minutes. The reaction is stopped by the addition of 2.75 ml. of 3.5% perchloric acid. The precipitated protein is removed by centrifugation, and the absorbancy of the solution at 290 mμ is determined in a quartz cuvette with a light path of 1 cm.

Definition of Unit and Specific Activity. One unit of enzyme is defined as that amount which causes the formation of 1 micromole of

GMP per minute under the above conditions. Specific activity is expressed as units per milligram of protein.

Purification Procedure

Growth of Cells. A purine-requiring mutant of *Aerobacter aerogenes*, strain PD-1, blocked at an early step of purine biosynthesis produces about four times as much xanthosine-5'-P aminase as the wild type when grown on limiting guanine or adenine. This mutant, like many purine-requiring strains, also requires thiamine for growth. The growth medium for this mutant contains, per liter, 13.6 g. of $Na_2HPO_4 \cdot 12H_2O$, 12.6 g. of KH_2PO_4, 0.2 g. of $MgSO_4 \cdot 7H_2O$, 0.02 g. of $CaCl_2$, 25 mg. of thiamine hydrochloride, 9 mg. of adenine, and 20 g. of $(NH_4)_2SO_4$; after sterilization by autoclaving at 120° for 15 minutes, 10 ml. of a solution containing 20 g. of glucose sterilized in the same manner are added. The inoculum is prepared by suspending the cells (*A. aerogenes*, strain PD-1) from a yeast extract–tryptone–agar slant culture grown at 37° for 24 hours in sterile distilled water; 0.5 ml. of this suspension (about 10^9 organisms) is added to 1 l. of medium contained in a 2-l. Erlenmeyer flask, and the culture is incubated with agitation on a rotary shaker at 37° for 16 hours.

Step 1. Extraction. The cells from 12 l. of medium are collected by centrifugation and washed three times with 100-ml. portions of 0.03 M potassium phosphate buffer, pH 7.40. All subsequent operations are carried out at temperatures between 0° and 4°; for centrifugation at 14,500 r.p.m., a high-speed Servall centrifuge was used. The cells are disrupted by subjecting 25-ml. batches of the suspension to sonic oscillation for 6 minutes in a 10-kc. magnetostrictive oscillator (Raytheon). The extract is clarified by centrifugation for 10 minutes.

Step 2. Protamine Fractionation. Nucleic acid and inactive protein are precipitated by slowly adding 9.6 ml. of 2% protamine sulfate. Nearly all the enzymatic activity is precipitated. The precipitate is washed with 158 ml. of distilled water and suspended in 15 ml. of distilled water. At this stage, storage of the enzyme at −17° for 24 hours facilitates subsequent purification.

Step 3. Fractional Elution of Enzyme from Protamine Precipitate. The suspension is thawed rapidly; the precipitate is collected by centrifugation and is resuspended in 0.10 M KCl. After 30 minutes of stirring, about half the enzymatic activity is found in the supernatant fluid. Most of the protein remains insoluble and is removed by centrifugation for 10 minutes.

Step 4. Ammonium Sulfate Fractionation. Fourteen milliliters of an ammonium sulfate solution saturated at 0° are added with stirring to

14 ml. of the enzyme solution. After 1 hour, the precipitate is removed by centrifugation, and 12 ml. of saturated ammonium sulfate solution are added to 28 ml. of the supernatant fluid. After 1 hour of stirring, the precipitate is collected by centrifugation and dissolved in 5 ml. of distilled water.

Further Purification. The purification achieved in the steps described so far and summarized in Table III results in a 90-fold increase in the specific activity of the enzyme. The purified fraction obtained after step 4 is contaminated with myokinase and ammonium salts. The removal of

TABLE III
PURIFICATION OF XANTHOSINE 5′ P AMINASE

Step	Volume, ml.	Activity, units	Protein, mg.	Specific activity, units/mg. protein
1. Extraction	160	30.5	2720	0.011
2. Protamine fractionation	15	25.8	512	0.051
3. Elution from protamine	14	14.0	48	0.28
4. $(NH_4)_2SO_4$ fractionation	4	11.8	11.5	1.03

these impurities can be accomplished by exhaustive dialysis against distilled water, followed by the adsorption of the enzyme to calcium phosphate gel. The gel is washed three times with distilled water, and finally the enzyme is eluted from the gel with $0.10\,M$ potassium phosphate buffer, pH 7.4, or with $0.25\,M$ Tris buffer, pH 7.80, depending on whether a phosphate-free preparation is desired. The exhaustive dialysis on several occasions caused the formation of an inactive precipitate, permitting an increase in specific activity to over 3.4. In these instances a 300-fold purification was obtained.

Properties

Amino Donor. Crude bacterial extracts can use glutamic acid, glutamine, or ammonium sulfate as the source of the amino group. The ability of glutamic acid and of glutamine to serve as amino donors decreases with purification and finally only an ammonium salt will serve. At pH 7.15, $111 \times 10^{-3}\,M$ ammonium acetate, and at pH 8.35, $8.8 \times 10^{-3}\,M$ ammonium acetate are required, respectively, for half-maximal rate. At these pH values and ammonium acetate concentrations the concentrations of NH_3 are $9.6 \times 10^{-4}\,M$ and 8.5×10^{-4}, respectively. The near equality of these values suggests that NH_3, rather than NH_4^+, is the amino donor.

Specificity. The enzyme requires xanthosine-5'-P, ATP, an ammonium salt, and Mg^{++} (K_s: 1.93×10^{-3}) for activity which is maximal at pH 8.5. The products are GMP, AMP, and PP.

Apparent Irreversibility. In the presence of excess ATP, xanthosine-5'-P is quantitatively converted to GMP. Neither AMP nor PP is incorporated into ATP in the complete system, nor in one from which xanthosine-5'-P or the ammonium salt has been omitted. The reaction catalyzed by the enzyme appears therefore to be irreversible.

Inhibition. Hydroxylamine is a potent inhibitor of the enzyme. Preincubation of the enzyme with 0.003 M hydroxylamine, ATP, xanthosine-5'-P, and $MgCl_2$ causes complete inhibition. Without preincubation the inhibited state is attained only gradually. Omission of xanthosine-5'-P and ATP, singly or together, from the reaction mixture during preincubation markedly reduces the extent of inhibition.

In contrast to IMP dehydrogenase this enzyme is not inhibited by GMP.

III. GMP Reductase from *Salmonella typhimurium*[3]

$$GMP + TPNH + H^+ \rightarrow IMP + TPN^+ + NH_3$$

Assay Method

Principle. The progress of the reaction is followed by measuring the decrease in absorbancy at 340 mμ due to the disappearance of TPNH.

Reagents

  0.2 M Tris buffer, pH 7.5.
  0.05 M cysteineHCl (freshly dissolved and neutralized).
  0.01 M GMP.
  0.001 M TPNH.
  0.1 M KCN.

Procedure. Into a quartz cell of 1-ml. capacity and with a light path of 1 cm. are pipetted 0.2 ml. of Tris buffer, 0.05 ml. of cysteine, 0.2 ml. of GMP, 0.1 ml. of TPNH, enzyme, and water to a volume of 1 ml. A blank control devoid of GMP is used to correct for the oxidation of TPNH by a TPNH oxidase present in crude extracts. It is possible to effect a considerable inhibition of this oxidase without affecting the GMP reductase by adding to the crude cell extract 0.2 ml. of 0.1 M KCN per milliliter and keeping the mixture for 5 minutes at 0° prior to the assay. The decrease in absorbancy at 340 mμ is measured at room temperature at 1-minute intervals over a period of 5 minutes.

Definition of Unit and Specific Activity. One unit of enzyme is defined as that amount which causes the disappearance of 1 micromole of TPNH per minute under the above conditions. Specific activity is expressed as units per milligram of protein.

Purification Procedure

Growth of Cells. Several strains of Enterobacteriaceae were found to have approximately the same levels of GMP reductase whether grown in the presence or in the absence of adenine or guanine.[3] On the other hand, the Harvard strain of *E. coli* contains a measurable level of this enzyme only when cultivated in a guanine-containing medium.[4] The enzyme of an adenine-requiring mutant of *S. typhimurium,* strain Ad-12, which lacks inosinicase has been purified. The cells are cultured in a minimal medium containing, per liter, 6 g. of $Na_2HPO_4 \cdot 7H_2O$, 7 g. of KH_2PO_4, 2 g. of $(NH_4)_2SO_4$, 0.2 g. of $MgSO_4$, and 0.01 g. of $CaCl_2$. The medium is supplemented with 20 mg. of adenine and 0.05 mg. of thiamine hydrochloride. The medium is brought to pH 6.5 and is sterilized by autoclaving at 115° for 15 minutes. Glucose is sterilized separately and added aseptically to the medium to give a concentration of 2 g./l. Each liter of complete growth medium (in a 2-l. Erlenmeyer flask) is inoculated with 0.5 ml. of an 8-hour slant culture suspended in 5 ml. of sterile water and vigorously shaken on a rotary shaker (New Brunswick) at 37° for 14 to 16 hours.

Step 1. Extraction. The cells are harvested by centrifugation, washed twice with H_2O and once with 0.025 M potassium phosphate buffer of pH 7.4, and then suspended at a concentration of 0.2 g. (wet weight) per milliliter in 0.025 M potassium phosphate buffer, pH 7.4, containing 0.005 M GSH. The suspended cells are disrupted in 20-ml. batches by sonic oscillation for 7 to 8 minutes in a cooled 10-kc. magnetostrictive oscillator (Raytheon). The cell debris is removed by centrifugation at $20,000 \times g$ for 20 minutes at 4°. The supernatant fluid is dialyzed against several changes of 0.008 M phosphate buffer, pH 7.5, containing 0.0025 M GSH. All subsequent steps are carried out at 4°, unless otherwise indicated.

Step 2. Removal of Nucleic Acid. The dialyzed extract is treated with one-tenth its volume of a 5% solution of streptomycin sulfate. The precipitate, consisting mainly of nucleic acids, is removed by centrifugation.

Step 3. Heating at 56°. The supernatant is heated at 54° to 56° with occasional stirring for 15 minutes, and the flocculent precipitate which appears is removed by centrifugation at $105,000 \times g$ for 15 minutes in a Spinco centrifuge, Model L. This procedure serves to remove the interfering TPNH oxidase. The supernatant is almost free of this activity but contains most of the GMP reductase.

Step 4. Treatment with Calcium Phosphate Gel. The supernatant is brought to pH 6.8, treated with aged calcium phosphate gel (5 mg./ml.), and centrifuged. The supernatant is then brought to pH 7.2 and again treated with calcium phosphate gel (20 mg./ml.). Most of the GMP reductase is adsorbed to the gel. The gel is collected by centrifugation, and the enzyme is eluted by stirring in 0.4 M potassium phosphate of pH 8.

Step 5. Ammonium Sulfate Fractionation. The eluate is fractionated with ammonium sulfate. The precipitate collected at 0.3 to 0.6 saturation

TABLE IV
PURIFICATION OF GMP REDUCTASE

Step	Volume, ml.	Activity, units	Protein, mg.	Specific activity, units/mg. protein
1. Dialyzed extract	50	14.5	1100	0.013
2. Streptomycin	51	14.3	870	0.016
3. Heated at 56°	45	11.3	230	0.049
4. Ca phosphate gel	6	6.3	11	0.58
5. (NH₄)₂SO₄ fractionation	4	5.2	4	1.16

of ammonium sulfate contains the bulk of the enzyme activity. A summary of the purification procedure of GMP reductase is given in Table IV.

Properties

Effect of pH. The enzyme is maximally active at pH 7.5 to 8.2.

Substrate Specificity and Affinity. The enzyme is highly specific for both GMP and TPNH; other derivatives of guanine, such as guanine itself, guanosine, 2'- and 3'-GMP, GDP, and GTP, are neither substrates nor inhibitors of the reaction, and DPNH cannot substitute for TPNH. The substrate constant for GMP is $9.6 \times 10^{-5}\ M$.

Purified preparations exhibit an absolute requirement for sulfhydryl compound, such as cysteine, glutathione, or mercaptoacetic acid.

Apparent Irreversibility. The products of GMP reduction are IMP and NH₃; for every molecule of GMP reduced, one molecule of TPNH is oxidized. No reduction of TPN⁺ could be detected when the enzyme was incubated in mixtures containing TPN⁺ in concentrations up to $2.5 \times 10^{-3}\ M$, IMP and ammonium sulfate in concentrations ranging from $5 \times 10^{-4} M$ to $2 \times 10^{-2} M$, and at pH values from 2.4 to 6.5. The reaction is therefore apparently irreversible.

Inhibition. The reduction of GMP is inhibited by ATP and, to a much lesser extent, by AMP and IMP. This effect can be overcome by increasing the concentration of GMP.

[14] Adenylic Acid Ribosidase

$$\text{Adenosine 5'-monophosphate} \xrightarrow{\text{(ATP)}} \text{Adenine} + \text{ribose 5-phosphate}$$

By Leon A. Heppel

This method is based on the procedure of Hurwitz *et al.*[1]

Assay Method

Principle. Two independent methods were used for the determination of AMP cleavage. The first procedure was based on the liberation of the free base, and the second on the appearance of reducing sugar.

Assay A. Liberation of Free Base

Reagents

$0.1 M$ MgCl$_2$.
$0.4 M$ Tris buffer, pH 7.95.
$0.02 M$ ATP.
$0.02 M$ AMP.
0.03 to 1.3 units of enzyme.

Procedure. The reaction mixture (0.2 ml.) contained 0.02 ml. of MgCl$_2$, 0.03 ml. of Tris buffer, 0.01 ml. of ATP, 0.01 ml. of AMP, and enzyme. Incubation was for 15 minutes at 37.5° in 12-ml. conical centrifuge tubes. After incubation, 0.03 ml. of 25% barium acetate solution was introduced, followed by 0.92 ml. of absolute ethanol. The tubes were briefly shaken, kept in ice for 20 minutes, and centrifuged at 2° for 10 minutes at about 2000 r.p.m. The optical density was determined at 250 mμ after the addition of 1.17 ml. of 0.15 N HCl to a 0.7-ml. aliquot of the supernatant. All of the ribose phosphate is precipitated by the alcoholic barium. It was found necessary to measure both adenine and

[1] J. Hurwitz, L. A. Heppel, and B. L. Horecker, *J. Biol. Chem.* **226**, 525 (1957).

hypoxanthine together because adenase was present at all stages of purification. At 250 mμ both adenine and hypoxanthine have identical molar extinction coefficients of 1.03×10^4.

Assay B. Appearance of Reducing Sugar

Reagents

> 0.02 M AMP.
> 0.02 M ATP.
> 0.1 M MgCl$_2$.
> 0.4 M Tris buffer, pH 7.95.

Procedure. The reaction mixture contained 0.02 ml. of AMP, 0.02 ml. of ATP, 0.01 ml. of MgCl$_2$, 0.025 ml. of Tris buffer, enzyme, and water to give a volume of 0.1 ml. After 10 minutes at 37.5°, an aliquot was removed, and the reducing power was measured with ribose 5-phosphate as a standard, by the procedure of Park and Johnson.[2]

In both assays, controls lacking enzyme were run. Further, the early enzyme fractions (crude extract to ammonium sulfate I) contained considerable ultraviolet spectrum-absorbing as well as reducing material, and additional controls containing enzyme, but lacking substrate, were included when these fractions were assayed. The reaction rate, with either assay, was linear with time and directly proportional to the enzyme concentration.

Definition of Unit and Specific Activity. A unit of enzyme activity was defined as equal to the amount required for the formation of 1 micromole of reducing sugar or adenine per hour at 38°, and specific activity as units per milligram of protein. Protein was determined as described by Sutherland *et al.*[3]

Purification Procedure

Cultivation and Harvest of Bacteria. This is based on the method of Hurwitz *et al.*[1] The enzyme was purified from extracts of *Azotobacter vinelandii* strain O. The organism was grown and harvested as described by Grunberg-Manago *et al.*[4] Cells were washed with cold H$_2$O and centrifuged at $13,000 \times g$, and the pellet was stored at $-15°$. Cells frozen up to 2 months showed no loss of ability to cleave AMP. A

[2] J. T. Park and M. J. Johnson, *J. Biol. Chem.* **181**, 149 (1949).
[3] E. W. Sutherland, C. F. Cori, R. Haynes, and N. S. Olsen, *J. Biol. Chem.* **180**, 825 (1949).
[4] M. Grunberg-Manago, P. J. Ortiz, and S. Ochoa, *Biochim. et Biophys. Acta* **20**, 269 (1956).

summary of the purification procedure is given in Table I. The details
follow.

TABLE I
PURIFICATION OF 5'-AMP RIBOSIDASE

Fraction	Units/ml[a]	Total units	Specific activity, units/mg. protein
Crude extract	20.7	3,300[b]	1.7[b]
Protamine fraction	34.2	10,800	8.1
Ammonium sulfate I, heated	85	9,350	15
Ammonium sulfate II	360	7,200	29
Ammonium sulfate III	970	6,900	130
Ammonium sulfate IV, heated	4400	4,200	250

[a] The assay conditions were those described for the reducing sugar procedure under Assay Method.

[b] The crude extract had a relatively low activity in this preparation. Such extracts usually contained a total of 10,000 units with a specific activity of 5.

Alumina Extract. Twenty-eight grams of cells were ground in a
mortar with 56 g. of Alumina A-301 (325 mesh, Aluminum Company of
America) for 10 minutes, and the paste was extracted with 112 ml. of
water. The turbid supernatant solution obtained by centrifugation for
15 minutes at 13,000 $\times$ g was decanted, and the sediment was re-extracted
with 56 ml. of water. The supernatant solutions were combined (alumina
extract, 160 ml.). The above operations were performed at room temperature. All subsequent steps were carried out at 2° except where
indicated.

Protamine Treatment. Alumina extract (160 ml.) was diluted with
160 ml. of water, and 12.8 ml. of a 2% solution of protamine sulfate
(Nutritional Biochemicals Corporation, Cleveland, Ohio) in 0.2 M sodium
acetate buffer, pH 5.0, were added. The solution was rapidly stirred
during the addition of protamine, kept standing for 10 minutes, and
centrifuged, yielding 315 ml. of supernatant solution (protamine fraction).

Ammonium Sulfate I, Heated. The protamine fraction was treated
with 150 g. of solid ammonium sulfate. After 15 minutes, the solution
was centrifuged, and the precipitate was dissolved in 89 ml. of water. The
entire solution (98 ml.) was heated to 55° in about 2.5 minutes by immersion in a water bath at 55° and held at this temperature for 5 more
minutes; the solution was then rapidly cooled to 3°. The voluminous
precipitate was removed by centrifugation and washed with about 15 ml.
of water. The supernatant and wash solutions were combined (ammonium sulfate I, heated, 110 ml.).

Ammonium Sulfate II. In all subsequent fractionations, ammonium sulfate solution saturated at room temperature and adjusted with concentrated NH_3 to pH 7.4 was used. The salt concentration of the ammonium sulfate I heated fraction was measured with the Barnstead purity meter and found to be about 0.10 saturated. This solution (110 ml.) was treated with 27.5 ml. of water and 27.5 ml. of 0.2 M Tris buffer, pH 7.7. To the mixture (165 ml.) were added 55 ml. of ammonium sulfate solution. After centrifugation, the supernatant solution (216 ml.) was treated with 33.2 ml. of saturated ammonium sulfate solution. Both precipitates were discarded. The supernatant solution (244 ml.) was treated with 73.2 ml. of saturated ammonium sulfate, and the heavy precipitate obtained was dissolved in water (ammonium sulfate II, 20.5 ml). This solution was found to be 0.05 saturated with ammonium sulfate.

Ammonium Sulfate III. The above fraction (20.5 ml.) was heated for 1 hour at 60°. The resulting suspension was cooled, diluted with 50 ml. of 0.04 M Tris buffer, pH 7.7, and treated with 30 ml. of ammonium sulfate solution. The heavy precipitate was removed by centrifugation. The supernatant solution (96 ml.) was treated with 16 ml. of ammonium sulfate solution, and the precipitate again was discarded. To the second supernatant solution (111 ml.) were added 79 ml. of ammonium sulfate solution. The flocculent precipitate was collected and dissolved in water (ammonium sulfate III, 7.1 ml.).

Ammonium Sulfate IV, Heated. The protein concentration was determined, and the solution was diluted with 0.1 M sodium acetate, pH 7.7, to a protein concentration of 4 mg./ml. To this solution (14.2 ml.), 12.9 ml. of acetone ($-10°$) were added rapidly, and the mixture was centrifuged for 3 minutes at $13,000 \times g$. Acetone (4.2 ml.) was added to the supernatant solution (25.2 ml.) in the same way. Both fractions were dissolved in water (2.5 ml.) and immediately assayed.[5]

The active acetone fraction (2.6 ml.) was treated with 6.5 ml. of 0.04 M Tris buffer, pH 7.7, and 6.13 ml. of ammonium sulfate solution. The precipitate was removed by centrifugation, and 10.3 ml. of ammonium sulfate solution were added to the supernatant solution (14.5 ml.). The precipitate was dissolved in water (1.0 ml.). This fraction was then heated for 1 hour at 60°, after which the small precipitate was removed by centrifugation at $13,000 \times g$ for 5 minutes (ammonium sulfate IV, heated, 0.95 ml.).

[5] This procedure is not reproducible, and the enzyme has been found to appear in either acetone fraction. The yield has varied from 40 to 100% of the activity initially present in ammonium sulfate III. The acetone fraction is unstable, and the enzyme cannot be stored at this stage.

The above procedure results in a purification of approximately fifty-fold (Table I).

Properties

Stability. The final preparation is stable when stored at −10°. After 2 months under such conditions, no loss of activity has been noted. Whereas the enzyme is stable to prolonged heating at 60° at pH 6.5, it is completely destroyed in 10 minutes at this temperature at pH 5.1. At 70° the enzyme is rapidly destroyed even at pH 6.5.

The final preparation is free of myokinase activity but is still contaminated with inorganic pyrophosphatase and adenase. The latter enzyme is very active in extracts of *A. vinelandii.*

Effect of pH and Metal Requirement. The reaction is most rapid at about pH 7.8. The rate drops sharply above this pH and below pH 7.0. The enzyme is completely inactive in the absence of Mg^{++}. The dissociation constant, K_m, for Mg^{++} calculated from the Lineweaver and Burk plot was about $1.1 \times 10^{-4} M$. Other divalent metals were able to substitute for Mg^{++}; Ca^{++}, Mn^{++}, and Co^{++} replaced Mg^{++} to varying degrees, but Fe^{++} and Zn^{++} were inactive. All the metals which activated purified 5′-AMP ribosidase stimulated the *A. vinelandii* myokinase activity. No means were found to avoid troublesome myokinase activity, except by selective destruction of this enzyme by prolonged heating.

It was observed that Mn^{++} markedly inhibits adenase, and this ion was useful for preventing the deamination of adenine to hypoxanthine, since the removal of adenase from 5′-AMP ribosidase was not complete.

Nature of Substrate. After removal of myokinase, ADP was inactive with 5′-AMP ribosidase. In order to obtain reducing sugar, or free base, both AMP and ATP are required. The role of ATP is catalytic. Thus, the yield of adenine was found to be equal to the amount of 5′-AMP added but greatly in excess of the ATP. Also, with C^{14}-labeled precursors it was found that only 5′-AMP gave rise to labeled base and labeled ribose-5-phosphate.

Specificity of Reactants. The following compounds did not replace AMP: adenosine, inosine, adenosine 3-phosphate, deoxyadenosine, deoxyuridine, deoxycytidylate, thymidylate, deoxyguanosine, deoxyadenylate, GMP, IMP, CMP, UMP, and AMP polymer. The dinucleotides 5′-phosphoadenosine 3′-adenosine 5′-phosphate and 3′-phosphouridine 5′-adenosine 3′-phosphate were also inactive. The following compounds did not replace ATP: GTP, ITP, CTP, UTP, 1-pyrophosphoryl ribose 5-phosphate, ribose triphosphate. In addition, various combinations such as GTP and GMP, UTP and UMP, and CTP and CMP did not yield either reducing sugar or free base.

It was found that inorganic pyrophosphate, linear tripolyphosphate, and adenosine tetraphosphate could partially or completely replace ATP.

Stoichiometry of Reaction. Starting with AMP and ATP, there was a disappearance of AMP with the appearance of equal amounts of reducing sugar and free base. In addition, ATP was quantitatively recovered, indicating that this compound was not utilized during the course of the reaction (Table II). Chromatography of reaction mixtures

TABLE II

Stoichiometry of Splitting of 5'-AMP by 5'-AMP Ribosidases[a]

Compound analyzed	With sodium pyrophosphate, micromoles/ml.	With sodium tripolyphosphate, micromoles/ml.	With ATP, micromoles/ml.	
			Expt. 1	Expt. 2
Δ AMP	−2.50	−3.55	−1.75	−3.40
Δ adenine	+2.4	+3.2	+2.16	+3.30
Δ reducing sugar	+2.75	+3.75	+2.07	+3.78
Δ ATP			0	0
Δ inorganic phosphate	+0.3	+0.1		
Δ acid-stable phosphate	0	0		
Δ acid-labile phosphate	0	0		

[a] The reaction mixture (total volume = 1.5 ml.) contained 90 micromoles of Tris buffer, pH 7.8, 60 micromoles of $MgCl_2$, 37.5 micromoles of NaF, 6 micromoles of 5'-AMP, 6.0 micromoles of either sodium pyrophosphate or sodium tripolyphosphate, adjusted to pH 7.8, and 6.4 μg. of enzyme (ammonium sulfate, IV, heated). The incubation was for 3.5 hours at 37.5°. In the reaction with ATP, 2.2 micromoles of ATP, 4.4 micromoles of AMP, 100 micromoles of Tris buffer, 5 micromoles of $MgCl_2$, and 5 μl. of ammonium sulfate IV, heated, in a total volume of 1 ml. were incubated for 20 and 55 minutes in experiments 1 and 2, respectively. All procedures are summarized under Assay Method. The results are given as change (Δ) in concentration of the various substances measured.

(Dowex 1, Cl⁻) after incubation with 5'-AMP ribosidase confirmed this finding. Quantitative elution of the ATP area as described by Cohn and Carter[6] yielded a compound with a base–pentose–labile P–organic P ratio of approximately 1:1:2:3. Ion-exchange chromatography gave no evidence for any intermediate reaction product derived from ATP and AMP. In other experiments, the reaction was halted after various time intervals, and the reaction mixture was examined by paper chromatography. The only changes detected by ultraviolet examination were AMP disappearance matched by the appearance of adenine and hypoxanthine. No ultraviolet spectrum-absorbing material of unique R_f, which might represent an intermediate of the reaction, was found.

[6] W. E. Cohn and C. E. Carter, *J. Am. Chem. Soc.* **72**, 4273 (1950).

Similar data for inorganic pyrophosphate and tripolyphosphate are also shown in Table II. In addition, there was no change in the 10-minute acid-labile P, suggesting that no utilization of these compounds occurred during the reaction.

Identification of Products. The evidence for ribose 5-phosphate as one of the products of AMP cleavage is summarized in Table III. The amount of reducing sugar formed with ribose 5-phosphate as the stand-

TABLE III
IDENTIFICATION OF RIBOSE 5-PHOSPHATE[a]

Method	Micromoles/ml.
Reducing sugar formed	5.02
Orcinol-reactive material	5.54
Phosphoribulokinase activity	5.14

[a] Twenty micromoles of AMP, 20 micromoles of ATP, 20 micromoles of Tris buffer, pH 7.8, 2 micromoles of Mg^{++}, and 23 μg. of ammonium sulfate IV, heated (30 units), in a total volume of 0.27 ml. were incubated for 90 minutes at 38°. The reaction was terminated with 0.015 ml. of 1 N $HClO_4$ and 0.1 ml. of 30% charcoal; after 1 hour at 0°, the charcoal was collected by centrifugation and washed with H_2O. The supernatant solutions were combined and neutralized with KOH; after 15 minutes at 0°, $KClO_4$ was removed by centrifugation. A total of 14 micromoles of orcinol-reactive material was present. The incubation with phosphoribulokinase was carried out as previously described [J. Hurwitz, A. Weissbach, B. L. Horecker, and P. Z. Smyrniotis, *J. Biol. Chem.* **218**, 769 (1956)], and the reaction was followed by ADP formation.

ard agreed well with that obtained by using the phosphoribulokinase assay. In the latter case, ribulose 5-phosphate, as well as ribose 5-phosphate, is active owing to the presence of phosphoriboisomerase in the kinase preparation. That the product is not ribulose 5-phosphate is indicated by the results in the orcinol test, since the ketopentose esters yield only 57% of the color of the aldopentose esters.[7] In addition, the product did not give the cysteine-carbazole test[8] which is given by ketopentoses. In view of the agreement between the methods used for determining the pentose phosphate, the reducing sugar is most likely ribose 5-phosphate.

The other product is adenine. This was established by paper chromatography, as well as by the action of the enzyme adenase which contaminates the enzyme preparation. Chromatographic identification, measurement of electrophoretic mobility, and examination of absorption spectrum were also carried out.

[7] J. Hurwitz and B. L. Horecker, *J. Biol. Chem.* **223**, 993 (1956).
[8] Z. Dische and E. Borenfreund, *J. Biol. Chem.* **192**, 583 (1951).

Inhibitors of Reaction. Orthophosphate markedly inhibited the cleavage of AMP when ATP was used as the cofactor. This inhibition appears to be due to a competition between ATP and orthophosphate for the enzyme. Phosphate $(7.5 \times 10^{-3} M)$ gave 100% inhibition when the ATP concentration was $0.5 \times 10^{-3} M$; with increasing concentrations of ATP, less inhibition was obtained. All attempts to demonstrate the uptake of phosphate were negative. No phosphate disappeared as measured with the Lowry and Lopez phosphate procedure[9] or by the use of P^{32}. Arsenate was also found to inhibit, but it was only approximately 50% as effective as phosphate under identical conditions.

IMP also inhibited the cleavage of AMP. With $2 \times 10^{-3} M$ IMP, $5 \times 10^{-3} M$ ATP, and $2 \times 10^{-3} M$ AMP, inhibition was 40%. This inhibition was overcome by increasing the concentration of either AMP or ATP. UMP was without effect under the conditions found to be inhibitory with IMP.

[9] O. H. Lowry, and J. A. Lopez, *J. Biol. Chem.* **162**, 421 (1946).

[15] Assay of Thymidylate Synthetase Activity

By Morris Friedkin

During the conversion of deoxyuridylate to thymidylate, shown in Eq. (1), a new carbon-to-carbon bond is formed, a one-carbon unit is reduced to a methyl group, and a hydrogenated form of folic acid is oxidized. Various assays for thymidylate synthetase activity are based on these chemical events.

$$\text{dUMP} + \text{5,10-methylenetetrahydrofolate} \xrightarrow[\text{enzyme}]{\text{Mg}^{++}} \text{dTMP} + \text{dihydrofolate}$$

$$(1)$$

The one-carbon unit can be supplied in the form of formaldehyde which reacts nonenzymatically with tetrahydrofolate to yield 5,10-methylenetetrahydrofolate. Thymidylate synthetase catalyzes the transfer of the one-carbon moiety to the 5-position of deoxyuridylate and its reduction to a methyl group. Thus far it has not been possible to fractionate thymidylate synthetase into separate enzymatic activities corresponding to these two steps.

Summary of Assay Methods

Assays that have been used in various laboratories are summarized below:

1. Conversion of P^{32}-labeled dUMP to P^{32}-labeled dTMP and the further conversion of the latter to thymidine triphosphate.[1]

2. Fixation of carbon 14-labeled formaldehyde in the methyl group of thymidylate.[2]

3. Same as procedure 2 but involves further conversion of the methyl group to volatile carbon 14-labeled iodoform.[3]

4. Fluorometric determination of thymine.[4,5]

5. Acid hydrolysis of tritium-labeled dUMP and dTMP followed by chromatographic separation of the free pyrimidine bases.[6]

6. Enzymatic hydrolysis of tritium-labeled dUMP and dTMP followed by chromatographic separation of the nucleosides.[7]

7. Separation of P^{32}-labeled dUMP from P^{32}-labeled dTMP by paper chromatography.[8,9]

COMPARISON OF VARIOUS THYMIDYLATE SYNTHETASE ASSAYS

Assay	Advantages	Disadvantages
1	Very high sensitivity No chromatography	Laborious preparation of P^{32}-labeled dUMP Requires extra enzymes
2	Moderately sensitive Simple No chromatography	High blanks
3	Moderately sensitive	Requires time-consuming manipulations
4	Sensitive	High blanks Requires time-consuming manipulations
5	High sensitivity	Chromatography
6	High sensitivity	Chromatography Requires extra enzyme
7	Very high sensitivity No further chemical or enzymatic treatment of nucleotides	Laborious preparation of P^{32}-labeled dUMP Chromatography
8	Very simple and fast	Not sensitive enough for assay of mammalian tissues
9	High sensitivity	Chromatography

[1] M. Friedkin and A. Kornberg, in "The Chemical Basis of Heredity" (W. D. McElroy and B. Glass, eds.), p. 609. Johns Hopkins Press, Baltimore, 1957.
[2] J. G. Flaks and S. S. Cohen, J. Biol. Chem. 234, 1501 (1959).
[3] B. M. McDougall and R. L. Blakley, J. Biol. Chem. 236, 832 (1961).
[4] D. Roberts and M. Friedkin, J. Biol. Chem. 233, 483 (1958).
[5] D. M. Greenberg, R. Nath, and G. K. Humphreys, J. Biol. Chem. 236, 2271 (1961).
[6] K.-U. Hartmann and C. Heidelberger, J. Biol. Chem. 236, 3006 (1961).
[7] A. Haggmark, Cancer Research 22, 568 (1962).
[8] A. J. Wahba and M. Friedkin, J. Biol. Chem. 236, PC 11 (1961).
[9] R. L. Blakley and B. M. McDougall, J. Biol. Chem. 237, 812 (1962).

8. Spectrophotometric method[8-10] based on the spectral change that occurs when 5,10-methylenetetrahydrofolate is converted to dihydrofolate.

9. Transfer of tritium from tritiated tetrahydrofolate to the methyl group of thymidylate.[11]

A comparison of the advantages and disadvantages of the various assays is given in the table. Unfortunately, the spectrophotometric assay (assay 8), which is excellent for microbial material, is not sensitive enough for use with mammalian tissues.

Details for assays 1, 7, and 8 are given below.

Three Assays of Thymidylate Synthetase

Reagents. PREPARATION OF dl,L-TETRAHYDROFOLIC ACID. Folic acid was reduced catalytically in glacial acetic acid according to the method of O'Dell *et al.*[12] as modified by Kisliuk.[13] The powder was stored *in vacuo* in the dark at room temperature and was flushed with hydrogen before each weighing. Tetrahydrofolate could also be stored as a frozen solution at −10° in the dark for several weeks: 5 mg. of dl,L-tetrahydrofolate per milliliter of $1 M$ mercaptoethanol adjusted to pH 7.3 with KOH.

PREPARATION OF DEOXYURIDYLIC ACID. To a solution of 175 mg. of deoxycytidylic acid (California Corp. for Biochemical Research) in 10 ml. of H_2O were added 2.12 g. of $NaNO_2$ and 2.5 ml. of glacial acetic acid. Twenty hours later barium deoxyuridylate was precipitated from this mixture by the addition of 0.9 g. of barium acetate, 5 ml. of $15 N$ NH_4OH, and 50 ml. of absolute ethanol. After centrifugation and suspension in 75% ethanol, absolute ethanol, and diethyl ether, the salt was dried *in vacuo* at room temperature. The nucleotide was freed of barium by passage over a Dowex 50 (H⁺) column and lyophilized as the free acid. Chromatography of dUMP with either solvent I or solvent II (see below) showed a single spot on examination with ultraviolet light. The molecular extinction coefficient of dUMP in $0.01 N$ HCl was 10,200 at 262 mμ based on the phosphorus content of the nucleotide.

PREPARATION OF P^{32}-LABELED DEOXYURIDYLIC ACID. DNA was isolated from *E. coli* B cells grown on a phosphate-deficient medium containing 20 mc. of P^{32}-labeled inorganic orthophosphate. The growth medium and isolation procedures were essentially those described by Lehman *et al.*[14]

[10] A. J. Wahba and M. Friedkin, *J. Biol. Chem.* **237**, 3794 (1962).

[11] E. J. Pastore and M. Friedkin, *J. Biol. Chem.* **237**, 3802 (1962).

[12] B. L. O'Dell, J. M. Vandenbelt, E. S. Bloom, and J. J. Pfiffner, *J. Am. Chem. Soc.* **69**, 250 (1947).

[13] R. L. Kisliuk, *J. Biol. Chem.* **227**, 805 (1957).

[14] I. R. Lehman, M. J. Bessman, E. S. Simms, and A. Kornberg, *J. Biol. Chem.* **233**, 163 (1958).

A mixture of P^{32}-labeled deoxyribonucleotides released from DNA by treatment with deoxyribonuclease and snake venom phosphodiesterase was adsorbed on a Dowex 1–Cl column. dCMP was separated from the other nucleotides by elution with $0.002 N$ HCl. To 34 ml. of pooled fractions of dCMP (1.72×10^8 c.p.m.) were added 2 ml. of glacial acetic acid and 1.5 g. of $NaNO_2$. This solution was kept at room temperature for 20 hours; then, after the addition of 1 g. of $Ba(OH)_2$, 4 ml. of $15 N$ NH_4OH, and 150 ml. of absolute ethanol, CO_2 was bubbled through the mixture until a gelatinous precipitate of $BaCO_3$ formed. The barium salt of $dUMP^{32}$ coprecipitated with the $BaCO_3$ gel under these conditions. The gel was centrifuged and resuspended in 40-ml. portions of 75% ethanol, absolute ethanol, and diethyl ether, and finally dried *in vacuo*. A solution of this powder dissolved in 10 ml. of H_2O and acidified with acetic acid was passed over a Dowex 50 (H^+) column (2×17 cm.) followed by a wash with H_2O. Two hundred milliliters of combined percolate and wash were lyophilized. The $dUMP^{32}$ thus obtained was used over a period of 3 months. Chromatography with either solvent I or solvent II (see below) gave a single radioactive spot.

SOLVENTS USED FOR SEPARATION OF dUMP AND dTMP BY PAPER CHROMATOGRAPHY. *Solvent I*, ascending: 600 g. of $(NH_4)_2SO_4$ dissolved in 1 l. of $0.1 M$ potassium phosphate buffer, pH 6.8, plus 20 ml. of *n*-propanol. R_f: dUMP, 0.55; dTMP, 0.43. *Solvent II*, descending ($15 N$ NH_4OH–H_2O–isobutyric acid; 1:29:50): This solvent was usually allowed to drip off the end of paper to provide better separation of the nucleotides. R_f: dUMP, 0.36; dTMP, 0.42.

Assay Method 1: Conversion of $dUMP^{32}$ to $dTP^{32}PP$.[1,10] The basis of the method (a modification of an assay for thymidylate kinase[14]) is as follows. Thymidylate kinase of *E. coli* catalyzes the phosphorylation of thymidylate by ATP but not of deoxyuridylate.[1] The nucleloside monophosphates are susceptible to dephosphorylation by phosphomonoesterase; the nucleoside triphosphates are not. Thymidine triphosphate is adsorbed on charcoal, whereas inorganic orthophosphate is not.

The assay mixture consists of 0.01 micromole of $dUMP^{32}$ (2000 c.p.m.), 0.1 micromole of tetrahydrofolate, 3 micromoles of HCHO, 2 micromoles of ATP, 5 micromoles of $MgCl_2$, 20 micromoles of Tris–HCl buffer, pH. 7.5, thymidylate kinase,[14] and thymidylate synthetase in a total volume of 350 μl. The mixture is incubated at 37° for 30 minutes, heated in a boiling-water bath for 3 minutes, then cooled in an ice bath. The following additions are made: 0.2 ml. of $1 M$ acetate buffer, pH 4.6; 0.2 micromole of dUMP in 0.02 ml.; and 0.05 ml. of acid phosphatase.[15] After an incubation for 30 minutes at 37°, the following additions are

[15] J. Wittenberg and A. Kornberg, *J. Biol. Chem.* **202**, 431 (1953).

made: 0.1 ml. of $2 N$ HCl and 0.2 ml. of a 20% suspension of acid-washed Norit. The activated charcoal with its adsorbed thymidine triphosphate is centrifuged, washed two times by resuspension in 2 ml. of H_2O, transferred to a planchet with $1.5 N$ NH_4OH, dried in an oven at 95° for 40 minutes, and assayed for radioactivity in an internal gas flow counter.

Assay Method 7: Chromatographic Separation of dUMP[32] and dTMP[32].[1,10] Mixture A (10 μl.) and enzyme (10 μl., containing 0.8 mg. of protein per ml. or less) are mixed at 0° in a 10×75-mm. test tube. (Mixture A, enough for twenty-five assays, consists of 0.025 micromole of dUMP[32] (100,000 c.p.m.), 0.25 micromole of *dl*,L-tetrahydrofolate, 7.5 micromoles of HCHO, 12.5 micromoles of $MgCl_2$, 25 micromoles of Tris–HCl buffer, pH 8.4, and 50 micromoles of 2-mercaptoethanol in a total volume of 275 μl.). A capillary tube is then inserted into the reaction mixture, whereupon most of the mixture usually rises into the tube. The test tube and capillary tube are immersed in a 37° bath. After 30 minutes the reaction mixture is transferred with the capillary tube to a spot on 3 MM Whatman paper where 0.1 micromole of dUMP and 0.1 micromole of dTMP have been applied previously.

dUMP and dTMP are separated by ascending chromatography (overnight at room temperature) with solvent I. The following morning the paper is dried at 70° to 80° and viewed with an ultraviolet light; the dTMP zone is cut out with a cork borer (22 mm.) and assayed for radioactivity with an end-window counter. The initial radioactivity of dUMP is determined by chromatographing 10 μl. of mixture A (minus enzyme), as above, and cutting out the dUMP zone.

Assay Method 8: Spectrophotometric Procedure.[8,10] A direct method has been devised for assay of thymidylate synthetase that depends on spectral changes associated with the oxidation of tetrahydrofolate to dihydrofolate.[8] The increase in molecular extinction during conversion of 5,10-methylene tetrathydrofolate to dihydrofolate amounts to 6400 at 340 mμ. An assay reaction mixture consists of enzyme in 0.9 ml. of buffer ($0.05 M$ Tris containing $0.01 M$ mercaptoethanol and $0.001 M$ disodium ethylenediaminetetraacetate adjusted to pH 7.4 with HCl) and 0.3 ml. of Plus Mix. This is read at 340 mμ against a reference cuvette containing 0.9 ml. of enzyme in buffer as above and 0.3 ml. of Minus Mix.

Plus Mix (enough for sixty assays): 6 ml. of $1 M$ mercaptoethanol adjusted to pH 7.4 with KOH, 1.8 ml. of tetrahydrofolate solution (5 mg. of *dl*,L-tetrahydrofolate per milliliter of $1 M$ mercaptoethanol adjusted to pH 7.4 with KOH), 3 ml. of $0.3 M$ HCHO, 3 ml. of $0.5 M$ $MgCl_2$, 3 ml.

of 0.001 M dUMP, and 1.2 ml. of H_2O. *Minus Mix:* same as Plus Mix, with dUMP omitted.

The reaction is run at room temperature or at 30° in a thermostated chamber of a Beckman DU spectrophotometer. Within the absorbance range 0 to 0.125 the increase of optical density at 340 mμ is usually linear with time. It is convenient to relate thymidylate synthetase activity to the change of absorbancy between minutes 2 to 12. Ten-minute increases of optical density are strictly proportional to increasing amounts of enzyme only after some fractionation of crude *E. coli* extracts.[10]

Purification of Thymidylate Synthetase of *Escherichia coli*[16]

Frozen *E. coli* B cells were obtained from the Grain Processing Corp., Muscatine, Iowa. Two buffers were used extensively: *Buffer A*—0.05 M Tris containing 0.01 M mercaptoethanol and 0.001 M disodium ethylenediaminetetraacetate adjusted to pH 7.4 with HCl. *Buffer B*—same components as in buffer A adjusted to pH 8.4 with HCl.

All operations were carried out in a cold room at 0° to 3°.

Fraction I. One hundred grams of *E. coli* B cells were ground with 200 g. of alumina (Alcoa A-301) and extracted with 300 ml. of buffer A. Centrifugation at 16,000 × g yielded a viscous supernatant fluid containing 20 mg. of protein per milliliter; specific activity, 0.02 micromole of dTMP formed per milligram of protein per hour at 30° under conditions of the spectrophotometric assay.

Fraction II. To 1 l. of fraction I enzyme were added 160 ml. of 5% streptomycin sulfate with constant stirring. After removal of the nucleic acid precipitate by centrifugation, the supernatant fluid was diluted with 18 l. of buffer B and percolated through a large DEAE-cellulose column[17] (diameter 17.5 cm., height 7.5 cm.) previously equilibrated with buffer B. This was followed by a wash of 8 l. of buffer A. A considerable amount of enzymatically inactive protein with eluted with 5 to 6 l. of 0.2 M NaCl in buffer A. Thymidylate synthetase was then eluted from the column with 0.4 M NaCl in buffer A and collected in 250-ml. Erlenmeyer flasks. The enzyme usually appeared in flasks 7 through 12. These fractions were combined (approximate volume 1500 ml.) and dialyzed overnight in 18 l. of buffer A. Enzyme at this stage of purification had a specific activity of 1 micromole of dTMP formed per milligram of protein per hour at 30°; recovery of activity from fraction I, 96%.

[16] M. Friedkin, E. J. Crawford, E. Donovan, and E. J. Pastore, *J. Biol. Chem.* **237**, 3811 (1962).
[17] The DEAE-cellulose column was regenerated after use by washing with 8 l. of 0.75 N NH$_4$OH, 8 l. of H_2O, and 18 l. of buffer B.

Fraction III. Approximately 1500 ml. of dialyzed fraction II enzyme were slowly poured into a column of DEAE-cellulose (1.4 × 15 cm.) previously equilibrated with buffer B. This usually required 12 to 20 hours. Concentrated thymidylate synthetase was eluted with 0.4 M NaCl in buffer A. Five-milliliter fractions were collected, and the enzyme usually appeared in tubes 5 through 8. The enzyme preparation at this stage was yellow and exhibited an intense blue fluorescence under a Shannon lamp emitting ultraviolet light with a wavelength of approximately 360 mμ. Its specific activity was 2.2 micromoles of dTMP formed per milligram of protein per hour at 30°; recovery of activity from fraction II, 78%.

Fraction IV. Ten milliliters of fraction III enzyme (approximately 60 mg. of protein) were layered on the surface of a Sephadex G-75 column (coarse grade; 3 × 81 cm., previously equilibrated with buffer A) and slowly displaced downward with buffer A. Percolate was collected in one 84-ml. fraction, followed by thirty-one separate 2.8-ml. fractions and thirty separate 13-ml. fractions at a rate of 0.45 ml./min. Tube numbering was started with the first 2.8-ml. fraction collected. The absorbancy of each fraction was determined at 280 mμ, and 0.05-ml. aliquots were taken for assay of thymidylate synthetase activity. Thymidylate synthetase was retarded on passage through the column, yielding a further threefold purification. The specific activity of thymidylate synthetase in tubes 16 to 24 (labeled fraction IV) was approximately 7 micromoles of dTMP formed per milligram of protein per hour at 30°.

The Sephadex G-75 column also separated an intensely blue fluorescent material from the main bulk of the protein. The fluorescent material finally emerged from the column in fractions 43 through 56, approximately 190 ml. after the thymidylate synthetase peak. This material inhibited thymidylate synthetase.[16]

Fraction V. Fraction IV enzyme (1.2 mg. of protein) was subjected to zone electrophoresis on a vertical column of Swedish cellulose (1 × 60 cm.) previously equilibrated with buffer A; 1000 volts, 8 ma., 29.5 hours. This technique was previously used for cruder preparations.[10] Thymidylate synthetase, despite extensive purification on DEAE-cellulose and Sephadex G-75, retained a strong negative charge at pH 7.4 and moved rapidly toward the positively charged lower end of the column. Its specific activity after electrophoresis was approximately 67 micromoles of dTMP formed per milligram of protein per hour at 30°.

[16] Deoxycytidylate Hydroxymethylase from Virus-Infected *E. coli*

Deoxycytidylate + formaldehyde → 5-Hydroxymethyldeoxycytidylate

By LEWIS I. PIZER and SEYMOUR S. COHEN

Occurrence

An enzyme which will convert deoxycytidylate (dCMP) to the 5-hydroxymethyldeoxycytidylate (dHMP) is induced in *Escherichia coli* strain B when the bacterium is infected with the bacterial viruses T2, T4, and T6.[1,2] The reaction requires the cofactor tetrahydrofolic acid (THFA). The T-even bacteriophages, at the present time, are unique in that their DNA contains 5-hydroxymethyl cytosine[3] and lacks cytosine. Since the dCMP hydroxymethylase is absent in normal cells, its induction is a prerequisite to viral multiplication, and the enzyme has been shown to appear in infected cells prior to synthesis of virus DNA.[2] A number of other enzymes involved in the production of viral DNA are also induced upon infection.

Assay Method

Principle. Three types of enzyme assay have been developed which are based on a determination of the C^{14}-formaldehyde converted into a form nonvolatile in acid or fixed to deoxycytidylate. Since formaldehyde reacts with numerous components in the assay system and in particular with the THFA, it is necessary, if total nonvolatile formaldehyde is being determined, to destroy the THFA–formaldehyde complex by heating with $FeCl_3$ in acid after the enzyme incubation.[1] It is also necessary in this assay to have a blank lacking substrate. Even with these precautions, this assay is not satisfactory for the most accurate work. A second assay developed for more exact studies makes use of added carrier 5-hydroxymethyldeoxycytidylate (dHMP) to enable the chromatographic isolation of C^{14}-labeled product formed during the reaction.[4] This assay has the disadvantages of requiring a stock of carrier dHMP and a lengthy purification of the dHMP or of hydroxymethycytosine generated. An accurate assay most useful for multiple samples, as in the study of enzyme purification or of properties of the purified enzyme, employs ion-exchange

[1] J. G. Flaks and S. S. Cohen, *J. Biol. Chem.* **234,** 1501 (1959).
[2] J. G. Flaks, J. Lichtenstein, and S. S. Cohen, *J. Biol. Chem.* **234,** 1507 (1959).
[3] G. R. Wyatt and S. S. Cohen, *Biochem. J.* **55,** 774 (1953).
[4] H. D. Barner and S. S. Cohen, *J. Biol. Chem.* **234,** 2987 (1959).

chromatography to obtain dHMP relatively free of C^{14}-containing contaminants and is described below.[5] This assay is applicable to the estimation of enzyme in both crude extracts and purified solutions.

Reagents

0.05 M deoxycytidylate.
0.05 M C^{14}-formaldehyde.[6]
1 M potassium phosphate buffer, pH 7.0.
0.01 M DL-THFA.[7]
Dowex 50 W (H⁺), 200 to 400 mesh, 8% X.

Procedure. The following solutions are mixed in a 12-ml. conical centrifuge tube stored in ice: 0.05 ml. of dCMP, 0.02 ml. of KPO_4, 0.05 ml. of CH_2O, and 0.05 ml. of THFA. Water and enzyme are added to give a final volume of 0.5 ml. After mixing of the contents, the tube is incubated in a 37° water bath for 20 minutes. The reaction is terminated by addition of 0.5 ml. of 10% trichloroacetic acid, and the precipitate is removed by centrifugation.

The reaction tube and pellet are washed with 2 ml. of 0.01 M formic acid, and the supernatant and wash fluids are combined. The mixed fluids are placed on a Dowex 50 (H⁺) column (1 × 8 cm) previously washed with 50 ml. of 0.01 M formic acid. The column is eluted with 0.01 M formic acid. Radioactive material passing directly through the column consists of uncharged compounds and negatively charged nucleotides. The dHMP is eluted in a 50-ml. fraction which starts after 50 ml. of eluate has passed through the column. The radioactivity appearing in this fraction is the measure of enzyme activity. The unreacted dCMP is eluted from the column in the 50-ml. fraction after that containing the dHMP.

Definition of Unit and Specific Activity. The unit of enzymatic activity is defined as the amount of enzyme which fixes 0.01 micromole of formaldehyde in dHMP in 20 minutes at 37°. Specific activity is expressed as units of enzyme activity per milligram of protein. Protein was determined by the method of Lowry *et al.*,[8] with bovine serum albumin as standard protein. Solutions containing streptomycin or high

---

[5] R. Somerville, K. Ebisuzaki, and G. R. Greenberg, *Proc. Natl. Acad. Sci. U.S.* **45**, 1240 (1959). The authors wish to thank Dr. Greenberg for making the details of this procedure available prior to publication.

[6] D. A. MacFadyen, *J. Biol. Chem.* **158**, 107 (1945).

[7] Prepared by the procedure of J. C. Rabinowitz and W. E. Pricer, Jr., *J. Biol. Chem.* **229**, 321 (1957); stored *in vacuo* at −20° and weighed just prior to use.

[8] O. H. Lowry, N. J. Rosebrough, A. L. Farr, and R. J. Randall, *J. Biol. Chem.* **193**, 265 (1951); see Vol. III [73].

concentrations of potassium phosphate were dialyzed against distilled water before assaying for protein.

Purification Procedure

The methods for cultivation and assay of phage are those described by Adams.[9] The methods for growing and infecting large batches of bacteria are those described by Kornberg et al.,[10] and the procedure for enzyme purification is substantially that described by the above workers.

Step 1. Growth and Infection. E. coli strain B is grown on a synthetic salts–glucose medium in 40-l. batches. A Biogen culture apparatus (available from the American Sterilizer Co., Erie, Pennsylvania) was used for this size preparation. The medium contains, per liter, 3 g. of KH_2PO_4, 6 g. of Na_2HPO_4, 0.24 g. of $MgSO_4 \cdot 7H_2O$, 0.5 mg. of $FeSO_4 \cdot 7H_2O$, 1 g. of NH_4Cl, and 5 g. of glucose. The glucose is autoclaved separately as a 10% salt-free solution.

An inoculum is prepared of 4 l. of a culture of bacteria grown aerobically for 16 hours, from a titer of 5×10^7 per to 10^9 per milliliter, with glucose as limiting substrate at 1 mg./ml. The inoculum is added to 36 l. of medium sterilized in the Biogen. The bacteria are aerated at 37° until the titer in the Biogen reaches 1×10^9. The relatively stable phage T6r$^+$ is used as the infecting agent, and a solution of 2 g. of DL-tryptophan is added as an adsorption cofactor at this point. Then sufficient purified phage to provide a phage-to-bacterium ratio of approximately 4 to 1 are introduced into the apparatus within 1 minute, and aeration with vigorous agitation is maintained for 15 minutes. To inhibit further maturation of phage within the infected cells, a solution of 2 g. of chloramphenicol (Parke-Davis Co.) is added 15 minutes after infection, and 2 minutes are allowed for mixing. The infected cells are then expelled by air pressure from the Biogen onto ice to chill them to 0°. The cells are harvested with a Sharples continuous-flow centrifuge and frozen at −20° for storage. The yield from 40 l. of infected culture is between 40 and 50 g. of frozen cells.

Where small quantities of bacterial extract containing hydroxymethylase are required, it is convenient to grow 1 l. of organisms to a titer of 2×10^8 per milliliter, to harvest in an angle centrifuge, and to resuspend at a titer of either 1 or 4×10^9 per milliliter. The simplest procedure is to warm the cells at 1×10^9 per milliliter to 37° with aeration and infect with phage. After 20 minutes the cells are chilled and

[9] M. H. Adams, *Methods in Med. Research* **2**, 1 (1950).

[10] A. Kornberg, S. B. Zimmerman, S. R. Kornberg, and J. Josse, *Proc. Natl. Acad. Sci. U.S.* **45**, 772 (1959). The authors wish to thank Dr. Kornberg for making the details of this procedure available prior to publication.

collected by centrifugation, and the extract is prepared by grinding with alumina or by sonication. If kinetic studies are planned on the development of the enzyme, it is useful to chill the cells at the titer of 4×10^9 per milliliter to 2° and to infect them in the cold. After 3 to 5 minutes allowed for adsorption, dilution to a titer of 2×10^8 in a warm (37°) aerated medium initiates enzyme production.

Step 2. Preparation of Extract. Infected cells may be disrupted by sonic vibrations or by grinding with alumina. All steps are carried out in a 5° cold room. For large batches of cells a simple procedure for obtaining an extract from frozen infected cells is the following: The beaker of cells is placed in ice, and the cells are allowed to thaw. The gelatinous material is suspended in $0.05 M$ glycylglycine buffer, pH 7.2 ($0.001 M$ with respect to glutathione), with 5 ml. of buffer per gram of cells. The suspension is then transferred to a cold Waring blendor and blended at full speed for 1 minute. The blendor cup should be full enough to avoid foaming of the protein. The mixture is centrifuged at 30,000 r.p.m. for 30 minutes in the Spinco Model L centrifuge, and the supernatant fluid is collected. Repetition of the blendor process with fresh buffer gives a protein solution with reduced enzyme activity. Both supernatant fluids are pooled and diluted with buffer to give a final protein concentration of 10 mg./ml. In a typical preparation a final volume of approximately 430 ml. is obtained.

Step 3. Nucleic Acid Precipitation. The nucleic acid is precipitated from the extract by the dropwise addition, with rapid stirring, of 0.3 vol. of a 5% solution of streptomycin sulfate. Ten minutes after the addition is completed, the solution is centrifuged at 10,000 r.p.m. for 10 minutes. The supernatant fluid is collected and adjusted to pH 8.0 to 8.2 by the careful addition, with stirring, of $0.1 M$ KOH. The supernatant fluid containing the hydroxymethylase may be stored in the cold at this pH.

Step 4. Column Fractionation. A column of DEAE-cellulose[11] (3.5×18 cm.) is thoroughly equilibrated at 4° with $0.02 M$ KPO$_4$ pH 8.0 ($0.01 M$ mercaptoethanol). The enzyme solution, at pH 8, is added to the column at about 20 ml. per 15 minutes; the same rate is used in elution, and 20-ml. fractions are collected. The column is subjected to two linear gradient elutions. The first is from $0.08 M$ KPO$_4$, pH 6.5, to $0.3 M$ KPO$_4$, pH 6.5. The reservoirs contain 500 ml. of buffer. The second gradient is from $0.3 M$ KPO$_4$, pH 6.5, to $1 M$ KPO$_4$, pH 6.1, and 1-l. reservoirs are used. All buffers contain $0.01 M$ mercaptoethanol. The enzyme appears near tube 90, reaches its peak near tube 100, and tails off to tube 120. The yields and enrichment obtained by this procedure are presented in the table.

[11] Type-selectacel DEAE obtained from Brown Company, Berlin, New Hampshire.

SUMMARY OF PURIFICATION PROCEDURE

Step	Volume solution, ml.	Total units	Protein, mg./ml.	Specific activity	Yield, %
1. Extract	430	127,000	10	30	100
2. pH 8 streptomycin supernatant	625	103,000	5.2	30	82
3. Fractions 95-107 from DEAE column	360	29,450	0.089	920[a]	23[a]

[a] The enrichments and yields reported in this table are probably low, owing to the presence of inactivated enzyme in the solutions. The specific activity recorded at step 3 was obtained 16 days after infection of the cells. The purified enzyme is somewhat unstable at 0°, as is shown by the fact that the assays on tube 101 with 4-day intervals gave values for specific activities of 2180, 1810, and 1200. Thus the specific activity of the purified enzyme is probably in excess of 2200.

Properties

Specificity. The enzyme is specific for deoxycytidine 5′-phosphate. The carrier assay has been used to show that cytidine 5′-phosphate and spongocytidine 5′-phosphate are very poor substrates.[12] The assay using the measurement of formaldehyde nonvolatile in acid showed that deoxycytidine and deoxycytidine 5′-methylphosphate[12, 13] are not substrates.

Ionic Effects. The activity does not alter with pH in the range 6.5 to 8.5. At pH 7.0 phosphate or maleate buffer results in the same activity, whereas phosphite reduces the activity to 80%, and imidazole reduces the activity to 37% of that found in phosphate. Magnesium ions at $10^{-2} M$ do not activate, but $10^{-2} M$ EDTA produces a 10% inhibition. Neither 5-fluorodeoxyuridylate, which powerfully inhibits thymidylate synthetase,[14] nor 5-fluorodeoxycytidylate[13] inhibits dCMP hydroxymethylase.[14]

Kinetic Properties. The production of dHMP is proportional to enzyme concentration but not to time. The dCMP has a K_m of $6 \times 10^{-4} M$. Formaldehyde has a K_m of $1.5 \times 10^{-3} M$, and THFA has a K_m of $1 \times 10^{-4} M$.

Properties of the Column-Purified Enzyme. The central tubes (99 to 103) in the column fraction containing the enzyme possessed the same specific activity of 1200, suggesting that only enzyme protein was present in these fractions. The fractions 95 to 107 were pooled to give the material for the physical studies. These analyses were carried out in $0.02 M$ KPO$_4$ buffer, pH 7.4, which was $0.1 M$ with respect to KCl. The

[12] L. I. Pizer and S. S. Cohen, *J. Biol. Chem.* **235**, 2387 (1960).
[13] J. Lichtenstein, H. D. Barner, and S. S. Cohen, *J. Biol. Chem.* **235**, 457 (1960).
[14] J. G. Flaks and S. S. Cohen, *J. Biol. Chem.* **234**, 2981 (1959).

ultracentrifuge pattern showed two components. The major component (90%) appeared to possess the enzymatic activity and had an $S_{w,20}$ of 4.4. The minor component had an $S_{w,20}$ of 25. The electrophoresis pattern showed two components; the minor one had a mobility of -9.27×10^{-5} cm.^{2} volt^{-1} sec.$^{-1}$, and the major component (90%) had a mobility of -5.35×10^{-5} cm.^{2} volt^{-1} sec.$^{-1}$.

[17] Nucleotide Synthesis from 5-Phosphoribosylpyrophosphate

Base + PRPP $\rightleftarrows$ Nucleoside 5'-phosphate + PP

By JOEL G. FLAKS

The synthesis of nucleotides from 5-phosphoribosylpyrophosphate (PRPP)[1] takes place in one step with the concomitant formation of pyrophosphate (PP), as shown in the general equation above. Enzymes catalyzing reactions of this type have been termed *nucleotide pyrophosphorylases*. The reaction is distinct from another widely distributed pathway of nucleotide synthesis which proceeds via the nucleoside, followed by phosphorylation to the nucleotide level. At the present writing four distinct nucleotide pyrophosphorylases have been described, and there are indications for the existence of other enzymes of this type.

The possible existence of a pathway of nucleotide synthesis involving ribose 1,5-diphosphate was indicated in a previous volume of this series,[2] but studies up to the present time have not substantiated this proposal.

I. Adenosine 5'-Phosphate Pyrophosphorylase

Adenine + PRPP $\rightleftarrows$ AMP + PP

AICA + PRPP $\rightleftarrows$ AICAR + PP

Assay Method

Principle. The enzyme catalyzes both the reactions shown above, and an assay method will be described for each.

AMP formation is determined by measuring the incorporation of C^{14}-

[1] In addition to the usual abbreviations, the following have been used: AICA, 5-amino-4-imidazolecarboxamide; AICAR, 5-amino-1-β-D-ribosyl-4-imidazolecarboxamide 5'-phosphate; IMP, inosine 5'-phosphate; OMP, orotidine 5'-phosphate; PRPP, 5-phosphoribosylpyrophosphate; XMP, xanthosine 5'-phosphate.
[2] M. Saffran, Vol. II [77].

adenine into a barium–ethanol insoluble fraction, obtained after the addition of carrier AMP. Residual C^{14}-adenine is removed, and the radioactivity of the insoluble barium adenylate is determined, the latter being a measure of the AMP formed.[3] Alternatively, the AMP formed may be determined spectrophotometrically with adenylic acid deaminase.[3]

The formation of AICAR is determined by separating the nucleotide from AICA by ion-exchange chromatography. The amount of AICAR formed is then estimated by assaying the diazotizable amine present in the nucleotide fraction.[4]

1. AMP Formation

Reagents

C^{14}-adenine, 0.01 M, with a specific activity of 1×10^5 counts/min./micromole.

PRPP,[5] Na or K salt, 0.001 M.

Tris–chloride buffer, pH 8.0, 1 M.

$MgCl_2$, 0.1 M.

AMP, 0.025 M.

Trichloroacetic acid, 10%.

Thymol blue, 0.04%.

KOH, 2 N.

Barium acetate, saturated solution.

Ethanol.

HCl, 0.1 N.

Procedure. The incubation mixture contains, in a final volume of 0.4 ml.: 0.03 ml. of C^{14}-adenine (0.3 micromole), 0.1 ml. of PRPP (0.1 micromole), 0.02 ml. of Tris–chloride buffer, pH 8.0 (20 micromoles), 0.02 ml. of $MgCl_2$ (2 micromoles), and less than 0.2 unit of the enzyme. The mixture is incubated at 36° for 15 minutes, and the reaction is terminated by the addition of 0.15 ml. of 10% trichloroacetic acid. The precipitate is removed by centrifugation, and the following are added to the supernatant fluid: 0.1 ml. of AMP (2.5 micromoles) as carrier, 1 drop of thymol blue, 2 drops of 2 N KOH, 0.1 ml. of saturated barium acetate, and 2.4 ml. of ethanol. The precipitate is collected by centrifugation, washed twice with 3-ml. portions of ethanol, and suspended in 0.5 ml. of 0.1 N HCl. An aliquot of 0.1 to 0.2 ml. of the suspension is

---

[3] A. Kornberg, I. Lieberman, and E. S. Simms, *J. Biol. Chem.* **215,** 417 (1955); see also Vol. VI [12].

[4] J. G. Flaks, M. J. Erwin, and J. M. Buchanan, *J. Biol. Chem.* **228,** 201 (1957).

[5] See Vol. VI [69].

plated onto a planchet; when it is dry, its radioactivity is determined and corrected for self-absorption.

Definition of Unit and Specific Activity. A unit of activity is defined as the amount of enzyme catalyzing the synthesis of 1 micromole of AMP per hour. Specific activity is expressed as units per milligram of protein. Protein is estimated by the procedure of Lowry *et al.*[6]

2. AICAR Formation

Reagents

AICA,[7] 0.025 M.
PRPP,[5] Na salt, 0.01 M.
Tris–chloride buffer, pH 8.0, 1 M.
$MgCl_2$, 0.1 M.
Dowex 1-X8, chloride form, 200 to 400 mesh.
HCl, 0.4 M.
H_2SO_4, 2 N.
Na nitrite, 0.1%.
Ammonium sulfamate, 0.5%.
N^1-naphthylethylenediamine dihydrochloride, 0.1% in 0.01 N HCl.

The latter three reagents for the diazotizable amine determination are stable for several months if stored in the refrigerator in amber bottles.

Procedure. The incubation mixture contains, in a final volume of 1.0 ml.: 0.2 ml. of AICA (5 micromoles), 0.1 ml. of PRPP (1 micromole), 0.1 ml. of Tris–chloride buffer, pH 8.0 (100 micromoles), 0.1 ml. of $MgCl_2$ (10 micromoles), and the enzyme. The mixture is incubated at 38° for 20 minutes, and the reaction is terminated by placing the vessels in a boiling-water bath for 2 minutes. After the denatured protein has been removed by centrifugation, the vessel contents are washed onto a Dowex 1–chloride column (3 cm. × 1-cm. diameter), and the column is rapidly washed (under pressure) with 100 ml. of water to remove the AICA. The AICAR is then removed with 5 ml. of 0.4 M HCl.

The diazotizable amine present in the eluted nucleotide fraction is determined with an aliquot of 2 ml. One milliliter of 2 N H_2SO_4 is added, followed by 0.5 ml. of 0.1% Na nitrite, and the contents are mixed. After 5 minutes, 0.5 ml. of 0.5% ammonium sulfamate is added with mixing. After 3 minutes, 0.5 ml. of 0.1% N^1-naphthylethylenediamine dihydrochloride is added; the contents are mixed and allowed to sit for 30

---

[6] O. H. Lowry, N. J. Rosebrough, A. L. Farr, and R. J. Randall, *J. Biol. Chem.* **193**, 265 (1951); see also Vol. III [73].
[7] E. Shaw and D. W. Woolley, *J. Biol. Chem.* **181**, 89 (1949).

minutes. The absorbancy at 540 mμ is determined against a blank that has been carried through the entire assay. The absorbancy at 540 mμ follows Beer's law over a limited concentration range, and reference should therefore be made to a calibration curve constructed with AICA as a standard.[8] Over that portion of the concentration range which follows Beer's law, the molar extinction coefficient of the product of the diazotization reaction is 26,400.

Definition of Unit and Specific Activity. A unit of activity is defined as the amount of enzyme required to synthesize 0.1 micromole of AICAR in 20 minutes. Specific activity is expressed as units per milligram of protein. Protein is determined by an absorbancy measurement at 280 mμ, with the assumption that a protein concentration of 1.0 mg./ml. has an absorbance of 1.6 at 280 mμ in a 1-cm. light path.[9]

Application of Assay Methods to Crude Tissue Preparations. Both the AMP and AICAR pyrophosphorylase assay procedures have been applied to several crude tissue preparations, and in each case a linear dependence of nucleotide formation on enzyme concentration was shown.

Purification Procedures

The enzyme appears to be widely distributed in nature and has at present been purified from brewer's yeast and beef liver. An extract of beef liver acetone powder has about 2% the specific activity of a brewer's yeast autolyzate, but the purified enzyme from liver is devoid of some contaminating activities present in the yeast preparation.

1. Purification of the Enzyme from Brewer's Yeast[3]

All operations are carried out at 0° to 3° unless specified.

Dried brewer's yeast is suspended in 3 vol. of 0.1 M KHCO$_3$ and incubated for 5 hours at 30° with occasional stirring. After the suspension has cooled to 5° to 10°, the supernatant fluid is obtained by centrifugation at 10,000 $\times$ g. The residue is washed with an amount of 0.1 M KHCO$_3$ equal to that originally used, and after centrifugation the supernatant fluids are combined. From 10 g. of dried yeast, 45 to 50 ml. of extract is obtained (yeast autolyzate).

To 20 ml. of the autolyzate, 40 ml. of 0.2 M K acetate buffer, pH 4.5, are added with stirring. With an ethanol–dry ice bath (—15°), the temperature of the mixture is lowered to 0°, and 11.5 ml. of cold ethanol (—15°) are added slowly with stirring. The precipitate is removed by centrifugation, and 15.0 ml. of cold ethanol (—15°) are slowly added to the supernatant fluid (now —6°) with constant stirring. The temperature

[8] See further in Vol. VI [9], Section I.B.
[9] J. M. Buchanan and C. B. Anfinsen, *J. Biol. Chem.* **180**, 47 (1949).

of the mixture is now —10°. The precipitate is collected by centrifugation, dissolved in 40 ml. of 0.025 M glycylglycine buffer, pH 6.8, and the pH adjusted to 6.8 with 0.1 N KOH (about 4 drops required) (ethanol fraction).

Forty milliliters of the ethanol fraction are stirred with 2.0 ml. of alumina C_γ gel[10] (15 mg. of solids per milliliter). After 5 minutes at 0°, the gel is collected by centrifugation and washed with 40 ml. of 0.1 M Na acetate buffer, pH 5.95. After centrifugation, the enzyme is eluted from the gel with 40 ml. of 0.01 M K phosphate buffer, pH 7.2 (alumina C_γ gel eluate).

Eight milliliters of water are added to 40 ml. of the alumina C_γ gel eluate, followed slowly, with stirring, with 12 ml. of calcium phosphate gel[11] (17 mg. of solids per milliliter). After 5 minutes at 0°, the gel is removed by centrifugation (calcium phosphate gel supernatant).

TABLE I

PURIFICATION OF AMP PYROPHOSPHORYLASE FROM YEAST

Fraction	Activity[a] units/ml.	Total activity, units	Protein, mg./ml.	Specific activity, units/mg. protein
1. Yeast autolyzate	18.8	376	15.6	1.2
2. Ethanol fraction	4.7	188	2.1	2.2
3. Alumina C_γ gel eluate	2.8	108	0.45	6.2
4. Calcium phosphate gel supernatant	1.2	72	0.05	24.0

[a] AMP formation determined enzymatically.

The enzyme at this point is purified 20-fold over the activity of the autolyzate (Table I). The ethanol- and gel-treated fractions retain 65 to 75% of their activity after 6 weeks at —15°.

2. Purification of the Enzyme from Beef Liver[4]

All operations are carried out at temperatures near 0° unless otherwise specified. Acetone powders of beef liver are prepared by the procedure used for pigeon liver (see Purification Procedure, Vol. VI [9], Section VI). The activity is stable for at least a year if the acetone powder is stored under desiccation at —15°. The purification procedure has also been carried out successfully on one-fourth the scale described below.

[10] R. Willstätter and H. Kraut, *Ber.* **56**, 1117 (1923); see also Vol. I [11].

[11] D. Keilin and E. F. Hartree, *Proc. Roy. Soc.* **B124**, 397 (1937–1938); see also Vol. I [11].

Four hundred grams of beef liver acetone powder are stirred at room temperature for 30 minutes with 4 l. of 0.05 M Tris–chloride buffer, pH 7.4. After centrifugation at 5400 $\times$ g for 30 minutes, 3500 ml. of extract are obtained (acetone powder extract).

To 3500 ml. of acetone powder extract, 350 ml. of 1 M Tris–chloride buffer, pH 8.0, are added, followed slowly, with stirring, with 862 g. of ammonium sulfate (0.38 saturation). After centrifugation at 5400 $\times$ g for 30 minutes, 485 g. of ammonium sulfate (0.55 saturation) are added, as above, to the supernatant fluid. The resulting precipitate is collected by centrifugation, as above, dissolved in 600 ml. of cold water, and dialyzed first against 20 l. of 0.005 M Tris–chloride buffer, pH 7.4, for 4 hours, followed by 20 l. of distilled water overnight (first ammonium sulfate step).

An amount of calcium phosphate gel[11] with a dry weight equivalent to one and one-half times the weight of protein in the ammonium sulfate fraction is centrifuged, and the water is discarded. The ammonium sulfate fraction is diluted to 4300 ml. with water, 43 ml. of 1 M K acetate buffer, pH 5.5, are added, and the enzyme solution is stirred with the packed gel for 10 minutes. After centrifugation, the supernatant fluid (4100 ml.) is retained (calcium phosphate gel supernatant).

The pH of the calcium phosphate gel supernatant fraction is adjusted to 5.5 with 5 M acetic acid. Then 1045 g. of ammonium sulfate (0.55 saturation) are slowly added with stirring, and the precipitate is collected by centrifugation and dissolved in 100 ml. of cold water. This solution is first dialyzed for 4 hours against 4 l. of 0.005 M Tris–chloride buffer, pH 7.4, and then against 4 l. of water overnight (second ammonium sulfate fraction).

An amount of alumina C_γ gel[10] with a dry weight equivalent to one and one-half times the weight of protein in the second ammonium sulfate fraction is centrifuged, and the water is discarded. The enzyme solution is adjusted to a protein concentration of 10 mg./ml. with water and stirred for 10 minutes with the packed gel. After centrifugation, the gel is successively eluted by stirring for 15 minutes with 120-ml. portions of K phosphate buffers, pH 7.4, of the following molarities: E-1, 0.02 M; E-2, 0.02 M; E-3, 0.025 M; E-4, 0.03 M. Eluate E-1, which has the highest specific activity, is further fractionated.

Next, 1.2 ml. of 1.0 M Tris–chloride buffer, pH 8.0, are added to 120 ml. of the alumina gel eluate E-1, and the temperature of the mixture is lowered to 0°. Then 190 ml. of 90% ethanol ($-15°$) are added dropwise with constant stirring while the temperature of the mixture is gradually lowered to $-15°$. The mixture is centrifuged at $-15°$, the precipitate is discarded, and 415 ml. of 90% ethanol ($-15°$) are added

TABLE II

PURIFICATION OF AMP AND AICAR PYROPHOSPHORYLASE FROM BEEF LIVER

Fraction	Volume, ml.	Protein, mg./ml.	AICA as substrate		Adenine as substrate	Adenine activity
			Yield,[a] %	Specific activity, units/mg. protein	Specific activity,[b] units/mg. protein	AICA activity
1. Acetone powder extract	3500	30.6	100	0.082	0.426	5.2
2. First ammonium sulfate fraction	840	43.4	71	0.171	0.992	5.8
3. Calcium phosphate gel supernatant	4100	2.79	51	0.390	2.13	5.3
4. Second ammonium sulfate fraction	115	15.5	28	1.38	7.60	5.5
5. Alumina C$_7$ gel eluate E-1	120	0.74	5.0	4.97	17.3	3.5
Alumina C$_7$ gel eluates E-2, E-3, E-4 (combined)	360	1.34	12.3	2.23	—	—
6. Ethanol fraction	10	1.56	1.6	9.0	27.9	3.1
7. Alumina C$_7$ gel supernatant	9.6	0.16	0.24	13.3	50.0	3.8

[a] The total number of units in the acetone powder extract was 8770.

[b] A unit is defined as the amount of enzyme required to synthesize 0.1 micromole of AMP in 20 minutes under the assay conditions of J. G. Flaks, M. J. Erwin, and J. M. Buchanan, *J. Biol. Chem.* **228**, 201 (1957).

as above. The mixture is allowed to stand overnight at $-15°$; the precipitate is then collected by centrifugation and extracted by stirring with 10 ml. of $0.005\,M$ Tris–chloride buffer, pH 7.4. The extract is clarified by centrifugation (ethanol fraction).

An amount of packed alumina C_γ gel equal to the amount of protein present is stirred for 10 minutes with the ethanol fraction. The mixture is centrifuged, and the supernatant fluid is retained (alumina C_γ gel supernatant).

The enzyme is purified approximately 156-fold over the acetone powder extract at the final step and loses only 10% of its activity after 3 months at $-15°$. The combined alumina gel eluates, E-1 through E-4, at step 5 in the purification procedure are adequate for the preparation of either AMP of AICAR, since this enzyme fraction is devoid of phosphatases and pyrophosphatases active on either PRPP or the products of the reaction. The purification of the beef liver is summarized in Table II.

Properties[3,4]

Specificity. The purified yeast enzyme is contaminated with inorganic pyrophosphatase and a pyrophosphorylase for IMP and GMP. This latter activity can be preferentially absorbed to calcium phosphate gel by increasing the amounts of gel used in the last step of the purification procedure.[3] The IMP–GMP pyrophosphorylase is completely removed from the beef liver enzyme at the third step (calcium phosphate gel) in the purification procedure.

Both enzyme preparations have a specific requirement for PRPP which cannot be replaced with ATP, ribose 5-phosphate, ribose 1-phosphate, or combinations of the latter two substances with ATP. Ribose 1,5-diphosphate is also inactive. Both enzymes are active with adenine and AICA. After 156-fold purification of the enzyme from beef liver there is little change in the ratio of activity with adenine to that with AICA (Table II), which suggests that a single enzyme may be responsible for both activities. The yeast enzyme is inactive with 2,6-diaminopurine. The liver enzyme is inactive with hypoxanthine, xanthine, 6-mercaptopurine, uric acid, guanine, 5-formamido-4-imidazolecarboxamide, and 5(4)-amino-1H-1,2,3-triazole-4(5)-carboxamide. The latter compound at $1 \times 10^{-3}\,M$ does not inhibit the reaction with AICA as substrate. Roy *et al.* [11a] have found the purified beef liver enzyme to be active with 8-azaadenine (between pH 6 and 7), 2-azaadenine, 4-aminopyrazolo(3,4-*d*)pyrimidine, and 2-fluoroadenine.

[11a] J. K. Roy, C. A. Haavik, and R. E. Parks, Jr., *Proc. Am. Assoc. Cancer Research* **3**, 146 (1960).

Pyrophosphorolysis of AICAR and AMP requires the presence of PP and does not occur with P_i or arsenate.

Activators and Inhibitors. Mg^{++} at a concentration of $2 \times 10^{-3} M$, or higher, is necessary for optimal activity with both enzyme preparations. The yeast enzyme shows 7% of the optimal activity in the absence of Mg^{++}.

The liver enzyme is not inhibited by p-chloromercuribenzoate, cyanide, iodoacetate, cysteine, or Cu^{++} at levels of $1 \times 10^{-3} M$, or by arsenate at $1 \times 10^{-2} M$. Fluoride at $5 \times 10^{-2} M$ inhibits the liver enzyme 44%, and the yeast enzyme from 25 to 40%. PP inhibits the yeast enzyme 50% at concentrations of $5 \times 10^{-3} M$, and 90% at $1 \times 10^{-2} M$.

pH Optimum. The optimal pH of the yeast enzyme with adenine as substrate is 7.6, with rates 67% of maximal at pH 6.5 and 8.4. The liver enzyme with AICA as substrate has a pH optimum from 7.6 to 8.4.

Equilibrium. With both enzyme preparations the equilibrium with adenine as substrate lies far in the direction of AMP formation. The equilibrium constant cannot be directly determined with the present assay procedures. Pyrophosphorolysis of AMP has, however, been demonstrated with both enzymes. With the liver enzyme and AICA as substrate, an equilibrium constant of 9.7 has been determined for the reaction in the direction of AICAR formation.

Kinetic Properties. The affinities of either enzyme for the substrates during AMP synthesis cannot be accurately determined with the present assay procedure. The K_m for adenine appears to be well below $4.5 \times 10^{-5} M$, and that for PRPP is somewhat below $4 \times 10^{-5} M$. For AICAR synthesis with the liver enzyme the K_m for AICA is $3.1 \times 10^{-4} M$, and that for PRPP is $1.4 \times 10^{-4} M$. AMP formation with the yeast enzyme was half-maximal at a Mg^{++} concentration of $2 \times 10^{-4} M$.

II. Guanosine 5′-Phosphate and Inosine 5′-Phosphate Pyrophosphorylase

$$\text{Guanine} + \text{PRPP} \rightleftarrows \text{GMP} + \text{PP}$$
$$\text{Hypoxanthine} + \text{PRPP} \rightleftarrows \text{IMP} + \text{PP}$$

Assay Method

Principle. The enzyme is most conveniently assayed with hypoxanthine as the purine substrate.[3, 12] Hypoxanthine is removed in an initial incubation with PRPP, and the residual quantity is determined spectrophotometrically in the presence of xanthine oxidase.[13] As an alternative, there are procedures of greater sensitivity which utilize incubations

---

[12] L. N. Lukens and K. A. Herrington, *Biochim. et Biophys. Acta* **24**, 432 (1957).
[13] H. M. Kalckar, *J. Biol. Chem.* **167**, 429 (1947); see also Vol. II [73].

containing C^{14}-purine substrates, followed by separation of the purine substrate from the nucleotide product on small ion-exchange columns.[3] A spectrophotometric assay is also available for the GMP activity.[13a]

Reagents for Assay A

Hypoxanthine, 0.0010 M.
PRPP,[5] Na salt, 0.01 M.
$MgCl_2$, 0.1 M.
K phosphate buffer, pH 7.4, 0.25 M.
Perchloric acid, 70%.
NaOH, 1 N.
Glycylglycine buffer, pH 7.5, 0.5 M.
Xanthine oxidase;[14] the alumina C_γ gel eluate of step 4 is diluted 5-fold just prior to use.

Reagents for Assay B

Guanine-8-C^{14}, 1.8 × 10^{-4} M, with a specific activity of 1.5 × 10^5 counts/min./μmole).
PRPP,[5] Na or K salt, 0.001 M.
Tris–chloride buffer, pH 8.0, 1 M.
$MgCl_2$, 0.1 M.
Dowex 1-X8, chloride form, 200 to 400 mesh.
HCl, 0.002 N.
HCl, 0.1 N.

Procedure for Assay A: IMP Pyrophosphorylase.[12] The incubation mixture contains, in a final volume of 2.5 ml.: 0.45 ml. of hypoxanthine (0.45 micromole), 0.2 ml. of PRPP (2 micromoles), 0.1 ml. of $MgCl_2$ (10 micromoles), 0.1 ml. of K phosphate buffer, pH 7.4 (25 micromoles), and the enzyme. The mixture is incubated at 38° for 15 minutes, and the reaction is terminated by the addition of 0.09 ml. of 70% perchloric acid. The denatured protein is sedimented by centrifugation, the supernatant fluid is brought to near neutrality (pH indicator paper) with 1 N NaOH, and the volume is made up to 5.0 ml.

Aliquots of 0.3 to 1.5 ml. of the above supernatant fluid are added to a 4-ml. cuvette containing 0.6 ml. of 0.5 M glycylglycine buffer, pH 7.5, and water to a final volume of 3.0 ml. The absorbance at 290 mμ is determined, and 0.1 ml. of the diluted xanthine oxidase is added, with mixing. The absorbance at 290 mμ is followed against a blank until the

[13a] C. E. Carter, *Biochem. Pharmacol.* **2**, 105 (1959).
[14] B. L. Horecker and L. A. Heppel, *J. Biol. Chem.* **178**, 683 (1949); see also Vol. II [73].

reaction is complete (about 30 minutes). The enzymatic oxidation of 1 μg. of hypoxanthine per milliliter results in an increase in absorbance at 290 mμ of 0.080.[13]

Procedure for Assay B: GMP Pyrophosphorylase.[3] The incubation mixture contains, in a final volume of 4.0 ml.: 2.0 ml. of guanine-8-C^{14} (0.36 micromole), 0.4 ml. of PRPP (0.4 micromole), 0.08 ml. of MgCl$_2$ (8 micromoles), 0.08 ml. of Tris–chloride buffer, pH 8.0 (80 micromoles), and the enzyme. The mixture is incubated for 20 minutes at 36°, and the reaction is terminated by holding the mixture at 100° for 1 minute. After removal of denatured protein by centrifugation, the supernatant fluid is passed through a Dowex 1 column, chloride form (3 cm. × 1-cm. diameter). Unreacted guanine is removed from the column by elution with 50 ml. of 0.002 N HCl. The GMP is then removed with 5 ml. of 0.1 N HCl, and aliquots are assayed for their radioactivity. A blank with the enzyme omitted should be carried through the entire procedure.

It should be pointed out that a more-sensitive assay procedure for the IMP pyrophosphorylase can be attained by using the incubation conditions of assay A, substituting C^{14}-hypoxanthine for the unlabeled purine, and terminating the incubation by heating the assay vessel at 100° for 1 to 2 minutes. After removal of denatured protein, the reaction mixture is subjected to the ion-exchange procedure exactly as described in assay B.

Definition of Unit and Specific Activity. For the enzyme purification described below, with assay A, a unit of activity is defined as the amount of enzyme effecting the removal of 0.1 micromole of hypoxanthine in 15 minutes. Specific activity is expressed as units per milligram of protein. Protein is estimated by absorbancy measurement at 280 mμ, with the assumption that a protein concentration of 1.0 mg./ml. has an absorbance of 1.6 at 280 mμ in a 1-cm. light path.[9]

Application of Assay Methods to Crude Tissue Preparations. Assay A, based on hypoxanthine removal, has been successfully used with liver acetone powder extracts of several species and with yeast autolyzates. Xanthine oxidase or dehydrogenase activities in these preparations do not appear to cause interference. Assay B, for the GMP pyrophosphorylase, is subject to interference by crude extracts from those animal tissues, notably liver, which contain guanase activity.

Purification Procedure[12]

The procedure is based on the observation that heating of extracts at 60° to 65° destroys AMP pyrophosphorylase but not the activity for hypoxanthine or guanine.[15]

[15] E. D. Korn, C. N. Remy, H. C. Wasilejko, and J. M. Buchanan, *J. Biol. Chem.* **217**, 875 (1955).

All operations are carried out at temperatures near 3° unless otherwise specified. Acetone powders of beef liver are prepared by the procedure used for chicken livers (see Purification Procedure, Vol. VI [9], Section IV). The activity is stable for over a year if the acetone powder is stored under desiccation at 4°.

Forty grams of the beef liver acetone powder are extracted by stirring for 1 hour with 400 ml. of 0.033 M K phosphate buffer, pH 7.4. The supernatant fluid is retained after centrifugation at 10,000 $\times$ g (acetone powder extract).

The acetone powder extract is placed in a water bath at 74° and kept there, with stirring, for 3 minutes after the temperature of the extract reaches 56°. The temperature is then rapidly lowered by chilling the extract in an ice bath, and the insoluble protein is removed by centrifugation (first heat step).

For each 100 ml. of enzyme solution from the first heat step, 31.2 g. of ammonium sulfate (0.50 saturation) are slowly added with stirring, and the precipitate obtained is removed by centrifugation. Then 11.8 g. of ammonium sulfate (0.65 saturation) are added, as above, for each 100 ml. of original enzyme solution. The precipitate obtained on centrifugation is dissolved in a minimal volume of 0.033 M K phosphate buffer, pH 7.4, and dialyzed against several changes of the same buffer (first ammonium sulfate fraction.)

The ammonium sulfate fraction (16 mg. of protein per milliliter) is subjected to a second heat step identical to that above, with the exception that the solution is held in the 74° bath for 5 minutes after the temperature reaches 56° (second heat step).

For each 100 ml. of solution from the second heat step, 27.8 g. of ammonium sulfate (0.45 saturation) are slowly added with stirring, and the precipitate collected by centrifugation is discarded. Then 15.2 g. of ammonium sulfate (0.65 saturation) are added, as above, for each 100 ml. of the second heat step solution, and the precipitate is collected by centrifugation and dissolved in a minimal volume of 0.003 M K phos-

TABLE III

PURIFICATION OF IMP–GMP PYROPHOSPHORYLASE OF BEEF LIVER

Fraction	Yield,[a] %	Specific activity, units/mg. protein
1. Acetone powder extract	100	1.6
2. First heat step	100	3.2
3. First ammonium sulfate fraction	76	19
4. Second heat step	76	32
5. Second ammonium sulfate fraction	59	58

[a] The acetone powder extract contained 38 units/ml.

phate buffer, pH 7.4. The solution is dialyzed against several changes of the same buffer.

The enzyme at this point is purified 36-fold (see Table III) over the acetone powder extract and is stable for at least 4 months when stored frozen. A similar enzyme has been purified from *Escherichia coli*.[13a]

Properties

Specificity. The enzyme is free from AMP and OMP pyrophosphorylase activity and enzymes that degrade either PRPP or the products of the reaction. It has a specific requirement for PRPP and either hypoxanthine, guanine, or 6-mercaptopurine. The following have been tested and found inactive for nucleotide synthesis: adenine, AICA, 5-formamido-4-imidazolecarboxamide, xanthine, uric acid, 8-azaguanine, 2,6-diaminopurine, and orotic acid. Similar specificity has been shown with a purified enzyme from *E. coli*.[13a]

Activators. Mg^{++} is required for the reaction, but the optimum level has not been reported.

Equilibrium. With hypoxanthine as substrate, the equilibrium lies far in the direction of IMP formation. The reaction is reversible, and pyrophosphorolysis of IMP has been demonstrated. No reversal of the reaction was noted with P_i.[3]

Kinetics. The rate of the reaction with guanine as substrate is about 42% that with hypoxanthine under similar reaction conditions.[3] The following K_s values have been obtained with the purified enzyme from *E. coli*:[13a] hypoxanthine, $1 \times 10^{-5} M$; 6-mercaptopurine, $1 \times 10^{-5} M$; guanine, $5 \times 10^{-4} M$. This greater affinity of the enzyme for 6-mercaptopurine as compared to guanine has been suggested as a possible site of action of 6-mercaptopurine.

III. Orotidine 5′-Phosphate Pyrophosphorylase

$$\text{Orotic acid} + \text{PRPP} \rightleftharpoons \text{OMP} + \text{PP}$$

Assay Method

Principle. The enzyme activity is assayed spectrophotometrically,[16] with the procedure based on the decrease in absorbance at 295 mμ on conversion of orotate to UMP according to:

$$\text{Orotate} + \text{PRPP} \rightleftharpoons \text{OMP} + \text{PP}$$
$$\text{OMP} \rightarrow \text{UMP} + CO_2$$

Excess OMP decarboxylase is added to the reaction components.

[16] I. Lieberman, A. Kornberg, and E. S. Simms, *J. Biol. Chem.* **215**, 403 (1955).

Alternatively, where low activity is encountered, the release of $C^{14}O_2$ from orotate-4,7-C^{14} on conversion to UMP is used to assay the enzyme.[16]

Reagents

Orotate, Na salt, 0.01 M.
PRPP,[5] Na salt, 0.001 M.
$MgCl_2$, 0.1 M.
Tris–chloride buffer, pH 8.0, 1 M.
OMP decarboxylase,[16] the ethanol fraction from step 2 of the purification procedure for inorganic pyrophosphatase from yeast,[17] contains OMP decarboxylase essentially free from OMP pyrophosphorylase. The decarboxylase is assayed by the procedure of Lieberman *et al.*[16]

Procedure. The reaction contains, in a final volume of 1.0 ml. in a quartz cuvette: 0.03 ml. of orotate (0.3 micromole), 0.1 ml. of PRPP (0.1 micromole), 0.02 ml. of $MgCl_2$ (2 micromoles), 0.02 ml. of Tris–chloride buffer, pH 8.0 (20 micromoles), 0.05 ml. of OMP decarboxylase (0.6 unit of decarboxylase), and the enzyme. The reaction is followed at room temperature by measurement of the decrease in absorbance at 295 mμ, with the removal of orotate calculated from the initial reaction rate (4 to 8 minutes). The molar absorption coefficient for the absorbancy decrease at 295 mμ on conversion of orotate to UMP is 3950.

Definition of Unit and Specific Activity. A unit of activity is defined as the amount of enzyme causing the removal of 1 micromole of orotate per hour as calculated from the initial reaction rate. Specific activity is expressed as units per milligram of protein. Protein is estimated by the procedure of Lowry *et al.*[6]

Application of Assay Method to Crude Tissue Preparations. The assay is applicable to crude tissue extracts with specific activities of about 0.1 or greater. Lieberman *et al.*[16] found that acetone powder extracts of the livers of pigeon, beef, and chicken have specific activities of 0.022, 0.020, and 0.018, respectively (about 1/35 that of brewer's yeast autolyzate). These estimations necessitated use of the radioactive assay procedure.

Purification Procedure[16]

All operations are carried out at 0° unless otherwise specified.
Dried brewer's yeast is suspended in 3 vol. of 0.1 M $KHCO_3$ and

[17] L. A. Heppel and R. J. Hilmoe, *J. Biol. Chem.* **192**, 87 (1951); see Vol. II [91].

incubated at 30° for 5 hours with occasional stirring. After the suspension has cooled to 5° to 10°, the supernatant fluid is collected by centrifugation at about 10,000 $\times$ g. The residue is washed with an amount of 0.1 M KHCO$_3$ equal to that originally used, and after centrifugation the supernatant fluids are combined. From 45 to 50 ml. of extract are obtained from 10 g. of dried yeast (yeast autolyzate).

To 48 ml. of the autolyzate, 96 ml. of 0.2 M Na acetate buffer, pH 4.5, are rapidly added with stirring. The mixture is cooled to $-2°$ in an alcohol–ice bath, and 48 ml. of ethanol ($-14°$) are added over the course of 4 minutes while the temperature is allowed to fall to $-6°$. Insoluble material is removed by centrifugation for 4 minutes at 10,000 $\times$ g at $-14°$. Then 144 ml. of ethanol are added to the supernatant fluid over the course of 5 minutes with the temperature maintained at $-10°$. The precipitate is collected by centrifugation at 10,000 $\times$ g for 3 minutes at $-14°$, dissolved in 0.025 M glycylglycine buffer, pH 7.0, and the pH adjusted to 7.0 with 1 N KOH. The volume is 48 ml. (ethanol fraction).

To 48 ml. of the ethanol fraction, sufficient calcium phosphate gel[11] (about 170 mg. of solids) is added to adsorb 85 to 90% of the OMP pyrophosphorylase activity. After 5 minutes the mixture is centrifuged, and the supernatant fluid is discarded. The enzyme is eluted from the packed gel with 24 ml. of 0.2 M Tris–chloride buffer, pH 8.0 (calcium phosphate gel eluate).

TABLE IV

PURIFICATION OF OMP PYROPHOSPHORYLASE FROM BREWER'S YEAST

| | OMP pyrophosphorylase | | | OMP decarboxylase, total activity, units |
Fraction	Total activity, units	Yield, %	Specific activity, units/mg. protein	
1. Yeast autolyzate	532		0.7	274
2. Ethanol fraction	240	45.1	1.7	240
3. Calcium phosphate gel eluate	141	26.5	6.7	2.9
4. Alumina C$_\gamma$ gel eluate	83	15.6	17.3	0

The calcium phosphate gel eluate (24 ml.) is adjusted to pH 7.0 with 1 N acetic acid, and 24 ml. of water are added. Sufficient alumina C$_\gamma$ gel[10] (about 29 mg. of solids) is added to adsorb all the pyrophosphorylase activity. The precipitate is collected by centrifugation after 5 minutes, washed with 48 ml. of 0.01 M K phosphate buffer, pH 7.2, and centrifuged. The enzyme is then eluted with 48 ml. of $0.06\,M$ K phosphate buffer, pH 7.2 (alumina C_γ gel eluate).

The enzyme at this point is purified approximately 25-fold over the autolyzate (Table IV).

Properties[16]

Specificity. With the purified enzyme no activity was found with adenine, uracil, DL-ureidosuccinate, L-dihydroörotate, or orotidine when substituted for orotate. Dahl *et al.*[18] have found the enzyme to be active with 5-fluoroörotate but inactive with the 5-bromo, 5-chloro, 5-amino, 5-nitro, and 5-methyl derivatives of orotate. The enzyme is also inactive with 5-fluorouracil. The 5-substituted orotates which are inactive as substrates, as well as uracil and 5-fluorouracil, do not inhibit the reaction with orotate and 5-fluoroörotate as substrates. The enzyme is also specific for PRPP and, when studied in the reverse direction, for PP.

Activators and Inhibitors. Mg^{++} is required for optimal activity with a reaction rate 25% of optimal in its absence. Optimal activity is shown with $0.002\,M$ Mg^{++}.

6-Uracilsulfonic acid, 6-uracilsulfonamide, and 6-uracil methyl sulfone competitively inhibit the conversion of orotate to OMP.[19] No evidence for the formation of nucleotides with these analogs was obtained.

An enzyme from *Lactobacillus bulgaricus* 09X was inhibited 80% in the presence of $0.15\,M$ phosphate or sulfate.[20] The reaction was inhibited 30% by $0.15\,M$ chloride.

pH Optimum. The optimal pH range for OMP formation lies between 7.5 and 8.5. For OMP pyrophosphorolysis, the optimal pH range is between 6.0 and 7.0.

Equilibrium. The equilibrium for the reaction lies in the direction of OMP pyrophosphorolysis. An equilibrium constant of 0.12 was found for the reaction as written in the direction of OMP formation. The reaction is not well suited for the synthesis of large amounts of OMP, as there is an early decline in the rate due to the unfavorable equilibrium.

Kinetics. The K_s values for both orotate and PRPP were found to be $2.0 \times 10^{-5}\,M$. The K_s for 5-fluoroörotate is also $2 \times 10^{-5}\,M$, but it reacts about twice as fast as orotate.[18] For the pyrophosphorolysis of OMP, a K_s of $1.2 \times 10^{-4}\,M$ was found for PP.

[18] J. L. Dahl, J. L. Way, and R. E. Parks, Jr., *J. Biol. Chem.* **234**, 2998 (1959).
[19] W. L. Holmes, *J. Biol. Chem.* **223**, 677 (1956).
[20] E. S. Canellakis, *J. Biol. Chem.* **227**, 329 (1957).

The K_i values were $7.0 \times 10^{-6} M$ for 6-uracilsulfonic acid, 3.9×10^{-4} M for 6-uracilsulfonamide, and $7.1 \times 10^{-4} M$ for 6-uracil methyl sulfone.[19]

IV. Uridine 5′-Phosphate Pyrophosphorylase

$$\text{Uracil} + \text{PRPP} \rightleftarrows \text{UMP} + \text{PP}$$

Assay Method

Principle. The method is based on the conversion of C^{14}-uracil to UMP.[21] The substrate and the nucleotide product are separated by anion-exchange chromatography on small resin columns, and the isotope incorporated into the UMP fraction is estimated.

An alternative spectrophotometric method is available, based on the difference in absorbance of uracil and UMP at alkaline pH's.[21] This procedure, however, is of much lower sensitivity.

Reagents

Uracil-2-C^{14}, $0.002 M$, with a specific activity of 6×10^5 counts/ min./micromole.
PRPP,[5] Na salt, $0.001 M$.
K phosphate buffer, pH 7.4, $1.0 M$.
$MgCl_2$, $0.1 M$.
Glutathione, pH 7.4, $0.1 M$.
Dowex 1-X2, chloride form, 200 to 400 mesh.
K acetate buffer, pH 5.0, $0.01 M$.
HCl, $0.1 M$.

Procedure. The reaction mixture contains, in a final volume of 1.0 ml.: 0.05 ml. of uracil-2-C^{14} (0.1 micromole), 0.1 ml. of PRPP (0.1 micromole), 0.03 ml. of $MgCl_2$ (3 micromoles), 0.05 ml. of K phosphate buffer, pH 7.4 (50 micromoles), 0.05 ml. of glutathione (5 micromoles), and the enzyme. The contents are mixed and incubated at 37° for 15 minutes, and the reaction is terminated by placing the mixture in a boiling-water bath for 30 seconds. After removal of insoluble material by centrifugation, the reaction mixture is passed through a Dowex 1 column, chloride form (1 cm. × 1-cm. diameter). Uracil is eluted from the column after the passage of 15 ml. of $0.01 M$ K acetate buffer, pH 5.0. The UMP fraction is then eluted with 10 ml. of $0.1 M$ HCl. Aliquots of the UMP fraction are then assayed for their radioactivity.

Definition of Unit and Specific Activity. A unit of activity is defined as the amount of enzyme causing the synthesis of 0.1 micromole of UMP

[21] I. Crawford, A. Kornberg, and E. S. Simms, *J. Biol. Chem.* **226**, 1093 (1957).

per hour. Specific activity is expressed as units per milligram of protein. Protein is estimated by the procedure of Lowry et al.[6]

Application of Assay Method to Crude Tissue Preparations. The assay method has been successfully applied to the crude extracts of a number of microorganisms.

Purification Procedure

The enzyme is present in fairly high activity in a number of strains of lactobacilli; the purification described below uses *L. bifidus* (ATCC 4963) as the source material.[21] The organism is grown on the semisynthetic media of Wright et al.,[22] supplemented with 0.03% of either orotate or uracil. Standing cultures after 20 to 24 hours at 37° yield about 4 g. wet weight of cells per liter. All operations are carried out at 0° to 3° unless otherwise specified.

The cells from a 2-l. culture are harvested by centrifugation and washed twice with cold distilled water by centrifugation. The pellet of cells is frozen in an ethanol–dry ice bath and ground in a chilled mortar with 16 g. of Alumina 301. The paste is suspended in 45 ml. of cold $0.05 M$ K phosphate buffer, pH 7.4, and centrifuged at $16,000 \times g$ to yield an extract containing from 2 to 7 mg. of protein per milliliter (cell-free extract).

To 40 ml. of the cell-free extract, 50 ml. of $0.1 M$ K acetate buffer, pH 4.0, are added with stirring to bring the pH to 4.5. After 5 minutes, the precipitate is collected by centrifugation and dissolved in 20 ml. of $0.05 M$ K phosphate buffer, pH 7.4. Four milliliters of 1% protamine sulfate (Eli Lilly) are added to the slightly turbid solution, and after 5 minutes the mixture is centrifuged and the precipitate is discarded (low pH protamine fraction).

To 23 ml. of the low pH protamine fraction, 4.6 ml. of $0.1 M$ K acetate buffer, pH 4.0, are added to bring the pH to 6.6. The solution is cooled to 0° in an ethanol–ice bath at −15°, and 12.4 ml. of ethanol (−15°) are added with stirring over the course of 2 minutes. The temperature falls to −2°. After 30 seconds the mixture is centrifuged at $10,000 \times g$ for 3 minutes, and the precipitate is discarded. Then 7.6 ml. of ethanol (−15°) are added with stirring over the course of 2 minutes during which time the temperature falls to −14°. The precipitate is collected by centrifugation at −15° and dissolved in 20 ml. of $0.01 M$ K phosphate buffer, pH 7.4 (ethanol fraction).

The ethanol fraction is brought to pH 6.8 with about 1 ml. of $0.1 N$ HCl and mixed with 2.4 ml of alumina C_γ gel[10] (15 mg. of solids per

[22] L. D. Wright, C. A. Driscoll, C. S. Miller, and H. R. Skeggs, *Proc. Soc. Exptl. Biol. Med.* **83**, 716 (1953).

milliliter). After 5 minutes, the suspension is centrifuged, and the gel is washed with 10 ml. of 0.05 M K phosphate buffer, pH 6.9, followed by centrifugation. The enzyme is eluted by extracting the gel with 0.1 M K phosphate buffer, pH 6.9 (alumina C_γ gel eluate).

The enzyme at this point is purified about 25-fold (see Table V) over the activity in the cell-free extract and loses less than 10% of its activity

TABLE V

PURIFICATION OF UMP PYROPHOSPHORYLASE FROM *Lactobacillus bifidus*

Fraction	Uracil activity		Orotate activity[a]	
	Total activity, units	Specific activity, units/mg. protein	Specific activity, units/mg. protein	Orotate activity / Uracil activity
1. Cell-free extract	63	0.6	0.24	0.4
2. Low pH protamine fraction	51	1.8	0.26	0.15
3. Ethanol fraction	18	8.5	0.10	0.01
4. Alumina C_γ gel eluate	4	15.0	0	0

[a] Assayed by the method described in Section C.

after 2 weeks at $-15°$. The cell-free extract and low pH–protamine fractions are equally stable, but the stability of the ethanol fraction is quite variable. Dialysis of the gel eluate against buffers of low ionic strength invariably results in complete inactivation, whereas after 3 hours dialysis against 0.5 M KCl only 40% of the activity is lost.

Properties

Distribution. Among several pyrimidine-requiring lactobacilli strains there is a correlation between those strains able to grow on uracil and the presence of the UMP pyrophosphorylase.[21] This does not extend to all uracil-requiring cells, since *E. coli* W_{c^-}, a cytidine- or uracil-requiring auxotroph of strain W, does not have an increased amount of the enzyme over that present in the parent, strain W. The levels of the enzyme in these strains is quite low (0.02 unit/mg. of protein), when compared to the level of the OMP pyrophosphorylase (about 0.7 unit/mg. of protein).[21] Strain W_{c^-} has an enhanced nucleoside phosphorylase activity and would seem to utilize the nucleoside pathway for its pyrimidine requirement. The UMP pyrophosphorylase was not detected in *E. coli* B,[21] but has been found in *E. coli* ATCC 9637.[23] Canellakis[20] has found the enzyme in a strain of *L. bulgaricus* 09X.

[23] R. W. Brockman, J. M. Davis, and P. Stutts, *Biochim. et Biophys. Acta* 40, 22 (1960).

The enzyme has not been found in yeast autolyzates or in acetone powder extracts of various mammalian and avian livers.[16] However, rapidly growing tissues (intestinal mucosa and regenerating liver) and tumor tissue may possess the enzyme.

Specificity. The specificity of the purified enzyme has not been reported, with the exception that it is not active with orotate. Brockman *et al.*[23] found both uracil and 5-fluorouracil to be active with an extract of *E. coli* ATCC 9637 and PRPP. Extracts from a 5-fluorouracil-resistant mutant derived from the parent strain would not convert either uracil or 5-fluorouracil to their respective nucleotides.

Activators and Inhibitors. Mg^{++} is required for the reaction with only 5% of the activity remaining in its absence.[21] Omission of glutathione reduced the rate of the reaction about 25%.[21]

The enzyme from *L. bulgaricus* 09X is inhibited 80% in the presence of $0.05\ M$ phosphate.[20]

Equilibrium. The equilibrium lies far in the direction of UMP formation, but some reversal of the reaction can be demonstrated with UMP and PP.[21]

Kinetics. The K_m for PRPP is $4.4 \times 10^{-5}\ M$. The lowest concentration of uracil used ($5.6 \times 10^{-5}\ M$) gave a maximal reaction rate. The K_m is thus well below this value.

V. Other Nucleotide Pyrophosphorylases

A large number of investigations have appeared suggesting nucleotide pyrophosphorylases active with various purine and pyrimidine analogs. Few of the studies have been carried out with purified or resolved enzyme preparations, and in many cases it is difficult to ascertain whether the system under study is the nucleotide pyrophosphorylase pathway or the pathway via the nucleoside. However, as more systems are studied and the assay procedures refined, it is becoming increasingly apparent that either a larger number of pyrophosphorylases exist in nature than those that have been resolved to date, or there exist enzymes of different specificity in various organisms. Some selected cases in point are discussed below.

Purines. The purified beef liver[12] and *E. coli*[13a] IMP-GMP pyrophosphorylases are inactive with xanthine as substrate, as are extracts of rat and pigeon liver.[24] Lagerkvist[24] has reported that extracts of lyophilized *E. coli* catalyze XMP formation from xanthine, ribose 5-phosphate, and ATP, and Brockman *et al.*[23] have found XMP formation from xanthine and PRPP with extracts of *E. coli* ATCC 9637. Studies with mutants,

[24] U. Lagerkvist, *J. Biol. Chem.* **233**, 138 (1958).

and extracts prepared from them, from *Streptococcus faecalis*[25] and *Salmonella typhimurium*[26] resistant to various purine analogs which are growth-inhibitory also suggest the presence of an XMP pyrophosphorylase activity. The work with *S. typhimurium* indicates the occurrence of at least three and possibly four distinct purine nucleotide pyrophosphorylases in this organism.[26]

Cross resistance to growth inhibition by 6-mercaptopurine, 8-azaguanine, and 6-thioguanine frequently occurs, and generally such mutants cannot convert hypoxanthine or guanine to their nucleotides. It has been found in *S. faecalis*,[25] *S. typhimurium*,[26] *L. casei*,[27] a leukemic cell line L1210,[28, 29] and in other organisms. Where studies have been carried out with cell-free extracts, the absence of, or a reduced, nucleotide pyrophosphorylase activity is correlated with the resistance. The purified beef liver[12] and *E. coli*[13a] IMP-GMP pyrophosphorylases are inactive with 8-azaguanine, but Way and Parks have reported the formation of 8-azaguanosine 5'-phosphate from hog and beef liver preparations and with yeast autolyzates.[30] An important observation was made that the pH optima for the reaction with various purines and their analogs is markedly different in some cases. Thus pyrophosphorylase activity with 8-azaguanine is hardly detectable at pH 8.0, the pH optimum for the IMP-GMP pyrophosphorylase, whereas it proceeds maximally at pH 7.0.[30] However, even at pH 7.2 the purified IMP-GMP pyrophosphorylase of *E. coli* is inactive with 8-azaguanine.[13a]

Nucleotide pyrophosphorylase activity for 2,6-diaminopurine has been found in *S. typhimurium*,[26] beef liver acetone powder extracts,[30] yeast autolyzates,[30] and *E. coli*.[31] Both the purified AMP pyrophosphorylase from yeast[3] and beef liver[4] and the purified IMP-GMP pyrophosphorylase[12] are inactive with 2,6-diaminopurine.

Although both the purified AMP pyrophosphorylases of yeast and beef liver are active with both adenine and AICA, extracts of *S. typhimurium* have considerable activity with adenine but relatively weak activity with AICA.[32]

The above examples suggest a re-examination of the activity of the

[25] R. W. Brockman, C. Sparks, D. J. Hutchinson, and H. E. Skipper, *Cancer Research* **19**, 177 (1959).

[26] G. P. Kalle, J. S. Gots, and C. Abramson, *Federation Proc.* **19**, 310 (1960).

[27] G. B. Elion, S. Singer, and G. H. Hitchings, *Federation Proc.* **15**, 248 (1956).

[28] R. W. Brockman, M. C. Sparks, and M. S. Simpson, *Biochim. et Biophys. Acta* **26**, 671 (1957).

[29] R. W. Brockman and P. Stutts, *Federation Proc.* **19**, 313 (1960).

[30] J. L. Way and R. E. Parks, Jr., *J. Biol. Chem.* **231**, 467 (1958).

[31] C. N. Remy and M. S. Smith, *J. Biol. Chem.* **228**, 325 (1957).

[32] G. P. Kalle and J. S. Gots, *Biochim. et Biophys. Acta* **51**, 130 (1961).

purified enzymes for various analogs with more-sensitive assay procedures and under different conditions, particularly pH, and the nature of the buffer (phosphate inhibits some pyrophosphorylase activities).

Pyrimidines. The formation of 2-thiouridine 5'-phosphate from ATP, ribose 5-phosphate, and an extract from *E. coli* K12 has been reported.[33] The same extract forms UMP from uracil, but the thiouracil nucleotide synthesis is inhibited 54% by the presence of 0.025 M K[+], whereas the uracil activity is not. Whether the reaction is proceeding via a nucleotide pyrophosphorylase is uncertain.

VI. General Assay for Nucleotide Pyrophosphorylases

In addition to the individual assay procedures described for each enzyme, general assay procedures can be devised for a large number of purines and pyrimidines. Incubations containing radioactive purine or pyrimidine substrates followed by anion-exchange chromatography to separate the substrate from the nucleotide product would provide the most-sensitive means of assaying a particular pyrophosphorylase. Assays of this type can be based on the procedures described above for the GMP and UMP pyrophosphorylases. An assay procedure of wide application, but lower sensitivity, has been described by Wyngaarden *et al.*[34] and is based on measurement of the residual PRPP as orcinol-reactive pentose after absorption of the nucleotide product on charcoal. The procedure described here is a modification of this method.[35]

The incubation mixture contains, in a final volume of 1.0 ml.: 1.3 micromole of purine or pyrimidine, 0.6 micromole of PRPP, 10 micromoles of phosphate buffer, pH 7.5, 5 micromoles of $MgCl_2$, and 0.2 to 0.4 unit of enzyme. After incubation at 37° for 30 minutes, the reaction is terminated by placing the vessels in a boiling-water bath for 2 minutes. The insoluble material is removed by centrifugation, and an aliquot of 0.5 ml. is removed and added to 0.5 ml. of water containing 20 mg. of Norit. The mixture is stirred for 3 minutes, centrifuged, and the Norit is washed with 2.0 ml. of water. The supernatant fluids are combined and assayed for pentose by the orcinol procedure.[36] Controls should be run with the purine or pyrimidine omitted, as well as a blank with PRPP omitted.

A unit of activity for this assay is defined as the amount of enzyme catalyzing the removal of 1 micromole of PRPP per hour. The method

[33] H. Amos, E. Vollmayer, and M. Korn, *Arch. Biochem. Biophys.* **77**, 236 (1958).
[34] J. B. Wyngaarden, H. R. Silberman, and J. H. Sadler, *Ann. N. Y. Acad. Sci.* **75**, 45 (1958).
[35] G. P. Kalle and J. S. Gots, *Biochim. et Biophys. Acta* **53**, 166 (1961).
[36] W. Mejbaum, *Z. physiol. Chem.* **258**, 117 (1939); see also Vol. III [12].

has been found to be applicable to all the purines and their analogs with a variety of crude tissue preparations. The pH and composition of the buffer should be modified for studies with some of the analogs, and in those cases where phosphate is inhibitory. Tris buffers should be avoided, as they appear to interfere with color development in the orcinol procedure.

[18] 5-Phosphoribose Pyrophosphokinase

Ribose 5-phosphate + ATP → PRPP + AMP

By JOEL G. FLAKS

Assay Method

Principle. The reaction is assayed by the convenient two-step procedure of Kornberg *et al.* [1] In step I the pyrophosphokinase reaction is carried out leading to the accumulation of 5-phosphoribosylpyrophosphate (PRPP), and in step II the accumulated PRPP is quantitatively utilized for the enzymatic conversion of orotate to a mixture of orotidylate and UMP according to the following equations:

$$\text{Orotate} + \text{PRPP} \rightleftarrows \text{orotidylate} + \text{PP}$$
$$\text{Orotidylate} \rightarrow \text{UMP} + CO_2$$

The removal of orotate in step II is determined spectrophotometrically.

Alternatively, two other two-step procedures of greater sensitivity (but less convenience) are available. These involve the enzymatic conversion of C^{14}-adenine[2] and 5-amino-4-imidazolecarboxamide[3] to their respective ribotides.

Reagents

ATP, K salt, 0.04 M.
Ribose 5-phosphate,[4] K salt, 0.025 M.
$MgCl_2$, 0.1 M.
Glutathione, 0.5 M.

[1] A. Kornberg, I. Lieberman, and E. S. Simms, *J. Biol. Chem.* **215**, 389 (1955).
[2] A. Kornberg, I. Lieberman, and E. S. Simms, *J. Biol. Chem.* **215**, 417 (1955); see also Vol. VI [17].
[3] J. G. Flaks, M. J. Erwin, and J. M. Buchanan, *J. Biol. Chem.* **228**, 201 (1957); see also Vol. VI [17].
[4] J. X. Khym, D. G. Doherty, and W. E. Cohn, *J. Am. Chem. Soc.* **76**, 5523 (1954).

KF, 1 M.

K phosphate buffer, pH 7.4, 1 M.

Orotate, 0.01 M.

Tris–chloride buffer, pH 8.0, 1 M.

Orotidylate pyrophosphorylase,[5] ethanol fraction from step 2 containing orotidylate decarboxylase.

Procedure. Step I. The incubation mixture contains, in a final volume of 1.0 ml.: 0.03 ml. of ATP (1.2 micromoles), 0.1 ml. of ribose 5-phosphate (2.5 micromoles), 0.02 ml. of $MgCl_2$ (2.0 micromoles), 0.02 ml. of glutathione (10 micromoles), 0.02 ml. of K phosphate buffer, pH 7.4 (20 micromoles), 0.05 ml. of KF (50 micromoles), and approximately 1 unit of the enzyme. The contents are mixed, incubated at 36° for 20 minutes, and the reaction is then terminated by heating the mixture at 100° for 1 minute.[6] The contents are immediately cooled and centrifuged if necessary.

Step II. The incubation contains, in a final volume of 1.0 ml. in a quartz cuvette: 0.02 ml. of orotate (0.2 micromole), 0.02 ml. of $MgCl_2$ (2 micromoles), 0.02 ml. of Tris–chloride buffer, pH 8.0 (20 micromoles), 0.2 ml. of orotidylate pyrophosphorylase (containing orotidylate decarboxylase), and an aliquot from the step I incubation containing about 0.03 micromole of PRPP. The contents are mixed, and the reaction is followed at room temperature in an ultraviolet spectrophotometer at 295 mμ against a suitable blank from step I. With the indicated amounts of orotidylate pyrophosphorylase, the utilization of 0.03 micromole of PRPP is usually completed in about 10 minutes. The enzymatic conversion of orotate to orotidylate and UMP results in a decrease in absorbancy at 295 mμ with a molar absorption coefficient of 3950.

Definition of Unit and Specific Activity. One unit of enzyme activity is defined as that amount catalyzing the synthesis of 1 micromole of PRPP (or the removal of 1 micromole of orotate) per hour. Specific activity is expressed as units per milligram of protein. Protein is determined by the procedure of Lowry *et al.*[7]

Application of Assay Method to Crude Tissue Preparations. The assay method has been successfully applied to a wide variety of crude

---

[5] I. Lieberman, A. Kornberg, and E. S. Simms, *J. Biol. Chem.* **215**, 403 (1955); see also Vol. VI [17].

[6] Prolonged heating of solutions of PRPP results in a marked, and eventually complete, destruction of the compound. The presence of divalent cations enhances this decomposition. Under the present assay conditions of heating for 1 minute at 100°, about 20% of the PRPP formed is destroyed. See further in Vol. VI [69].

[7] O. H. Lowry, N. J. Rosebrough, A. L. Farr, and R. J. Randall, *J. Biol. Chem.* **193**, 265 (1951); see also Vol. III [73].

tissue extracts. Where greater sensitivity is required, as in cases of low initial activity, one of the alternative assay procedures should be used. Some difficulties may be encountered with tissue extracts containing large endogenous amounts of purines or pyrimidines and their respective nucleotide pyrophosphorylases.

Purification Procedure

The enzyme is widely distributed; the highest concentration has thus far been found in the acetone powder extracts of pigeon liver. Similar acetone powder extracts of chicken liver and *Escherichia coli* possess roughly one-third the specific activity. The purification described below[1] uses pigeon liver as the enzyme source and has been successfully repeated in several laboratories. All operations are carried out at 0° to 3° unless otherwise noted, and without interruption, since the enzyme is not particularly stable.

Fresh pigeon livers are homogenized in a Waring blendor with 5 to 10 vol. of cold acetone (−10°) and filtered on a Büchner funnel. The residue is homogenized again in cold acetone, filtered as above, and dried in air. The acetone powders may be stored at −10° and have shown no loss in activity over a two-month period. Four grams of the acetone powder are suspended in 40 ml. of 0.02 M Tris–chloride buffer, pH 8.0, and extracted with occasional stirring over a 10-minute interval. The residue is removed by centrifugation at 8000 × g for 5 minutes, yielding a clear or opalescent supernatant fluid (acetone powder extract). To 34 ml. of the acetone powder extract are added 51 ml. of deionized water, followed slowly by 17 ml. of 1 M K acetate buffer, pH 5.4. After 5 minutes the precipitate is collected by centrifugation at 8000 × g for 5 minutes. The precipitate is dissolved in 7 ml. of 0.1 M Tris–chloride buffer, pH 8.0, and diluted to 30 ml. with deionized water. The pH is adjusted to 6.8 with 0.1 N HCl (about 1.3 ml. required), and the volume is made up to 34 ml. (low pH fraction).

For the last step enough alumina C_γ gel[8] is used to adsorb 85 to 90% of the enzyme. 3.75 ml. of the aged gel (15 mg. of dry solids per milliliter) are centrifuged, and the supernatant fluid is discarded. Thirty milliliters of the low pH fraction are thoroughly mixed with the gel, after which the mixture is held in an ice bath for 5 minutes and then centrifuged. The supernatant fluid is discarded, and the packed gel is washed by suspending it in 30 ml. of 0.05 M K phosphate buffer, pH 6.85, followed by centrifugation. The enzyme is then eluted by suspending the gel in 30 ml. of 0.10 M K phosphate buffer, pH 6.85, followed by centrifugation (alumina C_γ gel eluate).

[8] R. Willstätter and H. Kraut, *Ber.* **56**, 1117 (1923); see also Vol. I [11].

The enzyme at this point is purified about 27-fold over the acetone powder extract. It retains 65 to 70% of its activity after storage for 2 months at −10° provided glutathione (0.05 M) is present; in the absence of glutathione less than 5% of the activity remains.

PURIFICATION OF 5-PHOSPHORIBOSE PYROPHOSPHOKINASE

Step	Activity, units/ml.	Total activity, units	Protein, mg./ml.	Specific activity, units/mg. protein
1. Acetone powder extract	18.0	612	19.6	0.9
2. Low pH fraction	16.4	558	3.00	5.5
3. Alumina C_γ gel eluate	8.1	275	0.33	24.5

Tarr has purified the enzyme by an almost identical procedure from lingcod muscle.[9] The best preparation obtained from this source has a specific activity about one-thirtieth that of the purified pigeon liver enzyme. Attempts at fractionation with ammonium sulfate or DEAE-cellulose chromatography resulted in complete loss in activity. Further fractionation of this enzyme, or the one described from pigeon liver above, may have to include means for protecting what appears to be an exceedingly sensitive sulfhydryl group.

Properties

Specificity.[1] The enzyme appears to be specific for ATP and ribose 5-phosphate. ADP, ITP, and UTP, at concentrations equal to that of the ATP used in the assay, either failed to replace ATP or showed less than 2% of the activity of ATP. Similarly, ribose 1-phosphate, 2-deoxyribose 5-phosphate, and glucose 6-phosphate failed to replace ribose 5-phosphate. 2-Deoxyribose 5-phosphate in the presence of an equimolar concentration of ribose 5-phosphate did not inhibit the formation of PRPP.

Activators and Inhibitors. Mg^{++} is required for the reaction, with maximal activity shown at $3 \times 10^{-3} M$. In its absence there is no observable formation of PRPP. Excess Mg^{++} results in reduced yields of PRPP, particularly where the incubation time is extended. This appears to be due to a divalent cation activation of PRPP decomposition.[1, 6, 10, 11]

It has been reported that P_i is required for the reaction.[10] Whether this is an absolute requirement for enzymatic activity or is related to

[9] H. L. A. Tarr, *Can. J. Biochem. Physiol.* **38**, 683 (1960).
[10] C. N. Remy, W. T. Remy, and J. M. Buchanan, *J. Biol. Chem.* **217**, 885 (1955).
[11] H. G. Khorana, J. F. Fernandes, and A. Kornberg, *J. Biol. Chem.* **230**, 941 (1958).

the ability of P_i to inhibit the divalent cation-activated PRPP decomposition is not known. The ability of P_i to inhibit the Mg^{++}-activated PRPP decomposition has been reported.[1, 6]

Maximal activity of the enzyme, particularly where long incubation periods are used, and maximal stability on storage require the presence of glutathione $(0.05\,M)$. 2-Mercaptoethanol may be substituted during the incubation of the enzyme, but not when the enzyme is stored frozen.[12] Tarr has reported that the formation of PRPP with the cod muscle enzyme was maximal at the highest level of glutathione used $(0.05\,M)$.[9] These studies suggest that the enzyme has an exceedingly sensitive sulfhydryl group.

F^- has been included in the assay procedure to inhibit phosphatase activity. No inhibition of the reaction is detectable with F^- levels up to $0.1\,M$.[1, 9] However, $0.2\,M$ F^- is inhibitory.[9]

Optimum pH. The optimal pH is approximately 7.5, with 50% of the optimal activity at pH 6.7 and 8.2. No activity is detectable below pH 6.0 and above pH 9.0.

Equilibrium. The equilibrium lies well in the direction of PRPP formation, but the question of reversibility has not been adequately examined. Tarr has found a greater disappearance of PRPP than could be accounted for by the Mg^{++}-activated decomposition when AMP, PRPP, and the enzyme were incubated together. No accumulation of ATP could be demonstrated. However, the enzyme purified from cod muscle contains a strong adenylic acid deaminase activity which removes much of the AMP. It is also possibly contaminated with other activities which remove ATP, as indicated in the reported studies of the stoichiometry of the reaction.[9]

Kinetics. The reaction rate, with 0.3 unit of enzyme in step I of the assay procedure, is half-maximal at ATP levels of $4 \times 10^{-4}\,M$ and ribose 5-phosphate levels of $6 \times 10^{-4}\,M$. ATP at levels greater than $2 \times 10^{-3}\,M$ is inhibitory. Mg^{++} at $0.8 \times 10^{-3}\,M$ gives half-maximal activity.

[12] J. G. Flaks, unpublished results.

[19] Nucleoside Diphosphokinase from Rabbit Muscle[1]

ATP + nucleoside diphosphate $\rightleftarrows$ ADP + nucleoside triphosphate

By MAURICE J. BESSMAN

Assay Method[1]

Principle. The enzyme may be assayed with ATP and IDP as the substrates in the presence of an excess of adenylate kinase and adenylic acid deaminase.

$$2\ \text{ATP} + 2\ \text{IDP} \rightleftarrows 2\ \text{ITP} + 2\ \text{ADP}$$
$$2\ \text{ADP} \rightleftarrows \text{ATP} + \text{A-5-P}$$
$$\text{A-5-P} \rightarrow \text{I-5-P} + \text{NH}_3$$

$$\text{ATP} + 2\ \text{IDP} \rightarrow \text{I-5-P} + 2\ \text{ITP} + \text{NH}_3$$

The reaction is followed by the decrease in absorption at 265 mμ resulting from the deamination of A-5-P. Each molecule of A-5-P deaminated represents two molecules of IDP phosphorylated to form two molecules of ITP.

Reagents

Sodium succinate, 0.2 *M*, pH 6.0.
Magnesium chloride, 0.5 *M*.
ATP, 0.02 *M*.
IDP, 0.06 *M*.
Adenylate kinase.[2]
Adenylic acid deaminase.[3]

Procedure. The assay mixture contains sodium succinate, pH 6.0, (80 micromoles), magnesium chloride (5 micromoles), ATP (0.2 micromole), IDP (0.6 micromole), adenylate kinase (15 μg.), adenylic acid deaminase (40 μg.), and enzyme in a final volume of 0.5 ml. After incubation at 37° for 15 minutes, the reaction is terminated by the addition of 5.0 ml. of 0.1 *N* HCl, and the incubation mixture is read against a blank containing all the reagents except the enzyme. Under these conditions, the deamination of 1 micromole of A-5-P per milliliter leads to a decrease in

---

[1] P. Berg and W. K. Joklik, *J. Biol. Chem.* **210**, 657 (1954).
[2] See Vol. II [99] and Vol. VI [27].
[3] See Vol. II [68] and Vol. VI [12].

optical density measured at 265 mμ of 8.680 in a cell having a 1-cm. light path.

Definition of Unit and Specific Activity.[4] One unit of enzyme is that amount which catalyzes the transfer of 1 micromole of phosphate per minute under the above conditions. Specific activity is expressed as units per milligram of protein. Protein was determined nephelometrically according to the method of Bücher.[5]

Purification Procedure

Step 1. Preparation of Original Extract. The muscles from the legs and back of a rabbit are chilled and minced with a meat chopper. The ground tissue is extracted with 1 vol. of ice-cold distilled water for 30 minutes and squeezed through cheesecloth. The extract is dialyzed for 4 hours against tap water at about 10°, and any precipitate which forms is discarded.

Step 2. First Ammonium Sulfate Fractionation. To each volume of extract is added 0.55 vol. of saturated ammonium sulfate solution (saturated at room temperature), and the solution is allowed to remain in an ice bath for 30 minutes. The precipitate is removed by centrifugation and discarded. To each volume of the supernatant solution is added 0.45 vol. of saturated ammonium sulfate, and, after cooling as above, the precipitate is removed by centrifugation, dissolved in one-half the original volume with distilled water, and dialyzed overnight at 4° against distilled water.

Step 3. Acid Precipitation. The precipitate which forms during dialysis is discarded. The supernatant solution is made 0.06 M with respect to acetic acid by the addition of 1 M acetic acid, whereupon the pH drops to about 3.5. The solution is immediately neutralized to pH 6.8 with 1 N NaOH, and the heavy precipitate is centrifuged and discarded.

Step 4. Second Ammonium Sulfate Fractionation. To each volume of the supernatant solution from step 3 is added 1.2 vol. of saturated ammonium sulfate. The precipitate is removed and discarded. For every volume of this solution, 1.1 vol. of saturated ammonium sulfate are added. After centrifugation, the precipitate is dissolved in about one-third the volume used from step 3, and this solution is dialyzed overnight at 4°.

The data for a typical purification are given in the table.

A simple preparation[1] suitable for most purposes may be obtained by diluting the original extract with an equal volume of ice water, adding 0.05 vol. of 1 N acetic acid, heating 50-ml. aliquots at 55° for 1 minute,

[4] The unit of enzyme activity as defined originally by P. Berg and W. K. Joklik was one-sixtieth of that defined here.
[5] T. Bücher, *Biochim. et Biophys. Acta* 1, 292 (1947); see also Vol. III [73].

SUMMARY OF PURIFICATION PROCEDURE[a]

Step	Volume of solution, ml.	Units, micromoles/ min.	Protein, mg.	Specific activity, micromoles/ min./mg.	Yield, %
Original extract	150	384	2,960	0.13	(100)
First ammonium sulfate	60	284	677	0.42	74
Acid precipitation	70	157	99	1.58	41
Second ammonium sulfate	20	100	34	2.92	26

[a] The data for this table have been taken from P. Berg and W. K. Joklik, *J. Biol. Chem.* **210,** 657 (1954). The figures in columns headed "Yield" and "Protein" have been calculated by the author from the total units and the specific activity, respectively. (See footnote 4.)

and then cooling in an ice bath. The solution is adjusted to pH 6.8 with 1 N NaOH, and the heavy precipitate is discarded. This represents a purification of 10-fold over the original extract with a recovery of 92% of the activity.

Properties

Specificity. The enzyme requires a nucleoside triphosphate as donor, and nucleoside diphosphate as acceptor. Thus, whereas UDP and IDP accept a phosphoryl group from ATP, the corresponding monophosphates U-5-P and I-5-P are not phosphorylated. Inorganic pyrophosphate is not phosphorylated by ATP, and inorganic triphosphate cannot phosphorylate ADP. Sanadi *et al.*[6] reported that GTP can phosphorylate ADP in extracts of pig kidney cortex and ox heart muscle. Kirkland and Turner[7] reported that the relative rates of phosphorylation of ADP by ITP, UTP, GTP, and CTP were 100, 63, 39, and 31, respectively, in extracts of pea seeds.

Activators and Inhibitors. The enzyme requires a divalent cation for activity. Mg^{++} is optimally effective at a concentration of $5 \times 10^{-3} M$, and Ca^{++} or Mn^{++} are equally effective at this concentration.[1] Co^{++}, Zn^{++}, and Ni^{++} also activate the enzyme.[7]

Ethylenediaminetetraacetate (10mM) completely inhibits the enzyme, and arsenate (10 mM) inhibits the reaction 70%. Fluoride, iodoacetate, or inorganic pyrophosphate at 10 mM concentration, p-chloromercuribenzoate or molybdate at 1 mM concentration, or Hg^{++} at 0.1 mM concentration did not affect the activity.[7]

[6] D. R. Sanadi, D. M. Gibson, P. Ayengar, and M. Jacob, *J. Biol. Chem.* **218,** 505 (1956).

[7] R. J. A. Kirkland and J. F. Turner, *Biochem. J.* **72, 716** (1959).

pH Optimum. The enzyme has a pH optimum between pH 6 and 8. The rate at pH 5 and 9 is approximately 30% and 70%, respectively, of that at pH 6.

Equilibrium. The phosphorylation of IDP by ATP proceeds until 50% of the ATP has been converted to ADP. Likewise, phosphorylation of ADP by ITP proceeds until 50% of the ADP has been phosphorylated. Thus the equilibrium for the reaction is approximately 1. An equilibrium constant of 0.91 at pH 8.0 was found by Kirkland and Turner for the enzyme from pea seeds.[7]

Distribution. Nucleoside diphosphokinase activity has been found in the following sources: pigeon breast muscle and rat intestinal mucosa;[8] brewer's and baker's yeast, acetone powder extracts of beef brain;[1] extracts of chicken liver acetone powder;[9] the seeds and shoots of wheat, peas, broad beans, sugarcane, and barley, silver-beet leaves, and potato tubers.[7]

[8] H. A. Krebs, and R. Hems, *Biochim. et Biophys. Acta* **12**, 172 (1953).
[9] M. F. Utter, K. Kurahashi, and I. A. Rose, *J. Biol. Chem.* **207**, 803 (1954).

[20] Deoxynucleoside Monophosphate Kinases

Deoxynucleoside monophosphate + ATP
$$\rightleftarrows \text{Deoxynucleoside diphosphate} + \text{ADP}$$

By MAURICE J. BESSMAN

The name "deoxynucleotide kinase" has been loosely applied to tissue extracts capable of catalyzing the phosphorylation of deoxynucleotides to the corresponding polyphosphates. However, except for the enzyme which appears after infection of *Escherichia coli* by T2 bacteriophage, all extracts studied work as well or at a higher rate on the ribonucleotides.

This paper will describe the purification of deoxycytidylate kinase from *Azotobacter vinelandii* and the preparation of an extract from *E. coli* capable of catalyzing the phosphorylation of deoxyguanylate, deoxyadenylate, deoxycytidylate, and deoxythymidylate to the corresponding triphosphates. It is assumed that the formation of the triphosphate in the latter case is the result of a second enzyme or enzymes which catalyze the phosphorylation of the diphosphates as they are formed. Finally, the purification of T2-induced deoxynucleotide kinase will be described. This enzyme has a specificity directed toward deoxythymidylate, deoxyguanylate, and 5-hydroxymethyldeoxycytidylate.

I. Deoxycytidylate Kinase[1]

Assay Method

Principle. The method is based on the determination of the ADP formed during the reaction. The pyruvate formed during the phosphorylation of ADP by phosphopyruvate in the presence of pyruvate kinase is coupled with the lactic dehydrogenase reaction, and the disappearance of reduced DPN is measured spectrophotometrically. A summary of the

$$\text{dCMP} + \text{ATP} \rightleftarrows \text{ADP} + \text{dCDP}$$
$$\text{ADP} + \text{phosphopyruvate} \rightleftarrows \text{ATP} + \text{pyruvate}$$
$$\text{Pyruvate} + \text{DPNH} + \text{H}^+ \rightleftarrows \text{Lactate} + \text{DPN}^+$$

$$\text{dCMP} + \text{phosphopyruvate} + \text{DPNH} + \text{H}^+ \rightleftarrows \text{dCDP} + \text{lactate} + \text{DPN}^+$$

reactions involved is given here. This assay procedure is not applicable to crude extracts rich in ATPase or DPNH-oxidase.

Reagents

DCMP, 0.01 M.
ATP, 0.05 M.
DPNH, 0.01 M.
Disodium ethylenediaminetetraacetate (EDTA) 0.1 M.
Phosphopyruvate, 0.08 M.
KCl, 1 M.
Lactic dehydrogenase.[2]
Tris buffer, pH 7.5, 1 M.
MgSO$_4$, 0.1 M.

Procedure. The reaction mixture contains (in micromoles) Tris, 50; KCl, 80; MgSO$_4$, 8.0; EDTA, 2.0; ATP, 0.5; phosphopyruvate, 0.8; DPNH, 0.1; lactic dehydrogenase, 20 μg.; enzyme, 0.02 to 0.20 unit; and dCMP, 2.0. The final volume is 1.0 ml., and the incubation is done at 25°. The reaction is started by the addition of the dCMP, and the decrease in optical density at 340 mμ due to oxidation of DPNH is measured at 30-second intervals for a convenient time.

Definition of Unit and Specific Activity. A unit of activity is that amount causing a phosphorylation of one micromole of deoxynucleoside monophosphate to deoxynucleoside diphosphate per minute. Specific

[1] F. Maley and S. Ochoa, *J. Biol. Chem.* 233, 1538 (1958).
[2] The lactic dehydrogenase obtained from Sigma Chemical Co. (type 1:2X cryst) contains sufficient pyruvic kinase to supply both enzymes for the assay.

activity is defined as units per milligram of protein as determined by the procedure of Lowry et al.[3]

Purification Procedure

Growth of Cells. *Azotobacter vinelandii* cells are grown on the nitrogen-free medium of Burk and Lineweaver,[4] which contains, per liter, KH_2PO_4, 0.2 g.; K_2HPO_4, 0.8 g.; NaCl, 0.2 g.; $MgSO_4 \cdot 7H_2O$, 0.2 g.; $CaSO_4 \cdot 2H_2O$, 0.1 g.; $Fe_2(SO_4)_3$, 0.1 g.; and sucrose, 20 g. The culture is aerated vigorously at 30° and harvested by centrifugation after 15 hours. The yield is approximately 6 g. of cell paste.[5]

Crude Extract. Forty-three grams of cell paste are suspended in 100 ml. of 0.02 M phosphate buffer, pH 7.7. Thirty-milliliter aliquots at a time are subjected to sonic oscillation for 10 minutes in a Raytheon 10-kc. sonic oscillator. The combined suspensions are diluted to 200 ml. with water, made 0.14 M with NaCl, and centrifuged at 20,000 × g for 1 hour. The precipitate is discarded.

Ammonium Sulfate Fractionation. To the extract (184 ml.) are added 45 g. of finely powdered ammonium sulfate over a period of 10 minutes while the solution is stirred mechanically. The mixture is stirred for an additional 15 minutes and centrifuged at 20,000 × g for 10 minutes. The precipitate is discarded, and to the supernatant fluid are added 36 g. of ammonium sulfate as above. The precipitate is harvested by centrifugation and dissolved in 0.02 M phosphate buffer, pH 7.2, to give a final volume of 38 ml. This solution is dialyzed overnight against 2 l. of 0.02 M succinate buffer, pH 6.5.

Isoelectric Precipitation. To the dialyzed solution (54 ml.) are added 20 g. of sodium chloride. Enough 1.0 N HCl is then added dropwise to lower the pH of the solution to 4.3 to 4.1. The precipitate is removed by centrifugation and suspended in 0.02 M phosphate buffer, pH 7.2, to a volume of 55 ml. The suspension is stirred mechanically for 30 minutes, and the insoluble material is removed by centrifugation.

Heat Treatment. The supernatant solution from the previous step is immersed in a water bath at 55° to 60° and kept at a temperature of 50° for 2 minutes, after which it is rapidly cooled to 0°. The precipitate is centrifuged off and discarded.

Adsorption and Elution from Aluminum Hydroxide Gel. To the solu-

[3] O. H. Lowry, N. J. Rosebrough, A. L. Farr, and R. J. Randall, *J. Biol. Chem.* **193**, 265 (1951).

[4] D. Burk and M. Lineweaver, *J. Bacteriol.* **19**, 389 (1930).

[5] M. Grunberg-Manago, P. J. Ortiz, and S. Ochoa, *Biochim. et Biophys. Acta* **20**, 269 (1956).

tion from the previous step are added 150 ml. of alumina gel,[6] and after being stirred for 15 minutes the mixture is centrifuged and the supernatant solution is discarded. The gel is washed with 100 ml. of water and then eluted with 60 ml. of 0.1 M phosphate buffer, pH 6.4, by 15 minutes of stirring. After centrifugation, the gel is again eluted with 30 ml. of buffer, and the eluates are combined.

Removal of Inactive Protein at pH 4.0. In most instances, by decreasing the pH of the solution to 4.0 with 1 N HCl added dropwise, a precipitate of inactive protein forms. The precipitate is removed by centrifugation, and the clear supernatant solution is adjusted to pH 5.5 with NaOH.

Chromatography. The solution from the previous step is passed through a 2 × 2-cm. column of Amberlite IRC-50 at a flow rate of 1 ml./min. The resin has been purified according to the directions of Hirs[7] and equilibrated with 0.02 M sodium citrate of pH 5.3. Approxi-

TABLE I

PURIFICATION OF DEOXYCYTIDYLATE KINASE OF *A. vinelandii*[a]

Step	Volume, ml.	Units	Protein, mg.	Specific activity, micromoles/min./mg.	Yield, %
Sonic extract	184	172	3620	0.05	100
Ammonium sulfate	54	143	1240	0.12	81
pH 4.2 precipitate	50	135	535	0.25	76
Heat treatment	47	125	203	0.61	70
Alumina gel eluate	95	98	76.5	1.28	55
pH 4.0 supernatant fluid	90	77	42.3	1.82	43
First chromatography	21	47	4.7	10.0	26
Second chromatography	15	24	0.53	46.6	14

[a] F. Maley and S. Ochoa, *J. Biol. Chem.* **233**, 1538 (1958).

mately 20% of the enzyme and most of the protein pass through the column. The column is washed with four 5-ml. aliquots of 0.02 M sodium citrate, pH 5.3, and then eluted with 5-ml. aliquots of 0.2 M citrate at a flow rate of 0.5 ml./min. After the third fraction, the 280/260-mμ ratio

[6] The aluminum hydroxide gel used for this fractionation was prepared as follows:[1] 66 g. of Al$_2$(SO$_4$)$_3$·18 H$_2$O dissolved in 1200 ml. of water and 280 ml. of concentrated ammonium hydroxide were added with stirring. The suspension was diluted to 2 l. and stirred for 30 minutes. The gel was then centrifuged and washed repeatedly with distilled water until the supernatant fluid was neutral. Water was added to the thick gel paste to give a 500-ml. suspension, which was heated in a boiling-water bath for 4 hours. Longer heating was found to enhance the protein-binding power of the gel, and thus larger volumes of eluant are required.

[7] C. H. W. Hirs, Vol. I [13].

increases abruptly from less than 1.0 to 1.7 simultaneously with the appearance of the enzyme. This fourth fraction and the next three fractions are combined and dialyzed overnight against 2 l. of 0.02 M sodium citrate, pH 5.3.

Chromatography Repeated. The dialyzed enzyme solution may be rechromatographed as above, yielding about 15 ml. of a solution containing approximately 0.03 mg. of protein per milliliter. The enzyme may be stabilized by the addition of crystalline bovine serum albumin to a concentration of 1%, and under these conditions it is stable for at least 4 months even after repeated freezing and thawing.

A summary of the purification procedure is reported in Table I.

Properties

Specificity. The purified enzyme appears to be specific for deoxycytidylate and cytidylate as phosphate acceptors. The following compounds will not accept phosphate from ATP: deoxyguanylate, thymidylate, deoxyadenylate, adenylate, inosinate, and cytidine. Inosine triphosphate cannot substitute for ATP.

Activators and Inhibitors. There is an absolute requirement for a divalent cation which can be met with Mg^{++} or Mn^{++}. ADP ($1.3 \times 10^{-3} M$) inhibits the rate of deoxycytidylate phosphorylation by 40%.

The pH optimum lies between pH 7.5 and 8.1 for both deoxycytidylate and cytidylate.

Equilibrium. The apparent equilibrium constant approached from either direction, $K = (ADP)(dCDP)/(ATP)(dCMP)$, was determined as 1.49.

II. Deoxynucleoside Monophosphate Kinase of *E. coli*

The following is a description of the preparation of an extract from *E. coli* which phosphorylates thymidylate, deoxyadenylate, deoxycytidylate, and deoxyguanylate to the corresponding triphosphates at the expense of ATP.[8] Although it is very likely that this preparation represents a mixture of at least four enzymes specific for the corresponding deoxynucleotides (and possibly their ribo analogs),[9] the individual activities have not as yet been sufficiently purified to allow a clear description of the reaction pathways involved in the phosphorylations. Despite this relatively undefined status of the preparation, it has been very useful as a reagent for the preparation of deoxynucleoside triphosphates from all four of the above deoxynucleotides. This is especially true in the

[8] I. R. Lehman, M. J. Bessman, E. S. Simms, and A. Kornberg, *J. Biol. Chem.* **233**, 163 (1958).
[9] J. Hurwitz, *J. Biol. Chem.* **234**, 2351 (1959).

preparation of triphosphate derivatives from radioactive monophosphates, since virtually 100% yields based on the radioactive deoxynucleotide are easily obtained.

Assay Method

Principle. This assay takes advantage of the resistance of nucleoside diphosphates and triphosphates to semen phosphatase and the adsorption of purine and pyrimidine derivatives by Norit. The assay is divided into three parts. First, the P^{32}-labeled deoxynucleotide is reacted with the enzyme in the presence of ATP, and after a given time the reaction is stopped by heating. Next, the unreacted deoxynucleotide is hydrolyzed to inorganic P^{32} and nucleoside by semen phosphatase. The nucleotide phosphorylated to the diphosphate or triphosphate is not attacked by the phosphatase. In the third part of the assay, Norit is added which adsorbs the diphosphate and triphosphate but not the inorganic ortho-P^{32}, and hence the radioactivity associated with the Norit is a measure of deoxynucleotide phosphorylated.

Reagents

ATP, 0.1 M.
$MgCl_2$, 0.1 M.
P^{32}-Deoxynucleotide[8] (0.5 to 1×10^6 counts/min/micromole), 5×10^{-4} M.
Deoxynucleotide (unlabeled), 0.02 M.
Tris, pH 7.5, 1 M.
Sodium acetate, pH 5.0, 1 M.
Semen phosphatase.[10]
Norit A (acid-washed, 20% v/v).

Procedure. The incubation mixture contains (in micromoles) ATP, 2; $MgCl_2$, 4; Tris buffer, 100; P^{32}-labeled deoxynucleotide (15 millimicromoles); and enzyme (0.4 to 1.6×10^{-3} units), in a final volume of 0.25 ml. After incubation at 37° for 20 minutes, 0.5 ml. of water is added, and the tubes are placed in a boiling water bath for 2 minutes. The tubes are chilled, and the following additions are made (in micromoles): sodium acetate buffer, 100; unlabeled deoxynucleotide, 0.2; and semen phosphatase, 50 units, in a final volume of 1.0 ml. The incubation is for 15 minutes at 37°, after which the reaction mixture is chilled, and 0.1 ml. of cold 2 N HCl and 0.15 ml. of Norit are added. The tubes are shaken intermittently by hand for 5 minutes, after which the Norit is collected by centrifugation and washed three times with 2.5-ml. portions

[10] J. Wittenberg and A. Kornberg, *J. Biol. Chem.* **202**, 431 (1953).

of cold water. The final Norit precipitate is suspended in 0.5 ml. of 50% ethanol containing 0.3 ml. of concentrated ammonium hydroxide per 100 ml., and the entire suspension is plated and counted.

Definition of Unit and Specific Activity. A unit is defined as that amount catalyzing the formation of one micromole of nucleoside polyphosphate per minute from nucleoside monophosphate and ATP. Specific activity is defined as units per milligram of protein. Protein is determined according to Lowry *et al.*[3]

Purification Procedure

Growth and Harvest of Bacteria. Escherichia coli strain B is grown in a medium containing, per liter, K_2HPO_4, 11 g.; KH_2PO_4, 8.5 g.; Difco yeast extract, 6 g.; and glucose, 10 g. The cells are incubated at 37° with vigorous aeration and harvested after 15 hours by centrifugation (large quantities may be centrifuged in the Sharples). The packed cells are washed by suspension in 3 vol. of 0.5% NaCl–0.5% KCl and centrifuged. The washed cells (approximately 8 g./l. of culture) may be frozen and stored at −12° for at least 6 months.

Preparation of Extract. All manipulations are performed at 0°–5°. Cells are suspended in 0.05 M glycylglycine buffer, pH 7.0 (4 ml./g. of packed cells), and disrupted by treatment for 10 minutes in a Raytheon 10-kc. sonic oscillator. The suspension is centrifuged for 15 minutes at 12,000 × g, and the slightly turbid supernatant fluid is collected. The protein content is adjusted to a concentration of 10 mg./ml. by the addition of the same glycylglycine buffer.

Streptomycin Fractionation. Ninety milliliters of extract are stirred mechanically, while 27 ml. of 5% streptomycin sulfate are added dropwise. The suspension is allowed to stand for 5 minutes and then is centrifuged for 10 minutes at 10,000 × g. The supernatant solution is collected, and the precipitate is discarded.

Calcium Phosphate Gel Fractionation. Sixty-six milliliters of calcium phosphate gel[11] (15 mg. of solids per milliliter) are centrifuged at 5000 × g for 10 minutes, and the supernatant fluid is discarded. The streptomycin fraction (110 ml.) is mixed with the packed gel and allowed to stand for 5 minutes, after which the suspension is centrifuged as above and the gel is discarded. The supernatant fluid represents the calcium phosphate gel fraction (113 ml.).

Aluminum C_γ Fractionation. Forty-five milliliters of alumina C_γ (15 mg. of solids per milliliter) are packed by centrifugation as above, and the supernatant fluid is discarded. To the packed gel, 108 ml. of

[11] D. Keilin and E. F. Hartree, quoted in Vol. I [11].

calcium phosphate gel fraction are added, and the gel is thoroughly dispersed. After 5 minutes, the gel is collected by centrifugation and eluted with 54 ml. of 0.066 M potassium phosphate buffer, pH 7.4. This fraction showed no loss of activity against thymidylate after storage at $-12°$ for 6 months.

A summary of the purification is shown in Table II.

TABLE II

PURIFICATION OF DEOXYNUCLEOTIDE KINASE OF *E. coli*

Step	Volume, ml.	Units micromoles/min.	Protein, mg./ml.	Specific activity, $\times 10^3$	Yield, %
Extract	90	1.5	10.0	1.7	100
Streptomycin fraction	115	2.0	3.4	5.2	130
Calcium phosphate gel fraction	113	1.4	1.6	7.7	93
Alumina C$_\gamma$ gel eluate	52	1.1	1.8	11.3	73

[a] These values were obtained with thymidylate as the substrate. When assayed against deoxyadenylate, deoxycytidylate, and deoxyguanylate at the same concentration, the alumina C$_\gamma$ gel eluate had specific activities ($\times 10^3$) of 14.6, 6.6, and 11.8, respectively.

[b] These units are 600 times as great as those defined by I. R. Lehman, M. J. Bessman, E. S. Simms, and A. Kornberg, *J. Biol. Chem.* **233**, 163 (1958).

Properties

The activities reported here are not sufficiently purified to delineate the kinetics of the reaction, the pH optima, the effect of activators and inhibitors, etc.

III. Deoxynucleoside Monophosphate Kinase of T2-Infected *E. coli*

Assay

The same assay used for the *E. coli* enzyme has been employed during this purification, with two modifications. First, the substrate concentration has been raised tenfold, and, second, 20,000 units of lactic dehydrogenase have been added. The diphosphate formed during the reaction is immediately phosphorylated to the triphosphate by enzymes present in this lactic dehydrogenase preparation.

Purification Procedure

Preparation of Crude Extract. Escherichia coli cells are grown at 37° with vigorous aeration in a yeast extract medium.[8] When growth reaches a density of approximately 1×10^9 cells/ml., the culture is infected with enough T2 bacteriophage to give a phage-to-bacterium ratio of 4. Immediately after addition of the phage, the aeration is discontinued. Thirty minutes after addition of the phage, the culture is harvested in a Sharples centrifuge (50,000 r.p.m.) at a flow rate of 150 to 200 ml./min. The packed cells are suspended in 5 vol. of 0.05 M glycylglycine buffer, pH 7.4, and the mixture is exposed to sonic oscillation for 40 minutes at maximal output in a Raytheon 10-kc. magnetostrictive oscillator, Model DF101. The cell debris is sedimented at $15,000 \times g$ for 15 minutes, and the clear supernatant fluid is adjusted with 0.05 M glycylglycine buffer, pH 7.4, to give a final protein concentration of 10 mg./ml.

Fractionation with Streptomycin Sulfate. To the crude extract is added dropwise and with stirring 0.3 vol. of 5% streptomycin sulfate. After 10 minutes, the precipitate is removed by centrifugation. The supernatant fluid is designated as the streptomycin fraction.

First Fractionation on DEAE-cellulose. To a 4×12-cm. column of DEAE-cellulose washed with 320 ml. of 0.02 M potassium phosphate buffer, pH 6.2, are added 305 ml. of the streptomycin fraction. A linear gradient is established with 1600 ml. of 0.05 M NaCl in 0.02 M potassium phosphate buffer, pH 6.2, in the mixing vessel and 1600 ml. of 0.50 M NaCl in 0.02 M potassium phosphate buffer, pH 6.2, in the reservoir. Both eluents also contain 0.005 M 2-mercaptoethanol. Elution proceeds at a flow rate of 1.9 ml./min., and 47.0-ml. fractions are collected. The main peak of activity is preceded by two smaller peaks which constitute the deoxyguanylate kinase activity present in *E. coli* before infection. The six tubes in the last peak with the highest specific activity are pooled and constitute the DEAE I fraction.

Second Fractionation on DEAE-cellulose. The DEAE I fraction is dialyzed overnight against 50 vol. of 0.05 M potassium phosphate buffer, pH 8.1, containing 0.005 M 2-mercaptoethanol. The small precipitate which forms is discarded. The dialysis results in no significant loss of enzymatic activity. To a 2×12-cm. column of DEAE-cellulose previously washed with 80.0 ml. of 0.02 M potassium phosphate buffer, pH 8.1, are added 265 ml. of the dialyzed DEAE I fraction. A linear gradient is established with 400 ml. of 0.05 M NaCl in 0.02 M potassium phosphate buffer, pH 8.1, in the mixing vessel and 400 ml. of 0.50 M NaCl in 0.02 M potassium phosphate buffer, pH 8.1, in the reservoir. Both eluents contain 0.005 M 2-mercaptoethanol. Elution proceeds at a flow rate of 0.5 ml./min., and 10.8-ml. fractions are collected. The seven tubes with the highest specific activity are pooled and constitute the DEAE II fraction.

Fractionation with Calcium Phosphate Gel. The DEAE II fraction is dialyzed overnight against 80 vol. of 0.01 M potassium phosphate buffer, pH 7.0, containing 0.005 M 2-mercaptoethanol. There is no detectable loss in activity. For every milligram of protein in the dialyzed DEAE II fraction, 3.3 ml. of a calcium phosphate gel suspension (10.4 mg./ml.) are centrifuged to obtain the packed gel. To this pellet is added the dialyzed DEAE II fraction, and the suspension is stirred intermittently for 15 minutes. The suspension is centrifuged, and the supernatant fluid is discarded. The pellet contains 90 to 98% of the original enzymatic activity. The enzyme is eluted by suspension of the gel for 30 minutes in 5 ml. of 0.01 M potassium phosphate buffer, pH 8.1, per milligram of protein originally treated with gel. The suspension is centrifuged, and the packed gel is eluted once more in the same way. The two supernatant solutions are combined and constitute the calcium phosphate fraction.

Fractionation with Alumina C_γ Gel. The pH of the calcium phosphate fraction is lowered to about 7.0 by the addition of 0.01 vol. of M potassium phosphate buffer, pH 6.2. To this solution are added 1.7 ml. of an alumina C_γ suspension (4.6 mg./ml.) per milligram of protein. The suspension is stirred intermittently for 15 minutes, centrifuged, and the pellet is discarded. This procedure results in the loss of 10 to 15% of the enzymatic activity and a 1.5-fold increase in the specific activity of the supernatant fluid. The pH of the solution is lowered to about 6.5 by the addition of 0.0055 vol. (based on the original volume of calcium phosphate fraction) of 1 M HCl. To this solution are added 2.1 ml. of the alumina C_γ gel suspension per milligram of protein originally present. The suspension is stirred intermittently for 15 minutes, centrifuged, and the supernatant fluid which contains 5 to 15% of the enzymatic activity is discarded. The pellet is resuspended in 0.04 M potassium

TABLE III

PURIFICATION OF DEOXYNUCLEOSIDE MONOPHOSPHATE KINASE
OF T2-INFECTED *E. coli*

Fraction	Protein, mg./ml.	Enzyme activity,[a] units/ml.	Specific activity, units/mg. protein	Relative purity	Recovery, %
Crude enzyme	10.70	1.4	0.13	1	100
Streptomycin	3.37	1.0	0.29	2	83
DEAE I	0.34	0.7	2.1	16	55
DEAE II	0.41	2.0	5	38	43
Calcium phosphate	0.039	0.4	11	85	36
Alumina C_γ	0.054	1.3	24	185	24

[a] Activity was measured by the P^{32} assay, with $dGMP^{32}$ as the substrate.

phosphate buffer, pH 7.0 (5.8 ml. of buffer per milligram of protein originally present), stirred intermittently for 15 minutes, centrifuged, and the pellet is discarded. The supernatant fluid constitutes the alumina C_γ fraction.

The over-all results of the purification scheme are summarized in Table III.

Properties

Stability of Fractions. The crude extract and streptomycin fraction may be stored for at least a year at $-10°$ without any detectable loss in activity. The DEAE I and II fractions do not lose any activity for at least 2 months at 4°. The calcium phosphate fraction exhibits variable stability at 4°; it appears to be stable for at least 1 week, after which time some preparations lose activity rapidly. No attempt has been made to stabilize this fraction. The alumina C_γ fraction loses 20% or less of its activity in 6 months at 4° or $-10°$ if made 0.001 M with respect to ethylenediaminetetraacetate (EDTA). If EDTA is omitted, this fraction also shows variable stability.

Specificity. With ATP as the phosphate donor, dGMP, dTMP, and dHMP are phosphorylated at the relative rates of 2:2:1. No activity is detected toward dCMP, dAMP, GMP, or UMP. At equal concentrations, dUMP and 5-methyl dCMP are phosphorylated at 3% and 5% of the rate of dGMP. No other triphosphate tested (GTP, CTP, dGTP, dCTP, dTTP) except dATP can replace ATP as the phosphate donor.

Substrate Affinities. The K_m values for dGMP, dTMP, and dHMP are 0.85, 2.78, and 0.56 $\times 10^{-4} M$, respectively; the K_m values for ATP in the same order are 1.25, 0.82, and 5.0 $\times 10^{-3} M$.

Effect of Cations. With dGMP as substrate, the enzyme requires Mg^{++} or Mn^{++} for maximal activity. Ca^{++} at an equal concentration is 67% as effective as Mg^{++}. Mg^{++} is also required for the phosphorylation of dTMP and dHMP by this enzyme.

pH Optima. There is a broad pH optimum with all three substrates between pH 7.0 and 9.3. The pH of maximal activity with dHMP as substrate is about 8.6, and that of dGMP or dTMP is about 8.0.

Stoichiometry. In each case, the phosphorylation of one mole of deoxynucleoside monophosphate is accompanied by the formation of one molecule of the corresponding diphosphate at the expense of one molecule of ATP.

The reaction is freely reversible, and the disappearance of two moles of diphosphates (e.g., dGDP and ADP) is accompanied by the formation of one mole of the deoxynucleoside monophosphate (dGMP) and one molecule of ATP.

[21] Pyrimidine Synthesis and Breakdown

By PETER REICHARD and OLA SKÖLD

This article treats most of the enzymes which lead to the synthesis of the pyrimidine ring and those involved in pyrimidine catabolism. The enzymes catalyzing the formation of carbamyl phosphate and the transfer of the carbamyl group to aspartate are treated in Vol. V [124a] and [124b].

Included here also are uridine phosphorylase and uridine kinase. It might be considered that these two enzymes do not belong to the actual theme of this article, since they synthesize UMP from uracil and thus do not participate in the *de novo* synthesis of the pyrimidine ring. However, the general importance of this "uracil pathway of pyrimidine biosynthesis," especially in connection with the use of pyrimidine analogs in mammalian tissues, makes desirable the inclusion of these two enzyme reactions.

I. Ureidosuccinase (Carbamyl Aspartase)[1]

$$\text{Carbamyl aspartate} + H_2O \rightarrow CO_2 + NH_3 + \text{aspartate}$$

Assay Method

The measurement of $C^{14}O_2$ liberated from carbamyl-C^{14}–carbamyl aspartate is the basis for the estimation of enzyme activity.

Reagents

$MgCl_2$, 0.3 M.
Potassium phosphate buffer, pH 8.0, M.
$MnSO_4$, 0.01 M.
Cysteine, pH 7.0, 0.1 M.
DL-Carbamyl-C^{14}–carbamyl aspartate,[2] 0.02 M (8000 cpm/micromole).

Procedure. The incubation mixture (final volume 2.0 ml.), in a Thunberg tube, contains 0.02 ml. of $MgCl_2$, 0.1 ml. of potassium phosphate buffer, 0.1 ml. of $MnSO_4$, 0.1 ml. of cysteine, 0.1 ml. of labeled carbamyl aspartate, and 0.25 to 2.5 units of the enzyme. The bulb of the Thunberg tube contains 0.5 ml. of CO_2-free NaOH (0.2 M) and a drop

---

[1] I. Lieberman and A. Kornberg, *J. Biol. Chem.* **212**, 909 (1955).
[2] J. F. Nyc and H. K. Mitchell, *J. Am. Chem. Soc.* **69**, 1382 (1947).

of bromothymol blue (0.04%). Cysteine and enzyme are added last. The tube is evacuated and incubated at 30° for 30 minutes. The reaction is stopped by acidification. For this purpose approximately 0.3 ml. of a H_2SO_4–thymol blue solution (2 parts of 50% sulfuric acid + 1 part of 0.04% thymol blue solution) is cautiously admitted into the main space of the Thunberg tube through the side arm. After incubation at 60° to 65° for 30 minutes the bulb is removed, and 0.1 ml. of the NaOH solution is plated and counted on aluminum disks in a gas-flow counter. Self-absorption correction factors are obtained by diffusing and plating known samples of $NaHC^{14}O_3$ under assay conditions.

Definition of Unit and Specific Activity. One unit of enzyme is defined as the amount of protein causing the liberation of 1 micromole of CO_2 per hour under the conditions of the assay. Specific activity is defined as units of activity per milligram of protein, as determined by the method of Lowry *et al.*[3]

Purification Procedure

The enzyme was purified by Lieberman and Kornberg[1] from an extract of an anaerobic soil bacterium, *Zymobacterium oroticum.*

Growth of Bacteria. The growth medium consists of 2% tryptone, 0.05% Difco yeast extract, 0.2% orotic acid, and 0.05% sodium thioglycolate. Anaerobic conditions are maintained with a pyrogallol–Na_2CO_3 seal. Large cultures are grown in Erlenmeyer flasks (1 to 6 l.) without a seal. After 20 to 30 minutes of autoclaving at 15 pounds of pressure, the medium is cooled and neutralized with a sterile 50% K_2CO_3 solution, and the inoculum is added promptly. When growth is essentially complete (16 to 18 hours), the cells are harvested and resuspended in water (15 ml./l. of culture medium).

Preparation of Cell-Free Extract. Cell-free extracts are prepared by sonication of the cell suspension (45 to 75 ml.) for 8 to 10 minutes in a Raytheon 10-kc. oscillator, followed by centrifugation for 10 minutes at about 20,000 × g. The extract is stable at 0° for at least 10 days if kept *in vacuo* but rapidly loses activity when exposed to air. The whole procedure and all the following purification steps are carried out at 0°.

Step 1. Protamine Fractionation. To 30 ml. of cell-free extract are added, with stirring, 6 ml. of a 1% solution of protamine sulfate. After 5 minutes the precipitate is collected by centrifugation, and the supernatant solution is discarded. Potassium acetate buffer (30 ml., 0.1 M, pH 5.0) is then added, and the precipitate is suspended evenly. After 20 minutes, with occasional stirring, the precipitate is again collected by

[3] O. H. Lowry, N. J. Rosebrough, A. L. Farr, and R. J. Randall, *J. Biol. Chem.* **193**, 265 (1951).

centrifugation, and the supernatant solution is discarded. The precipitate is washed with water and then extracted with 30 ml. of sodium citrate buffer (0.075 M, pH 6.4). After 15 minutes the suspension is centrifuged, and the supernatant solution is collected.

Step 2. *Ammonium Sulfate Fractionation.* Fifteen milliliters of saturated ammonium sulfate (3°) are added to the protamine fraction with stirring, followed by 5.25 g. of the solid salt. After 5 minutes the precipitate is collected by centrifugation and dissolved in 12 ml. of water.

Step 3. *Isoelectric Fractionation.* Five milliliters of sodium acetate buffer (0.2 M, pH 4.2) are added to the ammonium sulfate fraction. After 3 minutes the precipitate is collected by centrifugation and dis-

TABLE I
PURIFICATION OF UREIDOSUCCINASE

Enzyme fraction	Units/ml.	Over-all recovery, %	Specific activity, units/mg. protein
Cell-free extract	5.1	100	0.7
Protamine	4.7	92	2.1
Ammonium sulfate	8.2	64	4.9
Isoelectric	10.0	65	7.4

solved in 10 ml. of potassium phosphate buffer (0.05 M, pH 8.0). See Table I for a summary of the purification procedure.

Properties

Stability. When *Z. oroticum* is disintegrated by shaking with glass beads instead of by sonication, complete loss of enzyme activity occurs. In all the partially purified fractions, enzyme activity decreases markedly during the first few days. *In vacuo* with cysteine, however, the loss of activity appears to be retarded.

Activators and Inhibitors. Cysteine stimulates the partially purified enzyme two- to sevenfold. Sodium thioglycolate is strongly inhibitory. The purified enzyme shows an absolute requirement for metal ions, Mn^{++} giving the highest effect. In the presence of Fe^{++} ($10^{-3} M$) and Cu^{++} ($10^{-4} M$) the rates are 69% and 3%, respectively, of that obtained with Mn^{++} ($10^{-3} M$). P_i and ADP do not stimulate the reaction.

Effect of pH. The optimal pH range for the reaction is between 7.8 and 8.5.

Equilibrium. The reaction is essentially irreversible, and no formation of carbamyl aspartate from aspartate $+ CO_2 + NH_3$ could be demonstrated.

Substrate Affinity. K_m (carbamyl aspartate) $= 2.8 \times 10^{-3} M$ with Fe^{++} and $1.3 \times 10^{-2} M$ with Mn^{++}.

II. Dihydroörotase[4]

L-Carbamyl aspartic acid $\rightleftarrows$ L-Dihydroörotic acid $+$ H$_2$O

This enzyme has not been freed from dihydroörotic dehydrogenase and carboxymethylhydantoinase.[4] The properties given below for the enzyme were obtained from studies with a crude protamine fraction from *Z. oroticum.*

Assay Method

Principle. In the presence of DPN and an excess of dihydroörotic dehydrogenase, the dihydroörotic acid formed during the reaction is quantitatively oxidized to orotic acid. The increase in optical density at 280 mμ accompanying orotic acid formation is the basis of the assay method.

Procedure. The incubation mixture contains 15 micromoles of MgCl$_2$, 100 micromoles of potassium phosphate buffer (pH 6.1), 0.05 micromoles of DPN, 30 micromoles of cysteine (pH 7.0), 20 micromoles of DL-carbamyl aspartate, potassium salt, and 2.0 ml. of enzyme (protamine fraction containing dihydroörotic dehydrogenase) in a final volume of 3.0 ml. The increase in optical density at 280 mμ is read against a blank containing the incubation mixture without added carbamyl aspartate.

Purification Procedure

Only a very limited purification of dihydroörotase has been performed. The enzyme source is *Z. oroticum* which is cultured as described earlier for ureidosuccinase.

Preparation of Cell-Free Extract. The cells are harvested in a Sharples centrifuge and resuspended in $0.01 M$ sodium orotate (7 ml./l. of original culture), potassium phosphate buffer (0.4 ml., pH $7.0, M$), and cysteine (0.4 ml., pH 7.0, $0.1 M$). The cell suspension is incubated *in vacuo* at 26° for 20 minutes. After centrifugation, the cells are suspended in ice-cold water (about 5 ml./l. of culture), and an aliquot of the suspension (about 6 ml.) is shaken with 6 g. of glass beads (0.10 to 0.15 mm. in diameter) in a Mickle vibrator for 15 minutes at 2°. The mixture is centrifuged at about $10,000 \times g$, and the precipitate is washed once with cold water. The volume of the extract (combined supernatant solutions) is adjusted to 10 ml./l. of culture.

Protamine Fractionation. Freshly prepared cell-free extract (100 ml.) is diluted with an equal volume of water, and 15 ml. of a 1% solution of

[4] I. Lieberman and A. Kornberg, *J. Biol. Chem.* **207**, 911 (1954).

protamine sulfate are added with stirring. After 5 minutes the precipitate is collected by centrifugation, and the supernatant solution is discarded. Citrate buffer (100 ml., $0.5\,M$, pH 6.0) is added to the hard and difficultly soluble precipitate. After 12 to 24 hours the softened precipitate is dissolved to a considerable extent by homogenization with a glass pestle. Water (200 ml.) is added with stirring, and the resultant stringy precipitate is discarded after centrifugation. The supernatant solution, which is essentially free of nucleic acid, is used as a source of enzyme.

Properties

Stability. As stated above, the enzyme preparation is contaminated with 5-carboxymethylhydantoinase and dihydroörotic dehydrogenase. Cell-free preparations stored at $-16°$ for several weeks loose hydantoinase activity, whereas dihydroörotase and dihydroörotic dehydrogenase do not decrease markedly.

Specificity. Only the L-isomers of carbamyl aspartate and dihydroörotate were active as substrates for the enzyme.

Substrate Affinity. K_m (carbamyl aspartate) $= 2.8 \times 10^{-4}\,M$ (rate of orotate synthesis from carbamyl aspartate).

Equilibrium. K_{eq} (carbamyl aspartate/dihydroörotate) $= 1.9$ at pH 6.1, the participation of water in the equilibrium being neglected.

III. Dihydroörotic Dehydrogenase[5]

$$\text{Orotate} + \text{DPNH} + \text{H}^+ \rightleftarrows \text{Dihydroörotate} + \text{DPN}$$

This enzyme was discovered by Lieberman and Kornberg[6] in *Z. oroticum;* a description of a partially purified enzyme is included in Vol. II [75]. Since then, the enzyme has been further purified and crystallized by Friedmann and Vennesland[5,7] and characterized as a flavoprotein. This fact justifies a renewed presentation of the enzyme based on the work of Friedmann and Vennesland[5] (for method, see Vol. VI [151a]).

IV. Hydropyrimidine Dehydrogenase[8]

$$\text{Hydropyrimidine} + \text{TPN} \rightleftarrows \text{Pyrimidine} + \text{TPNH} + \text{H}^+$$

Assay Method

Principle. The change in absorbancy at 340 mμ, accompanying the interconversion of TPN and TPNH, is the basis for the method.

[5] H. C. Friedmann and B. Vennesland, *J. Biol. Chem.* **235**, 1526 (1960).
[6] I. Lieberman and A. Kornberg, *Biochim. et Biophys. Acta* **12**, 223 (1953).
[7] H. C. Friedmann and B. Vennesland, *J. Biol. Chem.* **233**, 1398 (1958).
[8] S. Grisolia and S. S. Cardoso, *Biochim. et Biophys. Acta* **25**, 430 (1957).

Procedure. The enzyme is mixed with the following components in a 1-cm. Beckman cuvette in a final volume of 3 ml.: 200 micromoles of phosphate buffer (pH 7.35), 25 micromoles of hydrothymine or other substrate, and 0.5 micromole of TPN. Optical densities are measured at 340 mμ and 30° at suitable time intervals after mixing. When pyrimidine reduction is being measured, the hydropyrimidines are replaced by pyrimidines, and the TPN is replaced by 0.3 micromole of TPNH.

Definition of Unit and Specific Activity. One enzyme unit causes a change of optical density at 340 mμ and 30° of 0.001 per minute. Specific activity is defined as the number of enzyme units per milligram of protein.

Purification Procedure

It is important to carry out the fractionation procedure and any additional studies in a single day, since the enzyme is highly unstable.

Preparation of Crude Extract. Fresh calf liver is cut into approximately 1-inch cubes and homogenized for 2 to 3 minutes in a cooled Waring blendor with 5 vol. of cold (−20°) acetone. Five more volumes of acetone (−20°) are added with mixing, and the mixture is held at −20° for about 2 hours. The supernatant fluid is siphoned off as far as possible without disturbing the sediment, which is then filtered on a Büchner funnel. Just before the contents of the funnel are dry, suction is stopped, and a large piece of dental rubber dam is stretched over the top of the funnel and secured with rubber bands. After 40 to 66 minutes of suction with the water pump and an empty filter flask, a mechanical vacuum pump is used for 6 to 8 hours. The well-dried cake is broken, sifted, and transferred to a desiccator over alumina, evacuated for 30 minutes, and stored in the cold.

Fifty grams of acetone powder are extracted with 500 ml. of deionized water (0°) for 20 minutes and then centrifuged for 15 minutes (5000 × *g*).

All steps of the following fractionation are carried out at 0°. The volume of the reagents added during fractionation is referred to the volume at the beginning of the particular step. Saturated (at 0°) ammonium sulfate solutions of pH 7.4 are used.

Step 1. First Ammonium Sulfate Fraction. The crude extract from 50 g. of acetone powder is mixed with 1 vol. of ammonium sulfate solution and centrifuged for 15 minutes. The precipitate is dissolved in water to a concentration of 20 ± 1 mg. of protein per milliliter.

Step 2. Second Ammonium Sulfate Fraction. Ammonium sulfate solution (0.59 vol.) is added, followed by centrifugation. The precipitate is discarded, and the enzyme is precipitated by the addition of 0.465 vol. of ammonium sulfate solution to the supernatant fluid. The mixture is centrifuged for 20 minutes, and the precipitate is dissolved in water (80 ml.).

Step 3. pH. The pH is adjusted to 4.1 (pH measured at 5°) with 0.1 *M* acetic acid (about 15 ml. are required), and 0.1 vol. of 0.37 *M* sulfosalicylic acid solution (adjusted to pH 4.1) is added. The precipitate is discarded.

Step 4. Gel Eluate. The supernatant fluid, containing the enzyme, is mixed with 1 vol. of calcium phosphate gel (30 mg. dry weight per milliliter) and centrifuged for 5 minutes, and the supernatant is discarded. The gel is first eluted with 40 ml. of 0.2 *M* phosphate buffer, pH 7.35, and then with 50 ml. of 0.5 *M* phosphate buffer, pH 7.35. The second eluate contains most of the enzyme. The enzyme can be concentrated by precipitation with 3 vol. of a saturated ammonium sulfate

TABLE II
PURIFICATION OF HYDROPYRIMIDINE DEHYDROGENASE

Fraction	Volume, ml.	Units	Protein, mg.	Specific activity, units/mg. protein
Crude extract	420	1260	11,700	0.11
First ammonium sulfate	180	1230	3,600	0.34
Second ammonium sulfate	80	800	1,600	0.50
Gel eluate	50	375	168	2.2

solution. The purification of the hydropyrimidine dehydrogenase is summarized in Table II.

Properties

Specificity. This mammalian enzyme preparation was relatively unspecific with respect to the pyrimidine. Relative rates, expressed as per cent of values obtained with uracil as substrate, are 130, 57, 56, 55, 28, and 5 with 5-Br-uracil, 5-I-uracil, thymine, dihydrothymine, dihydrouracil, and dihydrouridylic acid. Cytosine, uridine, 5-Br-dihydrouracil, and 6-uracil methyl sulfone are inactive.

No activity is observed when TPN is replaced by DPN. Nothing is known about the possible participation of flavins in the reaction.

Activators. The enzyme is activated by sulfosalicylate and salicylate in a wide range of concentration, with maximum activation around 0.02 *M*.

Effect of pH. The pH optimum is between 7.0 and 7.8.

Substrate Affinities. K_m (uracil and thymine) $= <3 \times 10^{-6} M$; K_m (dihydrouracil and dihydrothymine) $= 6 \times 10^{-4} M$.

Other Preparations

A dihydrouracil dehydrogenase has been somewhat purified from rat liver by Fritzson.[9] The reported properties of this enzyme are in all respects similar to those of the beef liver enzyme.

Campbell[10] has purified a dihydrouracil dehydrogenase from *Clostridium uracilicum*. In contrast to the mammalian enzyme, the bacterial enzyme requires DPN instead of TPN. Besides, the preparation showed no activity toward thymine.

V. Hydropyrimidine Hydrase[11]

$$\text{Hydropyrimidine} + H_2O \rightleftarrows \text{Carbamyl amino acid}$$

Assay Method

Principle. The formation of the carbamyl amino acid is determined by the colorimetric method of Archibald.[12] In this method carbamyl compounds possess a higher chromogenicity than hydropyrimidines.

Procedure. The assay mixture in a conical 12-ml. centrifuge tube contains 25 micromoles of hydrouracil and enzyme in $0.25\,M$ Tris buffer, pH 9.2, in a final volume of 2 ml. Incubation is for 15 minutes at 30°. The incubation mixture is deproteinized by the addition of 5 ml. of 10% $HClO_4$, followed by centrifugation. Aliquots of the supernatant fluid are analyzed for carbamyl compounds according to Archibald[12] with a heating period of 15 minutes. Since the color production does not follow Beer's law, a calibration curve is necessary. In assays on enzyme preparations of higher purity than step 5 (see Purification Procedure below) 2.5 mg. of egg albumin (Difco Bacto egg albumin) should be added per milliliter of incubation mixture as an enzyme stabilizer.

Definition of Unit and Specific Activity. One unit of enzyme activity is defined as the amount of enzyme that converts 1 micromole of hydrouracil under the assay conditions. Specific activity is defined as units of activity per milligram of protein.

Purification Procedure

Steps 1 to 4 should be carried out in minimal time to ensure good recoveries. All centrifugations are carried out at 0° and at $4000 \times g$. The volumes of reagents added during the fractionation refer to the volume at the beginning of the particular step.

[9] P. Fritzson, *J. Biol. Chem.* **235**, 719 (1960).
[10] L. L. Campbell, *J. Biol. Chem.* **227**, 693 (1957).
[11] D. P. Wallach and S. Grisolia, *J. Biol. Chem.* **226**, 277 (1957).
[12] R. M. Archibald, *J. Biol. Chem.* **156**, 121 (1944).

Step 1. Acetone powder (80 g.) from calf liver prepared as described for hydropyrimidine dehydrogenase is extracted with 400 ml. of cold water, stirred gently for 10 minutes at 0°, and centrifuged. The precipitate is extracted with another 400 ml. of water and centrifuged. Combined supernatants (675 ml.) form the "crude extract" (see Table III).

Step 2. Acid Treatment. The crude extract is treated successively, and with vigorous stirring, with 0.1, 0.05, and 0.075 vol. of cold 0.1 M acetic acid. The precipitate is removed after each addition. Finally, 0.0375 vol. of 1 M acetic acid is added, followed immediately by 0.262 vol. of 1 M KHCO$_3$.

Step 3. Heating. The solution from step 2 is brought, with constant agitation, to 60° in a water bath maintained at 62° to 65° (it should take no more than 15 to 17 minutes to reach 60°), kept at this temperature for 5 minutes, and then cooled rapidly in an ice bath. The precipitate is discarded after centrifugation.

Step 4. First Ammonium Sulfate Fractionation. To the supernatant is added 0.8 vol. of a saturated (0°) ammonium sulfate solution. After 5 minutes the precipitate is centrifuged off and discarded. The enzyme is precipitated from the supernatant with 1.0 vol. of ammonium sulfate stirred in during 5 minutes, collected by centrifugation, and dissolved in cold water to half the initial volume of this step. This preparation is stable for several months at −20° and should be kept frozen at this temperature for at least 24 hours before further fractionation is attempted.

Step 5. First Acetone Fractionation. In this and subsequent solvent fractionations, the solvent concentration is calculated without correction for volumes of precipitates removed or volume changes due to solvent mixing.

The enzyme preparation (430 ml.) is thawed and diluted with water to 860 ml. It is cooled below 0° (cooling bath of −8° to −10°), and 624 ml. of cold (0°) acetone are added with stirring. The temperature should not rise above 3° during this and subsequent additions. The precipitate formed is discarded after centrifugation. Cold (0°) acetone (204 ml.) is added to the supernatant to give a second precipitate (42 to 49% acetone), which contains the bulk of the enzyme. It is collected by centrifugation, taken up in water to a final volume of 215 ml., and frozen at −20°. At this stage the enzyme is stable for 1 month and should be kept frozen for at least 24 hours before further fractionation is attempted.

Step 6. Second Ammonium Sulfate Fractionation. The enzyme preparation is thawed, and the insoluble material is discarded. A saturated (0°) ammonium sulfate solution (1.1 vol.) is stirred in during 5 minutes. The precipitate is discarded, and the enzyme is precipitated from the supernatant fluid with 0.7 vol. of ammonium sulfate solution. The precipitate is collected and dissolved in water to half the volume present at the initiation of this step.

Step 7. Second Acetone Fractionation. Cold (0°) acetone (72 ml.) is added to 104 ml. of the preceding fraction (same conditions as for step 5) to give 41% final concentration. The precipitate is centrifuged and taken up in water to a volume of 53 ml.; any insoluble material is discarded.

Step 8. Third Ammonium Sulfate Fractionation. The enzyme is concentrated by precipitation with 2 vol. of saturated ammonium sulfate and centrifugation at 18,000 × g for 10 minutes. The precipitate is taken up in water to a final volume of 2.5 ml., and any insoluble material is discarded after centrifugation (18,000 × g for 5 minutes). The ammonium sulfate present in the supernatant fluid is estimated. More saturated ammonium sulfate (0°) is added stepwise. There is no appreciable precipitation before 0.4 saturation is reached, and 90 to 95% of the enzyme has precipitated at 0.46 saturation. The precipitate is collected by centrifugation, dissolved in water, and made up to 2.0 ml. At this stage, the enzyme is stable for several months if kept frozen at

TABLE III
PURIFICATION OF HYDROPYRIMIDINE HYDRASE

Fraction	Total volume, ml.	Units/ml.	Total units, × 10³	Protein, mg./ml.	Specific activity, units/mg. protein	Yield, %
Crude extract	675	14.5	9.8	42.5	0.341	100
Acid-treated fraction	970	19.3	18.7	23.0	0.813	195
Heat-treated fraction	870	21.5	18.7	9.2	2.04	195
First ammonium sulfate	430	30.5	13.23	6.4	4.76	136
First acetone	215	47.38	10.2	5.2	9.13	104
Second ammonium sulfate	107	61.32	6.55	2.9	21.10	67
Second acetone	53	88.36	4.68	2.07	42.70	48
Third ammonium sulfate	2	1292	2.58	19.45	66.2	26

—20° and can be dialyzed at 4° against phosphate, pH 6.9, or Veronal buffer, pH 8.6, without loss of activity. A summary of the purification is given in Table III.

Properties

Specificity. The purified enzyme shows activity for hydrouracil and hydrothymine and for the corresponding carbamyl amino acids. Hydantoin is also very reactive, but carbamylglycine is not. Furthermore the purified enzyme is inactive with carbamyl β-alanine amide, barbituric acid, imidazole-4,5-dicarboxylic acid, 5-(carboxymethylidine) hydantoin, orotic acid, 4,5-aminoimidazole carboxamide, urocanic acid, and 5-(diphenyl)-hydantoin.

Activators and Inhibitors. Sn^{++} inhibits the reaction with hydrouracil completely at $1 \times 10^{-2} M$ concentration and to 50% at $5 \times 10^{-3} M$ concentration. Mg^{++} ions are inhibitory (10 to 20%) when carbamyl β-alanine or carbamyl β-aminoisobutyric acid is the substrate. Mg^{++} and Mn^{++} are marked accelerators, however, when hydrouracil is the substrate but are only slightly or not at all stimulatory with hydrothymine or hydantoin as substrates.

Effect of pH. The optimal pH for the hydration and opening of the ring is on the alkaline side (pH 8 to 9 for hydrothymine and hydantoin, and around pH 10 for hydrouracil). The optimal pH for ring closure is on the acidic side (pH 5 to 6).

Kinetic Properties. The rate of conversion of hydropyrimidines to carbamyl amino acids is ten to twenty times as fast as the conversion of carbamyl β-alanine or carbamyl β-aminoisobutyric acid to their corresponding hydropyrimidines. K_m values of $0.83 M$ for hydantoin, $11.75 \times 10^{-2} M$ for hydrouracil, and $2.1 \times 10^{-3} M$ for hydrothymine are obtained at pH 9.2. Turnover numbers (moles per 100,000 g. of protein per minute) at 30° and under optimal conditions for each substrate are 27,000 for hydantoin, 4300 for hydrouracil, and 420 for hydrothymine.

VI. Decarbamylation of Carbamyl β-Alanine[13]

Carbamyl β-alanine $+ H_2O \rightarrow CO_2 + NH_3 + \beta$-alanine

Assay Method

Principle. The sum of $NH_3 + \beta$-alanine formed during the reaction is determined with the ninhydrin reaction as described by Moore and Stein.[14]

Procedure. Tris buffer (300 micromoles, pH 7), 20 micromoles of carbamyl β-alanine, and enzyme are incubated in a final volume of 2 ml. at 38° for 60 minutes. The reaction is stopped by the addition of 2 ml. of $0.5 M$ perchloric acid, followed by centrifugation. Aliquots of the supernatant fluid are brought to pH 5 with $4 M$ NaOH and assayed with the ninhydrin reagent.[14] (The combination of NH_3 and β-alanine is additive under the conditions of the assay and analysis, and the chromogenicity is about eight times as high for NH_3 as for β-alanine.) Since the colorimetric method is markedly influenced by salts, it is necessary

[13] J. Caravaca and S. Grisolia, *J. Biol. Chem.* **231**, 357 (1958).
[14] S. Moore and W. H. Stein, *J. Biol. Chem.* **176**, 367 (1948); see also Vol. III [76].

to mix the ammonia standards with the components of the incubation mixture and subject them to all the manipulations, along with the experimental tubes.

Definition of Unit and Specific Activity. One enzyme unit is defined as the amount of enzyme that decomposes 1 micromole of carbamyl β-alanine in 60 minutes at 38° and pH 7. Specific activity is the number of units per milligram of protein.

Purification Procedure

Rat liver is the source of enzyme. The procedure should be carried out as fast as possible in one single day, since the enzyme is unstable in the earlier fractions. All centrifugations and manipulations are carried out at 0° and at 4000 $\times$ g for 15 minutes. Solvents are measured and added at 0°; the temperature should not rise above 5° during addition.

Step 1. Preparation of Crude Extract. Fresh rat livers (100 g.) are homogenized with an equal volume of isotonic KCl in a large-size Potter-Elvehjem homogenizer. The 50% homogenate is diluted four times and centrifuged.

Step 2. First Acetone Fractionation. To each 100 ml. of crude extract are added 42.5 ml. of acetone. The precipitate is discarded after centrifugation. The enzyme is then precipitated from the supernatant fluid by adding 24 ml. of acetone for each 100 ml. of the original crude extract. The 30 to 40% acetone fraction is taken up in water to a volume of approximately 0.2 of the volume of the original crude extract. Any insoluble protein is centrifuged off and discarded. The supernatant solution is adjusted to a protein concentration of 8 $\pm$ 1 mg./ml.

Step 3. Acid Treatment. Cold 0.1 M acetic acid is stirred in to bring the pH to 4.8 (about 0.2 ml. of acid per milliliter of first acetone fraction). The precipitate is discarded, and the supernatant solution is acidified to pH 4.4 with 0.1 M acetic acid (approximately 0.02 ml. of acid per milliliter of first acetone fraction). The supernatant solution after centrifugation is then brought to pH 6.5 with 0.1 M NaHCO$_3$ (approximately 0.38 ml./ml. of first acetone fraction). The precipitate is removed by centrifugation, and the supernatant forms the "acid-treated fraction" (see the table).

Step 4. Second Acetone Fractionation. To each milliliter of the acid-treated fraction is added 0.19 ml. of cold (0°) acetone, and the resulting precipitate is discarded after centrifugation. On the further addition of 0.24 ml. of acetone per milliliter of the starting acid-treated fraction to the supernatant fluid, the bulk of the enzyme is precipitated. Another fraction (30 to 35%), which is collected by adding 0.1 ml. of acetone per milliliter of the original acid-treated fraction, sometimes contains the bulk of the enzyme activity. The precipitates are taken up in water to 0.2 of the original volume of the acid-treated fraction and freed from all insoluble material by centrifugation. At this stage the enzyme is

TABLE IV

PURIFICATION OF CARBAMYL β-ALANINE DECARBAMYLASE

Fraction	Total volume, ml.	Units/ml.	Total units	Protein, mg./ml.	Specific activity, units/mg. protein	Yield, %
Crude extract	700	1.86	1302	20.5	0.09	100
First acetone fraction	135	8.30	1120	8.3	1.00	84
Acid-treated fraction	139	5.52	768	4.6	1.20	55
Second acetone fraction (15–30%)	28	11.98	332	5.0	2.36	25.5
Second acetone fraction (30–35%)	28	5.72	160	3.1	1.84	12.2

stable for over 3 weeks when frozen and kept at $-20°$. It can also be lyophilized and kept for longer periods without loss of activity. A summary of the purification of the enzyme is given in Table IV.

Properties

Specificity. The purified enzyme degrades carbamyl β-alanine and carbamyl β-aminoisobutyric acid but not carbamyl L-alanine, carbamyl γ-aminoisobutyric acid, carbamyl glycine, carbamyl L-glutamic acid, carbamyl L-aspartic acid, carbamyl L-proline, or L-citrulline.

Effects of pH. The optimal pH for the hydrolysis of carbamyl β-alanine and carbamyl β-aminoisobutyric acid is around 7.

Substrate Affinities. K_m (carbamyl β-alanine) $= 5 \times 10^{-4} M$; K_m (carbamyl β-aminoisobutyric acid) $= 1 \times 10^{-3} M$.

VII. Uridine Phosphorylase[15]

Uridine $+ P_i \rightleftarrows$ Uracil $+$ ribose-1-P

Assay Method

Principle. Ribose-1-P formation is measured by the orcinol method of Mejbaum.[16] The strong acidic conditions of the orcinol reaction cause a complete hydrolysis of ribose-1-P to ribose and some hydrolysis of

[15] H. Pontis, G. Degerstedt, and P. Reichard, *Biochim. et Biophys. Acta* **51**, 138 (1961).
[16] W. Mejbaum, *Z. physiol. Chem.* **258**, 117 (1939).

uridine. In order to avoid interference by the nucleoside, uridine is removed from the reaction mixture by the addition of charcoal.

Procedure. The incubation mixture contains 1 micromole of uridine, 10 micromoles of potassium phosphate buffer (pH 7.4), and enzyme in a final volume of 0.1 ml. After incubation for 15 minutes at 37°, the reaction is stopped by the addition of 0.1 ml. of 0.8 M perchloric acid, followed by 4 mg. of charcoal (Norit A). Aliquots of the supernatant after centrifugation are analyzed for ribose according to Mejbaum.[16] A blank is obtained by adding perchloric acid to a sample prior to incubation.

Definition of Unit. One unit of enzyme activity is defined as the amount of protein that produces 1 millimicromole of ribose-1-P during 15 minutes under the above conditions. Specific activity is expressed as units per milligram of protein. Protein is determined by the method of Bücher.[17]

Purification Procedure

Growth of Tumor. Ehrlich ascites tumor is the source of enzyme. Relatively large amounts of the tumor are obtained by intraperitoneal injection of 200 to 300 mice with 10^7 cells per mouse. This dose corresponds to 0.1 ml. of ascitic fluid diluted 1:10 with physiological saline. The tumor is harvested 10 to 11 days after inoculation. Two hundred mice usually give 750 ml. of blood-free, slightly viscous ascitic fluid. The cells are centrifuged at 1000 × g and 0° for 10 minutes. The sediment is washed twice with 750 ml. of physiological saline and finally suspended in enough physiological saline to give the original ascites volume.

Preparation of Crude Extracts. The suspension is mixed in a Waring blendor with 5 to 10 vol. of acetone (−15°) and filtered by suction. The filter cake is treated once more with the same amount of acetone and finally dried *in vacuo* in a desiccator. The yield of acetone powder from 750 ml. of ascites tumor is about 30 g. The powder could be stored in a desiccator at −20° without appreciable loss of activity for 1 to 2 months.

All the following operations are carried out in the cold. The acetone powder (21 g.) is extracted with 210 ml. of 0.02 M Tris–maleate buffer, pH 6.5, in 40-ml. Servall plastic centrifuge tubes by three treatments (10-minute intervals) with a tight-fitting plastic pestle. After centrifugation at 10,000 × g for 15 minutes, the residue is washed once with a small amount of the same buffer. The final volume of the pooled supernatants is 226 ml.

[17] T. Bücher, *Biochim. et Biophys. Acta* **1**, 292 (1947); see also Vol. III [73].

Step 1. Protamine Precipitation. To 225 ml. of crude extract are added 45 ml. of 10% protamine sulfate (pH 6.5) solution. The precipitate is allowed to settle overnight and finally centrifuged at 10,000 $\times$ g for 30 minutes. The precipitate is washed once with about 30 ml. of 0.02 M Tris–maleate buffer, pH 6.5. This protamine precipitate can be used for the preparation of uridine kinase. The volume of the pooled supernatants is 290 ml.

Step 2. Heat Denaturation. The pH of the supernatant from the step above is adjusted to 5.7 by the addition of 1 M acetic acid (3 to 4 ml.). The solution is heated in a water bath (56° to 58°) under constant stirring. When the temperature of the enzyme solution has reached 40°, the solution becomes cloudy and is heated for a further period of 3 minutes. The final temperature should be 49° to 50°. The solution is then cooled immediately in an ice bath and centrifuged. The precipitate is washed with about 20 ml. of 0.02 M Tris–maleate buffer, pH 5.7. The final volume of the pooled supernatants is 296 ml.

Step 3. Negative Adsorption on Aluminum Hydroxide. To 296 ml. of the above solution are added 89 ml. of aluminum hydroxide suspension (Cγ-gel, 20 mg./ml.). A ratio of milligrams of gel to milligrams of protein giving less than 10% adsorption of enzyme activity should be determined in a pilot experiment for each particular gel preparation. The suspension is stirred for 10 minutes and finally centrifuged. The supernatant is adjusted to pH 7.2 by the addition of 3 ml. of 1 M NaOH. The final volume is 355 ml.

Step 4. Protein Concentration. The supernatant solution from step 3 is concentrated by evaporation *in vacuo* at room temperature in a flash evaporator with a dry ice-cooled vapor trap. The final volume is 58 ml.

Step 5. Ammonium Sulfate Precipitation. To the enzyme solution (58 ml.) are added 8.35 g. of solid ammonium sulfate. The precipitate is centrifuged off and discarded. To the supernatant are added 5.9 g. of solid ammonium sulfate. The precipitate is allowed to settle overnight and is then collected by centrifugation. The precipitate is dissolved in 0.05 M potassium phosphate buffer, pH 7.5. The final volume is 2.75 ml.

Step 6. Zone Electrophoresis. If necessary, the enzyme solution from step 5 is concentrated to 15 to 20 mg. of protein per milliliter by ultra-filtration against M potassium phosphate buffer, pH 7.4. The concentrated protein solution is then dialyzed for 3 hours against two 1-l. changes of 0.025 M potassium phosphate buffer, pH 7.4, containing 0.025 M ammonium sulfate, and subsequently subjected to vertical zone electrophoresis in an agar gel suspension according to Hjertén.[18] The dialyzed enzyme solution (0.5 ml.) is applied to a column (length 124

[18] S. Hjertén, *Biochim. et Biophys. Acta* **53**, 514 (1961).

cm., diameter 1 cm.) of 0.15% agar in 0.025 M potassium phosphate buffer, pH 7.4, containing 0.025 M ammonium sulfate.[19] Electrophoresis is run for 15 hours at 750 volts and 15 ma. Ice water is circulated through the cooling jacket of the column. The enzyme moves downward toward the anode. After the electrophoresis, the agar gel suspension, containing the enzyme, is drained off the column. Small fractions (0.6 to 0.7 ml.) are collected with the aid of an automatic fraction collector. To each fraction 0.2 ml. of M phosphate buffer, pH 7.4, is added immediately in order to stabilize the enzyme. Analyses of protein (absorbancy at 280 mμ) and enzyme activity are carried out, and the most active fractions are pooled. At pH 7.4 the enzyme has a higher negative net charge than most of the contaminating proteins and thus moves relatively rapidly toward the anode. A considerable purification is therefore achieved in this step. The agar particles are centrifuged off at 20,000 $\times$ g (30 minutes), and the resulting precipitate is washed once with 0.2 ml. of M potassium phosphate buffer, pH 7.4. To the combined supernatants bovine serum albumin is added to give a final concentration of 1%. This enzyme preparation retains its activity for at least 2 weeks if kept frozen at $-20°$. It is inactivated almost completely on dialysis against 0.1 M phosphate buffer, pH 7.4. The purification method is given in Table V.

TABLE V
PURIFICATION PROCEDURE FOR URIDINE PHOSPHORYLASE

Fraction	Volume, ml.	Protein, mg./ml.	Units/ml.	Total units, $\times 10^3$	Specific activity, units/mg. protein	Yield, %
Extract	226	26.4	1300	294	49	100
Protamine supernatant	290	6.3	866	251	138	86
Heat-treated fraction	296	4.65	820	242	177	83
C$_\gamma$-treated fraction	355	2.95	900	320	306	108
After concentration	58	15.0	2700	157	180	53
Ammonium sulfate fraction	2.75	36	45000	124	1250	42
Before electrophoresis	0.5	16.6	22000	11	1325	100
After electrophoresis	5	0.06	900	4.5	15000	41

Properties

Stability. The rapid loss of enzyme activity during storage and purification represents one of the major difficulties when working with

---

[19] Difco Agar Noble (0.75 g.) is dissolved in 500 ml. of buffer solution by heating to the boiling point, with stirring. The solution is left to cool overnight at room temperature. Five hours before the preparation of the electrophoretic column, the agar gel suspension is placed in the cold room (0°).

uridine phosphorylase. This instability is not apparent in crude tissue extracts but is very pronounced with the purest enzyme fractions, especially at slightly acid pH values and in dilute solution. Addition of cysteine or glutathione did not increase the stability. Some stabilization could be obtained by phosphate or sulfate in relatively high concentration and by the addition of bovine serum albumin.

Specificity. The purest fractions catalyzed the phosphorolysis of uridine, deoxyuridine, and thymidine. Furthermore, the relative maximal velocities of phosphorolysis for all three nucleosides were the same in the crude extract and in all fractions obtained after agar electrophoresis. Nevertheless a considerable separation of the activities could be achieved by acid precipitation of the crude extract, but this treatment inactivated the enzymes to a large extent and could not be used as a purification step. From this and from other data it seems evident that there exists a specific protein, or at least a specific active center within a protein, which catalyzes the phosphorolysis of uridine, but not that of deoxyuridine and thymidine.

The enzyme preparation also cleaves 5-halogen nucleosides, but not cytidine, azauridine, azathymidine, orotidine, or purine nucleosides. On the other hand, synthesis of azauridine and azathymidine from the corresponding azapyrimidine and the pentose-1-P can be demonstrated.

Effect of pH. Optimal pH for the phosphorolysis of uridine is 8.1, for the phosphorolysis of deoxyuridine 6.5. In the crude extract the heat stability of uridine phosphorylase activity has a maximum at pH 7.5, that of deoxyuridine phosphorylase activity at 5.5.

Equilibrium. The reaction is readily reversible, and the equilibrium position favors synthesis (80% uridine, 20% uracil). The same holds true for the phosphorolysis of deoxyuridine.

Substrate Affinities. K_m (uridine) $= 7.6 \times 10^{-4}\ M$; K_m (deoxyuridine) $= 7.1 \times 10^{-4}\ M$; K_m (phosphate) $= 3.9 \times 10^{-3}\ M$ (for both uridine and deoxyuridine phosphorolysis).

Other Preparations

Uridine phosphorylase was first purified from *E. coli* by Paege and Schlenk.[20] They achieved a twentyfold purification with a yield of less than 5%. The microbial preparation did not cleave purine nucleosides, cytidine, or thymidine. Deoxyuridine was not tested.

Razzel and Khorana[21] have considerably purified a pyrimidine deoxyriboside phosphorylase from *E. coli* which did not split uridine but catalyzed the phosphorolysis of both thymidine and deoxyuridine.

[20] L. M. Paege and F. Schlenk, *Arch. Biochem.* **40**, 42 (1952).
[21] W. E. Razzell and H. G. Khorana, *Biochim. et Biophys. Acta* **28**, 562 (1958).

VIII. Uridine Kinase[22]

$$\text{Uridine} + \text{ATP} \rightarrow \text{UMP} + \text{ADP}$$

Assay Method

Principle. Measurement of the radioactivity adsorbed to an anion exchanger after incubation of uridine-2-C^{14} with ATP and enzyme is the basis of the isotopic assay method. A spectrophotometric assay can be used with the purified enzyme. This latter method is based on the determination of the ADP formed. In the presence of an excess of added lactic dehydrogenase, pyruvate kinase, phosphoenol pyruvate, and DPNH, ADP is quantitatively converted to ATP with the simultaneous stoichiometric oxidation of DPNH. Thus disappearance of DPNH, as measured by a decrease in absorbancy at 340 mμ, is a direct measure of uridine kinase activity.

Procedure (Isotope Method). Uridine-2-C^{14} [23] (0.5 micromole, about 80,000 c.p.m./micromole), 15.0 micromoles of ATP, 10.0 micromoles of MgCl$_2$, 10.0 micromoles of Tris–chloride buffer at pH 7.4, and enzyme are incubated in a final volume of 0.23 ml. for 15 minutes at 37°. The reaction is stopped by heating at 100° (3 minutes), and the incubation mixture is added directly to the top of a column of Dowex 2–acetate (diameter 0.4 cm., length 3 cm.). Uridine-2-C^{14} is eluted with water in a total volume of 15 ml. Labeled uridine phosphates are then eluted with 15 ml. of 1 M ammonium acetate–4 M acetic acid. Aliquots of both fractions are plated, and their radioactivities are determined at infinite thinness. From these values the percentage of phosphorylation of uridine can be calculated.

Definition of Unit and Specific Activity. One unit of activity is defined as that amount of enzyme producing 1 micromole of UMP under the above conditions. Specific activity is defined as enzyme units per milligram of protein.

Spectrophotometric Assay Method. The incubation mixture contains 0.04 ml. of DPNH (0.01 M), 0.10 ml. of phosphoenolpyruvic acid (0.01 M), 0.01 ml. of lactic dehydrogenase[24] containing pyruvate kinase (=0.1 mg. of protein), 0.03 ml. of KCl (1 M), 0.06 ml. of MgCl$_2$ (1 M), 1.70 ml. of Tris–chloride buffer (0.05 M, pH 7.4), 0.02 ml. of ATP (0.1 M), 0.02 ml. of uridine (0.1 M), and enzyme to a final volume of 2.00 ml. The whole test system is made up in a quartz cell. The reaction is initiated by the addition of uridine, and the decrease in optical density

[22] O. Sköld, *J. Biol. Chem.* **235**, 3273 (1960).
[23] O. Sköld, *Biochim. et Biophys. Acta* **44**, 1 (1960).
[24] Lactic dehydrogenase, type I, Sigma Chemical Company.

at 340 mμ is followed in a spectrophotometer. The average density decrease per minute during a 10-minute interval is determined. A blank run without substrate or enzyme is not necessary.

Purification Procedure

The purification procedure described below has been found highly reproducible. Ehrlich ascites tumor is the source of enzyme. The whole procedure is carried out in the cold (0° to 4°).

Growth of Tumor and Preparation of Crude Extract. This process is identical with that described under the same heading in the section on the purification of uridine phosphorylase. Both uridine kinase and phosphorylase can be prepared from the same tumor acetone powder.

Step 1. Protamine Fractionation. To 195 ml. of crude extract are added 39 ml. of a 10% solution of protamine sulfate (neutralized to pH 6.5 with M NaOH). The mixture is left overnight and finally centrifuged at 10,000 $\times$ g for 30 minutes. The tough, stringy precipitate is extracted with five 23-ml. portions of 0.5 M potassium phosphate buffer, pH 5.4. Each extraction is carried out in a Potter-Elvehjem homogenizer by two treatments with 10-minute intervals. Each portion of extract is centrifuged at 10,000 $\times$ g for 15 minutes. The final residue is discarded. Solid ammonium sulfate (62 g.) is added slowly to the pooled extracts (110 ml., 80% saturation), with stirring. The precipitate is centrifuged at 10,000 $\times$ g. It can be stored at $-15°$ for 1 month with no loss of activity.

Step 2. CM-Cellulose[25] Adsorption. The protamine contaminating the preparation of step 2 is removed by treatment with CM-cellulose. The ammonium sulfate precipitate (10.2 g. wet weight) from the previous step is dissolved in 62 ml. of 0.05 M ammonium acetate buffer, pH 8.0. The solution (68 ml.) is sometimes cloudy but turns clear if the temperature is raised to 13°. To the solution are added 5.8 ml. of a suspension of CM-cellulose (119 mg. dry weight per milliliter, capacity 0.8 meq./g.) in the same buffer. The CM-cellulose is previously equilibrated on a column with 0.1 M ammonium acetate, pH 8.0. The mixture is allowed to stand for 30 minutes with occasional stirring and is finally centrifuged at 10,000 $\times$ g for 15 minutes.

Step 3. Ammonium Sulfate Fractionation. The supernatant from step 2 (63 ml.) is analyzed for ammonium sulfate (=nitrogen volatile at pH 10). The difference between 14.1 g. and the amount in grams of ammonium sulfate found to be present in the solution is added slowly to give a final saturation of 40%. The precipitate is discarded after

[25] CM-cellulose = carboxymethylcellulose, DEAE-cellulose = diethylaminoethylcellulose, Brown Company, Berlin, New Hampshire.

centrifugation. Another 4.2 g. of solid ammonium sulfate are added to the supernatant (66 ml.), giving a final saturation of 50%.

Step 4. Chromatography on DEAE-Cellulose.[25] The precipitate obtained at 50% ammonium sulfate saturation is dissolved in 6 ml. of 0.05 M Tris–chloride buffer, pH 7.4, and desalted by gel filtration.[26] The protein solution is introduced to a Sephadex G-25[27] column (diameter 2.5 cm., length 16 cm.), previously equilibrated with 0.02 M potassium phosphate, pH 7.4. Protein emerges between 14.5 and 28.5 ml. after elution is started with 0.02 M potassium phosphate buffer, pH 7.4. The slightly opalescent solution (172 mg. of protein) is then added to the top of a DEAE-cellulose column (diameter 2 cm., length 8.6 cm., capacity 0.85 meq./g.) and allowed to drain into the column by gravity. The column was previously equilibrated with 2 l. of 0.02 M potassium phosphate, pH 7.4. The column is initially washed with 0.04 M potassium phosphate buffer (pH 7.4) until the absorbancy at 280 mμ is below 0.04. The enzyme is then eluted with 0.08 M potassium phosphate buffer (pH 7.4). Fractions with an absorbancy at 280 mμ greater than 0.04 are pooled, and the volume is reduced to about one-tenth by evaporation *in vacuo* at 0° to 4° in a flash evaporator with a dry ice-cooled vapor trap.

TABLE VI

SUMMARY OF PURIFICATION PROCEDURE FOR URIDINE KINASE

Step	Protein, mg.	Specific activity, units/mg. protein	Total units	Yield, %
Crude extract	6950	0.27	1890	100
Protamine fractionation	1520	0.62	967	51
Adsorption on CM-cellulose	830	1.26	1040	55
Ammonium sulfate precipitation	172	3.1	552	29
Chromatography on DEAE-cellulose	15	34.3	504	27

A further threefold purification is obtained[22] by zone electrophoresis on agarose gel suspension according to Hjertén.[18] The yield of this step was rather poor, however. The purification steps are summarized in Table VI.

Properties

Stability. The chromatographically purified enzyme can be stored frozen at −15° for 2 to 3 months with occasional thawing without loss of activity.

[26] J. Porath and P. Flodin, *Nature* **183**, 1657 (1959).
[27] Pharmacia, Uppsala, Sweden.

Specificity. The enzyme catalyzes the phosphorylation of uridine and cytidine. The activity toward uridine is twice as high as that toward cytidine in both the crude extract and the purified enzyme preparations indicating that the same enzyme is involved with both nucleosides. The enzyme preparation also phosphorylates 5-fluorouridine, 5-fluorocytidine, and 6-azauridine, but not 5-methyluridine, 4,5-dihydrouridine, 5-iodo-deoxyuridine, orotidine, 5-methylcytidine, 1-β-D-xylofuranosylthymine, or deoxynucleosides. GTP or ITP but not CTP or UTP could partly replace ATP.

Activators and Inhibitors. Mg^{++} is required for full enzyme activity. Mg^{++} could be partially replaced by Mn^{++} or Fe^{++}. The enzyme is inhibited by *p*-chloromercuribenzoate.

Effect of pH. Uridine kinase exhibits a broad pH optimum between pH 5.5 and 8.0.

Substrate Affinities. K_m (uridine) $= 4.8 \times 10^{-5} M$; K_m (cytidine) $= 2.3 \times 10^{-5} M$; K_m (5-fluorouridine) $= 3.8 \times 10^{-5} M$; K_m (ATP) $= 5.0 \times 10^{-4} M$.

[22] Dihydroörotic Dehydrogenase[1-3]

$$DPNH + H^+ + \text{orotate} \rightleftarrows DPN^+ + \text{dihydroörotate}$$

By HERBERT C. FRIEDMANN

Assay Method

Principle. The activity of the enzyme[3a] is determined spectrophotometrically by following the rate of decrease of the optical density at 340 mμ associated with the oxidation of DPNH to DPN^+ in the presence of excess orotate.

Reagents

It is advisable that all solutions except the bacterial culture medium be made with water redistilled from glass.

[1] I. Lieberman and A. Kornberg, *Biochim. et Biophys. Acta* **12**, 223 (1953); see Vol. II [75].
[2] H. C. Friedmann and B. Vennesland, *J. Biol. Chem.* **233**, 1398 (1958).
[3] H. C. Friedmann and B. Vennesland, *J. Biol. Chem.* **235**, 1526 (1960).
[3a] An application of this enzyme to assay orotic acid has been published. [H. C. Friedmann and G. Krakow, *in* "Methods of Enzymatic Analysis" (H. U. Bergmeyer, ed.), p. 508. Academic Press, New York, 1963.]

DPNH solution in water, aproximately 0.0035 M. Store at −15°.

Sodium orotate,[4] approximately 0.02 M. Dissolve an aqueous suspension of 0.312 g. of orotic acid, twice recrystallized from water, by adding 4 ml. of N NaOH, and make up volume to 100 ml. Store at 4°.

1 M sodium phosphate buffer,[4] pH 6.2, prepared by mixing requisite volumes of 1 M NaH$_2$PO$_4$ and 1 M Na$_2$HPO$_4$ solutions. Store at room temperature.

0.1 M cysteine hydrochloride. Use a freshly prepared solution of 79 mg. of cysteine hydrochloride in 5 ml. of water without neutralizing.

Enzyme, freshly diluted with ice-cold 0.2 M sodium phosphate buffer of pH 5.8, to obtain 120 to 240 units/ml. (See definition of unit below.)

Procedure. For maximal activity the enzyme, in the presence of orotate and buffer, must be preincubated with cysteine before addition of DPNH. The test system in a 3-ml. cuvette contains 0.4 ml. of the phosphate buffer, 0.2 ml. of the cysteine hydrochloride solution, 0.3 ml. of the sodium orotate solution, about 0.1 ml. of enzyme solution, and water to 2.9 ml. Part of the water is added before the enzyme, and the remaining volume of water is used to wash down the enzyme. After mixing, the contents of the cuvette are allowed to stand for 10 minutes at 20°. The reaction is started by adding 0.1 ml. of the DPNH solution, and as soon as possible after mixing the rate of DPNH oxidation at 340 mμ is followed in a spectrophotometer with the cuvette compartment maintained at 20°.

Definition of Unit and Specific Activity. A unit of enzyme is defined as that amount which causes the oxidation of 1 millimicromole of DPNH per minute in the assay system at 20°. No correction is made for the rate of oxidation of DPNH in the absence of added orotate. At the final pH of 6.5 obtained under the above conditions the difference between the rate of DPNH oxidation in the presence and in the absence of orotate is at its maximum. The specific activity is defined as the number of units per milligram of protein as determined by the method of Lowry *et al.*[5]

[4] Potassium salts are avoided, since potassium orotate is much less soluble than sodium orotate. Cf. G. Biscaro and E. Belloni, abstracted in *Chem. Zentr.* 1, 63 (1905).

[5] O. H. Lowry, H. J. Rosebrough, A. L. Farr, and R. J. Randall, *J. Biol. Chem.* 193, 265 (1951).

Purification Procedure

Growth of Bacteria. For preparation of the enzyme, the obligate anaerobic bacterium *Zymobacterium oroticum*[1,6,7] is grown on a medium containing, per 100 ml.: 2 g. of Bacto-tryptone (Difco), 0.4 g. of orotic acid, 0.1 g. of sodium thioglycolate, 0.15 g. of $Na_3PO_4 \cdot 12H_2O$, 0.1 g. of NaOH, and 1.5 mg. of riboflavin. The pH is 7.9. The tall flasks of 8.5-l. capacity, containing about 7.3 l. of medium, are plugged with nonabsorbing cotton. The amount of orotate is more than sufficient to saturate the solution, and a layer of solid orotic acid remains at the bottom of the flasks after autoclaving. Some of this can be brought into solution by gentle swirling of the hot flasks.

Each bottle is inoculated with 25 to 150 ml. of an actively growing culture. After about 12 hours at 30° the bacterial growth may be seen in the form of characteristic fluffy threads which settle gradually to the bottom of the flask. The contents of the bottle are mixed by gentle swirling two or three times during the course of the growth period, to ensure that the upper portion of the medium does not remain unutilized. The bacteria are harvested by centrifugation at about 10° as soon as growth is complete. The yield is about 4 to 5 g. wet weight of bacteria per liter of culture.

Preparation of Extract. The cells are disrupted by grinding with acid-washed glass beads[8] of 200 microns average diameter[9] in a Waring blendor immediately after harvesting. The paste from 22 l. of culture (100 g. wet weight) is washed into about 300 g. of chilled beads with approximately 50 ml. of 0.05 M sodium phosphate buffer of pH 6.5. The glass container of the blendor is surrounded by tubing through which ice water is circulated, and the temperature is not permitted to exceed 30°. Disruption of the cells may be detected by a marked change in the consistency of the suspension. This generally requires about 15 minutes of blending with the rheostat set at two-thirds of maximum. After disruption of the cells, an additional 100 ml. of phosphate buffer is added, the suspension is stirred gently for 2 minutes, the beads are allowed to

[6] J. T. Wachsman and H. A. Barker, *J. Bacteriol.* **68**, 400 (1954).
[7] The organism was kindly supplied by Dr. A. Kornberg; it was subcultured on the above freshly prepared medium at monthly intervals and stored under a pyrogallol seal at 4°. It is now available as ATCC 13619.
[8] C. Lamanna and M. F. Mallette, *J. Bacteriol.* **67**, 503 (1954).
[9] Available as 3M Superbrite glass beads, catalog number 100, from the Minnesota Mining and Manufacturing Company, St. Paul, Minnesota. Used beads may be washed by stirring for 3 to 4 hours with 0.05 N NaOH, followed by rinsing with tap and distilled water to neutrality.

settle, and the supernatant is decanted. It is of advantage to collect and centrifuge down foam which has formed during the disintegration. Fresh buffer is added, and the stirring and decanting are repeated three times with decreasing amounts of buffer, to give a final volume of about 350 ml. of extract. The combined extracts are cleared of solids by centrifugation, and the fractionation is begun as soon as possible. Unless otherwise noted, all steps are carried out with chilled reagents at about 0° to 5°.

Step 1. First Protamine Treatment and Ammonium Sulfate Precipitation. In this and in the following treatment with protamine sulfate, the optimal amount of the reagent is that which suffices to give near maximum precipitation. This amount is determined by the addition of increasing volumes of protamine solution to small aliquots of the enzyme solution. The samples are cleared by centrifugation, and the supernatants are treated with more protamine to determine whether more precipitate is formed.

If protamine is added to the first bacterial extract, the enzyme is precipitated along with nucleic acid.[1] In the presence of NaCl, however, nucleic acid is precipitated without coprecipitation of the enzyme. NaCl is added to bring the concentration to $0.4 M$. This is followed by addition of sufficient 1% protamine sulfate in $0.4 M$ NaCl until further addition gives little extra precipitate. About 0.4 to 0.5 vol. of the protamine sulfate solution per volume of crude extract is required. Excess is avoided. After 10 minutes, the white precipitate is removed by centrifugation, and 29.5 g. of ammonium sulfate is added for each 100 ml. of the yellow, slightly turbid supernatant solution. After 30 minutes, the slimy precipitate is completely removed by centrifugation. To the clear yellow supernatant solution, 9 g. of ammonium sulfate are added for each original portion of 29.5 g. A yellow precipitate containing the enzyme separates out in fine floccules after about 30 minutes. At this stage the material is usually allowed to stand overnight in the cold. The precipitate is collected by centrifugation and extracted with $0.2 M$ sodium phosphate buffer of pH 5.8. Several portions of buffer are employed to give a final total volume about one-tenth that of the original crude extract. The extraction removes all yellow pigment but leaves a white, less-soluble residue which is discarded after the centrifugation.

Step 2. Second Protamine Treatment and Precipitation by Dialysis. The clear yellow solution is treated with aqueous protamine sulfate to remove a small additional amount of nucleic acid. Although this treatment gives no detectable increase in purity, and the amount of precipitate obtained is often small, the step is necessary for success of subsequent procedures. When the solution is dialyzed overnight against several changes of cold distilled water, a bulky precipitate is formed.

The precipitate is collected by centrifugation, and the enzyme is extracted from it with successive portions of 0.2 M sodium phosphate buffer of pH 5.8. To achieve adequate extraction, the precipitate is repeatedly ground to a smooth paste with small portions of buffer (5 to 10 ml.). The yellow pigment is used as a guide. When the eluates become colorless, extraction is stopped, and the bulky green-gray residue is discarded.

Step 3. Third Protamine Sulfate Treatment and Crystallization. The solution obtained in step 2 has a volume about one-tenth that of the original extract. An equal volume of 1% protamine sulfate solution is added to precipitate the enzyme, which is collected by centrifugation in the form of an orange-yellow pellet. This precipitate is extracted with 5-ml. portions of 0.2 M NaH$_2$PO$_4$ until the eluate is only faintly colored. The combined eluates, decanted after centrifugation, are held at 20° for 1 hour and stored overnight at −15°. A precipitate which remains after thawing and warming to room temperature is removed by centrifugation in the cold. The enzyme crystallizes from the solution after 1 or more days at 4° in the form of fine blunt orange-yellow needles. If necessary, the solution is diluted to a point where the light absorbance at 450 mμ does not exceed 0.5, since the precipitate is often amorphous if crystallization is attempted from too concentrated a solution. About 85% of the yellow pigment is recovered in the form of crystals. These can be washed with cold water. A small further purification is achieved by extraction of the crystals with a small volume of 0.2 M phosphate buffer of pH 5.8. This extraction leaves a small amount of colorless residue which is removed by centrifugation. Recrystallization can be achieved by dialysis against water for 2 days or more. This results in a decrease (30 to 45%) in specific activity and is avoided. The first crop of crystals has been stored in the mother liquor at 4° for periods of over a month with little or no loss in activity.

The yellow supernatant from the third protamine sulfate precipitation contains considerable amounts of enzyme which may be precipitated by

SUMMARY OF PURIFICATION PROCEDURE

Stage of purification	Volume,[a] ml.	Protein, mg.	Total activity, units	Specific activity, units/mg.	Yield, %
Cell-free extract	700	4240	2,700,000	654	
Step 1	70	826	1,580,000	1,920	57
Step 2	83	270	1,110,000	4,190	40
Step 3 (crystals)		30.2	535,000	17,700	19

[a] In this particular preparation, cells were harvested from 44 l. of bacterial culture.

dialysis against water. A compact yellow pellet is obtained on centrifugation. Solution of the precipitate in $0.2\ M$ sodium phosphate buffer of pH 5.8, precipitation of the enzyme by protamine sulfate, and elution with $0.2\ M$ NaH_2PO_4 gives a solution from which a further crop of crystals separates. The heating to 20°, followed by freezing and thawing, is not needed in this case. This fraction may crystallize more readily than the bulk of the preparation.

A summary of the purification procedure is given in the table on p. 201.

Properties

Sedimentation Behavior. In the ultracentrifuge, solutions of the crystalline enzyme show a single peak with $S_{20,\,w} = 8.2\ S$.[10]

Spectrum and Flavin Content. The absorption spectrum with maxima at 273, 374, and 454 mμ, and minima at 310 and 403 mμ, is typical for a flavoprotein. The absorption band around 454 mμ has at least three well-defined shoulders. The low ratio of the absorbance at 273 mμ to that at 454 mμ (3.7) may in part be due to the iron content of the enzyme. Recent data[11] indicate that the iron participates in the reaction.

The crystalline enzyme contains both FMN and FAD in equimolar amounts. The minimum combining weight per two flavins is 62,000.

Turnover Number. With a sufficiently concentrated fresh enzyme solution, a turnover number based on flavin of the order of 1200 is obtained.[12]

Bleaching by Substrates. The yellow color of a solution of the enzyme is bleached instantaneously by excess DPNH or dithionite. In the absence of oxygen excess dihydroörotate bleaches slowly, less completely than DPNH, and only when cysteine is present. The rate of bleaching by dihydroörotate is greater when the enzyme is preincubated with cysteine than when cysteine is added with this substrate. Cysteine alone causes no bleaching.

Kinetics and Stability. The rate of DPNH oxidation by oxygen and by oxygen plus orotate is not linear, but decreases from the beginning. The oxidations of DPNH by orotate and by oxygen are additive. Activity is rapidly lost at moderately elevated temperatures, starting at about 20°.

Specificity, Activators, and Inhibitors. 5-Fluoroörotate can be used as a substrate in place of orotate. Its K_m (about $1.3 \times 10^{-4}\ M$) is the same as that for orotate, but the maximum velocity is greater. 5-Methyl

---

[10] These measurements were kindly made by Dr. E. Goldwasser.

[11] K. V. Rajagopalan, V. Aleman, P. Handler, W. Heinen, G. Palmer, and H. Beinert, *Biochem. Biophys. Res. Commun.* **8**, 220 (1962).

[12] G. Krakow and B. Vennesland, personal communication (1960).

orotate at $2 \times 10^{-3} M$ inhibits the over-all reaction and the "blank" by about 50%. This is the most potent inhibitor among various pyrimidine analogs tested. No activity and no inhibition of orotate reduction have been observed with uracil, cytosine, 5-methylcytosine, or thymine.[1] The 3-acetylpyridine analog of DPN can be substituted for DPN as a more effective oxidant for dihydroörotate than DPN itself. FAD causes irregular stimulation of the "blank" and over-all reaction, ranging from 11 to 37%. FMN alone or with FAD has no effect on the crystalline enzyme.

Cyanide $(3 \times 10^{-3} M)$, 8-hydroxyquinoline $(1 \times 10^{-3} M)$, Versene $(1.7 \times 10^{-2} M)$, and arsenite $(1 \times 10^{-2} M)$ do not inhibit; o-phenanthroline $(1 \times 10^{-3} M)$ inhibits 10%; p-chloromercuribenzoate $(1 \times 10^{-4} M)$ completely inhibits the "blank" reaction. Both $0.2 M$ NaCl and $0.2 M$ sodium phosphate (pH 6.5) inhibit the over-all reaction by about 55% and 20%, respectively.

Equilibrium Constant. The K for the system [dihydroörotate] [DPN⁺]/[orotate] [DPNH] [H⁺] has been determined as 2.3×10^9 at 20° over a wide pH range.[13] From this average value the E'_0 (pH 7) for the system dihydroörotate–orotate is -0.252 volt.

Effect of pH. The reaction in the presence of orotate has a maximum at pH 6.5, and the "oxidase" reaction a maximum at about pH 8.0. Orotate does not stimulate the rate of DPNH oxidation above a pH of about 8.

[13] G. Krakow and B. Vennesland, *J. Biol. Chem.* **236**, 142 (1961).

[23] Adenase from Yeast

Adenine + $H_2O \rightarrow$ Hypoxanthine + NH_3

By W. D. McELROY

Assay Method

Enzyme activity can be determined by measuring the decrease in optical density at 265 mμ according to the method of Kalckar.[1] Since the absorption spectra of adenine and hypoxanthine differ significantly, other wavelengths of light may be used if interfering substances are present at the longer wavelengths.

[1] H. M. Kalckar, *J. Biol. Chem.* **167**, 445 (1947).

Reagents

0.01 M sodium phosphate buffer, pH 7.0, plus 10^{-3} M Versene and 10^{-3} M glutathione.

Adenine. A Stock solution (80 μg./ml.) of adenine is prepared by dissolving 20 mg. of adenine in 250 ml. of the standard phosphate buffer. The standard adenine test solution was prepared from the stock solution by dilution using the phosphate–Versene buffer. A final concentration of 20 μg. of adenine per milliliter was used in the enzymatic assay.

Procedure. Place 3.0 ml. of the standard adenine solution (20 μg./ml.) into a silica cuvette (1-cm. light path), and measure the optical density at 265 mμ. Add 0.01 to 0.02 ml. of enzyme preparation that has been appropriately diluted with the phosphate–Versene buffer. Mix, and read the optical density change every minute or until the optical density has decreased by 0.2 to 0.3 of a unit. The first-order rate constant can be calculated from these data.

Definition of Unit. One enzyme unit is defined as the deamination of 1 μg. of adenine in 10 minutes. Specific activity is expressed as units per milligram of protein. The purest preparation of adenase obtained thus far has a specific activity of 350, which is approximately ten times as active as the adenase which has been partially purified from *Azotobacter vinelandii.*[2]

Growth of Torulopsis utilis, Strain 9926 (American Type Culture Collection). The existence of adenine deaminase or adenase in this yeast was first suggested by DiCarlo *et al.*[3] The enzyme has been prepared in crude extracts by Raush,[4] who first demonstrated that the enzyme concentration in the cells could be increased by adding adenine to the culture. The preparation of the partially purified adenase described in this paper has not been published previously. Joseph W. Berkow, Stanley Margulies, and Phillip A. Rierson made important contributions to the present procedures.

Growth Media for Torulopsis utilis (Modified after Schultz and Atkin[5]). The salt solution (16 ml.) contains KH_2PO_4, 17.6 g.; KCl, 13.6 g.; $CaCl_2 \cdot 2H_2O$, 4.0 g.; $MgSO_4 \cdot 7H_2O$, 4.0 g.; and water to make 1 l. The buffer solution (32 ml.) contains potassium citrate ($K_3C_6H_5O_7 \cdot H_2O$), 100 g.; citric acid ($C_6H_8O_7 \cdot H_2O$), 20 g.; and water to make 1 l.

[2] L. A. Heppel, J. Hurwitz, and B. L. Horecker, *J. Am. Chem. Soc.* **79**, 630 (1957).
[3] F. J. DiCarlo, A. S. Schultz, and D. K. McManus, *J. Biol. Chem.* **189**, 151 (1951).
[4] A. H. Raush, *Arch. Biochem. Biophys.* **50**, 510 (1954).
[5] A. S. Schultz and L. Atkin, *Arch. Biochem.* **14**, 369 (1947).

Glucose (50 g.) and ammonium sulfate (1 g.) are combined with enough tap water to make a total volume of 1 l.

Growth and Preparation of Acetone Powder. Eight liters of the modified growth medium is prepared in a 20-l. carboy and autoclaved for 20 minutes at 120°, 15 p.s.i. After cooling, the medium is inoculated by the yeast obtained from the washing of one slant. The yeast cells are grown at room temperature (25°) under filtered and humidified aeration for 3 to 4 days. Adenine (0.5 g./l.) is added daily after the first day of growth.

The yeast is harvested by centrifugation in the cold. Approximately 60 g. wet weight of cells is obtained from 8 l. of growth medium. Larger quantities of cells can be obtained in the same time period by using more ammonium sulfate as a nitrogen source; however, the adenase content is not so high.

The cells are resuspended in cold phosphate buffer (pH 7.0) to make a thin paste. This paste is then poured slowly into 1 l. of cold acetone (approximately −15°) with continuous stirring. The fluffy white precipitate is filtered and washed with 250 ml. of cold acetone. Acetone powder obtained can be stored at 0° for at least 3 years without detectable loss of adenase activity.

Purification Procedure

The following procedures were carried out at 0°. The pH of the preparations were maintained at 7.0 with NaOH, and all precipitates were dissolved in 0.01 M phosphate buffer, pH 7.0, containing 10^{-3} M glutathione and 10^{-3} M Versene.

Step 1. Crude Extract. Twenty grams of acetone powder is ground in a mortar until a very fine powder is obtained. To this is added 60 g. of sea sand (Merck), and the grinding is continued. The extraction seems to work best when the weight of the sand is three to four times the weight of the acetone powder. To this powder-sand mixture, 20 ml. of the phosphate buffer described above is added, and the grinding is continued for about 5 minutes. Additional buffer is added with continuous grinding until a total of 90 ml. is added. The solution is decanted into centrifuge tubes, and the sand is washed with 10 ml. of buffer. The combined decanted solution is centrifuged in the cold for 15 minutes at 18,000 $\times$ g. The supernatant fluid (crude extract) contains about 90% of the measurable adenase activity.

Step 2. Ammonium Sulfate Fractionation. The crude extract is brought to 40% saturation by the addition of solid ammonium sulfate. The pH is adjusted to 7.0, and the solution is allowed to stand in the cold for 20 minutes with occasional mixing. The precipitate is removed by centrifugation in the cold and discarded. Solid ammonium sulfate is added to the supernatant fluid to bring the per cent saturation to 80. The pH is adjusted to 7.0, and the solution is allowed to stand in the cold for 20 minutes before centrifugation. The precipitate is dissolved in 35 ml. of the cold phosphate buffer. Approximately 80% of the adenase in the crude extract is recovered in the precipitate.

Step 3. *Charcoal Treatment.* To 35 ml. of the ammonium sulfate fraction of step 2 is added 105 mg. of activated charcoal (Norit A). The pH is adjusted to 7.0, and the solution is allowed to stand for 20 minutes in the cold before centrifugation at $13,000 \times g$. Essentially all the adenase activity remains in the supernatant fluid. The amount of charcoal used in this step will vary depending on the amount of nucleic acid present (280/260 mμ). Usually, after step 2 the 280/260-mμ ratio is about 0.7, and a charcoal concentration of 3 mg./ml. is found to be satisfactory. If the 280/260 mμ ratio is between 0.5 and 0.65, higher amounts of charcoal should be used (5 mg./ml.).

Step 4. *Concentration of Enzyme by Ammonium Sulfate Precipitation.* Usually, several preparations of the enzyme are prepared through step 3 and then are concentrated by precipitation with ammonium sulfate. Solid ammonium sulfate is added to 70% saturation (pH 7.0), and, after 20 minutes, the precipitate collected by centrifugation is dissolved in phosphate buffer. The enzyme is dissolved in one-fifth the volume of buffer as the volume of enzyme solution used in step 3. This crude deaminase preparation appears to be highly specific for adenine. No deamination of adenosine 3'-phosphate, adenosine 5'-phosphate, adenosine, guanine, cytosine, ADP, or ATP was observed.

Step 5. *Calcium Phosphate Gel Column Chromatography.* Seven milliliters of the preparation from step 4 is placed onto a calcium phosphate gel column (3.3 cm. in diameter and 3.5 cm. high). After the enzyme solution has been adsorbed onto the gel it is eluted with 100 ml. of cold phosphate buffer. As the eluate is being collected, solid ammonium sulfate is added slowly in order to precipitate the protein. After all the eluate is collected, the final ammonium sulfate saturation should be approximately 75%. After the pH has been adjusted to 7.0, the precipitate is collected by centrifugation and dissolved in 5 ml. of phosphate buffer. The solution is then dialyzed overnight against 200 ml. of phosphate buffer.

Step 6. *Ammonium Sulfate Fractionation.* Ammonium sulfate is added to the dialyzed solution obtained from step 5 to 50% saturation. The pH is adjusted to 7.0, and the solution is allowed to stand in the cold for 20 minutes. The precipitate is removed by centrifugation and discarded, and solid ammonium sulfate is added to the supernatant fluid to obtain 70% saturation. The pH is adjusted to 7.0, and, after 20 minutes in the cold, the precipitate is collected by centrifugation, dissolved in 2 ml. of buffer, and dialyzed. Usually several batches collected in step 5 are combined for the final ammonium sulfate fractionation.

The results of the typical purification are shown in the table.

SUMMARY OF PURIFICATION STEPS

Fraction	Total units	Protein, mg.	Specific activity, units/mg. protein	Ratio, 280/260 mμ	Recovery, %
Step 1	16,600	545	31	0.39	100
Step 2	13,000	270	48	0.64	80
Step 3	12,300	210	59	0.72	74
Step 4	10,000	155	65	0.81	60
Step 5	6,700	32	210	0.80	40
Step 6	5,500	15	365	1.50	33

Properties

Stability. Various fractions of the enzyme have been kept in the frozen state without detectable loss of activity for over a year. Freezing and thawing every day for a period of a week led to a loss of approximately 10% of the activity. In the absence of substrate, over 50% of the activity is lost in 10 minutes at 35°.

Temperature Optimum. The temperature optimum for activity is approximately 30° and the energy of activation was found to be 18,500 cal.

pH Optimum. The pH optimum is fairly broad in both Tris and phosphate buffer, ranging from 6.2 to 7.5. No activity is observed above pH 9.0 or below pH 4.7.

Substrate Affinity. The substrate affinity is not so high as for the adenase isolated from *Azotobacter vinelandii*. The Michaelis-Menten constant, K_m, was found to be approximately $5 \times 10^{-5} M$.

Affect of Metals. Cobalt and manganese are inhibitors of adenase. A $10^{-3} M$ solution of $MnCl_2$ gives approximately 50% inhibition. None of a variety of metals tested stimulated activity.

Specificity. The enzyme has not been tested against a large number of nucleosides or nucleotides. However, no deamination of adenosine 3'-phosphate, adenosine 5'-phosphate, adenosine, guanine, cytosine, ADP, and ATP was observed.

Nature of the Reaction. The enzyme catalyzes the hydrolytic deamination of adenine to form ammonia and hypoxanthine. The latter has been identified spectroscopically and by microbiological assay.

Section II

Enzymes of Phosphate Metabolism

[24] Phosphoprotein Phosphatase

Protein-P + H_2O → Dephosphoprotein + P_i

By HELEN R. REVEL

Assay Method

Principle. The assay depends on measuring the formation of inorganic phosphate released from the phosphoprotein casein.

Reagents

1 M sodium acetate buffer, pH 5.8.

0.1 M ascorbic acid, freshly prepared and neutralized to pH 5.8 with NaOH.

Casein. Dissolve 10 g. of casein (Hammersten) in 84 ml. of water with the gradual addition of 16 ml. of 0.4 M NaOH to give a final pH of 7.0. Dialyze at 4° for about 20 hours against 20 vol. of 0.1 M sodium acetate buffer, pH 5.8. Assay the solution for organic phosphate and adjust to contain 20 micromoles of organic phosphate per milliliter.

Enzyme. Dilute enzyme with buffer to obtain a solution containing 0.05 to 0.5 unit/ml.

20% trichloroacetic acid.

Procedure. The following solutions are placed into a 12-ml. centrifuge tube: 0.1 ml. of acetate buffer, 0.05 ml. of ascorbic acid, 0.4 ml. of dialyzed casein, 0.35 ml. of water, and finally 0.1 ml. of enzyme. The tubes are incubated at 37° for 15 or 30 minutes. The reaction is stopped by the addition of 1 ml. of ice-cold 20% trichloroacetic acid. The tubes are centrifuged to remove denatured protein, and the supernatant solution is analyzed for inorganic phosphate.[1]

Definition of Unit and Specific Activity. The unit is defined as the amount of enzyme that catalyses the release of 1 micromole of inorganic phosphate from casein in 1 minute at 37°. Specific activity is defined as units per milligram of protein as measured by the biuret reaction.[2]

Application of the Assay Method to Crude Tissue Preparations. The presence of proteolytic enzymes, nonspecific phosphatases, and phosphorylated compounds other than casein in crude tissue preparations interferes with the assay for phosphoprotein phosphatase activity. With

[1] K. Lohmann and L. Jendrassik, *Biochem. Z.* **178**, 419 (1926).
[2] H. W. Robinson and C. G. Hogden, *J. Biol. Chem.* **135**, 727 (1940).

appropriate controls, however, a rough estimate of the enzyme activity can be made. A control tube without casein indicates the amount of inorganic phosphate released from phosphorylated compounds present in the crude extract. A second control tube in which α-glycerol phosphate replaces casein as substrate permits an estimate of nonspecific phosphatase activity.

Purification Procedure

Beef spleen is the most abundant source known for the enzyme.[3] Phosphoprotein phosphatase activity was originally observed in frog's eggs[4] and has been found in chick embryos,[5] mouse liver,[6] rat spleen, liver, and kidney,[7] and to a lesser extent in other tissues of the rat.[7]

The method described is the purification of phosphoprotein phosphatase from beef spleen by the author's modification[8] of the procedure of Sundararajan and Sarma[3,9] which results in a soluble stable enzyme preparation of high specific activity. A procedure for purification involving autolysis has been described by Singer and Fruton,[10] but the preparation has somewhat different properties.

All operations are carried out at 4° unless otherwise stated.

Step 1. Preparation of Crude Extract. Six iced beef spleens are freed of capsular material and ground in a meat grinder. Six batches of 400 g. of minced tissue are homogenized with 1 l. of a $0.5 M$ sodium chloride–$0.2 M$ acetate solution of pH 5.0 in a 2-l. Waring blendor. The homogenate is centrifuged immediately at $1000 \times g$ for 15 minutes, and the supernatant solution is filtered through several layers of gauze (4600 ml.).

Step 2. Ammonium Sulfate Fractionation. Solid ammonium sulfate (22.6 g. per 100 ml. of crude extract) is added, and after 30 minutes of stirring the mixture is centrifuged at $16,000 \times g$ for 20 minutes. The precipitate is discarded, and 26 g. of solid ammonium sulfate per 100 ml. of supernatant solution are added. After 30 minutes the mixture is centrifuged as before, and the precipitate is taken up in a minimum volume of water (about 300 ml.). Dialysis with stirring against 8 l. of distilled water (five to six changes) is continued (about 48 hours) until

[3] T. A. Sundararajan and P. S. Sarma, *Biochem. J.* **56**, 125 (1954).
[4] D. L. Harris, *J. Biol. Chem.* **165**, 541 (1946).
[5] M. Foote and C. A. Kind, *Arch. Biochem. Biophys.* **46**, 254 (1953).
[6] K. Paigen, *J. Biol. Chem.* **233**, 388 (1958).
[7] B. Norberg, *Acta Chem. Scand.* **4**, 1206 (1950).
[8] H. R. Revel and E. Racker, *Biochim. et Biophys. Acta* **43**, 465 (1960).
[9] T. A. Sundararajan and P. S. Sarma, *Biochem. J.* **65**, 261 (1957).
[10] M. F. Singer and J. S. Fruton, *J. Biol. Chem.* **229**, 111 (1957).

the dialyzate is free of sulfate ion. The precipitate formed during dialysis contains the major portion of the enzyme and is separated from the supernatant solution by centrifugation. The precipitate is extracted twice with 0.5 M sodium chloride–0.2 M acetate solution of pH 5.0, spinning after each extraction at 100,000 $\times$ g for 45 minutes in a Spinco centrifuge. The ratio of the ultraviolet extinction of the combined extracts (178 ml.) at 280 mμ to that at 260 mμ is 0.65 to 0.72, indicating the presence of some 11 to 14% nucleic acid.[11]

Step 3. Heat and Protamine Sulfate Treatment. The reddish solution is divided into 15-ml. aliquots and heated for 5 minutes in test tubes in a water bath maintained at 70°. The solutions are then quickly chilled in an ice bath and centrifuged to remove denatured protein. The clear amber supernatant solution contains over 90% of the total activity of the unheated enzyme and about 25% of the protein. To 164 ml. of the heated enzyme 15 ml. of a 2% protamine sulfate solution are added at room temperature, and after 10 minutes the mixture is centrifuged at 8000 $\times$ g for 15 minutes. From the supernatant solution, which retains all the activity, the enzyme is precipitated by the addition of 56 g. of solid ammonium sulfate per 100 ml. The precipitate is taken up in about 8 ml. of water and is dialyzed for about 20 hours against 3 l. of 0.005 M sodium chloride–0.002 M acetate solution of pH 6.0. A precipitate which forms contains no activity and is discarded. The clear yellow-brown supernatant solution (17.8 ml.) has a ratio of ultraviolet extinction at 280 mμ to that at 260 mμ of 1.35.

Step 4. Fractionation on a DEAE-Cellulose Column. The dialyzed enzyme solution is applied to DEAE-cellulose anion exchanger (5 $\times$ 3.5

SUMMARY OF PURIFICATION PROCEDURE[a]

Step	Volume, ml.	Protein, mg.	Activity, units	Specific activity, units/mg. protein
1	4600	165,000	3550	0.022
2	178	2,300	1000	0.44
3	17.8	178	600	3.4
4	33	96	600	6.3

[a] H. R. Revel and E. Racker, *Biochim. et Biophys. Acta* **43**, 465 (1960).

cm.) which has been equilibrated with 0.005 M sodium chloride–0.002 M acetate solution of pH 6.0. Fractions of 10 ml. are collected. The enzyme does not adhere to the cellulose and is eluted in tubes 4, 5 ,and 6 on

[11] O. Warburg and W. Christian, *Biochem. Z.* **310**, 384 (1941).

washing with the equilibrating buffer solution. The ratio of ultraviolet extinction at 280 mμ to that at 260 mμ of the purified enzyme at this stage is 1.45.

A summary of the purification procedure is given in the table.

Properties

Specificity. In addition to its action on the phosphoproteins such as casein and phosvitin, the purified beef spleen enzyme catalyzes the hydrolysis of pyrophosphates (ATP, ADP, inorganic pyrophosphate, etc.), aromatic monophosphate esters (phenyl phosphate, *p*-nitrophenyl phosphate, etc.), and certain other phosphate compounds such as phosphorylenol pyruvate, acetyl phosphate, and phosphoamide. The purified enzyme has little activity toward ordinary phosphomonoesters. For example, sugar phosphates, glycerol phosphate, serine phosphate, and small serine phosphate peptides are poor substrates.

Activators. Enzyme activity is considerably enhanced by the presence of reducing agents. Ascorbic acid and thioglycolic acid at the optimal concentration of $5 \times 10^{-3}\, M$ activate the enzyme about tenfold. At a concentration of $10^{-2}\, M$, thioglycerol, cysteine, or 2-mercaptoethanol increases enzyme activity about sixfold. Glutathione is not an effective activator.

Effect of Metals. Ferrous iron completely replaces the reducing agents and gives maximal stimulation at the optimal concentration of $10^{-3}\, M$. Higher concentrations of ferrous iron inhibit the enzyme. Ca^{++}, Mg^{++}, Mn^{++}, Cu^{++}, Co^{++}, Ni^{++}, Zn^{++}, and Fe^{+++}, tested in the range of 10^{-4} to $10^{-2}\, M$, did not stimulate phosphoprotein phosphatase activity and caused about 50% inhibition at $10^{-3}\, M$.

Inhibitors. Molybdate is the most effective agent, giving 50% inhibition at $4 \times 10^{-6}\, M$ and complete inhibition at $10^{-4}\, M$. Some metal chelators are potent inhibitors; 1,10-*o*-phenanthroline and α,α-dipyridyl inhibit the enzyme 50% at $10^{-5}\, M$ and $4 \times 10^{-5}\, M$, respectively. In contrast to molybdate, however, neither chelating agent inhibits the enzyme completely. Less effective inhibitors include *p*-chloromercuribenzoate (40% at $10^{-3}\, M$), fluoride (60% at $4 \times 10^{-3}\, M$), iodoacetate (30% at $10^{-2}\, M$), and phosphate (40% at $10^{-2}\, M$). Azide, cyanide, and ethylenediaminetetraacetate are without effect at $10^{-2}\, M$.

Stability. The purified enzyme is stable at $-18°$ for several months. Enzyme activity is preserved after exposure to temperatures up to 70° for prolonged periods of time, but the enzyme is rapidly inactivated at 37° in the presence of ascorbic acid or thioglycolic acid. Ferrous ions as well as substrate protect the enzyme against this inactivation.

K_m. The K_m for casein is about $6 \times 10^{-4}\, M$ for the enzyme preparation described.

[25] O-Phosphoserine Phosphatase

By Michael Schramm

Assay Method

Principle. P_i (inorganic phosphate) liberated from phosphoserine by enzymatic hydrolysis is measured colorimetrically.

Reagents

    0.02 M DL-O-phosphoserine, pH 7.4.
    0.02 M $MgCl_2$.
    0.30 M Tris buffer, pH 7.4.
    Enzyme. If the activity is too high, the enzyme may be diluted in
        0.02 M Tris buffer, pH 7.4.
    15% trichloroacetic acid.
    Reagents for determination of P_i.[1]

Procedure. Portions 0.5 ml. each of the phosphoserine, $MgCl_2$, and Tris buffer reagents are mixed in a test tube and placed in a 30° water bath. The reaction is started by addition of 0.5 ml. of enzyme solution. After incubation for 20 minutes, the reaction is stopped by addition of 1 ml. of trichloroacetic acid reagent. P_i is determined[1] after removal of denatured protein by centrifugation.

A control mixture, to which trichloroacetic acid is added before the enzyme, serves to correct for P_i present in the reagents. With crude extracts it is also necessary to set up an incubation mixture without phosphoserine to determine the amount of P_i liberated from substances present in the extract.

Definition of Unit and Specific Activity. One unit is defined as the amount of enzyme that catalyzes the formation of 1 micromole of P_i per minute. Specific activity is expressed as units per milligram of protein.

Application of Assay Method to Crude Extracts. Phosphatase activity toward phosphoserine in crude extracts may sometimes be due to nonspecific phosphatases. Since L-serine partially inhibits phosphoserine phosphatases,[2-4] inhibition by L-serine would indicate the presence of a specific phosphoserine phosphatase.

[1] C. H. Fiske and Y. SubbaRow, *J. Biol. Chem.* **66**, 375 (1925).
[2] M. Schramm, *J. Biol. Chem.* **233**, 1169 (1958).
[3] F. C. Neuhaus and W. L. Byrne, *J. Biol. Chem.* **234**, 113 (1959).
[4] L. F. Borkenhagen and E. P. Kennedy, *J. Biol. Chem.* **234**, 849 (1959).

Purification Procedure

In addition to the procedure given below, employing baker's yeast as the enzyme source,[2] procedures for isolation of phosphoserine phosphatase from chicken liver[3] and rat liver[4] have been described. The enzyme from baker's yeast is of higher specific activity, is less sensitive to inhibition by L-serine, and may therefore be used for the determination of phosphoserine.

Step 1. Preparation of Crude Yeast Extract. Three hundred grams of dried baker's yeast are suspended in 900 ml. of $0.1 M$ sodium bicarbonate and incubated at 41° for 2.5 hours with occasional shaking. The suspension is centrifuged, and the supernatant solution is cooled to 2°. The precipitate is re-extracted with 600 ml. of bicarbonate for 2.5 hours at 41°. After centrifugation the precipitate is discarded, and the supernatant solutions of the first and second extractions are combined (1030 ml.). All subsequent operations are carried out at 0° to 2°. Fifty milliliters of a 3% solution of protamine sulfate, adjusted to pH 6.5, are added with stirring to the crude yeast extract. After standing for 5 minutes the solution is centrifuged, and the small amount of precipitate is discarded.

Step 2. Fractionation with Ammonium Sulfate. Throughout the purification procedure, fractionation with ammonium sulfate is carried out at a pH of 6 to 7. The pH is maintained by addition of $2 M$ ammonia. For each 100 ml. of the yeast extract, 41.7 g. of ammonium sulfate are added. The precipitate is removed by centrifugation at $10,000 \times g$ for 10 minutes. An additional 15.2 g. of ammonium sulfate are added per 100 ml. of the supernatant solution, and stirring is continued for 20 minutes after the ammonium sulfate has dissolved. The precipitate obtained after centrifugation at $16,000 \times g$ for 30 minutes is taken up in a minimal amount of water and dialyzed overnight against 4 l. of $0.002 M$ Tris buffer, pH 7.7, containing $0.001 M$ $MgCl_2$.

Step 3. Removal of Inactive Protein by Alumina C_γ. Alumina C_γ (607 mg. in 4.4 ml.)[5] is added to the dialyzed protein solution (46 ml., 24.4 mg. of protein per milliliter). The suspension is centrifuged at $6000 \times g$ for 5 minutes, and the precipitate is discarded.

Step 4. Fractionation on DEAE-Cellulose Column. The supernatant solution (50 ml., 1000 mg. of protein) is applied to a DEAE-cellulose column (5×7 cm.²) which has been washed with 150 ml. of $0.005 M$ Tris buffer, pH 7.7, containing $0.001 M$ $MgCl_2$. The flow rate through the column averages 3 ml./min. Protein is eluted from the column with

[5] The exact amount of alumina C_γ required to remove the maximal amount of inactive protein without adsorbing the enzyme is determined on small samples.

solutions containing 0.05 M Tris, pH 7.7, 0.001 M $MgCl_2$, and the following different concentrations of KCl: solution I, no KCl; II, 0.04 M KCl; III, 0.1 M KCl; IV, 0.2 M KCl; and V, 0.4 M KCl. Forty-milliliter portions of each solution are successively applied to the column. The collection of eluate, in fractions of 10 ml., is started when solution III is applied to the column, since the enzyme is eluted by solutions III and IV. Enzyme of high specific activity is recovered in five to six fractions which contain 70 to 80% of the total activity. These fractions are pooled (50 ml.), and the protein is precipitated by addition of 30.5 g. of ammonium sulfate (61 g. per 100 ml. of the protein solution). The precipitate obtained after centrifugation at 12,000 $\times$ g for 25 minutes is drained thoroughly and taken up in 3 to 4 ml. of 0.02 M Tris buffer,

PURIFICATION OF PHOSPHOSERINE PHOSPHATASE

Fraction	Protein concentration, mg./ml.	Total units	Specific activity, units/mg. protein	Recovery, %
Crude yeast extract	8.4	520	0.06	100
Ammonium sulfate precipitate (68–90%)	24.4	208	0.18	40
Alumina C$_\gamma$	20.0	198	0.20	38
Cellulose column and precipitation by ammonium sulfate	11.7	108	1.68	21

pH 7.4, which contains 0.001 M $MgCl_2$. When stored at $-20°$ this enzyme preparation showed no loss of activity for at least a month. A summary of the purification and recovery of enzyme is given in the table.

Properties

Specificity. Both L- and D-phosphoserine are hydrolyzed by the enzyme. The following compounds do not yield P_i on incubation with the purified enzyme: sugar phosphates, adenosine tri-, di-, and monophosphate, phosphothreonine, phosphoethanolamine, phosphohydroxypyruvate, phosphovitin, and casein. However, the purified preparation still contains some pyrophosphatase.

Activators and Inhibitors. The enzyme is inactive in absence of a divalent cation. Although Mg^{++} is required for maximal activity, Co^{++}, Fe^{++}, and Ni^{++} could also activate the enzyme but not so efficiently. Sodium fluoride, calcium chloride, and manganese chloride are effective inhibitors.

Other Properties. The optimum pH is from 6.5 to 7.2. At pH 7.4, the

K_m for L-phosphoserine is 4×10^{-4}, and for D-phosphoserine it is 8×10^{-3}. The enzyme from liver shows strong transfer activity toward L-serine.[3, 4, 6]

[6] F. C. Neuhaus and W. L. Byrne, *J. Biol. Chem.* **234**, 109 (1959).

[26] Protein Phosphokinase from Brewer's Yeast

Protein + ATP $\rightleftarrows$ Protein-phosphate + ADP

By MURRAY RABINOWITZ

Transfer of the terminal phosphate from ATP to proteins was first demonstrated by Burnett and Kennedy[1] with soluble enzyme preparations from rat liver mitochondria. Subsequently, more highly purified enzymes have been obtained from rabbit mammary gland,[2] calf brain,[3, 4] and brewer's yeast.[3, 4] The procedure from brewer's yeast described here has proved superior to that previously reported[4] with regard to reproducibility and yield.

Assay Method

Principle. The reaction is followed by measuring the rate of transfer of P^{32} from terminally labeled ATP^{32} to a suitable protein such as phosvitin[5] or casein. Although phosphorylation occurs on the serine hydroxyl group in these proteins, other serine-containing proteins will not accept phosphate. Phosvitin, in which nearly every second amino acid is phosphoserine, must be partially dephosphorylated prior to use. Beef spleen protein phosphatase[6] has been used for this purpose; dephosphorylation with alkali, however, results in an inactive substrate. When the reaction is run at pH 7.5, the rapid exchange between ATP and protein phosphate, seen at a lower pH, is reduced so that a good agreement between P^{32} incorporation and phosphorylation of phosvitin is obtained (see Properties).

Preparation of Reagents. ATP^{32} is prepared by pyrophosphate exchange through the use of tryptophan-activating enzyme (Vol. V [97])

[1] G. Burnett and E. P. Kennedy, *J. Biol. Chem.* **211**, 969 (1954).
[2] T. A. Sundararajan, K. S. V. S. Kumar, and P. S. Sarma, *Biokhimiya* **22**, 135 (1957).
[3] M. Rabinowitz, *Federation Proc.* **16**, 236 (1957).
[4] M. Rabinowitz and F. Lipmann, *J. Biol. Chem.* **235**, 1043 (1960).
[5] D. K. Mecham and H. S. Olcott, *J. Am. Chem. Soc.* **71**, 3670 (1949).
[6] T. A. Sundararajan and P. S. Sarma, *Biochem. J.* **56**, 125 (1954).

P^{32}-Pyrophosphate was prepared according to Jones et al.[7]. The ATP^{32}, which contained small amounts of ADP, was isolated by charcoal adsorption and eluted with pyridine.[8] ATP^{32} prepared by a variety of other methods can also be used.

Twenty to forty per cent dephosphorylated phosvitin is prepared with beef spleen protein phosphatase (Vol. VI [24]). The substrate is isolated by precipitation with 0.03 M MnCl$_2$. The precipitate is dissolved in 1 × 10^{-2} M Versene and dialyzed extensively against 1 × 10^{-3} M Versene, pH 7.5, and against distilled water.

Casein may replace phosvitin in the assay but must be used in amounts five times as great as phosvitin.

Procedure. The 1-ml. assay mixture contains 2 mg. of 20 to 40% dephosphorylated phosvitin, 50 micromoles of Tris–chloride (pH 7.5), 5 micromoles of Mg^{++}, and 1 micromole of ATP^{32} having 10,000 to 20,000 c.p.m./micromole of acid-labile phosphate. After incubation for 15 minutes at 37°, the phosvitin is precipitated by the addition of TCA to a final concentration of 10%. It is washed three times with 10% TCA, and after drying on an aluminum planchette it is counted in a thin-window Geiger-Müller counter. The enzyme concentrations are adjusted to give less than 20% P^{32} incorporation into phosvitin during assay.

Definition of Unit and Specific Activity. A unit of enzyme is defined as that amount of enzyme which causes the transfer of 1 micromole of phosphate from ATP into phosvitin at 37° in an hour. Specific activity is expressed as units per milligram of protein. Protein may be determined by the TCA turbidity method[9] during early stages of purification (through step 3) and from the optical density at 280 mμ and 260 mμ[10] in later stages.

Application of Assay Method to Crude Tissue Preparation. The rapid breakdown of ATP and the presence of competing and diluting reactions in crude tissue prevent accurate quantitative assay of this reaction in crude tissue extract. In preparations with high ATPase activity the addition of 0.025 to 0.050 M fluoride is helpful.

Purification Procedure

The following procedure has given reproducible results in this laboratory. The addition of ATP and Mg^{++} stabilizes the enzyme during the later purification steps.

[7] M. E. Jones, F. Lipmann, H. Hilz, and F. Lynen, J. Am. Chem. Soc. 75, 3285 (1953).
[8] R. K. Crane and F. Lipmann, J. Biol. Chem. 201, 235 (1953).
[9] E. R. Stadtman, G. D. Novelli, and F. Lipmann, J. Biol. Chem. 191, 365 (1951).
[10] O. Warburg and W. Christian, Biochem. Z. 310, 384 (1942); see also Vol. III [73].

Step 1. Autolysis of Dried Brewer's Yeast. Five kilograms of dried brewer's yeast[11] is extracted by stirring overnight for 14 to 16 hours at 37° with 4 vol. of 0.1 N NaHCO$_3$. One milliliter of toluene is added per 250 ml. of extract. After the mixture has been cooled to below 10° by stirring in the cold room, the clear yellow extract is separated by centrifugation. (The VRA head of the Lourdes centrifuge at 12,000 r.p.m. was used in all separations. Other centrifuges providing similar centrifugal force may be substituted.)

Step 2. Ammonium Sulfate Precipitation. Solid ammonium sulfate (29.1 g. per 100 ml. of extract, 50% saturation) is added, the sodium Versenate concentration having been adjusted to $2 \times 10^{-3} M$, pH 7.5. The precipitate is collected by centrifugation for 20 minutes, dissolved in a minimal volume of water, and twice dialyzed for 24 hours against 20 vol. of $5 \times 10^{-3} M$ phosphate and $2 \times 10^{-3} M$ Versene, pH 7.5.

Step 3. Streptomycin and Protamine Precipitations. The protein concentration of the dialyzate is adjusted to 10 mg./ml. with $5 \times 10^{-3} M$ phosphate, pH 7.5, and 0.2 ml. of 5% sodium nucleate[12] is added per 10 ml. of extract. After adjustment of the pH to 5.8 with 1 N acetic acid, by means of a Beckman pH meter standardized with cold buffer, 1 ml. of 5% streptomycin sulfate is added slowly per 10 ml. of extract. The separated precipitate is dissolved in 0.03 M phosphate, pH 7.5, and the protein concentration is adjusted to about 8.0 mg./ml. Then 1.4 ml. of 2% protamine sulfate[13] is added slowly per 100 mg. of protein, and the heavy precipitate is discarded.

Step 4. Concentration of Enzyme by Alumina Adsorption and Ammonium Sulfate Precipitation. The dilute enzyme is adsorbed on 2 mg. of C$_\gamma$ alumina gel (Vol. I [11]) per milligram of enzyme protein. Elution from the gel is performed five times with 0.3 M phosphate, pH 7.5, the volume of the eluates being $\frac{1}{10}$, $\frac{1}{20}$, $\frac{1}{20}$, $\frac{1}{30}$, and $\frac{1}{40}$ the original volume adsorbed. Further concentration of enzyme is achieved by precipitation with ammonium sulfate added to 57.5 saturation in the presence of $2 \times 10^{-3} M$ Versene. After adjustment of the protein concentration to 12 to 14 mg./ml., 0.8 ml. of 2% protamine is added per 100 mg. of protein. MgCl$_2$ and ATP are also added to a final concentration of $1 \times 10^{-3} M$. The mixture is dialyzed for 8 hours against 25 vol. of $4 \times 10^{-2} M$ Tris–chloride (pH 7.5), $5 \times 10^{-4} M$ ATP, and $5 \times 10^{-4} M$ MgCl$_2$. The large inactive precipitate which forms after dialysis is discarded.

Step 5. Selective Acetone Denaturation. Ten micromoles each of ATP and MgCl$_2$ are added per milliliter, and the mixture is incubated at 37°

---

[11] Anheuser-Busch Inc., St. Louis, Missouri ("Special for enzyme work").
[12] Schwarz Laboratories Inc., Mount Vernon, New York.
[13] Nutritional Biochemicals Inc., Cleveland, Ohio.

SUMMARY OF PURIFICATION PROCEDURE

Step	Volume, ml.	Protein, mg.	Units	Specific activity, units/mg. protein	Yield	Purification
1. Autolyzate	27,580	967,000	145,000	0.15	100	
2. Ammonium sulfate (0–0.5 saturation)	10,200	127,500	129,500	1.02	89.0	6.7
3a. Streptomycin precipitation	13,700	104,000	121,000	1.16	83.3	7.5
3b. Protamine supernatant	14,200	9,950	107,500	10.8	74.0	72
4a. C_γ alumina eluate	4,000	7,000	79,100	11.3	54.5	75
4b. Ammonium sulfate-dialyzed (0–0.575 saturation)	465	4,050	58,000	14.3	40.0	95
5. Acetone denaturation	24.5	512	40,400	79.3	27.8	526
6. Carboxymethylcellulose chromatography	210	40	14,000	220–500 (mean = 350)	9.7	1470–3300

for 10 minutes. After cooling to 0°, an equal volume of cold acetone is added slowly, the temperature being kept near the freezing point in a salt-ice bath. At 50% acetone concentration the temperature is maintained at $-10°$.

The temperature is then rapidly elevated to 18° for 10 minutes and immediately cooled again to $-10°$. The precipitate is separated by centrifugation at $-10°$ and dispersed in a half volume of 0.1 M Tris–chloride, pH 7.5, containing 1 micromole of ATP and 1 micromole of Mg^{++} per milliliter. The mixture is then frozen at $-20°$ for 12 to 16 hours, and the heavy inactive precipitate is discarded. Acetone precipitation and denaturation were usually performed on 100-ml. aliquots of enzyme because of difficulty in accurate and rapid temperature adjustment when larger volumes were used.

Step 6. Carboxymethylcellulose Chromatography. The enzyme is concentrated to 25 to 50 mg./ml. by precipitation with 65% saturated ammonium sulfate. After dialysis for 12 hours against $6 \times 10^{-3} M$ phosphate (pH 7.2) and $5 \times 10^{-4} M$ ATP and Mg^{++}, it is adsorbed on a carboxymethylcellulose column (Vol. V [1]), 35×1.5 cm., which has been equilibrated with $6 \times 10^{-3} M$ phosphate, pH 7.2. Gradient elution is performed with a 250-ml. mixing chamber containing $6 \times 10^{-3} M$ phosphate, pH 7.2, and an upper reservoir containing $5 \times 10^{-2} M$ phosphate, also pH 7.2. Five-milliliter samples are collected. The enzyme, purified 1500- to 3500-fold, appears in two peaks, the first in about the fortieth fraction and the second in about the seventieth fraction. The enzyme may be concentrated by adsorption on a hydroxylapatite column (Vol. V [2]), equilibrated with 0.03 M phosphate, pH 7.2, and eluted with a small volume of 0.2 M phosphate, pH 7.2. A summary of a typical fractionation appears in the table on p. 221.

Properties

Specificity. Only phosvitin and casein, and phosphopeptides, containing twelve to twenty-four amino acids, derived from phosvitin or casein, were phosphate acceptors from ATP. Hexokinase, ribonuclease, trypsin, pepsin, beef serum albumin, and muscle phosphorylase b were inactive. Furthermore, phosphorylase b phosphokinase did not significantly catalyze a transfer from ATP to dephosvitin. Free serine is not phosphorylated by this system.

Activators and Inhibitors. As with other phosphokinases, this reaction is Mg^{++}-dependent. Other divalent ions such as Mn^{++}, Co^{++}, Fe^{++}, Zn^{++}, and Cu^{++} inhibited the reaction in concentrations of 1 to $2 \times 10^{-3} M$. Enzyme activity was unaffected by sulfhydryl reagents. In mammary gland phosphokinase, however, inhibition by sulfhydryl reagents has been reported.[2] Inorganic phosphate strongly inhibits the yeast protein phosphokinase reaction, half inhibition being obtained at $2.5 \times 10^{-2} M$ phosphate. In contrast, protein phosphokinase from calf brain is stimulated by inorganic phosphate.[4]

Equilibrium and Kinetics. The protein phosphokinase reaction is readily reversible, and enzymatic phosphate transfer from phosvitin to ADP can be followed analytically. The phosphate sites of rephosphorylated phosvitin show inhomogenous reactivity and therefore prevent accurate equilibrium determinations. However at pH 6.2, figures of the order of 20 to 50 were obtained for the equilibrium constant of the forward reaction. Although of a preliminary nature, they indicate the thermodynamic potential of phosphoryl in phosvitin to be considerably above that expected for a simple serine phosphate ester. In the forward reaction, i.e., ATP + dephosvitin, a broad pH optimum is present between 7 and 8. A fairly sharp pH optimum for transfer from phosvitin to ADP appears on the more acidic side at pH 6.5, the reaction becoming quite slow between pH 7.5 and 8.0.

[27] Myokinase, ATP-AMP Transphosphorylase

$$2\text{ADP} \rightleftarrows \text{ATP} + \text{AMP}$$

By LAFAYETTE NODA and STEPHEN A. KUBY

Determination of Enzymatic Activity

The activity of the enzyme has been determined by coupling the myokinase-catalyzed reaction with either the ATP–Cr transphosphorylase[1] or the adenylic deaminase-catalyzed reaction.[1,2] Two other methods employed by the authors are (1) separation of the adenine nucleotides on micro columns of Dowex 1[3] and (2) a rapid and sensitive pH-stat method employing crystalline hexokinase.[4] The procedure to be described here is basically the one employed during the isolation studies.

Dilutions of the enzyme are made (generally two successive dilutions, employing micropipets of 5- to 25-μl. capacity) in a solution containing freshly neutralized $0.01 M$ cysteine, $0.01 M$ recrystallized Tris (or

[1] L. Noda and S. A. Kuby, *J. Biol. Chem.* **226**, 541 (1957).
[2] H. M. Kalckar, *J. Biol. Chem.* **167**, 466 (1947).
[3] L. Noda, *J. Biol. Chem.* **232**, 237 (1958).
[4] T. A. Mahowald, E. A. Noltmann, and S. A. Kuby, *J. Biol. Chem.* **237**, 1535 (1962).

preferably 0.01 M glycylglycine), pH 8, and 0.80 mg. of ATP–Cr transphosphorylase per milliliter. (The ATP–Cr transphosphorylase and cysteine aid in stabilizing the highly diluted myokinase.)

For routine work a stock reaction mixture, adjusted to pH 8.0, is prepared, consisting of 0.002 M $MgSO_4$, 0.080 M creatine, and 0.080 M glycylglycine (or recrystallized Tris). In 10.0-ml. graduated Pyrex test tubes, the following aliquots are pipetted: 1.00 ml. of stock reaction mixture, 0.50 ml. of 0.008 M ADP (neutralized sodium salt), 0.20 ml. of 0.10 M freshly neutralized cysteine, and 0.20 ml. of ATP–Cr transphosphorylase (0.5 mg./ml.); the solutions are mixed and temperature-equilibrated at 30° for 10 minutes, during which any trace of ATP contaminant in the ADP is dephosphorylated and measured as Cr ~ P at zero time. The reaction is initiated by the addition of 0.10 ml. of properly diluted myokinase, and at suitable times (e.g., 0, 5, 10, 15 minutes) the reaction is stopped by the rapid addition of 1 ml. of 5% (w/v) ammonium molybdate·$4H_2O$ and 6 ml. of 8% (v/v) perchloric acid. After a 30-minute Cr ~ P hydrolysis period at room temperature, 0.40 ml. of reducing agent[1,5] is added, the sample is brought to 10.0 ml. with distilled water, and, after a 10-minute color development period, the absorbance at 660 mμ is determined. The samples and phosphorus standards are suitably spaced in time so that each tube (including the zero times) receives a 30-minute period of hydrolysis and a 10-minute color development. Thus the Cr ~ P formed in the zero-time and sample tubes is determined as inorganic phosphorus after hydrolysis by the acid molybdate.[5] The total protein added has been adjusted to the limit of its solubility in the 10.0-ml. final volume of acid molybdate, provided the molybdate is added before the perchloric acid. The initial velocity (measured in terms of Cr ~ P formed) is directly proportional to the myokinase concentration, if the extent of reaction does not exceed about 0.2 micromole/ml. of reaction mixture.

One unit of myokinase is defined as that amount of enzyme which catalyzes the transphosphorylytic reaction between 2 moles of ADP at an initial rate of 1 micromole of ATP formed (measured as Cr ~ P) per minute per milliliter of reaction mixture under the standard conditions stated above (30°, pH 8.0, 0.04 M glycylglycine, 0.001 M $MgSO_4$, 0.040 M Cr, 0.001 M cysteine, 0.002 M ADP, and 90 μg. of ATP–Cr transphosphorylase per milliliter). The specific activity is defined as units per milligram of protein; protein concentration is determined by the biuret method.[6]

[5] S. A. Kuby, L. Noda, and H. A. Lardy, *J. Biol. Chem.* **209**, 191 (1954).
[6] A. G. Gornall, C. J. Bardawill, and M. M. David, *J. Biol. Chem.* **177**, 751 (1949).

Isolation Procedure

The following procedure is essentially that previously reported,[1] except for the improved crystallization step.[4] All steps are conducted in a cold room (2° to 4°) or in an ice bath; protein-precipitating reagents (or pH adjustments) are added with good mechanical stirring, but foam is avoided; to facilitate centrifugation (or filtration) the suspensions are allowed to stand for 10 to 20 minutes after thorough mixing. All reagents employed are of analytical grade, and the filter aid (Celite 503, Johns-Manville), if employed, should be washed with 2 N HCl at 80°, exhaustively washed with distilled water, and oven-dried. A Beckman Model G pH meter with shielded extensible electrodes has been used for pH determinations.

Fraction I. Preparation of Crude Extract. Back and leg muscles from at least 8 to 10 adult white male rabbits are rapidly excised and chilled in ice. In the cold room, the muscle is ground and homogenized in Waring blendors (1 minute per batch) with a total of 3 l. of cold 0.01 M KCl per kilogram of ground muscle. After gentle stirring of the homogenate for 15 minutes, the suspension is centrifuged in large-capacity *refrigerated* centrifuges for 40 minutes at roughly 1500 × g (e.g., International 13 L and SR-3 centrifuges), and the supernatant liquid is decanted through bandage cloth to remove any floating debris. (An additional 5% in liquid volume may be recovered by further squeezing the sedimented residue through bandage cloth; however, if the initial volumes worked up exceed 20 to 30 l., this relatively small additional recovery in volume may be sacrificed.)

Fraction II. Acid Denaturation of Inert Protein. To fraction I, with good mechanical stirring, 2 N HCl is added to decrease *rapidly* the pH from about 6.1 to 2.0. After 5 to 10 minutes of stirring the pH is *slowly* increased to 7.0 by the addition of 2 N NaOH, resulting in a heavy precipitate which is stirred for an additional 5 to 10 minutes and allowed to stand for 20 minutes. The denatured protein is removed by centrifugation, as above. (Alternatively, filtration with 30 g. of Celite 503 per liter and two thicknesses of Whatman No. 1 may be employed.)

Fraction III. Precipitation with Zinc Acetate. Zinc acetate (1.0 M) is added, with constant mechanical stirring, to a final concentration of 0.0185 M, while the pH is maintained at 7.0 with 2 N NaOH, even during the addition of the zinc acetate. The suspension is centrifuged, as before, for ½ hour, and the precipitate is washed by suspending in 0.05 M acetate (Na⁺), pH 5.0 (⅕ vol. of fraction I), and recentrifuging. The washed precipitate is then resuspended by thorough mixing in 0.05 M acetate, pH 5.0 (ice bath and magnetic stirrer are convenient), with ⅛ vol. of fraction I, and dissolved by decreasing the pH to 3.5 with $2N$ HCl. The volume is measured.

Fraction IV. Ammonium Citrate Extraction. The pH of fraction III is readjusted to 4.65 with $2N$ NaOH; the resulting precipitate is collected by centrifugation and carefully resuspended in $0.25 M$ citrate (NH_4^+), pH 8, with $\frac{1}{4}$ vol. of fraction III. The thoroughly mixed suspension is stirred for $\frac{1}{2}$ hour, during which the pH is maintained at 7.6 by the addition of $5 N$ NH_4OH; the suspension is then gently stirred in an ice bath overnight. After centrifugation, the residue is re-extracted with $0.25 M$ ammonium citrate, pH 8.0; with 0.15 vol. of fraction III. After the residue is removed by centrifugation, the two extracts are combined.

Fraction V. Ammonium Sulfate Fractionation. The fraction precipitating between 0.58 and 0.88 saturation ammonium sulfate[7] is recovered by centrifugation (the 1.5-l. capacity rotors, GSA or VRA, of the refrigerated Servall or Lourdes centrifuges, respectively, are convenient) at $15,000 \times g$ for $\frac{3}{4}$ hour. (Alternatively, but less satisfactorily, filtration with 1 g. of Celite 503 per 100 ml. of solution may be used to collect the protein fraction; the final filter cake is extracted with the requisite $0.02 M$ ammonium citrate divided into three portions.) The precipitate is dissolved in $0.02 M$ ammonium citrate, pH 7.6, with $\frac{1}{5}$ vol. of fraction IV, and flow dialyzed for 5 hours against 12 l. of $0.02 M$ ammonium citrate, pH 7.6.

Fraction VI. Precipitation with Silver Nitrate. After determination of the protein concentration (the dialyzed solution is clarified, if necessary), fraction V is diluted to 5.0 mg./ml. with $0.02 M$ ammonium citrate, pH 7.6, and adjusted to 0.60 saturation with solid ammonium sulfate. The small precipitate is removed by centrifugation (or filtered, with 1 g. of Celite per 100 ml. of solution). To the clear supernatant liquid, $0.10 M$ $AgNO_3$ solution (freshly prepared) is added to a final concentration of $5 \times 10^{-4} M$, and the pH is adjusted to 6.0 with $1 N$ H_2SO_4. After thorough mixing, the suspension is allowed to stand overnight (shielded from light) in an ice bath. The precipitate is collected by centrifugation at $15,000 \times g$ for 20 minutes and suspended in a freshly prepared solution of $0.2 M$ cysteine–$0.05 M$ sodium cyanide–$0.4 M$ ammonium sulfate at pH 7.6, with a volume equal to one-tenth the volume of fraction V when diluted to 5 mg./ml. After flow dialysis overnight

---

[7] Ammonium sulfate to add for a given saturation at 0° is calculated by the formula $w = 0.515 V_1 (S_2 - S_1)/1 - S_2 (0.272)$, where w equals the weight of salts in grams, V_1 is the volume of solution in milliliters at saturation S_1, and S_2 represents the saturation desired.

against 12 l. of 0.005 M glycylglycine, pH 7.6, a trace of insoluble material is removed by centrifugation, and the protein concentration is determined.

Fraction VII. Adsorption on IRC 50 (XE 64). The resin (Rohm and Haas Co., Philadelphia, Pennsylvania) is washed and cycled according to Hirs[8] and stored as a moist solid (dry weight is determined). During the dialysis of the above step, the resin (as a suspension of 40 mg./ml. dry weight) is equilibrated with 1 N NaOH to pH 7.6 at 0° in the presence of 0.005 M glycylglycine. An amount of resin equal in weight to three and one-half times the total protein of fraction V is suspended in one-half the required volume of distilled water containing the calculated amount of freshly dissolved glycylglycine. Cautious equilibration to pH 7.6 with 1 N NaOH at ice-bath temperature is carried out with stirring during a period of 1 to 2 hours until the final pH (measured intermittently without stirring) is essentially constant for 10 minutes. The resin suspension is then brought to the required volume with distilled water and stored overnight for fraction VII.

The final conditions for adsorption of the enzyme are: 0°, 1 mg. of protein per milliliter, 0.005 M glycylglycine, pH 7.6, and equilibrated resin equal to 9 mg./ml. Thus, to the dialyzed protein solution, the required volume of 0.005 M glycylglycine, pH 7.6, is added, followed by the calculated amount of resin suspension (mixed thoroughly before taking aliquot) such that the total additive volumes result in the above stated conditions. The pH is maintained at 7.6 by the cautious addition of 1 N NaOH, with good stirring, over a 20-minute period (pH measurements are made intermittently without stirring). The resin is centrifuged off and transferred quantitatively to a small Büchner funnel containing two thicknesses of Whatman No. 1 filter paper and washed with about one-tenth the previous volume of 0.005 M glycylglycine, pH 7.6. (Filtration is conducted in the cold room.) The resin should not be allowed to dry, but suction may be applied until the pad just "cracks." The resin and one thickness of filter paper are suspended in a small beaker with 0.4 M ammonium citrate, pH 9.9, equal in milliliters to one-seventh the total milligrams of protein at the adsorption stage, and stirred intermittently for about 5 to 10 minutes in the cold room. The extract of enzyme is recovered from the resin with the same funnel (with a fresh upper layer of filter paper); suction is cautiously applied so as to avoid the formation of foam. The resin is then re-extracted three more times in the same manner with 0.4 M ammonium citrate, pH 8, the volume each time being equal to one-third the volume of the pH 9.9 ammonium

[8] C. H. W. Hirs, S. Moore, and W. H. Stein, *J. Biol. Chem.* **200**, 493 (1953).

citrate used. At the final filtration, the pad is pressed down tightly with the aid of a porcelain spoon, to collect all available liquid. The eluates are then combined.

Fraction VIII. Ammonium Sulfate Precipitation. To the combined eluates, solid $(NH_4)_2SO_4$ is added to 0.88 saturation, and the pH is adjusted to 6.1 with $1 N$ H_2SO_4. (A further addition of solid $(NH_4)_2SO_4$ is added to compensate for the dilution by the H_2SO_4.) The resulting precipitate is collected by centrifugation at about $15,000 \times g$ for $1\frac{1}{2}$ hours and dissolved in cold distilled water (approximately one-tenth the volume of fraction VII; the protein concentration should be at least 15 to 20 mg./ml.).

Fraction IX. Crystallization and Recrystallization. The enzyme solution is dialyzed against 2 l. of $2 \times 10^{-4} M$ Versene, pH 6.4, at 5°, adjusted to 0.55 saturation ammonium sulfate, for approximately 24 hours in the cold room. The small precipitate which forms during dialysis (about 5% of the total protein) contains mostly denatured myokinase and a very small amount of crystalline myokinase. (It is usually retained, combined with similar crops from a number of preparations, and after several successive crystallizations according to the procedure below may approach relatively high specific activities.)

The protein concentration of the dialyzed clarified solution is determined. If not at least 15 mg./ml., the protein is once again concentrated by bringing to 0.88 ammonium sulfate saturation and redissolving the precipitate in 0.01 M succinate (Na⁺), pH 6.4 at 5°, and solid $(NH_4)_2SO_4$ is cautiously added to 0.55 saturation. Over a period of several hours, small portions of solid $(NH_4)_2SO_4$ are added (generally to about 0.62 saturation). Initially, only a very slight opalescence may be observed, and after several hours there is a noticeable increase in viscosity. At this point the enzyme solution is allowed to stand in the refrigerator. When crystallization has been induced, the $(NH_4)_2SO_4$ concentration is then increased by 1% saturation to yield a fairly heavy crop of crystals. The crystal suspension is swirled occasionally during at least 2 weeks in the refrigerator. The enzyme is phenomenally stable as a crystalline suspension with essentially no loss in activity over more than a 6-month period.

Usually, at least a month has elapsed before several preparations have been set to crystallize, and they are centrifuged together in a refrigerated centrifuge at $30,000 \times g$, 0°, for 90 minutes. The supernatant is carefully removed with a capillary-tipped transfer pipet and retained. The crystalline pellet is washed with one-tenth the combined volumes, with a solution of 0.01 M Versene, pH 6.4 to 6.5 at 5°, adjusted to 0.64 saturation $(NH_4)_2SO_4$ (the Versene–$(NH_4)_2SO_4$ solution filtered if necessary), and the thick crystalline suspension is recentrifuged for 2 hours under above conditions.

The mother and wash liquors are combined and retained for further crops of crystals (see below).

The crystalline washed pellet (usually well packed and snow-white) is dissolved to a volume one-third to two-thirds that of the original volume, with 0.01 M Versene, pH 6.4 to 6.5 (adjusted with 1 N NaOH).

Dissolving the crystals may require an hour or so, and any trace of insoluble material (usually none) should be removed by centrifugation. The $(NH_4)_2SO_4$ concentration is conveniently estimated from the increase in volume over the 0.01 M Versene added (this increase in volume is assumed to be all due to 0.64 saturation $(NH_4)_2SO_4$). Then the solution is cautiously brought to about 0.60 to 0.62 saturation (as before, over a period of several hours). It is convenient to have about 30 to 35 mg. of initial protein concentration per milliliter at this point. Recrystallization is a far more rapid process, and overnight there often appears a thick crystalline suspension of beautifully snow-white crystals. The enzyme is conveniently stored as a crystalline suspension in ammonium sulfate. The yield from the first crystallization is often 70 to 80% (cf. fraction VIII), and from the second, 80 to 90% of the first crystals. Additional crops from the mother liquors may be obtained by reconcentrating the enzyme at 0.88 saturation $(NH_4)_2SO_4$, dissolving

FRACTIONATION OF ATP–AMP TRANSPHOSPHORYLASE[a]
(Initially, 1.0 kg. of rabbit skeletal muscle)

Fraction	Volume, ml.	Total protein	Total units	Units/mg.	Purification	Yield, %
I	3340	46,000	(206,000)	(4.5)		(100)
II	3280	5,520	167,000	30.2	6.7	81
III	566	2,670	154,000	57.8	13	75
IV	278	1,150	134,000	116	26	65
V	53	516	106,000	205	46	52
VI	14	186	86,700	465	103	42
VII	32	80.5	69,200	861	191	34
VIII	4.5	71.8	62,200	980	218	30
Crystals[b]		51.5	48,300	1100[b]	244	24

[a] From L. Noda and S. A. Kuby, *J. Biol. Chem.* **226**, 541 (1957).
[b] See text and footnote 9.

(at least 15 mg./ml.) in 0.01 M succinate, pH 6.4 (or 0.01 M Versene, pH 6.4, if it is to be stored as a final crop), and crystallization is conducted as described above. The protein may be lyophilized (after dialysis against 0.01 M sodium succinate pH 6.4), providing care is taken to exclude contact with any organic solvent vapors during the "shelling" process.

The specific activity of the first crystals is often about 1600 units/mg., and of the second crystals, about 1800 units/mg. A specific activity of 1100 units/mg. was previously reported.[3] The table summarizes the fractionation procedure previously reported.[1]

Properties

Purity. The crystalline enzyme prepared by the above procedure is homogeneous by the following tests: electrophoresis in various buffers, pH 5.5 to 8.3; sedimentation; chromatography on IRC 50; diffusion. It has been found to be free of 5'-adenylic deaminase, aldolase, hexokinase, fructokinase, ATP–Cr transphosphorylase, AMPase, ADPase, Mg^{++}- and Ca^{++}-activated ATPase and IDPase.

Specificity. With the possible exception of CDP, the enzyme appears to be highly specific for the adenine nucleotides. IDP, GDP, and UDP are inactive.

Activators, Inhibitors, and Stability Properties. Mg^{++}, or less suitably Mn^{++} (but not Ca^{++}), is required for enzymatic activity. The optimum concentration of Mg^{++} is equal to about half the initial concentration of ADP for the forward direction, or equal to about the initial concentration of ATP for the reverse direction. The enzymatic activity is inhibited by $AgNO_3$, zinc acetate, $CuSO_4$ and p-chloromercuribenzoate. Cysteine, EDTA, and inert protein (e.g., serum albumin or ATP–Cr transphosphorylase) stabilize the enzyme at high dilution. Myokinase is relatively unstable at low ionic strengths, but extraordinarily stable at high ionic strengths (e.g., 0.62 saturation $(NH_4)_2SO_4$).

Miscellaneous Kinetic Constants

The pH optimum for both the forward and reverse direction is approximately 8. At 25°, K_s (in moles per liter) is 3.3×10^{-4} for ATP and 2.6×10^{-4} for AMP; and K_i for ADP is about 3.3×10^{-4} M. V_{max} for the forward direction is about 46,000 moles of ATP formed per minute per mole of enzyme, and for the reverse direction about 41,000 moles of ATP utilized per minute per mole of enzyme.[9]

[9] In reference 3, the V_{max}'s are reported as 28,000 and 25,000, respectively; however, a preparation with a specific activity of about 1100 had been used, and therefore the results should be multiplied by 1800/1100.

[28] Inosine Diphosphatase (Nucleoside Diphosphatase) from Mammalian Liver

Nucleoside diphosphate + H_2O
$$\rightarrow \text{Nucleoside monophosphate} + \text{orthophosphate}$$

By G. W. E. PLAUT

Assay Method

Principle. The enzymatic liberation of orthophosphate from a susceptible nucleoside diphosphate is determined. Inosine diphosphate (IDP) is used as the substrate for routine work, since, at the moment, it is less expensive than other hydrolyzable nucleotides, such as uridine diphosphate (UDP) or guanosine diphosphate (GDP).

Reagents

0.2 ml. of 0.2 M tris(hydroxymethyl)aminomethane buffer at pH 7.4.
0.05 ml. of 0.1 M $MgSO_4$.
0.05 ml. of 0.1 M IDP.

The final volume is made up to 1.0 ml. with water.

Enzyme. Dilute the enzyme with 0.1 M tris(hydroxymethyl)aminomethane at pH 7.4, to a concentration which will lead to the liberation of no more than 2 micromoles of orthophosphate per milliliter of reaction mixture during the incubation period.

Procedure. The reaction components are incubated for 5 minutes in 15-ml. centrifuge tubes in a water bath at 20° prior to the addition of enzyme. The enzyme is then added, and incubation is continued for 10 minutes. The reaction is stopped by the addition of 1.0 ml. of ice-cold 10% trichloroacetic acid. Each incubated tube is accompanied by a nonincubated control, identical in composition, to which trichloroacetic acid is added prior to the enzyme. All values are corrected for zero-time orthophosphate content. After removal of coagulated protein, orthophosphate is determined by the method of Lowry and Lopez[1] or that of Fiske and SubbaRow.[2]

Definition of Unit and Specific Activity. No dilution effects have been observed so far with this enzyme, and the usual enzyme–time relationship is applicable. One unit of enzyme is taken as the amount

[1] O. H. Lowry and J. A. Lopez, *J. Biol. Chem.* **162**, 421 (1946).
[2] C. H. Fiske and Y. SubbaRow, *J. Biol. Chem.* **66**, 375 (1925).

catalyzing the liberation of 1.0 micromole orthophosphate per milliliter of reaction mixture per minute under the conditions of the assay. Specific activity is expressed as units per milligram of protein. Protein is determined by the method of Warburg and Christian (Vol. III [73]).

Purification Procedures

Enzyme from Beef Liver Particles

Preparation of Acetone Powder. One thousand grams of fresh beef liver are homogenized for 1 minute in a Waring blendor with 2.25 l. of a cold aqueous medium 0.25 M in sucrose and 0.03 M in dipotassium phosphate. The mixture is centrifuged at 1000 × g for 10 minutes. The supernatant solution is acidified to pH 5.9 with 3 N acetic acid and centrifuged at 4500 × g for 20 minutes. The clear supernatant layer is discarded, and the semifluid residue is recentrifuged at 18,000 × g for 30 minutes. The resulting residue is suspended in 450 ml. of 0.25 M sucrose and sedimented at 18,000 × g for 30 minutes. The washed residue is then converted to an acetone-dried powder, as previously described (Vol. I [119]). Yield, 30 to 35 g.

All subsequent operations are carried out at 2° to 5°.

Extraction. Ten grams of acetone powder are suspended with stirring in 100 ml. of glass-distilled water. The suspension is stirred for 15 minutes and centrifuged at 18,000 × g for 1 hour. The supernatant fluid is diluted to 100 ml. with distilled water.

First Calcium Phosphate Gel Fractionation. To the above extract are added 110 ml. of calcium phosphate gel (16.9 mg. of $Ca_3(PO_4)_2$ per milliliter)[3] (Vol. I [11]). After 20 minutes of stirring, the mixture is centrifuged at 5000 × g for 10 minutes. The supernatant fluid, which should contain 40 to 50% of the protein of the initial extract and no more than 5% of the enzyme activity, is discarded. The gel is suspended with vigorous stirring in 100 ml. of 0.125 M phosphate buffer, pH 5.4; after centrifugation, the residue is resuspended in 100 ml. of 0.125 M phosphate buffer, pH 5.9. The residue from this treatment is eluted by stirring with 100 ml. of 0.125 M phosphate buffer, pH 7.05. The eluate contains 10 to 20% of the protein and 50 to 90% of the activity of the initial extract with a 5- to 8-fold increase in specific activity. In order to concentrate the enzyme, 90 ml. of the eluate are treated with 54 g. of ammonium sulfate (0.8 saturation). The precipitate formed is collected by centrifugation and dissolved in 25 ml. of 0.01 M tris(hydroxymethyl)aminomethane buffer, pH 7.4.

Ammonium Sulfate Fractionation. Saturated ammonium sulfate (83

[3] D. Keilin and E. F. Hartree, *Proc. Roy. Soc.* **B124**, 397 (1938).

ml. of aqueous solution saturated with $(NH_4)_2SO_4$ at 25°) is added dropwise, with stirring, to every 100 ml. of the solution from the previous step. The resulting precipitate (0.45 saturation, containing about 8% of the enzyme; specific activity, 0.4) is discarded. A further 37.4 ml. of saturated ammonium sulfate are then slowly added to every 100 ml. of the supernatant solution, and the precipitate (between 0.45 and 0.6 saturation) is dissolved in 10 ml. of $0.2 M$ tris(hydroxymethyl)aminomethane buffer, pH 7.4.

Second Calcium Phosphate Gel Fractionation. The above solution is treated with 10 ml. of calcium phosphate gel. On centrifugation, the residue is washed with consecutive 10-ml. portions of $0.125 M$ phosphate buffer at pH 5.4 and 5.9, as previously. The major portion of the activity is then eluted with 10 ml. of $0.125 M$ phosphate buffer, pH 7.05. Practically all the activity of this eluate can be recovered by adding solid ammonium sulfate to 0.7 saturation and redissolving the precipitate in 5 ml. of $0.1 M$ tris(hydroxymethyl)aminomethane buffer at pH 7.4. This procedure achieves a 30- to 40-fold purification of the enzyme over that

TABLE I

PURIFICATION OF INOSINE DIPHOSPHATASE FROM BEEF LIVER PARTICLES[a]

(10 g. of acetone powder)

Step	Volume, ml.	Protein, mg.	Units	Specific activity, units/mg. protein	Yield, %
1. Original extract	97	2300	576	0.25	
2. First $Ca_3(PO_4)_2$ eluate[b]	26.5	242	315	1.31	55
3. $(NH_4)_2SO_4$ fractionation					
(0.45–0.6 saturation)	10	57	202	3.54	35
4. Second $Ca_3(PO_4)_2$	5.0	17	182	10.7	32

[a] Data from G. W. E. Plaut, *J. Biol. Chem.* **217**, 235 (1955).
[b] After concentration with ammonium sulfate (see the text).

of the acetone powder extract with a 20 to 30% yield. The results of a typical fractionation are summarized in Table I. This preparation is free of nucleoside diphosphokinase.

Enzyme from Whole Calf Liver

Extraction. Acetone powder from calf liver is prepared by the method of Horecker (Vol. II [123]). Except as noted, all steps are carried out at 0° to 3°. Thirty grams of powder are extracted with 600 ml. of cold distilled water for 30 minutes with gentle stirring. The mixture is centrifuged for 7 minutes at 13,000 × g, and the precipitate is discarded.

First Acid Ammonium Sulfate Step. The extract (535 ml.) is adjusted to pH 5.0 with 100 ml. of 0.1 N acetic acid under mechanical stirring. A precipitate forms which is separated from the supernatant solution (solution A, 605 ml.) by centrifugation. The precipitate is treated with 0.05 M sodium acetate, and the suspension is adjusted to pH 7.0 with NH_4OH. It dissolves only partly; the insoluble residue is removed by centrifugation and discarded. The supernatant fluid (fraction Ia) measures 104 ml.

The supernatant solution A (605 ml.) is mixed with 100 g. of ammonium sulfate (0.3 saturation). A precipitate forms which is collected by centrifugation and extracted with sodium acetate and NH_4OH, as described above, giving 72 ml. of extract (fraction Ib). Fractions Ia and Ib are combined (fraction I, 176 ml.).

Second acid ammonium sulfate step. Fraction I (176 ml.) is mixed with 29 g. of ammonium sulfate (0.3 saturation). An inactive precipitate is removed by centrifugation and discarded. The supernatant solution is adjusted to pH 4.6 with 13.8 ml. of 1 N acetic acid. After 10 minutes of standing, a precipitate is collected by centrifugation. This residue is treated with 0.05 M sodium acetate and NH_4OH as in step 2, yielding 30.4 ml. of extract (fraction II).

Alkaline Ammonium Sulfate Step. Fraction II is treated with 71 ml. of distilled water, and 16.7 g. of ammonium sulfate (0.3 saturation) are added. The acidity is adjusted to pH 8.0 by the addition of 1 ml. of a solution containing 1 N NH_4OH and 16.7 g. of ammonium sulfate per 100 ml. Saturated ammonium sulfate solution[4] (15.2 ml.) is then introduced (0.39 saturation). A precipitate is removed by centrifugation after 15 minutes and discarded. To the supernatant solution is added 165 ml. of saturated alkaline ammonium sulfate (0.75 saturation). This precipitate is collected by centrifugation, dissolved in 0.05 M acetate buffer, pH 6, and dialyzed against running distilled water for 6 hours (fraction III, 71 ml.).

Ethanol Step. Fraction III is diluted with distilled water to 113 ml. and adjusted to pH 5.2 with 0.5 ml. of 0.1 N acetic acid. A precipitate forms which is collected by centrifugation, dissolved in 0.05 M sodium acetate, and set aside (fraction IV, 71 ml.). The supernatant solution from acidified fraction III is mixed with 1.7 ml. of 2 M acetate buffer, pH 5.2, followed by 13.9 ml. of absolute ethanol (11%). The temperature is allowed to decline to −4° during the addition of ethanol. After 10 minutes the mixture is centrifuged at 13,000 × g for 3 minutes. The precipitate is allowed to drain at −8° and dissolved in 0.05 M sodium

[4] The saturated ammonium sulfate solution contains 3.7 ml. of concentrated NH_4OH per liter, and its reaction after 5-fold dilution with water is pH 8.0.

TABLE II
PURIFICATION OF INOSINE DIPHOSPHATASE FROM WHOLE CALF LIVER[a,b]
(30 g. of acetone powder)

Fraction	Total units	Over-all yield, %	Specific activity, units/mg. protein
Initial extract	5800	100	1.01
I. Ammonium sulfate, pH 5	2870	49	1.90
II. Ammonium sulfate, pH 4.6	1720	29	5.50
III. Alkaline ammonium sulfate	1125	19	8.66
IV. Precipitate, pH 5	505	8.5	9.50
V. 0–11% Ethanol	557	9.5	27.2

[a] Data from L. A. Heppel, J. L. Strominger, and E. S. Maxwell, *Biochim. et Biophys. Acta* **32**, 422 (1959).

[b] Assay of IDPase involves an incubation mixture containing 1.5 micromoles of Veronal-acetate buffer, pH 7.0, 1.6 micromoles of reduced glutathione, 0.8 micromole of $MgCl_2$, 0.01 micromole of EDTA, 0.03 mg. of crystalline serum bovine albumin, and 0.7 micromole of IDP in a total volume of 0.08 ml. This is incubated at 37° for 30 minutes, together with a control mixture lacking enzyme. After addition of 0.2 ml. of 2.5% perchloric acid, analysis for P_i is carried out.

acetate (fraction V, 71 ml.). The results of purification are summarized in Table II. Both fractions IV and V are free of nucleoside triphosphate-adenosine 5′-phosphate transphosphorylase activity.

Properties

Specificity. The enzyme from beef[5] or calf[6] liver does not catalyze the dephosphorylation of ATP, ITP, ribose 5-triphosphate, IMP, AMP, ADP, or CDP. The relative activities with IDP, UDP, and GDP are 1, 1.7, and 1.6 with purified enzyme from beef liver, and the relative rates of hydrolysis of IDP, UDP, GDP, and ribose 5-pyrophosphate are about 1.0, 0.6, 0.8, and 0.6 with the enzyme from calf liver. A preparation from lamb liver has been reported to hydrolyze thiamine pyrophosphate slowly.[7]

Other Sources. The enzyme has been detected in rat[5] and lamb[7,8] liver and in hog kidney.[9] No activity could be detected in aqueous

[5] G. W. E. Plaut, *J. Biol. Chem.* **217**, 235 (1955).

[6] L. A. Heppel, J. L. Strominger, and E. S. Maxwell, *Biochim. et Biophys. Acta* **32**, 424 (1959).

[7] J. D. Gregory, *Federation Proc.* **14**, 221 (1955).

[8] These studies[7] were done with lamb liver and not with beef liver as reported (J. D. Gregory, private communication, 1958).

[9] D. M. Gibson, P. Ayengar, and D. R. Sanadi, *Biochim. et Biophys. Acta* **16**, 536 (1955).

extracts from acetone powders of hearts from rat, guinea pig, hog, or cattle.[10]

Activators and Inhibitors. Inosine diphosphatase is activated by Mg^{++}. Mn^{++} and Ca^{++} are also reported to be active, but Ba^{++} and Sr^{++} are ineffective.[7] The enzyme from beef liver is inhibited by high concentrations of fluoride (75% inhibition by 0.1 M KF); ATP and ADP inhibit the hydrolysis of IDP.

pH Optimum and Affinity Constants. The optimal action of the enzyme from beef liver is at pH 6.9. The activity at pH 6.45 or 7.4 is only 10 to 15% lower than at the optimum. With IDP as substrate, K_s is approximately $5 \times 10^{-4} M$ at pH 7.4.[5]

[10] G. W. E. Plaut, unpublished observations (1955).

[29] Phosphodiesterases

By W. E. RAZZELL

The classification of enzymes that hydrolyze nucleic acids is in a state of flux.[1,2] This results partly from early attempts to bring order to a group of enzymes which has been enlarged greatly in recent years and partly from the results of studies on the substrate specificity of both well-known and newly discovered enzymes which have been made possible by advances in our understanding of the detailed structure of the nucleic acids, and in methods for their detection, separation, and synthesis. In general, phosphodiesterases can be considered to embrace all those enzymes which hydrolyze the bond between phosphate and one or the other of two molecules possessing alcoholic functional groups which it links. It is customary to assume that one is a nucleoside, but hydrolysis of phosphodiester bonds which do not involve nucleosides at all has been shown with at least one of this class of enzyme.

The particular phosphodiesterases described in this article are those which fall within the author's experience: (1) snake venom, (2) spleen, (3) enzymes in other tissues which appear to be related to the above,

[1] M. Laskowski, *Ann. N. Y. Acad. Sci.* **81**, 776 (1959).
[2] H. G. Khorana, Phosphodiesterases. In "The Enzymes" (P. D. Boyer, H. A. Lardy, and K. Myrbäck, eds.), 2nd ed., Vol. 5, pp. 79–94. Academic Press, New York, 1961.

(4) liver nuclei RNase I, (5) leukemic cell phosphodiesterase, and (6) 2',3'-cyclicphosphodiesterase.

The properties of a number of other enzymes have been recently reviewed,[2] and both pancreatic ribonuclease and deoxyribonuclease are discussed in Vol. II [62] and II [63].

Most phosphodiesterases have been studied to permit their use as tools for forming or characterizing oligonucleotides, and progress in this direction has been considerable. The problems of their distribution (in different animals, organs, or cell fractions) and dynamic function have received scant attention, however. It is reasonable to expect that, once the properties and specificities of the various phosphodiesterases, including those which rapidly depolymerize macromolecular nucleic acids, have been accurately evaluated, more emphasis will be placed on specific methods for their assay, permitting their distribution to be determined.

Assay Methods. Recently, assays based on the formation of acid-soluble fragments from naturally occurring nucleic acids have been replaced by those in which the hydrolysis of synthetic oligo- or polynucleotides has been followed. These, in turn, have in some cases (notably for venom, spleen, and related preparations) yielded to assays employing chromogenic diesters.

Substrates. Syntheses of p-nitrophenyl thymidine-5'- and 3'-phosphates, oligonucleotides of thymidylic acid, oligonucleotides of thymidylic acid terminated in a deoxycytidylic acid residue bearing a 5'-phosphate end group, and other compounds of interest in the study of phosphodiesterases are presented in Vol. VI [94].

The preparation and isolation of polyadenylic acid (poly A), polyuridylic acid (poly U), etc., are presented in Vol. VI [98].

Definition of Units and Specific Activity. A unit as used in this article refers to the hydrolysis of 1 micromole of substrate or the formation of 1 micromole of product, in 1 hour. Specific activity is defined as the number of such units per milligram of protein.

Phosphodiesterase I

It has become apparent that closely related enzymes from venom, kidney, and other tissues have the common property of hydrolyzing p-nitrophenyl thymidine-5'-phosphate and oligonucleotides (ribo- and deoxyribo-) with free 3'-hydroxyl end groups to form mononucleoside-5'-phosphates.[3] The purification and determination of properties of the venom and kidney enzymes are accomplished by the same assay procedures, and substrate specificities are almost identical.

[3] W. E. Razzell, *J. Biol. Chem.* **236**, 3031 (1961).

Assay Method

Principle. The hydrolysis of p-nitrophenyl thymidine-5'-phosphate results in the formation of thymidine-5'-phosphate and a yellow color due to the liberation of p-nitrophenylate at alkaline pH.[4]

Procedure. The assay is performed in a cuvette containing 100 micromoles of Tris buffer adjusted to pH 8.9 with hydrochloric acid, and 0.5 micromole of substrate in a total volume of 1.0 ml. Prior to the addition of enzyme the cuvette is equilibrated at 37° in a Cary recording spectrophotometer. On addition of enzyme the increase in absorbance is measured at 400 mμ. An increase in absorbance of 1.200 units is equivalent to the hydrolysis of 0.1 micromole of substrate.

Alternatively, the assay can be set up in small test tubes, in which case the quantities of reagents are scaled down by a factor of 5, and replicate tubes (containing identical quantities of buffer and substrate in volumes of 0.2 ml.) are preincubated at 37°. Identical quantities of enzyme are added to each tube in turn, and at subsequent intervals 1 ml. of 0.1 N NaOH is added to each tube. The contents of the tubes are then placed in 1-ml. cuvettes, and the absorbances at 400 mμ are determined in the spectrophotometer.

For nonchromogenic substrates such as oligonucleotides the hydrolysis is performed in small volumes from which aliquots are taken at intervals for chromatography. From a total volume of about 50 μl. containing 1 micromole of substrate, 20 micromoles of Tris, pH 8.9, and enzyme, aliquots are removed at intervals and mixed on a piece of parafilm with 5 μl. or more of glacial acetic acid to stop the reaction. These aliquots are then transferred directly to Whatman 3 MM paper and developed by descending chromatography in isopropanol–concentrated ammonium hydroxide–water (7:1:2, v/v). After chromatography the spots delineated under ultraviolet light are cut out and eluted either (1) by soaking the paper in 0.10 N hydrochloric acid overnight or (2) by allowing water to flow up the small piece of paper (thus bringing all the material on the paper to the top) and subsequently centrifuging off the solution in the paper into small test tubes. The eluates are made to volume, e.g., 3 ml., and the absorbance at suitable wavelengths (for example, 260 mμ) is determined in a spectrophotometer.

Purification Procedure, Phosphodiesterase I from Venom

Lyophilized Crotalus adamanteus venom is obtainable commercially. Purification procedures alternative to the one presented here are given

[4] W. E. Razzell and H. G. Khorana, J. Biol. Chem. **234**, 2105 (1959).

by Butler (Vol. II [89]), by Privat de Garilhe and Laskowski,[5] and by Felix et al.[6]

Step 1. Autolysis. The venom is dissolved in water, and the solution (pH 4.5) is incubated for 1 hour at room temperature. The mixture is centrifuged, and the supernatant is made to $0.2 M$ in ammonium acetate, pH 4.0. For example, 500 mg. of dry venom is dissolved in 30 ml. of water. To the 1-hour supernatant is added 18 ml. of $0.5 M$ ammonium acetate, pH 4, and the mixture is cooled to $0°$ in ice.

Step 2. Acetone Fractionation. To this mixture is added 35 ml. of acetone (previously cooled to $-20°$), the solution is swirled to mix the liquids, and the mixture is kept at $4°$ for 30 minutes. The supernatant from centrifugation at $8000 \times g$ for 10 minutes is poured off, and to this is added 6.8 ml. of cold acetone as before. This mixture is kept at $4°$ for 18 hours and recentrifuged. To the supernatant is added 8 ml. of acetone, and the mixture is allowed to stand for 30 minutes.[4, 7] The precipitate obtained by centrifuging this final mixture is dissolved in $0.002 M$ Tris, pH 8.9.

The diesterase at this stage should be free of contaminating 5'-nucleotidase, the absence of which may be determined conveniently by incubating a dinucleotide such as pTpT[8] with the enzyme in $M/20$ Tris, pH 8.9, and examining the products for the formation of mononucleoside. If the diesterase has not been precipitated, more acetone can be added to the supernatant, and the enzyme is recovered in this later precipitate. In general, complete removal of very minor quantities of 5'-nucleotidase together with some further purification of the diesterase is not inconvenient and can readily be accomplished on DEAE-cellulose columns.[4, 9]

Step 3. Chromatography. One-third of the precipitate obtained from 500 mg. of crude venom, dissolved in $0.002 M$ Tris, pH 8.9, is made to $0.02 M$ Tris, pH 8.9 (total volume approximately 10 ml.), and applied to the top of a column of DEAE-cellulose 30×3.0 cm. in diameter which has previously been equilibrated with $0.02 M$ Tris, pH 8.9. After the addition of the enzyme, 30 ml. of $0.02 M$ Tris, pH 8.9, is allowed to flow through the column, and elution is then begun with 100 ml. of $0.33 M$ Tris, pH

[5] M. Privat de Garilhe and M. Laskowski, *Biochim. et Biophys. Acta* **18**, 370 (1955).
[6] F. Felix, J. L. Potter, and M. Laskowski, *J. Biol. Chem.* **235**, 1150 (1960).
[7] J. F. Koerner and R. L. Sinsheimer, *J. Biol. Chem.* **228**, 1039 (1957). It is possible to obtain different protein forms of this enzyme, notably by *not* permitting the pH to remain low in the initial step (see refs. 4 and 9).
[8] In abbreviations for oligonucleotides, the letter "p" to the left of the initial for the nucleotide designates a 5'-phosphate; the letter "p" to the right, a 3'-phosphate.
[9] H. G. Boman and U. Kaletta, *Biochim. et Biophys. Acta* **24**, 619 (1957).

8.9, followed by 100 ml. of 0.6 M Tris, pH 8.9, and 100 ml. of 0.33 M Tris, pH 7.4. The enzyme emerges from the column just as the concentration of Tris rises from 0.02 to 0.33 M at pH 8.9. The enzyme as it is obtained from the column may be stored frozen in the buffer. A summary of the purification data is given in Table I. Dialysis has proved to inactivate

TABLE I

PURIFICATION OF VENOM PHOSPHODIESTERASE I

Step	Per cent of total units[a]	Specific activity[a]
1. Crude venom	100	600
1. Water supernatant at 20°[b]	100	950
2. Third acetone precipitate	36	7,400
3. Peak from DEAE-cellulose	10	34,000
3a. Total from DEAE-cellulose	32	—

[a] For definition, see page 237.

[b] The pH of the solution was about 4.5. Different chromatographic behavior, usually accompanied by greater purification but lower peak recovery, is obtained by adjusting the solution to pH 9 with NH₄OH at this stage.

large percentages of enzyme preparations at almost any stage of purification and has not been found useful as a routine procedure. It may be found, however, that some preparations of Tris buffer contain fluorescent materials which interfere with chromatographic identification of products of hydrolysis by these preparations. Therefore it has sometimes been necessary to dialyze the preparation briefly to eliminate this fluorescent material, and such dialysis should be performed against distilled water adjusted to pH 9 with ammonium hydroxide and maintained at 2°. Dialysis should not continue for more than 2 to 3 hours in any case.

Properties

Specificity. The acetone-precipitated preparation and the chromatographed preparations have essentially the same specificity. The substrate most rapidly hydrolyzed has been found to be p-nitrophenyl thymidine-5'-phosphate.[4] In general the deoxyribonucleotide diesters are hydrolyzed faster than the diesters of the ribonucleotides.

Other details of specificity toward compounds which have been tested in this laboratory are shown in Table II. In addition, it should be noted that uridine-2',3'-cyclic phosphate and cyclo-pTpT are hydrolyzed about 10^{-3} as fast as pTpT in crude venom, whereas only the activity toward cyclo-pTpT survives the acetone fractionation. It has been concluded that this activity is due to the phosphodiesterase itself.[4]

Oligonucleotides bearing 3'-phosphate end groups are hydrolyzed even more slowly than di(p-nitrophenyl) phosphate. No differences have been found between the purine or pyrimidine oligonucleotides. A 5'-5' phosphodiester linkage appears to be equivalent to a 5'-3' linkage.

These observations are consistent with a requirement by the enzyme for a nucleoside 5'-phosphoryl residue bearing a free or at least nonphosphorylated 3'-hydroxyl function.[10] Some other compounds hydrolyzed by the venom enzyme and by the analogous enzyme from kidney are shown in Table IV (see page 245).

TABLE II
SPECIFICITY OF VENOM PHOSPHODIESTERASE I

Compound	V_m (moles/hr./mg. protein)	K_m (M)
p-Nitrophenyl pT	36,500	5.0×10^{-4}
p-Nitrophenyl pU	1,275	5.4×10^{-4}
pTpT[a]	6,840	2.1×10^{-4}
TpT[b]	278	5.3×10^{-4}
Methyl p-nitrophenyl phosphate	714	1.2×10^{-2}
Benzyl p-nitrophenyl phosphate	315	6.8×10^{-3}
Di-p-Nitrophenyl pT (pH 8)	39	7.7×10^{-4}
p-Nitrophenyl pT (pH 8)	19,700	4.9×10^{-4}
3'-O-Acetyl p-nitrophenyl pT (pH 8)	7,250	1.5×10^{-4}

[a] Similar results were obtained with deoxy-pTpA.
[b] This is the 3',5' diester. Data for the 5',5' diester are essentially identical.

Inhibitors. The diesterase is inhibited by reducing agents such as cysteine, GSH, thioglycolate, and ascorbic acid, 50% inhibition being observed at $7.5 \times 10^{-3} M$, $1.5 \times 10^{-3} M$, $1.5 \times 10^{-3} M$, and $4 \times 10^{-3} M$, respectively.

Although citrate is not a strong inhibitor of the enzyme, EDTA is. Fifty per cent inhibition occurs at approximately $8 \times 10^{-5} M$ and is reversible by the addition of excess magnesium. Of a variety of cations tested, only copper inhibited the enzyme, this effect being easily demonstrated at $10^{-4} M$.

In spite of the fact that EDTA inhibits the enzyme and this inhibition can be reversed by magnesium, it has not been necessary in every case to add magnesium to the enzyme preparations. They are almost fully activated by the cations already present. In order to yield samples for chromatography free of as many interfering cations as possible, magnesium or other divalent cations are not normally added to the enzyme preparations.

[10] W. E. Razzell and H. G. Khorana, J. Biol. Chem. 234, 2114 (1959).

Competitive inhibition of p-nitrophenyl ester hydrolysis is readily observed on the addition of oligonucleotides such as pTpT, and this competitive inhibition presumably extends also to a variety of other oligonucleotides. The inhibition increases rapidly as the chain length increases, being more marked with oligonucleotides bearing 5′- (or 3′-) phosphate end groups. Similar observations have been made with p-nitrophenyl uridine-5′-phosphate.

Stepwise Degradation. The formation of intermediates during the hydrolysis of thymidine and other oligonucleotides is readily detected early in the course of the reaction.[10] The appearance of mononucleotide initially is accompanied by the appearance of the next lower homolog of the starting material, and subsequent aliquots contain a variety of homologs. Stepwise degradation by this enzyme has been observed with polymers bearing 5′-phosphate end groups, those without 5′-phosphate end groups, compounds with a 3′-acetyl group (with or without 5′-phosphate end groups), oligoadenylic acids, and mixed oligonucleotides such as pTpTpC. No intermediates have been observed in the hydrolysis of the macrocyclic oligonucleotides,[11] in which the limiting reaction appears to be the initial bond-breaking step in the opening of the macrocyclic ring, the linear compound thus formed being rapidly and preferentially hydrolyzed by the enzyme.[10]

Thus, attack on polymers at a point other than the end of a chain is possible, but Singer *et al.*[12] observed stepwise degradation in large riboöligonucleotides, and Adler *et al.*[13] observed the formation of mononucleotides at an early stage in the hydrolysis of DNA by the venom enzyme.

Purification Procedure, Phosphodiesterase I from Kidney

Step 1. Tissue Preparation and Homogenization. Hog kidneys obtained at the slaughterhouse are packed in ice and transported to the laboratory where they are immediately sliced lengthwise and the cortex is removed. The cortex is then cut into 1-inch cubes and homogenized in five times its volume of ice-cold 0.25 M sucrose for 4 minutes in the cold room. The mixture is centrifuged for 15 minutes at 5000 $\times$ g, and the supernatant is recovered.

Step 2. Ultracentrifugation and Washing of Microsomes. The supernatant from step 1 is centrifuged at 100,000 $\times$ g for 90 minutes. The

[11] G. M. Tener, H. G. Khorana, R. Markham, and E. H. Pol, *J. Am. Chem. Soc.* **80**, 6223 (1958).

[12] M. F. Singer, R. J. Hilmoe, and L. A. Heppel, *Federation Proc.* **17**, 312 (1958).

[13] J. Adler, I. R. Lehman, M. O. Bessman, E. S. Simms, and A. Kornberg, *Proc. Natl. Acad. Sci. U.S.* **44**, 641 (1958).

precipitate from high-speed centrifugation is resuspended in 0.01 M Tris buffer, pH 8.0, by brief sonic oscillation (2 minutes in a 10-kc. Raytheon oscillator). This resuspended precipitate is recentrifuged at 100,000 $\times$ g for 90 minutes and resuspended in Tris buffer as before by sonication.

Step 3. Solubilization. To the resuspended material tertiary amyl alcohol is added to a concentration of 33% (v/v) with rapid stirring in an ice bath. After standing for 18 hours at 0°, this mixture is centrifuged at 13,000 $\times$ g for 15 minutes, and the aqueous layer is recovered. The aqueous layer is transferred to a 500-ml. glass cylinder, and to it is added an equal volume of diethyl ether. This mixture is rotated and rocked carefully so as not to introduce too great a quantity of emulsion, while the ether extracts the tertiary amyl alcohol. The layers are allowed to separate, the ether layer is removed and replaced with a fresh ether layer, and the layers are rocked upon each other again. This process is repeated once more, and the ether which has dissolved in the water solution is taken off by vacuum evaporation in a round-bottomed flask connected to a water pump. The enzyme at this stage is no longer sedimentable by centrifugation at 100,000 $\times$ g for 90 minutes.

Step 4. Trypsin Digestion. The enzyme is dialyzed overnight at 2° against 10 volumes of 0.005 M Tris, pH 8.5. For every 250 mg. of protein in the solution remaining inside the dialysis sac, 1 mg. of crystalline trypsin is added together with a few crystals of thymol, the pH is adjusted to 8.5, and the mixture is incubated in a closed container at 50° for 24 hours. The resultant preparation should be clear, with a pH between 7.5 and 8.5.

Step 5. Acetone Fractionation. To each measure of 100 ml. of the above solution stirred at 2°, 20 ml. of 2 M potassium acetate and 66 ml. of acetone (−20°) are added and stirring is continued for 15 minutes. Insoluble material is then removed by centrifugation at 10,000 $\times$ g for 20 minutes and an additional amount of 58 ml. of acetone is stirred into the supernatant. After 15 minutes in ice the mixture is centrifuged as before and the precipitate is dissolved in 0.02 M Tris, pH 8, to one-fifth of the original volume.

Step 6. Chromatography on Ecteola Cellulose. A column of Ecteola cellulose, 35 cm. $\times$ 2 cm. diameter, is cleaned first by passing 100 ml. of 1 M NaCl–0.1 M NaOH through it, followed by passing through distilled water until the pH is less than 8. The acetone precipitate is diluted 5-fold with water, and 72 mg. of protein are applied to the column at 4°. Washing and elution are performed with 25 ml. of 0.05 M Tris, pH 8.0, 50 ml. of 0.05 M Tris–0.12 M KCl, pH 8.0, and 50 ml. of 0.05 M Tris–0.15 M KCl, pH 8.0. The enzyme is found in the last effluent. A summary of the purification procedure is given in Table III.

TABLE III
PURIFICATION OF PHOSPHODIESTERASE I FROM KIDNEY

Step	Per cent of total units[a]	Specific activity[a]
1. Homogenate	(100)	7.4
1. Supernatant, 5000 × g, 15 minutes	80	9.7
2. Washed particles	65	29
3. Soluble fraction	65	36
4. Trypsin digest	65	—
5. Acetone precipitate	54	362
6. Column peak	—	930–1740

[a] For definition, see page 237.

Properties

The enzyme is quite stable in the cold above pH 5.1, but is rapidly and apparently irreversibly inactivated at pH 4.5.

Competitive inhibition of the hydrolysis of p-nitrophenyl thymidine-5′-phosphate by various oligonucleotides has not been observed with this enzyme. Since higher concentrations of EDTA are required to inhibit this activity than in the case of venom phosphodiesterase, it is possible that the active sites are different.

The activity toward p-nitrophenyl thymidine-5′-phosphate is inhibited 40% by $10^{-3} M$ ADP or pT, 20% by $10^{-3} M$ ATP, and 100% by $10^{-3} M$ AMP. The polyphosphates are also substrates (see below).

Specificity. This enzyme is not completely inactive on oligonucleotides with 3′-phosphate end groups, presumably because it contains traces of alkaline phosphatase which is capable of removing the phosphate end group and thus providing a nonphosphorylated substrate.

The activity on p-nitrophenyl thymidine-3′-phosphate is less than 0.4% of the activity on the p-nitrophenyl thymidine-5′-phosphate at pH 6.0 to 9.5.

With the addition of small concentrations of inorganic phosphate to inhibit the alkaline phosphomonoesterase, stepwise degradation of oligonucleotides bearing 5′-phosphate (or hydroxyl) end groups can be demonstrated. With pTpTpC, it was shown that the degradation proceeds stepwise with the initial formation of pC and pTpT, so that the mode of action of this enzyme is quite analogous to that of venom phosphodiesterase itself.[3] It is likely that the exposed 3′-hydroxyl ends of nucleic acids in the cells in which the enzyme is found are also points of attack. Release of nucleoside-5′-phosphate occurs with a variety of compounds,[14] as is true for the enzyme from venom (Table IV).

[14] W. E. Razzell, unpublished observations (1962).

TABLE IV
SUBSTRATE SPECIFICITY OF VENOM AND KIDNEY PHOSPHODIESTERASE I

Compound	Rate relative to p-nitrophenyl thymidine-5'-phosphate
A. Venom	
ATP	0.2
ADP	0.015
Adenosine-5'-tetraphosphate	0.3
B. Kidney	
ATP	0.023
ADP	0.014
C. Venom or kidney	
UDP-glucose	0.25
GDP-mannose	0.25
Dephospho-coenzyme A	0.23
DPN$^+$	0.29
TPN$^+$	<0.03
Di(thymidine-5')pyrophosphate	0.23
Di(adenosine-5')pyrophosphate	0.23
Di(adenosine-5')triphosphate	0.22

Tissue Distribution

The specific activity of phosphodiesterase I acting on p-nitrophenyl thymidine-5'-phosphate is greatest in kidney and high in liver, intestinal mucosa, and lung compared with other tissues. The enzyme is found in the microsomal fraction (as is apparent from the purification procedure) but some appears in the nuclear fraction as well.[15]

Spleen Phosphodiesterase II

Assay Method

Principle. The spleen phosphodiesterase may be assayed by the formation of p-nitrophenol in an assay similar to the intermittent sample assay used for venom, but with p-nitrophenyl thymidine-3'-phosphate.[16]

Procedure. To small test tubes containing 1.2 micromoles of p-nitrophenyl thymidine-3'-phosphate, 50 micromoles of ammonium succinate, pH 5.9, or ammonium acetate, pH 5.7, and 0.01 ml. of 1% Tween 80, enzyme and water are added to a volume of 0.3 ml. The mixture, without enzyme, is preincubated for 2 minutes in a water bath at 37°, enzyme is added, and aliquots of 0.05 ml. are transferred at intervals

[15] W. E. Razzell, *J. Biol. Chem.* **236**, 3028 (1961).
[16] W. E. Razzell and H. G. Khorana, *J. Biol. Chem.* **236**, 1144 (1961).

to 1.0 ml. of 0.10 M NaOH. The absorbances of the resulting solutions are measured in a spectrophotometer at 400 mμ.

An increase in absorbance of 0.200 unit is equivalent to the hydrolysis of 0.1 micromole of substrate in the 0.3-ml. mixture.

Hydrolysis of RNA "core" may be determined by conventional spectrophotometric procedures as described by Heppel and Hilmoe.[17]

Paper chromatography may be performed in an incubation mixture similar to the above, with a total volume of 0.15 ml. containing 0.6 micromole of substrate. Aliquots are removed at intervals and mixed with acetic acid to stop the reaction, and the products are separated by descending chromatography in the solvent system noted above for venom diesterase.

Purification Procedure

Further purification steps beyond those reported already by Heppel and Hilmoe[17] have been added to the spleen fractionation technique[16, 18] and are included below. Their effect is to reduce the concentration of the enzyme responsible for the hydrolysis of nucleoside 2′,3′-cyclic phosphate esters and to increase the specific activity of the final preparation.

Step 6. Acetone Fractionation. To 246 ml. of Heppel and Hilmoe's fraction 5 (ammonium sulfate precipitate)[17] containing 3.6 mg. of protein per milliliter are added 2.2 ml. of M sodium acetate, pH 5.2, and M acetic acid to a final pH of 5.2. The solution is cooled in a dry ice–alcohol bath, and 19.4 ml. of acetone at −20° is added with continuous stirring over a period of 10 minutes.[18] During this time the temperature is allowed to drop to −8°. The mixture is left at −8° for 10 minutes and then centrifuged at 5000 × g for 5 minutes. To the supernatant 18 ml. of acetone is added slowly as before, and again after a 10-minute wait the mixture is centrifuged at −10°. The precipitate is suspended in 15 ml. of 0.01 M ammonium succinate, pH 6.5, and recentrifuged.

Step 7. Alumina Gel Adsorption and Elution. The supernatant is treated within 24 hours with alumina C_γ gel according to Heppel and Hilmoe,[17] with the additional provision that all solutions should contain 0.001% Tween 80.

Step 8. Calcium Phosphate Gel Adsorption and Elution. The active eluate from step 7 is directly adjusted to pH 5.75 with M potassium acetate, pH 5.2, and 4.37 ml. of calcium phosphate gel (20 mg./ml.) is stirred in.[16] As is usual with gel steps, the exact amount of gel to be

[17] See Vol. II [90].

[18] R. J. Hilmoe, *J. Biol. Chem.* **235**, 2117 (1960). Dr. Hilmoe kindly provided details of his method prior to their publication.

used is determined on small samples of the enzyme, at pH 5.75. The mixture is stirred occasionally during 10 minutes, and the gel is recovered by centrifugation and washed with 10 ml. of $0.02 M$ potassium

TABLE V
PURIFICATION OF SPLEEN PHOSPHODIESTERASE II

Step	Volume	Protein, mg./ml.	Specific activity, μmoles/hr./mg. protein	Total activity, μmoles/hr./kg. of spleen
5. See Vol. II [90]	46	3.6	116	17,000
6. Acetone precipitate	16	4.0	226	14,500
7. 0.01 M pyrophosphate eluate of alumina C$_\gamma$ (see Vol. II [90])	25	0.56	470	6,600
8. Eluate of calcium phosphate gel	10	0.28	2040	5,700

phosphate, pH 6. The enzyme is eluted with 10 ml. of 0.18 saturated ammonium sulfate in $0.04 M$ potassium phosphate, pH 7.2. A purification summary is given in Table V.

Properties

The preparation is stable in the refrigerator but loses considerable activity on freezing, presumably because of the ammonium sulfate present. Brief dialysis also causes some inactivation. The preparation at this stage of purification appears to have reached an unstable state.

Specificity. A comparison was made of the relative rates of hydrolysis of p-nitrophenyl thymidine-3'-phosphate versus RNA "core" to establish the identity of the enzyme hydrolyzing both substrates. The relative activities in these assays agreed well with the relative rates observed in assays at the acetone powder stage and suggest that the same enzyme is responsible for the hydrolysis of both substrates at both extremes of purification.[16] The rates of hydrolysis of a variety of substrates by the purified spleen diesterase are shown in Table VI. The relationships between activity, pH, and buffer concentrations are quite complex and are shown in Fig. 1. Although $0.167 M$ acetate buffer yields an activity considerably higher than the corresponding concentrations of succinate, at least at higher pH's, the buffering capacity of acetate above pH 6 is extremely limited.

The enzyme is quite specific for oligonucleotides with a free 5'-hydroxyl group. Thus the diesters of deoxyribonucleotides with 5'-phosphate end groups are resistant to hydrolysis, whereas the corresponding esters with 3'-phosphate end groups are hydrolyzed rapidly.

Stepwise Degradation. Hydrolysis of oligonucleotides begins from

TABLE VI

SUBSTRATE SPECIFICITY OF SPLEEN PHOSPHODIESTERASE II

(*Note:* Standard assay procedure, substrates at 3.6 micromoles/ml.)

Compound	Initial rate of mononucleotide liberation, μmoles/hr./mg.
1. *p*-Nitrophenyl thymidine-3′-phosphate	2040[a]
2. ApUp, ApApUp	3570
3. TpTp	2700
4. TpT	2600
5. pTpTpT	$<3 \times 10^{-4}$
6. Cyclo (Tp)$_2$	$<3 \times 10^{-4}$
7. Poly A[b]	43

[a] $V_m = 3020$; $K_m = 5 \times 10^{-3}\ M$.

[b] Substrate concentration = 13 micromoles of adenine residues per milliliter; thus end-nucleoside concentration is about 0.07 micromole/ml. At this substrate concentration, this rate of hydrolysis would be expected.

the end of the chain bearing the 5′-hydroxyl group, liberating mononucleotides with 3′-phosphate groups and successive lower homologs of the starting material.[16] This mode of action is consistent for the deoxy-

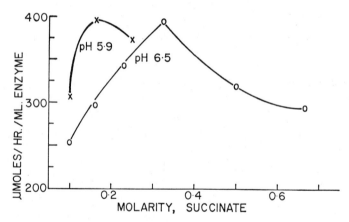

FIG. 1. Buffer and pH effects on spleen phosphodiesterase II.

ribose and ribose oligonucleotides, with or without 3′-phosphate end groups.[19] A comparison of the mechanisms of action of phosphodiesterases I and II is shown in Fig. 2.

[19] One of the important factors in a successful demonstration of stepwise degradation by the diesterases appears to be the use of the largest concentration of enzyme which will allow the necessary manipulations to be performed in the time available between aliquots, e.g., 2 to 5 minutes (see ref. 10).

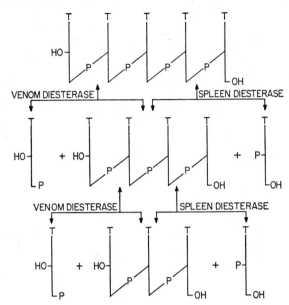

F<small>IG</small>. 2. Points of attack by venom phosphodiesterase I and spleen phosphodiesterase II.

Tissue Distribution

The enzyme is found at highest specific activity in spleen, thymus, lungs, and muscle. Lower activities in liver and kidney contrast with phosphodiesterase I. Although present in the mitochondrial-lysosomal fraction, considerable activity is found in the soluble proteins of the cells.[15]

Liver Nuclei RNase I

An enzyme from the nuclei fraction of guinea pig liver, reported by Heppel *et al.*[20] to form oligonucleotides and 5'-mononucleotides from polynucleotides, is very useful for the formation of substrates with 5'-phosphate end groups from poly A and other polymers.[21] Additional fractionation procedures have recently been applied to liver nuclei, and the enzyme has been purified further.[22] Products of the reaction may be purified by column chromatography (see Vol. VI [94] for methods).

[20] L. A. Heppel, P. J. Ortiz, and S. Ochoa, *Science* **123**, 415 (1956).
[21] M. F. Singer, L. A. Heppel, and R. J. Hilmoe, *Biochim. et Biophys. Acta* **26**, 447 (1957).
[22] L. A. Heppel, W. E. Razzell, and M. Lipsett, unpublished observations (1961).

Assay Method

Principle. The formation of oligonucleotides and mononucleotides from poly A is measured by the increase in optical density at 257 mμ of the perchloric acid-soluble fraction during the course of incubating poly A, inorganic phosphate, magnesium, and enzyme.

Procedure. In a small test tube are placed 0.05 ml. of a solution of poly A (10 mg./ml.), 0.01 ml. of 0.2 M magnesium chloride, 0.01 ml. of M potassium phosphate, pH 7, water, and enzyme to 0.15 ml. Aliquots of 0.04 ml. are removed at intervals to a small centrifuge tube containing 0.46 ml. of 3% perchloric acid, and the mixture is kept in ice for 10 minutes. This is then centrifuged, 0.1 ml. of the supernatant is made to 1.0 ml. with water, and the absorbance is read at 257 mμ. The absorbance of the perchloric acid blank is usually 0.010, and the poly A blank, 0.000. Typically, in 30 minutes 50 μl. of crude liver homogenate produces an increase in absorbance of 0.1. The units of activity are expressed as micromoles of adenylic acid residues liberated per hour.

Purification Procedure

Hog livers are obtained at the slaughterhouse, cut into 1-inch cubes, and kept in cold 0.25 M sucrose–0.0018 M calcium chloride.

Step 1. Homogenization, and Sedimentation of Nuclei. Several of the hog liver cubes are removed from the solution, blotted, and cut into 3-mm. cubes with scissors at 2°. The minced tissue is blended in fresh sucrose–calcium chloride solution (3000 cc. to 400 g. of mince) at 40 volts for 3 minutes in the Waring blendor or at the slowest speed available in a large Waring blendor. The homogenate is then filtered through cheesecloth and centrifuged in the cold with a 10-minute acceleration to 2500 r.p.m., 10 minutes at 2500 r.p.m., and a 10-minute deceleration from this speed. The supernatant is discarded, and the precipitate is suspended in 0.01 M potassium phosphate, pH 8.0 (approximately 650 ml.), containing 0.001 M mercaptoethanol and 0.001 M EDTA.

Step 2. Rupture of Nuclei and Solubilization of Enzyme. The nuclei suspension is stirred in an ice bath during the addition of one-fifth its volume of ice-cold n-butanol. Rapid stirring, without foaming, is continued for 5 minutes, and the resulting mixture is centrifuged at 6000 $\times$ g for 15 minutes. The butanol layer and the inactive protein gels are left behind, while the aqueous layer containing the enzyme is withdrawn with a long-tip pipet. This butanol-extracted solution is extracted further with 2 vol. of ether by means of a graduate cylinder, the formation of large quantities of emulsion being avoided. The ether extraction is repeated twice, and the final solution containing a small amount of ether and emulsion is centrifuged at $5000 \times g$ for 10 minutes. The aqueous layer is removed, and dissolved ether is evaporated off under vacuum.

Step 3. Ammonium Sulfate Fractionation. To every 100 ml. of the butanol–ether-treated solution is added 19.4 g. of ammonium sulfate over a period of 15 minutes. The mixture is centrifuged for 10 minutes at $5000 \times g$, and the precipitate is discarded. To the supernatant is added a further 11 g. of ammonium sulfate, and the precipitate is collected as before. Note: Sometimes, owing to incomplete removal of ether, the precipitate at this stage will float. It can be collected readily by carefully pouring the contents of the centrifuge tube through a folded filter paper with a small pinhole at the bottom. The precipitate is dissolved in $0.02\,M$ potassium phosphate, pH 7–$0.001\,M$ mercaptoethanol–$0.001\,M$ EDTA, and frozen for 2 days.

Step 4. Protamine Precipitation of Nucleic Acids. The inert precipitate which is found on thawing the above preparation is centrifuged down, and the supernatant is treated with protamine sulfate. The volume of the supernatant is about 12.8 ml., and to this is added an equal volume of cold water, plus a solution of 1% protamine sulfate until a precipitate no longer forms. This usually requires 0.32 ml. per 12.8 ml. of the solution, or 0.64 ml. per 25.6 ml. of the diluted material. The mixture is allowed to stand for 10 minutes in an ice bath and is centrifuged for 2 minutes at $10,000 \times g$.

Step 5. Alkaline Ammonium Sulfate Precipitation. Ammonium hydroxide $(0.3\,N)$ is added to the above supernatant with stirring until the pH reaches 8; then 0.73 vol. of saturated ammonium sulfate is added, and stirring is continued for a further 10 minutes. (The saturated ammonium sulfate is adjusted to pH 8.0 with concentrated ammonium hydroxide, the pH being determined on a fivefold dilution in water. The precipitate obtained by centrifuging the above mixture at $5000 \times g$ for 15 minutes is dissolved in the phosphate–mercaptoethanol–EDTA solution (approximately 10 ml.).

Step 6. Precipitation by Dialysis, and Extraction of Enzyme. The preparation is dialyzed against a solution of $0.001\,M$ potassium phosphate, $0.001\,M$ mercaptoethanol, and $5 \times 10^{-4}\,M$ EDTA, pH 7, for 4 hours at 2°. A precipitate develops, which is centrifuged down at $5000 \times g$ for 5 minutes and extracted with 1.5 ml of $0.2\,M$ phosphate–$0.001\,M$ mercaptoethanol, pH 7.0.

A summary of a representative purification is given in Table VII, from which it can be seen that, although the yields decrease, the purification is considerable. At this stage digestion of poly A results in the formation of little or no adenosine, indicating that monoesterases have been entirely removed. Little AMP is formed.

TABLE VII
PURIFICATION OF LIVER NUCLEI RNASE I

Step	Specific activity	Per cent of total units
1. Suspended nuclei	—	—
2. Solubilized enzyme	0.6	(100)[a]
3. Ammonium sulfate	0.9	33[b]
4. Protamine supernatant	1.1	46
5. Ammonium sulfate	4.4	65
6. Dialysis precipitation[c]		
a. Supernatant	1.9	25
b. Extract of precipitate	49.0	16

[a] The yield of enzyme is 150% of that obtained by rupture of the nuclei at pH 9,[20] and the specific activity is almost twice as high.
[b] An inhibitor is apparently concentrated here and removed later.
[c] These results are variable and require further study.

Properties

Prolonged storage may lead to inactivation at most stages of purification. This can be prevented by the addition of bovine plasma albumin to a concentration of 0.5% or of sucrose to 20%.

Phosphate in the assay may be replaced by 0.35 M Tris, provided the magnesium ion is $5 \times 10^{-4} M$ (final concentration).

Specificity. The enzyme hydrolyzes poly A and poly U to oligonucleotides of two to six units in length, terminated in 5′-phosphates. It solubilizes RNA[22] (as determined above with poly A, in a final concentration of 3% perchloric acid), and the products produced, unlike the original RNA, are rapidly hydrolyzed by venom phosphodiesterase to yield 5′-mononucleotides, indicating a consistent mechanism of action on the RNA and the biosynthetic polymers. The preparation at step 5 is free of phosphodiesterase I and II, but contains traces of monoesterase and of DNase I (viscometric assay). For the preparation of oligonucleotides with 5′-phosphate end groups, the monoesterase can be inhibited by including 0.01 M sodium fluoride in the reaction mixture.

Leukemic Cell Phosphodiesterase

Assay Method[23]

To 0.06 ml. of a mixture containing poly A (0.2 micromole of adenine residues), 0.15 micromole of magnesium chloride, 0.5 micromole of

[23] E. P. Anderson and L. A. Heppel, *Biochim. et Biophys. Acta* **43**, 79 (1960). The activity is found in a variety of leukemic cell tumors. The opportunity of reading this paper prior to its publication is gratefully acknowledged.

glutathione, 0.005 micromole of EDTA, and 4.0 micromoles of potassium phosphate, pH 7.6, is added 0.015 to 0.15 unit of enzyme. After 30 minutes at 37°, 0.34 ml. of 3% perchloric acid is added, and the suspension is kept in ice for 10 minutes, then centrifuged (1500 $\times$ g for 5 minutes). An aliquot (0.1 to 0.3 ml.) of the supernatant is made to 1.0 ml. with water, and the absorbance is measured at 250 mu (isosbestic point for AMP and IMP).

Purification Procedure

The tumor line employed by Anderson and Heppel[21] is carried in $(C_3H \times AKR)F_1$ hybrid mice. An intraperitoneal inoculum of 2×10^6 cells is allowed to proliferate in mice of the same strain for 6 days, the fluid is drained, and the cells are harvested by centrifugation. They are washed with 0.9% sodium chloride at 3°, resuspended in 0.1 M Tris, pH 7.3 (less than 3 vol.), and broken by vibration with stainless-steel beads. Other methods for cell rupture may also be applied.

Step 1. The homogenate is centrifuged to remove intact cells and debris (15,000 $\times$ g for 10 minutes). The supernatant is made to three times the original volume of packed cells with 0.1 M Tris, pH 7.3, plus mercaptoethanol to 0.005 M. After addition of 0.1 M sodium acetate, pH 3.5, sufficient to lower the pH to 6.9, the solution is heated in a water bath for 10 minutes at 53°, cooled to 3°, and centrifuged (15,000 $\times$ g for 3 minutes). This preparation is stable at −15°.

Step 2. A 1% protamine sulfate solution is added to the supernatant to a final concentration of 0.0433%, and the mixture is stirred for 15 minutes and centrifuged (15,000 $\times$ g for 1 minute).

Step 3. To the supernatant 5 vol. of cold distilled water, mercapto-ethanol to 0.0025 M, 0.2 N ammonium hydroxide to pH 7.5, and 1% protamine sulfate (0.56 ml. per 100 ml. of solution) are added. The mixture, stirred for 10 minutes at 3°, is centrifuged (15,000 $\times$ g for 1 minute).

Step 4. The precipitate is resuspended in 0.2 M potassium phosphate, pH 7.2, 0.005 M in mercaptoethanol and equal to 0.4 vol. of the initial protamine supernatant (step 2). This suspension is centrifuged, and the precipitate is discarded. The preparations are usually stable at −15°.

Step 2a. Purification by Chromatography. The supernatant from step 1 above is diluted by the addition of an equal volume of cold distilled water, and 9.5 ml. (28 mg. of protein) is allowed to flow into a column of DEAE-cellulose 2.2 cm. in diameter by 11 cm. long, equilibrated with 0.03 M Tris, pH 7.2, 0.001 M in EDTA and 0.005 M in mercaptoethanol. Elution is performed with a linear gradient by the introduction of the above solution containing, in addition, NaCl to 0.5 M. A flow rate of

1 to 2 ml./min. is maintained, and 5-ml. fractions are collected. The enzyme emerges between 100 and 150 ml. of effluent when the volumes of eluting solutions used for the gradient are 127 ml. each. For storage, the eluates are concentrated by ammonium sulfate precipitation, redissolved in one-fifteenth the pooled volume with a solution containing 0.04 M Tris, 0.004 M mercaptoethanol, 0.001 M EDTA, pH 7.2, and 25% glycerol.

Properties[23]

The degree of purification obtained by the protamine precipitation is less than that by the chromatographic procedure (Table VIII), but the yields of enzyme at the peak of highest specific activity from the column are about one-half of those obtained by the former method.

TABLE VIII

PURIFICATION RESULTS FOR LEUKEMIA CELL PHOSPHODIESTERASE

Step	Milliliters	Total units[a]	Specific activity[a]
Extract	100	700	1.5
1. Heated	93	575	3.6
2. Protamine	93	585	3.9
3–4. Protamine	37	520	16
2a. *Chromatographed after step 1*			
Initial	9.5	125	4.5
Tubes 15–21	50	75	16
Tube 20	5.2	19	51

[a] For definition, see page 237.

Contaminating phosphomonoesterase and AMP deaminase activities are absent from the final fractions, which permits detailed study of the substrate specificity and products formed, even though the over-all purification is not very large.

There is a broad pH optimum between 7.0 and 8.0.

Sulfhydryl reagents (mercaptoethanol, glutathione) stimulate the activity and are required for maximum stability of the preparations.

Mg^{++} stimulates the hydrolysis of poly A optimally at $3 \times 10^{-3} M$ but is inhibitory at $10^{-2} M$. It also stimulates the hydrolysis of oligonucleotides (pApA, etc.) at $8 \times 10^{-3} M$ but does *not* inhibit at higher concentrations. The oligonucleotides, however, liberate AMP more slowly than does poly A (one-tenth as fast). Inorganic pyrophosphate inhibits (40% at 0.01 M), as does fluoride (100% at 0.04 M).

The K_m for poly A hydrolysis was estimated to be $2.3 \times 10^{-3} M$ (as adenine), but after 20% hydrolysis the reaction ceased. There is some product inhibition by AMP, but this is insufficient to explain the above results. Oligonucleotides, however, are hydrolyzed completely to monomers, with the appearance of transient intermediates suggestive of stepwise degradation.

Deoxyriboöligonucleotides are hydrolyzed similarly, though more slowly, but pyrimidine oligonucleotides (pUpU, etc.) are hydrolyzed more rapidly than, say, pApApA.

The preparations also attack poly U, poly G, and poly AU (but more slowly than poly A). Poly C, poly AGUC, yeast RNA, TMV RNA, and *A. agile* RNA are very slowly hydrolyzed.

Benzyl *p*-nitrophenyl esters of 5'-mononucleotides, adenosine-2',3'-cyclic phosphate, and oligonucleotides with 3'-phosphate end groups are not split.

Oligonucleotides with no phosphate end groups are hydrolyzed more slowly than those with 5'-phosphate ends.

The enzyme is thus similar to the venom and venom-type phosphodiesterases except for its requirement for a nucleoside on either side of the phosphate bond.

Nucleoside 2',3'-Cyclic Phosphodiesterase

Assay Method

Principle. The rate of hydrolysis of adenosine-2',3'-cyclic phosphate to adenosine-2'-phosphate by a limiting amount of enzyme is followed by measuring the inorganic phosphate released in the presence of an excess of phosphomonoesterase.

Procedure. The assay mixture contains 10 μmoles of Tris buffer adjusted to pH 7.4 with acetic acid, 1.5 μmoles of ammonium, sodium, or potassium adenosine-2',3'-cyclic phosphate (see Vol. VI [94]), 25 μg. of purified *Escherichia coli* phosphomonoesterase[24, 25] (available commercially from Worthington Biochemicals, Freehold, New Jersey), and from 5 to 30 μg. dry weight, of the 2',3'-cyclic phosphodiesterase in a total volume of 0.10 ml. The mixture, without phosphodiesterase, is placed in a water bath at 37° for 2 minutes. The phosphodiesterase is added, and samples of 20 μl. are transferred directly to 0.38 ml. of 4% HClO$_4$ at 90-second intervals. Insoluble material (from very crude preparations) is removed by centrifugation at 1500 × *g* for 2 minutes and 0.3 ml. of the supernatant fraction from each sample is recovered for determination of inorganic phosphate by the method of Chen *et al.*,[26] as

[24] A. Garen and C. Levinthal, *Biochim. et Biophys. Acta* **38,** 470 (1960).

[25] M. Malamy and B. L. Horecker, *Biochem. Biophys. Research Commun.* **5,** 104 (1961).

[26] P. S. Chen, T. Y. Toribara, and H. Warner, *Anal. Chem.* **28,** 1786 (1956).

modified by Ames and Dubin.[27] A reaction mixture without adenosine-2',3'-cyclic phosphate is carried through the same procedure to establish the rate of liberation of inorganic phosphate from constituents in the 2',3'-cyclic phosphodiesterase preparation itself.

Purification Procedure

Concentrated, stable preparations from beef brain are obtained by the procedure of Drummond et al.[28]

Step 1. Tissue Preparation and Extraction with Acetone. Fresh beef brain is cut into one-inch slices and packed in ice immediately upon removal from the animal. Each slice is frozen (at $-20°$) in paraffin film and 100 g. of frozen brain are homogenized directly in 500 ml. of acetone, previously cooled to $-20°$, for 4 minutes in a high-speed blendor fitted with a variable transformer located in a cold room ($3°$). Care must be taken to accelerate the blendor gradually, to avoid a sudden overflow. The suspension is filtered on a Buchner funnel, the precipitate resuspended without delay in 500 ml. of acetone with the high-speed blendor, and again filtered. The suspension and filtration are repeated once more and the precipitate dried under vacuum at $3°$, using two dry ice–acetone traps in series with an oil pump.

Step 2. Butanol Extraction. The dried powder is suspended by blending for 4 minutes at $3°$ in 200 ml. of n-butanol ($-20°$), an additional 150 ml. of n-butanol are added with stirring at low speed, and the suspension is transferred to a stoppered flask containing a magnetic stirring bar. Stirring is continued for 5 hours and the precipitate is recovered by centrifugation in the cold at $10,000 \times g$ for 10 minutes.

Step 3. Extraction with Ether. The precipitate is suspended in 200 ml. of petroleum ether (b.p. $30°–60°$) previously cooled to $-20°$, homogenized at high speed, and stirred at low speed for 30 minutes. The suspension is filtered, the precipitate briefly stirred with 150 ml. of cold ether, and filtration repeated. The precipitate is dried under vacuum as before, in the presence of paraffin shavings.

Step 4. Extraction with Aqueous Buffer. The precipitate is homogenized in the cold at slow speed in 350 ml. of 0.02 M potassium phosphate, pH 7.5, for 3 minutes, and slowly stirred for 2 hours. The suspension and overlying froth are centrifuged at $30,000 \times g$ for 90 minutes and the supernatant is discarded. The precipitate is homogenized for 3 minutes at $3°$ in a mixture of 180 ml. of 5 M NaCl, 20 ml. of Tween 20 (Atlas Chemicals, Wilmington, Delaware), 0.5 ml. of 5 M KOH, and 100 ml. of 0.03 M potassium phosphate, pH 7.5. It is then stirred for 2 hours and

[27] B. N. Ames and D. T. Dubin, J. Biol. Chem. 235, 769 (1960).
[28] G. I. Drummond, N. T. Iyer, and J. Keith, J. Biol. Chem. 237, 3535 (1962).

centrifuged at 30,000 × *g* as before. The supernatant is kept in ice, the precipitate is suspended, stirred in one-half the previous volume of salt–Tween solution, centrifuged, and then the supernatant is pooled with the previous one. The combined supernatants are dialyzed until free of chloride, then lyophilized. The residue is freed of Tween by homogenization in 300 ml. of acetone (−20°) and centrifugation (−15°), three times, followed by drying under vacuum as before. About 4 g. of dry powder results. For use, the powder may be suspended in water or dilute buffers and subjected to sonic oscillation in order to effect maximum dispersion of the active particles.

Properties

The final preparation contains about 40% protein, with a specific activity of 104 μmoles per hour per mg. dry weight in the assay employing adenosine-2′,3′-cyclic phosphate. The dried preparation is stable at −20° for several months; the suspension in water loses activity slowly over the course of several weeks at the same temperature.

There is a broad pH optimum between pH 5.8 and 7.4, as determined by electrophoretic separation of the products of hydrolysis of adenosine-2′,3′-cyclic phosphate, and no divalent cation requirement.[28] Cupric, zinc, and mercuric ions are inhibitory.

Specificity. Although the preparation after step 4 shows no activity toward RNA when measured by the release of acid-soluble fragments, it does inactivate transfer-RNA. For example, an amount of enzyme which hydrolyzes 24 μmoles of adenosine-cyclic-p per hour (230 μg. dry weight) causes a 40% decrease in the amino acid acceptor activity of transfer-RNA in 5 minutes.[29] The particulate nature of the preparation precludes treatment on columns of ion-exchange resins or bentonite, or even with suspensions of bentonite, to remove ribonuclease activity. Treatment with the acetate form of Dowex 1 in suspension[28] did not decrease the rate of inactivation of transfer-RNA and is of questionable value.

The relative activity toward the cyclic phosphates is: adenosine, 100; guanosine, 55; cytidine, 31; and uridine, 18. In addition, the terminal 2′,3′-cyclic phosphate bonds of GpC-cyclic-p, ApC-cyclic-p, and CpC-cyclic-p may be hydrolyzed without detectable cleavage of the internucleotide linkage.[28] Similar results have been obtained using digests of RNA with commercial pancreatic ribonuclease II in which the rates of hydrolysis of the terminal cyclic phosphates of preparations with average chain lengths of 6–16 nucleotide residues were equivalent to the rate of hydrolysis of U-cyclic-p.[29] Thus, the diesterase may be employed in

[29] J. R. Beard and W. E. Razzell, in preparation.

concert with pancreatic ribonuclease II and alkaline phosphomonoesterase, to provide a direct estimate of the rate of phosphodiester bond
hydrolysis by the ribonuclease. Under appropriate conditions, the low
ribonuclease activity of the diesterase preparation does not interfere.[29]

[30] Deoxyctp- and Deoxycdp-Splitting Enzyme

$$dCTP + H_2O \rightarrow dCMP + PP$$
$$dCDP + H_2O \rightarrow dCMP + P$$

By STEVEN B. ZIMMERMAN

Assay Method

Principle. The release of inorganic pyrophosphate is measured by the
conversion of radioactive phosphorus in the terminal pyrophosphate
group of deoxycytidine triphosphate into a form not adsorbed by charcoal.[1]

Reagents

dCPP^{32}P^{32}, 0.3 micromole/ml., 10^6 c.p.m./micromole. Prepared by
phosphorylation of dCMP[2] with P^{32}-acetyl phosphate.[3] The
dCPP^{32}P^{32} is isolated by ion-exchange chromatography[2] and
concentrated with Norit.[4]
2-Mercaptoethanol, 0.125 M.
MgCl$_2$, 0.1 M.
Glycine buffer, pH 9.2, 0.5 M.
"Carrier." Crystalline bovine serum albumin (5 mg./ml.) in
0.025 M sodium pyrophosphate, pH 7, and 0.025 M potassium
phosphate buffer, pH 7.

[1] A. Kornberg, S. B. Zimmerman, S. R. Kornberg, and J. Josse, *Proc. Natl. Acad. Sci. U.S.* 45, 772 (1959).
[2] I. R. Lehman, M. J. Bessman, E. S. Simms, and A. Kornberg, *J. Biol. Chem.* 233, 163 (1958).
[3] A. Kornberg, S. R. Kornberg, and E. S. Simms, *Biochim. et Biophys. Acta* 20, 215 (1956).
[4] All steps at 0° to 5°. To the eluate at pH 2 is added 0.1 ml. of a 20% suspension of acid-washed Norit (prepared as in footnote 5) per micromole of nucleotide present. After 5 minutes of stirring, the Norit is collected by centrifugation, washed with 5 ml. of cold water, and eluted twice by stirring for 5 minutes with 2-ml. portions of 0.045 M NH$_4$OH in 50% ethanol and centrifuging. The volume of the combined eluants is reduced to 2 ml. by evaporation under an air stream. Colloidal charcoal is aggregated by freezing and thawing and removed by centrifugation.

Norit, 20% suspension.[5]

HCl, 0.1 N.

Enzyme. Dilute enzyme in 0.02 M Tris buffer, pH 7.5, containing 0.01 M 2-mercaptoethanol.

Procedure. To tubes containing 0.02 ml. each of the dCPP^{32}P^{32}, 2-mercaptoethanol, MgCl$_2$, and glycine buffer solutions and sufficient water to give a final volume of 0.25 ml. is added 1 to 20 × 10^{-5} unit of enzyme. The tubes are incubated for 20 minutes at 37°. The reaction is terminated by placing the tubes in an ice bath and immediately adding 0.5 ml. of 0.1 N HCl followed by 0.2 ml. of "carrier" and 0.1 ml. of the Norit suspension. The Norit is kept from settling by occasional mixing for 2 to 3 minutes. After centrifugation for 3 minutes at 5000 × g, a 0.5-ml. aliquot of the supernatant fluid is assayed for radioactivity.

Definition of Unit and Specific Activity. One unit of enzyme is defined as that amount releasing 1 micromole of PP per minute under the above conditions. Specific activity is expressed as units per milligram of protein. Protein is determined by the method of Lowry *et al.*[6]

Application of Assay Method to Crude Tissue Preparations. No interfering side reactions leading to error in applying the assay to crude extracts have been observed.

Purification Procedure

Preparation of T2-Infected E. coli Cells.[1] *Escherichia coli* B is grown at 37° with vigorous aeration in M-9 medium[7] modified to contain, per liter: KH$_2$PO$_4$, 3 g.; Na$_2$HPO$_4$, 6 g.; NH$_4$Cl, 1 g.; MgSO$_4$·7H$_2$O, 0.49 g.; glucose, 5 g.; and FeSO$_4$·7H$_2$O, 0.5 mg. At late exponential phase (2 × 10^9 cells/ml.), 3 to 4 T2r$^+$ per cell are added, and 10 minutes later 25 ml. of 2-mg./ml. chloramphenicol solution are added per liter of culture. The culture is chilled on ice and harvested by centrifugation. The yield is 2 g. of cells (wet weight) per liter. The infected cells may be used immediately or stored at −12° for at least 2 years with less than 20% loss of enzyme.

[5] Norit A (160 g.) is suspended in 2 l. of distilled water and allowed to settle for about 12 hours. The supernatant fluid is discarded. The Norit is suspended in 2 l. of 1 N HCl and held under vacuum until bubbling ceases (1 to 2 hours). The charcoal is washed with distilled water as before until the supernatant fluid is above pH 3, centrifuged briefly, and resuspended in four times its volume of distilled water.

[6] O. H. Lowry, N. J. Rosebrough, A. L. Farr, and R. J. Randall, *J. Biol. Chem.* **193**, 265 (1951). Protein in fractions containing interfering materials (streptomycin, 2-mercaptoethanol) is assayed after precipitation in the cold with trichloroacetic acid (10% final concentration).

[7] E. H. Anderson, *Proc. Natl. Acad. Sci. U.S.* **32**, 120 (1946).

All the following operations are carried out at 0° to 5°.

Step 1. Preparation of Extract. Thirteen grams of T2-infected cells are suspended in 60 ml. of $0.05\,M$ glycylglycine buffer, pH 7.0, and treated for 18 minutes in a Raytheon 10-kc. sonic oscillator. The suspension is centrifuged for 20 minutes at $12,000 \times g$, and the supernatant fluid is collected. The protein content is determined and adjusted to 10 mg./ml. by dilution with $0.05\,M$ glycylglycine buffer, pH 7.0 (fraction I).

Step 2. Streptomycin Precipitation of Inactive Materials. To 105 ml. of fraction I are added 32 ml. of 5% streptomycin sulfate with stirring over a 10-minute period. After an additional 10 minutes of stirring, the suspension is centrifuged for 15 minutes at $12,000 \times g$. The pellet is discarded. The supernatant fluid is adjusted to pH 8.0 to 8.2 by slow addition with constant stirring of $0.1\,N$ KOH (22 to 26 ml. required) (fraction II).

Step 3. First DEAE Chromatography. A column of DEAE[8] (8.8 cm.^{2} $\times$ 26 cm.) is equilibrated with $0.02\,M$ potassium phosphate buffer, pH 8.1, containing $0.01\,M$ 2-mercaptoethanol. Fraction II (150 ml.) is passed through the column at 3 ml./min. and washed into the column with 10 ml. of the equilibrating solution. A constant gradient of elution from $0.08\,M$ NaCl to $0.32\,M$ NaCl is applied, both limiting solutions (500 ml. each) containing $0.02\,M$ potassium phosphate buffer, pH 8.1, and $0.01\,M$ 2-mercaptoethanol. Fractions of 20 ml. are collected at a flow rate of 6 ml./min. Enzyme activity appears as a peak in the last quarter of the gradient. The fractions of highest specific activity are pooled (fraction III).

Step 4. Second DEAE Chromatography. Fraction III (108 ml.) is dialyzed for 4 hours against two 2-l. portions of $0.02\,M$ potassium phosphate buffer, pH 8.1, containing $0.01\,M$ 2-mercaptoethanol. A column (0.8 cm.$^2 \times$ 15 cm.) of DEAE is equilibrated with solution of the same composition. The dialyzed fraction III is passed through the column at a rate of 0.5 ml./min. and washed into the column with 5 ml. of the equilibrating solution. A constant gradient of elution from $0.04\,M$ to $0.28\,M$ potassium phosphate buffer, pH 6.5, is applied, both limiting solutions (130 ml. each) containing $0.01\,M$ 2-mercaptoethanol. Fractions of 6.5 ml. are collected at a flow rate of 1 ml./min. Enzyme activity appears as a peak in the middle third of the gradient. Peak fractions

[8] E. A. Peterson and H. A. Sober, *J. Am. Chem. Soc.* **78**, 751 (1956). Diethylaminoethylcellulose (Type 40, Brown Co., New Hampshire) is exposed to $0.25\,N$ NaOH for 15 minutes at room temperature and then exhaustively washed with water and buffer both before use and to regenerate used adsorbent.

may be pooled[9] (fraction IV) and are stable for at least 2 months at 5°. Fractions containing 2-mercaptoethanol are not frozen. Table I summarizes the yields and purification obtained at each step.

TABLE I
SUMMARY OF PURIFICATION PROCEDURE

Fraction	Step	Volume, ml.	Units	Protein, mg.	Specific activity, units/mg.	Yield, %
I	Sonic extraction	107	225	1070	0.21	100
II	Streptomycin	152	170	338	0.50	76
III	First DEAE chromatography	122	57	15.9	3.6	25
IV	Second DEAE chromatography	26	28	2.0	14.0	12

The activity in fraction IV may be concentrated in about 60% yield by addition of 0.8 g. of $(NH_4)_2SO_4$ per milliliter, equilibration for 30 minutes, and centrifugation for 40 minutes at 12,000 $\times$ g. The pellet is thoroughly drained and redissolved in a minimal volume of 0.05 M glycylglycine buffer, pH 7.0, containing 0.002 M glutathione. This solution may be stored at 12° for a year without loss of activity but loses 10 to 30% of its activity on a single freezing and thawing.

Properties

Specificity. The purified enzyme hydrolyzes both dCTP and dCDP. The relative activities on these two substrates remains constant throughout the purification, dCTP being cleaved several times as rapidly as dCDP.[10] There is little or no activity on dTTP, dGTP, dATP, 5-hydroxymethyl dCTP, 5-bromo dCTP, 5-bromo dUTP, ATP, GTP, CTP, or UTP.[1,10]

Activators and Inhibitors. Mg++ is required for activity. Zn++, Ca++, or Mn++ will not replace Mg++.[10]

dCMP is a competitive inhibitor of both dCDP and dCTP hydrolysis. dCDP inhibits competitively the cleavage of dCTP, as does dCTP the cleavage of dCDP (Table II).[10]

Fluoride is strongly inhibitory. Inhibitions of 60% and 85% are obtained at 2×10^{-4} M and 4×10^{-4} M, respectively. The inhibition is independent of phosphate.[10]

2-Mercaptoethanol in the assay stimulates two- to threefold the activity of enzyme fractionated in is absence.[10]

[9] Enzyme in the last half of the peak is significantly contaminated with inorganic pyrophosphatase.
[10] S. B. Zimmerman and A. Kornberg, *J. Biol. Chem.* in press.

TABLE II
AFFINITIES OF VARIOUS SUBSTRATES AND INHIBITORS

Substrate	Inhibitor	$K_m,$ $M \times 10^6$	$K_i,$ $M \times 10^6$
dCDP	dCTP	2.4	4.2
	dCMP		140
dCTP	dCDP	3.9	2.7
	dCMP		200

Kinetic Properties. The K_m and K_i values for several substrates and inhibitors are listed in Table II.[10]

pH optimum. Activity is maximal at about pH 9. Rates at pH 8 or 10 are about 40% of the value at pH 9.[10]

Distribution. The activity has been found in extracts of T2-,[1,10,11] T4-,[10] and T6-infected[10] cells at similar levels.[10] Less than 1% of this activity is found in extracts of noninfected or T5-infected *E. coli* B.[1,10]

[11] J. F. Koerner, M. S. Smith, and J. M. Buchanan, *J. Biol. Chem.* 235, 2691 (1960).

[31] Metaphosphate Synthesis by an Enzyme from *Escherichia coli*

$$n \text{ ATP} \rightleftarrows n \text{ ADP} + (PO_3^-)_n$$

By S. R. KORNBERG

Assay Method[1,2]

Principle. The assay measures the production of acid-insoluble labeled phosphate derived from terminally labeled ATP.

Reagents

ADP, 0.01 M.
P^{32}-acetyl phosphate, 0.015 M.
$MgCl_2$, 0.1 M.
Glycylglycine buffer, pH 7.0, potassium salt, 0.5 M.
Ammonium sulfate, 1 M.
Acetokinase,[3] 200 units/ml.

[1] A. Kornberg, S. R. Kornberg, and E. S. Simms, *Biochim. et Biophys. Acta* 20, 215 (1956).
[2] S. R. Kornberg, *Biochim. et Biophys. Acta* 26, 294 (1957).
[3] I. A. Rose, Vol. I [97], step 3.

Procedure. The incubation mixture contains 0.03 ml. of ADP, 0.03 ml. of acetyl phosphate, 0.01 ml. of $MgCl_2$, 0.025 ml. of glycylglycine buffer, 0.01 ml. of ammonium sulfate, 0.01 ml. of acetokinase, and water to a final volume of 0.25 ml. After incubation for 15 minutes at 37°, the mixture is treated with 0.25 ml. of cold 7% perchloric acid and then 0.5 ml. of bovine serum albumin (Armour Company, crystalline; 1.6 mg./ml.). The mixture is centrifuged at about 15,000 × g for 2 minutes. The precipitate is washed twice with 2.0-ml. portions of 3.5% perchloric acid and dissolved in 0.5 ml. of 0.5 M NH_4OH; an aliquot of 0.2 ml. is assayed for radioactivity.

Definition of Units and Specific Activity. One unit of enzyme is defined as the amount producing 0.01 micromole of acid-insoluble phosphate in 15 minutes. Specific activity is expressed as units per milligram of protein. Protein is determined by the method of Lowry *et al.*[4]

Purification Procedure

Growth of Cells. *Escherichia coli* strain B is grown in a medium containing 2% K_2HPO_4, 1.65% KH_2PO_4, 1% yeast extract (Difco, dehydrated), 1% glucose, and about 20 mg. of Antifoam A (Dow Corning Company) per liter. The pH is 7.0 to 7.2. The glucose is autoclaved separately and added to the cooled medium. Fifteen liters of the medium in a 20-l. Pyrex bottle is inoculated with 100 ml. of an 8-hour broth culture and incubated at 34° to 37° for 16 to 20 hours with vigorous forced aeration. The yield of cells, harvested in a Sharples supercentrifuge, is about 8 g./l. (wet weight). The cells are washed with 2 vol. of cold 0.5% NaCl–0.5% KCl and extracted immediately or stored at −13°; active extracts are obtained from cells stored as long as 6 weeks.

Preparation of Cell-Free Extract. Ten grams of cells, suspended in 50 ml. of glycylglycine buffer (0.02 M, pH 7.0), is treated in a Raytheon 10-kc. oscillator at 0° to 2° for 10 minutes. The suspension is centrifuged for 15 minutes at 10,000 × g, and the residue is discarded. The supernatant extract (fraction 1) is fractionated immediately or stored at −13°, with little loss in activity, for 6 weeks or longer.

Purification of the Enzyme. See the table. All operations are carried out at 0° to 3°.

Fraction 2. To 30 ml. of fraction 1 in each of five 50-ml. tubes is added, dropwise with stirring, an amount of streptomycin sulfate (5% solution) which leaves 5 to 10% of the enzyme in the supernatant fluid. The amount of streptomycin to be added is determined for each run by a small-scale titration; 3.2 ml. per tube is used for the preparation de-

[4] O. H. Lowry, N. J. Rosebrough, A. L. Farr, and R. J. Randall, *J. Biol. Chem.* **193**, 265 (1951).

scribed here. After 5 minutes the precipitate is separated by centrifugation, and the supernatant fluid is discarded. The precipitate in each tube is thoroughly dispersed in 30 ml. of potassium phosphate buffer (0.02 M, pH 6.5) with the aid of a pestle and centrifuged after 5 minutes. To the combined supernatant fluids of the five tubes are added 20 ml. of potassium phosphate buffer (0.1 M, pH 7.4) and water to bring the volume to 300 ml.

PURIFICATION OF THE ENZYME

Fraction	Units/ml.	Total units	Protein, mg./ml.	Specific activity, units/mg. protein
1. *E. coli* extract	215	32,200	15.0	14.3
2. Streptomycin eluate	95	28,500	1.82	52.1
3. Ammonium sulfate I	114	17,100	1.12	102
4. Ammonium sulfate II	262	9,850	0.98	265
5. Nuclease treatment	207	8,300		
6. Ammonium sulfate III	272	5,100	0.17	1600

Fraction 3. To 300 ml. of fraction 2 is added 60 g. of ammonium sulfate. After 5 minutes, the precipitate is removed by centrifugation, and to the supernatant fluid is added 31 g. of ammonium sulfate. After 5 minutes, the precipitate is collected by centrifugation and dissolved in 150 ml. of potassium phosphate buffer (0.02 M, pH 6.9).

Fraction 4. To 150 ml. of fraction 3 is added 33 g. of ammonium sulfate. The precipitate is discarded after separation by centrifugation, and 7 g. of ammonium sulfate is added to the supernatant fluid. This precipitate is collected and dissolved in 37.5 ml. of potassium phosphate buffer (0.02 M, pH 6.9).

Fractions 5 and 6. To 37.5 ml. of fraction 4 are added 1.87 ml. of potassium phosphate buffer (1 M, pH 7.4), 0.93 ml. of MgCl$_2$ (0.1 M), 0.375 ml. of a deoxyribonuclease solution (0.3 mg./ml.), and 0.375 ml. of a ribonuclease solution (0.5 mg./ml.). After incubation for 30 minutes at about 25°, the mixture is chilled in ice to 0° to 3° and treated with 8.4 g. of ammonium sulfate. The precipitate is separated by centrifugation and discarded, and to the supernatant fluid is added 2.25 g. of ammonium sulfate. This precipitate is collected and dissolved in 18.7 ml. of potassium phosphate buffer (0.02 M, pH 6.9). After about 1 hour, a fibrous precipitate develops which is separated by centrifugation and suspended in 18.7 ml. of the same buffer (fraction 6). Further purification of fraction 6, which is an insoluble fraction, is accomplished by adding 0.1 vol. of saturated ammonium sulfate (5.3 M) to the enzyme suspension. After centrifugation, the enzyme is recovered without loss in the supernatant fluid, whereas 70% of the protein is discarded in the precipitate.

Properties

Enzyme fractions stored for weeks at $-13°$, often with repeated thawing and refreezing, show no signs of deterioration. Fractions 2 and 3, retested after 6 to 8 weeks, show no detectable loss in activity. The optimum pH of the reaction, determined in Tris buffers in the presence of a severalfold excess of acetokinase, is found to be near 7.2. The rates at pH 6.2, 6.7, 7.5, and 8.2 have been found to be 45, 75, 83, and 73%, respectively, of the rate at pH 7.2. At pH 7.2 the rate in phosphate buffer is about 25% slower than in Tris buffer. Ammonium sulfate stimulates the reaction rate of enzyme fraction 6; the rate in the presence of $0.04\,M$ ammonium sulfate appears maximal and almost three times the rate in its absence. When the enzyme solubilized by ammonium sulfate was studied, the addition of ammonium sulfate did not stimulate the rate, although the concentration of the salt was only $0.002\,M$. Fluoride ($5 \times 10^{-4}\,M$) completely inhibits the enzyme; 2,4-dinitrophenol ($3 \times 10^{-4}\,M$) does not inhibit. Magnesium is essential to the reaction.

[32] Preparation and Assay of Phosphorylating Submitochondrial Particles

By C. L. WADKINS and A. L. LEHNINGER

Extraction of rat liver mitochondria with digitonin gives rise to submitochondrial fragments which possess intact respiratory chains and couple the phosphorylation of ADP with the oxidation of D-β-hydroxybutyrate.

Preparation of Submitochondrial Particles[1]

Twice-washed mitochondria prepared by the method described by Hogeboom[2] are suspended evenly in 6.0 ml of chilled $0.05\,M$ sucrose for each 100 g. of liver tissue used. After the total volume of this mitochondrial suspension is determined, 0.66 vol. of a chilled 2% solution of recrystallized[3] digitonin (Fisher Scientific Co.) in $0.05\,M$ sucrose is

[1] T. M. Devlin and A. L. Lehninger, *J. Biol. Chem.* 233, 1586 (1958).
[2] G. H. Hogeboom, Vol. I [3].
[3] Commercial digitonin should be recrystallized from hot ethanol; some batches of commercial preparations contain heavy metal ions.

added dropwise with continuous stirring at 0° to yield a final digitonin concentration of 0.8%. The suspension is then maintained at 0° for 20 minutes at which time sufficient chilled $0.25\,M$ sucrose is added to reduce the digitonin concentration to 0.25%. The mixture is transferred to chilled tubes and centrifuged at $20,000 \times g$ at 2° for 20 minutes to remove unfragmented mitochondria. The supernatant solution is removed so that the upper, loosely packed layer is also decanted; this supernatant is recentrifuged at $80,000 \times g$ for 25 minutes. The supernatant solution is discarded, and the pellet, which contains the active fragments, is suspended in cold distilled water. The pellet obtained from the mitochondria of 50 g. of rat liver is suspended to a final volume of 8 ml. and normally contains 0.8 to 1.2 mg. of nitrogen per milliliter. This suspension may be used directly in the phosphorylation assay system[1,4] or for studies of the ATP–$P_i{}^{32}$ exchange,[5] the ATPase activity,[6] or the ATP–ADP exchange activity.[7]

Oxidation Phosphorylation

Assay Method

Principle. In the presence of ADP and inorganic phosphate, the submitochondrial particles will oxidize D-β-hydroxybutyric acid quantitatively to acetoacetic acid with coupled production of ATP. Although P:O ratios of 1.5 to 2.0 are obtained with this system, the addition of a hexokinase–glucose phosphate "trap" has made it possible to obtain routinely P/O ratios above 2.0 (2.0 to 2.5). Acetoacetate is determined by a modification of the sensitive and specific method of Walker[8] which involves coupling with diazotized p-nitroaniline to produce the N,N'-di(-p-nitrophenyl-)C-acetylformazan, which shows characteristic absorption at 450 mμ. ATP production is estimated by employing P^{32}-labeled inorganic phosphate in the reaction medium and measuring the appearance of glucose-6-P^{32}. This is done by separation of the residual P^{32}-labeled inorganic phosphate and labeled glucose-6-P by reaction of the inorganic phosphate with ammonium molybdate followed by quantitative extraction of the phosphomolybdate complex with an isobutanol–benzene mixture.[9] The labeled glucose-6-P is present in the aqueous phase. If the hexokinase–glucose system is not used, labeled ATP will be present in the aqueous phase.

[4] C. Cooper and A. L. Lehninger, *J. Biol. Chem.* **219**, 489 (1956).
[5] C. Cooper and A. L. Lehninger, *J. Biol. Chem.* **224**, 561 (1957).
[6] C. Cooper and A. L. Lehninger, *J. Biol. Chem.* **224**, 547 (1957).
[7] C. L. Wadkins and A. L. Lehninger, *J. Biol. Chem.* **233**, 1589 (1958).
[8] P. G. Walker, *Biochem. J.* **58**, 699 (1954).
[9] S. O. Nielson and A. L. Lehninger, *J. Biol. Chem.* **215**, 555 (1955).

Reagents

0.2 M DL-β-hydroxybutyric acid, pH 6.5.
0.1 M histidine buffer, pH 6.5.
0.3 M potassium phosphate, pH 6.5.
0.025 M adenosine diphosphate, pH 6.5.
0.19 M trichloroacetic acid.
Hexokinase (Sigma Type V 500,000 K.M. units/g.), 10.0 mg./ml. in 0.25 M glucose (combined reagent).
Pyrophosphate-free P^{32}-labeled orthophosphate, 5,000,000 c.p.m./ml.
Digitonin particles, 0.8 to 1.2 mg. of nitrogen per milliliter.

Reaction Medium. The reaction medium contained the following components, in the final concentrations given: 0.1 ml. of DL-β-hydroxybutyric acid, 0.02 M; 0.1 ml. of histidine buffer, pH 6.5, 0.01 M; 0.1 ml. of K-phosphate, pH 6.5, 0.03 M; 0.1 ml. of P^{32}-orthophosphate, 500,000 c.p.m., containing 0.1 ml. of adenosine diphosphate, 0.0025 M; 0.1 ml. of glucose–hexokinase, 0.025 M, yielding 1 mg. of hexokinase per milliliter of reaction medium; water to a final volume of 1.0 ml.; and 0.2 ml. of digitonin particles, 150 to 200 μg. of nitrogen per milliliter.

Procedure. Enzyme is added last to start the reaction. The reaction is usually carried out for 20 minutes at 20° to 22° with shaking in air and is terminated by addition of 3.5 ml. of 0.19 M trichloroacetic acid. After removal of protein by centrifugation, aliquots of the supernatant liquid are used for ATP^{32} and acetoacetate analysis.

Acetoacetate Analysis

Reagents

p-Nitroaniline, 0.05% in 0.05 N HCl.
Sodium nitrite, 0.5% (must be prepared just before use).
Acetate buffer, 1.0 M, pH 5.7.
Hydrochloric acid, 5.0 N.
Sodium acetate, 0.2 M.
Diazo reagent. This reagent is prepared just before use by adding 15 ml. of sodium nitrite to 100 ml. of p-nitroaniline and then cooling to 2°. Then 35 ml. of 0.2 M sodium acetate is added.

Preparation of Samples. Deproteinized reaction system (1.5 ml.) containing 0.15 M trichloroacetic acid is mixed with 1.5 ml. of 1.0 M acetate buffer. The pH of the system at this point must be 4.9 to 5.0. Six milliliters of the freshly prepared diazo reagent is added, and the mixture is kept at 22° for 40 minutes followed by the addition of 2.0 ml.

of $5N$ HCl. The system is extracted with 8.0 ml. of ethyl acetate. Three milliliters of ethyl acetate phase is placed in a 1-cm. cuvette, and the extinction is determined at 450 mμ in a Beckman DU spectrophotometer. An optical density of 1.0 corresponds to 840 millimicromoles of aceto-acetate in the original enzymatic assay system.

It is unnecessary to use acetoacetate, an unstable compound, as working standard, since the extinction coefficient of the product of the reaction with diazotized P-nitroaniline has been found to be constant under the conditions used here.[8]

Analysis for Phosphate Uptake[9]

  Reagents

  Isobutanol–benzene reagent. This is prepared by shaking equal volumes of water-saturated isobutanol and water-saturated benzene and allowing the mixture to stand until the excess water separates.
  Ammonium molybdate–H_2SO_4 reagent, 5% ammonium molybdate in $4N$ sulfuric acid.
  Acetone.

Procedure. One milliliter of acetone and a 0.5-ml. sample of deproteinized reaction medium are mixed in a 40-ml. glass-stoppered shaking tube and allowed to stand for 10 minutes. Then 2.0 ml. of water and 7.0 ml. of isobutanol–benzene reagent are added, and the mixture is shaken vigorously for 30 seconds. Next 1.5 ml. of acid–molybdate reagent is added and mixed by gentle swirling. This mixture is allowed to stand for 5 minutes. The mixture is then shaken vigorously for 30 seconds and centrifuged at 1500 r.p.m. for 3 to 4 minutes. The bottom (aqueous) phase is removed and filtered through Whatman No. 50 paper into a second shaking tube containing 4.0 ml. of isobutanol–benzene reagent. This mixture is shaken vigorously and centrifuged at 1500 r.p.m. for 4 minutes. The aqueous phase is withdrawn and filtered through paper as before. The radioactivity of this sample is now determined by counting wet 1.0 ml. with constant geometry.

The net formation of ATP during the reaction is calculated from the radioactivity, A, of the extracted sample (counts per minute), the specific radioactivity, B, of the inorganic phosphate of the original reaction system (counts per minute per micromole), and the dilution introduced during the extraction procedure:

$$+ \ \Delta\mu\text{moles ATP} = \frac{A}{B} \times 4 \times \frac{1.0 + 3.5}{0.5}$$

The P/O ratio can then be expressed as

$$P/O = \frac{+ \; \Delta\mu\text{moles ATP}}{+ \; \mu\text{moles acetoacetate}}$$

Properties of the Phosphorylation System

Specificity. D-β-Hydroxybutyric acid, succinic acid, and ascorbate plus external cytochrome c are the only substrates whose oxidation is coupled to the synthesis of ATP in these preparations, but the efficiency with succinate is low (P/O = 0.2). ADP is the only nucleotide which will serve as a phosphate acceptor.

Stability. When aged at 30°, approximately 50% of the oxidation-coupled phosphorylation of ADP is lost in 60 minutes. At 2°, 50% loss occurs in 24 hours. Freezing and thawing completely destroy enzymatic activity. Addition of bovine serum albumin to the test medium (2 mg./ml.) will restore some of the activity of aged preparations.

Activators and Inhibitors. The addition of metal ions is not required for optimal P/O ratios, although the submitochondrial particles do contain bound Mg^{++}. Bovine serum albumin (2 mg./ml.) usually improves the P/O ratio and will prevent uncoupling by long-chain fatty acids. 2,4-Dinitrophenol, gramicidin, Dicumarol, arsenate, and azide uncouple at the same concentrations found to be effective with intact mitochondria, but thyroxine and its analogs and Ca^{++} do not act as uncoupling agents. Optimal P/O ratios are obtained at pH 6.5.

"Partial Reactions" of Oxidative Phosphorylation

The submitochondrial particles catalyze two exchange reactions which are considered to represent intermediate steps in the coupling process. The ATP-P_i[32] exchange is thought to be a manifestation of those steps involved in the uptake of inorganic phosphate with the ultimate synthesis of ATP,[5] and the ATP–ADP (C^{14}) exchange reaction is considered to represent the final transphosphorylation reaction whereby ADP reacts with a phosphorylated intermediate to form ATP.[7]

Assay Method for the ATP-P^{32} Exchange Reaction[5]

Principle. The assay of this reaction involves incubation of P^{32}-labeled inorganic orthophosphate, ATP, and the submitochondrial particles and, after termination of the reaction with trichloroacetic acid, estimation of the P^{32} content of the ATP of the reaction medium. This is accomplished by isobutanol–benzene extraction of the phosphomolybdate complex which leaves the labeled ATP in the aqueous phase. Determination of the radioactivity of the aqueous phase makes it possible to calculate the extent of incorporation of inorganic phosphate into ATP.

Procedure. Optimal conditions for assaying the ATP–$P_i{}^{32}$ exchange reaction consist of a reaction system containing 6 micromoles of ATP, 10 micromoles of sodium or potassium phosphate containing approximately 600,000 c.p.m. of $P_i{}^{32}$, 10 micromoles of histidine or imidazole buffer, pH 6.5, and water to a volume of 0.9 ml. The reaction is started by addition of 0.1 ml. of enzyme suspension. The system is incubated at 25° for 20 minutes and stopped by addition of 1.0 ml. of 13% trichloroacetic acid. The protein is removed by centrifugation, and 0.5 ml. of supernatant liquid is used for analysis of P^{32}-labeled ATP by the method described in the section on oxidative phosphorylation.

The extent of incorporation of $P_i{}^{32}$ into ATP is calculated from the determination of the radioactivity, A, of the extracted sample (counts per minute), the specific activity, B, of the inorganic phosphate of the original reaction system (counts per minute per micromole), and correction for the dilution imposed during the extraction procedure.

$$- \Delta P_i{}^{32} = \frac{A}{B} \times 4 \times \frac{1.0 + 1.0}{0.5} = \mu\text{moles } P^{32} \text{ exchanged}$$

Properties of the ATP–$P_i{}^{32}$ Exchange Reaction

Rate of Reaction. Under the respective optimal conditions, equivalent amounts of the enzyme catalyze the ATP-$P_i{}^{32}$ exchange reaction at a rate approximately 5% that of respiration-linked phosphorylation. However, the rate of the ATP–$P_i{}^{32}$ exchange reaction is markedly dependent on the ATP/ADP ratio; as this ratio becomes smaller, the exchange rate approaches zero. The extent to which the ATP-$P_i{}^{32}$ exchange occurs during respiration-linked phosphorylation has been found to be negligible.

In the exchange reaction the incorporation of P^{32}-labeled inorganic phosphate into ATP occurs almost linearly. When higher concentrations of enzyme are employed, approach to isotopic equilibrium can be observed. An accurate estimation of the initial reaction rate can be made when the conditions described above are employed.

Stability. The stability of the ATP-$P_i{}^{32}$ exchange system is quite similar to that of respiration-coupled phosphorylation. Approximately 50% loss of activity occurs in 24 hours at 2°. Freezing and thawing completely inactivate.

Inhibitors and Activators. The exchange activity is completely inhibited by 2,4-dinitrophenol, Dicumarol, gramicidin, arsenate, and azide at concentrations that also completely uncouple oxidative phosphorylation. The addition of metal ions is not required. Serum albumin (2 mg./ml.) will enhance the rate of exchange and will also prevent inhibition by long-chain fatty acids.

Assay Method for the ATP–ADP Exchange Reaction[7]

Principle. This reaction may be assayed by incubation of either P^{32}-labeled ADP[5] or C^{14}-labeled ADP,[10] unlabeled ATP, and the digitonin enzyme. After the reaction is stopped, the nucleotides are separated by paper chromatography and their specific activities determined after elution from the paper. Occurrence of the exchange is indicated by the amount of isotope found in the ATP.

Procedure. The ATP–ADP exchange reaction is carried out in a reaction system containing 3.0 micromoles of ATP, 3.0 micromoles of ADP containing 30,000 to 35,000 c.p.m. of C^{14}-ADP, 5.0 micromoles of histidine or imidazole buffer, pH 6.5, in 0.45-ml. volume. The enzyme suspension (0.05 ml.; 0.8 to 1.2 mg. of nitrogen per milliliter) is added to start the reaction which is allowed to proceed for 10 minutes at 25°. The reaction is stopped by addition of 0.05 ml. of 65% trichloroacetic acid, and the protein is removed by centrifugation.

Fifty microliters of the supernatant solution is placed on a sheet of acid-washed Whatman No. 1 paper (18 × 47 cm.), on a line approximately 8 cm. from the top. Four such spots can be placed on one sheet. The sheets are placed in jars for descending chromatography. The chromatogram is developed by a reagent consisting of 66 ml. of isobutyric acid, 1 ml. of concentrated ammonium hydroxide, and 33 ml. of water. Twenty hours is usually sufficient to produce quantitative separation of the ATP and ADP during which time the solvent front runs off the bottom of the paper. The nucleotides are located by scanning the dried sheets with an ultraviolet Mineralight.[11] In a typical chromatographic separation of a mixture containing ATP, ADP, and AMP, the leading edges of the spots corresponding to these nucleotides were 10 cm., 16 cm., and 24 cm., respectively, from the origin. The areas containing the nucleotides are marked, cut out, and sectioned in 1-cm. squares. The squares of paper are immersed in 3.0 ml. of water for 15 minutes with occasional swirling of the container. One milliliter is placed on a planchet, evaporated to dryness, and counted by means of a gas-flow windowless counter. The extent to which the exchange has occurred can be calculated by the relationship:

$$\frac{C.p.m._{(ATP)}}{C.p.m._{(ADP)}/micromoles\ ADP} \times 33 = Micromoles\ ATP\ exchanged$$

The dilution factor for this operation is 33 [3 × (0.50 + 0.05)/0.05].

[10] Adenosine diphosphate-8-C^{14}, Schwartz Laboratories, Inc., Mount Vernon, New York.
[11] Mineralight Model SL 2537, Ultra-Violet Products, Inc., San Gabriel, California.

Properties of the ATP–ADP Exchange Reaction

Specificity. The exchange reaction occurs only with ATP and ADP.

Stability. The exchange activity is quite stable; 50% loss occurs in 1 week when an aqueous suspension of the submitochondrial particles is stored at 2°.

Inhibitors. The ATP–ADP exchange reaction is inhibited 50% by $5 \times 10^{-5} M$ phenylmercuric acetate and by $1 \times 10^{-4} M$ p-chloromercuribenzoic acid. This inhibition is reversed by addition of cysteine or reduced glutathione. The exchange activity of *fresh* submitochondrial particles is inhibited by low concentrations of DNP and Dicumarol; these agents do not inhibit with *aged* enzyme preparations.

Other Reactions

The digitonin particles also catalyze DNP-stimulated ATPase activity[6] and the exchange of O^{18} between H_2O of the medium and inorganic phosphate and the terminal phosphate of ATP.

[33] Preparation and Assay of Phosphorylating Submitochondrial Particles: Sonicated Mitochondria[1]

By W. WAYNE KIELLEY

Assay Method

Principle. Although standard manometric and colorimetric techniques can be used for measuring oxygen and phosphate uptake with particles obtained by sonic disintegration of mitochondria,[2] it has been found convenient and somewhat more precise to utilize the oxygen electrode for measuring oxygen uptake with this material.

The Clark oxygen electrode[3] as used here measures the decrease in dissolved oxygen. As described, this involves amounts of the order of 0.2 to 0.3 μatom of oxygen. Therefore, an equally sensitive method for phosphate uptake, the incorporation of P^{32} into organic phosphate, is required.

Reagents

0.2 M PO_4, pH 7.0.
0.1 M $MgCl_2$.

[1] The methods described here are taken from W. W. Kielley and J. R. Bronk, *J. Biol. Chem.* **230**, 521 (1958).
[2] W. C. McMurray, G. F. Maley, and H. A. Lardy, *J. Biol. Chem.* **230**, 219 (1958).
[3] L. C. Clark, Jr., R. Wolf, G. Granger, and Z. Taylor, *J. Appl. Physiol.* **6**, 189 (1953).

0.1 M ADP, pH 7.0.
0.1 M AMP, pH 7.0.
0.05 M succinate, pH 7.0.
0.025 M DPNH.
0.01 M DPN$^+$, pH 7.0.
0.2 M βOH butyrate, pH 7.0.
0.5 M glucose.
Hexokinase solution (0.05 ml. to contain sufficient amount to transfer 5 micromoles of phosphate from ATP to glucose at pH 7.0 under otherwise optimal conditions).
P^{32} solution (to contain 1×10^6 c.p.m. in 0.05 ml.).
Enzyme solution as indicated in the preparative procedure.
Deoxygenated (boiled) water.
60% perchloric acid (PCA).

Procedure. One-tenth milliliter each of the phosphate, MgCl$_2$, ADP, AMP, and glucose solutions is added to the reaction vessel together with 0.05 ml. of the P^{32} solution. Either 0.1 ml. of the succinate or DPNH solution or 0.1 ml. of the βOH butyrate solution plus 0.1 ml. of the DPN$^+$ solution is then added. The volume is adjusted to 1.2 ml. with water, and just prior to incubation 0.05 ml. of hexokinase and 0.6 ml. of the boiled water are added. The incubation is carried out as described under oxygen uptake, and phosphate uptake is carried out by the method of Nielsen and Lehninger[4] (see Vol. VI [162]) after acidification with perchloric acid.

Oxygen Uptake. The essential features of the Clark oxygen electrode[5] and the experimental arrangement for its use are given in Fig. 1. The permeability properties of the polyethylene membrane make the electrode virtually indifferent to the composition of the medium, except as the composition influences the oxygen concentration. The small gap between electrode and reaction vessel (0.5 to 0.75 mm.) eliminates the problem of oxygen diffusion into the medium except in very protracted incubations. The reaction vessels are common shell vials cut off to an approximate length of 3 cm. and selected to slip into the water jacket without leakage around the O ring. Water from a constant-temperature bath is circulated through the jacket which is mounted on a magnetic stirrer of the type having an external rheostat. The electrode assembly in turn is mounted on a stand having a rack-and-pinion arrangement for raising and lowering the electrode into the reaction vessel. The electrical circuit for polarizing the electrode and coupling it to the recorder is given in Fig. 2. The recorder is a 10-mv. Varian, but a number of similar low-

[4] S. O. Nielsen and A. L. Lehninger, *J. Biol. Chem.* **215**, 555 (1955).
[5] Obtained from the Yellow Springs Instrument Company, Yellow Springs, Ohio.

input impedance recorders are available. Chart speeds of 1 or 2 inches per minute are convenient. The electrode is operated at a polarizing voltage of 0.6 volt (resistor A, Fig. 2). To set the scale of the recorder, 1.7 ml. of water and 0.2 ml. of the succinate solution are placed in a reaction vessel. Then 0.1 ml. of the original mitochondrial suspension is added (there is usually enough material left in the tube from the final

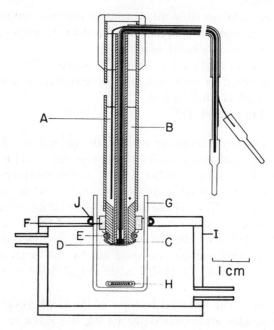

Fig. 1. Experimental arrangement for the Clark oxygen electrode. A, silver electrode; B, saturated KCl; C, platinum electrode; D, polyethylene membrane; E, rubber O ring; F, rubber gasket; G, reaction vessel; H, magnetic stirring bar; I, Lucite water jacket; J, rubber O ring.

resuspension prior to treatment in the sonic oscillator), the vessel is placed in the water jacket, and the electrode is immersed. Under these circumstances the oxidation of succinate will rapidly remove all the dissolved oxygen. By means of the zero adjustment on the recorder the minimum position is set at zero on the chart. The electrode is removed from the vessel, washed, and then placed in another vessel (in the water jacket) containing distilled water equilibrated with air at the temperature being used. After temperature equilibrium is reached, as determined by a stable reading on the recorder, the two resistors (B and C, Fig. 2) are adjusted to bring the pen of the recorder to 100%. The electrode is then raised and gently dried with facial tissue, and the reaction vessel containing water is replaced with one containing the incubation medium. The water needed to bring the latter to the proper volume was boiled to remove oxygen and kept stoppered. The medium when temperature-equilibrated usually showed about 80 to 90% of saturation under these circumstances. After reaching temperature (approximately 1 minute), the enzyme preparation is added with a micropipet (Carlsberg type), and when the oxygen has decreased about 25% the reaction is stopped by addition of the PCA (0.1 ml.). The response to oxygen is linear, so a 25% decrease represents 25% of the oxygen dissolved at this tempera-

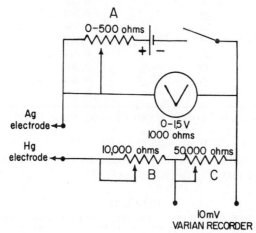

FIG. 2. Electrical circuit for oxygen electrode.

ture in equilibrium with air. Table I gives the oxygen concentration of air-saturated water at several temperatures. Phosphate uptake is then determined as indicated above by the procedure given in Vol. VI [32]. Some representative results for the three substrates are presented in Table II.

After being washed and dried, the electrode is ready for another incubation. Under the conditions described, incubation times vary from 0.5 to 3.0 minutes.

Preparation of Sonicated Particle. Mitochondria are prepared from a 0.25 M sucrose homogenate of about 20 g. of rat liver (two rats) by a modification of the isotonic sucrose fractionation procedure of Schneider[6] (see Vol. IV [34]). After the particles have been washed twice with 0.25 M sucrose, they are washed once by resuspension in about 20 ml. of 0.03 M PO$_4$, pH 7.0, and centrifuged at 20,000 $\times$ g for 10 minutes. After the supernatant fluid has been removed, the mitochondria are sus-

[6] W. C. Schneider, *J. Biol. Chem.* **176**, 259 (1948).

TABLE I

OXYGEN CONTENT OF AIR-SATURATED WATER AS A FUNCTION OF TEMPERATURE
AND AT A TOTAL AIR PRESSURE OF 760 MM.[a]

Temperature, °C	O_2 content, μmole/ml. H_2O
5	0.397
10	0.351
15	0.314
20	0.284
25	0.258
28	0.244
30	0.237
35	0.222
37	0.217
40	0.209
43	0.203
45	0.199

[a] Derived from data in *Handbook of Chemistry and Physics* (1949).

pended to a volume of 10 ml. with the 0.03 M PO_4 buffer and treated in a
9-kc. Raytheon sonic oscillator for about 25 seconds. This material is
centrifuged at 25,000 × g for 20 minutes. The supernatant fluid is then

TABLE II

OXIDATIVE PHOSPHORYLATION WITH PARTICLE PREPARATION

Substrate	Oxygen uptake, μatom/min.	Phosphate uptake, μmole PO_4/min.	P/O
Succinate	0.128	0.108	0.85
Succinate + 2 × 10^{-4} M dinitrophenol	0.116	0.000	0.00
DPNH	0.247	0.247	1.00
DPNH + 2 × 10^{-4} M dinitrophenol	0.235	0.020	0.09
β-Hydroxybutyrate + DPN^+	0.085	0.107	1.26
TPNH	0.010	0.003	0.30
TPNH + DPN^+	0.054	0.066	1.22

poured off and recentrifuged at 100,000 × g for 30 minutes. The sediment
is then resuspended in 0.03 M PO_4 buffer to a final volume of 2.0 to 3.0
ml. Aliquots (0.05 ml.) of this solution will give rapid rates of oxygen
uptake with any of the three substrates.

Properties

The submitochondrial particles are apparently deficient in most of
the mitochondrial dehydrogenases. As an index of this, neither isocitrate[2]
nor glutamate[1,2] is oxidized by these preparations. TPNH is slowly

oxidized by the particles with a very low P/O ratio. The rate of oxidation is markedly accelerated by addition of DPN+, indicating that the
particles possess transhydrogenase activity. In this circumstance the
P/O ratio observed is comparable to that obtained with β-hydroxybutyrate + DPN+. In the latter system DPN+ is an obligate component
for oxidation, indicating that the particles are deficient in bound DPN+.
Proline is also oxidized by the particles if DPN+ is added and, as in the
systems above, coupled phosphorylation occurs.[2] Deamino-DPNH and
the 3-acetylpyridine analog of DPNH are both oxidized (the latter quite
slowly) and exhibit P/O ratios comparable to that for DPNH.

The oxidation of ascorbate-reduced cytochrome c without phosphorylation and the recovery of the same P/2e ratios in the reduction
of cytochrome c as those obtained with oxygen as the electron acceptor
indicate that the phosphorylation observed with these particles is limited
to coupled processes in the steps leading to cytochrome reduction.

ADP and Mg++ are obligate components of the phosphorylation
mechanism, although Mg++ may be replaced by Co++, Mn++, or Fe++.[7]

The phosphorylation observed is dinitrophenol-sensitive, as demonstrated in Table II.

The use of hexokinase (plus glucose) is optional.

[7] J. R. Bronk, and W. W. Kielley, *Biochim. et Biophys. Acta* 24, 440 (1957).

[34] Preparation and Assay of Phosphorylating
Submitochondrial Systems: Mechanically
Ruptured Mitochondria

By MAYNARD E. PULLMAN and HARVEY S. PENEFSKY

General Principles

Mitochondria, isolated from beef heart, are subjected to high-speed
reciprocal shaking. Under appropriate conditions, the mitochondria are
fragmented, and the oxidative phosphorylation system may be separated
into a particulate and soluble fraction by differential centrifugation. The
particulate fraction catalyzes oxidation without concomitant phosphorylation. The addition of the soluble protein component to the particles restores oxidative phosphorylation.[1,2]

[1] M. E. Pullman, H. S. Penefsky, A. Datta, and E. Racker, *J. Biol. Chem.* 235, 3322
(1960).
[2] H. S. Penefsky, M. E. Pullman, A. Datta, and E. Racker, *J. Biol. Chem.* 235, 3330
(1960).

Assay for Oxidative Phosphorylation

Both the principle and the procedure for this assay have been previously described.[3]

Reagents

0.1 M MgSO$_4$ (0.0066 M).

0.2 M ATP, pH 7.4 (0.002 M).

1.0 M glucose (0.032 M).

1.0 M tris(hydroxymethyl)aminomethane (Tris), pH 7.4 (0.003 M).

3.0 mg. of yeast hexokinase[4] per milliliter (0.06 mg. per vessel).

1.0 M potassium phosphate buffer, pH 7.4 (0.012 M).

0.16 M ethylenediaminetetraacetic acid (EDTA), pH 7.4 (0.001 M)

0.25 M sucrose (0.024 M).

Mixed Medium. A mixed medium is prepared containing 0.6 ml. of MgSO$_4$, 0.15 ml. of ATP, 0.48 ml. of glucose, 0.075 ml. of Tris, 0.6 ml. of yeast hexokinase, 0.18 ml. of potassium phosphate buffer, 0.09 ml. of EDTA, and 0.84 ml. of sucrose. The concentrations of these components in the final reaction mixture (including the MgSO$_4$, sucrose, and Tris added with the enzyme) are indicated in parentheses in the reagent list.

Procedure. The particulate fraction[5] (1.5 mg.) is preincubated for 10 minutes at 30° with 1.2 to 1.5 mg. of the soluble coupling enzyme[6] and 0.04 ml. of 0.1 M MgSO$_4$. The mixture is adjusted to a final volume of 0.3 ml. with 0.25 M sucrose–0.01 M Tris, pH 7.4.

To the main compartment of chilled Warburg vessels are added 0.1 ml. of the mixed medium, 0.05 ml. of 0.5 M sodium succinate, 0.1 ml. of the preincubated enzyme mixture, and water to give a final volume of 0.5 ml. The enzyme mixture is added last. The vessels are immediately placed in the Warburg bath and equilibrated at 30° for 6 minutes, and oxygen uptake is measured for the next 18 minutes. The oxygen uptake during the 6-minute equilibration period is calculated by extrap-

---

[3] See Vol. II [101].

[4] The procedure described in Vol. V [25] for the purification of hexokinase was followed up to the bentonite step. At this point the enzyme is 20 to 30% pure and suitable for use in these experiments. Recently, crystalline hexokinase has become commercially available. To each vessel were added 5 units of a dialyzed enzyme solution, assayed in a system coupled to glucose-6-phosphate dehydrogenase (Chapman *et al.*, *J. Clin. Invest.* **41**, 1249, 1962).

[5] Particulate protein was measured by a biuret method modified for mitochondria; E. E. Jacobs, M. Jacob, D. R. Sanadi, and L. B. Bradley, *J. Biol. Chem.* **223**, 147 (1956). Soluble protein was measured by ultraviolet absorption; see Vol. III [73].

[6] For each 0.5 mg. of the particulate fraction, 0.4 to 0.5 mg. of the crude extract or 0.010 to 0.015 mg. of the purified enzyme is required for optimal P/O ratios.

olation, and this is added to the 18-minute value. Usually 5 to 7 micro-atoms of oxygen and 3 to 5 micromoles of P_i are taken up in the 24-minute period. The reaction is stopped by tipping in 0.05 ml. of 50% trichloroacetic acid from the side arm. After centrifugation of the precipitated protein, 0.1-ml. aliquots of the supernatant solution are removed for P_i determination.[7] This assay is also used to assay the coupling enzyme, in which case a control, in which the particulate fraction and $MgSO_4$ are preincubated without the coupling enzyme, is included.

Assay of the Soluble Coupling Enzyme

The soluble enzyme in addition to its coupling activity also catalyzes the hydrolysis of ATP. The latter activity affords a more rapid and convenient assay than does its coupling activity.

Principle. P_i liberated from regenerated ATP by ATPase, is measured colorimetrically. The liberation of P_i is in accord with the stoichiometry of the following equation:

$$ATP + H_2O \rightarrow ADP + P_i \qquad (1)$$

It is preferable to employ an ATP-generating system in the assay, since the ATPase activity of the coupling enzyme is inhibited by ADP. A most convenient one is the phosphoenolpyruvate–pyruvate kinase system.

Reagents

0.5 M Tris–acetate buffer,[8] pH 7.4.
0.2 M ATP, pH 7.4.
0.1 M $MgSO_4$.
0.05 M phosphoenolpyruvate, pH 7.4.
Pyruvate kinase,[9] 0.8 mg./ml. in 0.01 M Tris, pH 7.4.

Procedure. To 16 × 125-mm. test tubes are added 0.1 ml. of Tris–acetate buffer, 0.005 ml. of ATP, 0.01 ml. of $MgSO_4$, 0.1 ml. of phosphoenolpyruvate, 0.04 ml. of pyruvate kinase, and water to give a final volume of 1.0 ml. (including enzyme). After a 5-minute equilibration at 30°, an appropriate amount of enzyme is added, and the incubation is continued for 10 minutes. The reaction is stopped by the addition of 0.1 ml. of 50% trichloroacetic acid. After centrifugation of

[7] K. Lohmann and L. Jendrassik, *Biochem. Z.* **178**, 419 (1926).
[8] D. K. Meyers and E. C. Slater, *Biochem. J.* **67**, 558 (1957).
[9] Purchased from C. F. Boehringer and Soehne, Mannheim, Germany.

the precipitated protein, 0.5-ml. aliquots of the supernatant solution are removed for P_i determination by the method of Lohmann and Jendrassik.[7] Care must be taken to adjust the amount of enzyme used so that no more than 3.5 micromoles of P_i are split (70% of the added phosphoenolpyruvate). When the concentration of phosphoenolpyruvate drops below 1.5 micromoles/ml., the rate of P_i formation is no longer linear with time or proportional to enzyme concentration. With each new preparation of pyruvate kinase, the necessary excess to furnish maximal rates is determined. The recommended 32 μg. usually represents two or three times the amount necessary to give maximal rates.

Definition of Unit and Specific Activity. A unit of enzyme is defined as the amount of enzyme which will form 1 micromole of P_i per minute under the above conditions. Specific activity is defined as the units per milligram of protein.

Preparation of the Particulate Fraction

Attempts to disintegrate the mitochondria by a procedure which might be more easily adapted to large-scale preparations have, recently, been successful.[1,2] In the absence of a vacuum, breakage of the mitochondria was not so effective.

Step 1. Disintegration of the Mitochondria. Three hundred milligrams of "heavy-layer" mitochondria prepared according to the method of Green *et al.*[10] are suspended in 0.25 M sucrose containing 0.001 M EDTA, pH 7.4,[11] to give a final volume of 9.5 ml. The mitochondrial suspension is placed in a Nossal tube[12] containing 7.5 ml. of glass beads[13] which have been prechilled at 4° for at least 5 minutes. The suspension is gently mixed with the glass beads in order to remove as much of the trapped air as possible. This mixing procedure usually takes 3 to 4 minutes, and thoroughness is important to avoid bumping and loss of material during the subsequent evacuation procedure. A thin layer of vacuum grease is applied to the rim of the Nossal tube, and the entire tube, together with its cap resting slightly askew, is placed in a plastic tube slightly larger than the metal container and closed off by rubber stoppers at both ends. The entire container is evacuated for 3 to 4 minutes on a vacuum pump. After evacuation the cap is worked into

[10] D. E. Green, R. L. Lester, and D. M. Ziegler, *Biochim. et Biophys. Acta* 23, 516 (1957).

[11] This solution will be referred to as sucrose–EDTA.

[12] P. M. Nossal, *Australian J. Exptl. Biol. Med. Sci.* 31, 583 (1953). This is a reciprocal shaker with a ⅝-inch amplitude and a speed of 6000 c.p.m. It was purchased from the machine shop at Western Reserve University.

[13] Purchased from Minnesota Mining and Manufacturing Co., Ridgefield, New Jersey. 3M Superbrite, catalog number 090, 0.0110-inch diameter.

position and pressed onto the Nossal tube via a probe fitted into one of the rubber stoppers.

The evacuated Nossal tube is removed and placed in a Nossal shaker.[12] The shaking is carried out at 4° for 10 seconds. The tube is removed and cooled in ice for 1 minute. This process is repeated five times so that the total shaking time is 1 minute. If, on removal of the cap, no frothing of the mitochondrial suspension has occurred, the vacuum was as a rule sufficient to ensure effective breakage of the mitochondria.

Step 2. Separation of the Particulate Fraction and the Soluble Coupling Enzyme. The suspension is centrifuged for 2 minutes at $1800 \times g$, and the supernatant fluid is decanted from the glass beads. After the glass beads have been washed with 5 ml. of sucrose–EDTA, the combined supernatant solutions (approximately 10 ml.) are centrifuged at $26,000 \times g$ in the No. 40 rotor of the Spinco Model L centrifuge for 15 minutes. The firmly packed, brown residue is discarded, and the yellow turbid supernatant solution is decanted and recentrifuged at $105,000 \times g$ for 30 minutes. A red-brown, translucent, gelatinous residue (particulate fraction) and a faintly turbid, yellow supernatant solution are obtained. The supernatant solution is decanted and clarified by centrifugation for an additional 30 minutes at $105,000 \times g$, yielding a crude extract of the coupling enzyme. The particulate fraction is washed by homogenization in 5 ml. of sucrose–EDTA and centrifuged at $105,000 \times g$ for 30 minutes. A second wash with $0.25 M$ sucrose is carried out as above, and the residue is suspended in $0.25 M$ sucrose to give a final volume of approximately 2.5 ml.

The yield of the particulate fraction obtained under these conditions is usually about 50 mg. of protein.

Purification of the Soluble Coupling Enzyme

Step 1. Preparation of the Crude Extract. The crude extract prepared from 3.0 g. of heavy-layer mitochondrial protein is obtained as described in the previous section.

Step 2. pH Fractionation. The pH of the crude, clear[14] extract, containing 500 mg. of protein in 90 ml. of $0.25 M$ sucrose–$0.001 M$ EDTA, is adjusted with 0.4 to 0.5 ml. of $1 N$ acetic acid to pH 5.4 at 4° with mechanical stirring. The solution is centrifuged at $18,000 \times g$ for 15 minutes at 4°, and the precipitate is discarded. The pH of the super-

---

[14] Centrifugation of the crude extract at $105,000 \times g$ should be repeated until a clear supernatant solution is obtained. The presence of particulate material results in coprecipitation of the enzyme during the pH fractionation.

natant solution is adjusted to pH 6.7 by the addition of 0.1 ml. of $2 M$ Tris, pH 10.7.

Step 3. Protamine Fractionation. Two-tenths milliliter of 0.5% protamine sulfate solution for each 10 mg. of protein is added slowly at 4° with mechanical stirring to the step 2 fraction. Stirring is continued for an additional 15 minutes, after which the yellow protamine precipitate is centrifuged off at 10,000 $\times$ g for 15 minutes. *As a result of step 3 the enzyme becomes cold-labile. It is therefore necessary to carry out all further manipulations with solutions at room temperature unless otherwise indicated.* The supernatant solution is discarded, and the precipitate is dissolved in 6 ml. of a solution containing 0.4 M $(NH_4)_2SO_4$, pH 7.4, 0.25 M sucrose, 0.01 M Tris, pH 7.4, and 0.001 M EDTA, pH 7.4. The resulting yellow solution is centrifuged, and a small insoluble residue is discarded. To the clear yellow supernatant solution is added with gentle stirring an equal volume of saturated ammonium sulfate solution (pH 5.5). As an ammonium sulfate suspension the enzyme is not cold-labile. The suspension is kept at 4° for about 15 minutes to ensure complete precipitation of the enzyme and then centrifuged. The precipitate is dissolved in 4 ml. of sucrose (0.25 M)–Tris (0.01 M)–EDTA (0.001 M). The dissolved enzyme is again precipitated by the addition of an equal volume of saturated ammonium sulfate solution. The enzyme at this point (8 ml.) may be stored at 4° if necessary before proceeding to the next step.

Step 4. Temperature Fractionation. The ammonium sulfate suspension of the enzyme from step 3 is centrifuged and dissolved in sucrose–Tris–EDTA to give a final protein concentration of 10 mg./ml. To this solution is added 0.02 ml. of 0.2 M ATP for each milliliter of solution. The enzyme solution is placed in a 60° bath for 4 minutes. It is cooled to room temperature in a 25° bath and centrifuged at room temperature to remove the denatured protein. The enzyme is precipitated from the supernatant solution (1.7 ml.) by the addition of an equal volume of saturated ammonium sulfate. A second precipitation of the enzyme, carried out as described above, removes most of the added ATP.[15] The enzyme may be kept at 4° as a suspension in 50% ammonium sulfate for 2 months without appreciable loss in activity. For assay purposes, appropriate aliquots of the ammonium sulfate suspension are centrifuged, and, after the supernatant solution has been decanted and the walls of the centrifuge tube blotted with filter paper, the enzyme is dissolved in sucrose–Tris–EDTA solution. *The enzyme solution must now be kept*

[15] The enzyme may be precipitated a third time to remove more of the added ATP However, some losses in activity result from repeated precipitations.

at room temperature. A summary of the purification procedure is given in the table.

SUMMARY OF PURIFICATION PROCEDURE FOR COUPLING ENZYME

Fraction	Volume, ml.	Units[a]	Protein, mg.	Specific activity, units/mg. protein −DNP[a]	+DNP	Yield, %
Step 1: Crude extract	90	700	500	1.4	2.0	100
Step 2: pH fractionation	88	545	218	2.5	3.8	78
Step 3: Protamine fractionation	8	758	34	22.0	32.0	107
Step 4: Temperature fractionation	1.7	1100	10.0	110.0	160.0	157

[a] Activity measured in the absence of 2,4-dinitrophenol (DNP).

Properties of Enzymes of Oxidative Phosphorylation

Stability. The particulate fraction may be stored at −20° in 0.25 M sucrose for 10 days with little loss in activity.

The soluble coupling enzyme in the crude extract and the pH 5.4 supernatant solution is stable when stored at −20°. However, the purified enzyme has a half-life of 15 minutes at 4° but may be stored at 0° as a suspension in 50% ammonium sulfate for 2 months with little loss in activity. The purified enzyme may be dialyzed for 2 hours at room temperature against sucrose–Tris–EDTA containing 0.05 M salt, pH 7.4 (ammonium sulfate, ammonium phosphate, potassium sulfate, ammonium chloride), or 1 × 10^{-3} M ATP with only a 15 to 20% loss in activity. Dialysis in the absence of salts or ATP results in a loss of over 80% of the activity.

Specificity. The particulate fraction oxidizes succinate and will oxidize glutamate and β-hydroxybutyrate, providing the respective dehydrogenases and DPN are added. In all cases, coupled phosphorylation requires the addition of the coupling enzyme.

The purified coupling enzyme hydrolyzes ITP, GTP, and UTP at rates of 1.25, 0.75, and 0.6, respectively, of that of ATP; CTP is inactive. Neither the mono- nor the diphosphates of these compounds are split. Phosphorylated sugars and inorganic pyrophosphate are not attacked. The final preparation is free of myokinase, menadione reductase, and succinate–cytochrome c reductase.

Activators and Inhibitors. Succinate oxidation by the particulate fraction is inhibited completely by KCN (4 × 10^{-4} M), antimycin (0.5

μg./mg. protein), p-chloromercuribenzoate $(4 \times 10^{-5} M)$, and dihydro vitamin K_1 diphosphate $(5 \times 10^{-5} M)$. Azide $(5 \times 10^{-4} M)$ inhibits approximately 20 to 30%.

Azide $(2 \times 10^{-4} M)$, p-chloromercuribenzoate $(2 \times 10^{-5} M)$, and dihydro vitamin K_1 diphosphate $(1 \times 10^{-5} M)$ uncouple oxidative phosphorylation at the indicated concentrations. These inhibitors at these concentrations have little effect on respiration. In addition, 2,4-dinitrophenol $(5 \times 10^{-4} M)$, triiodothyronine $(2 \times 10^{-4} M)$, chlorpromazine $(2 \times 10^{-5} M)$, and atebrin $(3 \times 10^{-4} M)$ give 60 to 100% uncoupling.

The ATPase activity of the coupling enzyme is inhibited by all the above-mentioned compounds except KCN and antimycin. 2,4-Dinitrophenol and pentachlorophenol stimulate ATPase activity at low concentrations $(5 \times 10^{-4} M$ and $5 \times 10^{-5} M$, respectively) and inhibit at higher concentrations $(5 \times 10^{-3} M$ and $5 \times 10^{-4} M$, respectively). ADP, but not IDP, CDP, or GDP, inhibits the hydrolysis of all the triphosphates mentioned in the previous section.

pH Optimum. The pH optimum for oxidative phosphorylation of the combined system is rather narrow. At pH 6.5 and 8.0 the P/O ratio is 30% of the optimal ratio at pH 7.0. The ATPase activity gradually increases from almost zero at pH 5.5 to maximal activity at pH 9.0.

Kinetic Properties. The ATPase has a K_m for ATP (measured with the ATP-generating system) of $6 \times 10^{-4} M$.

[35] Oxidative Phosphorylation Systems: Microbial

By ARNOLD F. BRODIE

Introduction

Microorganisms, like other living matter, are dependent on electron-transport systems for the conversion of energy into a chemically utilizable form. In contrast to animal tissues which require organic compounds as sources of energy, various microorganisms obtain energy from the oxidation of a wide variety of compounds, either organic or inorganic. The cell-free bacterial preparations which have been studied appear to be similar to mammalian mitochondrial systems with respect to the conditions necessary for ATP formation, the lability of the particulate fraction, and effects of inhibitors, and in some systems with respect to the effects of uncoupling agents. In general, the P/O ratios observed with bacterial preparations tend to be lower than those found with mammalian mitochondria. The lower P/O ratios may be a reflection of the loss of a component in the electron-transport chain or may be due to other electron-transport pathways which are nonphosphorylative. The cell-free bacterial systems lend themselves to chemical fractionation.[1-4] Systems which can be fractionated and reconstituted facilitate analysis of the essential components involved in oxidative phosphorylation.

Manometric Assay Method

Principle. The crude cell-free system or fractionated components are incubated with a suitable electron donor, inorganic orthophosphate, and a phosphate acceptor system. The consumption of oxygen and esterification of inorganic phosphate are measured.

Reagents

0.1 M tris(hydroxymethyl)aminomethane buffer, pH 8.0.
0.025 M KF.
0.015 M MgCl$_2$.
0.15 M KCl.
0.005 M AMP, ADP, or ATP.
0.1 ml. of yeast hexokinase[5] (10 mg./ml.).
0.02 M glucose or mannose.
0.01 M orthophosphate.

Incubation Procedures. Respiration is measured by the conventional Warburg technique at 30°. The bacterial extract and hexokinase are added last to the manometric vessels. To ensure the same amount of inorganic phosphate in each vessel, the orthophosphate (10 to 15 micromoles per vessel) is added to the bacterial extract, and 0.7 ml. of extract (containing 15 to 20 mg. of protein) is added to the main compartment of the vessel.[6] In addition to the extract or fractionated components, Mg^{++}, glucose or mannose, and hexokinase are added to the main compartment. AMP, F$^-$, and substrate are added to the side arm of the vessel. The enzyme preparation is preincubated with Mg^{++} before the addition of fluoride to prevent precipitation and removal of the ions. In order to obtain a zero-time phosphate value, one or more tubes are

[1] A. Tissières and E. C. Slater, *Nature* 176, 736 (1955).
[2] A. F. Brodie and C. T. Gray, *Biochim. et Biophys. Acta* 19, 384 (1956).
[3] G. B. Pinchot, *J. Biol. Chem.* 205, 65 (1953).
[4] P. M. Nossal, D. B. Keech, and D. J. Morton, *Biochim. et Biophys. Acta* 22, 412 (1956).
[5] Yeast hexokinase obtained from Pabst Laboratories, Milwaukee 5, Wisconsin.
[6] A. F. Brodie and C. T. Gray, *J. Biol. Chem.* 219, 853 (1956).

set up containing the complete system, and 10% TCA is added before the addition of enzyme.

The reaction vessels are equilibrated at 30° for 10 minutes before the addition of substrate. Oxidation is then measured for 5 to 15 minutes, depending on the rate of oxidation of the particular extract. Although the rate of phosphorylation is linear with respect to time, higher P/O ratios are observed during the first 5 to 10 minutes after the addition of substrate.[6] The reaction is stopped by the addition of 1 ml. of 10% TCA. The mixture is poured into tubes, chilled, and centrifuged in order to remove the protein.

Measurement of Phosphate Esterification. Inorganic phosphate can be determined by any suitable procedure.[7,8] Esterification of phosphate can also be determined by measuring the formation of ATP or glucose-6-phosphate or the incorporation of P^{32} into these compounds. The ratio between inorganic phosphate disappearance and the atoms of oxygen consumed is the P/O ratio.

Spectrophotometric Assay Method

Principle. A spectrophotometric method for measuring oxidative phosphorylation in bacterial extracts has been described by Pinchot.[9] This method utilizes the spectrophotometer to measure the oxidation of DPNH by decrease in absorption at 340 mμ and simultaneously measures the formation of ATP formed from oxidation by coupling the reaction to a phosphate acceptor system containing hexokinase, glucose, glucose-6-phosphate dehydrogenase, and TPN. The accumulation of TPNH is therefore an index of the amount of ATP formed.

$$\text{DPNH} + \text{H}^+ + \text{ADP} + \text{P}_i + \tfrac{1}{2}\text{O}_2 \rightarrow \text{DPN}^+ + \text{ATP} + \text{H}_2\text{O} \quad (1)$$

$$\text{ATP} + \text{glucose} \xrightarrow{\text{hexokinase}} \text{G-6-P} + \text{ADP} + \text{H}^+ \quad (2)$$

$$\text{G-6-P} + \text{TPN}^+ \xrightarrow{\text{G-6-P dehydrogenase}} \text{TPNH} + \text{H}^+ + \text{6-PGA} \quad (3)$$

Reagents

0.05 M glycylglycine buffer, pH 7.2.
0.005 M orthophosphate.
0.006 M MgCl$_2$.
0.025 M glucose.
0.00005 M ADP.

---

[7] O. H. Lowry and J. A. Lopez, *J. Biol. Chem.* **162**, 421 (1946).
[8] C. E. Cardini and L. F. Leloir, Vol. III [114].
[9] G. B. Pinchot, *J. Biol. Chem.* **229**, 11 (1957).

Yeast hexokinase (43,000 to 97,000 units).
Glucose-6-phosphate dehydrogenase[10] (0.120 unit).

Procedure. At least two cuvettes are required for the spectrophotometric method. The first cuvette contains the complete system, consisting
of 0.05 ml. of the particulate fraction, 0.15 ml. of supernatant fraction,
50 micromoles of glycylglycine buffer (pH 7.2), 5.0 micromoles of inorganic phosphate, 6 micromoles of magnesium chloride, 25 micromoles of
glucose, 0.05 micromole of ADP, 0.2 micromole of TPN, 0.120 unit of
glucose-6-phosphate dehydrogenase, and hexokinase. The second vessel
lacks TPN, hexokinase, and glucose-6-phosphate dehydrogenase and is
used as a measure of the rate of oxidation of DPNH. A series of initial
readings are made for 3 minutes. After these readings an equal amount
of DPNH (0.3 micromoles) is added to each cuvette, and the readings
are again made at 1-minute intervals. The reactions are followed for
3 to 5 minutes. The initial reading plus the amount due to the addition
of DPNH is taken as the total initial reading. The change in optical
density at 340 mμ is calculated for each cuvette, and the difference
between the complete system and the control vessel is assumed to be the
accumulation of TPNH. The P/O ratio is calculated from the micromoles of TPNH accumulation per micromole of DPNH disappearance.

Properties of Phosphorylating Systems

Coupled phosphorylation with bacterial systems requires the presence
of the bacterial extract, a phosphate acceptor system, Mg^{++}, and fluoride
to inhibit ATPase. In contrast to mammalian mitochondria, the oxidation
observed with bacterial systems capable of coupled activity is usually
independent of the presence of a phosphate acceptor system. The P/O
ratios with *Mycobacterium phlei* extracts and various substrates are
shown in Table I. Evidence that this phosphorylation occurred above
the substrate level was provided by the demonstration of the formation
of ATP[6] from electron transport, by P/O ratios greater than 1.0, and by
the inhibition of phosphorylation by anaerobiosis and uncoupling agents.

Comparative P/O ratios obtained with different bacterial systems
are shown in Table II. The organisms are listed in the order of the
highest net phosphate disappearance (micromoles of P_i per milligram
of protein per minute) reported, since the P/O ratios do not indicate the
extent of phosphorylation. These values range from 0.3 to 0.4 micromole
of phosphate disappearing per milligram per 10 minutes with extracts of
M. phlei and *Azotobacter vinelandii* to 0.01 micromole with extracts of
Escherichia coli. The P/O ratios obtained with substrates representing

[10] A. Kornberg and B. L. Horecker, Vol. I [42].

two different pathways of electron transport (DPNH or malate and succinate) are shown in this table. With extracts of *Azotobacter* coupled phosphorylation has also been demonstrated with H_2,[11,12] TPNH,[13] β-hydroxybutyrate,[13,14] α-ketoglutarate, fumarate, and glutamate.[14,15] With extracts of *Proteus vulgaris*, coupled activity has also been demonstrated with formate,[4] whereas with yeast particles this activity has been demonstrated with isocitrate, ethanol, and α-ketoglutarate.[4,16,17]

TABLE I

P/O RATIO WITH DIFFERENT SUBSTRATES AND EXTRACTS OF *M. phlei*[a]

Substrate	Oxygen, μatoms	ΔP_i, μmoles	P/O
Succinate	3.1	5.5	1.78
β-Hydroxybutyrate	4.0	4.9	1.2
α-Ketoglutarate	2.9	6.9	2.4
Fumarate	3.8	6.5	1.7
Malate	4.0	7.2	1.8
Pyruvate	5.7	10.7	1.9
Glutamate	2.9	5.8	2.0
DPNH	2.6	3.9	1.5

[a] The vessels contained 15 micromoles of $MgCl_2$, 25 micromole of KF, 19.0 to 25 mg. of crude sonic extract containing 15 micromoles of P_i, and an acceptor system consisting of 2.5 micromoles of AMP, 20 micromoles of mannose, and 1 mg. of yeast hexokinase. The final volume of the vessels was 1.3 ml. The final concentration of substrate was as follows: 50 micromoles of succinate, β-hydroxybutyrate, and malate; 40 micromoles of pyruvate; 20 micromoles of α-ketoglutarate, glutamate, and fumarate; and 5.0 micromoles of enzymatically reduced DPN. The reactions were carried out at 30° for 10 minutes after the addition of substrate.

All the bacterial systems which have been studied extensively have lent themselves to fractionation and reconstruction.[1-4] Oxidative phosphorylation with fractionated systems is dependent on the presence of a highly organized particulate fraction which must be retained intact, and on the presence of soluble components found in the supernatant fraction (Table II).

Particulate Fraction (see Vol. V [4]). The particulate fraction con-

[11] L. Hyndman, R. Burris, and P. W. Wilson, *J. Bacteriol.* **65**, 522 (1953).
[12] E. H. Cota-Robles, A. G. Marr, and E. H. Nilson, *J. Bacteriol.* **75**, 243 (1958).
[13] I. A. Rose and S. Ochoa, *J. Biol. Chem.* **220**, 307 (1956).
[14] P. E. Hartman, A. F. Brodie, and C. T. Gray, *J. Bacteriol.* **74**, 319 (1957).
[15] A. Tissières, H. G. Hovenkamp, and E. C. Slater, *Biochim. et Biophys. Acta* **25**, 336 (1957).
[16] M. F. Utter, D. B. Keech, and P. M. Nossal, *Biochem. J.* **68**, 431 (1958).
[17] P. J. Russell, Ph.D. Thesis, Western Reserve University (1959).

TABLE II
COMPARATIVE PHOSPHORYLATION BY BACTERIAL EXTRACTS

Organism	Preparation	P/O ratio			Refs.
		Succinate	Malate	DPNH	
M. phlei	CE	1.78	1.8	1.5	[a, b]
	P	0.56	0.0	0.0	[b, c]
	P + S	0.98	1.6	1.1	[b, c]
A. vinelandii	CE	0.32	0.59[d]	0.43	[e]
	P	0.87	0.0	0.41	[e]
	P + S	0.82	0.41	0.32	[e]
A. aerogenes	CE	—	—	0.33[f]	[g]
	P	—	—	0.16	[g]
C. creatinovorans	CE	1.72	—	—	[a]
	P + S	0.79	—	—	[c]
P. vulgaris	CE	—	—	0.62[f]	[g]
	P	—	—	0.0	[g]
	P + S	—	—	0.52	[g]
Yeast	CE	0.74	—	—	[h, i]
	P	0.87	—	0.68[f]	[h, i]
	P + S	0.73	—	0.59	[h, i]
A. faecalis	CE	—	—	0.36	[j, k]
	P	—	—	0.03	[j]
	P + S	—	—	0.33–0.78	[j, k]
E. coli	CE	0.63	0.2	—	[m, n]
	P[l] + S	0.24	0.35	—	[n]

Note: CE = crude extract.
P = particulate fraction.
S = soluble fraction.

[a] A. F. Brodie and C. T. Gray, J. Biol. Chem. **219**, 853 (1956).
[b] A. F. Brodie and C. T. Gray, Biochim. et Biophys. Acta **19**, 384 (1956).
[c] A. F. Brodie, J. Biol. Chem. **234**, 398 (1959).
[d] Fumarate used as substrate.
[e] A. Tissières, H. G. Hovenkamp, and E. C. Slater, Biochim. et Biophys. Acta **25**, 336 (1957).
[f] Lactate used to generate DPNH.
[g] P. M. Nossal, D. B. Keech, and D. J. Morton, Biochim. et Biophys. Acta **22**, 412 (1956).
[h] M. F. Utter, D. B. Keech, and P. M. Nossal, Biochem. J. **68**, 431 (1958).
[i] P. J. Russell, Ph.D. Thesis, Western Reserve University (1959).
[j] G. B. Pinchot, J. Biol. Chem. **205**, 65 (1953).
[k] G. B. Pinchot, J. Biol. Chem. **229**, 11 (1957).
[l] Small particles obtained between 32,000 and 144,000 × g.
[m] D. F. Hersey and S. J. Ajl, J. Gen. Physiol. **34**, 295 (1951).
[n] E. R. Kashket and A. F. Brodie, Biochim. et Biophys. Acta, in press.

tains most of the dehydrogenase activities,[18] bound DPN, quinones, and cytochrome pigments. Prolonged sonic vibration of the crude extracts or of the isolated particulate fraction causes disruption of the particles and results in the loss of oxidative phosphorylation. Oxidation of certain substrates by the particulate fraction occurs to a limited extent. The particles from *M. phlei* are capable of a limited oxidative phosphorylation only with succinate; soluble components are required with all the other Krebs cycle intermediates and with DPNH. In contrast, particles isolated from *A. vinelandii*[15] carry out coupled phosphorylation with DPNH or succinate and require the addition of supernatant components with α-ketoglutarate or fumarate as electron donors. The particulate fraction from *Alcaligenes faecalis*[3,9,19] exhibits limited activity with DPNH and is stimulated by the addition of the supernatant fraction. Particles isolated from yeast can oxidize most of the Krebs cycle intermediates without the addition of soluble components. Thus among microorganisms there appears to be a spectrum of types of particles capable of oxidative phosphorylation. These particles differ with respect to the substrates oxidized, requirements for additional components, effects of inhibitors, and nature of their electron-transport systems. The extent to which a system is coupled may be a reflection of the nature of the electron-transport pathway. The respiratory pigments differ considerably among microbial species and are generally different from those found in mammalian tissues.[20]

Supernatant Fraction. In certain microbial systems stimulation or restoration of oxidative phosphorylation requires the addition of soluble components to the particulate fraction. The activity cannot be restored by the addition of known coenzymes, dyes, or boiled extracts of bacteria or yeast in place of the soluble components. The soluble components, found in the supernatant fraction after the removal of the particles from the crude extract, fall into two classes, one which is heat-labile and nondialyzable[1-3] and the other which is heat-stable.[2,3] These components can be fractionated by chemical means (see Vol. V [4]).

A summary of the purification procedures for the supernatant fraction from *M. phlei* is shown in Table III. Since the assay of the fractionated supernatant solution is dependent on particles which vary from day to day, it is difficult to describe the purification of essential components in terms of specific activity. Comparative activities are ascertained by testing a series of fractions simultaneously in the presence of a given suspension of particles, as shown in Table III. On a protein basis, however,

[18] A. F. Brodie and C. T. Gray, *Science* 125, 534 (1957).
[19] G. B. Pinchot, *J. Biol. Chem.* 229, 1 (1957).
[20] L. Smith, *Bacteriol. Revs.* 18, 106 (1954).

most preparations represent a ten- to fifteenfold purification of the factors necessary to restore both oxidation and esterification of orthophosphate. Regardless of the method or degree of fractionation, the increased oxidative activity is paralleled by a similar increase in phosphorylative activity.

TABLE III
SUMMARY OF PURIFICATION PROCEDURES
(Supernatant)

Step	Volume of solution, ml.	Units[a]	Protein, mg.	Specific activity, units/ml. protein	Yield, %
Crude supernatant fluid	98	460	3782	0.12	100
Ammonium sulfate fractionation	27	384	834	0.46	23
Acetone fractionation	20	234	450	0.53	12
Tricalcium phosphate-protamine fractionation	20	284	180	1.53	4.8

[a] The unit used in this table is the amount of fractionated supernatant which causes the disappearance of 1.0 micromole of orthophosphate in 10 minutes when combined with the particulate fraction (6.5 mg. of protein). The test system consists of 2.5 micromoles of ADP, 15 micromoles of $MgCl_2$, 20 micromoles of KF, 20 micromoles of glucose, 1 mg. of yeast hexokinase, 20 micromoles of succinate, and water to a final volume of 1.5 ml. The reactions are carried out at 30° for 10 minutes after the addition of substrate, and stopped by the addition of 1 ml. of 10% TCA. After centrifugation of precipitated protein, an aliquot of the supernatant is analyzed for inorganic phosphate by the method of Fiske and Subbarow; see A. F. Brodie and C. T. Gray, *J. Biol. Chem.* **219**, 853 (1956). The unit activity can also be defined as the amount of fractionated supernatant capable of consuming 1 microatom of oxygen in 10 minutes, since the increased oxidative activity is paralleled by an increase in phosphorylative activity.

The unfractionated supernatant from *M. phlei* contains the following enzymes known to be associated with electron transport: flavins, DPNH–cytochrome c reductase, menadione reductase,[2] a menadione-dependent cytochrome c reductase,[21] and a riboflavin phosphate-linked cytochrome c reductase.[21] It also contains myokinase and enzymes capable of hydrolyzing ADP or ATP. Many of these enzymes are lost or denatured during fractionation; however, fractionated supernatants capable of restoring oxidative phosphorylation contain menadione reductase and the menadione-dependent cytochrome c reductase. Further fractionation of the supernatant by the method of Warburg and Christian[22] results

[21] M. M. Weber and A. F. Brodie, *Biochim. et Biophys. Acta* **25**, 447 (1957).
[22] O. Warburg and W. Christian, *Biochem. Z.* **298**, 150 (1938).

in preparations which fail to promote oxidation or coupled phosphorylation with the particulate fraction. Restoration of both activities can be achieved by the addition of vitamin K_1 and FAD.[23]

Further purification of the supernatant components from *M. phlei* can be achieved by column chromatography on DEAE-cellulose or Sephadex (G-200). Following chromatography, different segments of the respiratory chain are restored by different protein components.[23a] The proteins are eluted with increasing concentrations of KCl. The material eluted between 0.13 and 0.14 N KCl is required for restoration of the DPN-linked pathway, while that obtained between 0.20 and 0.25 N is required for the succinoxidase pathway. The fractions containing adenylate kinase, ATPase, and the $P_i{}^{32}$-ATP exchange reaction are not required for restoration of activity whereas the ADP-C^{14}-ATP exchange reaction is associated with the fractions which reconstitute coupled phosphorylation.[23a] These results are similar to those obtained by Ishikawa and Lehninger with *Micrococcus lysodeikticus*.[23b]

Kinetic studies with limiting concentrations of particles or supernatant solution indicate that the supernatant fraction from *M. phlei* contains two components, one necessary for oxidation and the other for phosphorylation, whereas stimulation of activity with *Azotobacter* supernatant has been attributed to the removal of oxalacetate inhibition. Two soluble components have been described for *A. faecalis*. This system requires a heat-stable factor which has been identified as a polynucleotide of the RNA type and another component which is heat-labile. The heat-stable component in *M. phlei* supernatants has not been identified.[23] The supernatant requirement for *P. vulgaris* has not been extensively investigated. Coupled activity with a fractionated system from *M. lysodeikticus* is also dependent on a soluble component(s). This factor is necessary for restoration of phosphorylation only.[23a] Coupled phosphorylation can also be demonstrated with combinations of soluble components and particles from different bacterial species.[23]

Lability. Coupled phosphorylation with microbial systems, like that observed with mammalian preparations, is easily inactivated. The bacterial systems are sensitive to physical manipulation such as sonic vibration, dialysis, aging, freezing, and changes in tonicity. Of the two components in bacterial extracts required for oxidative phosphorylation, it is the particles which are sensitive to physical manipulation; however, the particles can be stored without loss of activity[14, 24] by quick freezing.

[23] A. F. Brodie, *J. Biol. Chem.* **234**, 398 (1959).
[23a] A. Asano and A. F. Brodie, unpublished observations (1962).
[23b] S. Ishikawa and A. L. Lehninger, *J. Biol. Chem.* **237**, 2401 (1962).
[24] P. J. Russell and A. F. Brodie, unpublished results; see Vol. V [4].

The supernatant factors are resistant to the treatments described above and can be lyophilized and stored indefinitely.

The rate of oxidation and phosphate esterification with extracts of M. phlei is pH-dependent. At pH 8.0 ATPase activity is increased and results in lower P/O ratios. A low pH (below 7.0) is also unfavorable because the endogenous activity is markedly increased and the substrate oxidation is lowered. The most satisfactory results are obtained within the pH range of 7.2 to 7.5.

Uncoupling. Bacterial systems vary greatly with respect to the effect of uncoupling agents. The effects of both chemical and biological uncoupling agents on oxidative phosphorylation with extracts from M. phlei are shown in Table IV. Uncoupling by low concentrations of DNP has

TABLE IV

EFFECTS OF UNCOUPLING AGENTS WITH EXTRACTS OF M. phlei[a]

Agent	Concentration, M	Oxygen, μatoms	P_i, μmoles	P/O
None	—	3.7	5.5	1.48
None	—	3.1	5.5	1.78
2,4-Dinitrophenol	8×10^{-5}	3.8	0.2	0.05
2-Amino-4-nitrophenol	8×10^{-4}	3.5	2.7	0.78
Na azide	1×10^{-4}	3.1	1.9	0.61
Na arsenate	1×10^{-2}	2.3	0.6	0.26
Methylene blue	8×10^{-5}	5.6	3.7	0.66
Lapachol[b]	5×10^{-4}	3.5	0.0	0.0
DL-Thyroxine	8×10^{-5}	3.6	1.2	0.33
Triiodo-L-thyronine	8×10^{-6}	4.2	1.5	0.35
3,5,3'-Triiodothyroacetic acid	3×10^{-4}	2.0	0.8	0.39
Gramicidin	8×10^{-6}	5.2	3.8	0.72
Dicumarol[b]	5×10^{-5}	2.1	0.7	0.32

[a] The conditions are similar to those of Table I except that succinate was used as an electron donor.
[b] DPNH (5 micromoles) was used instead of succinate.

been observed with all substrates oxidized by these preparations. The uncoupling agents generally neither stimulate nor inhibit oxidation with bacterial systems. KCN $(10^{-4}\,M)$ and arsenite $(10^{-2}\,M)$ inhibit both oxidation and phosphate esterification. Complete uncoupling by DNP $(10^{-3}\,M)$ has been described with particles from yeast[4, 16, 17] and for extracts of P. vulgaris.[4] Uncoupling has also been observed in A. faecalis[25] (73%) by $3.3 \times 10^{-4}\,M$ DNP, and 40% in Corynebacterium creatinovorans[6] by $10^{-4}\,M$ DNP. The systems described from A. vinelandii[1, 14, 18]

[25] G. B. Pinchot, J. Biol. Chem. 229, 25 (1957).

and *Aerobacter aerogenes*[4] are completely insensitive to DNP or other uncoupling agents.

Cofactor Requirements. Requirements for the addition of cofactors are dependent on the degree of fractionation, on special treatment (use of light at 360 mμ),[26] and on the nature of electron transport with the electron donor employed. The crude extract usually does not require the addition of cofactors, whereas stimulation of both oxidation and phosphate esterification occurs with the addition of DPN to supernatant and washed particles utilizing DPN-linked pathways. This cofactor appears to be lost from the particles during washing. Restoration of activity after light irradiation (360 mμ) requires the addition of vitamin K_1 or a closely related homolog.[27, 28] Restoration of activity with the particles and supernatant fraction from *A. faecalis* requires the addition of a polynucleotide of the RNA type.[19]

Comments

The extent of phosphorylation with bacterial systems is lower than that observed with mammalian mitochondria. A great deal of emphasis with bacterial preparations has been placed on the P/O ratios with little or no attention to the nature of the electron-transport pathways involved. Knowledge of these pathways is essential for an understanding of the basic mechanism of oxidative phosphorylation and is of particular importance with systems which are insensitive to uncoupling agents. The soluble oxidative enzymes, when present, may also contribute to the low P/O ratios. Some of the soluble nonphosphorylative oxidative enzymes can be removed from the supernatant fluid by fractionation. With certain systems the lower P/O ratios may be due in part to the high levels of ATPase found in some extracts.

An additional problem encountered with some microbial systems is the ability of the extracts to oxidize the glucose present in the acceptor system or to dephosphorylate G-6-P. Other carbohydrates which are not oxidized but are capable of being phosphorylated with ATP and hexokinase can be substituted for glucose. With extracts from organisms capable of fermentation, special controls (anaerobiosis, KCN, etc.) should be used to exclude these reactions.

[26] See Vol. VI [165a].
[27] A. F. Brodie and J. Ballantine, *J. Biol. Chem.* 235, 226 (1960).
[28] A. F. Brodie and J. Ballantine, *J. Biol. Chem.* 235, 232 (1960).

[36] Isolation and Photoinactivation of Quinone Coenzymes

By ARNOLD F. BRODIE

Introduction

Naturally occurring quinones have been isolated from animal tissues, plants, and bacteria. The benzo- and naphthoquinones are localized in subcellular structures such as the mitochondria and bacteria granules, which are essential for oxidative phosphorylation, and in chloroplasts, which carry out photosynthesis. The participation of quinones in oxidative metabolism has been demonstrated by the loss of activity which follows depletion of the naturally occurring compound and the restoration of activity on addition of the quinone or closely related homologs. Such depletion has been carried out successfully by selective photoinactivation[1-3] and by extraction with liquid solvents.[4-6]

Isolation of Quinones (Bacterial)

Quinones have been found in all microorganisms[1, 7-17] examined with the exception of the anaerobes.[10] Variation in the type and structure of

[1] A. F. Brodie, *Abstr. 132nd Meeting Am. Chem. Soc., New York*, p. 52C (1957).
[2] A. F. Brodie, M. M. Weber, and C. T. Gray, *Biochim. et Biophys. Acta* **25**, 448 (1957).
[3] D. R. Dallam and W. W. Anderson, *Biochim. et Biophys. Acta* **25**, 439 (1957).
[4] F. L. Crane, Y. Hatefi, R. L. Lester, and C. Widmer, *Biochim. et Biophys. Acta* **25**, 220 (1957).
[5] D. E. Green and R. L. Lester, *Federation Proc.* **18**, 987 (1959).
[6] F. L. Crane, C. Widmer, R. L. Lester, and Y. Hatefi, *Biochim. et Biophys. Acta* **31**, 476 (1959).
[7] M. Tishler and W. L. Sampson, *Proc. Soc. Exptl. Biol. Med.* **68**, 136 (1948).
[8] J. Francis, J. Madinaveitia, H. M. MacTurk, and G. A. Snow, *Nature* **163**, 365 (1949).
[9] A. F. Brodie, B. R. Davis, and L. F. Fieser, *J. Am. Chem. Soc.* **80**, 6454 (1958).
[10] H. Noll, *J. Biol. Chem.* **232**, 919 (1958).
[11] R. L. Lester and F. L. Crane, *J. Biol. Chem.* **234**, 2169 (1959).
[12] A. C. Page, Jr., P. H. Gale, H. Wallcik, R. B. Walton, L. E. McDaniel, H. B. Woodruff, and K. Folkers, *Arch. Biochem. Biophys.* **89**, 318 (1960).
[13] E. R. Kashket and A. F. Brodie, *Biochim. et Biophys. Acta* **40**, 550 (1960).
[14] B. K. Jacobsen and H. Dam, *Biochim. et Biophys. Acta* **40**, 211 (1960).
[15] R. A. Morton, *Ciba Foundation Symposium, Quinones in Electron Transport*, Ed. by G. E. W. Wolstenholme and C. M. O'Connor, J & A Churchill, London, p. 5 (1961).
[16] U. Gloor, O. Isler, R. A. Morton, R. Ruegg, and O. Wiss, *Helv. Chim. Acta* **41**, 2357 (1958).
[17] E. R. Kashket and A. F. Brodie, *Biochim. et Biophys. Acta*, in press.

TABLE I

PROPERTIES OF THE NAPHTHOQUINONES

(K_1 and K_2 homologs)

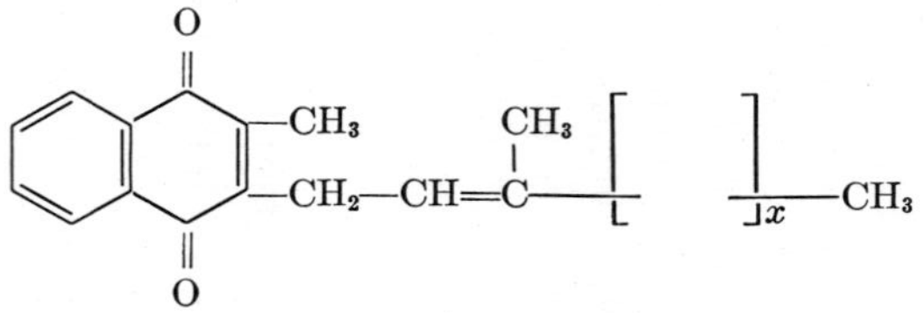

Chemical structure (naphthoquinone nucleus with CH_3 substituent and side chains):

K_1 series: $-CH_2-CH=C-CH_2-CH_2-CH_2-C-H$ (with CH_3 groups), $[\ ]_x$

K_2 series: $-CH_2-CH_2-CH=C-$ (with CH_3), $[\ ]_x-CH_3$

	K_1 series 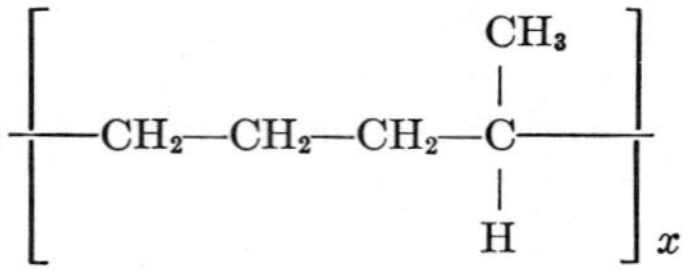				K_2 series 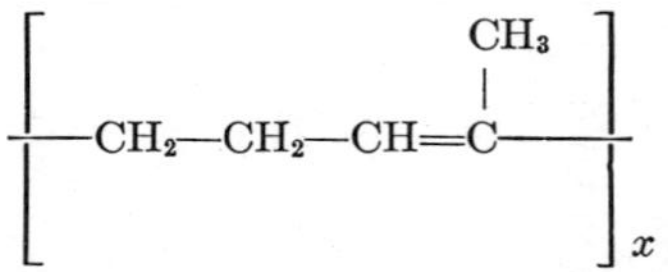				Refs.
	C_{10} $C_{21}H_{26}O_2$	C_{15} $C_{26}H_{36}O_2$	C_{20} $C_{31}H_{46}O_2$	C_{25} $C_{36}H_{56}O_2$	C_{10} $C_{21}H_{24}O_2$	C_{15} $C_{26}H_{32}O_2$	C_{25} $C_{36}H_{48}O_2$	C_{35} $C_{46}H_{64}O_2$	
Molecular weight	310.2	380.3	450.3	520.4	308.2	376.3	512.4	648.6	
Melting point	—	—	—	—	53°	—	39°	54°	a
$E_{1cm}^{1\%}$ 243 mμ	585[I]	—	418[I]	372[I]	609[I]	496	342[I]	278[P]	
248 mμ	626		441	398	646		365	292	
260 mμ	566		410	363	584		333	266	
269 mμ	544		411	360	582		330	267	
325 mμ	101		76.6	65	107		59.4	48	

R_f values							
Isoöctane–methanol (3:1)[V]	0.53	0.27	0.17	0.12	—	0.59	—[b]
Isoöctane–methanol–isopropanol (3:1:1)[V]	0.75	0.56	0.41	0.25	—	—	0.27[b,c]
Ethanol–acetic acid–H_2O (3:1:9)[S]	0.55	0.42	—	0.12	—	0.49	0.14[d]
Propanol–acetic acid–H_2O (25:1:15)[S]	0.73	0.54	—	0.26	—	0.68	0.30[a,d]

Note: I = spectral-grade isoöctane (Russell and Brodie, unpublished data, 1959).
P = petroleum ether [D. T. Ewing, J. M. Vanderbelt, and O. Kamm, *J. Biol. Chem.* **131,** 345 (1939)].
V = Vaseline-impregnated paper.
S = paper treated with silicone fluid (No. 1107).

[a] O. Isler, R. Ruegg, L. H. Chopard-dit-Jean, A. Winterstein, and O. Wiss, *Helv. Chim. Acta* **41,** 786 (1958).
[b] A. F. Brodie, B. R. Davis, and L. F. Fieser, *J. Am. Chem. Soc.* **80,** 6454 (1958).
[c] E. R. Kashket and A. F. Brodie, *Biochim. et Biophys. Acta* **40,** 550 (1960).
[d] J. P. Green and H. Dam, *Acta Chem. Scand.* **8,** 1341 (1954).

the quinones is found among the various microorganisms. Both benzo-
and naphthoquinones are found, and in some bacteria they can occur
together.[11,13] The benzoquinones which have been examined differ principally in the length of the isoprenoid side chain, whereas the naphthoquinones differ in the length and in the degree of saturation of the side
chain (vitamins K_1 and K_2). Marked variations in content and type of
quinone can be brought about by changes in cultural conditions.[11,13,17]

Step 1. Extraction of Bacterial Quinones. Quinones can be extracted
directly from packed cells and cell suspensions with isoöctane–isopropanol
(3:1),[9,13] methanol–ether (2:1),[10] ethanol–ether (3:1),[11] and, after
saponification with n-hexane[12] or methylene chloride.[16] Any of these procedures can be used successfully; only the method used to extract quinones from *Mycobacterium phlei* and *Escherichia coli* will be described in
detail. The best yields have been obtained from cells grown aerobically
and harvested in the exponential phase of growth. *M. phlei* cells are
grown and harvested from liquid medium (see Vol. V [4]) and washed
four times with distilled water. The quinones are extracted directly from
the packed cells (450 g. wet) or from cell-free extracts by refluxing with
3 to 5 vol. of isoöctane–isopropanol (3:1) for 30 minutes, cooling, and
filtering in order to remove the whole cells. The filtrate is concentrated
to dryness under vacuum, and the residue is resuspended in 1 to 3 ml. of
acetone. The waxes and other acetone-insoluble materials are removed,
and the material is again concentrated to dryness and further purified by
chromatography.

Step 2. Chromatographic Purification. The naphthoquinones can be
absorbed and eluted from alumino-silicate gel (DeCalso),[9,12,13] magnesium aluminum silicate (Florisil), magnesium silicate–Celite,[10] silicic
acid,[10] and aluminum oxide.[16] The brown or reddish oily residue from
step 1 is dissolved in petroleum ether (30° to 60° b.p.) and placed on a
column (2 × 30 cm.) containing packed DeCalso previously washed with
petroleum ether. After the absorption of the quinone the column is
washed two or three times with 250-ml. portions of petroleum ether, then
with 200 ml. of petroleum ether–diethyl ether (200:1), and the naphthoquinone is eluted with petroleum ether–diethyl ether (200:4). The
fractions containing the quinone are yellow and are assayed for purity
spectrophotometrically and by infrared spectroscopy. Since most of the
naphthoquinones in the K_1 series cannot be crystallized, further purification is achieved by repeated chromatography. Naphthoquinones belonging to the K_2 series, as in *M. tuberculosis*,[10] are eluted from magnesium
silicate–Celite with petroleum ether and purified further by crystallization. The material after chromatography on magnesium silicate–Celite
is concentrated *in vacuo*, and the yellow oil is allowed to stand for

several weeks at room temperature.[10] Bright yellow crystals separate from the oil and appear as small balls clumped together. The properties of the naphthoquinones of the vitamin K_1 and K_2 series are summarized in Table I. The benzoquinones of bacterial origin can be purified after extraction by similar chromatographic procedures described for mammalian systems (see step 2 below).

Isolation of Quinones (Mammalian)

The benzoquinone found in animal tissues (ubiquinone–coenzyme Q) is widely distributed in nature[11,18,19] and has been independently characterized by Morton and his colleagues[20,21] and by Green and his co-workers.[4,5,11] The compound was identified as a 2,3-dimethoxy-5-methyl-benzoquinone with an isoprenoid side chain consisting of 50 carbon atoms. Other benzoquinones found in nature differ in the length of the isoprenoid side chain.[11-16] The plastoquinones, 2,3-dimethyl benzoquinones with an unsaturated isoprenoid side chain in the 5-position, are another type of quinone found in plants.

Step 1. Extraction of Quinones (Mammalian). A number of procedures have been described for extraction of quinones from tissues and subcellular organelles.[16,19,22,23] The initial extraction after saponification results in an extensive purification; however, the quinones are easily destroyed by alkali and heat. To prevent destruction by alkali, pyrogallol is added to the saponification mixture, and the material is extracted under an atmosphere of nitrogen.

The fat and connective tissues are first removed from whole organs. The tissue is then passed through a grinder, and an ethanolic solution containing 10% sodium hydroxide (1.3 l./kg. of tissue) is added. The mixture is stirred under a nitrogen atmosphere, and pyrogallol (66 g./kg. of tissue) is added. The mixture is refluxed for 30 minutes, rapidly cooled by the addition of water, and extracted four times with *n*-hexane (Skellysolve B). The lipid extract is washed with water until free of alkali, dried over anhydrous sodium sulfate, and then concentrated under reduced pressure. The residue is dissolved in *n*-hexane (10 mg. per 150

[18] F. W. Heaton, J. S. Lowe, and R. A. Morton, *J. Chem. Soc.* **1956**, 4094.

[19] B. O. Linn, A. C. Page, Jr., E. L. Wong, P. H. Gale, C. H. Shunk, and K. Folkers, *J. Am. Chem. Soc.* **81**, 4007 (1959).

[20] G. N. Festenstein, F. W. Heaton, J. S. Lowe, and R. A. Morton, *Biochem. J.* **59**, 558 (1955).

[21] R. A. Morton, G. M. Wilson, J. S. Lowe, and W. M. F. Leat, *Biochem. J.* **68**, 16 P (1958).

[22] F. W. Heaton, J. S. Lowe, and R. A. Morton, *Biochem. J.* **67**, 208 (1957).

[23] F. L. Crane, R. L. Lester, C. Widmer, and Y. Hatefi, *Biochim. et Biophys. Acta* **32**, 73 (1959).

ml.) and filtered over a mat containing 100 g. of Supercel in order to remove insoluble material. The mat is washed with three 1-l. portions of n-hexane, and the filtrate is evaporated under reduced pressure. Other solvents such as isoöctane or low-boiling hydrocarbons can also be used in the initial extraction.

Mitochondrial preparations can be used for the isolation of ubiquinone–coenzyme Q, since this benzoquinone is localized in this subcellular structure. The quinone is extracted from mitochondrial preparations by direct extraction with ethanol–ether[23] (3:1, v/v) or after saponification as described for whole tissues.[19, 23, 24] Most of the impurities in the ethanol–ether extract appear to be phospholipids which are removed by acetone precipitation. Purification of small amounts can be achieved by paper chromatography.

Step 2. Chromatographic Purification.[19, 22, 23] The benzoquinones can be absorbed and eluted from alumina,[25] aluminum oxide,[22] DeCalso,[26] magnesium silicate (Florisil), fuller's earth and silicic acid (Mallinckrodt, 100 mesh),[23] or silica gel.[27] Crane et al.[23] have found that DeCalso is most efficient for purification after saponification procedures, whereas silicic acid is preferable for purification of the acetone-soluble material after ethanol–ether extraction. The material to be purified is dissolved in n-hexane, isoöctane, or petroleum ether and absorbed onto the packed column. The column containing the absorbed quinone is washed with about 1000 ml. of the original solvent to remove unabsorbed lipid materials, and the quinone is selectively eluted with more polar solvent mixtures. The elution pattern will vary depending on the absorbant, column size, packing, and solvent mixture. The benzoquinones are generally eluted with a mixture of 5% diethyl ether in either isoöctane,[23] n-hexane,[19] or petroleum benzene. The quinone distribution can be determined spectrophotometrically and by the yellow color of the eluates. Fractions containing the benzoquinones of less than 40 to 50% purity are pooled and rechromatographed. Fractions containing the quinone with greater than 50% purity are concentrated and purified by crystallization.

Step 3. Crystallization.[14, 22] Most of the benzoquinones, in contrast to the naphthoquinones of the vitamin K_1 series, are easily crystallized. The oily material obtained in step 2 is crystallized from absolute ethanol or methanol (40 ml./g. of oily residue). The mixture is warmed, filtered through a sintered-glass funnel, and allowed to cool slowly in the dark

[24] A. M. Pumphrey and E. R. Redfearn, *Biochem. J.* 76, 61 (1960).
[25] Merck (acid or basic form).
[26] DeCalso sodium aluminum silicate (50/80 mesh).
[27] Silica gel (20 to 65 mesh).

at room temperature.[19] The crystals formed are collected, washed three times with absolute ethanol, and dried *in vacuo*. Other solvents (10 mg. of oily residue per milliliter) used to obtain crystals[23] are amyl alcohol or ethyl acetate at −15° and acetone or acetic acid at 5°.

Crystals can also be obtained after direct extraction of tissues[23] or bacteria.[12,13] The concentrated residue after direct extraction is dissolved in either methanol or ethanol and stored at −15° for 24 hours. Occasionally a whitish contaminant precipitates out after 24 hours. This material is removed, and the mother liquor is stored at −15° until the yellow or orange benzoquinone crystals appear.

Identification of Benzo- and Naphthoquinones

Spectrophotometric Method. The benzo- and naphthoquinones have characteristic absorption spectra in the ultraviolet region. The benzoquinones have a characteristic absorption peak at 275 mμ in ethanol and 272 mμ in cyclohexane. In the reduced state the spectrum of the benzohydroquinone differs from that exhibited by the oxidized form in the position of the absorption peak (290 mμ reduced) and in intensity (at 275 mμ and 290 mμ). The $E_{1cm}^{1\%}$ for ubiquinone–coenzyme Q ($E_{1cm}^{1\%}$ quinone-hydroquinone) in ethanol (275 mμ) is 142.[23] The $E_{1cm}^{1\%}$ for other naturally occurring benzoquinones is given in Table II.

The sample to be assayed is dissolved in ethanol, and the absorption at 275 mμ is determined for the oxidized form. In order to ensure that the quinone is completely oxidized, the sample is first treated with solid silver oxide for 1 hour, and the salt is removed by filtration.[23] The benzohydroquinone is prepared by adding an excess of sodium borohydride (0.5 mg.) to the cuvette, the solution shaken, and absorption at 275 mμ determined. The concentration of benzoquinone can be determined from the difference in absorption of the oxidized and reduced forms.

The naphthoquinones belonging to the vitamin K$_1$ and K$_2$ series have similar characteristic ultraviolet absorption spectra which can be used for identification of the naphthoquinone and for quantitation. Maximum absorption bands in hexane and isoöctane occur at 243, 249, 260, 269, and 325 mμ, and the corresponding $E_1^{1\%}$ values in hexane for vitamin K$_1$ are 412, 435, 440, 390, 398, and 70.8.[28] The $E_{1cm}^{1\%}$ values for vitamin K$_2$ in petroleum ether are 278 (243 mμ), 295 (249 mμ), 266 (261 mμ), 267 (270 mμ), and 48 (325 mμ).[29] The naphthohydroquinones have characteristic absorption spectra which differ from those exhibited by the oxidized forms; however, this difference cannot be used for quantitation

[28] D. T. Ewing, J. M. Vandenbelt, and O. Kamm, *J. Biol. Chem.* 131, 345 (1939).
[29] O. Isler, R. Ruegg, L. H. Chopard-dit-Jean, A. Winterstein, and O. Wiss, *Helv. Chem. Acta* 41, 786 (1958).

TABLE II
PROPERTIES OF NATURALLY OCCURRING BENZOQUINONES

	Q_6 $C_{39}H_{58}O_4$	Q_7 $C_{43}H_{64}O_4$	Q_8 $C_{48}H_{72}O_4$	Q_9 $C_{53}H_{80}O_4$	Q_{10} $C_{58}H_{88}O_4$	Refs.
Molecular weight	590.9	644.9	713.1	781.2	849.3	
Melting point	—	30.5	37	45.2	49.9	a
$E^{1\%}_{1cm}$ max 275 mμ ethanol	—	221	206	185	165	a
$E^{1\%}_{1cm}$ max 272 mμ cyclohexane	246					b
$E^{1\%}_{1cm}$ max 290 mμ ethanol (reduced)		63.9	60.0	51.7	46.4	a
$\Delta E^{1\%}_{1cm}$ (oxid-red) 275 mμ ethanol		188	176	158	142	a
R_f values						
n-Propanol–H$_2$O (4:1)[S]	0.59	0.49	0.42	0.36	0.27	c
n-Propanol–H$_2$O (7:3)[S] (reduced quinones)		0.66	0.55	0.41	0.26	c
N,N-Dimethylformamide[V]				0.49	0.38	d
Isopropanol–acetic acid–H$_2$O (24:1:15)[S]					0.54	b

Note: S = silicone-treated paper (No. 550).

V = Vaseline-treated paper.

[a] R. L. Lester, Y. Hatefi, C. Widmer, and F. L. Crane, *Biochim. et Biophys. Acta* **33**, 169 (1959).

[b] R. A. Morton, U. Gloor, O. Schindler, G. M. Wilson, L. H. Chopart-dit-Jean, F. W. Hemming, O. Isler, W. M. F. Leat, J. F. Pennock, R. Ruegg, U. Schwieter, and O. Wiss, *Helv. Chem. Acta.* **41**, 2343 (1958).

[c] R. L. Lester and T. Ramasarma, *J. Biol. Chem.* **234**, 672 (1959).

[d] B. O. Linn, A. C. Page, E. L. Wong, P. H. Gale, C. H. Shunk, and K. Folkers, *J. Am. Chem. Soc.* **81**, 4007 (1959).

unless the hydroquinone is stabilized by acetylation, since the reduced forms are rapidly oxidized. Maximum absorption of the diacetate of the naphthohydroquinone occurs at 232 mμ ($E_{1cm}^{1\%}$ of the diacetate in isoöctane is 1640).[30] The naphthochromanol (naphthotocopherol) formed enzymatically[30] or chemically has a maximum absorption peak at 244 mμ ($E_m = 30,000$).[30]

The diacetate of the naphthohydroquinone is prepared[31] by refluxing 100 mg. of vitamin K$_1$ for 30 minutes with 100 mg. of fused sodium acetate dissolved in 10 ml. of acetic anhydride and 1 g. of finely powdered zinc dust. The mixture is filtered while hot, and the acetic anhydride is decomposed by the addition of water. The hydroquinone is extracted with ether, washed repeatedly with water, and concentrated *in vacuo.* The oily residue is dissolved and crystallized from methanol at $-5°$.

Paper Chromatography.[9, 16, 19, 32, 33] Paper chromatography with Vaseline- or silicone-impregnated filter paper can be used to separate and identify quinones. Migration of the benzo- and naphthoquinones depends largely on the length and degree of saturation of the isoprenoid side chain. The R_f values of the various homologs increase linearly with respect to the length of the side chain. The less polar members have lower solubilities in the mobile phase and corresponding lower R_f values (Tables I and II). The quinones can be separated by either ascending or descending chromatography or by the circular paper method using a reverse phase system.

Filter paper (Whatman No. 1) is impregnated with Vaseline by passing the paper through a 5% (w/v) solution of Vaseline in petroleum ether or with silicone by passing the paper through a 5% solution (v/v) of Dow Corning silicone fluid (No. 550 is usually used to separate benzoquinones, and No. 1107 is used to separate naphthoquinones) in chloroform or cyclohexane. The paper is allowed to dry in air at room temperature. The samples (5 to 50 μg.) are dissolved in ethanol, n-hexane, or isoöctane and applied to the impregnated paper. With amounts of quinones greater than 20 μg., the spot appears yellow.

The napthoquinones can be separated on Vaseline-impregnated paper with a mobile phase consisting of methanol–isoöctane (3:1) or methanol–isoöctane–isopropanol (3:1:1),[9] on silicone-treated paper (No. 550) with n-propanol–H$_2$O (4:1),[34] or on silicone-treated paper (No. 1107) with

[30] P. J. Russell and A. F. Brodie, *Biochim. et Biophys. Acta* **50**, 76 (1961).
[31] S. B. Brinkley, D. W. MacCorquodale, S. A. Thayer, and E. A. Doisey, *J. Biol. Chem.* **130**, 219 (1939).
[32] R. L. Lester and T. Ramasarma, *J. Biol. Chem.* **234**, 672 (1959).
[33] J. P. Green and H. Dam, *Acta Chem. Scand.* **8**, 1341 (1954).
[34] H. Noll, R. Ruegg, U. Gloor, G. Ryser, and O. Isler, *Helv. Chem. Acta* **43**, 433 (1960).

either ethanol–glacial acetic acid–H_2O (3:1:9) or isopropanol–glacial acetic acid–H_2O (24:1:15).[33]

Good separation of the benzoquinones on Vaseline-impregnated paper is achieved with a mobile phase consisting of a 3% (w/v) aqueous solution of N,N-dimethyl formamide saturated with Vaseline[19] or with a mixture of isopropanol–glacial acetic acid–H_2O (24:1:15).[16] The oxidized benzoquinones can be separated on silicone-treated paper with a mixture of n-propanol–H_2O (4:1, v/v), whereas separation of the benzohydroquinones on this paper is achieved with a mixture of n-propanol–H_2O (7:3, v/v).[32] The benzohydroquinones are chromatographed in an atmosphere of nitrogen to prevent reoxidation.

The quinone and hydroquinone derivatives can be separated by thin layer chromatography on silica gel G or cellulose with methanol-ether (9:1). The compounds are located by their absorption or fluorescence with light at 360 mμ and eluted with ether or methanol for spectrophotometric or chemical analysis. The silica gel or cellulose is washed with ether before use since they may contain ether-soluble materials which have absorption in the UV range. Some typical R_f values on cellulose are as follows: 6-chromanyl phosphate (0.82), 6-chromanyl acetate (0.76), vitamin K_1 diphosphate (0.61–0.66), vitamin K_1 diacetate (0.65).

Detection. The benzo- and naphthoquinones can be detected by ultraviolet absorption, by leucomethylene blue indicator,[19] or after treatment with $KMnO_4$,[32] neotetrazolium, or $FeCl_3$-α,α'-dipyridyl.[32]

With leucomethylene blue spray (prepared by mixing 100 mg. of methylene blue in 100 ml. of glacial acetic acid and 1 g. of zinc dust), the quinones appear as blue spots on a white or light-blue background. The spots should be outlined quickly, since the entire paper turns blue in about 5 minutes.

The quinones (as little as 5 μg.) appear as brown spots on an almost white background when treated with $KMnO_4$.[32] The spots are developed by dipping the paper into an 0.2% (w/v) solution of $KMnO_4$ (20 to 30 seconds) and washing with hot tap water until the unchanged $KMnO_4$ is removed.

Detection with neotetrazolium[32] or α,α'-dipyridyl[32] requires prior reduction of the quinone. Reduction is accomplished by placing the chromatogram in a freshly prepared solution of sodium borohydride (0.1% w/v) for 30 seconds or until the yellow spots fade. The paper is then drained and immersed in a 0.1 N HCl solution to destroy the borohydride. When the gas evolution has ceased, the excess HCl is drained off and the paper is placed in a 0.25% solution of neotetrazolium chloride (w/v) containing 0.25 M potassium phosphate, pH 7.0, drained, and heated at 80° to 100° for 30 to 60 seconds. The quinones appear as deep purple spots. The quinones can also be visualized after reduction by spraying with a solution consisting of $FeCl_3 \cdot 6H_2O$ (5% w/v in ethanol)–α,α'-dipyridyl (0.2% w/v in ethanol)–H_2O in a ratio of 1:1:2.[32]

The napthoquinones can be detected by the procedures used for the benzoquinones or by their characteristic fluorescence on filter paper before and after treatment with KOH.[33]

Catalytic Hydrogenation. The degree of unsaturation and index as to the length of the side chain of benzoquinones and naphthoquinones of the K_2 series can be determined by hydrogenation. Hydrogenation can also be used to differentiate between naphthoquinones in the vitamin K_1 and K_2 series. Lester et al.[35] have described a microscale manometric method for hydrogenation. The catalyst (3 mg. of 10% palladium on charcoal) is suspended in 3 ml. of ethanol and added to the main compartment of a Warburg vessel. The quinone (0.4 to 0.6 mg. total in ethanol) is added to the side arm, and the system is equilibrated in an atmosphere of hydrogen. The contents of the flask are mixed, and the uptake of hydrogen is followed until complete.

Infrared Spectroscopy.[35, 36] The benzo- and naphthoquinones have characteristic spectra in the infrared region. The benzoquinones Q_6 to Q_{10} have similar infrared spectra; however, the bands due to the isoprenoid side chain may vary in intensity. Q_{10} in carbon disulfide[36] has a prominent band at 6.05 μ due to the quinoid nucleus, two bands at 7.94 and 9.14 μ due to the ether linkage, and bands at 11.47, 12.61, and 13.40 μ due to the *trans*-isoprenoid side chain. The latter bands are similar to those found in the vitamin K_2 series.

The naphthoquinones in carbon disulfide have prominent bands at 6.04, 6.02, 7.55, 7.75, 11.25, and 12.75 μ.[9] Noll[37] has developed an infrared spectrophotometric method for qualitative and quantitative identification of vitamin K_1 and K_2 homologs. The absorption bands near 9.1 and 11.91 μ present in the K_2 series and not in the K_1 series permit differentiation.

Benzo-[37a] and naphthoquinones[37b] which differ from the known types of quinones in the nature of their side chains have been recently de-

[35] R. L. Lester, Y. Hatefi, C. Widmer, and F. L. Crane, *Biochim. et Biophys. Acta* 33, 169 (1959).

[36] R. A. Morton, U. Gloor, O. Schindler, G. M. Wilson, L. H. Chopart-dit-Jean, F. W. Hemming, O. Isler, W. M. F. Leat, J. F. Pennock, R. Ruegg, U. Schwieter, and O. Wiss, *Helv. Chem. Acta* 41, 2343 (1958).

[37] H. Noll, *J. Biol. Chem.* 235, 2207 (1960).

[37a] W. V. Lavate, J. R. Dyer, C. M. Springer, and R. Bentley, *J. Biol. Chem.* 237, P.C. 2715 (1962).

[37b] P. H. Gale, B. H. Arison, N. R. Trenner, A. C. Page, Jr., K. Folkers, and A. F. Brodie, *Biochemistry,* in press.

scribed. The isoprenoid side chain of the new quinones is more reduced than their corresponding homologs. The location of the reduced isoprene unit(s) is unknown; however, it is not located at either end of the chain. These quinones are difficult to distinguish from quinones containing typical isoprenoid side chains. They can be distinguished by nuclear magnetic resonance spectroscopy, catalytic hydrogenation, and chromatographically by comparison with known homologs before and after reduction.[37a,37b] Quinones with partially reduced side chains may be widely distributed in nature.

Colorimetric Determination. The Craven test has been modified for the quantitative determination of benzoquinones.[12] The material to be assayed is dissolved in 4 ml. of ethanol and mixed with 1 ml. of ethyl cyanoacetate and 1 ml. of 0.2 N KOH. A blue color develops in 10 minutes, and the tubes are read in a Lumitron colorimeter with a 620 filter.

A sensitive colorimetric test for naphthoquinones (K₁) has been developed by Schilling and Dam.[38] The naphthoquinone is dissolved in 3 ml. of ethanol and mixed with 1 ml. of xanthane hydride[39] solution (80% saturated xanthane hydride in ethanol) and 1 ml. of 1 N KOH. The mixture is incubated at 50° for 10 minutes, cooled immediately, and the absorption measured at 410 mμ.

Inactivation and Restoration of Enzymatic Activity

Depletion of the endogenous quinone has been achieved in bacterial extracts by selective photoinactivation (360 mμ).[40,41] Extracts or isolated particulate and supernatant fractions of *Mycobacterium phlei* (see Vol. VI [35]) are placed in petri dishes (in shallow layers 3 to 5 mm.) and treated in aluminum foil-lined ice cube trays containing crushed ice. A Gates lamp with black Raymaster tube B (maximum emission 360 mμ) is placed directly over the petri dish 3 cm. from the material to be irradiated. The length of exposure to light is generally 45 minutes to 1 hour. Such treatment results in the destruction of the bound K₁-like naphthoquinone[9] and in a loss of ability to conduct oxidative phosphorylation. Restoration of both activities is specifically dependent on the addition of the natural naphthoquinone from *M. phlei*,[9] vitamin K₁, or a closely related homolog.[40,41] Inactivation by light (260 mμ[3] and 360 mμ[42]) of oxidative phosphorylation and restoration by vitamin K₁ with mammalian mitochondrial preparations have been reported.

[38] K. Schilling and H. Dam, *Acta Chem. Scand.* 12, 347 (1958).
[39] Xanthane hydride is prepared by the method of F. D. Chattaway and H. P. Stevens, *J. Chem. Soc.* 71, 607 (1897).
[40] A. F. Brodie and J. Ballantine, *J. Biol. Chem.* 235, 226 (1960).
[41] A. F. Brodie and J. Ballantine, *J. Biol. Chem.* 235, 232 (1960).
[42] R. E. Beyer, *Biochim. et Biophys. Acta* 28, 663 (1958).

Destruction of the benzoquinones by light at 360 mμ occurs at a much slower rate than that observed with vitamins K_1 or K_2. The spectral changes which occur with irradiation of pure vitamin K_1 and ubiquinone–coenzyme Q are shown in Fig. 1. Attempts to reactivate

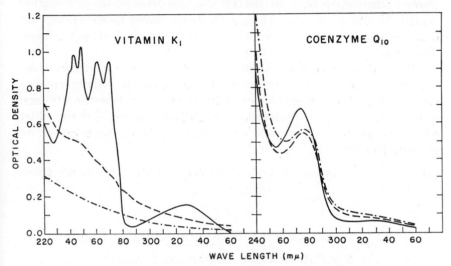

FIG. 1. The effects of 360-mμ light on pure vitamin K_1 and ubiquinone–coenzyme Q. The naphthoquinone (100 μg./ml.) was dissolved in spectral-grade isoöctane and exposed to light at 360 mμ. Samples were removed at 15 minutes (- - - - -) and 30 minutes (-·--·--·-), diluted, and assayed spectrophotometrically. The benzoquinone (100 μg./ml.) was dissolved in ethanol and treated in a similar manner.

oxidation and phosphorylation after light treatment (360 mμ) with benzoquinone systems have been unsuccessful owing to the length of time necessary to destroy this quinone.

Depletion of the endogenous quinone by extraction with lipid solvents has been found to be a suitable technique for the relatively light-insensitive benzoquinones of plant chloroplasts[43,44] and mammalian mitochondria. Mitochondria obtained from beef[45] or pig heart,[46] or from skeletal muscle,[47,48] as well as the electron-transport particle (ETP),[6] can be inactivated with lipid solvents such as isoöctane. The particles suspended in 0.25 M sucrose are extracted for 1 hour or more with cold

[43] N. I. Bishop, Proc. Natl. Acad. Sci. U.S. 45, 1696 (1959).
[44] D. W. Krogmann, Biochem. Biophys. Research Communs. 4, 275 (1961).
[45] J. Jarnefelt, R. E. Basford, H. D. Tisdale, and D. E. Green, Biochim. et Biophys. Acta 29, 123 (1958).
[46] E. R. Redfearn and A. M. Pumphrey, Biochim. et Biophys. Acta 30, 437 (1958).
[47] A. Nason and I. R. Lehman, J. Biol. Chem. 222, 511 (1956).
[48] J. Bouman and E. C. Slater, Biochim. et Biophys. Acta 26, 624 (1957).

isoöctane. The mixture is shaken at 15-minute intervals, the emulsion finally allowed to settle, and the aqueous layer withdrawn and centrifuged at $104,000 \times g$ in the Spinco preparative centrifuge. After removal of the supernatant fluid the particles are homogenized in 10 vol. of 0.25 M sucrose, centrifuged again at $104,000 \times g$, and resuspended in a smaller volume of 0.25 M sucrose. The suspension is placed under vacuum for 1 to 3 hours at room temperature to ensure removal of any residual isoöctane.

Other lipid components besides the benzoquinone are removed by isoöctane treatment.[5, 6, 45] Restoration of succinoxidase activity requires the addition of the specific benzoquinone (Q_{10}), lipoproteins,[6, 45] and cytochrome c.[4, 5, 6, 46, 47] Restoration of DPNH oxidase activity after extraction appears to require either lipid-soluble cytochrome c or α-tocopherol.[5, 49]

[49] D. Duel, E. C. Slater, and L. Veldstra, *Biochim. et Biophys. Acta* **22**, 133 (1958).

[37] Photosynthetic Phosphorylation in Plants

By F. R. WHATLEY and DANIEL I. ARNON

Photosynthetic phosphorylation[1, 2] (photophosphorylation for short) is a term coined in 1954 to describe a light-induced ATP formation by isolated chloroplasts, in accordance with Eq. (1):

$$n\text{ADP} + n \cdot \text{H}_3\text{PO}_4 \rightarrow n\text{ATP} \tag{1}$$

Reaction (1), in which the sole product is ATP, was subsequently[3] designated cyclic photophosphorylation to distinguish it from a second photophosphorylation reaction by isolated chloroplasts [Eq. (2)] which was found a few years later[4] and named noncyclic photophosphorylation.[3]

$$2\text{TPN} + 2\text{H}_2\text{O} + 2\text{ADP} + 2\text{H}_3\text{PO}_4 \xrightarrow{\text{light}} 2\text{TPNH}_2 + \text{O}_2 + 2\text{ATP} \tag{2}$$

The terms "cyclic" and "noncyclic" refer to the electron-flow mechanisms proposed for these two reactions. In cyclic photophosphorylation all the biochemically effective light energy is used for ATP

[1] D. I. Arnon, M. B. Allen, and F. R. Whatley, *Nature* **174**, 394 (1954).
[2] D. I. Arnon, F. R. Whatley, and M. B. Allen, *J. Am. Chem. Soc.* **76**, 6324 (1954).
[3] Compare review by D. I. Arnon, *in* "Handbuch der Pflanzenphysiologie" (W. Ruhland, ed.), Vol. V, p. 773. Springer, Heidelberg, 1960.
[4] D. I. Arnon, F. R. Whatley, and M. B. Allen, *Biochim. et Biophys. Acta* **32**, 47 (1959).

formation. In noncyclic photophosphorylation only a portion of the biochemically effective light energy is used for the formation of ATP; the remainder is used for the formation of a reductant, $TPNH_2$, and the excretion (evolution) of oxygen. A nonphysiological variant of noncyclic photophosphorylation (reaction 2) is reaction (3), in which TPN is replaced by ferricyanide[4] (represented here by Fe^{3+}).

$$4Fe^{3+} + 2H_2O + 2ADP + 2H_3PO_4 \rightarrow 4Fe^{2+} + O_2 + 2ATP + 4H^+ \quad (3)$$

Cyclic and noncyclic photophosphorylation supply ATP and $TPNH_2$, the two products formed at the expense of light energy, which suffice for converting CO_2 into carbohydrates in the dark.[3] Cyclic photophosphorylation [Eq. (1)] supplies only ATP, and the participation of this reaction in CO_2 assimilation is needed because the ATP formed in reaction (2) alone is insufficient for CO_2 assimilation to the level of carbohydrate.[3]

The discovery of photosynthetic phosphorylation by isolated chloroplasts, of both the cyclic and the noncyclic type, was confirmed and extended in other laboratories,[3] notably those of Jagendorf, Wessels, Vennesland, and Hill. Most of the work has been done with spinach chloroplasts, but Whatley et al.[3] have also demonstrated cyclic and noncyclic photophosphorylation in chloroplasts isolated from several other species of plants.

The methods described herein are those that are most frequently used in this laboratory. Variants of these and other procedures have been used with success in this and in other laboratories.[3]

Isolation of Chloroplasts

Reagents and Materials

Fresh spinach leaves.

0.35 M sodium chloride.

0.2 M Tris–HCl buffer, pH 8.0.

Washed sand, 18 to 25 mesh (particle diameter 0.7 to 1.0 mm.).

The highest rates of photophosphorylation are obtained with "broken" chloroplasts,[5] which have been prepared by suspending whole chloroplasts in a hypotonic salt solution, for example, 0.035 M NaCl. When whole chloroplasts are to be used, they are resuspended in 0.35 M NaCl.

Whole chloroplasts are most commonly isolated from spinach or Swiss chard leaves. We usually grow the plants in a greenhouse by a

[5] M. B. Allen, F. R. Whatley, and D. I. Arnon, *Biochim. et Biophys. Acta* 27, 16 (1958).

nutrient culture technique[6] and harvest the mature leaves prior to each experiment. However, we and others have obtained very active chloroplasts from fresh spinach leaves bought commercially.

Wash the leaves with distilled water, shake to remove excess water, and place in a plastic bag in the refrigerator for one to several hours to ensure their turgidity. Remove the midribs from the fully turgid leaves, and weigh the leaf blades. Slice quickly into pieces about 0.5 cm. square to facilitate grinding.

Grind 50 g. of sliced leaf blades by hand in a precooled mortar (20 cm. in diameter) with an "isotonic" solution consisting of 100 ml. of 0.35 M NaCl (ca. 2%) and 10 ml. of 0.2 M Tris buffer, pH 8, and about 50 g. of cold sand. Prolonged grinding (in excess of 2 minutes) should generally be avoided. Squeeze the slurry through a double layer of cheesecloth, and centrifuge the green juice at 0° for 1 minute at 200 × g to sediment sand, leaf debris, and whole cells. Decant the green supernatant liquid, and centrifuge it for 7 minutes at 1000 × g. In this centrifugation most of the intact chloroplasts are sedimented. Discard the supernatant fluid. Suspend the sedimented whole chloroplasts (P) in about 2 ml. of ice-cold 0.35 M NaCl by gently stirring them with the aid of a piece of absorbent cotton at the end of a stirring rod. Dilute to about 50 ml. with 0.35 M NaCl, centrifuge in the cold for 7 minutes at 1000 × g, and discard the pale supernatant liquid. Suspend, as above, the sedimented whole chloroplasts (P_1) in about 10 ml. of 0.035 M NaCl (final volume) to make a suspension of "broken" chloroplasts (P_{1s}). If further washing is desired, dilute the suspension of P_{1s} particles to about 40 ml. with 0.035 M NaCl, and centrifuge at 0° for 10 minutes at 18,000 × g. Discard the supernatant fluid, and resuspend the residue (P_{1s1}), as above, in about 10 ml. of 0.035 M NaCl.

Where greater stability of the chloroplasts is required and the presence of ascorbate does not interfere, add 0.01 M sodium ascorbate (200 mg. per 100 ml.) to all solutions used in the isolation, washing, and disruption of whole chloroplasts. The preparations made with ascorbate are designated C, C_1, C_{1s}, C_{1s1}, etc. (cf. Whatley et al.[7]).

In experiments where it is necessary to prepare chloride-free chloroplasts, proceed as above, using reagents free of chloride. Grind 50 g. of leaf blades in 100 ml. of a composite solution (with or without ascorbate) having a final concentration of 0.5 M sucrose, 0.05 M sodium acetate, and 0.02 M Tris–acetate buffer, pH 8. Wash the chloroplast preparation

---

[6] D. I. Arnon, M. B. Allen, and F. R. Whatley, *Biochim. et Biophys. Acta* **20**, 449 (1956).

[7] F. R. Whatley, M. B. Allen, and D. I. Arnon, *Biochim. et Biophys. Acta* **32**, 32 (1959).

P (without ascorbate) or C (with ascorbate) with 50 ml. of a composite solution of 0.5 M sucrose and 0.05 M sodium acetate. To obtain whole chloroplasts, suspend the sedimented P_1 or C_1 preparation in 10 ml. of a composite solution of 0.5 M sucrose and 0.05 M sodium acetate. To obtain broken chloroplasts (P_{1s} or C_{1s}), suspend the sedimented P_1 or C_1 preparation in 10 ml. of 0.05 M sodium acetate.

Chlorophyll Estimation.[8] Dilute 0.1 ml. of chloroplast suspension to 20 ml. with 80% acetone (20 ml. of H_2O made to 100 ml. with acetone), and filter through Whatman No. 1 filter paper. Read the optical density (against 80% acetone) at 652 mμ (1-cm. light path) in a spectrophotometer having a *narrow* band-width. Protect chlorophyll solutions in acetone from ambient light. Multiply the optical density by 5.8 to give milligrams of chlorophyll per milliliter of original suspension.

Assay of ATP Formed

Principle. The method commonly used to estimate the yield of ATP in cyclic photophosphorylation is to measure the "organic phosphate" formed during the light-dependent esterification of orthophosphate to ATP, with ADP as the phosphate acceptor. Orthophosphate is supplied at $K_2HP^{32}O_4$, the unesterified phosphate is precipitated as $MgNH_4PO_4$, and the radioactivity in the soluble "organic phosphate" fraction is determined. The blank value at $t = 0$ (or the dark control) is very low in this method.

If no P^{32} is used, photophosphorylation may be measured by other methods—for example, colorimetrically[9] by estimating the orthophosphate that remains after a period of illumination.

Procedure for Cyclic Photophosphorylation

The reaction is conveniently carried out in conical manometer vessels of about 18-ml. capacity. Add to the main compartment of the vessels, in micromoles: Tris–HCl buffer, pH 8.3, 80; $MgCl_2$, 5 or 10; sodium ascorbate, 10; adenosine diphosphate, neutralized to pH 8, 10; a cofactor[10] of photophosphorylation[7] and water to give a final volume of 3 ml. Place in the side arm 10 micromoles of K_2HPO_4, containing P^{32} (1×10^5 to 5×10^5 c.p.m.).

Chill the vessels in crushed ice, and add an aliquot of the P_{1s} or C_{1s}

---

[8] D. I. Arnon, *Plant Physiol.* **24**, 1 (1949).

[9] J. B. Summer and G. F. Somers, "Laboratory Experiments in Biological Chemistry." Academic Press, New York, 1949.

[10] The added cofactor of photophosphorylation may be *one* of these, in micromoles: phenazine methosulfate, 0.1; or flavin mononucleotide 0.3; or vitamin K_3 (in 0.1 ml. of methanol) 0.3.

preparation, containing 0.2 mg. of chlorophyll, to the main compartment. Attach the chilled vessels to Warburg manometers, shake at 15° in a refrigerated bath provided with a strong light source, and flush with nitrogen gas for 3 to 5 minutes. Start the reaction by pouring the radioactive phosphate from the side arm into the main compartment of the vessel, and turn on the light. Terminate the reaction after 15 minutes (or other suitable interval) by turning off the light and adding 0.3 ml. of 20% trichloroacetic acid to each vessel. Centrifuge the acidified reaction mixture in a 12-ml. Pyrex heavy-duty conical centrifuge tube in the cold. To measure the "organic phosphate" formed, mix a 1-ml. aliquot of the supernatant fluid with 1 ml. of magnesia mixture,[11] and add a drop of 0.2% phenolphthalein to check that the pH remains alkaline. We have found it desirable to add a small amount of inorganic phosphate to the stock magnesia mixture to ensure the presence of "seeds" of magnesium ammonium phosphate to start the precipitation. Allow the mixture to stand for 1 hour at room temperature, and then filter off the precipitate under suction. Collect the filtrate in a 10-ml. calibrated test tube, and wash the precipitate twice with a 1:10 dilution of the magnesia mixture. The filtrate now contains the radioactive ATP. Make the volume of the filtrate to 10 ml. with water, and, after thorough mixing, estimate radioactive phosphorus in a 1-ml. aliquot. Evaporate the aliquot to dryness on a planchet (an infrared lamp and a hair dryer may be used to accelerate drying). When it is dry, count the radioactivity under a thin-window Geiger counter. Compute the ATP formed by comparing the radioactivity in the filtrate with the total radioactive phosphate present in a 0.1-ml. aliquot of the original supernatant fluid, which was similarly evaporated but not precipitated with the magnesia mixture.

Procedure for Noncyclic Photophosphorylation

The procedure for determining ATP in cyclic photophosphorylation is also applicable to noncyclic photophosphorylation. Here oxygen evolution and $TPNH_2$ formation may be measured concurrently. When TPN is the electron acceptor, the reaction mixture contains, in micromoles: Tris–HCl buffer, pH 8, 80; $MgCl_2$, 5; adenosine diphosphate, neutralized to pH 8, 10; TPN, neutralized to pH 8, 4; $K_2H^{32}PO_4$, 10; also a solution of phosphopyridine nucleotide reductase,[11a] and water to a final volume of 3 ml. Prepare phosphopyridine nucleotide reductase by the method of

[11] See Vol. III [116], p. 850.
[11a] *Note added in proof:* Because of its similarities to ferredoxin in bacteria, Tagawa and Arnon [*Nature* **195**, 537 (1952)] have applied the name ferredoxin to "phosphopyridine nucleotide reductase."

Hill and Bendall,[12] omitting the crystallization step. The $K_2HP^{32}O_4$ and TPN are usually placed in the side arm of the manometer vessel. After chilling, add the P_{1s} preparation containing 0.2 mg. of chlorophyll, attach the vessels to manometers, and flush with nitrogen gas. Start the reaction by pouring the contents of the side arm into the main compartment of the vessel and immediately turning on the light. Measure oxygen evolution manometrically (with KOH and a strip of filter paper in the center well of the vessel). Determine ATP as described above. To measure the TPNH$_2$ formed, withdraw a 1-ml. sample prior to adding the trichloro-acetic acid. Centrifuge the sample at 0° at 18,000 × g for 10 minutes, dilute a 0.2-ml. aliquot of the supernatant liquid to 3 ml. with water, and read the optical density at 340 mμ, against a blank similarly prepared but without TPN.

Experiments with noncyclic photophosphorylation may also be made with ferricyanide as the electron acceptor [Eq. (3)]; in this case omit TPN and the reductase from the reaction mixture, and instead add 8 to 15 micromoles of $K_3Fe(CN)_6$ from the side arm. Since ascorbate, which is frequently present in the chloroplast preparation, reduces ferri-cyanide in the dark, enough ferricyanide must be added to leave a suitable excess for the photochemical reduction.

[12] R. Hill and F. Bendall, *Nature* **187**, 417 (1960).

[38] Photosynthetic Phosphorylation (Bacteria)

$$ADP + P_i \xrightarrow{h\nu} ATP$$

By Martin D. Kamen

Assay Method

This reaction, discovered by Frenkel,[1] is an anaerobic process cata-lyzed by extracts obtained from a variety of photosynthetic bacteria.[2-6] It can be assayed under aerobic conditions provided a reductant in small amounts is present. A number of procedures are useful. These rely on

[1] A. W. Frenkel, *J. Am. Chem. Soc.* **76**, 5568 (1954).
[2] A. M. Williams, *Biochim. et Biophys. Acta* **19**, 570 (1956).
[3] D. M. Geller, *Photophosphorylation by Rhodospirillum rubrum Preparations.* Ph.D. Thesis, Harvard University, 1957.
[4] J. W. Newton and M. D. Kamen, *Biochim. et Biophys. Acta* **25**, 462 (1957).
[5] A. W. Frenkel, *J. Biol. Chem.* **222**, 823 (1956).
[6] I. C. Anderson and R. C. Fuller, *Arch. Biochem. Biophys.* **76**, 168 (1958).

determination of either disappearance of inorganic phosphate[1-3] or appearance of ATP.[4-6] The former can be assayed by the usual Fiske-SubbaRow[3] or Lowry-Lopez[4,6] procedures. The latter can be determined by (1) isolation on absorbents specific for nucleotides, followed by elution and spectrophotometric determination,[4,5] or other means[4] (see below); and by (2) enzymatic assay with hexokinase and zwischen-ferment.[6]

A very simple procedure which appears to be generally applicable depends on the use of acid-washed Norit A to absorb P^{32}-labeled ATP produced by incubation of the enzyme system with ADP and P^{32}-labeled orthophosphate. Because the radioactivity assay equipment required is simple, readily procured, and in widespread use, this procedure will be assumed to be the one of choice. Details of other procedures will be found in the references cited.[3-6] (*Note:* See Addendum, p. 318.)

Reagents

All the following, with the exception of (c), (f), and (g), are made in 0.1 M Tris buffer, pH 7.8:

    (a) 0.01 M succinate.
    (b) 0.2 M $MgCl_2$.
    (c) 0.1 M Na_2HPO_4, containing P^{32} (specific activity $\sim 10^7$ c.p.m./ ml. based on an assay efficiency $\sim 5\%$).
    (d) 0.01 M ADP.
    (e) 0.001 M phenazine methyl sulfate.
    (f) Acid-washed Norit A.
    (g) 0.8 M perchloric acid.
    (h) Enzyme—cell-free extract (absorbancy at chlorophyll absorption maximum in infrared ~ 10 to 40).

Preparation of Reagents. Reagents (a), (b), and (d) require no comment.

REAGENT (c). Labeled orthophosphate can be obtained from the Isotopes Division, Atomic Energy Commission, Oak Ridge, Tennessee. It is supplied usually as H_3PO_4 in dilute HCl, specific radioactivity ~ 10 to 15 mc./ml. The aliquot to be used must be boiled for 10 minutes to ensure hydrolysis of labeled pyrophosphate sometimes present. As a rule, dilution to the specific activity required can be effected by addition of 100 vol. of carrier phosphate solution to 1 vol. of original. However, it is best to determine dilution empirically.

REAGENT (e). This dye may be synthesized by the method of Kehrmann and Havas.[7]

[7] F. Kehrmann and E. Havas, *Ber.* **46**, 343 (1913).

REAGENT (F). The absorbent is suspended in 10 vol. of HCl at 70° to 80° for 20 to 30 minutes and then washed with distilled water on a Büchner funnel until chloride-free. The final reagent is a suspension containing about 1 vol. of the charcoal absorbent for 5 to 10 vol. of water. One-tenth milliliter of suspension should have an absorptive capacity of ∼100 micromoles of nucleotide.

REAGENT (H). Cell-free extracts may be prepared from bacterial suspensions by sonication at 0° to 4° for varying lengths of time (2 to 20 minutes) by using a Raytheon 10-kc. magnetostriction oscillator. Alternatively, disruption by alumina grinding may be employed. It is best to begin by washing the fresh bacteria several times with distilled water, after which they are taken up in the buffer to give a suspension containing roughly 5 to 10 g. wet weight per 75 ml. of buffer.

After sonication, or disruption by other means, the suspension is centrifuged at $20,000 \times g$ for 10 minutes, and the supernatant fluid is used as enzyme. Enzyme concentration usually is referred to bacteriochlorophyll content. This is assayed by measuring absorbancy at one of the characteristic absorption bands in the infrared which vary in wavelength for the various bacteria; e.g., for *Rhodospirillum rubrum*, the band used is that at 880 to 890 mμ; for *Chromatium*, 800 mμ, etc.

Fractionation to obtain preparations with a variety of characteristics and invaried stages of purification has been described in the literature.[2-6, 8]

Assay Procedure

To a series of small centrifuge tubes (1.5 $\times$ 10 cm.) the following are added: 0.1 ml. each of reagents (a), (b), (c), (d), and (e), and 0.5 ml. of buffer. The smallest number of tubes is five—one for "zero" time and two each for the single time intervals in light and dark. Two of the tubes are wrapped with aluminum foil for the dark control. The tubes are mounted in a thermostat bath maintained at a convenient temperature (20° to 25°) and illuminated by a bank of four to six 60-watt Mazda bulbs suspended over the bath and about 1 to 2 feet from the tubes (intensity at tube position ∼200 to 400 foot-candles). The tube mount should be fabricated so that rapid shaking or swirling of the tube contents is possible. The support for a Warburg manometer flask can be used for this purpose. The reaction is started by addition of 0.1 to 0.2 ml. of enzyme (absorbancy at bacteriochlorophyll absorption maximum ∼10 to 40). Immediately after addition the reaction in the "zero-time" tube is terminated by addition of 1.0 ml. of 0.8 M perchloric acid. The reaction is allowed to continue in the assay tubes for a predetermined time

[8] A. W. Frenkel and D. D. Hickman, *J. Biophys. Biochem. Cytol.* 6, 285 (1959).

(10 to 20 minutes) and then terminated by rapid addition of 1.0 ml. of 0.8 M perchloric acid.

The tubes are transferred to a conventional table-top clinical centrifuge and spun at high speed to obtain a clear supernatant fluid which contains the P^{32}-labeled ATP formed. One milliliter of the clear supernatant fluid from each assay tube is transferred to fresh clean test tubes, and 0.1 ml. of well-mixed Norit suspension is added. The mixture is agitated well and centrifuged. The supernatant fluid is decanted, and the Norit is washed three times with distilled water, the wash liquid being centrifuged and decanted off each time. Finally the Norit is suspended in 1.0 ml. of water, mixed well, and two aliquots are transferred by pipet (0.3 ml. is a convenient volume) to planchettes for radioactivity assay.

The suspensions are pipetted carefully but rapidly in the center of the planchettes to ensure the radioactive sample lies in the central area of the G-M tube window when the planchette is placed in the counting rack in the position nearest the window. With a little practice with radioactive phosphate solutions, it is possible to obtain reproducible results by this procedure (deviations in duplicates <5%). The samples should be counted immediately to avoid losses by evaporation which can introduce errors owing to variations in self-absorption of the β-radiations because of inhomogeneities in sample thickness. Reliable assays are obtained when the final radioactivity is $\sim10^3$ to 10^4 c.p.m. with a thin-window G-M counting tube.

The usual assay geometry obtained by placement of a liquid P^{32} sample in the position closest to the G-M tube counter (within 1.0 cm. of the window) permits $\sim5\%$ counting efficiency. Thus, 1 μc. of P^{32} will give $0.05 \times 2.2 \times 10^6$ c.p.m. $= 1.1 \times 10^5$ c.p.m. The incubation mixture contains ~10 micromoles of phosphate with a radioactivity $\sim10^6$ c.p.m. The range of enzyme activity lies between ~1.0 and 10.0 micromoles of P_i esterified per hour for preparations with bacteriochlorophyll content like that specified in the assay procedure given above. For a 20-minute incubation, the least number of counts per minute appearing in the total ATP synthesized will be $\sim3 \times 10^4$. The aliquots taken usually assay at least $\sim1 \times 10^4$ c.p.m.

Properties

General Properties. Light saturation can be obtained at relatively low light intensities (~200 to 300 foot-candles). The pH optimum is broad and lies at $\sim$pH 8.0. Heat inactivation is effected by incubation for 1 to 2 minutes at 80°. The cell-free extracts in a crude state can be lyophilized and stored in an appropriate medium such as 40% ethylene glycol in the deep-freeze ($-15°$ to 20°) for months without appreciable deterioration. However, freezing and thawing will rapidly inactivate the enzyme. Intact bacterial cells can be frozen and stored at $-20°$ to $-30°$ for months and then thawed to give good yields of active cell-free extracts.

Specificity. When preparations consist of well-washed chromatophores, only ADP and IDP are esterified. Crude preparations may contain adenylate kinase, in which case AMP can serve as a substrate.

Rates of Photophosphorylation. Usually, the presence of phenazine methyl sulfate is necessary to ensure the greatest rates of photophosphorylation, which can be as high as 350 micromoles of orthophosphate esterified per milligram of bacteriochlorophyll per hour with *R. rubrum* preparations.[9] Rates observed with *Chromatium* preparations usually are no greater than one-fourth to one-third of this maximal rate.[6]

Effect of Reducing Agents. There is no substrate requirement for photophosphorylation. However, catalytic amounts of electron donors ensure good rates of reaction, even under strictly anaerobic conditions (gas, phase, helium). In the presence of air, appreciable amounts of a suitable reducing agent (succinate, DPNH) are required.[10]

Inhibitors. The usual inhibitors of mitochondrial oxidative phosphorylation such as dinitrophenol and cyanide are ineffective except at relatively high concentrations. Responses to inhibitors such as (1) Cu^{++}, (2) 8-hydroxyquinoline, (3) Dicumarol, (4) the butyl ester of 3,5-diiodo-4-hydroxybenzoic acid, (5) Antimycin-A, (6) 3-hydroxy, 1-heptylquinoline-*N*-oxime acid, and (7) alkylated napthoquinone (SN-5949) vary with purity and source of preparations. In general (1), (3), (5), (6), and (7) are effective at very low concentrations (10^{-4} to $10^{-6} M$). Dyes such as methylene blue, 2,6-dichlorophenolindophenol, and brilliant cresyl blue are inhibitors under aerobic conditions. Oxygen itself is an inhibitor. The extreme sensitivity toward Cu^{++} requires that care be taken in excluding this metal ion, and other heavy metal ions, from the reagents and water used.

In general, it appears that redox agents can act as either stimulators or inhibitors, depending on their degree of oxidation. Thus, phenazine methyl sulfate, which has been studied most intensively, appears to be effective as a stimulator when kept partially reduced.[3, 4, 11] There appears to be a difference in response of bacterial extracts to phenazine and its derivatives compared to that noted with green plant chloroplasts. Thus, whereas phenazine methyl sulfate stimulates all systems, pyocyanine

[9] L. Smith and M. Baltscheffsky, *J. Biol. Chem.* 234, 1575 (1959).
[10] A. W. Frenkel, *Brookhaven Symposia in Biol.* No. 11, 276 (1958).
[11] For a review of this and other aspects of the photophosphorylation reaction, see A. W. Frenkel, *Ann. Rev. Plant Physiol.* 10, 53 (1959).

seems to stimulate chloroplasts,[12] but it inhibits chromatophores as obtained from the photosynthetic anaerobe *Chromatium*.[4]

Concluding Remarks

The bacterial photophosphorylation reaction appears to be similar to the anaerobic reaction observed in green plant chloroplast preparations, discovered by Arnon *et al.*[13] and called by them "cyclic photophosphorylation" because of its lack of a substrate requirement. The occurrence of this reaction in cell-free extracts prepared from all types of photosynthetic system indicates a common mechanism in all photosynthetically active organisms for the light-induced storage of energy in the form of ATP. The simplest working hypothesis pictures phosphate esterification as occurring during recombination of the photoreductant and photo-oxidant systems formed by the photochemistry initiated by excitation of chlorophyll. This recombination is assumed to be mediated by an electron-transport chain of catalysts similar to those found in mitochondria. The substrate (succinate, DPNH, etc.) and H-acceptor (oxygen) needed in mitochondrial oxidation phosphorylation are replaced in the strictly anaerobic photophosphorylation system by the endogenous light-produced photoreductant and photo-oxidant.

Addendum

An alternative procedure, based on extraction of residual inorganic phosphate as phosphomolybdic acid using isobutanol-benzene, has been employed[14-16] in a number of laboratories. However, there appear to be no special advantages inherent in its use. A common disadvantage is that both methods are rather tedious. Methods based on disappearance of inorganic phosphate are more expedient but must be monitored carefully by occasional checks against results obtained based on determinations of esterified phosphate.

[12] R. Hill and D. A. Walker, *Plant Physiol.* **34**, 240 (1959).
[13] D. I. Arnon, F. R. Whatley, and M. B. Allen, *J. Am. Chem. Soc.* **76**, 6324 (1959).
[14] S. O. Nielsen and A. L. Lehninger, *J. Biol. Chem.* **215**, 555 (1955).
[15] M. Avron, *Biochim. et Biophys. Acta* **40**, 1 (1960).
[16] T. Horio and M. D. Kamen, *Biochemistry* **1**, 144 (1962).

[39] $P_i{}^{32}$-ATP Exchange Enzyme System

Adenosine triphosphate (A—R—P—P—P) + orthophosphate-P^{32} ($P_i{}^{32}$)
$\rightleftarrows$ Adenosine triphosphate-P^{32} (A—R—P—P—P^{32}) + orthophosphate (P_i)

$$(1)$$

Adenosine triphosphate (A—R—P—P—P)
    + adenosine diphosphate-P^{32} (A—R—P—P^{32})
    $\rightleftarrows$ Adenosine triphosphate-P^{32} (A—R—P—P^{32}P)
    + adenosine diphosphate (A—R—P—P)

$$(2)$$

By G. W. E. PLAUT

Assay Method

Principles. The purification of the $P_i{}^{32}$–ATP exchange activity is most conveniently followed by observing the incorporation of the label from orthophosphate into bound organic phosphate compounds by means of the molybdate extraction procedure. Since this method merely results in the removal of orthophosphate into the organic solvent phase, leaving other phosphate compounds in the aqueous layer, it is highly unspecific and cannot be used for the determination of the $P_i{}^{32}$–ATP exchange in very crude tissue preparations. The use of isolated mitochondria as the starting material for the enzyme extraction leaves behind most of the interfering enzymic activities. Even in this initial extract from mitochondrial acetone powder, radioactivity from $P_i{}^{32}$ is found in both ATP and ADP, probably because of the initial exchange reaction between $P_i{}^{32}$ and ATP which is followed by a redistribution of label from formed ATP^{32} into ADP via the adenylic kinase reaction. After the removal of adenylic kinase by the first protamine precipitation step of the purification procedure, practically all the label from $P_i{}^{32}$ is incorporated into the terminal phosphate group of ATP, and values obtained by the molybdate extraction method are in exact agreement with procedures involving the initial separation of ATP from the reaction mixture.

Reagents

    $4 \times 10^{-2} M$ imidazole (pH 6.7).
    $3 \times 10^{-4} M$ ATP (crystalline).
    $1.5 \times 10^{-3} M$ ADP.
    $2 \times 10^{-3} M$ $MnSO_4$.
    $1.5 \times 10^{-3} M$ $P_i{}^{32}$ (1×10^6 to 6×10^6 c.p.m.).
    Enzyme, diluted to a final volume of 2.0 ml. with water.

Procedure. The reaction is started by the addition of enzyme. After incubation for 30 minutes at 25°, 2 ml. of 10% trichloroacetic acid is added. It is advisable to include with the assays a zero-time sample, and two incubated blanks, one containing all components except ADP and ATP, and the other lacking only in enzyme. Precipitated protein is removed by centrifugation.

Two milliliters of the deproteinized supernatant is placed into a 16 × 125-mm. test tube and diluted to 3.0 ml. with water. Then 1.5 ml. of 5% ammonium molybdate [(NH$_4$)$_6$Mo$_7$O$_{24}$·4H$_2$O] and 0.3 ml. of 7 N sulfuric acid are added. Orthophosphate is removed from the aqueous phase by extraction with 5 ml. of isobutanol followed by 5 ml. of ether. Mixing of the phases is accomplished conveniently by aeration for 1 minute. After each separation, the solvent phase (upper) is removed by aspiration.

An aliquot (1.0 ml.) of the remaining aqueous phase is pipetted into a planchet of 25-mm. diameter, and the radioactivity of the solution is determined with a suitable detector, e.g., a Geiger-Müller end-window counter.

Definition of Units and Specific Activity. The counts obtained in the blank are subtracted from those in the incubated samples. Enzymic activity is estimated as micromoles of P$_i^{32}$ incorporated per milliliter of reaction mixture according to the formulation

$$\frac{\text{C.p.m./ml. reaction mixture (incorporated)}}{\text{C.p.m./micromole P}_i^{32} \text{ (initial)}}$$
$$= \text{Micromoles P}_i^{32} \text{ incorporated/ml. reaction mixture}$$

The relationship between enzyme concentration and time has been found to be linear in this assay up to a phosphate transfer of 0.15 micromole of P$_i^{32}$ per milliliter of reaction mixture.

The specific activity of the enzyme is expressed as micromoles of P$_i^{32}$ incorporated per milligram of protein in 30 minutes at 25°. Protein is determined by the method of Warburg and Christian (Vol. III [73]).

Purification Procedure

All manipulations are carried out at 2° to 5° unless specified otherwise.

Acetone Powder. Fresh pig liver is freed from connective tissue and cut into pieces of about 1 cm.³. The cubed tissue (1 kg.) is suspended in 500 ml. of 0.25 M sucrose, and the washing fluid is discarded. Portions (100 g.) of liver are homogenized in 400 ml. of 0.25 M sucrose for 1 minute in a Waring blendor at "slow" speed followed by 30 seconds at "full" speed. The suspension is centrifuged at 1000 × g for 10 minutes;

the residue is discarded, and the supernatant fluid is passed through a single layer of cheesecloth to remove large pieces of fat at the upper phase. The filtrate is centrifuged at 5000 $\times$ g for 30 minutes. On discarding of the supernatant layer, the residue is suspended in 1 l. of 0.25 M sucrose. The suspension is centrifuged at 5000 $\times$ g for 30 minutes, and the residue is washed twice more in this fashion with 500-ml. portions of 0.25 M sucrose. The residue is then treated three times with 400-ml. portions of acetone ($-20°$). The resulting residue is dried in a vacuum and stored in a tightly closed container at $-20°$.

Extraction of Enzyme. Acetone powder (10 g.) is mixed with 150 ml. of distilled water with mechanical stirring for 30 minutes. The suspension is centrifuged at 30,000 $\times$ g for 30 minutes, and the residue is discarded. The specific activity of the extract is between 0.02 and 0.04.

First Protamine Precipitation. Protamine sulfate solution (5%) [adjusted to pH 6.7 with KOH] is added to the clear supernatant solution from the previous step until no further precipitate is obtained. The resulting suspension is centrifuged at 30,000 $\times$ g for 15 minutes, and the supernatant fluid is discarded. The residue is washed with 100 ml. of water and then suspended in 100 ml. of 0.1 M ammonium acetate in a Potter-Elvehjem homogenizer. The mixture is mechanically stirred for 30 minutes and then settled at 30,000 $\times$ g for 15 minutes. The residue is discarded.

Second Protamine Precipitation. The supernatant solution is treated with protamine sulfate, and the resulting residue is washed with water as in the previous step. The precipitate is then suspended in 50 ml. of 0.3 saturated ammonium sulfate solution (the saturated solution at 25° is 4.10 M ammonium sulfate; see Vol. I [10]), homogenized, and stirred for 30 minutes. The mixture is centrifuged for 15 minutes at 30,000 $\times$ g, and the residue is discarded.

Ammonium Sulfate Precipitation. The ammonium sulfate concentration of the enzyme extract is adjusted from 0.3 to 0.5 saturation with saturated ammonium sulfate. The resulting precipitate is gathered by centrifugation for 30 minutes at 30,000 $\times$ g and dissolved in 10 ml. of 0.1 M ammonium acetate, pH 6.9. This solution usually has a protein concentration of about 3 mg./ml. and a specific activity of 1.2 to 1.8. It can be stored at 0° for several weeks without significant loss of activity.

Treatment with Anion Exchanger. One volume of the enzyme solution from the previous step is diluted with 4 vol. of water. One volume of wet Dowex 1-X4 chloride form (200 to 400 mesh), previously equilibrated with 0.02 M ammonium acetate to pH 6.4, is added to an equal volume of the diluted enzyme solution and suspended by stirring. The mixture is filtered, and the resin containing the enzyme is washed with an equal volume of $0.02 M$ ammonium acetate. The addition to the resin of an equal volume of 0.3 saturated ammonium sulfate at pH 6.5 leads to the elution of the activity. This treatment gives a threefold purification over that obtained in step 4 with a yield of 70%. However, the protein concentration of the eluate from the resin is only about 0.1 mg./ml., and the enzyme exhibits marked instability.

The specific activity of the final product should be 3 to 4.5 with an approximate activity yield from the aqueous extract from mitochondria of 30 to 50%.

A summary of the purification procedure is given in the table.

PURIFICATION OF EXCHANGE ENZYME[a]
(Ten grams of hog liver acetone powder was used.)

Steps	Volume, ml.	Activity yield, %	Total protein, mg.	Specific activity, micromoles $P_i{}^{32}$/mg. protein/30 min.
1. Aqueous extract	130	100	2600	0.027
2. Protamine precipitation				
Supernatant fluid	120	5.7	1800	0.002
Water wash	94	0.7	86	0.006
Ammonium acetate eluate	95	108	400	0.19
3. Second protamine precipitation				
Supernatant fluid	90	4	243	0.012
Water wash	48	0	0	0
Ammonium sulfate eluate				
(0.3 saturation)	48	128	101	0.9
4. Ammonium sulfate (0.3–0.5 saturation)				
Supernatant fluid	67	4.3	37	0.08
Precipitate	9.6	62	34.5	1.3
5. Resin step				
Supernatant fluid	50	0	0	0
Ammonium acetate wash	50	0	0	0
Ammonium sulfate eluate				
(0.3 saturation)	50	41	6.4	4.5

[a] Data from M. Chiga and G. W. E. Plaut, *J. Biol. Chem.* **234,** 3059 (1959).

Properties[1,2]

Specificity. The exchange between $P_i{}^{32}$ and ATP [reaction (1)] occurs specifically with the terminal phosphate group of ATP; however,

[1] G. W. E. Plaut, *Arch. Biochem. Biophys.* **69,** 320 (1957).
[2] M. Chiga and G. W. E. Plaut, *J. Biol. Chem.* **234,** 3059 (1959); *Biochim. et Biophys. Acta* **61,** 736 (1962).

this reaction is stimulated by the presence of ADP in the reaction mixture. ADP seems to be a participant in the P_i-ATP reaction, since purified enzyme preparations carry out an ADP-ATP exchange [reaction (2)] at the same rate as the P_i-ATP reaction. The ADP-ATP exchange is dependent on orthophosphate. The ATP and ADP requirements for reactions (1) or (2) cannot be replaced by other nucleoside monophosphates or nucleoside polyphosphates. In view of these interrelationships, it is possible that reactions (1) and (2) are catalyzed by the same protein molecule. However, in the absence of physiochemical data such a speculation is premature.

The purified enzyme does not contain adenylic kinase, nucleoside monophosphate kinase, or nucleoside diphosphokinase activities.

The exchange enzyme has been differentiated from a number of other enzymes possessing similar activities. One of these, a nucleoside diphosphokinase, also present in extracts from mitochondria, can carry out an ADP-ATP exchange but not the P_i-ATP reaction. This nucleoside diphosphokinase, which accounts for nearly 90% of the ADP-ATP exchange activity of the original extract from mitochondria (the exchange enzyme accounts for about 10% of the ADP-ATP exchange and practically all the P_i-ATP exchange of the extract), does not possess the strict adenosine nucleotide specificity exhibited by the exchange enzyme and differs furthermore by being activated about equally well by both Mg^{++} and Mn^{++}.

The ADP-ATP exchange enzyme of oxidative phosphorylation reported by Wadkins and Lehninger[3-5] also seems to differ from this exchange enzyme, since their enzyme shows no P_i^{32}-ATP exchange, the ADP-ATP exchange is not dependent on P_i, and it is activated equally well by Mg^{++} and Mn^{++}.

Distribution. The enzyme has been found in extracts of mitochondria from hog, beef, guinea pig, and rat liver, and from beef heart. Examination of various fractions of rat liver homogenate indicates that the enzyme is mainly in the mitochondria.

Activators and Inhibitors. The enzyme [reactions (1) and (2)] is activated by Mn^{++}, and to a lesser extent by Co^{++}. Mg^{++} is much less effective than Mn^{++}.

Both exchanges [reactions (1) and (2)] are inhibited by *p*-hydroxymercuribenzoate, mercuric chloride, and silver nitrate. The inhibition by these substances can be reversed by cysteine. Uncoupling agents of oxidative phosphorylation, e.g., 2,4-dinitrophenol, Dicumarol, sodium

[3] C. L. Wadkins and A. L. Lehninger, *J. Biol. Chem.* 233, 1589 (1958).
[4] C. L. Wadkins, *Federation Proc.* 18, 346 (1959).
[5] A. L. Lehninger, *Federation Proc.* 19, 952 (1960).

azide, and oligomycin, are ineffective. Arsenate is a potent inhibitor of both exchange reactions. The arsenate inhibition is partially reversed by orthophosphate, the reversal being of the competitive type in reaction (1).

Kinetic Properties. Optimal rates of P_i–ATP exchange are obtained in the presence of $3 \times 10^{-4} M$ ATP and $1.5 \times 10^{-3} M$ ADP. The Michaelis constant for P_i for reactions (1) and (2) is between 0.5×10^{-3} and $1.5 \times 10^{-3} M$.

[40] Acyl Phosphatase from Skeletal Muscle

Acetyl Phosphate → Acetate + Orthophosphate

By ISAAC HARARY

Assay Method

Principle. The assay of acyl phosphatase utilizes acetyl phosphate as the substrate because of its common availability and adequate stability. The method is based on following the hydrolysis of acetyl phosphate by the hydroxamic acid method of Lipmann and Tuttle[1] for measuring the residual substrate. Because of the high dissociation constant for acyl phosphatase and acetyl phosphate, the level of substrate used is somewhat below the concentration needed to saturate the enzyme for a long period of time. However, with the concentration indicated, zero-order kinetics prevail for 20 minutes, with activity independent of substrate concentration.

The enzyme concentration is chosen so that, of the 20 micromoles of acetyl phosphate used, close to one-half of the acetyl phosphate is hydrolyzed. Further hydrolysis resulting from higher enzyme activity lowers the concentration of substrate to a point which affects the rate of activity.

Reagents

1.0 M potassium acetate, pH 5.4.
0.1 M potassium acetyl phosphate.
4.0 M NH_2OH, 3.5 M NaOH, 3 N HCl, 12% TCA.
5% $FeCl_3 \cdot 6H_2O$ in 0.1 N HCl.
Enzyme. The enzyme is diluted with 0.1 M acetate buffer at pH 5.4 to give a solution containing approximately 50 units/ml.

[1] F. Lipmann and L. C. Tuttle, *J. Biol. Chem.* **159,** 21 (1945).

Procedure. To a solution containing 20 micromoles of acetyl phosphate and 100 micromoles of acetate buffer, pH 5.4, are added the enzyme and H_2O to a final volume of 1 ml. The reaction mixture is incubated at 37°, and the reaction is stopped after 20 minutes by addition of 0.2 ml. of the reaction mixture to 2.0 M freshly neutralized NH_2OH at pH 6.0. The acetyl hydroxamic acid formed is then measured by the standard procedure of Lipmann and Tuttle.[1] It is important to correct the value obtained by a blank containing everything but the enzyme because of the slow but measurable nonenzymatic hydrolysis of acetyl phosphate.

Definition of Unit and Specific Activity. One unit of enzyme activity is defined as that amount of enzyme which hydrolyzes 1.0 micromole of acetyl phosphate under the conditions described. Specific activity is expressed as units per milligram of protein. Protein may be determined spectrophotometrically after suitable dilution, by measurement of optical density at 280 mμ corrected for nucleic acid by the absorption at 260 mμ. Protein may also be determined by using sulfosalicylic acid precipitation of the protein and reading the turbidity developed in a Klett colorimeter.

Purification Procedure

Certain modifications have been made to the published procedure of Lipmann[2] and Koshland,[3] and a further purification has been achieved by fractionation with ammonium sulfate. A typical preparation is described below.

Step 1. Preparation of the Crude Extract. Ground horse skeletal muscle (665 g.) was added to 2 l. of boiling 1% KCl–0.067 M HCl for 10 minutes. The temperature was held between 90° and 94°. The mixture was filtered through gauze into a flask immersed in an ice bath, and the solution was brought as quickly as possible to room temperature. The pH was then brought from 4.0 to 6.0 by addition of 85 ml. of 1 M Na_2CO_3. The precipitate was centrifuged and discarded. The volume was 2080 ml.

Step 2. Trichloroacetic Acid and Ammonium Sulfate Precipitation. To 850 ml. of the heat-acid supernatant was added 42.5 ml. of 50% trichloroacetic acid at 0–2°C. to bring the trichloroacetic acid concentration to 2.5%. The precipitate was immediately centrifuged and extracted with ice-cold 0.05 M acetate, pH 5.4, and dialyzed against 0.02 M acetate, pH 5.4. Very little activity appeared in this fraction.

To the 2.5% trichloroacetic supernatant solution (volume 870 ml.) was added 487 g. of ammonium sulfate. The suspension was centrifuged,

[2] F. Lipmann, *Advances in Enzymol.* 6, 231 (1946).
[3] D. E. Koshland, Jr., Vol. II [87].

and the precipitate was dissolved in 0.05 M acetate, pH 5.4, and dialyzed against 0.02 M acetate, pH 5.4. This was repeated (using 850 ml. at a time) with the rest of the solution.

Step 3. Ammonium Sulfate Fractionation. The ammonium sulfate precipitate, dissolved and dialyzed against 0.02 M acetate, pH 5.4, was fractionated with solid ammonium sulfate. The fraction precipitating between 0 and 63% saturation yielded 10,000 units (specific activity 94); the fraction between 63 and 70% saturation yielded 4100 units (specific

SUMMARY OF PURIFICATION PROCEDURE

Step	Total units	Recovery, %	Specific activity
Water extract	60,000		2.0
Acid-heated supernatant	51,000	85	20.5
TCA supernatant	30,000	50	80.2
AmSO₄ precipitate (0–80%)	27,000	45	163
AmSO₄ fractionation (70–85%)	12,000	20	1270

activity 210); and the fraction between 70 and 85% saturation yielded 12,000 units (specific activity 1270). The lower fractions may be refractionated to yield high activity protein. The purification is summarized in the table.

Properties

Specificity. The enzyme catalyzes the hydrolyses of acetyl, butyryl, and palmityl phosphate.[1,4] 1,3-diphosphoglyceric acid,[5] and carbamyl phosphate.[6] It does not act on acetyl CoA, acetyl adenylate, pyrophosphate, ATP, ADP, AMP, glycerol phosphate, 3-phosphoglycerate, phosphoenol pyruvate, or succinyl choline.[5]

General Properties. The purified enzyme is stable at acid pH's and will withstand heating to 60° for 20 minutes without loss in activity. Heating under the same conditions at pH 8.6 in Tris buffer leads to a loss of approximately one-half to one-third of the activity. Paper electrophoresis studies of the purified enzyme preparations indicate one major and one minor component with isoelectric points in the neighborhood of pH 8.6. Its solubility in trichloroacetic acid solutions and the necessity for 80% saturation with ammonium sulfate for salting out indicate that the enzyme is a small protein. It withstands dialysis with

[4] A. L. Lehninger, *J. Biol. Chem.* **162**, 340 (1946).
[5] I. Harary, *Biochim. et Biophys. Acta* **26**, 434 (1957).
[6] S. Grisolia, J. Caravaca, and B. K. Joyce, *Biochim. et Biophys. Acta* **29**, 432 (1958).

a little loss in activity and retains its activity for months if kept frozen in solution.

The enzyme seems not to require a metal for activity as determined both by lack of effect of added metal ions and metal-binding inhibitors such as Versene or α,α-dipyridyl. Iodosobenzoate, iodoacetate, and PCMB do not inhibit, indicating that sulfhydryl groups are not necessary for activity. Acyl phosphatase will not catalyze the acetylation by acetyl phosphate of a group of possible acceptors, nor will it catalyze the phosphorylation of glucose or creatine. Moreover, it will not catalyze an exchange labeled acetate or phosphate with acetyl phosphate in the presence or in the absence of CoA or ATP. It does not contain transacetylase, acetothiokinase, myokinase, or ribonuclease activities.

The pH of maximum activity is in the region of 5.3. The Michaelis-Menton constant for acetyl phosphate is $8.0 \times 10^{-3} M$, and about $10^{-5} M$ for 1,3-diphosphoglyceric acid. The enzyme is competitively inhibited by ortho- and pyrophosphate and irreversibly inhibited by preincubation with $10^{-5} M$ thyroxine.[7,8]

Distribution. The enzyme is widely distributed. It is present in other tissues such as the brain,[6] kidney, liver, and leucocytes,[5] and also in many bacteria.[1,9] The enzyme in liver differs from that of the muscle enzyme. It is not acid-heat stable, and it is not inhibited by inorganic phosphate.

In both these tissues the enzymes are largely present in the so-called soluble portion of the cell.

[7] I. Harary, *Biochim. et Biophys. Acta* **25**, 193 (1957).
[8] I. Harary, *Biochim. et Biophys. Acta* **29**, 647 (1958).
[9] B. Shapiro and E. Wertheimer, *Nature* **156**, 690 (1945).

Section III
Enzymes of Coenzyme and Vitamin Metabolism

[41] Pyridoxamine 5-Phosphate Oxidase

Pyridoxamine 5-phosphate + $\frac{1}{2}O_2 \rightarrow$ Pyridoxal 5-phosphate + NH_3

By BURTON M. POGELL

Assay Method[1]

Principle. The method is based on the measurement of pyridoxal-5-P formation. Pyridoxal-5-P has a peak in its absorbancy spectrum at 415 mμ in Tris buffer of pH 8.0 and at 388 mμ in alkali. Under these conditions pyridoxamine-5-P has a negligible absorption in this region. Therefore the reaction may be followed by measuring the increase in absorbancy at either of these wavelengths.

Molar Extinction Coefficients for Pyridoxal-5-P. E_{415} (Tris, pH 8) = 6480; E_{388} (0.1 M NaOH) = 6550.

Reagents

0.01 M pyridoxamine-5-P.
0.1 M tris(hydroxymethyl)aminomethane, pH 8.0.
0.05 M sodium carbonate, pH 10.4.
5 N NaOH.
Enzyme. Dilute with water if necessary.

Procedure. Since pyridoxal-5-P is easily decolorized by light, carry out the enzyme assays in minimal light.

A. Place 0.15 ml. of pyridoxamine-5-P and 2.75 ml. of Tris buffer in 25-ml. Erlenmeyer flasks, and cool in an ice bath. Add 0.2 ml. of enzyme, and incubate in a constant-temperature shaker for 30 minutes at 38°. Cool the flasks in an ice bath, transfer the contents to cold small test tubes, and place them in a boiling-water bath for 2 to 3 minutes. Cool the tubes, filter the contents, and read the absorbancy at 415 mμ in a Beckman Model DU spectrophotometer, using microcells with water as a blank. Include controls for enzyme alone and substrate alone, and calculate the results by using the appropriate extinction coefficient. This procedure is applicable with crude extracts.

B. The following procedure can be used when a partially purified, transparent enzyme fraction is available. Place 0.1 ml. of pyridoxamine-5-P and 1.7 ml. of carbonate buffer in flasks, and cool as above. Add 0.2 ml. of enzyme, and incubate as above. Stop the reaction by the addition of 0.2 ml. of 5 N NaOH, and read the absorbancy at 388 mμ.

[1] B. M. Pogell, *J. Biol. Chem.* **232**, 761 (1958).

The oxidase reaction with purified fractions may be followed directly in cuvettes by using an appropriate spectrophotometer. The complete reaction mixture minus pyridoxamine-5-P serves as a blank. Enzyme is added last to start the reaction. Linearity of product formation with time is found in both buffers at 30°, but not in carbonate buffer at 38°. A decrease in specific activity of the oxidase is found with increasing protein concentration at both 38° and room temperature. Enzyme activity also may be followed by measurement of the specific activation by pyridoxal-5-P of the tyrosine apodecarboxylase of vitamin B_6-deficient *Streptococcus faecalis* cells.[1,2]

Purification Procedure[1]

All steps are carried out at 0° to 4° unless otherwise indicated. Protein is determined spectrophotometrically.[3]

Step 1. Preparation of Crude Extract. Place rabbit livers immediately after exsanguination in cold isotonic KCl. Drain, and homogenize in a Waring blendor for 2 minutes in 2 ml. of isotonic KCl per gram of tissue. Centrifuge the homogenate for 100 minutes at $18,000 \times g$, and discard the precipitate.

Step 2. Acid Precipitation. Adjust the supernatant fluid to pH 5.0 by the dropwise addition of $1.0\,N$ acetic acid, continue stirring for at least 15 minutes, and remove the precipitate by centrifugation.

Step 3. Alcohol Fractionation. Place the supernatant, whose temperature is 0°, in a deep-freeze at −23°, and add dropwise with constant stirring 0.3 vol. of 95% ethanol, previously cooled to −23°. Collect the precipitate by centrifugation at −2° to −3°, suspend to 0.2 of the original volume in water, and dialyze overnight against distilled water. Centrifuge the dialyzed enzyme at $18,000 \times g$, and discard any insoluble matter.

Specific activities of 0.0040 to 0.0045 micromole of pyridoxal-5-P formed per hour per milligram of protein were found in initial $18,000 \times g$ supernatant fluids by using assay A. A final preparation with specific activity of 0.063 was obtained with assay procedure B, with an over-all recovery of 52%.

Properties[1]

Substrate Concentration and Specificity. For each mole of pyridoxamine-5-P oxidized, 1 mole each of pyridoxal-5-P and NH_3 are formed.

[2] I. C. Gunsalus and R. A. Smith, Vol. III [142].
[3] B. M. Pogell and R. W. McGilvery, *J. Biol. Chem.* **208**, 149 (1954); E. Layne, Vol. III [73].

Saturation with pyridoxamine-5-P is found at 0.2 mM at pH 10.4. The partially purified enzyme fraction also catalyzes the following reactions:

$$\text{Pyridoxamine} + \tfrac{1}{2}O_2 \rightarrow \text{Pyridoxal} + NH_3$$
$$\text{Pyridoxamine-5-P} + H_2O \rightarrow \text{Pyridoxamine} + P$$
$$\text{Pyridoxal-5-P} + H_2O \rightarrow \text{Pyridoxal} + P$$

It is not known whether pyridoxamine-5-P and pyridoxamine are oxidized by the same or different enzymes. Much higher concentrations of pyridoxamine are needed for maximal activity (about 7.5 mM), but the rates of oxidation are of the same order of magnitude.

No oxidation of benzylamine or histamine is detectable with the purified oxidase. The enzyme is different from the known monoamine, diamine, and D-amino acid oxidases.

Activators and Inhibitors. There is an oxygen requirement for this reaction; it will not proceed anaerobically. Azide ion at 0.0034 M does not inhibit the oxidase when tested in Tris buffer. The remaining inhibitor studies were performed at pH 10.4. Pyridoxamine inhibits 34% at 0.005 M, but this compound also is deaminated by the enzyme preparation. NH$_4$Cl inhibits 15% at 0.0025 M and 34% at 0.01 M. *p*-Chloromercuribenzoate inhibits 27% at 0.0001 M and 83% at 0.001 M. Neither pyridoxine nor pyridoxal produces any inhibition at 0.0002 M. Pyridoxine, oxaloacetate, and α-ketoglutarate inhibit from 26 to 31% at 0.0033 M. MgSO$_4$ has no effect over the concentration range of 0.0005 to 0.05 M.

The enzyme is reactivated by either flavin adenine dinucleotide or riboflavin 5'-phosphate after partial resolution of the oxidase by acid–ammonium sulfate precipitation.

Effect of pH. The activity rises as the pH is raised from 6 to 8, and a plateau is reached at pH 8 to 9. Further increase in pH causes a rise in activity, and the maximum is found at pH 10.1 to 10.4.

Stability. The partially purified enzyme is stable for several weeks when stored either frozen or lyophilized at −20°. After 5 minutes at pH 5, 50% of the activity is destroyed at 43°, and 100% at 60°.

Distribution. Highest oxidase activity is found in rabbit and rat liver, and smaller amounts are present in rat kidney and brain and beef liver. Activity also can be detected in whole cells of *Streptococcus faecalis* (vitamin B$_6$-deficient) and baker's yeast.

ADDENDUM

Wada and Snell[4] have reported a 65-fold purification of the rabbit liver oxidase. They developed a sensitive colorimetric procedure for assay of the enzyme by reac-

[4] H. Wada and E. E. Snell, *J. Biol. Chem.* **236**, 2089 (1961).

tion of pyridoxal-5-P with phenylhydrazine. Oxidation of both pyridoxine-5-P and pyridoxamine-5-P appear to be catalyzed by the same enzyme. Pyridoxal-5-P and phosphorylated vitamin B_6 analogs were potent inhibitors of the oxidase. Riboflavin-5'-P was the most effective reactivator of the apoenzyme (K_m of $3.1 \times 10^{-8} M$), but flavin adenine dinucleotide worked equally well at higher concentrations.

Turner and Happold[5] have reported the presence of pyridoxamine-5-P oxidase in *Escherichia coli* extracts. The presence of this enzyme and a pyridoxal-5-P phosphatase apparently completely explain the earlier claims of interconversion of pyridoxamine-5-P and pyridoxal-5-P by a transaminase type reaction.

[5] J. M. Turner and F. C. Happold, *Biochem. J.* **78**, 364 (1961).

[42] Ascorbic Acid Synthesis in Animal Tissues

By CLARK BUBLITZ and ALBERT L. LEHNINGER

It is now well established by the work of several laboratories that the formation of ascorbic acid from D-glucuronate in animal tissues involves the reactions catalyzed by TPN-L-gulonate dehydrogenase, aldonolactonase, and L-gulonolactone oxidase. The assay and purification of TPN-L-gulonate dehydrogenase and aldolactonase and an assay of L-gulono-γ-lactone oxidase and dehydrogenase will be described, since neither the oxidase nor the dehydrogenase has been purified to a significant extent.

I. TPN-L-Gulonate Dehydrogenase

$$\text{D-Glucuronate} + \text{TPNH} + \text{H}^+ \rightleftarrows \text{L-Gulonate} + \text{TPN}^+$$

This enzyme has been purified from pig kidney[1] and rat liver.[2] The purification procedure of the two enzymes is similar. The purification of the pig kidney enzyme will be described.

Assay Method

Principle. The enzyme is most conveniently assayed spectrophotometrically by measuring the rate of change of absorbancy at 340 mμ of TPNH formed during the oxidation of L-gulonate.

[1] J. L. York, A. P. Grollman, and C. Bublitz, *Biochim. et Biophys. Acta* **47**, 298 (1961).

[2] Y. Mano, K. Yamada, K. Suzuki, and N. Shimazono, *Biochim. et Biophys. Acta* **34**, 563 (1959).

Reagents

0.1 M potassium L-gulonate.

0.01 M TPN.

0.2 M tris(hydroxymethyl)aminomethane–glycine buffer, pH 9.0.

Procedure. To a cuvette with a 1-cm. light path are added 0.3 ml. of L-gulonate, 0.2 ml. of TPN, 1.0 ml. of buffer, enzyme, and water to give a final volume of 3.0 ml. The increase in absorbancy at 340 mμ is recorded. When assaying crude, undialyzed preparations, it is convenient to use a blank which contains the complete system except for L-gulonate and to record the difference in the rate of change of absorbancy between the cuvette containing L-gulonate and the blank.

Definition of Unit and Specific Activity. A unit of dehydrogenase activity has been defined as that amount of enzyme which is necessary to produce a rate of increase in absorbancy at 340 mμ of 0.001 per minute under the conditions described above. Specific activity is expressed in units per milligram of protein as determined spectrophotometrically.[3]

Purification Procedure

All operations described in this section are carried out at temperatures between 0° and 5°.

Pig kidneys, obtained immediately after slaughter, are placed on cracked ice, brought to the laboratory, and frozen at −25°. The frozen kidneys can be stored for at least 3 months without loss of enzyme activity.

Homogenization and Differential Centrifugation. A total of 1000 ml. of a 25% homogenate of pig kidney in 0.15 M KCl is prepared in a Waring blendor; each batch is homogenized for 1 minute. Most of the particulate matter is then removed by centrifugation at 21,000 × g for 2 hours.

Ammonium Sulfate Precipitation. To 760 ml. of the supernatant fluid obtained after centrifugation (clarified homogenate) are added slowly, with mechanical stirring, 172 g. of solid ammonium sulfate. After 30 minutes the precipitate that forms is removed by centrifugation. To 800 ml. of the supernatant fluid are added 96 g. of solid ammonium sulfate. The precipitate that forms is gathered by centrifugation, suspended in 20 ml. of distilled water, and dialyzed against 4 l. of distilled water with mechanical stirring for 4 hours. The dialyzing solution is changed, and the dialysis is carried out for another 2 hours. The precipitate which forms during dialysis is removed by centrifugation, and

[3] See Vol. III [73].

the supernatant fluid (ammonium sulfate fraction) is used in the further purification. The enzyme can be stored at this stage for at least a month at —5° without loss of activity.

Fractionation on Diethylaminoethylcellulose.[4] The ammonium sulfate fraction, diluted with 15 vol. of a solution containing $0.001 M$ ethylenediaminetetraacetate and $0.005 M$ tris(hydroxymethyl)aminomethane–HCl, pH 8.0, is adsorbed on a diethylaminoethylcellulose column (12×2 cm.) previously equilibrated with a solution containing $0.002 M$ ethylenediaminetetraacetate and $0.01 M$ tris(hydroxymethyl)aminomethane–HCl, pH 8.0, and subsequently eluted by a linear gradient from a 500-ml. mixing chamber containing initially a solution of $0.002 M$ ethylenediaminetetraacetate in $0.01 M$ tris(hydroxymethyl)aminomethane–HCl, pH 8.0, and a reservoir containing 500 ml. of a solution of $0.2 M$ KCl and $0.002 M$ ethylenediaminetetraacetate in $0.01 M$ tris(hydroxymethyl)aminomethane–HCl, pH 8.0. The enzyme is eluted between 120 and 140 ml. of eluent. It is essential to add ethylenediaminetetraacetate during this step, since its omission results in complete loss of enzyme.

A summary of this purification procedure is given in Table I.

TABLE I

PURIFICATION OF TPN-L-GULONATE DEHYDROGENASE FROM PIG KIDNEY

Enzyme preparation	Total units	Specific activity units/mg. protein
Clarified homogenate	4.48×10^5	55.7
Ammonium sulfate fraction	5.40×10^5	335
Eluates from diethylaminoethylcellulose:		
Fraction 8	3,200	334
Fraction 9	17,600	1290
Fraction 10	16,000	1820
Fraction 11	12,000	1500
Fraction 12	9,300	605
Fraction 13	3,200	228

Properties

Stability. The purified enzyme is stable when stored at —5° for 2 weeks, when it begins to show some loss in activity.

Activators and Inhibitors. The rate of the reaction is maximum between pH 8.6 and 9.3. The enzyme is reversibly inactivated by *p*-chloromercuribenzoate ($0.001 M$).

Specificity. The enzyme reduces D-glucuronate at about the same rate

[4] See Vol. V [1].

as D-galacturonate. The enzyme oxidizes L-galactonate at a rate about 60% that of L-gulonate. DPN is 1% as active as TPN.

Kinetic Properties. The K_m is $7.3 \times 10^{-3} M$ for L-gulonate at $0.01 M$ TPN, and $2 \times 10^{-5} M$ for TPN at $0.02 M$ L-gulonate.

II. Aldonolactonase

$$\text{Aldonolactone} + H_2O \rightleftarrows \text{Aldonate}^- + H^+$$

Aldonolactonase has been purified from beef[5] and rat liver.[6,7] The purification of the enzyme from rat liver will be described here.[6]

Assay Method

Principle. The hydrolysis of a lactone at neutral pH involves the production of an equivalent of acid and thus can be measured manometrically by following the release of CO_2 from a bicarbonate buffer.

Reagents

$0.1 M$ KHCO$_3$ (freshly prepared).
$0.005 M$ MnCl$_2$.
$0.2 M$ D-Galactono-γ-lactone.

Procedure. To the main compartment of a Warburg vessel are added 0.3 ml. of KHCO$_3$, 0.6 ml. of MnCl$_2$, water, and enzyme. To the side arm is added 0.6 ml. of lactone. The total volume is 2.0 ml. After the system is equilibrated at 25° under an atmosphere of 95% N$_2$–5% CO$_2$, substrate is tipped in from the side arm. Manometric readings are taken every 5 minutes for 20 minutes. Readings are corrected for retention of carbon dioxide and for nonenzymatic hydrolysis of lactone.

Definition of Unit and Specific Activity. A unit is defined as the amount of enzyme that causes the release of 1 micromole of CO$_2$ per 15 minutes under the conditions described above. Specific activity refers to units per milligram of protein, as determined by the biuret method or method of Lowry.[3] Since MnCl$_2$ which is added to enzyme preparation interferes with both methods of protein estimation, it is necessary to precipitate the protein with trichloroacetic acid (final concentration 5%, w/v, trichloroacetic acid) and wash the precipitated protein with trichloroacetic acid prior to protein estimation.

[5] K. Yamada, *J. Biochem.* **46**, 361 (1959).
[6] C. Bublitz and A. L. Lehninger, *Biochim. et Biophys. Acta* **47**, 288 (1961).
[7] F. A. Isherwood, L. W. Mapson, and Y. T. Chen, *Biochem. J.* **76**, 157 (1960).

Purification Procedure

All operations are carried out at temperatures between 0° and 5° unless otherwise indicated.

Homogenization. A total of 980 ml. of a 15% homogenate of rat liver in 0.25 M sucrose is prepared with a Teflon homogenizer.

Fractional Centrifugation and Treatment with $MnCl_2$. To 770 ml. of the supernatant fraction (8000 $\times$ g supernatant fraction) obtained after removal of the nuclei and mitochondria[8] are slowly added, with stirring, 85.7 ml. of M $MnCl_2$ to give a final concentration of 0.1 M $MnCl_2$.

Isoelectric Precipitation. To 855 ml. of the Mn^{++}-treated supernatant fraction is added cautiously, with stirring, 1.0 M acetic acid to give a final pH of 5.2. The precipitate which forms is removed 15 minutes later by centrifugation.

Heat Treatment. The supernatant fraction (800 ml.) from the preceding step is distributed among seventy test tubes, which are then placed in a 65° water bath equipped with a mechanical stirrer. When the temperature inside the tubes reaches 55°, the temperature of the bath is lowered to 55° by the addition of ice water. After an incubation period of 30 minutes at 55° to 56°, the test tubes are immersed in ice water until the temperature inside the tubes drops to 2°. The precipitate which forms during this procedure is removed by centrifugation at 21,000 $\times$ g for 2 hours.

Acetone Fractionation. To 674 ml. of the heat-treated supernatant fluid are added 290 ml. of cold (−15°) acetone. The temperature is gradually lowered to −7° during this step. The precipitate which forms is removed by centrifugation at −7°, 15 minutes after the last addition of acetone. To 925 ml. of the supernatant fluid thus obtained are added an additional 697 ml. of acetone. The temperature is maintained between −5° and −7° during this step. The precipitate which forms is collected by centrifugation at −7° and suspended in a solution containing 0.01 M potassium acetate, pH 4.9, and 0.1 mM $MnCl_2$. This suspension is clarified by centrifugation (acetone fraction).

Chromatography on Carboxymethylcellulose.[4] The acetone fraction is adsorbed on a carboxymethylcellulose column (29.5 $\times$ 2.24 cm.), packed by gravity, and equilibrated with 0.01 M potassium acetate, pH 5.2. The enzyme is eluted by a linear gradient of KCl in a mixing chamber containing initially 500 ml. of a solution of 0.01 M potassium succinate, pH 6.0, and 0.1 mM $MnCl_2$, and a reservoir of 500 ml. of 0.01 M potassium succinate, pH 6.0, 0.1 mM $MgCl_2$, plus 0.15 M KCl. The aldonolactonase forms a colored band on the column which begins to appear after 216 ml. of eluent has run through the column.

[8] See Vol. I [3].

A summary of the purification procedure is given in Table II.

TABLE II

PURIFICATION OF ALDONOLACTONASE FROM RAT LIVER

Enzyme	Total units	Specific activity units/mg. protein
Homogenate	581,000	19.1
8000 × g supernatant fraction	594,000	35.8
pH 5.2 supernatant fluid	541,000	90.7
Heat-treated supernatant fluid	400,000	652
Acetone fraction	391,000	1410
Eluates from carboxymethylcellulose:		
Fraction 13, 216–234 ml.	4,850	583[a]
Fraction 14, 234–252 ml.	195,000	2120
Fraction 15, 252–270 ml.	59,400	645[a]
Fraction 16, 270–288 ml.	5,830	375[a]
Fraction 17, 288–306 ml.	2,790	555[a]
Fraction 18, 306–324 ml.	1,480	226[a]

[a] Protein determined by method of Lowry et al., Vol. III [73].

Properties

Stability. The enzyme is protected from heat denaturation by $MnCl_2$ (0.5 mM). The enzyme is most stable at pH 7.4 to 8.0. The purified enzyme is stable for months at 0°.

Activators and Inhibitors. The enzyme requires metal ions for optimal activity and is inhibited completely by 0.001 M ethylenediaminetetraacetate.

Specificity. The enzyme hydrolyzes the following lactones at the relative rates shown: D-galactono-γ-lactone, 100; L-gulono-γ-lactone, 6.4; D-glucurono-γ-lactone, 0.9; D-gulono-γ-lactone, 64; α-D-glucoheptono-γ-lactone, 2.8; D-arabono-γ-lactone, 15.9; D-ribono-γ-lactone, 5.5; L-galactono-γ-lactone, 25; D-mannono-γ-lactone, 10.8; D-mannurono-γ-lactone, 4.0; D-glucono-γ-lactone, 62. The enzyme also catalyzes formation of lactone from anion at slightly acid pH.

Kinetic Properties.[9] The K_m for D-gulono-γ-lactone is $1.65 \times 10^{-2}\,M$; for L-gulono-γ-lactone, $6.0 \times 10^{-2}\,M$; and for L-galactono-γ-lactone, approximately $2.1 \times 10^{-2}\,M$.

III. L-Gulono-γ-lactone Oxidase

L-Gulono-γ-lactone + $\frac{1}{2}O_2$ → L-Ascorbic acid + H_2O

This enzyme is found in liver microsomes obtained from all animals except those requiring dietary ascorbic acid such as primates and guinea

[9] J. Winkelman and A. L. Lehninger, *J. Biol. Chem.* **233**, 794 (1958).

pigs. Since this enzyme has not been extensively purified, only an assay for the oxidase is described.[10] Similar assays of the oxidase have been described.[7,11,12] In a later section an aerobic assay for the dehydrogenase component is described.

Assay Method for Oxidase

Principle. The oxidase (with oxygen as electron acceptor) is most conveniently assayed by measuring the formation of ascorbic acid by the method of Roe and Kuether.[13] When assaying crude tissues, it is useful to measure ascorbic acid destruction by including a vessel incubated with an appropriate amount of ascorbic acid (1 to 2 micromoles) in place of L-gulono-γ-lactone.

Reagents

0.1 M imidazole–HCl, pH 6.8.

0.1 M L-gulono-γ-lactone (freshly prepared).

Procedure. To a 20-ml. beaker are added 0.4 ml. of imidazole–HCl, 0.1 ml. of L-gulono-γ-lactone, water, and enzyme to a final volume of 2.0 ml. The reaction is carried out with shaking in air at 37°, and the reaction rate is linear with time and enzyme concentration until 4 micromoles of ascorbic acid are formed. The reaction is conveniently carried out for 1 to 2 hours. The reaction is stopped by the addition of 8.0 ml. of 6% (w/v) trichloroacetic acid, and, after the precipitated protein is removed, ascorbic acid is determined in 2.0-ml. aliquots by the method of Roe and Kuether.[13]

Properties

Activators and Inhibitors. The enzyme is reversibly inhibited by *p*-chloromercuribenzoate. The enzyme is not inhibited by azide, cyanide, ethylenediaminetetraacetate, *o*-phenanthroline, Amytal, or antimycin A.

Specificity.[11] The enzyme catalyzes the oxidation of the following compounds at the relative rates shown: L-gulono-γ-lactone, 100; L-galactono-γ-lactone, 87; D-mannono-γ-lactone, 68; D-altrono-γ-lactone, 47; D-talono-γ-lactone, 50; ethyl-D-idonate, 65.

Kinetic Measurements. In air, the K_m for L-gulono-γ-lactone is 4 × 10^{-4} M. The K_m for oxygen is 0.2 mM.[7]

[10] C. Bublitz, *Biochim. et Biophys. Acta* **48**, 61 (1961).

[11] J. Kanfer, J. J. Burns, and G. Ashwell, *Biochem. et Biophys. Acta* **31**, 556 (1959).

[12] I. B. Chatterjee, G. C. Chatterjee, N. C. Ghosh, J. J. Ghosh, and B. C. Guha, *Biochem. J.* **74**, 193 (1960).

[13] J. H. Roe and C. A. Keuther, *J. Biol. Chem.* **147**, 399 (1943).

Assay Method for Dehydrogenase

Principle. The assay method is based on the reduction of phenazine methosulfate by the dehydrogenase and subsequent reduction of 2,6-dichlorophenolindophenol by the reduced phenazine methosulfate. The reduction of the latter dye is measured spectrophotometrically at 600 or 610 mμ. Provided the proper control is used, this assay may be done under aerobic conditions.

Reagents

0.001 M 2,6-dichlorophenolindophenol in 0.02 M tris(hydroxymethyl)aminomethane–HCl, pH 7.4.

0.01 M KCN.

Freshly prepared phenazine methosulfate (10 mg./ml.) in 0.2 M tris(hydroxymethyl)aminomethane–HCl, pH 7.4.

0.2 M tris(hydroxymethyl)aminomethane–HCl, pH 7.4.

0.1 M L-gulono-γ-lactone (freshly prepared).

Bovine serum albumin, 4% (w/v).

Procedure. Exactly 0.1 ml. of the above solutions with the exception of L-gulono-γ-lactone is added to two or more cuvettes with a 1-cm. light path. To the reaction cuvettes are added 0.1 ml. of gulonolactone and enzyme. The volumes of the control cuvette, i.e., the cuvette not containing enzyme or substrate, and the reaction cuvettes are brought to 1.0 ml. with water, and the contents of each cuvette are mixed by inversion. A Beckman B spectrophotometer is adjusted to 100% transmission at 610 mμ with a cuvette containing the enzyme. Exactly 15 seconds after the instrument has been adjusted to 100% transmission, the absorbancy of the control cuvette is determined. Readings are taken every minute for 3 minutes. The reaction rate in three cuvettes can be conveniently followed by arranging the cuvettes in such a way that the instrument is brought to 100% transmission with the cuvette containing the most rapid reaction and the other cuvettes arranged in order of descending reaction rates. When assaying crude preparations, it is necessary to add serum albumin and enzyme to the control cuvette.

Definition of Unit. A unit of dehydrogenase activity has been defined as the amount of enzyme required to produce a change in absorbancy at 610 mμ of 0.01 per minute under the conditions described above after corrections have been made for changes in the control cuvette.

[43] FAD-Forming Enzyme (Liver)

$$FMN + ATP \xrightarrow{Mg^{++}} FAD + PP$$

By CHESTER DE LUCA

Assay Method

Principle. The activity of the FAD-forming enzyme is measured by determining the initial rate of formation of the dinucleotide from the substrate mononucleotides. The most common method of assay for FAD is the manometric method of Warburg and Christian (described in Vol. II [23]) using the D-amino acid oxidase which specifically requires it as its coenzyme. For convenience, the rapid spectrophotometric adaptation[1] of this classical method was used throughout these studies.

Reagents

0.01 M FMN, neutralized.
0.01 M ATP, neutralized.
0.01 M MgCl₂.
0.10 M tris(hydroxymethyl)aminomethane (Sigma-121) buffer, pH 7.5.

Procedure. Appropriate dilutions of the enzyme preparation are added to a reaction mixture containing 1.0 ml. of Tris (Sigma-121) buffer, water, 0.3 ml. of MgCl₂, 0.2 ml. of FMN, and 0.2 ml. of ATP, mixed in that order, to give a total volume of 3 ml. The reaction components are then incubated at 37°. At zero time and again after 30 to 60 minutes, two 0.25- to 1.0-ml. aliquots are removed and assayed directly after addition to cuvettes containing blank and assay medium, respectively.

FAD in the reaction aliquot taken above is quantitatively determined as the rate-limiting component during the oxidation of DPNH (0.55 micromole) in the coupled enzyme assay system. The assay cuvette includes, besides the reduced pyridine nucleotide, 0.2 ml. of D-amino acid apoöxidase (about 0.25 mg. of protein prepared as described in ref. 1), 0.1 ml. of rabbit skeletal muscle lactate dehydrogenase (20-fold dilution of Worthington Corp. preparation), 112 micromoles of DL-alanine, and 250 micromoles of Tris buffer, pH 8.2, in a total volume of 3 ml. The aliquot used as a blank is diluted with water and buffer.

The assay is run over the next 3½ minutes at 340 mμ in the Beckman

[1] C. DeLuca, M. M. Weber, and N. O. Kaplan, *J. Biol. Chem.* **223**, 559 (1956).

Model DU spectrophotometer with a photomultiplier attachment when necessary. The reaction aliquot in the blank cuvette serves to null the high absorption due to the presence of FMN. The increase in FAD concentration between the time of initiation of the synthesis reaction, and the 30- to 60-minute incubation period is a measure of the initial rate of dinucleotide synthesis.

Application of Assay Method to Crude Tissue Homogenates. Owing to the presence of competing activities in whole crude homogenates, little or no FAD synthetic ability can be detected. In fact, such preparations, incubated with FAD, cause a decrease in the concentration of the dinucleotide, presumably by destroying it. The first point at which a reasonable synthesis assay may be made is after step 2 of the purification procedure given below.

Purification Procedure

The procedure described here has been followed repeatedly with rat liver.[2] The FAD-synthetic activity, however, is widespread throughout mammalian tissues, being found in the soluble fraction of all tissues examined, both normal and neoplastic. The cell particulates contain the FAD-destructive ability.

Step 1. Preparation of Crude Homogenates. Rat tissue homogenates are prepared in the cold usually with 5 vol. of 0.25 M sucrose solution in a glass Ten Broeck homogenizer with a relatively loose-fitting pestle.

Step 2. Separation of Active Cell Fraction. Initial studies with rat liver were performed on major cell fractions isolated by the method of differential centrifugation. In all cases synthetic activity is found to be localized in the soluble fraction not sedimented by centrifugation at approximately 100,000 $\times$ g for a period of 1 hour in the Spinco Model L preparative ultracentrifuge.

Step 3. Ammonium Sulfate Precipitation. Closer examination reveals that in even the soluble fraction the synthetic activity is accompanied by some destructive activity. Forty per cent ammonium sulfate is capable of precipitating the synthesizing enzyme; yet, as in the case of the yeast enzyme,[3] the precipitate is still contaminated by some of the FAD-inactivating system.

Step 4. Chromatographic Purification. Column chromatographic separation of the FAD-synthetic and FAD-destructive activities is carried out at 2° to 10° with Whatman Ashless powdered cellulose, Standard Grade, previously washed with 0.5 N HCl. Columns are packed with

[2] C. DeLuca and N. O. Kaplan, *Biochim. et Biophys. Acta* **30**, 6 (1958).
[3] A. W. Schrecker and A. Kornberg, *J. Biol. Chem.* **182**, 795 (1950).

positive pressure applied to cellulose poured as a slurry in 40% ammonium sulfate solution.

Proteins to be separated are reprecipitated with ammonium sulfate, and enough dry cellulose is added to form a loose suspension; the entire mixture is then applied to the prepared column. The over-all dimensions of the columns are about 50 mm. high by 33 mm. wide for 100 mg. of total protein applied.

Columns are eluted at a rate of about 1 ml./min. Five-milliliter fractions are collected. A system of gradient elution with nonlinearly decreasing concentrations of ammonium sulfate is employed to recover the synthetic enzyme free of the destructive activity.

The over-all recovery of total protein is poor, only 10 to 20% being removed from the column in the manner given; not all the enzyme is recovered. Two major peaks are eluted, the second of which is active for the synthesis of FAD. The remaining protein can be removed by elution with alkaline agents; this ordinarily shows no synthetic activity. The active protein is recovered from combined eluates by ammonium sulfate precipitation and dialyzed before used any further.

A summary of over-all purification achieved is not included here because purification on the free synthetic activity was not carried out. Before the separation of the two competing activities, no accurate determination of purity can be made directly. For this reason no unit has been defined.

Properties of the Partially Purified Enzyme

Effect of Nucleotides. Concentration curves for the effects of FMN and ATP on the rate of the synthetic reaction show both compounds to be inhibitory at levels above $5 \times 10^{-4} M$.

No synthesis of FAD was observed when riboflavin was substituted for FMN in the purified system.

With ADP, 70% of the maximum rate, obtained in the presence of ATP, could be observed. The presence of myokinase activity in the purified enzyme has not been ruled out.

5'-AMP is inactive in this system. It does not itself condense with FMN to give FAD, nor will it interfere with the synthesis of the dinucleotide.

Activators and Inhibitors. For its maximum rate the system requires the addition of Mg^{++} to a concentration of $1 \times 10^{-3} M$. Mn^{++} can also serve as cofactor giving 30% of the maximum rate at an optimum concentration of $1 \times 10^{-4} M$.

Cysteine, though ordinarily not required for activity, was found to sustain the synthetic activity during extended incubation periods.

p-Chloromercuribenzoate inhibited the system completely at a concentration of $1 \times 10^{-4} M$.

pH Effect. The optimum pH for the FAD-forming system is at approximately 7.5.

Reversibility. It has not been possible to demonstrate the breakdown of FAD when the dinucleotide is incubated with inorganic pyrophosphate and magnesium in the presence of the purified enzyme. The synthetic enzyme has been freed of pyrophosphatase activity by aging or by the addition of $6.7 \times 10^{-3} M$ sodium fluoride.

[44] Synthesis of Diphosphopyridine Nucleotide from Nicotinic Acid

By John Imsande, Jack Preiss, and Philip Handler

Synthesis of DPN from nicotinic acid in yeast, mammalian liver, and human erythrocytes has been shown to proceed by the following three consecutive reactions:[1]

Nicotinic acid + PRPP
$$\overset{Mg^{++} \; P_i}{\rightleftarrows} \quad \text{Nicotinic acid mononucleotide} + PP_i \qquad (1)$$

Nicotinic acid mononucleotide + ATP
$$\overset{Mg^{++}}{\rightleftarrows} \quad \text{Nicotinic acid adenine dinucleotide} + PP_i \qquad (2)$$

Nicotinic adenine dinucleotide + ATP + glutamine
$$\overset{Mg^{++}}{\underset{K^+}{\rightarrow}} \quad DPN + \text{glutamic acid} + AMP + PP_i \qquad (3)$$

This article will describe the partial purification of the enzymes involved in these steps. The intermediates, nicotinic acid mononucleotide (deamido-NMN) and nicotinic acid adenine dinucleotide (deamido-DPN), have been isolated from human erythrocytes after incubation with nicotinic acid.[1] Deamido-DPN has also been isolated from *Penicillin chrysogenum*,[2] mouse liver,[3] and rat liver,[1] Deamido-NMN has been isolated from baker's yeast.[4]

[1] J. Preiss and P. Handler, *J. Biol. Chem.* **233**, 488, 493 (1958).
[2] G. Serlupi-crescenzi and A. Ballio, *Nature* **180**, 1204 (1957).
[3] T. A. Langan, Jr., N. O. Kaplan, and L. Shuster, *J. Biol. Chem.* **234**, 2161 (1959).
[4] R. W. Wheat, *Arch. Biochem. and Biophys.* **82**, 83 (1959).

I. Nicotinic Acid Mononucleotide Pyrophosphorylase

Nicotinic acid $+$ PRPP $\rightleftarrows$ Nicotinic acid mononucleotide $+$ PP$_i$

Assay Method

Principle. Nicotinic acid mononucleotide pyrophosphorylase activity is measured by following the formation of nicotinic acid-C^{14} mononucleotide from nicotinic acid-7-C^{14} with a strip counter after separation of reaction components by paper chromatography.

Reagents. The reaction mixture contains the following (in micromoles): MgCl$_2$, 5.0; ATP, 2.0; PRPP, 0.2; phosphate, 40, pH 7.3; nicotinic acid-7-C^{14}, 0.1; and approximately 9 units of enzyme in a volume of 0.8 ml. Nicotinic acid-7-C^{14} (specific activity 3.09 mc./millimole) may be obtained from the New England Nuclear Corporation. PRPP is prepared by the procedure of Kornberg *et al.*[5] or obtained commercially from the Pabst Laboratories.

Procedure. Reaction mixtures are incubated for 5 hours at 37°, then deproteinized by heating in a boiling-water bath for 1 minute. An aliquot (25 μl.) of the incubation is spotted on Whatman No. 1 paper and chromatographed overnight. The solvent system consists of 7 parts 95% ethanol and 3 parts M ammonium acetate at pH 5.0. Chromatograms thus obtained are analyzed for the per cent conversion of nicotinic acid-7-C^{14} to nicotinic acid-C^{14} mononucleotide by means of a chromatogram strip counter.

Definition of Units and Specific Activity. A unit of activity is defined as the amount of enzyme required for the synthesis of 1 millimicromole of nicotinic acid mononucleotide per hour; specific activity is expressed as units per milligram of protein.

Purification Procedure

Beef liver is obtained immediately after slaughter and chilled in ice. All subsequent purification steps are performed in the cold except where otherwise specified.

Acetone Powder Preparation. Fresh chilled beef liver (100-g. portions) are homogenized for 1 to 2 minutes in a Waring blendor with 500 ml. of acetone ($-12°$), then filtered rapidly with suction on a 25-cm. Büchner funnel with Whatman No. 1 paper. The moist liver residue is rehomogenized with 500 ml. of cold acetone, filtered, and quickly dried at room temperature. Drying is accelerated and connective tissue is removed by passing the filter pad through a wire screen. When the smell of acetone can no longer be detected, the dry powder is stored at $-12°$. Storage for 6 months results in approximately 30% loss in activity.

---

[5] A. Kornberg, I. Liebermann, and E. S. Simms, *J. Biol. Chem.* **215**, 389 (1955).

Step 1. Extraction. Beef liver acetone powder (30 g.) is extracted for 30 minutes at room temperature with 300 ml. of 0.05 M Tris, pH 7.4, with occasional stirring. The suspension is then centrifuged at 10,000 × g for 10 minutes, and the residue is discarded.

Step 2. Ammonium Sulfate Fractionation. Ammonium sulfate (61.3 g.) is added slowly to 250 ml. of the extract with continuous stirring. Stirring is continued for an additional 10 minutes, the mixture is centrifuged as in step 1, and the supernatant is discarded. The precipitate is taken up in 200 ml. of 0.01 M Tris, pH 7.4.

Step 3. pH and Ammonium Sulfate Fractionation. The solution obtained in step 2 is adjusted to pH 4.8 with approximately 3.8 ml. of 0.4 M HCl. Stirring is continued for 1 hour, the mixture is centrifuged as in step 1, and the precipitate is discarded. The supernatant is adjusted to pH 7.4 with approximately 1.5 ml. of 0.5 M Tris, pH 10.7, brought to 32% saturation with 45.7 g. of ammonium sulfate as in step 2, centrifuged, and the supernatant is discarded. The precipitate is taken up in 70 ml. of water and dialyzed for 1 hour against 6 l. of water. Centrifugation is repeated if turbidity develops during dialysis.

Step 4. Calcium Phosphate Gel Adsorption. Seventy milliliters of the above preparation are diluted to 490 ml., and 10 ml. of 0.5 M phosphate, pH 7.3, are added. The solution is combined with 2 g. (dry weight) of aged calcium phosphate gel, stirred for 10 minutes, centrifuged, and the gel is discarded. The supernatant is brought to 35% saturation with 122.5 g. of ammonium sulfate as in step 2, centrifuged, and the supernatant discarded. The precipitate is taken up in 25 ml. of water and dialyzed for 2 hours against 4 l. of water.

Step 5. DEAE-Cellulose Fractionation. The aqueous solution obtained in step 4 is added to a packed 1.6 × 6-cm.^{2} column prepared from a 0.005 M phosphate suspension of DEAE-cellulose, pH 7.4. Contaminating materials which appear as a dark band are washed from the column with approximately 40 ml. of 0.005 M phosphate, pH 7.4. The column is eluted batchwise with 40 ml. of 0.025 M phosphate, pH 7.4, followed by 40 ml. of 0.05 M phosphate, pH 7.4. The two fractions are collected separately, dialyzed against 4 l. of water, and assayed. Activity is usually associated with the 0.025 M phosphate eluate.

Step 6. Alumina C$_\gamma$ Adsorption. Alumina C$_\gamma$ (100 mg. dry weight) is added with stirring to the active preparation obtained in step 5. Stirring is continued for 10 minutes, and the suspension is centrifuged at 2000 r.p.m. for 5 minutes. The supernatant is set aside, and the gel is eluted with 10 ml. of 0.025 M phosphate, pH 7.4. The suspension is centrifuged, and the gel is discarded. Supernatant and eluate fractions are assayed; the majority of the activity is usually associated with the latter fraction.

The procedure for purification of nicotinic acid mononucleotide pyrophosphorylase from beef liver acetone powder is summarized in Table I.

TABLE I

SUMMARY OF PURIFICATION PROCEDURE
FOR NICOTINIC ACID MONONUCLEOTIDE PYROPHOSPHORYLASE

Fraction	Volume, ml.	Activity, units/ml.	Protein, mg./ml.	Purification, -fold
Acetone powder extract	250	3.3	49.0	1.0
0–35% $(NH_4)_2SO_4$	200	4.0	20.4	3.0
pH 4.8 $(NH_4)_2SO_4$[a]	70	48.0	13.9	12.6
$CaPO_4$ gel supernatant	25	80.0	4.50	65.5
DEAE-cellulose eluate	40	34.0	.75	170.0
Alumina C_γ gel eluate	10	52.0	.71	270.0

[a] A 4-fold increase in total activity accompanies this step.

Properties

Stability. Enzyme preparations representing the various steps of purification retain at least 50% of the activity for 3 months when stored at −12°, with the exception of the dilute DEAE-cellulose eluate.

Specificity. Only a limited attempt has been made to determine the substrate specificity of enzyme; however, the enzyme may be specific for nicotinic acid, since nicotinamide in 100-fold excess over nicotinic acid-7-C^{14} does not reduce the rate of formation of nicotinic acid-C^{14} mononucleotide, nor does it result in formation of nicotinamide mononucleotide.

Activators and Inhibitors. Maximal activity of the enzyme has been observed only in the presence of magnesium, orthophosphate, and ATP. There is an absolute requirement for magnesium, with an optimal concentration of approximately $5 \times 10^{-3}\,M$. Whether or not the enzyme has an absolute requirement for orthophosphate could not be determined, since PRPP breakdown and contaminating enzymes produce phosphate during the relatively prolonged incubation; however, a 2.5-fold stimulation occurs in the presence of $3 \times 10^{-2}\,M$ phosphate. Phosphate stimulation is suppressed, but not eliminated, by $1.2 \times 10^{-2}\,M$ arsenate. Stimulation by ATP has been found to vary from 60 to 300% with different enzyme preparations. The maximal effect is observed with aged preparations of the enzyme. Other nucleoside triphosphates do not effectively replace ATP. Step 3 of the purification procedure is accompanied by a 4-fold increase in total activity.

Effect of pH. Nicotinic acid mononucleotide pyrophosphorylase exhibits a rather broad pH optimum in phosphate buffer, with optimal activity at approximately pH 7.2.

Kinetic Properties. The apparent Michaelis constants are $1 \times 10^{-6} M$ for nicotinic acid and $5 \times 10^{-5} M$, for PRPP. Nicotinic acid mononucleotide, the product of the reaction, inhibits competitively against both nicotinic acid and PRPP, showing apparent K_i's of 4×10^{-5} and $3.5 \times 10^{-5} M$, respectively.

Reversibility. The equilibrium of the reaction greatly favors synthesis. Consequently, exchange data obtained by incubating nicotinic acid-C^{14} mononucleotide in the presence of PP_i and a small amount of unlabeled nicotinic acid are the only evidence for reversibility.

II. Nicotinic Acid Adenine Dinucleotide Pyrophosphorylase

$$\text{Deamido-NMN} + \text{ATP} \rightleftarrows \text{Deamido-DPN} + PP_i$$

Assay Method

Principle. Deamido-DPN pyrophosphorylase is assayed by determining the amount of deamido-DPN-C^{14} formed from deamido-NMN-C^{14} after separation of the reaction components by paper chromatography.

Reagents. The incubation mixture contains the following, in micromoles: deamido-NMN-C^{14}, 0.2 $(1.4 \times 10^5$ c.p.m.); ATP, 1; Tris, 15, pH 7.4; $MgCl_2$, 5; plus enzyme in a total volume of 0.5 ml. Deamido-NMN-C^{14} is prepared from nicotinic acid-7-C^{14} with nicotinic acid mononucleotide pyrophosphorylase.

Procedure. Varying amounts of enzyme are added to the reaction mixture and incubated for 20 minutes at 37°; the reaction is terminated by heating for 1 minute in a boiling-water bath. Aliquots of the incubation are spotted on Whatman No. 1 paper and chromatographed for 12 to 18 hours with a solvent system containing 7 parts of 95% ethanol and 3 parts of M ammonium acetate adjusted to pH 5.0 with HCl. After drying, the deamido-NMN and deamido-DPN spots are counted with a chromatogram strip counter. Since, in this solvent system, NMN and DPN have almost the same R_f's as deamido-NMN and deamido-DPN, respectively, they can be used as "markers." The per cent conversion of deamido-NMN to deamido-DPN permits calculation of the amount of deamido-DPN synthesized.

Definition of Unit and Specific Activity. A unit of enzyme activity is defined as the amount catalyzing synthesis of 1 micromole of deamido-DPN per hour, and specific activity as units per milligram of protein. Proportionality to enzyme concentration is observed with crude as well as with purified preparations when 1 unit or less is present in the test system.

Purification Procedure

Deamido-DPN and DPN pyrophosphorylase activities are closely associated throughout fractionation of an extract of acetone-dried hog liver. The ratios of both activities are constant during fractionation.[1] The distribution of both activities in subcellular fractions of rat liver homogenates prepared according to Schneider and Hogeboom[6] is identical. It appears, therefore, that in hog liver synthesis of DPN and deamido-DPN from their respective mononucleotides is catalyzed by the same enzyme. The most purified preparation of deamido-DPN pyrophosphorylase has been obtained by following the procedure of Kornberg[7] for purifying DPN pyrophosphorylase from hog liver extracts. Deamido-DPN pyrophosphorylase activity has also been demonstrated in extracts of human erythrocyte and in yeast autolyzates.[1]

III. DPN Synthetase

Deamido-DPN + ATP + glutamine
$$\rightarrow DPN + AMP + PP_i + \text{glutamic acid}$$

Assay Method

Principle. This enzyme is assayed by determining the amount of DPN formed from deamido-DPN by reduction of the DPN with alcohol dehydrogenase. DPN-C^{14} formation from deamido-DPN-C^{14} can also be estimated after separating the dinucleotides by paper electrophoresis at pH 3.5 or 7.4 and then assaying the radioactive spots with a chromatogram strip counter.

Reagents. The incubation mixture contains the following components, in micromoles: DPN, 0.5; ATP, 1; MgCl$_2$, 2.5; glutamine, 10; phosphate or Tris, pH 7.4, 15; NaF, 5; KCl, 28; and 0.3 to 0.5 unit of enzyme in a volume of 0.5 ml. Deamido-DPN is prepared by the procedure of Lamborg *et al.*[8] or by incubation of nicotinic acid-7-C^{14} with erythrocyte actone powder extracts and isolation of deamido-DPN-C^{14} on a Dowex 1–formate column.[1]

Procedure. After addition of the enzyme, the reaction mixture is incubated for 30 minutes at 37°. The reaction is terminated by heating in a boiling-water bath for 30 to 60 seconds. The suspension is diluted with 1.0 ml. of water and cleared by centrifugation (if necessary), and a 1.0-ml. aliquot is taken for DPN analysis.

Definition of Unit and Specific Activity. One unit of enzyme is

[6] G. H. Hogeboom and C. W. Schneider, *J. Biol. Chem.* 197, 611 (1952).
[7] A. Kornberg, Vol. II [116].
[8] M. Lamborg, F. E. Stolzenbach, and N. O. Kaplan, *J. Biol. Chem.* 231, 685 (1958).

defined as that amount necessary to catalyze synthesis of 1 micromole of DPN per hour under standard conditions. Specific activity is expressed as units per milligram of protein.

Purification Procedure

Step 1. Autolysis. Fleischmann 20–40 yeast (60 g.) is autolyzed for 5 hours at 37° in 180 ml of 0.1 M KHCO$_3$. The autolyzate is chilled, and all subsequent operations are performed in the cold. The autolyzate is centrifuged at 10,000 r.p.m. in a Lourdes Model AB centrifuge, and the supernatant fluid is retained. The insoluble residue is washed with 180 ml. of cold 0.1 M KHCO$_3$. The combined washings and supernatant fluid (288 ml.) are dialyzed for 4 hours against 4.5 l. of 0.025 M Tris, pH 7.4, containing 0.001 M glutathione and Versene.

Step 2. Ammonium Sulfate Fractionation I. $(NH_4)SO_4$ is added to 285 ml. of autolyzate to 0.4 saturation. The relatively small precipitate is dissolved in a minimal volume of 0.05 M Tris, pH 7.4, dialyzed against this solvent for 2 hours, and then diluted with 0.05 M Tris, pH 7.4, to 40 ml.

Step 3. Calcium Phosphate Gel Fractionation. Solid calcium phosphate gel (1 mg./mg. of protein) is added, and the suspension is stirred for 10 minutes. The gel is removed by centrifugation for 5 minutes at 3000 r.p.m. The supernatant fluid usually contains 90% of the enzyme activity. (If more than 30% of the activity has been absorbed, the gel is eluted with 40 ml. of 0.05 M phosphate, pH 7.4, and the eluate and previous supernatant fluid are combined for the next step.)

Step 4. Ammonium Sulfate Fractionation II. $(NH_4)_2SO_4$ is added to the supernatant fluid to 0.34 saturation. The resulting precipitate is collected and redissolved in 20 ml. of 0.05 M Tris, pH 7.4, containing 0.001 M glutathione and Versene.

Step 5. Alcohol Fractionation. After addition of 1.25 ml. of M sodium acetate, the solution is cooled to 0°, and 9.1 ml. of 95% ethanol at −12° are slowly added. After 5 minutes of stirring at −12°, the precipitate

TABLE II
SUMMARY OF DPN SYNTHETASE PURIFICATION PROCEDURE

Fraction	Volume, ml.	Activity, units/ml.	Protein, mg./ml.	Purification, -fold
Yeast autolyzate	286	0.86	16	1
First AmSO$_4$ (0–0.4) fraction	40	9.7	16.7	11
Supernatant from calcium phosphate gel	40	5.6	4.9	21
Second AmSO$_4$ (0–0.34) fraction	20	13.0	5.4	44
EtOH precipitate	10	16.1	2.8	106

is removed by centrifugation at $-12°$ in a high-speed centrifuge. The precipitate is taken up in 10 ml. of $0.025 M$ Tris, pH 7.4, containing $0.001 M$ glutathione and Versene.

Table II summarizes the course of this fractionation. The enzyme has also been purified 7-fold from the supernatant of rat liver homogenates.[1]

Properties

Specificity of Amide Donor. Glutamine and NH_4^+ salts are the only known amide donors effective for formation of DPN. Glutamate, aspartate, and asparagine at concentrations up to $0.02 M$ do not serve as amide donors.

Specificity of the Triphosphate. CTP, UTP, ITP, and GTP do not replace ATP at concentrations up to $10^{-2} M$.

Enzymatic Impurities. The 100-fold purified yeast enzyme contains virtually no ATPase activity and exhibits only slight pyrophosphatase activity when incubated in $0.05 M$ pyrophosphate. This activity is inhibited by inclusion of $0.01 M$ fluoride without appreciably affecting the synthetase reaction. There is also a trace of adenylate kinase.

Activators and Inhibitors. Omission of K^+ from the reaction mixture results in a 75% diminution in DPN synthesis; there is an absolute requirement for Mg^{++}.

The synthetase is inhibited by azaserine which appears to act as a competitive inhibitor with respect to the amide donors. The K_i of azaserine versus glutamine is $1.3 \times 10^{-3} M$, and versus NH_3, $2.7 \times 10^{-3} M$. However, preincubation studies indicate that glutamine, rather than reversibly competing with azaserine for the enzyme site, only delays essentially irreversible binding of azaserine to the enzyme.[1]

Kinetic Properties. The K_m values found for the yeast enzyme were: glutamine, $3.5 \times 10^{-3} M$; NH_4^+, $1.4 \times 10^{-1} M$ (at pH 7.4); deamido-DPN, $1.4 \times 10^{-4} M$; ATP, $6 \times 10^{-4} M$; and Mg^{++}, $1.3 \times 10^{-3} M$.

ADDENDUM

Since this article was written, it has been found that cell-free extracts of *Escherichia coli*,[9,10] *Bacillus subtilis*,[11] and yeast[11] contain highly active pyridine nucleotide synthesizing systems. The specific activity of nicotinic acid mononucleotide pyrophosphorylase in some of these crude preparations is approximately equal to that of the most highly purified beef liver preparation described above.

[9] J. Imsande, *J. Biol. Chem.* **236**, 1494 (1961).
[10] J. Imsande and A. B. Pardee, *J. Biol. Chem.* **237**, 1305 (1962).
[11] J. Imsande, unpublished results.

[45] DPNH Synthesis from DPNH-X

$$\text{DPNH-X} + \text{ATP} \xrightarrow{\text{Mg}^{++}} \text{DPNH} + \text{ADP} + \text{P} + \text{H}_2\text{O}$$

By EDWIN G. KREBS

Glyceraldehyde 3-phosphate dehydrogenase from yeast or muscle catalyzes the conversion of DPNH to a derivative of uncertain structure which has been designated as DPNH-X.[1,2] Certain polybasic anions such as pyrophosphate, phosphate, or citrate accelerate the reaction, which has a pH optimum slightly above 5.0. Even under the best conditions the turnover number for the enzyme in this reaction is very low. The formation of DPNH-X from DPNH is accompanied by an increase in absorption at 290 mμ and loss of the absorption peak at 340 mμ. DPNH-X can be converted back to DPNH by an enzyme which has been partially purified from yeast.[3] This reaction requires ATP, which is split to ADP and P$_i$ in the process; water may be a product of the reaction.

Preparation of DPNH-X

Reagents

DPNH (Sigma Chemical Company, St. Louis, Missouri).
0.03 M maleate–0.005 M cysteine buffer, pH 6.0.
Yeast glyceraldehyde 3-phosphate dehydrogenase.[4]
Pyrophosphate buffer, 0.15 M, pH 6.0.
NaHCO$_3$, 0.05 M.
NaOH, 5 N.
BaBr$_2$, 2.0 M.
Ethanol, 95%.
Ethanol, absolute.

Procedure. To 500 mg. of DPNH dissolved in 10 ml. of H$_2$O are added 2.0 g. of twice-recrystallized yeast glyceraldehyde 3-phosphate dehydrogenase dissolved in 75 ml. of maleate–cysteine buffer. To this solution are added 25 ml. of pyrophosphate buffer to start the reaction which is carried out at 25°. Small portions of acid or base are added to adjust the final pH to 6.0. At intervals aliquots are removed and diluted

[1] G. W. Rafter, S. Chaykin, and E. G. Krebs, *J. Biol. Chem.* **208**, 799 (1954).
[2] S. Chaykin, J. O. Meinhart, and E. G. Krebs, *J. Biol. Chem.* **220**, 811 (1956).
[3] J. O. Meinhart, S. Chaykin, and E. G. Krebs, *J. Biol. Chem.* **220**, 821 (1956).
[4] E. G. Krebs, G. W. Rafter, and J. McBroom Junge, *J. Biol. Chem.* **200**, 479 (1953).

with the bicarbonate solution for measurement of absorption at 340 mμ. After 20 minutes an additional 0.5 g. of enzyme dissolved in 15 ml. of maleate–cysteine buffer are added, and the reaction is allowed to continue for an additional 30 minutes.[5] The reaction is stopped by addition of sufficient 5 N NaOH to adjust the pH to 8.5. BaBr$_2$ solution (2.5 ml.) is added, and the resulting precipitate is removed by centrifugation and discarded. To the supernatant solution is added 1 vol. of 95% ethanol, and the mixture is cooled to 0°. The precipitate which forms is again removed by centrifugation and discarded. Additional BaBr$_2$ solution (about 2.0 ml.) is added to the supernatant solution until no additional precipitation occurs, and the mixture is again centrifuged. To the clear supernatant solution are added 5 vol. of absolute ethanol at −15°, and a flocculant precipitate of the barium salt of DPNH-X appears. The suspension is stored at −20° for 24 hours, and the precipitate is collected by centrifugation at this temperature. The DPNH-X is washed with cold absolute ethanol, alcohol–ether (1:1), and ether, dried *in vacuo* over paraffin and CaCl$_2$, and stored in a desiccator at 4°. Approximately 200 mg. of material are obtained. For use, the barium salt is converted to the sodium salt by using Na$_2$SO$_4$ in the customary manner to remove the barium.

Characteristics of DPNH-X. The millimolar extinction coefficients[2] of DPNH-X are: 28 at 260 mμ, 29 at 265 mμ, 14 at 290 mμ, and 0.2 at 340 mμ. In preparations made on a large scale as described above, 2 to 10% of the total pyridine nucleotide isolated may be unchanged DPNH. If the reaction is run at a lower pH nearer the optimum, all the DPNH will react, but under these conditions appreciable amounts of the primary acid product of DPNH are formed. This latter compound resembles DPNH-X closely in spectrum,[1] but is nonreactive in the reaction described below.

Enzymatic Conversion of DPNH-X to DPNH

Enzyme Preparation. Ground dried baker's yeast (100 g.) is incubated with 300 ml. of 0.17 M (NH$_4$)$_2$HPO$_4$ for 4 hours at 38°, and the mixture is centrifuged for 60 minutes in a Servall angle centrifuge at 7000 r.p.m. at 3°. To the supernatant solution are added 4 vol. of saturated (NH$_4$)$_2$SO$_4$ solution at 25°. The precipitated proteins are collected by centrifugation in the cold and suspended in water to a final volume of 75 ml. Then 10-ml. portions of the suspension are heated for 3 minutes

---

[5] The second addition of enzyme and the additional time are unnecessary, if without these measures the reaction has proceeded to near completion as evidenced by almost complete disappearance of absorption at 340 mμ. (Appropriate corrections should be applied for absorption of other components in the mixture.)

at 60° and centrifuged after cooling. The clear pale-yellow supernatant solution, after dialysis against cold distilled water for 16 hours, is centrifuged at 100,000 $\times$ g for 2 hours in the Spinco Model L preparative ultracentrifuge. The supernatant solution is decanted, and the transparent sediment remaining is suspended evenly in 5 ml. of water. The enzyme is stored at $+3°$ and is stable for several weeks. The enzyme can be further purified[3] with alumina C_γ.

Reaction. The DPNH-X to DPNH reaction is conveniently followed by using the decrease in absorption at 290 mμ or the increase at 340 mμ. $\Delta E_{290} = 11.4$ and $-\Delta E_{340} = 6.0$ for a DPNH-X concentration of 1×10^{-3} M. If any DPNH oxidase is present, reduced DPNH formed will be oxidized to DPN, which can be readily determined. A typical reaction mixture for carrying out this reaction is made up in a 3.0-ml. volume to contain 5×10^{-5} M DPNH-X, 1.3×10^{-2} M $MgSO_4$, 5×10^{-4} M ATP, and Tris–acetate buffer, 2.5×10^{-3} to 2.5×10^{-3} M, pH 6.5. Initial absorption readings are made, and enzyme diluted in water is added to start the reaction. Suitable corrections are applied for enzyme absorption.

Observations on the Structure of DPNH-X

DPNH-X differs from DPNH through changes in the reduced nicotinamide portion of the molecule.[2,3] From spectral evidence[6,7] and results of catalytic hydrogenation[2] it appears that DPNH-X is a derivative of the 1,4,5,6-tetrahydronicotinamide analog of DPN. DPNH-X formed in the presence of D_2O incorporates D in the 5-position.[8,9] Thus DPNH-X may be the 6-hydroxy derivative formed by the addition of water to DPNH. The conversion of DPNH-X to DPNH would consist of an energy-requiring dehydration reaction.

[6] M. Von Marti, M. Viscontini, and P. Karrer, *Helv. Chim. Acta* **39**, 1451 (1956).
[7] K. Wallenfels and H. Schüly, *Biochem. Z.* **329**, 75 (1957).
[8] J. O. Meinhart and M. C. Hines, *Federation Proc.* **16**, 425 (1957).
[9] J. O. Meinhart, M. C. Hines, A. F. Fluharty, and E. G. Krebs, in preparation.

[46] UDPG Pyrophosphorylase from Muscle

$$UTP + G\text{-}1\text{-}P \rightleftharpoons UDPG + PP$$

By C. VILLAR-PALASI and J. LARNER

Assay Method

Principle. The most commonly used method of determination of the enzymatic activity is based on the spectrophotometric measurement of the glucose 1-phosphate formed from UDPG on addition of inorganic pyrophosphate (see Vol. II [118]). Isolation by paper chromatography of the nucleotides formed in the reaction has been used in studies of specificity.[1]

Reagents

a. Prepare daily a mixture of: 0.025 M Tris, pH 7.45, 0.002 M $MgCl_2$, 0.0004 M UDPG, 2 × 10^{-6} M glucose 1,6-diphosphate, 0.0005 M TPN, 0.05 mg. of lyophilized glucose 6-phosphate dehydrogenase[2] per milliliter, and 0.02 ml. of freshly dialyzed phosphoglucomutase[3] per milliliter.

b. 0.02 M potassium pyrophosphate.

Procedure. To 0.95 ml. of the reaction mixture (*a*), in a quartz cell with a 1-cm. light path, the sample of enzyme is added (about 0.01 to 0.05 ml.), and the reaction is started by addition of 0.05 ml. of pyrophosphate solution (*b*). Optical density readings at 340 mμ are recorded every 30 to 60 seconds for 5 to 10 minutes.

Definition of Unit and Specific Activity. One unit of enzymatic activity is defined as that amount which will form 1 micromole of glucose 1-phosphate per minute at 30°, an increase of optical density of 6.200 being equivalent under the conditions of assay to the split of 1 micromole of UDPG. Specific activity is expressed as units per milligram of protein, as measured by the biuret method.[4]

Application of Assay Method to Crude Tissue Preparations. This method appears to be valid for the assay of crude muscle extracts. As muscle contains about ten times as much phosphoglucomutase as UDPG pyrophosporylase,[5] the addition of phosphoglucomutase to the assay system appears to be necessary only when highly purified samples of the enzyme are tested. Glucose 6-phosphate dehydrogenase, as usually prepared,[2] is contaminated with trace amounts of UDPG pyrophosphorylase; this contaminating activity, however, disappears slowly after the lyophilized enzyme has dissolved.

Purification Procedure

Step 1. Preparation of Crude Extract. Rabbit muscle is used as starting material. The animals are killed by a blow on the head and decapi-

[1] C. Villar-Palasi and J. Larner, *Arch. Biochem. Biophys.* **86**, 61 (1960).
[2] A. Kornberg and B. L. Horecker, Vol. I [42].
[3] V. A. Najjar, Vol. I [36].
[4] H. W. Robinson and C. G. Hogden, *J. Biol. Chem.* **135**, 727 (1940).
[5] C. Villar-Palasi and J. Larner, *Arch. Biochem. Biophys.* **86**, 270 (1960).

tated; the muscles of the back and hind legs are removed, ground twice in a prechilled meat grinder, and weighed. All subsequent manipulations take place in the cold room at 3°. The muscle is extracted with 2 vol. (w/v) of cold 0.03 N KOH–0.005 M EDTA for 30 minutes, with occasional stirring. The suspension is filtered through four layers of gauze. The gauze bag is squeezed, such that the recovery of fluid is over 90% of the added extraction medium. About 50% of the UDPG pyrophosphorylase activity is extracted in this way. The enzyme is unstable at this stage, and it is necessary to proceed to the next step without delay.

Step 2. Ammonium Sulfate Fractionation. The pH of the extract is adjusted to 6.8 measured at 15° with N acetic acid (about 10 ml./l. of extract). Solid ammonium sulfate, 277 g./l. of extract, is slowly added, the suspension is stirred, and after 3 hours at 0° the precipitate formed is collected by centrifugation (10 minutes at 8000 × g). The fatty layer floating at the top is separated by filtration through glass wool. Ammonium sulfate, 134 g./l., is then added to the clear filtrate, and the precipitate formed after 3 hours at 0° is collected by centrifugation, as above. The precipitate is dissolved with 0.03 N KOH–0.005 M EDTA and dialyzed overnight against 100 vol. of 0.005 N KOH–0.001 M EDTA. The dialyzed enzyme is stable for at least a week at 3°.

Step 3. Acetone Fractionation. The dialyzed solution is made 0.005 M in pyrophosphate with neutral 0.1 M $K_4P_2O_7$, and the pH is adjusted to 8.0. The solution is cooled to 0°, and acetone, prechilled at −20°, is added dropwise to a 37% final concentration, the mixture being stirred in an ice-salt bath at −10°. After the addition is completed, the suspension is kept at −20° for 30 minutes and centrifuged in prechilled tubes. The precipitate is rapidly suspended in 0.01 M $K_4P_2O_7$–0.005 M EDTA, pH 9.6, and dialyzed for several hours, with mechanical stirring, against 200 vol. of 0.005 M $K_4P_2O_7$–0.001 M EDTA, pH 9.6. The dialyzed fluid is adjusted to pH 6.0 with N acetic acid and centrifuged at 8000 × g for 10 minutes; after the protein content of the supernatant fluid has been measured, the following step is carried out without delay.

Step 4. Alumina C_γ Treatment. To the solution from step 3, a suspension of alumina C_γ (20 mg./ml.) and distilled water are added to bring the protein concentration to 7.5 mg./ml. and the alumina C_γ to 5.0 mg./ml. After 10 to 15 minutes the suspension is centrifuged, and the precipitate is washed with distilled water and centrifuged. The enzyme is eluted with several portions of 250 to 300 ml. of 0.01 M $K_4P_2O_7$, pH 9.6, and the eluates with higher specific activity are collected. Solid ammonium sulfate is added (351 g./l. of pooled eluates), the suspension is kept at 0° overnight, and the precipitate is collected by centrifugation at 8000 × g for 10 minutes. The precipitate is dissolved in 0.01 M $K_4P_2O_7$–

0.005 M EDTA, pH 9.6, and dialyzed against the same solution for 3 hours; an inactive precipitate that forms is removed by centrifugation. The supernatant fluid is adjusted to pH 8.0 and dialyzed overnight against 140 vol. of 0.005 M $K_4P_2O_7$, pH 8.0.

Step 5. DEAE-Cellulose Chromatography. The columns are prepared by allowing a slurry of 10 g. of DEAE-cellulose (Type 20, capacity 0.86 meq./g.), equilibrated with 0.005 M $K_4P_2O_7$, pH 8.0, to pack in a column 2.5 × 14 cm. first by gravity, and then, to obtain the desired flow rate (2 ml./min.), with pressure of N_2. The columns are washed with several liters of 0.005 M $K_4P_2O_7$, pH 8.0, the dialyzed enzyme of step 4 (about 100 ml.) is applied, and the effluent liquid is collected in 25-ml. fractions. The columns are then first washed with 300 ml. of 0.005 M $K_4P_2O_7$, pH 8.0, and the enzyme is eluted by applying a gradient increase in the concentration of pyrophosphate (to 0.1 M). The fractions with higher specific activity are pooled; practically all the enzymatic activity is recovered in one peak. The collected fractions contain very diluted protein, and the recovery of the enzyme in a concentrated form appears difficult. The enzyme loses activity after some weeks at −20° in this dilute solution.

A summary of the purification procedure is given in the table.

SUMMARY OF PURIFICATION OF UDPG PYROPHOSPHORYLASE

Step	Volume, ml.	Protein, mg.	Units, μmoles/min.	Specific activity, units/mg.
1. Extract	2250	60,400	3800	0.06
2. (NH₄)₂SO₄ fraction	650	21,200	3800	0.18
3. Acetone fraction	250	4,000	2420	0.55
4. Alumina C$_\gamma$ treatment	752	3,000	2580	0.86
5. DEAE-cellulose chromatography	50	16	1320	82.50

Properties

Specificity. Several hexose and pentose phosphates have been tested as substrates for the purified enzyme.[1] Only α-D-glucose-1-phosphate appears to be a reasonable substrate. β-D-glucose 1-phosphate, α- and β-L-arabinose 1-phosphate, α- and β-D-galactose 1-phosphate, α- and β-D-xylose 1-phosphate, and α-D-ribose 1-phosphate did not react with UTP when incubated in the presence of the purified enzyme.

Activators and Inhibitors. The enzyme has a total requirement of Mg^{++} for activity. On the other hand, excess Mg^{++} is inhibitory.

Equilibrium. Preliminary determinations of the K_{eq} of the reaction appear to be close to 1.

Kinetic Properties. The K_m for UDPG is $4.5 \times 10^{-5}\,M$. The K_m for both Mg^{++} and $P_2O_7^{-4}$ are of the order of $1 \times 10^{-4}\,M$.

Effect of pH. The pH of maximal stability is 9.8. The pH of optimal activity in short incubation periods is 7.45.

[47] A Bacterial Pterin Deaminase[1]

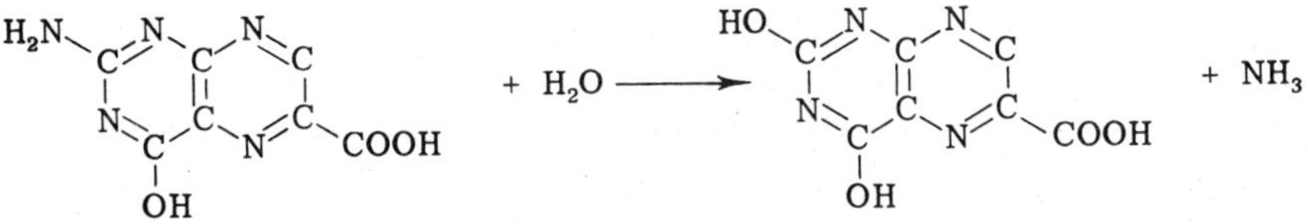

By OSAMU HAYAISHI

Assay Method

Principle. The reaction can be followed either by measuring the formation of ammonia or by the spectral change of pterin carboxylic acid[2] accompanying the conversion of the 2-amino group to the 2-hydroxyl group at neutral pH.

Reagents

Pterin carboxylic acid, $0.002\,M$.
$0.1\,M$ potassium phosphate buffer, pH 6.3.
Pterin deaminase preparation.

Procedure. In the standard assay, the mixture contains 0.05 ml. of pterin carboxylic acid, 0.8 ml. of phosphate buffer, and 0.02 to 0.15 ml. of enzyme in a final volume of 1.0 ml. Incubation is carried out at 23° in a 1-ml. Beckman quartz cuvette with a light path of 1.00 cm. Reaction is begun by the addition of enzyme, and readings of optical density are taken every 3 minutes thereafter. (The control cuvette contains buffer and enzyme, but no pterin carboxylic acid.) With crude preparations, light measurement at 360 mμ is employed. However, with more purified fractions of the deaminase, it is possible to follow the course of the reaction at 290 mμ, the wavelength at which there is observed the largest ΔE value between substrate and product.

Definition of Unit and Specific Activity. A unit of activity is defined as that quantity of enzyme which, under these conditions, will catalyze

[1] B. Levenberg and O. Hayaishi, *J. Biol. Chem.* **234**, 955 (1959).
[2] Abbreviation: Pterin carboxylic acid = 2-amino-4-hydroxypteridine-6-carboxylic acid.

the deamination of 1 micromole of pterin carboxylic acid in 10 minutes ($\Delta E_{290} = 9700$, and $\Delta E_{360} = 4400$). Specific activity is expressed as units per milligram of protein. Protein is determined by the method of Warburg and Christian.[3]

Purification Procedure

Culture Conditions. *Alcaligenes metalcaligenes* (ATCC No. 13270) is the source of this enzyme. The medium was prepared in the following manner; K_2HPO_4 (1.5 g.), KH_2PO_4 (0.5 g.), $MgSO_4 \cdot 7H_2O$ (0.2 g.), NH_4NO_3 (2.0 g.), Difco nutrient broth (5.0 g.), and Difco yeast extract (0.1 g.) were dissolved in 500 ml. of tap water plus 450 ml. of glass-distilled water. Dextrose (0.2 g. in 50 ml. of distilled water) was sterilized separately and subsequently added aseptically to the above autoclaved mixture just before the introduction of a few milliliters of an inoculum of the organism. The bacteria were grown at 25° for approximately 55 hours with moderate aeration in 20-l. carboys each containing 10 l. of media. Cells were harvested with the aid of a Sharples supercentrifuge and washed several times with cold distilled water. The packed cells could be stored in this condition at −60° for at least 4 months without appreciable loss of pterin deaminase activity. Extracts were prepared by suspending 1 part, by weight, of cells in 1.8 parts of 0.05 M potassium phosphate buffer, pH 7.0, and disrupting the bacterial mass, in batches of 30 ml. for 25 minutes at 2°, in a Raytheon 10-kc. sonic oscillator. The resulting material was centrifuged for 30 minutes at 2° in the Servall SS-1 centrifuge at 20,000 × *g*.

Step 1. Treatment of Extracts with Protamine Sulfate. Crude extracts (127 ml., 8900 units, specific activity = 1.0) were treated, dropwise, with 60 ml. of a 1% solution of protamine sulfate (Eli Lilly) in 0.05 M potassium phosphate buffer, pH 7.0, with stirring over a period of 10 minutes. The suspension was stirred for an additional 10 minutes and then centrifuged for 20 minutes at 20,000 × *g*. The supernatant solution (170 ml.) contained the major portion of the deaminase (7300 units, specific activity = 1.3).

Step 2. Heat Treatment. The enzyme solution from step 1 was placed in a 750-ml. Erlenmyer flask, rapidly warmed to room temperature, and treated with 34 ml. of 1.3 M potassium chloride. The flask and contents were immersed in a water bath maintained at 60° and held at this temperature with gentle agitation of the solution for 7 minutes. The mixture was chilled in ice, and denatured protein was removed by centrifugation. The resulting pale-yellow supernatant solution (170 ml.) contained 6670 units of enzyme with a specific activity of 3.0.

[3] O. Warburg and W. Christian, *Biochem. Z.* **310**, 384 (1941–1942).

Step 3. Treatment with Activated Charcoal. The preparation from step 2 was warmed to 27° and treated with 19 ml. of a 20% suspension of Nuchar for 8 minutes with gentle stirring. The mixture was then chilled in ice and clarified by centrifugation at 20,000 × g for 20 minutes. The major portion of the pterin deaminase was found in the resulting supernatant solution (175 ml., 6230 units, specific activity = 5.1).

Step 4. Alkali Ammonium Sulfate Fractionation. The pH of the above solution was adjusted to 8.4 by the careful addition of a few milliliters of 1 N NH$_4$OH. Solid ammonium sulfate (28 g.) was then added to each 100 ml. of this solution. Addition of the salt required 15 minutes, after which the suspension was equilibrated for another 10 minutes and centrifuged at 15,000 × g. The inactive precipitate (0 to 40% fraction) was discarded, and the supernatant solution was treated, as described above, with an additional 19 g. of ammonium sulfate per 100 ml. of original volume. If further purification was not sought, the paste obtained by centrifugation of this fraction was dissolved in an amount of 0.1 M Tris buffer, pH 9.0, equal to about one-fifth the volume of the original solution of step 3. After aging in ice for an hour, this 40 to 67% fraction was freed from a small amount of inactive material by centrifugation. The resulting clear, very pale-yellow supernatant solution (4000 units, specific activity = 13) could be stored at 0° for approximately 3 weeks with no appreciable loss of activity. For most of the enzymatic studies, pterin deaminase preparations were employed which had been purified as far as this stage. A further step was included only when it was desirable to obtain a fraction as free as possible from residual guanase activity.

Step 5. Treatment with Alumina C$_\gamma$ Gel. The ammonium sulfate 40 to 67% paste from step 4 was dissolved in 50 ml. of 0.005 M potassium phosphate buffer, pH 7.0, and stirred for 15 minutes with an equal volume of alumina C$_\gamma$ gel suspension (16 mg. of gel per milliliter). After centrifugation of this mixture, the supernatant was discarded, and the gel

TABLE I
PURIFICATION OF PTERIN DEAMINASE

Step	Yield, %	Pterin deaminase, units/mg. protein
Sonic extract	(100)	1.0
1. Protamine sulfate	82	1.3
2. Heat treatment	75	3.0
3. Nuchar	70	5.1
4. Alkali ammonium sulfate	45	13.0
5. Alumina C$_\gamma$ gel	27	21.0

was washed twice with small volumes of cold water. Elution of the activity was accomplished by stirring the gel for 45 minutes with 30 ml. of 0.085 M phosphate buffer, pH 7.5. On centrifugation, the supernatant solution contained 2400 units of deaminase, with a specific activity of 21 (Table I).

Properties

Specificity. Some of the structural features required for a pteridine compound to serve as a substrate were investigated by testing a limited number of available derivatives.

1. Only those pteridines possessing the pterin structure (i.e., the 2-amino and 4-hydroxyl functional groups) are deaminated. Exchange of these groups (compound *e* in Table II), removal of the 4-hydroxy (compound *a*), or its substitution by an amino group (compounds *f, g, h*) produces pteridines not attacked by the enzyme (Table II).

TABLE II
SUBSTRATE SPECIFICITY OF PTERIN DEAMINASE

Pteridines attacked		Pteridines not attacked
	⎡—H	(*a*) 2-Amino-
	—CH$_3$	(*b*) 2-Amino-4-hydroxy-7-methyl-
	—CH$_2$OH	(*c*) 2-Amino-4-hydroxy-7-carboxy-
2-Amino-4-	⎨—CHO	
hydroxy-6-	—COOH	(*d*) 2-Amino-4-hydroxy-6,7-dimethyl-
	—CH(OH)—CH(OH)—	(*e*) 2-Hydroxy-4-amino-6-methyl-
	—CH$_2$OH	
	⎣—CH$_2$COOH	(*f*) 2,4-Diamino-
Pteroic acid		(*g*) Aminopterin
	⎡(Glutamic) 1,2,3	(*h*) Amethopterin
Pteroyl-	⎨Aspartic	(*i*) Xanthopterin
	⎣Glycine	(*j*) Leucovorin

2. The nature of the substitution at carbon 6 is relatively unimportant. However, a hydroxyl group at this position (compound *i*) results in an inactive substance.

3. Carbon 7 must be unsubstituted, since blocking of this position by

a methyl (compounds *b* and *d*) or carboxyl function (compound *c*) destroys the ability of the pterin to serve as a substrate.

4. An *N*-5-formylated and reduced pterin (compound *j*) is not deaminated.

Stability and Activity at Various pH Values. The most satisfactory conditions for storage of enzyme fractions were found to be at pH 9.0 in 0.1 *M* Tris buffer at 0°. Under such conditions pterin deaminase activity remained almost unchanged for several weeks. The pH optimum for pterin deaminase activity occurs between pH 6.3 and 6.7. Although the reaction proceeds in the complete absence of phosphate, the rate of deamination is slightly faster in phosphate than in Tris buffer.

Nonidentity of Pterin Deaminase and Guanase. Extracts of *A. metalcaligenes* contain a highly active guanase. However, the ratio of the specific activity of guanase to that of pterin deaminase decreased markedly throughout the fractionation procedures. Additional evidence indicating the nonidentity of the two deaminase activities was obtained from studies of the effect of guanine and pterin carboxylic acid on the rate of deamination of another pterin substrate, 2-amino-4-hydroxy-6-methyl pteridine. These data, together with those on fluoride sensitivity, have permitted the conclusion that the pterin deaminase and guanase activities of this strain of *Alcaligenes* are probably distinct enzymatic entities.

Inhibitors. Pterin deaminase activity was not significantly affected by extensive dialysis of enzyme fractions against 0.1 *M* Tris buffer, pH 9.0. Preincubation with Versene, KCN, or 8-hydroxyquinoline sulfate likewise did not result in a decrease of activity. However, *p*-chloromercuribenzoate and sodium or potassium fluoride each caused severe inhibition when either was preincubated with the deaminase or added directly to the experimental cuvette containing pterin carboxylic acid and enzyme. The inhibitory action of *p*-chloromercuribenzoate could be effectively prevented with glutathione. The fluoride inhibition could not be overcome by extended incubation or the addition of a large excess of Mg^{++} but was completely removed on dialysis of a fluoride-treated enzyme preparation against two changes of a large volume of 0.01 *M* Tris buffer, pH 9.0. A 50% inhibition of the initial rate of pterin carboxylic acid deamination was observed at fluoride ion concentrations of approximately 3×10^{-5} *M*. This is in contrast to an inhibitor concentration of over 10^{-3} *M* required to effect a similar reduction in the guanase activity of these preparations. None of the other halide ions showed significant inhibitory properties.

Irreversibility of Deamination. Evidence for the reversibility of the reaction could not be obtained by the use of the sensitive spectrophotometric method.

[48] Dihydrofolic Reductase

Dihydrofolate $+$ TPNH $+$ H$^+$ $\rightleftarrows$ Tetrahydrofolate $+$ TPN$^+$ (1)

By C. K. MATHEWS, K. G. SCRIMGEOUR, and F. M. HUENNEKENS

Assay Method

Principle. Dihydrofolic reductase (also called folic reductase[1]) catalyzes the reversible, TPNH-dependent reduction of dihydrofolate to tetrahydrofolate, as shown in Eq. (1). At pH 7.5, the equilibrium lies relatively far to the right, and the reaction goes essentially to completion in the forward direction. The decrease in absorbancy at 340 mμ is used to follow the disappearance of TPNH.

Reagents

0.5 M potassium phosphate buffer, pH 7.5.
0.1 M 2-mercaptoethanol.
0.004 M TPNH.
Dihydrofolate. Dihydrofolic acid is prepared by the reduction of folic acid with hydrosulfite according to the method of Futterman (Vol. VI [112]); the lyophilized material is stored in evacuated, sealed ampules until needed. Dihydrofolate at a concentration of 0.002 M is prepared just prior to use by dissolving the solid material in water with the addition of 0.5 M potassium phosphate buffer in order to adjust the pH to 7.5; the solution is also 0.05 M in 2-mercaptoethanol.

Procedure. The following components are added to a silica cuvette having an optical path of 1.0 cm. and a volume of approximately 1.5 ml.: 0.1 ml. of phosphate buffer, 0.1 ml. of mercaptoethanol, 0.025 ml. of TPNH, 0.05 ml. of dihydrofolate, and water to make a total volume of 1.25 ml. The optical blank contains all components except TPNH. The reaction is initiated by the addition of 0.01 to 0.05 ml. of enzyme, and the decrease in absorbancy at 340 mμ is measured at 1-minute intervals for 5 minutes. The rate is corrected for a small blank reaction in which dihydrofolate is omitted. One unit of enzyme activity is defined as that amount which causes a decrease in absorbancy of 0.010 per minute under the above conditions.[2] Protein is determined from absorbancy measure-

[1] S. F. Zakrzewski, *J. Biol. Chem.* **235**, 1776 (1960).
[2] Units of enzyme activity may be converted to micromoles of substrate reacting per minute. In converting changes in absorbancy to turnover values for substrate,

ments at 260 and 280 mμ with crystalline bovine serum albumin as the standard.[3] Specific activity is defined as units of activity per milligram of protein.

Application of Assay Method to Crude Tissue Preparations. The spectrophotometric method described above is quite satisfactory for measuring dihydrofolic reductase in crude preparations of chicken, pigeon, beef, and rabbit liver. However, with leukemic cells[4] and ascites tumor cells,[5] reductase activity by this optical method can be observed only after fractionation of the crude preparation with ammonium sulfate, pH 8 (the 60 to 80% fraction contains most of the activity). With the latter two tissues where the level of reductase is low, crude preparations may be assayed, nevertheless, by an alternative procedure[4] in which the tetrahydrofolate formed in the reductase reaction is converted to N^{10}-formyl tetrahydrofolate via reaction (2) catalyzed by the formate-activating enzyme.

Tetrahydrofolate + formate + ATP

$$\rightleftarrows N^{10}\text{-Formyl tetrahydrofolate} + \text{ADP} + \text{P} \quad (2)$$

Purification Procedure

Dihydrofolic reductase has been prepared previously from chicken liver,[6] sheep liver,[7] *Streptococcus faecalis*,[8] and calf thymus.[9] The present procedure,[10] which represents an extension of our previous method,[6] leads to a highly purified, stable preparation of the enzyme. Fresh chicken

---

it is necessary to correct for the fact that dihydrofolate and TPNH have almost the same molar extinction coefficients at 340 mμ. Thus, the observed change in optical density is the sum of contributions from TPNH oxidation and dihydrofolate reduction. The value 6.2×10^3 l./mole-cm. is used as the extinction coefficient for TPNH at 340 mμ, and 5.8×10^3 l./mole-cm. is taken as the corresponding value for dihydrofolate. The extinction coefficient of tetrahydrofolate at this wavelength is negligible. For dihydrofolate reduction, if one unit gives an absorbancy change of 0.010 per minute and that fraction of the change resulting from TPNH oxidation is (6.2/6.2 + 5.8), then one unit will give an absorbancy change of 0.010 (6.2/6.2 + 5.8), or 0.0052 per minute due to pyridine nucleotide disappearance. This corresponds to the oxidation of 1.05 mμmoles of TPNH per minute in the 1.25-ml. volume used for the standard assay.

[3] See Vol. III [73].
[4] J. R. Bertino, B. W. Gabrio, and F. M. Huennekens, *Biochem. Biophys. Research Communs.* 3, 461 (1960).
[5] Unpublished observations.
[6] M. J. Osborn and F. M. Huennekens, *J. Biol. Chem.* 233, 969 (1958).
[7] J. M. Peters and D. M. Greenberg, *Biochim. et Biophys. Acta* 32, 273 (1959).
[8] R. L. Blakley and B. M. McDougall, *J. Biol. Chem.* 236, 1163 (1961).
[9] R. Nath and D. M. Greenberg, *Biochemistry* 1, 435 (1962).
[10] C. K. Mathews, Ph.D. Thesis, University of Washington, 1962.

livers obtained from a commercial packing plant are either used immediately or stored frozen. The subsequent operations are carried out at 0° to 5°, and, unless otherwise specified, centrifugations are performed for 10 minutes at 4000 r.p.m. in the International refrigerated centrifuge, PR-1.

Step 1. Preparation of Crude Extract. One hundred grams of either fresh or frozen chicken liver is cut into small pieces and homogenized for 1 minute in a Waring blendor with 150 ml. of 0.1 M potassium phosphate buffer, pH 7.5. The resultant homogenate is then centrifuged at 30,000 r.p.m. for 1 hour in the Spinco Model L preparative ultracentrifuge (No. 30 rotor). The supernatant fluid is strained through cheesecloth to remove lipid material.

Step 2. Fractionation by Precipitation at pH 5.4. The lipid-free supernatant fraction is treated with 0.3 vol. of 2% protamine sulfate solution and centrifuged again for 15 minutes at 30,000 r.p.m. The pH of the supernatant fluid is then adjusted carefully to 5.4 with N acetic acid, the precipitated protein is removed by centrifugation, and the pH of the supernatant fraction is readjusted to 7.5 with N NaOH.

Step 3. Fractionation with Ethanol–Chloroform. The protein content of the pH 5.4 supernatant fraction is adjusted to 20 mg./ml. by the addition of water. To every 10 ml. of this solution is added 2.1 ml. of absolute ethanol at —20° followed by 1.3 ml. of chloroform also at —20°. The mixture is shaken vigorously for 1 minute and then centrifuged. The clear, aqueous phase, after being separated carefully from the denatured protein and the organic layer, is dialyzed for 4 hours against two 6-l. portions of cold water.

Step 4. Fractionation with Ammonium Sulfate. The above preparation, centrifuged if necessary to remove denatured protein, is fractionated with solid ammonium sulfate, and the protein precipitating between 0 and 60% of saturation is collected by centrifugation and discarded. The ammonium sulfate concentration in the supernatant fraction is raised to 85% of saturation, and the resulting 60 to 85% precipitate is dissolved in 20 ml. of 0.05 M phosphate buffer, pH 6.5, and dialyzed for 4 hours against 3 l. of the same buffer. At this stage the preparation can be stored frozen for several weeks with little loss in activity.

Step 5. Chromatography on Hydroxylapatite. Hydroxylapatite is prepared according to Tiselius et al.[11] and stored in the cold as a suspension in $10^{-3} M$ sodium phosphate buffer, pH 6.8. The column is prepared as follows. A suspension of hydroxylapatite is poured into a 3×25-cm. glass column to give a final bed height of 16 cm. The column is washed with 1 l. of 0.05 M phosphate buffer, pH 6.5, under gentle pressure (1 to 2 p.s.i.) of nitrogen, if necessary. The above 60 to 85% ammonium sulfate

[11] A. Tiselius, S. Hjertén, and Ö. Levin, *Arch. Biochem. Biophys.* **65**, 132 (1956).

fraction (about 20 ml.) is adsorbed on the column, and gradient elution is carried out with 500 ml. of 0.05 M phosphate buffer, pH 6.5, in the mixing chamber and 500 ml. of 0.3 M phosphate buffer, pH 6.5, in the reservoir. Nitrogen pressure is applied during chromatography to give a flow rate of about 0.5 ml./min. Fractions of 7.5 ml. are collected with an automatic fraction collector. The enzyme appears in the effluent between tubes 35 and 45. The active fractions can be used in this form, or they may be concentrated by lyophilization. It was also possible to rechromatograph the pooled, active fractions in order to achieve higher specific activities.

The purification procedure is summarized in the table. After step 5, the enzyme is purified about 250-fold and is recovered in an over-all yield of about 37%. Rechromatography of the preparation on hydroxylapatite increases the activity another threefold (specific activity about

PURIFICATION OF DIHYDROFOLIC REDUCTASE FROM CHICKEN LIVER

Step	Volume, ml.	Reductase activity, units[a]	Protein concentration[b] mg./ml.	Specific activity, units/mg.	Recovery, %
1. Crude extract	155	15,000	70	1.38	100
2. Fractionation by precipitation	165	14,600	46	1.92	97
3. Fractionation with ethanol–chloroform	345	14,000	5.4	7.5	93
4. Fractionation with ammonium sulfate (60–85% fraction)	40	9,600	13.0	18.4	64
5. Chromatography on hydroxylapatite	45	5,600	0.34	336	37

[a] One unit of activity is defined as that amount of enzyme which catalyzes an absorbancy change of 0.010 per minute under conditions of the assay.

[b] Protein concentration is determined by the ultraviolet absorption method described in Vol. III [73].

1000 units/mg. of protein) with little loss in total activity. At the final stage of purification, the enzyme is colorless and shows only a single absorption maximum at 280 mμ. After being concentrated by lyophilization, the purified enzyme is stable to dialysis against dilute phosphate buffers at neutral pH. In this form, it may also be stored frozen for several months with negligible loss of activity, whereas the more dilute fractions from the column lose up to 40% of their activity after 2 to 3 days' storage in the frozen state. The concentrated, purified enzyme also retains its activity when stored for several days at 5°.

Properties[10]

Contamination with Other Enzymes. The purified enzyme is free from formate-activating enzyme, serine hydroxymethylase, N^5,N^{10}-methylene tetrahydrofolic dehydrogenase, and cyclohydrolase.

Specificity. The specificity of the enzyme for the two substrates in reaction (1) is somewhat complicated by the pH-activity behavior of the enzyme. With dihydrofolate and TPNH, the enzyme exhibits two pH optima, one at about pH 4.5 and the other near 7.5. At neutral pH values neither folate nor DPNH can substitute for dihydrofolate or TPNH, but at the acidic pH values DPNH is about one-half as active as TPNH, and folate is roughly one-fifth as active as dihydrofolate.

Activators. The enzyme is activated by K^+.[12]

Inhibitors. Dihydrofolic reductase from chicken liver is insensitive to SH-inhibitors and to metal-binding agents. It is inhibited, however, by very low levels (about $10^{-8}\,M$) of aminopterin and amethopterin.[13, 14]

Thermodynamic and Kinetic Properties. At pH 7.0, the equilibrium constant for reaction (1) is 5.6×10^4. The K_m value for dihydrofolate is $5.0 \times 10^{-7}\,M$ with the dihydrofolic reductase from chicken liver.

[12] J. R. Bertino, *Biochim. Biophys. Acta* **58**, 377 (1962).
[13] S. Futterman and M. Silverman, *J. Biol. Chem.* **224**, 31 (1957).
[14] M. J. Osborn, M. Freeman, and F. M. Huennekens, *Proc. Soc. Exptl. Biol. Med.* **97**, 429 (1958).

[49] N^5, N^{10}-Methylenetetrahydrofolic Dehydrogenase

(Hydroxymethyltetrahydrofolic Dehydrogenase)

N^5,N^{10}-Methylenetetrahydrofolate + TPN$^+$
$$\rightleftarrows (N^5,N^{10}\text{-Methenyltetrahydrofolate})^+ + \text{TPNH} \quad (1)$$

By K. G. SCRIMGEOUR and F. M. HUENNEKENS

It has been demonstrated that many tissues[1-3] contain a TPN-linked N^5,N^{10}-methylenetetrahydrofolic dehydrogenase (formerly called hydroxymethyltetrahydrofolic dehydrogenase[1]). Ehrlich ascites cells contain both the TPN enzyme and a DPN-linked, metal ion-requiring N^5,

[1] Y. Hatefi, M. J. Osborn, L. D. Kay, and F. M. Huennekens, *J. Biol. Chem.* **227**, 637 (1957).
[2] M. J. Osborn and F. M. Huennekens, *Biochim. et Biophys. Acta* **26**, 646 (1957).
[3] B. V. Ramasastri and R. L. Blakley, *J. Biol. Chem.* **237**, 1982 (1962).

N^{10}-methylenetetrahydrofolic dehydrogenase.[4] The DPN-linked dehydrogenase, however, appears to be of limited distribution compared to the TPN enzyme.

Assay Method

Principle. The assay for N^5,N^{10}-methylenetetrahydrofolic dehydrogenase is based on the measurement of one, or both, of the products (TPNH and N^5,N^{10}-methenyltetrahydrofolate) of reaction (1). The substrate, N^5,N^{10}-methylenetetrahydrofolate, is generated *in situ* by the chemical interaction of HCHO with tetrahydrofolate. The continuous rate of formation of TPNH is followed by the increase in absorbancy at 340 mμ. The other product, N^5,N^{10}-methenyltetrahydrofolate, may be measured at the end of the reaction by its absorbancy at 355 mμ in acid solution. In the latter case, the addition of acid stops the enzymatic reaction, destroys TPNH which would interfere in this spectral assay, and reconverts to the N^5,N^{10}-methenyl derivative [cf. Eq. (2)] any

$(N^5,N^{10}$-Methenyltetrahydrofolate$)^+ +$ H$_2$O
$$\rightleftarrows N^{10}\text{-Formyltetrahydrofolate} + \text{H}^+ \quad (2)$$

N^{10}-formyltetrahydrofolate which may have formed enzymatically (via cyclohydrolase) or chemically. Measurement of N^5,N^{10}-methenyltetrahydrofolate formation, rather than TPNH formation, does not allow the rate of the reaction to be followed continuously, but less enzyme is required, owing to the fact that the molal extinction coefficient of the former material is approximately four times that of TPNH. For this reason, N^5,N^{10}-methenyltetrahydrofolate formation is quite feasible for assay of the enzyme in crude tissue preparations.

Reagents

0.5 M potassium phosphate, pH 7.5.
0.1 M 2-mercaptoethanol.
0.006 M TPN.
0.006 M DPN.
0.1 M MgCl$_2$.
0.025 M formaldehyde, prepared fresh daily.
0.002 M DL-tetrahydrofolate. Tetrahydrofolic acid (243-B) (2.67 mg.) is suspended in 3.0 ml. of 0.05 M 2-mercaptoethanol and then dissolved by the addition of a few drops of 1 M phosphate buffer, pH 7.5.
1.0 N hydrochloric acid.

[4] K. G. Scrimgeour and F. M. Huennekens, *Biochem. Biophys. Research Communs.* **2**, 230 (1960).

Procedure. The following components are added to a 1-cm. Corex cell of 3.5-ml. capacity: 0.3 ml. of phosphate buffer, 0.2 ml. of mercaptoethanol, 0.1 ml. of TPN, 0.4 ml. of formaldehyde, 0.3 ml. of tetrahydrofolate, and water to make 3.0 ml. The optical blank contains water. The enzyme (0.02 to 0.10 ml.) is added to start the reaction, and the production of TPNH is determined by the increase in absorbancy at 340 mμ. A small correction is made for the absorbancy change in a control cuvette containing the identical components except for the omission of HCHO.

If N^5,N^{10}-methenyltetrahydrofolate is to be measured, 0.3 ml. of 1.0 N HCl is added after the reaction has proceeded for 10 to 15 minutes. After removal of denatured protein by centrifugation, the mixture is allowed to stand for 10 minutes before the absorbancy is measured at 355 mμ. Control and blank vessels are the same as above.

For the DPN-linked N^5,N^{10}-methylenetetrahydrofolic dehydrogenase, TPN is replaced by 0.1 ml. of DPN, and 0.1 ml. of MgCl$_2$ is added. All other procedures are identical to those used for the assay of the TPN-linked dehydrogenase.

Application of the Assay Method to Crude Tissue Preparations. In turbid suspensions it is usually difficult to follow the appearance of TPNH at 340 mμ, owing to endogenous TPNH-oxidizing systems. For this reason, the enzyme is assayed by measuring the production of N^5, N^{10}-methenyltetrahydrofolate.

Purification Procedure

The TPN-linked N^5,N^{10}-methylenetetrahydrofolic dehydrogenase has been partially purified from beef liver,[1] pigeon liver,[1] and chicken liver.[2] Ramasastri and Blakley[3] have obtained a 125-fold purified preparation of the enzyme from baker's yeast. The present procedure is an adaptation of our previous method with chicken liver acetone powder as the starting material. All operations are carried out at 0° to 3°, and centrifugations are performed for 10 minutes at 4000 r.p.m. in the International refrigerated centrifuge, PR-2.

Step 1. Extraction. Five grams of chicken liver acetone powder are extracted with 30 ml. of 10^{-2} M phosphate buffer, pH 7.5, for 1 hour with mechanical stirring. After centrifugation, the extract is treated with 10 ml. of 2% protamine sulfate, and the mixture is stirred for 5 minutes and again centrifuged.

Step 2. Ammonium Sulfate Fractionation. Saturated ammonium sulfate, pH 8, is added to the supernatant fraction until 30% saturation is reached. After standing for 10 minutes, the precipitate is recovered by centrifugation and discarded. After the ammonium sulfate concentration in the supernatant fluid is increased to 45% saturation, the precipitate is collected by centrifugation, dissolved in 10 ml. of cold water, and dialyzed for 4 hours against $10^{-3}\,M$ phosphate buffer, pH 7.5. After dialysis, any precipitate is removed by centrifugation.

Step 3. Zinc Precipitation. Sufficient $0.05\,M$ ZnCl$_2$ is added to the preparation to bring the final concentration of Zn^{++} to $0.005\,M$, and the pH is maintained at 6.0. After 5 minutes the mixture is centrifuged. The precipitate is collected, dissolved in 10 ml. of $0.01\,M$ EDTA, pH 7.5, and dialyzed for 3 hours against water.

Step 4. Chromatography. Traces of cyclohydrolase and serine hydroxymethylase, which are still present as contaminants, are removed readily[5] by chromatography of the preparation on a 1.3 × 25-cm. column of DEAE-cellulose.[6] Thirty to 40 milligrams of the enzyme preparation (in about 10 ml.) are adsorbed on the column and eluted by a gradient technique, with 250 ml. of 2 × $10^{-3}\,M$ phosphate buffer, pH 7.3, in the mixing vessel and 250 ml. of $0.5\,M$ phosphate buffer, pH 7.3, in the reservoir. Fractions of 2.5 ml. are collected with an automatic fraction collector. The dehydrogenase activity is concentrated sharply in the initial tubes containing protein.

After chromatography on DEAE-cellulose, the enzyme is purified about 100-fold and recovered in about a 50% yield. The DPN-linked

TABLE I
SUMMARY OF PURIFICATION PROCEDURE

Fraction	Volume	Units[a]	Protein, mg.	Specific activity, units/mg. × 10^{-3}	Recovery, %
Acetone powder extract	22	7.4	1240	6	100
Ammonium sulfate fractionation (30–45%)	12	5.3	252	21	72
ZnCl$_2$ precipitation	12	4.2	33	126	58
DEAE-cellulose chromatography (pooled fractions)	15	3.8	6.5	585	51

[a] A unit is defined as the amount of enzyme that catalyzes the conversion of 1 μmole of substrate per minute under the given assay conditions.

[5] Adsorption and elution from calcium phosphate gel has also been used to free the dehydrogenase from cyclohydrolase (M. J. Osborn, Ph.D. Thesis, University of Washington, 1958).

[6] Commercial DEAE-cellulose is washed exhaustively first with $1\,M$ K$_2$HPO$_4$ followed by water, and the pH of the suspension is brought to 7.3. The resin is packed by gravity into the column and washed with 250 ml. of 2 × $10^{-3}\,M$ phosphate buffer, pH 7.3.

N^5,N^{10}-methylenetetrahydrofolic dehydrogenase has been partially purified[6] from phosphate buffer extracts of acetone-dried Ehrlich ascites
tumor cells by removal of nucleic acids with protamine, followed by
precipitation of the enzyme with saturated ammonium sulfate, pH 8
(0 to 48% fraction), and with solid ammonium sulfate (0 to 35%
fraction).

A summary of the purification procedure is given in Table I.

Properties

Activators. The DPN enzyme, but not the TPN enzyme, is stimulated
by Mg^{++} or Mn^{++}.

Equilibrium. The equilibrium constant[7] for reaction (1) is 1.7×10^{-2}
at pH 7.

Kinetic Properties. The K_m values for the substrates of the various
N^5,N^{10}-methylenetetrahydrofolic dehydrogenases are summarized in
Table II.

TABLE II
K_m VALUES (M)

Enzyme	Substrate				
	DL-Tetra-hydrofolate	TPN	DPN	Mg^{++}	Ref.
Beef liver	2.7×10^{-5}				a
Baker's yeast		3.7×10^{-5}			b
Ascites cells	3.0×10^{-5}	4.0×10^{-5}			c
Ascites cells	4.2×10^{-5}		1.2×10^{-4}	3×10^{-3}	c

a Y. Hatefi, M. J. Osborn, L. D. Kay, and F. M. Huennekens, *J. Biol. Chem.* **227,**
637 (1957).

b B. V. Ramasastri and R. L. Blakley, *J. Biol. Chem.* **237,** 1982 (1962).

c K. G. Scrimgeour and F. M. Huennekens, *Biochem. Biophys. Research Communs.*
2, 230 (1960).

[7] F. M. Huennekens and M. J. Osborn, *Advances in Enzymol.* **21,** 369 (1959).

[50] N^{10}-Formyltetrahydrofolic Deacylase

N^{10}-Formyltetrahydrofolate $+ H_2O \rightarrow$ Formate $+$ tetrahydrofolate   (1)

By F. M. HUENNEKENS and K. G. SCRIMGEOUR

Assay Method

Principle. N^{10}-Formyltetrahydrofolic deacylase[1] catalyzes the hydrolysis of N^{10}-formyltetrahydrofolate into formate and tetrahydrofolate as shown by Eq. (1).[2] For as-yet-unexplained reasons, a *catalytic* amount of TPN or TPNH is required in the deacylase reaction. Reaction (1) may be followed: (1) by combining it with the serine hydroxymethylase–N^5,N^{10}-methylenetetrahydrofolic dehydrogenase system [reaction (2)] and cyclohydrolase [reaction (3)] (in the presence of a *limiting amount*

Serine $+$ tetrahydrofolate $+$ TPN$^+$
$\rightleftharpoons$ Glycine $+$ (N^5,N^{10}-methenyltetrahydrofolate)$^+$ $+$ TPNH   (2)

(N^5,N^{10}-Methenyltetrahydrofolate)$^+$ $+ H_2O$
$\rightleftharpoons N^{10}$-Formyltetrahydrofolate $+ H^+$ (3)

of tetrahydrofolate, the continuous production of TPNH measured spectrophotometrically[3] or manometrically[4] is dependent on the presence of deacylase); (2) by measuring directly the disappearance of N^{10}-formyltetrahydrofolate in reaction (1) after stopping the reaction with perchloric acid[5] (acidification converts any residual N^{10}-formyltetrahydrofolate to N^5,N^{10}-methenyltetrahydrofolate, via reversal of reaction (3), and the latter compound is estimated by its absorbancy at 355 mμ); and (3) by the continuous spectrophotometric method[1] described below which is based on the increase in absorbancy at 300 mμ accompanying the conversion of N^{10}-formyltetrahydrofolate ($\epsilon = 8.5 \times 10^6$ cm.^{2}/mole) to tetrahydrofolate ($\epsilon = 22 \times 10^6$ cm.^{2}/mole).

Reagents

10^{-3} M N^{10}-formyltetrahydrofolate. This solution is prepared just prior to each assay. Crystalline N^5,N^{10}-methenyltetrahydrofolate

---

[1] M. J. Osborn, Y. Hatefi, L. D. Kay, and F. M. Huennekens, *Biochim. et Biophys. Acta* **26**, 208 (1957).

[2] This reaction is comparable to the enzymatic deacylation of acetyl or succinyl coenzyme A [J. Gergeley, P. Hele, and C. V. Ramakrishnan, *J. Biol. Chem.* **198**, 323 (1952)].

[3] M. J. Osborn, Ph.D. Thesis, University of Washington, 1958.

[4] Y. Hatefi, Ph.D. Thesis, University of Washington, 1956.

[5] H. R. Whiteley, *Comparative Biochem. Physiol.* **1**, 227 (1960).

(0.9 mg.) is dissolved in 1.0 ml. of 0.1 M mercaptoethanol, 1 N KOH is added dropwise to adjust the pH to 8 to 9, and the solution is diluted to 2.0 ml.

1 M Tris buffer, pH 8.5.

0.1 M 2-mercaptoethanol.

10^{-4} M TPNH.

Procedure. The following components are added to a silica cuvette: 0.3 ml. of N^{10}-formyltetrahydrofolate, 0.1 ml. of Tris buffer, pH 8.5, 0.2 ml. of mercaptoethanol, 0.05 to 0.2 ml. of enzyme, and water to make 3.0 ml. The blank cuvette is identical except for the omission of N^{10}-formyltetrahydrofolate. After the initial absorbancy is measured at 300 mμ, 0.15 ml. of TPNH is added to start the reaction, and the increase in absorbancy at 300 mμ is followed over a 15 to 30-minute period. Since the substrate, N^{10}-formyltetrahydrofolate, is a *dl*-mixture with respect to the C-6 position, only 50% will have been deacylated at the completion of the reaction.

Definition of Unit. One unit of enzyme is defined as the amount that catalyzes the deacylation of 1 μmole of N^{10}-formyltetrahydrofolate per minute under these assay conditions.

Application of Assay Method to Crude Tissue Preparations. The above assay can be used to measure the deacylase activity in crude tissue extracts which have been clarified by centrifugation. For turbid preparations, the assay devised by Whiteley[2] is preferable.

Purification Procedure[3]

All operations are carried out at about 5°, and centrifugations are performed for 10 minutes at 4000 r.p.m. in the International refrigerated centrifuge, PR-2. Five grams of beef liver[6] acetone powder are extracted with 30 ml. of 10^{-2} M phosphate buffer, pH 7.5, for 1 hour with stirring. After centrifugation, the residue is discarded. The supernatant fraction (about 25 ml.) is treated with 9 g. of dry Dowex 1 resin (X-10, chloride form, 200 to 400 mesh). The mixture is stirred mechanically for 20 minutes, after which the resin is removed by centrifugation. The supernatant fraction is diluted with 1 vol. of water, the pH is taken to 5.4 by the dropwise addition of 5% acetic acid, and then without centrifugation immediately readjusted to pH 7.2 with 1 N NaOH. Six-tenths volume of acetone (previously cooled to −70° by immersion in dry ice) is added with stirring over a 10-minute period, during which the temperature of

---

[6] N^{10}-Formyltetrahydrofolic deacylase is found in beef liver but is absent in pigeon and chicken liver. The levels of this enzyme in a number of other tissues have been reported by Whiteley.[5]

the solution is gradually lowered to $-10°$. After standing at $-10°$ for 10 minutes, the mixture is centrifuged at -5 to $-10°$, and the supernatant solution is discarded. The pellet is taken up in 20 ml. of $10^{-2} M$ phosphate buffer, pH 7.5, and saturated ammonium sulfate (pH 8) is added until 30% saturation is reached; after centrifugation, the residue is discarded. The ammonium sulfate concentration is raised to 45% of saturation, and the precipitate is collected by centrifugation and dissolved in 10 ml. of $10^{-2} M$ phosphate buffer, pH 7.5.

Properties

After the above procedure, the enzyme is purified about tenfold and contains no N^5,N^{10}-methylenetetrahydrofolic dehydrogenase. The enzyme is stable both to dialysis for 12 hours against $10^{-3} M$ phosphate buffer, pH 7.5, and to storage in the frozen state; the activity is destroyed, however, by heating the preparation to 50° for 2 minutes. The enzyme is activated by catalytic amounts of TPN or TPNH[1] and is inhibited 70% by $6.7 \times 10^{-4} M$ p-chloromercuribenzoate, The enzyme is also inhibited by excess amounts of the reaction product, tetrahydrofolate.

[51] Formyltetrahydrofolate Synthetase[1]

By JESSE C. RABINOWITZ and W. E. PRICER, JR.

HCOOH + ATP + (tetrahydrofolate) ⇌ (10-formyltetrahydrofolate) + ADP + P_i

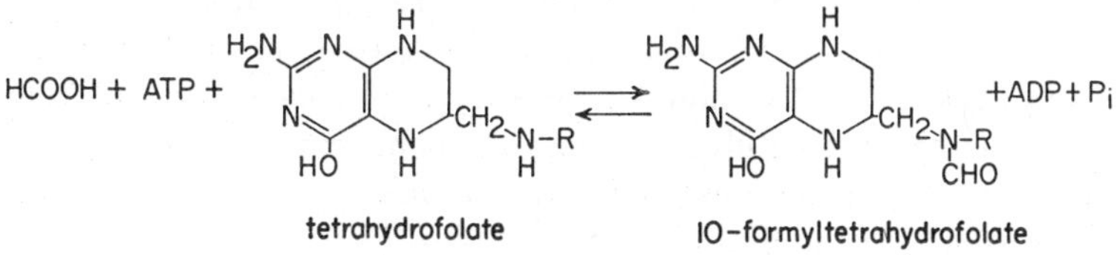

Assay Method

Principle. The 10-formyltetrahydrofolic acid formed in the enzymatic reaction shown above is quantitatively converted to 5,10-methenyltetra-

[1] *The enzyme:* This name is suggested by the Commission on Enzymes of the International Union of Biochemistry (Pergamon Press, 1961). The enzyme has also been referred to as tetrahydrofolic acid formylase,[2-4] and formate-activating enzyme.[5]

[2] G. R. Greenberg, L. Jaenicke, and M. Silverman, *Biochim. et Biophys. Acta* **17,** 589 (1955).

[3] J. C. Rabinowitz and W. E. Pricer, Jr., *J. Biol. Chem.* **229,** 321 (1957).

[4] L. Jaenicke and E. Brode, *Biochem. Z.* **334,** 108 (1961).

[5] H. R. Whiteley, M. J. Osborn, and F. M. Huennekens, *J. Biol. Chem.* **234,** 1538 (1959).

hydrofolic acid by the addition of acid (see equation below). This product is determined spectrophotometrically by its characteristic absorption maximum at 350 mμ.

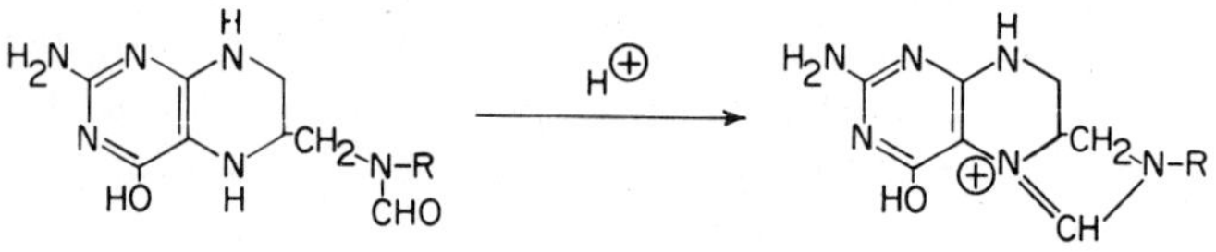

10-formyltetrahydrofolate   5,10-methenyltetrahydrofolate

Reagents

1.0 M triethanolamine buffer, pH 8.0.

0.01 M *dl*-tetrahydrofolic acid.[3,6] Dissolve 28 mg. of tetrahydro-folic acid diacetate in 5 ml. of 1.0 M 2-mercaptoethanol, and neutralize with about 4 drops of 1 N KOH.

0.1 M MgCl$_2$.

0.05 M ATP.

0.2 M sodium formate, pH 8.0.

Enzyme. Dilute the enzyme solution with 0.05 M maleate buffer at pH 7.0, 0.1 M with respect to 2-mercaptoethanol.

Procedure. Add the following amounts of reagents to 13 × 100-mm. test tubes: triethanolamine buffer, 0.1 ml.; MgCl$_2$, 0.1 ml.; ATP, 0.1 ml.; HCOONa, 0.2 ml.; *dl*-tetrahydrofolic acid, 0.2 ml.; and water, 0.3 ml. At 30-second intervals, add 0 to 40 μl. of the diluted enzyme. Incubate the solutions at 37° for 10.0 minutes. Stop the reaction by adding 2.0 ml. of 0.36 N HCl. Allow the tubes to stand at room temperature for 10 to 30 minutes. Determine the absorbance at 350 mμ.

Definition of Unit and Specific Activity. One unit of enzyme is the amount which causes an absorbance change of 1.000 per 10 minutes under the conditions described. Specific activity is expressed as units per milligram of protein. Protein is determined by using the Folin-Ciocalteu reagent[7] with crystalline serum albumin containing 14.6% nitrogen as a standard.

Application of Assay Method to Crude Tissue Preparations. Several enzymes occur in crude tissue extracts which might lead to an error in assay of formyltetrahydrofolate synthetase. The product of the reaction,

---

[6] See Vol. VI [113].

[7] E. W. Sutherland, C. F. Cori, R. Haynes, and N. S. Olsen, *J. Biol. Chem.* **180**, 825 (1949).

10-formyltetrahydrofolic acid, may be destroyed by the action of a specific deacylase.[8] 10-Formyltetrahydrofolic acid may also be formed from a variety of formyl donors by enzymatic reactions. Among these is the conversion of histidine to glutamic acid and formiminotetrahydrofolic acid.[9] The latter product is converted to 10-formyltetrahydrofolic acid by a series of enzymic reactions which have been shown to occur in mammalian species.[10] Purines[11] and serine[12] may also serve as sources of the formyl group in a series of enzymatic reactions which do not involve the participation of formyltetrahydrofolate synthetase.

Purification Procedure[13]

Step 1. Cell Autolyzate. Suspend 1 g. of lyophilized cells of *Clostridium cylindrosporum*[14] in 20 ml. of 0.05 M maleate buffer at pH 7.0, 0.1 M with respect to 2-mercaptoethanol. Incubate the mixture for 30 minutes at 37° in an evacuated flask. Centrifuge the autolyzate for 10 minutes at 105,000 $\times$ g, and discard the residue.

Step 2. Protamine Treatment. To 18 ml. of the autolyzate, add 5 ml. or protamine sulfate (10 mg./ml., neutralized with ammonium hydroxide). Allow the mixture to stand at room temperature for 5 minutes. Centrifuge at 30,000 $\times$ g, and discard the precipitate.

Step 3. First Ammonium Sulfate Fractionation. To 23 ml. of the cold protamine supernatant solution, add 6.25 g. of ammonium sulfate, 0.025 ml. of 2-mercaptoethanol, and concentrated ammonium hydroxide to bring the solution to pH 6.35. Remove the precipitate by centrifugation at 30,000 $\times$ g for 5 minutes. Adjust the pH of the supernatant solution to 6.9 with ammonium hydroxide.

Step 4. Heat Treatment. Transfer the solution to a large test tube, and place the tube in a water bath at 55°. Stir the contents with a thermometer for 15 minutes. The internal temperature was 53° after 10 minutes and reached 55° after 15 minutes. Cool the contents of the tube in an ice bath, and remove the precipitate by centrifugation at 30,000 $\times$ g for 5 minutes.

[8] M. J. Osborn, Y. Hatefi, L. D. Kay, and F. M. Huennekens, *Biochim. et Biophys. Acta* **26**, 208 (1957).
[9] A. Miller and H. Waelsch, *Arch. Biochem. Biophys.* **63**, 263 (1956).
[10] H. Tabor and J. C. Rabinowitz, *J. Am. Chem. Soc.* **78**, 5705 (1956); see also Vol. V [105].
[11] S. C. Hartman and J. M. Buchanan, *J. Biol. Chem.* **234**, 1812 (1959).
[12] D. A. Goldthwait and G. R. Greenberg, Vol. II [78].
[13] J. C. Rabinowitz and W. E. Pricer, Jr., *J. Biol. Chem.* **237**, 2898 (1962).
[14] See Vol. VI [97].

Step 5. Second Ammonium Sulfate Fractionation. To 23 ml. of the previous fraction, add 0.025 ml. of 1 *M* MgCl$_2$ and 2.5 g. of ammonium sulfate and concentrated ammonium hydroxide to bring the solution to pH 7.0. Allow the mixture to stand at 2° for 75 minutes or longer, and centrifuge. Dissolve the precipitate in 5.0 ml. of 0.05 *M* maleate buffer at pH 7.0, 0.1 *M* with respect to 2-mercaptoethanol.

Step 6. First Crystallization. Add 1.5 g. of ammonium sulfate to 5.0 ml. of the previous fraction. Crystals form in several minutes. Adjust the pH of the solution to 7 with ammonium hydroxide, and incubate the mixture for 3 hours or more at 0°. Remove the crystals by centrifugation. Dissolve the precipitate in 5.0 ml. of 0.05 *M* maleate buffer at pH 7.0, 0.1 *M* with respect to 2-mercaptoethanol.

Step 7. Second Crystallization. Add 1.48 g. of ammonium sulfate to 5 ml. of the solution obtained in the previous step, and bring the pH to 7.0 with ammonium hydroxide, if necessary. Incubate the mixture in an ice bath for 3 hours or longer.

Step 8. Recrystallization. Dissolve the crystals of the enzyme in 5 ml. of 0.05 *M* maleate buffer at pH 7.0, 0.1 *M* with respect to 2-mercaptoethanol. Add 1.5 g. of ammonium sulfate with 15 μl. of concentrated ammonium hydroxide per 5 ml. Allow the mixture to stand at 0° for several hours.

A summary of the purification procedure is given in the table.

SUMMARY OF PURIFICATION PROCEDURE

Step	Volume, ml.	Units	Protein, mg./ml.	Specific[a] activity, units/mg. protein	Yield, %
1. Autolyzate	39	1,760,000	17.6	2560	100
2. Protamine treatment	45	1,633,500	13.1	2770	93
3. First ammonium sulfate precipitate	48	1,670,000	10.3	3400	95
4. Heat treatment	45	1,395,000	5.7	5410	79
5. Second ammonium sulfate precipitate	10	1,420,000	12.7	11,200	81
6. First crystallization	10	1,120,000	9.37	12,000	64
7. Second crystallization	11.5	1,080,000	3.65	25,700	61
8. Third crystallization	12.0	909,000	3.21	23,500	52
9. Fourth crystallization	10.0	892,000	2.86	31,200	51
10. Fifth crystallization	10.0	834,000	2.57	32,400	47

[a] The average final specific activity obtained with various crystalline preparations is about 35,000, although values have ranged from 28,000 to 40,000.

Properties

Physical Properties. The crystalline enzyme is homogeneous by sedimentation. The molecular weight based on sedimentation and diffusion measurements is 230,000.[13, 15]

Specificity. The crystalline enzyme is entirely free of ATPase activity, adenylate kinase activity, and diphosphonucleotide kinase activity.[13, 16] The enzyme is specific for the *l*-isomer of tetrahydrofolic acid. Higher homologs of formic acid are completely without activity, and no other formyl donors have been detected. Although crude preparations obtained from *C. cylindrosporum* show activity with a variety of nucleotide tri- and diphosphates,[3] the crystalline enzyme is specific for ATP and ADP shows no activity.

Activators and Inhibitors. The crystalline enzyme requires the addition of either magnesium, manganese, or calcium ions for activity. The enzyme is strongly inhibited by sulfhydryl group reagents such as *p*-chloromercuribenzoate and iodoacetamide. Optimal activity is exhibited at 41°. The activity falls off to 75% of this value at 33° and at 46°. Optimal activity is noted in triethanolamine buffer at pH 8.0. The activity observed in triethanolamine at pH 7.0 is 74% of the optimal value.

Equilibrium. The equilibrium of the reaction lies far to the right. However, in the presence of an excess of enzyme and substrates shown on the right side of the equation, the reaction can be shown to be reversible. In the presence of an excess of any of the substrates shown on the left, the reaction goes to completion, and the enzyme is useful for the quantitative determination of ATP, tetrahydrofolic acid, and formic acid.[3, 17]

Kinetic Properties.[16] The following constants were obtained with the crystalline enzyme incubated at pH 8.0: K_m (*l*-tetrahydrofolic acid) = $2.6 \times 10^{-4} M$; K_m (ATP) = $2.9 \times 10^{-4} M$; K_m (formate) = $5.6 \times 10^{-3} M$; and $K_m(Mg^{++})$ = $2.1 \times 10^{-3} M$.

[15] J. C. Rabinowitz and W. E. Pricer, Jr., *Federation Proc.* **17**, 293 (1958).
[16] J. C. Rabinowitz and R. H. Himes, *Federation Proc.* **19**, 963 (1960).
[17] D. H. Rammler and J. C. Rabinowitz, *Anal. Biochem.* **4**, 116 (1962).

[52] Formiminotetrahydrofolic Acid Cyclodeaminase[1]

5-FORMIMINO-THF   5,10-METHENYL-THF

$$ (1) $$

By KOSAKU UYEDA and JESSE C. RABINOWITZ

Assay Method

Principle. 5,10-Methenyltetrahydrofolic acid formed by the enzymatic reaction shown by Eq. (1) is determined from its absorption at 356 mμ. The rate of its formation from 5-formiminotetrahydrofolic acid is a function of the concentration of formiminotetrahydrofolic acid cyclodeaminase.

Reagents

0.001 M 5-formiminotetrahydrofolic acid. One milligram of 5-formiminotetrahydrofolic acid[2] is dissolved in 2 ml. of water with the addition of 2 drops of ethanol and 1 drop of 0.1 N HCl. Undissolved material, presumably 5,10-methenyltetrahydrofolic acid, is removed by centrifugation. The solution can be stored at —20° for a week. The concentration of 5-formiminotetrahydrofolic acid is determined by converting it to 5,10-methenyltetrahydrofolic acid by treatment with 0.25 N HCl; the increase in absorbancy is determined at 350 mμ, and the concentration of 5-formiminotetrahydrofolic acid is calculated by using a value of 24,900 for the molar extinction coefficient of the 5,10-methenyltetrahydrofolic acid in acid solution.

0.1 M potassium maleate buffer, pH 7.0.

Enzyme. The enzyme is diluted in 0.5 M potassium maleate buffer, pH 7.0, for assay.[3]

Procedure. An assay mixture is prepared by mixing 2.9 ml. of maleate buffer, 0.1 micromole of 5-formiminotetrahydrofolic acid, and water to

---

[1] J. C. Rabinowitz and W. E Pricer, Jr., *J. Am. Chem. Soc.* 78, 5702 (1956).

[2] J. C. Rabinowitz and W. E. Pricer, Jr., *Federation Proc.* 16, 236 (1957); see Vol. VI [115].

[3] Fractions obtained by elution from the DEAE columns with phosphate buffer as described in step 4 of the purification are diluted in cold water for assay.

make the volume to 3.0 ml. The reaction is initiated by the addition of enzyme to the sample cuvette, and the change in absorbancy at 356 mμ is determined with a recording spectrophotometer at 25°.

Definition of Unit and Specific Activity. One unit of activity is defined as that amount of enzyme which causes a change in absorbancy of 1.0 per minute under the above conditions and is equivalent to the formation of 0.12 micromole of 5,10-methenyltetrahydrofolic acid per milliliter per minute. Specific activity is expressed as units per milligram protein. Protein is determined by using the Folin-Ciocalteu reagent with crystalline serum albumin containing 14.6% nitrogen as a standard.

Application of Assay Method to Crude Tissue Preparations. The presence of cyclohydrolase interferes with the assay, since it hydrolyzes the product to 10-formyltetrahydrofolic acid, which has a decreased absorbancy at 356 mμ. The amount of interference encountered in natural products, however, will depend on the relative amounts of cyclodeaminase and cyclohydrolase. In cases so far investigated, the cyclohydrolase is relatively inactive and does not seriously interfere with the assay of cyclodeaminase by this method.

Purification Procedure[4]

Step 1. Cell Autolyzate. Two grams of lyophilized cells of *Clostridium cylindrosporum*[5] are suspended in 40 ml. of 0.05 M potassium maleate buffer, pH 7.0. The suspension is incubated for 30 minutes at 37° and is then centrifuged with a Spinco Model L ultracentrifuge for 15 minutes at 144,000 $\times$ g.

Step 2. Acid Supernatant Solution. The autolyzate (35 ml.) is added with stirring to 160 ml. of 0.1 M potassium acetate buffer at pH 5.0 at 2°. After 5 minutes, the pH is adjusted to 4.15 by the addition of approximately 5 ml. of glacial acetic acid. After an additional 5 minutes, the mixture is centrifuged at 30,000 $\times$ g for 5 minutes. The precipitate is discarded. The pH of the supernatant is adjusted to 7.0 with 5 N KOH.

Step 3. Ammonium Sulfate–TCA. To the supernatant solution from step 2 are added 80 g. of powdered ammonium sulfate, and the mixture is held at 2°. After 10 minutes, the mixture is centrifuged at 30,000 $\times$ g for 5 minutes. Then 15 ml. of 50% (w/v) trichloroacetic acid solution

---

[4] The 5-formiminotetrahydrofolic acid cyclodeaminase has been purified both from the cell autolyzate and from the supernatant solution obtained in step 2 of the purification of formiminoglycine formimino-transferase (Vol. VI [10]) which had been stored frozen for 6 months. Cyclodeaminase prepared from the latter source had a specific activity almost three times as high as could be obtained from the freshly prepared cell autolyzate described here. However, the procedure has not been studied in detail.

[5] H. A. Barker and J. V. Beck, *J. Bacteriol.* **43**, 291 (1942); see Vol. VI [97].

is added to the 300 ml. of supernatant solution obtained. After 20 minutes at 2°, the mixture is centrifuged at 30,000 × g for 10 minutes. The precipitate is dissolved in 20 ml. of 0.1 M potassium phosphate at pH 7.0. The solution is then dialyzed overnight against 2 l. of 0.005 M potassium phosphate at pH 6.0.

Step 4. Chromatography on DEAE-cellulose. The dialyzed enzyme solution from step 3 is chromatographed on a DEAE-cellulose column (2 × 27 cm.) at 2° to 4°C which has been equilibrated with 0.005 M potassium phosphate buffer at pH 6.0. The buffer solutions used in the chromatography are varying concentrations of potassium phosphate, pH 6.0. After the column has been charged with the enzyme solution, it is washed with 100 ml. of 0.03 M buffer solution. The enzyme is then eluted from the column with a linear gradient of phosphate obtained by adding 200 ml. of 0.18 M buffer solution from an open vessel into an open mixing chamber containing 200 ml. of 0.03 M buffer. Five milliliter samples are collected on an automatic fraction collector. The protein and enzyme contents of each fifth tube are determined. The enzyme is found in the fractions numbers 64 to 73.

Step 5. Concentration. The pooled fractions obtained in step 4 are frozen in an ethanol–dry ice bath and lyophilized until about 5 ml. of solution remains. The lyophilization takes about 4 hours. The concentrated enzyme solution is then dialyzed at 2° for 10 hours against 1 l. of 0.05 M potassium phosphate at pH 7.0. The enzyme may be stored at −20°.

A summary of the purification procedure is given in the table.

SUMMARY OF PURIFICATION PROCEDURE

Fraction	Volume, ml.	Units/ml.	Total units	Protein, mg./ml.	Specific activity, units/mg.	Recovery, %
1. Cell autolyzate	35	3500	122,000	17.6	200	100
2. Acid supernatant solution	200	668	133,600	0.9	743	110
3. Ammonium sulfate TCA	38	3680	140,000	1.6	2300	114
4. Chromatography on DEAE-cellulose	50	1110	55,500	0.2	5550	46
5. Concentration	9.2	7160	66,000	1.0	7160	54

Properties

Physical Properties. The molecular weight based on sedimentation measurements is approximately 38,000.

Specificity. Formiminotetrahydrofolic acid can be replaced by formiminotetrahydropteroyl-(L)-aspartic acid, formiminotetrahydropteroyltri-(L)-glutamic acid-(α), and formiminotetrahydropteroyltriglutamate-(γ).[6]

Activators and Inhibitors. The optimal activity was observed in maleate buffer at pH 7.2. The activity in diethanolamine and triethanolamine buffers at this pH was decreased by approximately 20%. The following compounds were found to be competitive inhibitors of formiminotetrahydrofolic acid cyclodeaminase: 4-amino-4-deoxyfolic acid, 4-amino-4-deoxy-10-methylfolic acid, *dl*-4-amino-4-deoxytetrahydrofolic acid, *dl*-4-amino-4-deoxy-10-methyltetrahydrofolic acid, *dl*-tetrahydrofolic acid, *dl*-10-methyltetrahydrofolic acid, and 2-hydroxy-2-deaminotetrahydrofolic acid. The K_i of the most active inhibitor, 4-amino-4-deoxytetrahydrofolic acid, is $1.4 \times 10^{-7} M$.

Equilibrium. Spectral evidence suggests that the reaction may be reversible at pH 8.3, but not at pH 6.8 or 9.3. However, the equilibrium constant could not be determined because of the instability of the reactants under the conditions of the experiment.

Kinetic Properties. The K_m (*l*-5-formiminotetrahydrofolic acid) is $3.1 \times 10^{-5} M$.

---

[6] These compounds were prepared by subjecting the chemically reduced folic acid derivatives to the action of formiminoglycine formimino-transferase (see Vol. VI [10]) in the presence of formiminoglycine. The formimino derivatives of the reduced folic acid compounds were isolated by chromatography on DEAE-cellulose according to the procedure described in Vol. VI [115].

---

[53] Synthesis and Transformations of Folic Coenzymes
I. Cyclodehydrase (5-Formyl to 5,10-Methenyl)

$$\text{5-Formyltetrahydrofolic acid} \xrightarrow{\text{ATP, Mg}^{++}} \text{5,10-Methenyltetrahydrofolic acid}$$

By DAVID M. GREENBERG

This enzyme has been isolated from sheep liver acetone powders[1,2] and from extracts of *Micrococcus aerogenes*.[1,3]

Assay Method

Principle. This is based on the accumulation of a product with an adsorption maximum at 343 mμ[1,2] or 355 mμ[3] on incubation of *dl*-calcium leucovorin with enzyme and ATP.

---

[1] J. M. Peters and D. M. Greenberg, *J. Am. Chem. Soc.* **80**, 2719 (1958).

[2] L. K. Wynston and D. M. Greenberg, *Biochemistry*, publ. pending.

[3] L. D. Kay, M. J. Osborn, Y. Hatefi, and F. M. Huennekens, *J. Biol. Chem.* **235**, 195 (1960).

Reagents

Buffer solutions prepared as described in Vol. I [16].

Stock solutions: 0.1 M aqueous solutions stored at $-5°$.

Procedure. The assay was performed at 30° in Corex cuvettes with a 1-cm. light path in a total volume of 3.0 ml. Components of the reaction mixture were 100 micromoles of sodium citrate buffer, pH 6.0, 0.2 micromole of calcium leucovorin, 2 micromoles $MgSO_4$, 2 micromoles of ATP, and enzyme protein in amounts of about 1 mg. The reaction was started by addition of calcium leucovorin to the medium. Readings at 343 mμ are taken frequently over a 5- to 10-minute period.

Definition of Unit and Specific Activity. An enzyme unit was defined as the amount of enzyme which brought about a change in optical density of 0.001 unit in 30 minutes at 30°. The specific activity is represented by enzyme units per milligram of protein.

Purification Procedure

Preparation of Acetone Powder Extracts. The enzyme activity was extracted from the acetone powder by stirring for 2 hours with cold 0.05 M sodium citrate, pH 6.5, using 1500 ml. of buffer per 100 g. of powder. To the supernatant liquid obtained after centrifugation of the suspension at 10,000 $\times$ g for 30 minutes, solid $(NH_4)_2SO_4$ was added to reach 30% saturation (16.0 g. per 100 ml.). The pH was maintained at 6.0 to 6.5 by addition of dilute NH_4OH. The precipitated protein was removed by centrifugation and more $(NH_4)_2SO_4$ added to 50% saturation (12.1 g. per 100 ml.). After stirring overnight at 4°, the suspension was centrifuged at 10,000 $\times$ g for 30 minutes and the supernatant liquid discarded. Preliminary to further purification, the salts were removed by dialysis against 4 changes of 4 l. each of distilled water. The enzyme could then be lyophilized and stored.

Chromatography on CM- and DEAE-Cellulose. A higher degree of purification was achieved by first chromatographing on CM-cellulose followed by DEAE-cellulose, rather than the reverse. The cellulose ion exchangers were first purified by washing with NaOH and HCl.[4] About 4 g. of protein in a volume of 200 ml. was placed on a 2.5 $\times$ 21 cm. column of CM-cellulose which had been equilibrated at pH 5.4 with 0.005 M sodium citrate buffer. Chromatography was carried out in the cold room at 4° and fractions of 15 ml. were collected at a rate of 3 ml. per minute on elution with 0.005 M sodium citrate buffer at pH 5.4. The cyclodehydrase activity was highest in the leading edge of the peak which emerged with the solvent front.

[4] E. A. Peterson and H. A. Sober, *J. Am. Chem. Soc.* **78**, 751 (1956).

The active material from the CM-cellulose chromatography; after dialysis against distilled water and 0.001 M potassium phosphate buffer, pH 7.5, was concentrated on a rotary evaporator and chromatographed on a 2.0 × 30 cm. column of DEAE-cellulose, previously equilibrated at pH 7.5 with 0.005 M potassium phosphate buffer. The column was loaded with 1–1.5 g. of protein and eluted first with 0.005 M phosphate buffer, and later with 0.035 M buffer, pH 7.5. Fractions of 10 ml. volume were collected. The purification achieved in the different steps of the procedure is shown in Table I.

TABLE I
SUMMARY OF PURIFICATION OF SHEEP LIVER CYCLODEHYDRASE

Fraction	Specific activity units/mg.
Aqueous acetone powder extract	2.5
30 to 50% $(NH_4)_2SO_4$ fraction	22.5
CM-cellulose chromatography eluate	50 2
First DEAE-cellulose chromatography eluate	340
Second DEAE-cellulose chromatography eluate	1142

Properties

Nature of Reaction Product. The product has been shown to be identical with 5,10-methenyltetrahydrofolate. The previous report that it was not[1] has been shown to be erroneous.[2]

pH Optimum. This was determined to be at pH 4.8.[2]

Activators and Inhibitors. The cyclodehydrase is an SH enzyme. Its activity is inhibited by the usual SH reagents, and this inhibition can be completely reversed by glutathione. ATP and Mg^{++} are required in the reaction.

Kinetic Properties. The Michaelis constant of N^5-formyltetrahydrofolate was estimated to be $1.4 \times 10^{-4} M$.[2]

[54] Synthesis and Transformations of Folic Coenzymes
II. Cyclohydrolase (5,10-Methenyl to 10-Formyl)

5,10-Methenyltetrahydrofolic acid + H_2O
→ 10-Formyltetrahydrofolic acid

By DAVID M. GREENBERG

This enzyme was observed in lyophilized cells of *Clostridium cylindrosporum* by Rabinowitz and Pricer,[1] and a method for its estimation was reported. It has been slightly purified from rabbit liver by Tabor and Wyngarden.[2]

Assay Method

Reagents. dl-5,10-Methenyltetrahydrofolic acid (anhydroleucovorin) was prepared by dissolving approximately 15 micromoles of calcium leucovorin (American Cyanamide Company) in 10 ml. of 0.1 N HCl containing 600 micromoles of mercaptoethanol. This was incubated in a vacuum in the dark at 25° for 24 hours and then stored in a vacuum at 0°.

Procedure. The assay was performed by measuring the decrease in absorption at 355 mμ in an incubation mixture containing 0.2 ml. of 1 M potassium maleate buffer, pH 6.5, 280 micromoles of mercaptoethanol, enzyme, and 0.075 micromole of dl-5,10-methenyltetrahydrofolic acid in a total volume of 1 ml. The HCl in the substrate was first neutralized with KOH. Incubations were performed at room temperature ($\sim$25°).

For a satisfactory assay it was necessary to obtain initial reaction rates. These were determined with a Cary recording spectrophotometer. In an alternative procedure, first-order reaction constants were calculated, after the substrate concentration had been suitably corrected for the 50% which is enzymatically inactive.

Definition of Unit and Specific Activity. An enzyme unit is defined as the amount of enzyme causing an optical density change of 1.0 per minute at 355 mμ calculated from the optical density change during the initial phase of the reaction. An optical density of 1.0 is equivalent to 0.04 micromole of 5,10-methenyltetrahydrofolic acid per milliliter. Specific activity is expressed as the ratio of enzyme units per milligram of protein.

[1] J. C. Rabinowitz and W. E. Pricer, Jr., *J. Am. Chem. Soc.* **78**, 4176, 5702 (1958).
[2] H. Tabor and L. Wyngarden, *J. Biol. Chem.* **234**, 1830 (1959).

Purification Procedure

Acetone Powder Extracts. Acetone powders were prepared from fresh rabbit livers by homogenizing the liver with 5 vol. of cold acetone for 1 minute in a Waring blendor. The mixture was rapidly filtered through a Buchner funnel, and the moist filter cake was again homogenized with acetone and refiltered. The material was then passed through a coarse screen, air-dried at room temperature, and stored at −10° in a vacuum. The enzyme activity of this material was stable for at least 3 months.

Acetone powder (2 g.) was extracted with 20 ml. of water, with occasional stirring at room temperature for 15 minutes. The suspension was then centrifuged at 20,000 × g for 10 minutes, and the supernatant liquid collected.

Ammonium Sulfate Fractionation. To the supernatant fluid (14 ml.) was added 3 g. of ammonium sulfate. After standing for some time, the precipitate was centrifuged down and discarded. An additional 2 g. of ammonium sulfate was added to the supernatant fluid. The precipitate was collected and dissolved in 7 ml. of 0.1 M potassium maleate buffer, pH 6.5. This fractionated enzyme was unstable in aqueous media at 0°, losing most of its activity in 4 days, but could be stabilized by adding glycerol to a 50% concentration.

The acetone powder extract had a specific activity of 1.35, and the ammonium sulfate precipitation brought this up to 2.2.

Properties

Nonenzymatic Reaction. In neutral and alkaline solutions, the reaction catalyzed by this enzyme proceeds spontaneously. This nonenzymatic hydrolysis is markedly affected by the nature and concentration of the buffers used. Phosphate and imidazole accelerate the reaction at pH 6.5, whereas the rate is relatively slow with triethanolamine sulfate and potassium maleate. The equilibrium constant for the reaction:

$$\frac{[\text{10-Formyl-folate} \cdot \text{H}_4][\text{H}^+]}{[\text{5,10-Methenyl-folate} \cdot \text{H}_4][\text{H}_2\text{O}]} = K'$$

has been estimated to be 2.4×10^{-8}. At pH 7 the degree of hydrolysis at equilibrium is calculated to be 93%.

Enzymatic Reaction. The enzymatic hydrolysis has been shown to follow the same course with enzymatically formed and chemically formed substrate, when a correction is applied for the inactive isomer present in the chemical product. The ultraviolet spectrum of 5,10-methenyltetrahydrofolic acid is the same when formed by either method.

Kinetics. The K_m for the enzymatically active isomer of the substrate was estimated to be $7.5 \times 10^{-5} M$.

The reverse reaction can also be catalyzed by the enzyme.

Section IV

Respiratory Enzymes

[55] Some Bacterial Cytochromes

MARTIN D. KAMEN, ROBERT G. BARTSCH, TAKEKAZU HORIO,
and HENK DE KLERK

Sources

Heme proteins of the C type, B type, and certain variants occur in bacteria.[1,2] Notable exceptions are found among fermentative anaerobes, such as the clostridia, and among some species of streptococci. No single procedure suffices for extraction and isolation.[3] Despite the general impression that C-type cytochromes are stable to extremes of pH and temperature, the fact is that they exhibit variable degrees of lability. Hence, it is best to avoid the drastic conditions sometimes prescribed in the older literature for preparation of mammalian cytochrome c.

We shall present methods for isolation and purification of two types of bacterial heme protein—the C type and the variant form, known as RHP[3,4] the latter so-called because the first example was isolated from the facultative photoheterotroph, *Rhodospirillum rubrum,* hence Rhodospirillum heme protein, or RHP. We shall describe three examples of the C type—*R. rubrum* cytochrome c (cytochrome c_2),[5] *Chromatium* cytochrome 552,[6] and *Desulfovibrio* cytochrome 552 (cytochrome c_3)[7-9]—and two examples of the RHP type—*R. rubrum* RHP[3,4,10] and *Chromatium* RHP.[3,6]

The culture media for the strains of *R. rubrum* and the obligate photoanaerobe, *Chromatium,* that we have employed are described by Newton (Vol. V [7]). Mass cultures are grown in stoppered 5-gallon carboys under tungsten lamp illumination. Cultures are started with a heavy inoculum (5 to 10%) and cooled by forced-air circulation with fans. Usually yields are in the range 200 to 500 mg. dry weight of cells per liter, after 5 to 7 days.

We have isolated cytochrome c_3 from dried powders of the obligately

[1] L. Smith, *Bacteriol. Revs.* 18, 106 (1954).
[2] M. D. Kamen, *Bacteriol. Revs.* 19, 250 (1955).
[3] M. D. Kamen and L. P. Vernon, *Biochim. et Biophys. Acta* 17, 10 (1955).
[4] R. G. Bartsch and M. D. Kamen, *J. Biol. Chem.* 230, 41 (1958).
[5] S. Elsden, M. D. Kamen, and L. P. Vernon, *J. Am. Chem. Soc.* 75, 6347 (1953).
[6] R. G. Bartsch and M. D. Kamen, *J. Biol. Chem.* 235, 825 (1960).
[7] J. R. Postgate, *Biochem. J.* 58, IX (1954).
[8] J. R. Postgate, *J. Gen. Microbiol.* 14, 545 (1956).
[9] M. Ishimoto, J. Koyama, T. Ohmura, and Y. Nagai, *J. Biochem. (Tokyo)* 41, 527 (1954).
[10] T. Horio and M. D. Kamen, *Biochim. et Biophys. Acta* 48, 266 (1961).

anaerobic bacterium, *Desulfovibrio desulfuricans* (Hildenborough strain) donated to us by Dr. J. R. Postgate, who grew this microorganism in the following medium: KH_2PO_4, 0.5 g.; NH_4Cl, 1 g.; Na_2SO_4, 2.6 g.; $CaCl_2 \cdot 6H_2O$, 0.1 g.; $MgSO_4 \cdot 7H_2O$, 2 g.; Na lactate (as 70% w/v solution), 6 g.; yeast extract, 1 g.; distilled H_2O to 1 l. The mixture is brought to pH 7.8, boiled for 15 minutes, filtered, and the pH readjusted to 7.4. It is then autoclaved for 15 minutes at 15 p.s.i. The mixture is cooled, and $Fe(NH_4)_2(SO_4)_2$ is added to 10 p.p.m. from a sterile solution (1%).

Cultures are maintained under 1 atm. H_2 and approximately 30% v/v CO_2. Cells are collected with a Sharples centrifuge under N_2, deposited on large sheets of Whatman No. 1 filter paper, and dried in *vacuo* in a vacuum desiccator over $CaCl_2$ at 4°. Peck[11] has reported a procedure in which he uses a Biogen fermentor (American Sterilizer Co.) for preparation of large quantities.

Stock Solutions. One to five liters of the following reagents (C.P. grade) should be prepared and available for dilution as required:

1. 1.0 M solutions of the acids: acetic, hydrochloric, and citric.
2. 1.0 M solutions of the bases: sodium hydroxide, ammonia, and tris (hydroxymethyl)aminomethane (Tris).
3. 4.0 M sodium chloride.
4. Buffers: 3 M sodium acetate, pH 5.0 to 5.1, 5.5, 5.65, and 7.0; 2 M ammonium phosphate, pH 7.0; 1.0 M sodium citrate, pH 6.0; 1 M potassium phosphate, pH 7.0; 2 M Tris, pH 8.0.

Solids: Sodium dithionite (reagent grade), ammonium sulfate (C.P.).

Resins

1. Amberlite CG-50, 100 to 200 mesh, Type I; Amberlite CG-50, 200 to 400 mesh, Type II. This resin is prepared for use by threefold washing with distilled H_2O (10 vol.), followed successively by threefold washing with 0.2 N NaOH (3 vol.), twofold washing with distilled H_2O (10 vol.), threefold washing with 0.2 N HCl (3 vol.), and, finally, threefold washing with distilled H_2O (3 vol.). The pH is adjusted to 5.0 with dilute NaOH, and the resin is stored in distilled H_2O. Batches of this treated resin (Type II) are equilibrated with 0.1 N ammonium phosphate buffer, pH 7.0 (called pH 7 resin). Others (Type II) are equilibrated with 0.1 N sodium acetate buffer at pH 5.1, 5.5, 5.65, and 5.75.
2. DEAE-cellulose of two grades is used, coarse-grade Standard

[11] H. D. Peck, Jr., *J. Biol. Chem.* **235**, 2734 (1959).

Selectacel and Type 40 Selectacel, obtained from Brown Paper Co., Berlin, New Hampshire. Pretreatment follows procedures given by Peterson and Sober (Vol. V [1]).

I. Soluble Heme Proteins of *R. rubrum*—Cytochrome c_2 and RHP

(Procedure of H. de Klerk and M. D. Kamen.
Scale: 10 to 1000-g. dry weight of cells)

Assay

For cytochrome c_2, the purity index is defined as the ratio of absorbancy for the reduced form at the absorption maximum (α-peak), 550 mμ, to that for the oxidized form at 280 mμ; symbol: $A_{550}{}^{red}/A_{280}{}^{oxid}$. $A_{550}{}^{red}$ is measured in the presence of a slight excess of sodium dithionite, obtained by adding a few crystals of the solid; $A_{208}{}^{oxid}$ is measured in the presence of a slight excess of potassium ferricyanide (addition of a trace quantity of solid). The purity index for the best preparations (greater than 95% pure by spectroscopic and electrophoretic criteria) ranges from 1.12 to 1.20.

For RHP, the purity index is defined as the ratio of absorbancies for the oxidized form at the characteristic maxima, 638 mμ (hematin peak) and 282 mμ; symbol: $A_{282}{}^{638}$. The value of $A_{282}{}^{638}$ for the best preparations (greater than 95% pure by spectroscopic and electrophoretic criteria) is in the range 0.13 to 0.14.

Preparation

Harvest the fresh cells by centrifugation, and wash them free of medium by suspending them twice in 3 vol. of distilled H_2O. Two alternative procedures—lyophilization and sonication—can be employed for cell rupture; see also Newton (Vol. V [7]). If the former method is used, transfer the washed cells in a minimal volume of distilled H_2O to the lyophilizer apparatus, freeze-dry, and then suspend in 5 vol. of 0.1 M acetate buffer, pH 5.0. Stir in the cold ($4°$) overnight. Allow the solids to settle, decant the supernatant fluid, and repeat the extraction with buffer for several hours. Combine the supernatant fluids. Recover the residual fluid by centrifugation at approximately $10,000 \times g$ to $20,000 \times g$ for 30 minutes, and add it to the combined extracts. The residues can be stored in the deep-freeze ($-20°$) for possible further examination. The extract contains cytochrome c_2 (purity approximately 1%; total, approximately 1 g. per kilogram dry weight of cells) and RHP (purity approximately 1%; total, approximately 1.2 g. per kilogram dry weight of cells).

If sonication is used, centrifuge the cells, and suspend in 0.1 M acetate buffer, pH 5.0; use 300 ml. for 100 g. wet weight of cells. Mix thoroughly for at least 30 minutes with a mechanical stirrer; then process in an appropriate sonication apparatus. Whatever apparatus is used, it is important to cool the extracts so that temperatures at the transducer do not rise above 10°. Separate the sonicate by centrifugation for 20 minutes at approximately 10,000 $\times g$ in the cold. Repeat until the supernatant fluid is freed of cellular debris. Treat the extracts obtained either by sonication or by lyophilization in the following manner. All procedures from this point are performed in the cold (0° to 4°), unless specified otherwise.

Test the pH of the extract. If it is not 5.0, bring it to that value by careful addition of 0.1 M acetic acid, or 0.1 N NaOH, whichever is appropriate. Stir the extract for 1 hour to promote precipitation of residual photoactive pigments, small particles, and some contaminant protein. Let stand for a few minutes, then decant, or centrifuge the mixture to obtain a clarified solution.

To remove the bulk of protein contaminants, effect a preliminary precipitation by adding solid ammonium sulfate, final concentration 30% w/v. Check the pH again, and, if necessary, adjust to 5.0. The supernatant fluid, after centrifugation in the usual manner, should appear reddish-orange and should be completely freed of purple material. If it is not, increase the concentration of the ammonium sulfate *cautiously* stepwise to 40% w/v, and repeat the centrifugation. It is advisable to save the residue obtained between 30% and 40% w/v ammonium sulfate for later reworking.

Occasionally, some lipid-containing material collects at the top of the supernatant fluid. It can be removed by filtration through Whatman No. 1 filter paper or by careful manipulation with a flat spoon. To minimize loss, it is best to collect most of the fluid first by aspiration, starting from the bottom of the solution, and then submit only the top layers to the filtration or scoop procedure.

If large amounts of protein appear as a precipitate when the supernatant fluid is allowed to stand in the cold, these should be removed by centrifugation. Such precipitates can also be redissolved, if in small amount, by reestablishment of the pH at 5.0. Dialyze the supernatant fluid for 6 to 8 hours against 0.01 M acetate buffer, pH 5.0, in dialysis tubing (previously rinsed with Tris buffer) with one or two changes of outside buffer (total dialysis time, ~20 to 24 hours). The final salt content of the solution should be less than 0.1% to avoid interference with the chromatographic procedures which follow later. If a precipitate forms during dialysis, it should be removed by centrifugation. The clear supernatant solution contains the two soluble heme proteins. The procedure for both of these proteins remains identical until after the initial chromatographic adsorption step (see below).

Before proceeding with adsorption, pass an aliquot through a pilot column of Type I Amberlite to determine the quantity of resin needed. Stop adding solution when about three-fourths of the resin is colored. Now adsorb the main batch of salt-free preparation on the appropriate amount of resin. Rinse the column with 10 vol. of $0.1 N$ acetate buffer, pH 5, or until the absorbancy of the effluent at 280 mμ is negligible. Then elute the heme proteins as a sharp band with $3 N$ sodium acetate, pH 7.0.

Collect the colored eluate, and dialyze against distilled H_2O. Then adjust the pH to 5.65 with $0.1 N$ acetic acid, and dialyze overnight against the $0.1 N$ acetate buffer, pH 5.65, twice. Addition of a little dithionite to bring the cytochrome c_2 into its reduced form is helpful here because, in subsequent chromatography, the ferrocytochrome is more easily separated from the oxidized RHP than is the ferricytochrome.

Place the dialyzed solution on the Type II Amberlite column, equilibrated at pH 5.65 with $0.1 N$ acetate buffer, pH 5.65. The cytochrome c_2 moves through the column, while the RHP is absorbed. Rinse the column with the same buffer until all the cytochrome is collected. Wash the column, with its adherent RHP, several times with buffer, than elute the RHP with $3 N$ acetate buffer, pH 7.0. Collect the RHP, dialyze against distilled H_2O to remove all salt, and store for further treatment (see below).

Adjust the pH of the cytochrome c_2 solution from the pH 5.65 column to 5.5 with dilute acetic acid, and run the solution through a column equilibrated against $0.1 N$ acetate buffer, pH 5.5. Residual RHP is adsorbed completely, while the cytochrome c_2 runs through somewhat more slowly than in the previous chromatography. In this step, be certain to load the resin only to the point at which half the resin is colored. A small amount of flavoprotein and other yellowish material moves ahead of the cytochrome and can be discarded before the cytochrome is eluted.

Collect the cytochrome in 10- to 15-ml. aliquots, and monitor each aliquot by determination of the purity index (see Assay).

Combine the purest fractions, absorb these on $0.1 N$ Type II Amberlite, pH 5.1, wash several times with $0.1 N$ sodium acetate buffer, pH 5.1, elute with $3 N$ sodium acetate, pH 7.0, and dialyze against $0.01 M$ sodium phosphate buffer, pH 7.0. If any insoluble material is seen, centrifuge, discard the residue, and collect the clear supernatant fluid. Store in the cold. For long-term storage, add sufficient finely powdered ammonium sulfate to bring the solution to approximately 50% saturation in ammonium sulfate.

Yields vary, depending on the age of the culture and the scale of operations. The procedure described above usually involves approximately 100 g. or more dry weight of cells from 4- to 5-day cultures, and yields approximately 700 to 800 mg. of cytochrome c_2 ($A_{280\ (oxid)}^{550\ red}$ = 1.2) per kilogram dry weight of cells. This represents a purification of 100-fold over that of the cytochrome c_2 in the initial extract and a recovery of approximately 70 to 80% of the total cytochrome extracted. By reworking tailings and residues on columns, up to 10% more protein may be recovered. Horio and Kamen,[10] using smaller quantities of starting material, have reported similar yields for crystallized pure cytochrome c_2.

Adsorb the salt-free RHP solution from the pH 5.65 chromatography (above) on a pH 5.75 column. Wash several times by passage of 0.1 M acetate buffer, pH 5.75, and then elute with 3 M acetate, pH 7.0. Dialyze the eluate against 0.01 M phosphate buffer, pH 7.0, and store in the cold. For long-term storage, it is best, as with cytochrome c_2, to add sufficient ammonium sulfate (fine powder) to bring the solution to approximately 50% saturation in ammonium sulfate.

Yields are similar to those found for cytochrome c_2. Thus, approximately 100-fold purification relative to the initial extract is obtained with a recovery of approximately 80%.

Comments. The chromatographic procedures described are not necessarily the best or simplest. Many variations are possible; for instance, Horio and Kamen[10] have used alumina, at pH values 7.0 and 8.0, and obtained similarly purified preparations in yields much like those reported here. We use Amberlite resin because it is easier to obtain in the United States and more economical for large-scale work. This may not be true elsewhere; those who prefer alumina will find details for its use in the literature cited.

Crystallization is simple to accomplish, once samples of the high purity described above have been obtained; Horio and Kamen[10] give detailed instructions for crystallization procedures. However, no significant change in purity is brought about by crystallization.

Properties

Cytochrome c_2 is a monoheme protein with molecular weight 12,000, on the basis of heme content, iron content, and sedimentation studies, all of which determinations are in good agreement.[10] The amino acid composition, as given by Coval *et al.*,[12] reveals the characteristic high content of the basic amino acid, lysine. Unlike the mammalian forms of cytochrome c, the bacterial preparation contains large amounts of the

[12] M. Coval, T. Horio, and M. D. Kamen, *Biochim. et Biophys. Acta* **51**, 246 (1961).

dicarboxylic residues, aspartic and glutamic acids, so that the isoelectric point[10] is at pH 6.4, whereas, as is well-known, that for mammalian cytochrome c is at pH 10.5. In this respect, cytochrome c_2 is like other bacterial C-type cytochromes, practically all of which are acidic proteins.[3] Another important difference between cytochrome c_2 and mammalian cytochrome c is that, like most bacterial C-type proteins, cytochrome c_2 does not react with oxygen in the presence of cytochrome oxidase.[5]

The RHP of *R. rubrum* is a diheme protein with molecular weight approximately 28,000.[10] It is similar in its spectrochemical properties to the *Chromatium* RHP[6]; that is, it resembles "open-type" hematin compounds, like myoglobin and peroxidase. However, it appears also to be closely related to C-type cytochromes because its heme groups are attached by covalent bonds, presumably thioether linkages as in cytochrome c (see article by K. G. Paul, Vol. II [133]). It is remarkable in that in its reduced form it reacts only with the simple ligand CO, and with NO both when oxidized and reduced.[13] Further information is available in a review by Kamen and Bartsch.[14]

TABLE I

MOLECULAR EXTINCTION COEFFICIENTS AT ABSORPTION MAXIMA
FOR RHP AND CYTOCHROME c_2 (*R. rubrum*, pH 7.0)[a]
$(A \times 10^3 \text{ l./mol. cm.})$

RHP[b]				Cytochrome c_2[b]			
Oxidized		Reduced		Oxidized		Reduced	
$m\mu$	A	$m\mu$	A	$m\mu$	A	$m\mu$	A
638	5.9	550	22	525	10.5	550	28.1
497	21.5	423	175	410	115	521	17.0
390	159			357	30	415	143
282	44			275	25	316	37
						272	34

[a] T. Horio and M. D. Kamen, *Biochim. et Biophys. Acta* **51**, 246 (1961).
[b] RHP contains two heme moieties per molecule (mol. wt. approximately 28,000), whereas cytochrome c_2 contains one per molecule (mol. wt. approximately 12,000).

In Table I, data are presented for molecular extinctions at pH 7.0. Data at other pH values are given by Horio and Kamen.[10]

[13] S. Taniguchi and M. D. Kamen, *Biochim. et Biophys. Acta* in press (1963).
[14] M. D. Kamen and R. G. Bartsch, *in* "Symposium on Hematin Enzymes" (J. E. Falk, R. Lemberg, and R. K. Morton, eds.), pp. 419 *et seq.*, IUB Symposium, Canberra, 1959. Pergamon Press, New York, 1961.

II. *Desulfovibrio desulfuricans* Cytochrome-552 (Cytochrome c_3)

(Procedure of Horio and Kamen.[10] Scale: 5 to 50 g. dry weight of cells)

Assay

The purity index is defined as the absorbancy at the α-peak of the reduced form, 552 mμ, to that of the oxidized form at 280 mμ; symbol: $A_{280\,(oxid)}^{552\,(red)}$. For the purest preparations (greater than 95% by spectroscopic and electrophoretic criteria), $A_{280\,(oxid)}^{550\,(red)} = 2.9$ to 3.1.

Procedure

Lyophilize fresh washed cells. Suspend 5 g. of the resultant lyophilized powder in 400 ml. of 0.1 M citrate buffer, pH 6.0, and mix thoroughly in a blendor. Adjust the homogenate to pH 6.0 with 0.1 M citric acid, and stir overnight in the cold room (4°). Perform all succeeding operations in the cold unless otherwise noted.

Centrifuge the mixture at 10,000 $\times$ g for 20 minutes. Decant the supernatant fluid, and extract the residue once more. Combine the extracts, and dialyze them overnight against distilled water. Pass the dialyzed solution through a column packed with pH 7 resin (Amberlite Type II). A small column (2.5 cm. in diameter $\times$ 3.0 cm. in height) suffices for this quantity of heme protein which can be seen to adsorb completely at the top of the column. Use a spatula to remove the portion of the column which contains the cytochrome, and wash this material by low-speed centrifugation three times with distilled H_2O. Repack the washed resin in a column (1 cm. in diameter), and elute with 2 M ammonium phosphate buffer, pH 7.0. The cytochrome can be obtained in a very small volume by this means; e.g., the quantity encountered in extraction of 5 g. of bacterial powder is easily concentrated in 1.5 ml.

Dilute this eluate with 0.2 M ammonium phosphate buffer, pH 7.0, until the absorbancy of the reduced form (add dithionite for assay), A_{552}^{red}, is approximately 1.0. Add sufficient ammonium sulfate (finely powdered) to give approximately 95% saturation (about 71 g. per 100 ml.). Allow to stand at room temperature for 3 hours. The solution may be slightly turbid. It can be clarified by high-speed centrifugation (10,000 $\times$ g) for 30 minutes. The brown precipitate is set aside for possible re-examination. If a thin film of white material (probably lipid) appears, remove it by careful manipulation with a flat spoon or spatula. Filtration is a more positive procedure (see above).

Dialyze the clarified solution overnight against distilled H_2O and then against 10^{-5} M ammonium phosphate buffer, pH 7, for 12 hours. It is essential to introduce this low-salt solution because the cytochrome in a salt-free state has a marked tendency to adsorb completely on the cellophane dialysis tubing. Pass the dialyzed cytochrome solution through the pH 7 resin column (1 cm. in diameter $\times$ 15 cm. in height), and wash with buffer (10^{-5} M).

Elute with 2 M ammonium phosphate buffer, pH 7.0, so that the cytochrome is obtained in a very concentrated solution. The solubility of cytochrome c_3 in ammonium sulfate is quite high; solutions with A_{552}^{red} < 15 fail to develop turbidity even when saturated with ammonium sulfate. It also appears that the amount of the protein required to produce turbidity in the presence of ammonium sulfate is greater at low temperature than at room temperature.

The cytochrome c_3 so obtained can be stored in saturated ammonium sulfate, or subjected to crystallization, by following the procedure of Horio and Kamen.[10] Thus, the final eluate (pH 7 to 8, as attained by addition of ammonia water at room temperature) is carefully supplemented with powdered ammonium sulfate until a slight turbidity develops (approximately 80 to 100% saturation, depending on the protein concentration). Centrifuge immediately at 10,000 $\times$ g for 30 minutes (3° to 5°). Decant the supernatant fluid, treat it with ammonium sulfate until a heavy turbidity develops, and let stand in the refrigerator. Small needle-like crystals form accompanied by appearance of flat plates after several hours.

Yields of cytochrome c_3 are high. The initial eluate from 5 g. dry weight of bacterial powder usually contains approximately 20 mg. of the protein; most of this (approximately 90%) can be recovered in the solution, just prior to chromatography, with a purity index approximately 2.5 to 2.7 (approximately 95% pure). However, only approximately 30% is in the "native" form. Chromatography separates the native form from the rest to give almost quantitative recovery of protein with purity index about 2.9 or greater. Crystallization effects very little further purification.

Comments. Cytochrome c_3 can be obtained essentially as a single homogeneous compound when the starting material is fresh. However, stored powders which have stood at varying temperatures appear to offer opportunities for degradation or alteration of the cytochrome so that a variety of fractions is obtained. Thus, Horio and Kamen[10] found that on careful elution at least five fractions resulted. The major fraction was the native protein, and the four other fractions represented the protein in various states of alteration and denaturation. Hence, it is wise to monitor the preparation, as described by Horio and Kamen.[10] They begin with 0.12 M ammonium phosphate buffer, pH 7.0 (flow rates approximately

10 ml./hour), and collect the cytochrome in 5-ml. fractions. All but a small amount of material absorbed at the top of the column can be recovered in this way.

Properties

Cytochrome c_3 was discovered by Postgate[7,8] and independently by Ishimoto et al.[9] It is a diheme protein with molecular weight of 12,000, which appears to have its prosthetic groups bound covalently in the same manner as in mammalian cytochrome.[7,8] The heme groups present seem to be identical with that of cytochrome c, in that the same mesoheme is released by fission in dilute acetic acid in the presence of silver salts.[8] The mid-point potential $(E_m \equiv E'_0)$ is at $\sim$ -205 mv., so that this cytochrome has its central iron atoms poised some 500 mv. below the potential of the iron in mammalian cytochrome c. In its amino acid composition[12] it is much like mammalian cytochrome c; i.e., it exhibits the usual high lysine content, together with a relatively low content of dicarboxylic amino acids, so that its isoelectric point is approximately 10.5. Thus, it is a basic protein, unlike most other bacterial C-type cytochromes.

Cytochrome c_3 is unusual in that it exhibits a very low absorbancy in the ultraviolet. Thus, the purity index, $A_{280 \text{(oxid)}}^{552 \text{(red)}}$, is 3.0, as compared with values of approximately 1.2 for cytochrome c_2. Of course, one expects a doubled absorption of the α-band maximum (552 mμ) compared to the protein absorption maximum (280 mμ), because there are two heme groups present instead of just one. However, the increase is by a factor of 3.

The molecular extinction coefficients (ϵ_M) at the Soret maxima (oxidized, 409 mμ; reduced, 418 mμ) and at the other characteristic maxima in the visible region (α-band of reduced form, 552 mμ, and β-band of reduced form, 552 mμ) are essentially double those for the corresponding absorption maxima of mammalian cytochrome c. Thus, the molar absorbancy for cytochrome c at the α-peak (550 mμ) is 28.0×10^3 l./mol. cm., and that for cytochrome c_3 at its α-peak (552 mμ) is 56.1×10^3 l./ mol. cm. A discussion of physicochemical properties of cytochrome c_3 will be found in the literature cited.[7,8,10]

III. *Chromatium* Heme Proteins

(Method of R. G. Bartsch. Scale: 1 to 500 g. dry weight of cells)

Assay

The heme proteins which can be extracted are a C-type cytochrome, *Chromatium* cytochrome 552, and an RHP type, *Chromatium* RHP.

Two wavelengths (635 mμ and 524 mμ) are used to monitor approximate concentrations of both heme proteins in crude extracts. These wavelengths are at, or near, the maxima for non-overlapping absorption bands in the reduced-minus-oxidized difference spectra. The two proteins are autoxidizable, appear in the oxidized form in dialyzed extracts, and remain oxidized throughout purification.

Prepare an aliquot, by dilution of a sample of crude extract with 0.05 M phosphate buffer, pH 7.0, and add a few crystals of sodium dithionite. Measure the reduced-minus-oxidized absorbancies at 635 mμ and 524 mμ; for *Chromatium* RHP, the molar absorbancy, $A_{(red-oxid)}{}^{635}$, is approximately -4.3×10^3 l./mol. cm., and for *Chromatium* cytochrome 552, $A_{(red-oxid)}{}^{524}$ is approximately 15.5×10^3 l./mol. cm. These measurements can be performed in open cuvettes, provided an excess of dithionite is present.

After preliminary purification by chromatography, it is convenient to use the Soret absorption maxima. Molar absorbancies for the various

TABLE II
MOLAR ABSORBANCIES OF *Chromatium* HEME PROTEINS[a]
($A \times 10^3$ l./ml. cm.)

State of cytochrome	Cytochrome c		RHP	
	Wavelength of absorption peak, mμ	A	Wavelength of absorption peak, mμ	A
Oxidized	278	175	280	63
	410	320	400	192
	525	26	495	24.5
			635	7.2
Reduced	416	364	426	210
	523	41.6	547	22.7
	552	61.1	565	20.9
Difference spectrum (reduced minus oxidized)	406	−10.7	398	−115
	422	165	429	158
	524	15.5	495	−14.4
	553	44.6	565	5.4
			635	−4.3
Carbon monoxide complex (reduced)	414	500	418	583
	533	38.8	535	28
	555	50.5	563	23.8
Difference spectrum (reduced carbon monoxide minus reduced)	414	165	418	404
	535	10.7	535	7.2
	565	17.5	567	3.6

[a] The spectra were measured in 0.05 M phosphate buffer, pH 6.8, $d = 1.0$ cm., with the Cary Model 14 spectrophotometer.

absorption peaks are listed in Table II. The ratio of absorbancies at 275 to 280 mμ to that at the Soret maximum (400 mμ for RHP, 410 mμ for cytochrome 552), symbols, A_{400}^{280} and A_{410}^{280}, respectively, is used as a purity index.

Procedures

Step 1. Suspend 1 kg. wet weight of fresh, washed cells in 1.5 l. of phosphate buffer, pH 7.0. Sonicate 75-ml. portions (cool the transducer vessel with ice water). Alternatively, as with *R. rubrum* heme proteins, lyophilize and extract. However, repeated trials appear to indicate that recoveries, especially of the *Chromatium* RHP, are doubled by use of sonication. Centrifuge for 10 minutes at 25,000 $\times$ g to remove cell debris. Resuspend, and repeat the sonication procedure. Centrifuge, and combine extracts. Add 30 g. of finely powdered ammonium sulfate to each 100 ml., let stand in the cold overnight, and then centrifuge at 25,000 $\times$ g for 30 minutes to recover the extract and remove the voluminous precipitate (mostly small particles).

Resuspend the precipitate in approximately 2 vol. of ammonium sulfate (30 g./per 100 ml). Let stand for several hours in the cold, and centrifuge. Combine extracts, and add 30 g. of finely powdered ammonium sulfate for each 100 ml. Let stand overnight in the cold, and centrifuge for 10 minutes at 25,000 $\times$ g. Dissolve the precipitate in a minimal volume of distilled H$_2$O, to which Tris (free base) is added so that the pH is 7.0. Then, dialyze against several changes of distilled H$_2$O and finally against 0.02 M Tris buffer. The heme proteins will remain in solution. Remove precipitated material by centrifugation at 25,000 $\times$ g for 15 minutes.

Step 2. Chromatography is performed in two steps. In the first, pack a 4.0 $\times$ 40-cm. column with coarse-grade DEAE-cellulose (Standard Selectacel), equilibrated with 0.02 M Tris buffer, pH 8.0. Adjust the pH of the crude extract from step 1 to pH 8.0 (use free Tris base), and apply to the column. Elute with 0.02 M Tris buffer, pH 8.0, to which NaCl is added stepwise to give elution at increasing values of ionic strength.

A typical elution schedule is as follows: 2 l. of buffer; 2 l. of buffer 0.02 M NaCl; 4 l. of buffer 0.04 M NaCl, which elutes a greenish-brown protein;[15] 4 l. of buffer 0.08 M NaCl, which elutes *Chromatium* RHP; 2 l. of buffer 0.1 M NaCl; 4 l. of buffer 0.12 M NaCl, which elutes the main portion of cytochrome 552; 3 l. of buffer 0.2 M NaCl, which removes residual heme protein (mainly altered cytochrome 552). The cytochrome 552 occurs in two bands, which overlap. One is orange-colored and contains flavin; the other is brown and is free of flavin.

[15] R. G. Bartsch, *Federation Proc.* **21**, 47 (1962).

At each step, continue passage of buffer until the protein concentration in the effluent becomes negligible. The relation between salt and buffer concentrations required for elution of different heme protein fractions may vary with different batches of adsorbent and different bacterial extracts. Hence, it is advisable to make a pilot run on a small column to establish conditions for the preparative chromatography. Collect 20- to 50-ml. fractions from the preparative column (fraction collector). Measure the purity index of each colored fraction, and combine fractions of similar purity.

To concentrate the various fractions, dilute with 2 vol. of distilled H_2O, and apply to a column 2×20 cm. to 2×30 cm., packed with the coarse-grade DEAE-cellulose, equilibrated with $0.02 M$ Tris buffer, pH 8.0. The column should be sufficiently large so that it is approximately half-saturated with the heme protein applied. Elute with buffer $0.5 M$ NaCl to displace the protein as a compact zone. The leading two-thirds of the zone (30 to 40 ml. of eluent) is usually purer than the initial solution; hence, collect 5-ml. fractions, measure the purity index of each, and pool comparable fractions. Desalt either by dialysis or by treatment with a Sephadex G-25 (medium) column (see E. A. Peterson and H. A. Sober, Vol. V [1]).

Step 3. The second chromatographic procedure utilizes a finer grade of absorbent type 40 Selectacel DEAE-cellulose equilibrated in a $2 \times$ 30-cm. column with $0.02 M$ Tris buffer, pH 8.0. Apply the RHP fraction, and elute stepwise with 1- to 2-l. portions of buffer, which contain 0.05, 0.06, 0.07, 0.08, and $0.1 M$ NaCl, successively. The purest fraction of RHP appears in the eluate between $0.07 M$ and $0.08 M$ NaCl. At lower salt concentrations, the RHP migrates slowly, while contaminant protein is eluted more rapidly. Once the colored zone appears in the effluent, increase the salt concentration immediately by $0.02 M$ to elute the bulk of RHP in 1 to 2 l. Collect 20- to 30-ml. fractions, assay purities, pool, and concentrate as above. The best fraction should be essentially homogeneous, with purity index 0.3 to 0.35.

Similarly, rechromatograph the cytochrome 552. Elute the contaminant protein with buffer 0.08 to $0.12 M$ NaCl. Elute the cytochrome with buffer 0.12 to $0.15 M$ NaCl; collect 20 to 30-ml. fractions, assay purities, pool, and concentrate. The best fractions show purity indices in the range 0.4 to 0.5.

Step 4. Cytochrome 552 which has lost flavin (FMN) is not completely separated from the flavin-containing form, which is presumed to be native. Further separation can be effected by starch-bed electrophoresis (see M. Bier, Vol. V [3]). Use $0.02 M$ Tris buffer, pH 8.0. The orange component (flavin-heme form) migrates more rapidly toward the positive electrode. Elute each zone separately with buffer, dialyze against distilled H_2O, and lyophilize. Flavin removal from cytochrome 552 is progressive and can be accelerated by incubation of the protein at alkaline pH (greater than 9.0).

In Table III, yields and purities of heme protein fractions, as obtained

TABLE III
SUMMARY OF PURIFICATION PROCEDURE; *Chromatium* HEME PROTEINS

	Yield	
Fraction	RHP	Cytochrome 552
Step 2. Dialyzed NH₄SO₄ precipitate	50–100 micromoles	10–20 micromoles
Step 3. First DEAE-cellulose chromatogram	35–75 micromole purity index ≅0.8	7–15 micromole purity index ≅1.0
Step 4. Second DEAE-cellulose chromatogram	10–25 micromoles, purity index ≅0.30–0.33	2–5 micromoles, purity index ≅0.5–0.6

from 1 kg. of packed wet cells, are shown. Both heme proteins have been stored as desalted solutions at −20° for several months without appreciable deterioration.

Properties

Some physicochemical properties are shown in Table IV. The values given are preliminary.

TABLE IV
SOME PROPERTIES OF *Chromatium* HEME PROTEINS

Protein	Molecular weight	pI	E'_0 (pH 7)	Purity index	Heme	FMN
RHP	36,000	5.5	−0.005 volt	0.30	2	—
Cytochrome 552 I	97,000	5.4	+0.01 volt	0.59	3	1
Cytochrome 552 II	—	—	—	0.43	3	0

[56] Cytochrome Photoöxidase

$$2Fe^{++} \text{ (cytochrome c)} + \tfrac{1}{2}O_2 + 2H^+ \rightarrow 2Fe^{+++} \text{ (cytochrome c)} + H_2O$$

By BIRGIT VENNESLAND

Cytochrome photoöxidase is an enzyme complex which catalyzes the oxidation of ferrocytochrome c to ferricytochrome c by O_2 in the light. It is a characteristic component of chloroplasts, and protein-bound chlorophyll is one of its components. In order to demonstrate the photoöxidase, it is necessary to destroy or inhibit the photoreducing capacity of the chloroplasts. This is done by disintegration with digitonin. The enzyme can be demonstrated in a digitonin-treated chloroplast suspension, or in the clear green "solution" obtained by triturating the chloroplasts with aqueous digitonin and removing the solids by centrifugation. The reaction is followed by spectrophotometric measurement at 550 mμ of the decrease in light absorbance which accompanies the oxidation of added reduced cytochrome c. Cyanide is added to inhibit the cytochrome oxidase which is active in the dark.[1]

The cytochrome photoöxidase can be separated into two components, Factor 1 and Factor 2, both of which are necessary for activity.[2] Factor 1 contains protein–chlorophyll complex and is insoluble when the digitonin concentration is low. Factor 2 is a water-soluble, heat-labile component, presumably protein in nature, which has been extensively purified. Factor 2 is assayed by determination of its activating effect on a preparation of Factor 1. Factor 1 may be assayed by determination of its activity in the presence of an excess of Factor 2. Factor 1 has not been extensively fractionated but may be obtained from spinach leaves by a standardized procedure which gives preparations of approximately equivalent activity. The amount of Factor 1 may be conveniently expressed in terms of its chlorophyll content. The assay procedure described below is designed to follow the purification of Factor 2 with a given preparation of Factor 1 as a standard.

Assay Method

Principle. Suitable amounts of reduced cytochrome c together with the two components of cytochrome photoöxidase are illuminated under conditions which inhibit the dark cytochrome oxidase activity of the Factor 1 preparation. The rate of oxidation of reduced cytochrome c is followed by measurement of the decrease in light absorbance at 550 mμ.

[1] R. H. Nieman and B. Vennesland, *Plant Physiol.* **34**, 255 (1959).
[2] R. H. Nieman, H. Nakamura, and B. Vennesland, *Plant Physiol.* **34**, 262 (1959).

Reagents

Cytochrome c,[3] reduced, 0.4 mM.

$(NH_4)_2SO_4$, 4 M.

Enzyme, Factor 1 and Factor 2, prepared as described below.

Procedure. An assay mixture is made up with 0.1 ml. of the cytochrome c and 1 ml. of the ammonium sulfate solution, the amount of Factor 2 to be assayed, and water, to make a final volume of 3 ml. The reaction mixture is allowed to equilibrate to 20°. A solution of Factor 1 containing about 15 to 30 μg. of chlorophyll[4] is added in the dark. The optical density at 550 mμ is recorded, and the light is switched on. The rate of decrease of the optical density is recorded. The procedure employed to illuminate the reaction mixture may vary with circumstances. In the author's laboratory, measurements are made with a Beckman DU spectrophotometer in cells of 1-cm. light path, against water as a blank. The samples are illuminated by a beam of white light focused from above into the second position of the cuvette carrier of the Beckman spectrophotometer. The intensity of the light at the surface of the liquid is about 40 foot-candles. To take a reading, the cell is pulled out of the light beam into the position in front of the photocell. Units are defined in the table. It should be understood that these units will vary in magnitude with the activity of Factor 1 used. The problem of cross standardization has not been satisfactorily solved.

Properties of the Reaction.[1, 2, 5] The conditions of illumination described for the assay are not optimal. For light saturation, about 500 foot-candles of white light are required. The effective wavelengths are those absorbed by the chlorophyll. The rate of photoöxidation of reduced cytochrome decreases as the substrate is oxidized, but not as much as for a first-order reaction. It is important for accurate assay that the reaction should be no more than about half complete in the time interval used for rate measurement.

At saturating light intensity, the log of the rate is linearly proportional to the reciprocal of the absolute temperature below about 40°. The experimental activation energy (ΔE_{exp}) is 4.3 kcal./mole.

The reaction rate is strongly accelerated by high salt concentrations, divalent and trivalent anions being more effective than monovalent

[3] Commercial cytochrome c preparations may be employed. See Vol. II [130] and [133] for methods for preparing and assaying reduced cytochrome c.
[4] The desired amount of factor 1 is determined empirically and will vary with the preparation.
[5] N. I. Bishop, H. Nakamura, J. Blatt, and B. Vennesland, *Plant Physiol.* 34, 551 (1959).

anions. At $1.3\,M$, $(NH_4)_2SO_4$ gives an optimal effect. At this sulfate concentration, the rate is constant over a pH range of 5.2 to 8.7. If the reaction is carried out in $0.67\,M$ phosphate buffer, the rate at pH 9 is more than twice that at pH 5, an effect probably related to the increase in the concentration of divalent anion at the higher pH.

Definition of Unit and Specific Activity. One unit of Factor 2 is defined as that amount which causes an increase of 0.001 per minute in the rate of decrease of optical density, as measured in the standard ammonium sulfate assay system, with an amount of Factor 1 containing 25 μg. of chlorophyll. The specific activity is defined as the number of units per microgram of protein, determined according to Lowry *et al.*[6] The results in the table are only representative. The activity of Factor 1 is variable, and the standard assay system does not give optimal rates.

Purification Procedure

Preparation of Digitonin Extracts of Chloroplasts. A medium of $0.35\,M$ NaCl and $10^{-4}\,M$ disodium ethylenediaminetetraacetate (Versene) is used to grind an equal weight of spinach leaves to a fine suspension at $0°$. Large particles are removed by straining through cloth followed by very brief centrifugation. The intact chloroplasts are collected from the suspension by centrifugation at $2500 \times g$ for 10 minutes. After being washed in the NaCl–Versene medium, the chloroplasts are extracted with 0.05 ml. of 1% aqueous digitonin per gram of fresh leaf. The suspension is stirred for about 20 minutes while the solution is allowed to come to room temperature, and then it is centrifuged for 30 minutes at 25,000 $\times g$ and $0°$. The clear green supernatant contains both Factors 1 and 2. Re-extraction of the green sediment with 1% aqueous digitonin gives a larger yield of Factor 1.

Separation of Factor 1. The combined chloroplast extracts are cooled in an ice bath, and ethanol, precooled to $-15°$, is added dropwise, with stirring, until 0.3 ml. of absolute ethanol has been added per milliliter of extract. After 15 minutes, the mixture is centrifuged for 20 minutes at 25,000 $\times g$ and $0°$. The bulk of the chlorophyll is sedimented by this procedure. The supernatant is decanted. The sediment is dissolved in 1% digitonin (10 ml./kg. fresh leaf). This preparation is Factor 1. Its chlorophyll content ranges from 1 to 3 mg. of chlorophyll per milliliter.[7, 8]

Separation and Purification of Factor 2: Step 1. To the supernatant from the ethanol precipitation is added with stirring an equal volume of acetone, precooled to $-15°$. After 15 minutes, the precipitate is removed

[6] O. H. Lowry, N. J. Rosebrough, and A. L. Farr, *J. Biol. Chem.* **193**, 265 (1951).
[7] For the assay of chlorophyll, see Vol. IV [15].
[8] D. I. Arnon, *Plant Physiol.* **24**, 1 (1949).

by centrifugation for 20 minutes at 2500 $\times$ g and 0°. An additional 2 vol. of acetone are added in the cold, and after 30 minutes the precipitate is collected by centrifugation. The supernatant is discarded, and the well-drained precipitate is dissolved in 0.01 M potassium phosphate buffer of pH 7.0 (10 ml./kg. fresh spinach).

Step 2. For further purification, the above solution is diluted with 40 ml. of H_2O, and 23.5 g. of $(NH_4)_2SO_4$ are added with stirring in the cold. The precipitate is removed by centrifugation, and an additional 9 g. of $(NH_4)_2SO_4$ are added. After 2 hours, the precipitate is recovered by centrifugation. This precipitate has about the same specific activity and about one-third of the total activity of the preparation designated in step 1. The supernatant, which contains most of the activity, is close to saturation with $(NH_4)_2SO_4$ at 4°. Acetic acid is added carefully, with rapid stirring in the cold, to bring the pH to 3.6. Precipitation begins at pH 3.9. Care must be taken to add no more acid than the amount just sufficient to give a maximum amount of precipitate. After standing at 0° for 2 hours, the precipitate, which is almost white in color, is recovered by centrifugation. It is dissolved in 5 ml. of 0.01 M phosphate buffer of pH 7.0 and dialyzed against the same buffer for 6 hours at 0°.

Step 3. Acetone, cooled to $-15°$, is added dropwise with stirring until 3.5 ml. has been added for each milliliter of solution from step 2. The precipitate represents the highest purification achieved. The supernatant from this precipitation still contains a considerable amount of Factor 2,

FRACTIONATION OF FACTOR 2

Step	Volume of solution, ml.	Units	Specific activity, units/μg. protein	Yield, %
Whole leaf[a]	1000	17,000	0.0005	100
1	10	3,500	0.21	21
2	5	1,150	0.58	7
3	As desired	520	1.00	3

[a] The values for whole leaf are crude approximations estimated from the total protein present and the units of activity which can be recovered from the various fractions. Values given are calculated for 1 kg. of spinach. About two-thirds of the total activity is in the cytoplasm and only one-third in the chloroplasts, but the Factor 2 activity of the cytoplasm has not been successfully purified to the levels achieved with the material from the chloroplasts.

which may be recovered by further addition of a volume of acetone equal in amount to that used previously. The specific activity of this material is about the same as that obtained in step 2. A summary of the results of the fractionation is given in the table.

Properties of Factor 1[2]

Factor 1 is a chlorophyll–lipoprotein complex containing nucleic acid and phospholipid. There are about 6 mg. of protein per milligram of chlorophyll. The complex contains chlorophyllase, which is not active unless the chlorophyll is partially solubilized by addition of organic solvents. Under these circumstances, Factor 1 is inactivated. Solutions of chlorophyll or of chlorophyllide have no Factor 1 activity under the conditions of the assay for cytochrome photoöxidase.

At neutral pH about a week's standing at room temperature is required for inactivation of Factor 1, but at 50° inactivation is complete in 5 minutes. Inactivation is also rapid at room temperature above pH 11.5 and below pH 4.

A considerable amount of dark cytochrome oxidase activity is present in Factor 1. Unlike the photoöxidase, this dark oxidase is unaffected by Factor 2 and is almost completely inhibited by $10^{-3} M$ cyanide and by high concentrations of ammonium sulfate.

Properties of Factor 2[2, 5]

Factor 2 is inactivated completely in 10 minutes at 90° and is unstable at room temperature above pH 12, but half of the activity is retained after 30 minutes of exposure at pH 0 at room temperature in the absence of $(NH_4)_2SO_4$. Solutions of Factor 2 show protein absorption in the ultraviolet. No absorption bands in the visible light range have been associated with the activity. It is stable to dialysis and sediments on ultracentrifugation with an S_{20} of about 0.7 S.

[57] Lipid-Dependent DPNH–Cytochrome c Reductase from Mammalian Skeletal and Heart Muscle*

$$DPNH + 2 \text{ ferricytochrome } c \rightarrow DPN^+ + 2 \text{ ferrocytochrome } c + H^+$$

By ALVIN NASON and FRANK D. VASINGTON

Assay Method

Principle. Enzymatic activity is best measured spectrophotometrically by determining the rate of reduction of cytochrome c as indicated

* Contribution No. 389 of The McCollum-Pratt Institute, The Johns Hopkins University, Baltimore, Maryland. The information in this article represents a portion of a study supported in part by a research grant (No. 2332) from The National Institutes of Health, U. S. Public Health Service.

by the increase in absorption at 550 mμ. It is based on the original observation by Keilin[1] and is similar to that described by Haas[2] for measuring TPNH–cytochrome c reductase from yeast and by Mahler,[3] Brodie,[4] and Lehman and Nason[5] for measuring DPNH–cytochrome c reductase from mammalian and bacterial cells. Cyanide is used in the reaction mixture to inhibit any cytochrome oxidase activity that may be present. The enzyme may also be assayed for by the less-sensitive procedure of measuring the rate of oxidation of DPNH as indicated by the decrease in absorption at 340 mμ.

Reagents

0.1 M phosphate buffer, pH 7.5.
0.05 M KCN, prepared as a fresh solution every 3 or 4 days.
2% aqueous horse-heart ferricytochrome c.[6]
0.0012 M DPNH.[6]

Procedure. The reaction mixture is routinely composed of the following: 0.05 ml. of suitably diluted enzyme, 0.1 ml. of KCN, 0.1 ml. of cytochrome c solution, and 0.70 ml. of phosphate buffer in a 1.5-ml. Beckman Corex or silica cuvette (1.0-cm. light path). At zero time 0.05 ml. of DPNH solution is added, and the increase in optical density at 550 mμ is measured at 30-second intervals for 3 minutes in a Beckman DU spectrophotometer.

Definition of Unit and Specific Activity. One unit of enzyme activity is that amount which will cause a linear increase in log I_0/I of 0.001 in the interval from 1 to 3 minutes. A few preliminary assays are made with different concentrations of each enzyme preparation in order to be certain that the rate of cytochrome reduction is directly proportional to the protein concentration under these conditions of measurement. Specific activity is defined as units of enzymatic activity per milligram of protein. Protein is determined by the method of Lowry *et al.*[7]

Application of Assay Method to Crude Tissue Preparations. The procedure is applicable to the determination of cytochrome c reductase activity in crude cell-free preparations or homogenates provided that the

[1] D. Keilin, *Proc. Roy. Soc.* **B98**, 312 (1925).
[2] See Vol. II [122].
[3] See Vol. II [120].
[4] See Vol. II [121].
[5] I. R. Lehman and A. Nason, *J. Biol. Chem.* **222**, 497 (1956).
[6] Both horse-heart cytochrome c and DPNH, of 60% and 90% purity, respectively, are usually used and can be obtained commercially. For the preparation of these substances see Vol. II [133] and Vol. III [127].
[7] O. H. Lowry, N. J. Rosebrough, A. L. Farr, and R. J. Randall, *J. Biol. Chem.* **193**, 265 (1951); also see Vol. III [73].

turbidity of the enzyme is not too great and that no appreciable settling out of the enzyme occurs during the assay. Endogenous reductase activity (in the absence of added DPNH) should also be determined and corrected for. KCN is used in the reaction mixture, as indicated above, to inhibit cytochrome oxidase activity which is present at all stages of purification.

Purification Procedure

All steps in the purification procedure, as summarized in the accompanying table for rat skeletal muscle, were carried out at 0° to 4°. The method of purification with indicated minor modifications is essentially similar for both rat skeletal muscle[5] and bovine heart muscle.[8]

Step 1. Preparation of Crude Extract. Fresh rat skeletal muscle obtained from the hind legs of adult male Wistar rats, or frozen bovine heart muscle from which most of the exterior fat connective tissues, interior valves, and gristle have been removed, is finely minced in a meat grinder and then homogenized with ten times its weight of 0.1 M phosphate buffer, pH 7.5, in a TenBroeck tissue grinder. The resulting homogenate is centrifuged at approximately 3000 $\times$ g for 15 minutes to yield a turbid active supernatant solution (fraction 1).

Step 2. Preparation of Dialyzed Supernatant Solution. Dialysis of fraction 1 for 2 hours against 10 vol. of 0.01 M phosphate buffer, pH 7.5, results in the formation of a heavy gelatinous precipitate in the case of the rat skeletal muscle preparation, and a considerably smaller precipitate in the case of the heart muscle. The precipitate is removed in both cases by centrifugation for 15 minutes at 3000 $\times$ g to give an active supernatant solution (fraction 2).

Step 3. Preparation of the Pellet. Fraction 2 is centrifuged for 30 minutes at 140,000 $\times$ g in the Spinco preparative ultracentrifuge. The resulting pellet, which contains most of the total activity of fraction 2, is then resuspended in one-tenth its original volume of 0.1 M phosphate buffer, pH 7.5 (fraction 3).

Step 4. Preparation of Digitonin-Solubilized Enzyme. Fraction 3 is ground in a TenBroeck homogenizer with an equal volume of 4% aqueous digitonin suspension in the case of the skeletal muscle preparation, or with digitonin powder to give a final concentration of 3% in the case of the heart muscle preparation, followed by centrifugation for 30 minutes at 140,000 $\times$ g. The resulting clear supernatant solution (fraction 4) usually contains 50 to 100% of the total activity of the resuspended pellet. In some cases the total activity is significantly greater than 100%, ranging as high as 300%.

[8] F. D. Vasington and A. Nason, in preparation.

Step 5. Adsorption of Calcium Phosphate Gel. To fraction 4 is added a one-third to one-quarter volume of calcium phosphate gel,[9] aged 9 months or longer (21 mg. dry weight per milliliter), and the mixture is stirred intermittently for 10 minutes.

Step 6. Elution of the Enzyme from the Calcium Phosphate Gel. The above calcium phosphate gel is collected by centrifugation at 3000 $\times$ g for 5 minutes and eluted by stirring for 5 minutes with a quantity of 0.1 M pyrophosphate buffer, pH 7.5, equal to one-half the volume of fraction 4 originally treated with gel in the previous step. Centrifugation of the mixture for 5 minutes at about 3000 $\times$ g yields a clear supernatant solution, designated as the first eluate, which contains DPNH–cytochrome c reductase activity. Four or five more elutions are performed in the same manner, with the last three or four eluates usually possessing a high specific activity as indicated in the table. Further elution often results in little or no activity. Usually the third, fourth, and fifth eluates which exhibit comparably high specific activities are combined (fraction 5) and mixed with crystalline bovine serum albumin (final concentration 0.2%) to enhance stability. Fraction 5 is dialyzed for 1 hour against 0.1 M phosphate buffer, pH 7.5, to remove some of the digitonin.

PURIFICATION OF DPN–CYTOCHROME c REDUCTASE FROM RAT SKELETAL MUSCLE

Fraction	Total units[a]	Total, protein, mg.	Specific activity, units/mg. protein	Recovery, %
1. Crude extract	228,000	820	277	100
2. Dialyzed supernatant solution	134,000	294	455	59
3. Resuspended high-speed pellet	84,000	24	3,500	37
4. Digitonin-treated	48,000	17	2,820	21
First Ca$_3$(PO$_4$)$_2$ gel eluate	5,040	3.8	1,330	
Second Ca$_3$(PO$_4$)$_2$ gel eluate	1,920	2.1	914	
Third Ca$_3$(PO$_4$)$_2$ gel eluate	17,400	1.15	15,100	
Fourth Ca$_3$(PO$_4$)$_2$ gel eluate	4,050	0.25	16,200	
Fifth Ca$_3$(PO$_4$)$_2$ gel eluate	3,900	0.12	32,500	
5. Combined third, fourth, fifth gel eluates	25,350	1.52	16,700	11

[a] $\Delta E_{550\ m\mu} \times 10^3$ per 2 minutes.

The range of purification obtained by this procedure varies from 50- to 300-fold with yields ranging from 10 to 30% for the skeletal muscle system and from 1 to 5% for the heart muscle system. The specific activity of the final heart muscle enzyme preparation (not shown) is usually ten to twenty times as great as that of the corresponding rat

[9] For the preparation of calcium phosphate gel, see Vol. I [11].

skeletal muscle enzyme. This is attributed largely to the fact that the specific activity of the crude extract of heart muscle initially is considerably higher than that of the rat skeletal muscle.

In terms of turnover number, the most highly purified preparations of rat skeletal muscle and bovine heart muscle catalyze the reduction of 85 moles and 1600 moles, respectively, of cytochrome c per mole of protein per minute, if a molecular weight of 100,000 is assumed for the enzyme.

A summary of the purification procedure is given in the table.

Properties

Stability. The resuspended pellet (fraction 3) of rat skeletal muscle is quite stable, retaining over 70% of its activity on storage at −15° for 1 month, whereas the final purified preparation (fraction 5) loses 65% of its activity within 4 days at −15°. Fraction 3 also stores well at 0° to 4°, losing little or none of its activity after a week, whereas the more purified fractions are considerably less stable. The corresponding heart muscle preparations exhibit somewhat similar stability characteristics. The enzymes at all stages of purity appear to withstand dialysis against $0.01 M$ phosphate buffer, pH 7.5, for periods of 1 to 2 hours. Exposure of fractions 3 and 5 of the rat skeletal muscle preparation for 10 minutes at 40° causes a 60 to 75% loss in activity.

Other Enzymes Present. Cytochrome c oxidase and DPNH oxidase activities are present at all stages of purification of the DPNH–cytochrome c reductase, although the levels of these activities vary markedly. The succinate–cytochrome c reductase activity of the skeletal muscle preparation is usually present in fractions 1 through 3 but is not detectable in the gel eluate fractions. In the heart muscle system succinate–cytochrome c reductase usually extends into the second gel eluate. Other electron acceptors such as 2,6-dichloroindophenol and ferricyanide may be used in place of cytochrome c.

pH Optimum. The rat skeletal enzyme displays a broad maximum activity at pH 7.5 in phosphate buffer. The activity is lower in Tris buffer despite recrystallization and extraction with 8-hydroxyquinoline. Pyrophosphate buffer $(0.01 M)$ causes an almost complete inhibition at pH levels ranging from 6.5 to 9.0. Although the heart muscle enzyme also displays a broad pH optimum in the region of 7.5, the enzyme is hardly inhibited by Tris buffer and is inhibited only about 35% by pyrophosphate buffer.

Specificity. There is a marked specificity for DPNH as the electron donor, with TPNH showing only a small effect. The dissociation constant (K_m) for the rat skeletal enzyme–DPNH complex is $2.1 \times 10^{-6} M$, and

414 RESPIRATORY ENZYMES [57]

that for the enzyme–cytochrome c complex is $1.5 \times 10^{-5} M$. The dissociation constant for the heart muscle enzyme–DPNH complex is $3.5 \times 10^{-6} M$, but that for the cytochrome c complex is difficult to assess accurately because of the rapid reduction of low concentrations of cytochrome c.

Components of the Enzyme Complex. Examination of the concentrated resuspended skeletal muscle pellet preparation (fraction 3) in the sensitive split-beam recording spectrophotometer or in a Cary recording spectrophotometer (Model 14) shows the presence of cytochromes $c + c_1$, b, and a ($+$cytochrome a_3). Similar components were observed in the concentrated fourth gel eluate of the heart muscle enzyme. Although there is some indication of a possible flavin component, this aspect has not yet been sufficiently studied to warrant a conclusive decision. Thus far there is no evidence of a flavin requirement by the enzyme. Preincubation of the enzyme with FAD, FMN, or boiled pig heart extract fails to stimulate enzyme activity. Precipitation of the enzyme with ammonium sulfate at acid pH as well as prolonged dialysis under various conditions does not result in a subsequent stimulation by added flavin.

α-Tocopherol is found to be present as a component of the bovine heart muscle preparation in variable quantities in all stages of purification, with the more purified fractions usually, but not always, showing higher concentrations.[10] The presence of tocopherol in a highly purified but differently prepared particulate DPN–cytochrome c reductase from bovine heart muscle has also been reported.[11] Extraction of the enzyme with isoöctane (2,2,4-trimethyl-pentane), or aging, especially under conditions of freezing and thawing, results in a marked loss of enzyme activity which can be restored by the addition of α-tocopherol.[12] Successive isoöctane extractions of the enzyme are accompanied by a progressive removal of vitamin E as well as a corresponding decrease in DPNH–cytochrome c reductase activity. Restoration of activity of relatively freshly extracted enzyme may also be effected by vitamin K_1 and by a number of other natural and synthetic products including butter, oleomargarine, and n-butyl stearate. By aging the enzyme preparations in order to dissociate more of the endogenous tocopherol, it is possible to show that enzymatic restoration is accomplished specifically by the tocopherols. The aging process is not a well-defined one, however, so that

[10] A. Nason and F. D. Vasington, *in* "Cell, Organism and Milieu" (D. Rudnick, ed.), p. 63. Roland Press, New York, 1959.

[11] Personal communication from Y. Hatefi, Institute for Enzyme Research, University of Wisconsin, Madison, Wisconsin.

[12] A. Nason and I. R. Lehman, *J. Biol. Chem.* **222**, 511 (1956); K. O. Donaldson, A. Nason, and R. H. Garrett, *ibid.* **233**, 572 (1958).

the attainment of enzymatic reactivation specifically by tocopherol occurs for some preparations but not for others. Little or no inhibition by the isoöctane treatment as described with the above prepared enzyme fractions has been observed in the authors' laboratory.

The lipid content of the rat skeletal particulate preparation (fraction 3) represents 60 to 65% of the enzyme on a dry-weight basis. One of the lipid components from the heart muscle enzyme has been isolated, purified, and identified as a mixed triglyceride with stearate, palmitate, and oleate fatty acid components.[13] It will also restore the activity of relatively fresh enzyme recently extracted with isoöctane.

The presence of coenzyme Q in these enzyme fractions has not yet been investigated.

Inhibitors. Antimycin A is a powerful inhibitor of all the fractions of DPN–cytochrome c reductase, with as little as 0.001 μg./ml. causing approximately 50% inhibition, depending on the protein concentration. In a number of preparations, antimycin A inhibition can be reversed by the addition of α-tocopherol and in nearly all cases can be specifically prevented by the prior addition of the vitamin to the enzyme preparation. When 2,6-dichloroindophenol serves as the electron acceptor in place of cytochrome c, antimycin A has no effect. Amytal (0.12 mg./ml.) or 2n-heptyl-4-hydroxyquinoline-N-oxide (0.1 μg./ml.) results in a 50% or greater inhibition of heart muscle enzyme activity. Fe^{++} and Cu^{++} ions at $5 \times 10^{-4} M$ cause a virtually 100% inhibition of the skeletal muscle enzyme whereas Zn^{++} and Co^{++} at this same concentration give 63% and 48% inhibition, respectively. However, the heart muscle enzyme is not affected by Cu^{++} and Zn^{++} ions at this concentration. Mn^{++}, Ca^{++}, BO_3^{---}, MoO_4^{--}, Fe^{+++}, and Mg^{++} have no effect. At a final concentration of $5 \times 10^{-3} M$, α,α'-dipyridyl and 8-hydroxyquinoline inhibit the skeletal muscle enzyme to the extent of 40 to 50%, but azide has no effect. In the case of the heart muscle enzyme, o-phenanthroline or 8-hydroxyquinoline each causes an 80% inhibition at $10^{-3} M$. At final concentrations of $10^{-4} M$ and $5 \times 10^{-4} M$ of p-chloromercuribenzoate, a 60% and 100% inhibition, respectively, of the skeletal muscle enzyme occurs. No reversal of this inhibition is observed by the addition of either glutathione or cysteine at from two to five times the level of p-chloromercuribenzoate employed.

[13] K. O. Donaldson, A. Nason, I. R. Lehman, and A. Nickon, *J. Biol. Chem.* **233**, 566 (1958).

[58] Electron Transport Particles

By D. E. GREEN and D. M. ZIEGLER

Source

The electron transport particle (ETP) may be defined as a unit which catalyzes the oxidation of both succinate and DPNH by oxygen, and which retains the structural and functional characteristics of the segment of the mitochondrial electron transport chain that intervenes between succinate and DPNH at one end and oxygen at the other. The purified particle is free from dehydrogenases other than the succinic and DPNH dehydrogenases and contains a higher concentration of the components of the terminal electron transport system than does the parent mitochondrion. Heart muscle has many advantages over other mammalian tissues as a source of the mitochondrial electron transport particle. The technical difficulties of liberating the mitochondria from the muscle matrix and the consequent low yield of particles are more than offset by the great stability of these mitochondria and their relatively high concentration of cytochromes. All the procedures described below have been developed for beef heart mitochondria. Electron transport particles have also been isolated from hearts of other species,[1] from other mammalian tissues,[1,2] and from microorganisms.[3]

Assays

Assays only for measuring DPNH oxidase, succinoxidase, and cytochrome c reductase activity will be described. Methods for measuring oxidative phosphorylation in mitochondria and in the electron transport particle have been described elsewhere.[4,5]

Reagents

0.5 M potassium phosphate buffer, pH 7.4.
1.0 M potassium succinate.
0.1 M potassium cyanide.
0.01 M Versene (ethylenediaminetetraacetate), pH 7.4.
0.2% (w/v) DPNH.

[1] D. Keilin and E. F. Hartree, Biochem. J. 44, 205 (1949).
[2] W. C. McMurray, G. F. Maley, and H. A. Lardy, J. Biol. Chem. 230, 219 (1958).
[3] J. H. Bruemmer, P. W. Wilson, J. L. Glenn, and F. L. Crane, J. Bacteriol. 73, 113 (1957).
[4] H. A. Lardy and H. Wellman, J. Biol. Chem. 195, 215 (1952).
[5] A. W. Linnane and D. M. Ziegler, Biochim. et Biophys. Acta 29, 638 (1958).

1.0% cytochrome c. In most commercial preparations cytochrome c is partially reduced. The following method can be applied to convert cytochrome c entirely to the oxidized form. Five milliliters of 1% cytochrome c solution are incubated at 10° for 4 hours with 0.5 mg. of a suspension of well-washed ETP (see preparations). The particles are then removed from the solution of oxidized cytochrome c by centrifugation at 110,000 × g for 40 minutes.

10% BPA (w/v). Ten grams of crystalline bovine plasma albumin (Armour Pharmaceutical Comp.) are dissolved in 60 ml. of water, and the solution is dialyzed for 48 hours against water with four changes. The dialyzed solution of BPA is made up to a final volume of 100 ml. with distilled water.

Assay of DPNH Oxidase Activity

The oxidation of DPNH is measured spectrophotometrically at 38° by the decrease in absorbance at 340 mμ. In a quartz cell the following solutions (see above) are combined: 0.6 ml. of phosphate buffer, 0.03 ml. of Versene, 0.03 ml. of cytochrome c, 0.15 ml. of BPA, and 0.12 ml. of DPNH. The volume is made up to 3 ml. with water. The reaction is started by adding 1 to 20 μg. of enzyme in a volume of 1 to 5 μl. Readings are taken against a water blank every 15 or 30 seconds for 2 to 5 minutes. The decrease in optical density at 340 mμ is a linear function of time until approximately 85% of the DPNH is oxidized. The extinction coefficient for DPNH is taken to be 6.22 × 10^6 cm.^{2}/mole. The specific activity of the preparation is expressed as micromoles of DPNH oxidized per minute per milligram of protein.

Assay of Succinoxidase Activity

The rate of oxidation is measured manometrically at 38° in a Warburg apparatus. The main compartment of the vessel contains 0.1 ml. of phosphate buffer, 0.1 ml. of cytochrome c, 0.5 to 2 mg. of enzyme, and water to bring the final volume in the main compartment to 1.5 ml. Potassium succinate (0.1 ml.) is added to the side arm, and 0.2 ml. of 6 N KOH is added to the center well. The manometers are gassed with oxygen for 3 minutes. After temperature equilibration (approximately 6 minutes) the succinate is tipped into the main compartment of the flask from the side arm to start the reaction. Manometers are read every 5 minutes for 30 minutes. The rate of oxygen uptake is linear with time for at least 15 minutes, but a decline in the rate of oxidation is usually observed after 30 minutes. The rate of oxidation is calculated from the linear portion of the curve, and the succinoxidase activity of the preparation is expressed as microatoms of oxygen taken up per minute per milligram of protein. The product of oxidation, fumarate, is not oxidized by ETP. Therefore, the number of microatoms of oxygen consumed is equivalent to the number of micromoles of succinate oxidized.

Assays of Cytochrome c Reductase Activity

The reduction of cytochrome c is measured spectrophotometrically at 38° by the increase in optical density at 550 mμ. The assay system is essentially the same as that for the assay of DPNH oxidase activity, except that the final concentration of cytochrome c is 1 mg./ml., and sodium cyanide (0.05 ml.) is added to the cuvette to block the reoxidation of cytochrome c. Either the enzyme or the substrate (3 micromoles of DPNH or 20 micromoles of succinate) may be added last to start the reaction. Readings are taken every 15 seconds for 2 to 3 minutes. The rate of cytochrome c reduction is usually linear for the first minute but declines rapidly after more than 40% of the cytochrome c has been reduced. The rate of cytochrome c reduction is calculated from the linear portion of the curve, and the specific activity of the preparation is expressed as micromoles of cytochrome c reduced per minute per milligram of protein. The difference between the extinction coefficients of oxidized and reduced cytochrome c at 550 mμ is taken to be 18.5 $\times$ 10^6 cm.^{2}/mole. Just prior to the assay, an aliquot of the enzyme preparation is diluted to a final protein concentration of not less than 1 mg./ml. This dilution is made with a solution of 0.25 M sucrose which contains 10 mg. of bovine plasma albumin (BPA) per milliliter. The cytochrome c reductase activity of submitochondrial particles is more stable in the presence of BPA, but even in the presence of BPA dilute solutions (less than 1 mg. of protein per milliliter) are unstable.

Preparations

The procedures that have been developed for isolating electron transport particles depend on the method by which the mitochondria are liberated from heart tissue. Mitochondria isolated from beef heart muscle either by the method of Crane et al.[6] or by the method given below can be used for the preparation of the electron transport particles described in this report.

Isolation of Mitochondria

The hearts are collected at the slaughterhouse and routinely packed in ice a few minutes after the animals are killed, although for most purposes the hearts can be left at room temperature for 4 to 5 hours

[6] F. L. Crane, J. L. Glenn, and D. E. Green, Biochim. et Biophys. Acta 22, 475 (1956).

without seriously altering the enzymatic properties of the mitochondria. For example, mitochondria isolated from beef hearts stored for 4 hours at room temperature and then for 24 hours at 5° still manifest the theoretical P/O ratio for the oxidation of citric cycle substrates. If the hearts are left at room temperature longer than 6 hours, fragmentation of mitochondria is almost complete. These fragmented mitochondria are, however, quite satisfactory for the isolation of the nonphosphorylating ETP or other nonphosphorylating subunits of the electron transport system.

The hearts are trimmed of excess fat and connective tissue and passed through a cold meat grinder (plate holes of 4-mm. diameter). The minced muscle is washed four times with 5 vol. of $0.25\,M$ sucrose containing $10^{-3}\,M$ Versene. The pH is maintained at 7.0 by the addition of $1.0\,N$ potassium hydroxide whenever necessary. Two hundred grams (wet weight) of the washed mince are suspended in 500 ml. of $0.25\,M$ sucrose and homogenized in a Waring blendor at top speed for 20 seconds. The homogenate is centrifuged at $1000 \times g$ for 10 minutes. The turbid supernatant is carefully decanted from the loosely packed cell debris and centrifuged at $15,000 \times g$ for 10 minutes. The pellet which is obtained contains only mitochondria and mitochondrial fragments.[7] The pellet is thoroughly homogenized with a Potter-Elvejhem homogenizer (glass-Teflon) in 10 vol. of $0.25\,M$ sucrose containing $10^{-3}\,M$ Tris buffer of pH 7.8, and the suspension is again centrifuged at $12,000 \times g$ for 10 minutes. During centrifugation the pellet separates into two layers—the upper, light mitochondrial fraction (L-BHM), which consists of mitochondrial fragments and swollen mitochondria, and the lower, well-packed, heavy mitochondrial fraction (H-BHM), which contains the more intact mitochondria.[7] The upper layer is separated from the well-packed lower layer by covering the pellet with a small volume of sucrose solution and gently swirling the tube. Both residues are washed separately in 10 vol. of $0.25\,M$ sucrose and then resuspended in 4 vol. of the sucrose solution. Two hundred grams (wet weight) of the minced heart tissue yield about 200 mg. (protein) of the H-BHM fraction and 300 mg. (protein) of the L-BHM fraction.

Isolation of the Nonphosphorylating Electron Transport Particle (ETP)

ETP can be isolated in high yield only from the L-BHM fraction.[7] The L-BHM fraction is adjusted to a protein concentration of 30 mg./ml. with $0.25\,M$ sucrose, and then potassium phosphate ($1.0\,M$, pH 7.6) is added to bring the final phosphate concentration to $0.15\,M$. Cold ($-10°$) 95% ethanol is slowly added, to a final concentration of 15% (v/v),

[7] D. M. Ziegler, A. W. Linnane, D. E. Green, C. M. S. Dass, and H. Ris, *Biochim. et Biophys. Acta* **28**, 524 (1958).

and the suspension is homogenized with a Potter-Elvejhem (glass-Teflon) homogenizer. The suspension is centrifuged at $17,000 \times g$ in a No. 30 rotor of the Spinco Model L ultracentrifuge for 5 minutes (excluding the time required for acceleration and deceleration). The turbid supernatant is carefully decanted from the loosely packed pellet (residue 1) and centrifuged at $110,000 \times g$ for 40 minutes. The deep-red residue (ETP) is washed twice with 10 vol. of $0.25\,M$ sucrose to remove most of the phosphate and ethanol, and then suspended in 3 vol. of $0.25\,M$ sucrose.

The ETP can be stored at $-20°$ for several months with very little loss of either DPNH oxidase or succinoxidase activity. Table I shows

TABLE I

YIELD AND ACTIVITIES OF ETP ISOLATED FROM THE L-BHM FRACTION

Fraction	Per cent protein[a]	Specific activity[b]			
		DPNH oxidase		Succinic oxidase	
		+ cyt. c	− cyt. c	+ cyt. c	− cyt. c
L-BHM	100	1.3	0.8	0.8	0.3
Residue #1	56	0.7	0.3	0.4	0.1
ETP	36	3.9	4.0	2.1	1.4

[a] The protein not accounted for in the two particulate fractions is soluble protein discarded in the final supernatant fraction.

[b] The increase in specific activity of the ETP over the L-BHM is due in large measure to an activation effect rather than an increase in purity. The concentration of the DPNH- and succinic flavoproteins is only 30% greater in ETP than in the L-BHM fraction.

the yield of ETP from the L-BHM fraction in a typical preparation and lists the enzymatic activities. A nonphosphorylating particle very similar in composition and properties to ETP can be isolated in high yield from the H-BHM by exposing the mitochondria to sonic vibrations for 5 minutes in a 10-kc. Raytheon sonic oscillator. The disrupted mitochondria are then fractionated by the method described above for the isolation of ETP from the L-BHM fraction. The nonphosphorylating ETP derived from the sonicated H-BHM has the same composition and oxidase activity as does the phosphorylating ETP_H (cf. Table II), the preparation of which is described below.

Isolation of the Phosphorylating Electron Transport Particle (ETP_H)

An electron transport particle capable of coupling the oxidation of DPNH and succinate to the esterification of phosphate can be prepared

TABLE II
COMPOSITION OF ETP AND ETP$_H$ PREPARATIONS^a

	Micromoles $\times$ 10^{-3} per mg. protein		
Component	L-BHM	ETP	ETP$_H$
Cytochrome a^b	0.59	0.76	0.72
Cytochrome b^b	0.58	0.72	0.65
Cytochrome c + c$_1$^{b}	0.50	0.60	0.32
Coenzyme Q	2.8	3.5	3.4
Flavin	0.42	0.64	0.60
Lipid (% total dry weight)	27	34	32

^{a} The preparations were analyzed by the methods described by D. E. Green, S. Mii, and P. M. Kohout, *J. Biol. Chem.* **217**, 551 (1955) and by F. L. Crane and J. L. Glenn, *Biochim. et Biophys. Acta* **24**, 100 1957).

^{b} Later studies in our laboratory have shown that the correct molar ratios of cytochrome a:b:c$_1$ in BHM, ETP, and ETP$_H$ are 6:3:1, and that there are two molecules of flavin in the chain for each molecule of cytochrome c$_1$ [Blair, Oda, Green and Fernandez-Moran, *Biochem.*, in press (1963)]. The values shown in the table for the three cytochromes (a, b, and c$_1$) are underestimated. The comparison of the three particles is still valid nonetheless.

from the H-BHM fraction by exposing mitochondria to sonic oscillations for a short period.[2, 5, 8] The H-BHM fraction must be frozen for at least 16 hours in order to obtain a good yield of ETP$_H$. Just prior to sonic oscillation the suspension of H-BHM is thawed, adjusted to 20 mg. of protein per milliliter with 0.25 M sucrose, and then made 0.015 M with respect to MgCl$_2$ and 0.001 M with respect to ATP. Finally the pH is adjusted to 7.3 with 0.1 N potassium hydroxide. The suspension (in 26-ml. aliquots) is then exposed for 30 seconds to sonic vibration in a 10-kc. Raytheon sonic oscillator operating at an energy output of 1.0 amp. The temperature inside the oscillator chamber is kept between 0° and 5° by regulating the flow of the cooling fluid through the outer jacket.

The sonicated suspension is centrifuged for 6 minutes at 15,000 $\times$ g to remove the larger mitochondrial fragments (residue 1). The turbid supernatant is decanted and centrifuged for 40 minutes at 110,000 $\times$ g. The sedimented particles (ETP$_H$) are washed once with a 0.25 M sucrose solution containing 10 micromoles of MgCl$_2$ and 1 micromole of ATP per milliliter and then resuspended in another aliquot of the same solution. The particles manifest a P/O ratio of 1.6 to 2.0 for the oxidation of DPNH and 0.8 to 1.0 for oxidation of succinate. The phosphorylative activity of ETP$_H$ is quite stable. The particles can be frozen in 0.25 M

[8] W. W. Kielley and J. R. Bronk, *J. Biol. Chem.* **230**, 521 (1958).

sucrose or lyophilized in $0.25\,M$ sucrose and stored at $-20°$ for several weeks with very little loss of activity.

The composition of ETP and ETP_H is shown in Table II. On the whole, the two particles are very similar in composition. The main difference lies in the lower cytochrome c content of the ETP_H preparation, and this is reflected in a requirement for cytochrome c in order to achieve a rapid oxidation of DPNH (Table III).

TABLE III
YIELD AND DPNH AND SUCCINOXIDASE ACTIVITIES OF ETP_H
ISOLATED FROM H-BHM

		Specific activity[a]			
		DPNH oxidase[b]		Succinoxidase	
Fraction	Per cent protein	+ cyt. c	− cyt. c	+ cyt. c	− cyt. c
H-BHM	100	0.43	0.15	0.73	0.41
Residue #1	73	0.47	0.23	0.61	0.38
ETP_H	12	3.8	1.5	1.7	0.7

[a] The specific activities reported here were measured by the methods described under Assays. In the assays of phosphorylative activity the rates of oxidation of both succinate and DPNH are somewhat lower; see A. W. Linnane, and D. M. Ziegler, *Biochim. et Biophys. Acta* **29**, 638 (1958).

[b] Intact mitochondria do not oxidize exogenous DPNH; the increase in DPNH oxidase activity in the fractions is due primarily to the disruption of mitochondria achieved by sonic vibrations.

Isolation of DPNH *and Succinic–Cytochrome c Reductase Particles*

The electron transport particle can be fractionated into a series of simpler enzymatically active preparations. By treating the particle with lipid-extracting alcohols, or by dispersing it with bile salts and then fractionating with ammonium sulfate, several different enzyme complexes have been prepared. These include: succinic–cytochrome c reductase (SDC) ;[9,10] DPNH–cytochrome c reductase;[11] cytochrome oxidase;[12] and a DPNH oxidase which is free from succinoxidase activity.[13] With the exception of the succinic dehydrogenase complex (SDC),[9] all the other fractions have been obtained by fragmenting the electron transport particles with bile salts followed by ammonium sulfate fractionation. A

[9] D. E. Green, S. Mii, and P. M. Kohout, *J. Biol. Chem.* **217**, 551 (1955).
[10] H. W. Clark, H. A. Neufeld, C. Widmer, and E. Stotz, *J. Biol. Chem.* **210**, 851 (1954).
[11] B. Mackler and N. Penn, *Biochim. et Biophys. Acta* **24**, 294 (1957).
[12] L. Smith and E. Stotz, *J. Biol. Chem.* **209**, 819 (1954).
[13] F. L. Crane and J. L. Glenn, *Biochim. et Biophys. Acta* **24**, 100 (1957).

modification of the method of Rabinowitz and de Bernard[14] for the isolation of a particle with DPNH and succinic–cytochrome c reductase activity is described below. This method illustrates the general principles involved in the preparation of enzymatically active subunits of the electron transport particle by bile salts fractionation.

Beef heart mitochondria, either the light or heavy fractions, suspended in 0.25 M sucrose at a protein concentration of 35 mg./ml., are exposed to sonic vibrations in a Raytheon sonic oscillator for 5 minutes at 0° to 5°. The suspension of mitochondrial fragments is centrifuged at 110,000 $\times$ g for 40 minutes. The soluble supernatant fraction is discarded, and the particles are washed once with 10 vol. of 0.25 M sucrose and then resuspended in a sufficient quantity of 0.25 M sucrose to achieve a final protein concentration of 40 $\pm$ 1 mg./ml. To each 57 ml. of the suspension are added 3.2 ml. of neutral potassium cholate (200 mg. cholic acid per milliliter) and 40 ml. of cold neutral saturated ammonium sulfate. The ratio of cholate to protein, as well as the absolute concentration of cholate, is critical. In the subsequent steps of the fractionation the temperature must be kept at 0° to 2°, and the intervals between steps must be timed accurately. Fifteen minutes after addition of the ammonium sulfate, the precipitated protein is removed by centrifugation at 50,000 $\times$ g for 10 minutes and discarded. The deep-red supernatant fluid is fractionated with solid ammonium sulfate into four fractions. The samples are left at 0° for 10 minutes after each addition of ammonium sulfate, and the precipitated fractions are collected by centrifugation at 60,000 $\times$ g for 10 minutes. The following ammonium sulfate fractions are collected (the amount of solid ammonium sulfate per 100 ml. of solution required to precipitate each fraction is given in parentheses). Fraction I (2.0 g. per 100 ml.) contains substantial quantities of cytochrome oxidase but in all other respects is quite similar to fraction II. Fraction II (3.5 g. per 100 ml.) has high DPNH– and succinic–cytochrome c reductase activities and is essentially free from cytochrome oxidase activity. Fraction III (5.8 g. per 100 ml.) contains only succinic–cytochrome c reductase activity. The final supernatant fraction contains less than 5% of the original protein, and cytochrome c is the only hemoprotein found in it.

The ammonium sulfate precipitates are resuspended in 4 vol. of 0.25 M sucrose. If the fractions are to be stored for periods longer than a week, they should be dialyzed for 5 hours against 0.25 M sucrose to remove excess ammonium sulfate. Preparations repeatedly frozen and thawed over a period of a week in the presence of ammonium sulfate are rapidly inactivated. Dialyzed preparations, however, can be stored for 2

[14] M. Rabinowitz and B. de Bernard, *Biochim. et Biophys. Acta* **26**, 29 (1957).

weeks at $-10°$ with very little loss of cytochrome c reductase activity. The yield and cytochrome c reductase activities of the three fractions are shown in Table IV.

TABLE IV

Yield and Activities of the Cholate-Ammonium Sulfate Fractions

Fraction	Per cent protein	Cytochrome c reductase activity (specific activity)	
		Succinate	DPNH
Mitochondria[a]	100	1.0	2.1
Fraction I	12	4.5	7.4
Fraction II	15	6.0	9.5
Fraction III	8	12.0	0

[a] Intact mitochondria react very slowly with exogenous cytochrome c. The activities reported here were obtained by treating the mitochondria with 0.4 mg. of deoxycholate per milligram of protein in the presence of 0.1 M phosphate buffer; see B. Mackler and D. E. Green, Biochim. et Biophys. Acta 21, 1 (1956).

[59] Electron-Transferring Flavoproteins from Pig Liver and Beef Heart

By HELMUT BEINERT and WILLIAM LEE

Introduction

The so-called electron-transferring flavoproteins (ETF) belong to a new class of flavoproteins, which have the function of mediating electron (or hydrogen) transport between certain flavoproteins and electron acceptors. The only known representatives of this class are the enzymes involved in the dehydrogenation of saturated fatty acyl derivatives of CoA and of sarcosine. It is likely that the electron-transferring flavoproteins of both systems are identical.[1] The flavoproteins concerned with fatty acid metabolism have been found in beef, pig,[1a] and sheep liver,[2] in beef heart,[3] and also in mycobacteria.[4] The ETF of the sarcosine pathway has been found in rat liver.[4a] Only the enzymes from pig liver and beef heart have been obtained in high purity. The preparations

[1] H. Beinert and W. R. Frisell, J. Biol. Chem. 237, 2988 (1962).
[1a] F. L. Crane and H. Beinert, J. Biol. Chem. 218, 717 (1956).
[2] F. Lynen and K. Decker, Ergeb. Physiol. biol. Chem. u. exptl. Pharmakol. 49, 327 (1957).
[3] H. Beinert and W. Lee, Federation Proc. 18, 189 (1959).
[4] D. S. Goldmann, personal communication.
[4a] W. R. Frisell, J. R. Cronin, and C. G. MacKenzie, J. Biol. Chem. 237, 2975 (1962).

obtained to date will interact with all acyl dehydrogenases isolated from beef heart and pig liver. ETF from pig liver also functions with the beef and sheep liver dehydrogenases, those from mycobacteria, and the sarcosine dehydrogenase of rat liver.

Assay Method

The assay for ETF activity is identical with the assay described for the acyl dehydrogenases[5] except that in the ETF assay an excess of dehydrogenase has to be present and ETF should be limiting. It is therefore advisable to perform assays at more than one level of dehydrogenase and to choose assay conditions such that addition of more dehydrogenase does not lead to an increased rate of dye reduction except for what can be attributed to the blank rate of the dehydrogenase. Blank runs on the dehydrogenase preparation are therefore necessary. The specific activity depends on the particular dehydrogenase and substrate used.[6] It is therefore recommended that one combination of substrate and dehydrogenase be used throughout, if activities are to be compared. Otherwise all remarks made previously concerning the assay of the acyl dehydrogenases[5] apply also to the assay of ETF.

ETF will also interact with acceptors other than indophenol, such as ferricyanide or cytochrome c. However, the rate of reduction of cytochrome c is slow and depends on a variety of factors.[1a] ETF of pig liver and of beef heart will interact efficiently with the cytochromes of particulate mitochondrial subfractions,[3] and an assay can be devised on this basis.

Definition of Unit and Specific Activity. These are identical to those proposed for the acyl dehydrogenases.[5]

Preparation of Pig Liver ETF

Extraction:[7] Three hundred grams of an acetone powder of pig liver mitochondria[5] are stirred for 30 minutes with 2.5 l. of $0.02 M$ Tris acetate of pH 7.2. The mixture is centrifuged for 30 minutes at 2500 r.p.m. in the International 4-l. centrifuge $(2000 \times g)$.[8] About 2 l. of a reddish-brown clear supernatant are decanted through two layers of cheesecloth, which removes floating particles.

---

[5] Vol. V [73].
[6] It has been observed during the purification of ETF that there may be a change in the ratio of specific activities with different dehydrogenases or substrates. This may indicate the existence of different kinds of ETF which differ in their substrate specificity. Definite information on this point is not available, as clear-cut separations have not been observed.
[7] All operations during extraction and fractionation are carried out at ice-bath temperature unless otherwise stated.
[8] For more details on centrifugation see Vol. V [73].

Fractionation: Step 1. Zn⁺⁺ Treatment. Twenty milliliters of 0.05 M zinc lactate, pH 5.5, are slowly added with stirring for every 100 ml. of extract. Stirring is continued for 10 minutes, and the mixture is clarified by centrifugation as above.

Step 2. Ammonium Sulfate Fractionation. The greenish-yellow supernatant of step 1 is adjusted to pH 7.2 to 7.5 with a small amount of 6 N KOH and is fractionated with solid ammonium sulfate. After an equilibration period of 30 minutes after each addition of salt, fractions are collected at 35%, 50%, 65%, and 85% saturation. Centrifugation is carried out as described in Vol. V [73], step 1. The fractions are dissolved in 0.02 M Tris acetate of pH 7.2.

Step 3. Zn⁺⁺–Ethanol Fractionation. The fractions obtained at 65% and 85% saturation are dialyzed for 6 hours against a twenty fold excess of 0.02 M Tris acetate of pH 7.2. Since removal of salt is essential for a successful fractionation, both the samples and the dialysis fluid are stirred during this dialysis, and the fluid is replaced twice. A precipitate which occasionally forms during dialysis is removed by high-speed centrifugation. After dialysis, the protein concentration of the samples is determined and adjusted to ⩽20 mg./ml. by addition of 0.02 M Tris acetate. Twenty milliliters of 0.05 M zinc lactate are then slowly added with stirring for every 100 ml. of the dialyzed solution. Stirring is continued for another 5 to 10 minutes. The precipitate is sedimented, and the supernatant is fractionated with cold ethanol in a manner analogous to that described previously.[5] Fractions are collected at 10%, 15%, and 40% ethanol concentration, calculated on the basis of volumes added. The fractions are dissolved in 0.4 M citrate and then dialyzed for 2 to 3 hours against 0.02 M Tris acetate of pH 7.2 containing 0.001 M Versene. The fraction obtained at 15 to 40% ethanol contains most of the ETF activity and is generally free of acyl dehydrogenase activity.

Step 4. Ammonium Sulfate Fractionation. This fraction is adjusted to a protein concentration of ⩽20 mg./ml. with 0.02 M Tris acetate of pH 7.2 and fractionated as above with solid ammonium sulfate. Fractions are collected at 50%, 55%, 65%, 90%, and 100% saturation. The fractions are dissolved in 0.02 M Tris acetate of pH 7.2 and are assayed. The fraction obtained between 65 and 90% saturation usually contains the bulk of the ETF and the material of highest specific activity. This fraction is suitable for most work requiring ETF.

Step 5. Ammonium Sulfate Fractionation at pH 8.1. Some further purification is possible by fractionation at alkaline pH. The fraction obtained at 90% saturation in step 4 is dialyzed for 3 hours with internal stirring against 0.2 M Tris acetate of pH 8.1. The protein concentration is adjusted to 15 to 20 mg./ml. with the same buffer, and solid ammonium sulfate is added as above. Fractions are obtained at 70%, 75%, 80%, and 100% saturation. The fractions are dissolved in 0.02 M Tris acetate of pH 7.2 and are assayed. The highest activity is generally found in the 75 to 80% fraction. Occasionally, however, high activity appears in the 80 to 100% fraction. A summary of the purification is given in the table.

SUMMARY OF PURIFICATION PROCEDURE FOR PIG LIVER ETF

Step	Protein, mg.	Specific activity (1000×)	Units	Recovery, %
Extraction of 300 g. acetone powder	57,000	~3	171	100
Zn++–ammonium sulfate precipitation (50–65% and 65–85% fractions)	6,450	~10	64.5	38
Zn++–ethanol treatment	900	25	22.5	13
Ammonium sulfate precipitation at pH 7 (65–90%)	124	80	9.9	5.8
Ammonium sulfate precipitation at pH 8.1: 0–70%	102	57	5.8	3.4
70–75%	5	140	0.70	0.4
75–80%	8	190	1.5	0.9
80–95%	2.6	90	0.2	0.1

Preparation of Beef Heart ETF

Step 1. Extraction and DEAE Treatment. These initial steps of the preparation have been described.[5] To the red filtrate from the cellulose which retains the acyl dehydrogenases, 100 ml. of 1 M phosphate of pH 7.9 are added.

Step 2. Ammonium Sulfate Fractionation. Solid ammonium sulfate is slowly added with mechanical stirring to the buffered solution of step 1. Fractions are collected at 40% and 85% saturation with respect to ammonium sulfate. The precipitate formed at 40% is sedimented in the International 4-l. centrifuge (2000 × g) for 40 minutes, and that formed at 85% is sedimented in the 1.7-l. Servall rotor (8000 × g) for 30 minutes. The precipitate obtained at 85% saturation is dissolved in the minimum volume of 0.02 M Tris acetate of pH 7.4 and dialyzed with internal and external stirring against 5 l. of the same buffer. The dialysis fluid is changed twice.

Step 3. Second DEAE Treatment. The pH of the dialyzed material is adjusted to 8.5 by adding about 2 ml. of 2 M Tris hydroxide. The solution is then stirred for 5 minutes with 100 ml. of packed damp DEAE-cellulose previously equilibrated with 0.02 M Tris chloride of pH 8.5. The cellulose is filtered off with suction on a coarse sintered-glass funnel in the cold room. Any imbibed material is washed out of the cellulose with 200 ml. of 0.02 M Tris chloride of pH 8.5.

Step 4. Zn^{++}-Ethanol Fractionation. The filtrate and washings are neutralized with about 2.2 ml. of 1 N acetic acid. Ten milliliters of 0.05 M zinc lactate are then added slowly with stirring to every 100 ml. of protein solution. Thereafter 5 ml. of cold ethanol are added in the same way for every 100 ml. of the resulting mixture. The precipitate is sedimented in an angle head of the International centrifuge (3200 $\times g$) for 15 minutes. The supernatant is dialyzed for 3 hours with internal and external stirring against 5 l. of 0.02 M Tris acetate, pH 7.4, which contains Versene at a concentration of 0.003 M.

Step 5. Ammonium Sulfate Fractionation. To the dialyzed solution is added 0.03 vol. of 1 M phosphate of pH 7.9. A slight turbidity at this point may be disregarded. Fractionation with solid ammonium sulfate is carried out on this material in the same way as described in step 2. The precipitate obtained at 40% saturation may be spun in this case in a 1-l. angle head of the International centrifuge. The sediment is flushed together with the minimum volume of 0.02 M Tris acetate or chloride, pH 7.5, and is dialyzed against this buffer.

Step 6. Column Electrophoresis. The dialyzed material is left in the dialyzing bag and concentrated by pervaporation or in a concentrated solution of polyvinylpyrrolidone, which is buffered with Tris chloride (0.1 M, pH 7.5). After concentrating the fraction is once more dialyzed for 3 hours against 0.1 M Tris chloride of pH 8.5 and is then placed on a cellulose (or starch) column as described previously.[5] At this pH, ETF moves slowly down the cellulose column as an anion. Essentially three major colored bands separate. One red fraction containing hemoproteins moves upward toward the cathode, and another red fraction moves down ahead of the fraction containing ETF. On starch, because of strong endosmotic flow, an upward movement of all components is superimposed on this pattern, so that, as a net effect, ETF moves slowly in the opposite direction. However, since on both supporting media the migration of ETF is slow, minor differences in pH, type of supporting medium, or impurities present may change the direction of migration from that expected. It is therefore advisable to follow the progress of the separation. The eluted fractions are monitored by observing their spectral properties (cf. Vol. V [73]) and by assays of enzymatic activity on the fractions of the lowest ratio of E_{415}/E_{450}. The most active fractions are combined and precipitated with ammonium sulfate at 90% saturation.

Step 7. Ammonium Sulfate Fractionation. Specific activity may be substantially increased by an additional ammonium sulfate fractionation. The protein concentration of the most active fraction from electrophoresis is adjusted to about 10 mg./ml., and 0.03 vol. of 1 M phosphate of pH 7.9 is added. The initial ammonium sulfate concentration is calculated, and the material is fractionated with solid ammonium sulfate. Fractions are collected at 50%, 60%, 65%, 85%, and 95% saturation. The 65 to 85% fraction is generally most active.

Remarks. In this procedure ETF is free of acyl dehydrogenases and thus suitable for assay of these enzymes at the end of step 3. When one starts with 100 g. of acetone powder (specific activity of extract ~0.004), 400 mg. of specific activity 0.012 are obtained after electrophoresis, and about 80 mg. of specific activity 0.036 after the final ammonium sulfate step. Additional electrophoresis on cellulose at pH 7 has yielded ETF of specific activity 0.12.

Properties

Stability. ETF activity measured with indophenol as acceptor declines noticeably during storage even at $-15°$. Within a month about 50% of the activity is lost. Similarly losses of activity during fractionations were often found. Thus according to criteria of flavoprotein chemistry definite purification may be achieved in some steps while activity may be unchanged or diminished. This instability of ETF from pig liver and from beef heart has prevented attempts at further purification. The activity with cytochrome c as acceptor is subject to even greater variability.[1a]

Specificity. Until recently no substrate other than reduced acyl dehydrogenases had been found for ETF (see, however, next paragraph). It has now been found, however, that reduced sarcosine dehydrogenase of rat liver is an additional substrate.[1] The ETF preparations obtained from beef heart and pig liver will react with all acyl dehydrogenases from either source. ETF from pig liver will also interact with the dehydrogenases of beef and sheep liver and those from mycobacteria. Octanoyl CoA with the acyl dehydrogenase (C_4 to C_{16}) will sustain the highest rate of ETF turnover. It thus appears that the kind of substrate used for the dehydrogenase exerts an influence on the function of ETF.

Contamination. ETF of highest purity is free of acyl dehydrogenases. All ETF preparations obtained to date exhibit pyridine nucleotide dehydrogenase activity[1a] at a level of specific activity similar to that shown with the reduced acyl dehydrogenases as substrates. TPNH and DPNH are dehydrogenated at about equal rates. Reduction of flavin, however, could not be demonstrated with reduced pyridine nucleotides. Since considerable variation was found in the ratio of pyridine nucleotide and fatty acyl CoA dehydrogenase activities during purification and since the former activity was 50% inhibited by $10^{-8}\,M$ dicoumarol, it is concluded that the pyridine nucleotide dehydrogenase activity is due to contamination with the highly active "DT-diaphorase" (phylloquinone reductase).[9,10] Otherwise remarks made concerning the dehydrogenases[5] are applicable here also. ETF of beef heart is not easily freed of hemoprotein contaminants.

Spectral Properties. ETF of pig liver or of beef heart shows a flavoprotein spectrum, which, however, exhibits some peculiarities. In addition to the usual flavoprotein peaks, located with ETF at 270, 375, and 437.5 mμ, there is an additional minor peak at 460 mμ. A shoulder at 410 mμ may indicate a minor hemoprotein contaminant. For the purest preparations the absorbancy ratios have been: $E_{270}/E_{310}/E_{375}/E_{437.5}/E_{460} = 6.5:0.3:0.9:1.0:0.86$. There is one mole of FAD present per mole of enzyme (mol. wt. ~80,000).

Kinetic Properties. The remarks made previously[5] on the kinetics of the acyl dehydrogenases are pertinent here. The turnover of ETF with indophenol is of the order of a few hundred per minute at 30°.

[9] F. Märki and C. Martius, *Biochem. Z.* 333, 111 (1960).
[10] L. Ernster, M. Ljunggren, and L. Danielson, *Biochem. Biophys. Research Communs.* 2, 88 (1960).

[60] Chloroplast TPNH Diaphorase

TPNH + 2,3′,6-trichlorophenolindophenol
$\rightleftarrows$ TPN+ + leuco-2,3′,6-trichlorophenolindophenol

By ANDRÉ T. JAGENDORF

Assay Method

Principle. As with other diaphorase enzymes,[1] the reaction may be observed by decolorization of a blue indophenol dye as it becomes reduced to the leuco form. The most convenient and sensitive assay is based on the red absorption band of 2,3′,6-trichlorophenolindophenol which has a maximum at 645 mμ (millimolar extinction coefficient found to be 27 at pH 7.3, by titration with a standard concentration of ascorbate). The dye and TPNH are mixed in a buffered solution, and the decrease in optical density at 645 mμ is measured at 15-second intervals as the dye is reduced. Many variations of this procedure using other electron acceptors are feasible; however, care should be taken either to avoid or to

[1] H. R. Mahler, Vol. II [124].

correct for simultaneous changes due to TPNH and the electron acceptor if the wavelength is one where the two absorption spectra overlap. Alternatively, if a rapidly autoxidizable dye is used (such as methylene blue), one can measure simply the decrease in optical density at 340 mμ.

Reagents

0.00165 *M* 2,3′,6-trichlorophenolindophenol (the optical density of a 0.1-ml. aliquot diluted with 2.9 ml. of buffer should be 1.50).

0.15 *M* Tris–chloride buffer, pH 7.5.

0.0015 *M* TPNH, prepared enzymatically from TPN.[2]

Enzyme, diluted with Tris at pH 7.5 to give a solution containing approximately 0.25 unit/ml. (See definition of a unit below.)

Procedure. Mix 1.0 ml. of Tris buffer, 0.1 ml. of indophenol dye, and 0.1 ml. of water to give a final volume of 2.9 ml. Add 0.1 ml. of the enzyme solution, and observe the optical density at 645 mμ at intervals of 15 seconds for 1 to 2 minutes.

Definition of Unit and Specific Activity. One unit of enzyme is defined here as that amount which causes an initial rate of dye reduction of 1.0 micromole/min. under the above conditions. Specific activity is expressed as units per milligram of protein.

Application of Assay Method to Crude Tissue Preparations. A crude homogenate of spinach or other leaves will contain diaphorases other than the one described here, including some that catalyze a reaction with DPNH as well as with TPNH, and probably other enzymes as well (e.g., nitrate reductase[3]) which function in different reactions entirely but nevertheless possess diaphorase activity. After separation of the chloroplasts, however, practically all the diaphorase activity associated with the particles will be the TPNH-specific enzyme described below.

The recommended pH for the assay procedure is not the optimum one for activity with the purified enzyme (pH 9.0). One reason for the lower pH is to avoid possible difficulties with reoxidation of the leuco dye. If the enzyme is measured at pH 9.0, approximately double the activity will be found compared to pH 7.5.

Purification Procedure[4,5]

The preliminary steps in purification, up to step 3, have been carried out successfully in at least two other laboratories. A very similar en-

[2] H. J. Evans and A. Nason, *Plant Physiol.* **28**, 233 (1953).

[3] D. J. D. Nicholas and A. Nason, *J. Biol. Chem.* **211**, 183 (1954).

[4] M. Avron and A. T. Jagendorf, *Arch. Biochem. Biophys.* **65**, 475 (1956).

[5] M. Avron and A. T. Jagendorf, *Arch. Biochem. Biophys.* **72**, 17 (1957).

zyme, if not the same one, has been partially purified by Marrè and Servettaz.[6] The enzyme appears to be a relatively stable one and is not easily damaged by exposure to room temperature. All operations were nevertheless carried out at 0° to 5°, unless otherwise indicated.

Step 1. Preparation of Leaf Homogenate. After removal of the petioles, spinach leaves purchased from a grocery are sliced finely and divided into 100-g. lots. Each lot is ground in a Waring blendor for 15 seconds at 75 volts (or in an Omnimixer at 45 volts for 45 seconds), together with 200 ml. of $0.05 M$ phosphate buffer at pH 7.5 containing $0.40 M$ sucrose and $0.01 M$ NaCl. The homogenate is strained through a pad of cheesecloth and glass wool.

Step 2. Isolation of Chloroplasts and Extraction of Enzyme. The homogenate is centrifuged for 5 minutes at $2500 \times g$. The supernatant solution is discarded, and the chloroplast pellet is resuspended in a small volume of the phosphate–sucrose–NaCl solution and washed once by centrifugation. After the wash the chloroplasts are resuspended in $0.01 M$ Tris–HCl buffer at pH 8.0 and allowed to stand at room temperature for 1 hour. They are then centrifuged at $140,000 \times g$ for 1 hour. The supernatant is collected and used as the crude enzyme preparation.

Step 3. Acetone Fractionation. Three volumes of acetone at $-10°$ are added slowly with constant stirring to 1 vol. of enzyme. The protein suspension is centrifuged at $5000 \times g$ for 5 minutes, and the pellet is resuspended thoroughly in $0.005 M$ Tris buffer at pH 8.0. The extracted enzyme is centrifuged at $17,000 \times g$ for 20 minutes, and the pellet is discarded.

Step 4. Phosphate Gel Fractionation. Tricalcium phosphate gel suspension is added to the supernatant at a ratio of approximately 14 mg. of gel to the enzyme solution from 300 g. of leaves. The suspension is allowed to stand for 15 minutes with occasional stirring, then centrifuged at $5000 \times g$ for 10 minutes, and the supernatant is discarded. The pellet is extracted with $0.1 M$ sodium pyrophosphate buffer at pH 9.0 and allowed to stand for 15 minutes with occasional mixing. The extracted gel is centrifuged at $17,000 \times g$ for 20 minutes, and the pellet is discarded.

Step 5. Chromatography on Dowex 50. Five milliliters at a time of the pyrophosphate enzyme solution is passed through a Dowex 50 (2% cross-linkage) 2×10-cm. column, followed by $0.05 M$ Tris buffer at pH 8.0. The enzyme is obtained in the effluent between 7 and 14 ml. after addition of the enzyme solution at the top of the column. The column procedure is carried out at room temperature.

Step 6. Ammonium Sulfate Fractionation. Solid ammonium sulfate is

[6] E. Marrè and O. Servettaz, *Arch. Biochem. Biophys.* **75**, 309 (1958).

added to the enzyme solution until 50% saturation is reached. The solution is centrifuged, and the pellet is discarded. More ammonium sulfate is added until 62% saturation is reached, and the majority of the enzyme precipitates. After centrifugation the pellet must be redissolved in 0.1 M pyrophosphate buffer at pH 9.0 and dialyzed against pyrophosphate buffer to remove residual ammonium sulfate. The final ammonium sulfate step can give a specific activity 60% higher than that achieved in step 5, with a loss of about 40% of the enzyme activity.

A summary of the purification procedure is given in the table.

SUMMARY OF PURIFICATION PROCEDURE[a]

Step	Units micromoles/min.	Specific activity, units/mg. protein	Yield, %
Homogenate	—	0.24	—
Chloroplast extract	481	1.09	100
Acetone fractionation	528	4.35	110
Phosphate gel fractionation	243	19.2	51
Chromatography on Dowex 50	241	28.1	50

[a] The starting material consisted of 1.76 kg. of spinach leaves.

Properties

Specificity. The purified enzyme is completely inactive with DPNH as electron donor. A high degree of specificity for TPNH is found even in the crude chloroplast extract (step 2). Electron acceptors include trichlorophenolindophenol, ferricyanide, methylene blue, menadione, FMN, benzoquinone, ferric chloride with o-phenanthroline, and probably many other dyes. Since acetylpyridine analog of TPN is also an acceptor[7] the enzyme could also be called a transhydrogenase. Oxygen, nitrate ions, cytochrome c, dehydroascorbate, a mixture of ascorbic acid and ascorbic acid oxidase, and glutathione do not act as electron acceptors.

Inhibitors. Fifty per cent inhibition of the enzyme occurs with $2 \times 10^{-5} M$ mercuric chloride, $1 \times 10^{-4} M$ p-chloromercuribenzoate, $5 \times 10^{-4} M$ zinc or cobalt ions, or $3 \times 10^{-3} M$ iodoacetate. 2′-AMP is an inhibitor, as is TPN.

Coenzyme. The most highly purified preparations are yellow. The content of FAD bound to the enzyme remains in constant ratio to activity throughout purification, and the flavin absorption spectrum can be seen to be bleached by added TPNH. There appears to be one FAD molecule per molecule of protein with molecular weight 35,000.

Physical Constants. The K_m for TPNH is $6 \times 10^{-6} M$. Affinity for

[7] M. M. Weber and N. O. Kaplan, *J. Biol. Chem.* **225,** 27 (1957).

various oxidants has not been determined. The purified protein has a single symmetrical peak on ultracentrifugation with a sedimentation constant of 3.5 Svedberg units, suggesting a molecular weight between 30,000 and 40,000. A major peak with a smaller fast-moving component is seen in electrophoresis; the major peak has a mobility of 8.0×10^{-5} cm./(sec. $\times$ volts/cm.) at pH 8.0 and 7.6, and 2.7×10^{-5} cm./(sec. $\times$ volts/cm.) at pH 7.0.

Note added in proof: What is most probably the same enzyme has also been isolated directly from spinach leaf homogenates by Keister, San Pietro, and Stolzenbach [*J. Biol. Chem.* **235**, 2989 (1960)] measuring primarily transhydrogenase activity.

[61] Pyridine Nucleotide Transhydrogenase from Spinach

$$\text{TPNH} + \text{DPN} \rightarrow \text{TPN} + \text{DPNH} \tag{1}$$

$$\text{TPNH} + \text{deamino-TPN} \rightarrow \text{TPN} + \text{deamino-TPNH} \tag{2}$$

By DONALD L. KEISTER and ANTHONY SAN PIETRO

Assay Method

Principle. The usual assay [reaction (1)] involves the generation of DPNH mediated by the transhydrogenase in the presence of the isocitric dehydrogenase system and a catalytic amount of TPN as described by Kaplan.[1] The reaction is followed at 340 mμ. Under the assay conditions described, however, the reduction of deamino-TPN by TPNH [reaction (2)] is a much more sensitive method for measuring the transhydrogenase activity in spinach preparations.

Reagents

DPN, 0.01 M.
TPN, 0.001 M.
Deamino-TPN, 0.005 M.
Sodium DL-isocitrate, 0.1 M.
MgCl$_2$, 0.1 M.
Tris buffer (pH 8.7), 0.5 M.
Pig heart TPN isocitric dehydrogenase.[2]

Procedure. The reaction mixture in a Beckman cuvette contains 0.3 ml. of buffer, 0.1 ml. of MgCl$_2$, 0.1 ml. of isocitrate, 0.1 ml. of TPN,

[1] See N. O. Kaplan, Vol. II [119].
[2] See S. Ochoa, Vol. I [116].

0.1 ml. of DPN, an excess of isocitric dehydrogenase, and water to 3 ml. When all the TPN has been reduced, the transhydrogenase preparation is added, and the increase in optical density at 340 mμ is followed for 3 minutes.

The reaction mixture and the procedure for assay of reaction (2) are identical to that above with the exception that 0.2 ml. of deamino-TPN is substituted for DPN.

Definition of Unit and Specific Activity. A unit of activity is defined as the amount of transhydrogenase which will produce an increase in optical density of 0.01 in 3 minutes. Specific activity is expressed as units per milligram of protein.

Application of Assay Method to Crude Extracts. The reduction of DPN by TPNH [reaction (1)] is very difficult to measure in crude extracts of spinach for two reasons. First, this enzymatic activity is low, and, second, the amount of crude extract which can be used in the assay is small and is limited by the high light absorption, at 340 mμ, of the crude extract. On the other hand, it is possible to demonstrate transhydrogenase activity in crude extracts with the deamino-TPN assay [reaction (2)]. The results obtained by this latter assay are reproducible but consistently low. In general, the total units found in the crude extract is about one-half that demonstrable in the extract of acetone precipitate (Table I). It is suggested, therefore, that the enzyme be partially purified (through the extract of acetone precipitate) prior to assay. At this stage, either assay procedure is applicable.

Purification Procedure

The procedure described below has been repeated numerous times with consistent and reproducible results.

Steps 1–3. Preparation of Crude Homogenate; Extract of Acetone Precipitate; Treatment with Protamine Sulfate. These three steps are carried out as described by San Pietro[3] for the purification of photosynthetic pyridine nucleotide reductase. The transhydrogenase activity is not precipitated with protamine sulfate and remains in the supernatant fluid from the precipitation of photosynthetic pyridine nucleotide reductase with protamine sulfate.

Step 4. Adsorption on Bentonite. Solid bentonite[4] is added to the

[3] See Vol. VI [62]. A preliminary report describing the presence of transhydrogenase in spinach and certain of its properties has been published [D. L. Keister and A. San Pietro, *Biochem. Biophys. Research Communs.* 1, 110 (1959)].
[4] The amount of bentonite necessary for optimum purification and recovery of transhydrogenase in this step varies with each preparation. It is best, therefore, to do a trial run of this step with a small volume of enzyme. In general, the ratio (w/w) of bentonite to protein used varies between 0.5 and 1. Under these condi-

enzyme solution (protamine sulfate supernatant fluid from step 3), and the suspension[5] is centrifuged at 16,000 $\times$ g. The supernatant is discarded, and the bentonite residue is washed once, by centrifugation, with a volume of 0.1 M Tris, pH 8.7, equal to the volume of enzyme used for this step. The transhydrogenase activity is then eluted from the washed bentonite residue, first with 1 vol., then with 0.5 vol., of 0.1 M potassium phosphate, pH 8.1. The majority of the activity which can be eluted is usually present in the first phosphate eluate. In some cases, the second phosphate eluate contains very little activity and is discarded.

Step 5. Concentration of Enzyme. The protein content in the phosphate eluate(s) from step 4 is quite low (about 0.3 mg. of protein per milliliter). Concentration has been accomplished by first using the water-absorbing properties of Carbowax[6] and then precipitating the enzyme with acetone.

Powdered Carbowax is placed in dialysis tubing, and the dialysis tubing is immersed in the dilute enzyme solution from step 4. Water is absorbed by the Carbowax, and a five- to tenfold concentration is accomplished in a few hours with very little loss in activity. After treatment with Carbowax, the protein concentration is sufficiently high to permit additional concentration, with good recovery, by precipitation with acetone.[7] The enzymatic activity is readily precipitated at an acetone concentration of 75%. After solution in water, the enzyme solution is dialyzed against 20 vol. of distilled water, in the cold, for a total of 3 hours; the dialysis solution is changed at the end of the first and second hour of dialysis.

Further purification of the enzyme has been obtained with Dowex 50-W (200 to 400 mesh, $\times$2, sodium form). The dialyzed enzyme solution from step 5 is placed on a Dowex 50-W column. The column is then washed with water until most of the enzymatic activity is recovered. If

---

tions, between 40 and 65% of the enzymatic activity is adsorbed on the bentonite. Approximately 30 to 40% of the enzyme initially present in the protamine sulfate supernatant fluid can then be eluted from the bentonite. At higher ratios of bentonite to protein, more activity is adsorbed on the bentonite. However, there is no significant increase in the amount of enzyme which can be eluted from the bentonite, and the specific activity of bentonite eluates is decreased. This behavior of bentonite is not understood.

[5] The bentonite is difficult to suspend evenly. Suspension is best accomplished either with a TenBroeck glass homogenizer or a Waring blendor, depending on the volume of the enzyme solution used.

[6] Carbowax (Polyethylene Glycol Compound 20-M), Union Carbide Chemicals Co., Charleston, West Virginia.

[7] Acetone precipitation serves also to remove any Carbowax that may have gotten into the enzyme solution (evidenced by a higher E_{280}/E_{260} ratio than before the treatment with Carbowax).

the volume of the effluent and water washings is equal to the volume of enzyme placed on the column, the enzymatic activity is found in water washings 3 to 5. By this procedure, it is possible to obtain approximately a twofold purification with 70% recovery.

A summary of the yields and purifications obtained at each step in the preparation of the enzyme is presented in Table I.

TABLE I
SUMMARY OF TYPICAL PURIFICATION[a]

Fraction	Total volume, ml.	Total protein, mg.	Total units	Specific activity, units/mg. of protein	Yield, %
1. Crude homogenate	7800	133,380	(340,800)[b]	(2.55)	100
2. Extract of acetone precipitate	945	25,326	645,000	24.6	190
3. Supernatant of protamine precipitation	1235	11,140	553,000	49.8	159
4. Eluate of bentonite	1819	664	216,500	326	66
5. Concentrate of step 4	68	390	163,900	420	48

[a] Starting material, 10.8 kg. of whole spinach leaves. Enzyme assays carried out according to reaction (1).

[b] This is a calculated value. The activity in the crude homogenate was measured according to reaction (2) by using deamino-TPN. The value obtained by this assay was used to estimate the rate expected with DPN.

Properties

Specificity. The enzyme is capable of promoting the following reactions:

$$TPNH + DPN \rightarrow TPN + DPNH \tag{3}$$

$$TPNH + deamino\text{-}TPN \rightarrow TPN + deamino\text{-}TPNH \tag{4}$$

$$TPNH + acetylpyridine\text{-}TPN \rightarrow TPN + acetylpyridine\text{-}TPNH \tag{5}$$

$$TPNH + deamino\text{-}DPN \rightarrow TPN + deamino\text{-}DPNH \tag{6}$$

$$TPNH + acetylpyridine\text{-}DPN \rightarrow TPN + acetylpyridine\text{-}DPNH \tag{7}$$

The maximal velocities and the relative ratios of the maximal velocities of these reactions are shown in Table II. The maximal velocity of reaction (4) is about twelve times that of reaction (3). This observation, together with the low substrate concentration of TPNH and deamino-TPN necessary for half-maximal velocity, as contrasted with DPN, may serve to explain the apparent irreversibility of reaction (3).[3] It is tacitly assumed that the DPNH concentration necessary for half-maximal velocity is very much larger than that determined for TPNH. Under

TABLE II
MAXIMAL VELOCITIES WITH VARIOUS ACCEPTORS

Acceptor	Maximal velocity[a]	Relative maximal velocity
DPN	9.1	1
Deamino-TPN	108	12
Acetylpyridine-TPN	45.6[b]	5
Acetylpyridine-DPN	2.6	0.3
Deamino-DPN	1.8	0.2

[a] The maximal velocities are presented as micromoles of acceptor reduced per milligram of protein in 3 minutes and have been calculated from Lineweaver-Burk plots.
[b] The assay system contained 150 micromoles of Tris buffer, pH 8.7, 0.3 micromole of TPNH, 2.7 μg. of enzyme protein, and a saturating amount of acetylpyridine-TPN in a volume of 3 ml. The reaction is initiated with enzyme, and the reduction of acetylpyridine-TPN is followed at 375 mμ at which wavelength the difference in extinction coefficients between TPNH and acetylpyridine-TPNH is 5.1 [A. M. Stein, N. O. Kaplan, and M. M. Ciotti, *J. Biol. Chem.* **234**, 979 (1959)]. The reaction is measured for only 1 minute and corrected to 3 minutes, since the rate is not linear beyond the first minute, probably owing to the accumulation of TPN which is inhibitory.

these conditions, as soon as a small amount of TPNH is formed by reversal of reaction (3), it will interact with the TPN present and thereby inhibit the interaction between DPNH and TPN.

Effect of Nucleotide Concentration. The concentrations of the nucleotides necessary for half-maximal velocity, which have been measured, are $2.3 \times 10^{-6} M$ for TPNH; $3.3 \times 10^{-5} M$ for deamino-TPN; $7 \times 10^{-5} M$ for acetylpyridine-TPN; $1.5 \times 10^{-4} M$ for deamino-DPN; $1.1 \times 10^{-3} M$ for acetylpyridine-DPN; and $1.8 \times 10^{-3} M$ for DPN.

pH Optimum. The pH optimum of spinach transhydrogenase is approximately 8.8.

Inhibitors.[8] In contrast to its effect on transhydrogenase from bacteria and animal tissues,[1,9] 2′-adenylic acid inhibits the spinach enzyme. 3′-Adenylic acid, 5′-adenylic acid, ADP, and ATP also inhibit, but to a somewhat lesser degree than the 2′-isomer. 2′-Adenylic acid causes an inhibition of 43% at $1.7 \times 10^{-3} M$, and this inhibition has been shown to be competitive with DPN.

Other inhibitors are the α-isomer of DPN (33% at $10^{-3} M$), thyroxine

[8] In some of the inhibitor studies, acetylpyridine-TPN was used as the acceptor [reaction (5); see also footnote b to Table II] to eliminate any possible effect of the inhibitor on the isocitric dehydrogenase system. In all others, the standard assay with DPN was employed.
[9] N. O. Kaplan, S. P. Colowick, and E. F. Neufeld, *J. Biol. Chem.* **205**, 1 (1953).

(34% at $3.3 \times 10^{-5} M$), TPN (33% at $5 \times 10^{-5} M$), p-chloromercuribenzoate (40% at $10^{-3} M$), and zinc sulfate (45% at $10^{-4} M$).

Compounds of interest that have no significant effect include cyanide ($10^{-3} M$), azide ($2 \times 10^{-3} M$), fluoride ($5 \times 10^{-2} M$), Versene ($3 \times 10^{-3} M$), magnesium chloride ($3.3 \times 10^{-2} M$), copper sulfate ($3 \times 10^{-3} M$), estradiol-17β ($3.3 \times 10^{-5} M$), atabrine ($5 \times 10^{-4} M$), nicotinamide ($2 \times 10^{-2} M$), and quinine ($5 \times 10^{-4} M$). Manganese stimulates slightly.

FAD Content of Enzyme. Analysis of the enzyme (specific activity of 543) for FAD content by the method of DeLuca *et al.*[10] indicates the presence of 8.5×10^{-3} micromole of FAD per milligram of protein.[11] In addition, the spectrum of a more purified enzyme preparation (specific activity of 712) indicates the presence of a flavin.

Other Enzymatic Activities. The purified enzyme exhibits TPNH-diaphorase and TPNH-menadione reductase activity. It is interesting that the purification of these activities parallels that for the transhydrogenase.

Distribution. The enzyme has been found in plants other than spinach. These include pea leaves, turnip greens, parsley, and watercress. It has also been found in *Euglena gracilis.*

[10] C. DeLuca, M. M. Weber, and N. O. Kaplan, *J. Biol. Chem.* **223**, 559 (1956).
[11] By assuming that the enzyme contains one mole of flavin, it is possible to calculate a molecular weight of approximately 118,000 for the enzyme. However, preliminary measurements of the molecular weight of the enzyme in the Spinco ultracentrifuge gave a value of about 36,000. Thus it would appear that this preparation is about 30% pure and that the specific activity of the pure enzyme would be close to 1700. To date, the highest specific activity we have attained is 1300.

[62] Photosynthetic Pyridine Nucleotide Reductase

$$\text{TPN} + \text{H}_2\text{O} \xrightarrow[\text{light}]{\text{chloroplasts}} \text{TPNH} + \text{H}^+ + \tfrac{1}{2}\text{O}_2$$

By ANTHONY SAN PIETRO

Assay Method

Principle. Enzymatic activity is determined by measuring the initial rate of formation of TPNH in the Hill reaction.[1] In this procedure, the photochemical reduction process is followed directly by measuring the increase in optical density at 340 mμ as described below.

[1] See Vol. IV [15].

Reagents

Chloroplast suspension,[2,3] approximately 1 mg. of chlorophyll per milliliter.

TPN, 0.01 M.

Tris buffer, pH 7.2, 0.5 M.

Enzyme. Dilute the enzyme with 0.005 M Tris, pH 8, to give a solution containing approximately 3 to 5 units/ml. (See definition of unit below.)

Procedure. The reaction mixture in a modified glass cuvette[4] contains 0.5 micromole of TPN, 0.3 ml. of Tris buffer, 0.3 to 0.5 unit of enzyme, and 0.1 ml. of chloroplast suspension in a final volume of 3 ml. The optical density of the reaction mixture at 340 mμ is measured against a blank which contains everything except enzyme. Both the blank and the reaction cuvette are then placed in a water bath, kept at room temperature, and illuminated for 5 minutes. The light source is a 75-watt bulb at a distance of approximately 2 inches. After illumination, the optical density of the reaction mixture is again measured as indicated above. The increase in optical density, calculated from the two measurements, serves as a measure of the enzymatic activity.

It should be noted that, since the blank contains everything except enzyme, the data obtained in this manner are corrected for the slow endogenous rate of TPN reduction by chloroplasts alone in the absence of added enzyme.

Definition of Unit and Specific Activity. One unit of enzyme is defined as that amount which produces a change in optical density of 1.0 in 10 minutes at 340 mμ when the reaction mixture contains 0.1 mg. of chlorophyll per 3 ml. This unit of enzyme corresponds to the reduction of 4.8 micromoles of TPN per milligram of chlorophyll in 10 minutes. Specific activity is expressed as units per milligram of protein. The

[2] D. I. Arnon, M. B. Allen, and F. R. Whatley, *Biochim. et Biophys. Acta* **20**, 449 (1956).

[3] Chloroplasts are prepared according to Arnon *et al.*[2] as follows: After removal of the midribs, 25 g. of spinach leaves are ground in a mortar and pestle, in the cold, with 4 g. of sand and 35 ml. of 0.35 M NaCl. The slurry is filtered through a double layer of cheesecloth, and the filtrate is centrifuged in the cold for 1 minute at 200 × g to remove sand, leaf debris, and whole cells. The residue is discarded, and the supernatant is centrifuged in the cold for 8 minutes at 1350 × g. The supernatant is discarded, and the sedimented chloroplasts are suspended in 10 ml. of 0.35 M NaCl. In general, the chloroplast suspension contains approximately 0.7 to 1.0 mg. of chlorophyll per milliliter. Chlorophyll concentration is determined by the modification of Arnon [*Plant Physiol.* **24**, 1 (1949)].

[4] Purchased from the Kontes Glass Company, Vineland, New Jersey. These cells have a light path of 1 cm. and are 9 cm. in length.

protein content of the enzyme preparations is determined either by the Folin phenol reagent method of Lowry et al.[5] or by the ultraviolet absorption procedure of Warburg and Christian.[6]

Application of Assay Method to Crude Homogenate. It is difficult to determine accurately the enzymatic activity in a crude homogenate of spinach for several reasons. First, the specific activity of the enzyme in the crude homogenate is rather low. Second, the amount of crude homogenate which can be used in the assay is small and is limited by the high light absorption, at 340 mμ, of the crude homogenate.

Purification Procedure

The enzyme purification presented below has been described, in part, by San Pietro and Lang.[7] All steps in the purification are carried out at 0° to 4°; centrifugations are carried out in the cold.

Step 1. Preparation of Crude Homogenate. Two kilograms of spinach leaves, purchased at a local grocery, are freed from veins and ground in 2.6 l. of cold distilled water in the Waring blendor for 3 minutes at 100 volts. The dark-green homogenate is filtered through a double layer of cheesecloth and glass wool, and the residue is discarded. To the dark-green filtrate is added sufficient 0.5 M Tris-HCl buffer, pH 8, to give a final concentration of 0.05 M Tris.

Step 2. Extract of Acetone Precipitate. Acetone, previously cooled in the deep-freeze, is added to 3362 ml. of the adjusted filtrate slowly and with mechanical stirring to a final concentration of 35%. The preparation is centrifuged at 1000 × g for 15 minutes, and the clear yellow-green supernatant fluid, which contains the majority of the activity, is decanted and saved. The dark-green residue contains very little of the activity and is discarded.

The activity is precipitated from the supernatant solution by the slow addition of acetone, precooled in the deep-freeze, to a final concentration of 75%. During the addition, the solution is stirred mechanically. The precipitate is flocculent and settles rapidly when stirring is discontinued. After 15 minutes, the greater part of the supernatant fluid can be decanted before collection of the precipitate by centrifugation for 5 minutes at 1000 × g. The resulting clear yellow-green supernatant fluid is discarded, and the light brown-colored residue is suspended thoroughly in ice-cold 0.005 M Tris, pH 8. The suspension is centrifuged at

---

[5] O. H. Lowry, N. J. Rosebrough, A. L. Farr, and R. J. Randall, *J. Biol. Chem.* **193**, 265 (1951). See also Vol. III [73].

[6] O. Warburg and W. Christian, *Biochem. Z.* **310**, 384 (1941–42). See also Vol. III [73].

[7] A. San Pietro and H. M. Lang, *J. Biol. Chem.* **231**, 211 (1958).

18,000 × g for 20 minutes, and the residue is discarded. The clear brown supernatant solution is dialyzed against 10 to 20 vol. of 0.005 M Tris, pH 8, in the cold. To ensure complete removal of residual acetone, the dialysis solution is changed once at the end of 3 hours; the total time of dialysis is 15 hours. The volume of the enzyme solution after dialysis is 370 ml.

Step 3. Precipitation with Protamine Sulfate. The enzyme is precipitated by the addition of 1% protamine sulfate, pH 6, at a ratio of approximately 9 mg. of protamine sulfate to 100 mg. of protein. The solution is centrifuged for 15 minutes at about 1500 × g, and the brown supernatant fluid is decanted and saved for the purification of transhydrogenase.[8] The residue is extracted thoroughly with 75 ml. of 0.5 M Tris, pH 8, and after centrifugation, as above, the residue is re-extracted two times with the same volume of buffer. The volume of the combined extracts was 215 ml.

Step 4. Treatment with Dowex–Bentonite[9] and Concentration with Acetone. A suspension of Dowex 1 formate (volume after settling, 215 ml.) and 21.5 g. of bentonite is centrifuged for 15 minutes at 10,000 × g

SUMMARY OF PURIFICATION PROCEDURE

Fraction	Total volume, ml.	Total units	Total protein, mg.	Specific activity, units/mg. protein	Yield, %
Crude homogenate	3362	9,840	57,826	0.17	
Extract of acetone precipitate	370	10,100	6,711	1.5	103
Protamine sulfate	215	5,880	735	8.0	60
Dowex–bentonite concentrate	11.3	2,320	69	33.6	24

and the supernatant is decanted and discarded. The residue of bentonite and Dowex 1 formate is suspended in the 215 ml. of combined extracts[10] from the previous step, and the resulting suspension is centrifuged as

[8] See D. L. Keister and A. San Pietro, Vol. VI [61].
[9] A given volume of a suspension of Dowex 1 formate (200 to 400 mesh, ×10) is allowed to settle for 15 minutes, and the volume of the settled Dowex 1 formate is measured. For every milliliter of settled Dowex 1 formate, 100 mg. of solid bentonite are added. To facilitate suspension of the bentonite and Dowex 1 formate, it is convenient to add water at this point. The amounts of settled Dowex 1 formate and bentonite used per milligram of protein (in the combined extracts from step 3) are 0.3 ml. and 30 mg., respectively.
[10] When the protein concentration of the combined extracts from step 3 is greater than 3 to 4 mg./ml., it is adjusted to this concentration prior to treatment with the Dowex 1 formate and bentonite.

above. The residue is discarded,[11] and the enzyme is precipitated from the supernatant by the addition of 3 vol. of acetone, precooled in the deep-freeze. The residue is extracted thoroughly with $0.05 M$ Tris, pH 8; any insoluble material is removed by centrifugation as above. The resulting reddish-brown supernatant is referred to as the Dowex–bentonite concentrate.

A summary of the yields and purifications obtained at each step in the preparation of the enzyme is presented in the table.

Properties

Homogeneity and Physical Properties.[12] The purified enzyme is homogeneous in the Spinco ultracentrifuge and Tiselius electrophoresis. The molecular weight of the enzyme calculated from sedimentation data is approximately 14,000.

Amino acid analysis indicates a total of ninety-five amino acid residues (not including the tryptophan content of the enzyme), and there is no methionine. The minimum molecular weight calculated from amino acid analysis is 11,780, but this value does not include the contribution of tryptophan.

The purified enzyme is reddish-brown in color. The absorption spectrum of the enzyme exhibits peaks at 278, 320, 413, and 457 mμ and shoulders at 284 and 502 mμ. To date it has not been possible to associate these spectral characteristics with any of the usual electron transport compounds. Although the nature of the redox group, if any is present, in the protein is unknown, it does not appear to be either a heme or a flavin. The ratio of the absorption at 280 and 260 mμ is only 1.2.

Specificity. In the presence of illuminated chloroplasts or grana, the purified enzyme catalyzes the reduction of TPN, but not DPN, under the assay conditions described.[13]

With the "extract of acetone precipitate" used as the enzyme source and illuminated grana, the reduction of both TPN and DPN is observed. When chloroplasts are used in place of grana, only the reduction of TPN is observed. In this latter system, it is possible to demonstrate reduction of DPN, provided a high concentration of DPN is used.[13]

Acetylpyridine-TPN and deamino-TPN are both reduced in the pres-

---

[11] Additional enzyme has been recovered from this residue by extraction with water.

[12] A. San Pietro and F. E. Stolzenbach, unpublished observations, 1959.

[13] It is possible to demonstrate DPN reduction in this system provided the DPN concentration is high, approximately $5 \times 10^{-3} M$. It appears, at present, that the photochemical reduction of DPN requires both the enzyme described herein and transhydrogenase.[14]

[14] D. L. Keister and A. San Pietro, *Biochem. Biophys. Research Communs.* 1, 110 (1959). See also Vol. VI [61].

ence of chloroplasts or grana and the purified enzyme. The corresponding analogs of DPN are inactive in this system, as are NMN (nicotinamide mononucleotide), NR (nicotinamide riboside), and the α-isomer of DPN.

It should be noted that the enzyme is without effect on the photolytic activity of chloroplasts or grana when measured spectrophotometrically with 2,3,6-trichlorophenolindophenol or manometrically with ferricyanide.[7, 15]

Photoinactivation.[16] The enzyme is irreversibly photoinactivated in the presence of chloroplasts. The photoinactivation is prevented by a number of reducing compounds, e.g., cysteine, thioglycolate, and glutathione, as well as ferricyanide. However, none of these compounds is capable of reactivating the photoinactivated enzyme.

Inhibitors.[17] The enzyme is completely inhibited by p-chloromercuribenzoate at a concentration of $10^{-5} M$; this same concentration of p-chloromercuribenzoate is without effect on the Hill activity of the chloroplasts.

Both benzyl and methyl viologen inhibit the enzymatic activity. At a concentration of $8.5 \times 10^{-6} M$, benzyl and methyl viologen cause an inhibition of 78% and 59%, respectively.[18] This same concentration of either viologen is without effect on the Hill activity of the chloroplasts.

FMN is inhibitory and causes an inhibition of 100% at $10^{-4} M$, 55% at $10^{-5} M$, and 18% at $10^{-6} M$.

Phenazine methosulfate causes an inhibition of 70% at $3.3 \times 10^{-6} M$ and 93% at $10^{-5} M$.

A variety of compounds are without effect on the enzymatic activity at concentrations as high as $10^{-2} M$. These include sodium azide, iodoacetic acid, potassium cyanide, and sodium arsenite.

pH Optimum.[18] The photochemical reduction of TPN by illuminated chloroplasts requires at least two component systems: the photolytic system and a system for transferring hydrogen from the photolytic system to the TPN. The pH optimum for the complete system is about 7. At present, it is not possible to measure the pH-dependence of the transferring system which is felt to be the site of action of the enzyme described here.

Stability. The purified enzyme can be stored at $-15°$ for 3 months

[15] A. San Pietro, *Brookhaven Symposia in Biol.*, No. 11, p. 262 (1958).

[16] J. Giovanelli and A. San Pietro, *Arch. Biochem. Biophys.* 84, 471 (1959).

[17] Any inhibitor which will inhibit the Hill activity of the chloroplasts will also inhibit the photochemical reduction of TPN in the presence of illuminated chloroplasts. Examples of this type of inhibitor are o-phenanthroline and 3-(p-chlorophenyl) 1,1-dimethylurea.

[18] An enzyme fraction equivalent to that designated as "protamine sulfate" in the table was used in these experiments.

with little or no loss in activity. The activity is completely lost by keeping the enzyme at 100° for 2 minutes.

Under acid conditions, the enzyme is irreversibly bleached, and the activity is completely lost. In addition it appears that hydrogen sulfide is released from the enzyme under these conditions. The bleaching of the enzyme is accompanied by a loss of, at least, the absorption at 413 and 457 mμ. A similar loss in absorption in this region is observed under alkaline conditions.

Other Enzymatic Activities. The purified enzyme is devoid of ATPase activity, alcohol dehydrogenase activity with either TPN or DPN, and lipoic dehydrogenase activity with either DPN or TPN.

[63] Crystalline Firefly Luciferase

$$LH_2 + ATP \rightleftarrows LH_2\text{—}AMP + PP$$
$$LH_2\text{—}AMP + O_2 \rightarrow L\text{—}AMP + light + H_2O$$

By W. D. McElroy

The production of light by extracts from fireflies depends on an oxidative reaction catalyzed by an enzyme, luciferase. The initial step is the reaction of ATP with the carboxyl group of luciferin to form luciferyl adenylate and pyrophosphate. The enzyme-bound luciferyl adenylate reacts with oxygen to create an excited state which subsequently emits light.

Assay Method

Since the intensity of the light emitted depends on both enzyme and substrate concentration (ATP and LH$_2$), the quantitative determination of these substances can be made by measuring the light intensity by a suitable photocell arrangement. The light units are arbitrary. The reaction mixture consists of enzyme, 0.1 ml. of 0.1 M MgSO$_4$, 0.2 ml. of luciferin solution, 0.2 ml. of ATP (1 mg./ml.), and 0.025 M glycylglycine buffer, pH 7.5, to make the final volume to 2.5 ml. The reaction is started by injecting the ATP with a syringe into the reaction mixture held in a small test tube in front of the photocell. The initial flash of light is proportional to the enzyme concentration. One unit of enzyme is defined as that amount which gives one arbitrary light unit.

Preparation of Crude Enzyme

The following procedure is based on the original report of McElroy and Coulombre.[1] The live fireflies are dehydrated in a continuous vacuum over calcium chloride for 24 hours. The dried lanterns are stable for several years provided they are kept at low temperatures (deep-freeze). Thirty grams of dried lanterns are ground in a mortar with 2 to 3 g. of sand. One hundred milliliters of cold acetone are added, mixed with the lanterns, and ground thoroughly. The suspension is filtered on a Büchner funnel and washed twice with 200-ml. aliquots of cold acetone and finally air-dried. All subsequent processes are carried out in the cold, and all centrifugation is at 5° or less.

The acetone powder is ground in a mortar and then extracted with 100 ml. of $10^{-3} M$ Versene plus 10 ml. of $1 N$ NaOH. The pH of the crude suspension should be near 8.0 as measured by the glass electrode standardized at room temperature with the solution cold. If the pH is below 7.5 to 7.6, additional alkali should be added. The suspension after standing for 10 to 15 minutes is centrifuged at 3000 r.p.m. for 10 minutes, and the residue is extracted twice with 60 ml. of $10^{-3} M$ Versene plus 1 ml. of $1 N$ NaOH. The combined extracts are then centrifuged at $20,000 \times g$ in a refrigerated centrifuge for 30 minutes. The enzyme plus luciferin remain in the supernatant. (This preparation is the one recommended for ATP assay, since the light response depends only on the addition of ATP and magnesium; see Vol. III [122]).

Three parts of the crude enzyme extract are mixed with one part of calcium phosphate gel containing 16 to 17 mg. of the gel per milliliter. The gel, at pH 7.5, is measured and then centrifuged, and the supernatant is discarded. The gel is thoroughly mixed with the extract and recentrifuged. The supernatant from the first gel treatment is mixed, as above, with two and one-half times the volume of gel. The luciferase is absorbed, and the luciferin remains in the supernatant. Luciferase is eluted from the gel with cold 20% saturated $(0.82 M)$ $(NH_4)_2SO_4$ in $10^{-3} M$ Versene at pH 8.0. Three elutions of 100 ml. each are made.

The pH of the solution is adjusted to 8.0, and solid ammonium sulfate is added. Fractions are taken at 40%, 50%, 60%, and 70% saturated ammonium sulfate (pH 7.5). The precipitates are separated by centrifuging at $20,000 \times g$ for 15 minutes and dissolved in $10^{-3} M$ Versene at pH 8.0. The fraction collected between 50 and 60% saturated $(NH_4)_2SO_4$ contains over 60% of the luciferase at approximately 50% purity and can be used for crystallization. The other fractions are combined and can be used in a later purification or immediately refractionated and added to the 50 to 60% $(NH_4)_2SO_4$ fraction described above.

[1] W. D. McElroy and J. Coulombre, *J. Cellular Comp. Physiol.* **39**, 475 (1952).

Crystallization of Luciferase

The crystallization procedure is based on an original report by Green and McElroy.[2] The first crystals of luciferase were observed when the partially purified enzyme described above was subjected to electrophoresis on a sponge system devised by Dr. Herschel K. Mitchell (personnel communication, 1955). As the enzyme moved from the sponge containing the high salt content in which it was dissolved to one of low ionic strength, crystallization rapidly occurred. This observation led to the realization that luciferase was a euglobulin and could be crystallized on dialysis against a solution of low ionic strength.

The 50 to 60% $(NH_4)_2SO_4$ fraction is dialyzed against a solution containing $10^{-3} M$ Versene, $0.01 M$ NaCl, and $0.002 M$ Na_2HPO_4, pH 7.2 to 7.4. After dialysis overnight the crystals are centrifuged in the cold and dissolved in $0.4 M$ $(NH_4)_2SO_4$ and $0.001 M$ Versene, pH 8.0. Recrystallization is effected by redialyzing against the same solution as described above. Unusually large crystals may be obtained by dialyzing against $0.15 M$ NaCl and $0.02 M$ phosphate buffer, but the process is slow. One series of crystallization using a number of ammonium sulfate fractions is given in the table.

Fraction	Total light units	Light units/mg. protein
Crude extract	6.1×10^7	3,000
Supernatant of first gel	5.2×10^7	3,500
Elution of second gel	3.8×10^7	6,000
50–60% $(NH_4)_2SO_4$ fraction	2×10^7	11,000
Frist crystallization	1.8×10^7	22,000
Second crystallization	1.5×10^7	26,000
Third crystallization	1.3×10^7	27,500
Fourth crystallization	1.2×10^7	28,000
Fifth crystallization	1.2×10^7	28,000

Recovery of Luciferin

Most of the luciferin remains in the supernatant after the calcium phosphate gel treatment. The solution is adjusted to pH 3.5 and extracted two times with an equal volume of redistilled ethyl acetate. All the active luciferin passes into the ethyl acetate. The ethyl acetate is removed by vacuum distillation in the presence of a small amount of water. After all the ethyl acetate is removed, the pH is adjusted to approximately 7.0, and water is added to give a volume of 25 ml. This crude luciferin can be used for enzyme assay.

[2] A. A. Green and W. D. McElroy, *Biochim. et Biophys. Acta* **20**, 170 (1956).

Stability of the Crystalline Enzyme

The crystalline enzyme is precipitated by 60% ammonium sulfate in the presence of 0.001 M Versene at a pH between 7.5 and 8.0 and stored in this state at 4°. Under such conditions the enzyme is stable for months. To prepare for use, a portion of the suspension is centrifuged, and the precipitate then is dissolved in Versene at pH 8.0.

Physicochemical Characteristics

The isoelectric point of firefly luciferase is between pH 6.2 and 6.3. The sedimentation constant is S_{25} (0.15 M NaCl) = 5.6, and this evidence plus the combination with oxyluciferin suggests a molecular weight of 100,000. The five-times-crystallized enzyme is homogeneous as judged by electrophoretic, ultracentrifugation, and enzymatic tests. Inorganic pyrophosphatase and luciferyl adenylate hydrolase are common contaminants. These are completely absent after three crystallizations.

Properties

The temperature optimum of the enzyme-catalyzed light-emitting reaction is 23° to 25°. The pH optimum is 7.8. Phosphate is an inhibitor, as are other salts at high concentration such as calcium and potassium. The enzyme is not affected by arsenate, cyanide, and azide but is strongly inhibited by various amines, copper, and p-chloromercuribenzoate. Pyrophosphate inhibits, owing to the reversal of the initial activating step. Benzimidazole, benztriazole, and substituted derivatives act as competitive inhibitors of luciferin.

The enzyme catalyzes the formation of oxyluciferyl and luciferyl adenylate in the presence of ATP and magnesium. Luciferyl adenylate remains tightly bound to the enzyme which seems to be a characteristic of most enzymes catalyzing the formation of an acyl adenylate. The enzyme-bound oxyluciferyl adenylate readily reacts with two compounds: pyrophosphate to form ATP, enzyme, and oxyluciferin; and coenzyme A to form luciferyl CoA, enzyme, and adenylic acid.

[64] Glutathione Homocystine Transhydrogenase

GSH + homocystine $\rightleftarrows$ GSSG + homocysteine

By E. RACKER

Assay Method

Principle. The assay system is dependent on the reduction of homocystine by a system regenerating reduced glutathione.[1] The latter is provided by alcohol, alcohol dehydrogenase, yeast glutathione reductase, and catalytic amounts of DPN and GSSG. Under these conditions, the rate of spontaneous hydrogen transfer from reduced glutathione to homocystine is negligibly small and markedly accelerated by transhydrogenase. Homocysteine which accumulates is determined colorimetrically with nitroprusside.[2]

Reagents

1 M dibasic ammonium phosphate.
10^{-2} M DPN.
0.08% GSSG.
5 M ethanol.
1% alcohol dehydrogenase.
0.3% GSSG reductase.[3]
10^{-3} M homocystine.

Procedure. In a final volume of 1 ml. the following solutions are pipetted: 0.03 ml. of ammonium phosphate, 0.02 ml. of DPN, 0.05 ml. of GSSG, 0.02 ml. of ethanol, 0.02 ml. of alcohol dehydrogenase, 0.02 ml. of glutathione reductase, 0.7 ml. of homocystine, and finally the sample of transhydrogenase to be tested. After 30 minutes at 37°, the mixture is deproteinized and analyzed for SH as described by Grunert and Phillips.[2] One control vessel without homocystine and one without transhydrogenase should always be run to correct for the small amounts of GSH formed and for the nonenzymatic hydrogen transfer.

Application of Assay Method to Crude Tissue Preparations. The assay procedure as described is employed for measurements of transhydrogenase in crude extracts of liver and yeast. If added homocystine disappears under the test conditions, it is advisable to run the reaction in a nitrogen atmosphere.

[1] E. Racker, *J. Biol. Chem.* **217**, 867 (1955).
[2] R. R. Grunert and P. H. Phillips, *Arch. Biochem. Biophys.* **30**, 217 (1951).
[3] E. Racker, Vol. II [127] p. 722.

Purification Procedure

The liver enzyme is difficult to purify and is readily inactivated by contact with ammonium sulfate. A partially purified preparation is prepared as follows: Beef liver is mixed with 2 to 3 vol. of ice-cold acetone in a Waring blendor, and the mixture is poured into 8 vol. of acetone and then rapidly centrifuged in the cold room. The precipitate is washed once with 8 vol. of acetone, then pressed out between paper towels and crumbled into a fine powder which dries rapidly when distributed on large filter papers. The powder is extracted with 8 vol. of distilled water containing 600 μg. of neutralized EDTA per milliliter at 2° for 45 minutes with slow mechanical stirring. A crude extract is obtained by centrifuging this mixture at 8000 $\times$ g for 20 minutes. The activity in this extract from the acetone powder is approximately 10 μg. of homocysteine formed per milligram of protein in 30 minutes. To the crude extract, an equal volume of 3 M dibasic ammonium phosphate is added, the mixture is centrifuged for 20 minutes at 2° at 13,000 $\times$ g, and the precipitate is discarded. For each 100 ml. of supernatant solution, 15 g. of solid dibasic ammonium phosphate is added, and the mixture is centrifuged as above. The precipitate is collected in a small volume of water to yield a solution of a final protein concentration of about 50 mg./ml. This preparation forms about 32 μg. of homocysteine per milligram of protein in 30 minutes, thus representing approximately a three-fold purification from the crude extract of acetone powder.

Properties

The enzyme preparations obtained thus far are rather crude and relatively unstable. Preparations of greater stability have recently been obtained from baker's yeast, but a satisfactory purification procedure has not as yet been worked out. Crude preparations of the glutathione homocystine transhydrogenase transfer hydrogen from glutathione to cystine as well as to homocystine. Transhydrogenase also accelerates the reverse reaction, namely the reduction of GSSG by homocysteine.

PREPARATION AND ASSAY OF SUBSTRATES

Section I

Carbohydrates

[65] Sialic Acids and Derivatives: Preparation

By LARS SVENNERHOLM

Nomenclature and Occurrence

The term "sialic acids" is a group name for acylated nonulosaminic acids, which exist in nature. For the basic unsubstituted structure, $C_9H_{17}O_8N$, common to all, Blix *et al.*[1] have suggested the name neuraminic acid. The naturally occurring acids are all N-acylated, and part of them in addition are O-substituted. They are designated according to the nature and linkage of the substituent(s), e.g., *N*-acetylneuraminic acid, *N*-glycolylneuraminic acid, and *O,N*-diacetylneuraminic acid.

N-Acetylneuraminic acid is the most commonly occurring acid and exists alone in all the tissues of man, birds, reptiles, and fishes.[2] In all other mammals investigated, *N*-acetyl- and *N*-glycolylneuraminic acids occur together in the tissues, except in the central nervous system, where there is *N*-acetylneuraminic acid only.[2, 3] In many sources the sialic acids also contain in the native state one or two *O*-acetyl groups. The *O*-acetyl groups are very labile, and some of them are lost from both the diacetyl- and triacetyl-neuraminic acids during preparation and purification. Because of the large difference in stability, the methods elaborated for the preparation of only N-acylated neuraminic acids are quite distinct from those for the sialic acids containing *O*-acetyl groups as well. Therefore it has been considered suitable to give a detailed description of the isolation procedure for both *N*-acetylneuraminic acid and *O*-acylated neuraminic acids. The procedure for the preparation of methoxyneuraminic acid (methylglycoside of neuraminic acid) is also described.

Preparation of *N*-Acetylneuraminic Acid

Principle. *N*-Acetylneuraminic acid free from *N*-glycolylneuraminic acid is suitably prepared from human material, e.g., serum. The acid is liberated by mild acid hydrolysis from its natural source and purified by ion-exchange chromatography.

Reagents

0.05 *M* sulfuric acid.
Barium hydroxide solution, saturated.

---

[1] G. Blix, A. Gottschalk, and E. Klenk, *Nature* **179**, 1088 (1957).
[2] L. Svennerholm, unpublished results.
[3] E. Mårtensson, A. Raal, and L. Svennerholm, *Biochim. et Biophys. Acta* **30**, 124 (1958).

Ethanol.
Methanol, redistilled.
Diethyl ether, redistilled.
Dowex 50-X8, 50 to 100 mesh (H+ form).
Dowex 1-X8, 100 to 200 mesh (formate form).

Procedure. Five liters of serum (containing about 600 mg. of *N*-acetylneuraminic acid per liter) is poured into 4 vol. of hot ethanol and boiled for about 30 minutes. The ethanol is filtered off on a large Büchner funnel, and the protein precipitate is washed on the funnel with 1 vol. of ethanol. Cations and low-molecular-weight organic compounds are removed by stirring the protein precipitate at 4° with 10 l. of 0.01 M sulfuric acid. The procedure is repeated until the pH of the supernatant is about pH 2.5. The precipitate is then heated with 5 l. of 0.05 M sulfuric acid at 80° for 1 hour. The sulfuric acid solution is filtered off with suction, and the hydrolysis is repeated twice with the same amount of 0.05 M sulfuric acid. The combined filtrates are neutralized with saturated barium hydroxide to pH 5 to 6. The precipitate of barium sulfate is removed by filtration, and the solution is poured into a column of Dowex 50-X8, 50 to 100 mesh (30 × 300 mm.), in H+ form. The outlet from this column is coupled to a second column with Dowex 1-X8, 100 to 200 mesh (30 × 300 mm.), in formate form. After the columns have been rinsed with 2 l. of distilled water, the Dowex 1 column is coupled to a large Mariotte flask containing 0.3 N formic acid. The eluate is collected in 25-ml. fractions on an automatic collector and tested for sialic acid by the resorcinol reagent.[4] About 98% of the sialic acid is eluted in a main peak. Only the contents of the tubes belonging to this main peak are pooled.[5] An equal volume of distilled water is added, and then the solution is concentrated with a rotating evaporator in a water bath, the temperature not exceeding 40°, to one-fourth its original volume. To avoid destruction of the sialic acid by the increasing concentration of formic acid, further water is added, and evaporation is continued. The addition of water and renewed evaporation are repeated until no or only a faint smell of formic acid is recognized; then lyophilization can take place. The lyophilized white powder is placed in a vacuum desiccator over KOH pellets and left there for at least 48 hours. Yield, 2.7 to 2.8 g.

Purification. The lyophilized substance is dissolved in 8 ml. of distilled water (3 ml. of water to 1 g. of substance). Eighty milliliters of methanol is added with warming to about 40°. Then 100 ml. of ether is added. An amorphous precipitate appears, which is immediately filtered off on a

[4] L. Svennerholm, *Biochim. et Biophys. Acta* **24**, 604 (1957).
[5] L. Svennerholm, *Acta Soc. Med. Upsaliensis* **61**, 75 (1956).

sintered-glass filter. The filtrate is transferred to a flask with a ground stopper, and further ether is added until a faint opalescence appears. The flask is left at room temperature, and further ether is added during the following 2 days to obtain complete crystallization. After 48 hours the crystals are collected on a sintered-glass filter, washed with methanol–ether (1:1, v/v) and dried over P_2O_5. Yield, 2.5 g.; purity 97 to 100%. Recrystallization is performed as described above. As the sialic acids have a great tendency to form methyl esters in methanol, the crystallization has also been performed in glacial acetic acid.[6] This procedure does not remove impurities as efficiently as the described technique. By the use of water as suggested here, the tendency to ester formation is very small. The crystallization time, however, should not be prolonged.

Preparation of N-Glycolylneuraminic Acid. A reasonably pure *N*-glycolylneuraminic acid (85 to 90%) can be obtained from hog submaxillary mucin. The mucin is prepared as described below for the isolation of bovine mucin. The sialic acid is liberated from the mucin by heating to 80° with 0.05 *M* sulfuric acid and purified by chromatography on the anion-exchange resin as described for *N*-acetylneuraminic acid.

The admixture of *N*-acetylneuraminic acid can be determined by glycolic acid analyses[7] and quantitative paper partition chromatography.[8]

Preparation of O,N-Diacetylneuraminic Acid and N-Acetyl-O-Diacetyl-neuraminic Acid

Principle. Bovine submaxillary mucin is isolated and purified by isoelectric precipitation. The sialic acids are liberated by boiling water and separated by cellulose column chromatography.[9] The final purification is achieved by crystallization.

Reagents

4 *M* hydrochloric acid.
Methanol, redistilled.
Diethyl ether, A.R.
1-Butanol, redistilled.
Acetic acid, glacial A.R.
Light petroleum, b.p. 40°.

Preparation. The method to be described has been published by Blix and Lindberg.[9] Fresh submaxillary glands are freed from adherent

[6] F. Zilliken, G. A. Braun, and P. György, *Arch. Biochem. Biophys.* 63, 394 (1956).
[7] E. Klenk and G. Uhlenbruck, *Z. physiol. Chem.* 307, 266 (1957).
[8] E. Svennerholm and L. Svennerholm, *Nature* 181, 1154 (1958).
[9] G. Blix and E. Lindberg, *Acta Chem. Scand.* 14, 1809 (1960).

connective tissue and lymph glands and are then frozen. They are cut into cubes with about 1 cm:s side. Three kilograms of cut glands is covered with distilled water at 4° for 24 hours with occasional stirring. The mucin extract is collected by "filtration" on a stainless-steel Büchner funnel with holes about 4 mm. in diameter. It is freed from tissue fragments by centrifugation. The extraction of the glands with water may be repeated four or five times. Hydrochloric acid is added to the extract until the mucin just dissolves on the acid side of its isoelectric point. The mucin is precipitated by dilution with a few volumes of water. By stirring the solution, the mucin precipitate adheres as an elastic clot to the stirring rod. The mucin is dehydrated by ethanol and torn into small pieces during this procedure. It is ground to powder and defatted by ether extraction.

Forty grams of mucin is dissolved in a 2-l. flask by addition of small portions of distilled water to a final volume of 500 ml. under vigorous shaking. The suspension should then have a pH below 3.5; otherwise dilute hydrochloric acid is added drop by drop until this pH is reached. The suspension is heated for 1 hour in a boiling-water bath, then cooled to room temperature and centrifuged. The undissolved residue is washed with 200 ml. of water, and the two supernatants are combined and lyophilized.

The freeze-dried material is extracted twice with 0.5 l. of methanol for 4 to 5 hours at 4°. The extracts are combined and evaporated to dryness below 0° in order to minimize methyl ester formation. The residue is dissolved in 2 ml. of water, and 30 ml. of methanol and 90 ml. of ether are added. The amorphous precipitate found is immediately filtered off, and the filtrate is evaporated to dryness. The residue is dissolved in 5 ml. of the upper phase from 1-butanol–acetic acid–water (4:1:5, v/v) and chromatographed on the LKB Chromax pressurized paper chromatography column system No. 3502. The eluate is collected in fractions of about 20 ml. One-tenth milliliter of each fraction is taken for sialic acid determination with the resorcinol reagent. The fastest-moving peak contains the triacetylneuraminic acid. Then there appears a small peak of a sialic acid with unknown structure, suggested to be a labile isomerization product of diacetylneuraminic acid.[9] The next large peak contains O,N-diacetylneuraminic acid.

The fractions containing the di- and triacetylneuraminic acids are pooled separately and evaporated below 0° to dryness. In order to remove the last traces of butanol, the residues are dissolved in about 100 ml. of methanol and evaporated in the same manner.

For crystallization the residue is dissolved in 0.5 to 1.0 ml. of water, and 30 ml. of methanol and 90 ml. of diethyl ether are added. After filtration, light petroleum is added drop by drop to the filtrate until faint turbidity appears. Crystallization takes place at room temperature or at 4°. More substance crystallizes on each addition of a few drops of light petroleum during the next days. Recrystallization is performed in the same way. Yield, 150 mg. of diacetylneuraminic acid; 400 to 500 mg. of triacetylneuraminic acid.

The purity of the compounds is suitably investigated by paper partition chromatography with 1-butanol–pyridine–water (6:4:3, v/v) as solvent.

Preparation of Methoxyneuraminic Acid

Principle. Submaxillary mucin is hydrolyzed with methanolic hydrochloric acid by which the methyl ester of methoxyneuraminic acid is formed. After saponification with ammonia, the methoxyneuraminic acid is chromatographed on a cation-exchange resin.[10]

Preparation. Fifty grams of bovine submaxillary mucin is treated with 5% methanolic hydrogen chloride at 105° to 110° in an autoclave for 3 hours. The hydrolyzate is centrifuged, and the residue is washed with methanol. The supernatants are combined and evaporated *in vacuo* to a sirup. The neuraminic acid is saponified on a boiling water bath for 2 hours with an excess of 100 ml. of 2 N ammonia.

After filtration, 80 to 90% of the chloride ions is removed on Dowex 2-X10, 20 to 50 mesh (OH⁻ form), and the ammonia in a desiccator over concentrated sulfuric acid. The methoxyneuraminic acid is freed from amino acids by chromatography on a column with Dowex 50-X4, 20 to 50 mesh. The methoxyneuraminic acid is eluted with distilled water. The fractions containing neuraminic acid free from amino acids as ascertained by paper partition chromatography are pooled, evaporated to 50 ml., and precipitated with 300 ml. of ethanol in the cold. The precipitate is dissolved in water and decolorized on a small column with alumina. The neuraminic acid is crystallized by addition of ethanol and recrystallized from water. Yield, 3.0 to 3.5 g.

Properties

Pure sialic acids are colorless. Except for the triacetylneuraminic acid, which melts at 130° to 131° with gas evolution and without discoloration, they do not melt but decompose over a range over several degrees under discoloration (see the table). They do not show any mutarotation in aqueous solution. The sialic acids and methoxyneuraminic acid are easily soluble in water. The triacetylneuraminic acid dissolves readily in

[10] F. Weygand and H. Rinno, *Z. physiol. Chem.* **306**, 173 (1957).

methanol; the other acids, especially N-glycolylneuraminic acid, are sparingly soluble in methanol. All the acids are insoluble in diethyl ether and light petroleum. The sialic acids are relatively strong acids with a pK_a value of 2.6 for N-acetylneuraminic acid[4] and diacetylneuraminic acid[11] and 2.75 for N-glycolylneuraminic acid.[11] Pure N-acetylneuraminic

DECOMPOSITION POINTS AND SPECIFIC ROTATIONS OF SIALIC ACIDS

Compound	Formula	Decomposition point	$[\alpha]_D^{22°}$ (in water)
N-Acetylneuraminic acid	$C_{11}H_{19}O_9N$	185°–187°	−32°
N-Glycolylneuraminic acid	$C_{11}H_{19}O_{10}N$	185°–187°	−32°
O,N-Diacetylneuraminic acid (bovine)	$C_{13}H_{21}O_{10}N$	138°–140°	+6°
N-Acetyl-di-O-acetylneuraminic acid	$C_{15}H_{23}O_{11}N$	130°–131°	+9°
Methoxyneuraminic acid	$C_{10}H_{19}O_8N$	About 200°	−55°

acid and N-glycolylneuraminic acid are very stable in water solution and can be stored for months at 4° without measurable changes. If the solution contains only very small amounts of mineral acids, the stability is much less. The di- and triacetylneuraminic acids are not stable. When kept at 37° they will be transformed rather soon to N-acetylneuraminic acid. They are all unstable in alkaline milieu except methoxyneuraminic acid, which, being a normal O-glycoside, is stable even to a strong alkali at 100°.

All sialic acids, but not methoxyneuraminic acid, reduce Fehling's solution. N-Acetylneuraminic and N-glycolylneuraminic acids consume 2 moles of periodate and form 1 mole of formic acid and about 1 mole of formaldehyde. Bovine diacetylneuraminic acid consumes only 1 mole of periodate, and the triacetylneuraminic acid does not consume any periodate at all.

By the action of aldolase from *Clostridium perfringens* and *Escherichia coli* K-235, N-acetylneuraminic acid is split into pyruvic acid and N-acetylmannosamine.[12]

Analysis

The sialic acids can be identified by several chemical and physicochemical methods. Of the chemical methods, elemental analyses of nitrogen, total acetyl, and O-acetyl are valuable, further glycolic acid analysis, and periodate oxidation. The optical rotation, the X-ray powder diagram, and the infrared spectra are the most important physico-

[11] G. Blix, E. Lindberg, L. Odin, and I. Werner, *Acta Soc. Med. Upsaliensis* **61**, 1 (1956).

[12] D. G. Comb and S. Roseman, *J. Biol. Chem.* **235**, 2529 (1960).

chemical methods. The X-ray powder diagrams were used by Blix and co-workers[11] for the original division of the sialic acids; but, as the acids form solid solutions, the method cannot be adopted for the estimation of the homogeneity of isolated sialic acids. For that purpose the infrared spectra and paper partition chromatography are convenient. Suitable solvents for chromatography are 1-butanol–acetic acid–water (4:1:5, v/v), 1-butanol–pyridine–water (6:4:3, v/v), and 1-butanol–1-propanol– 0.1 N HCl (1:2:1, v/v). For detection, orcinol, Ehrlich, or thiobarbituric acid reagents can be applied. Many different colorimetric methods have been elaborated for the determination of sialic acids; some of them are discussed in Vol. VI [66].

[66] Sialic Acids and Derivatives: Estimation by the Ion-Exchange Method

By LARS SVENNERHOLM

Sialic acids have been estimated by many colorimetric methods, such as the orcinol,[1] resorcinol,[2] diphenylamine,[3] direct Ehrlich,[4] tryptophan-perchloric acid,[5] hydrochloric acid,[6] sulfuric–acetic acid,[7] and thiobarbituric acid[8] procedures. None of the color reactions can claim absolute specificity, and none can be applied to the direct measurements of sialic acids in tissues. Most of the methods have previously been used in some modified form for the estimation of other carbohydrates. Because of that, mainly the influence of other carbohydrates has been investigated when applying a new procedure for sialic acid assay. As the direct Ehrlich reagent gives a negligible color with most carbohydrates, the Ehrlich reaction has been considered to be the most characteristic color reaction for sialic acids.[9] However, the direct Ehrlich reaction is essentially a pyrrole reaction and cannot be used for the assay of sialic acids

[1] E. Klenk and H. Langerbeins, Z. physiol. Chem. 270, 185 (1941).
[2] L. Svennerholm, Biochim. et Biophys. Acta 24, 604 (1957).
[3] W. Ayala, L. W. Moore, and E. L. Hess, J. Clin. Invest. 30, 781 (1951).
[4] I. Werner and L. Odin, Acta Soc. Med. Upsaliensis 57, 230 (1952).
[5] F. B. Siebert, M. L. Pfaff, and M. V. Siebert, Arch. Biochem. Biophys. 18, 279 (1948).
[6] J. Folch, S. Arsove, and J. A. Meath, J. Biol. Chem. 191, 819 (1951).
[7] E. L. Hess, A. F. Coburn, R. C. Bates, and P. Murphy, J. Clin. Invest. 36, 449 (1957).
[8] L. Warren, J. Biol. Chem. 234, 1971 (1959).
[9] A. Gottschalk, "The Chemistry and Biology of Sialic Acids and Related Substances." Cambridge University Press, Cambridge, 1960.

in body fluids and tissues, which may contain preformed pyrroles. In our experience the Ehrlich reaction gives the largest error of all the common sialic acid methods, when applied to quantitative estimations of sialic acid in different tissues. Similar data have been reported by Warren.[8] Reliable results can be obtained when the Ehrlich reaction is performed on isolated glycoproteins or fairly pure gangliosides, but under these circumstances many of the other methods are as reliable. A very sensitive and rather specific method for the assay of sialic acids, the thiobarbituric acid method, has been described by Warren.[8] A further advantage of the method is that it can be used for determination of free sialic acids. The present experience with the method is still too limited to allow any definite evaluation of the method. However, it cannot be used for the determination of lipid-bound sialic acids. In our hands 50% of the sialic acid in monosialogangliosides and 70% of the sialic acid in disialogangliosides were recovered. In lipid extracts still lower recoveries were obtained.

The specificity of the color reactions can be greatly enhanced by the isolation of the sialic acids from the biological material to be tested. As the sialic acids probably always have a terminal position in animal material, they are easily released by mild acid hydrolysis. A satisfactory purification of the hydrolyzate is achieved in a simple way on anion exchangers.[10]

Principle. The sialic acids are liberated with weak sulfuric acid and retained on a column with Dowex 2 in acetate form. Most of the contaminants are eluted with distilled water, and the sialic acids are displaced with an acetate buffer of pH 4.6. The concentration of sialic acid in the eluates is analyzed with a resorcinol method.[2]

Reagents

Dowex 2-X8, 200 to 400 mesh, wet-screened.
1 M acetic acid–sodium acetate buffer, pH 4.6.
1 mM N-acetylneuraminic acid. N-Acetylneuraminic acid (30.9 mg.)
   is dissolved in 100 ml. of distilled water; 1 ml. contains 1 micromole of N-acetylneuraminic acid.
0.05 M sulfuric acid.
Resorcinol reagent. Resorcinol A.R. (0.2 g.) is dissolved in 10 ml. of distilled water. Concentrated hydrochloric acid (80 ml.) and 0.1 M copper sulfate (0.25 ml.) are added, and the volume is made up to 100 ml. with distilled water.
1-butylacetate–1-butanol (85:15, v/v).[11]

[10] L. Svennerholm, *Acta Chem. Scand.* 12, 547 (1958).
[11] T. Miettinen and I.-T. Takki-Luukkainen, *Acta Chem. Scand.* 13, 856 (1959).

Preparation of Resin Columns. Dowex 2-X8, 200 to 400 mesh wet-screened through 150 mesh, is heated on a water bath with three times the volume of 2 N hydrochloric acid. After the resin has settled for 30 minutes, the supernatant is decanted to remove the fines. After undergoing this treatment four times, the resin is stirred up in water and poured into a large tube fitted with a glass filter. The resin is converted to acetate form by passing 2 N sodium acetate through the column until the effluent gives a negative test for chloride. The resin is then slurried up in 1 N acetic acid, transferred to a Büchner funnel, and excessive moisture is removed by suction.

For the preparation of the columns, about 3 g. of the moist resin is suspended in about 2 vol. of distilled water and poured into a chromatographic tube (inner dimensions 0.7 × 20.0 cm.), fitted with a glass filter plate (G 2) at the bottom and widened at the top (2 × 3 cm.). The resin is allowed to settle to a height of 6.0 cm., and excess resin is removed by suction. A small filter paper is placed over the resin, and above it sand to a height of 0.5 cm. After the resin has been washed with 10 ml. of 0.1 N acetic acid, the column is ready for use.

Regeneration of the Columns. After elution the columns are regenerated in the following order: 2 N sodium hydroxide, distilled water, 2 N sodium acetate, and 0.1 N acetic acid. Ten milliliters of each solution is used.

Hydrolysis of the Samples and Chromatography. Samples containing 0.1 to 2.0 micromoles (about 30 to 600 µg.) of sialic acid are hydrolyzed in small tubes having glass stoppers in 5 ml. of 0.1 N sulfuric acid at 80°. Having been heated, the tubes are cooled to room temperature, and the hydrolyzates are filtered into the column with resin. After filtration the hydrolysis tubes and the columns are washed twice with 5 ml. of water.

Standards of 0, 0.5, and 1.0 micromole of N-acetylneuraminic acid are treated under the same condition as the samples in each experiment.

The sialic acids are eluted from the resin columns with 8 ml. of 1 M acetic acid–sodium acetate buffer (pH 4.6). The effluent is collected in a 10-ml. cylinder. Distilled water is added to the 10-ml. mark.

Color Reaction. Two milliliters of effluent in duplicates is pipetted into a centrifuge tube (16 × 120 mm.), and 2 ml. of resorcinol reagent is added. The tubes are heated for 15 minutes in a bath of boiling water. After the tubes have cooled in tap water, 4 ml. of butylacetate–butanol is added, and the tubes are vigorously shaken and chilled for 10 minutes in a bath of ice water. After centrifugation, the organic solvent phase is transferred to 1.0-cm. absorption cells and read against pure organic solvent in a Beckman spectrophotometer Model B or DU at 450 mµ and 580 mµ. In the eluates of hydrolyzed tissue, sugar phosphates also occur.

Their influence on the absorbancy values at 580 mμ is calculated from the reading at 450 mμ. If greater sensitivity is wanted (for samples containing 0.05 to 0.5 micromole of sialic acid), the organic solvent phase may be read in 50-mm. microcells.

Calculations. The amount of total sialic acid is calculated from the standards of N-acetylneuraminic acid. The molar absorbancy indices (extinction coefficients) of di- and triacetylneuraminic acids are the same as that of N-acetylneuraminic acid. As the molar absorbancy index of N-glycolylneuraminic acid is 30% greater than that of N-acetylneuraminic acid, the found amount has to be multiplied by 0.77, if the sample contains N-glycolylneuraminic acid only.

If no standards are available, the amount of N-acetylneuraminic acid present in the samples taken for hydrolysis can be determined from the following equation:

$$\text{Micromoles } N\text{-acetylneuraminic acid} = \frac{V \times A_{580} \times 5}{8} = 2.5 \times A_{580}$$

where V is the volume of the organic solvent phase, 8 is the molar absorbancy index of N-acetylneuraminic acid divided by 1000, and 5 is the volume factor (2 ml. of a 10-ml. eluate are analyzed). If microcells are used, the equation is divided by 5.

Interfering Substances. Only negligible amounts of interfering carbohydrates are eluted together with the sialic acids in the determination of sialic acid in body fluids such as serum, lymph, milk, synovial fluid, and transudate. When tissues are subjected to assay, sugar phosphates, mainly ribose phosphates, are eluted together with the sialic acids. The occurrence of interfering substances is detected from the reading at 450 mμ. A rather accurate calculation of the interference from other sugars can be made from the dichromatic readings.[2] When the amount of interfering substances is great, as judged from the reading at 450 mμ, it is desirable to analyze the eluates by a second method also, e.g., the thiobarbituric acid assay.[8]

[67] Thiobarbituric Acid Assay of Sialic Acids[1,2]

By LEONARD WARREN

Assay Method

Principle. Periodate oxidation of the neuraminic acid backbone of sialic acids results in the formation of β-formylpyruvic acid from carbon atoms 1 to 4. The N-acetyl or N-glycolyl group of sialic acids apparently does not interfere with periodate oxidation. β-Formylpyruvic acid is then coupled with 2-thiobarbituric acid to form a red chromophore with a maximum absorption at 549 mμ. Since only free sialic acids are reactive in the assay, hydrolysis of sialic acid-containing material must be carried out for the measurement of total sialic acids. The assay is suitable for measuring the release of bound sialic acid by sialidase (neuraminidase).

Reagents

Sodium metaperiodate (0.2 M) in 9 M phosphoric acid (53%).
Sodium arsenite (10%) in 0.5 M sodium sulfate solution.
2-Thiobarbituric acid[3] (0.6%) in 0.5 M sodium sulfate solution.
Cyclohexanone.

All solutions are stable for several months at room temperature.

Procedure. The assay is usually carried out in 130 $\times$ 10-mm. test tubes. To the sample of sialic acid (0.01 to 0.05 micromole) in a volume of 0.2 ml. is added 0.1 ml. of the solution of sodium periodate. These are thoroughly mixed and allowed to stand at room temperature. After 20 minutes, 1.0 ml. of the sodium arsenite solution is added, and the tubes are shaken vigorously. After 2 minutes the tubes are shaken again to assure the complete discharge of yellow-brown color. Three milliliters of 2-thiobarbituric acid solution are added, the contents are mixed by inversion of the tubes, and the tubes are inserted into a vigorously boiling-water bath for exactly 15 minutes. The tubes are cooled in tap water for 5 minutes, and 4 ml. of cyclohexanone are added to each tube.

---

[1] L. Warren, *J. Biol. Chem.* **234**, 1971 (1959). A graphic method for correction can also be used. This is presented in: R. K. Jakoby and L. Warren, *Neurology* **11, 232** (1961).

[2] D. Aminoff, *Virology* **7**, 355 (1959); *Biochem. J.* **81**, 384 (1961).

[3] Several of the commercial preparations labeled "2-thiobarbituric" acid contain approximately 2 moles of base per mole of 2-thiobarbituric acid. The solubility of the free acid is approximately 0.9 g. per 100 ml. of water at room temperature, and that of its salt is 6 g. per 100 ml.

The tubes are shaken vigorously twice and centrifuged briefly. The top clear cyclohexanone phase is transferred to cuvettes of 1-cm. light path, and optical densities are determined at 532 mμ and 549 mμ.

Calculations. The optical density reading at 549 mμ is a measure of the sialic acid present. This may be calculated from Eq. (1), which is based on an observed molecular extinction coefficient of 57,000 for N-acetylneuraminic acid.

$$\text{Micromoles} = 0.070 \times OD_{549} \qquad (1)$$

There are however, nonsialic acid materials in many biological preparations which give rise to a chromophore(s) with an absorption maximum at 532 mμ. Correction may be made for this interfering material by inserting the observed optical density values at 532 and 549 mμ into Eq. (2). The derivation of this equation depends on the molecular extinction coefficients at 532 mμ and 549 mμ of sialic acids and the interfering material (considered 2-deoxypentose[4]) observed by the author[1] (see footnote 1 for derivation).

$$\text{Micromoles} = (0.084 \times OD_{549}) - (0.031 \times OD_{532}) \qquad (2)$$

Comments. The thiobarbituric acid assay measures only unbound sialic acids—N-acetylneuraminic acid, N-glycolylneuraminic acid ($\epsilon =$ 46,000), and equine diacetylneuraminic acid. Pure bovine diacetylneuraminic acid and triacetylneuraminic acids are not reactive in the assay.[5, 6] Since these forms are reactive in the orcinol assay for sialic acids, the amount of bovine di- and triacetylneuraminic acids in a sample can be estimated by subtracting the thiobarbituric acid value from the orcinol value.

A series of 2-keto, 3-deoxy sugar acids (pentonic, hexonic, heptonic, and octonic), found in bacteria, also react in the thiobarbituric acid assay. These produce a chromogen with a peak at 549 mμ. They can be readily distinguished from sialic acids because they are not reactive in the orcinol or direct Ehrlich assays[7] for sialic acids.

The hydrolysis (0.1 N H$_2$SO$_4$, 80°, 1 hour[8]) frees all the sialic acids from several mucoproteins that have been tested except for brain tissues where release of sialic acids takes place for several hours.[9] Neutralization is not required before the thiobarbituric acid assay.

[4] V. S. Waravdekar and L. D. Saslaw, *J. Biol. Chem.* **234**, 1945 (1959).
[5] Personal communications from Drs. G. Blix and D. Aminoff.
[6] We wish to thank Dr. G. Blix for samples of equine diacetylneuraminic acid and triacetylneuraminic acid.
[7] L. Werner and L. Odin, *Acta Soc. Med. Upsaliensis* **57**, 230 (1952).
[8] L. Svennerholm, *Acta Chem. Scand.* **12**, 547 (1958).
[9] Personal communication from Dr. W. E. van Heyningen.

Thiobarbituric Acid Spray Reagent for Sialic Acids.[10] The paper is sprayed with 0.05 M sodium metaperiodate in 0.05 N H_2SO_4 and allowed to stand at room temperature for 15 minutes. A spray of ethylene glycol, acetone, and concentrated H_2SO_4 (50:50:0.3) is applied, and after 10 minutes the paper is sprayed with an aqueous solution of 6% sodium 2-thiobarbiturate. Red spots appear after heating at 100° for 5 minutes. Approximately 3 μg. of sialic acids can be detected. Chromatograms run in basic solvent systems should be sprayed first with 0.1 N H_2SO_4.

[10] L. Warren, *Nature* **168**, 237 (1960).

[68] Enzymatic Determination of Sialic Acids

N-Acylneuraminic acid $\rightleftarrows$ N-Acyl-D-mannosamine + pyruvate

By PAOLO BRUNETTI, ANN SWANSON, and SAUL ROSEMAN

Principle. N-Acetylneuraminic and N-glycolylneuraminic acids are cleaved by N-AN-aldolase (N-acylneuraminic acid aldolase) to pyruvate and the corresponding N-acyl-D-mannosamine.[1,2] Either of the products can be used to determine the concentration of these sialic acids in the sample. The other known sialic acids contain O-acetyl groups[3] and are resistant to N-AN-aldolase. However, the O-acetyl groups are readily removed by dilute alkali or acid,[4] and the residual N-acetylneuraminic acid can then be determined.

Two procedures can be used for the determination of the N-acylneuraminic acids. The first involves a two-step method similar to that employed for assay of N-AN-aldolase as previously described.[5] The second procedure utilizes a more highly purified preparation of N-AN-aldolase, free of DPNH oxidase, and couples this enzymatic reaction with lactic dehydrogenase and DPNH for the continuous spectrophotometric determination of pyruvate. A modification of the two-step spectrophotometric method is also described below, with fluorometry used for the determination of pyruvate.

Reagents

0.1 M potassium phosphate buffer, pH 7.2.

0.10 M N-acetylneuraminic or N-glycolylneuraminic acid. These

[1] D. G. Comb and S. Roseman, *J. Biol. Chem.* **235**, 2529 (1960).
[2] P. Brunetti, G. W. Jourdian, and S. Roseman, *J. Biol. Chem.* **237**, 2447 (1962).
[3] G. Blix and E. Lindberg, *Acta Chem. Scand.* **14**, 1809 (1960).
[4] G. Blix, *Proc. 4th Intern. Congr. Biochem., Vienna, 1958* **1**, 94 (1959).
[5] D. G. Comb and S. Roseman, Vol. V [391].

solutions are unstable; the compounds are dissolved before use, and the solution is adjusted to pH 4 to 5 with NaOH and diluted to volume. After several days at −16°, the standard solutions yield somewhat less than theoretical DPNH oxidation values, despite the fact that they still give essentially the same optical density values with the resorcinol reagent.[6]

0.003 M DPNH (stored in dilute alkali at −16°).

Crystalline lactic dehydrogenase, commercial preparation. A solution is prepared before use by dilution of the crystalline suspension in water.

Assay Method. Purification of N-AN-aldolase is followed by the assay procedure previously described;[5] the definitions of the unit and specific activity of the enzyme preparations are also the same.

Purification Procedure

Growth of Clostridium perfringens. The medium is prepared by autoclaving 630 g. of Todd Hewitt broth (Difco) in 17.7 l. of water for 30 minutes at 15 pounds pressure and is kept anaerobic by adding sterile mineral oil to the carboy while the solution is hot. After cooling, a solution of sterile-filtered glucose (27 g. in 100 ml. of water) and sterile-filtered salts (45 g. of NaCl, 32.4 g. of K_2HPO_4, and 0.9 g. of cysteine in 180 ml. of water) is added. The 18 l. of medium is inoculated with 1 l. of similar broth containing the organism grown for 10 hours in still culture at 37°. The contents of the carboy are gently stirred with a magnetic stirrer, the temperature is maintained at 37°, and growth is followed by removing aliquots and determining turbidity. The top of the log phase is usually reached in 3 to 4 hours. If necessary, Dow-Corning Antifoam A is used to depress foaming. Immediately after reaching the top of the log phase, the mixture is cooled to 10°; the cells are harvested, washed twice with cold water by stirring in a blendor, and finally stored at −20°. The yield of cells obtained under these conditions approximates 50 g. The frozen cells were kept for periods as long as 4 months and seemed to retain full activity.

A single lot of 80 l. was grown as described above but with an open, steam-jacketed vat as the container. Here, the medium was brought to a boil, cooled under mineral oil, and the glucose and salts were added, followed by 6 l. of inoculum grown in sterile medium. Gentle stirring was performed with a mechanical stirrer. The yield of washed cells obtained here was 450 g.; this preparation of cells contained more enzyme and

[6] L. Svennerholm, *Biochim. et Biophys. Acta* **24**, 604 (1957).

gave better results in the fractionation procedure than did the cells obtained from the carboys.

Unless specified otherwise, the following operations are conducted at temperatures between 0° and 4°, and centrifugations are performed for 20 minutes at 32,000 $\times g$ in a cold room.

Crude Extract. The washed cells (20 g.) are suspended in 40 ml. of 0.15 M KCl and disrupted by treatment for 15 to 30 minutes in a 10-kc. Raytheon sonic oscillator. After centrifugation for 20 minutes, the residue is discarded.

Bentonite Step. Bentonite is washed with water and is then added to the crude extract (75 mg. dry weight per milliliter). The mixture is stirred intermittently for 10 minutes and centrifuged. The residue is washed once with 0.5 vol. of 0.15 M KCl, and the supernatant fluids are pooled.

Polymyxin Step. The bulk of the nucleic acids is removed by adding a 10% solution of Polymyxin sulfate (Polymyxin B, Chas. Pfizer Co.) dropwise with stirring to the bentonite supernatant fluid; 0.25 ml. of Polymyxin solution is added per milliliter of the fluid obtained above. After 10 minutes of stirring, the mixture is centrifuged and the residue is discarded.

Protamine Step. The enzyme is precipitated by adding a 2% solution of protamine sulfate (Eli Lilly Co.) dropwise to the supernatant fluid obtained above. In this case, 0.40 ml. of protamine solution is added per milliliter of the enzyme preparation. After 10 minutes of intermittent stirring, the mixture is centrifuged at temperatures between 4° and 10°, and the supernatant fluid is discarded. The precipitate is washed once with an equal volume of 0.01 M phosphate buffer, pH 6.5, and then extracted by stirring repeatedly with 10-ml. portions of 0.10 M phosphate buffer, pH 7.2, for 10 minutes each. The extraction process is repeated until enzymatic activity is no longer detectable in the supernatant fluid; generally, five extractions are required.

Heat Step. DPNH oxidase is removed by heating 10-ml. portions of the solution at 60° to 61° for 8.5 minutes. After chilling, the mixture is centrifuged, and the residue discarded.

Ammonium Sulfate Step. The enzyme solution obtained in the previous step is adjusted to 80% of saturation by adding solid ammonium sulfate, stirred for 30 minutes, and centrifuged. The supernatant fluid is then brought to 90% of saturation with ammonium sulfate, centrifuged, and the final supernatant fluid is assayed. Generally, more than 70% of the activity is found in the final, 90% saturated supernatant fraction. This solution is dialyzed overnight, with mechanical stirring, against at least two changes of 50 vol. of water each. A conductivity meter is used to determine completion of dialysis. As the salt concentration inside the dialysis bag decreases, N-AN-aldolase slowly precipitates from solution. It is collected by centrifugation, and the bulk of the precipitate is dissolved in a minimum quantity of 0.10 M phosphate buffer, pH 7.2. A slight sediment is removed by centrifugation. The results of the purification procedure are summarized in the table.

SUMMARY OF PURIFICATION OF N-ACYLNEURAMINIC ACID ALDOLASE

Fraction	Volume, ml.	Protein, mg./ml.	Activity, units/ml.	Total units	Specific activity, units/mg. protein
Original					
Expt. 1[a]	58	17	217	12,600	12.8
Expt. 2[a]	88	12	30	2,640	2.5
Bentonite					
Expt. 1	47	4.0	200	9,400	50
Expt. 2	82	3.6	23	1,890	6.4
Polymyxin					
Expt. 1	58	1.0	150	8,700	150
Expt. 2	100	0.48	16	1,600	33
Protamine[b]					
Expt. 1	40	0.68	116	4,640	171
Expt. 2	50	0.35	8.3	414	24
Heat					
Expt. 1	38	0.56	153	5,810	273
Expt. 2	48	0.25	11	528	44
90% ammonium sulfate supernatant					
Expt. 1	45		89	4,000	
Expt. 2	61		14	854	
Dialyzed					
Expt. 1	4.5	0.76	592	2,660	779
Expt. 2	4.7	0.25	122	573	488

[a] Experiment 1 represents typical results obtained with cells grown in 80-l. quantities in a vat as described in the text. Experiment 2 represents the poorest results obtained with carboy-grown cells. In each of these experiments, 20 g. of cells was extracted.

[b] Protamine inhibits enzyme activity.

Properties

Stability. The purified enzyme is stable for at least several months when stored at −20°; it is also stable when stored in the presence of bovine serum albumin, or in 90% saturated ammonium sulfate. The purified enzyme loses most of its activity when the solution is lyophilized.

Other Properties. The purified enzyme exhibits essentially the same properties as those previously reported[1,5] for the less pure preparations, with the important exception that it contained no DPNH oxidase.

Enzymatic Determination of *N*-Acetylneuraminic and *N*-Glycolylneuraminic Acids

The equilibrium constant for the reactions *N*-acetylneuraminic acid $\rightleftharpoons$ Pyruvate $+$ *N*-acetyl-D-mannosamine is $K_{eq} = 0.096\,M$.[2] At sialic acid concentrations below $0.001\,M$, essentially all the sialic acid is cleaved. At higher concentrations, however, significant quantities remain in the equilibrium mixture. *N*-AN-aldolase alone can be used at the lower concentrations; the reaction is stopped by heating the incubation mixture, and either pyruvate or *N*-acylmannosamine is determined. When *N*-AN-aldolase is coupled with DPNH and lactic dehydrogenase, the reaction is "pulled" to completion, and the sialic acid concentration is estimated by one or more of the following methods: (A) determination of the quantity of DPNH oxidized at the end of the reaction; (B) determination of the initial rate of DPNH oxidation (which is a function of sialic acid concentration); (C) determination of acylhexosamine by a modified Morgan-Elson method[2] by using the corresponding compound as a standard. The coupled lactic dehydrogenase procedures (A and B) are described below.

Procedure A. Spectrophotometric. The following components are added to a 1-ml. cuvette with a 1.0-cm. light path: 0.50 ml. of $0.1\,M$ potassium phosphate buffer, pH 7.2; 0.01 to 0.25 μmole of *N*-acetyl- or *N*-glycolylneuraminic acid; 0.30 μmole of DPNH; and 0.01 ml. of lactic dehydrogenase solution. The volume is adjusted to 0.98 ml. with water, and the reference cell contains all components except the sialic acid. After determination of the optical density of the solution, 30 units of *N*-AN-aldolase is added to each cuvette to a final volume of 1.0 ml. The reaction is allowed to proceed to completion, the change in optical density indicating the quantity of DPNH oxidized. As indicated in Fig. 1, with either of the two sialic acids, the DPNH oxidized is essentially equivalent to the amount of sialic acid added.

Procedure A. Fluorometric. The fluorescence assay offers the most sensitive and specific method available for determining two of the sialic acids.[7] In this case, the quantity of DPNH oxidized is determined by measuring the decrease in fluorescence.[8,9] The range and sensitivity of

[7] This method was originally suggested by Dr. Eli Robins, Washington University School of Medicine, Department of Psychiatry and Neurology, St. Louis, Missouri.

[8] S. Udenfriend, "Fluorescence Assay in Biology and Medicine." Academic Press, New York, 1962.

[9] O. H. Lowry, N. R. Roberts, and J. I. Kapphahn, *J. Biol. Chem.* **224**, 1047 (1957).

the procedure is dependent on the initial concentrations of DPNH and of the sialic acids. A Farrand photoelectric fluorometer is used with primary filter No. 7-37 and a secondary filter combination of Nos. 5-61, 4-70, and 3-72.[9]

The following components are added to a standard 10-mm. borosilicate glass tube (final volume, 0.10 ml.) : 10 μmoles of phosphate buffer, pH 7.2; 0.05 to 0.4 mμmole of the sialic acid sample; 0.5 unit of N-AN-aldolase; 0.5 mμmole of DPNH; and 5 μl. of commercial crystalline

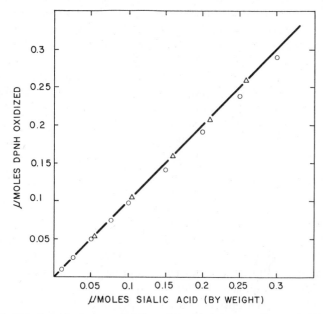

Fig. 1. Relationship between sialic acid added to cuvettes and total DPNH oxidized. Incubation mixtures as described in the text (procedure A); reactions were generally complete within 20 minutes at room temperature when 30 units of N-AN-aldolase was added. ○, N-acetylneuraminic acid; △, N-glycolylneuraminic acid.

lactic dehydrogenase (diluted 1 to 20). The tubes are incubated for 15 minutes at 37°, 1.0 ml. of 0.10 M phosphate buffer, pH 7.8, is added to each tube, and the fluorescence is determined. Appropriate controls include standard pyruvate in place of N-acetylneuraminic acid, known concentrations of one of the sialic acids, and tubes lacking either standards, unknowns, N-AN-aldolase, or lactic dehydrogenase. The method described here involves measurement of the oxidation of DPNH in each sample by comparison with control tubes. This procedure may seem less accurate and reliable than measuring DPNH changes directly on each sample. However, standard curves obtained by the described procedure were highly satisfactory and agreed well with those obtained with known concentrations of pyruvate. The fluorometer range is extremely broad, and the sensitivity of the procedure can be increased even further by varying the concentrations of the sialic acid and DPNH used.

Procedure B. The initial rate of the N-AN-aldolase reaction is directly proportional to the sialic acid concentration under the conditions described above for the spectrophotometric method (Figs. 2 and 3). In this case, 10 rather than 30 units of N-AN-aldolase is used.

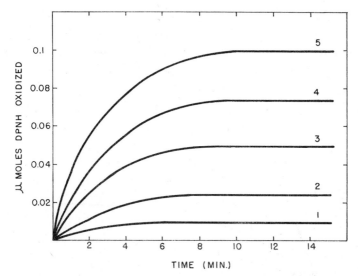

TIME (MIN.)

Fig. 2. Rate of cleavage of N-acetylneuraminic acid. Incubation mixtures as described under procedure B in the text. The following concentrations of N-acetyl-neuraminic acid were used in final volumes of 1.0 ml.: (1) 0.01 micromole; (2) 0.025 micromole; (3) 0.05 micromole; (4) 0.075 micromole; (5) 0.10 micromole. The curves are tracings from the Cary spectrophotometer.

Comments on the Procedure. The methods described above, particularly procedure B (spectrophotometric), have been tested with extracts obtained from submaxillary glands and lipid fractions from brain. The methods appear satisfactory in the samples per se, and in equivalent samples hydrolyzed with 0.1 N sulfuric acid at 80° for 1 hour (followed by neutralization). To obviate errors due to the presence of pyruvate or of enzyme inhibitors, the procedure was modified as follows: All the constituents were added to the cuvette except N-AN-aldolase; when the optical densities were constant (i.e., 2 to 3 minutes), N-AN-aldolase was added; after completion of the reaction, a known quantity of

N-acetylneuraminic acid was added to the cuvette. If the change in optical density obtained after the last addition corresponded to the amount of sialic acid added, it was assumed that the coupled enzyme reaction was not inhibited and that the change in optical density obtained after the addition of N-AN-aldolase to the cuvette was due to the sialic acid present in the incubation mixture.

The enzyme preparation appears to be free of neuraminidase, so that only free N-acetylneuraminic and N-glycolylneuraminic acids are determined.

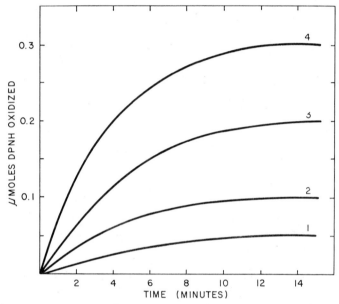

Fig. 3. Rate of cleavage of N-glycolylneuraminic acid. Incubation mixtures as described under procedure B in the text. The following concentrations of N-glycolylneuraminic acid were used in final volumes of 1.0 ml.: (1) 0.05 micromole; (2) 0.10 micromole; (3) 0.20 micromole; (4) 0.30 micromole. The curves are tracings from the Cary spectrophotometer.

The specificity of the enzyme for different sialic acids is shown in Fig. 4. In these studied, excess sialic acids were added to the cuvette in order to determine relative rates of cleavage. The two samples of N,O-diacetylneuraminic acids were highly purified B- and E-sialic acids, provided through the kindness of Dr. G. Blix. As shown in Fig. 4, these materials are apparently not cleaved by N-AN-aldolase; the slight activities obtained (less than 3% of the total added to the cuvettes) are thought to be due to contamination with N-acetylneuraminic acid. The

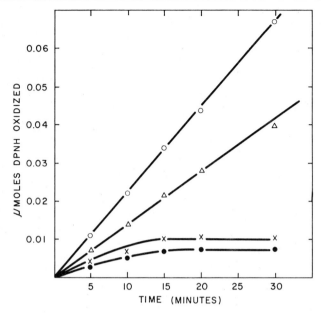

Fig. 4. Relative rates of cleavage of the sialic acids. In this experiment, each incubation mixture contained in a final volume of 0.10 ml.: 5 micromoles of phosphate buffer, pH 7.2; 2 micromoles of the sialic acid; 0.5 unit of N-AN-aldolase. ○, N-acetylneuraminic acid; △, N-glycolylneuraminic acid; ×, N,O-diacetylneuraminic acid (E sialic acid); ●, N,O-diacetylneuraminic acid (B sialic acid). At the indicated times, 0.020-ml. aliquots of the incubation mixtures were transferred to tubes containing 0.50 ml. of phosphate buffer and 0.46 ml. of water at 100°. The tubes were maintained at 100° for 2 minutes, cooled, and the pyruvic acid determined enzymatically.

O-acetyl group is notoriously labile; in fact, if solutions of the N,O-diacetyl compounds are neutralized, stored at −18°, and assayed after several days, approximately 20 to 30% of total samples are cleaved by N-AN-aldolase.

[69] 5-Phosphoribosylpyrophosphate

By JOEL G. FLAKS

Preparation

Principle. 5-Phosphoribosylpyrophosphate (α-D-ribofuranose 1-pyrophosphate 5-phosphate; PRPP) is obtained in good yield by the enzymatic transfer of pyrophosphate from ATP to ribose 5-phosphate.[1-3] Isolation of the phosphate ester is accomplished by adsorbing nucleotides and protein onto Norit and separating the remaining mixture of phosphate compounds by anion-exchange chromatography. PRPP is then isolated as the ethanol-insoluble magnesium salt. Further purification is obtained by repeating the chromatography and magnesium salt isolation steps.[3,4]

Enzymatic Synthesis. 5-Phosphoribose pyrophosphokinase, which catalyzes the pyrophosphoryl transfer from ATP, has been partially purified from pigeon liver actone powder extracts and may be used as the source of the enzyme.[1] However, a more convenient and stable source is available with a lyophilized preparation obtained by ethanol precipitation of the soluble fraction of pigeon liver homogenates. Details for the preparation of this lyophilized fraction, which precipitates between 0 and 15% ethanol, are given in Vol. VI [9], Section I.A. The following comments regarding the procedure should be noted. The pyrophosphokinase appears to have an exceedingly sensitive sulfhydryl group.[1] It has been found advisable to include $0.005\,M$ 2-mercaptoethanol in the buffer–salts medium used for homogenization of the pigeon livers. The protein fraction precipitating after the addition of 0.2 vol. of 90% ethanol (0 to 15% fraction), with temperatures maintained as indicated, is taken up in a minimal volume of ice-cold water containing $0.003\,M$ glutathione, pH 7.5, and lyophilized as described. This protein fraction is not completely soluble in water. The lyophilized powder stored *in vacuo* at $-15°$ over P_2O_5 shows no loss in activity for at least 6 months. About 2 g. of lyophilized powder are obtained from the livers of 24 pigeons.

---

[1] A. Kornberg, I. Lieberman, and E. S. Simms, *J. Biol. Chem.* **215**, 389 (1955); see also Vol. VI [18].

[2] C. N. Remy, W. T. Remy, and J. M. Buchanan, *J. Biol. Chem.* **217**, 885 (1955).

[3] J. G. Flaks, M. J. Erwin, and J. M. Buchanan, *J. Biol. Chem.* **228**, 201 (1957).

[4] The procedure described here represents several modifications of the procedures of Kornberg et al.[1] and Remy et al.,[2] which considerably improve both the yield and the purity of the product. Termination of the enzymatic incubation by the addition of Norit, rather than by acid addition or heating, minimizes the losses of PRPP by hydrolysis and conversion to the cyclic 1,2-phosphate. It also serves to remove adenine nucleotides, which otherwise result in contamination of the PRPP, after chromatography, with ATP. It has not been our experience that excessive losses in PRPP result from adsorption to the Norit; at most, the final yield is reduced by 10%. Isolation of PRPP as the magnesium salt is preferred to the barium salt. The latter is difficultly soluble in water, and large losses of PRPP occur during the removal of the barium. The modified procedure represents the collaborative efforts of A. J. Guarino, S. C. Hartman, and J. G. Flaks.

Reagents

ATP, K or Na salt, pH 7.5, 0.05 M.
Ribose 5-phosphate,[5] Na salt, 0.1 M.
Potassium phosphate buffer, pH 7.5, 1.0 M.
$MgCl_2$, 0.5 M.
2-Mercaptoethanol, 1 M.
Potassium fluoride, 1 M.
0 to 15% lyophilized ethanol fraction (see above). Just prior to its use, 1.2 g. of the lyophilized powder are taken up in 50 ml. of 0.1 M potassium phosphate buffer, pH 7.5, containing 0.01 M 2-mercaptoethanol.

Procedure. The procedure described below has some recent modifications over that previously published.[3]

A reaction mixture is prepared in a 2-l. Erlenmeyer flask by the addition of 440 ml. of water, 16 ml. of ribose 5-phosphate, 7 ml. of potassium phosphate buffer, pH 7.5, 12 ml. of $MgCl_2$, and 18 ml. of 2-mercaptoethanol. The flask is placed in a 37° water bath; when the temperature of the solution reaches 37°, 24 ml. of ATP are added, followed by 30 ml. of potassium fluoride and the enzyme. The contents are mixed and incubated at 37° for 30 minutes at which time the yield of PRPP is maximal (70 to 80% based on the ATP which is limiting). The flask is rapidly chilled in an ice bath, and, after the temperature of the solution falls to 10°, 20 g. of Norit are added and stirred with the solution for a few minutes. The mixture is filtered at 0° to 3°, with suction, on a Büchner funnel containing a thin pad of Celite, and the Norit is then washed with 400 ml. of ice-cold water. PRPP is isolated and purified from the combined filtrate and washings as described below. It is advisable to carry out the purification immediately.

Purification Procedure

The anion-exchange chromatography is carried out at 0° to 3°. The combined filtrate and washings from the incubation mixture are rapidly (3 to 4 ml./min.) passed through a Dowex 1 column, formate form (15 cm. × 2.5 cm. diameter, 200 to 400 mesh). A linear gradient elution of the column is carried out, with 1.5 M ammonium formate, pH 5.0, entering into a mixer containing 500 ml. of water. The elution is rapid (about 3 ml./min), and fractions of 20 ml. are collected. Aliquots of 0.1

[5] J. X. Khym, D. G. Doherty, and W. E. Cohn, *J. Am. Chem. Soc.* **76**, 5523 (1954).

CARBOHYDRATES

[69]

ml. are assayed for pentose by the orcinol procedure,[6] with a heating time of 20 minutes. Two major pentose-containing peaks are obtained. The first, eluted after several resin bed volumes, is ribose 5-phosphate; the second, after 300 to 400 ml. have passed through the column, is PRPP. Occasionally a small pentose peak, consisting of 5-phosphoribose 1,2-cyclic phosphate (see Properties), immediately precedes the PRPP peak.

Fractions containing PRPP are combined, and 10 ml. of 0.5 M $MgCl_2$ are added, followed by 3 vol. of ethanol ($-15°$) to precipitate the magnesium salt. The mixture is allowed to stand overnight at $-15°$, after which much of the supernatant fluid can be removed by decantation. The magnesium salt of PRPP is collected by centrifugation in the cold ($-15°$) and washed twice at $-5°$ to $-10°$ with cold ethanol. This is followed by two washes with acetone and two with ether at room temperature.[7] The material is quickly dried at room temperature[7] and then *in vacuo* over P_2O_5. Storage of the dry magnesium salt at $-15°$ *in vacuo* over P_2O_5, results in little loss after a few months. However, since PRPP is subject to decomposition in the presence of divalent cations, it is recommended that the material be converted to the Na or K salt as described below. On the basis of $C_5H_8O_{14}Mg_2$, the product is approximately 75% pure at this point, with the major contaminants being the magnesium salts of phosphate (about 10%) and pyrophosphate (3 to 5%). The preparation is free from adenine nucleotides. The yield is 310 to 390 mg. (55 to 70% based on the ATP).

The preparation at this point is suitable for most purposes after removal of the magnesium. This is accomplished by dissolving the salt in 20 ml. of cold water and passing the solution through a column of Dowex 50, Na or K form (6 cm. $\times$ 1 cm. diameter, 200 to 400 mesh), at $0°$ to $3°$. The column is washed with 10 ml. of cold water, and the effluent fractions are combined. Hartman[8] suggests the use of Chelex 100 (Bio-Rad Laboratories), Na or K form, since this resin is more selective for divalent cations than is Dowex 50. Solutions of the Na or K salt have been stored frozen for over 2 years without detectable loss of PRPP.

Further purification can be accomplished in two ways. By repeating the anion-exchange chromatography and magnesium salt isolation, a preparation of 85 to 90% purity is obtained, which is free from inorganic

[6] W. Mejbaum, *Z. physiol. Chem.* **258,** 117 (1939); see also Vol. III [12].

[7] The magnesium salt of PRPP is exceedingly hygroscopic. Unless it is thoroughly dried and stored under desiccation, it forms a sirupy glass which is difficult to solubilize further, resulting in large losses of PRPP.

[8] S. C. Hartman, unpublished results.

phosphate and pyrophosphate. The loss in yield is about 10%. Alternatively, the inorganic pyrophosphate can be removed as the manganous salt by the procedure of Kornberg.[9] The preparation is then freed of excess manganous ion by the Dowex 50 treatment described above.

Hartman[8] has scaled up the preparation 3.5-fold (based on the ATP) with yields of PRPP of 50%. The procedure is essentially that described here, with the following modifications: The incubation mixture, in a volume of 660 ml., contains 3.5 times the amount of ATP and ribose 5-phosphate, 3 times the amount of phosphate buffer, and 1.5 times the amount of $MgCl_2$ and enzyme, with reference to that described above. All other conditions were the same with the chromatography carried out with one-third of the preparation on the column described here.

PRPP has also been synthesized chemically.[10]

Properties

The Na, K, NH_4, and Mg salts of PRPP are readily soluble in water; the barium salt is difficult to solubilize, and almost inevitably losses of PRPP will result.

Elevated temperatures and both low and high pH all lead to rapid decomposition of PRPP (see table). Owing to the effects of temperature, solutions of PRPP should not be kept at room temperature. At acidic pH the compound cleaves to ribose 5-phosphate and inorganic pyrophosphate. At alkaline pH values, particularly in the presence of divalent cations, the pyrophosphoryl bond is cleaved, with the elimination of inorganic phosphate and the concomitant formation of 5-phosphoribose cyclic 1,2-phosphate.[2,11] This structure is suggested by the following facts: (1) the compound does not take up 1 mole of $NaIO_4$, as do PRPP and ribose 1,5-diphosphate; (2) titration of the compound reveals only one secondary phosphate; (3) the compound possesses a single acid-labile phosphate. The cyclic diphosphate can also be chromatographically separated from either PRPP or ribose 1,5-diphosphate.[2,11] Under further alkaline treatment the cyclic phosphate cleaves to a mixture of ribose diphosphates and ribose monophosphates. The formation of the cyclic diphosphate indicates that the pyrophosphoryl attachment at carbon 1 of the pentose is of the α-configuration.[2,10,11] Complete conversion of PRPP to the cyclic phosphate occurs on incubation for 60 minutes at 25° and pH 10.5 in the presence of Ba^{++} $(0.005\ M)$.[11]

PRPP is not cleaved by inorganic pyrophosphatase or by nucleotide pyrophosphatase.[1] It acts as a 5-phosphoribosyl donor in *de novo* purine

[9] A. Kornberg, *J. Biol. Chem.* **182**, 779 (1950).
[10] G. M. Tener and H. G. Khorana, *J. Am. Chem. Soc.* **80**, 1999 (1958).
[11] H. G. Khorana, J. F. Fernandes, and A. Kornberg, *J. Biol. Chem.* **230**, 941 (1958).

DECOMPOSITION OF PRPP[a]

System	Decomposition, %	Conditions
PRPP at pH 3.1 (0.02 M formate)[b]	100	65° for 20 minutes
PRPP at pH 4.0 (0.02 M acetate)	70	
PRPP at pH 6.7 (0.02 M glycylglycine)	50	
PRPP at pH 9.0 (0.02 M glycine)	63	
PRPP[c]	50–55	100° for 1 minute
PRPP + Mg^{++} (0.02 M)	78–80	
PRPP + K phosphate, pH 7.0 (0.02 M)	29	
PRPP + Mg^{++} + K phosphate	18–22	
PRPP + Mg^{++} + Na pyrophosphate, pH 8.0 (0.02 M)	10	
PRPP + Mg^{++} + EDTA, pH 7.0 (0.02 M)	17	
PRPP[c] (pH 7.0) + Mg^{++} (0.002 M)	51	36° for 60 minutes

[a] A. Kornberg, I. Lieberman, and E. S. Simms, *J. Biol. Chem.* **215**, 389 (1955); C. N. Remy, W. T. Remy, and J. M. Buchanan, *J. Biol. Chem.* **217**, 885 (1955); see also Vol. VI [18].

[b] Decomposition is over 95% complete in 10 minutes at 65°.

[c] 1×10^{-4} M PRPP adjusted to pH 7.0.

biosynthesis,[12] pyrimidine biosynthesis,[13] nucleotide synthesis,[13] nicotinic acid mononucleotide formation,[14] tryptophan biosynthesis,[15] and the initial step in histidine biosynthesis.[16]

Analysis

Analysis of a sample of the purified magnesium salt (twice chromatographed on Dowex 1) showed: orcinol-reactive pentose (with AMP or adenosine as standard),[6] 1.00; inorganic phosphate,[17] <0.02; inorganic pyrophosphate,[18] <0.02; acid-labile phosphate,[17] 2.05; total phosphate,[16] 3.03. A sample also showed an uptake of 1 mole of NaIO$_4$ per mole of pentose at pH 5.0.[2] Enzymatic assay (see below) indicated 0.98 mole of PRPP per mole of pentose.

The usual assay values obtained are the ratio of pentose:total phosphate:enzymatically active material.

Almost any of the several reactions in which PRPP functions may

[12] See Vol. VI [9].

[13] See Vol. VI [17] and VI [21].

[14] See Vol. VI [44].

[15] See Vol. V [107].

[16] B. N. Ames, R. G. Martin, and B. J. Garry, *J. Biol. Chem.* **236**, 2019 (1961).

[17] C. H. Fiske and Y. SubbaRow, *J. Biol. Chem.* **81**, 629 (1929); see Vol. III [115].

[18] Assayed as inorganic phosphate after incubation with yeast inorganic pyrophosphatase.

be used to assay the compound enzymatically, but the most convenient is the procedure of Kornberg *et al.*[1] It is based on the conversion of orotate to a mixture of orotidylate and UMP, which results in a decrease in absorbance at 295 mμ. The procedure is exactly that described in step 2 of the assay method, Vol. VI [18].

[70] Preparation and Properties of D-Erythrose 4-Phosphate

By CLINTON E. BALLOU

Preparation

Principles. Procedure A involves the phosphorylation of a suitable blocked derivative of D-erythrose, leading to D-erythrose 4-phosphate dimethyl acetal. Free D-erythrose 4-phosphate is obtained by mild acid hydrolysis of the acetal.

Procedure B is a one-step synthesis carried out by the lead tetraacetate oxidation of D-glucose 6-phosphate. Although far simpler than procedure A, the product is not so pure.

Procedure A.[1] *Preparation of 4-Trityl-2,3-diacetyl-D-erythrose Diethyl Mercaptal.* Twenty grams of 4,6-ethylidene-D-glucose (m.p. 175° to 180°)[2] is oxidized with sodium metaperiodate to 2,4-ethylidene-D-erythrose, as has been reported for its enantiomorph.[3] The colorless sirup resulting from the oxidation is dissolved with swirling in 50 ml. of ethanethiol, and to the ice-cold solution is added 15 ml. of concentrated hydrochloric acid. The mixture is shaken at 0° for 20 minutes and then made slightly basic by cautious addition of concentrated ammonium hydroxide. The reaction mixture is concentrated to dryness *in vacuo,* and the residue is dried by distilling absolute ethanol from it two or three times. Absolute ethanol is then added, and the insoluble ammonium chloride is removed by filtration. Removal of the alcohol at reduced pressure gives a mixture of mercaptal and a considerable amount of ammonium chloride; the product is then further dried by azeotropic distillation after the addition of benzene.

The resulting mixture is dissolved in 150 ml. of anhydrous pyridine,

[1] C. E. Ballou, H. O. L. Fischer, and D. L. MacDonald, *J. Am. Chem. Soc.* **77,** 5967 (1955).

[2] R. C. Hockett, D. V. Collins, and A. Scattergood, *J. Am. Chem. Soc.* **73,** 599 (1951).

[3] D. A. Rappoport and W. Z. Hassid, *J. Am. Chem. Soc.* **73,** 5524 (1951).

and 30 g. of triphenylchloromethane is added. After standing for about 18 hours, the solution is cooled in ice, and 100 ml. of acetic anhydride is added. After one-half hour at 0°, the mixture is set at 20° to 25° overnight. The solution is then cooled in ice, and the excess acetic anhydride is decomposed by the addition of 50 ml. of water. After one-half hour, the solution is concentrated *in vacuo*, and the residue is taken up in 250 ml. of chloroform, washed with 1 *N* sulfuric acid, 1 *N* potassium carbonate, and water, and dried (sodium sulfate). The solvent is removed at reduced pressure, and the residual sirup is taken up in 500 ml. of hot methanol, treated with charcoal, and filtered hot. 4-Trityl-2,3-diacetyl-D-erythrose diethyl mercaptal crystallizes from the dark-red solution. It is dissolved in hot methanol, decolorized with charcoal, and from the cooled solution 26 g. of almost colorless material is obtained (m.p. 105° to 106°).

4-Trityl-2,3-diacetyl-D-erythrose Dimethyl Acetal. Five grams of the acetylated mercaptal is dissolved in 75 ml. of warm, dry methanol in a three-necked round-bottomed flask. The solution is cooled rapidly to room temperature, and 2 ml. of 0.5 *N* barium methoxide solution is added. After one hour, when deacetylation is complete, the vessel is fitted with a mercury-sealed glass stirrer and a reflux condenser. After the addition of 7.5 g. of mercuric oxide, the stirrer is adjusted to such a speed that the oxide is kept well suspended, and a solution of 7.5 g. of mercuric chloride in warm dry methanol is added. The mixture is stirred at room temperature for 10 minutes and then under reflux in a water bath for 20 minutes.

The cooled solution is filtered, and the filtrate is concentrated *in vacuo* to dryness in the presence of a little mercuric oxide. The solid residue is extracted with two 50-ml. portions of chloroform, and the combined chloroform filtrate is washed in a separatory funnel with 10% aqueous potassium iodide solution to remove the orange color, then twice with 100-ml. portions of water. The organic layer, dried over anhydrous sodium sulfate, is concentrated *in vacuo* to a stiff sirup that weighs 3.7 g.

This sirup is acetylated in 20 ml. of dry pyridine with 5 ml. of acetic anhydride. After 18 hours at room temperature, a little water is added to destroy the excess acetic anhydride, and the solution is concentrated *in vacuo* to remove most of the pyridine. The residue is taken up in 100 ml. of chloroform, and the solution is washed with 100-ml. portions of water, cold 1 *N* hydrochloric acid, cold 1 *M* potassium bicarbonate, and finally water. The dry chloroform solution (sodium sulfate) is concentrated *in vacuo* to a sirup (5.0 g.) which crystallizes on addition of 5 ml. of methanol. After several hours at 5° the crystals are filtered off and dried in air. The yield is 3.5 g. (79%). Recrystallization from a small volume of methanol gives 3.1 g. of granular crystals with m.p. 99° to 101°. The substance shows $[\alpha]_D^{25} + 10.8°$ (c 2.4, chloroform).

4-Trityl-2,3-dibenzoyl-D-erythrose Dimethyl Acetal. Three grams of the acetylated acetal is deacetylated in 50 ml. of dry methanol with 1 ml. of 0.5 N barium methoxide. After one hour the solution is concentrated *in vacuo* to a thick sirup and benzoylated in 15 ml. of dry pyridine with 3 ml. of benzoyl chloride. After 18 hours at room temperature, the reaction is worked up as described for the acetylation of the acetal above. The product crystallizes from methanol in a yield of 3.5 g. (93%). After recrystallization from methanol, the dibenzoate melts at 122° to 124° and shows $[\alpha]_D^{25} + 18.3°$ (c 3, chloroform).

This dibenzoate can be prepared directly from the demercaptalation reaction, but the over-all yield is better when the preparation is carried out via the diacetate.

Detritylation and Phosphorylation. Three grams of 4-trityl-2,3-dibenzoyl-D-erythrose dimethyl acetal in 100 ml. of absolute ethanol is shaken with 3 g. of reduced and washed 5% palladium chloride-on-carbon catalyst[4] in the presence of hydrogen gas at atmospheric pressure for 16 hours. The hydrogen uptake (170 ml.) is in excess of the theoretical amount (110 ml.). The catalyst is centrifuged off, and the ethanol solution is concentrated to dryness *in vacuo*. Crystals of triphenylmethane separate.

Without separation, the mixture is dissolved in 10 ml. of dry pyridine, cooled in ice water, and 2.5 g. of diphenyl phosphorochloridate is added dropwise. After 18 hours at 5°, the reaction is worked up as described for the acetylation of the acetal above. The yield of the phosphorylated product (contaminated with triphenylmethane) is 3.7 g.

D-*Erythrose 4-Phosphate Dimethyl Acetal.* The sirup from the above condensation is hydrogenated at atmospheric pressure in 250 ml. of absolute ethanol with 1 g. of platinum oxide catalyst. The hydrogen uptake is 1340 ml. in 10 hours. The catalyst is removed by centrifugation, and 30 ml. of 1 N sodium hydroxide is added to the ethanol solution. After 18 hours to allow saponification, the alcohol is removed by distillation *in vacuo*, and the residue, in 100 ml. of water, is extracted with ether to remove some water-insoluble material. The water layer is treated batchwise with 50 ml. of Dowex 50 (H⁺ form, 2 meq./ml.) to remove the cations and is again extracted with ether to remove the cyclohexylcarboxylic acid. The water layer is immediately adjusted to about pH 9

⁴ H. Gilman and A. H. Blatt, *Organic Syntheses* **26**, 77 (1946); C. E. Ballou and H. O. L. Fischer, *J. Am. Chem. Soc.* **76**, 3188 (1954).

(indicator paper) with cyclohexylamine, and the solution is concentrated *in vacuo* to dryness. The residue is dissolved in 5 ml. of absolute ethanol, and ether is added to turbidity. Crystallization occurs, and, after 18 hours at 5°, the mixture is filtered by suction on a hardened filter paper. The product is washed with ether on the funnel, then dried in air, and finally for an hour at room temperature in a high vacuum over phosphorus pentoxide. The yield is 0.6 g. (31%) of a product with m.p. 160° to 165°, $[\alpha]_{589}^{25}$ 0°($\pm$)0.2° (c 5, water or 1 N hydrochloric acid).

D-*Erythrose 4-Phosphate.* A solution of 100 mg. of the cyclohexylamine salt of the acetal in 5 ml. of water is swirled for a minute with 2 ml. of Dowex 50 (H$^+$ form, 2 meq./ml.). The resin is filtered off, and the filtrate is left in a tightly stoppered container at 40° for 18 hours. An aliquot (0.2 ml.), removed and analyzed for aldehyde content by the Willstätter-Schudel titration,[5] is found to consume about 0.018 meq. of oxidant, or 90% of the calculated requirement. No further increase in reducing power occurs. The above solution of the free acid of D-erythrose 4-phosphate may be neutralized by adding sodium bicarbonate and the neutral solution stored frozen. It decomposes slowly over a period of several months.

Procedure B.[6] *Preparation of* D-*Erythrose 4-Phosphate from* D-*Glucose 6-Phosphate.* Barium D-glucose 6-phosphate heptahydrate (0.53 g., 1.0 millimole[7]), moistened with 1 ml. of water, is taken up in 5 ml. of glacial acetic acid (reagent grade), and the suspension is warmed for a few minutes at 50° to 60° to effect solution of the salt. Sulfuric acid (6 N, 0.35 ml.) is added to the solution with stirring, followed by 245 ml. of glacial acetic acid.[8]

A solution of lead tetraacetate[9] in acetic acid (0.85 g., 1.9 millimoles in about 50 ml. of glacial acetic acid[10]) is measured into a separatory

[5] F. Auerbach and E. Bodländer, *Z. angew. Chem.* **36**, 602 (1923).

[6] J. N. Baxter, A. S. Perlin, and F. J. Simpson, *Can. J. Physiol.* **37**, 199 (1959). This procedure is an unpublished modification by the above authors, which allows the preparation to be done on a larger scale and gives a product less highly contaminated with D-glyceraldehyde 3-phosphate. The author is indebted to Dr. Perlin for supplying this modification.

[7] Satisfactory preparations of barium D-glucose 6-phosphate may be obtained commercially. The proportion of water of crystallization present varies somewhat, and the salt may decompose on storage. It is advisable, therefore, to assay the salt before use and choose the appropriate sample weight accordingly.

[8] It is not necessary to remove the precipitate of barium sulfate at this stage.

[9] Lead tetraacetate is available commercially or may be prepared conveniently on the laboratory scale [see, e.g., R. C. Hockett and W. S. McClenahan, *J. Am. Chem. Soc.* **61**, 1667 (1939)].

[10] The required volume is determined by iodimetric titration of an aliquot of the solution (see note 9).

funnel, followed by sulfuric acid (6 N, 0.7 ml.) just prior to use.[11] The mixture is then added from the funnel dropwise into the rapidly stirred solution of D-glucose 6-phosphate at room temperature (about 23° to 25°), its rate of addition being regulated so that the reaction mixture contains no more than a slight excess of lead tetraacetate at any given time.[12] This may be checked by spot-testing the reaction mixture periodically with moistened starch-iodine paper; if a positive test is given, addition of the oxidizing solution is halted temporarily, or its rate of addition is slowed accordingly.

The reaction mixture is filtered through a thin layer of Filter-aid, and the filtrate is concentrated *in vacuo* at 35° to 40° to a volume of 15 to 20 ml. The filter is washed with water (100 ml.), the washings are combined with the concentrate, and evaporation of solvent is continued to remove most of the acetic acid, the final volume being 15 to 20 ml. The concentrate is diluted with water to a volume of 75 to 100 ml., extracted continuously with ether for 15 to 20 hours, and the aqueous layer is then filtered with washing through a short column holding 1 to 2 g. of Amberlite IR-120 (H⁺). Concentration of the solution to a volume of 20 to 25 ml. removes the ether still present, and the volume and pH of the solution may then be adjusted as required. The final solution is found by enzymatic assay to contain 0.75 to 0.8 millimole (75% to 80% yield) of D-erythrose 4-phosphate, about 0.02 millimole of D-glyceraldehyde 3-phosphate, and about 0.05 millimole of unoxidized D-glucose 6-phosphate.

Properties

D-Erythrose 4-phosphate is similar to glyceraldehyde 3-phosphate in its sensitivity to acid, the half-time for hydrolysis with the formation of inorganic phosphate in 1 N acid at 100° being 20 minutes.[1] On the other hand, treatment with 1 N sodium hydroxide at room temperature results in the rapid but incomplete (about 50%) elimination of phosphate as inorganic phosphate. Longer treatment is without effect.

---

[11] A small white precipitate may form, owing to the presence of divalent lead as a contaminant; the sulfuric acid has little effect on the lead tetraacetate during the reaction period required.

[12] Divalent lead formed in the reaction is removed preferentially as the sulfate and thus is prevented from precipitating the sugar phosphates; also, the concentration of sulfuric acid is maintained at a low level in the reaction mixture by this procedure, minimizing overoxidation. A rate of about 1 ml./min. is satisfactory for addition of about one-half of the oxidizing solution, but the rate is gradually diminished as the end of the reaction is approached.

The specific rotation of D-erythrose 4-phosphate in water varies with wavelength as follows:[13]

	420 mμ	500 mμ	589 mμ	700 mμ
Free acid	− 0.9°	+0.3°	+0.5°	+0.6° ± 0.1°
K salt	−14.8°	−7.8°	−4.7°	−3.8° ± 0.1°

D-Erythrose 4-phosphate is oxidized slowly by D-glyceraldehyde 3-phosphate dehydrogenase, and a method for its estimation is based on this fact.[14] It is also rapidly condensed with dihydroxyacetone phosphate in the presence of rabbit muscle aldolase to yield sedoheptulose diphosphate,[1] and with phosphoenolypyruvate in the presence of enzymes from *Escherichia coli* to yield dehydroshikimic acid.[15]

[13] C. E. Ballou, unpublished observations, 1957.
[14] E. Racker, V. Klybas, and M. Schramm, *J. Biol. Chem.* **234**, 2510 (1959).
[15] P. R. Srinivasan, M. Katagiri, and D. B. Sprinson, *J. Am. Chem. Soc.* **77**, 4943 (1955).

[71] Microestimation, Biosynthesis, and Isolation of 2,3-Diphosphoglycerate (2,3-PGA)

By Santiago Grisolia

Microestimation

The catalytic effect of 2,3-PGA on phosphoglyceric acid mutase from several sources (see Vol. V [26]) and the estimation of the extent of the reaction by colorimetric or spectrophotometric methods provide sensitive and highly specific methods to estimate 2,3-PGA. We present here the two more commonly used in our laboratory.[1]

Reagents

1.0 M stock Tris–HCl buffer, pH 9.0.
1.0 M stock Tris–phosphate buffer,[1] pH 7.3.
1.0 M stock MgSO$_4$.
0.04 M ADP.
Enolase, second ethanol fraction (see Vol. V [26]).
Pyruvic kinase (see Vol. I [66]). It is sufficient to use 50% pure enzyme.
0.1 M 3-PGA (free of 2,3-PGA).
Reagents for pyruvic estimation (see Vol. III [66]).

[1] J. C. Towne, V. W. Rodwell, and S. Grisolia, *J. Biol. Chem.* **226**, 777 (1957).

Procedure. SPECTROPHOTOMETRIC ASSAY 2.[1] The components are as follows: purified 3-PGA, 10 micromoles; Tris, pH 9.0, 200 micromoles; MgSO$_4$, 30 micromoles; enolase, 40 units (see Vol. V [26]); and water to 2.9 ml. At zero time, 0.10 ml. of the 2,3-PGA sample is added, and the increase in optical density at 240 mμ and 30° is recorded. The experimental values are corrected for the endogenous rate.

At pH 9, mutase activity is decreased some 25-fold; however, at this pH, zero-order kinetics are observed for the first minutes, and the 2,3-PGA effect can be measured. The increase in the initial reaction rate caused by added 2,3-PGA is nearly linear up to about 0.1 micromole.

COLORIMETRIC METHOD 4.[1] A simple method is to use a coupled assay system of mutase, enolase, and pyruvic kinase, starting with purified 3-PGA and sufficient ADP to act as a phosphate acceptor for the phosphoenolpyruvic acid formed on the addition of 2,3-PGA sample.

The conditions are: 2,3-PGA sample to 2.0 ml.; and a solution containing, per 2.0 ml., ADP, 20 micromoles; purified 3-PGA, 50 micromoles; Tris–phosphate buffer, pH 7.3, 400 micromoles; MgSO$_4$, 30 micromoles; enolase, 1.0 mg.; pyruvic kinase, 0.5 mg. An endogenous control is included. All the components are mixed at 0° and then placed in a water bath at 38° for 30 minutes. After the incubation period, the reaction mixture is cooled in an ice bath for 5 minutes and then deproteinized with 2.0 ml. of 10% HClO$_4$. After centrifugation, aliquots are taken for pyruvate analysis.

When it is necessary to find the correction value for endogenous pyruvic, 2-PGA, PEP, and 3-PGA (when 2,3-PGA is present), a sample is analyzed as above, the 50 micromoles of purified 3-PGA being omitted.

Preparation of 3-PGA Free of 2,3-PGA[1]

Dowex 1-X8, 200 to 400 mesh, is washed twice successively with acetone, distilled water, 2 N NaOH, distilled water, acetone, distilled water, and 2 N HCl. After the resin is washed free from the HCl, a column is prepared (1 cm.^{2} $\times$ 6 cm.); 28.6 g. of the acid barium salt of 3-PGA are mixed in 40 ml. of H$_2$O, 16.0 ml. of 10 N H$_2$SO$_4$ are added with stirring, and the mixture is centrifuged at 5000 $\times$ g for 10 minutes. The supernatant solution is retained, and the BaSO$_4$ residue is washed twice with 50-ml. portions of H$_2$O. The combined supernatant solutions are diluted to 4.0 l. to give 0.02 M 3-PGA at pH 1.85 to 1.88, and this solution is percolated through the column. The effluent is collected, concentrated ten times *in vacuo*, neutralized to pH 4.0 with dilute KOH, and an excess of BaCl$_2$·2H$_2$O is added. After addition of 2 vol. of 95% ethanol, the acid barium salt of 3-PGA crystallizes on standing at −20°.

The crystals are washed with 65% ethanol, then with 95% ethanol, and then dried *in vacuo* over $CaCl_2$. The yield is 24.2 g., containing less than 1 part of 2,3-PGA per 10,000 to 50,000 parts of 3-PGA.

Biosynthesis and Isolation[2]

The high activity for the biosynthesis of 2,3-PGA shown by extracts of chicken breast muscle–acetone powders[3] provides the basis for a simple method which permits the isolation of large quantities of 2,3-PGA with little effort. The following components are mixed and made up to a total volume of 830 ml.: 0.8 millimole of ATP potassium salt at pH 8.5; 8 millimoles of $MgCl_2$; and 80 millimoles of 3-PGA potassium salt at pH 8.5. The incubation mixture is brought to 38° and then completed with 170 ml. of extract of chicken breast muscle–acetone powder, (Each gram of chicken breast muscle–acetone powder is extracted for 10 minutes at 0° with 10 ml. of water and then centrifuged; the supernatant fluid is filtered through paper and used as such.) After 2 hours of incubation[5] at 38°, the pH is adjusted to 1.85 with $2 N$ H_2SO_4 (approximately 100 ml. required), and the mixture is heated to the boiling point, cooled in ice water, centrifuged, and filtered through coarse filter paper. Analysis at this stage shows a total of 37.3 millimoles of 2,3-PGA. A small drop of Dow Corning Antifoam B is added to the almost protein-free filtrate, which is then boiled for 30 minutes. After cooling at 0°, the hydrolyzed preparation is filtered through Whatman No. 1 filter paper, diluted to approximately 5000 ml. with water, and percolated through a column (32 × 225 mm.) of Dowex 1-X8 anion-exchange resin (200 to 400 mesh, chloride form) with an overhead pressure of approximately 5 feet of water. The effluent is discarded, and the column is washed with 150 ml. of deionized water and then eluted with a $0.5 M$ NaCl–$0.5 M$ HCl solution. The first 183 ml. of eluate are discarded, and the next 175 ml. are retained, since they contain the bulk of the 2,3-PGA. Analysis of this fraction shows 32.4 millimoles of 2,3-PGA.

Sixty milliliters of $2 M$ barium perchlorate are then added to the eluate, and the insoluble material formed is removed by centrifugation; the excess of barium is added because sulfate or the other barium acid-insoluble materials are still present in the eluate. The precipitate is

[2] S. Grisolia and B. K. Joyce, *J. Biol. Chem.* 233, 18 (1958).
[3] The high activity for diphosphoglyceric acid mutase found in chicken breast muscle led to the purification of the enzyme from this source.[4]
[4] B. K. Joyce and S. Grisolia, *J. Biol. Chem.* 234, 1330 (1959).
[5] The quantity of muscle extract and length of incubation for optimal synthesis should be determined in a small incubation mixture for each batch of acetone powder.

removed by centrifugation, washed once with about 30 ml. of deionized water, and discarded. The pH of the combined supernatant fluid and washing is adjusted to pH 4 by the addition of 2.5 M LiOH (approximately 70 ml. required, added with constant stirring to prevent local precipitation of $BaCO_3$). The barium salt of the 2,3-PGA is precipitated by the addition of 2 vol. of 95% ethanol and dried over $CaCl_2$ in a vacuum desiccator. The dry material yields 10.8 g. The isolated 2,3-PGA is over 99% pure. The only impurities present in this preparation are traces of inorganic phosphate and of adenine.

The earlier method of isolation (see Vol. III [37]) or modifications thereof[6] require handling large volumes, whereas the method for synthesis and isolation presented above can be conducted easily with five to ten times the quantities described without special equipment, particularly in view of the concentration of the 2,3-PGA at the resin adsorption step. Since commercial 3-PGA is relatively cheap, the biosynthetic procedure appears to be the choice, particularly for large preparations. Also, the whole procedure can be conducted in one day.

[6] J. C. Towne, V. W. Rodwell, and S. Grisolia, *Biochem. Preparations* **6**, 12 (1958).

[72] Preparation and Determination of Diacetylmethylcarbinol

$$CH_3—CO—C(—CH_3)(—OH)—CO—CH_3$$

By Elliot Juni

Chemical Synthesis

Principle. Methylacetylacetone is oxidized with lead tetraacetate, and the resulting acetoxymethylacetylacetone is hydrolyzed to diacetylmethylcarbinol (DAMC) and acetic acid by dilute H_2SO_4.[1] DAMC is separated from acetic acid and a small amount of acetoin, formed as a partial breakdown product during the hydrolysis, by a series of distillations.[2]

Reagents

Absolute ethanol.
Metallic sodium.
Acetylacetone.
Iodomethane.

[1] E. Juni and G. A. Heym, *J. Biol. Chem.* **218**, 365 (1956).
[2] E. Juni and G. A. Heym, *Arch. Biochem. Biophys.* **67**, 410 (1957).

Sodium-dried benzene (thiopene-free).
Lead tetraacetate.
Anhydrous MgSO₄.
Drierite.
Concentrated H₂SO₄.

Procedure. METHYLACETYLACETONE. The synthesis is performed in a
three-necked round-bottomed 2-l. flask fitted with a mercury-sealed
stirrer, a reflux condenser, the open end of which is covered with a drying
tube containing magnesium perchlorate, and a dropping funnel. To 500
ml. of absolute ethanol are added 32.2 g. of finely divided sodium (1.4
moles), and mixing is continued until all the sodium is dissolved. This is
followed by the slow addition, with stirring, of 140 ml. of distilled acetylacetone. The solution becomes reddish-brown, and a brown precipitate
forms. To this mixture 200 g. of iodomethane (1.4 moles) are added
slowly, with stirring, followed by refluxing for 12 hours. The liquid is now
slightly acid to moist litmus paper.

The liquid phase is decanted from the NaI. The precipitate is washed
twice with 50-ml. portions of benzene, and the washes are added to the
decanted liquid. On removal of most of the ethanol by distillation, a
large precipitate of NaI forms. Once again the supernatant is decanted
and combined with six 50-ml. benzene washes of the precipitate. Remaining traces of NaI are removed by suction filtration. Ethanol and benzene
are then removed by distillation under reduced pressure, and methylacetylacetone is fractionally distilled at 70° to 72° (24 mm.). A yield of
74.4 g. (0.65 mole) is obtained.

ACETOXYMETHYLACETYLACETONE. To a solution of 74.4 g. of methylacetylacetone in 250 ml. of dry benzene are slowly added 288 g. (0.65
mole) of lead tetraacetate with vigorous stirring. The temperature is
maintained at about 40° by the use of an external ice bath during addition of lead tetraacetate. The ice bath is then removed, and the mixture
is stirred at room temperature for 24 hours. Precipitated lead diacetate is
separated from the supernatant by suction filtration, washed four times
with 100-ml. portions of benzene, and the washes are combined with the
supernatant. The supernatant is then washed ten times with 100-ml.
portions of water to remove acetic acid. The pH of the last wash is
approximately equal to that of the water used for washing. The benzene
solution is filtered through several thicknesses of filter paper, dried with
anhydrous MgSO₄ and Drierite in succession, filtered, and the benzene
removed by distillation. Acetoxymethylacetylacetone (25.2 g., 0.146
mole) is distilled at 80° (5 to 6 mm.).

Hydrolysis to DAMC and acetic acid is performed by adding 0.4 ml.

of concentrated H_2SO_4 to a solution of 1 ml. (5.8 millimoles) of acetoxy-methylacetylacetone in 49 ml. of water and permitting the solution to stand at room temperature for 1 week. The brownish solution is then vacuum-distilled to dryness (30° at 20 mm.), and the distillate is collected in an ice-cooled flask (first distillate). This distillate contains 90.5% of the theoretical hydrolysis products, 17.1% being acetoin formed by partial hydrolysis of DAMC. In order to remove acetic acid, the distillate is carefully adjusted to pH 7.0 with 1.0 M tris(hydroxymethyl)-aminomethane and vacuum-distilled to dryness as above (Tris distillate). The small amount of acetoin remaining in the Tris distillate is removed by two successive vacuum distillations from sodium bisulfite (10 g./ml.). A summary of the purification procedure is given below. The entire procedure from the initial synthesis to the final purification has been successfully reproduced several times.

Solutions of DAMC are best preserved by acidification with dilute acid and storage in the frozen state. Such preparations show little or no decomposition for several months.

A summary of the purification procedure is given in the table.

SUMMARY OF PURIFICATION PROCEDURE FOR CHEMICALLY SYNTHESIZED DIACETYLMETHYLCARBINOL[a]

Step	DMAC, micromoles/ml.	Acetoin, micromoles/ml.	Diacetyl, micromoles/ml.	Acetic acid, micromoles/ml.	DAMC recovered, % of theory
Theoretical	115	0	0	115	
First distillate	85.5	17.8	0.69	110	74.4
Tris distillate	83.7	21.3		0	72.7
First NaHSO₃ distillate	69.5	5.5			60.4
Second NaHSO₃ distillate	51.4	0.9	0.22	0	44.7

[a] From E. Juni and G. A. Heym, *Arch. Biochem. Biophys.* **67**, 410 (1957).

Enzymatic Synthesis

Principle. DAMC is synthesized enzymatically from diacetyl according to the following equation:[1,3]

$$2CH_3—CO—CO—CH_3 + H_2O \rightarrow$$
$$CH_3—CO—C(—CH_3)(—OH)—CO—CH_3 + CH_3—COOH$$

Bacteria grown with 2,3-butanediol or acetoin as the carbon source contain a thiamine pyrophosphate-requiring enzyme which catalyzes the

[3] E. Juni and G. A. Heym, *J. Bacteriol.* **72**, 746 (1956).

above reaction. Pyruvic oxidase preparations also convert diacetyl to DAMC.[1]

Reagents

$NaHCO_3$.
Diacetyl.
TPP.
$MgSO_4$.
CO_2 gas (100%).
Bacterial enzyme. Alumina C_γ eluate which is first dialyzed overnight against 0.025 M Veronal buffer, pH 7.6, containing 0.01% glutathione, to remove ammonium sulfate.[3]
Pigeon breast muscle enzyme. The enzyme fraction obtained in step 2 of the purification of Jagannathan and Schweet[4] is used.

Procedure. To a 125-ml. Warburg vessel are added 2.0 millimoles of $NaHCO_3$, 3.0 millimoles of diacetyl, 0.0015 micromole of TPP, 0.015 millimole of $MgSO_4$, and 10 ml. of enzyme in a total volume of 25 ml. The gas phase is 100% CO_2, and the reaction mixture is incubated at 30° with shaking for 7 hours. The reaction is usually complete in this time as evidenced by the lack of further CO_2 evolution from the bicarbonate buffer.

With the bacterial enzyme preparation, DAMC and acetic acid are the only end products of the reaction. With the pigeon breast muscle preparation, diacetyl is converted to a mixture of DAMC, acetoin (about 20% of the amount of DAMC), and acetic acid. DAMC free of these other products is obtained by treating acidified mixtures according to the purification scheme outlined above. If the reaction is not carried to completion, excess diacetyl can be removed by heating to 60° and gassing with a stream of helium, or some other inert gas, for about an hour.[5] Purified solutions of DAMC should be acidified with dilute acid and stored in the frozen state.

Determination

Principle. DAMC reduces acid solutions of ammonium molybdate with the formation of a blue color characteristic of lower molybdenum oxides.[2] The absorbance of the color obtained is proportional to the concentration of DAMC present. Since the presence of as little as 1 μg. of inorganic phosphate results in inhibition of the maximum color obtained in this test, samples of DAMC are distilled before the determination is

[4] V. Jagannathan and R. S. Schweet, *J. Biol. Chem.* **196**, 551 (1952).
[5] M. Dolin, *J. Bacteriol.* **69**, 51 (1955).

made. This distillation also leaves behind protein and other nonvolatile products.

Reagents

Glass-distilled water. All reagents must be prepared in glass-distilled water. The use of demineralized water has been shown to result in depressed and abnormal colors.

A Stotz still[6] or similar distillation apparatus.

NaCl.

H_2SO_4, 1 N.

16-mm. Pyrex colorimeter tubes.

Klett-Summerson colorimeter with a 660-mμ filter.

Acid–molybdate reagent. Prepared by mixing equal volumes of $\frac{2}{3}$ N H_2SO_4 and freshly prepared 5% ammonium molybdate solution.

Procedure. DAMC is first distilled from solutions in which it is contained by adding 13 g. of NaCl to 20 ml. of solution, acidifying with one or two drops of 1 N H_2SO_4, and heating with a free flame. Collection of 10 ml. of distillate results in quantitative recovery of DAMC.

To a 16-mm. Pyrex colorimeter tube containing 5.0 ml. of distilled sample of DAMC (0.05 to 0.5 micromole) is added 1 ml. of acid–molybdate reagent, and the contents are thoroughly mixed by shaking. A colorimeter blank is prepared by using 5.0 ml. of distilled water instead of sample. The colorimeter is set to zero absorbance with the blank, which is completely colorless. A blue color develops rapidly in tubes containing DAMC and reaches a maximum value in less than 10 minutes. At ordinary room temperatures, the maximum color is stable for approximately 30 to 40 minutes, after which time it starts to fade slowly. The maximum absorbance obtained is proportional to the amount of DAMC in the colorimeter tube and is independent of the temperature. For unusually high room temperatures, however, the color may start to fade several minutes after the maximum intensity has been obtained. A Klett reading (optical density $\times$ 500) of 100 is obtained for 0.1 micromole of DAMC. When the absorbance of this colored solution for 0.1 micromole of DAMC is measured in a 1-cm. cell in a Beckman Model DU spectrophotometer at 660 mμ, a maximum optical density of 0.150 is obtained. Reproduction of the same color intensity for a given concentration of DAMC has been observed consistently.

The following substances (100 micromoles) neither react nor interfere in the colorimetric test for DAMC: ethanol, acetaldehyde, acetic acid,

[6] E. Stotz, *J. Biol. Chem.* **148**, 585 (1943).

2,3-butanediol, sodium lactate and sodium pyruvate. Diacetyl and acetoin in concentrations fifty to one hundred times as great as the DAMC concentration do produce a color which, unlike the color for DAMC, increases linearly with time. When present in concentrations as much as ten times that of the DAMC concentration, diacetyl and acetoin do not interfere to any appreciable extent. With interfering concentrations of diacetyl and/or acetoin it is still possible to determine the concentration of DAMC by a procedure involving extrapolation of the time curve for color formation.[2] In general, it has been found that any compound that produces color in this test must be present in relatively large concentrations and gives rise to a color the intensity of which increases linearly with time. DAMC is the only substance tested which results in the rapid formation of a color of maximum intensity which then proceeds to fade.

DAMC may also be determined by the Voges-Proskauer test[7] and the diacetyl test of White et al.[7] In each of these tests DAMC gives the same reaction as an equimolar quantity of diacetyl.

Properties

DAMC resembles diacetyl and acetoin in many ways,[1,2] such as codistillation with water. Distillation at 100° of 50% of a solution containing DAMC, acetoin, and diacetyl removes 86% of the DAMC, 63% of the acetoin, and 100% of the diacetyl. In the presence of salts, such as NaCl, distillation of DAMC is facilitated, making it possible to achieve a ninefold concentration by two successive distillations, each of one-third of the volume treated, with better than 95% recovery. Under alkaline conditions, in the absence of oxygen, DAMC is hydrolyzed to acetoin and and acetic acid. There is considerable destruction of the acetoin formed on alkaline hydrolysis of DAMC if this reaction takes place in the presence of air. Periodate oxidation of DAMC yields 3 moles of acetic acid per mole of DAMC. DAMC is oxidized to diacetyl by ferric chloride in a manner similar to the oxidation of acetoin by this reagent. Solutions of DAMC form a derivative with 2,4-dinitrophenylhydrazine in the cold; melting point, 234° to 238° (uncorrected).

DAMC is reduced by DPNH in the presence of a suitable secondary alcohol dehydrogenase.[8]

[7] See Vol. III [50].
[8] E. Juni and G. A. Heym, J. Bacteriol. 74, 757 (1957).

[73] The Preparation of 3-Deoxy-D-*arabino*-heptulosonic Acid 7-Phosphate[1]

By D. B. SPRINSON, J. ROTHSCHILD, AND M. SPRECHER

Principle

Starting with 2-deoxy-D-*arabino*-hexose[2] (2-dexoy-D-glucose), the cyanohydrin is prepared and hydrolyzed, and a crystalline methyl 3-deoxy-D-*gluco*-heptonate[3] (I) is isolated. Tritylation and then benzoylation of (I) gives a crystalline tetra-*O*-benzoyl-7-*O*-trityl methyl ester (II), which is detritylated by catalytic hydrogenolysis and then esterified with diphenylphosphorochloridate to yield (III). Unmasking of the phosphate group by catalytic hydrogenolysis (the benzene rings are reduced in this process) and saponification of the cyclohexylcarboxylic and methyl ester groups affords 3-deoxy-D-*gluco*-heptonic acid 7-phosphate (IV), which is isolated as a crystalline cyclohexylammonium salt. The latter is oxidized by a pyridine–vanadium pentoxide reagent, and the resulting 3-deoxy-D-*arabino*-heptulosonic acid 7-phosphate (V) is purified by ion-exchange chromatography and isolated as a noncrystalline barium salt (Scheme 1).

Preparation

Methyl 3-Deoxy-D-*gluco-heptonate (I).* To 25.7 g. (0.16 mole) of 2-deoxy-D-*arabino*-hexose dissolved in 80 ml. of water and cooled in ice are added cold solutions of 40.0 g. (0.23 mole) of calcium acetate monohydrate in 80 ml. of water, and 10.0 g. (0.21 mole) of NaCN in 80 ml. of water, and the reaction mixture is allowed to stand overnight in the refrigerator.[4] After the addition of 11 g. (0.15 mole) of Ca(OH)$_2$, the mixture is heated for 4 hours on the steam bath with occasional shaking, and cooled.

The above mixture is combined with an identical second preparation and filtered through Celite. A column 26 × 7.6 cm. is prepared from 1 kg. of Dowex 50-X8 (H$^+$, 50 to 100 mesh) and washed with water. The

---

[1] Abbreviated as DAHP. For the enzymatic formation of DAHP and its conversion to 5-dehydroquinate see Vol. V [52a] and [52b].

[2] Purchased from Aldrich Chemical Co., Milwaukee, Wisconsin. All other reagents (and solvents) were reagent grade.

[3] The configuration of the hydroxyl group on C-2 was determined by oxidation of I with periodate, and then with Ag$_2$O to yield 3-deoxy-D-*glycero*-tetraric (D-malic) acid.

[4] Titration of an aliquot with AgNO$_3$ showed disappearance of cyanide equivalent to the deoxyglucose added. This was essentially the case also after 4 hours.

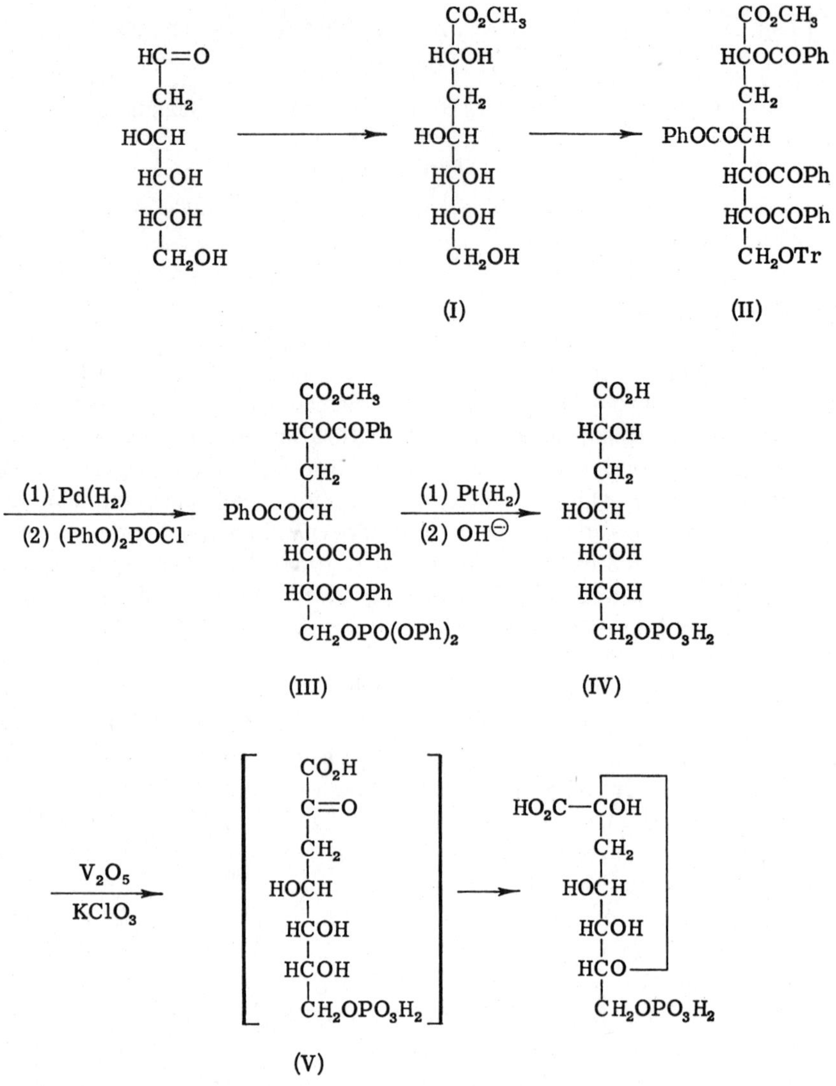

$$
\begin{array}{ccc}
\text{HC}=\text{O} & & \text{CO}_2\text{CH}_3 \\
| & & | \\
\text{CH}_2 & & \text{HCOH} \\
| & & | \\
\text{HOCH} & \longrightarrow & \text{CH}_2 \\
| & & | \\
\text{HCOH} & & \text{HOCH} \\
| & & | \\
\text{HCOH} & & \text{HCOH} \\
| & & | \\
\text{CH}_2\text{OH} & & \text{HCOH} \\
& & | \\
& & \text{CH}_2\text{OH} \\
& & (\text{I})
\end{array}
$$

(I) → (II)

(II)
$$
\begin{array}{c}
\text{CO}_2\text{CH}_3 \\
| \\
\text{HCOCOPh} \\
| \\
\text{CH}_2 \\
| \\
\text{PhOCOCH} \\
| \\
\text{HCOCOPh} \\
| \\
\text{HCOCOPh} \\
| \\
\text{CH}_2\text{OTr}
\end{array}
$$

$$\xrightarrow[\text{(2) (PhO)}_2\text{POCl}]{\text{(1) Pd(H}_2)}$$

(III)
$$
\begin{array}{c}
\text{CO}_2\text{CH}_3 \\
| \\
\text{HCOCOPh} \\
| \\
\text{CH}_2 \\
| \\
\text{PhOCOCH} \\
| \\
\text{HCOCOPh} \\
| \\
\text{HCOCOPh} \\
| \\
\text{CH}_2\text{OPO(OPh)}_2
\end{array}
$$

$$\xrightarrow[\text{(2) OH}^\ominus]{\text{(1) Pt(H}_2)}$$

(IV)
$$
\begin{array}{c}
\text{CO}_2\text{H} \\
| \\
\text{HCOH} \\
| \\
\text{CH}_2 \\
| \\
\text{HOCH} \\
| \\
\text{HCOH} \\
| \\
\text{HCOH} \\
| \\
\text{CH}_2\text{OPO}_3\text{H}_2
\end{array}
$$

$$\xrightarrow[\text{KClO}_3]{\text{V}_2\text{O}_5}$$

(V)
$$
\left[
\begin{array}{c}
\text{CO}_2\text{H} \\
| \\
\text{C}=\text{O} \\
| \\
\text{CH}_2 \\
| \\
\text{HOCH} \\
| \\
\text{HCOH} \\
| \\
\text{HCOH} \\
| \\
\text{CH}_2\text{OPO}_3\text{H}_2
\end{array}
\right]
\longrightarrow
\begin{array}{c}
\text{HO}_2\text{C}-\text{COH} \\
| \\
\text{CH}_2 \\
| \\
\text{HOCH} \\
| \\
\text{HCOH} \\
| \\
\text{HCO}- \\
| \\
\text{CH}_2\text{OPO}_3\text{H}_2
\end{array}
$$

Scheme 1

Ph , Phenyl

Tr , Trityl

filtrate from the hydrolysis of the cyanohydrin is diluted to 2200 ml., the solution is percolated through the column, and the resin is washed with at least 1 l. of water. The washings are combined with the main fraction, and the solution is evaporated to dryness *in vacuo*.
The residue is taken up in 75 ml. of dry methanol, and 0.7 g. of anhydrous HCl in 10 ml. of dry methanol is added. Crystallization begins after several minutes, and, after chilling for 3 hours in an ice bath, the crystals are collected (41 g., m.p.[5] 172° to 175°). On standing overnight at 0°, another 2.3 g. is deposited (m.p. 169° to 172°; yield 62%). This material can be used in the next step. Analytical sample, recrystallized from methanol, m.p. 175° to 176°, $[\alpha]_D^{25} + 11.0°$ (5% in H_2O).

Analysis. Calculated for $C_8H_{16}O_7$ (224.2): C, 42.9; H, 7.2. Found: C, 42.9; H, 7.3.

Methyl 3-Deoxy-2,4,5,6-tetrabenzoyl-7-trityl-D-gluco-heptonate (II). To 8 g. (0.036 mole) of the thoroughly dried methyl ester in 200 ml. of pyridine (dried over BaO) is added 10.8 g. (0.039 mole) of tri-phenylmethyl chloride. The reaction mixture is shaken gently at room temperature for 24 hours, the resulting clear solution is cooled in an ice bath, and 20.3 g. (0.145 mole) of benzoyl chloride is added slowly. The mixture is allowed to stand for 48 hours, 2 ml. of water is added to hydrolyze any remaining reagent, and most of the pyridine is removed by distillation at 40° *in vacuo*. The residue is dissolved in 150 ml. of chloroform and washed with successive 150-ml. portions of cold water, 1 M HCl, 1 M NaHCO₃, and water. The chloroform solution is dried over Na_2SO_4 and reduced to an oil *in vacuo*, and the oil is triturated with 200 ml. of methanol to yield white crystals. After chilling in an ice bath the material is filtered off and washed with cold methanol; yield 20.6 g., m.p. 145° to 150° with softening at 120°. Recrystallization from ethyl acetate–methanol yields 15.5 g. (49%), m.p. 167° to 168°, $[\alpha]_D^{24°} + 73.4°$ (5.07% in ethyl acetate).

Analysis. Calculated for $C_{55}H_{46}O_{11}$ (883): C, 74.8; H, 5.2. Found: C, 75.2; H, 5.5.

Tricyclohexylammonium 3-Deoxy-D-gluco-heptonate 7-phosphate (IV). Seventeen grams of 5% palladium chloride on charcoal (Darco G-60),[6] suspended in absolute ethanol, is stirred magnetically in an atmosphere of hydrogen until no more gas is taken up (preferably overnight). The catalyst is filtered off, washed with methanol and methyl acetate, and added to 16.9 g. (0.019 mole) of the trityl tetrabenzoyl compound (II) in 100 ml. of methyl acetate. After 18 hours of magnetic stirring in hydrogen at 1 atm. 610 ml. of hydrogen is taken up [1.3

---

[5] Melting points on Fischer-Johns block, uncorrected.
[6] *Org. Syntheses, Coll. Vol.* **III**, 686 (1955).

moles/mole of (II).[7] The catalyst is removed by filtration and washed with methyl acetate. The filtrate and washings are evaporated to an oil *in vacuo*, and the residue is dried over P_2O_5 in a desiccator evacuated with a mechanical pump to yield a glassy solid (15 g.).

The product is dissolved in 40 ml. of dry pyridine, 10.2 g. (0.038 mole) of diphenylphosphochloridate is added slowly with chilling, and the mixture is place in a refrigerator overnight. After the addition of 0.5 ml. of water to decompose excess reagent, most of the pyridine is removed *in vacuo*, and 200 ml. of ether is added to the residue. The resulting solution is extracted with successive 200-ml. portions of cold water, 1 M HCl, 1 M $NaHCO_3$, and water. The solvent is removed *in vacuo*, and ethanol is added several times and evaporated *in vacuo*.

The residual oil (containing crystals of triphenylmethane) is dissolved in 250 ml. of absolute ethanol,[8] 3.5 g. of platinum oxide catalyst is added, and hydrogenation is continued (overnight) until no additional hydrogen is taken up.[9] The catalyst is filtered off and washed with ethanol, 200 ml. of 1 N NaOH is added to the alcoholic solution (600 ml.), and the mixture is allowed to stand at room temperature overnight.[10] The solution is reduced at the water pump to 100 ml., 100 ml. of water is added, and the mixture is extracted with ether to remove triphenylmethane. The aqueous solution is treated with 100 ml. of Dowex 50 (H^+) and stirred mechanically for 5 minutes. The resin is filtered off, washed with water, and the filtrate and washings are extracted three times with ether to remove cyclohexylcarboxylic acid. The aqueous solution is immediately brought to pH 9 with redistilled cyclohexylamine, reduced to dryness at the water pump, and the residue dried over P_2O_5 under high vacuum. The white crystals (9.2 g., m.p. 152° to 160°) are dissolved in approximately 20 ml. of water and crystallized by the addition of acetone to a slight turbidity, seeding, and chilling in the refrigerator. Once crystallization

---

[7] Excess H_2 uptake has been observed in the hydrogenolytic cleavage of trityl ethers [C. E. Ballou, H. O. L. Fischer, and D. L. MacDonald, *J. Am. Chem. Soc.* **77**, 5967 (1955)].

[8] By extracting with smaller volumes of alcohol it is possible to remove some insoluble triphenylmethane. This was usually not done.

[9] Approximately 8.0 l. of hydrogen was used. On the assumption that all the benzene rings are reduced except those of triphenylmethane, the theoretical uptake for 0.019 mole of III is 8.5 l. A gas burette was used for small batches. A Parr shaker was more convenient for large batches of material and gave the same uptake of H_2.

[10] Titration of aliquots with acid, with thymolphthalein as indicator, showed that 80% to 95% of the theoretically required amount of alkali (seven acidic groups per mole of trityl benzoyl compound) had been neutralized during the saponification. The maximum value was reached after 2 hours.

has started, acetone is added slowly until about 100 ml. is used. Yield
6.1 g., second crop 0.8 g. (62%), m.p. 153° to 157°. This material is sufficiently pure for the next step. Recrystallization from aqueous acetone
gives 4.8 g. (43%), m.p. 155° to 158° $[\alpha]_D^{25} + 9.4°$ (4.9% in water).

Analysis. Calculated for $C_{25}H_{54}N_3O_{10}P$ (588): C, 51.1; H, 9.3; N,
7.15; P, 5.27. Found: C, 50.4; H, 9.3; N, 7.13; P, 5.46.

3-Deoxy-D-arabino-*heptulosonic Acid 7-Phosphate* (*DAHP, V*).[11]
The catalyst for oxidation is prepared by dissolving 750 mg. of V_2O_5
in 45 ml. of concentrated HCl. To this solution is rapidly added 45 ml.
of pyridine with cooling. The resulting suspension is adjusted to pH 3.2
with HCl and pyridine. It can be stored without special precautions for
several months.

One gram (1.7 millimoles), of the tricyclohexylammonium salt of
(IV) and 0.10 g. (0.82 millimole) of $KClO_3$ are dissolved in 2 ml. of
water in a small beaker, and 3 ml. of the catalyst suspension is added.
The reaction mixture is adjusted to pH 3.6 (pH meter) with HCl and
pyridine, transferred to a small glass-stoppered Erlenmeyer flask with
1 ml. of H_2O, and the suspension is stirred with a magnetic stirrer at
room temperature (22° to 25°) for exactly 16.5 hours. The dark gray-
green clear solution is passed through a column (2 cm. in diameter)
of 15 ml. of Dowex 50 (H^+), and the green filtrate is passed through a
second 15 ml. of the resin, yielding a white, clear solution. The columns
are washed thoroughly with water, the combined filtrate and washings
are brought to pH 6 with NH_4OH, and 680 mg. (2.7 millimoles) of
barium acetate is added. The solution is evaporated *in vacuo* to 15 ml.
at 30°, more NH_4OH is added to pH 8, and the barium salts are precipitated by the addition of 2 vol. of absolute alcohol. After chilling in an
ice bath for several hours, the precipitate is centrifuged off, washed
three times with 5 ml. of cold 70% ethanol and once with 5 ml. of absolute ethanol, and dried *in vacuo* at room temperature. Yield of crude
barium salt 800 to 900 mg., containing approximately 40% DAHP by
periodate–thiobarbiturate assay.[12]

The crude barium salt is suspended in water and dissolved by the
addition of 5 ml. of Dowex 50 (H^+). After removal of the resin by filtration, the filtrate is passed through a column of 5 ml. of Dowex 50, and
the resins are washed thoroughly with water. The combined filtrate and
washings are brought to pH 8 with NH_4OH, and the volume is adjusted

---

[11] The method described here is a modification of one previously described for the
    oxidation of aldonic acids to ulosonic acids: P. P. Regna and B. P. Caldwell, *J.
    Am. Chem. Soc.* **66**, 243 (1944).
[12] Vol. V [52a]. The modification of this method by L. Warren [*J. Biol. Chem.* **234**,
    1971 (1959)] is more convenient.

to 300 ml. This solution is loaded at a rate of 1.5 to 2.0 ml./min. on a
10-ml. column (1.3 cm. in diameter and 7 cm. high) of Dowex 1 chloride-
X8 (200 to 400 mesh, fined and washed thoroughly with 3 N HCl and
water). The column is washed with 100 ml. of water and eluted with
0.01 N HCl on an automatic fraction collector. Fourteen 50-ml. fractions are collected and neutralized with NH_4OH. The acid breaks through
after three fractions, and fractions 4 and 5 contain small amounts of
inorganic phosphate and unidentified impurities. Unoxidized (IV) (approximately 0.4 millimole of organic phosphate) is eluted in fractions 6
to 10, followed by four fractions devoid of any phosphate. Elution is
now begun with 0.02 N HCl, 25-ml. fractions being collected, neutralized
with NH_4OH, and tested by the periodate–thiobarbiturate assay. After
two fractions devoid of activity, 0.8 to 0.9 millimoles of DAHP is
obtained in the following nine fractions. The neutralized solutions containing (IV) and DAHP are treated with 165 and 290 mg., respectively,
of barium acetate, and reduced *in vacuo* to 15 and 20 ml., respectively.
The pH is adjusted to 8 with NH_4OH, and absolute ethanol is added to
60% concentration (v/v). The solutions are then chilled in an ice bath,
after which the Ba salts are collected by centrifugation, washed three
times with cold 60% ethanol and once with absolute ethanol, and dried
in vacuo at room temperature. Yield of Ba salt of (IV), 155 mg., which
can be used for the recovery of its cyclohexylammonium salt. Yield of
the tetrahydrate of BaDAHP, 400 to 600 mg. $[\alpha]_D^{25} + 15.7°$ (K salt,
3.8% in water). For the free acid, $[\alpha]_D^{25} + 42°$ (1.5% in water). After
the barium salt is dried *in vacuo* at 100° the dihydrate is obtained. Loss
in weight 6.3% (calculated 6.4%).

Analysis. Calculated for $C_7H_{10}O_{10}PBa_{1.5} \cdot 2H_2O$ (527.1): C, 15.94; H,
2.66; P, 5.88; Ba, 39.09. Found: C, 15.94; H, 2.57; P, 5.90; Ba, 39.45.

[74] The Preparation and Identification of 5-Dehydroquinic and 5-Dehydroshikimic Acids

By E. HASLAM, R. D. HAWORTH, and P. F. KNOWLES

Principle. 5-Dehydroquinic and 5-dehydroshikimic acids can be prepared by the nitric acid oxidation of quinic acid.[1] A more convenient
procedure for the preparation of 5-dehydroquinic acid is, however, the

[1] R. Grewe and J. P. Jeschke, *Ber.* **89**, 2080 (1956).

platinum-catalyzed dehydrogenation of quinic acid; the stereospecificity of this reaction is dependent on quinic acid existing predominantly in a conformation in which the 5-dehydroxyl group is axially disposed. The method has been described by two groups of authors,[2,3] that suggested by Haslam et al.[3] being given in the present account. 5-Dehydroshikimic acid may be prepared by a similar selective oxidation of shikimic acid[2] or by the dehydration of 5-dehydroquinic acid under the conditions outlined by Davis et al.[4] Convenient modifications of both these procedures for the preparation of 5-dehydroshikimic acid are described here.

I. 5-Dehydroquinic Acid

Procedure. Quinic acid (0.5 g.) is dissolved in water (10 ml.) and the pH is adjusted to 6.0 with solid sodium bicarbonate. Meanwhile platinum oxide (0.5 g.) is placed in a hydrogenation flask, covered with water (10 ml.), and shaken with hydrogen at atmospheric pressure until uptake of hydrogen ceases. The platinum, after decantation of the water, is added to the sodium quinate solution, and a stream of oxygen is passed through the suspension, maintained at 40°, for 6 hours. Additional agitation of the reaction mixture is achieved by magnetic stirring. The catalyst is then filtered off and washed with 5 × 10-ml. aliquots of water, the combined filtrate and washings are decationized by passage through a column of Dowex 50W-X8 (50 to 100 mesh, 20 × 2 cm., hydrogen form[5]), and the eluate is concentrated to a volume of 20 ml. by rotary evaporation before adsorption on a column of Dowex 1-X8 (100 to 200 mesh,[6] 20 × 2 cm., acetate form).[7] The acids are removed by gradient elution[7] by using a reservoir containing 6 N acetic acid (2000 ml.) and a mixing chamber containing 0.5 N acetic acid (500 ml.). Eighty fractions (7.5 ml. each) are collected at a flow rate of 75 ml./hr. and analyzed by paper chromatography (see final section). Concentration of the relevant fractions (in the region of fraction 50) by rotary evaporation and high-vacuum drying overnight gives 5-dehydroquinic acid (0.4 g.) as an oil which is dissolved in dry acetone (10 ml.) and filtered. The filtrate is allowed to evaporate slowly in a desiccator under reduced pressure (20 cm. Hg). A colorless crystalline solid is obtained which is washed free

[2] K. Heyns and H. Gottschalck, *Ber.* **94**, 343 (1961).
[3] E. Haslam, R. D. Haworth, and P. F. Knowles, *J. Chem. Soc.* 1857 (1961).
[4] U. Weiss, B. D. Davis, and E. S. Mingioli, *J. Am. Chem. Soc.* **75**, 5572 (1953).
[5] See Vol. III [16B].
[6] Amberlite CG 400 anion-exchange resin (100 to 200 mesh) can replace Dowex 1-X8 with equal success.
[7] See Vol. III [70]; gravity alone is sufficient to force acetic acid from the reservoir into the mixing flask during gradient elution.

from residual oil with acetone (cooled to −20°) and recrystallized by being dissolved in the minimum volume of hot acetone with chloroform being added dropwise until a faint turbidity appears. The solution is chilled overnight, giving white needles (0.25 g.), melting point 136° to 138°.

II. 5-Dehydroshikimic Acid

Procedure A. From Shikimic Acid. Shikimic acid (1 g.) is dissolved in water (50 ml.), platinum, freshly prepared by the hydrogenation of platinum oxide (1 g.) (see procedure I), is added, and a stream of oxygen is passed through the suspension at room temperature for 5 hours. Additional agitation of the reaction mixture is achieved by magnetic stirring. The platinum is filtered off, washed with water (5 × 10-ml. aliquots), and the combined filtrate and washings are adsorbed directly on a column of Dowex 1-X8 (100 to 200 mesh, 30 × 2 cm., acetate form).[7] Gradient elution of the acid is as described previously for the preparation of 5-dehydroquinic acid. Fractions (100 × 10 ml.) are collected at a flow rate of 75 ml./hr. and analyzed by paper chromatography (see final section). 5-Dehydroshikimic acid is present in the region of fraction 70, and concentration of the appropriate fractions by rotary evaporation followed by drying at high vacuum overnight yields rosette-shaped crystals (0.17 g.). Recrystallization from hot ethyl acetate (which need not be dry) gives white needles (0.12 g.); melting point 146° to 147°.

Shikimic acid (0.74 g.) is recovered as a crystalline solid by concentration of the appropriate fractions (in the region of fraction 20). Heyns and Gottschalck[2] with this method obtained 35% yields of 5-dehydroshikimic acid. An important factor in this oxidation is the degree of dispersion of the platinum; finely colloidal platinum, freshly prepared makes the best catalyst.

Procedure B. From 5-Dehydroquinic Acid. 5-Dehydroquinic acid (0.4 g.) is refluxed for 1 hour with 0.1 N hydrochloric acid (20 ml.). The solution is concentrated to dryness by rotary evaporation taken up in distilled water (10 ml.) and adsorbed on a column of Amberlite CG-400 ion-exchange resin (100 to 200 mesh, 20 × 2 cm., acetate form). The acids are removed by gradient elution as described above in the preparation of 5-dehydroquinic acid. Fractions (100 × 10 ml.) are collected and analyzed by paper chromatography (see final section). Concentration of the relevant fractions (in the region of fraction 70) gives 5-dehydroshikimic acid (0.2 g.) as an oil which is crystallized as described in procedure II.A above. 5-Dehydroquinic acid (0.1 g.) is recovered as an oil by concentration of the relevant fractions (in the region of fraction 50) and crystallized as described previously in procedure I.

Analysis of Reaction Products by Paper Chromatography

Test samples (5 drops) of every second fraction from the above elutions are spotted onto Whatman No. 1 chromatography paper and developed overnight (descending) in the solvent system benzyl alcohol–butan-3-ol–propan-2-ol–water (3:1:1:1 w/v) containing 2% formic acid (90%). The dried chromatogram is sprayed with a freshly prepared solution of sodium metaperiodate (160 mg.) in a mixture of N acetic acid (12.5 ml.) and N sodium acetate (12.5 ml.) followed 20 minutes later by a spray of 3% alcoholic aniline, according to the procedure described by Hasegawa and Yoshida.[8] The chromatogram is studied in the visible and under ultraviolet light (Hanovia fluorescence 11 lamp). A summary of the results is given in the table.

CHROMATOGRAPHIC PROPERTIES[a]

Compound	Visible	Ultraviolet	R_f
5-Dehydroquinic acid	Yellow	Yellow fluorescence	0.25
5-Dehydroshikimic acid	Yellow	Yellow fluorescence	0.55
Quinic acid	Pink	Adsorption	0.20
Shikimic acid	Red	Adsorption	0.40

[a] Considerable economy in time is achieved in the analysis of column eluates if the fractions are spotted onto filter paper and sprayed directly. Chromatograms are then prepared of the relevant fractions.

[8] S. Yoshida and M. Hasegawa, *Arch. Biochem. Biophys.* **70**, 377 (1957).

Section II
Lipids and Steroids

[75] Preparation and Determination of Intermediates in Cholesterol Synthesis

By T. T. TCHEN

Introduction

Among the many intermediates between mevalonic acid and cholesterol, only one compound, squalene, is available commercially in reasonably pure form, and then only in nonradioactive form and therefore of little use to biochemists. Mevalonic acid is commercially available both in unlabeled form and labeled with C^{14} in C-2. In this chapter, the procedures for the preparation of some labeled intermediates are described. The methods are primarily those in use in the laboratory of Professor Konrad Bloch at Harvard University.[1] Unfortunately, for most of the intermediates there is no convenient method of preparation. The compounds to be included in this chapter are, therefore, only a small fraction of the intermediates involved in the biogenesis of cholesterol.

Mevalonic Acid

Since this compound is commercially available, only the assay method will be described here. Purified mevalonic kinase[2] free from phosphomevalonic kinase (see Vol. V [66]) is coupled to pyruvic kinase and lactic dehydrogenase to give a spectrophotometric determination of mevalonic acid.[3] The reaction mixture (3 ml.) at the start of the experi-

MVA + ATP → P-MVA + ADP

ADP + PEP → ATP + pyruvate

Pyruvate + DPNH → Lactate + DPN+

Sum: MVA + pyruvate + DPNH → P-MVA + lactate + DPN+

[1] The procedures described in this chapter include only those used personally by the author or by other members in the laboratory of Professor Konrad Bloch. So much has now been published on the metabolism of MVA that it is not possible to cite all the literature in this chapter, which is not meant as a review article. Hence, only those papers which describe the experimental procedure used in this chapter are cited.

[2] T. T. Tchen, J. Biol. Chem. 233, 1100 (1958).

[3] The following abbreviations are used: mevalonic acid, MVA; 5-phosphomevalonic acid, P-MVA; 5-pyrophosphomevalonic acid, PP-MVA; 3-methyl-3-butenylpyrophosphate, isopentenyl-PP; adenosine diphosphate and triphosphate, ADP and ATP; phosphoenolpyruvate, PEP; inorganic phosphate, P_i; oxidized and reduced diphosphopyridine and triphosphopyridine nucleotides, DPN+, DPNH, TPN+, and TPNH; and all phosphate groups, P.

ment should contain the following: DPNH, 0.3 micromole; ATP, 30 micromoles; PEP, 10 micromoles; Mg^{++}, 30 micromoles; KF, 100 micromoles; phosphate buffer (pH 6.8), 300 micromoles; lactic dehydrogenase, 50 μg.; pyruvic kinase, 50 μg.; and mevalonic kinase, 1 unit.[4] The mevalonic kinase prepared according to Vol. V [66] still contains some ATPase activity. This activity is first determined by following the oxidation of DPNH in a spectrophotometer (disappearance of absorption at 340 mμ). After the rate of this blank DPNH oxidation is established, the MVA sample is added (approximately 0.3 micromole of the racemic mixture or 0.15 micromole of the active isomer). The rate of DPNH oxidation is followed until it has returned to the blank rate prior to the addition of MVA. The amount of the active isomer of MVA added is equal to the amount of DPNH oxidized after subtraction of the blank oxidation. This value is determined graphically.

Phosphomevalonic Acid

Preparation. Although this compound has been synthesized chemically, the yield is very low. It is, therefore, more convenient to prepare it enzymatically with purified mevalonic kinase (free from phosphomevalonic kinase).[2] (See Vol. V [66]). For this conversion, the reaction mixture should contain 2-C^{14}-MVA ($5 \times 10^{-3} M$), ATP ($2 \times 10^{-2} M$), Mg^{++} or Mn^{++} ($10^{-2} M$), phosphate buffer (pH 6.8, $10^{-2} M$), and an appropriate amount of enzyme (see Vol. V [66]. At the end of incubation, the entire reaction mixture is poured onto a Dowex 1–formate column (four times excess of resin). Unreacted mevalonic acid is readily eluted by 1 N HCOOH (complete with less than five times the hold-up volume of the column). Phosphomevalonic acid is then eluted with 6 N HCOOH, and the elution is followed by the appearance of radioactivity. With a column of 10 g. wet weight of the resin, the elution is complete with about 100 ml. of formic acid. The strongly acidic eluate thus obtained contains also inorganic phosphate and ADP. Formic acid is removed by vacuum distillation to near dryness. Water is added, and the distillation is repeated. The residue is dissolved with a small volume of water, neutralized with K$_2$CO$_3$, and stored at $-20°$. If one wishes to remove ADP from this preparation, the above solution is applied to Whatman No. 1 filter paper and chromatographed with t-butanol–HCOOH–water (20:5:8 by volume).[5] In this system, P-MVA has an R_f of 0.55 and is readily separated from ADP ($R_f \sim 0.10$).

Some of the properties of this compound are given in the tables.

[4] One unit is defined as that amount of enzyme which will convert 1 micromole of mevalonic acid per hour. For purification of the enzyme, see Vol. V [66].

[5] K. Bloch, S. Chaykin, A. H. Phillips, and A. deWaard, *J. Biol. Chem.* **234**, 2595 (1959).

Determination. With labeled samples, the easiest way to determine P-MVA is by chromatography followed by counting of the radioactivity. The chromatographic behavior of this compound in various systems is compared to that of other intermediates in Table II.

With unlabeled samples, a coupled spectrophotometric assay could be used. This is identical to that described for MVA with the exception that the substrate and enzyme should be changed to P-MVA and P-MVA kinase.

Pyrophosphomevalonic Acid

Preparation. Phosphomevalonic acid (10^{-4} to $10^{-3} M$) is incubated with purified phosphomevalonic kinase[5] (Vol. V [66]) and ATP ($10^{-2} M$), Mg^{++} ($10^{-2} M$), F^- ($0.03 M$), and phosphate buffer ($0.02 M$, pH 7.0). If the reaction volume is small, the enzyme is heat-inactivated (boiling-water bath, 2 minutes), and the mixture is chromatographed on paper. With *t*-butanol–formic acid–H_2O (20:5:8) as solvent, pyrophosphomevalonic acid (R_f 0.39) can be readily separated from phosphomevalonic acid (R_f 0.55). With a relatively large reaction mixture (10 ml. or more), it is more convenient to separate pyrophosphomevalonic acid from phosphomevalonic acid by column chromatography on Dowex 1–formate.[5] Phosphomevalonic acid is first completely eluted with $4 N$ formic acid (approximately 200 ml. for a 10×1-cm. column). Pyrophosphomevalonic acid is then eluted with $4 N$ formic acid $+ 1 N$ ammonium formate. After removal of formic acid by vacuum distillation at room temperature, the ammonium formate is removed by sublimation at 45° under vacuum. The pyrophosphomevalonic acid (ammonium salt) thus obtained is contaminated with ADP and ATP from which it can be separated by chromatographing twice on paper with *t*-butanol–formic acid–H_2O. The properties of this compound are given in the tables.

Determination. As with P-MVA, labeled PP-MVA is most readily determined by chromatography followed by counting. The R_f values of this compound are given in Table II. Unlabeled PP-MVA can be determined by the coupled spectrophotometric method as described for MVA with the modification that ATP concentration be reduced to $10^{-3} M$. The substrate and enzyme used, of course, should be changed to PP-MVA and PP-MVA decarboxylase.

PP-MVA can also be determined chemically by the liberation of inorganic phosphate when PP-MVA is heated with acid. (For the methods of determination of acid-labile phosphate, see Vol. III [115]).

Isopentenylpyrophosphate (3-Methyl-3-butenyl-1-pyrophosphate)

Preparation. This compound has been synthesized chemically with relatively good yield. However, for ordinary laboratory use, it is more convenient to prepare it enzymatically according to Agranoff *et al.*[6] This method is based on the fact that iodoacetamide inhibits the isomerase that converts isopentenylpyrophosphate to dimethylallylpyrophosphate. The soluble fraction of yeast autolyzate, prepared according to either Lynen *et al.*[7] or Bloch and co-workers (Vol. V [66]), is incubated with 2-C^{14}-mevalonic acid ($10^{-4} M$), ATP ($10^{-2} M$), Mg^{++} ($10^{-2} M$), iodoacetamide ($10^{-2} M$), and phosphate buffer ($10^{-2} M$, pH 6.8). The protein concentration in the reaction mixture should be approximately 5 to 10 mg./ml. After 1 to 2 hours of incubation at 30°, the reaction mixture is transferred to a Dowex 1–formate column and chromatographed by gradient elution. With a 12 × 0.7-cm. column, and after gradient elution according to Hurlbert *et al.*,[8] the elution pattern is as shown in Fig. 1. The mixing flask contains 150 ml. of H_2O, and the

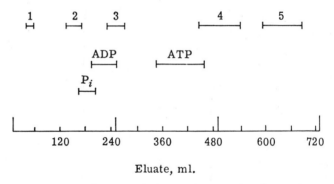

Eluate, ml.

Fig. 1. A 12 × 0.7-cm. Dowex 1–formate column is eluted according to Hurlbert *et al.*[8] The mixing flask contains 150 ml. of water, and the reservoir is charged successively with 250-ml. quantities of $4 N$ formic acid, $4 N$ formic acid + $0.4 M$ ammonium formate, and finally $4 N$ formic acid + $0.8 M$ ammonium formate. The brackets 1 to 5 represent MVA, decomposition product of isopentenyl-PP, P-MVA, PP-MVA and isopentenyl-PP respectively.

reservoir is charged successively with 250 ml. of formic acid and ammonium formate as indicated. Formic acid and ammonium formate are removed by vacuum distillation and sublimation at 45°.

Determination. Labeled samples can be determined by chromatography (see Table II) and counting. Since the pyrophosphate linkage is completely hydrolyzed by heating with $1 N$ HCl for 10 minutes at 100°,

[6] B. W. Agranoff, H. Eggerer, U. Henning, and F. Lynen, *J. Biol. Chem.* **235**, 326 (1960).
[7] F. Lynen, H. Eggerer, U. Henning, and I. Kessel, *Angew. Chem.* **70**, 738 (1958).
[8] R. B. Hurlbert, H. Schmitz, A. F. Brumm, and V. R. Potter, *J. Biol. Chem.* **209**, 23 (1954).

this liberation of inorganic phosphate can be used for the determination of nonlabeled samples after the complete removal of other pyrophosphates by paper chromatography with t-butanol–HCOOH–H$_2$O (see Table II).

Dimethylallylpyrophosphate

This compound is extremely labile to acid and cannot be isolated by column or paper chromatography with formic acid. An isomerase that catalyzes the conversion of isopentenyl-PP to dimethylallyl-PP has been partially purified by Agranoff et al.[6] However, this enzyme has not been purified from biological sources that are commercially available in this country. Hence, no procedure for the preparation of dimethylallyl-PP can be given here. However, the acid lability serves as a convenient method for the determination of labeled dimethylallyl-PP. The sample (neutral pH) is dried in an aluminum planchet over a steam bath, and the radioactivity is determined. Four drops of dilute formic acid are now added to the planchet and again evaporated to dryness. Dimethylallyl-PP is completely converted to a steam-volatile compound and is lost from the planchet. It should be pointed out, however, that geranyl-PP and farnesyl-PP share this property with dimethylallyl-PP. The loss in radioactivity thus represents the total amount of the three allyl pyrophosphates.

Some Properties of the Above Compounds

The stabilities of these compounds are listed in Table I. The R_f values in two paper chromatography systems are shown in Table II. Their behavior on a Dowex 1–formate column is shown in Fig. 1.

TABLE I
STABILITY OF DERIVATIVES OF MVA

Derivative	1 N KOH 100°, 10 minutes	1 N HCl 100°, 10 minutes
MVA	Stable	Stable
P-MVA	Stable	Stable
PP-MVA	Stable	Labile
Isopentenyl-PP	—	Labile
Dimethylallyl-PP	—	Labile

Squalene

Preparation. Nonlabeled squalene is commercially available. Labeled squalene can be synthesized chemically, but it is much easier to prepare

TABLE II
RANGE OF R_f VALUES OF MVA AND ITS DERIVATIVES
AT ROOM TEMPERATURE[a]

	n-Butanol–HCOOH–H_2O	t-Butanol–HCOOH–H_2O
MVA	0.75	0.8
P-MVA	0.15	0.53–0.61
PP-MVA	0.1	0.29–0.35
Isopentenyl-PP	0.1	0.53–0.61
Orthophosphate	0.15	0.50–0.58
AMP	—	0.26–0.30
ADP	0.0	0.08–0.16
ATP	0.0	0.0–0.16

[a] From T. T. Tchen, *J. Biol. Chem.* **233**, 1100 (1958).

it enzymatically. One part of rat liver is homogenized at 0° in a Potter-Elvehjem homogenizer with two parts of phosphate buffer (pH 7.4, 0.1 M) containing 4 mg. of nicotinamide per milliliter.[9] The soluble proteins and the microsomes are decanted off after centrifugation for 10 minutes at 5000 $\times$ g in a refrigerated centrifuge. To 2 ml. of this preparation are added 30 micromoles of ATP, 30 micromoles of $MgSO_4$, 6 micromoles of DPN, and 2 micromoles of 2-C^{14}-mevalonic acid (final volume, 3 ml.). After incubation under anaerobic conditions at 37° for 2 hours, six to ten pellets of KOH, 3 vol. of methanol, and two boiling stones are added, and the solution is brought to a gentle boil for 10 minutes. After cooling, the saponified solution is extracted twice with petroleum ether. The extract is washed twice with water and dried over Na_2SO_4. All but 3 to 4 ml. of the petroleum ether is removed on a steam bath under a stream of nitrogen. The residual solution is transferred to a 5 $\times$ 1-cm. column of alumina (Merck alumina for chromatography). Squalene is eluted quantitatively with 50 ml. of petroleum ether. The squalene thus obtained is radiochemically pure.

Small amounts of squalene (a few milligrams or less) are sensitive to air oxidation. Radioactive squalene should, therefore, be kept in petroleum ether solution in the cold and in the absence of light. For use in experiments, the petroleum ether is evaporated, and the radioactive squalene is dissolved in acetone containing small quantities of Tween-80 (from Atlas Powder Co.). The acetone is then evaporated with the aid of a stream of nitrogen. On addition of water, an emulsion of squalene is obtained. The amount of Tween-80 should be at least ten times the weight of squalene.

Determination. A modification of the Liebermann-Burchard reaction

[9] N. L. R. Bucher, *J. Am. Chem. Soc.* **75**, 498 (1953).

can be used for the determination of nonlabeled squalene.[10] With labeled squalene obtained from most sources, the petroleum ether eluate from the Al_2O_3 column (see above) is radiochemically pure. This radiopurity is determined as the hexachloride.[11] Carrier squalene is added to the labeled sample and dissolved in petroleum ether. Approximately 100 mg. are transferred to a small flask, and the solvent is evaporated. Then 1.0 ml. of cold HCl-saturated acetone is added. Dry HCl is bubbled through the solution while the temperature is kept at $-5°$ with a KCl–ice bath. After crystals begin to appear, the stream of HCl is continued for 30 minutes. The crystals (hexachloride of squalene) are filtered, washed three times with cold etther, and dried *in vacuo* overnight. The radiopurity of the squalene sample is determined from the observed and calculated specific activities of the hexachloride.

Lanosterol

Preparation. Pure lanosterol is not commercially available but can be prepared from "isocholesterol" (wool fat sterol) which contains 20 to 45% lanosterol as obtained commercially. This procedure has been published in "Biochemical Preparations"[12] and therefore will not be repeated here.

C^{14}-Labeled lanosterol can be prepared enzymatically from $2\text{-}C^{14}$-MVA.[13] The same reaction mixture that was discussed earlier for the preparation of labeled squalene can be used here with the following modification. The homogenate is first incubated at $0°$ for 20 minutes with 0.002 M arsenate. The final incubation, after addition of cofactors and $2\text{-}C^{14}$-MVA (1 µmole/5 ml. of homogenate), is carried out for 3 hours at $37°$ under aerobic conditions and in 0.002 M arsenate. A typical yield of sterols from 1 µmole of racemic MVA under these conditions is 40% of theoretical (20% of added racemic MVA). Over 90% of the label is in lanosterol.

For the isolation of radiochemically pure lanosterol, the reaction mixture is saponified and extracted with petroleum ether as discussed before (preparation of labeled squalene). The volume is reduced under a stream of nitrogen to about 10 ml., and the solution is transferred to an alumina column (5 g. of Merck alumina) for chromatography. Squalene is eluted with 100 ml. of petroleum ether. Some unidentified compounds are then eluted with 100 ml. of benzene. Finally, the sterol mixture is eluted with 100 ml. of ether. The ether is evaporated under a stream of

[10] V. R. Wheatley, *Biochem. J.* **55**, 637 (1953).
[11] I. M. Heilbron, E. D. Kamm, and W. M. Owens, *J. Chem. Soc.* 1630 (1926).
[12] K. Bloch and J. Urech, *Biochem. Prep.* **6**, 32 (1958).
[13] M. L. Moller and T. T. Tchen, *J. Lipid Res.* **2**, 342 (1961).

nitrogen. The residue is redissolved in petroleum ether and chromatographed on a column of three hundred times excess of deactivated alumina. (One hundred parts of Merck alumina in Skellysolve are shaken for 6 hours with seven parts of 10% acetic acid. The alumina is then filtered and dried in air.) Most of the lanosterol is eluted from a 5-g. Al_2O_3 column with approximately 100 ml. of 5% benzene in petroleum ether. Cholesterol is not eluted until much later.[14] The lanosterol thus obtained is radiochemically pure.

Determination. The radiopurity of labeled lanosterol can be determined by chromatography with deactivated alumina as described above. Quantitative determination of lanosterol by the Liebermann-Burchard reaction has been reported. The method, however, is not applicable to crude mixtures of sterols.[15]

Other Sterol Intermediates

Many other sterols have been reported to be intermediates between lanosterol and cholesterol. Unfortunately, no convenient method is available for the preparation of these labeled sterols. The reader is therefore referred to the numerous original publications which have been summarized in a review article by the author.[16]

Comments on the Use of Labeled Sterols (and Squalene)

Stability. Although these compounds are normally considered as stable chemicals, small amounts (a few milligrams or less) are very sensitive to air oxidation. These biologically prepared labeled intermediates should therefore always be kept in petroleum ether solution, in the cold, and in the absence of light.

Use as Substrate. These compounds are all insoluble in water. The best way of emulsifying these compounds is by the aid of Tween-80 (Atlas Powder Co.). An aliquot of a petroleum ether solution of these compounds is pipetted to an Erlenmeyer, and the solvent is evaporated under a stream of nitrogen. An acetone solution of Tween-80 is added. The amount of Tween-80 should be at least ten times the weight of the sterol (or squalene) but less than 1 mg. per 2 ml. of final reaction mixture. Acetone is also evaporated under a stream of nitrogen. The residue gives a fine emulsion when water is added. This emulsion should be prepared fresh each time.

[14] The exact elution patterns of lanosterol and cholesterol vary with different batches of deactivated alumina and should be determined when a new batch of deactivated alumina is used.
[15] For this and other methods, see R. P. Cook and J. B. M. Rattray, in "Cholesterol" (R. P. Cook, ed.), Chapter 3. Academic Press, New York, 1958.
[16] T. T. Tchen, in "Chemical Pathways of Metabolism" (D. M. Greenberg, ed.), 2nd ed., Vol. 1, pp. 389–429. Academic Press, New York, 1960.

[76] The Identification of Fatty Acids by Gas Chromatography

By S. R. LIPSKY and R. A. LANDOWNE

Principle

The technique of gas–liquid chromatography[1,2] involves the application of a small quantity of a sample mixture capable of being volatilized to a tube containing a column of supporting particles, the *solid support*, coated with a specific chemical substance, *the stationary phase*. The chromatographic column, which has a constant flow of an eluting or *carrier gas* passing into it, is maintained at an optimal operating temperature for the specific analysis. The sample components in the form of a vapor distribute themselves between the gas phase and the stationary liquid. Differences in the affinity of the stationary liquid for the various component vapors cause them to be moved by the carrier gas down the length of the column at varying rates of speed. The substances thereby separated from one another emerge from the column as individual "bands" which on passing through a suitable detector give rise to a signal that is amplified and fed into an automatic recording device.

Preparation of Methyl Esters of Fatty Acids

The analysis of fatty acids by the method of gas chromatography is most advantageously carried out when these substances are present in the form of methyl esters. The methyl ester is very conveniently prepared and generally has a much lower boiling point than the free acid. Since it is also less polar and considerably less reactive, there is less likelihood of interaction with the stationary phase, the solid support, or the walls of the chromatographic column when metal tubing is used for either packed or capillary columns.

Transesterification

A sample of lipid which is taken up in a small volume of absolute methanol–benzene (1:1 v/v) is placed in a graduated centrifuge tube with a ground-glass joint. To the tube is added 0.5 ml. of 2,2-dimethoxy-propane and 5% concentrated H_2SO_4 in absolute methanol (v/v). Although approximately 1 to 2 ml. of the acid alcohol is sufficient for the hydrolysis of 10 mg. of material, a minimum of 5 ml. of solution should

[1] A. T. James and A. J. P. Martin, *Biochem. J.* **50**, 679 (1952).
[2] A. T. James and A. J. P. Martin, *Biochem. J.* **63**, 144 (1956).

be used in order to avoid excessive evaporation. The mixture is refluxed for 2 hours. If complex phospholipids are present, the reaction should proceed for an additional 6 hours in order to obtain the complete release of fatty acids beyond the C-20 range.

For convenience or when the quantity of available material is limited, the sample may be placed in a 10-ml. ampule, 5 ml. of 5% H_2SO_4 in methanol is added, and the vial is sealed and stored in an oven at 70° for 8 to 12 hours.

The contents are cooled, and a half volume of distilled water is added. The solution is then extracted by shaking with 1 vol. of petroleum ether (30° to 60°), centrifuging for 5 minutes if necessary, and carefully removing the petroleum ether layer with a long needle attached to a small hypodermic syringe. The extraction is repeated two more times. The combined petroleum ether extracts are washed twice with distilled water. Anhydrous sodium sulfate is then added to dry the solution. The petroleum ether is removed by evaporation either under nitrogen or at reduced pressure, and the dried residue is quickly taken up in a few milliliters of warm petroleum ether and transferred to a ground-glass-stoppered conical centrifuge tube. The esters, now ready for analysis, can be safely stored as a *dilute* solution in the refrigerator for periods up to 3 months.

In instances where the lipid extract has not been subjected to absorption chromatography (silicic acid or Florisil) for separation into various classes of lipid complexes, high-molecular-weight (above 200) hydrocarbons, alcohols, or aldehydes may be present which interfere with the gas chromatographic assay of the fatty acid esters. Under these circumstances a microsaponification procedure can be used to obtain the free fatty acids. A sample of lipid material is hydrolyzed by refluxing with 0.5 N KOH in ethanol for 4 to 8 hours. The length of time required for hydrolysis is again dependent on the nature of the fatty acids present in the phospholipid moieties and the extent to which they are cleaved. Five milliliters of alcoholic KOH is a convenient volume to use, although 1 to 2 ml. is sufficient for a 10-mg. sample. The mixture is allowed to cool, and 5 to 10 ml. of a 2:1 ethanol–H_2O solution is added. The mixture is extracted three times with petroleum ether (30° to 60°) to remove the unsaponifiable material. The petroleum ether washings are combined and extracted once with alcoholic KOH. This ethanol wash is added to the original aqueous material remaining in the flask. The mixture is then made acid to phenolphthalein with 6 N H_2SO_4. The free fatty acids are then removed from the mixture by extracting them three times with small volumes of petroleum ether. The combined petroleum ether extracts are washed once with 5% sodium bicarbonate, twice with distilled water, and then dried over anhydrous sodium sulfate. The free fatty acids are readily converted to the methyl esters by the use of diazomethane.

Preparation of Diazomethane

The apparatus for this purpose (Fig. 1) is assembled in a hood. Twenty-five milliliters of 95% ethanol is added to 8 ml. of a 62.5% aqueous KOH in the 100-ml distilling flask. The receiving flasks connected to the condenser should be cooled to 0°. A solution of 3.0 g. of

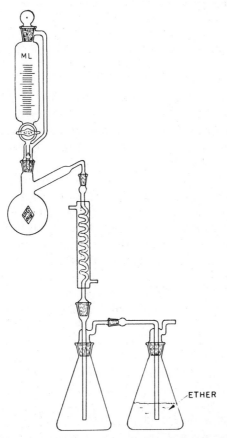

FIG. 1. Apparatus for the generation of diazomethane.

N-methyl-N-nitroso-p-toluenesulfonamide (Diazald, Aldrich Chem., Milwaukee, Wisconsin) in 50 ml. of ether is added through the dropping funnel in a period of 10 to 15 minutes. The flask is heated in a water bath at 65°, and distillation is allowed to progress toward completion.

Another 20 ml. or more of ether is then added slowly to the flask until the distillate is colorless. Several milliliters of the yellow ethereal solution of diazomethane is slowly added to a conical centrifuge tube containing the free fatty acid. This addition should continue until the yellow color persists. The sample container is then placed in a hood until the solution has become colorless and the diazomethane is liberated. Although diazomethane can be stored for periods up to 2 weeks, it is desirable to use fresh solutions for methylating fatty acids. When diazomethane is prepared it is advisable to use rubber stoppers instead of ground-glass joints in the distillation apparatus.

Analysis of the Methyl Esters of Fatty Acids by Gas Chromatography

Preparation of the Solid Support

Great care should be used in the preparation of the chromatographic support in an effort to reduce the number of active sites in these diatomaceous earth materials which can react with the sample component vapors (particularly polyunsaturates). Chromosorb-W, Celite 545, firebrick, and Chromosorb are known to have different physical properties such as bulk density, surface area, and pore structure which influence their activity toward polar compounds. The latter two substances should not be employed as solid supports, since a relatively high incidence of absorption phenomena, asymmetrical peaks, and "tailing" has been noted coincident with their use in the gas chromatographic analysis of fatty acid esters.

Batches of several hundred grams of untreated Celite 545 or Chromosorb-W are placed on No. 170 mesh screens and sieved for 20 minutes. This operation is most effectively accomplished by using an automatic device which vigorously rotates and pounds the metal screens. All material passing through the screen is discarded. After 2 to 3 pounds of the powder remaining on top of the screens is collected, it is placed in a large glass cylindrical chromatography tank half filled with water. The slurry is stirred with a thick glass rod and allowed to settle for 8 minutes. The supernatant containing dust and fine particles is quickly removed by aspiration. This operation is repeated twice, after which time excess water is taken off and concentrated HCl is carefully added in quantities sufficient to cover the surface of the powder. The slurry is allowed to stand for 15 minutes with intermittent stirring. Much of the acid is then removed by several washings with water. An aqueous solution containing 5% KOH is added to the tank, and the slurry is again stirred for an additional 10 minutes. The mixture is made neutral to pH paper by repeated rinsings with water. Water is then removed from the powder by suction filtration on a large Büchner funnel. The wet powder is spread on aluminum foil and set in an oven at 200° to dry overnight. The dried powder is very thoroughly and repeatedly sieved to various narrow particle sizes (60 to 80 mesh, 80 to 100 mesh, 100 to 140 mesh, and 140 to 170 mesh) and stored in tightly capped bottles.

The Stationary Phase

The nonpolar liquid phase[2] most widely employed for the analysis of fatty acid esters by gas chromatography is the high-vacuum stopcock grease, Apiezon L (James Biddle Co., Philadelphia, Pennsylvania). The most popular and effective polar materials are the adipate, glutarate, or succinate polyesters of ethylene or diethylene glycol.[3-5]

Apiezon L. A coated support containing 15% Apiezon L is produced by weighing 30 g. of this material in a 3-l. stainless-steel beaker. Several hundred milliliters of chloroform is then added with stirring, and the beaker is gently warmed on a hot plate until the grease is completely dissolved. At this point, 170 g. of 140- to 170-mesh solid support (Celite 545 or Chromosorb-W) is added to the solution. Sufficient chloroform should be present in the beaker to cover the surface of the slurry. The temperature of the hot plate is increased, and the chloroform is carefully evaporated off by continuous stirring of the mixture with a stout glass rod. To prevent overheating, as the last traces of the solvent are removed, the temperature of the hot plate is reduced. Agitation is continued until a dry free-flowing powder is obtained. The material is stored in a dry atmosphere in screw-capped glass bottles until ready for use.

Polyesters. The starting materials employed in the formation of these high-molecular-weight polymers (3000 to 5000+) should be of the highest purity. All experimental conditions should be rigidly controlled in order to avoid batch variations. For research laboratories engaged in the analysis of a wide variety of fatty acid esters, it is recommended that three to six different polyesters of varying polarity should be available (see Analysis). These include the adipate polyester of diethylene glycol, the glutarate polyester of diethylene glycol, the succinate polyester of diethylene glycol, the adipate polyester of ethylene glycol, the glutarate polyester of ethylene glycol, and the succinate polyester of ethylene glycol.

The glycol and dicarboxylic acid are mixed in equimolar ratios in a three-necked reaction flask heated to 180° under a stream of nitrogen. A condenser is attached to the flask to trap out water vapor released into

[3] S. R. Lipsky and R. A. Landowne, *Biochim. et Biophys. Acta* **27**, 666 (1958).
[4] S. R. Lipsky, R. A. Landowne, and M. R. Godet, *Biochim. et Biophys. Acta* **31**, 336 (1959).
[5] S. R. Lipsky, R. A. Landowne, and J. E. Lovelock, *Anal. Chem.* **31**, 852 (1959).

the nitrogen stream during polymerization. For best results (1) the quantities of starting materials should be chosen to yield no more than 100 g. of polymer in one batch and (2) a 3 to 5% excess of glycol should be employed in the starting mixture to bind any free carboxyl groups that may remain. When the reaction temperature is reached, the catalyst, 25 mg. of methane sulfonic acid or p-toluenesulfonic acid, is added, and heating is continued for exactly 2 hours. The nitrogen flow is then discontinued, and the flask is evacuated at 5 mm. Hg for an additional hour with the temperature still maintained at 180°. The contents are then cooled. In order to minimize the "bleeding" of decomposition products derived from these materials during their use as stationary liquids at elevated temperatures, the catalyst and any free residual carboxyl groups are removed by passage of the polymer through an anion-exchange column.[6] A chromatographic column, $1\frac{1}{2}$ inches in diameter and 25 inches high, consisting of 100- to 200-mesh IR-4B (OH), is slowly washed with three to four column charges of 5% aqueous KOH. It is then made neutral to pH paper with repeated washings of distilled H_2O. The water is removed from the column with acetone. One hundred grams of polymer is then dissolved in 1 l. of acetone and passed through the column at a rate of 5 to 10 ml./min. A 12 to 15% concentration of polyester is then applied to 60- to 80-mesh or 80- to 100-mesh solid support in a manner similar to that described for Apiezon L.

Another very satisfactory but more sophisticated method of coating the solid support is that of solution loading.[7] This requires closer scrutiny of details than the other technique, but it is less time-consuming and avoids fragmentation of the solid support and the production of "fines" which occur during vigorous stirring.

Polyester (37.5 g.) weighed to 0.1 g. is dissolved in 250 ml. of acetone or chloroform (15% solution w/v) in a 1-l. beaker. To this is added 150 g. (also accurately weighed) of 60- to 80-mesh or 80- to 100-mesh Celite 545 or Chromosorb-W. Sufficient solution should be present to cover the surface of the powder. The mixture is allowed to stand for 10 minutes with occasional gentle stirring. It is then carefully and quickly transferred with continuous swirling to a coarse fritted-disk Büchner funnel, and much of the solution is removed by suction filtration. The solvent in the recovered solution may be removed, and the polyester reclaimed for future use. The coated material on the Büchner funnel is then quantitatively placed on aluminum foil and put in an oven to dry overnight at 50°. The dried powder is reweighed, and the percentage of stationary

[6] J. Corse, and R. Teranishi, *J. Lipid Research* **1**, 191 (1960).
[7] E. C. Horning, E. A. Moscatelli, and C. C. Sweeley, *Chem. & Ind. (London)* p. 751 (1959).

liquid present is calculated. This step is essential, since the pore density of Celite 545 and Chromosorb-W differs considerably. The concentration of polyester in solution may be increased or decreased accordingly in order to procure an optimal column coating under ideal operating conditions. Best results are obtained with columns containing 12 to 15% liquid phase. At these levels the separation factor, α, for oleate–stearate should be as follows: adipate polyester of ethylene or diethylene glycol, 1.12 to 1.14; glutarate polyester of ethylene or diethylene glycol, 1.16 to 1.18; and succinate polyester of ethylene or diethylene glycol, 1.16 to 1.18. An increase in the concentration of liquid phase tends to prolong relative retention times and decrease the resolving power of the column.

For the quantitative analysis of fatty acid esters by gas chromatography it is essential that the proportion of any stationary phase should not be reduced below 10%, a point where the absorptive properties of Celite 545 or Chromosorb-W become readily manifest. On the other hand, under circumstances where it is desirable to use larger sample loads (10 to 20 mg.) for semipreparative work, the concentration of liquid phase should be increased to 25 to 30%. This results in some loss in column efficiency. For samples up to 50 mg., another alternative is to increase the internal diameter of the column to 12 mm.

The Packed Column

The coated solid support is usually contained in a U-shaped column 4 to 9 feet in over-all length with an internal diameter of 5 mm. and made of double-thickness Pyrex glass (Fig. 2). This type of column is much preferred to the coiled metal column for ease and uniformity of packing. Straight 4-foot glass columns have also been successfully used but have several disadvantages: (1) they do not provide the same magnitude of resolution of closely related components as that noted with 7- to 9-foot columns; (2) the detector housing by necessity sits at the bottom of the column, making sample collection awkward; and (3) under these circumstances, it is inconvenient to achieve a very desirable operational feature, that of maintaining the detector in a separate temperature-control system.

A few grams of dried, carefully coated, free-flowing powder is first added to the inlet leg of the column through a funnel. The column is then vibrated by quickly moving a standard Black and Decker drill containing a 2-inch-long brass bit, flattened on one side, which rotates at a speed of 2250 r.p.m., up and down the extent of the leg. The procedure is repeated several times until one side of the column is filled. The same technique is then applied to the other leg of the column. Overvigorous vibration of the column is to be avoided, since this tends to pack the coated support too tightly and makes optimal gas-flow rates difficult to achieve. A small tuft of glass wool is then placed on top of the outlet side of the powdered column. Appropriate high-temperature silicone rubber "through hole" and "diaphragm" seals are then placed in position (see Fig. 2). Nitrogen at 35 p.s.i. is introduced through the gas inlet side of the column for approximately 10 minutes. The flow

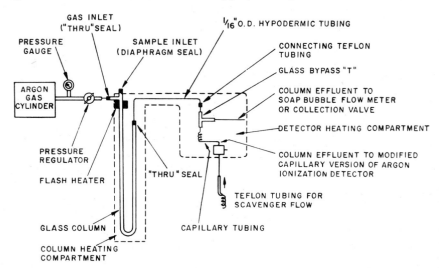

Fig. 2. Schematic diagram of a gas chromatographic apparatus showing the column, the ionization detector, and a bypass system.

through a typical 8-foot column containing a 15% polyester on 60- to 80-mesh solid support should be at least 120 ml./min. at this pressure when tested with a soap bubble flowmeter. After this time, with the gas still flowing, the column is once again quickly vibrated. The gas flow is then discontinued, and the pressure is allowed to drop to zero before the inlet side of the column is opened. The void on this side of the column is filled with packing material and glass wool in order to minimize the dead volume.

The column is now "conditioned" by placing it in a carefully controlled heating compartment maintained at 5° to 10° above the usual operating temperatures normally employed for a particular stationary liquid (range 170 to 210°). A high rate of carrier gas (125 to 250 ml./min.) is allowed to flow through the column for 12 to 24 hours. During this phase of the operation the column is vented to the atmosphere in order to prevent contamination of the detector by small quantities of residual volatile materials which are continuously stripped from the stationary phase. Resolution of the common saturated and unsaturated fatty acid ester components is dependent on: (1) the type and per cent of liquid phase employed; (2) the mesh size of the solid support; (3) the length of the column; (4) the pressure drop across the column; (5) the operating temperature of the column; (6) the type and rate of flow of carrier gas; (7) the "age" of the column; (8) the over-all characteristics of the detector, i.e., sensitivity, dead volume, time constant, etc.; and (9) the size of the sample applied to the column.

Typical examples of experimental conditions which have provided very satisfactory results are as follows:

Nonpolar Columns

Column: 4- to 8-foot U glass column, 5 mm. i.d.
Phase: 12 to 15% Apiezon L on 140- to 170-mesh Celite 545
Column temperature: 193° to 225°
Sensing device: Argon ionization detector (modified capillary version)
Gas: Argon, 10 to 45 p.s.i. column inlet pressure
Column flow rate: 40 to 170 ml. argon per minute
Cell bypass ratio: 60 to 1
Scavenger flow rate: 30 to 80 ml. argon per minute

Polar Columns

Column: 6- to 9-foot U glass column, 5 mm. i.d.
Phase: 12 to 15% adipate, glutarate, or succinate polyester of ethylene or diethylene glycol on 60- to 80-mesh or 80- to 100-mesh Chromosorb-W
Column temperatures: 170° to 210°
Sensing device: Argon ionization detector (modified capillary version)
Gas: Argon, 20 to 50 p.s.i. column inlet pressure
Column flow rate: 110 to 275 ml. argon per minute
Cell bypass ratio: 60 to 1
Scavenger flow rate: 30 to 80 ml. argon per minute.

After the preconditioning period, the column is coupled to the detector and is ready for use. At this point, several types of standard mixture should be analyzed in an effort to establish optimal column operating conditions and relative retention times (or volumes) for typical saturated and unsaturated acid esters (see Analysis).

The life of a properly prepared column is strictly dependent on the afore-mentioned factors involved in its operation. In many instances, it is possible to use a column for 2 to 4 months before appreciable loss in resolving power is noted. Temperature cycling and frequent changes in inlet pressures should be avoided if possible. Carrier gas should flow at all times through any column maintained above room temperature. At the end of a working day the inlet pressure may be dropped to 5 p.s.i. Prior to its use again, the usual operating pressure is applied to the column, and approximately 30 minutes is allowed to pass for the column to equilibrate. The flow rate is recorded, and the pressure is readjusted if necessary. This maneuver usually brings about a slight but temporary base-line drift due to further bleeding of small quantities of liquid phase from the column. Quantitative analysis should not be attempted if the noise level due to "bleeding" is extraordinarily high. This is particularly true when high-sensitivity ionization detectors are used, since their response to a sample vapor under these circumstances would tend to be nonlinear.

The U-shaped glass columns are easily cleaned by removing the silicone rubber seals and glass wool with a "barbed" 18-gage hypodermic needle, inverting and clamping the column to a stand at the edge of a laboratory bench, inserting a long, straight, $1/16$-inch o.d. metal rod into one leg of the column to break up the packing, and, with the rod in place, vigorously vibrating the column with a drill.

If the quantity of sample is extremely limited or an additional slight increase in speed of analysis or resolving power of the column is desired without resorting to capillary columns, a 2-mm. i.d. packed column may be conveniently utilized. Here, the amount of stationary liquid applied to the solid support is of necessity 10% or less in order to avoid large re-strictions in the flow of carrier gas through the narrow tube. If the solid support shows little or no evidence of interaction with the sample com-ponents, optimal resolution of very small sample loads (5 to 25 μg.) may be achieved with a 3 to 7% coating. Several additional comments should be made about the 2-mm. column: (1) It should not be tightly packed when it is prepared; (2) it may be necessary to coat larger parti-cles of solid support i.e., 40 to 60 mesh, in order to obtain optimal gas flows; and (3) the column should be cleaned immediately on its removal from the heated compartment. This is best accomplished by inverting the column and placing a long piece of Nichrome wire in the outlet leg. Gas pressure is then applied to column while the packing is removed by continuous vibrations.

Introduction of Sample (Packed Columns). The two most common methods of introducing the sample onto the chromatographic column are the *open*[8] and *closed* loading techniques. The former system can be utilized only with detection devices which are relatively insensitive to changes in pressure or gas flow. The pressure applied to the column is interrupted, and the column is allowed to reach atmospheric pressure. This usually takes several minutes. A 10/30 ground-glass joint on the inlet side of the column is quickly opened, and the sample is inserted onto the glass wool by means of a self-filling calibrated glass micropipet. In order to facilitate complete emptying, the rubber tubing attached to the

[8] J. W. Farquhar, W. Insull, P. Rosen, W. Stoffel, and E. H. Ahrens, *Nutrition Revs.* **17**, Part II (1959).

pipet is stripped downward two or three times. The column is quickly closed, and pressure is reapplied.

Advantage: (1) Good reproducibility is achieved in applying sample loads of similar size onto column. This may be of particular importance in the analysis of radioactive samples.

Disadvantages: (1) Volatile or semivolatile sample components are lost to the atmosphere in varying amounts. (2) Since it takes several minutes after loading for the column to come to equilibrium and give a steady base-line recording, components emerging during this period may be distorted and thus provide difficulties in accurate quantitative analysis. (3) The continuous introduction of air or moisture onto polyester columns shortens the life of the column. (4) Each loading operation takes from 5 to 10 minutes. (5) The calculation of absolute retention times is relatively laborious, since the position of the air peak is often preceded by the carrier gas front. (6) High back pressures occurring in columns longer than 5 feet tend to produce gas leaks around the ground-glass joint.

The preferred system is the closed loading technique in which the sample is taken up into a calibrated 1-μl. Hamilton microsyringe and injected through a self-sealing silicone rubber seal onto the column. The syringe is held in place for 2 to 3 seconds in order that the ejected sample is completely removed from the needle tip by the rapid flow of carrier gas. Loading in this manner is greatly facilitated by the presence of a "flash heater" (see Fig. 2) maintained at 75° to 100° above the operating temperature of the column in order to provide instantaneous volatilization of components with a wide range of boiling points.

Advantage: (1) This is a rapid, efficient, and convenient method of introducing a small quantity of sample as a "concentrated band" onto the column, which avoids many of the pitfalls of an open loading system.

Disadvantage: (1) There is difficulty in reproducibly applying very small samples to the column without resorting to dilution with a solvent.

Optimal sample size is dependent on the following factors: (1) the number of components present in the mixture; (2) the relative ratios of each component; (3) the relative retention time of each component which in turn is determined by column length, the ratio of liquid phase to solid support, particle size of solid support, column temperature, and pressure; (4) the type of detection system used.

A 5-mm. i.d. column containing 15% polar or nonpolar liquid phase and operated under favorable conditions can accept a sample load of 5 to 10 mg. without appreciable loss of resolution. For analytical purposes, however, smaller sample sizes are preferable for optimal column efficiency and resolving power. The average load of a seven-component mixture of fatty acid esters extending from C-14 to C-20, which is required to provide an easily detectable signal from a variety of sensing systems, is as follows: gas density balance, 2 to 3 mg.; thermal conductivity cell, 2 to 5 mg.; argon ionization detector (modified capillary version with bypass system), 0.05 μg. to 5.0 mg.; flame ionization detector, 0.050 μg. to 5.0 mg.

Capillary Columns

Principle.[9] Long capillary tubes (usually 50 to 300 feet) whose inner surfaces have been uniformly coated with a thin layer of stationary liquid have been shown to have efficiencies far exceeding those of packed columns. Since the internal diameter of such tubing is small (0.010 to 0.020-inch), there is a large resistance to gas flow which results in a substantial reduction in outlet flow. The over-all effect is a significant increase in linear velocity, very high column efficiencies, and a decrease in the time required for analysis. In general within the limitations of the technique, the longer the length of the column and the smaller its cross section, the greater is the efficiency of the column. Speed of analysis may be further increased with only slight loss in efficiency by using shorter columns or columns containing wider bore areas.

Preparation of Capillary Columns for Gas Chromatography. Coated capillary columns are usually made from stainless-steel, copper, glass, or nylon tubing. Since the oxide which form readily on the surfaces of copper at 200° may interact with sample vapors, this type of column should not be routinely employed in the analysis of fatty acid esters. Treated nylon tubing, which is inexpensive and easy to coat, unfortunately cannot tolerate continuous use at temperatures above 175°. Under these circumstances this material can be best utilized at this time when coated with polar polyester phases. Since the nature of the surface of the bore is one of the major factors which determines the manner in which a thin (3000 to 5000 A.) uniform layer of stationary liquid is laid down on the inner walls, slightly different procedures are employed for the coating of stainless-steel, glass, or nylon tubing.

The capillary column is thoroughly cleaned by passing 20 to 30 ml. of chloroform through it under a nitrogen pressure of 60 p.s.i. (30 p.s.i. per 100 feet of tubing). A schematic representation of the device used for filling and coating capillary columns is seen in Fig. 3. During the cleaning procedure the outlet end of the capillary tube is placed in a small beaker half filled with water. After the chloroform has passed through the tubing and settled at the bottom of the beaker, gas bubbles begin to

[9] M. J. E. Golay, in "Gas Chromatography" (D. H. Desty, ed.), p. 36. Academic Press, New York, 1958.

appear. At this point, the gas flow through the column is maintained for an additional hour in order to remove the last traces of solvent.

Stainless-Steel Capillaries. One end of the capillary tube (50 to 300 feet in length, 0.010-inch i.d.) is fastened to the loading chamber. A 2-inch piece of Teflon tubing, to which is fastened a 30-foot length of 0.010-inch i.d. uncoated capillary tubing which serves as a pressure

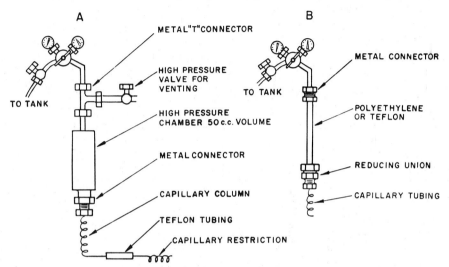

FIG. 3. Devices for coating capillary columns. *A*, high pressure. *B*, low pressure.

throttle, is snugly attached to the other end of the column. A filtered 10% solution of Apiezon L in pentane is placed in the loading chamber. The volume should be equivalent to six times the capacity of the capillary column to be coated. Nitrogen at 600 p.s.i. N_2 is applied to the chamber until the last of the solution is noted to pass through the Teflon connection. The gas flow is interrupted, and the chamber and column are brought to atmospheric pressure by careful venting of the gas lines. At this point, the Teflon connector and the capillary throttle are removed, and the capillary column unit is attached to a low-pressure gas valve. Nitrogen at 6 p.s.i. is allowed to flow through the column for an additional 12 hours. The column is then placed in the gas chromatographic unit, and 5 to 8 p.s.i. of carrier gas (per 100 feet of 0.010-inch i.d. tubing) is applied while the column is brought up to operating temperature. The column is "baked out" under conditions of analysis for an additional 12 to 24 hours. A nonpolar column 100 to 200 feet long and 0.010 inch i.d. that yields less than 300 theoretical plates per foot for a component that emerges in three times the time required for the appearance of the air or solvent peak is considered to be a poor column.

Another successful method of loading 0.010- to 0.020-inch i.d. stainless-steel capillaries can also be used. A similar volume of a 10% solution of Apiezon L is added to a polyethylene or Teflon loading tube attached to a length of cleaned capillary tubing. The tube is then fastened to a nitrogen cylinder, and 30 p.s.i. of pressure is applied for each 100 feet of capillary tubing. As the loading tube is emptied of solution, the pressure is dropped one-fifth of the starting pressure. The gas flow is maintained for 12 hours, and the column is then conditioned. The latter procedure has been more successful. In addition a polar capillary column can be fabricated with glutarate ethylene glycol polyester using a 10% solution in acetone.[9a, b]

Glass and Nylon Capillaries. Because of the relatively smooth bore surfaces of these materials, best results are achieved by loading at low pressures. Under these circumstances an inlet pressure of 2 to 6 p.s.i. is used for every 100 feet of 0.010- to 0.020-inch i.d. tubing to be coated. It is necessary to condition nylon capillary tubing for 12 hours at 175° before it is cleaned and coated with the liquid phase.

A properly prepared capillary column can last from 2 to 4 months if temperature and pressure programming are minimized. Variations in technique, differences in viscosities and vapor pressures of various liquid phases, and batch differences in the porosity and smoothness of the capillary bore account for failures in approximately 30% of the attempts to coat capillary columns for use in the analysis of fatty acid esters at high temperatures (170° to 240°).

Introduction of Sample (Capillary Columns). One of the most arduous problems facing the analyst who seeks effectively to employ high-efficiency capillary columns is that of placing a *small representative* sample onto the column. Since columns of 0.010 to 0.020-inch i.d. usually contain a coating of stationary liquid which is several microns thick, an optimal sample load (dependent on the factors listed under The Packed Column) would be in the range of 0.01 to 3.0 μg. A simple but effective bypass system was first used to introduce accurately this very small quantity of sample onto the column. At a fixed inlet pressure, due to its high resistance there is a low flow of gas through the capillary column. A relatively larger flow passes through the low variable resistance represented by the needle valve. Further investigation soon demonstrated that fractionation of the sample vapor was occurring within the bypass "T." This phenomenon existed despite the presence of a flash heater maintained at 100° above the column temperature which surrounded the bypass in order to provide instantaneous volatilization of the fatty acid

[9a] R. A. Landowne and S. R. Lipsky, *Biochim. et Biophys. Acta* **46**, 1 (1961).
[9b] R. A. Landowne and S. R. Lipsky, *Biochim. et Biophys. Acta* **47**, 589 (1961).

esters extending from C-8 to C-26. The lower-boiling components apparently expanded more rapidly in a confined area during their rapid passage down the "T" and entered onto the capillary column in a proportion greater than their actual representation in the sample mixture. Possible solutions to the problem of nonlinear stream splitting are now being tested. Thus far the most promising approach is that in which the liquid sample is introduced by means of a thin hypodermic needle (Hamilton microliter syringe) directly into bore of a 20- to 30-foot 0.020- to 0.025-inch i.d. uncoated length of capillary tubing. The capillary tubing is surrounded by a flash heater maintained at 75° to 100° above column temperature. The sample components emerge from the capillary outlet as a concentrated band. An appropriate portion of this band is then split and fed into the coated capillary column by means of the bypass system.

Typical experimental conditions which have provided very satisfactory results with capillary columns are as follows:

NONPOLAR CAPILLARY COLUMNS[5, 10]

Column: 60- to 250-foot lengths of 0.010-inch i.d. stainless-steel or glass capillary tubing.
Stationary phase: Inner surface of tubing coated with a 10% solution of Apiezon L or polybutene in pentane or any other appropriate light solvent.
Column temperature: 190° to 240°.
Sensing device: Argon ionization detector (modified capillary version).
Gas: Argon, 5 to 25 p.s.i.
Average total sample load: 0.1 to 1.0 µg.
Average column outlet flow: 0.5 to 2.0 ml./min.
Bypass to column split: 50 to 500: 1 ml./min.
Scavenger flow rate: 30 to 80 ml./min.
Theoretical plates: 400 to 1000 per foot of column for a component that has a retention time approximately three times that of the air or solvent peak.

POLAR CAPILLARY COLUMNS

Column: 60- to 250-foot lengths of 0.010-inch i.d. stainless-steel, glass, or nylon capillary tubing.
Stationary phase: Inner surface of tubing coated with a 5 to 15% solution of the adipate, glutarate, or succinate polyester of diethylene or ethylene glycol.
Column temperature: 160° to 225° (175° limit for nylon tubing).
Other operating parameters: As noted for nonpolar capillary columns.
Theoretical plates: 300 to 600 per foot of column for a component that has a retention time approximately three times that of the air or solvent peak.

Properly conditioned columns may be conveniently temperature- or pressure-"programmed." When longer lengths of wider-bore tubing are

[10] S. R. Lipsky, J. E. Lovelock, and R. A. Landowne, *J. Am. Chem. Soc.* 81, 1010 (1959).

used, it is possible to analyze considerably larger sample loads and still maintain relatively high orders of resolution. Thus:

Column length	Internal diameter	Sample size	Outlet flow	Optimum theoretical plates
1000 feet	0.020 inch	1.0 mg.	50–70 ml./min.	250,000
1000 feet	0.034 inch	2.0 mg.	75–100 ml./min.	250,000
5000 feet	0.066 inch	20.0 mg.	250 ml./min.	1,200,000

In these instances the sample is placed directly into the bore of the coated column. Since the amount of material applied to such columns is in the milligram range, small-volume bulk property devices (gas density balance, thermal conductivity cells, cross-section detectors, etc.) may at times be employed as sensing systems. Sample collection from the comparatively large cross-sectional capillary column is now feasible. The wide-bore column is preferably coated by gravity loading. A volume of a 10 to 15% solution of stationary phase equivalent to six times the capacity of the column is placed into the loading tube and allowed to flow slowly into and out of the column by gravity. When very long columns are to be coated, slight pressure may be applied to the loading chamber. The coating procedure when properly carried out should take 8 to 24 hours. The column is then gently conditioned in the usual manner.

Detection of Sample Vapor

The simplest, most sensitive, and versatile detector available to date is a modified capillary version of the argon ionization detector (Fig. 4). This model can be most effectively used with either capillary or packed

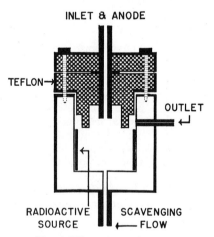

INLET & ANODE

TEFLON→

OUTLET

RADIOACTIVE SOURCE

SCAVENGING
←— FLOW

FIG. 4. Micro argon detector.

columns. In the former the capillary column is connected directly to the detector by means of a short piece of snug-fitting Teflon tubing. It is important to minimize the dead volume. In the latter instance a bypass system with a variable restriction is employed whereby 0.1 to 10% of the packed column effluent passes into the detector (Fig. 5). An average

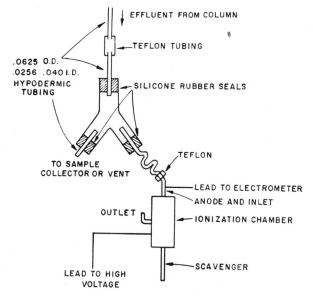

FIG. 5. Bypass system for micro argon detector.

bypass-to-cell flow ratio of approximately 75:1 provides the analyst with a wide range of operating parameters. This split in effluent flow is accomplished by placing approximately 7 inches of 0.020-inch i.d. uncoated capillary tubing in the leg of the bypass leading to the detector. The 1- to 2-inch piece of 0.025 to 0.040-inch i.d. tubing which is placed in the other leg to create slight back pressure is connected to a sample collection device or may be vented to the atmosphere. A variable restriction in the form of a high-quality needle valve externally mounted across this flow path permits the user to alter bypass-to-cell ratios conveniently. To prevent overloading of the detector, usually no more than 3 to 6 ml. of column effluent per minute should flow through the anode. The counter flow or scavenger flow ranges between 30 and 80 ml./min. If it is desirable to have a larger proportion of column effluent pass into the capillary detector, an alternative method consists in bringing a part of the effluent into the scavenging stream which then enters the detector.

The argon ionization detector consists of a small metal chamber to which a high potential is applied. It has an internal volume of 1 to 3 ml.

The effective volume, i.e., the partially enclosed area surrounding the anode, is approximately 10 μl. A thin foil impregnated with an alpha or beta emitter provides the source of ionizing radiation. Depending on the type used, slightly different operational characteristics will be noted, owing to differences in noise levels derived from the stochastic nature of emission of particles by a particular radioactive material. Thus:

Alpha sources: Radium 226, or radium D.
Strength: That concentration which provides a current of 1 to 2 $\times$ 10^{-8} amp. at 1000 volts at 2000° with argon flowing through detector.
Temperature limitations: To 400°+.
Half-life: Radium 226, 1612 years; radium D, 22.2 years.
Noise: Random fluctuations associated with radium D sources are three times those of radium 226.

Beta sources: Sr^{90} or tritium impregnated in titanium.
Strength: Sr^{90}, 10 mc.; tritium, 100 mc.; to yield a similar range of standing currents as that required of alpha sources.
Temperature limitations: Sr^{90}, 400°+; tritium, 225°.
Half-life: Sr^{90}, 28 years; tritium, 12.5 years.
Noise: Random fluctuations associated with radium 226 are fourteen times those of Sr^{90} and forty-five times those of T^3.

Principle of Operation of Argon Ionization Detector.[11-13] When argon, specifically employed as a carrier or scavenger gas in the gas chromatographic technique, enters an ionization chamber (Fig. 4) to which a high voltage is applied, it is continuously irradiated by the radioactive source. Under these conditions the gas is partly ionized and a small (1 to 3 $\times$ 10^{-8} amp.) but constant ionization current, i.e., the "standing" or "baseline" current, is produced. Simultaneously, additional reactions occur within the device. Other primary electrons set free in the argon are accelerated by the applied potential to velocities sufficient to excite large numbers of argon atoms to their metastable state. The over-all sensitivity of the detector is directly dependent on the concentration of these highly excitable but non-ionized atoms within the chamber, since, within a given range, increases in voltage to the cell provide proportionately larger increments in their number without the coincident production of a significant quantity of argon ions.

When a trace amount of a sample in the form of an organic vapor whose ionization potential is lower than that of the excitation potential of the metastable argon atoms (11.6 e.v.) emerges from the chromatographic column and enters the argon-filled ionization chamber, collisions

[11] J. E. Lovelock, *J. Chromatog.* 1, 35 (1958).
[12] J. E. Lovelock, *Nature* 182, 1663 (1958).
[13] J. E. Lovelock, A. T. James, and E. A. Piper, *Ann. N.Y. Acad. Sci.* 72, 720 (1959).

between the vapor and the metastable atoms occur. The transfer of energy from the metastable atoms to the organic vapor molecules readily takes place and results in the ionization of this material. The secondary electrons produced in the process, when collected at the anode, give rise to an increase in the ionization current which is directly related to the vapor concentration. They also generate more metastable atoms, replacing those lost on collision with vapor molecules. The reactions may be depicted as follows:

$$\text{Argon}° + \xrightarrow{\text{α- or β-rays}} \text{Argon}^+ + e^- \text{ (primary)}$$

$\text{Argon}° + e^- + \text{high voltage} \to \text{Argon*} \text{ (metastable state, 11.6 e.v.)}$

$\text{Argon*} + \text{organic vapor} \to \text{Organic vapor}^+ + e^- \text{ (secondary)} + \text{argon}$

$e^- \text{ (secondary)} + \text{argon}° + \text{high voltage} \to \text{Argon*}$

$$e^- \text{ (secondary)} \xrightarrow{\text{to anode}} \text{current (amperes)}$$

Operational Characteristics of the Detector. Operational characteristics are governed largely by the cell geometry, the source of radioactivity (preferably T^3 or Sr^{90}), and the associated electronics. The detector is relatively insensitive to changes in flow and temperature.

Dynamic Range. The modified version of the argon detector has a 10^5 range, drawing currents from 10^{-6} amp. down to 10^{-11} amp. Its sensitivity is such that it will readily detect vapors emerging from the column at levels as low as 10^{-13} mole/sec.

Operating Voltage Range. The operating voltage range usually extends from 600 to 2000 volts under optimal conditions. For fatty acid analysis at 200° it is customary to operate at 800 to 1500 volts, since the formation and multiplication of metastable atoms occurs at a lower potential when the density of the gas is less at elevated temperatures. Space charges within the chamber usually limit the upper level of operating voltages.

Electrometer Settings. With ionization chambers containing T^3 or Sr^{90} sources this instrument should be operated in the range 3×10^{-8} to 1×10^{-9} amp. The electrometer-recorder system should be so constructed that when a current results within the chamber that is three times that of the standing current a full-scale recorder deflection is produced.

Current-Limiting Resistor. Usually a resistor is not required. However, it is preferable to include a 50- to 100-megohm resistor in series with the input to the electrometer in order to prevent "overloading" of this device.

Calibration Factors. No calibration factor is required for fatty acid esters with molecular weights over 150. The peak areas of the chromatogram are directly proportional to the mass of the substance separated.

Per Cent of Sample Destroyed in Chamber. The proportion of molecules ionized at full sensitivity ranged between 5 and 10%. Under ordinary operating conditions, about 0.5 to 1.5% of the sample *entering the detector* is destroyed.

Quantitative Analysis[14]

Several suitable methods may be employed to calculate the areas under a peak. These include triangulation, automatic electromechanical integration, planimetry, and weighing of the paper. The first two techniques are most widely used. An additional order of reliability in quantitative analysis may be obtained by adding a known amount of a select standard substance to a given volume of the unknown mixture. For the triangulation of symmetrical peaks, the peak height (in millimeters) is multiplied by the peak width (in millimeters) at half height. A properly operated modified argon detector provides a linear response to fatty acid esters over a wide dynamic range and does not require the use of calibration factors, since the peak area is proportional to the amount (mass) of substance present. The area represented by an individual component when divided by the sum of the areas for all components provides the fraction of each present in the sample. Accurate calculation of peak areas is best carried out when the recorder chart speed is at least 30 inches/hour.

In certain instances where overlapping of bands occurs, good quantitative estimation of a component may be obtained by using either the method of peak ratios (area of the peak/total area of both peaks) or the system of approximating triangles. If feasible, the sample may be rechromatographed by using another stationary liquid which provides complete resolution of the overlapping bands.

Qualitative Analysis

Straight-Chain Saturated Fatty Acid Esters. The corrected retention times (t_{R_1}) or volumes (t_{V_1}) of at least three standard commonly occurring esters are plotted against the number of carbon atoms on semilogarithmic paper. A straight-line relationship should be observed.[2,4] The chain length of an unknown saturated straight-chain ester is usually determined by noting that point on the graph which corresponds to the time required for the unidentified component to emerge from the column. It is common practice to equate the slight operational variations noted in

[14] R. L. Pecsok, "Principles and Practice of Gas Chromatography." Wiley, New York, 1959.

the use of different columns by calculating the separation factor (α) of all the components relative to methyl stearate. Thus:

$$\alpha = \frac{\text{Corrected retention time of } x}{\text{Corrected retention time of methyl stearate}}$$

Straight-Chain Unsaturated Fatty Acid Esters

1. *Nonpolar Columns* (*Apiezon M or L, Polybutene*). The unsaturated esters are, without exception, eluted from the column *before* the corresponding saturated ester. The comparative rate of movement of these substances from a nonpolar column is governed by the number of double

TABLE I[a]

SEPARATION OF THE POSITIONAL ISOMERS OF CERTAIN C-18 FATTY ACID ESTERS
ON POLAR AND NONPOLAR COLUMNS

Time	Time (minutes)	Separation factor (Sf) (relative to stearate)
Polar column: 8 feet, 15% adipate ethylene glycol polyester on 80- to 100-mesh Chromosorb-W. Temperature 196°. Gas flow, argon 230 ml./min. Inlet pressure, 38 p.s.i.		
Methyl stearate	16.3	1.00
Methyl oleate	18.0	1.10
Methyl $\Delta^{8,11}$-linoleate	21.3	1.31
Methyl $\Delta^{9,12}$-linoleate	21.6	1.32
Methyl $\Delta^{10,13}$-linoleate	22.0	1.35
Methyl $\Delta^{11,14}$-linoleate	23.3	1.43
Nonpolar column: 125 feet, 0.010-inch i.d. hypodermic tubing coated with polybutene. Temperature 200°. Gas flow: 0.75 ml./min. Inlet pressure 16 p.s.i. Scavenger flow, 40 ml./min.		
Methyl $\Delta^{8,11}$-linoleate	42.0	0.81
Methyl $\Delta^{9,12}$-linoleate	42.4	0.82
Methyl $\Delta^{10,13}$-linoleate	44.2	0.86
Methyl oleate	45.2	0.88
Methyl $\Delta^{11,14}$-linoleate	46.0	0.89
Methyl elaidate	47.1	0.92
Methyl stearate	51.5	1.00

[a] Taken in part from R. A. Landowne and S. R. Lipsky, *Biochim. et Biophys. Acta* **46**, 2 (1961).

bonds, their position and configuration within the carbon chain, and the nonpolar London dispersion forces.[15] Theoretically each additional double bond present in the carbon chain beyond the 9,10-position increases the polarity and lowers the boiling point of the molecule. However,

[15] A. T. James, *J. Chromatog.* **2**, 555 (1959).

despite the use of high-efficiency capillary columns,[5,10] it has been impossible to separate 18:2 from the 18:3 or the 20:4 from the 20:5 esters (the number after the colon indicates the number of double bonds in an ester of stated chain length, i.e., 18:2 = octadecadienoate).

CONFIGURATIONAL ISOMERS. The presence of *trans* double bonds tends to decrease polarity (minor effect) and greatly increase the boiling point (major effect) of an unsaturated ester. As the net result the *trans* isomers usually emerge from the nonpolar column after the *cis* isomers, i.e., methyl oleate—*cis*-9,10-octadecenoate—is followed by methyl elaidate—*trans*-9,10-octadecenoate.[5,10] These materials are conveniently and rapidly separated from one another on capillary columns which yield 20,000 to 30,000 theoretical plates (as calculated for methyl stearate).

POSITIONAL ISOMERS. From the analysis of several C-18 diunsaturated isomers on a nonpolar capillary column, it is noted (Table I) that, the closer the double bonds are to the carboxyl carbon, the more rapid is the relative movement of the component from the column.

BRANCHED-CHAIN FATTY ACID ESTERS.[2,9b,15] The saturated simple branched-chain esters emerge before the corresponding straight-chain ester on both polar and nonpolar columns. The sequence of elution is as follows:

$$
\text{Neo compounds (R}\underset{\underset{\textstyle C}{|}}{\overset{\overset{\textstyle C}{|}}{-C}}\text{—C)}
$$

$$
\text{Iso compounds (R}\text{—C}\overset{\overset{\textstyle C}{|}}{-C}\text{—C)}
$$

$$
\text{Anteo-iso compounds (R}\text{—C}\overset{\overset{\textstyle C}{\diagup}}{-C}\text{—C)}
$$

Straight-chain saturated compounds (R—C—C—C—C)

A confirmatory chemical method for location of a branch in a saturated carbon chain consists in oxidative degradation of the isolated component by chromic acid in glacial acetic acid followed by separation of neutral and acidic degradation products and analysis of each fraction by gas chromatography.

2. *Polar Columns (Various Polyesters).*[3-5,16] The unsaturated fatty acids, without exception, progressively emerge *after* their saturated homologs on polyester columns. The speed and facility with which the

[16] C. H. Orr and J. E. Callen, *J. Am. Chem. Soc.* **80**, 249 (1958).

separation of closely related components is achieved are directly related
to the chemical composition of the stationary liquid. In certain circumstances, depending on the degree of polarity of the polyester phase, the
length and age of the column, the flow rate, the temperature, the mesh
size, and the ratio of liquid phase to inert support, some polyunsaturated
acid esters may be retained to a point where they emerge close to, coincident with, or after the next higher saturated component. Thus with a
properly prepared column containing the adipate polyester of ethylene
or diethylene glycol, 18:3 appears before 20:0, and 20:4 before 22:0.
However, if more polar columns containing the glutarate or succinate
polyester of ethylene or diethylene glycol are used, 18:3 is usually eluted
after 20:0, and 20:4 *after* 22:0 (Fig. 6).

POSITIONAL ISOMERS. Although the magnitude of retardation of components usually increases with the degree of unsaturation, this effect is
offset to some extent by the position of the double bonds within the chain.
If the double bonds are close to the carboxyl carbon, there is a tendency
for the component to have a relatively short retention time (Table I).

CONFIGURATIONAL ISOMERS. On polar columns the increase in the boiling point of a *trans* component usually outweighs the effect of a decrease
in the polarity of the molecule. This tends to lengthen the relative retention times. This effect becomes more apparent as the number of *trans*
double bonds increases.

Although reference to published relative retention factors may provide information concerning the *tentative* nature of unknown component
bands, each laboratory should establish its own factors by the use of
standards under conditions by which the analyses are being routinely
performed. This is particularly important, since slight differences in the
operating parameters involved in the gas chromatographic technique
have a direct bearing on the absolute and relative retention times of the
various fatty acid esters. Despite the advent of high-efficiency capillary
columns and the effective use of a number of different stationary phases,
it is usually very difficult to separate and positively identify *all* the
components present in a complex mixture on a single chromatographic
column. In many instances definitive proof of the chemical structure of a
component peak is obtained only after an aliquot of the sample has been
subjected to ancillary chemical procedures such as hydrogenation or
bromination followed by chromatographic analysis of the reaction mixture. Since the polybrominated esters are relatively nonvolatile, the
peaks that appear in these instances represent the saturated acids. Caution should be used in the interpretation of the chromatogram when
brominated products are rerun on polyester columns, since the dibromo
compounds derived from monosaturated esters tend to dehydrobrominate

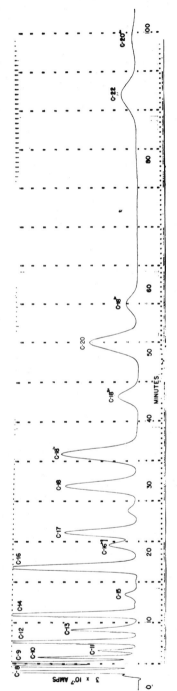

Fig. 6. The separation of the methyl esters of long-chain fatty acids. Column: 8 feet, 15% succinate ethylene glycol on 60- to 80-mesh Chromosorb-W. Temperature, 171°. Inlet pressure, 40 p.s.i. Flow rate, argon 250 ml./min. Separation factor, stearate-oleate, 1.18.

at temperatures above 190° and produce peaks which represent unsaturated artifacts.[17]

A third method that has proved essential for the unequivocal identification of an unsaturated band is oxidative ozonolysis. Information concerning the position and number of double bonds in an isolated peak is provided by the nature of the major products of the fragmentation procedure. The monocarboxylic and dicarboxylic acids which are formed are esterified with diazomethane and identified by gas chromatography by using an 8-foot column coated with a succinate or glutarate polyester at 130° to 150°.

A flow sheet for fatty acid chromatography is given in Table II.

TABLE II

FLOW SHEET FOR THE IDENTIFICATION BY GAS CHROMATOGRAPHY OF THE
INDIVIDUAL COMPONENTS PRESENT IN A MIXTURE OF FATTY ACID ESTERS

Sample unknown

↙   ↘

"Screening Column"	Aliquot for Bromination and/or Hydrogenation
4 to 6 feet long, 2 to 6 mm. i.d., 10 to 15% succinate or glutarate polyester of ethylene or diethylene glycol on 60- to 80-mesh solid support, 145° to 190°, gas flow 200 to 300 ml./min. a. Determine carbon chain length of the saturated esters. b. Determine the presence of polyunsaturated acid esters in the C-20 to C-24 range.	a. Differentiate between those peaks that represent either saturated or unsaturated fatty acid esters. ↓ Packed Columns ⎯→ (nonpolar and polar) a. Isolate and collect those peaks which may represent more than one component and rerun on an appropriate capillary or packed column.
Capillary column (nonpolar and polar) ←⎯⎯ a. Tentatively determine the general chemical structure of the majority of components in the sample. b. Tentatively determine the general chemical structure of certain additional peaks by comparing relative retention times on polar and nonpolar capillary columns. Relative ratios of several components on each column are used to correlate the position of an unidentified band. c. Tentatively determine the presence of certain configurational and positional isomers and simple branched acids.	b. Isolate and collect those peaks where the position and number of double bands require confirmation by oxidative ozonolysis. c. The products of oxidative ozonolysis (the mono- and dicarboxylic acids) are esterified with diazomethane and identified by analysis on either packed or capillary polar or nonpolar columns.

[17] R. A. Landowne and S. R. Lipsky, *Nature* **182**, 1731 (1958).

[77] Methylmalonyl Coenzyme A

By MARTIN FLAVIN

Preparation

Two methods have been reported for the preparation of the mono-thioester of CoA and methylmalonate: the mixed anhydride method,[1] and the reaction catalyzed by propionyl CoA carboxylase.[2] The former will be described here.

The mixed anhydride of methylmalonic and ethylchlorocarbonic acids is prepared as follows[3]: To a small centrifuge tube containing 1.2 ml. of tetrahydrofuran, which has been redistilled within the preceding hour and collected over sodium, is added 118 mg. of dry methylmalonic acid. The latter compound is prepared by saponification of the diethylester (obtainable from Sapon Laboratories) and purified by ether extraction and recrystallization from acetone or acetone–petroleum ether.[4] The tube is immediately stoppered and, as soon as the methylmalonic acid has dissolved, is placed in a −5° bath, supported by a clamp attached to a vibrating motor. While the tube is agitated, 0.081 ml. of anhydrous pyridine is added, followed by 0.094 ml. of ethylchlorocarbonate. The tube is at once restoppered and briefly centrifuged at −5°. The clear supernatant solution containing the mixed anhydride is decanted into a dry tube which is immediately stoppered and chilled. Yield: 20 to 30% as determined by the hydroxylamine assay,[1] and with the assumption that the absorbancy of the hydroxamate of methylmalonate is the same as that of succinate. Strictly anhydrous conditions are essential to the success of this preparation.

An aliquot of the tetrahydrofuran solution containing 38 micromoles of the mixed anhydride is added to an ice-cold test tube containing 30 micromoles of CoA, 60 mg. of potassium bicarbonate, and 3.0 ml. of water. After 5 minutes of shaking in the cold, the mixture is allowed to stand at room temperature for 10 minutes. If excess mixed anhydride is present, it is decomposed hydrolytically. The solution, which will contain 18 to 24 micromoles of methylmalonyl CoA by hydroxylamine assay, can be stored for 1 week at pH 6.0 at −18°, without decline in hydroxylamine titer.

---

[1] W. S. Beck, M. Flavin, and S. Ochoa, *J. Biol. Chem.* **229**, 997 (1957).
[2] See Vol. V [77].
[3] T. Wieland and L. Rueff, *Angew. Chem.* **65**, 186 (1953).
[4] M. Flavin and S. Ochoa, *J. Biol. Chem.* **229**, 965 (1957).

It appears likely that this preparation does not contain appreciable amounts of dithioester of methylmalonic acid, since chromatography after treatment with hydroxylamine reveals only one component, having the same R_f value as an authentic preparation of methylmalonmonohydroxamic acid.[4] However, a satisfactory preparation of methylmalondihydroxamic acid has not been available for comparison, to ensure that it does not decompose to yield the former compound during the chromatographic procedure.

Methylmalonyl pantetheine has been prepared by the same procedure[5] and is inactive in the methylmalonyl isomerase reaction.[6]

Properties

The properties of methylmalonyl CoA have not been extensively investigated. The enzymatically prepared compound has been reported to be more completely converted to succinyl CoA by methylmalonyl isomerase than that prepared by the mixed anhydride method.[7] Material active in the isomerase reaction may be obtained by elution of methylmalonyl CoA from paper chromatograms developed in the ascending direction for 24 hours at $+2°$ with a 1:1 mixture of ethanol and 0.1 N sodium acetate, pH 4.5.[5] Methylmalonyl CoA is alkali-labile, and relatively stable to acid.[7]

Note added in proof: A number of additional procedures for the nonenzymatic preparation of malonyl coenzyme A have recently been reported.[8-10] In several cases these have also been successfully applied, without modification, to the preparation of methylmalonyl coenzyme A.[9-11] Procedures such as that of Trams and Brady[9] are superior to the one previously described, in terms of reproducibility, yield, and unequivocal isolation of the monothioester. The mixed anhydride method may still be useful under special circumstances, as for the preparation of radioactive methylmalonyl coenzyme A on a small scale.[11]

[5] W. S. Beck, personal communication, 1960.
[6] See Vol. V [79].
[7] W. S. Beck and S. Ochoa, *J. Biol. Chem.* 232, 931 (1958).
[8] P. R. Vagelos, *J. Biol. Chem.* 235, 346 (1960).
[9] E. G. Trams and R. O. Brady, *J. Am. Chem. Soc.* 82, 2972 (1960).
[10] H. Eggerer and F. Lynen, *Biochem. Z.* 335, 540 (1962).
[11] M. Flavin and C. Slaughter, *J. Am. Chem. Soc.* 83, 397 (1961).

[78] Malonyl Coenzyme A

By S. J. WAKIL

Enzymatic Synthesis

Principle. The first intermediate in the synthesis of long-chain fatty acids from acetyl CoA is the formation of malonyl CoA by the carboxylation of acetyl CoA by CO_2 in the presence of ATP, Mn^{++}, and a biotin-containing enzyme known as acetyl CoA carboxylase.[1,2] The CO_2 appears to be activated by ATP with the simultaneous formation of CO_2–biotin–enzyme intermediate which transfers the CO_2 to acetyl CoA to form malonyl CoA according to the following equations:

$$CO_2 + ATP + \text{biotin–enzyme} \xrightarrow{Mn^{++}} CO_2\text{–biotin–enzyme} + ADP + P_i$$
$$CO_2\text{–biotin–enzyme} + CH_2COSCOA \rightarrow$$
$$HOOCCH_2COSCoA + \text{biotin–enzyme}$$

Avidin inhibits the carboxylation of acetyl CoA by binding to the enzyme-bound biotin. This inhibition can be relieved by pretreatment of avidin with free biotin.

Reagents

1.0 M potassium phosphate buffer, pH 6.5.
0.1 M ATP, potassium or sodium salt, pH 6.5.
0.1 M $MnCl_2$.
0.1 M $KHCO_3$.
Acetyl coenzyme A.
Acetyl CoA carboxylase.

Preparation of Acetyl CoA Carboxylase. Acetyl CoA carboxylase was prepared by the following procedure: One volume of fresh chicken livers was blended with 2 vol. of 0.1 M potassium phosphate buffer, pH 7.0, in a Waring blendor at full speed for 1 minute. The homogenate was centrifuged for 30 minutes at $3000 \times g$. The supernatant fluid was decanted through a cheesecloth and was further centrifuged at $100,000 \times g$. For each 100 ml. of the clear supernatant solution, 13.4 g. of solid ammonium sulfate were added slowly with constant stirring. The precipitate was separated by centrifugation, taken up in minimum amounts of 0.005 M phosphate buffer, pH 7.0, and the solution was dialyzed against 4 l. of the

[1] S. J. Wakil, *J. Am. Chem. Soc.* **80**, 6465 (1958).
[2] R. O. Brady, *Proc. Natl. Acad. Sci. U.S.* **44**, 993 (1958).

same buffer for 3 hours. The enzyme solution was adjusted to a protein concentration of 30 to 40 mg./ml. by the addition of 0.005 M potassium phosphate buffer, pH 7.0. To this solution, 3 parts (v/v) of precooled calcium phosphate gel[3] (30 mg./ml.) were then added slowly with gentle stirring. The mixture was allowed to stand for 3 to 4 minutes with occasional stirring and was then centrifuged for 3 minutes at 3000 $\times$ g. The supernatant was discarded, the precipitated gel was resuspended in 3 vol. of 0.1 M potassium buffer, pH 7.0, and the mixture was centrifuged as quickly as possible. The eluate was saved, and the precipitate was again suspended in 2 vol. of the same phosphate buffer and recentrifuged. This procedure was repeated four more times. The eluates were combined, and solid ammonium sulfate (20 g. per 100 ml.) was added. Then the mixture was centrifuged at 3000 $\times$ g for 20 minutes. The precipitate was redissolved in a minimum amount of 0.005 M potassium phosphate buffer, pH 7.0. The resulting solution was centrifuged at 20,000 $\times$ g to remove any insoluble residue, and the clear supernatant fluid was dialyzed for 3 hours against 0.005 M potassium phosphate buffer, pH 7.0.

At this level of purity the enzyme may be used for the synthesis of malonyl CoA with satisfactory results. For additional purification the following procedure was used: DEAE-cellulose column (2 $\times$ 20 cm.) was prepared according to the procedure of Sober and Peterson.[4] Then 200 to 300 mg. of protein were added, and the inactive protein was washed with 150 to 200 ml. of 0.01 phosphate buffer, pH 7.4, and 0.1 M NaCl. The enzyme was eluted with 150 to 200 ml. of 0.01 M phosphate buffer, pH 7.4, and 0.25 M NaCl. Fractions of 6.0 ml. each were collected, and fractions No. 6 to 12, containing active protein, were pooled. Then 20 g. of $(NH_4)SO_4$ were added per 100 ml. of solution. The protein was separated by centrifugation, redissolved in 0.005 M phosphate buffer, pH 7.0, and stored in $-15°$.

Procedure. The reaction mixture was prepared by mixing 3.0 ml. of Tris buffer, 1.0 ml. of ATP, 0.3 ml. of $MnCl_2$, 4.0 ml. of $KHCO_3$, 30 micromoles of acetyl CoA, 3.0 mg. of acetyl CoA carboxylase, and H_2O to a final volume of 20 ml. After incubation for 20 minutes at 38°, the reaction was stopped either by heating in a boiling-water bath for 5 minutes or by careful acidification of the mixture with 60% $HClO_4$ to pH 3.0. The denatured protein was separated by centrifugation and washed once with 2.5 ml. of H_2O. The supernatant fluid and washings were pooled and lyophilized to dryness. The white powder was dissolved in a minimum amount of water (2 to 3 ml.). Then the solution was streaked on What-

[3] D. Keilin and E. F. Hartree, *Proc. Roy. Soc.* **B124**, 397 (1938).
[4] H. A. Sober and E. A. Peterson, *Federation Proc.* **17**, 1116 (1958); see also Vol. V [1].

man filter paper, No. 17 mm. The chromatogram was developed by descending chromatography in a isobutyric–concentrated NH_4OH–H_2O (66:1:33) system. Malonyl CoA separates from ATP, ADP, and acetyl CoA with an R_f of 0.47 to 0.52 and was spotted by quenching the ultraviolet light. The acyl CoA was eluted by water in the usual manner. The eluate was lyophilized, and the residue was dissolved in a minimum amount of water. The yield was approximately 15 to 20 micromoles according to hydroxamic acid assay. Rechromatography of malonyl CoA may be necessary in order completely to separate traces of ADP and acetyl CoA from the malonyl CoA.

Chemical Synthesis (for an alternative procedure, see Vol. VI [79])

Principle. The chemical synthesis of malonyl CoA is based essentially on the procedure of Wieland and Köppe for the formation of acyl coenzyme A via the corresponding thiophenol derivatives. The procedure involves two steps: The first requires the formation of the monomalonyl thiophenol from the mixed anhydride, and the second is the displacement of thiophenol by coenzyme A. The chemical reactions are indicated in reactions (1) and (2).

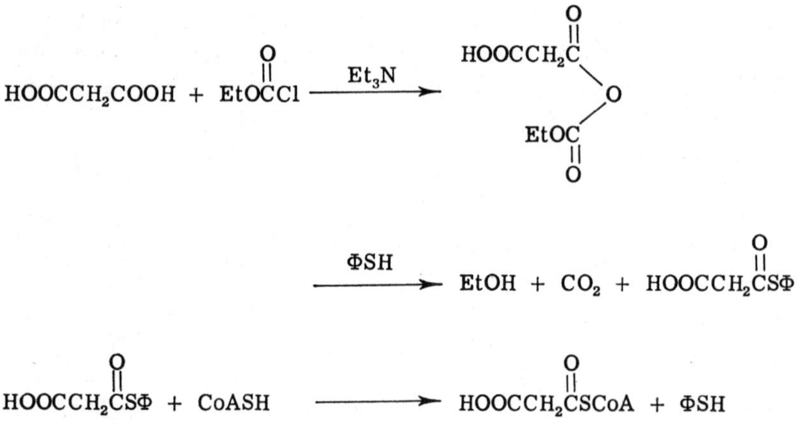

$$HOOCCH_2COOH + EtOCCl \xrightarrow{Et_3N}$$

(1)

$$\xrightarrow{\Phi SH} EtOH + CO_2 + HOOCCH_2CS\Phi$$

$$HOOCCH_2CS\Phi + CoASH \longrightarrow HOOCCH_2CSCoA + \Phi SH \quad (2)$$

Reagents

Triethylamine.
Tetrahydrofuran.
Ethyl chlorocarbonate.
Thiophenol.
Malonic acid.
CoASH.
Dowex 50.
$NaHCO_3$.

Preparation of the Malonyl Thiophenol. Fifty millimoles of anhydrous malonic acid were dissolved in 25 ml. of anhydrous tetrahydrofuran at −20° followed by the addition of 25 millimoles of triethylamine. To this a solution of 25 millimoles of ethyl chlorocarbonate dissolved in 5 ml. of tetrahydrofuran was added dropwise with constant mixing. Precipitation of triethylamine hydrochloride began after several minutes. Standing for 20 minutes ensured completion of the reaction. Samples were then withdrawn and added to a solution of hydroxylamine in water; the resultant mixture was incubated for 10 minutes at 38°. The violet color due to the formation of the hydroxamine acid was developed by the addition of HCl and $FeCl_3$ solutions in the usual manner.[5]

The main portion of the mixed anhydride was treated with 25 millimoles of thiophenol and allowed to stand at room temperature for 24 hours. The solvent was then removed *in vacuo*, and the residue was suspended in a minimum amount of water. The oily product was extracted in diethyl ether and dried over sodium sulfate. Evaporation removed the ether, with the malonyl thiophenol remaining as a viscous oily residue.

Reaction with CoASH. A solution of reduced CoA was adjusted to a concentration of 5 to 10 micromoles/ml. Sufficient amount of $NaHCO_3$ was added to the CoASH solution to attain a pH of 8.2. The oily acyl thiophenol was then added in small amounts with vigorous agitation by bubbling through nitrogen until a small test sample (0.01 to 0.02 ml.) gave a very low nitroprusside test.[6] At this end point agitation of the mixture was continued for 10 minutes at room temperature. Then the reaction mixture was acidified with Dowex 50 to pH 3 to 4, and the Dowex was removed by centrifugation and washed once with water. The supernatant fluid and the wash were combined, and the thiophenol was extracted with diethyl ether. The aqueous solution, containing malonyl CoA, was then lyophilized. Recovery was 40 to 50%.

The preparation of malonyl CoA thus obtained would contain a considerable amount of unreacted CoA. Further purification of malonyl CoA may be achieved by the afore-mentioned chromatographic procedure.

Assay of Malonyl CoA

Malonyl CoA may be assayed either by the hydroxanic acid formation and subsequent development of color according to the general method of Lipmann and Tuttle[5] or enzymatically by the oxidation of TPNH in the presence of acetyl CoA and the R_{2gc} enzyme fraction. The latter resulted in the formation of palmitic acid.[7] The enzymatic assay

---

[5] F. Lipmann and L. C. Tuttle, *J. Biol. Chem.* **161**, 415 (1945); see also Vol. III [39].
[6] R. R. Grunert and P. H. Phillips, *Arch. Biochem.* **30**, 217 (1951).
[7] S. J. Wakil and J. J. Guanguly, *J. Am. Chem. Soc.* **81**, 2597 (1959).

mixture contained 30 micromoles of potassium phosphate buffer, pH 6.5, 10 millimicromoles of acetyl CoA, and about 10 millimicromoles of malonyl CoA in a total volume of 0.4 ml. The reaction was initiated by the addition of 100 to 200 μg. of R_{2gc} enzyme fraction. Under these conditions malonyl CoA was quantitatively converted to palmitic acid. Here the amount of TPNH oxidized was about twice the amount of malonyl CoA consumed.

[79] Preparation of Thiolesters of Coenzyme A

By P. ROY VAGELOS

The chemical synthesis of lactyl CoA, malonyl CoA, and β-ketoöctanoyl CoA will be described. These three compounds are examples of CoA thiolesters that are difficult or impossible to synthesize by the usual thiolester preparative procedures[1] because of interference by α-hydroxy, α-carboxy, or β-keto functional groups, respectively.

Lactyl CoA

Principle. This method, previously described by Wieland and Koppe,[2] is based on the fact that an α-hydroxy acid reacts with phosgene to form the acid chloride. The acid chloride readily reacts with thiophenol to yield the acyl thiophenyl ester, which is used to acylate CoA.

Reagents

Sodium L($+$)-lactate (thoroughly dried over P_2O_5).
Phosgene.
Thiophenol.
Pyridine (anhydrous).
CoASH.

Procedure. The procedure is carried out in the absence of moisture. To 5.6 g. of sodium lactate covered with 50 ml. of dry ether at 0° is added dropwise a solution of 25 g. of phosgene in 20 ml. of ether with constant stirring. The stirring is continued for 24 hours at room temperature, and the sodium chloride that forms is then removed by centrifugation. After evaporation of the ether *in vacuo*, a pale-yellow, thick, acrid-smelling oil remains which can react with hydroxylamine to

[1] E. R. Stadtman, Vol. III [137].
[2] T. Wieland and H. Köppe, *Ann. Chem. Liebigs* **588**, 15 (1954).

form lactyl hydroxamic acid. The lactyl chloride is dissolved in ether or benzene, and a slight excess of thiophenol is added. The solution is cooled, and to it is added a slight excess of dry pyridine. A precipitate of pyridine hydrochloride forms. After a few hours, the solvent is removed *in vacuo*, and the oily residue is mixed several times with petroleum ether to remove thiophenol. Lactyl-thiophenyl ester is freed of pyridine salt by washing with a little water and then dried *in vacuo*.

To a 5% aqueous solution of CoASH containing 25% pyridine at room temperature is added a slight excess of the lactyl-thiophenyl ester dissolved in pyridine while nitrogen is constantly bubbled through. Completeness of the CoASH esterification can be determined by the nitroprusside test for —SH,[3] since thiophenol gives a negative reaction. Thiophenol and excess thiophenyl ester are removed by extracting the solution with ether. Lactyl CoA is purified by paper chromatography.[4] The yield of thiolester is 75 to 80%, based on CoASH. Alkaline or enzymatic hydrolysis of this lactyl CoA gives rise to stoichiometric amounts of L(+)-lactate and CoASH.[4]

Malonyl CoA via Mixed Anhydride of Malonic and Acetic Acids
(for an alternative procedure, see Vol. VI [78])

Principle. This method[5] is based on the fact that ketene reacts with malonic acid in ether to form a mixed anhydride of malonic and acetic acids. The mixed anhydride reacts with thiophenol to give malonyl-thiophenyl ester, which is used to acylate CoASH.

Reagents

Malonic acid.
Ketene (delivered directly from a ketene generator).
Thiophenol.
Pyridine (anhydrous).
CoASH.
Potassium bicarbonate, 1.0 M, pH 7.

Procedure. Dry malonic acid (5.2 g., 50 millimoles) and 100 ml. of dry ether are stirred continuously in a rubber-stoppered flask fitted with glass tubing for delivery of ketene until the malonic acid is almost completely dissolved. Ketene is delivered into the mixture at the rate of 0.6 to 0.8 millimole/min. for 1 hour at room temperature, at which point the mixture becomes yellow or amber. The total amount of hydroxamate-

[3] G. Toennies and J. J. Kolb, *Anal. Chem.* **23**, 823 (1951).
[4] P. R. Vagelos, J. M. Earl, and E. R. Stadtman, *J. Biol. Chem.* **234**, 765 (1959).
[5] P. R. Vagelos, *J. Biol. Chem.* **235**, 346 (1960).

forming material is measured by the method of Lipmann and Tuttle.[6] One equivalent of thiophenol (based on hydroxamate assay) is added; the solution is cooled to 0°; and 50 millimoles of pyridine are then added. A reddish-brown tar forms in the ether solution. The mixture at this point contains malonyl-mono- and dithiophenyl esters as well as acetyl-thiophenyl ester. After 30 minutes, 1.0 M potassium bicarbonate, freshly adjusted to pH 7, is added dropwise (50 to 70 ml. total) to the ether mixture at 0° (stirred constantly) until all the tarry material dissolves and the aqueous layer remains at pH 7. The aqueous layer is removed, and the ether layer is extracted twice with 20-ml. aliquots of the bicarbonate solution. The combined aqueous layers contain essentially all the malonyl-monothiophenyl ester. This is acidified to pH 1 by dropwise addition of 5 N sulfuric acid and then extracted three times with 15-ml. aliquots of ether, thereby extracting the thiophenyl ester back into ether. The combined ether extracts are then treated with bicarbonate solution followed by acidification of the aqueous layer and extraction with ether as above except that consecutively smaller volumes are used. This is repeated twice more, the final amber ether extract (5 to 10 ml.) being saved. This procedure effectively separates malonyl-mono-thiophenyl ester from the other products. The usual yield is 2 to 3.5 millimoles. The ether solution of the malonyl-monothiophenyl ester is stable at −20° for at least 9 months.

Acyl transfer to CoASH is accomplished by addition of 1 equivalent of the thiophenyl ester in ether to a solution of 70 micromoles of CoASH in 2 ml. of 0.2 M potassium bicarbonate, pH 8.0, at room temperature, with helium bubbling through the mixture. The pH is maintained by dropwise addition of 1 N potassium hydroxide if necessary. When the CoASH is completely esterified,[3] the solution is acidified to pH 1 and extracted three times with 3 ml. of ether to remove excess thiophenyl ester and thiophenol. The aqueous solution is adjusted to pH 5, and the malonyl CoA is purified by paper chromatography. The yield is usually 55 to 60 micromoles.

β-Ketoöctanoyl CoA

Principle. This method[7] is based on the fact that the keto group of an ethyl ester of a β-keto acid can be blocked by converting it to an ethyleneketal. The ethyleneketal derivative is converted to a thiolester of thioglycolic acid. The keto function is regenerated by treating the thiolester with sulfuric acid. β-Ketoacyl CoA is then formed by an ester interchange reaction between β-ketoacyl-S-thioglycolate and CoASH.

[6] F. Lipmann and L. C. Tuttle, *J. Biol. Chem.* **159**, 21 (1945).
[7] P. R. Vagelos and A. W. Alberts, *Anal. Biochem.* **1**, 8 (1960).

Reagents

Ethyl β-ketoöctanoate.
Ethylene glycol.
p-Toluenesulfonic acid.
Potassium bicarbonate, 1.0 M.
Potassium hydroxide, 1.0 M.
Pyridine (anhydrous).
Ethyl chloroformate.
Potassium thioglycolate, 1.0 M.
Concentrated sulfuric acid.
CoASH.
Tris(hydroxymethyl)aminomethane hydrochloride, 1.0 M, pH 8.75.

Procedure. A mixture of ethyl β-ketoöctanoate (18.6 g., 0.1 mole), ethylene glycol (6.8 g.), p-toluenesulfonic acid (190 mg.), and 120 ml. of toluene is refluxed under a take-off condenser for removal of water for about 3 hours or until water stops accumulating. The reaction mixture is cooled in ice, and 1.0 M potassium bicarbonate is added cautiously to neutralize the acid. The toluene layer is separated and washed twice with water and then dried over anhydrous magnesium sulfate. The toluene is removed by distillation, and the residue is subjected to vacuum distillation. The fraction containing the ethyleneketal of ethyl β-ketoöctanoate distills at 81° to 95° (0.5 to 0.6 mm.). The usual yield is about 6.0 g. Although this product is usually contaminated with a little unreacted keto acid (as noted by infrared spectrum), it need not be further purified.

To 5.0 g. of this fraction are added 50 ml. of ethanol and 50 ml. of 1.0 M potassium hydroxide. This solution is heated over a steam bath for 1½ hours to hydrolyze the ethyl ester and then concentrated under a stream of air to about 10 ml. The remaining solution is cooled in ice and acidified to pH 1 by dropwise addition of 6 N sulfuric acid with continuous stirring. The mixture, from which potassium sulfate separates, is extracted three times with 5-ml. portions of ether. The combined ether layers, containing ethyleneketal β-ketoöctanoic acid, are dried over anhydrous magnesium sulfate. The free acid present in the ether is determined by titration of an aliquot with standard alkali. The usual yield of acid is 12 to 13 millimoles. The mixed anhydride is then formed essentially by the method of Wieland and Rueff.[8] To the ether solution, cooled in ice, are added 1.1 equivalents of anhydrous pyridine and 1.1 equivalents of ethyl chloroformate. After 2 hours at 0°, the white precipi-

---

[8] T. Wieland and L. Rueff, *Angew. Chem.* **65**, 186 (1953).

tate that forms is removed by centrifugation and discarded. The total amount of hydroxamate-forming material, the mixed anhydride of ethyleneketal β-ketoöctanoate and ethyl hydrogen carbonate, is measured by the method of Lipmann and Tuttle.[6] The usual yield is 5 to 6 millimoles.

After the ether is blown off with a stream of dry helium, the mixed anhydride preparation is dissolved in 20 ml. of tetrahydrofuran (freshly distilled over lithium aluminum hydride). This solution (cooled to 0°) is added to an ice-cold solution containing 1 equivalent of potassium thioglycolate, 2.5 ml. of 1 M potassium bicarbonate, and 5 ml. of water in 10 ml. of tetrahydrofuran. The pH of the resulting solution is maintained at 7.5 by dropwise addition of 5 N potassium hydroxide. Esterification of the mercaptan is complete in about 10 minutes, as indicated by a negative nitroprusside spot test.[3] To this thiolester solution are added 5.7 ml. of concentrated sulfuric acid for the purpose of splitting the ethyleneketal, and the potassium sulfate which forms is removed by centrifugation. The solution is incubated at 30° in a stoppered flask for 14 hours. At the end of that time the tetrahydrofuran is removed by blowing a jet of air onto the solution at 30°. The remaining aqueous solution is extracted three times with 5-ml. portions of ether, which are combined and dried over anhydrous magnesium sulfate. When the ether is evaporated, a yellowish sirup containing some crystalline material remains. The crystalline mixture is washed six times with 0.5-ml. portions of n-hexane (cooled to 0°). The addition of cold n-hexane causes many more crystals to form in the sirup. The crystals are collected by filtration and recrystallized from hot n-hexane. Two additional crystallizations yield white, thin plates of β-ketoöctanoyl-S-thioglycolic acid (30 to 35% yield based on mixed anhydride). This product has a melting point of 73.5° to 74.5° and the following elementary analysis for $C_{10}H_{16}O_4S$: Calculated: C, 51.70; H, 6.95; S, 13.80. Found: C, 51.59; H, 6.88; S, 13.70.

β-Ketoöctanoyl-S-thioglycolate is converted to β-ketoöctanoyl CoA by an ester interchange reaction. Sixty micromoles of β-ketoöctanoyl-S-thioglycolate and 3.0 micromoles of CoASH in 1.0 ml. of solution, buffered at pH 8.75 with Tris–hydrochloride, are incubated for 10 minutes at 24°. The yield of β-ketoöctanoyl CoA is 95% or better, based on CoASH.

Other Acyl CoA Preparations Described Recently

Enzymatic synthesis of malonyl CoA has been reported by Formica and Brady[9] and by Wakil.[10] Chemical synthesis of β-hydroxy-β-methyl

[9] J. V. Formica and R. O. Brady, *J. Am. Chem. Soc.* **81**, 752 (1959).
[10] S. J. Wakil, *J. Am. Chem. Soc.* **80**, 6465 (1958).

glutaryl CoA has been described by Hilz *et al.*[11] Methylmalonyl CoA has been prepared by Beck *et al.*[12] Malonyl semialdehyde CoA and β-hydroxypropionyl CoA have been prepared by Vagelos and Earl.[13]

[11] H. Hilz, J. Knappe, E. Ringelmann, and F. Lynen, *Biochem. Z.* **329**, 476 (1958).
[12] W. S. Beck, M. Flavin, and S. Ochoa, *J. Biol. Chem.* **229**, 997 (1957).
[13] P. R. Vagelos and J. M. Earl, *J. Biol. Chem.* **234**, 2272 (1959).

[80] Synthesis of Malonic Semialdehyde, β-Hydroxy-propionate, and β-Hydroxyisobutyrate

By WILLIAM G. ROBINSON and MINOR J. COON[1]

Malonic Semialdehyde[2]

Preparation

Principle. Ethyl β,β-diethoxypropionate, made by the procedure of Sorm and Smrt,[3] is converted to malonic semialdehyde by a slight modification of a method reported previously.[4] The ester acetal is obtained in

$$CH(OC_2H_5)_3 + CH_2CO \xrightarrow{ZnCl_2} CH(OC_2H_5)_2CH_2CO_2C_2H_5 \xrightarrow{NaOH}$$

$$CH(OC_2H_5)_2CH_2CO_2Na \xrightarrow{HCl} OCHCH_2CO_2H$$

considerably higher yield by this procedure than by the conversion of ethyl formate and ethyl acetate to the sodium derivative of malonic semialdehyde ethyl ester[5,6] and subsequent treatment with hydrogen chloride in absolute ethanol.[7]

Reagents

Ethyl orthoformate (redistilled).
Ketene, generated from acetone.
Zinc chloride.

[1] We wish to acknowledge the assistance of Bruce A. Brown, LaVerne G. Schirch, and Roman R. Lorenz in checking these procedures.
[2] A detailed procedure for the synthesis of methylmalonic semialdehyde, the product of the enzymatic dehydrogenation of β-hydroxyisobutyrate, has been published elsewhere [F. P. Kupiecki and M. J. Coon, *Biochem. Prep.* **7**, 69 (1960)].
[3] F. Sorm and J. Smrt, *Chem. listy* **47**, 413 (1943).
[4] H. Den, W. G. Robinson, and M. J. Coon, *J. Biol. Chem.* **234**, 1666 (1959).
[5] M. Cogan, *Bull. soc. chim.* (5) **8**, 125 (1941).
[6] W. Deuschel, *Helv. Chim. Acta* **35**, 1587 (1952).
[7] E. Dyer and T. B. Johnson, *J. Am. Chem. Soc.* **56**, 222 (1934).

Sodium.
Absolute ethanol.
0.5 N sodium hydroxide.
4.5 N hydrochloric acid.
M potassium phosphate buffer, pH 7.4.
M tripotassium phosphate.

Procedure. Zinc chloride (6.8 g.) is suspended in ethyl orthoformate (74.0 g.) in a 250-ml. gas washing bottle equipped with a coarse sintered-glass disk. The bottle is cooled in an ice bath, and ketene, generated according to the method of Williams and Hurd,[8] is bubbled through the suspension at the rate of 0.5 mole/hr. The bottle is opened several times, and the solution is stirred to resuspend the zinc chloride. At the end of 2½ hours the bottle is tightly stoppered and allowed to stand at −18° for 20 hours. The light-brown solution is decanted into a flask, and a solution of sodium ethoxide, prepared by dissolving 0.75 g. of sodium in 15 ml. of absolute ethanol, is added. The resulting solution is freed of ethyl orthoformate by distillation at about 15 mm. (water aspirator) and at a bath temperature not exceeding 60°. The product, ethyl β,β-diethoxypropionate, distills at 65° at 2 mm. The yield is 70.7 g. (74% based on ethyl orthoformate).[9]

One gram of the redistilled ester acetal is added to 15 ml. of 0.5 N sodium hydroxide, and the mixture is shaken vigorously in a mechanical shaker at room temperature until solution is complete (about an hour). The aldehyde acid is obtained by allowing a mixture of equal volumes of this solution and 4.5 N hydrochloric acid to stand for 1 hour at room temperature. The malonic semialdehyde solution is chilled in an ice bath, 5 ml. of M phosphate buffer, pH 7.4, are added, and the pH is adjusted to 6 by the addition of M tripotassium phosphate. The yield of aldehyde acid (based on the ester acetal) is about 80%, as determined by enzymatic assay. Paper chromatography of the semialdehyde and its dinitrophenylhydrazone are decsribed elsewhere.[4]

Stability. Solutions of malonic semialdehyde in phosphate buffer made as described are not stable on prolonged storage. A loss of 10 to 20% occurs in 24 hours at 0° or in the frozen state, and about 50% in 24 hours at room temperature. On the other hand, the alkaline solution of sodium β,β-diethoxypropionate is stable for several weeks when stored in the frozen state.

[8] J. W. Williams and C. D. Hurd, *J. Org. Chem.* **5**, 122 (1940).
[9] The purity of a redistilled sample of the ester acetal was established by elementary analysis.

Assay Procedure

Enzymatic Assay. The concentration of malonic semialdehyde solutions may be determined by measuring the extent of DPNH oxidation in the presence of β-hydroxypropionic dehydrogenase (Vol. V [61]). A reaction mixture containing 50 micromoles of phosphate buffer, pH 7.4, approximately 10 to 15 micromoles of the aldehyde acid, 16 micromoles of DPNH, an excess of β-hydroxypropionic dehydrogenase free of DPNH oxidase activity (e.g., 40 mg. of protein after step 4 of the purification procedure), and water to a final volume of 3.0 ml. is incubated at room temperature. The semialdehyde is omitted from a control tube. When the reaction is complete (at about 20 minutes), 0.1-ml. aliquots are diluted to 3.0 ml. with 0.01 M phosphate buffer, pH 7.4, and the absorbancy at 340 mμ is determined in the spectrophotometer. The total amount of DPNH oxidized (difference between complete and control experiments) is equal to the amount of aldehyde acid. Since the K_m for malonic semialdehyde is relatively high, this method is not applicable to small amounts of the compound. β,β-Diethoxypropionate and acetaldehyde do not interfere with the enzymatic assay and give no color in the procedure described below.

Colorimetric Assay. Since pure, anhydrous preparations of the aldehyde acid or its salts have not been obtained, the concentration of a standard solution for the colorimetric assay must be determined by the enzymatic assay. When treated with diazotized *p*-nitroaniline according to the procedure employed by Kalnitsky and Tapley,[10] malonic semialdehyde gives a yellow product with maximum absorption at 440 mμ. When the colored product is extracted into 4 ml. of ethyl acetate and read in a cuvette having a light path of 1 cm., an absorbancy of about 0.600 corresponds to 0.1 micromole of aldehyde acid in the sample assayed. The preparation of a standard curve is advisable.

β-Hydroxypropionic Acid[3]

Principle. β-Hydroxypropionic acid preparations obtained commercially or made from ethylene cyanohydrin according to published directions[11] contain varying amounts of impurities, chiefly a material which may be a polyester acid of the type described by Gresham *et al.*[12] Furthermore, concentrated solutions of β-hydroxypropionic acid develop these impurities on storage. As described below, β-hydroxypropionic acid

---

[10] G. Kalnitsky and D. F. Tapley, *Biochem. J.* **70**, 28 (1958).
[11] R. R. Read, *Org. Syntheses, Coll. Vol.* **1**, (2nd ed.), 321 (1941).
[12] T. L. Gresham, J. E. Jansen, and F. W. Shaver, *J. Am. Chem. Soc.* **70**, 998 (1948).

generated from propiolactone is purified as the crystalline calcium-zinc salt.[13, 14] Solutions of potassium β-hydroxypropionate are then generated as needed by treating the calcium-zinc salt with Dowex 50 (K⁺ form).

Reagents

Propiolactone, freshly distilled.
Barium hydroxide octahydrate.
Calcium carbonate.
12 N sulfuric acid.
50% zinc chloride solution.
Dowex 50.

Procedure. Propiolactone (14.4 g.) is dissolved in about 10 ml. of water, 34 g. of solid barium hydroxide are added cautiously with stirring, and the mixture is heated for 30 minutes on a steam bath. The preparation is cooled to 0°, concentrated sulfuric acid is added dropwise to about pH 1, and 12 g. of solid calcium carbonate are then added with care to avoid vigorous foaming. The preparation is heated for 30 minutes on a steam bath, cooled, and the bulky precipitate of barium sulfate obtained on centrifugation is washed twice with about 100 ml. of water. The combined supernatant solution and washes are concentrated in a rotary flash evaporator to a thick sirup, and 10 ml. of 50% zinc chloride solution are added. The preparation is chilled in an ice bath, and the precipitate of the calcium-zinc salt is collected by filtration. When recrystallized from a minimal amount of hot water and dried *in vacuo* the pure salt is obtained in 45% yield.

A solution of the salt (1.15 g. in a minimal amount of water) is passed through a 1.1 × 15-cm. column of Dowex 50 (K⁺ form). Water is passed through the column, and 150 ml. of eluate are collected and concentrated to a volume of 10 ml. The yield of potassium β-hydroxypropionate, as determined by titration with hydrochloric acid, is 93% in this step. Paper chromatography of β-hydroxypropionic acid is described elsewhere.[4]

β-Hydroxyisobutyric Acid

Principle. dl-β-Hydroxyisobutyric acid is prepared from formaldehyde and ethyl α-bromopropionate by a Reformatsky reaction essentially according to the procedure of Blaise and Herman.[15]

[13] W. Heintz, *Ann. Chem. Liebigs* **157**, 291 (1871).
[14] H. Johannson, *Lund Univ. Ann.,* New Series, Division 2, **12**, No. 8 (1915).
[15] E. E. Blaise and I. Herman, *Ann. chim. et phys.* **17**, 371 (1909).

Reagents

Paraformaldehyde.
Zinc dust.
Ethyl α-bromopropionate.
Ethyl acetate (redistilled).

Procedure. Paraformaldehyde (7 g.), zinc dust (17.5 g.), and ethyl acetate (67 ml.) are placed in a three-necked round-bottomed flask equipped with a reflux condenser, dropping funnel, and sealed stirrer. The mixture is heated to reflux temperature, and 37.5 g. of ethyl α-bromopropionate are added dropwise. If necessary, a crystal of iodine is added to initiate the reaction. If the reaction becomes too vigorous, the flask is chilled. After the addition of ethyl α-bromopropionate is complete, the mixture is heated at reflux temperature for 30 minutes, cooled, and then poured into a mixture of crushed ice and sodium chloride. The preparation is centrifuged, and the ethyl acetate layer is removed and saved. A 20% solution of sulfuric acid is added slowly to the combined aqueous and solid layers until solution is complete. This solution is extracted twice with 50 ml. of ether, and the combined ether and ethyl acetate layers are washed three times with saturated ammonium sulfate solution, once with saturated potassium bicarbonate solution, once with water, and dried over anhydrous sodium sulfate. The solvents are removed in a rotary flash evaporator, and ethyl β-hydroxy-isobutyrate is obtained by distillation at 14 mm. (b.p. 84° to 90°). The yield is 10.0 g. (37% based on ethyl α-bromopropionate).

The ester (10.0 g.) is added slowly to 8.6 ml. of 12.5 N sodium hydroxide solution and stirred at room temperature for 3 hours. The ethanol is largely removed by distillation in a flash evaporator, the solution is cooled, and 13.5 ml. of 12 N sulfuric acid are added slowly. The solution is saturated with ammonium sulfate and extracted six times with 20-ml. portions of ether. The combined ether solutions are washed with 4 ml. of saturated ammonium sulfate solution and dried over anhydrous sodium sulfate. The ether is removed by distillation, and the residue is dissolved in 4 ml. of water and neutralized to pH 9 with sodium hydroxide solution. The crude sodium β-hydroxyisobutyrate obtained by removal of the water in a flash evaporator is crystallized twice from a minimal volume of boiling ethanol. The yield of purified sodium β-hydroxyisobutyrate is 3.45 g. (36% based on the ethyl ester). The dry salt is stable on storage. Paper chromatography of the product is described elsewhere.[16]

[16] W. G. Robinson, R. Nagle, B. K. Bachhawat, F. P. Kupiecki, and M. J. Coon, *J. Biol. Chem.* **224**, 1 (1957).

Section III
Proteins and Derivatives

[81] Acetylglutamic Acid

By Leonard Spector and M. E. Jones

N-Acetyl-L-glutamic acid[1] is a metabolite in the biosynthesis of ornithine in a number of microorganisms,[2] and it appears to be the "natural" catalyst required for the synthesis of citrulline with ureotelic vertebrate livers.[3-6]

Synthesis

Enzymatic synthesis of AGA from acetyl coenzyme A and L-glutamic acid is catalyzed by *Escherichia coli* extracts.[7] A good procedure for the chemical synthesis of AGA from acetic anhydride and L-glutamic acid has been reported.[8,9] These are not given in detail here, since commercial preparations of good purity are available.

Isolation

AGA has been isolated from beef liver and yeast.[6] The isolation was carried out by separation of AGA from the other anions present in hot-water extracts of liver and yeast on a column of Dowex 2 resin by a modification of the procedure for carbamyl-L-glutamic acid.[10] The isolation on the Dowex 2 resin was repeated several times. Cations were then removed by treatment with Dowex 50, and the AGA was finally crystallized from water and then n-butanol or subjected to paper chromatography followed by crystallization from the concentrated water eluates.[11] The procedure described here is a purely chemical method for the isola-

---

[1] The abbreviation AGA will be used to denote N-acetyl-L-glutamic acid.
[2] H. J. Vogel, *in* "Amino Acid Metabolism," (W. D. McElroy and B. Glass, eds.), p. 335. Johns Hopkins Press, Baltimore, 1955.
[3] S. Grisolia and P. P. Cohen, *J. Biol. Chem.* **204**, 753 (1953).
[4] A. D. Anderson and M. E. Jones, *Abstr. 135th Meeting Am. Chem. Soc., Boston, April, 1959*, p. 63c.
[5] G. W. Brown, Jr., and P. P. Cohen, *Biochem. J.* **75**, 182 (1960).
[6] L. M. Hall, R. L. Metzenberg, and P. P. Cohen, *J. Biol. Chem.* **230**, 1013 (1958).
[7] W. K. Maas, G. D. Novelli, and F. Lipmann, *Proc. Natl. Acad. Sci. U.S.* **39**, 1004 (1953).
[8] B. H. Nicolet, *J. Am. Chem. Soc.* **52**, 1194 (1930).
[9] M. E. Jones and L. Spector, *J. Biol. Chem.* **235**, 2897 (1960).
[10] See Vol. III [94], p. 647.
[11] Whole liver contains 1 to 3 parts per million; yeast, 2 parts per 10,000 of the total solids in an 80° water extract.[6]

tion of isotopic AGA from incubation mixtures.[9] It yields crystalline material of high purity and in reasonably good yield. Advantages of the method are its rapidity and avoidance of all chromatography.

Procedure. The denatured incubation mixture[12] (2 l.) is chilled and centrifuged rapidly in an International centrifuge, size 2, to remove the bulk of protein. The supernatant is filtered[13] and concentrated *in vacuo*[14] at 50° to 60° to a volume of about 50 ml. The yellowish, turbid solution is charcoaled[15] twice, and the filtrate is adjusted to pH 3.5 to 4.0 with 6 N HCl. The resultant solution is charcoaled twice more, and the clear, colorless filtrate is concentrated in a 100-ml. flask *in vacuo* at 30° to 40° to a moist solid. This is dissolved in 3.0 ml. of water, and the pH is adjusted, if necessary, to 2.0 to 2.5 with a few drops of 6 N HCl. Freshly distilled 1,2-dimethoxyethane (DME)[16] is added to a volume of 100 ml., and the mixture is thoroughly shaken. After the mixture has stood at room temperature for 20 minutes, the supernatant is decanted, charcoaled, and filtered through a medium-porosity glass filter. The slightly turbid filtrate is charcoaled again and filtered through a fine-porosity filter. The clear filtrate is concentrated in three portions, in a 100-ml. flask, *in vacuo* at 30° to 40°. The residue is dissolved in 6 drops of water (warmed gently on the steam bath, if necessary) and diluted with 50 ml. of DME. The turbid solution is charcoaled and filtered into a 100-ml. flask through a fine-porosity filter. The clear filtrate is concentrated *in vacuo* at 30° to 40° to a film which is *immediately* treated with 50 ml. of DME. The flask walls are carefully scratched down with a spatula. The mixture is charcoaled and filtered in three portions through a fine-porosity filter into a 50-ml. flask. Solvent is removed *in vacuo* at 30° to 40°. The crystalline residue is dissolved in 15 ml. of water, filtered from dust, and concentrated *in vacuo* at 40° to 50° to a volume of 0.5 ml. or less. On release of vacuum, tiny clusters of prisms usually separate in

[12] The following reagents (in micromoles) are incubated at 37° for 30 minutes in a total volume of 300 ml. at pH 8.1 or 9.0: ATP (3000), MgCl₂ (3000), NH₄Cl (3000), glutathione (2000), L-ornithine (3000), KHCO₃ (12,000), AGA (200), and enzyme protein, 110 mg. The reaction is terminated by the addition of 1700 ml. of absolute ethanol.

[13] All filtrations are performed on sintered-glass filters.

[14] All concentration operations are performed in a rotating still at the water pump.

[15] About 40 mg. of acid (HCl)-washed Norit A (neutral) is used throughout in each charcoaling operation. The charcoal mixture is shaken by hand for 3 to 5 minutes at room temperature and filtered.

[16] Only freshly distilled DME is used in this procedure. This is most easily accomplished by maintaining a small still in constant readiness; a few minutes before the DME is required, the heat under the distilling flask is turned on.

the body of the solution.[17,18] The amount of crystallization is augmented by overnight refrigeration. The crystals are slurried out onto a coarse filter with 2.0 ml. of chilled water, washed, and air-dried. Yield, 10 to 12 mg. Melting point, 191.0° to 192.5°. Mixed melting point with starting material, unchanged.

Physical Properties

The melting point of AGA is 191° to 193°. It is soluble in water and ethanol, insoluble in benzene, chloroform, acetone, or 1,2-dimethoxyethane. AGA can be identified by ion-exchange resin chromatography,[6,10] by paper chromatography (Table I), and by paper ionophoresis (Table II). Localization of the AGA can be effected by: (1) direct visualization

TABLE I
PAPER CHROMATOGRAPHY OF AGA

Solvent	R_f
3:1 (v/v) butanol, 1.0 N acetic acid	0.80[a]
8:1:1 (v/v/v) methanol, H_2O, concentrated NH_4OH	0.52[a]
2:1 (v/v) isopropanol, 0.1 N NH_4OH	0.46[a]
20:4:1 (v/v/v) n-butanol, H_2O, glacial acetic acid	0.65[b,c]

[a] L. M. Hall, R. L. Metzenberg, and P. P. Cohen, *J. Biol. Chem.* **230**, 1013 (1958).
[b] W. K. Maas, G. D. Novelli, and F. Lipmann, *Proc. Natl. Acad. Sci. U. S.* **39**, 1004 (1953).
[c] Glutamate and glutamine had an R_f of 0 to 0.1.

on paper by spraying with an indicator, such as bromcresol green, after removal of volatile acidic solvents; (2) the enzymatic assay given below; or (3) the ability to permit the growth of bacterial mutants.[7,19]

Enzymatic Assay

Principle. The method utilizes the catalytic action of AGA in citrulline synthesis.[3] It is a modification of one already published.[6] The major modifications, apart from the use of a different enzyme source, are the reduction of the ammonia concentration below inhibitory levels, and the inclusion of glutathione to preserve maximal enzyme activity. This

---

[17] Sometimes crystals separate on the flask walls instead of from solution. For maximum purity it is advisable to redissolve the solid on the steam bath and allow crystallization to proceed from solution.
[18] To arrive at this point in the isolation procedure generally requires only 6 to 7 hours of steady work.
[19] B. D. Davis, *J. Biol. Chem.* **200**, 417 (1953).

TABLE II
Electrophoretic Migration of AGA

Conditions	Compound	Distance (D) from origin toward (+) pole	$\dfrac{D \text{ for compound}}{D \text{ for aspartic acid}}$
0.05 M potassium phosphate buffer, pH 6.8, 5.5 volts/cm. for 4 hours	Alanine	− 0.6	—
	Aspartate	6.8	[1.00]
	Carbamyl aspartate	11.5	1.68
	Glutamate	6.0	0.88
	Carbamyl glutamate	10.7	1.57
	Acetyl glutamate	10.1	1.51
0.1 M sodium formate buffer, pH 3.8, 5.5 volts/cm. for 4 hours	Alanine	−0.4	—
	Aspartate	3.9	[1.00]
	Carbamyl aspartate	5.7	1.46
	Glutamate	1.9	0.49
	Carbamyl glutamate	4.9	1.26
	Acetyl glutamate	4.8	1.23

method is more sensitive than the earlier methods[20] and is linear for AGA concentrations from 2×10^{-5} to $1.2 \times 10^{-4} M$.

Reagents

1.0 M Tris buffer, pH 8.0.
0.2 M ATP, pH 7.5.
1.0 M MgCl$_2$.
0.2 M L-ornithine–HCl.
1.0 M KHCO$_3$.
0.2 M NH$_4$Cl.
0.25 M reduced glutathione.

Procedure. The vessels required for the assay are (1) assay mixture and enzyme; (2) assay mixture, enzyme, and standard AGA (0.01, 0.02, 0.03, 0.04, 0.06, 0.08, and 0.1 micromole); (3) assay mixture, enzyme, and an aliquot of unknown in duplicate, one sample incubated and the other deproteinized at zero time to check on noncitrulline chromogens.

Each vessel contains 0.05 ml. of ATP, 0.05 ml. of Tris buffer, 0.01 ml. of MgCl$_2$, 0.025 ml. of L-ornithine–HCl, 0.03 ml. of KHCO$_3$, 0.025 ml. of NH$_4$Cl, and 0.01 ml. of glutathione. Generally, a mixture of the com-

---

[20] A chemical assay has been reported.[7] It depends on the hydrolysis of AGA in basic hydroxylamine to generate acethydroxamate which is quantitatively determined through its color formation with ferric ion. This method is lengthy and relatively insensitive.

ponents is prepared, and 0.2 ml. of it is added to each tube. AGA or sample is then added along with enough water so that the final volume will be 0.5 ml. after all additions are made. The tubes are placed in a 37° bath, and enzyme is added at zero time. After 30 minutes of incubation, 1 ml. of 5% trichloroacetic acid is added, denatured protein is removed by centrifugation, and citrulline is estimated on a 0.5-ml. aliquot of the reaction mixture by a modification of the Archibald method.[21] The tubes are heated for 30 minutes in a boiling-water bath in the dark, then transferred to a covered cold-water bath for 10 minutes to cool to room temperature (shaking is not necessary), and finally read at 490 mμ.[22, 23]

The AGA standards are plotted on graph paper, and the values of the unknowns are read from the graph. The AGA content of the unknown aliquot is obtained by subtracting the value of the zero-time tube from that of the incubated tube. If the unknown is in an unusual solvent or a buffer whose pH is somewhat far removed from that of the enzyme assay, it is wise to check its effect on the AGA standard curve.

Enzyme. The enzyme can be prepared from a number of ureoteles. In this procedure an extract of hog liver mitochondria[24] (0.5 to 1.0 mg. of protein) or an extract of frog liver mitochondria[25] (0.04 to 0.08 mg. of protein) is used. The hog liver mitochondria are prepared by following the Jencks procedure through the centrifugation step in the Sharples centrifuge[24] or the 13,000 $\times$ g centrifugation in the Servall SS2.[26] The paste is then suspended in an equal volume of 0.15 M KCl and frozen. Just before use, the hog mitochondria are extracted with 1 vol. of 0.05 M glycylglycine buffer, pH 8.0, and centrifuged at 100,000 $\times$ g for 30 minutes. This extract contains approximately 10 mg. of protein per milliliter. With the 0.1-micromole standard AGA tube, 0.05 to 0.1 ml. of this extract should produce 0.5 to 1.0 micromole of citruilline.[27] The paste, when held at −20°, has been stable for over 2 years. The glycylglycine

[21] See Vol. III [92], p. 642.
[22] A heating period of 30 minutes was necessary for full color development. After only 15 minutes (the heating period suggested by Archibald), the color development is still proceeding at a linear rate.
[23] The color production has been measured in either the Klett or the Beckman spectrophotometer. The useful range in the former is 0.1 to 0.5 micromole, and in the latter 0.025 to 0.3 micromole.
[24] See Vol. V [63], p. 469.
[25] See Vol. V [124a], p. 919.
[26] See Vol. V [63], pp. 470–471.
[27] For AGA assay a volume of extract should be used which produces 1 ± 0.2 micromole of citrulline.

extract is stable for only 2 days if it is kept ice-cold when in use and frozen at $-20°$ when not in use.[28]

[28] Glutathione is in the assay mixture to ensure reduced enzyme; however, it is not necessary until the paste is approximately a year old. Glutathione decreases the color yield of a standard amount of citrulline. If it is necessary to determine the amount of citrulline produced with a standard amount of AGA, one must add an equal amount of zero-time reaction mixture to each citrulline standard.

[82] Preparation and Analysis of N-Acetyl-L-aspartic Acid

By HARRIS H. TALLAN

Preparation

Principle. The acetylation of aspartic acid is best effected chemically, with acetic anhydride; purification of the product is carried out by preparation of the readily crystallized acetylaspartic anhydride, which is rehydrated by dissolving in water. Since crystallization of acetylaspartic acid is rather difficult, the anhydride would seem to be the compound of choice for most applications (cf. Jacobson[1]).

By use of ion-exchange chromatography on an anion exchanger, natural acetylaspartic acid has been isolated from cat brain tissue, but only in small amounts.[2] A number of enzymes have been described that bring about the synthesis of acetylaspartic acid; however, their use for the preparation of the compound has not yet been reported. These enzymes include the specific aspartic acetylase of rat brain, which catalyzes the formation of N-acetyl-L-aspartic acid from L-aspartic acid and acetyl coenzyme A,[3] and enzyme systems in extracts of *Escherichia coli*[4] and *Bacterium cadaveris*.[5]

Chemical Synthesis. The procedure of Barker[6] is described here.

Reagents

L-Aspartic acid.
Acetic anhydride.

[1] K. B. Jacobson, *J. Gen. Physiol.* **43**, 323 (1959).
[2] H. H. Tallan, S. Moore, and W. H. Stein, *J. Biol. Chem.* **219**, 257 (1956).
[3] F. B. Goldstein, *J. Biol. Chem.* **234**, 2702 (1959).
[4] W. K. Maas, G. D. Novelli, and F. Lipmann, *Proc. Natl. Acad. Sci. U.S.* **39**, 1004 (1953).
[5] S. R. Mardashev, Lu-Ju-shan, and J. A. Romakov, *Mikrobiologiya* **28**, 641 (1959).
[6] C. C. Barker, *J. Chem. Soc.* 453 (1953).

Procedure. L-Aspartic acid (5 g.) is dissolved in 100 ml. of boiling water, 25 ml. of acetic anhydride is added, the reaction mixture is cooled rapidly to room temperature, another 75 ml. of acetic anhydride is added, and the solution is stirred at 20° (with cooling) for 6 hours, or until an aliquot of the reaction mixture, adjusted to pH 5, gives no color with ninhydrin. The solvents are removed on a rotary evaporator to give a heavy oil, which will crystallize on long standing (several months), especially if seeded. Crystallization is facilitated if the anhydride of acetylaspartic acid is first prepared and purified. Acetic anhydride (50 ml.) is added to the oil obtained above, and the mixture is heated at 95° to 105° for 20 minutes. Any insoluble material is removed by filtration, and the filtrate is concentrated to half its original volume. After being kept at 0° for 1 hour, the crystals that form are filtered off, washed on the filter with acetic anhydride, ethyl acetate, and petroleum ether, and then dried. Recrystallization from hot acetic anhydride (10 to 15 ml.) will raise the melting point to 173°. The yield is about 3 g. The anhydride is dissolved in 10 ml. of warm water to form acetylaspartic acid, and the solution is taken to dryness in a rotary evaporator. The resulting oil crystallizes readily on seeding. Lacking seed crystals, crystallization may be hastened by rigorous drying, followed by exposure of a thin film of the oil to the atmosphere at room temperature. Yield, 2.5 g.

Properties

N-Acetyl-L-aspartic acid melts at 140° to 141°.[6] The di-*p*-nitrobenzyl ester has been described.[2]

Hydrolysis of acetylaspartic acid is 90% complete after heating in a boiling-water bath for 30 minutes with 2 *N* HCl.[2] It has also been reported that a heating time of 60 minutes is required to achieve the same extent of hydrolysis.[1]

Analysis

In pure solution, acetylaspartic acid may be determined either by measurement with ninhydrin of the aspartic acid liberated by hydrolysis, or by direct reaction with hydroxylamine according to the procedure developed by Katz *et al.*[7] for acetyl amino acids. Neither method is satisfactory with a tissue extract, because of interfering substances, from which the acetylaspartic acid must be separated prior to analysis. Chromatographic systems for this purpose have been described, using paper, cation-exchange resins, and anion-exchange resins. The separation of acetylaspartic acid from pyrrolidonecarboxylic acid, which

[7] J. Katz, I. Lieberman, and H. A. Barker, *J. Biol. Chem.* **200**, 417 (1953).

arises from glutamine in tissue extracts and which has many properties similar to those of acetylaspartic acid, should be established before any method, but particularly paper chromatography, is applied.

The chromatography of acetylated amino acids on Whatman No. 1 paper, with butanol–acetic acid as developer, was described by Katz et al.[7] This method has been applied to acetylaspartic acid by Mardashev et al.[5] The compound (minimal amount, 0.1 micromole) is detected by spraying with bromcresol green solution. Tsukada et al.[8] report the position of acetylaspartic acid on Toyo No. 51 paper after two-dimensional chromatography (n-butanol–ethanol in the first direction, phenol–water in the second); radioautographs were prepared.

On a 0.9×150-cm. column of Dowex 50-X4 equilibrated with buffer at pH 2.2, and developed with buffer of pH 3.1,[9] acetylaspartic acid emerges at 62 ml., between glycerophosphoethanolamine and phosphoethanolamine.[2] Effluent fractions 1 ml. in size are collected, subjected to hydrolysis as described below, and analyzed with ninhydrin.[2,10]

Because of the acidic nature of acetylaspartic acid, chromatography on an anion-exchange resin is particularly suitable. In what follows are the details of the modification, only briefly noted earlier,[2] of the published Dowex 2-X4 procedure.[2] A similar system, using, however, ammonium acetate buffers to which one-tenth volume of 95% ethanol was added just before use, has been employed by Goldstein.[3] In this application, the effluent fractions were collected directly on stainless-steel planchets, dried by infrared radiation, and assayed by radioactivity.

Procedure. Dowex 1-X8 (−400 mesh) in the chloride form is screened wet through a 200-mesh sieve with a strong jet of water. The material that passes through the sieve is heated with twice the volume of 4 N HCl for 2 hours on a steam bath, with occasional stirring. After the resin has settled for 1 hour, the supernatant fluid is decanted to remove the fine particles. The resin is treated twice more in this manner, residual fine particles are removed by repeated suspension and settling of the resin and decantation of the supernatant fluid, and finally the resin is filtered, washed with water, and then with 4 N NaOAc until the filtrate (diluted, and acidified with HNO_3) gives a negative test for Cl⁻ with $AgNO_3$ (cf. Tallan et al.[11]).

The resin in the acetate form is washed with detergent-free 0.2 N

[8] Y. Tsukada, Y. Nagata, S. Hirano, and G. Takagaki, *J. Biochem.* (*Tokyo*) **45,** 979 (1958).

[9] S. Moore and W. H. Stein, *J. Biol. Chem.* **211,** 893 (1954); see Vol. VI [117].

[10] N. Okumura, S. Otsuki, and H. Nasu, *J. Biochem.* (*Tokyo*) **46,** 247 (1959).

[11] H. H. Tallan, S. T. Bella, W. H. Stein, and S. Moore, *J. Biol. Chem.* **217,** 703 (1955).

NaOAc buffer (prepared by diluting the 2 N buffer described below), and a column 0.9 × 15 cm. is poured. Buffer containing a detergent (5 ml. of a 50% solution of BRIJ 35 per liter) is then percolated through the column until detergent is present in the effluent. The sample to be analyzed (0.5 to 2.0 ml. of solution, containing 0.05 to 0.5 mg. of acetylaspartic acid) is adjusted to pH 5.5, added to the column, and washed in with three small portions of 0.2 N buffer. Enough 0.2 N buffer is added to make a layer 3 cm. high above the resin, and elution is begun with NaOAc buffer of gradually increasing ionic strength, prepared by allowing 2 N buffer (272 g. of NaOAc·3H$_2$O and 50 ml. of glacial acetic acid, made up to 2 l. with distilled water) to flow into 140 ml. of 0.2 N buffer in a mixing device, such as that described by Moore and Stein.[9] The buffers used for chromatography contain 5 ml. of BRIJ 35 per liter. The effluent is collected in 1-ml. fractions at a rate of 5 to 7 ml./hr. Acetylaspartic acid emerges at about 75 ml. Fractions 46 to 95 (the preceding 45 ml. may be collected in a single forerun, if convenient) are heated for 30 minutes in a boiling-water bath with 0.5 ml. of 6 N HCl, cooled, neutralized with 0.5 ml. of 6 N NaOH, and analyzed with ninhydrin.[12] A recovery factor for the hydrolysis should be included in the calculations.

Since the column of resin shrinks as the molarity of the buffer increases, a new column is prepared for each analysis. The resin should be thoroughly washed with 2 N NaOAc before reuse.

Note added in proof: Du Ruisseau[13] has reported use of the two-dimensional paper chromatographic system of Nordmann *et al.*[14] for the detection of acetylaspartic acid. The acidic components of the tissue extract are absorbed on a small column of Dowex 2 and eluted with 12 N formic acid. After removal of the formic acid by evaporation, the sample, redissolved in water, is applied to paper. Solvent 1 is absolute ethanol–21.6% ammonia–water (80:5:15); solvent 2 is n-propanol-eucalyptol–98% formic acid (50:50:20), plus water to turbidity. Acetylaspartic acid moves with R_f values of 0.35 and 0.39 in solvents 1 and 2, respectively. There is good separation from pyrrolidonecarboxylic acid. After location of acidic compounds by dipping the dried chromatogram into a 0.1% acetone solution of bromcresol green, the acetylaspartic acid spot is eluted with 5 ml. of 20% methanol and measured by titration.

Several ion-exchange procedures have been reported. Margolis *et al.*[15] make a preliminary separation on Dowex 50-X4 (200–400 mesh). The dried residue of a tissue extract prepared with perchloric acid and neutralized with KOH is dissolved in 2.5 ml. of 0.1 N HCl, applied to a 0.9 × 7.5-cm. column, and eluted with 30 ml. of water. The acetylaspartic acid in the water effluent may then be hydrolyzed (in

[12] S. Moore and W. H. Stein, *J. Biol. Chem.* **211**, 907 (1954).
[13] J.-P. Du Ruisseau, *Can. J. Biochem. Physiol.* **38**, 763 (1960).
[14] R. Nordmann, O. Gauchery, J.-P. Du Ruisseau, Y. Thomas, and J. Nordmann. *Bull. soc. chim. biol.* **36**, 1461 (1954).
[15] R. U. Margolis, S. S. Barkulis, and A. Geiger, *J. Neurochem.* **5**, 379 (1960).

2 N HCl at 100° for 30 min.) and determined as aspartic acid by chromatography on a column of Dowex 1-X8 (0.9 × 30 cm.). Aspartic acid emerges in 75 to 90 ml. with 0.5 N sodium acetate as eluant. Alternatively, the water effluent is concentrated, the residue redissolved in 2.5 ml. of 0.2 N sodium acetate buffer (pH 5.0), and the solution chromatographed on Dowex 50-X4[16] with gradient elution (2 N sodium acetate flowing into a 140-ml. mixing chamber containing 0.2 N sodium acetate buffer of pH 5.0). Fractions are collected and subjected to hydrolysis prior to ninhydrin analysis.

Tallan[17] makes a preliminary separation of a perchloric acid extract on Dowex 1-X8 (minus 200 mesh). The resin column, 0.55 × 5.5 cm., is in the hydroxide form. The sample solution, usually 1 ml., is applied and then eluted successively with 2 ml. of water, 5 ml. of 2 N acetic acid, and 5 ml. of 2 N HCl. The HCl fraction, which contains acetylaspartic acid, is hydrolyzed, and aspartic acid is determined on an Amino Acid Analyzer.[18]

[16] S. Moore and W. H. Stein, J. Biol. Chem. 211, 893 (1954).
[17] H. H. Tallan, in "Amino Acid Pools" (J. T. Holden, ed.), p. 465. Elsevier, Amsterdam, 1962.
[18] D. H. Spackman, W. H. Stein, and S. Moore, Anal. Chem. 30, 1190 (1958).

[83] Newer Methods for Preparation of S-Adenosyl-methionine and Derivatives

By JAKOB A. STEKOL

Like the mammalian liver, growing or metabolizing yeast synthesizes the corresponding S-adenosyl derivatives from methionine or ethionine in amounts ranging from 7 to 20 micromoles per gram of yeast, depending on the strain of yeast employed.[1] From the perchloric acid extracts of yeast, pretreated with methionine or ethionine, the corresponding S-adenosyl derivatives are isolated by precipitation of the sulfonium compound with ammonium reineckate,[2] followed by the decomposition of the salt with methyl ethyl ketone, removal of the ketone by ether, and the concentration of the extracted aqueous layer containing the S-adenosyl derivative, which is then assayed, without isolation, spectrophotometrically, and used as is. In another procedure,[1] the S-adenosyl derivative is also precipitated from the perchloric acid extracts of yeast as the reineckate, but the reineckate is passed through a Dowex 50 column, and the retained S-adenosyl derivative is eluted off the column with sulfuric acid. Further purification of the S-adenosyl derivative is

[1] F. Schlenk, J. L. Dainko, and S. M. Stanford, Arch. Biochem. Biophys. 83, 28 (1959).
[2] See Vol. III [85].

accomplished by precipitation with phosphotungstic acid, decomposition of the acetone–water solution of the phosphotungstate with amyl alcohol and ether, and removal of the sulfate with $BaCO_3$; the solution of the S-adenosyl derivative is assayed spectrophotometrically and used as is.

Neither the reineckate nor the phosphotungstic acid is a specific precipitant of the sulfonium bases from the perchloric acid extracts of yeast or mammalian tissues. In the procedure described in detail below, the precipitation of the S-adenosyl derivatives of methionine or ethionine from the perchloric acid extracts of yeast or mammalian tissues (or "methionine-activating enzyme" digests containing ATP and methionine or ethionine[2]) is accomplished with phosphotungstic acid. The phosphotungstate of the sulfonium derivative is extracted with aqueous acetone in the same manner as was described for the isolation of methionine methyl sulfonium from mammalian tissue preparations.[3] Most, if not all, of the impurities which are precipitated along with the phosphotungstate of the sulfonium base are insoluble in aqueous acetone. The phosphotungstate of the S-adenosyl derivative, soluble in aqueous acetone, is decomposed with tetraethyl ammonium bromide or iodide, a step which completely removes the phosphotungstic acid. The bromide salt of the S-adenosyl derivative, after the adjustment of the pH to 5.8 with ammonia, is precipitated from methanolic solution with ethanol, washed with ethanol to remove all traces of tetraethyl ammonium bromide and NH_4Br, then with ether, and dried *in vacuo*. The advantages of this procedure are that the salts of the S-adenosyl derivatives of methionine or ethionine can be prepared in any desired quantity in crystalline form, and that the salts are stable for indefinite periods when stored *in vacuo* over silica gel in the cold. The impurities which are still present in the solutions of the S-adenosyl derivatives employed without isolation of the compounds are absent or undetectable by chromatographic methods in the crystalline preparations of the salts of the S-adenosyl derivatives.

Preparation of the Bromide Salts of S-Adenosylmethionine or S-Adenosylethionine from Yeast[4]

A 3-l. round-bottomed flask was fitted with a two-hole rubber stopper with sintered-glass aerator and vacuum line lead. Air was drawn into the culture vessel after passage through a gas-bubbling train consisting of a $6\,N$ H_2SO_4 scrubber and two water scrubbers. The medium was made up as follows: (1) $MgCl_2 \cdot 6H_2O$, 0.429 g., and $CaCl_2$, 0.143 g.;

[3] S. Weiss, E. I. Anderson, Peng Tung Hsu, and J. A. Stekol, *J. Biol. Chem.* **214,** 239 (1955).

[4] J. A. Stekol, E. I. Anderson, and S. Weiss, *J. Biol. Chem.* **233,** 425 (1958).

(2) $MnSO_4 \cdot H_2O$, 0.087 g., and $ZnSO_4 \cdot 7H_2O$, 0.143 g.; (3) Na·citrate, 1.428 g., and $(NH_4)_2SO_4$, 2.857 g.; (4) KH_2PO_4, 2.857 g., and K_2HPO_4, 1.428 g.; glucose, 21.42 g.; methionine or ethionine, 6.7 millimoles. The first pair of ingredients was dissolved and diluted to 400 ml. Each of the remaining three pairs of ingredients was dissolved and added, with rinsing, in a volume of 200 ml. By using such large dilutions during mixing, the formation of insoluble salts was avoided. Glucose and the radioactive or nonradioactive substrates (methionine or ethionine) were added separately in a solution to bring the final volume of the medium to 1430 ml. A drop of Dow Antifoam A was added, and 65 g. of either National or Fleishmann brand baker's yeast were crumbled into the medium. Aeration was continued for 24 hours at room temperature (24° to 27°) with one interruption after 7 to 8 hours to recharge the culture with an additional 14.29 g. of glucose. The yeast, after being harvested in the centrifuge at 1800 r.p.m., was washed twice with 3 vol. of ice water. Decantation of the washes was easily effected if the suspension was centrifuged for 20 minutes at 1800 r.p.m. The yeast may be either stored in the frozen state or immediately extracted with 4 vol. of 1.5 N perchloric acid for 1 hour at room temperature. After removal of the yeast residue by centrifugation, the slightly opaque yellow extract was chilled overnight and filtered through a fine glass filter to remove turbidity and precipitated perchlorates.

From the measured volume, a 25-ml. aliquot was taken to ascertain the amount of freshly prepared 50% phosphotungstic acid solution necessary for complete precipitation. Generally, 20 ± 5 ml. were required. After 2 hours of chilling, the yellow phosphotungstic precipitate was removed by centrifugation and thoroughly washed with 4 to 5 vol. of water four times. This was followed by washing four times with 2 vol. each of 95% ethanol, once with absolute ethanol, and finally with ethyl ether before drying *in vacuo*. The third and fourth water wash and the last 95% ethanol wash were turbid.

The weighed dry phosphotungstate (3.2 ± 0.7 g.) was then triturated four or five times with approximately 10 ml. of 90% acetone. After each trituration the suspension was centrifuged and the yellow acetone extract was filtered through a fine sintered-glass filter. Completion of the extraction was arbitrarily judged by a progressive decrease in color of the extracts. The insoluble residue, after it was dried with ether, amounted to about 30% of the original weight of the phosphotungstate. To the combined aqueous acetone extracts 1 M tetraethyl ammonium bromide was added until precipitation of the phosphotungstate was complete. The extract was diluted with an equal volume of cold water, chilled, and centrifuged. After adjustment of the pH to 5.8 ± 0.1 with 1 M NH_4OH, the supernatant solution was decolorized with Darco (activated carbon, Atlas Powder Company) at room temperature and filtered. Decolorization with Darco can be repeated if necessary to obtain a colorless solution. The filtrate was evaporated *in vacuo* at room temperature to a thin layer of yellowish oil. On addition of 10 ml. of absolute methanol, a white granular material separated. The methanol was distilled off *in vacuo* to facilitate the removal of water, and another portion of methanol was added. The solid, slightly hygroscopic material was rapidly transferred with absolute ethanol to a centrifuge tube. After the sulfonium salt was centrifuged, washed with absolute ethanol three times for complete removal of NH_4Br, then with ether, it was dried *in vacuo* at room temperature. The yields of the sulfonium salts prepared from L- or D-methionine or ethionine, with various strains of yeast, and their optical rotations in 1% solution in water are summarized in the table.

SYNTHESIS OF *S*-ADENOSYLMETHIONINE AND *S*-ADENOSYLETHIONINE BY YEAST

Substrate	Metabolizing yeast[a]		Baker's yeast[d] (Fleischmann or National)	D^e $[\alpha]_{24}$
	Torulopsis[b]	*Saccharomyces*[c]		
L-Methionine	6.2	16.0	5.7	$+16.8°$
D-Methionine	5.0	2.4	1.2	$- 8.2$
L-Ethionine	7.8	17.0	5.3	$+23.9$
D-Ethionine	6.9	0.5	0.9	$- 9.6$

[a] Assayed spectrophotometrically in the solution containing the *S*-adenosyl derivative; F. Schlenk, J. L. Dainko, and S. M. Stanford, *Arch. Biochem. Biophys.* **83**, 28 (1959). The yields are calculated as micromoles per gram of moist cell centrifugate containing $20 \pm 2\%$ of solids after drying at 110°.

[b] Commercial *Torulopsis*.

[c] Activated dry yeast.

[d] By direct isolation as the bromide salt in micromoles per gram of yeast cake used; J. A. Stekol, E. I. Anderson, and S. Weiss, *J. Biol. Chem.* **233**, 425 (1958).

[e] For a solution of 1% in water; J. A. Stekol, E. I. Anderson, and S. Weiss, *J. Biol. Chem.* **233**, 425 (1958).

It will be noted that the yields of *S*-adenosylmethionine and *S*-adenosylethionine from the D-isomer of methionine or ethionine, respectively, are considerably lower than those from the corresponding L-isomer. In contrast to the mammalian "methionine-activating enzyme" prepared from rat or rabbit liver,[5] the yeast can utilize the D-isomer of methionine or ethionine for activation with ATP. However, as judged from the optical rotation data in the table, the *S*-adenosyl derivative of methionine or ethionine, obtained from the D-isomer of methionine or

[5] G. L. Cantoni and J. Durell, *J. Biol. Chem.* **225**, 1033 (1957).

ethionine, respectively, is a mixture of about 70% of the D- and about 30% of the L-isomer. It would appear that not only can the yeast invert the D-isomer of methionine or ethionine to the L-form, but it can directly activate the D-form of methionine or ethionine in the presence of ATP to yield the D-isomers of the corresponding S-adenosyl derivatives.

During the 24-hour incubation employed in the described procedure, the yeast synthesizes very small amounts, if any at all, of methionine. This was indicated by the fact that the specific activity of either S-adenosyl-L-methionine or S-adenosyl-L-ethionine formed by yeast from L-methionine or L-ethionine, respectively (labeled with S^{35} or with C^{14} in the methyl, ethyl group, or in carbons 1 or 2 of the chain), was almost identical with that of the labeled respective precursors. Furthermore, S-adenosyl-L-ethionine did not reveal the presence of S-adenosylmethionine in the crystalline preparation on chromatographic examination of the overloaded spots obtained from digests of S-adenosyl-L-ethionine with 0.18 N NaOH for 2 hours. Under these conditions, S-adenosylmethionine is decomposed to methionine, homoserine, and adenine, and S-adenosylethionine is decomposed to ethionine, homoserine, and adenine.[4]

Methionine- or Ethionine-Activating Enzyme from Yeast[6]

Procedure A.

Step 1. Drying. Fresh Fleischmann's baker's yeast was passed through a screen (openings about 2×2 mm.) and spread between two sheets of heavy paper in a layer about 10 mm. thick. It was allowed to dry at room temperature for 6 days and then stored at $-15°$.

Step 2. Extraction and Acetone Fractionation. Dried yeast (600 g.) was suspended in 1800 ml. of a 0.067 M solution of K_2HPO_4 containing 3.6 g. of DL-methionine and very gently stirred at 32° for 4 hours. The residue was removed by centrifugation (0°, $2500 \times g$, 20 minutes). The extracts thus obtained were immediately carried through the next procedure, since the enzyme was not uniformly stable at this stage. Time and temperature of incubation are relatively critical, longer incubation leading to lower yields.

To 805 ml. of the cloudy supernatant fluid, 254 ml. of acetone were added dropwise to reach a final concentration of 24%, when calculated on the basis of additive volumes. During the addition, which took about 45 minutes, the extract was stirred in an alcohol–dry ice bath, and the temperature was lowered to $-8°$ as rapidly as possible without freezing the solution, and thereafter maintained at that temperature. After the addition was complete, the mixture was allowed to stand for 15 minutes,

[6] S. Harvey Mudd and G. L. Cantoni, *J. Biol. Chem.* **231**, 481 (1958).

and the residue was removed by centrifugation (—8°, 2500 × g, 10 minutes) and discarded. The clear yellow supernatant fluid was brought to 45% acetone by dropwise addition of 405 ml. of acetone over 45 minutes, allowed to stand for 15 minutes, and the precipitate was collected by centrifugation (—8°, 800 × g, 10 minutes). The precipitate was suspended in 130 ml. of potassium phosphate buffer, 0.02 M, pH 6.6, and dialyzed overnight at 3° in a rocking dialyzer against 10 vol. of the same buffer. The dialyzed enzyme preparation was clarified by centrifugation (3°, 18,000 × g, 5 minutes), the small residue discarded, and the supernatant fluid (acetone I fraction) (213 ml.) stored at —15°. At this stage the enzyme is stable on storage at —15° for many months. This step was quite reproducible as to the specific activity of the product. The total yield in units varied between about 400 and 1400 in different batches of yeast. Usually several batches were prepared as above and combined for the next step.

Step 3. Calcium Phosphate Gel Adsorption and Elution. All operations were carried out at 0° to 3° during this and subsequent steps. A 400-ml. aliquot of the combined acetone I fractions was diluted with an equal volume of cold glass-distilled water and slowly brought to pH 5.4 to 5.2 by addition of 30 ml. of 0.25 N acetic acid. The suspension was kept at 0° for 10 minutes, and, after removal of the precipitate by centrifugation (800 × g, 10 minutes), the clear supernatant fluid was treated with 415 ml. of calcium phosphate gel (20.5 mg./ml.). After 10 minutes of gentle stirring, the gel was collected by centrifugation (800 × g, 10 minutes), washed in 620 ml. of sodium acetate (0.03 N, pH 5.5), collected again by centrifugation, and eluted by stirring with 135 ml. of potassium phosphate, 0.04 M, pH 6.6, for 60 minutes. The gel was removed by centrifugation (18,000 × g, 3 minutes), and the elution was repeated with 65 ml. of buffer. Usually the first eluate (gel eluate I) contained the bulk of the enzyme, which was stable for months on storage at —15°.

Step 4. Negative Bentonite Adsorption. Bentonite (30 mg./ml. of enzyme) was suspended as evenly as possible in the enzyme solution. The suspension was stirred in an ice bath for about 20 minutes, and the bentonite was removed by centrifugation (11,000 × g, 6 minutes). Without delay, an additional 30 mg. of bentonite were added to each milliliter of the supernatant fluid, and after 20 minutes the bentonite was removed as before. Sometimes a second centrifugation was required at this stage to achieve complete removal of the lighter particles. At this stage the enzyme shows variable stability and therefore is usually carried on quickly to the next step. To each 100 ml. of very faintly cloudy solution (bentonite supernatant fluid) were added 58 g. of solid ammonium sulfate recrystallized in the presence of ethylenediaminetetraacetate. The pH was kept at 6 to 6.5 by addition of 0.1 N KOH as needed. The precipitated enzyme was collected by centrifugation (8000 $\times$ g, 10 minutes), resuspended in a small volume of Tris buffer, 0.02 M, pH 7.6, and dialyzed overnight against the buffer. At this stage, the enzyme has been stored only at $-60°$, at which temperature one preparation lost little activity during 2 months. The effect of storage at $-15°$ is not known.

Procedure B.

Steps 1 through 3 of procedure A were followed.

Step 4. Acetone Fractionation. Gel eluate I was fractionated with acetone at $-8°$ as in step 2. The 38% precipitable fraction was discarded, and the 38 to 55% fraction was collected by centrifugation at $-8°$, suspended in 0.02 M potassium phosphate buffer, pH 6.6, containing 0.2 M KCl (with use of a volume of buffer equal to about one-eighth the original volume of eluate I), and dialyzed against the same buffer for 2.5 hours (25 vol., changed once).

Step 5. The dialyzed enzyme (acetone B) was diluted with 1.6 vol. of potassium phosphate–KCl buffer, and solid ammonium sulfate (39 g. per 100 ml.) was added with stirring at 0°. The precipitate was removed by centrifugation and discarded. The enzyme was precipitated by adding further ammonium sulfate (6 g. per 100 ml. of original solution), collected by centrifugation, and resuspended in a small volume of potassium phosphate–KCl buffer (ammonium sulfate B).

Properties. GSH protects the enzyme against the effects of heating at 55° for 13 minutes (56% of the activity is retained). The optimum pH is around 7.6. The yeast methionine-activating enzyme, in contrast to the liver methionine-activating enzyme,[5] is not sensitive to fluoride ions even at concentrations up to 0.01 M. Whereas the liver enzyme shows a specific requirement for Mg^{++}, the yeast enzyme responds well when Mn^{++} or other divalent cations (less well, however) replace Mg^{++}. Liver- or yeast-activating enzyme show about the same concentration-activity relationship, $K_{methionine}$ being 2.6×10^{-3} M. Either enzyme reacts with methionine or ethionine, and only the liver enzyme is specific for the L-form of either methionine or ethionine, whereas the yeast enzyme reacts with either the L- or the D-form of methionine or ethionine (although not with the same efficiency, depending on the strain of yeast). The liver enzyme is inactive with homocysteine, whereas the yeast enzyme reacts with L-homocysteine at about 2% of the rate induced by L-methionine, the product being S-adenosyl-L-homocysteine. Neither the liver nor the yeast enzyme reacts with S-methyl or S-ethyl cysteine, and the N-acetylation of methionine or ethionine does not yield the corresponding sulfonium with either enzyme. The —NH$_2$ group of methionine can be replaced by —OH, and the yeast enzyme reacts with the hydroxy analog of methionine.

Synthesis of S-Adenosylmethionine or S-Adenosylethionine by Employing the Yeast-Activating Enzyme

The conditions are the same as those previously described for the mammalian enzyme.[2, 5, 6]

Enzymatic Synthesis of S-Adenosyl-L-homocysteine

S-Adenosyl-L-homocysteine is the product formed from S-adenosyl-L-methionine in the course of enzymatic synthesis of creatine from guanidinoacetic acid, N'-methyl nicotinamide from nicotinamide, etc.[7]

Preparation of Adenosine-Homocysteine Condensing Enzyme[8]

See Vol. V [10] for the preparation of the enzyme.

Properties. None of the following mercaptans could replace L-homocysteine as a substrate when used in the concentration of 0.05 M: D-homocysteine, L-cysteine, glutathione, coenzyme A, 3-mercaptoethylamine, 2-mercaptoethanol, and thioglycolic acid. The following nucleosides could not replace adenosine: inosine, 2-deoxyadenosine, guanosine, 2-deoxyguanosine, xanthosine, cytidine, 2-deoxycytidine, uridine, and thymidine. The 2'-, 3'-, and 5'-adenylic acids were also ineffective, as was ribose itself. In the presence of the purified enzyme, serine will not condense with homocysteine to yield cystathionine. No requirement for pyridoxal phosphate or any other cofactor could be demonstrated for the adenosine-homocysteine condensing enzyme.

Preparation of S-Adenosyl-L-homocysteine[8]

In a final volume of 100 ml. the following components were present: 1 millimole of adenosine, 4 millimoles of DL-homocysteine, 0.5 millimole of potassium phosphate buffer (pH 6.9), and 800 units of condensing enzyme (40 to 60% ethanol fraction). The reaction mixture was then incubated at 37°. The reaction was followed by measuring —SH groups until an amount equivalent to the amount of adenosine used had been utilized. For this purpose, a sample was withdrawn and deproteinized with metaphosphoric acid for a zero-time determination of —SH groups immediately after addition of enzyme, and at 60 and 120 minutes. After 120 minutes, the reaction was stopped with 5 ml. of 70% perchloric acid, and the mixture was cooled for approximately 1 hour in an ice bath. It

[7] G. L. Cantoni and E. Scarano, *J. Am. Chem. Soc.* **76**, 4744 (1954).
[8] G. de la Haba and G. L. Cantoni, *J. Biol. Chem.* **234**, 603 (1959); see Vol. V [102].

was then centrifuged, and to the supernatant fluid 100 ml. of 20%
phosphotungstic acid were added. After being stored in a cold room
overnight, the precipitate was centrifuged and then suspended in 50 ml.
of 0.1 N sulfuric acid. To decompose the phosphotungstate, this suspension was shaken at room temperature with a 1:1 mixture of n-butanol
and ether. It was then centrifuged, the organic layer was siphoned off
and discarded, the acid layer was decanted, and the remaining precipitate
was resuspended in 30 ml. of 0.1 N sulfuric acid and again treated as
above. After centrifugation and removal of the organic layer as before,
the acid layer was combined with the first fraction. After addition of
50 micromoles of potassium phosphate buffer (pH 6.5), the solution was
adjusted with 10 N KOH to pH 6.7 and lyophilized. The resulting dry
powder was dissolved in water to a final volume of 15 ml., and the sulfate
was removed by addition of barium iodide (0.5 M). The barium sulfate
precipitate was centrifuged, washed once with water, centrifuged, and the
wash solution was combined with the main fraction. Recovery of S-
adenosylhomocysteine was almost quantitative, with negligible loss due
to adsorption. The solution was again lyophilized, and the powder was
dissolved in water to a final concentration of about 0.05 M as judged by
the absorption at 260 mμ ($E = 16 \times 10^3$). Absolute ethanol was then
added to a concentration of 50%, and the mixture was placed in the
cold room for a few hours. It was then centrifuged, and the small gelatinous precipitate which contained only traces of ultraviolet-absorbing
material was discarded. Enough ethanol was added to the supernatant
fluid to reach 94% saturation, and the suspension was kept at 4° overnight. The resultant amorphous precipitate was centrifuged, washed
twice with cold absolute ethanol, and dried in vacuo. At this stage,
S-adenosylhomocysteine is about 70 to 80% pure on a weight basis.
Crystallization can be achieved by dissolving the powder in water to a
final concentration of 0.045 M, freezing the solution overnight at −20°,
and placing the solution in ice the next day. Crystallization begins after
a day or two. The product was collected by centrifugation, washed once
with a small volume of cold water and twice with cold absolute ethanol,
and then dried in vacuo at room temperature. The yield was about 70%.
After drying in vacuo at 100° for 5 hours, the product lost 4.63% in
weight. The compound has the correct molecular weight of 384.42 and
elementary composition of S-adenosylhomocysteine; the melting point
with decomposition is 202° (synthetic compound melted at 190° to
193°[9]). The picrate melted at 170°, corresponding to that of a synthetic
product.[9] S-Adenosylhomocysteine reacts in the modified McCarthy-

[9] J. Baddiley and G. A. Jamieson, J. Chem. Soc. 1085 (1955).

Sullivan reaction,[10] has an absorption peak at 500 mμ, and a molar extinction coefficient of 290.

Synthesis of S-Adenosyl-L-methionine from S-Adenosyl-L-homocysteine by Methylation[8]

The procedure is that of Baddiley and Jamieson.[9] The method is equally applicable for the synthesis of S-adenosyl-L-ethionine. Since another center of asymmetry around the sulfur atom arises in the course of methylation or ethylation, a mixture of two diastereoisomers of S-adenosylmethionine and S-adenosylethionine is produced by this procedure. Forty milligrams of S-adenosyl-L-homocysteine of 66% purity was dissolved in 2 to 3 ml. of formic acid, 0.15 ml. of methyl iodide (about 30-fold excess) was added, and the tube was sealed and placed in the dark for 5 days. This step is identical with that employed by Toennies and Kolb[11] in the methylation of methionine with methyl iodide to produce methionine methyl sulfonium. Some insoluble material which appeared was removed by filtration, the solvent was removed by lyophilization, and the residue was dissolved in 2 ml. of 0.1 N HCl and extracted twice with ether. Eight milliliters of a 1.5% solution of ammonium reineckate in 5% trichloroacetic acid were added to the aqueous fraction, and the mixture was placed in a cold room overnight. The precipitate was then centrifuged in the cold, washed twice with cold 0.75% ammonium reineckate in 2.5% trichloroacetic acid, and finally dissolved in about 10 ml. of methyl ethyl ketone. This solution was extracted twice with 5-ml. portions of 0.1 N sulfuric acid; the acid extracts were combined and extracted once with methyl ethyl ketone and once with ether, and then neutralized to pH 5.2 with NaOH.

Ionophoresis of 0.4 micromole of the product in 0.2 M acetate buffer of pH 5.2 on Whatman No. 1 paper strips with constant circulation of ice water through the chamber and with operation at 8 volts/cm. produced a migration of about 50% of the total ultraviolet-absorbing material toward the cathode with the same mobility as authentic S-adenosyl-L-methionine. The remainder of the product stayed at the origin. The areas containing the moving component were cut out and eluted overnight with 3 ml. of water. The solution was then assayed for S-adenosylmethionine as the methyl donor in the enzymatic synthesis of creatine. The synthetic product was about 50% as active as the natural product, owing to the racemic nature of the preparation around the sulfur atom. The resolution

[10] J. Durell, D. G. Anderson, and G. L. Cantoni, *Biochim. et Biophys. Acta* **26**, 270 (1957).

[11] G. Toennies and J. J. Kolb, *J. Am. Chem. Soc.* **67**, 849 (1945).

of this mixture of diastereoisomers of S-adenosylmethionine was accomplished by reacting the synthetic product with a methyl pherase and a suitable acceptor of the methyl group of the natural component of the mixture of the diastereoisomeric forms of S-adenosylmethionine, and isolation of the unreactive diastereoisomer from the mixture.[12]

Enzymatic Synthesis of S-Adenosyl-L-selenomethionine[13]

When synthetic selenomethionine is incubated with methionine-activating enzyme prepared from rabbit liver or yeast (under the conditions of incubation described for methionine), S-adenosyl-L-selenomethionine is formed which can be isolated and identified by the procedures described for S-adenosyl-L-methionine. The selenium analog proved to be available in the enzymatic synthesis of creatine by creatine methyl pherase from guanidinoacetic acid.

Decomposition Products of S-Adenosylmethionine and S-Adenosylethionine

5'-Methylthioadenosine and α-amino-γ-butyrolactone are formed from S-adenosylmethionine after a pH 4 solution of the compound is heated for 20 minutes at 100°.[14] The α-amino-γ-butyrolactone and 5'-ethylthioadenosine are formed from S-adenosylethionine under similar conditions. Under these conditions, S-ribosylmethionine yields only α-amino-γ-butyrolactone. The products formed from S-adenosylmethionine under acid hydrolysis were also found on metabolic breakdown of S-adenosylmethionine by cell-free preparations of $Aerobacter$[15] or yeast preparations.[16]

Preparation of S-Ribosylmethionine[14]

Hydrolysis of S-adenosylmethionine in 0.1 N NaOH in an ice bath for 5 hours, followed by removal of adenine with Ag_2SO_4, and removal of Ag with H_2S and of H_2SO_4 by Dowex 2–HCO_3, gives a solution of S-ribosylmethionine (and occasionally a few per cent of methionine and homoserine as well). With butanol–water–acetic acid (60:25:15) as the solvent, the R_f of ribosylmethionine is 0.1; with ethanol–water–acetic acid (63:34:1) as the solvent, the R_f is 0.35. The compound proved refractory to isolation in the dry or crystallized state.

[12] G. de la Haba, G. A. Jamieson, S. Harvey Mudd, and H. H. Richards, $J. Am. Chem. Soc.$ **81**, 3975 (1959).
[13] S. Harvey Mudd and G. L. Cantoni, $Nature$ **180**, 1052 (1957).
[14] L. W. Parks and F. Schlenk, $Arch. Biochem. Biophys.$ **75**, 291 (1958).
[15] S. K. Shapiro and A. N. Mather, $J. Biol. Chem.$ **233**, 631 (1958).
[16] S. Harvey Mudd, $J. Biol. Chem.$ **234**, 87 (1959).

Preparation of Methylthioribose[14]

Hydrolysis of S-ribosylmethionine at pH 5.3 for 30 minutes at 100° yields methylthioribose and homoserine in practically quantitative amounts.

Chemical Synthesis of S-Adenosylhomocysteine[9]

Metallic sodium (0.46 g.) is added to a solution of 1.36 g. of homo-cysteine in 50 ml. of liquid ammonia, and then a solution of freshly prepared 2′,3′-O-isopropylidene-5′-O-toluene-p-sulfoadenosine[17] in 50 ml. of liquid ammonia is added. Moisture is excluded by a sodium hydroxide tube, and solvent is allowed to evaporate. The residue is dissolved in water, and unchanged homocysteine (0.1 g.) is removed by filtration. The filtrate is passed through a column of Amberlite IR-120 (ammonium form) to remove sodium ions, and the eluate is concentrated *in vacuo* and passed through a column of Amberlite IR-4B (hydroxyl form) to remove toluene-p-sulfonate ions. The eluate from the second column is evaporated to dryness *in vacuo*, leaving a solid residue of S-adenosyl-homocysteine, R_f 0.3 in butanol–acetic–water, and R_f 0.61 in n-propanol–ammonia–water. The isopropylidene residue is removed by hydrolysis in 16 ml. of N H_2SO_4 at room temperature for 36 hours. The sulfate ions are removed by calculated amounts of $Ba(OH)_2$, and $BaSO_4$ is removed by filtration through Supercel silica and washed with water. The com-bined filtrate and washings are evaporated to dryness *in vacuo*, and the residue is recrystallized from aqueous alcohol. The yield is 520 mg., small prisms, m.p. 204°, after several recrystallizations.

[17] P. A. Levene and R. S. Tipson, *J. Biol. Chem.* **106**, 113 (1934), prepared from 4.6 g. of 2′,3′-O-isopropylidene adenosine.

[84] Intermediates in Histidine Synthesis: The Formation of Imidazole Glycerolphosphate

By H. S. MOYED

Imidazole glycerolphosphate (IGP) has been shown by Ames and Mitchell[1] to be an early precursor of both the imidazole ring and the side chain of histidine. The synthesis of this histidine precursor has been observed in extracts of several bacteria.[2] The reaction requires ATP,

[1] B. N. Ames and H. K. Mitchell, *J. Biol. Chem.* **212**, 687 (1955).
[2] H. S. Moyed and B. Magasanik, *J. Biol. Chem.* **235**, 149 (1960).

ribose-5-P, glutamine or ammonium salts, and $MgCl_2$. Acetyl-P and glutathione are stimulatory. In addition to IGP, 4-amino-5-imidazole-carboxamide ribonucleotide (AICAR) is found as a reaction product. In the absence of an amino donor an intermediate, "compound III," accumulates which yields AICAR on mild acid hydrolysis. These observations and incorporation studies have led to the suggestion that the adenine ring of ATP is cleaved into two moieties by way of the intermediate, "compound III." The N-1, C-2 portion of the adenine ring combines with ribose-5-P, and with the introduction of an amino group this moiety cyclizes to form IGP. The remainder of the adenine ring appears as AICAR.[2]

This article describes methods for (1) routine measurement of the enzymatic synthesis of IGP, AICAR, and the intermediate, "compound III"; and (2) separation of the enzymes necessary for the synthesis of "compound III" from the enzymes which convert "compound III" to AICAR and IGP.

Assay Method

Principle. IGP is estimated by its reaction with diazosulfanilic acid by using the method of Ames[3] (see Ames and Mitchell[1]). AICAR is estimated by the test of Bratton and Marshall[4] for diazotizable amines. The identity of these products had been previously established by analyses of the isolated compounds.[2] "Compound III" is estimated by the increase in AICAR after mild acid hydrolysis of samples of the reaction mixture.

Reagents

Tris buffer, 1 M, pH 8.0.
$MgCl_2$, M.
ATP, 0.05 M, pH 7.0 to 7.5.
Ribose-5-P, 0.10 M, pH 7.0 to 7.5.
Reduced glutathione, 0.075 M, pH 7.0 to 7.5.
Acetyl-P, 0.20 M, pH 6.8 to 7.2.
Glutamine, 0.10 M.

Procedure. Reagents were added in the following order: 0.05 ml. of Tris buffer, 0.01 ml. of $MgCl_2$, 0.02 ml. of ATP, 0.02 ml. of ribose-5-P, 0.10 ml. of glutathione, 0.05 ml. of acetyl-P, 0.10 ml. of glutamine, and enzyme(s). The final volume was 0.6 ml. ATP was omitted from the control. After 15 minutes of incubation at 37°, two aliquots of the reac-

[3] B. N. Ames and H. K. Mitchell, *J. Am. Chem. Soc.* **74**, 252 (1952).
[4] A. C. Bratton and E. K. Marshall, Jr., *J. Biol. Chem.* **128**, 537 (1939).

tion mixture were made $0.2\,N$ in HCl. One aliquot was used for the estimation of AICAR synthesis by determining the diazotizable amine present.[4] "Compound III" synthesis was estimated by determining the increase in diazotizable amine after heating the other acidified aliquot for 5 minutes at 100°. 4-Amino-5-imidazolecarboxamide was used as a standard in the test for diazotizable amines. A third aliquot of the reaction mixture was deproteinized by adding an equal volume of 10% trichloroacetic acid prior to the determination of IGP by its reaction with diazosulfanilic acid (see Ames and Mitchell[1]). The amounts of AICAR and IGP formed are proportional to enzyme concentration provided that no more than 0.10 micromole of each are produced. In the absence of an amino donor, which can be either glutamine or an ammonium salt, considerably reduced amounts of AICAR and only traces of IGP are formed; however, there is a corresponding increased formation of "compound III" (Table I).

TABLE I

EFFECT OF GLUTAMINE ON IGP SYNTHESIS[a]

Glutamine added[b]	IGP	AICAR, micromoles/20 min.	"Compound III"
1. 20 micromoles	0.05	0.10	0.04
2. None	0.01	0.05	0.09
Difference (1 − 2)	+0.04	+0.05	−0.05

[a] The enzyme preparation was an extract of *Salmonella typhimurium*, strain hi B-12, which had been partially freed of nucleic acids (step 2 in the fractionation procedure).

[b] Ammonium sulfate also stimulates the formation of IGP and AICAR but is much less effective than glutamine; however, glutamine and ammonium sulfate serve equally well for the conversion of preformed "compound III" to AICAR and IGP.

Separation of Enzymes. Strains of *Escherichia coli* were grown with aeration in basal medium containing, per liter, 18.9 g. of $Na_2HPO_4 \cdot 7H_2O$, 6.3 g. of KH_2PO_4, 0.2 g. of $MgSO_4 \cdot 7H_2O$, 0.01 g. of $CaCl_2$, 2 g. of $(NH_4)_2SO_4$, and 2 g. of glucose. The bacteria have been grown under conditions of histidine deficiency in order to relieve the repressive effect of this amino acid on the formation of enzymes required for its own biosynthesis. Cells grown in this manner are found to contain from ten to twenty times the normal amounts of the enzymes necessary both for the synthesis of "compound III" and for its conversion to AICAR and IGP. Histidine deficiency during growth has been created by forcing a histidine auxotroph which excretes histidinol, *E. coli* strain H-4, to obtain the amino acid at a low rate from β-alanylhistidine. For this purpose the basal medium was supplemented with 15 mg. of β-alanylhistidine per liter.

A histidine deficiency can be induced in the wild-type parent organism, *E. coli* W, by growth in the presence of a partially inhibitory level of 2-thiazolealanine. This analog, a false feedback inhibitor, mimics the normal inhibitory effect of histidine on the action of an enzyme necessary for the synthesis of "compound III" and thus reduces the rate of histidine biosynthesis.[5] Cells grown in the basal medium supplemented with 15 mg. of 2-thiazolealanine contain ten times as much of the early enzymes of histidine synthesis as when grown in the unsupplemented medium.[6]

The following procedure was employed for the partial purification of the enzymes necessary for the synthesis of "compound III": *Step 1.* Extracts were prepared by treating suspensions of *E. coli* W grown in the presence of 2-thiazolealanine (4 g. wet weight of cells in 25 ml. of phosphate buffer at pH 7.4) for 5 minutes in a 10-kc. magnetostrictive oscillator (Raytheon). The extracts were clarified by centrifugation at 20,000 × *g* for 15 minutes. *Step 2.* Inactive protein and nucleic acids were precipitated by the addition of 12 ml. of a 2% solution of protamine sulfate per 100 ml. of extract. *Step 3.* The further addition of 4 ml. of

TABLE II

PARTIAL PURIFICATION OF THE ENZYMES REQUIRED
FOR "COMPOUND III" SYNTHESIS

Step	Volume, ml.	Protein[a]	Units[b]	S.A.[c]
1. Extraction	90	1032	15.4	0.015
2. Removal of nucleic acids and inactive proteins	91	455	13.7	0.030
3. Precipitation of "compound III"-synthesizing enzymes	7.5	84	12.6	0.150

[a] Protein was determined by the Biuret method (Vol. III [73]).

[b] One unit is equivalent to the formation of 1 micromole of "compound III" per minute.

[c] S.A. = specific activity (units per milligram of protein).

2% protamine sulfate resulted in the precipitation of the enzymes necessary for the synthesis of "compound III." The precipitate was dissolved in 10 ml. of *M* KCl containing 0.05 *M* reduced glutathione (Table II). The enzyme(s) necessary for the conversion of "compound III" to

[5] H. S. Moyed and M. Friedman, *Science* **129**, 968 (1959).
[6] H. S. Moyed, *Science* **131**, 1449 (1960).

AICAR and IGP are not precipitated by protamine and remain in the supernatant fluid.

Comments. The partially purified preparation can utilize either ribose-5-P or 5-phosphoribosyl-1-pyrophosphate (PRPP) for the synthesis of "compound III." In either case ATP is necessary. The preparation at this stage contains the enzyme which catalyzes the reaction ATP + ribose-5-P ⇌ PRPP + AMP. A "compound III"-synthesizing preparation can be obtained free of this enzyme by several stepwise precipitations with small amounts of protamine sulfate rather than by two precipitations as described above. This separation of "compound III"-synthesizing activity from PRPP kinase cannot always be accomplished with the same amounts of protamine sulfate, so that it is necessary to determine the correct amount of protamine sulfate for each extract by performing a pilot experiment. Preparations freed in this manner of PRPP kinase cannot utilize ribose-5-P for "compound III" synthesis but instead require PRPP. The requirement for an adenine nucleotide is met only by ATP. ADP and AMP both are inactive.

[85] Intermediates in Histidine Breakdown

Urocanic Acid, Formimino-L-glutamic Acid, and Formyl-L-glutamic Acid

By HERBERT TABOR

I. Urocanic Acid[1]

$$\begin{array}{c} \diagup^{\text{CH}}\diagdown \\ \text{N} \qquad \text{NH} \\ | \qquad\qquad | \\ \text{HC} =\!\!=\!\!= \text{CH}-\text{CH}=\text{CHCOOH} \end{array}$$

Urocanic acid can be determined by its absorption in the ultraviolet region ($\epsilon_{max} = 1.88 \times 10^4$ at 277 mμ at pH 7.4; $\epsilon_{max} = 2.00 \times 10^4$ at 306 mμ in 1 N KOH); or colorimetrically after coupling with diazotized 4-nitroaniline.[1] With most biological materials, such as urine, however, numerous interfering materials are present which have to be eliminated by a preliminary purification. The procedure listed here involves a preliminary rapid purification with Dowex 1–acetate; if needed, further specificity is obtained by the use of urocanase.

[1] The preparation, spectra, and chromatography of urocanic acid have been presented in Vol. III [90].

Materials

0.01 N acetic acid.

0.06 N acetic acid.

Dowex 1–acetate (200 to 400 mesh, 8 or 10% crosslinked). Commercial Dowex 1–chloride is washed in a column with saturated sodium acetate until essentially chloride-free. The resin is then washed thoroughly with water and suspended in an equal volume of water.

Chromatographic column. Each column consists of a glass tube, 0.7 × 8 cm., slightly narrowed at the lower end. A larger glass tube, 2.3 × 11 cm., is sealed to the upper end as a reservoir. A small amount of glass wool is inserted into the bottom of each column, and the column is filled with 3.3 ml. of the Dowex 1 suspension (0.6 × 5.5 cm. after packing). During the chromatographic procedure a relatively rapid flow rate is obtained by the use of air pressure (2 pounds). With an appropriate manifold ten columns can be used simultaneously.

Urocanase. A convenient preparation for this purpose is a twelve-fold purified enzyme obtained from hog liver.[2]

---

[2] An unpublished procedure of R. Yankee, H. Tabor, and V. Childs is used for the preparation of liver urocanase. The assay method has been presented in Vol. II [29]. One unit of enzyme is defined as that amount of enzyme that causes a decrease in the optical density at 277 mμ of 0.001 per min. at 25°; the entire procedure can be carried out in a 5-hour period.

One hundred grams of commercial frozen hog liver is thawed and homogenized with 300 ml. of water for 1.5 minutes in a Waring blendor. An additional 100 ml. of water and 50 g. of bentonite are added while the blendor is still running. The mixture is centrifuged for 20 minutes at 20,000 × g in an angle centrifuge. The supernatant is collected and subjected to DEAE chromatography.

Diethylaminoethyl cellulose (DEAE; 100 to 200 mesh) is washed successively with disodium ethylenediaminetetraacetate, 0.5 N NaOH, and water [E. A. Peterson and H. A. Sober, *J. Am. Chem. Soc.* **78**, 751 (1956); Vol. V [1], p. 3). The DEAE suspension is then adjusted to pH 7.2 with acetic acid and tris(hydroxymethyl)aminomethane (Tris), packed in a column (3 × 11 cm.), and thoroughly washed with 0.01 M Tris acetate, pH 7.2.

The above supernatant (300 ml.) is mixed with 3 ml. of 1 M Tris acetate, pH 7.2, and passed through the DEAE column. The column is washed with 200 ml. of a solution of 0.12 M NaCl in 0.01 M Tris acetate, pH 7.0. The urocanase is then eluted with 100 ml. of a solution of 0.36 M NaCl in 0.01 M Tris acetate, pH 7.0. The most active fractions (70 to 200 units/ml.; 25 units/mg. of protein) are pooled; this preparation is stable for >1 months when stored at −15°. The chromatographic procedures are all carried out rapidly with the aid of air pressure (2 pounds).

Further purification can be attained by the use of steps involving calcium phosphate and alumina C$_\gamma$ gel adsorption and elution, but this is not necessary for

Procedure. One milliliter of urine[3,4] is placed in a conical centrifuge tube and thoroughly mixed with 5 ml. of absolute ethanol. After >10 minutes at room temperature the mixture is centrifuged for 10 minutes in a clinical-type centrifuge. The supernatant is decanted, and alkalinized (pH 7.5 to 9.5) with NH_4OH (usually 0.1 to 0.2 ml. of $0.4 N$ NH_4OH). The resultant mixture is passed through the Dowex 1–acetate column. The column is then washed with 10 ml. of water and two 12-ml. aliquots of 0.01 N acetic acid; the washings are discarded. The urocanic acid is then eluted with two 5-ml. portions of 0.06 N acetic acid, and the eluates are pooled.

In most instances this material is sufficiently purified for direct spectrophotometric analyses. (1) To 1 ml. of the combined eluates 0.2 ml. of M K_2HPO_4 is added, and the optical density read at 277 mμ. Under these conditions a reading of 0.157 represents a urocanic acid concentration of 0.01 micromole/ml. in the eluant, or of 0.1 micromole/ml. in the original solution. If the optical density reading is >1.5, the mixture should be diluted with $0.2 M$ K_2HPO_4. (2) Alternatively, 0.2 ml. of 6 N KOH is added to 1 ml. of the eluant, and the optical density is read at 306 mμ. Under these conditions a reading of 0.167 represents a urocanic acid concentration of 0.01 micromole/ml. in the eluant, or of 0.1 micromole/ml. in the original solution. If the optical density reading is >1.5, the mixture should be diluted with 1 N KOH. If the optical density readings are very low, the sensitivity of the determination can be increased by evaporating the combined eluates on the steam bath in an evaporating dish and redissolving the residue in 1.5 ml. of water prior to the spectrophotometric measurements.

To increase the specificity of the method, aliquots of the eluate can

use in the above assay procedure. Purification procedures for urocanase from liver have also been reported by R. H. Feinberg and D. M. Greenberg, *J. Biol. Chem.* **234**, 2670 (1959); M. Takeuchi, *J. Biochem. (Japan)* **34**, 1 (1941); Y. Sera and D. Aihara, *J. Osaka Med. Soc.* **41**, 745 (1942); and D. D. Brown and M. W. Kies, *J. Biol. Chem.* **234**, 3182 (1959). Extracts of histidine-adapted *Pseudomonas* cells [H. Tabor and O. Hayaishi, *J. Biol. Chem.* **194**, 171 (1952)] can also be used for this purpose, particularly after treatment with protamine sulfate. These extracts have a considerable higher activity than the liver preparations (approximately 2800 units/ml.; 140 units/mg. of protein).

[3] The procedure presented here has been satisfactory for a number of samples of normal human urine. Further studies are necessary to test the general applicability of this method.

[4] This procedure can also be used for the determination of urocanic acid in other solutions. For the determination of urocanic acid in tissues a perchloric acid extract is used. This is neutralized with KOH, and the potassium perchlorate is removed before the supernatant is applied to the column.

be treated with urocanase, and the decrease in optical density measured. The incubation mixture contains 1 ml. of eluate, 0.06 ml. of M K_2HPO_4, and 30 λ of urocanase (5 units).[5] After 1 hour at 37° 0.2 ml. of 6 N KOH is added, and the optical density is determined at 306 mμ. A control mixture is also prepared in which the 6 N KOH is added at zero time. The amount of urocanic acid is calculated from the decrease in optical density in the incubated sample. Further support for the presence of urocanic acid can be obtained by comparing the spectra of the incubated and the nonincubated samples; the difference spectra should be similar to that of urocanic acid.

II. Formimino-L-glutamic Acid[6-9]

$$
\begin{array}{c}
NH \\
\parallel \\
CH \\
| \\
NH \\
| \\
HOOC-CH_2-CH_2-CH-COOH
\end{array}
$$

Preparation

Principle. L-Glutamic acid is treated with formamidine hydrochloride and silver carbonate in formamide. The product, formiminoglutamic acid, is purified by adsorption onto Dowex 1–acetate, elution with acetic acid, and crystallization as the barium salt.

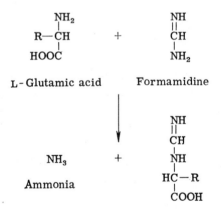

Formiminoglutamic acid

---

[5] Under these conditions this amount of enzyme will degrade 0.15 to 0.2 micromole of urocanic acid in 1 hour. If the eluate contains more than 0.1 micromole/ml., the enzymatic incubation should be carried out with a smaller aliquot.

[6] H. Tabor and J. C. Rabinowitz, *Biochem. Preparations* **5**, 100 (1957). This refer-

Reagents

L-Glutamic acid.
Ag_2CO_3.
Formamidine hydrochloride.[10]
Formamide.

Procedure.[6] L-Glutamic acid (1.47 g., 10 millimoles), Ag_2CO_3 (3.0 g., 11 millimoles), and formamidine hydrochloride (1.6 g., 20 millimoles) are suspended in 5 ml. of redistilled formamide in a 50-ml. two-neck round-bottom flask. This is fitted with an inlet tube for helium, and an outlet tube containing $CaCl_2$. The mixture is incubated at room temperature for 36 hours with vigorous magnetic stirring; during this period helium is slowly passed through the vessel to remove the NH_3 formed. The flask contents are then cooled to 0°, cautiously treated with 100 ml. of 0.06 N HCl, filtered, and placed under reduced pressure for 30 minutes to remove dissolved CO_2. The solution is then passed through a Dowex 1–acetate column (2.4 × 18 cm.; 200 to 400 mesh; 8 to 10% crosslinked) and eluted by gradient elution. The reservoir vessel contains 0.3 N acetic acid, and the mixing vessel contains 250 ml. of water. The fractions containing most of the formiminoglutamic acid[11] are pooled and evaporated to dryness over KOH. The glassy residue is dissolved in water (10 to 20 ml.), adjusted to pH 7 to 8 with saturated barium

---

ence also contains directions for the synthesis of formiminoaspartic acid and formiminoglycine (cf. also Vol. VI [97]).
[7] H. Tabor and L. Wyngarden, *J. Clin. Invest.* **37**, 824 (1958); *J. Biol. Chem.* **234**, 1830 (1959).
[8] H. Tabor and A. H. Mehler, *J. Biol. Chem.* **210**, 559 (1954).
[9] Formiminoglutamic acid has also been prepared: (*a*) by the reaction of formamidine with benzylglutamic acid and reduction of the product [A. Miller and H. Waelsch, *J. Am. Chem. Soc.* **76**, 6194 (1954)]; (*b*) by the reaction of formimino ether with glutamic acid [J. E. Seegmiller, M. Silverman, H. Tabor, and A. H. Mehler, *J. Am. Chem. Soc.* **76**, 6205 (1954)]; (*c*) by isolation from the urine of folic acid-deficient rats [M. Silverman, R. C. Gardiner, and H. A. Bakerman, *J. Biol. Chem.* **194**, 815 (1952)]; and (*d*) from the enzymatic degradation products of histidine and urocanic acid [B. Borek and H. Waelsch, *J. Biol. Chem.* **205**, 459 (1953); H. Tabor and A. H. Mehler, ref. 8].
[10] Formamidine hydrochloride can be purchased commercially. Radioactive formamidine can be prepared by the reductive desulfurization of C^{14}-thiourea with Raney nickel (cf. ref. 6).
[11] The formiminoglutamic acid and glutamic acid content of each fraction can be determined qualitatively by the paper chromatographic procedure described below. Glutamic acid is usually eluted from the Dowex 1 column immediately after the formiminoglutamic acid; any of the formiminoglutamic acid fractions that contain substantial quantities of glutamic acid are discarded.

hydroxide, treated with ethanol until turbid (approximately 40 ml.), and cooled. The crystals are collected by filtration, washed with cold 50% ethanol, and dried *in vacuo*. Additional material is obtained by the addition of ethanol (about 100 ml.) to the mother liquor. The yield is 900 to 1300 mg. (3.5 to 5 millimoles) of $Ba_{1/2}$ $C_6H_9N_2O_4 \cdot H_2O$. The barium salt[12] can be stored over silica gel at room temperature for long periods (>12 months) without deterioration.

Properties

Formiminoglutamic acid is labile to alkali and produces NH_3, HCOOH, and glutamic acid on hydrolysis. It has a half-life at 25° of approximately 10 minutes in $1 N$ KOH, 20 minutes at pH 11.5, 70 minutes at pH 10.8, 40 hours at pH 9.2. It is stable for long periods (at least several months) when stored under sterile conditions at pH 7.2 and pH 4, and in $0.1 N$ acetic acid, $0.1 N$ HCl, and $6 N$ HCl.

Upon titration in 0.05 to 0.1 M solutions the pK'_1, pK'_2, and pK'_3 are 2.0 to 2.7, 4.0 to 4.7, and 11.1 to 11.9. The optical rotation for barium formimino-L-glutamate is $[\alpha]_D^{20°} = -9.8°$ (2.5% in H_2O); for formimino-L-glutamic acid[9a, 9d] $[\alpha]_D^{28°} = -10.3°$ (0.8% in $1 N$ HCl).

Analysis

Paper Chromatography. A convenient method for the detection of formiminoglutamic acid involves descending chromatography with Whatman No. 1 paper in *t*-butanol (70), formic acid (15), H_2O (15). The R_f for formiminoglutamic acid is 0.47; for glutamic acid the R_f is 0.39.

Formiminoglutamic acid only gives a faint spot with ninhydrin[13] even when heated at 110° for 1 hour. To detect formiminoglutamic acid, therefore, the papers are placed in an atmosphere of NH_4OH for 2 hours. This serves to hydrolyze the formiminoglutamic acid. The NH_4OH is

[12] The barium salt is easily soluble in water and is readily converted to the sodium salt by the addition of a stoichiometric amount of sodium sulfate.

[13] Although formiminoglutamic acid gives only a faint spot with ninhydrin when tested in this manner, it reacts quantitatively when heated in the quantitative reduced ninhydrin procedure of S. Moore and W. H. Stein (Vol. III [76], p. 468). Although the reaction is delayed, essentially full color development (compared to glutamic acid) is observed at the end of the 20-minute heating period.

Formiminoglutamic acid also gives a positive reaction when heated in the reduced ninhydrin method of W. Troll and R. K. Cannan, *J. Biol. Chem.* **200**, 803 (1953). However, only a faint reaction ($<3\%$ of the corresponding glutamic acid color) is found when this method is modified by being carried out at room temperature.

then removed by aeration for an hour, the ninhydrin spray is applied, and the color is allowed to develop at room temperature. The color obtained with formiminoglutamic acid spots is markedly intensified by this treatment; the NH_4OH treatment, on the other hand, does not affect the color of a glutamic acid spot. Therefore, we have found it useful to carry out the paper chromatography in duplicate, and to compare the intensity of the spots obtained with and without the NH_4OH treatment.

Formiminoglutamic acid can also be visualized by the Cl_2–starch-iodine procedure of Rydon and Smith.[8, 14]

Column Chromatography. Formiminoglutamic acid can be chromatographed on Dowex 1–acetate, as described above in the "procedure" section. Formiminoglutamic acid can also be chromatographed on Dowex 50[8, 9c] (H form) and eluted with either HCl or H_2SO_4. The HCl can be removed by evaporation in a vacuum desiccator (over KOH) without any destruction of the formiminoglutamic acid.

Quantitative Assays

Enzymatic Assay.[7, 15] The quantitative assay of formiminoglutamic acid can be carried out with the enzyme solution described in Vol. V [105]. The formiminoglutamic acid-containing solution is incubated with 0.25 ml. of the enzyme solution (preparation 3 of Vol. V [105]), 0.1 ml. of 0.007 M sodium dl-tetrahydrofolate in 1 M mercaptoethanol, 0.1 ml. of 1 M potassium phosphate buffer (pH 7.2), and water in a final volume of 1 ml. in a narrow 3-ml. test tube at 25° for 30 minutes. At the end of the incubation period 0.3 ml. of 10% perchloric acid is added. The

[14] H. N. Rydon and P. W. G. Smith, *Nature* **169**, 922 (1952).
[15] Another method for the determination of formiminoglutamic acid has been reported by M. Silverman, R. C. Gardiner, and P. T. Condit [*J. Natl. Cancer Inst.* **20**, 71 (1958)]. In this procedure formiminoglutamic acid is incubated with folic acid, TPN, and a dialyzed chicken liver extract. The final product is converted to 5-formyltetrahydrofolic acid by autoclaving at pH 6, and this is assayed microbiologically. A modification of this method, using a spectrophotometric assay for the final determination, has recently been reported by I. Chanarin and M. C. Bennett, *Brit. Med. J.*, p. 27 (1962).

Methods using paper electrophoretic separation of the formiminoglutamic acid have been reported by J. P. Knowles, T. A. J. Prankerd, and R. G. Westall, *Lancet* **II**, 347 (1960), and by J. Kohn, D. L. Mollin, and L. M. Rosenbach, *Lancet* **I**, 112 (1961), *J. Clin. Pathol.* **14**, 345 (1961). Formiminoglutamic acid has also been determined by microbiological determinations of free and bound glutamic acid [M. Silverman, R. C. Gardiner, and H. A. Bakerman, *J. Biol. Chem.* **194**, 815 (1952)].

mixture is heated for 55 seconds in a boiling-water bath and cooled immediately in ice water. After centrifugation the optical density (O.D.$_1$) of the supernatant is read at 350 mμ against a control solution, by using semimicrocuvettes (1.2-ml. volume) with a 1-cm. light path. This control solution is prepared in the same way except for the use of a 1 M mercaptoethanol solution instead of the tetrahydrofolate–mercaptoethanol mixture.

Since the above procedure does not correct for the small blank absorption due to the tetrahydrofolate solution, a correction value is obtained daily by incubating the buffer, enzyme, and tetrahydrofolate-mercaptoethanol solutions as above; after the addition of perchloric acid the optical density (O.D.$_2$) is read against a similar mixture containing mercaptoethanol solution instead of the tetrahydrofolate–mercaptoethanol mixture. This reading should be <0.1; if it exceeds this value, the tetrahydrofolate solution is discarded and replaced with a fresh dilution.

The quantity of formiminoglutamic acid in micromoles in the aliquot used is calculated by multiplying the corrected optical density reading (O.D.$_1$ minus O.D.$_2$) by 0.052.[16] The sensitivity of the procedure is approximately 0.001 micromole.

The specificity of the reaction can be checked by repeating the assay on a sample of urine that had been hydrolyzed at an alkaline pH, since this treatment destroys formiminoglutamic acid.[7]

Colorimetric Assay. This assay is less sensitive than the enzymatic assay and is essentially the same procedure as that developed by Rabinowitz and Pricer for formiminoglycine determinations.[17]

The formiminoglutamic acid solution (in 0.5-ml. volume) is mixed with 2 ml. of saturated sodium borate and 0.5 ml. of the ferricyanide-nitroprusside reagent (4 g. each of NaOH, potassium ferricyanide, and sodium nitroprusside in 120 ml. of water). The optical density at 485 mμ is read after 30 minutes at room temperature against a corresponding blank. The blank is prepared by preincubating 0.3 ml. of urine with 0.1 ml. of 2.5 N KOH for 2 hours at 25° to destroy the formiminoglutamic acid present; the mixture is then neutralized with 0.1 ml. of 2.5 N HCl, and treated with the above reagents.

Under the above conditions an optical density of 0.60 is obtained with 1 micromole of formiminoglutamic acid.

---

[16] If the optical density reading at 350 mμ is too high, readings can be made at 365 mμ or at 380 mμ. The corresponding factors are 0.059 and 0.088, respectively.

[17] J. C. Rabinowitz and W. Pricer, Jr., *J. Biol. Chem.* **222**, 537 (1956). See also Vol. VI [97].

III. Formyl-L-glutamic Acid[8, 18]

$$\underset{\underset{\underset{COOH-CH-CH_2CH_2COOH}{|}}{NH}}{CHO}$$

Preparation

Principle. Formyl-L-glutamic acid is synthesized by treating L-glutamic acid with a mixture of formic acid and acetic anhydride.

Reagents

L-Glutamic acid.
Formic acid (98%).
Acetic anhydride.

Procedure. Sixty grams of L-glutamic acid is placed in a round-bottom flask with 900 ml. of formic acid; 300 ml. of acetic anhydride is then added slowly. The temperature of the mixture rises to >50° owing to the exothermic reaction.

After the mixture begins to cool (at least 30 minutes after mixing), the solvents are removed by vacuum distillation. The residual oil is dissolved in a minimum volume of absolute ethanol, and any insoluble solid (free glutamic acid) is removed by filtration. Benzene (10 to 15 vol.) is then added to the ethanol. The solution is cooled to 5°, and rosettes of large fragile crystals slowly form. These are collected by filtration (52 g.) and recrystallized from ethanol–benzene as above to yield 44 g., melting point 112° to 113°.

Formylglutamic acid must be stored in an anhydrous atmosphere, as otherwise it is slowly hydrolyzed. When stored over a drying agent, such as silica gel, at room temperature, it is stable for long periods (>12 months).

Properties

Formylglutamic acid is more stable than formiminoglutamic acid in alkali but is relatively labile in acid. At room temperature (19° to 26°) formylglutamic acid has a half-life (as measured by the appearance of ninhydrin-reactive material) of approximately 24 hours in 1 N HCl, 5 days in 0.1 N HCl, 35 days in 0.01 N HCl, 35 days in 1 N acetic acid, and 3 days in 0.54 N KOH; no ninhydrin-reactive material is liberated in 10 days at pH 7.2.

[18] A comparable procedure has also been used for the synthesis of formylaspartic acid [O. Hayaishi, H. Tabor, and T. Hayaishi, *J. Biol. Chem.* **227**, 161 (1957)].

The optical rotation of formyl-L-glutamic acid is $[\alpha]_D^{20} = -7.25°$ (2% solution in water).

Analysis

Paper Chromatography. The R_f of formylglutamic acid in *t*-butanol (70), formic acid (15), H_2O (15) (Whatman No. 1 paper, descending chromatography) is 0.74. The formylglutamic acid spot cannot be visualized directly with a ninhydrin spray. To hydrolyze the formylglutamic acid the papers are first placed in an atmosphere of HCl for 2 hours; this is conveniently carried out by placing the paper in a closed glass cylinder containing a beaker of concentrated hydrochloric acid. The HCl is then largely removed by aeration for several hours, and the papers are sprayed with a 0.2% ninhydrin solution in *n*-propanol, containing 0.5% pyridine. Formylglutamic acid can also be visualized by the Cl_2–starch–iodine procedure of Rydon and Smith.[14]

Column Chromatography. Formylglutamic acid (adjusted to pH 7) is adsorbed on Dowex 1–acetate and can be eluted with acetic acid.

Quantitative Analysis. The quantitative analysis of formylglutamic acid can be carried out by the reduced ninhydrin method of Moore and Stein[19] after hydrolysis of the formylglutamic acid in 1 *N* HCl at 100° for 10 minutes.

[19] Vol. III [76], p. 468. Formylglutamic acid does not give a positive reaction in the ninhydrin procedure without prior hydrolysis.

[86] Intermediates in the Biosynthesis of Tryptophan

Anthranilic acid + PP-ribose-P → Anthranilic deoxyribulotide →

By OLIVER H. SMITH and CHARLES YANOFSKY

The reaction sequence indicated above has been shown to be the mechanism of tryptophan biosynthesis in several species of enteric bacteria and in *Neurospora crassa*. Shikimic acid[1] and shikimic acid-5-phosphate[2] are known precursors of anthranilic acid, but the mechanism of

[1] B. D. Davis, *J. Biol. Chem.* 191, 315 (1951).
[2] P. R. Srinivasan, *J. Am. Chem. Soc.* 81, 1772 (1959).

their conversion to this compound has not yet been elucidated. Indole is accumulated in culture filtrates of some tryptophan auxotrophs and supports growth of other tryptophan auxotrophs, but enzymatic experiments indicate that it does not participate in tryptophan biosynthesis as a free intermediate.

Commercial preparations of anthranilic acid, indole, and tryptophan are generally of sufficient purity to serve as substrates or standards in enzymatic assays. The routine assays for these compounds have been described in a previous volume (see Vol. V [107]) and will not be further considered here.

To alleviate the necessity for extensive fractionation to separate the different enzymes involved in tryptophan formation, bacterial mutants which lack one of the various enzymes are employed as a source of the other enzymes. Thus a mutant which accumulates indoleglycerol in its culture filtrates lacks the enzymatic activity necessary to convert indole-3-glycerolphosphate (InGP) to tryptophan. An extract of this mutant can then be used to study the enzymatic synthesis of InGP from anthranilic acid, since it is unable to metabolize the InGP to indole or tryptophan. Similarly, extracts of a mutant lacking the enzyme which converts anthranilic deoxyribulotide to InGP can be used for the enzymatic synthesis of anthranilic deoxyribulotide from anthranilic acid. The characteristics of these different mutant types are described in Vol. V.

I. Anthranilic Deoxyribulotide

Preparation

Principle. Anthranilic acid and 5-phosphoribosyl 1-pyrophosphate (PP-ribose-P) react with the splitting out of pyrophosphate to form N-(o-carboxyphenylamino)-1-deoxyribulose-5-phosphate (anthranilic deoxyribulotide). The reaction appears to involve a dehydration and Amadori-type rearrangement.[3, 4]

Enzymatic Synthesis. The enzymatic synthesis of anthranilic deoxyribulotide is readily accomplished by using extracts of *Escherichia coli* or *Salmonella typhimurium* tryptophan auxotrophs grown under conditions of low tryptophan supplementation. These extracts contain considerable amounts of 5-phosphoribose pyrophosphokinase; thus ATP and ribose-5-P can serve as substrates for PP-ribose-P synthesis.

Preparation of Enzyme Extract. Since the enzyme responsible for the synthesis of anthranilic deoxyribulotide has not been fractionated free of

[3] C. Yanofsky, *J. Biol. Chem.* **223**, 171 (1956).
[4] O. H. Smith and C. Yanofsky, *J. Biol. Chem.* **235**, 2051 (1960).

the enzymes which further metabolize this intermediate, demonstrable synthesis of this compound is possible only if extracts of Tryp-3a mutants are used. These strains lack the enzyme which converts anthranilic deoxyribulotide to InGP. The bacteria are grown in a glucose–mineral salts medium[5] supplemented with indole (2.5 μg./ml.). When the indole in the medium is exhausted, the cells are harvested by centrifugation, washed once with cold 0.9% NaCl, and finally resuspended in 0.1 M phosphate buffer, pH 7.8. The cells are disrupted by sonication, and cell debris is removed by high-speed centrifugation. The supernatant solution is treated with solid ammonium sulfate to 26% of saturation. The precipitate is collected by centrifugation, dissolved in a minimum quantity of 0.1 M phosphate buffer, pH 7.8, and dialyzed in the cold for 3 hours against the same buffer (0.05 M) with one change of buffer. Such preparations usually contain 10 to 15 mg. of protein per milliliter.

Reagents

0.5 M phosphate buffer, pH 8.2.
0.054 M ATP, K salt.
0.05 M ribose-5-P, Na salt.
0.1 M MgCl$_2$.
0.0044 M anthranilic acid.
0.1 M phosphate buffer, pH 6.0.

Procedure. An incubation mixture containing 20 micromoles of phosphate buffer, pH 8.2, 0.8 micromole of ATP, 0.5 micromole of ribose-5-P, 1 micromole of MgCl$_2$, 0.22 micromole of anthranilic acid, enzyme, and distilled water to a final volume of 0.5 ml. is usually sufficient to demonstrate the synthesis of anthranilic deoxyribulotide. The reaction mixture can be scaled up to provide larger quantities. The course of the reaction is followed by diluting 0.05 ml. of the reaction mixture with 1.0 ml. of phosphate buffer, pH 6.0, and measuring the disappearance of anthranilic acid by a fluorometric assay (see Vol. V [107]). The reaction is usually completed in 20 to 40 minutes, and a 3-minute heat treatment at 100° serves to inactivate the enzyme. After filtration, this preparation may be used as a source of the deoxyribulotide for enzymatic and chemical studies. These preparations spontaneously decompose to anthranilic acid and should be stored frozen and reassayed frequently.

Chemical Synthesis. Two millimoles of the sodium salt of ribose-5-P dissolved in 17.3 ml. of water are mixed with 6.0 millimoles of anthranilic acid in 10 ml. of methanol. The mixture is heated to 70° in a water bath,

[5] H. J. Vogel and D. M. Bonner, *Microbial Genetics Bull.* **13**, 43 (1956).

and 2.5 ml. of ethyl malonate are added. Heating at 70° to 80° is continued for 30 minutes, during which time the reaction mixture becomes progressively darker yellow-colored and may become slightly orange. The reaction is stopped by chilling the flask in an ice bath. After storage for 24 hours at 4°, a small amount of an orange oil may settle out and is discarded. The residual yellow solution is adjusted to pH 4.5 with 1 N HCl and extracted twice with 25-ml. portions of ethyl acetate. Residual ethyl acetate is aerated off from the aqueous layer. The yield of anthranilic deoxyribulotide is approximately 10%, based on the limiting amount of ribose-5-P used in the synthesis.

Purification

No satisfactory method has been found to purify anthranilic deoxyribulotide. Anion exchangers such as Dowex 1 and Dowex 2 adsorb anthranilic deoxyribulotide; however, it has not yet been possible to develop a satisfactory elution procedure, presumably owing to the instability of the compound. Some purification has been achieved by chromatography on DEAE-cellulose and elution with a NaCl gradient. The presence of unreacted ribose-5-P in the chemically synthesized preparations of anthranilic deoxyribulotide exerts no demonstrable inhibition of the enzymatic reaction over a wide range of substrate concentrations.

Properties

Anthranilic deoxyribulotide shows strong reducing activity in alkaline solutions as evidenced by its rapid reaction with triphenyltetrazolium, and methylene blue and its positive reaction in the Folin-Malmros reducing test and the cysteine–carbazole test. Since alkaline solutions of the compound are capable of reducing o-dinitrobenzene in the cold, it seems likely that the intermediate exists, at least in part, in the enol form. As would be expected of a 1,2-enol, anthranilic deoxyribulotide is rapidly inactivated by alkali (82% by 10 minutes at room temperature in 0.1 N NaOH) and is relatively stable to acid treatment (2% inactivation after 10 minutes at room temperature in 0.1 N HCl). Alkali treatment leads to hydrolysis of the compound and liberation of free anthranilic acid.

In 0.05 M phosphate buffer, pH 7.7, anthranilic deoxyribulotide exhibits a clearly defined absorption peak at 320 mμ and a shoulder at 250 mμ. These shifts from the anthranilic acid peaks at 310 mμ and 240 mμ are characteristic of N-substituted anthranilate derivatives. Solutions of the dephosphorylated form of the intermediate, anthranilic deoxyribuloside, have an absorption maximum at 346 mμ in ethyl acetate as contrasted with the maximum at 336 mμ for anthranilic acid.[6]

Analysis

Anthranilic deoxyribulotide at levels of 0.05 to 0.5 micromole can be assayed enzymatically by conversion to InGP with extracts of *E. coli* or *S. typhimurium* mutants. The enzyme responsible for this conversion, indole-3-glycerolphosphate synthetase, is contained in the ammonium sulfate precipitate obtained from an *E. coli* crude extract at 26 to 40% saturation. The incubation mixture contains only substrate, enzyme, and Tris buffer, pH 8.8.[7] The InGP formed is oxidized to indole-3-aldehyde by metaperiodate, and indole-3-aldehyde is measured spectrophotometrically as described in a later section.

The rapid and quantitative hydrolysis of anthranilic deoxyribulotide to anthranilic acid by alkali can also be utilized to assay the compound. A preliminary ethyl acetate extraction is performed on the acidified solution (pH 4.5) to remove any free anthranilic acid. The aqueous layer is then hydrolyzed in 0.1 N NaOH at 100° for 5 minutes. The liberated anthranilic acid can be quantitatively extracted from the aqueous solution at pH 4.5 by several extractions with 2 to 3 vol. of ethyl acetate. The anthranilic acid concentration can be determined spectrophotometrically at 336 mμ (extinction coefficient of 4900[6]).

II. Indole-3-glycerolphosphate

Preparation

Chemical Synthesis. A brief description of the chemical synthesis of indole-3-glycerolphosphate has been published.[8] It involves reaction of 3-diazopyruvoylindole with diphenyl hydrogen phosphate and mild alkaline hydrolysis to remove the phenyl groups. An improved method utilizing phosphoric acid in place of diphenyl hydrogen phosphate has been reported to be more reliable.[9]

Enzymatic Synthesis. Indole-3-glycerolphosphate (InGP) is formed from anthranilic acid and PP-ribose-P through the intermediary formation of anthranilic deoxyribulotide.[3]

InGP can also be formed enzymatically from indole and triose phosphate. Triose phosphate is generated from fructose diphosphate by the action of added aldolase. Subsequent purification of InGP is accom-

[6] C. H. Doy and F. Gibson, *Biochem. J.* **72,** 586 (1959).

[7] F. Gibson and C. Yanofsky, *Biochim. et Biophys. Acta* **43,** 489 (1960).

[8] A. A. P. G. Archer and J. Harley-Mason, *Proc. Chem. Soc.* **285,** (October, 1958).

[9] A. A. P. G. Archer, J. H. New, and J. Harley-Mason, manuscript in preparation.

plished by the same procedures as described below for the InGP synthesized from anthranilic acid.

Preparation of Enzyme Extract. Wild-type strains of *E. coli* form only 5 to 10% as much of the enzymes involved in tryptophan biosynthesis as do tryptophan-requiring mutants grown under conditions where tryptophan repression is released, i.e., with a growth-limiting supply of tryptophan. Thus any tryptophan auxotroph which forms the enzymes required to convert anthranilic acid to InGP can be used to synthesize the latter compound. A Tryp-4 mutant, lacking an enzyme involved in the synthesis of anthranilic acid, has been used for this purpose and is treated in the following manner. The cells are grown on glucose–salts medium, and the crude extract is prepared as described in the previous section. Solid ammonium sulfate (18.7 g.) is added to 100 ml. of crude centrifuged extract. The mixture is allowed to stand in an ice bath for 20 minutes and then centrifuged in the cold. The precipitate is dissolved in 0.1 M phosphate buffer at pH 7.0 (30 ml. for every 100 ml. of extract) and dialyzed for 3 hours against 0.02 M phosphate buffer, pH 7.8. The dialyzed preparation is treated with ammonium sulfate (4.3 g. for each 40 ml.) and, after 20 minutes, centrifuged in the cold. The precipitate is discarded, and 3 g. of ammonium sulfate are added to the supernatant solution. After 20 minutes the precipitate is collected by centrifugation, dissolved in 15 ml. of 0.1 M phosphate buffer, pH 7.8, and dialyzed for 3 hours against 0.02 M phosphate buffer at pH 7.8. This preparation catalyzes the conversion of anthranilic acid to InGP and is unable to convert the latter compound to indole.

Reagents. The reagents used in the preparation of anthranilic deoxyribulotide are also used in the synthesis of InGP.

Procedure. A reaction mixture is prepared containing 0.75 millimole of anthranilic acid, 2 millimoles of ATP, 0.8 millimole of ribose-5-P, 1.7 millimoles of $MgSO_4$, 17 millimoles of phosphate buffer at pH 8.2, and approximately 30 to 40 ml. of enzymes in a final volume of 1080 ml. The mixture is incubated at 37° for 30 to 40 minutes. Aliquots are removed every 5 to 10 minutes to follow the disappearance of anthranilic acid fluorometrically. At the end of the incubation period the mixture is chilled rapidly.

Purification Procedure

The pH of the solution is adjusted to 6.5 to 7.0 with 1 N acetic acid. Several portions of Darco G-60 (acid- and alkali-washed) are then added to adsorb the InGP. After each addition, the Darco is removed by filtration, and a portion of the filtrate is assayed for InGP with ferric chloride reagent (1 ml. of 0.5 M $FeCl_3$ plus 50 ml. of water plus 30 ml. of concentrated H_2SO_4). A pink color signifies the presence of InGP. The Darco treatment is discontinued when a negative test for InGP is obtained. The charcoal can then be washed once with water, and the InGP is eluted by stirring with 40% alcohol containing 5 ml. of concentrated NH_4OH per liter. The eluate is freed from charcoal by filtration and concentrated *in vacuo* to a small volume (20 to 30 ml.). This solution is adjusted to pH 8.0 to 8.5 with saturated $Ba(OH)_2$, and a 25% solution of barium acetate is added until further additions no longer produce a precipitate. The precipitate is removed by centrifugation and, before being discarded, washed three times with small volumes (3.5 ml.) of water containing a few drops of $Ba(OH)_2$. The washings are combined with the original supernatant solution, and 1 ml. of barium acetate solution is added, followed by 2 vol. of acetone. The precipitate is collected by centrifugation and washed twice with acetone. The final precipitate contains more than 70% of the InGP initially present. This precipitate is dissolved in water and applied to a 2×32-cm. column of Dowex 1–chloride (2% cross-linked, 200 to 400 mesh). The resin is prepared by treatment of the column with 200 ml. of $1\,M$ NaCl (this and all subsequent solutions added to the column should contain 0.1 ml. of $1\,M$ NaOH per 100 ml.) and then with 100 ml. of $0.01\,M$ NaOH. The InGP solution is applied, and the column is again treated with 100 ml. of $0.01\,M$ NaOH. It is essential that all solutions applied to the column be alkaline; otherwise the InGP is destroyed during isolation. The column is washed with 600 ml. of $0.1\,N$ NaCl. Gradient elution is then performed to remove the InGP, with 400 ml. of $0.1\,M$ NaCl in the mixing flask and $0.5\,M$ NaCl in the reservoir flask. Fractions of 20 ml. are collected, and a sample from each is tested for InGP with ferric chloride reagent. The InGP is usually present in fractions 18 to 30. The purity of these fractions is determined by measuring their absorption at 240, 260, and 280 mμ. Of the ten to twelve fractions containing InGP, the first three to five usually contain appreciable amounts of adenylic acid, and the last two fractions occasionally contain small amounts of ADP. The InGP fractions containing impurities are combined, and the Darco and Dowex steps are repeated. InGP fractions free of absorbing impurities are adjusted to pH 6.5 to 7 and passed through a 2×5-cm. (diameter) column of Darco G-60. The InGP is adsorbed completely. The column is washed with water, and finally the InGP is eluted with 40% alcohol containing 1 ml. of concentrated NH_4OH per 100 ml. The eluate is concentrated *in vacuo* to about 5 ml., and $1\,N$ acetic acid is added until the pH is lowered to about 6. This treatment precipitates a small amount of charcoal which always contaminates the eluate. The precipitate is removed by centrifugation, and the supernatant solution is adjusted to pH 8 to 8.5 with saturated

Ba(OH)$_2$. One milliliter of a 25% solution of barium acetate is added, and the barium salt of InGP is precipitated by the addition of 4 vol. of alcohol and 2 vol. of acetone. The precipitated barium salt is washed twice with acetone and air-dried on a Büchner funnel.

Properties

When heated with dilute alkali, InGP is cleaved to form indole. In dilute acid solution at room temperature the intermediate is rapidly destroyed. The barium salt of InGP has a well-defined ultraviolet absorption maximum at 280 mμ in aqueous solutions. Its absorption spectrum is similar but not identical with that of indole.

Analysis

Isolated InGP yields 0.9 to 0.95 mole of indole per mole of phosphorus on treatment with an appropriate enzyme fraction. Triose phosphate would also be expected from the enzymatic hydrolysis of InGP, and the presence of a compound which gives a 2,4-dinitrophenylosazone with an absorption spectrum identical with that of the 2,4-dinitrophenylosazone formed by the triose phosphates can be detected. Alkali-labile phosphate also appears, thus further suggesting that triose phosphate is a reaction product.[3]

Neutral or alkaline aqueous solutions of InGP, when free of other absorbing materials, can be analyzed spectrophotometrically at 280 mμ. The 280/260-mμ and 280/240-mμ ratios are 1.32 and 3.1, respectively, for purified InGP.

InGP can be conveniently measured in enzymatic reaction mixtures by oxidation to indole-3-aldehyde as follows: 0.1 ml. of 1 M acetate buffer at pH 5.0 and 0.5 ml. of 0.1 N sodium metaperiodate are added to 0.4 ml. of a solution of InGP, and the mixture is incubated at room temperature for 20 minutes. Then 0.25 ml. of 1 N NaOH is added, and the indole-3-aldehyde formed is extracted with 5 ml. of ethyl acetate. The ethyl acetate layer can be clarified by a 1-minute centrifugation, and its indole aldehyde content determined spectrophotometrically at 290 mμ (extinction coefficient of 11,400).

[87] 5-Hydroxytryptophan and Derivatives

By Sidney Udenfriend and Herbert Weissbach

Although 5-hydroxyindoles have now been identified in animals, plants, and microorganisms,[1-5] most of the biochemical studies have employed animal tissues. 5-Hydroxytryptamine has been found in many tissues including the gastrointestinal tract, brain, blood platelets, skin, and lungs, and it is the hydroxyindole found in the largest quantities in animal tissues. Biosynthesis proceeds from tryptophan via 5-hydroxytryptophan, and the major route of metabolism is oxidative deamination by monoamine oxidase. This yields 5-hydroxyindoleacetaldehyde, which is rapidly converted to 5-hydroxyindoleacetic acid.[6] The latter compound is a normal constituent of urine; in man the average daily excretion ranges from 2 to 9 mg.

Some properties and assay procedures for 5-hydroxytryptophan, 5-hydroxytryptamine, and 5-hydroxyindoleacetic acid will be presented below. These compounds are now all commercially available and can be obtained from the following companies: California Corp. for Biochemical Research, Los Angeles, California; Nutritional Biochemicals Corp., Cleveland, Ohio; Mann Research Laboratories, Inc., New York, New York; and Regis Chemical Company, Chicago, Illinois.

General Methods of Detection and Quantitation

Absorption Spectra

5-Hydroxyindole compounds have characteristic absorption spectra in both acid and alkali. In dilute acid, absorption maxima are observed at 275 and 295 mμ, whereas in alkali the 295-mμ maximum is replaced by one at 325 mμ. The molar extinction in acid at 275 mμ is 5.3×10^{-3}.

Colorimetric Reactions

Hydroxyindoles react with nonspecific indole color reagents such as p-dimethylaminobenzaldehyde and xanthydrol,[7] as well as with phenolic

[1] V. Erspamer, *Rend. sci. farmital.* **1**, 1 (1954).
[2] M. M. Rapport, *J. Biol. Chem.* **180**, 961 (1949).
[3] H. Wieland, W. Konz, and H. Mittasch, *Ann. Chem.* **513**, 1 (1934).
[4] V. L. Stromberg, *J. Am. Chem. Soc.* **76**, 1707 (1954).
[5] C. Mitoma, H. Weissbach, and S. Udenfriend, *Arch. Biochem. Biophys.* **63**, 122 (1956).
[6] S. Udenfriend, E. Titus, H. Weissbach, and R. E. Peterson, *J. Biol. Chem.* **219**, 335 (1956).
[7] S. R. Dickman and A. L. Crockett, *J. Biol. Chem.* **220**, 957 (1956).

reagents such as diazotized sulfanilic acid and 1-nitroso-2-naphthol.[8] Under certain conditions 1-nitroso-2-naphthol will react almost exclusively with 5-hydroxyindoles, and this reaction is commonly employed for the assay of 5-hydroxytryptamine and its derivatives.[8] The following procedure can be employed for the colorimetric determination of all the 5-hydroxyindole compounds when present in amounts greater than 5 μg./ml.

Reagents

1-Nitroso-2-naphthol reagent. 0.1% 1-nitroso-2-naphthol in 95% ethanol.

Nitrous acid reagent. To 5 ml. of 2 N H_2SO_4 or HCl is added 0.2 ml. of 2.5% $NaNO_2$. This reagent should be prepared fresh daily.

To 2 ml. of an acid extract containing 0.03 to 0.8 μmole of 5-hydroxyindole material in a glass-stoppered shaking tube are added 1 ml. of the nitroso-naphthol reagent and, after thorough mixing, 1 ml. of the nitrous acid reagent. The tube is stoppered, shaken again, and allowed to stand at room temperature for 10 minutes. Ten milliliters of ethylenedichloride are then added, and the tube is shaken for a sufficient time so as to extract all the excess nitrosonaphthol. After centrifugation at low speed, the supernatant aqueous layer is transferred to a cuvette, and the optical density of the stable purple chromophore is measured at 540 mμ in an appropriate photometer. When measured in a Beckman spectrophotometer, optical density is proportional to concentration up to 0.8 micromole of compound. The colorimetric procedure has been employed in the determination of 5-hydroxyindoleacetic acid in urine, of 5-hydroxytryptamine in carcinoid tumors, and in enzymatic experiments involving the formation of 5-hydroxytryptamine from 5-hydroxytryptophan, or the metabolism of 5-hydroxytryptamine.

Fluorescent Procedure

In dilute acid 5-hydroxyindoles emit a maximum fluorescence at 350 mμ when activated at 295 mμ. However, in 3 N HCl the 350-mμ fluorescence peak is depressed, and a new peak appears in the visible range at 550 mμ.[9] Until the development of the spectrophotofluorometer, it was not possible to assay for 5-hydroxyindoles fluorometrically, since prior instruments were not able to produce exciting light of sufficient

[8] S. Udenfriend, H. Weissbach, and C. T. Clark, *J. Biol. Chem.* **215**, 337 (1955).
[9] S. Udenfriend, D. F. Bogdanski, and H. Weissbach, *Science* **122**, 972 (1955).

intensity below 330 mμ. However, commercial spectrophotofluorometers are now capable of delivering high-intensity monochromatic radiation at all wavelengths from 240 to 800 mμ, and they can analyze the resulting fluorescence throughout this entire range.[10]

For most tissues fluorescence is the method of choice for the assay of 5-hydroxytryptamine and its derivatives. The fluorescence at 550 mμ, although less sensitive than the 350-mμ fluorescence in weak acid, is more specific and is generally employed. As little as 0.1 μg. of 5-hydroxyindole material can be assayed by the fluorescent techniques.

Paper Chromatography

The various 5-hydroxyindoles can be conveniently separated and detected on paper chromatograms. Table I summarizes the R_f values of

TABLE I
CHROMATOGRAPHIC BEHAVIOR OF 5-HYDROXYINDOLE COMPOUNDS AND TRYPTOPHAN[a,b]

Substance	Mean R_f, propanol–NH$_3$[c]	Mean R_f, butanol–acetic acid–water	Mean R_f, 20% KCl[d]
5-Hydroxytryptamine (free base)	0.65	0.42	0.39
5-Hydroxytryptophan	0.10	0.17	0.44
5-Hydroxyindoleacetic acid	0.30	0.76	0.49
N,N'-Dimethyl-5-hydroxytryptamine (free base)	0.90	0.50	0.51
Tryptophan	0.55	0.41	0.59

[a] Taken from S. Udenfriend, H. Weissbach, and B. B. Brodie, *Methods of Biochem. Anal.* **6**, 95 (1958).
[b] Whatman No. 1 paper used in all cases.
[c] Ascending.
[d] Descending.

tryptophan and several 5-hydroxyindole compounds in three solvent systems.

The hydroxyindoles can be detected on the developed chromatograms by spraying with a number of reagents. First, these compounds emit pink fluorescence on paper when sprayed with 0.1 N HCl and examined under a short-wave ultraviolet light (Mineralite B 41). This fluorescence is characteristic of 5-hydroxyindoles and is sufficiently sensitive to detect 5 to 10 μg. Ehrlich's reagent (p-dimethylaminobenzaldehyde), a common spray for indoles, also gives blue colors with hydroxyindoles which with 5-hydroxytryptamine and 5-hydroxytryptophan gradually turn green.

[10] S. Udenfriend, "Fluorescence Assay in Biology and Medicine." Academic Press, New York, 1962.

This reagent is prepared by diluting 1% p-dimethylaminobenzaldehyde in 12 N HCl with an equal volume of ethanol. The test can be made more sensitive by respraying the dry chromatograms, following the initial spray with Ehrlich's reagent, with a 0.2% nitrite solution in dilute HCl. With this procedure as little as 1 μg. can be detected. The nitrosonaphthol reagent can also be applied as a spray. The paper is first sprayed with the same nitrosonaphthol reagent as is used in the colorimetric procedure. After drying, the paper is sprayed with the nitrite reagent prepared in HCl. The hydroxyindoles appear as violet spots on a faint yellow background. At least 5 to 10 μg. of compound are needed with this spray.

Erspamer[1] has described many spray reagents, for 5-hydroxytryptamine and related compounds, which depend on the reactivity of the phenolic group or amino group.

Determination of Total 5-Hydroxyindoles

Direct Procedure

In tissues in which the 5-hydroxyindole content is greater than 1 μg./g., it is possible to use fluorescence assay after precipitation of the proteins with zinc hydroxide. The following procedure is recommended. Tissues are homogenized in 2 vol. of 0.1 N HCl, and an aliquot of not more than 1 g. is diluted to 8 ml. with water. One milliliter of 10% zinc sulfate is added and, after swirling, is followed by 0.5 ml. of 1 N NaOH. Five minutes after mixing, the precipitated proteins are removed by centrifugation for 15 minutes at 10,000 × g. When the 5-hydroxyindole level of blood is above 1 μg./ml., 1 ml. is diluted to 8 ml. with water, and

TABLE II
TISSUE LEVELS OF 5-HYDROXYTRYPTAMINE

Tissue	5-Hydroxytryptamine content, μg./g. or ml.
Rat lung	3
Rabbit intestine	10
Guinea pig intestine	6
Rat intestine	1.5
Mouse intestine	5.3
Guinea pig lung	0.2
Rabbit blood	4–6
Chicken blood	1–3
Carcinoid blood	1–3
Carcinoid tumor	1000
Brain (most species)	0.3–0.8

the same procedure is followed. After centrifugation, 1 ml. of the supernatant layer is transferred to a test tube containing 0.3 ml. of 12 N HCl, and the fluorescence is measured at 550 mμ after activation at 295 mμ. Standards, internal recoveries, and a reagent blank are all treated in the same manner.

In most tissues 5-hydroxytryptamine is the only 5-hydroxyindole compound present, and the direct procedure is useful for its determination in intestine, lungs, and spleen of most species and in the blood of rabbits, chickens, and patients with malignant carcinoid (Table II). The major limitations of this rapid procedure are its lack of sensitivity (levels higher than 1 μg./g. are needed) and its lack of specificity (any 5-hydroxyindole present will react).

Measurement of Specific 5-Hydroxyindoles

Assay for 5-Hydroxytryptophan

The amounts of this amino acid normally present in animal tissues are too small to be detected by any of the methods now available. It is formed in large quantities from tryptophan by resting-cell suspensions of *Chromobacterium violaceum,* and its formation can be quantitated by using the nitrosonaphthol color reaction (see Vol. V [110]). Where large amounts of 5-hydroxytryptophan are present, as in experiments in which this amino acid has been administered to animals, the following procedure can be employed. Tissues are homogenized in 2 vol. of 0.1 N HCl, and 1 ml. of the homogenate is diluted with 2 ml. of water. The proteins are precipitated by the addition of 1 ml. of 40% trichloroacetic acid. After centrifugation a 3-ml. aliquot of the supernatant solution is extracted twice with 10 ml. of ether to remove both the trichloroacetic acid and any 5-hydroxyindoleacetic acid present. The aqueous phase is then adjusted to about pH 8 to 10 with 0.5 ml. of 20% Na_2CO_3 and extracted with 15 ml. of n-butanol to remove 5-hydroxytryptamine. The remaining unextractable 5-hydroxyindole is taken to represent 5-hydroxytryptophan. To 1 ml. of the residual aqueous layer is added 0.6 ml. of 12 N HCl, and the fluorescence is determined in the spectrophotofluorometer (activation wavelength, 295 mμ; fluorescent emission wavelength, 550 mμ). Standards, reagent blank, and internal recoveries are carried through the entire procedure. Tissues from control animals contain no detectable 5-hydroxytryptophan. However, they do contain small amounts of fluorescent material having other fluorescent characteristics, and there is some overlapping at 550 mμ. This nonspecific fluorescence in the control tissues, which may amount to 1 to 5 μg./g. of tissue, calculated as 5-hydroxytryptophan, should be subtracted from the values obtained after 5-hydroxytryptophan administration. Each application of this procedure should be checked for specificity by paper chromatography. The assay yields best results when the 5-hydroxytryptophan levels are greater than 15 μg./g.

An assay for 5-hydroxytryptophan which was recently found applicable to carcinoid urine involves removal of 5-hydroxytryptamine on a base exchanger, conversion of 5-hydroxytryptophan to 5-hydroxytryptamine by purified kidney aromatic amino acid decarboxylase, and fluorometric assay of the enzymatically generated 5-hydroxytryptamine.[11]

Determination of 5-Hydroxytryptamine in Tissues by Extraction and Fluorometric Assay

In some tissues, such as brain, the quantities of 5-hydroxytryptamine are too small to permit direct assay after precipitation of the proteins with zinc hydroxide. Extraction with solvent is therefore used to isolate and concentrate it before assay. The extraction procedure which must be used in such instances also adds specificity, since only 5-hydroxyindole-amines (5-hydroxytryptamine and its N-methylated derivatives) are extracted by this procedure. Therefore, in experiments in which other 5-hydroxyindoles (5-hydroxytryptophan and 5-hydroxyindoleacetic acid) are present, the direct procedure can be used to determine total 5-hydroxyindoles, but the extraction procedure must be used to determine 5-hydroxytryptamine specifically. The extraction procedure for 5-hydroxytryptamine in brain,[12] which is presented here, can be applied to other tissues as well.

Reagents

Borate buffer. To 94.2 g. of boric acid dissolved in 3 l. of water, 165 ml. of 10 N NaOH are added. The buffer solution is then saturated with purified n-butanol and NaCl by adding these substances in excess and shaking. Excess n-butanol is removed by aspiration, and excess salt is permitted to settle. The final pH should be approximately 10.

n-Butanol. Reagent-grade butanol is purified by shaking first with an equal volume of 0.1 M NaOH, then with an equal volume of 0.1 N HCl, and finally twice with distilled water.

Heptane. Practical grade of heptane is treated in the same manner as the n-butanol.

[11] A. Sjoerdsma, J. A. Oates, P. Zaltzman, and S. Udenfriend, *J. Pharmacol. Exptl. Therap.* **126**, 217 (1959).

[12] D. F. Bogdanski, A. Pletscher, B. B. Brodie, and S. Udenfriend, *J. Pharmacol. Exptl. Therap.* **117**, 82 (1956).

Procedure. One part of brain tissue is homogenized in 2 parts of 0.1 N HCl. An aliquot of homogenate containing 0.5 to 5.0 μg. of 5-hydroxy-tryptamine is transferred to a 60-ml. glass-stoppered bottle and adjusted to approximately pH 10 by the addition of anhydrous sodium carbonate. Five milliliters of borate buffer, pH 10, are added, and the solution is diluted with water to a volume of 15 ml. Then 5 g. of NaCl and 15 ml. of *n*-butanol are added. After 10 minutes of shaking, the bottle is centrifuged, and the fluid is decanted from the solid material into another bottle. The aqueous layer is removed by aspiration, and the butanol phase is washed by shaking it with an equal volume of borate buffer.[13] Ten milliliters of the butanol phase are transferred to another bottle containing 20 ml. of heptane and 1.5 ml. of 0.1 N HCl. The bottle is shaken and centrifuged, and the supernatant solvent is removed. Then 1 ml. of the acid layer is added to 0.3 ml. of concentrated HCl in a quartz cuvette. The solution is activated at 295 mμ in the spectrophoto-fluorometer, and the resultant fluorescence at 550 mμ is measured. Standards and reagent blanks and internal recoveries are carried through the entire extraction procedure. The specificity of this procedure has been discussed elsewhere.[12]

5-Hydroxytryptamine Determination in Blood Platelets

In most species the 5-hydroxytryptamine levels in blood are too small to be assayed by the direct method. The following procedure for the isolation of platelets and determination of their 5-hydroxytryptamine content is recommended. About 10 ml. of blood are collected in siliconized glassware by using 0.1 vol. of 1% Versene–0.7% saline as anticoagulant. All subsequent steps in the isolation of platelets (carried out at 3° with siliconized glassware) are performed according to the procedure of Dillard *et al.*[14] The blood is centrifuged at 600 r.p.m. in an International No. 1 centrifuge for 45 minutes. The platelet-rich plasma is transferred to another centrifuge tube and centrifuged at 2000 r.p.m. for 30 minutes. The plasma is decanted, and the sedimented platelets are carefully suspended in 5 ml. of cold saline to wash out adhering blood proteins. The tube is centrifuged at 2500 r.p.m. for 20 minutes, and the saline is

---

[13] Washing of the butanol phase with borate buffer is essential when other 5-hydroxy-indoles are present. This would be likely in experiments where 5-hydroxytrypto-phan is administered or used as an enzyme substrate. In the presence of 5-hydroxy-tryptophan it is necessary to wash the butanol three times with borate buffer. Under normal conditions, omitting the buffer wash does not appreciably influence the analysis. In this laboratory the butanol layer is washed once with borate buffer as a precautionary measure.

[14] G. H. L. Dillard, G. Brecher, and E. P. Cronkite, *Proc. Soc. Exptl. Biol. Med.* **78**, 853 (1951).

discarded. This washing procedure is repeated once more, and then the platelets are suspended in about 3.5 ml. of 0.02 N HCl. Gentle agitation causes lysing of the platelets. One milliliter is then removed for 5-hydroxytryptamine assay. It is diluted to 2 ml. with water, and the proteins are precipitated by the addition of 0.2 ml. of 10% zinc sulfate and 0.1 ml. of 1 N NaOH. After centrifugation the supernatant fluid is assayed fluorometrically (activation, 295 mμ; fluorescence, 350 mμ). Standards and a reagent blank are treated in the same manner. A second aliquot of the lysed platelet solution is assayed for protein according to the method of Sutherland *et al.*[15] Results are expressed as micrograms of 5-hydroxytryptamine per milligram of platelet protein.

5-Hydroxyindoleacetic Acid in Tissues

Normally the amounts of this acid in tissues are too small to be determined. However, 5-hydroxyindoleacetic acid is formed during the oxidation of serotonin by monoamine oxidase in crude tissue preparations. The following procedure can be used to assay for enzymatically formed 5-hydroxyindoleacetic acid.

An aliquot of the incubation mixture (containing about 2 μg. of 5-hydroxyindoleacetic acid) is transferred to a 40-ml. glass-stoppered shaking tube, acidified by the addition of 1 ml. of 1 N HCl, and diluted to 4 ml. Sufficient NaCl is added to saturate the aqueous phase, and 15 ml. of peroxide-free ether are added. The tube is shaken well and centrifuged. Ten milliliters of the ether layer are transferred to another tube containing 1.5 ml. of 0.5 M PO$_4$ buffer, pH 7.0. After shaking and centrifuging the buffer layer is assayed spectrophotofluorometrically (activation, 295 mμ; fluorescence, 340 mμ). Standard and reagent blanks are carried through the entire procedure.

Crude incubation mixtures may contain colored materials that interfere with the assay, but these usually can be removed by a preliminary precipitation of proteins with zinc hydroxide (see above for total 5-hydroxyindoles) followed by the ether extraction.

A modification of this extraction procedure has been used to determine 5-hydroxyindoleacetic acid in urine.[16]

[15] E. W. Sutherland, C. F. Cori, R. Haynes, and N. S. Olsen, *J. Biol. Chem.* **180**, 825 (1949).
[16] S. Udenfriend, E. Titus, and H. Weissbach, *J. Biol. Chem.* **216**, 499 (1955).

[88] Acyl Imidazoles

By Thomas C. Bruice

Preparation

The great susceptibility of N-acetylimidazole to hydrolysis[1-6] delayed its preparation until 1952, when Boyer[7] reported the first successful synthesis of the compound by treatment of imidazole with isopropenyl acetate. Under Schotten-Baumann conditions imidazoles generally undergo ring opening; e.g., the reaction of imidazole with benzoyl chloride in sodium hydroxide solution yields dibenzoyl diaminoethylene and formic acid.[8-10] However, in the instance of a few substituted imidazoles (an example is 4-phenylimidazole[11, 12]) the reaction has been noted to yield the N-acyl derivative. This cleavage reaction, known as the Bamberger reaction, is remarkable in view of the general stability of the imidazole nucleus. Imidazoles may be acylated by the reaction of one mole of an acid chloride with two moles of the imidazole compound (Gerngross synthesis[13, 14]), in which case, however, one mole of the imidazole is squandered as the hydrochloride. Employing the Gerngross method, Wieland and Schneider[1] were able to effect the synthesis of N-acetylimidazole. These workers also prepared 1-carbobenzoxyglycyl-L-histidine methyl ester, which is considerably more stable than N-acetylimidazole. The greater instability of the aliphatic N-acetylimidazoles as compared to the aromatic N-acyl derivatives is illustrated by the fact that 1-acetyl-N-acetylhistamine evaded preparation until 1956,[15]

[1] T. Wieland and G. Schneider, Ann. **580**, 159 (1953).

[2] E. R. Stadtman, in "The Mechanism of Enzyme Action" (W. D. McElroy and B. Glass, eds.), p. 581. Johns Hopkins Press, Baltimore, 1954.

[3] E. R. Stadtman and F. H. White, J. Am. Chem. Soc. **75**, 2022 (1953).

[4] H. A. Staab, Ber. **89**, 1927 (1956).

[5] H. A. Staab, Ber. **89**, 2088 (1956).

[6] W. P. Jencks and J. Carriuolo, J. Biol. Chem. **234**, 1273, 1280 (1958).

[7] J. H. Boyer, J. Am. Chem. Soc. **74**, 6274 (1952); see also Biochem. Preparations **4**, 54 (1955).

[8] E. Bamberger and B. Berle, Ann. **273**, 342 (1893).

[9] G. Heller, Ber. **37**, 3112 (1904).

[10] P. Ruggli, R. Ratti, and E. Henzi, Helv. Chim. Acta **12**, 332 (1929).

[11] R. L. Grant and F. L. Pyman, J. Chem. Soc. **119**, 1893 (1921).

[12] R. Ruggli, Helv. Chim. Acta **3**, 559 (1920).

[13] O. Gerngross, Ber. **46**, 1908 (1913).

[14] O. Gerngross, Z. physiol. Chem. **108**, 50 (1919).

[15] H. A. Staab, Angew. Chem. **68**, 616 (1956).

whereas dibenzoylhistamine[16] has been known for some time. N,N'-Di-imidazolylterephthalic acid and N,N'-diimidazolyladipic acid are examples of symmetrical bis(N-carboimidazolyl) compounds; other interesting acyl imidazoles include the highly reactive N,N'-carbonyl-diimidazole (from phosgene and imidazole[17]), N-carboethoxyimidazole (from imidazole magnesium bromide and ethyl chlorocarbonate[18, 19]), and N-phenylcarbamylimidazole (from phenylisocyanate and imidazole[20-22]).

Benzimidazole is also subject to the Bamberger reaction (yielding N,N'-dibenzoyl-o-phenylenediamine[8, 14, 23, 24]) but can be acetylated by the Gerngross method[13] or by the reaction of benzimidazole with acetic anhydride in the absence of base[25] (for other N-acylbenzimidazoles, see refs. 6, 26, and 27; and for N-benzoylbenzimidazole, see ref. 6). Examples of N-acylnaphthimidazoles are N-acetylnaphth-(1,2)-imidazole and N-benzoylnaphth-(1,2)-imidazole.[13, 28] Naphthimidazole does not undergo the Bamberger reaction, and its N-acyl derivatives are thus easily available. In no case is it known which nitrogen bears the acyl group in unsymmetrical N-acyl imidazoles.

The recent finding that imidazole hydrochlorides are readily acylated by anhydrides to yield the corresponding N-acyl imidazole hydrochlorides has led to the most satisfactory synthesis of the unstable N-acyl imidazoles. This synthetic technique has been employed by Katchalski and co-workers[29] in the synthesis of 1,N-dicarbobenzoxy-L-histidine methyl ester hydrochloride and by Bruice and Sturtevant[30] in the preparation of N,O-diacetyl-4-(2'-hydroxyethyl)imidazole. The preparation of N-acetylimidazole hydrochloride by this procedure follows.

N-Acetylimidazole Hydrochloride. Imidazole (1.36 g.; 0.02 mole) was dissolved in 6 ml. of acetic anhydride, and the solution was saturated with hydrogen chloride under anhydrous conditions to precipitate imidazole hydrochloride. The suspension was next brought just to reflux under

[16] H. A. Staab, *Ber.* **90**, 1326 (1957).
[17] H. A. Staab, *Ann.* **609**, 75 (1957).
[18] B. Oddo and Q. Mingois, *Gazz. chim. ital.* **58**, 584 (1928).
[19] W. John, *Ber.* **68**, 2283 (1935).
[20] R. A. Henry and W. M. Dehn, *J. Am. Chem. Soc.* **71**, 2297 (1949).
[21] K. Schlögl and H. Woidich, *Monatsh. Chem.* **87**, 679 (1956).
[22] H. A. Staab, *Ann.* **609**, 83 (1957).
[23] B. Oddo and L. Raffa, *Gazz. chim. ital.* **67**, 537 (1937).
[24] B. Oddo and L. Raffa, *Gazz. chim. ital.* **68**, 199 (1938).
[25] A. Bistrzycki and G. Prezeworski, *Ber.* **45**, 3483 (1912).
[26] G. Heller, *Ber.* **37**, 3112 (1901).
[27] B. Oddo and F. Ingraffia, *Gazz. chim. ital.* **62**, 1092 (1932).
[28] O. Fischer, *Ber.* **34**, 930 (1901).
[29] A. Patchornik, A. Berger, and E. Katchalski, *J. Am. Chem. Soc.* **79**, 6417 (1957).
[30] T. C. Bruice and J. M. Sturtevant, *J. Am. Chem. Soc.* **81**, 2860 (1959).

anhydrous conditions, charcoaled, and the filtrate refrigerated for 12 hours. The mother liquor was then allowed to drain from the crystalline product and discarded. The white crystals of N-acetylimidazole hydrochloride were washed repeatedly with absolutely dry and alcohol-free ether (over Na) until the odor of acetic anhydride was no longer detectable. In this manner the product was obtained in 65% yield; m.p. 118° to 123°.

Chemistry

The anhydride nature of acyl imidazoles was applied by Bergmann and Zervas[31] in the synthesis of peptides (1-acetyl-N-benzoylhistidine methyl ester plus glycine or arginine in aqueous solution yields N-acetylglycine and N-acetylarginine, respectively; the 1-hippuryl derivative of N-benzoylhistidine methyl ester behaved similarly). Wieland and Schneider[1] reported that N-acetylimidazole and 1-carbobenzoxyglycyl-L-histidine methyl ester react with hydroxylamine to yield acethydroxamic acid and carbobenzoxyglycylhydroxamic acid. The latter also reacted with aniline and alanine methyl ester to form carbobenzoxyglycyl anilide and carbobenzoxyglycyl alanine methyl ester.

The first quantitative studies on the hydrolysis of N-acetylimidazole were made by Stadtman.[2,3] He found that at 26° the half-life of this compound was 60 minutes at pH 6.0 and 7.5, and 2.5 minutes at pH 5.0 and 8.8, indicating strong catalysis of the hydrolysis by hydronium and hydroxide ions. In addition, Stadtman reported the acetylation of amines, alcohols, thiols, hydroxylamine, and inorganic phosphate by N-acetylimidazole. Staab[4,5] provides additional kinetic data on the hydrolysis and aminolysis of N-acetylimidazole and several other aliphatic N-acyl imidazoles. Increasing the steric requirements of the acyl group results in a proportionate enhancement in rates of hydrolysis at neutral pH, while the rates of basic hydrolysis and aminolysis are markedly decreased. Staab proposed that in neutral aqueous solution the hydrolysis of N-acyl imidazole involved initial heterolytic dissociation of the compound into imidazole anion and acylium ion, whereas it was suggested that the basic hydrolysis and aminolysis occur by a nucleophilic attack of the hydroxide ion or amine on the carbonyl group of N-acetylimidazole. Jencks and Carriuolo[6] concur with Staab's mechanistic suggestions for basic hydrolysis but have shown that at pH below 2.5 the rate of hydrolysis is constant and therefore appears to be due to the attack of water on the N-acetylimidazolium ion (pK'_a 3.6). In addition, the latter workers have also studied the reaction of N-acetylimidazole with acetate, succinate, formate, phosphate, arsenate, mercaptans, water, and amines.

[31] M. Bergmann and L. Zervas, Z. physiol. Chem. 175, 145 (1928).

The reaction with the latter three reagents was found to be catalyzed by imidazole, acting, presumably, as a general base by abstracting a proton from the attacking reagent in the transition state.

The high reactivity of N-acetylimidazoles extends to the N-carbo-alkoxy- and the N-carbaminoimidazoles.[19] A more stable member of this group is 1,2-N-dicarbobenzoxyhistidine, of potential utility for the synthesis of histidine peptides.[32] The N-carbaminoimidazoles react rapidly with hydroxide ion, amines, and alcohols.[22, 33, 34]

N-Acetylbenzimidazole has a half-life of 21 hours[35] at room temperature in water, as compared to a half-life of 40 minutes for N-acetylimidazole under the same conditions.[4] N-Benzoylbenzimidazole is stable for days in cold carbonate solution[13]; its greater stability relative to N-acetylbenzimidazole parallels the similar relationship which holds for the corresponding imidazoles. The N-acyl naphthimidazoles are even more stable than the N-acetylbenzimidazoles and are only slowly hydrolyzed in cold dilute NaOH.

As an Intermediate in the Reaction of Imidazoles with Esters

Imidazoles catalyze the hydrolysis of esters formed from alcohols that are reasonably strong acids.[36, 37] Both the neutral imidazole and the imidazole anion are catalysts, but the positive imidazolium ion is not.[37] For the imidazole-catalyzed hydrolysis of p-nitrophenyl acetate, it has been shown that N-acetylimidazole accounts quantitatively, as an intermediate, for the conversion of the ester to hydrolytic products.[38-40] In addition, N-acetylimidazole has been isolated as an intermediate in this reaction.[41] In the reaction of benzimidazole with p-acetoxybenzoic acid, the path through N-acetylbenzimidazole has been shown to account for at least 90% of the conversion of ester to hydrolytic products.[42] The

---

[32] A. Robertson and R. Robinson, *J. Chem. Soc.* 1460 (1928).

[33] R. A. Henry and W. M. Dehn, *J. Am. Chem. Soc.* **71**, 2297 (1949).

[34] K. Schlögl and H. Woidich, *Monatsh. Chem.* **87**, 679 (1956).

[35] H. A. Staab, *Ber.* **90**, 1320 (1957).

[36] For the initial studies of this phenomenon, see M. L. Bender and B. W. Turnquest, *J. Am. Chem. Soc.* **79**, 1656 (1957); T. C. Bruice and G. L. Schmir, *ibid.* **79**, 1663 (1957); and subsequent papers in this series.

[37] T. C. Bruice and G. L. Schmir, *J. Am. Chem. Soc.* **80**, 148 (1958).

[38] S. A. Bernhard and H. Gutfreund, *in* "Proceedings of the International Symposium on Enzyme Chemistry, Tokyo-Kyoto (1957)" (K. Ichihara, ed.), p. 124. Academic Press, New York, 1958.

[39] D. M. Brower, Doctoral Thesis, University of Leiden, s'Gravenhage, 1957.

[40] D. M. Brower, M. J. van der Vlugt, and E. Havinga, *Koninkl. Ned. Akad. Wetenschap. Proc.* **60B**, 275 (1957).

[41] W. Langenbeck and R. Mahrwald, *Chem. Ber.* **90**, 2423 (1957).

[42] G. L. Schmir and T. C. Bruice, unpublished data.

latter experiment is particularly noteworthy, since the stability of
N-acetylbenzimidazole ($t_{1/2}$ 21 hours) allows for its accurate determination. In addition, an acyl imidazole has been established as an intermediate in the catalysis of the hydrolysis of an acyl thiol.[43]

Determination

The anhydride nature of the N-acyl imidazoles allows their ready
determination by the quantitative hydroxamic acid method of Lipmann
and Tuttle.[44] The determination of the rate of formation or disappearance of the N-acyl bond is most easily carried out by observation of the
absorbance at 254 mμ (characteristic λ_{max} for N-acetylimidazoles and at
which wavelength aromatic imidazoles show little absorption).[2]

[43] T. C. Bruice, *J. Am. Chem. Soc.* **81**, 5444 (1959).
[44] F. Lipmann and L. C. Tuttle, *J. Biol. Chem.* **159**, 21 (1945).

[89] γ-Aminobutyric Acid

By EUGENE ROBERTS

γ-Aminobutyric acid (GABA), which is widely distributed in plants
and bacteria, first became of interest in mammalian physiology because
it was found in an easily extractable form in large amounts in the brain
and spinal cord of various species.[1] Since that time work in a number
of laboratories has shown that GABA has inhibitory effects in tests with
invertebrate and vertebrate nervous systems and has indicated that
GABA might play an important role in regulation of activity in the
nervous system.[2] Studies on the distribution of GABA and GABA-
metabolizing systems in a wide variety of organisms are being carried
on in many laboratories.

Availability of GABA from Commercial Sources. GABA now is available in pure form at reasonable cost from California Corporation for
Biochemical Research, 3625 Medford Street, Los Angeles, California,
and from General Mills, Inc., 2010 East Hennepin Ave., Minneapolis,
Minnesota. GABA-1-C^{14} in activities of 1 to 2 mc./millimole is also
listed in the catalogues of various companies which sell labeled chemicals.

[1] E. Roberts, *Progr. in Neurobiol.* **1**, 11 (1956).
[2] K. A. C. Elliott and H. H. Jasper, *Physiol. Revs.* **39**, 383 (1959).

Synthesis of GABA

By far the simplest and cheapest method of preparing GABA is by hydrolysis of 2-pyrrolidone (butyrolactam).[3] Pyrrolidone has been obtained by the author in 1- to 2-pound lots at no cost from Cliffs Dow Chemical Company, Marquette, Michigan. The pyrrolidone is refluxed with two to three times its weight of barium hydroxide in approximately 10 vol. of water for 2 hours. After cooling, the barium is precipitated by passing in carbon dioxide. The barium carbonate is removed by filtration or centrifugation, and the precipitate is washed. The combined filtrate and washing is treated with a slight excess of sulfuric acid to liberate the amino acid from its barium salt, and, after digestion on the steam bath with an excess of barium carbonate to remove sulfate, the precipitate is removed. The filtrate is concentrated and frozen. On thawing, a precipitate causing a slight turbidity is removed by centrifugation, and the filtrate is concentrated to a sirup. After the addition of at least 10 vol. of alcohol, crystallization is allowed to take place in the cold. The amino acid can be recrystallized in the cold by dissolving it in a minimal quantity of water and adding a tenfold excess of absolute alcohol.

The synthesis of GABA from potassium phthalimide and γ-chlorobutyronitrile has been described.[4] GABA-1-C^{14} has been prepared (Tracerlab, Inc.) by condensation of potassium cyanide-C^{14} and γ-iodopropylphthalimide:

The above intermediate was hydrolyzed with sulfuric acid to give GABA-

[3] J. Tafel and M. Stern, Ber. 33, 2224 (1900).
[4] C. C. de Witt, Org. Syntheses Coll. Vol. 2, 25 (1943).

1-C^{14}. After addition of barium carbonate and treatment with charcoal, the GABA was crystallized from 90% ethanol.

GABA was prepared from L-glutamic acid by quantitative decarboxylation with lyophilized washed cells of *Clostridium perfringens* and crystallized as the hydrochloride.[5] GABA-4-C^{14} has been prepared by enzymatic decarboxylation of DL-glutamic acid-2-C^{14},[6] by employing a suspension of lyophilized cells of *Clostridium welchii*,[7] followed by ion-exchange separation of D-glutamic acid and GABA, and in a similar fashion we have made this preparation employing an acetone powder of glutamate-adapted *Escherichia coli*.[8] The enzymatic methods can be employed for the production of GABA-U-C^{14} of high specific activity from L-glutamic-U-C^{14}. Small quantities can be prepared at one time and can be isolated easily from the reaction mixture by one-dimensional paper chromatography of streak preparations followed by elution.

Properties

GABA is a white crystalline solid. Various melting points have been found for different analytically pure batches prepared by hydrolysis of 2-pyrrolidone: 185° to 190°, 185° to 188°, 185° to 208°, 189° to 204°, 190° to 210°, 192° to 194°. It is highly soluble in water (130 g. per 100 g. of water),[9] aqueous solutions showing no characteristic spectral properties in the ultraviolet. It is stable to acid and alkali under conditions ordinarily employed for hydrolysis of proteins. In contrast to α-amino acids, GABA does not chelate copper under alkaline conditions.[10]

Determination of GABA

Enzymatic Assay of GABA in Tissue Extracts. The most rapid, specific, and sensitive determination of GABA in biological extracts is based on the use of an enzyme system found in the bacterium *Pseudomonas fluorescens* EC grown on pyrrolidine which converts GABA to succinate via transamination and oxidation coupled to TPN reduction[11-13] (see Vol. V [104a]). The procedure described below is employed for the

[5] M. N. Camien, L. E. McClure, A. Lepp, and M. S. Dunn, *Arch. Biochem. Biophys.* **43**, 378 (1953).

[6] W. E. Wilson, R. J. Hill, and R. E. Koeppe, *J. Biol. Chem.* **234**, 347 (1959).

[7] A. Meister, H. A. Sober, and S. V. Tice, *J. Biol. Chem.* **189**, 577 (1951); see Vol. III [75A].

[8] P. Ayengar, E. Roberts, and G. B. Ramasarma, *J. Biol. Chem.* **193**, 781 (1951).

[9] L. S. Mason, *J. Am. Chem. Soc.* **69**, 3000 (1947).

[10] H. Ley, *Ber. deut. chem. Ges.* **42**, 354 (1909).

[11] E. M. Scott and W. B. Jakoby, *Science* **128**, 361 (1958).

[12] W. B. Jakoby and E. M. Scott, *J. Biol. Chem.* **234**, 937 (1959).

[13] E. M. Scott and W. B. Jakoby, *J. Biol. Chem.* **234**, 932 (1959).

determination of GABA in extracts of brain[14] but obviously can be modified for extracts from other types of material. Brain areas are removed immediately after decapitation, weighed, and dropped into separate vials containing 4 to 6 ml. of ice-cold 80% ethanol. Samples are homogenized in a ground-glass homogenizer and transferred quantitatively to 12-ml. centrifuge tubes. The homogenate is spun at $2000 \times g$ for 15 minutes, and the supernatant is decanted. The precipitate is reextracted twice with 3 ml. of 75% ethanol, and the pooled extracts are evaporated to dryness in 20-ml. beakers under an infrared lamp with the aid of a fan. One milliliter of distilled water is added to dried extract for every 100 mg. of original fresh weight of tissue. The extract is triturated thoroughly, and the suspension is spun in a Spinco ultracentrifuge at $20,000 \times g$ for 30 minutes. Aliquots of the supernatant (usually 0.2 ml.) are pipetted into numbered optical tubes, taken to dryness in a vacuum oven at 60°, and stored in vacuum desiccator until use. For analysis the extract in optical tubes is resuspended in 0.1 ml. of water. The final assay system contains the extract and 0.3 ml. of $0.1 M$ pyrophosphate buffer, pH 8.3; 0.1 ml. of mercaptoethanol, 2 mg./ml. in pyrophosphate buffer; 0.1 ml. of TPN, 8 mg./ml.; 0.1 ml. of enzyme; 0.1 ml. of α-ketoglutaric acid neutralized with sodium hydroxide, 14.6 mg./ml. Final volume of assay system is 0.8 ml. The mercaptoethanol, TPN, enzyme, and α-ketoglutarate are premixed and stored on ice. Fraction 3, the first dialyzed acetone fraction of the bacterial enzyme preparation,[13] is used for all the determinations. The reaction is started by addition of 0.4 ml. of the mixture to 0.4 ml. of buffer and sample contained in the optical tube. Optical density at 340 mμ is read immediately after the reaction mixture has been stirred. All spectrophotometric readings are made with the Beckman spectrophotometer Model DU, equipped with tube adapter.[15] Enzyme concentration is so adjusted that 0.1 ml. of enzyme in incubation mixture of 0.8 ml. produces maximal $\Delta O.D._{340}$ 15 to 20 minutes after initiation of the reaction which is proportional to amount of GABA in a sample. One microgram of GABA in the system corresponds to $\Delta O.D._{340}$ of 0.055. This represents close to an isomolar relationship between GABA in a sample and formation of TPNH. Samples containing no GABA show no changes in $O.D._{340}$.

Paper Chromatographic Method.[16] Spots corresponding to known amounts of GABA and accurately measured amounts of unknown samples are placed along the narrow edge of sheets of Whatman No. 1

[14] C. F. Baxter and E. Roberts, *Proc. Soc. Exptl. Biol. Med.* **101**, 811 (1959).
[15] O. H. Lowry, N. R. Roberts, K. Y. Leiner, M. Wu, and A. L. Farr, *J. Biol. Chem.* **207**, 1 (1954).
[16] E. Roberts and S. Frankel, *J. Biol. Chem.* **187, 55** (1950).

filter paper and run in water-saturated phenol for 24 hours. After thorough removal of the phenol by directing a fan at the sheets for 12 to 18 hours, the sheets are sprayed on both sides with a 0.1% solution of ninhydrin in butanol, and the color development is allowed to take place for 24 hours in the dark at room temperature or in 30 minutes at 90° in a moist atmosphere. The developed spots are cut out, care being taken to include the same total area of paper in all the known and unknown samples in any set of determinations. Suitably chosen paper blanks always are included. The pieces of paper containing the spots are cut into small strips with minimal handling and put into test tubes. Five milliliters of glass-distilled water are then added, and the tubes are shaken vigorously, to achieve quantitative elution of the color. The paper fiber is then removed by centrifugation, and the color is read at 570 mμ in the Beckman spectrophotometer. The optical density is proportional to the concentration in the range studied (1 to 15 μg.). A series of standards is run with each set of determinations. By using concentrations of protein-free brain extract corresponding to 20 mg. of original fresh weight of tissue or less, the quantities of the interfering substances are reduced below a level at which positive reactions with ninhydrin are given so that the above method is applicable. In the case of extracts which contain substances which may overlap with GABA in the one-dimensional system, a similar procedure may be employed in which the unknown and the standard are run on two-dimensional chromatograms in phenol and lutidine. An accuracy of ±8% can be achieved with these methods. There is a marked destruction of GABA on the paper chromatograms if the sheets are dried after removal from the phenol solvent in smog-laden air. The relative rates of destruction of GABA are greater than those of the other ninhydrin-active constituents, and the only solution to this problem is to perform the operation in smog-proof quarters. Many variations of the above procedure are possible, and the choice of chromatographic solvents may depend on the nature of the constituents found in the extracts.

Other Procedures

The inhibition of crayfish stretch receptor by GABA has been used as the basis for an assay for GABA in extracts of brain.[17] This assay system responds not only to GABA, but also to other substances, so that analytical values obtained in this manner are probably a composite of several substances with different specific activities in the test system. Column chromatographic procedures have been employed for the meas-

---

[17] E. Florey and E. Florey, *J. Physiol. (London)* 144, 220 (1958).

urement of GABA,[18,19] but the amount of effort required to perform the analyses is warranted only if quantitative information about constituents other than GABA also is desired.

[18] S. Berl and H. Waelsch, *J. Neurochem.* 3, 161 (1958).
[19] H. H. Tallan, S. Moore, and W. H. Stein, *J. Biol. Chem.* 211, 927 (1954).

[90] Determination of Spermine, Spermidine, Putrescine, and Related Compounds

By CELIA WHITE TABOR and SANFORD M. ROSENTHAL

Principle

Diamines and polyamines can be separated by ion-exchange chromatography.[1,2] Quantitative determination of these compounds after separation is carried out by colorimetric determination of the dinitrophenyl derivatives[3] or by isotopic measurements when C^{14} or N^{15} materials are used.

Columnar Chromatography

Several different procedures using different resins have been developed to meet the needs of specific determinations. To separate spermine, spermidine, and putrescine from each other, and from basic amino acids or acetylated polyamines, Amberlite XE-64 in the K[+] form has been found most suitable.[1,2,4]

Method I (for General Analyses)

Reagents. Amberlite XE-64 is first washed with $4 N$ HCl, then converted to the K[+] form by equilibration with $1 N$ KOH, and finally thoroughly washed with water. The eluting salt solution contains 1.65 moles of Na_2SO_4 and 0.1 mole of sodium phosphate buffer, pH 7.2, per liter.

Procedure. The material to be analyzed is extracted with $0.3 N$

[1] S. M. Rosenthal and C. W. Tabor, *J. Pharmacol. Exptl. Therap.* 116, 131 (1956).
[2] H. Tabor, S. M. Rosenthal, and C. W. Tabor, *J. Biol. Chem.* 233, 907 (1958).
[3] Modification of the method of F. Sanger, *Biochem. J.* 39, 507 (1945).
[4] D. Dubin and S. M. Rosenthal, *J. Biol. Chem.* 235, 776 (1960).

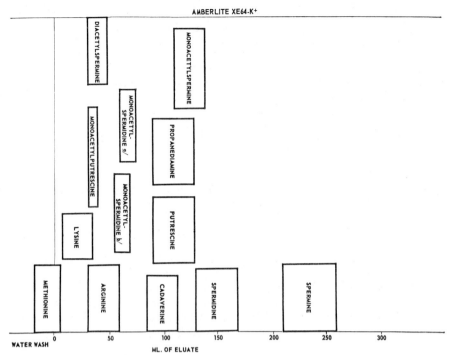

Fig. 1. Gradient elution [H. Busch, R. B. Hurlbert, and V. R. Potter, *J. Biol. Chem.* **196,** 717 (1952)] was carried out with a solution containing 0.1 mole of sodium phosphate buffer, pH 7.2, and 1.65 moles of sodium sulfate per liter. The mixing vessel contained 200 ml. of water. Air pressure (2 pounds) was used for the elution. Flow rate was maintained at approximately 20 ml./hr. The data in this and the following figures are taken from the paper of H. Tabor, S. M. Rosenthal, and C. W. Tabor, *J. Biol. Chem.* **233,** 907 (1958), supplemented by data of D. Dubin and S. M. Rosenthal, *J. Biol. Chem.* **235,** 776 (1960). (*a*) $NH_2CH_2CH_2CH_2NHCH_2$-$CH_2CH_2CH_2NHCOCH_3$. (*b*) $CH_3CONHCH_2CH_2CH_2NHCH_2CH_2CH_2CH_2NH_2$.

trichloroacetic acid; the excess acid is removed by three extractions with 3 vol. of ether. The water layer is adjusted to pH 7 with KOH.[5] The extract is chromatographed on a column 0.6 × 8 cm. (inside diameter × height). The fractions can then be assayed directly for dinitrofluorobenzene reacting material as described below. The details and elution pattern

[5] For analysis of urine, it is convenient to remove amino acids by extraction of the polyamines into tertiary butanol [F. C. McIntire, L. W. Roth, and J. L. Shaw, *J. Biol. Chem.* **170,** 537 (1947)]. Urine plus sufficient water to adjust the volume to 5 ml., 1.4 g. of the alkaline salt mixture (3 parts Na_2SO_4, 1 part $Na_3PO_4·12 H_2O$) and 5 ml. of tertiary butanol are shaken for 20 minutes. The butanol layer is mixed with 1 ml. of 1 N hydrochloric acid and evaporated to dryness on the steam bath. The residue is taken up in water, adjusted to pH 7, and chromatographed. This extraction procedure can also be used with trichloroacetic acid extracts of tissues after removal of excess trichloroacetic acid.

are shown in Fig. 1. In this method, the eluates contain large amounts of salts. Therefore, for isotopic studies, method II is preferable.

Method II (for Radioactive Analyses)

   Reagents

   Dowex 50 (H⁺ form), 2% crosslinked, 100 to 200 mesh.
   2.5 N HCl.

   Procedure. The material is prepared as for method I[6] and adsorbed onto a 0.6 × 8-cm. column (inside diameter × height). The details and elution pattern are shown in Fig. **2**. This method does not separate

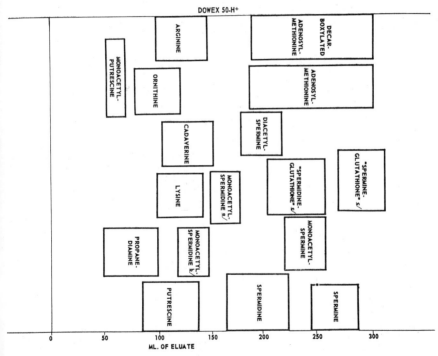

FIG. 2. Gradient elution was carried out with 2.5 N HCl; 300 ml. of H₂O were in the mixing vessel. Flow rate was maintained at approximately 20 ml./hr. In many experiments, Dowex 50 (H⁺ form), 2% crosslinked, 200 to 400 mesh, was used instead of 100 to 200 mesh. With the same elution conditions, the elution volumes were found to be as follows: putrescine, 120 to 144 ml.; spermidine, 233 to 277 ml.; spermine, 343 to 390 ml.; and S-adenosylmethionine, 262 to 452 ml. (a) NH₂CH₂-CH₂CH₂NHCH₂CH₂CH₂CH₂NHCOCH₃. (b) CH₃CONHCH₂CH₂CH₂NHCH₂CH₂-CH₂CH₂NH₂. (c) D. Dubin, *Biochem. Biophys. Research Communs.* **1**, 262 (1960).

⁶ In method II, it is unnecessary to remove and neutralize excess trichloroacetic acid before adsorbing the extract onto the column.

putrescine from several contaminating amino acids[5]; however, in most protein-free tissue extracts, the concentrations of these amino acids are low. Aliquots can be plated (preferably on Monel planchets), dried *in vacuo*, and counted. Determination by the dinitrofluorobenzene procedure is also possible if the acid is either removed[7] or neutralized.[8]

Method III

This method is a modification of method II which has been found suitable for rapid routine assays in studies of the enzymatic biosynthesis of C^{14}-spermidine from C^{14}-putrescine.

Reagents

Dowex 50 (H+ form), 2% crosslinked, 100 to 200 mesh.
0.5 N HCl.
2.5 N HCl.

Procedure. To the incubation mixture 2 micromoles of unlabeled spermidine (as spermidine phosphate) and 1 ml. of water are added. Sufficient trichloroacetic acid is added to bring the pH to approximately 1 (a thymol blue indicator is used), and the precipitate is removed by centrifugation. The supernatant is put directly onto a 0.5 × 3.5-cm. column of the Dowex 50. Under 2 pounds of air pressure, the column is washed twice with 5 ml. of water, and then with 150 ml. of 0.5 N HCl to remove the C^{14}-putrescine. The spermidine is then eluted with 10 ml. of 2.5 N HCl. An aliquot is plated, dried, and counted.

Method IV

In enzymatic studies on the biosynthesis of C^{14}-spermidine from C^{14}-decarboxylated adenosylmethionine,[9] it was necessary to separate these two compounds. Since method III does not accomplish the separation, the following procedure was developed.

Reagents

XE-64 (H+ form). The resin is prepared by washing it with 4 N HCl, followed by thorough washing with water.
1 N acetic acid.

---

[7] The acid can be removed by evaporating the aliquots to be assayed in a dessicator over KOH.
[8] When the material is in acid solution, neutralization of the acid is necessary. A drop of phenolphthalein is added to the solution. Saturated K_2CO_3 is added cautiously until the first appearance of pink color.
[9] See Vol. V [103a], p. 759.

Procedure. The material to be assayed is prepared as in method I. The neutralized extract is put onto a 0.6 × 8-cm. column. Figure 3 gives the elution pattern. Aliquots can be plated, dried, and counted directly.

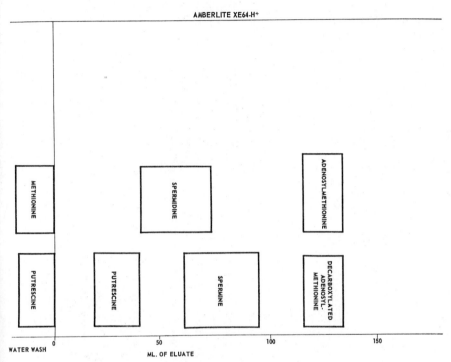

Fig. 3. Gradient elution was carried out with 1 *N* acetic acid; 200 ml. of water were in the mixing vessel. Flow rate was maintained at approximately 40 ml./hr.

Dinitrofluorobenzene Procedure[1-3]

Reagents

5 *N* HCl.
Saturated sodium borate solution.
1.3% dinitrofluorobenzene in acetone.
Saturated K$_2$CO$_3$.
4-Methyl-2-pentanone.
Cyclohexanone (redistilled).

Procedure. An aliquot of the material to be tested, containing 0.02 to 0.1 micromole of amine, is diluted to 2 ml. with water.[8,10] Then 0.5 ml.

[10] To determine the presence of conjugated derivatives, a portion of the sample can be analyzed before and after hydrolysis.[4]

SOLVENT SYSTEMS AND R_f VALUES

Solvent system[a]	1,3 Propane-diamine	Putrescine	Spermidine	Spermine	Monoacetyl putrescine	Monoacetyl-spermidine A[b]	Monoacetyl-spermidine B[c]	Monoacetyl-spermine	Diacetyl-spermine
I[d]	0.41	0.45	0.22	0.11	—	—	—	—	—
II[d]	0.40	0.48	0.27	0.10	—	—	—	—	—
III[e]	0.35	0.89	—	0.33	—	—	—	—	—
IV[f,g]	—	0.41	0.31	0.24	0.67	0.54	0.53	0.39	0.66
V[g,h]	—	0.45	0.33	0.23	0.79	0.65	0.65	0.53	0.78
VI[g,i]	—	0.15	0.28	0.48	0.50	0.73	0.33	0.50	0.43

[a] Solvent I: diethylene glycol monoethyl ether–propionic acid–water (70:15:15) saturated with NaCl. Whatman No. 1 paper. Solvent II: ethylene glycol monomethyl ether–propionic acid–water (70:15:15) saturated with NaCl. Whatman No. 1 paper. Solvent III: phenol, in an atmosphere saturated with 50% aqueous acetic acid. Whatman No. 4 paper. Solvent IV: 1-butanol–acetic acid–pyridine–water (4:1:1:2) Schleicher and Schuell No. 598. Solvent V: 1-propanol–concentrated HCl–H$_2$O (3:1:1) Schleicher and Schuell No. 598. Solvent VI: 1-propanol–triethylamine–water (85:3:15) Schleicher and Schuell No. 597.

[b] $NH_2CH_2CH_2CH_2NHCH_2CH_2CH_2CH_2NHCOCH_3$.

[c] $CH_3CONHCH_2CH_2CH_2NHCH_2CH_2CH_2CH_2NH_2$.

[d] E. J. Herbst, D. L. Keister, and R. H. Weaver, Arch. Biochem. Biophys. 75, 178 (1958).

[e] J. M. Bremner and R. H. Kenten, Biochem. J. 49, 651 (1951).

[f] Solvent of S. Friedman, unpublished data.

[g] D. Dubin and S. M. Rosenthal, J. Biol. Chem. 235, 776 (1960).

[h] Modification of solvent used by A. D. Hershey, Virology 4, 237 (1957).

[i] Modification of solvent used by R. R. Redfield, Biochim. et Biophys. Acta 10, 344 (1953).

of saturated sodium borate solution and 0.25 ml. of the dinitrofluorobenzene solution are added; after thorough mixing, the tube is heated in a
60° bath for 10 minutes. Next 0.5 ml. of 5 N HCl is added, and after
cooling to room temperature, the dinitrophenyl derivative is extracted
into 2 ml. of cyclohexanone[11] or 4-methyl-2-pentanone. The absorption is
determined at 420 mμ. A positive test is obtained with all primary and
secondary amines, but not with acetylated amino groups. Standards of
the polyamines concerned are included in each run. Under these conditions, the optical density of 0.1 micromole of putrescine is 0.44, of 0.1
micromole of spermidine is 0.71, and of 0.1 micromole of spermine is 0.90.

Paper Chromatography and Electrophoresis

Various solvent systems have been described for the separation of
spermine and related compounds. The table gives the solvent systems
and R_f values. The compounds were located by a ninhydrin spray.
Biautographic techniques, using *Hemophilus parainfluenzae*, have also
been described.[12, 13]

Paper electrophoresis has been used by Herbst *et al.*[14] for separation
of these amines.

Miscellaneous Procedures

Spermine, spermidine, and putrescine react in the reduced ninhydrin
procedure of Moore and Stein.[15] These amines are steam-distillable and
have been isolated by this technique.[12] Solid derivatives, as the picrates,
flavianates, chloroaurates, phosphates, and chloroplatinates, have been
prepared in large-scale isolations.[16, 17]

Spermine hydrochloride, spermine phosphate, spermidine phosphate,
and putrescine hydrochloride are available commercially.

We have synthesized C^{14}-putrescine and N^{15}-putrescine[2]; C^{14}-putres-
cine is now available commercially. Spermidine and spermine, labeled

---

[11] Cyclohexanone is used for extraction of the putrescine dinitrophenyl derivative,
since the latter is very insoluble in 4-methyl-2-pentanone. Either cyclohexanone or
4-methyl-2-pentanone can be used for the other amines.

[12] E. J. Herbst, R. H. Weaver, and D. L. Keister, *Arch. Biochem. Biophys.* **75,** 171
(1958).

[13] P. H. A. Sneath [*Nature* **175,** 818 (1955)] described a putrescine-requiring mutant
strain of *Aspergillus nidulans*. This organism can also be used for the bioassay of
putrescine.

[14] E. J. Herbst, D. L. Keister, and R. H. Weaver, *Arch. Biochem. Biophys.* **75,** 178
(1958).

[15] See Vol. III [76], p. 468.

[16] H. W. Dudley, M. C. Rosenheim, and O. Rosenheim, *Biochem. J.* **18,** 1263 (1924).

[17] M. Guggenheim, "Die biogenen Amine," p. 325. S. Karger, Basel, 1951.

with C^{14} in the 3-carbon or 4-carbon moiety, have also been synthesized.[18] The following acetylated derivatives have been prepared synthetically: monoacetylputrescine,[19] monoacetylspermine and diacetylspermine,[4, 20] and two isomers of monoacetylspermidine.[4, 19]

[18] E. L. Jackson and S. M. Rosenthal, *J. Org. Chem.* **25**, 1055 (1960).
[19] E. L. Jackson, *J. Org. Chem.* **21**, 1374 (1956).
[20] H. Bauer, unpublished results.

[91] β-Aspartyl Phosphate and Aspartic-β-semialdehyde

By SIMON BLACK

Preparation of β-Aspartyl Phosphate

Principle. The chloride group of N-carbobenzoxy-L-aspartyl α-benzyl ester β-chloride is replaced with a phosphate group by treatment with AgH_2PO_4. The carbobenzoxy and benzyl groups are then removed by catalytic hydrogenation under especially mild conditions.[1] An alternative method for preparing this substance has been described by Katchalsky and Paecht.[2]

Method. One-half gram (1.4 millimoles) of N-carbobenzoxy-L-aspartic acid α-benzyl ester[3] is dissolved in 2 ml. of dry ether in a 15 × 125-mm. test tube. The solution is cooled in an ice bath, and 0.4 g. of PCl_5 is added; the tube is then removed from the ice bath, and the mixture is stirred with a glass rod until the PCl_5 is largely dissolved. Ten milliliters of petroleum ether are added, and the product is allowed to crystallize for about 30 minutes. The walls of the tube may have to be scratched with the glass rod to induce crystallization. The crystals are centrifuged to the bottom of the tube, which is capped during this procedure, and washed once with 10 ml. of petroleum ether. The compound is then immediately transferred with 25 ml. of dry ether, in which it is partially dissolved, to a 125-ml. Erlenmeyer flask which contains an intimate mixture of 0.2 ml. of H_3PO_4 (85%) plus 0.4 g. of Ag_3PO_4. The silver phosphate mixture under the ether solution is broken into small fragments with a stirring rod. A $CaCl_2$ tube is then attached to the flask, and the flask is mechanically shaken for 30 minutes at room temperature. The suspension is briefly centrifuged in capped tubes to give a clear ether solution of N-carbobenzoxy-L-aspartyl α-benzyl ester β-phosphate. The yield of this substance is about 80%.

[1] S. Black and N. G. Wright, *J. Biol. Chem.* **213**, 27 (1955).
[2] A. Katchalsky and M. Paecht, *J. Am. Chem. Soc.* **76**, 6042 (1954).
[3] M. Bergmann, L. Zervas, and L. Salzmann, *Ber. chem. G.* **66B**, 1288 (1933).

The ether solution is transferred to a glass "gas absorption" tube into which H_2 can be passed from the bottom through a fritted disk. This tube is immersed in an ice bath and contains, in addition to the ether solution, 5 ml. of water, 2.0 ml. of $2\,M$ $KHCO_3$, and 1 g. of palladium black. Hydrogen is passed through the apparatus at a rate sufficient to mix the ether and aqueous layers vigorously and keep the palladium suspended. This is continued until all the hydroxamic acid-forming substance in the aqueous solution is reactive in the enzyme test described below, usually about 4 hours. By this time all the ether has completely evaporated. The solution is filtered by suction through the fritted disk of the hydrogenation vessel and, if necessary, separated from any toluene not carried away by the hydrogen stream. The pH should be determined at intervals during the hydrogenation. If it falls below 6.5, it should be brought to 6.5 to 7.0 with small amounts of $5\,N$ KOH. The over-all yield of β-aspartyl phosphate is 20 to 30%. It should be stored as this impure solution at $-20°$ at which temperature it deteriorates slowly over a period of weeks.

Assay. β-Aspartyl phosphate can be estimated by the hydroxamic acid method.[4] It can be more positively identified by its disappearance in a reverse test of the β-aspartokinase reaction.[4] One to two micromoles of β-aspartyl phosphate and 5 micromoles of ADP are substituted for aspartate and ATP in this test, and hydroxylamine is omitted from the incubation. After 20 minutes of incubation at 15°, 0.1 ml. of $0.1\,M$ p-chloromercuribenzoate (a suspension) is added, and 5 minutes later 0.4 ml. of $2.0\,M$ hydroxylamine (pH 8.0). After 20 minutes 1.5 ml. of the $FeCl_3$ solution is added, and the optical density at 540 mμ, which is proportional to the hydroxamic acid concentration, is determined.

Stability. β-Aspartyl phosphate hydrolyzes at a substantial rate in solutions at room temperature; hence enzyme tests with this substance are made at 15° where it is reasonably stable for several hours. Its stability is not greatly affected by changes in pH between pH values 4.5 and 10.5, but in more acid or alkaline solutions it decomposes rapidly. It has not been successfully isolated from the reaction medium in which it is formed.

Preparation of Aspartic-β-semialdehyde

Principle. Allylglycine is treated with ozone to yield aspartic-β-semialdehyde and formaldehyde. The method described employs DL-allyl-glycine which is available commercially. It is equally suitable for the L-isomer.[5]

[4] See Vol. V [111].
[5] S. Black and N. G. Wright, *J. Biol. Chem.* **213**, 39 (1955).

Method. Allylglycine (2.3 g., 20 millimoles) is dissolved in 20 ml. of 1 N HCl. Ozone is passed through this solution, held at 0°, at a rate of 0.4 millimole/min. for 100 minutes, or until absorption is no longer measurable. The yield at this point is practically 100%. To free it of other reaction products, 6 millimoles are placed on a column of Dowex 50 (hydrogen form, 200 to 400 mesh, 2.4 × 14 cm.) and washed with a large volume of water. The product is then eluted with 4 N HCl. One 6-ml. fraction is collected every 30 minutes; 80% of the compound has been found in fractions 6 to 9. The unpurified product of ozonization is suitable for most enzymatic tests.

Assay. L-Aspartic-β-semialdehyde is readily determined by use of the homoserine dehydrogenase test.[4] With limiting amounts of substrate, DPNH disappearance in this test is equivalent to the amount of L-aspartic-β-semialdehyde added.

Stability. In acid solutions this compound is quite stable, and at —20° it has been stored successfully for several years. On neutralization, however, it deteriorates within a few hours. Therefore solutions must be neutralized immediately before use. The compound has not been successfully isolated from solutions.

[92] α,ε-Diaminopimelic Acid

$$COOH \cdot CH(NH_2) \cdot (CH_2)_3 \cdot CH(NH_2) \cdot COOH \text{ (mol. wt. 190)}$$

By ELIZABETH WORK

α,ε-Diaminopimelic acid, a symmetrical α,α'-diaminodicarboxylic acid, exists in three stereoisomeric forms; the LL-, the DD-, and the *meso*- (internally compensated) isomers. Since each isomer behaves differently toward enzymes, it is essential that pure isomers be used as substrates. Diaminopimelic acid can be produced synthetically or microbiologically —in the latter case only the *meso*- and LL-isomers are formed.

Synthesis

The general method[1] for synthesis of amino acids through the diethyl-phthalimido derivative is used.[2] Another method[3] through the diphos-phiniminoester is said to give 100% yield.

[1] J. C. Sheehan and W. A. Bolhofer, *J. Am. Chem. Soc.* **72**, 2786 (1950).
[2] E. Work, S. M. Birnbaum, M. Winitz, and J. P. Greenstein, *J. Am. Chem. Soc.* **77**, 1916 (1955).
[3] F. Lingens, *Z. Naturforsch.* **15b**, 811 (1960).

To 203 g. of diethyl α,ε-dibrompimelate dissolved in 920 ml. of dimethyl formamide is added 296 g. of potassium phthalimide. The reaction mixture is heated over a steam bath for 2 to 3 hours with occasional shaking. After cooling, 1040 ml. of chloroform is added, and the mixture is poured into 4 l. of water. The aqueous layer is separated and extracted twice with 800-ml. portions of chloroform. The combined chloroform extracts are washed once with 0.1 N NaOH and twice with water. After the extracts have been dried with anhydrous sodium sulfate and the chloroform has been removed under reduced pressure, 275.5 g. of a clear oil is obtained. The oil is dissolved in 2.5 l. of absolute methanol, 34.9 ml. of anhydrous hydrazine is added, and the solution is refluxed over a steam bath for 3 hours. The suspension is concentrated and mixed with 1.3 l. of water, and the remaining methanol is removed under reduced pressure. Then 1.3 l. of 10 N HCl is added, and the mixture is refluxed for 4 hours, after which it is cooled to 0° and the precipitated phthalyl hydrazide is filtered off. Excess HCl is removed by concentration in vacuo, addition of water, and reconcentration. The residue, dissolved in a minimal amount of water, is adjusted to pH 6.2 with LiOH, treated with charcoal, and filtered (total volume about 1 l.). Five volumes of ethanol is added, and the mixture is left at −10° overnight. The precipitated oil is separated by decantation and dissolved in 600 ml. of hot water; on precipitation with ethanol it yields a solid precipitate. This is recrystallized from ethanol–water: yield, 95.6 g.

Microbiological Preparation

Principle. Lysine-requiring mutants of *Escherichia coli* which lack diaminopimelic decarboxylase accumulate diaminopimelic acid in the culture filtrate when grown in suboptimal concentrations of lysine. Diaminopimelic acid is prepared from the culture filtrate by ion-exchange chromatography.[4-6] It is a mixture of *meso-* and LL-isomers.[5,6]

Growth of Organism.[7-9] Lysine-requiring mutant strains **26-26** or ATCC **12408** of *E. coli*,[7,10] are used. These are unstable mutants which revert easily to the wild forms after exhaustion of lysine from the medium. They should therefore be maintained in the presence of ade-

[4] E. Work and R. F. Denman, *Biochim. et Biophys. Acta* **10**, 183 (1953).
[5] L. D. Wright and E. L. Cresson, *Proc. Soc. Exptl. Biol. Med.* **82**, 354 (1953).
[6] D. S. Hoare and E. Work, *Biochem. J.* **61**, 562 (1955).
[7] L. E. Casida, U.S. Patent 2,771,396 (1955).
[8] D. A. Kita and H. T. Huang, U.S. Patent 2,841,532 (1956).
[9] J. Angulo, T. D. G.-Mauriño, A. M. Municio, and W. Rivero, *Anales real. soc. españ. fís y quím. (Madrid)* **56B**, 413 (1960).
[10] B. D. Davis, *Nature* **169**, 534 (1952).

quate lysine (at least 200 μM), and on each subculture a test for reversion should be carried out by inoculating a tube of minimal medium + glucose. *E. coli* 12408 produces very much higher amounts of diaminopimelic acid than does strain 26-26, particularly under high aeration rates,[7-9] which are, however, difficult to attain under laboratory conditions. The requisite amount of lysine in the growth medium is also dependent on growth conditions and should be determined in each laboratory; if it is exceeded, the yield of diaminopimelic acid decreases. The medium used by the author for *E. coli* 12408 consists of the following: $(NH_4)_2HPO_4$, 15 g.; KH_2PO_4, 2 g.; Mg $SO_4 \cdot 7$ H_2O, 50 mg.; $Na_2SO_4 \cdot 10$ H_2O, 0.1 g.; lysine hydrochloride, 80 mg.; pH adjusted to 7.4; water to 1 l. The energy source can be either glucose (2%) or preferably a mixture of sucrose (1.5%) and glycerol (1.5%); sterile sucrose or glucose is added after autoclaving. In a typical experiment designed to produce C^{14}-diaminopimelic acid, 150 ml. of medium in two 2-l. flasks is inoculated with 0.1-ml. portions of an overnight culture of *E. coli* 12408 grown in the same medium. The flasks are shaken at 37° on a rotary shaker until the concentration of diaminopimelic acid in the culture filtrate reaches a maximum (40 to 50 hours). If growth is continued beyond this point, there is danger that the culture will revert to a nonexacting organism containing diaminopimelic decarboxylase,[8] and the diaminopimelic acid level may drop. Final diaminopimelic acid concentration, estimated colorimetrically by the ninhydrin reaction (see later) on 0.1 ml. of culture filtrate after removal of cells by centrifugation, was 2.1 mg./ml.

Treatment of Culture Filtrate. The cells are removed by centrifugation, and the culture filtrate is dialyzed to separate diaminopimelic acid from high-molecular-weight extracellular materials. The dialyzate is concentrated and then treated with an excess of charcoal, filtered, and evaporated to dryness *in vacuo*.

Ion-Exchange Column. In earlier work on this preparation, the sulfonated polystyrene resin Zeo-Karb 225, 8% crosslinked ($\times$8), was used;[11] a column 4.8 × 50 cm. would separate diaminopimelic acid from 4 l. of culture filtrate from *E. coli* 26-26.[4, 6] Later, Zeo-Karb 225, 4.5% crosslinked (100 to 150 mesh), was used to prepare C^{14}-labeled diaminopimelic acid.[12] This form has an advantage over the $\times$8 resin in that diaminopimelic acid is more easily removed.[6] The resin ($\times$4.5) is prepared by cycling with 2 N NaOH, 4 N HCl, and 2 N HCl and finally is equilibrated on the column with 1.25 N HCl. The dried dialyzed culture filtrate is dissolved in the minimum quantity of 1.25 N HCl and applied

[11] From Permutit Co. Ltd., Gunnersbury Ave., London, England. This resin is similar to Dowex 50.
[12] P. Meadow and E. Work, *Biochem. J.* 72, 396 (1959).

to the column (1 × 60 cm.); elution is carried out with acid of the same strength at room temperature. Samples (0.2 ml.) of eluate fractions are evaporated to dryness in a vacuum desiccator over KOH and examined for amino acids by paper chromatography[13]; 0.1-ml. samples are also examined colorimetrically for diaminopimelic acid by reaction with ninhydrin in acid solution (see later).

Small amounts of aspartic acid, glutamic acid, glycine, alanine, and valine will have been eluted by 1000 ml. of HCl; diaminopimelic acid is found in the next 400 ml. The column should then be washed with stronger HCl (2.5 and 4.0 N) to remove basic substances; after equilibration with 1.25 N HCl, it will be ready for further use.

The eluate fractions containing diaminopimelic acid are combined and concentrated to dryness *in vacuo;* water is added, and the solution is decolorized with charcoal. The crude free amino acid is obtained, after the pH is adjusted to 6.2 with LiOH, by precipitation with 2 vol. of ethanol. Yield, 260 mg.

Separation of Isomers

Principle. meso-Diaminopimelic acid differs sufficiently in solubility from the other two isomers to enable separation to be carried out by fractional crystallization either as the free amino acid or as a salt. Derivatives may also be similarly fractionated. The racemic mixture of LL- and DD-isomers must be resolved by enzymatic means. Hog kidney amidase has been successfully used; this enzyme specifically hydrolyzes amide groups in the L-configuration.[14] If the mixture of diamides of all three isomers of diaminopimelic acid is digested with amidase, the final reaction mixture consists of free LL-diaminopimelic acid, *meso*-diaminopimelic acid D-monoamide, and DD-diaminopimelic acid diamide [reaction (1c)]. These compounds can be separated by ion-exchange chromatography, and the amino acids subsequently recovered from the amides.

Fractional Crystallization. meso-Diaminopimelic acid can'be separated from the other isomers by three or four crystallizations from aqueous ethanol.[6,15] If the original diaminopimelic acid is of microbiological origin, and so contains only *meso*- and LL-isomers, LL-isomer may be separated from the aqueous–ethanol mother liquors, left after

[13] Methanol–water–10 N HCl–pyridine (80:17.5:2.5:10) (see Vol. V [117]) is a useful solvent for examining acidic solutions, as the chromatograms are unaffected by small amounts of acid or salts. Phenol (NH₃ atmosphere) may also be used, but acid must be removed more thoroughly.

[14] J. P. Greenstein, Vol. III [82].

[15] For these fractional crystallizations, make a strong aqueous solution, cool to room temperature, add organic solvent until a faint permanent turbidity appears, and leave at room temperature for several hours.

crystallization of the *meso*-isomer, by crystallization as the monohydrochloride from dilute HCl with acetone.[6,15] The progress of the separations can be followed by chromatography on Whatman No. 1 paper with methanol–water–pyridine (80:20:10) or methanol–water–10 N HCl–pyridine (80:17.5:2.5:10).[6,16] These solvents separate LL-diaminopimelic acid from the *meso*- and DD-isomers.

Another method, used by Gilvarg[17] for the microbiologically produced material, is to form the ternary salt with β-naphthalenesulfonic acid; two crystallizations from 1 N HCl were reported to produce the pure salt of LL-diaminopimelic acid. The mother liquor, already enriched in *meso*-isomer, is said to yield the pure *meso*-form by two crystallizations as the dihydrochloride from 10 N HCl. Wide variations may occur in the ratio of the two isomers in commercially available diaminopimelic acid. It has been reported[17a] that this ratio determines which isomer crystallizes out first; if the *meso*-isomer predominates in the mixture, then its salt occurs in the first crop of crystals, whereas with an original mixture rich in LL-isomer, crystallization proceeds as described by Gilvarg.[17] Fractionation of synthetic carbobenzoxydiaminopimelic acid is reported to separate the *meso*-isomer from the other isomers[18]; crystallization from ethyl acetate produced dicarbobenzoxy-DD,LL-diaminopimelic acid (m.p. 164° to 165°); the dried mother liquors, on crystallization from chloroform, produced dicarbobenzoxy-*meso*-diaminopimelic acid (m.p. 123° to 125°). This method should also be applicable to microbiologically produced diaminopimelic acid, but here again the isomer composition of the original mixture may affect the relative solubility of the derivatives.

Enzymatic Resolution. The steps used in the resolution of synthetic diaminopimelic acid[2] are summarized in reactions (1a) to (1f), where diaminopimelic acid is abbreviated to DAP.

$$\text{LL,DD,}meso\text{-DAP} \xrightarrow[\text{HCl}]{\text{methanol}} \text{LL,DD,}meso\text{-DAP-dimethyl ester·2HCl} \qquad (1a)$$

$$\text{LL,DD,}meso\text{-DAP-dimethyl ester·2HCl} \xrightarrow[\text{methanol, NH}_3]{\text{anhydrous}} \qquad (1b)$$
$$\text{LL,DD,}meso\text{-DAP-diamide·2HCl}$$

[16] L. E. Rhuland, E. Work, R. F. Denman, and D. S. Hoare, *J. Am. Chem. Soc.* **77**, 4844 (1955).

[17] C. Gilvarg, *J. Biol. Chem.* **234**, 2955 (1959).

[17a] D. Jusic, C. Roy, A. J. Schocher, and R. W. Watson, *Can. J. Biochem. Physiol.* **41**, 817 (1963).

[18] R. Wade, S. M. Birnbaum, M. Winitz, R. J. Koegel, and J. P. Greenstein, *J. Am. Chem. Soc.* **79**, 648 (1957).

$$\text{LL,DD,}meso\text{-DAP-diamide} \xrightarrow{\text{amidase}} \begin{cases} \text{LL-DAP} + 2NH_3 \\ meso\text{-DAP-D-monoamide} + NH_3 \\ \text{DD-DAP-diamide} \end{cases} \quad (1c)$$

Separate three components by ion exchange.   (1d)

$$meso\text{-DAP-D-monoamide} \xrightarrow{\text{HCl}} meso\text{-DAP} \quad (1e)$$

$$\text{DD-DAP-diamide} \xrightarrow{\text{HCl}} \text{DD-DAP} \quad (1f)$$

Reagents

Hog kidney amidase,[19] freshly dialyzed.

Diaminopimelic acid diamide dihydrochloride, prepared by steps (1a) and (1b).

Amberlite X-64 ion-exchange column (3.5 × 90 cm.) in the Li^+ form, buffered with lithium acetate to pH 6.5, and washed with water.

Procedure. The digest (570 ml.) containing 14 g. of diaminopimelic acid diamide dihydrochloride in water, dialyzed amidase preparation (285 mg. of nitrogen), and 0.01 M $MnCl_2$, adjusted to pH 8.0 with LiOH, is incubated for 7 hours at 37°, or until the level of α-COOH amino groups, determined by the Van Slyke ninhydrin–CO_2 method, shows no change; further enzyme is added, and the incubation is continued for 2 hours, during which there is no further change in amino nitrogen content. The digest is dialyzed at 2° against three changes of water, and the dialyzates are concentrated *in vacuo* to 50 ml. The solution is adjusted to pH 9.2 with LiOH and immediately placed on the column. The column is washed with water (1.1 l.) until the effluent fractions give no ninhydrin reaction for amino acids. The ninhydrin-positive fractions of the effluent are examined by paper chromatography (phenol–ammonia, see also footnote 13); the main band consists of LL-diaminopimelic acid ($R_f =$ 0.27), and there may be small bands of other materials ($R_f = 0.15, 0.55,$ 0.62). The column is then irrigated with 1% (v/v) acetic acid; *meso*-diaminopimelic acid D-monoamide ($R_f = 0.69$) with DD-diaminopimelic acid diamide ($R_f = 0.92$) are eluted successively. In the resolution carried out by the author, some overlapping of these two components occurred. A buffered acidic eluant would probably have avoided this, but time did not permit further investigations. The overlap was insufficient to interfere with resolution, although it reduced the final yield of pure *meso*- and DD-isomers.

[19] S. M. Birnbaum, Vol. II [55].

The fractions containing the pure amides are concentrated separately *in vacuo*, each dissolved in 1 l. of 3 N HCl and hydrolyzed under reflux for 5 hours. The amino acid hydrochlorides can be purified by chromatography on Zeo-Karb 225 or on Dowex 50 with HCl (1.25 N), as previously described. Free amino acids can be then obtained by ethanol precipitation at pH 6.2, after neutralization with LiOH.

Modified Enzymic Resolution. A repeat resolution has been reported,[18] in which *meso*-diaminopimelic acid is first removed from the racemic mixture as the carbobenzoxy derivative. The steps used are indicated in reactions (2a) to (2h), where the carbobenzoxy (carbobenzoxyloxy) group ($C_6H_5CH_2OCO$) is indicated by the abbreviation Cbz.

$$\text{LL,DD,}meso\text{-DAP} \xrightarrow{\text{CbzCl}} \text{DiCbz-LL,DD,}meso\text{-DAP} \qquad (2a)$$

$$\text{DiCbz-LL,DD,}meso\text{-DAP} \xrightarrow[\text{ethylacetate}]{\text{crystallize}} \text{DiCbz-LL,DD-DAP} \qquad (2b)$$

$$\text{Residue} \xrightarrow[\text{chloroform}]{\text{crystallize}} \text{DiCbz-}meso\text{-DAP}$$

$$\text{DiCbz-}meso\text{-DAP} \xrightarrow{\text{H}_2 \text{ palladium}} meso\text{-DAP} \qquad (2c)$$

$$\text{DiCbz-LL,DD-DAP} \xrightarrow[\text{+ isovaleryl chloride + NH}_3]{\text{triethylamine}} \qquad (2d)$$
$$\text{DiCbz-LL,DD-DAP-diamide}$$

$$\text{DiCbz-LL,DD-DAP-diamide} \xrightarrow{\text{H}_2 \text{ palladium}} \text{LL,DD-DAP-diamide-diacetate} \quad (2e)$$

$$\text{LL,DD-DAP-diamide} \xrightarrow{\text{amidase}} \begin{cases} \text{LL-DAP} + 2\text{NH}_3 \\ \text{DD-DAP-diamide} \end{cases} \qquad (2f)$$

Separate two components by ion exchange   (2g)

$$\text{DD-DAP-diamide} \xrightarrow{\text{HCl}} \text{DD-DAP} \qquad (2h)$$

The method is claimed to be advantageous, since it avoids any possibility of overlap of mono- and diamides on the column; it does, however, involve several additional steps. In the opinion of the author, the preliminary removal of *meso*-diaminopimelic acid as the dicarbobenzoxy derivative [step (2b)] should not be necessary, provided conditions are found for the separation of mono- and diamides on the column. On the other hand, the preparation of diaminopimelic acid diamide through the crystallizable dicarbobenzoxy derivatives [steps (2a) and (2d)] may be an easier process than that through the diester hydrochloride [steps (1a) and (1b)], which is not crystallizable.

Properties

meso-Diaminopimelic acid is sparingly soluble in water (0.92%, w/v, at 21°).[20] The LL- and DD-isomers are extremely soluble in water and are precipitated by ethanol from aqueous solution as gels.[2, 6] Both the crystals from water and the powder obtained by dehydrating the gels contain 1 mole of water of crystallization in the case of the LL- and DD-isomers. *meso*-Diaminopimelic acid crystallizes from aqueous solutions in the anhydrous form.

The optical rotations are given in the table.

OPTICAL ROTATIONS OF ISOMERIC COMPONENTS OF α,ε-DIAMINOPIMELIC ACID[a]

Compound	$[\alpha]_D^{24}$ in:		
	H_2O (c5)	1 N HCl (c1)	5 N HCl (c2.6)[b]
LL-Diaminopimelic acid monohydrate	+8.14°	+45.0°	+45.1°
LL-Diaminopimelic acid monohydrochloride	—	—	+38.5°
DD-Diaminopimelic acid monohydrate	−8.45°	−45.5°	−44.6°
DD-Diaminopimelic acid monohydrochloride	—	—	−38.0°
meso-Diaminopimelic acid	—	0	0

[a] From E. Work, S. M. Birnbaum, M. Winitz, and J. P. Greenstein, *J. Am. Chem. Soc.* **77**, 1916 (1955); R. Wade, S. M. Birnbaum, M. Winitz, R. J. Koegel, and J. P. Greenstein, *J. Am. Chem. Soc.* **79**, 648 (1957).

[b] c = g./100 ml. of solution.

The dissociation constants[18] of the *meso*- and LL-isomers are identical; they are $pK'_1 = 1.8$; $pK'_2 = 2.2$; $pK'_3 = 8.8$; $pK'_4 = 9.9$. The calculated isoelectric point, pI, is close to 5.5.

The solid-state infrared spectra (3 to 15 μ) have been published.[18] Those of LL- and DD-diaminopimelic acids are identical; the spectrum of the *meso*-isomer has a number of sharp bands in the region of 6 to 8 μ, in contrast to the more diffuse spectrum of the other isomers in this region.

Paper Chromatography. In acidic solvents all isomers behave the same, but in certain basic solvents the LL-isomer travels slightly faster than the *meso*- or DD-isomers, the ratio R^{LL}/R^{meso} being up to 1.4:1.[16] The R_f values vary with the pH of the mixture under investigation, as well as with the solvent mixture, and cannot be regarded as physical

[20] E. Work, *Biochem. J.* **49**, 17 (1951).

constants worthy of quotation. In the solvent system[6, 16, 21] methanol–water–10 N HCl–pyridine (80:17.5:2.5:10), on Whatman No. 1 paper, the LL-isomer separates well, and sometimes the *meso*-isomer has been observed[22] to travel slightly faster than the DD-isomer. With the exception of cystine, diaminopimelic acid travels more slowly than all other natural amino acids in this solvent, and when the chromatograms are developed by ninhydrin in acetone, diaminopimelic gives characteristic olive-green spots fading to permanent yellow[21]: other amino acids, except lysine, give transitory spots. On two-dimensional chromatograms, in phenol–ammonia, and collidine–water or butanol–acetic acid–water, diaminopimelic acid separates well from all other common amino acids, except cystine (from which it can be distinguished by its stability to hydrogen peroxide).[20, 23]

Assay

Both enzymatic and colorimetric methods are available; the former differentiate between the isomers, the latter do not.

Enzymatic Assay. Diaminopimelic decarboxylase specifically removes 1 mole of CO_2 from *meso*-diaminopimelic acid[24] and can be used for quantitative estimation provided certain precautions are taken. The enzyme frequently occurs with diaminopimelic racemase, which interconverts the LL- and the *meso*-isomers[21]; consequently, enzymatic assay is easier to carry out if differentiation between these isomers is not required. If, however, each isomer is to be assayed separately, the preparation of decarboxylase must be completely freed from racemase; any CO_2 produced will then be derived only from *meso*-diaminopimelic acid. In either case, a good source of enzyme is provided by washed cells of *E. coli* ATCC 9637 grown with aeration at 37° in glucose salts medium[25] and harvested a short time before the end of logarithmic growth. Total *meso*- and LL-diaminopimelic acid can be estimated manometrically[24] by using a crude enzyme preparation consisting of these cells which contain no lysine decarboxylase, but which require more added pyridoxal phosphate for maximum enzyme activity than the amount previously quoted; i.e., stock solution of pyridoxal phosphate contains 1 mg./ml., 0.1 ml. used in each Warburg flask. Acetone-dried preparations of these cells are convenient, as they can be stored at −15°, but their decarboxylase activity is lower than that of fresh cell suspensions containing cetyl-trimethylammonium bromide (60 μg./mg. dry weight) where

[21] E. Work, Vol. V [117].
[22] P. Meadow and E. Work, *Biochim. et Biophys. Acta* **28**, 596 (1958) (see figure).
[23] E. Work and D. L. Dewey, *J. Gen. Microbiol.* **9**, 394 (1953).
[24] E. Work, Vol. V [118].
[25] B. D. Davis and E. S. Mingioli, *J. Bacteriol.* **60**, 17 (1950).

activities are in the region of $Q_{CO_2}^{DAP} = 30$. Diaminopimelic racemase may be removed from cell-free extracts of *E. coli* 9637 grown as above by co-precipitation with the nucleic acids.[21] All steps are carried out at 2°. The washed cells are suspended (about 30 mg. dry weight per ml.) in 0.1 *M* phosphate buffer, pH 6.8, containing 10^{-4} *M* dimercaptopropanol, and disrupted by passage through a needle valve at 12,000 pounds per square inch.[26] The debris is removed by centrifugation at 20,000 × *g* for 20 minutes. To each 5 ml. of supernatant liquid are added, successively, with stirring, 1 ml. of 1% (w/v) protamine sulfate and 0.6 ml. of 25% (w/v) streptomycin sulfate. After 2 hours, the precipitate is removed by centrifuging. The supernatant liquid contains no diaminopimelic racemase; it has all the diaminopimelic decarboxylase activity of the washed organisms, if 10^{-3} *M* dimercaptopropanol and 1.6 × 10^{-3} *M* pyridoxal phosphate are present in the manometers. This enzyme preparation may be stored at −15°. Diaminopimelic decarboxylase is easily inhibited by acidic amino acids and salts and should be used only on solutions from which these substances have been removed, for example by ion exchange.

A less specific enzyme which attacks diaminopimelic acid is L-amino acid oxidase from *Neurospora*. The *meso*-isomer is oxidized with uptake of 1 equivalent of oxygen; the LL-isomer takes up 2 equivalents.[27]

Colorimetric Assay. The ninhydrin reaction in solution can be rendered selective for amino acids containing two amino groups or for cyclic imino acids, if it is carried out in acid solution.[28-30] The modification (a) described here can estimate diaminopimelic acid (2.5 to 220 μg.), provided that lysine, ornithine, proline, cystine, or tryptophan is not present in excess.[29] In the presence of these amino acids, suitable reaction conditions can be used to eliminate interference.

Reagents

Acetic acid (AnalaR).
Ninhydrin reagent[31]: 250 mg. of ninhydrin (recrystallized or AnalaR) dissolved in 6 ml. of acetic acid and 4 ml. of aqueous 0.6 *M* phosphoric acid.

Procedure: (a) SAMPLE IN PURE AQUEOUS OR ACIDIC SOLUTION. One-half milliliter of the test solution, containing 5 to 220 μg. of diamino-

[26] H. W. Milner, N. S. Lawrence, and C. S. French, *Science* **111**, 633 (1950).
[27] E. Work, *Biochim. et Biophys. Acta* **17**, 410 (1955).
[28] F. P. Chinard, *J. Biol. Chem.* **199**, 91 (1952).
[29] E. Work, *Biochem. J.* **67**, 416 (1957).
[30] C. Gilvarg, *J. Biol. Chem.* **233**, 1501 (1958).
[31] Stable for a few days at 2°.

pimelic acid, is mixed with 0.5 ml. of acetic acid and 0.5 ml. of ninhydrin reagent. The solutions are immediately covered and heated at 100° in a water bath for 2 to 5 minutes, then cooled rapidly to room temperature and diluted to 5 ml. with 3.5 ml. of acetic acid. The absorption of the yellow solutions is read at either 345 mμ or 440 mμ against a reagent blank treated identically. A plot of optical density against concentration is linear within the range specified, and the readings are repeatable. HCl does not interfere. Smaller amounts may be estimated by halving the quantities of all solutions.

(b) LYSINE ALSO PRESENT, NOT IN EXCESS. The above procedure is followed except that the heating period must be only 2 minutes, during which lysine does not react.[32] Method (c) may also be used.

(c) LYSINE, ORNITHINE, PROLINE, CYSTINE, OR TRYPTOPHAN ALSO PRESENT. The reaction is carried out in a volume of 5 ml. (i.e., 4 ml. of acetic acid is added before the ninhydrin reagent); after the solution is heated at 37° for 1.5 hours, the optical density is measured at 440 mμ. Standards should be included with each estimation. This procedure is applicable in the presence of 4 moles excess of lysine but cannot be used in the presence of HCl: Gilvarg's method should then be used.[30]

(d) IN PROTEIN-CONTAINING SOLUTIONS (E.G., ENZYMATIC DIGESTS). The solution is deproteinized by adding at least 1 vol. of acetic acid, the amount of acid being chosen so that the final solution contains the appropriate concentration of diaminopimelic acid for estimation. After centrifugation, water is added to the required volume of supernatant, if necessary, to bring the acetic acid concentration to 50% (v/v) before the ninhydrin reagent is added; then procedure (b) is followed, or (c), if other amino acids, particularly lysine, are present. The standard curve is made by adding known amounts of diaminopimelic acid to the deproteinized contents of the blank enzymic digest.

(e) SPOTS ON PAPER CHROMATOGRAMS.[12] The piece of filter paper covered by diaminopimelic acid is cut into small pieces and immersed in 1.0 ml. of 50% (v/v) acetic acid and 0.5 ml. of ninhydrin reagent. After 5 minutes of heating, and dilution with acetic acid to 5 ml., the filter paper is removed by centrifugation,[33] and the optical density of the supernatant is read against a blank similarly prepared from adjacent areas of amino acid-free filter paper. A standard curve is constructed from known amounts of diaminopimelic acid subjected to the same chromatographic procedure.

[32] Lysine can be estimated by this method, after a heating time of 1 hour.
[33] Clear supernatants are often obtained without centrifugation.

[93] Preparation and Properties of Soluble Collagens

By Paul M. Gallop and Sam Seifter

Principle. Suitable substrates for detection or measurement of collagenase activity may be prepared by extraction of various tissues with dilute acid or neutral salt solutions. Preparations thus obtained have the advantages of constancy of properties, freedom from nonconstituent carbohydrates and unrelated protein, and solubility in reagents which permit examination of enzymatic activity at neutral pH values.

Zachariades[1] and Nageotte[2] demonstrated the extractability of collagens by dilute acid solutions. Orekhovich *et al.*[3] showed that, in particular, citrate buffers of pH 3.5 to 4.0 could be used for extraction of a portion of the collagen in a given tissue. The soluble "procollagen" in such an extract could be "reconstituted" into needlelike, "crystalline" fibrils by dialysis against water or dilute salt solutions. Gallop[4] extended this method of preparation to collagen of carp swim bladder which, being one of many fish collagens, has been designated "ichthyocol."

The procedures employed for preparation of ichthyocol and of soluble collagen of calfskin are outlined below.

Reagents

0.5 *M* sodium acetate.

0.1 *M* sodium citrate buffer, pH 4.3, made by mixing equal volumes of 0.1 *M* trisodium citrate and 0.1 *M* citric acid (for extraction of ichthyocol).

0.075 *M* sodium citrate buffer, pH 3.7, made by mixing equal volumes of 0.05 *M* trisodium citrate and 0.1 *M* citric acid (for extraction of calfskin collagen).

0.02 *M* disodium hydrogen phosphate.

Preparation of Ichthyocol from Carp Swim Bladders

The tunica externa, or simply tunic, of fresh carp swim bladder is a tissue of choice for preparation of ichthyocol. It is obtained from the whole bladder by separation from the inner muscular tissue. The tunic consists of an outer, tough, lamellated section of collagenous fibers and

---

[1] P. A. Zachariades, *Compt. rend. soc. biol.* **52**, 182, 251, 1127 (1900).

[2] J. Nageotte, *Compt. rend. soc. biol.* **96**, 172 (1927).

[3] V. N. Orekhovich, A. A. Tustanovski, K. D. Orekhovich, and N. E. Plotnikova, *Biokhimiya* **13**, 55 (1948).

[4] P. M. Gallop, *Arch. Biochem. Biophys.* **54**, 486 (1955).

elastic tissue which merges into an inner, thicker section of finer collagenous fibers. Before use the tissue is washed thoroughly with running cold tap water. A large number of tunics may be collected at one time, and those not used immediately may be stored at −20°; from the standpoint of preparation of ichthyocol, no significant deterioration occurs in tunics at this temperature for several years.

Because collagen readily becomes converted to gelatin (i.e., becomes denatured), special precautions must be taken that tissues and derived extracts be maintained at temperatures between 0° and 5°. Thus, tissues must not be permitted to become unduly warm during grinding or blending procedures; and solutions used for extraction, washing, or dialysis must be prechilled.

Approximately 500 g. of swim bladder tunics are blended in a Waring blendor for 1 to 2 minutes at 5° with prechilled 0.5 M sodium acetate solution; the volume of homogenate is then brought to 2 l. with the same reagent. The mixture is stirred mechanically for 15 to 18 hours at 5° and then centrifuged at 2000 r.p.m. at 5° for 1 hour. The supernatant is discarded. The pasty residue is transferred to a clean towel and squeezed until most of the liquid is removed. The material is then extracted once more with 2 l. of prechilled 0.5 M sodium acetate at 5° for 15 to 18 hours; the mixture is centrifuged as before, the supernatant is discarded, and the residue is squeezed free of gross liquid in a clean towel. The process of extraction with sodium acetate solution is performed two times more, yielding finally a residue from which noncollagenous soluble proteins and polysaccharides have been removed; a fraction of the soluble collagen is also removed by the slightly alkaline medium.

The residue is next suspended in 2 l. of cold distilled water and centrifuged at 2000 r.p.m. at 5° for 1 hour. The supernatant is discarded, and the precipitate is again suspended in 2 l. of water and centrifuged. A third washing with water is also carried out in an identical manner. The precipitate is then transferred to a clean towel and squeezed free of gross liquid.

The residue is next suspended in 2 l. of cold 0.1 M sodium citrate buffer, pH 4.3, and the mixture is stirred for 15 to 18 hours at 5°. It is then centrifuged at 5°, and the supernatant is removed and stored on ice. Extraction of the residue with citrate buffer is carried out two times more; in each case the supernatant after centrifugation is combined with the first citrate extract. The combined extracts are centrifuged in a Model L Spinco centrifuge (No. 21 rotor) at 18,000 r.p.m. at 2° to 5° for 1 hour. To remove residual particles, the supernatant is filtered with suction through glass wool at 5°.

The filtrate (approximately 5 to 6 l.) is transferred to several dialysis sacks made of Visking tubing approximately 3 inches in diameter. It is dialyzed with stirring at 5° against large volumes of prechilled $0.02\,M$ disodium hydrogen phosphate solution. The dialysis medium is replaced frequently by fresh solution so that its pH continues to be alkaline. When the pH of the protein solution within the sacks becomes neutral or slightly alkaline, thick, rigid, needlelike fibrils of ichthyocol appear. This process usually occurs after 8 hours, but the time varies with efficiency of dialysis. Dialysis is then continued for 48 hours, during which time the external medium is replaced several times by fresh solution. The ichthyocol is then harvested by centrifugation at 2000 r.p.m. at 5°; the supernatant is discarded, and the precipitate is resuspended in cold distilled water and collected once more by centrifugation. The ichthyocol is washed two times more in this manner. It is uniformly distributed in a small amount of cold distilled water, and the resulting suspension is brought to a volume of approximately 250 ml. by addition of cold water. The suspension is then lyophilized. The dry ichthyocol is stored at 5° in stoppered bottles in a desiccator containing calcium chloride. Under these conditions it is stable for several years, although material kept for approximately 1 year or more occasionally appears to be less readily dissolved in $0.5\,M$ calcium chloride solution than is newly prepared ichthyocol. The yield of ichthyocol is approximately 5 g. for each 500 g. of tunic (wet weight).

Collagen obtained in this manner can be redissolved in citrate buffer of pH 4.3 and reconstituted into fibrils once more by dialysis against $0.02\,M$ disodium hydrogen phosphate solution.

Preparation of Soluble Collagen of Calfskin

A skin obtained from a freshly killed calf is washed thoroughly with cold water and pinned to a board in a cold room at 5°. The hair on the upper side and the fat and loose tissue on the lower are trimmed with a straight razor. The hide is then cut with scissors into rough squares of approximately 1 inch side. The material is ground with a sturdy electrical tissue grinder at 0° to 2°, extreme care being exercised that the tissue does not become unduly warm. Cooling may be promoted by addition of ice cubes to the grinder together with the pieces of tissue. The ground skin is extracted then with $0.5\,M$ sodium acetate (four times) as described for the preparation of ichthyocol. The residue thus obtained is washed with cold distilled water (three times), also as described under preparation of ichthyocol, and then extracted with $0.075\,M$ citrate buffer of pH 3.7; note that the buffer is different from that used for extraction of ichthyocol. The citrate extract of calfskin is clarified by centrifugation and dialyzed, as described above, against $0.02\,M$ disodium hydrogen phosphate solution; the resulting fibrils are harvested, washed with cold distilled water, resuspended in water, and lyophilized. The calfskin collagen is stored in the same manner as ichthyocol.

Properties

Although the collagen extracted from carp swim bladder or calfskin by acidic buffer solutions constitutes only a small portion of the total collagen in the tissue, it is representative of the class of collagenous proteins as determined by appearance under the electron microscope and by X-ray diffraction. Fibrils show characteristic bands with a macroperiod of approximately 640 A. The amino acid composition is also characteristic of the class of collagens.

In addition to solubility in mildly acidic solution, the preparations described here are soluble in various salt solutions.[5] Of particular importance for their use as substrates is a capacity for solution without denaturation in neutralized calcium chloride solution. Thus, 0.4 g. of ichthyocol distributed by homogenization in 50 ml. of ice-cold distilled water dissolves immediately when the suspension is added to an equal volume of cold 1 M calcium chloride buffered at pH 7. The same amount of calfskin collagen distributed in 50 ml. of cold water dissolves somewhat more slowly when the mixture is added to an equal volume of cold 2 M calcium chloride at pH 7.

TABLE I
CONTENTS OF NITROGEN, AMINO GROUPS, AND CARBOHYDRATE
OF SOLUBLE COLLAGENS

	Ichthyocol	Calfskin collagen
Total nitrogen (Kjeldahl), %	18.9	18.7
Amino groups (ninhydrin[a]), micromoles leucine equivalents per 100 mg. protein	29	29
Carbohydrate (anthrone[b]), micromoles glucose equivalents per 100 mg. protein	2.3	2.3

[a] H. Rosen, Arch. Biochem. Biophys. **67**, 10 (1957).
[b] S. Seifter, S. Dayton, B. Novic, and E. Muntwyler, Arch. Biochem. Biophys. **25**, 191 (1950); see also Vol. III [7].

Table I presents some general features of chemical composition of ichthyocol and calfskin soluble collagen which may be of particular interest to enzymologists. In addition the following chemical facts are of interest: Both ichthyocol and calfskin collagen exhibit about half the

[5] P. M. Gallop, S. Seifter, and E. Meilman, J. Biophys. Biochem. Cytol. **3**, 545 (1957).

color yield of an equal weight of serum albumin in the protein method of Lowry *et al.*[6]; this is due to the large number of bonds involving imino acids. Both collagens contain bonds (described as "ester-like"[7]) which are cleaved by aqueous 1 M hydroxylamine or hydrazine at pH values between 8 and 10; 6 micromoles of hydroxamic acid or of acid hydrazide are found per 100 mg. of protein. Both collagens do not contain significant quantities of hexosamine, hexuronic acid, nor sialic acid by any of the currently used methods.

TABLE II

AMINO ACID COMPOSITIONS OF SOLUBLE COLLAGENS[a]

(Residues per 1000 total residues)

Component[b]	Ichthyocol	Calfskin collagen
Glycine	340	350
Alanine	128	111
Valine	17	19
Isoleucine	11	11
Leucine	22	24
Proline	118	124
Hydroxyproline	70	91
Phenylalanine	14	12
Tyrosine	2–3	2–3
Serine	32	30
Threonine	27	16
Methionine and methionine sulfoxide[c]	12	—
Hydroxylysine	7	7
Lysine	27	25
Histidine	4	5
Arginine	51	48
Aspartic acid	46	42
Glutamic acid	71	70
Amide	42	42

[a] Analyses were actually performed on gelatins derived as follows. Ichthyocol is suspended in distilled water, and the mixture is heated at 50° for 15 minutes; solution occurs as gelatin forms. Calfskin collagen is suspended in water and heated at 60° for 15 minutes.

[b] Neither cystine nor tryptophan was detected by methods employed.

[c] Formed during hydrolysis of protein with 6 N HCl.

Table II contains the amino acid analyses (uncorrected for losses during hydrolysis) of typical preparations of collagen obtained in our laboratory; the automatic chromatographic procedure of Spackman

[6] O. H. Lowry, N. J. Rosebrough, A. L. Farr, and R. J. Randall, *J. Biol. Chem.* **193**, 265 (1951); see also Vol. III [73].

[7] P. M. Gallop, S. Seifter, and E. Meilman, *Nature* **183**, 1659 (1959).

TABLE III
PHYSICOCHEMICAL PROPERTIES OF SOLUBLE COLLAGENS AND THEIR GELATINS[a]

Property	Ichthyocol	Calfskin collagen	Gelatin from ichthyocol	Gelatin from calfskin collagen
Denaturation temperature, T_d (transition to gelatin), °C.	29^b	36^b	—	—
Intrinsic viscosity, $[\eta]$, 100 ml./g.	13.2^c	13.5^b	0.34^d 0.44^e	0.56^b
In neutral $CaCl_2$ solution	$15-17^f$	$16-18$		
Partial specific volume, v, ml./g.	$0.72-0.73$		0.705^d	
Sedimentation coefficient, $S^\circ_{20,w}$, Svedberg units	2.85^c 2.96^e	3.28^b	$3.31^{d,g}$ 3.77^e	3.85^b
Refractive index increment, dn/dc, ml./g.	0.192^b			
Molecular weight, M, weight average, g./mole	$345,000^e$	$352,000^b$		
Length of molecule, L, angstroms	$3,000^e$	$3,000^h$		
Diameter of molecule, D, angstroms	13.6^e	15^h		
Optical rotation, $[\alpha]_D$, degrees	-350^i	-415^b	-110^i -124.8^i	
λ_c, $m\mu$	205^i		205^i 213^i	

[a] Except for those cases in which dimensions were determined by electron microscopy, all measurements were made in solution in citrate buffer of pH 3.7 or, as indicated in one instance, in neutral calcium chloride solution.
[b] P. Doty and T. Nishihara, in "Recent Advances in Gelatin and Glue Research" (G. Stainsby, ed.), p. 92. Pergamon Press, London, 1958.
[c] P. M. Gallop, Arch. Biochem. Biophys. 54, 486 (1955).
[d] P. M. Gallop, Arch. Biochem. Biophys. 54, 501 (1955).
[e] H. Boedtker and P. Doty, J. Am. Chem. Soc. 78, 4267 (1956).
[f] P. H. von Hippel, P. M. Gallop, S. Seifter, and R. S. Cunningham, J. Am. Chem. Soc. 82, 2774 (1960).
[g] Presumably for the α component.
[h] R. V. Rice, Proc. Natl. Acad. Sci. U. S. 46, 1187 (1960).
[i] C. Cohen, J. Biophys. Biochem. Cytol. 1, 203 (1955).
[j] P. H. von Hippel, personal communication, 1963.

et al.[8] was employed. The analyses compare well with those reported by Piez and Gross.[9]

Harrington and von Hippel[10] have summarized the known physico-chemical properties of ichthyocol and calfskin soluble collagens. In Table III we present selected parameters of immediate interest.

It is beyond the scope of this article to describe the α and β components of the collagen molecule[11-15] and other fractions obtained by chromatography.[16] Because these may prove of great significance for the biosynthesis and structure of collagen as it occurs in tissues, the reader may desire to consult the noted references.

[8] D. H. Spackman, W. H. Stein, and S. Moore, *Anal. Chem.* **30**, 1190 (1958).

[9] K. A. Piez and J. Gross, *J. Biol. Chem.* **235**, 995 (1960).

[10] W. F. Harrington and P. H. von Hippel, *Advances in Protein Chem.* **16**, 1 (1961).

[11] V. N. Orekhovich and V. O. Shpikiter, *Doklady Akad. Nauk S.S.S.R.* **101**, 529 (1955).

[12] V. N. Orekhovich and V. O. Shpikiter, *Biokhimiya* **23**, 286 (1958).

[13] P. M. Gallop, *Arch. Biochem. Biophys.* **54**, 501 (1955).

[14] K. A. Piez, E. Weiss, and M. S. Lewis, *J. Biol. Chem.* **235**, 1987 (1960).

[15] K. A. Piez, M. S. Lewis, G. R. Martin, and J. Gross, *Biochim. et Biophys. Acta* **53**, 596 (1961).

[16] A. Kessler, H. Rosen, and S. M. Levenson, *J. Biol. Chem.* **234**, 989 (1960).

Nucleic Acids, Coenzymes and Derivatives

[94] Preparation of Nucleotides and Derivatives

By Michael Smith *and* H. G. Khorana

Introduction

During the last decade, chemical methods for the preparation of nucleotides and their derivatives have been developed to the point where they now often provide the easiest and most efficient routes to these compounds. The scope of the chemical synthetic approach is illustrated by the preparations described in this article. These have been selected on the basis of their general utility as methods in nucleotide chemistry and because of their reliability in use in the authors' laboratory.

The preparations of simpler mononucleotides and derivatives include those of the *p*-nitrophenyl esters of thymidine 3'-phosphate and thymidine 5'-phosphate (convenient assay substrates for the spleen and snake venom types of phosphodiesterases; see Vol. VI [29], ribonucleoside cyclic 2',3'-phosphates, and ribonucleoside cyclic 3',5'-phosphates. β-Cyanoethyl phosphate on activation with dicyclohexylcarbodiimide gives a new and exceedingly powerful phosphorylating agent. Its preparation and examples of its use in the synthesis of mononucleotides are described. Of the alternative methods for phosphorylating alcoholic hydroxyl functions, that employing polyphosphoric acid in the preparation of uridine and cytidine 5'-phosphates is included.

Improved methods for the synthesis of nucleoside 5'-pyrophosphate derivatives have been developed recently. An example is the preparation of nucleoside 5'-triphosphates by reaction of nucleoside 5'-phosphates with orthophosphoric acid and dicyclohexylcarbodiimide. Nucleoside 5'-diphosphates and unsymmetrical diesters of pyrophosphoric acid (nucleoside 5'-diphosphate coenzymes) are readily obtained through the use of nucleoside 5'-phosphoramidates and, more especially, their substituted analogs, the nucleoside 5'-phosphoromorpholidates. This latter type of compound is used in the described syntheses of cytidine 5'-diphosphate and uridine diphosphate glucose.

Finally, in the polynucleotide field, the preparation of low-molecular-weight polymers from mononucleotides is illustrated by chemical polymerization and ion-exchange separation of pure members of the resulting homologous series of oligonucleotides derived from thymidine and deoxycytidine 5'-phosphates.

General Techniques and Materials

All liquid reagents should be distilled before use. Pyridine (reagent grade) is kept over calcium hydride for several days before use. Distilled water is used in all experiments requiring water, and solutions are concentrated in a rotary evaporator at low temperature (usually 30° or less) under reduced pressure afforded by a water aspirator.

The use of cellulose anion exchangers (e.g., diethylaminoethylcellulose, DEAE-cellulose) represents a marked improvement over the use of conventional ion-exchange resins for the separation of nucleotides and their derivatives. Many of the procedures described below include the use of such cellulose anion-exchangers. The columns are prepared by packing down a thin aqueous slurry of the anion exchanger under 5 to 6 pounds of air pressure. Usually the exchanger is employed in the carbonate form by using the volatile buffer triethylammonium bicarbonate (pH 7.5) as eluting agent.[1]

The increasing commercial availability of many nucleotides is welcome, and these products can be used as starting materials in the preparations described in this article. However, it is very important to check carefully the purity of commercially procured materials. The paper chromatographic homogeneity of nucleotides is checked in this laboratory by applying approximately 2 micromoles (no less) of the substance as a spot on Whatman No. 40 or 44 (doubly acid-washed) paper strips and developing the chromatograms in suitable solvent systems. The latter are noted at appropriate places in the text. Complete ultraviolet absorption spectra should always be determined, and the detailed spectral characteristics compared with published data.[2]

p-Nitrophenyl Thymidine 3′-Phosphate[3]

Principle. 5′-*O*-Tritylthymidine is reacted with *p*-nitrophenyl phosphorodichloridate, the trityl group is hydrolyzed with aqueous acetic acid, and the resultant *p*-nitrophenyl thymidine 3′-phosphate is isolated as its ammonium salt in 94% yield.

Sodium p-Nitrophenoxide.[4] Aqueous sodium hydroxide (10 M, 110 ml.) is added with stirring to a suspension of *p*-nitrophenol (150 g.) in boiling water (700 ml.). To the resulting clear solution is added a further 100 ml. of 10 M aqueous sodium hydroxide, and the mixture is cooled rapidly. The yellow sodium *p*-nitrophenoxide which separates is collected by filtration and washed three times with 50-ml. portions of ice-cold water. After being dried thoroughly at 110° and then pulverized, the anhydrous salt is obtained as a red powder (165 g., 95%).

[1] J. Porath, *Nature* **175**, 478 (1955).
[2] G. H. Beaven, E. R. Holiday, and E. A. Johnson, *in* "The Nucleic Acids" (E. Chargaff and J. N. Davidson, eds.), Vol. I, p. 493. Academic Press, New York, 1955.
[3] A. F. Turner and H. G. Khorana, *J. Am. Chem. Soc.* **81**, 4651 (1959).
[4] J. G. Moffatt and H. G. Khorana, *J. Am. Chem. Soc.* **79**, 3741 (1957).

p-Nitrophenyl Phosphorodichloridate. Finely powdered anhydrous sodium *p*-nitrophenoxide (41 g.) is added slowly with stirring to phosphorus oxychloride (250 ml.) contained in a flask equipped with a reflux condenser and cooled in an ice-salt bath. A vigorous reaction takes place, and the color of the added salt rapidly disappears. When the addition is complete (1 hour), sodium chloride is removed by filtration, the excess of phosphorus oxychloride is removed under reduced pressure, and the residue is kept at 0.01 mm. of pressure for 1 hour at room temperature. The oil is distilled in a short-path apparatus (b.p. 128°, 0.02 mm.); the distillate crystallizes to give *p*-nitrophenyl phosphorodichloridate (m.p. 43.5° to 44.5°). The yields vary but are better than 40%.

p-Nitrophenyl Thymidine 3'-Phosphate. A solution of 5'-O-tritylthymidine[5] (1.12 g., 1.98 millimoles) in anhydrous dioxane (5 ml.) is added dropwise, with exclusion of moisture, to a magnetically stirred solution of *p*-nitrophenyl phosphorodichloridate (1.06 g., 3.92 millimoles) in anhydrous dioxane (5 ml.) and dry pyridine (0.63 ml., 7.84 millimoles). After the addition is complete (1 hour), the mixture is kept at room temperature for 2 hours. A solution of pyridine (0.65 ml.) in water (3.0 ml.) is then added rapidly with stirring, and the total solution is evaporated to a gum under reduced pressure. The residue is taken up in a mixture of chloroform and water, and the chloroform layer is separated after gentle agitation. The organic layer is washed with aqueous pyridine hydrochloride (1.0 M, pH 5.5). The chloroform solution is then taken to dryness, and the residual pale-yellow gum is dissolved in 80% acetic acid (25 ml.) and heated at 100° for 20 minutes. The solvent is removed under reduced pressure, and the residue is diluted with water. After 18 hours at 4°, the crystalline triphenylcarbinol is removed, and the aqueous solution is lyophilized. The solid product is dissolved in water, and the lyophilization is repeated. The product is converted to the free acid by passing through a column of cation exchanger (Amberlite IR-120) in the acid form. A small amount of *p*-nitrophenol which is present in the product is removed by repeated ether extraction of the acidic solution, which is then neutralized with aqueous ammonium hydroxide. Lyophilization yields the anhydrous ammonium salt of *p*-nitrophenyl thymidine 3'-phosphate (0.86 g.). The product has a λ_{max} of 270.5 mμ and ϵ_{max} of 15,400 and is chromatographically homogeneous in *n*-butanol–water (86: 14), isopropanol–ammonia–water (7:1:2), and *n*-butanol–acetic acid–

---

[5] A. R. Todd, Vol. III [113]. It should be noted that, in repetitions of this preparation, the product has always been found to contain one molecule of benzene per molecule of 5'-O-tritylthymidine; P. T. Gilham and H. G. Khorana, *J. Am. Chem. Soc.* **81**, 4647 (1959). Material containing solvent of crystallization is used directly in the present preparation.

water (5:2:3) and is also homogeneous when examined electrophoretically under neutral and acidic conditions.

p-Nitrophenyl Thymidine 5'-Phosphate[6]

Principle. Thymidine 5'-phosphate is esterified with p-nitrophenol on treatment with dicyclohexylcarbodiimide in the presence of triethylamine. The p-nitrophenyl ester is isolated as its sodium salt in 70% yield after partition chromatography on cellulose.

p-Nitrophenyl Thymidine 5'-Phosphate. One millimole of thymidine 5'-phosphate,[7] as the free acid, in water is converted to its triethylammonium salt by addition of 0.28 ml. (2.0 millimoles) of the base. The solution is evaporated to dryness under reduced pressure, and the residual gum is dissolved in anhydrous pyridine (10 ml.). p-Nitrophenol (1.39 g., 10.0 millimoles) is added to the solution, and the pyridine is removed under reduced pressure. The mixture is rendered anhydrous by dissolving it in anhydrous pyridine (10 ml.) and evaporating the pyridine under reduced pressure (this process is repeated twice). Finally the mixture of p-nitrophenol and thymidine 5'-phosphate is dissolved in anhydrous pyridine (10 ml.) and reacted with dicyclohexylcarbodiimide (2.06 g., 10.0 millimoles) for 3 days at 25°. The solvent is removed by evaporation under reduced pressure (last traces at 0.1 mm.) at 25°. The residual gum is dissolved in water (25 ml.), and the solution is extracted twice with ether. A small amount of solid which separates is filtered off, and then the aqueous filtrate is passed through a column (5 cm. × 1 cm. in diameter) of a cation exchanger (Dowex 50) in the hydrogen form. The column is washed with water until the eluate is neutral. The combined eluate is adjusted to pH 3.5 with M aqueous sodium hydroxide and then washed repeatedly with ether until the ethereal washings do not give a yellow color on shaking with dilute aqueous ammonium hydroxide. The washed solution is evaporated to dryness under reduced pressure, and the residue is dissolved in the minimum volume of isopropanol–water (3:1). This solution is applied to a column of cellulose powder (Whatman standard grade, 30 cm. × 4 cm. in diameter) previously packed in isopropanol–water (3:1). Elution of this column with the same solvent mixture gives p-nitrophenyl thymidine 5'-phosphate (fractions 29 to

---

[6] J. G. Moffatt, see W. E. Razzell and H. G. Khorana, *J. Biol. Chem.* **234**, 2105 (1959).

[7] Contaminating deoxyribonucleoside 5'-phsophates (e.g. deoxyuridine 5'-phosphate in thymidine 5'-phosphate) can be detected by chromatography in the isobutyric acid (100 ml.)–1.0 M ammonium hydroxide (60 ml.)–0.1 M ethylenediaminetetraacetic acid (1.6 ml.) system. Contaminating ribonucleoside 5'-phosphates appear as slow-moving spots in the isopropanol–concentrated ammonia–0.1 M boric acid (7:1:2) system.

42; 10-ml. fractions) and a smaller amount of thymidine 5′-phosphate (fractions 46 to 58). The fractions containing the diester are combined and concentrated to dryness under reduced pressure (bath temperature, 30°). The residual gum is dissolved in methanol (approximately 3 ml.), and the solution is diluted with anhydrous ether (30 ml.). The trihydrate of sodium p-nitrophenyl thymidine 5′-phosphate which is precipitated is collected by centrifugation (360 mg.). It is rendered anhydrous by drying at 100° under reduced pressure. The nucleotide has a λ_{max} of 270 mμ and ϵ_{max} of 16,250 and is chromatographically homogeneous in the isopropanol–concentrated ammonium hydroxide–water (7:1:2) and the ethanol–M ammonium acetate (5:2) systems.

Ribonucleoside Cyclic 2′,3′-Phosphates[8]

Principle. The ribonucleoside 2′(3′)-phosphate is reacted with dicyclohexylcarbodiimide in the presence of triethylamine. The resultant cyclic phosphate is isolated as its anhydrous calcium salt (100% yield). The conditions used below in the preparation of adenosine cyclic 2′,3′-phosphate are equally applicable to the preparation of all other ribonucleoside cyclic 2′,3′-phosphates.

Adenosine Cyclic 2′,3′-Phosphate. Adenosine 2′(3′)-phosphate as the free acid (1.0 millimole) is dissolved in water (2 ml.) containing redistilled triethylamine (0.15 ml., 1.0 millimole). Methanol (15 ml.) and dicyclohexylcarbodiimide (1.03 g., 5.0 millimoles) are added to give a homogeneous solution which is kept at 25°. Dicyclohexylurea begins to separate after a few minutes. After 8 hours the solution is evaporated to dryness under reduced pressure. Water (100 ml.) and ether (50 ml.) are added to the residue, and the mixture is shaken. The insoluble dicyclohexylurea is removed by filtration, and the aqueous layer is then separated and evaporated to dryness. The resulting gum is dissolved in 95% ethanol (10 ml.) containing triethylamine (0.15 ml., 1.0 millimole). To this is added M calcium chloride in 95% ethanol (10 ml.). Precipitation of the nucleotide is completed by addition of anhydrous ether (20 ml.). The white precipitate is collected by centrifugation. It is resuspended in 95% ethanol (10 ml.), ether (20 ml.) is added, and the precipitate is collected again as above. This washing is repeated to ensure that all the calcium chloride is removed. Finally the precipitate is washed with anhydrous ether and dried at 50° at 0.1 mm. overnight to yield the anhydrous calcium salt of adenosine cyclic 2′,3′-phosphate quantitatively. The product (as its ammonium salt, obtained by treating the calcium salt with the ammonium form of a cation exchanger) is homogeneous both

[8] The method is based on the procedure of M. Smith, J. G. Moffatt, and H. G. Khorana, *J. Am. Chem. Soc.* **80**, 6204 (1958).

chromatographically in the isopropanol–concentrated ammonia–water (7:1:2) system and electrophoretically at pH 7.5.

Nucleoside Cyclic 3′,5′-Phosphates[9]

Principle. The 4-morpholine-*N,N′*-dicyclohexylcarboxamidine (I) salt of adenosine 5′-phosphate is reacted with dicyclohexylcarbodiimide in

$$C_6H_{11}N{=}C\Big\langle\begin{array}{l} NHC_6H_{11} \\[4pt] N\text{(morpholine ring)} \end{array}$$

(I)

anhydrous pyridine at reflux temperature under dilute conditions. The nucleoside cyclic 3′,5′-phosphate is obtained in over 80% yield. This same procedure can be used for the preparation of uridine cyclic 3′,5′-phosphate. Because of their insolubility in anhydrous pyridine, guanosine and cytidine 5′-phosphates are first converted to their more soluble *N*-benzoyl derivatives. Cyclization under standard conditions, followed by treatment with concentrated aqueous ammonia to remove the *N*-benzoyl group, affords the corresponding nucleoside cyclic 3′,5′-phosphates.

4-Morpholine-N,N′-dicyclohexylcarboxamidine.[10] Dicyclohexylcarbodiimide (10.3 g., 50 millimoles) and morpholine (8.7 g., 100 millimoles) in *t*-butanol (10 ml.) are heated under reflux on a steam bath for 5 hours. The solution is cooled to 0°, and the product separates as thick white needles which are collected by filtration and washed with low-boiling petroleum ether (30° to 60°). Recrystallization from aqueous methanol gives the carboxamidine (I) as thick needles which are then dried at 50° at 0.1 mm. over phosphorus pentoxide (m.p. 105° to 106°; yield 10 g.).

Adenosine Cyclic 3′,5′-Phosphate. Adenosine 5′-phosphate[11] (free acid, 1.0 millimole) and 4-morpholine-*N,N′*-diclohexylcarboxamidine (293 mg., 1.0 millimole) are dissolved in pyridine (25 ml.) containing water (5 ml.), and the solution is evaporated to dryness, under reduced pressure. The resulting gum is dissolved in anhydrous pyridine (50 ml.), and the solution is again evaporated to dryness. This procedure is repeated twice to ensure that water is completely removed by coevaporation. The

---

[9] M. Smith, G. I. Drummond, and H. G. Khorana, *J. Am. Chem. Soc.* **83**, 698 (1961).
[10] J. G. Moffatt and H. G. Khorana, *J. Am. Chem. Soc.* **83**, 649 (1961).
[11] Examined for purity in the isobutyric acid (100 ml.)–1.0 *M* ammonium hydroxide (60 ml.)–0.1 *M* ethylenediaminetetraacetic acid (1.6 ml.) system.

residual gum is finally dissolved in anhydrous pyridine (100 ml.). This solution is run dropwise (down the condenser) into a solution of dicyclohexylcarbodiimide (412 mg., 2.0 millimoles) in anhydrous pyridine (100 ml.) which has previously been raised to reflux temperature. The whole system is protected from moisture by a silica gel tube. The time taken for the addition is 3 hours. (On occasion some of the nucleotide may crystallize in the dropping funnel. This is rinsed into the reaction flask with a little anhydrous pyridine, after the dropwise addition is completed.) The mixture is then refluxed for a further 2 hours. Next, the solution is concentrated to dryness under reduced pressure, and water (100 ml.) and ether (50 ml.) are added to the residue with vigorous shaking. The insoluble dicyclohexylurea is filtered off, and the aqueous layer in the filtrate is concentrated to a small volume (10 ml.) which is adsorbed onto a column of DEAE-cellulose (30 cm. $\times$ 3.5 cm. in diameter) in the carbonate form. Elution is carried out by using a linear salt gradient, the mixing chamber containing $0.002\,M$ aqueous triethylammonium bicarbonate (2.0 l.), and the reservoir a $0.10\,M$ solution of the same salt (2.0 l.). (A stock solution of M triethylammonium bicarbonate is prepared by bubbling carbon dioxide through a mixture of 500 ml. of distilled water and 140 ml. of redistilled triethylamine, cooled in an ice bath, until the pH drops to 7.5. The solution is then diluted to 1.0 l. with water.) The flow rate through the column is maintained at 4 ml./min., 20-ml. fractions being collected. The elution of products is followed spectrophotometrically at 260 mμ. Adenosine cyclic 3',5'-phosphate, the major product, is eluted between fractions 100 and 150. The fractions containing the cyclic phosphate are combined and evaporated to dryness under reduced pressure. The residual solid is dissolved in a small amount of water, and the solution is re-evaporated to dryness. This is repeated twice to remove completely triethylammonium bicarbonate. The residual gum is dissolved in water (2 ml.) and ethanol (2 ml.), and the solution is adjusted to pH 1.0 with M hydrochloric acid. Adenosine cyclic 3',5'-phosphate separates as the free acid, and after cooling to 0° it is collected, washed with ethanol, and dried over active alumina to give the monohydrate, 280 mg. The product is chromatographically and electrophoretically homogeneous in acid, neutral, and basic systems. It is completely hydrolyzed to a mixture of adenosine 5'-phosphate and adenosine 3'-phosphate on treatment with $0.4\,M$ barium hydroxide at 100° for 30 minutes.

N⁶-Benzoyl Cytidine 5'-phosphate. Cytidine 5'-phosphate[11] (free acid, 1.0 millimole) is dissolved in water (100 ml.) containing a little pyridine. The solution is lyophilized to give the nucleotide as a finely dispersed white powder. This is dissolved, by shaking, in anhydrous pyridine (15 ml.) containing redistilled benzoyl chloride (2.5 ml.). The solution is kept at 25° for 1 hour with exclusion of light. Fifty milliliters of ice-cold water are then added, and after 5 minutes at 25° the solution is extracted with chloroform (3 × 50 ml.). The combined chloroform extracts, which, presumably, contain the tetrabenzoyl nucleotide (II) and benzoic acid as pyridine salts, are washed with water (2 × 10 ml.) and then concen-

(II)

trated to a gum under reduced pressure. Pyridine (20 ml.) and water (10 ml.) are added, followed by 2 M aqueous sodium hydroxide (30 ml.). Initially a cloudy pale-yellow suspension is obtained, but, on shaking, a clear orange-colored solution results in a few seconds. After 4 minutes, the freshly prepared pyridinium form of a cation exchanger (Amberlite IR-120, 50 ml. of wet resin) is added to the solution, which turns pale yellow, the pH dropping to around 7. The aqueous solution is filtered free of resin, the latter is washed well with water, and the combined aqueous solutions are concentrated to about 25 ml. at low temperature. Benzoic acid separates during the evaporation and is removed by extraction with ether. The aqueous solution is passed through a column of pyridinium cation exchanger (Amberlite IR-120, 50 ml. of wet resin) to complete the removal of sodium ions. The column is washed well with water, and the eluate and washings are concentrated to 25 ml. (At this stage the pH of the solution should be about 3.5. If it is lower than 3.0, pyridine should be added immediately to prevent N-debenzoylation.) Removal of benzoic acid is completed by exhaustive extraction with ether, the pH being checked at each extraction. Finally pyridine (10 ml.) is added to the aqueous solution. The solution, when examined chromatographically in 95% ethanol–M ammonium acetate (pH 7.5) (7.5:3), should contain only N^6-benzoyl cytidine 5′-phosphate, which gives a blue fluorescent spot ($R_f = 0.41$). In this system cytidine 5′-phosphate and benzoic acid have R_f values of 0.18 and 0.83, respectively.

Cytidine Cyclic 3',5'-Phosphate. To the above aqueous pyridine solution of N^6-benzoyl cytidine 5'-phosphate is added 293 mg. (1.0 millimole) of 4-morpholine-N,N'-dicyclohexylcarboxamidine, and the mixture is concentrated to a gum. This is redissolved in anhydrous pyridine, and the solution is re-evaporated to dryness. Finally, the solution of the nucleotide salt in 100 ml. of dry pyridine is run dropwise over a 2-hour period into a refluxing solution of dicyclohexylcarbodiimide (412 mg., 2.0 millimoles) in the same solvent (100 ml.). Heating is continued for a further 2 hours, and then the product is worked up as in the preparation of adenosine cyclic 3',5'-phosphate to give the crude N^6-benzoyl cytidine cyclic 3',5'-phosphate as a solution in water (20 ml.). To this is added concentrated ammonia (20 ml.), and the mixture is kept for 15 hours at 25°. Ammonia is removed under reduced pressure, and the aqueous solution is passed through a column of pyridinium cation exchanger (Amberlite IR-120). Benzoic acid and benzamide are removed by ether extraction, and then the product is purified by anion-exchange chromatography as in the preparation of adenosine cyclic 3',5'-phosphate. The major peak is concentrated to dryness, and the cytidine cyclic 3',5'-phosphate is crystallized by dissolving the residue in water (2 ml.) and adjusting the solution to pH 1.0. The yield is 220 mg. (73%).

The procedure for the preparation of guanosine cyclic 3',5'-phosphate is identical, except that treatment with concentrated ammonia for 2 hours at 100° is required to remove the N-benzoyl group.

Phosphorylation with 2-Cyanoethyl Phosphate[12]

Principle. 2-Cyanoethyl phosphate is prepared as the crystalline barium salt by phosphorylation of β-hydroxypropionitrile with a stoichiometric amount of phosphorus oxychloride. The reaction of a mixture of the cyanoethyl phosphate and dicyclohexylcarbodiimide with 2',3'-O-isopropylidene guanosine followed by alkaline treatment to remove the cyanoethyl group (elimination of acrylonitrile) and acidic treatment to remove the isopropylidene group affords guanosine 5'-phosphate in 72% yield. A similar procedure has been used to phosphorylate in high yield a variety of other protected nucleosides[12] as in the preparation of thymidine 3'-phosphate from 5'-O-tritylthymidine which is also described below.

2-Cyanoethyl Phosphate. Phosphorus oxychloride (18.4 ml., 0.2 mole) is mixed with anhydrous ether (200 ml.) in a 1-l. three-necked flask equipped with thermometer, stirrer, dropping funnel, and drying tube. The solution is cooled to —13° in an ice-salt bath, and a mixture of β-hydroxypropionitrile (hydracrylonitrile, 14.2 g., 0.2 mole) and anhy-

[12] G. M. Tener, *J. Am. Chem. Soc.* **83**, 159 (1961).

drous pyridine (16.1 ml., 0.2 mole) is added dropwise with stirring, the temperature not being allowed to rise above —10°. The addition takes about 1 hour. After a further hour at —10° the whole mixture, from which some pyridine hydrochloride separates, is slowly poured, with stirring, into a mixture of water (750 ml.), pyridine (80 ml.), and ice (300 g.). To this is added a solution of barium acetate (100 g.) in water (300 ml.). Barium phosphate, which separates, is removed after 2 hours by filtration on a Büchner funnel. (Any residual ether evaporates at this stage.) Barium 2-cyanoethyl phosphate is crystallized by addition of 2 vol. of 95% ethanol to the filtrate. The product, which separates as white platelets, is collected after an hour at 0°. After a washing with 50% ethanol and pure ethanol followed by air drying, barium 2-cyanoethyl phosphate is obtained as the dihydrate (41 g.). A standard M-solution of 2-cyanoethyl phosphate for use in phosphorylation reaction is prepared by dissolving the barium salt (16.1 g.) in water containing a suspension of acidic cation exchanger (Dowex 50, H^+ form). The solution is passed through a column of cation exchanger in the acid form, which is washed until neutral. Pyridine (20 ml.) is added to the eluate, and the solution is concentrated to a small volume, after which it is made up to 50 ml. with pyridine in a volumetric flask. This solution can be kept for a month without detectable decomposition.

2′,3′-O-Isopropylidene Guanosine.[13] Zinc chloride (56 g. of a freshly fused anhydrous sample) is dissolved in anhydrous acetone (350 ml.). Twenty-two grams of guanosine (dried at 100° *in vacuo* over phosphorus pentoxide) is added, and the mixture is heated under reflux overnight with exclusion of moisture. The resulting clear solution is concentrated to a clear sirup, and 8 M ammonium hydroxide is added until the initially formed precipitate redissolves and a clear solution results. A cation exchanger (Amberlite IR-120) in the ammonium form (100 ml. of resin) is added, and if a precipitate forms it is redissolved by gradual addition of concentrated ammonia. The mixture is stirred for 15 minutes. The resin is then removed by filtration and washed twice with 50-ml. portions of very dilute aqueous ammonia. This whole process is repeated four times to ensure that zinc ions are completely removed. The final volume of filtrate is about 1 l. Ammonia is removed by heating the solution on a water bath. On cooling, 2′,3′-O-isopropylidene guanosine crystallizes. It is collected by filtration, washed with water, and air-dried to yield 15.2 g. of pure material. A further amount (4.2 g.) is recovered by concentrating the mother liquor and retreating the solution with cation exchanger (two 100-ml. portions). The total yield is 19.4 g. with a

[13] R. W. Chambers, J. G. Moffatt, and H. G. Khorana, *J. Am. Chem. Soc.* **79**, 3747 (1957).

melting point of 296° (resolidification). 2',3'-O-Isopropylidene guanosine can be recrystallized from water. It has an R_f value of 0.60 in the isopropanol–concentrated ammonia–water (7:1:2) system in which guanosine has an R_f of 0.27.

   Guanosine 5'-Phosphate. 2',3'-O-Isopropylidene guanosine (323 mg., 1.0 millimole) in 50% aqueous pyridine (10 ml.) and 4 ml. of the stock M solution of 2-cyanoethyl phosphate are combined, and the mixture is evaporated to dryness under reduced pressure (bath temperature 30°). The residue is made anhydrous by dissolving it in dry pyridine and evaporating off the solvent, the process being repeated several times. Finally, the oily residue is dissolved in anhydrous pyridine (10 ml.), dicyclohexylcarbodiimide (2 g.) is added, and the reaction mixture is kept under anhydrous conditions at 25° for 18 hours. Then 1.0 ml. of water is added to the reaction mixture. After 30 minutes at 25°, the solvent is removed under reduced pressure. Last traces of pyridine are removed by adding water (20 ml.) and repeating the evaporation. Lithium hydroxide (40 ml. of 0.5 M) is then added to the residue, and the mixture is heated under reflux for 1 hour. After cooling to room temperature, the mixture is filtered, and the filtrate is passed through a column (4.5 cm. × 3 cm. in diameter) of cation exchanger in the acid form (Dowex 50, H⁺). The column is washed with water until the eluate is neutral. The total eluate (320 ml.) is kept at 25° for 2 hours and then concentrated under reduced pressure (bath temperature 30°) to 50 ml. On neutralization to pH 7.5 with saturated aqueous barium hydroxide, barium phosphate precipitates and is removed by centrifugation. The precipitate is repeatedly washed with water. The aqueous solution and washings are combined (180 ml.) and diluted with 2 vol. of 95% ethanol to precipitate the barium salt of guanosine 5'-phosphate which is collected by centrifugation and washed with ethanol, acetone and, ether. The disodium salt is obtained by treating a suspension of the barium salt in water (10 ml.) with excess of a cation exchanger in the acid form. The resin is washed thoroughly with water to remove completely guanosine 5'-phosphate, after which the combined washings (200 ml.) are neutralized to pH 7.5 with M sodium hydroxide. The solution is concentrated to dryness, the residue is dissolved in water (5 ml.), and acetone (15 ml.) is added. Disodium guanosine 5'-phosphate precipitates, is collected by centrifugation, and is washed with alcohol, acetone, and ether. After drying over phosphorus pentoxide at 25° at 0.1 mm. for 24 hours, the product (318 mg., 72%) is obtained as the dihydrate.

   Thymidine 3'-Phosphate. 5'-O-Tritylthymidine[5] (568 mg., 1.0 millimole) is dissolved in dry pyridine (10 ml.) containing 2 ml. of the stock M solution of 2-cyanoethyl phosphate, and the solution is evaporated to dryness *in vacuo*. The residual gum is rendered anhydrous by dissolving it in dry pyridine (10 ml.) and concentrating the solution to dryness under reduced pressure, the process being repeated three times. Finally the residue is dissolved in dry pyridine (5 ml.), dicyclohexylcarbodiimide (1.67 g.) is added, and the mixture is kept under anhydrous conditions for 48 hours at 25°. Water (5 ml.) is then added, the mixture is kept at 25° for 1 hour, and the solvent is removed by evaporation *in vacuo*. Last traces of pyridine are removed by coevaporation with water (10 ml.). The residue, in 80% acetic acid (20 ml.), is heated under reflux for 30 minutes, and then the solvent is removed under reduced pressure. The residue is suspended in water (25 ml.), and insoluble triphenylcarbinol and dicyclohexylurea are removed by filtration, the latter being washed with water (15 ml.). The combined filtrate and washing is treated with *M* lithium hydroxide (25 ml.) under reflux for 15 minutes. After the mixture has been cooled to 0°, insoluble trilithium phosphate is removed by centrifugation. The supernatant solution is passed through a column (30 cm. $\times$ 2 cm. in diameter) of cation exchanger (Dowex 50) in the acid form, the column being washed with water until the emerging solution is neutral. The eluate is concentrated to a small volume (20 ml.) *in vacuo* at 30° and is then adjusted to pH 7.5 with aqueous barium hydroxide. A small amount of barium hydroxide separates and is removed by centrifugation. Addition of ethanol (40 ml.) to the supernatant solution induces the precipitation of the barium salt of thymidine 3'-phosphate. The nucleotide is collected by centrifugation and washed with 50% ethanol, pure ethanol, and ether to yield the trihydrate, 450 mg. (88%).

Pyrimidine Ribonucleoside 5'-Phosphates with Polyphosphoric Acid as the Phosphorylating Agent

Principle. The 2',3'-*O*-isopropylidine derivatives of uridine or cytidine are reacted with a mixture of orthophosphoric acid and phosphorus pentoxide. After acidic hydrolysis of the "polyphosphoric acid," inorganic phosphate is removed as its insoluble lithium salt, and the nucleotide is purified by ion-exchange chromatography.

Uridine 5'-Phosphate.[14] Five grams of finely powdered 2',3'-*O*-isopropylidine uridine is stirred into a warm (60°) solution[15] (25 ml.) of phosphorus pentoxide in 85% phosphoric acid (1:1.3, w/w). The sirupy mixture is stirred at intervals with exclusion of moisture. The 2',3'-

---

[14] R. H. Hall and H. G. Khorana, *J. Am. Chem. Soc.* **77**, 1871 (1955).

[15] This solution is prepared by adding the required amount of 85% orthophosphoric acid to a weighed sample of phosphorus pentoxide and agitating the mixture, with washing, until it is homogeneous.

isopropylidene uridine gradually dissolves to give a light-yellow solution. After 2 hours, cold water (100 ml.) is added to the reaction mixture which is then heated at 100° for 30 minutes. The solution is then brought to pH 10 with 4 M lithium hydroxide, and the precipitate of trilithium phosphate is removed by centrifugation. This precipitate is washed with three portions of water (40 ml.).[16] The combined supernatant solutions are concentrated under reduced pressure to a volume of 50 ml. A little lithium phosphate which separates is removed by centrifugation. The solution is applied to a column (14 cm. $\times$ 4 cm. in diameter) of anion exchanger (Dowex 2) in the chloride form. The column is washed with water (1 l.) to remove uridine. Subsequent elution with 0.015 M hydrochloric acid (5 l.) gives the uridine 5'-phosphate. The acidic eluate is concentrated to a small volume (20 ml.) under reduced pressure at 30° to 35°. The last traces of solvent are removed by suction on an oil pump followed by desiccation over potassium hydroxide. The residue is washed twice with dry ether (50-ml. portions), and then dissolved in anhydrous ethanol (10 ml.). Uridine 5'-phosphate precipitates on addition of anhydrous ether (100 ml.). This operation is repeated twice, the nucleotide being collected by centrifugation each time. After drying in a desiccator over phosphorus pentoxide, the residual amorphous resin is dissolved in water (40 ml.) and neutralized to pH 9.0 with 4 M lithium hydroxide. Aqueous barium acetate (2 M, 10 ml.) is added to this solution, which is set aside for a few hours. A small amount of barium phosphate which separates is removed by centrifugation. Addition of an equal volume of ethanol results in the precipitation of the barium salt of uridine 5'-phosphate. The nucleotide is collected by centrifugation and washed with 50% ethanol, pure ethanol, and ether. The yield of the hydrated barium salt is 65%. An aqueous solution of the nucleotide is passed through a cation exchanger (Dowex 50) in the hydrogen form, the column being washed with water until the solution emerging from the column is neutral. The eluate is neutralized to pH 8.0 with M sodium hydroxide and then concentrated to a small volume (20 ml.). Addition of a small amount of acetone induces the crystallization of disodium uridine 5'-phosphate as the dihydrate.

Cytidine 5'-Phosphate.[17] The procedure, starting from 2',3'-O-isopropylidene cytidine is identical with the above preparation of uridine 5'-phosphate to the point where the nucleotide is eluted from the anion exchanger. Formic acid (0.02 M) is used to elute cytidine 5'-phosphate.

[16] If the water is adjusted to pH 10 to 11 with lithium hydroxide, precipitation of lithium phosphate is complete, and the subsequent removal of residual inorganic phosphate as its barium salt is unnecessary.

[17] E. H. Pol and H. G. Khorana, unpublished results.

The eluate is concentrated to a small volume (20 ml.) under reduced pressure. Addition of acetone to the residue induces crystallization of cytidine 5'-phosphate.

Nucleoside 5'-Triphosphates[18]

Principle. The nucleoside 5'-phosphate is treated with acetic anhydride in pyridine to acylate the amino group on the purine or pyrimidine ring (this step is not necessary in the preparation of thymidine and uridine 5'-triphosphates). The sugar hydroxyl groups are also acetylated by this procedure, which serves both to increase the solubility of the nucleotide in pyridine and to prevent formation of N-phosphorylated derivatives during the subsequent preparation of the nucleoside 5'-triphosphate. This latter reaction is brought about by reacting the nucleoside 5'-phosphate with orthophosphoric acid and dicyclohexylcarbodiimide. Aqueous ammonia removes the acetyl groups from the product, which is freed from inorganic polyphosphates on charcoal and purified by chromatography on DEAE-cellulose.

Cytidine 5'-Triphosphate. Cytidine 5'-phosphate[11] (1.0 millimole, free acid) is dissolved in water (100 ml.) containing pyridine (5 ml.), and the solution is freeze-dried to give the nucleotide as a finely dispersed white powder. To this are added anhydrous pyridine (10 ml.) and acetic anhydride (5 ml.). The suspension is shaken at 25° for 12 hours, with exclusion of light, to give a clear, colorless solution. After a further 12 hours at 25°, water (50 ml.) is added, with cooling in ice water, and the mixture is kept at 25° for 1 hour. The solution is then evaporated to dryness *in vacuo* (bath temperature no higher than 30°). Water (25 ml.) is added to the residual gum, and the evaporation is repeated. After another such evaporation, the residual gum is dissolved in water (100 ml.), and the solution is lyophilized to give triacetylcytidine 5'-phosphate as a fluffy white powder (the lyophilization should be repeated if any gummy material is obtained). The nucleotide is dissolved in pyridine (20 ml.) containing 85% orthophosphoric acid (1.16 g., 10 millimoles) and tri-n-butylamine (5 ml., 21 millimoles). Dicyclohexylcarbodiimide (10.3 g., 50 millimoles) is added, and the solution is kept at 25° for 24 hours. (Dicyclohexylurea starts to separate after 1 hour.) The solvent is removed *in vacuo*, and the residual mixture is suspended in water (100

---

[18] Modified from the method of M. Smith and H. G. Khorana, *J. Am. Chem. Soc.* **80,** 1141 (1958). In this earlier preparation, no acetylation step was included. This led to difficulties in the preparation of cytidine and deoxycytidine 5'-triphosphates. In the preparation of adenosine 5'-triphosphate described in that work, any N-phosphoryl derivatives formed would be hydrolyzed under the acid conditions employed during work-up.

ml.) and ether (100 ml.). After filtration to remove the insoluble dicyclo-hexylurea, the aqueous layer, combined with water washings of the urea and the ether layer, is evaporated to dryness under reduced pressure (bath temperature, 30°). The residual gum is dissolved in water (25 ml.), concentrated ammonia (25 ml.) is added (tri-n-butylamine separates), and the mixture is kept at 25° for 1 hour. The mixture is diluted with water (50 ml.) and extracted with ether (50 ml.). A little dicyclohexyl-urea which separates is removed by filtration, and the aqueous solution is concentrated to dryness to remove ammonia. The residue is dissolved in water (100 ml.) and adjusted to pH 3.5 with M formic acid. Twenty grams of activated charcoal[19] are then stirred into the solution. After filtration on a Büchner funnel under gentle suction, the charcoal is washed with water (500 ml.) adjusted to pH 3.5 with M formic acid. The nucleotide is then eluted with 50% ethyl alcohol–water (2 l.) con-taining concentrated ammonia (20 ml.). This eluate is evaporated to dryness in $vacuo$. The residue is dissolved in water (100 ml.), and the solution is centrifuged at 15,000 $\times$ g for 10 minutes at 0° to remove some colloidal charcoal. The supernatant liquid is then adsorbed onto a column (30 cm. $\times$ 3.5 cm. in diameter) of DEAE-cellulose (carbonate form). The nucleotides are eluted from the column by using a salt gradient, 0.002 M triethylammonium bicarbonate (4 l.) in the mixing chamber, and 0.5 M triethylammonium bicarbonate (4 l.) in the reservoir (for preparation of triethylammonium bicarbonate, see the preparation of uridine cyclic 3′,5′-phosphate). Twenty-milliliter fractions are col-lected, the flow rate being 4 ml./min. The elution is followed spectro-photometrically at 271 mμ. Cytidine 5′-phosphate and 5′-diphosphate are eluted first, followed by cytidine triphosphate, which is eluted when the salt concentration reaches 0.2 M (fractions 150 to 170). The combined fractions containing the nucleoside 5′-triphosphate are evaporated to dryness under reduced pressure. The residue is freed from residual tri-ethylammonium carbonate by dissolving in water and re-evaporating the solution. The nucleotide is obtained as a colorless gum which is dissolved in water (25 ml.) and passed through a column (10 cm. $\times$ 1.5 cm. in diameter) of cation exchanger (Amberlite IR-120) in the pyridinium form. The eluate is concentrated in $vacuo$ (bath temperature below 30°) to a small volume (about 10 ml.). The pH at this stage is 3 to 3.5. Lithium hydroxide (M) is added to bring the solution to pH 7.5. Addition of acetone (30 ml.) precipitates the lithium salt of the nucleoside 5′-triphosphate as a gum. After collection by centrifugation, the gum is

---

[19] Prepared by washing commercially available Norit A (100 g.) successively with M hydrochloric acid (2 l.), M ammonium hydroxide (2 l.), and then water until the washings are neutral, and finally by drying at 110° for 18 hours.

triturated with acetone (20 ml.) and yields the nucleotide as a white powder. After a further wash with acetone and drying at 25° at 0.1 mm. over phosphorus pentoxide for 3 hours, tetralithium cytidine 5'-triphosphate (150 mg.) is obtained as the nonahydrate (22% yield). The product is free of inorganic phosphate and polyphosphates and is homogeneous chromatographically in the isobutyric acid–concentrated ammonia–water (66:1:33) system and on electrophoresis under acid and neutral conditions.

The above procedure can be used for the preparation of adenosine 5'-triphosphate. In the preparation of guanosine 5'-triphosphate by this method, treatment with 50% concentrated ammonia for 36 hours is required to ensure that N-deacetylation is complete.

Nucleoside 5'-Phosphoramidates[20]

Principle. A nucleoside 5'-phosphate and aqueous ammonia are condensed together by reaction with dicyclohexylcarbodiimide. The carbodiimide also reacts with ammonia to form 1,3-dicyclohexylguanidine, so that, as in the preparation of adenosine 5'-phosphoramidate described below, the nucleoside 5'-phosphoramidate is obtained as its 1,3-dicyclohexylguanidinium salt. These conditions are equally applicable to the preparation of all nucleoside 5'-phosphoramidates.

Adenosine 5'-Phosphoramidates. Adenosine 5'-phosphate[11] (free acid, 3.0 millimoles) is dissolved in a mixture of $2 M$ ammonium hydroxide (7.5 ml.) and formamide (5 ml.). To this is added dicyclohexylcarbodiimide (3.09 g., 15 millimoles) in t-butanol (20 ml.). The two-phase reaction mixture is heated at 80° in a well-stoppered bottle (or alternatively under very gentle reflux) for 7 hours (after 2 to 3 hours the mixture becomes homogeneous). After cooling, the dicyclohexylurea which separates is removed by filtration and washed with water. The combined filtrate and washings are concentrated under reduced pressure to remove t-butanol. After an ether extraction, the aqueous solution is concentrated further to remove water, the last traces being removed on an oil pump. On dropwise addition of acetone to the residual formamide solution, 1,3-dicyclohexylguanidinium adenosine 5'-phosphoramidate crystallizes as the monohydrate; it is obtained anhydrous by drying at 90° over phosphorus pentoxide for 3 hours. The yield is 1.43 g., 87%; the melting point is 239° to 241° (dec.). The product is homogeneous on chromatography in isopropanol–ammonia–water (7:1:2), $R_f = 0.23$, and in ethanol–ammonium acetate (0.5 M, pH 3.8) (5:2), $R_f = 0.27$. On electrophoresis at pH 7.5, the mobility of the phosphoramidate relative to that of adenosine 5'-phosphate is 0.58.

[20] R. W. Chambers and J. G. Moffatt, *J. Am. Chem. Soc.* **80**, 3752 (1958).

Nucleoside 5'-Phosphoromorpholidates[10]

Principle. The method is analogous to that used for the preparation of phosphoramidates, a nucleoside 5'-phosphate being condensed with morpholine by reaction with dicyclohexylcarbodiimide. 4-Morpholine-N,N'-dicyclohexylcarboxamidine (I) is formed concurrently, and consequently the phosphoromorpholidate is isolated as its carboxamidinium salt. The method is completely general for the preparation of nucleoside 5'-phosphoromorpholidates.

Cytidine 5'-Phosphoromorpholidate. To a refluxing solution of cytidine 5'-phosphate[11] (1.0 millimole of the free acid) in water (10 ml.) and *t*-butanol (10 ml.) containing redistilled morpholine (0.35 ml., 1.0 millimole) is added a solution of dicyclohexylcarbodiimide (824 mg., 4.0 millimoles) in *t*-butanol (15 ml.), over a period of 2 to 3 hours. Heating is continued for several hours, until electrophoresis at pH 7.5 of a small aliquot shows that cytidine 5'-phosphate has disappeared. *t*-Butanol is removed *in vacuo*, and the residual aqueous solution is washed three times with ether and then evaporated to dryness, the last traces of water being removed by suction on an oil pump at 25°. The glassy residue is dissolved in the minimum amount of methanol (3 ml.) and transferred to a centrifuge tube. On addition of ether (30 ml.), the nucleotide derivative separates as a sticky solid. After centrifugation and removal of the supernatant liquid, the precipitate is triturated with ether, where it changes to a white powder. This is washed once with ether and then is dried *in vacuo* at 25° to give the 4-morpholine-N,N'-dicyclohexylcarboxamidinium salt of cytidine 5'-phosphoromorpholidate as the dihydrate (680 mg., 95%). The anhydrous salt is obtained on drying at 100° *in vacuo*.

Nucleoside 5'-Diphosphates[10]

Principle. A nucleoside 5'-phosphoromorpholidate is condensed with orthophosphoric acid as its tri-*n*-butylammonium salt in anhydrous pyridine. The method is quite general for the preparation of nucleoside 5'-diphosphates and is well suited to the preparation of these compounds labeled with P^{32} in the terminal (β) phosphate group.

Cytidine 5'-Diphosphate. One millimole of the 4-morpholine-N,N'-dicyclohexylcarboxamidinium salt of cytidine 5'-phosphoromorpholidate (716 mg. of the dihydrate) is dissolved in anhydrous pyridine (10 ml.), and the solvent is removed *in vacuo*. This procedure is repeated three times to ensure that water is completely removed, and the nucleotide derivative is finally dissolved in anhydrous pyridine (10 ml.). Separately an anhydrous solution of the mono-(tri-*n*-butylammonium) salt of orthophosphoric acid (3.0 millimoles) in pyridine (10 ml.) is prepared in the same way. The two solutions are mixed, and the solvent is removed *in vacuo* at 25°. The residue is dissolved in anhydrous pyridine (10 ml.), and the solution is evaporated to dryness. This dissolution and evaporation are repeated twice, and the residue is finally dissolved in anhydrous pyridine (10 ml.). The mixture is shaken mechanically for 1 hour at 25° to redissolve a small amount of material which separates initially. After 50 hours at 25°, the solvent is evaporated under reduced pressure, last traces of pyridine being removed by coevaporation with water. The product is then dissolved in water (15 ml.) containing lithium acetate (409 mg., 4.0 millimoles), and the solution is washed with ether. Molar lithium hydroxide is added to bring the aqueous solution to pH 12. After standing for 30 minutes at 0°, the precipitate which separates is removed by filtration and washed with 0.01 M lithium hydroxide. The combined filtrate and washings are adjusted to pH 8.0 by addition of a cation exchanger (Dowex 50) in the acid form and then applied to a column (10 cm. $\times$ 2 cm. in diameter) of anion exchanger (Dowex 2) in the chloride form. After a water wash, the column is eluted by using a linear salt gradient with 0.003 M hydrochloric acid (2 l.) in the mixing chamber and 0.05 M lithium chloride in 0.003 M hydrochloric acid (2 l.) in the reservoir. The flow rate should be 2.5 ml./min., 20-ml. fractions being collected. The elution of nucleotides is followed spectrophotometrically at 280 mμ. A small amount of cytidine 5′-phosphate (12%, estimated spectrophotometrically) is eluted first, followed by cytidine 5′-diphosphate (82%). The combined fractions containing the diphosphate are adjusted to pH 7.5 with lithium hydroxide and then evaporated to dryness *in vacuo*. Last traces of water are removed from the residue by suction on an oil pump at 25° to yield a white solid. This is suspended in methanol (10 ml.), and the mixture is stirred thoroughly. On addition of acetone (60 ml.) the nucleotide separates and is collected by centrifugation. This treatment with methanol and acetone is repeated until the supernatant liquid is free of chloride ion (absence of a precipitate when an aliquot is evaporated to dryness, and then aqueous silver nitrate is added). The nucleotide is then dried under reduced pressure at 25° to give the dihydrate or trilithium cytidine 5′-diphosphate as a white powder (361 mg., 78%). The product is completely free of cytidine 5′-phosphate and inorganic phosphate.

Nucleoside 5′-Diphosphate Coenzymes[21]

Principle. The method is completely analogous to that used in the synthesis of cytidine 5′-diphosphate. In the preparation of the pyrophos-

[21] S. Roseman, J. J. Distler, J. G. Moffatt, and H. G. Khorana, *J. Am. Chem. Soc.* **83**, 659 (1961).

phate coenzymes, a nucleoside 5′-phosphoromorpholidate is condensed with an appropriate phosphate monoester. The latter are often used as their trialkylammonium salts, since they are soluble in anhydrous pyridine. The product is then isolated by ion-exchange chromatography. The method is illustrated by the synthesis of uridine diphosphate glucose described below.

Uridine Diphosphate Glucose. The 4-morpholine-N,N'-dicyclohexyl-carboxamidinium salt of uridine 5′-phosphoromorpholidate (227 mg., 0.33 millimole) is dissolved in anhydrous pyridine (10 ml.), and the solution is evaporated to dryness under reduced pressure. This procedure is repeated, twice, to ensure that the nucleotide is anhydrous. (Only dry air is allowed to enter the flask during additions of pyridine.) Separately, an aqueous solution of dipotassium α-D-glucose 1-phosphate monohydrate (365 mg., 1.0 millimole) is passed through a column (5 cm. $\times$ 1 cm. in diameter) of cation exchanger (Dowex 50) in the pyridinium form, and the resin is washed with water (25 ml.). The combined eluate is evaporated to dryness, and the residue is dissolved in water (5 ml.). To this is added a solution of tri-n-octylamine (355 mg., 1.0 millimole) in pyridine (15 ml.). The mixture is shaken until homogeneous and then evaporated to dryness. The residue is rendered anhydrous by repeatedly dissolving it in anhydrous pyridine (10 ml.) followed by evaporation of the solvent (four times). The anhydrous tri-n-octylammonium salt of α-D-glucose 1-phosphate in dry pyridine (15 ml.) is transferred to the flask containing the uridine 5′-phosphoromorpholidate. This solution is evaporated to dryness, and the residue is dissolved in dry pyridine (5 ml.). This solution is kept at 25° for 3 days. Then, the solvent is removed *in vacuo*, and the oily residue is suspended in water (15 ml.). Lithium acetate (150 mg.) is added with stirring, and the mixture is extracted with ether (15 ml.). The ether layer is washed with water (15 ml.) containing lithium acetate (50 mg.). The combined aqueous solutions are diluted to 100 ml. with water and then passed through a column (10 cm. $\times$ 2 cm. in diameter) of anion exchanger (Dowex 2) in the chloride form. The column is washed with water to remove pyridine, and then uridine 5′-phosphate is eluted with 0.02 M lithium chloride in 0.003 M hydrochloric acid, elution being followed spectrophotometrically at 260 mμ. Uridine diphosphate glucose is next eluted with 0.06 M lithium chloride in 0.003 M hydrochloric acid (frequently a small amount of P^1,P^2-diuridine 5′-pyrophosphate is eluted immediately after the coenzyme). The solution containing uridine diphosphate glucose is adjusted to pH 4.0 with M lithium hydroxide and then carefully evaporated to dryness under reduced pressure (bath temperature 30°). Last traces of water are removed from the residue by evaporation on an oil pump at 25° to leave a dry white solid. This is dissolved in methanol (5 ml.), and the nucleotide is precipitated by addition of acetone (35 ml.) and ether (5 ml.). The precipitate is collected by centrifugation, dissolved in 3 ml. of methanol, and reprecipitated by addition of acetone (30 ml.) and ether (5 ml.). The reprecipitation procedure is repeated until the supernatant liquid is free of chloride ions. The final precipitate is washed with ether and dried *in vacuo* to give the dilithium salt of uridine diphosphate glucose as the hexahydrate (142 mg.) which can be converted to the dihydrate by drying *in vacuo* at 100°. The product is completely identical with authentic uridine diphosphate glucose.

Polymerization of Deoxyribonucleotides[22]

Principle. The deoxyribonucleotide is polymerized by treatment with dicyclohexylcarbodiimide in anhydrous pyridine. The products, homologous series of linear and cyclic oligonucleotides, are separated chromatographically. In the experiment described below, thymidine 5'-phosphate is polymerized in the presence of $N^6,O^{3'}$-diacetyldeoxycytidine 5'-phosphate. After removal of the acetyl groups, three series of products are

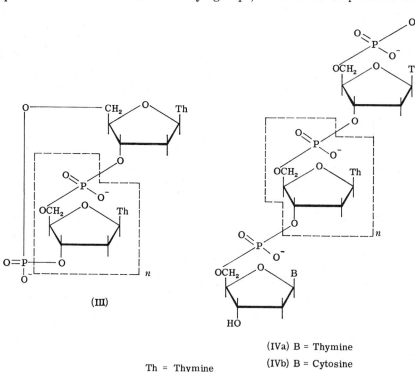

(III)

Th = Thymine

(IVa) B = Thymine

(IVb) B = Cytosine

[22] H. G. Khorana and J. P. Vizsolyi, *J. Am. Chem. Soc.* **83**, 675 (1961); H. G. Khorana, A. F. Turner, and J. P. Vizsolyi, *ibid.* p. 686 and subsequent papers.

obtained: the homologous cyclic oligonucleotides derived from thymidine 5'-phosphate [(III), $n = 1$, 2, 3], corresponding linear polymers of thymidine 5'-phosphate [(IVa), $n = 0$ to 9], and the linear polymers terminated with a deoxycytidine group [(IVb), $n = 0$ to 9]. The $N^6,O^{3'}$-diacetyldeoxycytidine 5'-phosphate, as well as giving rise to the series (IVb) polymers, reduces the amount of cyclic oligonucleotides, since any oligonucleotide produced with it as a terminating group cannot cyclize.

$N^6,O^{3'}$-Diacetyldeoxycytidine 5'-Phosphate. Deoxycytidine 5'-phosphate[7] (0.5 millimole of the free acid) is dissolved in water (10 ml.) containing pyridine (1 ml.), and the solution is lyophilized. The resulting finely divided material is suspended in anhydrous pyridine (5 ml.) containing acetic anhydride (1.5 ml.). On shaking for a few hours at 25° in the dark, a clear solution is obtained. After a total of 20 hours at 25°, the mixture is diluted with cold water (20 ml.) and kept for 90 minutes at the same temperature. The solution is then evaporated to dryness *in vacuo* (bath temperature, less than 20°). Water is added to the sirupy residue, and the solution is re-evaporated. This process is repeated twice to remove most of the pyridinium acetate. Finally the residue is dissolved in water, and the solution is lyophilized to give a white powder. Electrophoresis at pH 3.5 and chromatography in the systems ethyl alcohol–0.5 M ammonium acetate, pH 3.8 (5:2), and ethyl alcohol–M ammonium acetate, pH 7.5 (5:2), should indicate the presence of only one nucleotidic product. The lyophilized material is dissolved in anhydrous pyridine (2 ml.) and used directly in the polymerization reaction.

Polymerization. To the above solution of $N^6,O^{3'}$-diacetyldeoxycytidine 5'-phosphate is added a solution of the pyridinium salt of thymidine 5'-phosphate[7] (1.5 millimoles, prepared from the diammonium salt by treatment with a cation exchanger in the pyridinium form) in anhydrous pyridine (10 ml.). The solution is concentrated to dryness *in vacuo* and is rendered anhydrous by dissolving it in dry pyridine (about 10 ml.) and evaporating the solvent. This coevaporation step is repeated four times. The resultant gum is dissolved in anhydrous pyridine (1 ml.) with exclusion of moisture. Dicyclohexylcarbodiimide (825 mg., 4.0 millimoles) in dry pyridine (1 ml.) is added rapidly to this solution by using a pressure-equalizing dropping funnel. The mixture separates into two liquid phases, the lower phase hardening to a gum on being shaken vigorously by hand for 5 minutes. The mixture is then shaken mechanically with exclusion of light for a total of 5.5 days. (At this stage the gum has been replaced by solid material.) Sodium hydroxide (2 M, 3 ml.) is added with shaking to the mixture, together with water (5 ml.). The alkaline solution is extracted with ether to remove unreacted dicyclo-

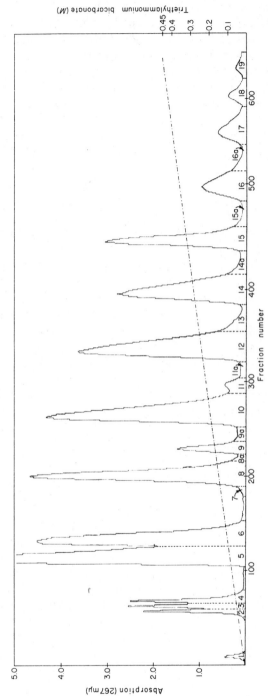

Fig. 1. Chromatography of total products obtained on polymerization of a mixture of thymidine 5'-phosphate and $N^6,O^{3'}$-diacetyldeoxycytidine 5'-phosphate. For details of procedure, see text.

hexylcarbodiimide and then filtered through a cotton plug to remove dicyclohexylurea. Forty-five minutes after the addition of alkali, an excess of cation exchanger (Amberlite IR-120) in the ammonium form is added. The resin is removed by filtration and washed thoroughly with water. The combined filtrates are evaporated to dryness, and the residual gum is dissolved in concentrated ammonium hydroxide to remove the N^6-acetyl group from the deoxycytidine nucleotides. The ammonia is removed by evaporating the solution to dryness, and the residue is dissolved in water (10 ml.). For convenience, one-half of this mixture of polymers is used in the separation that follows.

Separation of Oligonucleotides. Five milliliters of the above solution of polynucleotides (corresponding to 1.0 millimole of starting material) is applied to a column (60 cm.$\times$ 4 cm. in diameter) of DEAE-cellulose (carbonate form—prepared by washing the column with 2 M ammonium carbonate) followed by a water wash (300 ml.). The column is eluted by using a linear salt gradient with water (6 l.) in the mixing chamber and 0.45 M triethylammonium bicarbonate (pH 7.5, 6 l.; see synthesis of adenosine cyclic 3',5'-phosphate for preparation). The flow rate should be about 2.5 ml./min. Fractions of approximately 20 ml. are collected. The elution of nucleotides is followed spectrophotometrically at 267 mμ. A typical elution pattern is shown in Fig. 1. The fractions constituting the peaks (numbered in Fig. 1) are combined, the amount of product present is estimated spectrophotometrically, and then the solutions are evaporated to dryness *in vacuo*. To ensure that all the triethylammonium carbonate is removed, the residue is dissolved in water, and the solution is evaporated to dryness several times. Peaks 1 to 4, 7, 8a, 9a, 11a, 13, 14a, 15a, 16a, and 19 are discarded. Peaks 6, 9, and 11 contain the cyclic di-, tri-, and tetrathymidylic acids, respectively. Trace impurities in each peak are removed by chromatography on Whatman 3 MM paper in the isopropanol–concentrated ammonia–water (7:1:2) system; the tetranucleotide (peak 11) requires several days' chromatography for complete resolution from its contaminants. The major ultraviolet-adsorbing band on each chromatogram is eluted with water, and the solution is passed through a small column of cation exchanger (Dowex 50) in the ammonium form to give the pure ammonium salts of the cyclic oligonucleotides. These can be stored as the frozen solutions.

Peaks 8, 10, 12, 14, 15, 16, 17, and 18 contain the homologous series of linear oligonucleotides [(IVa) and (IVb)]. Thus peak 8 contains the dinucleotides [(IVa), $n = 0$; and (IVb), $n = 0$], peak 10 contains the trinucleotides [(IVa), $n = 1$; and (IVb), $n = 1$], etc., up to peak 18, which contains the nonanucleotides [(IVa), $n = 7$; and (IVb), $n = 7$]. Separation of the pairs of oligonucleotides and removal of some trace impurities is achieved by chromatography on a DEAE-cellulose column 15 cm. $\times$ 2 cm. in diameter (16 cm. $\times$ 1 cm. in diameter for peaks 17 and 18) by using a linear lithium chloride gradient in acid solution. The salt concentration in the mixing vessel (1 l.) and in the reservoir (1 l.) used for the separation of each pair of oligonucleotides is listed in the table. The flow rate through the column is 1.5 ml./min.; 15-ml. fractions

SALT CONCENTRATIONS USED FOR SEPARATION OF PAIRS OF OLIGONUCLEOTIDES

Peak	Nucleotides	Mixing vessel, $0.003\ M$ HCl $+ x\ M$ LiCl	Reservoir, $0.003\ M$ HCl $+ y\ M$ LiCl
8	Di-	$x = 0$	$y = 0.05$
10	Tri-	0.05	0.1
12	Tetra-	0.1	0.2
14	Penta-	0.1	0.2
15	Hexa-	0.15	0.25
16	Hepta-	0.175	0.3
17	Octa-	0.2	0.3
18	Nona-	0.2	0.3

are collected. The elution is followed spectrophotometrically at 267 mμ. The deoxycytidine-ended member of a pair of oligonucleotides is eluted first from the column; it is the major component, the average ratio of the (IVb) series to the (IVa) series being 3:2. Occasionally small peaks are eluted from the column (in addition to the two major peaks); these are discarded. The fractions containing the oligonucleotides are combined, neutralized to pH 7.0 with M lithium hydroxide, and the solution is concentrated to dryness in vacuo. The last traces of water are removed from the residue by suction on an oil pump at 25°. The residual white solid (lithium chloride and oligonucleotide) is dissolved in the minimum amount of methanol (2 to 4 ml.), and then the nucleotide is precipitated by addition of acetone (25 to 30 ml.) and ether (5 to 10 ml.). The mixture is cooled to 0° for 2 to 3 hours, and then the nucleotide is collected by centrifugation. The precipitate is stirred with methanol (1 ml.), and acetone (15 ml.) is added. After collection by centrifugation, the nucleotide is washed with acetone and dried in vacuo at 25°. The lithium salt so obtained is dissolved in a little water, and the solution is passed through a cation exchanger (Dowex 50) in the ammonium form. The combined eluate and washings are concentrated to a small volume. This solution of the oligonucleotide ammonium salt is frozen and stored. The yields (%) of the various oligonucleotides, estimated spectrophotometrically, are as follows: (1) Cyclic di-, tri-, and tetranucleotides, 10%, 1.5%, and 0.5%, respectively. (2) Linear thymidine-ended oligonucleotides [series (IVa)], di-, tri-, and tetranucleotides, 3 to 4%. The higher oligonucleotides are obtained in decreasing amounts down to the nonanucleotide, 0.5%. (3) Deoxycytidine-ended oligonucleotides [series (IVb)], di-, tri-, and tetranucleotides, about 6% each, then decreasing amounts of the higher oligonucleotides down to the nonanucleotide, 0.6%.

[95] Preparation of ATP Free of Metal Ions

By John M. Lowenstein

Principle. The barium salt of ATP is dissolved in the presence of Dowex 50 resin (H^+ form), the resin is removed by filtration, residual barium ions are precipitated as barium sulfate, and the solution is neutralized to the desired pH with a base such as diethanolamine (pK = 9.0) or tris(hydroxymethyl)aminomethane ($pK = 8.0$).[1] The total amount of base added depends on the required pH but is usually between 3 and 4 equivalents per mole of ATP. If barium ATP is dissolved with acid instead of Dowex 50 resin, 4 equivalents of HCl are required per mole of ATP. A further 4 equivalents of H_2SO_4 are then required to precipitate the barium ions. Between 11 and 12 equivalents of base are usually required to bring the resulting solution to the desired pH. Such large quantities of salt are generally undesirable, and the method employing Dowex 50 resin is to be preferred.

Reagents

Barium ATP.
Dowex 50 resin, H^+ form, 8% crosslinked, 200 to 400 mesh. (The resin should be cycled through the sodium and acid forms and washed thoroughly with distilled water before use.)
Base: Diethanolamine (redistilled under reduced pressure), tris-(hydroxymethyl)aminomethane, and tetramethyl- or tetraethyl-ammonium hydroxide are all suitable.

Procedure. Two grams of barium ATP are suspended in 6 ml. of water in a 25-ml. beaker. A graduated centrifuge tube is filled with 14 ml. of a 50% (v/v) suspension of Dowex 50 resin (H^+ form), and the suspension is centrifuged. The volume of the resin is measured and adjusted to 7.0 ml. by the addition or subtraction of small amounts of resin and repeated centrifugation. When the desired volume of packed

[1] J. M. Lowenstein, *Biochem. J.* **75**, 269 (1960).

resin has been obtained, the water is decanted. The following operations are performed on ice or in a cold room. The resin (7.0 ml.) is added slowly to the suspension of barium ATP with thorough stirring. The ATP dissolves very slowly at first, but, after a certain amount of resin has been added, a point is reached where the characteristic white slurry of barium ATP suddenly disappears. This is usually accompanied by the conversion of the well-dispersed resin into a sticky mass. The remainder of the resin is added slowly with thorough stirring. After the addition of the resin is complete, stirring is continued for 5 minutes. The resin is filtered off on a sintered funnel and resuspended in 3 ml. of water. The suspension is stirred thoroughly, and the washing liquid is filtered off on a sintered funnel. The washing of the resin is repeated twice more. The combined filtrates are neutralized to between pH 3 and 5 by the addition of (say) diethanolamine. This is most conveniently done by using the liquid base (m.p. 23°). A check of the total amount of diethanolamine added is made by weighing out an excess of the base and weighing the remainder at the end of the preparation. The pH is raised at this point in order to minimize the hydrolysis of ATP in the very acid solution obtained from the resin treatment. Residual barium is then precipitated from the solution by the dropwise addition of 2 N H_2SO_4. If the solution becomes too cloudy to judge the completeness of precipitation, it is centrifuged, and the addition of H_2SO_4 is continued until no further precipitate forms. This point is harder to judge than is the case in, say, the precipitation of barium from solutions of its chloride. It is best to add a drop of H_2SO_4 to the clear solution and to wait a few minutes to see if the solution remains clear, before deciding that precipitation of barium is complete. The clear, barium-free solution is neutralized to the desired pH, say pH 9.0, by the addition of diethanolamine, the pH being checked first on indicator papers, and near the desired pH on a meter. The solution is then freeze-dried. The number of equivalents of base per mole of ATP which results in the stablest dry preparation must be determined for each base.

If a dry preparation is not required, the concentration of ATP is determined optically and is adjusted to the desired value, usually 0.1 M, with water. If the solution contains less than the desired concentration of ATP, the cold solution of ATP is placed in a round-bottomed flask, and some water is removed from the cold solution in a rotary drier. No appreciable breakdown of ATP occurs in the course of this treatment. The concentration of ATP is redetermined and adjusted to the desired value.

[96] Intermediates in Purine Nucleotide Synthesis

By LEWIS LUKENS and JOEL FLAKS[1,2]

The individual enzymatic reactions which are responsible for the biosynthesis of the purine nucleotides have been extensively reviewed in a recent article.[3] The purification procedures for these enzymes are presented in Vol. VI [9], to which article the reader is also referred for an illustration (Fig. 1) of the structural formulas of the intermediates in purine biosynthesis along with their enzymatic reactions. Preparations of several of these intermediates have been described in Vol. II [78]. The preparation of 5-phosphoribosylpyrophosphate is described in Vol. VI [69].

A. General Procedures

Materials. Deionized water is used in making up all solutions which are added to enzymatic incubations. Solutions are adjusted to approximately the pH of the incubation before their addition—thus disodium ATP refers to the composition of the solid and not to the solution, which is neutralized before use with KOH or NaOH. For approximate adjustments of this sort, including those made on solutions prior to ion-exchange chromatography or precipitation of ribotides, a single, multiply-indicating pH paper may be used. Where necessary to convert barium salts to their sodium, potassium, or ammonium counterparts, the

---

[1] We wish to express our indebtedness to Drs. J. M. Buchanan, T. C. French, S. C. Hartman, and R. W. Miller for making available information prior to its publication and for providing the details of recent modifications of a number of published procedures.

[2] The following are the trivial and corresponding systematic names of the intermediates of purine biosynthesis: glycinamide ribotide, 2-amino-*N*-ribosylacetamide-5′-phosphate; formylglycinamide ribotide, 2-formamido-*N*-ribosylacetamide-5′-phosphate; formylglycinamidine ribotide, 2-formamido-*N*-ribosylacetamidine-5′-phosphate; 5-aminoimidazole ribotide, 5-amino-1-ribosylimidazole-5′-phosphate; 5-amino-4-imidazolecarboxylic acid ribotide, 5-amino-1-ribosyl-4-imidazolecarboxylic acid 5′-phosphate; 5-amino-4-imidazole-*N*-succinocarboxamide ribotide, *N*-(5-amino-1-ribosyl-4-imidazolecarbonyl)-L-aspartic acid 5′-phosphate; 5-amino-4-imidazolecarboxamide ribotide, 5-amino-1-ribosyl-4-imidazolecarboxamide-5′-phosphate; 5-formamido-4-imidazolecarboxamide ribotide, 5-formamido-1-ribosyl-4-imidazolecarboxamide-5′-phosphate. All these compounds, as well as 5-phosphoribosylamine, are assumed to have the β-configuration at carbon 1 of ribose, since they are converted to inosinic acid which is known to be the β-anomer without apparent opportunity for change of configuration.

[3] J. M. Buchanan and S. C. Hartman, *Advances in Enzymol.* **21**, 199 (1959).

conversion is generally accomplished by batchwise treatment of the barium salt with an excess of Dowex 50 of the appropriate form. Per cent, when referring to trichloroacetic acid solutions, means grams of acid per 100 ml. of final solution.

Enzyme Preparations. Pigeon liver acetone powder is prepared as described in Section F of Vol. VI [9]. The yield of powder is approximately 2 g. per pigeon liver. All enzyme purification procedures are performed at 3° unless otherwise noted. The calculations of the per cent of saturation of solutions with $(NH_4)_2SO_4$ are based on the figures in Table I of Vol. I [10]. When referring to ethanol concentration, per cent means milliliters of ethanol per 100 ml. of solution. The protein concentration of enzyme solutions, where mentioned, is approximate and is obtained from the optical density of the solution at 280 mμ on the assumption that 1 mg. of protein per milliliter gives an optical density of 1.6 in a cell of 1-cm. light path.

Ion-Exchange Chromatography. The ion-exchange resins employed are Dowex 1 and Dowex 50 (or the newer Dowex 50-W), 200 to 400 mesh and 8% crosslinked, in both cases. The resins are washed and converted to the desired forms by conventional procedures[4] and when used for column chromatography are present in at least 100-fold excess, based on their ion-exchange capacity (approximately 1.2 and 1.7 meq./wet ml. for Dowex 1 and 50, respectively). The height of the columns is ten to fifteen times their diameter. Dimensions, where cited, are invariably diameter by height. For columns 2.2 cm. in diameter, flow rates of the order of 1 to 2 ml./min. are generally used. Unless otherwise noted, 10- to 15-ml. fractions of the eluate are collected automatically with a fraction collector, after the initial effluent and any washes have been collected.

The elution of unlabeled aminoimidazole ribotide and later intermediates is easily followed by means of the Bratton and Marshall procedure described below; in addition, intermediates later than aminoimidazole ribotide are conveniently detected by their ultraviolet absorption. There is, however, no equally convenient method for following the elution of the early acyclic ribotides (of glycinamide, formylglycinamide, and formylglycinamidine, respectively). It is therefore convenient to carry out the enzymatic synthesis of these ribotides in the presence of a C^{14}-labeled precursor. The specific activity of the latter may in general be as low as 3000 c.p.m./micromole. The radioactivity of solutions obtained during the isolation procedure is determined in an approximate way by pipetting measured aliquots onto planchets. The latter are counted after the liquid has been evaporated off under infrared lamps.

[4] See, for example, Vol. III [82], p. 564, and Vol. III [107], p. 732.

In following the effluents from column chromatography, aliquots (0.05 to 0.40 ml.) of the initial effluent, washes, and every second or third subsequent fraction are plated. In special cases where a ribotide devoid of radioactivity is desired, the elution of the early acyclic ribotides may be followed by the orcinol assay for pentose (see below) or by performing, as a guide, a duplicate elution, under identical conditions, with radioactive ribotide.

Precipitation of Ribotides as Salts. In general, all effluent fractions from ion-exchange columns containing 0.2 micromole or more of product per milliliter are combined prior to concentration. The combined eluates are concentrated on a rotary flash evaporator attached to a vacuum pump. The bath in which the receiver rotates contains dry ice suspended in methyl Cellosolve; the other bath, containing water, is heated at a sufficient rate to maintain the temperature of the bath at 30° to 40°. This procedure gives rapid evaporation and minimizes the thermal destruction of unstable compounds. The concentrated solution, usually 2 to 10 ml., should contain at least 2 micromoles of ribotide per milliliter. The ribotide is then precipitated in the cold at neutral pH as the ammonium, barium, or other salt by the addition of ethanol and the appropriate cation as described below for specific preparations. The salt is collected by centrifugation and is washed successively with 10 to 20 vol. of ethanol (or acetone) and ether. These solvents are also removed by centrifugation. The original supernatant solution from the precipitation step should never be discarded until the recovery of the product in the precipitate has been confirmed. In preparations of radioactive ribotides, the completeness of the precipitation may be conveniently checked by noting the decrease in radioactivity of the supernatant solution.

Analytical Methods. In addition to the more generally applicable methods described here, specialized methods are described under the individual ribotides. Total nitrogen may be determined by the Kjeldahl procedure, and total ammonia colorimetrically with Nessler's reagent.[5] Total phosphate is determined by the procedure of Gomori[6] after digestion of the samples in $10 N$ H_2SO_4 for 90 minutes at 150°. Inorganic phosphate is measured by the same procedure on undigested samples, and organic phosphate is obtained by difference. Pentose is analyzed, except where otherwise noted, by the first orcinol procedure described in Vol. III [12], with 5'-AMP as a standard. Glycine is determined by the method of Alexander et al.[7] after hydrolysis of the ribotide in 2.5 N H_2SO_4 for 90 minutes at 150°. The 5-amino group, which is present in all

[5] See Vol. III [145].
[6] G. Gomori, *J. Lab. Clin. Med.* **27**, 955 (1942).
[7] B. Alexander, G. Landwehr, and A. M. Seligman, *J. Biol. Chem.* **160**, 51 (1945).

the imidazole ribotides that are intermediates in purine biosynthesis, enables these compounds to be determined by the colorimetric assay described by Bratton and Marshall for aromatic amines.[8] The following detailed procedure has been employed except where otherwise noted. The sample, containing from 0.005 to 0.04 micromole of ribotide in a final volume of 0.40 ml., is mixed with 0.15 ml. of 0.2 N HCl. Then 0.05 ml. of 0.1% sodium nitrite is added, followed after 5 minutes with 0.05 ml. of 0.5% ammonium sulfamate and after 3 additional minutes with 0.05 ml. of 0.1% N-(1-naphthyl)ethylenediamine dihydrochloride. The optical density of the solution may then be measured any time after 10 minutes in a Beckman DU spectrophotometer fitted with an attachment for the use of microcells. The wavelengths of maximum absorbance and the molar extinction coefficients of the colored derivatives produced in the Bratton and Marshall procedure are given below under the preparation of the individual ribotides. It should be mentioned here, however, that the four imidazole ribotides that contain an unsubstituted 5-amino group, namely the ribotides of 5-aminoimidazole, 5-amino-4-imidazolecarboxylic acid, 5-amino-4-imidazole-N-succinocarboxamide, and 5-amino-4-imidazolecarboxamide, may each be distinguished from the others by means of this procedure. Thus, only the first ribotide yields an orange product, in contrast to the red-purple color given by the other three ribotides, whereas the succinocarboxamide ribotide uniquely fails to yield any color under the usual conditions of the procedure at room temperature (although it does so at 0°). Although the two remaining ribotides give quite similar colors when treated according to this procedure, they may be readily distinguished by placing the samples, after adjusting them to pH 1 or below with HCl, in a boiling-water bath for 10 minutes, prior to performing the usual procedure. This treatment quantitatively destroys the aminoimidazolecarboxylic acid structure while the full yield of color is realized from 5-amino-4-imidazolecarboxamide ribotide. A similar method may be used to measure the latter ribotide quantitatively in the presence of 5-aminoimidazole ribotide (see Section I).

B. Preparation of 5-Phosphoribosylamine

Since 5-phosphoribosylamine has not as yet been isolated from an enzymatic incubation mixture, presumably because of its instability, the only method presently available for its preparation is the chemical synthesis reported by Goldthwait,[9] whose procedure, with slight modifications successfully employed by Hartman,[10] is described below.

[8] A. C. Bratton and E. K. Marshall, Jr., J. Biol. Chem. 128, 537 (1939).
[9] D. A. Goldthwait, J. Biol. Chem. 222, 1051 (1956).
[10] S. C. Hartman, unpublished results, 1957.

Principle. The potassium salt of ribose-5-phosphate reacts with liquid ammonia under anhydrous conditions to produce 5-phosphoribosylamine, which remains, together with unreacted ribose-5-phosphate, as a powder when the ammonia is allowed to evaporate. The yield is from 10 to 30% based on total pentose. A lower yield is obtained if the lithium or calcium salts of ribose-5-phosphate are used.

Reagents

Potassium R-5-P dried *in vacuo* over P_2O_5. Conveniently prepared by acid hydrolysis of 5'-AMP.[11]

Tank ammonia (anhydrous).

Procedure. Approximately 100 mg. of powdered R-5-P are placed with minimal exposure to air in a closed flask equipped with a stirrer and glass inlet and outlet tubes. The inlet tube should reach to within a couple of inches of the bottom of the flask, whereas the outlet tube should be above the level of the ammonia which is to be condensed. The outlet tube is attached to a tower of NaOH pellets to ensure anhydrous conditions. The flask is placed in a bath of dry ice suspended in acetone, in a hood, and ammonia from a cylinder is passed through the flask until 30 ml. of ammonia have condensed. During this time the contents of the flask are mechanically stirred. The inlet tube is then closed, and the reaction mixture is allowed to stand at room temperature without further stirring until all the ammonia has evaporated. The dry powder which remains is placed in a vacuum desiccator which is evacuated on an oil pump (pressure equals approximately $50\,\mu$) for one-half hour at room temperature. The solid may be stored in the dry state in the cold, or it may be dissolved in a small amount of cold $0.02\,N$ KOH (e.g., 3 ml.) and distributed to a number of small tubes which are then stored at $-15°$. Some purification of 5-phosphoribosylamine may be achieved by fractional precipitation of its barium salt which is less soluble at low ethanol concentrations (up to 50% v/v) than the barium salt of R-5-P.[10]

Properties. 5-Phosphoribosylamine is very unstable, especially in acid, but can be kept frozen in $0.02\,M$ KOH for several weeks with little loss of activity.[9] It gives the same color value in the Elson-Morgan reaction as glucosamine.[9] Synthetic 5-phosphoribosylamine reacts with glycine in the presence of ATP and the enzyme glycinamide ribotide kinosynthase to yield glycinamide ribotide, ADP, and inorganic phosphate.[9, 12]

Analysis. The amount of 5-phosphoribosylamine present is determined by analysis for ammonia, which is released from 5-phosphoribo-

[11] J. X. Khym, D. G. Doherty, and W. E. Cohn, *J. Am. Chem. Soc.* **76**, 5523 (1954).
[12] S. C. Hartman and J. M. Buchanan, *J. Biol. Chem.* **233**, 456 (1958).

sylamine by heating for 6 minutes at 100° in 1 N HCl.[9] Owing to the presence of extraneous NH_3, this assay procedure has been found to be somewhat unreliable. A very accurate estimation of the amount of 5-phosphoribosylamine is not presently available. The amount of contaminating R-5-P may be obtained as the difference between total pentose and 5-phosphoribosylamine.

C. Preparation of Glycinamide Ribotide

Principle. Although methods for the isolation of glycinamide ribotide from enzymatic incubations have been reported,[13, 14] the best method from the standpoint of yield involves the isolation and acid hydrolysis of formylglycinamide ribotide.[12] The latter compound is much easier to obtain because of its greater accumulation in enzymatic incubations due to more favorable equilibrium conditions. Although glycinamide ribotide prepared by this method contains 25 to 35% of the enzymatically inactive α-form, in contrast to the pure β-isomer obtained by the earlier methods, the inactive compound does not interfere in any of the enzymatic reactions which have been observed. Since the hydrolysis of formylglycinamide ribotide to glycinamide ribotide is practically quantitative, the yield of the desired β-isomer, before isolation, is approximately 70%. The ribotide is chromatographed on Dowex 1–acetate and is finally isolated as the barium salt.[13] Losses during isolation should not exceed 10%.

Reagents

Ammonium salt of C^{14}-labeled formylglycinamide ribotide. See Section D.

Concentrated hydrobromic acid (48%, specific gravity 1.48).

Procedure. The ammonium salt of formylglycinamide ribotide is dissolved in the minimum amount of water, and concentrated HBr is added to a final concentration of 0.1 N. The solution is heated for 15 minutes in a boiling-water bath, cooled to 0°, and adjusted to pH 8 with NH_4OH. A twofold excess of saturated barium bromide solution and 4 vol. of absolute ethanol are added to precipitate glycinamide ribotide as the barium salt. The precipitate is collected by centrifugation and dissolved in a minimum amount of water. This solution is adjusted to pH 8 with NH_4OH and is then placed on a column of Dowex 1–acetate. For a

[13] S. C. Hartman, B. Levenberg, and J. M. Buchanan, *J. Biol. Chem.* **221**, 1057 (1956).

[14] R. A. Peabody, D. A. Goldthwait, and G. R. Greenberg, *J. Biol. Chem.* **221**, 1071 (1956).

preparation containing 50 micromoles of ribotide, a column 1 × 15 cm. is sufficient. The column is eluted with 0.025 M ammonium acetate buffer, pH 5.35. The glycinamide ribotide, which is followed by means of its radioactivity, begins to be eluted after approximately 4 resin-bed volumes. The pooled fractions containing the product are evaporated to a small volume (the final concentration of ribotide should be 2 to 5 micromoles/ml.), and the barium salt is precipitated as described above. Glycinamide ribotide exhibits no exceptional chemical instability, but it is routinely stored as the dry barium salt at 0° or below, or as a frozen solution. When prepared from formylglycinamide ribotide purified by a second passage through Dowex 1–acetate, as described in Section D, the glycinamide ribotide should be essentially free from organic contaminants (except for acetate derived from the elutriant) and is satisfactory for enzymatic studies. The product may be further purified by reprecipitation as the barium salt from ethanol and, if necessary, by further chromatography on Dowex 1–bromide.[13]

Properties.[13] The pK's of the secondary phosphate hydrogen and of the protonated amino group are 6.05 and 8.15, respectively The results of the pH titration also indicate that, in the barium salt, the amino group of glycinamide ribotide is not protonated and that the phosphate group is doubly ionized. The calculated molecular weight of the barium salt ($C_7H_{13}N_2O_8$ PBa), with possible water of hydration neglected, is 422. The ribotide does not absorb light throughout the range from 230 to 700 mμ and exhibits only weak end absorption below 230 mμ.

Analysis. The most specific and sensitive method for determining glycinamide ribotide depends on its participation in the enzymatic reaction illustrated in Eq. (1).[15] The carboxamide ribotide produced, which

Inosinic acid + glycinamide ribotide + H_2O → 5-Amino-4-imidazole-
carboxamide ribotide + formylglycinamide ribotide (1)

is equal to the glycinamide ribotide initially present, is measured colorimetrically by the Bratton and Marshall procedure (see Section A). The assay is performed exactly as described in Vol. VI [9] for glycinamide ribotide kinosynthase, except that the first incubation with kinosynthase is omitted and the neutralized ethylenediaminetetraacetate is added directly to the sample of glycinamide ribotide. The sample should contain from 0.01 to 0.05 micromole of ribotide in a final volume of 0.25 ml.

The organic phosphate, total nitrogen, glycine, and pentose determinations as described in Section A may be performed on glycinamide ribotide with the exception that it is necessary in the pentose determina-

[15] S. C. Hartman and J. M. Buchanan, *J. Biol. Chem.* **233,** 451 (1958).

tion, in order to get a color development equivalent to that from 5'-AMP, to hydrolyze all samples in 1.3 N HCl for 40 minutes at 100° immediately prior to color development.[13] The ribose may also be determined by periodate titration.[14] The amide nitrogen is acid-labile and may be determined as ammonia after the sample has been heated in 1 N HCl for 2 hours at 100°.[14] The ribotide gives a color value with ninhydrin[16] approximately equal to that of glycine.

D. Preparation of Formylglycinamide Ribotide

Principle. An extract of pigeon liver acetone powder contains all the enzymes required for the synthesis of this ribotide from ribose-5-phosphate, glutamine, glycine, and formate. The accumulation of the ribotide is increased by adding to the incubation medium the antibiotic azaserine which specifically inhibits the conversion of formylglycinamide ribotide to formylglycinamidine ribotide. The isolation of the ribotide is accomplished by ion-exchange chromatography.[13] A somewhat different preparative procedure has been briefly described in a previous volume.[17] It is also possible to obtain formylglycinamide ribotide by chemical formylation of glycinamide ribotide.[14]

Reagents

L-Azaserine[17a] (o-diazoacetyl-L-serine), chemically synthesized.[18] The product of a single chromatographic purification on a charcoal–Celite column is sufficiently pure.

Glycine-1-C^{14} (specific activity 2000 c.p.m./micromole) is generally used to facilitate following the ribotide during its isolation.

Ribose-5-phosphate prepared as in Section B.

Pigeon liver acetone powder extract. The powder is prepared as described in Section F of Vol. VI [9]. The pigeons should be sacrificed 2 to 5 hours after feeding. The powder·is extracted by stirring it for 30 minutes at 3° with 0.1 M potassium phosphate buffer, pH 7.4 (100 ml. per 10 g. of powder). Insoluble material is removed by centrifugation, and the supernatant solution is used.

Other reagents are as described in Vol. II [78].

[16] See Vol. III [76].
[17] See Vol. II [78].
[17a] Sample of compound a gift from Drs. Alexander Moore and John R. Dice of Parke, Davis and Co.
[18] E. D. Nicolaides, R. D. Westland, and E. L. Wittle, *J. Am. Chem. Soc.* **76**, 2887 (1954).

Procedure.[19] The incubation mixture contains the following quantities of materials, expressed in micromoles: Tris–chloride buffer, pH 8.0, 7000; sodium formate, 1080; magnesium chloride, 600; L-glutamine, 1440; disodium 3-phosphoglycerate, 1800; disodium ribose-5-phosphate, 1400; disodium ATP, 400[20]; azaserine, 400; glycine-1-C^{14}, 1200; and the extract from 5.0 g. of pigeon liver acetone powder. The final volume is brought to 200 ml., and the solution is incubated at 38° with gentle shaking for 90 minutes. The proteins present are precipitated by the addition of 10 ml. of 60% trichloroacetic acid and are removed by centrifugation. The precipitate is washed once with 40 ml. of cold 5% trichloroacetic acid, and the washings are added to the first supernatant solution.

The combined supernatant solutions, at room temperature, are promptly passed through a column of Dowex 50–ammonium, 2.2 × 16 cm., at 2 to 3 ml./min. The effluent is collected in a receiver cooled in an ice bath and is saved. The column is then washed with 100 ml. of 0.05 M ammonium formate buffer, pH 3.3 (4.7 ml. of 88% formic acid and 3.16 g. of ammonium formate per liter). The first 40 ml. of buffer effluent and any subsequent 10-ml. fractions which contain more than 0.2 micromole of ribotide per milliliter, as judged by radioactivity, are combined with the initial effluent. The combined solutions are immediately adjusted to pH 8 with several drops of 10 N KOH. Since glycine is retained by the column, whereas the ribotide passes through, the radioactivity in the combined effluents furnishes an approximate measure of the yield of formylglycinamide ribotide. The neutralized solution, as well as the eluate from the Dowex 1–acetate column (see below), may be stored overnight at 3°.

The pooled and neutralized fractions from the Dowex 50 column are percolated at room temperature through a column of Dowex 1–acetate, 2.2 × 19 cm. The column is washed with 1 resin-bed volume of water (75 ml.) and is then eluted with 0.1 M ammonium acetate buffer, pH 5.35 (7.68 g. of ammonium acetate and 1.44 ml. of glacial acetic acid per liter). The ribotide is eluted after approximately 3 resin-bed volumes of buffer in a total volume of approximately 180 ml. If a small radioactive peak is observed just before the main ribotide-containing peak, it consists of the inactive α-isomer and is discarded. The fractions containing more than 0.2 micromole of ribotide per milliliter are combined and concentrated down to 2 to 3 ml. on a rotary evaporator. The ammonium

---

[19] Includes unpublished modifications of T. C. French.

[20] The experimental observation that higher levels of ATP decrease the yield of formylglycinamide ribotide is probably explained by the finding of J. B. Wyngaarden and D. M. Ashton [*J. Biol. Chem.* **234**, 1492 (1959)] that ATP and ADP inhibit 5-phosphoribosylpyrophosphate amidotransferase.

salt of the ribotide is precipitated from this solution by the addition of 50 vol. of absolute ethanol. After the solution has stood at −20° for 1 hour or more, the precipitate is collected by centrifugation and washed successively with absolute ethanol and then with ether. The product is stored in a desiccator at 3° or below, or in frozen solution. It is only about 30% pure on a weight basis (the calculated molecular weight of ammonium formylglycinamide ribotide is 301) but is adequate for use in enzymatic incubations. It may be conveniently purified by repeating the above chromatographic procedure with a smaller column, 0.8 × 22 cm., of Dowex 1–acetate. The final yield of ammonium salt, precipitated as before, is 100 to 200 micromoles. A sample prepared by a similar procedure was found to be 85% pure on a weight basis and gave the expected analyses for phosphate, etc. (see below).[13]

The above preparation may be scaled up five times, in which case columns of Dowex 50–ammonium (4.5 × 20 cm.) and Dowex 1–acetate (4.5 × 24 cm.) are employed. Flow rates of 10 to 15 ml./min. may be used with such columns.

Properties.[13] The pK of the secondary phosphate hydrogen of formylglycinamide ribotide is 6.40. The compound resembles glycinamide ribotide in exhibiting only weak end absorption below 230 mμ. The failure of both compounds to react with the Pauly diazo reagent,[21] which couples with phenols, aromatic amines, histidine, or imidazoles to produce highly colored azo compounds, is taken as evidence that these ribotides are not imidazole derivatives but are acyclic structures. The ease with which the formyl group of formylglycinamide ribotide is hydrolyzed under acidic conditions is illustrated by the preparation of glycinamide ribotide described in Section C.

Analysis. Formylglycinamide ribotide may be specifically and accurately determined by its enzymatic conversion to 5-aminoimidazole ribotide which is then estimated colorimetrically by the procedure of Bratton and Marshall. The incubation mixture is similar to that described for the assay of FGAR-amidotransferase in Vol. VI [9] except that formylglycinamide ribotide is made limiting. The two enzymes required are obtained in a single ammonium sulfate fraction (0 to 60% saturation) from pigeon liver acetone powder extract.[10] The latter is prepared by stirring 10 g. of the acetone powder for 30 minutes at 3° with 100 ml. of 0.05 M potassium phosphate buffer, pH 7.0, containing L-glutamine (0.001 M) and ethylenediaminetetraacetate (0.001 M). The suspension is centrifuged, and solid $(NH_4)_2SO_4$ is added slowly with stirring to the supernatant solution (39 g. per 100 ml.). The precipitate obtained after centrifugation is dissolved in 10 ml. of 0.05 M potassium

[21] K. K. Koessler and M. T. Hanke, *J. Biol. Chem.* **39**, 497 (1919).

phosphate buffer, pH 7.4, containing glutamine $(0.001 M)$ and ethylene-diaminetetraacetate $(0.001 M)$ and is dialyzed against 2 l. of the same buffer for 3 hours. The preparation is stable for 2 weeks if kept frozen. The incubation mixture contains, in addition to the sample of formylglycinamide ribotide to be assayed $(0.003$ to 0.04 micromole), the following quantities of materials, expressed as micromoles, in a final volume of 0.6 ml.: ATP, 2; L-glutamine, 1; $MgCl_2$, 2; KCl, 50; Tris–chloride, pH 7.4, 10; and 0.10 ml. of enzyme preparation. The vessels should be incubated at 38° until no further aminoimidazole ribotide is formed (30 minutes is usually sufficient). The incubation mixtures are deproteinized by the addition of 0.10 ml. of 30% trichloroacetic acid solution, followed by centrifugation. The Bratton and Marshall procedure (Section A) is then performed directly on the supernatant solutions or on aliquots (0.4 ml.) transferred to other tubes. In either case the total amount of aminoimidazole ribotide produced is calculated from the molar extinction coefficient $(24,600)$ of its colored derivative at 500 mμ. The amount of formylglycinamide ribotide originally present is equal to the aminoimidazole ribotide that is formed.

The analyses for organic phosphate, total nitrogen, acid-labile (amide) nitrogen, pentose, and glycine may be performed by the same methods employed for glycinamide ribotide. In addition, formic acid may be determined by the method of Grant[22] after hydrolysis of the ribotide in 1 N NaOH for 50 minutes at 100°.[13]

E. Preparation of Formylglycinamidine Ribotide

Principle. Formylglycinamide ribotide is converted to formylglycinamidine ribotide by incubation with a partially purified preparation of formylglycinamide ribotide amidotransferase from chicken liver in the presence of ATP, Mg++, and glutamine.[23, 24] The enzyme fraction used in this preparation must be free from the enzyme, also present in avian liver, which catalyzes the conversion of the amidino-ribotide to 5-aminoimidazole ribotide in the presence of ATP. The formylglycinamidine ribotide is obtained as the ammonium salt after chromatography on Dowex 1–acetate.

Reagents

Ammonium salt of C^{14}-labeled formylglycinamide ribotide. The product from rechromatography on Dowex 1–acetate should be used. See Section D.

[22] W. M. Grant, *Anal. Chem.* **20**, 267 (1948).
[23] B. Levenberg and J. M. Buchanan, *J. Biol. Chem.* **224**, 1019 (1957).
[24] T. C. French and S. C. Hartman, unpublished results, 1957.

The enzyme preparation is a dialyzed 0 to 43% $(NH_4)_2SO_4$ fraction derived from chicken liver acetone powder extract (step 4 in Section E, Vol. VI [9], where its preparation is described). The activity of such a preparation may be checked in a preliminary incubation of the same composition as the large-scale reaction mixture described below, but of 0.3-ml. final volume. After incubation for 30 minutes at 38°, the formylglycinamide ribotide amidotransferase is inactivated by placing the reaction mixture in a boiling-water bath for 1 minute. The yield of the amidino-ribotide is then measured by adding to the cooled reaction mixture an excess of the enzyme formylglycinamidine ribotide kinocyclodehydrase, which catalyzes the quantitative ATP-dependent conversion of the latter ribotide to 5-aminoimidazole ribotide. The kinocyclodehydrase, which must be free from the amidotransferase, is easily obtained by performing a single ammonium sulfate fractionation of pigeon liver acetone powder extract, as described in steps 1 and 2 of Section F, Vol. VI [9]. If the protein concentration of the enzyme solution is approximately 25 mg./ml., 0.05 ml. should be sufficient. After a second incubation at 38° for 30 minutes, the reaction mixture is deproteinized by chilling in an ice bath, followed by the addition of 0.10 ml. of 30% trichloroacetic acid. After sedimentation of the denatured protein, the aminoimidazole ribotide is measured colorimetrically according to the procedure of Bratton and Marshall (Section A). The molar extinction coefficient of the colored derivative of this ribotide is 24,600 at 500 mμ.[25]

Procedure.[23, 24] The incubation mixture contains, in a final volume of 240 ml., the following quantities of materials expressed in micromoles: formylglycinamide ribotide, 50; disodium ATP, 400; L-glutamine, 400; $MgCl_2$, 1000; Tris–chloride, pH 8.0, 5000; and 166 mg. (10 ml.) of the dialyzed 0 to 43% $(NH_4)_2SO_4$ fraction.

After incubation for 30 minutes at 38°, the mixture is chilled to 0° and adjusted to pH 10 by addition of a few drops of 10 N KOH. The mixture is immediately passed, under sufficient pressure to maintain a flow rate of 12 to 14 ml./min., through a column of Dowex 1–acetate, 2.2 × 10 cm., at 3°. The column is previously adjusted to pH 8 by washing with 0.001 M Tris–acetate buffer, pH 8. After the incubation mixture has passed through, the column is washed with 40 ml. of 0.001 M Tris–acetate buffer, pH 8.0, and is then eluted with 200 ml. of 0.1 M

[25] B. Levenberg and J. M. Buchanan, *J. Biol. Chem.* 224, 1005 (1957).

ammonium acetate buffer, pH 5.35. Ten-milliliter fractions are collected every 3 minutes. The initial effluents and the subsequent even-numbered fractions are assayed for radioactivity by plating 0.1-ml. aliquots. The fractions (extending approximately from the first 40 to 100 ml. of effluent), which contain a total of more than 0.5 micromole of ribotide, are combined and adjusted, if necessary, to between pH 6 and 7. If unreacted formylglycinamide ribotide is present, it can be recovered in the later fractions. The combined fractions containing formylglycinamidine ribotide are evaporated down to 4 to 8 ml. on a rotary evaporator at an internal flask temperature below 10°. This solution is transferred to a chilled 40-ml. centrifuge tube and is further evaporated to a slurry. Absolute ethanol (20 ml.) and ether (20 ml.) are added to the slurry, and the resulting suspension is centrifuged. The supernatant solution is discarded, and the precipitated ammonium salt of the ribotide is washed once by suspending it in 15 ml. of anhydrous ether and centrifuging again. It is then dried in a vacuum over P_2O_5 at 3°. The formylglycinamidine ribotide may be stored dry as the ammonium salt or in frozen aqueous solution. The yield of ribotide is approximately 30 micromoles, or 60% based on the formylglycinamide ribotide.

Properties.[23] Formylglycinamidine ribotide contains two titratable groups between pH 3 and 11, one with a pK of 6.0, assigned to the secondary phosphate hydrogen, and the other with a pK of 9.2, assigned to the amidinium group. The compound is similar to the two ribotides described immediately above, in having only weak end absorption (below 240 mμ), and in its lack of reaction in the Bratton and Marshall procedure for aromatic amines or in the Pauly procedure[21] for imidazoles. It is more unstable than the ribotides of glycinamide or formylglycinamide, and some destruction to the latter occurs even in the cold.

Analysis. The most specific analysis for formylglycinamidine ribotide depends on its enzymatic conversion to aminoimidazole ribotide, which is estimated colorimetrically. The procedure is identical with the analysis for formylglycinamide ribotide described in Section D, except that glutamine is omitted from the incubation mixture.[10] The conversion to aminoimidazole ribotide is quantitative. When the amidino-ribotide sample does not contain glutamine (as, for instance, after the ion-exchange chromatography described above), the crude enzyme preparation described in Section D under Analysis [0 to 60% $(NH_4)_2SO_4$ fraction] may be used. In cases, such as that described under Reagents, where it is desired to measure formylglycinamidine ribotide in the presence of both glutamine and formylglycinamide ribotide, it is necessary, in order to avoid extraneous synthesis of the amidino-ribotide during the assay itself, to use a preparation of the kinocyclodehydrase free from the amidotransferase. The simple preparation of the kinocyclodehydrase described in steps 1 and 2 of Section F, Vol. VI [9], is suitable for this purpose.

Formylglycinamidine ribotide may be assayed for glycine, etc., by the same methods employed for formylglycinamide ribotide and with identical results except that, as expected, two acid-labile nitrogen atoms, rather than one, are present per molecule.[23]

F. Preparation of 5-Aminoimidazole Ribotide

In addition to the incubation described in detail below for the enzymatic synthesis of aminoimidazole ribotide, two alternative incubation procedures have been successfully employed. In one of these,[25] formylglycinamide ribotide is replaced by a combination of the more readily prepared 5-phosphoribosylpyrophosphate[26] and glycine. Since repeated attempts to isolate aminoimidazole ribotide from such an incubation were unsuccessful,[25] the usefulness of this procedure is at present limited to cases in which the ribotide may be used *in situ*. In the other procedure,[27] the ribotide is produced by a reversal of reactions (2) and (3). By replacing phosphate with arsenate, the equilibrium of

5-Aminoimidazole ribotide $+ CO_2 \rightleftarrows$
$$5\text{-Amino-4-imidazolecarboxylic acid ribotide} \quad (2)$$
5-Amino-4-imidazolecarboxylic acid ribotide $+ ATP +$ L-aspartate $\rightleftarrows$
5-Amino-4-imidazole-N-succinocarboxamide ribotide $+ ADP + HPO_4{}^{--}$
$$(3)$$

reaction (3) is shifted to the left. Since the isolation of aminoimidazole ribotide from such an incubation has not been attempted (although it is expected that the isolation procedure described below would be applicable), and since approximately as much labor is required to obtain the succinocarboxamide ribotide as to obtain formylglycinamide ribotide, this procedure is not described in detail. The following quantities of materials, expressed in micromoles, are incubated together for 4 hours at 38° in a final volume of 10.5 ml.: succinocarboxamide ribotide, 6.8; disodium ADP, 63; disodium arsenate, 250; $MgCl_2$, 100; K_2HPO_4, 150; and approximately 12 units, based on its activity in catalyzing reaction (3), of the 30 to 50% fraction obtained from the $(NH_4)_2SO_4$ step described in Section H, Vol. VI [9]. The yield of aminoimidazole ribotide is 4.5 micromoles, or 66% based on the succinocarboxamide ribotide.

The preparation of aminoimidazole ribotide from formylglycinamide ribotide is described in detail below.[25]

[26] See Vol. VI [69].
[27] R. W. Miller, unpublished results, 1958.

Principle. In the presence of a crude enzyme preparation from avian liver, formylglycinamide ribotide reacts with ATP and glutamine to yield formylglycinamidine ribotide which is cyclized in a further ATP-requiring reaction to yield 5-aminoimidazole ribotide. For maximum synthesis of the latter, the concentration of potassium ions in the incubation mixture must be 0.06 M or higher, owing to a requirement for this ion in the cyclization reaction. Because of the instability of the 5-aminoimidazole structure,[25, 28] the isolation is performed rapidly and in the cold.

Reagents

Formylglycinamide ribotide, prepared as described in Section D. Since aminoimidazole ribotide may be followed during chromatography by means of the Bratton and Marshall procedure, the presence of radioactivity in the formylglycinamide ribotide is not so important as in the previously described preparations.

Ethanol fraction (13 to 33%) from pigeon liver extract, prepared as described for the 15 to 30% fraction in Section A, Vol. VI [9]. The optimum amount of enzyme is usually tested in incubations identical to that described below but of 0.55-ml. final volume. After deproteinizing by the addition of 0.10 ml. of 30% trichloroacetic acid, the aminoimidazole ribotide in the supernatant solution is measured by means of the Bratton and Marshall procedure (Section A). The molar extinction coefficient of the colored derivative of this ribotide is 24,600 at 500 mμ.[25]

An extract of pigeon liver acetone powder and the 0 to 60% $(NH_4)_2SO_4$ fraction prepared from the extract, both of which are described under Analysis in Section D, both containing the two required enzymes, and both being more convenient to prepare than the 13 to 33% ethanol fraction. Neither fraction has yet been tested, however, by actual use in large-scale preparations.

Procedure. The incubation mixture contains, in a final volume of 440 ml., the following quantities of materials, expressed in micromoles: sodium salt of formylglycinamide ribotide, 64; glutamine, 640; disodium ATP, 720; magnesium acetate, 2400; potassium acetate, 48,000; Tris–bromide buffer, pH 8.0, 8000; and approximately 400 mg. of the 13 to 33% ethanol fraction from pigeon liver. After incubation for 65 minutes at 38°, the reaction mixture is deproteinized by placing it in a boiling-water bath. Since it is desirable to keep the exposure to heat as short as is consistent with precipitation of the bulk of the protein (3 to 5 min-

[28] J. C. Rabinowitz, *J. Biol. Chem.* **218,** 175 (1956).

utes), one-third of the reaction mixture is heated at a time, with stirring, and is then rapidly cooled in a −10° bath. After removal of the precipitated protein by centrifugation at 3°, the amount of aminoimidazole ribotide in the supernatant solution may be determined by performing the Bratton and Marshall assay on a 0.20-ml. aliquot to which is added 0.20 ml. of water.

The supernatant solution is chromatographed at 3°, with a flow rate of 3 to 4 ml./min., on a column of Dowex 1–acetate, 2.2 × 12 cm. The column is eluted first with a resin-bed volume of water and then with 0.04 M ammonium acetate buffer, pH 5.1. Two fractions of approximately 50 ml. are collected by hand, and 0.20-ml. aliquots are analyzed for product as described above. It is convenient to follow the pH of the eluate with pH paper, and to collect a new fraction of approximately 100 ml. as soon as the pH, which is at first more basic, reaches that of the elutriant, since this change coincides with the elution of the ribotide. Subsequent fractions of 10 to 20 ml. are collected until they are found to contain negligible product. The combined ribotide-containing fractions (approximately 150 ml.) are evaporated to 2 or 3 ml. during which time the temperature of the solution is not allowed to exceed 20°. Four volumes of ethanol and a threefold excess of saturated barium bromide solution are added to precipitate the ribotide as the barium salt. After the solution has stood for 1 to 2 hours at −17°, the precipitate is collected by centrifugation in the cold. An aliquot of the supernatant solution is analyzed by the Bratton and Marshall procedure or by radioactivity, if C^{14}-formylglycinamide ribotide was the starting material, to assure adequate precipitation of the product. The barium salt is immediately placed in a vacuum over P_2O_5, under which conditions it is stable for at least a month (and probably longer) if stored at −17°. The yield of ribotide is seldom more than 60%, based on formylglycinamide ribotide, and its purity is 35% or less, based on a calculated molecular weight for the monobarium salt of 432. Although in general the above operations should be performed with the minimum of delay, the eluate from the Dowex 1–acetate column may be stored overnight at 3°, since the ribotide is relatively stable at pH 5.

Material obtained by the foregoing procedure has reacted satisfactorily in all observed enzymatic reactions. A further purification may be achieved, however, by dissolving 80 mg. of the barium salt in 28 ml. of 0.02 N HCl and passing the solution through a column of Dowex 50–sodium, 2.2 × 3.8 cm., at 3°.[25] The column is washed with water after the ribotide-containing solution has passed through. The initial effluent and the first 10 ml. of water wash, which contain the ribotide, are combined and adjusted to pH 8 with a few drops of 4 N KOH. The above procedure removes any contaminating glutamine or glutamic acid. The aminoimidazole ribotide is then reisolated as before from a column of Dowex 1–acetate, 0.8×14 cm. Both of the above chromatographic procedures should be performed at 3° without interruption because of the instability of this ribotide. The best sample of ribotide prepared by this procedure, after several washings of the barium salt with 95% ethanol and a final wash with ether, was 60% pure on a dry-weight basis and gave, on hydrolysis, the expected molar ratios of glycine, formate, pentose, organic phosphate, acid-labile nitrogen, and total nitrogen of 1:1:1:1:2:3.[25]

Properties.[25] 5-Aminoimidazole ribotide, in contrast to all the foregoing intermediates, gives a positive reaction with the Pauly diazo reagent for imidazoles[21] and with the Bratton and Marshall reagents (Section A) for aromatic amines. The product in both cases has a bright orange color which in the former procedure rapidly fades but in the latter procedure is stable. The molar extinction coefficient of the derivative produced in the Bratton and Marshall procedure at pH 1.5 is 24,600 at 500 mμ (λ_{max}). The absorption spectrum, which is similar to that of the corresponding derivative obtained from 5-aminoimidazole itself, is strongly affected by the pH and in stronger acid has its maximum shifted to longer wavelengths.[29] 5-Aminoimidazole ribotide is distinguished from all later purine precursors by the absence of a peak in its ultraviolet absorption spectrum, which exhibits only increasing absorption down to 210 mμ. The stability of the ribotide under various conditions has not been studied quantitatively but appears similar to that reported for 5-aminoimidazole.[28]

Analysis. The most convenient method for the determination of 5-aminoimidazole is by the Bratton and Marshall procedure (Section A). If the sample is contained in 0.4 ml. or less, as little as 0.004 micromole can be conveniently determined. The analyses for the different components of the ribotide (e.g., pentose) are performed in the same way as described for formylglycinamidine ribotide.

G. Preparation of 5-Amino-4-Imidazolecarboxylic Acid Ribotide

Principle. A carboxylation reaction occurs at carbon 4 of 5-aminoimidazole ribotide to yield 5-amino-4-imidazolecarboxylic acid ribotide in the presence of bicarbonate and an enzyme from avian liver [reaction (2), Section F]. Although the equilibrium favors decarboxylation at low levels of bicarbonate, in the presence of 0.3 M bicarbonate up to 50%

---

[29] This fact explains why the absorption maximum of the colored product from 4(5)-aminoimidazole occurs at 514 mμ in 4% $HClO_4$,[28] and at 500 mμ in 0.04 N HCl or H_2SO_4 or in 0.8 M sodium phosphate buffer, pH 1.5.

of the aminoimidazole ribotide may be converted to the carboxylic acid ribotide. The ribotide is isolated by chromatography at 3° on a column of Dowex 1–acetate at pH 9 and is precipitated as the barium salt. The alkaline conditions of chromatography are chosen to minimize the acid-catalyzed decarboxylation reaction but have the disadvantage that a considerable salt concentration is required to elute the ribotide at this pH. Owing to the instability of the carboxylic acid ribotide, the loss during isolation is usually 50% or more.

Reagents

Potassium salt of 5-aminoimidazole ribotide, prepared from the barium salt (obtained as described in Section F) by batchwise treatment with Dowex 50-K+.

Lithium bromide, commercially available.

Diethanolamine, commercially available.

As a source of the carboxylation enzyme, the 30 to 55% $(NH_4)_2SO_4$ fraction of chicken liver extract or any of the subsequent fractions described in Vol. VI [9], Section G, may be used. For a preparation of the carboxylic acid ribotide of the size described below, a single liver provides a surplus of enzyme. Extracts of chicken liver acetone powder also contain the carboxylation enzyme and are presumably a satisfactory source of the enzyme, although such extracts have not been tested by use in an actual preparation.

Procedure.[30] Fifty micromoles of the potassium salt of 5-amino-imidazole ribotide and 9500 micromoles of freshly dissolved $KHCO_3$ are combined in a final volume of 250 ml. with sufficient enzyme to cause equilibrium to be reached in 20 to 30 minutes at room temperature. The synthesis of 5-amino-4-imidazolecarboxylic acid ribotide is followed by observing, with a Beckman DU spectrophotometer, the increase in optical density at 265 mμ, at which wavelength the carboxylic acid ribotide, in contrast to the starting materials, has considerable absorption. The optimum amount of a particular enzyme preparation is conveniently determined in a preliminary incubation in which different amounts of enzyme are incubated in silica cuvettes, under the same conditions as above, in a final volume of 2 to 3 ml. For 50 micromoles of aminoimidazole ribotide, approximately 20 mg. of the 30 to 55% $(NH_4)_2SO_4$ fraction or 10 mg. of the fraction obtained from the dialysis step are required. It is necessary to use freshly prepared solutions of

[30] L. N. Lukens and J. M. Buchanan, *J. Biol. Chem.* **234**, 1799 (1959).

$KHCO_3$ (pH approximately 8.3), since solutions stored at 3° for several days result in lower yields of the carboxylic acid ribotide. When no further increase in optical density at 265 mμ is observed, the incubation mixture is cooled in a −10° bath to just above its freezing point, and approximately 7 ml. of 70% perchloric acid are added to bring the solution to pH 2 or below. After nitrogen gas has bubbled through the solution for 1 minute to aid in the removal of CO_2, the precipitated protein is removed by centrifugation for 2 minutes at 3°. The supernatant solution is immediately adjusted to approximately pH 9 by the addition of 10 N KOH (1.3 ml.), after which 5 ml. of 0.1 M diethanolamine hydrobromide buffer, pH 9, are added. Since the carboxylic acid ribotide is especially unstable in acid solution, it is advisable to work as rapidly as possible and to keep the solution as near 0° as feasible between the addition of the acid and the neutralization.

The above solution, at pH 9, is placed on a column (2.2 × 20 cm.) of Dowex 1-acetate at 3°, at which temperature the subsequent elution is performed. The unreacted aminoimidazole ribotide is first eluted with 0.04 M ammonium acetate buffer, pH 5.1, as described in Section F under the preparation of this ribotide. The column is then adjusted to pH 9 by the rapid addition of 1 to 2 resin-bed volumes of 0.01 M diethanolamine hydrobromide buffer, pH 9. Because of the instability of the carboxylic acid ribotide in acid solution, this point in the procedure should be reached within a maximum of 4 to 5 hours after the first addition of the pH 5.1 buffer. A gradient elution is then performed, in which the reservoir contains a solution of 0.2 M LiBr and 0.01 M diethanolamine hydrobromide buffer, pH 9, and the mixer contains 400 ml. of a solution of the same buffer and 0.08 M LiBr. The 5-amino-4-imidazolecarboxylic acid ribotide is eluted after approximately 450 ml. in a total volume of 1 to 2 resin-bed volumes. The ribotide is most conveniently located in the effluent fractions (generally of approximately 15 ml.) by performing the Bratton and Marshall procedure (Section A) on 0.20-ml. aliquots. The combined fractions which contain product are evaporated in the usual way to approximately 5 ml., with care that the temperature of the solution does not exceed 10°. The concentrated solution, which contains some undissolved salts precipitated during evaporation, is transferred to a centrifuge tube with the aid of an additional milliliter of H_2O as wash. The ribotide at this point, although usually uncontaminated with aminoimidazole ribotide, is heavily contaminated with salts derived from the elutriant. An effective and reproducible procedure for obtaining the ribotide free from these salts has not been developed, so the following manipulations, included as a guide, can probably be improved upon.

Five volumes of ethanol at 3° are added to the centrifuge tube, and any undissolved salts are removed by centrifugation. The supernatant solution is saved. The precipitate, which contains some product, is washed once with 80% ethanol and then extracted with 1-ml. portions of H_2O until no further ribotide appears in the extracts. The presence of the carboxylic acid ribotide is detected most conveniently by its radio-activity if radioactive aminoimidazole ribotide is used for its synthesis, or alternatively by means of the Bratton and Marshall procedure. The ethanol wash is discarded after it has been confirmed that it does not contain product, and the combined extracts (2 to 4 ml.) which contain ribotide are added to the original supernatant solution. The ribotide is then precipitated from this ethanolic solution as the barium salt by the addition of 100 mg. of solid $BaBr_2 \cdot 2H_2O$. The solution is allowed to stand at $-17°$ for 1 hour and is then centrifuged at $-10°$. The super-natant solution is discarded after it has been determined that it contains negligible product. The precipitate, which always contains less than 50% of the 5-amino-4-imidazolecarboxylic acid ribotide present before chromatography, is dried in a vacuum at room temperature and is stored in the dry state at 3° or below. This salt, although still contaminated with salts from the elutriant and with material which absorbs ultraviolet light below 250 mμ, is satisfactory for use in enzymatic reactions. When required, some further purification of the ribotide may be obtained by reprecipitations of the barium salt from ethanol. If the ribotide present in a barium precipitate fails to be extracted by H_2O, the precipitate is mixed for a minute with an equal volume of Dowex 50–potassium, and the extraction is repeated. A sample subjected to such reprecipita-tions was still apparently contaminated slightly with organic phos-phate.[30] It should be mentioned that there is always the possibility of fractions of 5-amino-4-imidazolecarboxylic acid ribotide becoming con-taminated with 5-aminoimidazole ribotide, owing to the ease of decar-boxylation of the former. The presence of 5-aminoimidazole ribotide as a contaminant is qualitatively revealed by means of the Bratton and Marshall procedure (see below).

Properties.[30] 5-Amino-4-imidazolecarboxylic acid ribotide is readily distinguished from 5-aminoimidazole ribotide both by the color produced in the Bratton and Marshall procedure and by the presence of a peak in the ultraviolet absorption spectrum of the carboxylic acid ribotide. The latter ribotide yields a red product in the Bratton and Marshall proce-dure with λ_{max} at 520 mμ, compared to the orange material produced from 5-aminoimidazole ribotide with λ_{max} at 500 mμ. Contamination of the carboxylic acid ribotide by aminoimidazole ribotide is revealed by a shift of the absorption maximum of the solution resulting from the

Bratton and Marshall procedure toward 500 mμ. More strongly acid conditions than those described for the Bratton and Marshall procedure in Section A result in a marked shift of the absorption of the colored derivatives of the aminoimidazole compounds to longer wavelengths (as described in Section F), with little effect on the derivatives of the aminoimidazolecarboxylic acid compounds. Since the colored product formed from aminoimidazole ribotide might be confused with that produced from the carboxylic acid ribotide under such strongly acidic conditions, it is desirable, when performing the Bratton and Marshall procedure on the carboxylic acid ribotide, to substitute 4 M phosphate buffer, pH 1.5, for the 0.2 N HCl usually employed (Section A). Owing to the difficulty of isolation of this ribotide, occasioned by its instability, a sample has not as yet been sufficiently analyzed for phosphate, glycine, and the other entities released on hydrolysis to allow a precise determination of the molar extinction coefficients of the ribotide itself (in the ultraviolet region) or of the colored derivative produced from it in the Bratton and Marshall procedure. The latter value may be estimated, however, from preliminary analyses to be of the order of 1.5 $\times$ 10^4 at 520 mμ. The colored derivative from the corresponding synthetic aglycone,[28] 5-amino-4-imidazolecarboxylic acid, exhibits a similar absorption spectrum under the same conditions.

The peak in the ultraviolet absorption spectrum of 5-amino-4-imidazolecarboxylic acid ribotide occurs at 249 mμ from pH 8 through 12, with an *estimated* (see above) molar extinction coefficient of 1 $\times$ 10^4. In 0.25 N H$_2$SO$_4$ the spectrum is two-peaked (λ_{max} at 243 mμ and 264 mμ) and resembles in this respect the spectrum reported for 5-amino-4-imidazolecarboxylic acid itself at pH 5.[28] In acid solution the carboxylic acid ribotide rapidly decarboxylates, with the result that, after 2 to 3 hours at room temperature in 0.25 N H$_2$SO$_4$, solutions of the ribotide exhibit only the end absorption characteristic of 5-aminoimidazole ribotide. The rate of the decarboxylation may be measured by following the decrease of absorption at 265 mμ and is identical for the ribotide and the aglycone, 5-amino-4-imidazolecarboxylic acid. The ribotide is also decarboxylated when it is incubated with the enzyme which catalyzes reaction (2) in the absence of added bicarbonate.

Analysis. Although the methods of analysis which have been employed for the fragments released from 5-aminoimidazole ribotide by hydrolysis are presumed to apply to 5-amino-4-imidazolecarboxylic acid ribotide, sufficient pure material for such a complete analysis has not as yet been obtained. In addition to such analytical methods, the carboxyl group may be determined, if the ribotide is synthesized from bicarbonate of known specific activity, by trapping and counting the C^{14}O$_2$ released on enzymatic or acid-catalyzed decarboxylation. The ribotide may also be determined by either the Bratton and Marshall procedure or by its ultraviolet absorption, although it is not possible by either method to get a reliable figure for the absolute concentration of the ribotide, since accurate values for the requisite extinction coefficients are not available. In the presence of 5-aminoimidazole ribotide the latter method is preferable and has been used for the assay of the enzyme which catalyzes reaction 2.

H. Preparation of 5-Amino-4-Imidazole-N-succinocarboxamide Ribotide

Principle. The succinocarboxamide ribotide may be synthesized from 5-aminoimidazole ribotide according to reactions (2) and (3), Section F.[31] It may also be synthesized from 5-amino-4-imidazolecarboxamide ribotide according to reaction 4 in reverse.[32] The latter method has the

5-Amino-4-imidazole-N-succinocarboxamide ribotide $\rightleftarrows$ Fumaric acid +
5-amino-4-imidazolecarboxamide ribotide   (4)

advantages that the carboxamide ribotide is more readily prepared than 5-aminoimidazole ribotide and that ADP, which is separated with difficulty from the succinocarboxamide ribotide, is not present in the reaction mixture. The enzyme which catalyzes reaction (4) has been shown[32] to be apparently identical with adenylosuccinase, which catalyzes the reversible cleavage of adenylosuccinic acid to AMP and fumaric acid.[33] Although this enzyme may be obtained from avian liver, a yeast enzyme is employed in the procedure described below.

Reagents

Disodium fumarate. Fumaric acid is neutralized to pH 7, and the sodium salt is crystallized from ethanol–water.

Sodium salt of 5-amino-4-imidazolecarboxamide ribotide, prepared from the barium salt (Section I) by treatment with Na_2SO_4.

Enzyme. Adenylosuccinase, prepared from yeast as described in Vol. VI [9], Section I. The enzyme preparation need not be purified 180-fold, but the purification must be carried at least through the heat step, in order to remove fumarase. The assay for the enzyme and the definition of a unit of activity are also given in Section I of Vol. VI [9].

Hydrobromic acid for ion-exchange chromatography. It has been

[31] L. N. Lukens and J. M. Buchanan, *J. Biol. Chem.* **234**, 1791 (1959).
[32] R. W. Miller, L. N. Lukens, and J. M. Buchanan, *J. Biol. Chem.* **234**, 1806 (1959).
[33] C. E. Carter and L. H. Cohen, *J. Biol. Chem.* **222**, 17 (1956).

found advisable[27] to purify reagent-grade concentrated HBr (48%, $d = 1.48$) by passing it through a column of Dowex 1–bromide, (2.2 $\times$ 4 cm. for 50 ml. of acid), which has been first thoroughly washed with the same acid. This treatment removes all yellow material and may prevent some loss of the succinocarboxamide ribotide on chromatography.

Procedure.[27] Sodium fumarate (250 mg.) and the sodium salt of 5-amino-4-imidazolecarboxamide ribotide (143 micromoles) are dissolved in 10 to 11 ml. of water, and the pH of the solution is adjusted, if necessary, to 7.5 with KOH. To this solution 34 units of adenylosuccinase are added (approximately 1 ml. of the supernatant solution from the heat step or 4 ml. of the final, 180-fold purified fraction from the second DEAE column). The volume is made up to 15 ml., and the mixture is incubated for 5 hours at 38°. The protein is then coagulated by heating the incubation mixture, with stirring, in a boiling-water bath for 4 minutes. After the mixture has been cooled, the coagulated protein is removed by centrifugation. The supernatant solution, at room temperature, is adjusted to pH 10 with KOH and is diluted to 750 ml. with water.

The diluted solution is percolated through a column of Dowex 1–bromide, 2.2 $\times$ 20 cm., at room temperature and is followed by 700 ml. of water. The column is then developed with 0.007 N HBr which begins to elute the carboxamide ribotide and fumaric acid after approximately 250 ml. All the ribotide and almost all the fumarate are eluted in the next 500 ml. The elution of the succinocarboxamide ribotide begins when approximately 800 ml. (total) of 0.007 N HBr have passed through the column and is completed during the next 2000 ml. The fractions containing the product (detected most conveniently by its ultraviolet absorption or by its behavior in a modified Bratton and Marshall procedure; see below) are combined and concentrated to approximately 5 ml. in the usual way. The pH of the concentrated solution is adjusted to 4 or 5 with NaOH, and 5 vol. of ethanol at $-17°$ are added. The precipitated ribotide is collected by centrifugation in the cold and is then washed once by stirring with 3 ml. of cold ethanol. The solid is again collected by centrifugation. This method of precipitation succeeds in largely removing a contamination with fumarate, caused by some trailing of this compound in the chromatographic procedure. The sodium salt of the succinocarboxamide ribotide is dried in a vacuum over P_2O_5. It has been stored as the dry salt for some months at 3° or below without loss. If appreciable product fails to precipitate as the sodium salt, it should be recovered from the supernatant ethanolic solution by the addition of a calculated twofold excess of a saturated barium bromide solution. The barium salt may subsequently be converted to the sodium salt by treatment with Na_2SO_4. The yield of the succinocarboxamide ribotide is approximately 124 micromoles before isolation and 100 micromoles after, the latter figure representing a yield of 70% based on the 5-amino-4-imidazolecarboxamide ribotide.

Samples of ribotide isolated by the above procedure are suitable for most purposes, although they may be contaminated with small amounts of fumaric acid and even with ADP, if the carboxamide ribotide employed contains an appreciable amount of ADP as contaminant. Further purification of the ribotide may be accomplished by chromatography on a second, smaller Dowex 1–bromide column, which is eluted with 0.006 N HBr. Material purified in this way has been found to be 95% pure, based on a calculated molecular weight of 496 for the sodium salt $C_{13}H_{15}O_{12}N_4$-PNa_2, possible water of hydration being neglected.[27] Such samples may still be contaminated with traces of ADP, as judged by the very sensitive test based on the requirement for ADP in the arsenolysis of the succinocarboxamide ribotide [reaction (3) in reverse]. A reproducible procedure for removing this contamination consists in streaking 10 micromoles of the succinocarboxamide ribotide on a strip of Whatman 3 MM paper, 7 inches in width.[27] The paper is developed by the descending technique in a solvent system composed of isobutyric acid–concentrated ammonia–water in a volume ratio of 100:2:58. The area of the paper which contains the succinocarboxamide ribotide is readily identified by its blue fluorescence under ultraviolet light. This strip is cut out, and the ribotide is eluted with water.

Properties.[31] The ultraviolet absorption spectrum of 5-amino-4-imidazole-N-succinocarboxamide ribotide does not vary appreciably between pH 0 and 1, between pH 5 and 6, and between pH 8 and 14. The spectra at pH 5 and 8 differ only slightly, and both possess a single maximum at 269 ± 1 mμ, with a molar extinction coefficient of 13.3×10^3. The spectrum at pH 1 is markedly different and exhibits, in addition to a lower maximum at 269 mμ ($\epsilon = 11.0 \times 10^3$), a shoulder extending from 244 to 255 mμ, with a molar extinction coefficient of 8.9×10^3.

The ribotide unexpectedly gives no color in the usual Bratton and Marshall procedure, owing to the rapid disappearance of its diazonium salt.[34] A reddish purple product is obtained, however, if the procedure is carried out in an ice bath or if the coupling agent is added within 1 minute of the nitrite (see Analysis).

Four titratable hydrogen ions are released by the succinocarboxamide ribotide in the pH range 2 to 8. These protons are derived in all proba-

[34] J. S. Gots and E. G. Gollub, *Proc. Natl. Acad. Sci. U.S.* **43**, 826 (1957).

bility from the two carboxyl groups, the protonated amino group, and a secondary phosphate dissociation. Owing to overlapping of the pK's, only one pK has been determined, which is approximately equal to 6.6, and is tentatively assigned to the secondary phosphate dissociation. The ribotide is strongly bound by Dowex 1 and is eluted from columns of Dowex 1–bromide or chloride under approximately the same conditions as is ADP.

Analysis. In addition to the determination of the ribotide by its ultraviolet absorption, it may be conveniently determined by the Bratton and Marshall procedure described in Section A, with the following modifications: (1) The sample and reagents are kept at 0°; (2) 10 N H_2SO_4 is substituted for the usual 0.2 N acid; (3) the sulfamate is added 3 minutes after the nitrite; and (4) the dye is added 1 minute after the sulfamate. After this last addition the samples are removed from the ice bath. The optical density at 560 mμ should be measured between 10 and 30 minutes after the addition of the dye, since a slow decrease in optical density occurs on standing. The molar extinction coefficient, based on the concentration of the succinocarboxamide ribotide, is 17.7 $\times$ 10^3 at 560 mμ, the absorption maximum.[31]

A convenient and specific enzymatic method for determining the succinocarboxamide ribotide consists in incubating it in 0.05 M phosphate buffer, pH 7.4, with an excess of adenylosuccinase (see Reagents) in a convenient volume (e.g., 0.55 ml.). The succinocarboxamide ribotide is quantitatively cleaved to 5-amino-4-imidazolecarboxamide ribotide, which is determined by the usual Bratton and Marshall procedure (see Analysis, Section I). For an undetermined reason, the values obtained for the concentration of the succinocarboxamide ribotide by calculations based on the accepted value of 26,400 at 540 mμ for the molar extinction coefficient of the colored derivative of the carboxamide ribotide are low by approximately 10%, as judged by other criteria. If it is desired to obtain accurate values for the absolute concentration of the succinocarboxamide ribotide with this assay, a calibration curve must first be established by measuring the absorbancies at 540 mμ that are given by known amounts of the succinocarboxamide ribotide.

The previously employed methods (Section A) may be used for the analysis of organic phosphate and nitrogen. The pentose determination, however, is performed according to the Dische modification of the orcinol reaction,[35] except that the heating time is extended from 20 to 70 minutes. Such conditions were found to be necessary to get the maximum yield of color from the ribotide, since its N-ribosyl bond is hydrolyzed much less readily than the corresponding bond of AMP, which is employed as the

[35] See Vol. III [12], p. 88.

pentose standard. There is, however, no assurance that even under these conditions the yield of colored product from AMP and from the succinocarboxamide ribotide is identical. Aspartic acid is quantitatively released from the ribotide by digestion of the latter in sealed tubes with 3 N HCl at 121° for 8½ hours in an autoclave. The aspartic acid can be assayed with high precision by its enzymatic conversion to oxalacetate on incubation with α-ketoglutarate in the presence of glutamic-aspartic transaminase. The oxalacetate produced is measured by following spectrophotometrically the oxidation of DPNH which occurs during the quantitative reduction of oxalacetate to malate in the presence of excess DPNH and malic dehydrogenase.[31] It is of interest that the method used to release glycine quantitatively from the previously described ribotides (see Section A) gives consistently low yields of glycine from the succinocarboxamide ribotide. Various more vigorous conditions of hydrolysis failed to raise the yield above 82% of the theoretical value.[31]

I. Preparation of 5-Amino-4-Imidazolecarboxamide Ribotide

Principle. In addition to the preparation described below, a different preparation of the carboxamide ribotide is described in Vol. II [78], in which the riboside is isolated and then enzymatically phosphorylated with ATP. A chemical procedure for obtaining this riboside in gram quantities from inosine has been described.[36]

In the present preparation, the ribotide is made from the corresponding free base by enzymatic condensation with 5-phosphoribosylpyrophosphate according to reaction (5). This reaction is catalyzed by the same

5-Phosphoribosylpyrophosphate + 5-amino-4-imidazole-carboxamide $\rightleftarrows$

Pyrophosphate + 5-amino-4-imidazolecarboxamide ribotide   (5)

enzyme, AMP-pyrophosphorylase, which catalyzes the reversible condensation of adenine with phosphoribosylpyrophosphate to yield 5'-AMP.[37] In the procedure described below, the enzymatic synthesis of phosphoribosylpyrophosphate from ATP and ribose-5-phosphate and its condensation with 5-amino-4-imidazolecarboxamide are accomplished in a single incubation. Since chemically synthesized 5-amino-4-imidazolecarboxamide is available commercially (see Reagents), the procedure based on reaction (5) provides an extremely convenient synthetic method for this ribotide. Preliminary attempts to find similar enzymes which would catalyze the synthesis of the ribotides of 5-aminoimidazole or 5-amino-4-imidazole-N-succinocarboxamide from these bases and phos-

[36] E. Shaw, *J. Am. Chem. Soc.* **81**, 6021 (1959).
[37] J. G. Flaks, M. J. Erwin, and J. M. Buchanan, *J. Biol. Chem.* **228**, 201 (1957).

phoribosylpyrophosphate have been unsuccessful.[38] The procedure which follows is that of Flaks et al.,[37] including unpublished modifications of R. W. Miller. In particular, the substitution of Dowex 1–bromide for Dowex 1–chloride in the chromatography procedure is advantageous, because the acid concentration in the elutriant can be kept lower. The carboxamide ribotide tends to decompose to a red pigment in the presence of a hydrogen ion concentration greater than 0.04 N.

Reagents

Ribose-5-phosphate, see Reagents, Section B.

5-Amino-4-imidazolecarboxamide. Purchased from the California Corporation for Biochemical Research.

Pigeon liver acetone powder extract, prepared as described under Reagents, Section D. This extract contains the enzyme that catalyzes the synthesis of 5-phosphoribosylpyrophosphate from ATP and ribose-5-phosphate.

Beef liver AMP-pyrophosphorylase. Since use of a crude acetone powder extract does not give good yields of ribotide, the following partial purification[37] is recommended. Beef liver acetone powder (100 g.) is stirred with 1 l. of 0.05 M Tris buffer, pH 7.4, at room temperature for one-half hour. Subsequent operations are performed at 3°. Insoluble material is removed by centrifugation (e.g., $5300 \times g$ for an hour). The supernatant solution is mixed with 1 M Tris buffer, pH 8 (1 ml. of buffer per 10 ml. of solution), and is fractionated with solid $(NH_4)_2SO_4$. The fraction precipitating between 38 and 55% of saturation is collected by centrifugation, dissolved in water (approximately 150 ml.), and dialyzed against two 5-l. batches of 0.005 M Tris–chloride buffer, pH 7.4, for, respectively, 4 hours and 8 or more hours. The weight of protein present in the dialyzed solution is calculated from its optical density at 280 mμ, as described in Section A. A suspension of calcium phosphate gel, equivalent in dry weight to one and one-half times the calculated weight of the protein, is centrifuged, and the water is discarded. The protein solution is diluted with water to approximately 1100 ml. and, after addition of 11 ml. of 1.0 M potassium acetate buffer, pH 5.5, is stirred for 10 minutes with the gel. After removal of the gel with its adsorbed protein by centrifugation, the supernatant solution is adjusted to pH 5.5 with 5 N acetic acid and is then fractionated with solid $(NH_4)_2$-SO_4. The fraction which precipitates between 0 and 55% of satu-

[38] Unpublished results from laboratory of J. M. Buchanan, 1955–1958.

ration is collected, dissolved in approximately 25 ml. of water, and dialyzed as before, except that the volume of the Tris buffer is reduced to 1 l.

Although the above procedure has been found to be quite reproducible, the activity of any fraction may be checked by means of the following assay.[37] The enzyme is incubated for 20 minutes at 38° in a final volume of 1 ml. with the following materials, expressed in micromoles: 5-amino-4-imidazolecarboxamide, 5; 5-phosphoribosylpyrophosphate,[26] 1; Tris–chloride buffer, pH 8, 100; and $MgCl_2$, 10. The reaction is stopped by placing the incubation vessel in a boiling-water bath for 2 minutes, the precipitated protein is removed by centrifugation, and the supernatant solution is placed on a column of Dowex 1–chloride, 1×3 cm. The imidazolecarboxamide is rapidly washed from the column with 100 ml. of water under pressure, and the imidazolecarboxamide ribotide is then quantitatively eluted with 5 ml. of $0.4 N$ HCl. The ribotide in the eluate is measured by performing the Bratton and Marshall procedure on an aliquot (see under Analysis, below). With limiting amounts of enzyme, the synthesis of ribotide is proportional to the concentration of enzyme. The original acetone powder extract (about 870 ml.) should contain a total amount of enzyme capable of synthesizing approximately 220 micromoles of ribotide under the assay conditions. The fraction from the second $(NH_4)_2SO_4$ step should contain about 28% of the above activity. An amount of enzyme capable of synthesizing 30 micromoles of ribotide under the assay conditions should be used in the preparation described below.

Procedure. The incubation mixture contains the following materials, expressed in millimoles, in a final volume of 500 ml.: ribose-5-phosphate, 2.4; ATP, 1.6; $MgCl_2$, 10; 5-amino-4-imidazolecarboxamide, 1.54; potassium phosphate buffer, pH 7.4, 25; one-half of the second $(NH_4)_2SO_4$ fraction prepared from 100 g. of beef liver acetone powder, as described above; and the extract obtained from 10 g. of pigeon liver acetone powder. After incubation for 4 hours at 38° with gentle agitation, the reaction mixture is heated for 4 minutes in a boiling-water bath and is then cooled to 3°, at which temperature all subsequent operations are performed. The solution is adjusted to pH 9 with KOH. Precipitated protein and inorganic salts are removed by filtration through a Celite pad (or through filter paper) on a Büchner funnel. The pad is washed once with 100 ml. of water, and the combined wash and original filtrate are diluted to 1 l. with water. The diluted filtrate is passed overnight through a column of Dowex 1-bromide, 2.2 × 40 cm. The column is then washed with 500 ml. of water which are followed by 700 ml. of 0.007 N HBr. The carboxamide ribotide is eluted with 0.014 N HBr after about 300 ml., in a volume of approximately 1 l. The combined ribotide-containing fractions are adjusted to pH 4 with NH_4OH and are then evaporated to 10 ml. Some impurities may be precipitated from this solution by the addition of 0.3 ml. of 1 M barium acetate and 25 ml. of 95% ethanol. After standing for an hour at −15°, the solution is centrifuged at the same temperature, and the precipitate is discarded. It is advisable to confirm by the Bratton and Marshall procedure the absence of ribotide in any fraction which is to be discarded. The barium salt of the carboxamide ribotide is obtained from the supernatant solution by a procedure identical with that above, after the addition of 15 ml. of 95% ethanol, 20 ml. of ether, and 0.5 ml. of 1 M barium acetate. The salt is washed twice with cold acetone and once with ether. After drying briefly in air, it is suspended in 3 ml. of water, and any insoluble material is removed by centrifugation. The barium salt is reprecipitated from the supernatant solution by the addition of 15 ml. of cold acetone. After standing for an hour at −15° it is collected by centrifugation and washed and dried as above. The yield of the carboxamide ribotide is 700 micromoles at the end of the incubation and 500 micromoles at the end of the above isolation procedure, or 30% based on the 5-amino-4-imidazolecarboxamide. The barium salt is approximately 75% pure, being contaminated mainly with ribose-5-phosphate, and is satisfactory for routine metabolic work. It may be stored indefinitely in the dry state at 3° or below. If it is desired to purify the ribotide further, a second chromatography on Dowex 1-bromide may be employed to obtain a product free from ribose-5-phosphate or ADP.[27] The precipitation of the product is performed as described above.

It should be possible to scale up the entire procedure, in which case larger columns should of course be employed (see, for example, Section D).

Properties.[37] The ultraviolet absorption spectrum of 5-amino-4-imidazolecarboxamide ribotide resembles closely the spectrum of the corresponding riboside[39] and the spectrum of 5-amino-4-imidazole-N-succinocarboxamide ribotide. At pH 7 there is a single peak at 269 mμ with a molar extinction coefficient of 12,600, and at pH 1 there is a lower peak at 269 mμ (ϵ = 9400) with a shoulder at 245 to 255 mμ (ϵ = 7500). The ribotide, like the free base and riboside, produces a purple derivative in the Bratton and Marshall procedure with maximum absorption at 540 mμ, at which wavelength the molar extinction coefficient is 26,400.

[39] G. R. Greenberg and E. L. Spilman, *J. Biol. Chem.* **219**, 411 (1956).

The 5-amino-4-imidazolecarboxamide structure, whether unsubstituted on N-1 or present as the riboside or ribotide, may be distinguished from many other aromatic amines by its failure to be easily acetylated with acetic anhydride. The ribotide, for instance, may be measured in the presence of p-aminobenzoylglutamate by adding to the samples, in 5% (wt./ml.) trichloroacetic acid, one-tenth volume of acetic anhydride, and allowing the samples to stand at room temperature for 20 minutes, before performing the Bratton and Marshall procedure.[40] The carboxamide ribotide gives its full yield of color under these conditions, whereas the p-aminobenzoylglutamate is completely acetylated and gives no color.

The pK of the secondary phosphate dissociation of the ribotide is 6.3. The pK of the amino group has not been determined. The glycosidic bond of the carboxamide ribotide is, as in the case of the N-succinocarboxamide ribotide, considerably more stable to acid hydrolysis than that of the purine ribotides, but less so than that of the pyrimidine ribotides.

Analysis. The most generally useful method for determining the carboxamide ribotide is by the Bratton and Marshall procedure, described in Section A. The succinocarboxamide ribotide does not interfere in this procedure, since it gives no color under the usual room-temperature conditions. The removal of interference from anilines by acetylation is described above. The ribotide may be measured in the presence of the ribotides of aminoimidazole and aminoimidazolecarboxylic acid by adding 0.1 ml. of 5 N H_2SO_4 to the sample contained in 0.4 ml. The tube is capped with a glass marble and is placed in a boiling-water bath for 15 minutes. After the tube has cooled to room temperature, the Bratton and Marshall procedure is performed without the addition of further acid. The aminoimidazole rings of the latter two ribotides are completely destroyed under these conditions, whereas the aminoimidazolecarboxamide structure remains intact and gives a quantitative yield of colored product.

The ribotide has been analyzed for total phosphate and for pentose by the usual procedures. For the latter analysis, the samples and the standard 5'-AMP were first hydrolyzed in 3 N HCl for one-half hour at 100°.[7]

J. Preparation of 5-Formamido-4-Imidazolecarboxamide Ribotide

Principle. 5-Formamido-4-imidazolecarboxamide ribotide has not yet been isolated from an enzymatic reaction mixture because all the enzyme preparations obtained to date which catalyze its synthesis from 5-amino-4-imidazolecarboxamide ribotide and N^{10}-formyltetrahydrofolic acid also

[40] J. G. Flaks, L. Warren, and J. M. Buchanan, *J. Biol. Chem.* **228**, 215 (1957).

catalyze its cyclization to inosinic acid.[41] It is not feasible to obtain sufficient amounts of the formamido-ribotide for isolation purposes from the reversal of the latter reaction because the equilibrium lies so far in favor of inosinic acid.[42] The formamido compound may, however, be conveniently prepared by the chemical formylation of 5-amino-4-imidazolecarboxamide ribotide under the same conditions as were originally employed by Shaw for the formylation of the corresponding aglycone, 5-amino-4-imidazolecarboxamide.[43]

Reagents

5-Amino-4-imidazolecarboxamide ribotide. Prepared as described in Section I.

Procedure.[41] Ten to twenty micromoles of the barium salt of 5-amino-4-imidazolecarboxamide ribotide are added to 37 mg. of sodium formate in 0.25 ml. of 98% formic acid. After the addition of 0.5 ml. of acetic anhydride, the mixture is warmed gently to initiate the reaction. The reaction mixture is allowed to stand at room temperature for 10 minutes, after which it is kept at 50° for 20 minutes. It is then cooled and lyophilized to dryness. The residue is stirred with 2 ml. of water, and the insoluble material is removed by centrifugation. The supernatant solution, which contains 5-formamido-4-imidazolecarboxamide ribotide in essentially quantitative yield, is adjusted to pH 6 with 3 N ammonium hydroxide.

For purposes of storage, the ribotide is precipitated from the above solution as the barium salt by adding 3 vol. of 95% ethanol and 300 micromoles of barium acetate. After standing for an hour at −15°, the salt is collected by centrifugation and may be washed successively with acetone and ether before drying in a vacuum at room temperature. It is stable indefinitely in the dry state at 3° or below.

This crude barium salt has been found to be satisfactory for use in enzymatic reactions.[41] Further purification may be achieved, however, by applying a concentrated aqueous solution of the barium salt to a sheet of Whatman No. 1 paper as a thin band. The chromatogram is then developed at 3° for 36 hours by the descending technique, with a solvent system of isobutyric acid–water–concentrated ammonia (66:33:1, by volume). The product, which is the main ultraviolet-absorbing band, has an R_f of approximately 0.18. It is eluted from the paper with 40 ml. of water and, after lyophilization of the solution to 5 to 10 ml., is placed on

[41] J. G. Flaks, M. J. Erwin, and J. M. Buchanan, *J. Biol. Chem.* **229**, 603 (1957).
[42] L. Warren, J. G. Flaks, and J. M. Buchanan, *J. Biol. Chem.* **229**, 627 (1957).
[43] E. Shaw, *J. Biol. Chem.* **185**, 439 (1950).

a column of Dowex 1–chloride, 0.9 × 2.5 cm. The column is eluted at 3° with 0.005 N HCl. The ribotide, which is detected by its ultraviolet absorption, begins to be eluted after approximately 70 ml. The combined product-containing fractions are immediately neutralized with 3 N NH$_4$OH and lyophilized to a volume of 3 to 6 ml. The barium salt is then precipitated, washed, and dried in the same manner as described above. A sample purified by this method was found to be 80% pure on a dry-weight basis and gave the expected ratios on analysis for pentose, phosphate, formate, and diazotizable amine released after acid hydrolysis. It was uncontaminated with 5-amino-4-imidazolecarboxamide ribotide.[41]

Properties.[41] The ultraviolet absorption spectrum at pH 7 of 5-formamido-4-imidazolecarboxamide ribotide resembles closely that reported for the aglycone[43] and has a single peak at 270 mμ with a molar extinction coefficient of 11,300. The ribotide also resembles the free base in its ability to cyclize in alkaline solution and in the ease of hydrolysis of its formyl group. The ribotide is quantitatively converted to inosinic acid by treatment with 0.1 N NaOH for 30 minutes at 38° and is quantitatively deformylated to 5-amino-4-imidazolecarboxamide ribotide by treatment with 0.2 N H$_2$SO$_4$ for 3 minutes at 100°. It is also converted enzymatically to inosinic acid in the presence of the enzyme inosinicase. As expected from its role as an intermediate in the enzymatic conversion of 5-amino-4-imidazolecarboxamide ribotide to inosinic acid, the formamido-ribotide can give rise to either of these two ribotides on incubation with enzymes of avian liver under the appropriate conditions.[41, 42]

Analysis.[41] In addition to the determination of the formamido-ribotide, in the absence of interfering substances, by its ultraviolet absorption, it may be more specifically determined by its quantitative release of 5-amino-4-imidazolecarboxamide ribotide, on acid hydrolysis. A convenient procedure consists in heating 0.003 to 0.03 micromole of the formamido-ribotide with 1.6 N H$_2$SO$_4$ in a final volume of 0.5 ml. for 5 minutes at 100°. After cooling to room temperature, the 5-amino-4-imidazolecarboxamide ribotide is measured by the Bratton and Marshall procedure, which is begun with the addition of the nitrite reagent, without the addition of further acid (see Section A).

The ribotide may be analyzed for organic phosphate, pentose, and formate by the usual methods, and for aromatic amine produced on acid hydrolysis, by the procedure described above. In the pentose determination, the prehydrolysis employed in the case of the aminoimidazolecarboxamide ribotide should be performed.

[97] Intermediates in Purine Breakdown

By JESSE C. RABINOWITZ

Principle. Xanthine is quantitatively degraded to formiminoglycine by extracts of lyophilized cells of *Clostridium cylindrosporum.* The compounds shown in Fig. 1 occur as intermediates in the degradation. Each

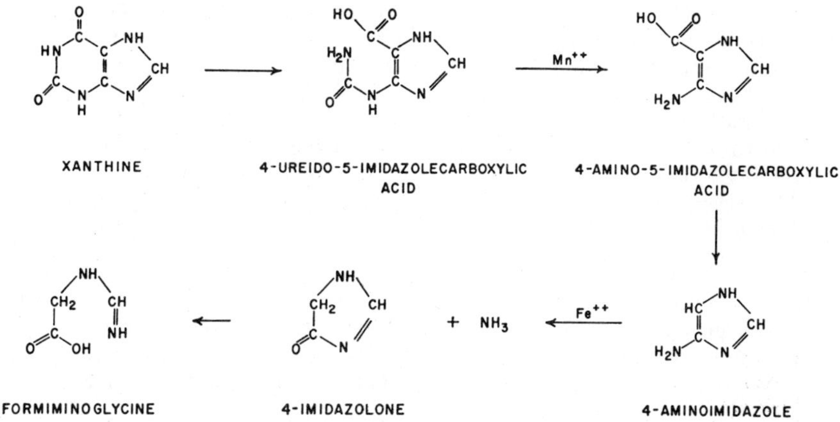

XANTHINE   4-UREIDO-5-IMIDAZOLECARBOXYLIC   4-AMINO-5-IMIDAZOLECARBOXYLIC
                            ACID   ACID

FORMIMINOGLYCINE   4-IMIDAZOLONE   4-AMINOIMIDAZOLE

FIG. 1.

compound may be formed from the appropriate substrate in reaction mixtures containing an unfractionated cell extract under particular incubation conditions which prevent the further degradation of the product. Each of the compounds described may also be prepared by chemical synthesis.

I. Enzyme Preparation

Growth of Clostridium cylindrosporum.[1] Stock cultures of the organism[2] are maintained on 10 ml. of medium A (Table I) supplemented with 2% agar in test tubes made anaerobic with pyrogallol seals.[3]

[1] H. A. Barker and J. V. Beck, *J. Bacteriol.* 43, 291 (1942).
[2] Available from the American Type Culture Collection, 2112 M Street, N.W., Washington 7, D.C.
[3] The cotton plug extending beyond the test tube is cut off, flamed, and pushed into the tube. The remaining portion of the tube is filled with absorbent cotton, and 4 drops of a 60% pyrogallol solution are dropped onto the cotton. Four drops of 10% sodium carbonate are then dropped onto the cotton, and the tube is quickly

Freshly inoculated media are incubated at 37° for 24 to 48 hours, or until growth is visible. Organisms may be maintained on the agar stabs kept at 0° for 6 months.

TABLE I

MEDIA FOR CULTURE OF *Clostridium cylindrosporum*[a]

Component	A	B
1. Uric acid	2 g.	30 g.
2. KOH (10 N)	1.2 ml.	18 ml.
3. Distilled water	750 ml.	3000 ml.
4. $K_2HPO_4\cdot3H_2O$, (700 mg./ml.)	1.3 ml.	20 ml.
5. $MgSo_4\cdot7H_2O$ (50 mg./ml.)	0.7 ml.	10 ml.
6. $FeSO_4\cdot7H_2O$ (2.5 mg./ml.)	0.7 ml.	10 ml.
7. $CaCl_2\cdot2H_2O$ (6.0 mg./ml.)	0.7 ml.	10 ml.
8. Difco yeast extract	1.0 g.	10 g.
9. Mercaptoacetic acid (80%)	2.0 ml.	—
10. Methylene blue (0.1%)	—	2.0 ml.
11. Distilled water to bring volume to:	1.0 l.	15.0 l.

[a] *Procedure:* Mix 1, 2, and 3 and boil. Add 4, and continue boiling until all the uric acid is dissolved. Discontinue heating, and add 5, 6, and 7. Add 8, after first dissolving it in a small amount of water. Add 9, and adjust the pH of the medium to 7.0 to 7.2 with KOH or H_2SO_4. Add 10. Sterilize by autoclaving at 15 pounds pressure. Sterilized media should be stored at 37° to minimize the precipitation of uric acid.

To grow 15-l. or larger liquid cultures, 10 ml. of medium A are inoculated with the organism from the agar stab. The liquid medium is sealed with pyrogallol[3] and incubated at 37° for 24 hours. This culture is used to inoculate a 1.5-l. culture contained in a 2-l. volumetric flask. The flask is sealed with pyrogallol[3] and incubated at 37° for 16 to 24 hours. The 1.5-l. culture is then used to inoculate 15 l. of medium B contained in a 20-l. carboy equipped with a rubber stopper and two glass tubes in an arrangement like that on a wash bottle, to permit mixing of the contents by a stream of helium or nitrogen after each addition described. Prior to inoculation, the 15 l. of medium are brought to 37°, and 50 g. of sodium carbonate dissolved in 250 ml. of water are added to each carboy. The pH of the medium is adjusted to 7.4 to 7.6 by adding about 13 ml. of concentrated sulfuric acid to each carboy. Small amounts of solid sodium hydrosulfite are added until the green color of the medium disappears. The carboy is then inoculated with 2 l. of culture, and small amounts of sodium hydrosulfite are added as required to keep the medium reduced. When growing larger amounts of organism, the

sealed with a rubber stopper. Fresh seals are made whenever tubes are opened for use.

contents of one carboy may be used to inoculate four to eight additional carboys.

The absorbance of uric acid in the growing culture is determined at 290 mμ. The initial absorbance of a 1:100 dilution is about 1.0. When this value has decreased to 0.025 or less, the cells are harvested with a Sharples centrifuge. The time required to achieve full utilization of the substrate varies but is about 6 hours under the growth conditions described.[4] The harvested cells are suspended in cold distilled water and lyophilized. The lyophilized cells are stored *in vacuo* at −10°. The yield of dry cells is 2 to 2.5 g. per 15 l. of medium.

Cell Extract.[5] Five hundred milligrams of lyophilized cells are incubated with 20 ml. of 0.1 M potassium phosphate buffer at pH 7.0, 0.01 M with respect to cysteine, in an evacuated vessel at 37° for 1 hour. The suspension is then centrifuged at 30,000 × g for 10 minutes at 0°, and the clear supernatant solution obtained is stored in an evacuated tube at 0° to 4°.

II. Intermediates

4-Ureido-5-imidazolecarboxylic acid[6]

Principle. The enzymatic degradation of 4-ureido-5-imidazolecarboxylic acid requires either Mn^{++} or Fe^{++}. This intermediate can, therefore, be formed quantitatively from xanthine by unpurified cell extracts incubated with xanthine in the presence of ethylenediaminetetraacetic acid. The product is isolated by successsive chromatography on Dowex 50-H$^+$ and Dowex 1–formate columns.

Procedure. A reaction mixture is prepared by mixing 2 g. of xanthine, 6 millimoles of ethylenediaminetetraacetic acid adjusted to pH 9.0, 12 millimoles of cysteine adjusted to pH 9.0, 120 millimoles of diethanolamine buffer at pH 9.0, an extract of *C. cylindrosporum* equivalent to 200 mg. of protein, and water to make the volume 900 ml. Aliquots (0.05 ml.) are removed at hourly intervals, and the amount of 4-ureido-5-imidazolecarboxylic acid formed is determined by the modified imidazole test.[7] The maximum amount of product is formed after about 4 hours. The reaction is stopped by the addition of perchloric acid to a final

---

[4] The culture sporulates and lyses on prolonged incubation in the absence of substrate.

[5] J. C. Rabinowitz and H. A. Barker, *J. Biol. Chem.* **218**, 161 (1956).

[6] J. C. Rabinowitz and W. E. Pricer, Jr., *J. Biol. Chem.* **218**, 189 (1956).

[7] Determinations of 4-ureido-5-imidazolecarboxylic acid based on absorbance in the ultraviolet region are not satisfactory because of interference caused by components of the cell extract.

concentration of 1%. The precipitated protein is removed by centrifugation. The acidified reaction mixture is stored at 2° overnight, and the precipitated material is removed by centrifugation and discarded. The supernatant solution is passed over a column of Dowex 50-H$^+$ (12 by 8.3 cm.^{2}). The column is developed by gradient elution by using a reservoir containing 0.1 M ammonium hydroxide and a mixing chamber containing 2 l. of water. After 30 column volumes have been eluted,[8] the eluting solutions are replaced with 0.3 M ammonium hydroxide. The intermediate is eluted with 30 to 43 column volumes.[9] These fractions are combined, and the volume is reduced to 750 ml. in a rotary flash evaporator. The eluate is then placed on a column of Dowex 1–formate (10 by 30 cm.^{2}). The column is eluted with 12 column volumes of 0.003 M formic acid, followed by 2 column volumes of 0.01 M formic acid, and finally 0.05 M formic acid. Fractions eluted with 18 to 42 column volumes[9] are concentrated under reduced pressure to about 20 ml. A crystalline material is separated by centrifugation, washed with cold water, and recrystallized from water.

Chemical Synthesis.[10] 4-Ureido-5-imidazolecarboxylic acid may be synthesized by treating 4-amino-5-imidazolecarboxylic acid[11] with potassium cyanide. A solution containing 7 millimoles of 4-amino-5-imidazolecarboxylic acid, 14 millimoles of potassium cyanide, and 14 millimoles of hydrochloric acid in 420 ml. of water is stored at room temperature overnight. The reaction mixture contains both 4-ureidoimidazole and the desired product, 4-ureido-5-imidazolecarboxylic acid, when examined by paper chromatography. The product may be isolated by chromatography on Dowex, as described for the enzymatic product.

Properties. The isolated material contains a variable amount of water of crystallization but may be dried at 100° to the anhydrous form. The molecular extinction coefficients at the wavelength of maximal absorption were found to be 13,800 at 255 mμ in 0.2 N HCl; 14,300 at 242 mμ, pH 3 to 4; 11,700 at 256 mμ, pH 6 to 9; and 12,600 at 270 mμ in 6 N KOH. The pK'_a values calculated from the absorption spectra, which were not obtained at constant ionic strengths, are 2.0, 4.9, and 12.2. The solubility of the isolated material in water is 1.5 millimoles/ml. at 35.5°, 1.0 millimole/ml. at 23°, and 0.5 millimole/ml. at 0°. The chromatographic behavior of the compound is indicated in Table II.

[8] The fractions collected were approximately 75 ml.
[9] The compound may be detected by its absorbance at 250 mμ in 0.1 M sodium pyrophosphate buffer at pH 9.0.
[10] G. Hunter and I. Hlynka, *Biochem. J.* 31, 488 (1937).
[11] The synthesis is described in Section II.

TABLE II
CHROMATOGRAPHY OF IMIDAZOLE DERIVATIVES[a]

Compound	R_f in Solvent:					Color in Pauly test
	1	2	3	4	5	
4-Aminoimidazole	0.43	0.35	—	—	—	Blue
4-Amino-5-imidazolecarboxylic acid	0.42	0.13	0.19	0.62	0.21	Blue
4-Ureido-5-imidazolecarboxylic acid	0.29			0.51	0.31	Magenta
4-Ureidoimidazole	0.44			0.61	0.50	Magenta
4-Amino-5-imidazolecarboxylic acid methyl ester	0.72	0.61	0.84			Orange
4-Amino-5-imidazolecarboxamide	0.49	0.33	0.57			Blue
4-Hydroxy-5-imidazolecarboxamide	0.30	0.21				Red
5-Imidazolecarboxylic acid	0.30	0.15	0.49			Yellow
Xanthine	0.40	0.25				Orange

[a] See B. N. Ames and H. K. Mitchell, *J. Am. Chem. Soc.* **74**, 252 (1952) and H. Tabor, Vol. III [90] for chromatographic properties of other imidazoles, and a description of the spray reagent.

[b] Solvent 1, n-propanol–1 N acetic acid (3:1). Solvent 2, n-propanol–water (3:1). Solvent 3, acetone–2 M triethylamine–water (160:1:40). Solvent 4, methanol–chloroform–10% formic acid (3:3:1). Solvent 5, n-propanol–1 M ammonium hydroxide (3:1).

4-Amino-5-imidazolecarboxylic Acid[12]

Principle. 4-Amino-5-imidazolecarboxylic acid is not degraded enzymatically in media of pH above 8.0. This compound, therefore, accumulates when xanthine is incubated with an extract of *C. cylindrosporum* at pH 9.0 and is isolated by chromatography on Dowex 50.

Procedure. A reaction mixture is prepared by mixing 5 ml. of 0.02 M xanthine, 1 ml. of 0.10 M cysteine at pH 9.0, 1 ml. of 1 M diethanolamine buffer at pH 9, and 1.0 ml. of the cell extract. The volume is adjusted to 10.0 ml. with water, and the mixture is incubated at 37°. When a maximum amount of diazotizable amine is formed,[13] the reaction mixture, without deproteination or pH adjustment, is placed on a column of

[12] J. C. Rabinowitz, *J. Biol. Chem.* **218**, 175 (1956).

[13] Aliquots of 0.1 ml. (or less as the reaction proceeds) are added to 5 ml. of 3% perchloric acid, and the diazotizable amine content is determined as described in Section III. Under the conditions described, maximum amounts of diazotizable amine were formed in 2 hours, but, because of the variability in the activity of the cell extracts, it is necessary to follow the course of the reaction by the colorimetric test.

Dowex 50-H+ (12 by 1 cm.²). Fractions of 5 ml. are collected. The column is washed with 15 ml. of water and 50 ml. of 0.2 M NaCl. These fractions may all be discarded. The column is then eluted with 0.1 M ammonium formate. The 4-amino-5-imidazolecarboxylic acid is eluted in the first 50 ml. of eluate.[14] The compound has not been isolated as a solid.

Chemical Synthesis.[15] Solutions of 4-amino-5-imidazolecarboxylic acid are prepared by catalytic reduction of 4-nitro-5-imidazolecarboxylic acid. The latter compound is synthesized from pyruvic aldehyde,[16] which is converted successively to 5-methylimidazole,[17] 4-nitro-5-methylimidazole,[18] and 4-nitro-5-styrylimidazole.[18] The latter compound is finally oxidized with permanganate to the desired product.[18] The procedures for reduction of the nitro group and for storage of a product are similar to those described in the following section for the preparation of 4-amino-imidazole, except that the reduction is carried out in phosphate buffer at pH 8.5, and the product is stored at this pH. Just before use, the solution is passed over a column of Dowex 50-NH₄+ to remove 4-aminoimidazole which is formed by spontaneous decarboxylation.

Properties. The chromatographic properties of 4-amino-5-imidazole-carboxylic acid are shown in Table II. The molecular extinction coefficients of the compound and the wavelength of maximum absorption are 9300 at 260 mμ at pH 9.0 and 7260 mμ in 0.1 M KOH. The spectrum of the product formed in the Bratton-Marshall test under the conditions described in Section III has a maximum absorption of 502 mμ.

4-Aminoimidazole[12]

Principle. 4-Aminoimidazole accumulates briefly as the main diazotizable amine incubated at pH 7 containing xanthine and extracts of *C. cylindrosporum*. It is isolated from these mixtures by chromatography on Dowex 50-H+.[19]

Procedure. A reaction mixture is prepared by mixing 5 ml. of 0.02 M

---

[14] The compound may be detected in the eluted fractions by measuring the absorbance at 240 and 260 mμ. The ratio of absorbance of 4-amino-5-imidazolecarboxylic acid at 260 to 240 mμ at pH 5.0 is 0.88. The ratio for 4-aminoimidazole under these conditions is 0.37. Both compounds may also be detected in the eluted fractions by the diazotizable amine test (see Section III).
[15] A. Windaus and W. Langenbeck, *Ber.* **56**, 683 (1923).
[16] Pyruvic aldehyde satisfactory for this synthesis can be obtained from Carbide and Carbon Chemicals Co., 30 East 42nd Street, New York 17, New York.
[17] J. M. Gulland and T. F. Macrae, *J. Chem. Soc.* **662** (1933).
[18] W. E. Allsebrook, J. M. Gulland, and L. F. Story, *J. Chem. Soc.* **232** (1942).
[19] The further degradation of 4-aminoimidazole requires metal ions, and therefore it also can be prepared by incubation from 4-amino-5-imidazolecarboxylic acid with the cell extract in the presence of ethylenediaminetetraacetic acid.

xanthine, 1 ml. of 0.10 M cysteine at pH 7, 1 ml. of 1 M phosphate buffer at pH 7, 1 ml. of triethanolamine buffer at pH 7, and 1.0 ml. of the cell extract in a total volume of 10 ml. The reaction mixture is incubated at 37°. The course of the reaction is followed by measuring a formation of diazotizable amine (see Section III). When the absorbance of a 0.1-ml. aliquot is above 1.0 in the diazotizable amine test, the reaction mixture, without deproteinization or pH adjustment, is placed on a column of Dowex 50-H[+] as described in the preceding section. The elution is carried out as described there, and, after elution of the 4-amino-5-imidazolecar-boxylic acid with 125 ml. of 0.1 M ammonium formate, the column is eluted with 0.5 M ammonium formate. The 4-aminoimidazole is eluted in 100 ml. and detected in individual fractions by the diazotizable amine test.

Chemical Synthesis. 4-Aminoimidazole is synthesized by the catalytic reduction of 4-nitroimidazole.

4-Nitroimidazole.[20] Twenty grams of imidazole are treated with 40 ml. of nitric acid and 40 ml. of sulfuric acid. The mixture is refluxed for 2 hours and poured into ice water. The precipitate is collected by filtration, washed with water and acetone, and recrystallized from glacial acetic acid. The recrystallized product is obtained in 33% yield.

4-Aminoimidazole. A suspension of 114 mg. of 4-nitroimidazole, 0.8 ml. of 1 M monopotassium phosphate, 14.0 ml. of water, and 100 mg. of 5% palladium–charcoal (Baker and Co., Inc., Newark, New Jersey) is shaken under 1 atmosphere of hydrogen until the gas uptake is complete. The rate of uptake is linear and is complete after 30 to 45 minutes with the apparatus used. Then 1.2 ml. of 1 M dipotassium phosphate, 0.08 ml. of 2 M sodium sulfide, 0.04 ml. of 4 N HCl, and 3.9 ml. of water are added. The solution is filtered with filter aid and stored at 0° in a syringe closed with a sealed hypodermic needle. Samples can be withdrawn without exposing the solution to oxygen. Exposure of neutral solutions of 4-aminoimidazole to oxygen results in the formation of blue pigments and a partial loss of diazotizable amine.

Properties. The chromatographic properties of 4-aminoimidazole are shown in Table II. The approximate molecular extinction coefficient is 3800 at 238 mμ in 0.1 M HCl or at pH 5.0. The spectrum of the product formed from 4-aminoimidazole in the diazotizable amine test under the conditions described has a maximum at 514 mμ. The compound shows maximum stability at room temperature when stored at pH 5 in in 0.1 M acetate buffer. The half-life of 4-aminoimidazole at pH 7 at room temperature is about 75 hours. There is a marked decrease of stability in acid or alkaline solution.

[20] R. G. Fargher and F. L. Pyman, *J. Chem. Soc.* **115**, 217 (1919).

4-Imidazolone[21]

Principle. 4-Imidazolone is formed from 4-aminoimidazole by a partially purified enzyme preparation,[22] but it is most conveniently prepared as a solution by the addition of an equivalent amount of NaOH to chemically synthesized formiminoglycine-β-phenyl ester hydrochloride.

Chemical Synthesis. FORMIMINOGLYCINE-β-PHENYLETHYL ESTER HYDROCHLORIDE. A mixture of 350 mg. of formiminoglycine[23] and 20 ml. of β-phenylethyl alcohol saturated with dry HCl gas is heated to 90° for 1 hour. The light-brown oil formed on cooling is extracted with 500 ml. of ether. After cooling for several minutes in an ice bath, crystals appear, which can be recrystallized from methanol–ether. The product has a melting point of 120°. The yield is 500 mg. (60% theory).

4-IMIDAZOLONE. An equivalent amount of NaOH is added to an aqueous solution of the formiminoglycine-β-phenylethyl ester hydrochloride. Solutions containing 0.2 micromole of the ester per milliliter are converted to the imidazolone within 5 minutes after addition of alkali.[24]

Properties. 4-Imidazolone has an absorption maximum in neutral solution at 255 mμ. The extinction coefficient is roughly 7000. The dimer formed when solutions of the ester containing 50 micromoles of the ester per milliliter are neutralized has an absorption maximum at 325 mμ, with an extinction coefficient similar to that of the imidazolone.

Formiminoglycine[22]

Principle. 4-Aminoimidazole is quantitatively converted to formiminoglycine by extracts of *C. cylindrosporum* at pH 7.0. The product is isolated by chromatography on Dowex 50-H+.[25]

Procedure. A reaction mixture containing the following material is incubated under anaerobic conditions at 37°: 10.0 millimoles of 4-aminoimidazole, 26.8 millimoles of potassium phosphate buffer at pH 7, 0.34 millimole of ferrous sulfate, 0.09 millimole of sodium sulfide, 10 ml. of the cell extract, and water to make the volume to 180 ml. After 100 minutes, 3 ml. of 60% perchloric acid is added, and the precipitated protein is removed by centrifugation. The supernatant solution is neutralized with potassium hydroxide and centrifuged at 0° to remove the precipitated potassium perchlorate. The solution is lyophilized. The lyophilized powder is dissolved in 100 ml. of water and adjusted to pH

[21] K. Freter, J. C. Rabinowitz, and B. Witkop, *Ann.* **607**, 174 (1957).
[22] J. C. Rabinowitz and W. E. Pricer, Jr., *J. Biol. Chem.* **222**, 537 (1956).
[23] The synthesis is described in the next section.
[24] Solutions containing higher amounts of the ester form a dimer when neutralized.
[25] Xanthine is also converted to formiminoglycine quantitatively under these conditions, but the product is formed in low concentration because of the low solubility of the purine.

2.5. Insoluble material is removed by filtration. The solution is placed on a column of Dowex 50-H$^+$ (9.25 $\times$ 32.5 cm.^{2}). The column is washed with 2 l. of water. Formiminoglycine is eluted with 1 M ammonium acetate (0.1 M with respect to acetic acid). Ten 150-ml. fractions are collected. The formiminoglycine is detected by using the ferricyanide-nitroprusside reagent (see Section III). Fractions 7 and 8, which contain the formiminoglycine, are lyophilized to remove the ammonium acetate. The solid obtained is crystallized from an ethanol–water mixture and recrystallized from 50% ethanol with the use of charcoal.

Chemical Synthesis.[26] Twenty grams (0.27 mole) of glycine, 20 g. (0.25 mole) of formamidine hydrochloride,[27] 30 g. of redistilled pyridine, and 200 ml. of redistilled formamide are placed in a three-necked flask equipped with a seal stirrer and a condenser stoppered with a calcium chloride tube. The mixture is stirred for 24 hours in a water bath at 50°. The mixture is cooled at 0°, and 200 ml. of absolute ethanol are added. The precipitate is removed by filtration on a Büchner funnel, washed with ethanol, and dried under reduced pressure over KOH. The precipitate weighs 23.7 g. (93% yield based on formamidine hydrochloride). For recrystallization, 5 g. of formiminoglycine are dissolved with heating in 40 ml. of 50% ethanol. On cooling, a 70% yield of the material is obtained.

Properties. The compound is extremely labile to alkali. Formiminoglycine has a half-life at 25° of approximately 10 minutes in 1 N KOH, 20 minutes at pH 11.5, 70 minutes at pH 10.8, and 40 hours at pH 9.2, and is stable for long periods (at least several months) at pH 7.2 and pH 4, and in 0.1 N acetic acid, 0.1 N HCl, and 6 N HCl. The pK'_a values for formiminoglycine are approximately 2.2 and 11.6. Formiminoglycine can be determined with the ferricyanide–nitroprusside reagent described in Section III. The compound may also be detected on paper chromatograms sprayed with this reagent.[28]

III. Color Tests

Diazotizable Amine Test

4-Aminoimidazole and 4-amino-5-imidazolecarboxylic acid are determined as diazotizable amines by using a modification of the procedures described by Bratton and Marshall.[29]

[26] H. Tabor and J. C. Rabinowitz, *Biochem. Preparation* **5**, 100 (1957).
[27] Formamidine hydrochloride may be obtained commercially from Fluka Inc., Buchs SG, Switzerland, or from Mann Research Laboratories, Inc., 136 Liberty St., New York 6, N.Y. The compound may also be synthesized as described.[26]
[28] The R_f of formiminoglycine in methanol–chloroform–90% formic acid (3:3:1) is 0.54, and in t-butanol–90% formic acid–water (70:15:15) it is 0.44.
[29] A. C. Bratton and E. K. Marshall, Jr., *J. Biol. Chem.* **128**, 537 (1939).

Reagents

5% perchloric acid.
0.1% sodium nitrite.
0.5% ammonium sulfamate.
0.1% N-1-naphthylethylenediamine dihydrochloride.

Procedure. The sample, deproteinized when necessary with perchloric acid and diluted with water to a volume of 0.2 ml., is added to 5 ml. of perchloric acid. Then 0.5 ml. of sodium nitrite is added. After 3 minutes, 0.5 ml. of ammonium sulfamate is added, followed 2 minutes later by 0.5 ml. of N-1-naphthylethylenediamine dihydrochloride. After 10 minutes, the optical density at 510 mμ is determined with a Coleman Jr. spectrophotometer in 16-mm. test tubes. One micromole of 4-aminoimidazole, 4-amino-5-imidazolecarboxylic acid, or 4-amino-5-imidazolecarboxamide has an absorbance of 4.5, 4.5, 4.6, respectively.

Imidazole Test[30]

Imidazole derivatives are determined quantitatively by this modification of the Pauly test. Imidazole derivatives substituted with primary amino groups may be destroyed by treatment with nitrous acid.

Reagents

Sulfanilic acid: 900 mg. + 9 ml. of HCl diluted to 100 ml.
5% sodium nitrite.
Diazotized sulfanilic acid: 10 ml. of sulfanilic acid solution are
· added to 15 ml. of sodium nitrite.
1% sodium carbonate.
Tertiary butanol.
5% perchloric acid.

Procedure. To a colorimeter tube containing 5 ml. of sodium carbonate, 0.1 ml. of the diazotized sulfanilic acid reagent, and 1 ml. of tertiary butanol is added 0.5 ml. of sample. The solution is mixed, and the absorbance at 520 mμ is determined 0.5 to 2 minutes after mixing of the sample and reagent.

4-Ureido-5-imidazolecarboxylic acid may be determined in mixtures containing 4-aminoimidazole after the latter compound has been destroyed with nitrous acid.[6] The deproteinized sample is diluted to 0.7 ml. with water and treated with 0.1 ml. of 5% perchloric acid and 0.1 ml. of 0.1% sodium nitrite. After 2 minutes, 0.1 ml. of 0.5% ammonium

[30] K. K. Koessler and M. F. Hanke, *J. Biol. Chem.* **39**, 497 (1919).

sulfamate is added. The imidazole is then determined by the imidazole test, as described.

Formimino Test[22]

Formiminoglycine is determined with the modification of the alkaline ferricyanide–nitroprusside test first described for the detection of guanidine and its methyl derivatives.[31]

Reagents

Ferricyanide–nitroprusside reagent.[32] Four grams of sodium hydroxide, 4.0 g. of sodium nitroprusside, and 4.0 g. of potassium ferricyanide are dissolved in water and diluted to 120 ml. The solution is filtered before use and stored in a refrigerator. The reagent can be kept under these conditions for at least a month. 0.16 M sodium tetraborate (saturated).

Procedure. To a sample containing 0 to 1.0 micromole of formiminoglycine and water to make the volume to 0.5 ml. are added 2.0 ml. of sodium tetraborate, followed immediately by 0.5 ml. of the ferricyanide-nitroprusside reagent. After 30 minutes the optical density is determined at 485 mμ by using 1-cm. cells in the Beckman DU spectrophotometer, with a reagent blank of the proper dilution in the reference cell. The absorbance of 1 micromole of formiminoglycine under these test conditions is 1.14.

[31] O. W. Tiegs, *Australian J. Exptl. Biol. Med.* **24,** 712 (1927).
[32] This reagent may also be used to detect formiminoglycine on paper chromatograms. After being thoroughly dried in air, the paper is resprayed with the reagent and then with saturated sodium borate. Color development is relatively slow and may take 5 to 10 minutes. A very faint reaction can be detected at 0.05 micromoles of formiminoglycine, but it requires 0.5 micromole to give a strong reaction. The colors formed are generally not stable for more than a day.

[98] Enzymatic Synthesis of Polyribonucleotides

By CARLOS BASILIO and SEVERO OCHOA

The preparation of polymers as reported here largely represents the experience so far accumulated with polynucleotide phosphorylase[1] from *Azotobacter vinelandii.*

[1] See Vol. VI [1].

Assay Method

The extent of polymerization can be determined by measuring the orthophosphate liberated during the reaction. A slight modification of the method of Lohmann and Jendrassik,[2] described below, is very convenient for this purpose.

Reagents

0.125 M ribonucleoside diphosphate. Dissolve in distilled water, and neutralize with KOH, if necessary, with phenol red as indicator. Determine the absorbancy at λ_{max} for exact determination of concentration.[3]

0.1 M MgCl$_2$.

1.0 M Tris buffer, pH 8.1, containing 0.0066 M EDTA.

40% TCA.

2.5% ammonium molybdate in 5 N H$_2$SO$_4$.

Reducing agent. Dissolve 0.1 g. recrystallized 1-amino-2-naphthol-4-sulfonic acid, 6.0 g. of anhydrous NaHSO$_3$, and 1.2 g. of anhydrous Na$_2$SO$_3$ in 50 ml. of distilled water, and filter; keep the solution cool and dark.

Standard phosphate, 0.001 M KH$_2$PO$_4$.

85% phenol made with redistilled phenol.

Dialysis fluid, 0.04 M NaCl containing 0.005 M sodium citrate.

Procedure for Phosphate Determination. To the tube containing the sample from the reaction mixture, add 0.1 ml. of 40% TCA, allow it to stand for 10 minutes in an ice-water bath, and centrifuge. Transfer 0.5 ml. of the clear supernatant to another tube, and add 0.4 ml. of distilled water, 0.24 ml. of ammonium molybdate, and 0.06 ml. of reducing agent. Heat the mixture at 50° for 5 minutes, and immediately chill it in an ice-water bath. After 10 minutes, determine the absorbancy at 670 mμ in a Beckman spectrophotometer. Run a blank containing water and a standard containing 0.1 micromole of KH$_2$PO$_4$ in the same manner.

Synthesis of Polyribonucleotides

The conditions described below can be used for preparing any homo- or copolymer with the exception of poly G. The concentration of the components, in micromoles per milliliter, is as follows: total nucleoside diphosphate, 60.0; MgCl$_2$, 5.0; Tris buffer, pH 8.1, 150.0; and EDTA, 1.0. Start the reaction by adding 10 to 16 "exchange" units of enzyme (mini-

---

[2] K. Lohmann and L. Jendrassik, *Biochem. Z.* **178**, 419 (1926).

[3] See Vol. III [111], p. 804, for λ_{max} and ϵ_{max} of nucleotides.

mum specific activity of 50), and incubate at 30°. At zero time and at 0.5- or 1-hour intervals, remove a 0.01 ml. aliquot of the reaction mixture, and pipet it into 1.0 ml. of ice-cold distilled water. These are used for determination of orthophosphate liberated during the course of the polymerization reaction and are taken until the formation of orthophosphate ceases. This usually occurs after 2 to 4 hours of incubation for synthesis of homopolymers and for copolymers containing AMP, CMP, and UMP. For copolymers containing GMP, orthophosphate liberation stops after approximately 5 to 8 hours. This time can be shortened by using twice as much enzyme. In general, the higher the GDP concentration in the reaction mixture, the longer is the time required for the reaction to come to completion. The addition of a drop of toluene is recommended for long incubations in order to prevent bacterial growth.

Synthesis of Polyguanylic Acid and Priming Effect

Under the conditions described above, GDP reacts very slowly. Singer et al.[4] have found that polymerization of GDP takes place only when oligonucleotides having an unesterified 3'-hydroxyl end group such as pApA, ApA, pApApA, or pUpUpU are used as primers. Under these conditions, the guanosine monophosphate units are added to the 3'-hydroxyl end group of the oligonucleotide. The same priming reaction occurs when ADP, UDP, or thymine riboside pyrophosphate is used for polymerization.[5]

Polyguanylic acid obtained by this method has an average chain length of no more than 30 residues. The following reaction mixture is representative of an oligonucleotide-primed synthesis of poly G (concentration in micromoles per milliliter): GDP, 54.8; Tris buffer, pH 8.2, 150.0; MgCl$_2$, 10; EDTA, 0.4; pUpUpU, 1.8 (as mononucleotide); and enzyme (specific activity 160), 100 "exchange" units. The incubation is carried out at 37° for 6.5 hours.

Ochoa and Mii[6] have obtained polyguanylic acid with an average chain length of 200 residues by using highly purified polynucleotide phosphorylase from *Azotobacter vinelandii*. The following reaction mixture is used in this case (concentration in micromoles per milliliter): GDP, 5.0; Tris buffer, pH 8.0, 150.0; MgCl$_2$, 2.0; EDTA, 1.0; and enzyme (specific activity 300), 25 exchange units. The mixture is incubated at 30° for 24 hours. Essentially the same results are obtained when 0.25 micromole of poly C (as CMP) are added as primer.

[4] M. F. Singer, R. J. Hilmoe, and L. A. Heppel, *J. Biol. Chem.* 235, 751 (1960).
[5] M. F. Singer, L. A. Heppel, and R. J. Hilmoe, *J. Biol. Chem.* 235, 738 (1960).
[6] S. Ochoa and S. Mii, unpublished results.

Isolation of the Polymers

The polymers are isolated by treating the reaction mixture with phenol according to Gierer and Schramm,[7] followed by dialysis and lyophilization. This method, which differs from the one previously reported[8] (ethanol precipitation), has the advantage that the polymers are obtained essentially free of protein. After the reaction is completed, chill the tube in an ice-water bath. Add an equal volume of 85% phenol, and shake the mixture in the cold room for 8 minutes. Centrifuge, and separate the aqueous layer. Wash the phenol layer by shaking with an equal volume of cold distilled water. Add the water wash to the initial aqueous layer, and repeat the phenol treatment on the combined aqueous

TABLE I

SUBSTRATE SPECIFICITY OF POLYNUCLEOTIDE PHOSPHORYLASE

Substrates

Adenosine-5'-diphosphate[a]	Cytidine-5'-diphosphate[a]
Guanosine-5'-diphosphate[a]	Uridine-5'-diphosphate[a]
Inosine-5'-diphosphate[a]	Xanthosine-5'-diphosphate[b]
Ribothymidine-5'-diphosphate[c]	2-Thiouridine-5'-diphosphate[d]
5-Fluorouridine-5'-diphosphate[b]	5-Bromouridine-5'-diphosphate[e]
5-Chlorouridine-5'-diphosphate[e]	5-Iodouridine-5'-diphosphate[e]
N-Methyluridine-5'-diphosphate[f]	8-Azaguanosine-5'-diphosphate[g]

Nonsubstrates

Azauridine-5'-diphosphate[b,h,i]	5,6-Dihydrouridine-5'-diphosphate[f,i]
Adenyl-5'-methylphosphonate[k]	Adenyl-5'-methylenediphosphonate[i,k]
6-Mercaptopurineriboside-5'-diphosphate[i,l]	Ribose-5-pyrophosphate[j]

[a] M. Grunberg-Manago, P. J. Ortiz, and S. Ochoa, *Biochim. et Biophys. Acta* **20**, 269 (1956).

[b] P. Lengyel, and C. Basilio, to be published.

[c] B. E. Griffin, A. R. Todd, and A. Rich, *Proc. Natl. Acad. Sci. U. S.* **44**, 1123 (1958).

[d] P. Lengyel, and R. W. Chambers, *J. Am. Chem. Soc.* **82**, 752 (1960).

[e] A. M. Michelson, J. Dondon, and M. Grunberg-Manago, *Biochim. et Biophys. Acta* **55**, 529 (1962).

[f] W. Szer, and D. Shugar, *Acta Biochim. Polon.* **8**, 235 (1961).

[g] D. H. Levin, *Biochim. et Biophys. Acta* **61**, 75 (1962).

[h] J. Skoda, J. Kara, Z. Sormova, and F. Sorm, *Biochim. et Biophys. Acta* **33**, 579 (1959).

[i] Also inhibitors.

[j] S. Ochoa and L. A. Heppel, *in* "The Chemical Basis of Heredity" (W. McElroy and B. Glass, eds.), p. 615. Johns Hopkins Press, Baltimore, 1957.

[k] L. N. Simon and T. C. Myers, *Biochim. et Biophys. Acta* **51**, 178 (1961).

[l] J. A. Carlson, *Biochem. Biophys. Research Communs.* **7**, 366 (1962).

[7] A. Gierer and G. Schramm, *Nature* **177**, 702 (1956).

[8] M. Grunberg-Manago, P. J. Ortiz, and S. Ochoa, *Biochim. et Biophys. Acta* **20**, 269 (1956).

layers by shaking for 2 minutes. Separate the resultant aqueous layer, and wash the phenol layer as before. Extract the phenol from the last combined water wash and aqueous layer by repeated extractions with ether. Then remove the ether by bubbling N_2 through the solution. After the ether has been removed, dialyze the polymer solution twice against 1 l. of 0.04 M NaCl–0.005 M sodium citrate, and then against several changes of cold distilled water (1 l. per change) until no more ultraviolet-absorbing material is detected in the dialysis fluid. Then lyophilize the polymer, and store it over silica gel at −15°.

Substrate Specificity. Table I indicates various ribonucleoside diphosphates shown thus far to be substrates for polymer and copolymer synthesis. Also included are several related compounds which are inactive or which inhibit polynucleotide phosphorylase.

Properties

Table II shows actual experimental data for some representative polymers synthesized by polynucleotide phosphorylase. A detailed discussion of their chemical and physical properties can be found in recent

TABLE II
Data on Synthesis of Various Polymers

Polymer	Polymerization, %	Yield mg./ml.	Sedimentation coefficients	Actual base ratio
Poly A	68	12	17.7	
Poly C	65	11	8.6	
Poly U (sample 1)	63	8	9.7	
Poly U (sample 2)	65	12	10.3	
Poly I	58	15.5	16.9	
Poly UA, 5:1[a]	54	14	9.0	4.7:1
Poly UC, 5:1	53	13	9.5	4.8:1
Poly UG, 5:1	66	11.5	4.1	5.2:1
Poly UCA, 6:1:1	46	12	2.8	6:0.9:1.2
Poly UCG, 6:1:1	50	10	4.9	6:0.6:1
Poly UGA, 6:1:1	50	11.3	4.8	6:1.0:1.3

[a] Ratio of ribonucleoside diphosphates in the reaction mixture.

reviews.[9-11] The primary structure of synthetic polymers is analogous to ribonucleic acid, consisting of nucleoside residues linked in a chain by

[9] M. F. Singer, L. A. Heppel, R. J. Hilmoe, S. Ochoa, and S. Mii, *Proc. 3rd Can. Cancer Research Conf.* **3**, 41 (1959).
[10] H. G. Khorana, *in* "The Nucleic Acids" (E. Chargaff and J. N. Davidson, eds.), Vol. 3, p. 105. Academic Press, New York, 1960.
[11] R. F. Steiner and R. F. Beers, Jr., "Polynucleotides." Elsevier Publishing Co., Amsterdam, 1961.

3′,5′-phosphodiester bonds.[12,13] Alkaline, acid, and enzymatic hydrolysis with specific phosphodiesterases gives rise to the same products as result from naturally occurring RNA.[12,13] The base composition of the co-polymers, as shown in Table II, largely reflects the relative concentration of the ribonucleoside diphosphates present during the polymerization. Other investigators have reported results at variance with ours, with substantial divergence between base composition of the polymers and nucleoside diphosphate input.[14] Contamination of the polynucleotide phosphorylase preparations with nucleases and/or nucleoside phospho-kinases, ATPase, etc., might account for these results. Nearest-neighbor frequency studies with poly AU (1:1)[13] and poly AGUC (1:1:1:1)[15] have shown that the nucleotide distribution is essentially random. The molecular weight of the polymers, based on sedimentation data,[16] may vary from 10^4 to 10^6, depending on the base composition of the polymer and the conditions of polymerization.[17]

The study of synthetic polymers has greatly contributed to our present understanding of the physical chemistry of nucleic acids. The finding that copolymers of different base composition promote the incorporation of specific amino acids into polypeptide chains in cell-free systems has led to the deciphering of the amino acid code.[18-20]

[12] L. A. Heppel, P. J. Ortiz, and S. Ochoa, *J. Biol. Chem.* **229**, 679 (1957).
[13] L. A. Heppel, P. J. Ortiz, and S. Ochoa, *J. Biol. Chem.* **229**, 695 (1957).
[14] M. S. Bretscher and M. Grunberg-Manago, *Nature* **195**, 282 (1963).
[15] P. J. Ortiz and S. Ochoa, *J. Biol. Chem.* **234**, 1208 (1959).
[16] R. C. Warner, unpublished data.
[17] J. R. Fresco and P. Doty, *J. Am. Chem. Soc.* **79**, 3928 (1957).
[18] P. Lengyel, J. Speyer, and S. Ochoa, *Proc. Natl. Acad. Sci. U.S.* **47**, 1936 (1961).
[19] J. Speyer, P. Lengyel, C. Basilio, and S. Ochoa, *Proc. Natl. Acad. Sci. U.S.* **48**, 441 (1962).
[20] J. H. Matthaei, O. W. Jones, R. G. Martin, and M. W. Nirenberg, *Proc. Natl. Acad. Sci. U.S.* **48**, 666 (1962).

[99] Enzymatic Preparation and Properties of Deoxy-ribonucleic Acid and Polydeoxyribonucleotides

By MAURICE J. BESSMAN

This report describes procedures for the enzymatic synthesis of DNA and polydeoxyribonucleotides by a highly purified enzyme from *Escher-*

ichia coli.[1] The synthesis of DNA by purified extracts of calf thymus has been described by Bollum.[2]

DNA will refer to the product synthesized in a complete reaction mixture. Polydeoxyribonucleotide refers to a product synthesized in a reaction mixture lacking one or more of the deoxyribonucleoside triphosphates of adenine, cytosine, guanine, and thymine and/or primer DNA.

Materials

Polymerase, fraction VII[3] or any fraction of equivalent purity.
dATP, dCTP, dGTP, dTTP.[4]
Primer DNA.[5]
Carrier DNA.[6]

I. Synthesis of DNA

Synthesis of Tracer Amounts of DNA

For many purposes it is desirable to measure properties of DNA during initial phases of its synthesis under conditions in which the total DNA synthesized represents only a small fraction of the primer DNA present in the reaction mixture. For this purpose, radioactive triphosphates must be employed. A typical reaction mixture contains (in 0.3 ml.) 20 micromoles of glycine (pH 9.2), 2 micromoles of $MgCl_2$, 0.3 micromole of 2-mercaptoethanol, 20 millimicromoles of DNA,[7] 5 millimicromoles each of dATP, dCTP, dGTP, and dTTP (any one may be labeled), and 0.05 unit of enzyme. After 30 minutes at 37°, the reaction mixture is chilled, and carrier is added (0.2 ml. of thymus DNA, 5 micromoles/ml.). The DNA is precipitated with 0.5 ml of 1 N $HClO_4$, and, after 5 minutes at 0°, 2.5 ml. of water are added, and the precipitate is dispersed. The precipitate is collected by centrifugation at 20,000 $\times$ g for 5 minutes, redissolved in 0.3 ml. of 0.2 N NaOH, and then reprecipitated with 0.4 ml.

---

[1] I. R. Lehman, M. J. Bessman, E. J. Simms, and A. Kornberg, *J. Biol. Chem.* **233**, 163 (1958).
[2] F. J. Bollum, *J. Biol. Chem.* **234**, 2733 (1959).
[3] See Vol. VI [5].
[4] See Vol. III [111] for preparation of unlabeled deoxyribonucleoside triphosphates. These are also available commercially. For the preparation of P^{32}-labeled substrates, see Lehman *et al.*[1]
[5] All preparations of DNA isolated under conditions preventing depolymerization tested so far serve as primers. Articles [102] and [103] in Vol. III describe suitable preparations from several sources.
[6] We have routinely used calf thymus DNA as carrier; E. R. M. Kay, N. S. Simmons, and A. L. Dounce, *J. Am. Chem. Soc.* **74**, 1724 (1952).
[7] Quantities of DNA are expressed as micromoles of DNA phosphorus.

of 1 N HClO$_4$. Three milliliters of water are added, and the precipitated DNA is dispersed and centrifuged as before. The precipitate is treated once more in the same way and then may be dissolved in water with the aid of 0.1 N NaOH. After this wash procedure, less than 0.1% of the unreacted triphosphates remains in the final precipitate. It should be noted that DNA harvested in this manner cannot be expected to retain its secondary structural features (double-strand character, helicity, hydrogen bonding). To preserve these characteristics in the synthetic material, alternative isolation procedures must be employed (see below).

Properties. Although only a very small part of the total DNA in the final product represents newly synthesized material, it is possible to measure several of its properties by taking advantage of the incorporated radioactivity. The newly synthesized DNA is resistant to alkali and is hydrolyzed by acid (Table I). The internucleotide linkages are 3',5'-

TABLE I[a]

CHARACTERISTICS OF THE PRODUCT

Product	Counts per minute
Acid-insoluble product	274
DNase-treated	13
RNase-treated	240
NaOH (1 N), 18 hours at 37°	214
Perchloric acid (0.01 N), 5 minutes at 100°	232
Perchloric acid (0.04 N), 5 minutes at 100°	82

[a] From A. Kornberg, in "The Chemical Basis of Heredity" (W. D. McElroy and B. Glass, eds.), p. 600. Johns Hopkins Press, Baltimore, 1957.

phosphodiester bonds,[8] which are susceptible to deoxyribonucleases I and II and completely resistant to pancreatic ribonuclease. The material has sedimentation constants ranging from 20 to 30 S, with a molecular weight in the neighborhood of 6 million.[9]

Perhaps the most striking feature of the newly synthesized DNA is its reflection of the compositional details of the primer DNA used in the reaction even during the early stages of the synthesis (Table II). Of special interest is the recent "nearest-neighbor" sequence analysis which suggests that each DNA species has a unique arrangement of its nucleotides in respect to each other.[10] Also deducible from the data presented

[8] M. J. Bessman, I. R. Lehman, E. S. Simms, and A. Kornberg, J. Biol. Chem. 233, 171 (1958).
[9] H. K. Schachman, I. R. Lehman, M. J. Bessman, J. Adler, E. S. Simms, and A. Kornberg, Federation Proc. 17, 304 (1958).
[10] J. Josse, A. D. Kaiser, and A. Kornberg, J. Biol. Chem. 236, 864 (1961); see Vol. VI [101].

TABLE II[a,b]

BASE RATIOS OF THE PRODUCT ISOLATED EARLY IN THE REACTION
AS DETERMINED WITH ISOTOPICALLY MARKED SUBSTRATES

Primer DNA	Increase in DNA, %	T incorporated, millimicromoles	C incorporated, millimicromoles	T/C ratios Product	T/C ratios Primer
M. phlei	2	0.047	0.11	0.42	0.49
M. phlei	35	0.97	2.05	0.47	
Calf thymus	8	0.40	0.28	1.43	1.24
Calf thymus	63	3.08	2.43	1.27	
A. aerogenes	18	0.74	0.93	0.79	0.82

[a] From I. R. Lehman, *Ann. N. Y. Acad. Sci.* **81**, 752 (1959).

[b] Duplicate reaction mixtures, as described in the text, were used in each experiment, with the exception that C^{14}-dTTP was used in one vessel and C^{14}-dCTP was used in the second. After the appropriate incubation period, aliquots were removed, and incorporation of isotope into DNA was measured.

in this paper is the fact that the two strands of the double helix have opposite polarity, that is, run in opposite directions.

Net Synthesis of DNA

The synthesis of products containing more than 90% newly synthesized DNA may be accomplished in incubation mixtures containing (in 3.0 ml.) 1 micromole each of dATP, dGTP, dCTP, and dTTP, 200 micromoles of potassium phosphate buffer (pH 7.4), 20 micromoles of $MgCl_2$, 7 units of polymerase fraction VII, and 0.3 micromole of DNA primer. After 4 hours at 37°, NaCl is added to a final concentration of 0.2 M, and the reaction mixture is heated to 70° for 5 minutes. The product may be freed of unreacted deoxyribonucleoside triphosphates by dialysis against 0.2 M NaCl. Some DNA preparations successfully used as primers to prepare synthetic DNA were obtained from calf thymus, *E. coli*, *Aerobacter aerogenes*, *Mycobacterium phlei*, and T2 bacteriophage.

Properties.[9] The DNA preparations used to make the following observations were products containing over 90% newly synthesized material; that is, less than 10% of the total DNA analyzed represented primer DNA originally present in the incubation mixture.

Analysis revealed that the newly synthesized material has a sedimentation constant in the neighborhood of 25 and a reduced viscosity of 40 dl/g. Heating the DNA in solution or hydrolyzing it with pancreatic DNase leads to an increase in the optical density at 260 mμ (hyperchromic effect) identical with similarly tested thymus DNA.[11] These

[11] A. Kornberg, *in* "The Chemical Basis of Heredity" (W. D. McElroy and B. Glass, eds.), p. 600. Johns Hopkins Press, Baltimore, 1957.

properties indicate that the DNA is a highly ordered stiff rod with an effective volume greater than would be expected from single strands.

One property of the synthetic DNA which is still puzzling is its ability to recover its original structure after heating and cooling. That is, it does not display the hysteresis of similarly treated DNA preparations from natural sources.

Analysis of the products for molar ratios of the purine and pyrimidine bases revealed a striking similarity between the newly synthesized DNA and the primer used in its synthesis (Table III). Thus the replication of

TABLE III[a,b]
CHEMICAL COMPOSITION OF ENZYMATICALLY SYNTHESIZED DNA

	A, moles	T, moles	G, moles	C, moles	$\dfrac{A + G}{T + C}$	$\dfrac{A + T}{G + C}$
Mycobacterium phlei						
Primer	0.65	0.66	1.35	1.34	1.01	0.49
Product	0.66	0.65	1.34	1.37	0.99	0.48
Escherichia coli						
Primer	1.00	0.97	0.98	1.05	0.98	0.97
Product	1.04	1.00	0.97	0.98	1.01	1.02
Bacteriophage T2						
Primer	1.31	1.32	0.67	0.70	0.98	1.92
Product	1.33	1.29	0.69	0.70	1.02	1.90
Calf thymus						
Primer	1.14	1.05	0.90	0.85	1.05	1.25
Product	1.12	1.08	0.85	0.85	1.02	1.29
A-T copolymer	1.99	1.93	<0.05	<0.05	1.03	40

[a] From A. Kornberg, *Science* **131**, 1507 (1960).
[b] A = adenine, T = thymine, G = guanine, C = cytosine.

this compositional feature in preparations containing over 90% new material bears out the similar observations based on tracer techniques mentioned earlier.

II. Synthesis of Polydeoxyribonucleotides

Synthesis of a Copolymer of Deoxyadenylate and Thymidylate and a Homopolymer of Deoxyguanylate and Deoxycytidylate

A copolymer of deoxyadenylate and thymidylate can be synthesized by the polymerase from *E. coli*. A suitable reaction mixture contains (in 1.0 ml.) 0.3 micromole of dATP, 0.3 micromole of dTTP, 6 micromoles of $MgCl_2$, 60 micromoles of potassium phosphate buffer (pH 7.4), and 2 units of polymerase fraction VII. Under these conditions, a considerable lag (2 to 4 hours) occurs before synthesis of polymer can be measured. The lag may be eliminated by priming the reaction with deoxyadenylate-thymidylate (d-AT) copolymer. No other DNA or polynucleotide preparations tested can substitute for d-AT in shortening the lag period of this reaction. The synthesis may be followed by incorporation of P^{32}-labeled dATP or dTTP into an acid-insoluble product, by viscosity increase, or by measuring the decrease in optical density (hypochromicity) during the polymerization. The reaction may be terminated in either of the two ways described above. For studies of the physical properties of the d-AT polymer, the second method (0.2 M NaCl, followed by heating to 70°, followed by dialysis) was used to isolate the product.

Properties. The d-AT polymer is composed of equimolar amounts of dAMP and dTMP linked together by 3′,5′-phosphodiester bridges. The nucleotides are not assembled in a random manner, since nearest-neighbor sequence analysis has shown that dAMP and dTMP are present in strictly alternating sequence (Table IV).

TABLE IV[a]

HYDROLYSIS OF d-AT COPOLYMER TO 3′-DEOXYRIBONUCLEOTIDES

		Products of hydrolysis			
		3′-Deoxyadenylate		3′-Deoxythymidylate	
Experiment	Substrates for preparation of dAT copolymer	C.p.m.	C.p.m. in polymer, %	C.p.m.	C.p.m. in polymer, %
1	dAPPP and dTP³²PP	9670	100	23	<0.5
2	dAP³²PP and dTPPP	46	<0.5	8510	100

[a] For details of this experiment and interpretation of results, see H. K. Schachman, J. Adler, C. M. Radding, I. R. Lehman, and A. Kornberg, *J. Biol. Chem.* **235**, 3245 (1960).

Physicochemical properties are dependent on the length of the lag period, longer lags favoring larger polymers (Table V). Thus variations from 1.8 to 8.8 million have been calculated for preparations obtained after different lag periods. Small-molecular-weight intermediates have not been detected during the lag period. The molecules formed in the early stages of the reaction have properties similar to those formed later (Table VI).

The polymer shows a 30 to 40% hyperchromic effect on heating, which unlike DNA is completely reversible on cooling. This, together with sedimentation and viscosity data, indicates that the d-AT copolymer is a highly organized, double-stranded, rigid rod with a molecular weight

TABLE V[a,b]

SUMMARY OF PHYSICOCHEMICAL DATA FOR d-AT COPOLYMER

Preparation	Time to maximum synthesis, minutes	Reduced viscosity, (g./100 ml.)$^{-1}$	Sedimentation coefficient, S	Molecular weight, $\times 10^{-6}$
1	274	9	14	1.8
2	500	17	15	2.8
3	730	27	17	4.2
4	377	25	20	5.2
5	760	26	28	8.8

[a] From H. K. Schachman, J. Adler, C. M. Radding, I. R. Lehman, and A. Kornberg, *J. Biol. Chem.* **235**, 3245 (1960).

[b] The length of the lag period is given as the time between the addition of the enzyme and the attainment of the maximal viscosity.

in the millions. Recently, a new product has been described[12] composed entirely of deoxyguanylate and deoxycytidylate. It forms in the absence of primer and the other two deoxynucleotides under conditions similar to

TABLE VI[a]

RELATIONSHIP BETWEEN EXTENT OF REACTION AND SIZE OF PRODUCT

	Extent of reaction		Properties of product	
Experiment	Viscosity, %	Hypochromicity	Sedimentation coefficient, S	Reduced viscosity, (g./100 ml.)$^{-1}$
1	8	5	29	
	37	46	28	26
	100	100	27	24
2	36	28	19	23
	80	91	19	25
	100	100	20	26

[a] From H. K. Schachman, J. Adler, C. M. Radding, I. R. Lehman, and A. Kornberg, *J. Biol. Chem.* **235**, 3245 (1960).

those described for the d-AT reaction. A striking difference, however, is that the d-GC product is a homopolymer made up of poly dGMP and poly dCMP in separate chains linked through hydrogen bonding.

Addition of Single Nucleotides to the Ends of DNA Molecules[13]

Omission of one or more of the deoxyribonucleoside triphosphates from the normal reaction mixture reduces the synthetic reaction to very

[12] C. M. Radding, J. Josse, and A. Kornberg, *J. Biol. Chem.* **237**, 2869 (1962).

[13] J. Adler, I. R. Lehman, M. J. Bessman, E. S. Simms, and A. Kornberg, *Proc. Natl. Acad. Sci. U.S.* **44**, 641 (1958).

low but significant values.[8] A typical reaction mixture contains 5 milli-micromoles of deoxyribonucleoside triphosphate (0.5 to 1×10^8 c.p.m./micromole), 6 micromoles of $MgCl_2$ 60 micromoles of potassium phosphate buffer of pH 7.4, 1.6 micromole of DNA, and 4 units of polymerase fraction VII. Any one, two, or three of the deoxyribonucleoside triphosphates may be used in this reaction. The high specific radioactivity is necessary to increase the sensitivity of measurement, since the incorporation of nucleotides is of the order of micromicromoles. It will be noted that a 25-fold increase in primer concentration over that in the usual incubation mixture is used, since this increases the extent of the limited reaction. Under these conditions, incorporation of nucleotides is complete by 2.5 hours, and the product may be harvested after addition of carrier by precipitation with perchloric acid or by dialysis as described above. Since high levels of radioactivity are used in this procedure, it is well to wash the precipitate an extra time in order to minimize the background radioactivity due to unreacted nucleotides contaminating the product.

Properties. It can be shown that the nucleotides incorporated into acid-insoluble products in this limited reaction add to the nucleoside end of the primer molecules.[13] No preference for DNA chains ending with specific nucleotides is evident from analysis of the products, since each of the deoxyribonucleoside triphosphates can be shown to add to DNA chains ending with any of the four deoxyribonucleotides. For example, if dCTP is used as the labeled substrate, it adds to DNA chains ending in dAMP, dGMP, dTMP, or dCMP. The same is true for the other deoxyribonucleoside triphosphates. The incorporated nucleotides are present in covalent linkage and sediment with the high-molecular-weight fraction in the ultracentrifuge. Since, on the average, one deoxyribonucleotide is incorporated per chain of approximately 20,000 deoxyribonucleotides, it would seem unlikely that the properties of the polymer would be altered to any appreciable extent. Although the physical properties of these products have not been studied in extenso, no unusual properties have been observed.

Synthesis of DNA Containing Base Analogs[14]

Several purine and pyrimidine bases not normally found in calf thymus DNA may be incorporated into DNA when supplied as their respective deoxyribonucleoside triphosphates. The incorporation of uracil and of hypoxanthine has been shown directly with $dUP^{32}PP$ and $dIP^{32}PP$. Incorporation of other bases has been inferred by their ability

[14] M. J. Bessman, I. R. Lehman, J. Adler, S. B. Zimmerman, E. S. Simms, and A. Kornberg, Proc. Natl. Acad. Sci. U.S. 44, 633 (1958).

to support DNA synthesis in the absence of one of the required deoxy-ribonucleoside triphosphates. The evidence suggests that thymine may be replaced by uracil and 5-bromouracil, cytosine by 5-methyl- and 5-bromocytosine, and guanine by hypoxanthine. The reaction mixtures for the synthesis of analog-containing DNA are exactly the same as the reaction mixtures for synthesis of normal DNA except that the analog (as the triphosphate) replaces one of the normal triphosphates.

Properties. The properties of the analog-containing DNA prepara-tions have not been studied extensively. It has been shown that, in the case of uracil-containing DNA, the deoxyuridylate is linked to each of the other three nucleotides and to itself in typical 3′,5′-phosphodiester linkages. It is not known whether any of the analog-containing products can act as primer for the synthesis of more DNA.

ACKNOWLEDGMENT

This report was written during the tenure of a grant (C4088) from the National Institutes of Health, U. S. Public Health Service.

[100] A Procedure for the Isolation of Deoxyribonucleic Acid from Microorganisms

By J. MARMUR

Introduction

A method is described for the isolation of DNA from microorganisms which yields stable, biologically active, highly polymerized preparations relatively free from protein and RNA. Alternative methods of cell dis-ruption and DNA isolation have been described and compared. DNA capable of transforming homologous strains has been used to test various steps in the procedure, and preparations have been obtained possessing high specific activities. Several procedures have described the isolation of DNA from selected groups of microorganisms.[1-3] However, no detailed account is available for the isolation of DNA from a diverse group of

[1] R. D. Hotchkiss, Vol. III [102].
[2] S. Zamenhof, B. Reiner, R. DeGiovanni, and K. Rich, J. Biol. Chem. 219, 165 (1956).
[3] E. Chargaff, in "The Nucleic Acids" (E. Chargaff and J. N. Davidson, eds.), Vol. 1. p. 308. Academic Press, New York, 1955.

microorganisms. The reason for this is that microorganisms vary greatly in the ease with which their cell walls can be disrupted, in their content of capsular polysaccharides (which are difficult to separate from DNA), and in the association of DNA to protein which influences the ease of DNA purification.[4] Most of these difficulties have been overcome in the present procedure which has been applied successfully to approximately two hundred different species of microorganisms. Included in this number are those organisms whose DNA can transform homologous and closely related strains and which have thus provided a very useful tool in determining the efficacy of many of the steps outlined in the procedure.

In general, the method to be described can be outlined as follows: The cells are first disrupted, the cell debris and protein are removed by denaturation and centrifugation, and the RNA is removed by RNase and selective precipitation of the DNA with isopropanol. Degradation by DNase and divalent metal ion contamination is prevented by the presence of chelating agents and by the action of sodium lauryl sulfate. The products obtained, although they may not reflect the *in vivo* molecular weight, have molecular weights in excess of 10×10^6 and a high specific transforming activity, where this is present.

Materials and Methods

Reagents

Saline–EDTA. $0.15\,M$ NaCl plus $0.1\,M$ ethylenediaminetetraacetate (EDTA), pH 8. The EDTA, and/or high pH, inhibit DNase activity.

Sodium lauryl sulfate, 25%. The anionic detergent ($NaC_{12}H_{26}SO_4$) will lyse most nonmetabolizing cells, inhibit enzyme action, and denature some proteins.[5,6]

Lysozyme, crystalline (Armour), used to lyse cells resistant to detergent action. Cells lysed with lysozyme are then subjected to sodium lauryl sulfate as well.

Sodium perchlorate, $5\,M$. The high salt concentration provided by the perchlorate[7] helps to dissociate protein from nucleic acid.

Chloroform–isoamyl alcohol, 24:1 (v/v), used to deproteinize, according to the method of Sevag *et al.*[8] The chloroform causes

[4] K. S. Kirby, *Biochem. J.* **66**, 495 (1957).
[5] M. Bayliss, *J. Lab. Clin. Med.* **22**, 700 (1937).
[6] A. Bolle and E. Kellenberger, *Schweiz. Z. allgem. Pathol. u. Bakteriol.* **21**, 714 (1958).
[7] L. S. Lerman and L. J. Tolmach, *Biochim. et Biophys. Acta* **26**, 68 (1957).
[8] M. G. Sevag, D. B. Lackman, and J. Smolens, *J. Biol. Chem.* **124**, 425 (1938).

surface denaturation of proteins. The isoamyl alcohol reduces foaming, aids the separation, and maintains the stability of the layers of the centrifuged, deproteinized solution.

Ethyl alcohol, 95%, used to precipitate nucleic acids after deproteinization. Denatured ethyl alcohol may also be used.

Saline–citrate, 0.15 M NaCl plus 0.015 M trisodium citrate, pH 7.0 $\pm$ 0.2. Maintains ionic strength of dissolved DNA, and chelates divalent ions.

Dilute saline–citrate, 0.015 M NaCl plus 0.0015 M trisodium citrate. DNA dissolves more readily in dilute salt solutions but should *never* be dissolved in pure water.

Concentrated saline–citrate, 1.5 M NaCl plus 0.15 M trisodium citrite. The concentrated solution is used to bring the dilute saline–citrate solute, in which the nucleic acid is dissolved, up to saline–citrate concentration. The volume added need only be approximate until the final pure product is obtained.

Ribonuclease, 0.2% (crystalline, Armour) in 0.15 M NaCl, pH 5.0. The solution is heated at 80° for 10 minutes to inactivate any contaminating DNase. The RNase digests the RNA and facilitates its separation from DNA.

Acetate–EDTA, 3.0 M sodium acetate plus 0.001 M EDTA, pH 7.0. This provides the proper ionic environment in the isopropanol step for the separation of DNA from RNA or its digestion products (N. Simmons, personal communication).

Isopropanol, used to precipitate DNA selectively; RNA remains in solution. In some cases it will selectively precipitate and separate DNA from polysaccharides, but this is not always the case.

Equipment for DNA Isolation

Centrifuges, Servall SS-1 operating at 5000 to 10,000 r.p.m. (3000 to 13,000 $\times$ g) and a clinical swinging-bucket centrifuge capable of spinning at 2000 to 3000 r.p.m. (300 to 600 $\times$ g).

Glass-stoppered flasks, for deproteinization. A "Vortex" mixer may also be used.

Shaker, wrist action or reciprocal for deproteinization; several hundred strokes per minute.

Volumetric pipet, 10 to 15 ml., fitted with an 18-inch (approximately) rubber tube attached to the upper end, used to remove the aqueous layer from the deproteinized, centrifuged mixture.

Stirring motor, fitted with a glass stirring rod with a screw taper, used to stir the solution (500 to 1000 r.p.m.) during the isopropanol addition. A "Vortex" mixer may also be used.

Physical and Biological Measurements

Determination of T_m. The method has been previously described.[9]

Determination of Sedimentation Coefficient and Molecular Weight. The sedimentation coefficient, $S_{20,w}$, of DNA dissolved in standard saline–citrate is determined in the Spinco ultracentrifuge Model E at a concentration of 20 μg./ml. at a speed of 35,600 r.p.m. by using ultraviolet optics. The centrifuge cell is fitted with a Kel-F centerpiece. The molecular weight of the sample can then be estimated by using the relationship established by Doty *et al.*[10]:

$$S_{20,w} = 0.0063\ M_w^{0.37}$$

Transformation. The transformation of *Diplococcus pneumoniae* is carried out by the method of Fox and Hotchkiss[11] with transformable glycerol-treated cells stored at −20°. When it is necessary to examine the biological properties of *Bacillus subtilis* DNA, the method of Spizizen[12] is used to transform this organism.

Isolation Procedure

The procedure is designed for 2 to 3 g. of wet packed cells. The volumes are only approximate unless otherwise stated. All operations can be performed at room temperature except the RNase treatment, which is carried out at 37°.

Bacteria grown to the logarithmic phase of their growth cycle are harvested by centrifugation and washed once with 50 ml. of saline–EDTA. After being collected by centrifugation, the cells are suspended in a total volume of 30 to 35 ml. of saline–EDTA. Lysis[a,b,c,12a] is effected by the addition of 2.0 ml. of sodium lauryl sulfate, and the mixture is placed in a 60°[d] water bath for 10 minutes and then cooled to room temperature. Lysis of the culture results in a dramatic increase in viscosity[e] accompanying the release of the nucleic acid components and some clearing. If, on preliminary testing, the cells are insensitive to the detergent but sensitive to lysozyme, approximately 10 mg. of lysozyme are added to the cells suspended in saline–EDTA. The mixture is then incubated at 37° with occasional shaking, and the lysis is followed by noting the increase in viscosity. In some cases 30 to 60 minutes may be required for optimum results. When lysozyme is used, sodium lauryl sulfate is

[9] J. Marmur and P. Doty, *J. Mol. Biol.* **5**, 109 (1962).
[10] P. Doty, B. B. McGill, and S. A. Rice, *Proc. Natl. Acad. Sci. U.S.* **44**, 432 (1958).
[11] M. S. Fox and R. D. Hotchkiss, *Nature* **179**, 1322 (1957).
[12] J. Spizizen, *Federation Proc.* **18**, 957 (1959).
[12a] Italic letters refer to the paragraphs in the following section on Procedure notes.

added as well, *after* the cells have lysed, followed by the 60° exposure and cooling.

Perchlorate is added to a final concentration of $1 M$ to the viscous, lysed suspension, and the whole mixture is shaken with an equal volume of chloroform–isoamyl alcohol in a ground-glass-stoppered flask for 20 minutes.[f] The resulting emulsion[g] is separated into three layers by a 5-minute centrifugation at 5000 to 10,000 r.p.m. in the Servall. The upper aqueous phase contains the nucleic acids and is carefully pipetted off into a tube or narrow flask. The nucleic acids are precipitated by gently layering approximately 2 vol. of ethyl alcohol on the aqueous phase. When these layers are gently mixed with a stirring rod, the nucleic acids "spool" on the rod as a threadlike precipitate[h] and are easily removed. The precipitate is drained free of excess alcohol by pressing the spooled rod against the vessel. The precipitate is then transferred to approximately 10 to 15 ml. of dilute saline–citrate[i] and gently removed from the stirring rod by swirling it back and forth. The solution is gently shaken or pipetted until dispersion is complete (lumps can be recognized by adhering air bubbles when the solution is shaken). The solution is adjusted approximately to standard saline–citrate concentration by adding concentrated saline–citrate, shaken as before with an equal volume of chloroform–isoamyl alcohol for 15 minutes, centrifuged,[j] and the supernatant removed. It is then deproteinized repeatedly with chloroform–isoamyl alcohol,[k] as described, until very little protein is seen at the interface.

The supernatant obtained after the last in the series of deproteinizations is precipitated with ethyl alcohol and dispersed in saline–citrate (about 0.5 to 0.75 the supernatant volume) in the manner already described. Ribonuclease[l] is added to a final concentration of 50 μg./ml., and the mixture is incubated for 30 minutes at 37°. After the digestion of the RNA, it becomes possible to remove protein which has resisted earlier chloroform deproteinizations. The digest is again subjected to a series of deproteinizations until there is little or no denatured protein visible at the interface after centrifugation. The supernatant, after the last such treatment, is again precipitated with ethyl alcohol, and the drained nucleic acid is dissolved in 9.0 ml. of dilute saline–citrate. When solution has occurred, 1.0 ml. of acetate–EDTA is added, and, while the solution is rapidly stirred, 0.54 vol.[m] of isopropyl alcohol is added dropwise into the vortex. The DNA usually precipitates in a fibrous form after first going through a gel phase at about 0.5 vol. of isopropyl alcohol. RNA or oligoribonucleotides and cellular or capsular polysaccharides remain behind, while the DNA threads precipitate out of solution. If the yield is good, the DNA is redissolved and precipitated once more with isopropanol in the manner described. The final precipitate is washed free of acetate and salt by gently stirring the adhered precipitate in 70% ethyl alcohol and is then dissolved (immediately after precipitation) in the solvent of choice. If the solution is not clear, it can be clarified by centrifugation in the Servall centrifuge for 10 minutes at 5000 r.p.m.[n]

By using caution and recovery steps, up to 50% of the DNA from the cell is obtained; in general 1 to 2 mg. of DNA is obtained from 1 g. of wet packed cells.[o] The DNA can be stored in solution at 5° in the presence of several drops of chloroform.[p, q, r] If it is to be stored for over a month, the DNA solution should either be frozen or precipitated with ethyl alcohol. If it is so desired, the DNA (free of spores) can be sterilized by exposure to 75% ethyl alcohol for several hours and then transferred to a sterile solvent.

Notes on the Procedure for the Isolation of DNA

It would be very difficult to describe a definitive technique for the efficient isolation of DNA from a wide variety of microorganisms. The method described can undoubtedly be modified or improved to eliminate difficulties encountered with specific strains. In general, the gram-negative organisms yield themselves readily to the procedure, resulting in good recoveries of highly polymerized DNA. Several suggestions are offered to improve yields and to eliminate some of the difficulties that may arise.

a. A spot test on a centrifuged portion of the culture should be made to determine whether the organisms are susceptible to sodium lauryl sulfate, lysozyme, or neither of the two. If lysed by both detergent and enzyme, the former is preferable, since DNase is inactivated in its presence. If lysozyme is used, the detergent is added *after* maximum enzymatic lysis is attained.

b. Organisms readily lysed by sodium lauryl sulfate include all gram-negative strains thus far encountered as well as *D. pneumoniae* (which also lyses readily with deoxycholate), *Bacillus stearothermophilus, B. macerans, B. licheniformis, Clostridium madisoni, Cl. chauvei, Cl. butylicum,* and *Mycoplasma* (PPLO). *Euglena gracilis, Chlamydomonas reinhardii,* and the slime mold *Dictyostelium discoideum* are readily lysed by the detergent. Some strains of *Streptococcus* and of *Staphylococcus aureus* lyse slowly with sodium lauryl sulfate, and cell disruption is sometimes facilitated by raising the temperature of the lysing mixtures to between 70° and 75°. The former genus can also be lysed by extracts from *Streptomyces albus.*[13]

The organisms lysed with lysozyme include *B. subtilis, B. natto, B. cereus, B. megaterium,* and *Cl. perfringens,* as well as the Actinomycetes

[13] M. McCarty, *J. Exptl. Med.* **96**, 555 (1952).

(*S. albus* and *S. viridochromogenes*), which lyse very slowly with the enzyme,[14] depending on the state of growth when harvested.

It has been found that some strains of *Streptococcus* and of *Staphylococcus aureus*, *Lactobacillus acidophilus*, and baker's yeast are insensitive to lysis by either lysozyme or detergent. Yeast cells can be lysed by an extract from the snail *Helix pomatia*.[15] If no means of enzyme or detergent lysis is available, the cells can be disrupted by grinding with alumina or glass powder (see below). The product, however, usually has a lower molecular weight than that obtained by the other methods described. (For other means of cell wall disruption, see review by Weibull.[16])

We have also used mercaptoethanol (ME) for isolation of DNA from *Bacillus* spores, *B. brevis*, *B. cereus*, and yeast. Harvested saline-EDTA-washed cells are suspended in a small volume of saline-EDTA and an equal volume of 14 M ME added. The mixture is incubated for 30 min. at 37°, the cells freed of ME by dialysis or centrifugation and then treated with lysozyme, detergent, frozen and thawed several times and the procedure continued as outlined.

c. It has been noted that some cells grown in the presence of 5-bromo-deoxyuridine have altered their susceptibility to lysis. Thus, *D. pneumoniae* grown in the presence of this analog will lyse slowly if at all with sodium lauryl sulfate but remains susceptible to deoxycholate. This situation has not arisen in the case of *E. coli*.

d. If potent nuclease action is anticipated (e.g., *Serratia marcescens*), the detergent-lysed suspension should be heated to within 10° of the DNA T_m to eliminate its activity. For the isolation of protozoan DNA, the heating step should be omitted. Immediately after lysis, the protozoan lysate should be treated with phenol (see notes *l* and *q*). Experience with several protozoans indicates a potent DNase which resists inactivation by $CHCl_3$ and detergent as well as the exposure to heat (M. Mandel, personal communication).

e. Some organisms (e.g., *Klebsiella pneumoniae*) are difficult to harvest and when lysed give rise to extremely viscous solutions. This is due to polysaccharide which is usually eliminated in the final stages of the DNA isolation but may also be removed earlier by initial use of the isopropanol step.

f. In many instances the addition of perchlorate may be omitted or replaced by the same concentration of NaCl.

g. A very critical step in the procedure is the concentration of cells

[14] A. Sohler, A. H. Romano, and W. J. Nickerson, *J. Bacteriol.* **75**, 283 (1958).
[15] A. A. Eddy and D. H. Williamson, *Nature* **179**, 1252 (1957).
[16] C. Weibull, *Ann. Rev. Microbiol.* **12**, 1 (1958).

being lysed. Too low a cell concentration will give rise to losses in subsequent alcohol precipitation, whereas, if the cell suspension is too thick, the first chloroform deproteinization will result in a "lumpy" emulsion. The lumps consist of denatured protein with large amounts of occluded DNA and form a voluminous middle layer when the mixture is centrifuged. This difficulty is easily remedied in subsequent preparations by lysing a more dilute suspension of cells. In order to recover DNA from the denatured protein layer, because of excessive occlusion or when a high yield of DNA is desired under normal circumstances, the residue is shaken with a 10-ml. portion of dilute saline–citrate for 15 minutes and centrifuged; the precipitate obtained from the ethyl alcohol addition is combined with the remainder of the preparation.

h. If difficulty is encountered in collecting a majority of the nucleic acid threads after ethyl alcohol addition, or if no threads appear because of DNA degradation or if its concentration is low, the solution is subjected to a short centrifugation, the sediment is dissolved in a small volume of dilute saline–citrate, and the alcohol precipitate is added to the remaining portion of the preparation. In this case, it is best to treat the suspension with RNase at this time and continue to deproteinize.

i. When dissolving the fibrous nucleic acid precipitate it is well to keep the concentration of DNA at a level of about 0.2 to 0.8 mg./ml. Too low a concentration results in degradation[17] and loss of biological activity during handling. Higher concentrations are highly viscous and difficult to handle and disperse. In the early stages of the preparation, the redissolved precipitate gives rise to turbid suspensions; as the purification proceeds, the nucleic acid takes on a glassy appearance when being dispersed and results in clear solutions.

j. If the emulsion separates readily into two phases after shaking, centrifugation in a swinging-bucket clinical centrifuge at 2000 to 3000 r.p.m. is sufficient to clear the aqueous layer of chloroform and most of the denatured protein.

k. The best time to interrupt the procedure, if it should be necessary, is after any of the deproteinization steps. In this case the uncentrifuged emulsion should be stored until the procedure can be resumed. The average length of time required to isolate purified DNA, starting from the lysis of the cells, is approximately 5 to 8 hours.

l. It is also possible to isolate RNA by omitting the RNase treatment, recovering the ethyl alcohol precipitates by centrifugation, and saving the solution after the DNA is removed by isopropanol step. It would, however, be best to carry out the isopropanol step earlier in the procedure and to use sodium dodecyl sulfate for lysis. The ribonuclease step should

[17] A. D. Hershey and E. Burgi, *J. Mol. Biol.* **2**, 143 (1960).

be omitted if the DNA is to be used in *in vitro* studies as a primer for RNA synthesis; alternatively, the enzyme can be inactivated by treating with warm (45°), buffer-saturated (pH 8–9) phenol.

m. Several cases have been encountered where the isopropanol step does not precipitate the DNA when 0.54 vol. has been added (*C. reinhardii*). When this occurs, larger volumes of isopropanol should be added. At times the DNA precipitate is granular and can be collected by centrifugation. The isopropanol step should be omitted if the yield of DNA fibers in the last ethyl alcohol step is low.

n. It has been repeatedly observed that the purified DNA isolated from some spore formers (e.g., *B. brevis, B. subtilis*) is contaminated with viable spores. The number of spores can be reduced or eliminated by harvesting the cells when spore formation is at a minimum, centrifuging the purified DNA for about 10 minutes at 10,000 r.p.m. in the Servall, and/or treatment of the DNA with phenol (see notes *l* and *q*).

o. The method can be applied to bacterial harvests of the order of 100 g., but, to avoid the awkwardness of handling large volumes, the amount of reagents used can be scaled up less than proportionately.

p. If protein removal is incomplete, storage of the DNA solution in the presence of chloroform will sometimes leave a halo of denatured protein surrounding the interface of the solutions. The protein can be removed by centrifugation.

q. The introduction of the phenol method[4,18] to isolate nucleic acids prompted its use as a protein denaturant in some preliminary experiments. After cell lysis, omitting the addition of perchlorate, an equal volume of distilled phenol (saturated with buffer, pH 8–9) is added, and the mixture is shaken gently for 10 minutes and centrifuged. The DNA in the supernatant is precipitated with ethyl alcohol, dissolved as described, RNased, and the chloroform-isoamyl treatment is continued as above (following procedure note "*l*").

r. The amino acid content (other than glycine) of the purified DNA on acid hydrolysis is approximately 0.3 to 0.5% (determined by Dr. H. Van Vunakis on the Spackman *et al.*[19] automatic amino acid analyzer). Typical ratios for absorption of the DNA at 260:230:280 mμ are 1.0:0.450:0.515.

Alternative Methods of Cell Disruption and Lyophilization

Mechanical Disruption. Various techniques have been described[20] for the mechanical disruption of cells for the isolation of enzymes, nucleic

[18] K. S. Kirby, *Biochim. et Biophys. Acta* **36**, 117 (1959).
[19] D. H. Spackman, W. H. Stein, and S. Moore, *Anal. Chem.* **30**, 1190 (1958).
[20] I. C. Gunsalus, Vol. I [7].

acids, etc. An attempt was made to disrupt *D. pneumoniae* cells by grinding with glass powder as well as by sonic treatment and to isolate the DNA from the disrupted cells according to the method described above. Since these and other methods might be applied to cells that resist enzyme and detergent action, we have applied them to *D. pneumoniae* as a basis of comparison (using transforming activity) to the detergent method of cell lysis.

Grinding with glass or alumina (particle size 500 mesh) is carried out at 5°. The harvested, washed cells (with saline EDTA) are placed in a precooled mortar, an equal weight of glass powder (Fisher Scientific Co.) is added, and the mixture is ground with a pestle for 5 to 10 minutes. Ten volumes of cold saline–EDTA, containing 2% sodium lauryl sulfate, are added, and the suspension is centrifuged to remove glass and large cell debris. The supernatant is then treated in the manner described above for the isolation of DNA.

Sonic treatment is also carried out in the cold at 5°. Cells washed with saline–EDTA are suspended in 10 vol. of saline–EDTA and placed in the cup of the sonic apparatus (Raytheon). The suspension is exposed to 9-kc. (50-watt) sound waves for 5 to 30 minutes. Sodium lauryl sulfate is then added, and the suspension is centrifuged to remove any cell debris and then subjected to the isolation procedure. If the DNA does not precipitate as threads on ethyl alcohol addition, the nucleic acids are collected by centrifugation.

Lyophilization of Cells. Lyophilization is a commonly used laboratory technique for the preservation of cells. It was thought of interest to examine the DNA isolated from *D. pneumoniae* which had been subjected to freeze-drying. The method employed for the isolation of DNA from the lyophilized cells is the same as that described with sodium lauryl sulfate as the lysing agent.

Isolation of DNA by Cesium Chloride Density Gradient Centrifugation

The introduction of the cesium chloride density gradient technique by Meselson *et al.*[21] has made it possible to separate RNA, DNA, and protein under mild conditions. By selecting the proper density and conditions of centrifugation, the RNA collects at the bottom of the centrifuge tube, the protein floats on the top, and the DNA bands at or near the center. Biologically active DNA can readily be located by assaying the transforming activity of various fractions. The technique can also be adapted to separate DNA samples differing in base composition (and thus in buoyant density).

---

[21] M. Meselson, F. W. Stahl, and J. Vinograd, *Proc. Natl. Acad. Sci. U.S.* **42**, 581 (1957).

As applied to the isolation of DNA from cell lysates, the following outline has been used. A concentrated lysed cell suspension is mixed with a concentrated solution or solid CsCl containing $0.005\,M$ Tris buffer (2-amino-2-hydroxymethylpropane-1,3-diol) plus $0.005\,M$ EDTA, pH 8.0. After the density has been adjusted to 1.706 ± 0.002 g./ml., the mixture is centrifuged in Lusteroid tubes in the Model L Spinco preparative centrifuge at 35,000 r.p.m. in the SW-39 rotor at room temperature for 3 days. When the rotor head has coasted (unbraked) to a stop, the tubes are removed and secured firmly in a vertical position, and a small hole is bored in the bottom of the tube with a small-gage needle. The collected fractions are then monitored either by their absorbance, radioactivity, biological activity, etc.

Results

A representative group of DNA samples isolated from bacteria by either lysozyme or detergent cell rupture is listed in Table I. The sedi-

TABLE I

PROPERTIES OF REPRESENTATIVE DNA SAMPLES ISOLATED FROM CELLS
SENSITIVE TO LYSOZYME AND/OR SODIUM LAURYL SULFATE[a]

Organism	$S_{20,w}$	T_m
Aerobacter aerogenes	22.4	94.0
Bacillus brevis	25.3	87.1
B. cereus	24.2	82.7
B. megaterium	29.0	85.2
B. subtilis	26.0	87.4
Diplococcus pneumoniae	24.0	85.2
Escherichia coli (K12)	29.0	90.0
Hemophilus influenzae	28.5	85.5
Klebsiella pneumoniae	25.6	92.0
Micrococcus lysodeikticus	24.0	98.6
Staphylococcus aureus	28.0	83.2
Pseudomonas aeruginosa	28.2	96.7
Salmonella typhimurium	24.3	91.1
Serratia marcescens	23.7	93.6
Shigella dysenteriae	22.7	90.0
Streptococcus salivarius	24.4	85.2
Streptomyces albus	25.0	100.1
Calf thymus	22.1	86.0
Salmon sperm	23.2	87.0

[a] The sedimentation coefficient and the T_m were determined by using saline–citrate ($0.15\,M$ NaCl plus $0.015\,M$ Na citrate) as the solvent. The T_m for S. albus was obtained at a lower ionic strength and corrected to that for saline–citrate solvent. The sedimentation coefficients, determined immediately or within a day after the DNA preparation, did not alter on prolonged storage.

mentation coefficient (at 20 μg./ml.) and T_m values have been determined and recorded. For careful comparisons, the DNA samples should be dialyzed against the same solvent. (The sedimentation coefficient was the same regardless of whether the centrifuge cells—regular or synthetic boundary—were filled slowly with a syringe or a wide-tipped pipet.) The purified DNA samples have approximately the same sedimentation coefficients with calculated molecular weights of the order of 8 to 12 million. Calf thymus and salmon sperm DNA's, isolated and supplied by Dr. N. Simmons, are also listed for comparison. In determining the T_m values, the temperature–absorbance curves showed that all the DNA samples isolated were predominantly in the native configuration. Variations of the sedimentation coefficient and the T_m of different preparations of DNA from the same organism were of the order of $\pm 1\,S$ and $\pm 0.5°$, respectively.

The effects of mechanical disruption and lyophilization of cells on the transforming activity and molecular weight of *D. pneumoniae* DNA are compared to detergent lysis in Table II. Cells subjected to freeze-

TABLE II

EFFECT OF METHOD OF CELL DISRUPTION AND LYOPHILIZATION ON THE
TRANSFORMING ACTIVITY AND MOLECULAR WEIGHT OF *Diplococcus pneumoniae* DNA

Method of cell disruption	Relative transforming activity	Molecular weight, $\times 10^{-6}$ [a]
SLS[b]	100%[c]	9.6
Grind with glass	36	5.0
Sonic, 5 minutes	4.1	1.0
Sonic, 30 minutes	0.42	0.38
Lyophilize, SLS	108	10.5

[a] Estimated from the sedimentation coefficient.
[b] SLS = sodium lauryl sulfate.
[c] 100% represents 3.6×10^6 transformants to streptomycin resistance per microgram of transforming factor.

drying yield DNA with no detectable differences from that isolated from freshly harvested cells. However, as might be expected, sonic treatment[22] has a deleterious effect on DNA. Grinding with glass is less destructive than sonic treatment and would thus be the preferred method of mechanical disruption. Other methods applying a shear force[17, 23, 24] would probably result in degradation of the DNA.

[22] M. Litt, J. Marmur, H. Ephrussi-Taylor, and P. Doty, *Proc. Natl. Acad. Sci. U.S.* **44**, 144 (1958).
[23] P. F. Davison, *Proc. Natl. Acad. Sci. U.S.* **45**, 1560 (1959).
[24] L. F. Cavalieri, and B. H. Rosenberg, *J. Am. Chem. Soc.* **81**, 5136 (1959).

Discussion

The method described for the isolation of DNA has anticipated most of the difficulties that may be encountered with various microorganisms. In general, the method described has yielded DNA from a variety of microorganisms which is native, is highly polymerized, and possesses a fairly uniform molecular-weight distribution. It might be argued that the method used for the isolation of DNA results in a degradation of the molecules and thus does not represent the true *in vivo* value. Degradation could have taken place during the shaking[17] used for deproteinization. The initial lysate is much more viscous than the final product; if the viscosity is due to a continuous DNA structure without protein links, then degradation has most likely taken place. The genetic evidence does indicate that the genome of *E. coli* behaves as one linkage group[25]; however, definitive evidence is still lacking as to whether this "chromosome" consists of large subunits or molecules.

The present method yields DNA from microorganisms with a molecular weight of about 8 to 12×10^6, which is adequate for most experimental purposes. Those DNA's that possess transforming activity (*D. pneumoniae, B. subtilis,* etc.) can transform their homologous strains with an efficiency of 1 to 5% at saturating levels of DNA.

DNA isolated by selective buoyancy in a CsCl density gradient has physical, chemical, and biological properties similar to the product obtained by the method described, which involves selective denaturation with chloroform and alcohol precipitations. However, unless great care is taken, the DNA is still subjected to mild shear forces during the CsCl method, and it is again possible that degradation may have taken place.

The methods of DNA isolation described in the literature, and here,[26] are continually being improved, and undoubtedly agents will be uncovered that will selectively precipitate or fractionate DNA under milder conditions than heretofore employed. The method presented in this report has been useful in providing samples of DNA from microorganisms with widely varying base ratios and has proved of great value in studying their physicochemical and genetic properties.

Acknowledgment

This report was written during the tenure of grants from the NIH, NSF, and NASA.

[25] F. Jacob and E. L. Wollman, *Symposia Soc. Exptl. Biol.* No. 12, 75 (1958).
[26] J. Marmur, *J. Mol. Biol.* 3, 208 (1961).

[101] Determination of Frequencies of Nearest-Neighbor Base Sequences in Deoxyribonucleic Acid

By JOHN JOSSE and MORTON SWARTZ

I. Introduction

At the present time only limited information is available about the sequential arrangement of bases in the polynucleotide chains of DNA. Two general lines of approach to this problem have been employed. The first of these is systematic analysis of the fragments resulting from either partial enzymatic or controlled chemical degradation of the DNA. The nature of the products of enzymatic degradation of DNA is a function of the specificity of the enzyme employed as well as of the structure of the DNA being digested.[1-3] Chemical degradation studies of DNA, carried out in the presence of acid, have yielded information on the relative frequencies of solitary pyrimidine nucleotides flanked on both sides by purine nucleotides and on the frequencies of short runs of pyrimidine nucleotides in the DNA chains.[4, 5] The second line of approach, known as nearest-neighbor analysis and to be described in this section, yields additional information and allows one to estimate the relative frequency with which a given nucleotide (purine or pyrimidine) is adjacent to any other nucleotide in the polynucleotide chains.[6, 7]

II. General Principles of Procedure

The four commonly occurring bases of DNA can align in $(4)^2$ or sixteen possible nearest-neighbor sequences. The relative frequencies of these sixteen possible sequences in a given DNA sample are determined by using the sample as "primer" in the enzymatic synthesis of DNA from the four deoxyribonucleoside triphosphates in the presence of DNA polymerase. The method clearly depends on the synthesis by DNA polymerase of a faithful copy of the nucleotide sequence of the primer

[1] R. L. Sinsheimer, *J. Biol. Chem.* **208**, 445 (1954); **215**, 579 (1955).
[2] M. P. de Garilhe, L. Cunningham, U. Laurila, and M. Laskowski, *J. Biol. Chem.* **224**, 751 (1957); J. L. Potter and M. Laskowski, *ibid.* **234**, 1263 (1959).
[3] G. W. Rushizky, C. A. Knight, W. K. Roberts, and C. A. Dekker, *Biochem. Biophys. Research Communs.* **2**, 153 (1960).
[4] H. S. Shapiro and E. Chargaff, *Biochim. et Biophys. Acta* **26**, 608 (1957); **39**, 62 (1960).
[5] K. Burton and G. B. Peterson, *Biochem. J.* **75**, 17 (1960).
[6] J. Josse, A. D. Kaiser, and A. Kornberg, *J. Biol. Chem.* **236**, 864 (1961).
[7] M. Swartz, T. Trautner, and A. Kornberg, *J. Biol. Chem.* **237**, 1961 (1962).

DNA, which is considered to act as a template.[8] Incorporation of P^{32}-labeled 5'-nucleotide substrates into the synthetic DNA is followed by quantitative degradation of this DNA into 3'-nucleotides, which are then isolated and analyzed for their P^{32} content.

Synthesis of P^{32}-Labeled DNA's. The primer DNA is reacted enzymatically with labeled substrates in four parallel reactions according to the following pattern:

Reaction 1: dATP^{32}, dTTP, dGTP, dCTP
Reaction 2: dATP, dTTP^{32}, dGTP, dCTP
Reaction 3: dATP, dTTP, dGTP^{32}, dCTP
Reaction 4: dATP, dTTP, dGTP, dCTP^{32}

As shown in Fig. 1, the P^{32}, which is attached to carbon 5 of the deoxyribose of the reacting nucleoside triphosphate (Y), becomes esterified with carbon 3 of the deoxyribose of the nucleotide (X) at the growing

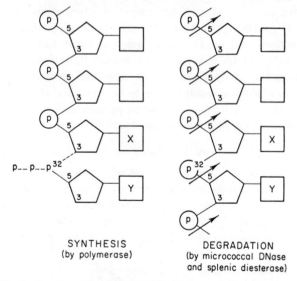

SYNTHESIS
(by polymerase)

DEGRADATION
(by micrococcal DNase
and splenic diesterase)

FIG. 1. Synthesis of a P^{32}-labeled DNA chain and its subsequent enzymatic degradation to 3'-deoxyribonucleotides. The arrows indicate the linkages cleaved by micrococcal DNase and calf spleen phosphodiesterase, yielding a digest composed exclusively of 3'-deoxyribonucleotides. Reproduced with the permission of the *Journal of Biological Chemistry.*

end of the chain. The extent of the DNA synthesis generally represents a 20% increment over the amount of primer DNA added, although it is not essential that the reactions in each of the four mixtures proceed to

[8] A. Kornberg, *Science* 131, 1503 (1960).

the same extent; the nearest-neighbor pattern does not vary when the net increase in DNA varies, for example, from 10 to 30%.

Degradation of P^{32}-Labeled DNA's. The DNA, isolated and thoroughly washed free of unreacted substrates, is hydrolyzed by the consecutive actions of micrococcal DNase and calf spleen phosphodiesterase. These enzymes cleave the bonds between phosphate and carbon 5 of deoxyribose, producing as products 3'-mononucleotides. Thus, in effect, the P^{32} is "transferred" from the deoxyribose carbon 5 of the reacting triphosphate substrate to the deoxyribose carbon 3 of that nucleotide in the chain with which the labeled triphosphate substrate reacted, i.e., its nearest-neighbor nucleotide (Fig. 1). By determining the relative P^{32} content of each of the four 3'-mononucleotides isolated from each reaction digest, the frequency with which any nucleotide is linked to any other in the newly synthesized chains can be ascertained.

III. Detailed Procedure

Substrates and Enzymes

1. Unlabeled deoxyribonucleoside triphosphates synthesized by the method of Smith and Khorana.[9]

2. P^{32}-Labeled deoxyribonucleoside triphosphates, containing P^{32} in the phosphate esterified to the sugar, prepared from P^{32}-labeled *Escherichia coli* DNA.[10]

3. *Escherichia coli* DNA polymerase, prepared from fraction VII by refractionation on DEAE-cellulose.[10] The specific activity was 500 units/mg. of protein.

4. DNA primers isolated and purified by a variety of procedures,[6, 7] with molar absorbancies (based on deoxypentose) ranging between 6.3 and 7.5×10^3. It is important that the DNA preparations be free of nuclease activity and that during their isolation they are not discernibly exposed to DNases, since this may introduce distortion of the nearest-neighbor pattern.

5. Micrococcal DNase, with specific activity of 7500 units/mg. of protein, prepared according to Cunningham *et al.*[11]

6. Calf spleen phosphodiesterase, with specific activity of 35 units/mg. of protein, isolated according to the procedure of Hilmoe.[12] It is

[9] M. Smith and H. G. Khorana, *J. Am. Chem. Soc.* **80**, 1141 (1958).
[10] I. R. Lehman, M. J. Bessman, E. S. Simms, and A. Kornberg, *J. Biol. Chem.* **233**, 163 (1958).
[11] L. Cunningham, B. W. Catlin, and M. P. de Garilhe, *J. Am. Chem. Soc.* **78**, 4642 (1956). For the assay of micrococcal DNase see footnote 2 of Josse *et al.*[6]
[12] R. J. Hilmoe, *J. Biol. Chem.* **235**, 2117 (1960).

essential that both this enzyme and the micrococcal DNase be entirely free of phosphomonoesterase activity.

7. Crude human semen phosphomonoesterase prepared according to Wittenberg and Kornberg.[13] This preparation had a specific activity of 1200 units/mg. of protein and was free of diesterase activity. Alternatively, chromatographically purified *Escherichia coli* alkaline phosphatase may be used[14]; this should be assayed to verify the absence of contaminating diesterase activity.

1. Enzymatic Synthesis of P³²-Labeled DNA

Each incubation mixture (0.3 ml.) has the following composition: 20 micromoles of glycine buffer (pH 9.2), 2 micromoles of $MgCl_2$, 0.3 micromole of 2-mercaptoethanol, 20 millimicromoles of a DNA primer (expressed as nucleotide phosphorus equivalents), 5 millimicromoles each of dATP, dTTP, dGTP, and dCTP, only one of which is labeled with P³² (specific activity of 0.5 to 1.0×10^8 c.p.m./micromole), and DNA polymerase. The amount of enzyme employed should be sufficient to incorporate approximately 1 millimicromole of the labeled nucleotide into DNA in 30 minutes. The amount of enzyme required varies with the different DNA primers used and is determined by preliminary assays with each type of primer; 0.25 to 2.5 units of enzyme are usually required. For each primer, four such incubation mixtures are prepared, each containing the same amount of enzyme but a different labeled triphosphate, as described under General Principles of Procedure. Each incubation is carried out at 37° for 30 minutes.

2. Isolation of Enzymatically Synthesized DNA

At the end of the incubation, each mixture is chilled and separately treated as follows: All steps in the isolation are carried out at 2°. Calf thymus DNA (0.2 ml. of a solution containing 5 micromoles of nucleotide phosphorus per milliliter) is added as "carrier," followed immediately by the addition of 0.5 ml. of 7% perchloric acid. After 5 minutes for complete precipitation of DNA, 2.5 ml. of cold water are added, and the precipitate is dispersed. The precipitate is then collected by centrifugation, dissolved in 0.3 ml. of 0.2 N NaOH, and reprecipitated with 0.4 ml. of 7% perchloric acid. After 5 minutes, 3 ml. of water are added, and the suspension is stirred and centrifuged. The precipitate is treated again in the same fashion, then collected and dispersed in 0.4 ml. of water, and

[13] J. Wittenberg and A. Kornberg, *J. Biol. Chem.* **202**, 431 (1953). For the assay of phosphomonoesterase activity see footnote 3 of Josse *et al.*⁶

[14] A commercial preparation available from the Worthington Biochemical Corporation has proved satisfactory.

dissolved by the addition of 0.1 N NaOH. The pH is then adjusted to 8.6, and the volume to 0.5 ml. With such repeated reprecipitations, less than 0.1% of the unreacted triphosphates remains in the final precipitate.

3. Enzymatic Digestion of DNA to 3'-Nucleotides

A. Micrococcal Nuclease Digestion. To each of the four parallel DNA solutions at pH 8.6 are added 2 micromoles of Tris buffer (pH 8.6), 1 micromole of $CaCl_2$, and 180 units of micrococcal DNase, in a final volume of 0.54 ml. The incubation is carried out at 37° for 2 hours, at the end of which time all the DNA will have been converted to acid-soluble fragments.

B. Calf Spleen Phosphodiesterase Digestion. After the pH of each digest has been reduced to 7.0 with 0.1 N HCl, 0.2 unit of calf spleen phosphodiesterase is added, and the mixture is incubated at 37°. At the end of 1 hour and again at the end of 2 hours, a like amount of enzyme is added. The total incubation time is 3 hours. At the end of this time, aliquots should be removed from each of the mixtures and assayed for completeness of digestion, as judged by the fraction of P^{32} which has become susceptible to the action of a phosphomonoesterase. From 95 to 100% of the radioactivity should have been converted to a phosphomono-esterase-sensitive form which in this case represents 3'-mononucleotides. If less than 95% has been converted to the phosphomonoesterase-sensitive form, the digestion has been inadequate, and additional treatments with calf spleen phosphodiesterase are indicated.

Assay of Completeness of Digestion

a. EMPLOYING HUMAN SEMEN PHOSPHOMONOESTERASE. Two 0.03-ml. aliquots are removed from each digest and added to separate incubation mixtures containing 50 micromoles of sodium acetate buffer (pH 5.0) and 3 micromoles of $MgCl_2$ in a final volume of 0.2 ml. To one of the tubes is added 70 units of human semen phosphomonoesterase. After 20 minutes of incubation at 37°, the mixtures are chilled, and the following additions are made to each tube: 2.25 ml. of cold water, 0.1 ml. of N HCl, 0.20 ml. of Norit-adsorbing solution,[15] and 0.15 ml. of a Norit suspension (20% packed volume). The tubes are mixed and incubated for 5 minutes at 0°. The Norit precipitates are collected by centrifugation, washed three times with 3-ml. portions of 0.005 M potassium phosphate buffer (pH 2.0), and then suspended in 0.5 ml. of 50% ethanol containing 0.3 ml. of concentrated ammonium hydroxide per 100 ml. The entire Norit suspension is then plated and counted.

[15] Bovine serum albumin (5 mg./ml.) in 0.025 M potassium phosphate buffer, pH 2.

The radioactivity adsorbed to Norit in the tube which contained phosphomonoesterase should be less than 5% that adsorbed to Norit in the enzyme-free tube.

b. EMPLOYING CHROMATOGRAPHICALLY PURIFIED ALKALINE PHOSPHATASE OF *E. coli.* The 0.03-ml. aliquots from each digest are added to two incubation mixtures each containing 30 micromoles of Tris buffer (pH 8.0) and 5 micromoles of $MgCl_2$ in a final volume of 0.3 ml. To one of the tubes is added 0.01 mg. of *E. coli* alkaline phosphatase. After a 20-minute incubation at 37°, the mixtures are handled in exactly the same fashion as described above for the semen phosphomonoesterase assay.

4. Separation of 3'-Nucleotides

Each digest is now centrifuged to remove any protein or other precipitate that may have come out of solution. (If a precipitate is present, it should contain no radioactivity.) The supernatant is pipetted off and then taken to dryness under a fine air stream at room temperature; the residue is dissolved in 0.08 ml. of distilled water. Each preparation is then subjected to paper electrophoresis by the technique of Markham and Smith[16] (Whatman No. 3 MM paper, 0.05 M ammonium formate buffer at pH 3.5, 1200 volts, 25° to 35°, 2 to 2½ hours). The four 3'-nucleotide bands are identified under an ultraviolet lamp, cut out, and eluted overnight at room temperature into 3 ml. of 0.01 N HCl. Strips of paper between the sharply resolved nucleotide bands and in the forward area where inorganic phosphate travels should also be cut out and eluted.

5. Determination of P^{32} Content of 3'-Nucleotides

The eluates (1.0-ml. aliquots) are assayed for radioactivity in any appropriate counter. It is important that essentially all the radioactivity (>95%) applied to the paper be recovered in the nucleotide bands. Little or no radioactivity (<2%) should be found in the strips between the nucleotide bands or in the inorganic phosphate band (which travels ahead of thymidylate), confirming that digestion of the DNA has been complete and that there has been no hydrolysis of the mononucleotides.

IV. Calculation of Relative Frequencies of Nearest-Neighbor Nucleotide Sequences

The results of an experiment with the DNA from *Mycobacterium phlei* as primer will be used as an illustration (Table I). The P^{32} content of each of the four 3'-mononucleotides in a given reaction digest is con-

[16] R. Markham and J. D. Smith, *Biochem. J.* **52**, 552 (1952).

TABLE I[a]

RADIOACTIVITY MEASUREMENTS IN EXPERIMENT WITH *M. phlei* AS PRIMER

Isolated 3'-deoxyribonucleotide	Labeled triphosphate											
	Reaction 1, dATP^{32}			Reaction 2, dTTP^{32}			Reaction 3, dGTP^{32}			Reaction 4, dCTP^{32}		
	Sequence	C.p.m.	Fraction	Sequence	C.p.m.	Fraction	Sequence	C.p.m.	Fraction	Sequence	C.p.m.	Fraction
Tp	TpA	873	0.075	TpT	1,665	0.157	TpG	3,490	0.187	TpC	4,130	0.182
Ap	ApA	1,710	0.146	ApT	2,065	0.194	ApG	2,500	0.134	ApC	4,300	0.189
Cp	CpA	4,430	0.378	CpT	2,980	0.279	CpG	7,730	0.414	CpC	6,070	0.268
Gp	GpA	4,690	0.401	GpT	3,945	0.370	GpG	4,960	0.265	GpC	8,200	0.361
Sum		11,703	1.000		10,655	1.000		18,680	1.000		22,700	1.000

[a] Reproduced with the permission of the *Journal of Biological Chemistry.*

verted to the decimal fraction of the total. For example, in the tube where dATP[32] was the labeled substrate, 873 counts (of a total of 11,703 counts incorporated into DNA) were found in 3'-deoxythymidylate. Thus 0.075 of the incorporated deoxyadenylate residues are located next to deoxythymidylate in the newly synthesized DNA, representing the relative fraction of the dinucleotide sequence TpA. The fractional values in a given reaction do not depend on the particular synthetic conditions in that reaction; identical fractional values have been obtained for a given DNA primer with varying amounts of substrates, enzyme, and incubation times. However, the relative proportions of each of the four bases incorporated into the enzymatically synthesized DNA are known to vary, depending on the base composition of the particular DNA primer. Thus, it is necessary to multiply the fractional values of each reaction by a base-incorporation factor, which expresses the relative frequency with which the particular base (originally labeled with P[32] as a 5'-nucleotide) is incorporated into the synthesized DNA. These four base-incorporation factors, designated a, t, g, and c for the labeled deoxynucleoside triphosphates of adenine, thymine, guanine, and cytosine, respectively, may be derived by either of two methods:

1. *From values obtained by independent chemical determination of base composition of primer DNA.* In the case of *M. phlei* primer DNA, the molar proportions of adenine, thymine, guanine, and cytosine are known to be 0.162, 0.165, 0.338, and 0.335, respectively,[17] and therefore the base-incorporation factors are: $a = 0.162$, $t = 0.165$, $g = 0.338$, and $c = 0.335$.

2. *From data of nearest-neighbor analysis itself.* This is the preferred method, since it requires no independent knowledge of the base composition of the particular primer DNA. Instead, the base-incorporation factors are derived from the radioactivity measurements in the experiment. In nearest-neighbor experiments, digestions of the DNA and subsequent recoveries of 3'-nucleotides must be complete; consequently, the amount of a given base recovered from the synthesized DNA as a 3'-nucleotide should be equal to the amount of that base incorporated into the DNA during synthesis as a 5'-nucleotide. Four equations involving a, t, g, and c and the sixteen fractional values of Table I can be written expressing this equivalence. To illustrate, the total quantity of adenine incorporated as a 5'-nucleotide (TpA + ApA + CpA + GpA) must be equal to the total amount of adenine recovered as a 3'-nucleotide (ApA + ApT + ApG + ApC). For *M. phlei* DNA (Table I) this equivalence is expressed in the following equation:

[17] I. R. Lehman, S. B. Zimmerman, J. Adler, M. J. Bessman, E. S. Simms, and A. Kornberg, *Proc. Natl. Acad. Sci. U.S.* **44**, 1191 (1958).

$$0.075a + 0.146a + 0.378a + 0.401a = 0.146a + 0.194t + 0.134g + 0.189c$$
$$a = 0.146a + 0.194t + 0.134g + 0.189c$$

Similarly for thymine, guanine, and cytosine nucleotides:

$$t = 0.075a + 0.157t + 0.187g + 0.182c$$
$$g = 0.401a + 0.370t + 0.265g + 0.361c$$
$$c = 0.378a + 0.279t + 0.414g + 0.268c$$

The solution of these four equations is

$$a = 0.489c; \; t = 0.483c; \; g = 1.000c$$

TABLE II[a]

NEAREST NEIGHBOR FREQUENCIES OF *M. phlei* DNA

(Identical Roman numerals designate those sequence frequencies which should be equivalent in a Watson and Crick DNA model with strands of opposite polarity; identical lower-case letters designate sequence frequencies which should be equivalent in a model with strands of similar polarity.)

Reaction	Labeled triphosphate	Isolated 3'-deoxyribonucleotide			
		Tp	Ap	Cp	Gp
1	dATP³²	*a* TpA 0.012 I	*b* ApA 0.024 II	*c* CpA 0.063 III	*d* GpA 0.065
2	I dTTP³²	*b* TpT 0.026 I	*a* ApT 0.031	*d* CpT 0.045 IV	*c* GpT 0.060 V
3	dGTP³²	*e* TpG 0.063 II	*f* ApG 0.045 IV	*g* CpG 0.139	*h* GpG 0.090 VI
4	dCTP³²	*f* TpC 0.061 III	*e* ApC 0.064 V	*h* CpC 0.090 VI	*g* GpC 0.122
	Sums	0.162	0.164	0.337	0.337

These data:

$$\frac{Ap + Tp}{Gp + Cp} = 0.48$$

Independent chemical analysis of primer DNA:[b]

$$\frac{A + T}{G + C} = 0.49$$

[a] Reproduced with the permission of the *Journal of Biological Chemistry.*
[b] I. R. Lehman, S. B. Zimmerman, J. Adler, M. J. Bessman, E. S. Simms, and A. Kornberg, *Proc. Natl. Acad. Sci. U.S.* **44,** 1191 (1958).

These data show an incorporation of deoxyadenylate equal to deoxythymidylate and an incorporation of deoxyguanylate equal to deoxycytidylate. The $(a + t)/(g + c)$ ratio of 0.48 is the same as was obtained by independent chemical analysis of the *M. phlei* DNA primer.[17]

For convenience it has been found most suitable to express the frequencies of each of the nearest-neighbor sequences as a decimal proportion of 1. Thus, we may arbitrarily set $a + t + g + c = 1.000$. By substituting the values previously derived, the following are obtained: $a = 0.164$; $t = 0.162$; $g = 0.337$; $c = 0.337$.

These algebraically determined base-incorporation factors, derived from the radioactivity measurements, are identical to those obtained from independent chemical determination of base composition. These results indicate that the derived base-incorporation factors are accurate expressions of the nucleotide composition of both primer and product DNA's.

As a final step in the calculation, the decimal fractions given in Table I are now properly weighted for base frequencies, each being multiplied by its base-incorporation factor. The final nearest-neighbor frequencies, expressed as decimal proportions of 1.0, are shown in Table II. As can be seen, the sums of the nearest-neighbor frequencies for each nucleotide are the respective base-incorporation factors.

V. Applications of the Method[18]

Determination of Base Composition of Enzymatically Synthesized DNA

The over-all parity of the adenine with the thymine content, and the guanine with the cytosine content, of the newly synthesized DNA is found in all analyses regardless of the primer DNA (except in the case of the single-stranded ϕX174 viral DNA). This can be seen for *M. phlei* DNA (Table II) on comparing the sums of the four vertical columns, representing the fractional values of each of the four isolated nucleotides. Furthermore, the $(Ap + Tp)/(Gp + Cp)$ ratio from this analysis (as well as all nearest-neighbor analyses thus far performed with double-stranded DNA primers) is in close agreement with the $(A + T)/(G + C)$ ratio obtained on independent chemical analysis of the primer DNA (Table II), suggesting that in the experiment faithful replication of the over-all composition of the primer DNA has been achieved.

Confirmation of Base-Pairing and of Opposite Polarity of Strands in Enzymatically Synthesized DNA

Two different sets of predictions concerning nucleotide frequencies can be made, depending on whether the strands of the DNA double helix

[18] For a more complete discussion, see Josse *et al.*[6] and Swartz *et al.*[7]

are of similar or opposite polarity. In the model with opposite polarity there are six matching sequences, and they are indicated in Table II by the same Roman numeral. In each instance the agreement is good. Each of the four values along the diagonal represents its own matching pair and cannot be checked; i.e., every TpA sequence would be matched by a TpA sequence in the complementary strand of opposite polarity, and the same would apply for ApT, CpG, and GpC sequences. In the model with strands of similar polarity, eight pairs of matching values would be predicted among the sixteen nearest-neighbor sequence frequencies, and they are indicated in Table II by the same lower-case letter. In only a few instances are the values reasonably close. Statistical analysis of data from sixteen different DNA samples has established good agreement with the model of opposite polarity but significant deviation from the model of similar polarity.

Detection of Nonrandomness of Nearest-Neighbor Frequencies

A random assembly of bases in the DNA chain would be reflected by a definite ordering of the nearest-neighbor patterns. Randomness would predict that the frequency of a given nearest-neighbor or dinucleotide sequence (e.g., f_{TpA}) in a particular DNA would be equal to the product of the frequencies of the two constituent mononucleotides (e.g., $f_{Tp} \times f_{Ap}$) in that DNA. Thus, the frequencies of sequences ApT and TpA would be equal and predicted by the product $f_{Tp} \times f_{Ap}$. Inasmuch as Ap and Tp occur equally often in DNA, the ApA and TpT sequences would also be predicted to be equal to this product. It can be seen from Table II that, in *M. phlei* DNA, TpA and ApT differ markedly in frequency and that the ApA and TpT frequencies are in turn different from either of these. (ApA is, of course, equal to TpT because of base-pairing, as noted above.) Deviation from randomness can also be observed for the other *M. phlei* sequences in Table II. In every DNA thus far examined there is significant deviation from random prediction in the frequencies of nearly every nearest-neighbor sequence, although some approximate conformity to random ordering is often observed and is more marked in certain of the DNA's (for example, in the DNA's of the T-even bacteriophages).

Study of Sequence Patterns in DNA's from a Variety of Bacterial, Viral, Plant, and Animal Sources

Such studies have shown that each different DNA has a unique and nonrandom pattern of the sixteen nearest-neighbor sequence frequencies. Certain general trends of phylogenetic interest have been observed. For example, in the case of DNA's isolated from tissues of animals, the sequence of CpG has a remarkably low frequency, which deviates from a random distribution prediction more markedly than any other sequence in any DNA tested. Bacterial DNA's, ranging in base composition from $(A + T)/(G + C) = 0.41$ to 1.62, have been analyzed and show some over-all influence by the base composition, although there is always significant deviation from purely random ordering of the nucleotides. DNA's of several bacteriophages have been examined; among the temperate viruses the DNA of phage λ was not distinguishable in nearest-neighbor frequencies from the host cell DNA (*Escherichia coli*). By contrast, the nearest-neighbor frequencies of DNA's of the virulent T-series of phages were very different from those of *E. coli* DNA.

This procedure has also been applied to a study of the mechanisms of enzymatic replication of the single-stranded DNA of the bacterial virus ϕX174; it was found that replication proceeds initially through the formation of a new strand with base composition complementary to that in the original ϕX174 primer strand. The method has been used to establish the composition of minor DNA components (available only in small quantities), such as that described by Sueoka[19] in the DNA obtained from crab testes (*Cancer borealis*). That this unique material consisted of alternating A and T residues to the extent of 93% of the sequences has been established through the use of nearest-neighbor analysis.

Nearest-Neighbor Sequence Analysis of RNA

The RNA (labeled with P^{32}) synthesized by the DNA-directed RNA polymerase has been subjected to a "nearest-neighbor" type of analysis. Parallel synthetic reactions were carried out, each containing a different one of the four ribonucleoside triphosphates bearing a P^{32} in the sugar-esterified phosphate. On alkaline hydrolysis of the product to 2'- and 3'-ribonucleotides, the P^{32} label is "transferred" to the neighboring nucleotide, just as occurs on enzymatic digestion of the newly synthesized DNA in the nearest-neighbor analysis. Analyses employing many different DNA primers have shown good agreement between the nearest-neighbor sequence frequency patterns of the primer DNA and that of the newly synthesized RNA.[20-22]

VI. Limitations of the Method

1. Inability to provide information on sequences longer than dinucleotides. Since complete degradation of the synthesized DNA to 3'-nucleo-

[19] N. Sueoka, *J. Mol. Biol.* 3, 31 (1961).
[20] S. B. Weiss and T. Nakamoto, *Proc. Natl. Acad. Sci. U.S.* 47, 1400 (1961).
[21] J. J. Furth, J. Hurwitz, and M. Goldmann, *Biochem. Biophys. Research Communs.* 4, 431 (1961).
[22] M. Chamberlin and P. Berg, *Proc. Natl. Acad. Sci. U.S.* 48, 81 (1962).

tides is carried out, any probing at trinucleotide or longer sequences is precluded.

2. Failure to distinguish sequences involving the uncommon nucleotides, such as 6-methyl deoxyadenylate in bacterial DNA or 5-methyl deoxycytidylate in animal or plant DNA's. For example, when calf thymus primer is used, deoxycytidylate always replaces the 5-methyl derivative, obscuring the important information on the arrangement of 5-methyl deoxycytidine nucleotides and at the same time giving erroneous values for deoxycytidine nucleotide sequences.

3. Limitations in the accuracy of analysis. An analytic error of as little as 1% in the method would actually permit a relatively large number of errors to be made in determining the sequence frequencies in a DNA molecule. Even the small chromosome of λ phage, comprising about 10^5 nearest-neighbor sequences, could differ from a related DNA in one thousand nucleotide sequences without any difference being detected by this technique.

[102] Preparation and Properties of Amino Acyl Adenylates

By ALTON MEISTER

Amino acyl adenylates are formed as enzyme-bound products according to reaction (1), in which E represents an amino acid-specific

$$\text{Amino acid} + \text{ATP} + \text{E} \rightarrow \text{E–amino acyl adenylate} + \text{pyrophosphate} \quad (1)$$

activating enzyme. The amino acyl moiety of the enzyme-bound anhydride may be transferred to a suitable acceptor, as in reaction (2).

$$\text{E–amino acyl adenylate} + \text{acceptor} \rightarrow$$
$$\text{E} + \text{amino acyl acceptor} + \text{AMP} \quad (2)$$

Equations (1) and (2) are therefore analogous to the reactions catalyzed by the acetate-activating enzyme in which the acceptor is coenzyme A.[1] Similar reactions are involved in the activation of the carboxyl groups of other molecules; thus, it is probable that enzyme-bound acyl adenylates are formed as activated intermediates in the synthesis of hippuric acid,[2] pantothenic acid,[3] phenylacetyl-L-glutamine,[2] fatty acyl coen-

[1] P. Berg, *J. Am. Chem. Soc.* **77**, 3163 (1955); also see Vol. V [62].

[2] K. Moldave and A. Meister, *J. Biol. Chem.* **229**, 463 (1957).

[3] W. K. Maas and G. D. Novelli, *Arch. Biochem. Biophys.* **43**, 336 (1953).

zyme A derivatives,[4,5] acetyl-L-aspartic acid,[6] taurocholic acid,[7] and carnosine.[8]

In the enzymatic synthesis of carnosine (and related peptides), β-alanyl adenylate appears to be formed by reaction of β-alanine with ATP according to equation (1); the available evidence[8,9] indicates that L-histidine is the acceptor in equation (2). There is evidence that the amino acyl moieties of enzyme-bound α-amino acyl adenylates are transferred to soluble ribonucleic acid acceptors; such RNA–amino acid complexes are generally believed to be intermediates in the incorporation of amino acids into proteins.[10,11] The amino acyl moieties of enzyme-bound amino acyl adenylates may also be transferred to hydroxylamine to yield the corresponding amino acyl hydroxamates. Under certain experimental conditions, enzymatically synthesized amino acyl adenylates may acylate the ribose hydroxyl groups of nucleotides (e.g., ATP) and also the free amino groups of proteins.[12-14]

Amino acid-activating enzymes catalyze pyrophosphate–ATP exchange in the presence of the respective amino acid.[10] This reaction and the synthesis of the amino acyl adenylate appear to be very specific. On the other hand, the synthesis of ATP from amino acyl adenylate and pyrophosphate catalyzed by amino acid-activating enzymes [reversal of equation (1)] has been observed to be less specific.[15-19] For example, tryptophan-activating enzyme catalyzes ATP synthesis from pyrophos-

[4] W. P. Jencks and F. Lipmann, *J. Biol. Chem.* **225**, 207 (1957).
[5] H. S. Moyed and F. Lipmann, *J. Bacteriol.* **73**, 117 (1957).
[6] F. B. Goldstein, *J. Biol. Chem.* **234**, 2702 (1959).
[7] W. H. Elliott, *Biochem. J.* **62**, 427, 433 (1956).
[8] G. D. Kalyankar and A. Meister, *J. Biol. Chem.* **234**, 3210 (1959).
[9] G. D. Kalyankar and A. Meister, *J. Am. Chem. Soc.* **81**, 1515 (1959).
[10] Symposium on Amino Acid Activation, *Proc. Natl. Acad. Sci. U.S.* **44**, 67 (1958).
[11] See reviews by R. B. Loftfield, *Progr. in Biophys. and Biophys. Chem.* **8**, 347 (1957); H. Chantrenne, *Ann. Rev. Biochem.* **27**, 35 (1958); A. Meister, *Revs. Modern Phys.* **31**, 210 (1959).
[12] P. Castelfranco, K. Moldave, and A. Meister, *J. Am. Chem. Soc.* **80**, 2335 (1958).
[13] P. Castelfranco, A. Meister, and K. Moldave, Symposium on Microsomal Particles and Protein Synthesis, p. 115. Biophysical Society, Washington Academy of Science, 1958.
[14] C. Zioudrou, S. Fuji, and J. F. Fruton, *Proc. Natl. Acad. Sci. U.S.* **44**, 439 (1958).
[15] P. Berg, *J. Biol. Chem.* **233**, 601 (1958).
[16] G. D. Novelli, *Proc. Natl. Acad. Sci. U.S.* **44**, 86 (1958).
[17] M. Karasek, P. Castelfranco, P. R. Krishnaswamy, and A. Meister, *J. Am. Chem. Soc.* **80**, 2335 (1958).
[18] M. Karasek, P. Castelfranco, P. R. Krishnaswamy, and A. Meister, Symposium on Microsomal Particles and Protein Synthesis, p. 109. Biophysical Society, Washington Academy of Science, 1958.
[19] P. R. Krishnaswamy and A. Meister, *J. Biol. Chem.* **235**, 408 (1960).

phate and a wide variety of amino acyl adenylates.[16-19] However, the affinity of the activating enzyme has been shown to be much greater for L-tryptophanyl adenylate than for the other amino acyl adenylates.[19] Thus, the formation of enzyme-bound amino acyl adenylate occurs with much lower concentrations of L-tryptophanyl adenylate than with other amino acyl adenylates.

Although amino acyl adenylates prepared by organic synthesis hydrolyze readily and exhibit great tendency to acylate nonenzymatically, they are useful in enzymatic studies carried out under appropriate conditions. Thus, incubation of tryptophan-activating enzyme with synthetic L-tryptophanyl adenylate has been shown to yield enzyme-bound L-tryptophanyl adenylate, which is capable of reacting with pyrophosphate to give ATP[19] or with ribonucleic acid to yield ribonucleic acid–tryptophan complex.[20, 21] Although the participation of amino acyl adenylates in the synthesis of peptide bonds of proteins has not yet been proved by experiment, there is evidence that at least one amino acyl adenylate (β-alanyl adenylate) can serve in the enzymatic synthesis of a peptide bond of a naturally occurring compound (carnosine).[8, 9] Incubation of β-alanyl adenylate with DL-histidine yields carnosine nonenzymatically; however, this carnosine has been shown to be racemic. Carnosine formed in the presence of enzyme from β-alanyl adenylate and DL-histidine is predominantly of the L-configuration.[9] The ability of specific activating enzymes to bind amino acyl adenylates prepared by organic synthesis makes it possible to study the transfer function of such enzymes without utilizing their amino acyl adenylate-synthesizing activity.

Methods of Preparation

Amino acyl adenylates have been prepared by condensing amino acid chlorides with silver adenosine 5′-phosphate.[22] Amino acyl adenylates have also been prepared by condensing free amino acids with adenosine 5′-phosphate in the presence of N,N′-dicyclohexylcarbodiimide.[23] The preparation of glycyl adenylate and leucyl adenylate via the corresponding benzyl mercaptoformyl derivatives has been reported.[24] The en-

[20] K. K. Wong, A. Meister, and K. Moldave, *Biochim. et Biophys. Acta* **36**, 1777 (1959).

[21] K. K. Wong and K. Moldave, *J. Biol. Chem.* **235**, 694 (1960).

[22] J. A. DeMoss, S. M. Genuth, and G. D. Novelli, *Proc. Natl. Acad. Sci. U.S.* **42**, 325 (1956).

[23] P. Berg, *J. Biol. Chem.* **233**, 608 (1958).

[24] D. J. McCorquodale and G. C. Mueller, *Arch. Biochem. Biophys.* **77**, 13 (1958).

zymatic synthesis of L-tryptophanyl adenylate has been demonstrated in two laboratories.[17-19, 25]

Principle. In the present method[26] for the preparation of amino acyl adenylates, N-carbobenzoxy amino acids[27] are condensed with adenosine 5'-phosphate in the presence of N,N'-dicyclohexylcarbodiimide[28] in aqueous pyridine. The carbobenzoxy amino acyl adenylate is isolated from the reaction mixture and converted to the free amino acyl adenylate by catalytic hydrogenation with palladium. The present method has been successfully applied to the preparation of sixteen amino acyl adenylates including those of β-alanine, α-alanine, asparagine, glutamine, glycine, isoleucine, leucine, methionine, phenylalanine, proline, serine, threonine, tryptophan, tyrosine, valine, and α-aminoisobutyric acid. A similar method has been employed by Zioudrou et al.[14, 29] for the synthesis of tyrosinyl adenylate and glycyl tyrosinyl adenylate.

Reagents

N-Carbobenzoxy amino acid.
Adenosine 5'-phosphate.
N,N'-dicyclohexylcarbodiimide.
Pyridine, anhydrous.
75% pyridine (aqueous).
Palladium catalyst (see below).
Ethylene glycol monomethyl ether.
Acetone.
Diethyl ether, anhydrous.
90% acetic acid (aqueous).
2 M hydroxylamine hydrochloride (adjusted to pH 6.5).
Ferric chloride reagent (see below).

Preparation of Palladium Catalyst. Palladium catalyst is prepared by dissolving 10 g. of palladium black (Fisher Scientific Company) in aqua regia and evaporating to dryness at 128°. The residue is dissolved in concentrated hydrochloric acid and then evaporated again; this procedure is repeated twice. The residue is dissolved in the minimal volume of hot concentrated hydrochloric acid and poured rapidly into

[25] H. S. Kingdon, L. D. Webster, Jr., and E. W. Davie, *Proc. Natl. Acad. Sci. U.S.* **44**, 757 (1958).
[26] K. Moldave, P. Castelfranco, and A. Meister, *J. Biol. Chem.* **234**, 841 (1959).
[27] M. Bergmann and L. Zervas, *Chem. Ber.* **67**, 1192 (1932); N-Carbobenzoxy amino acids may also be obtained from Mann Research Laboratories.
[28] H. G. Khorana, *J. Am. Chem. Soc.* **76**, 3517 (1954); may be obtained from Schwarz Laboratories, Inc.
[29] C. Zioudrou and J. S. Fruton, *J. Biol. Chem.* **234**, 583 (1959).

boiling water; 10 ml. of 88% formic acid are added to the boiling suspension, followed by sufficient 5 N potassium hydroxide to make the solution alkaline to litmus paper. After addition of 0.2 ml. of 88% formic acid, the solution is cooled, and the catalyst is washed several times with water by decantation.

Procedure. N-Carbobenzoxy amino acid (0.002 mole) and adenosine 5′-phosphate (684 mg., 0.002 mole) are dissolved in 10 ml. of 75% aqueous pyridine, and the solution is cooled to 0°. A solution of N,N'-dicyclohexylcarbodiimide prepared at 0° by adding 8 g. of N,N'-dicyclohexylcarbodiimide to 8 ml. of anhydrous pyridine is added rapidly to the solution of carbobenzoxy amino acid and adenosine 5′-phosphate. The mixture is shaken or stirred continuously with a magnetic stirrer at 0° to 5°. A precipitate of N,N'-dicyclohexyl urea forms gradually during the reaction.

The reaction is followed by carrying out a hydroxamate reaction on suitable aliquots of the reaction mixture as follows. The sample (0.01 to 0.5 ml.) is treated with 0.5 ml. of 2 M hydroxylamine hydrochloride (adjusted to pH 6.5 with potassium hydroxide). After 5 minutes, 1.5 ml. of ferric chloride reagent (0.2 N trichloroacetic acid, 0.67 N hydrochloric acid) is added, and the final volume is brought to 2.5 ml. with water. The colors are compared with those obtained with authentic amino acyl hydroxamates in a Beckman Model DU spectrophotometer at 535 mμ.

The reaction is considered to be complete when there is no further increase in hydroxamate-forming material (usually about 2 to 3 hours). The mixture is then filtered through a sintered disk at 5°, and the precipitate is washed with 3 ml. of ice-cold 75% aqueous pyridine. On standing at 5° for 10 minutes, the combined filtrate and washing separate into two layers. The upper layer is carefully removed, and the lower layer, which contains more than 90% of the product, is added to 200 ml. of acetone previously cooled to −15°. The mixture is allowed to stand at −15° for 12 to 18 hours, and the precipitate is collected by centrifugation for about 1 hour at 800 × g at −15°. The precipitate is washed by centrifugation with two 100-ml. portions of −15° acetone. The precipitate is drained for a few minutes and then stirred for 5 minutes with 50 ml. of ethylene glycol monomethyl ether.[30] The precipitate is removed by centrifugation, and the extraction with ethyl glycol monomethyl ether is repeated.[31] The extracts are combined and mixed with 500 ml. of

---

[30] May be obtained from Howe and French, Inc.; it is advisable to fractionate the material through a short vigreux column (b.p. 123° to 124°).

[31] The carbobenzoxy amino acyl adenylate is soluble in ethylene glycol monomethyl ether, whereas significant quantities of adenosine 5′-phosphate are not extracted. Two extractions are usually sufficient; occasionally there is no residue

diethyl ether. The N-carbobenzoxy amino acyl adenylate precipitates immediately and is separated by centrifugation at $-15°$. The product is washed twice with 100 ml. of ether by centrifugation in the cold and then allowed to drain at room temperature for about 10 minutes. The product is placed in a vacuum desiccator over P_2O_5 and stored *in vacuo* at $-15°$.

The N-carbobenzoxy amino acyl adenylate (150 mg.) is dissolved in 5 ml. of 90% acetic acid in a small Erlenmeyer flask equipped with a magnetic stirrer and a two-hole stopper. The exit tube is connected with rubber tubing to a test-tube trap containing 5 ml. of freshly filtered saturated barium hydroxide. The solution is cooled to $0°$, and approximately 1 g. (wet weight) of catalyst is added. The vessel is flushed with nitrogen for 3 minutes, and then hydrogen is admitted into the system. The reaction is considered to be complete when there is no further formation of barium carbonate in a freshly prepared barium hydroxide trap for 5 minutes. This usually takes about 10 to 30 minutes, depending on the quantity of catalyst added. The solution is flushed with nitrogen for 3 minutes and then is transferred with 1.5 ml. of cold 90% acetic acid to a conical centrifuge tube. After removal of the catalyst by brief centrifugation in the cold, the supernatant solution is transferred to a small flask with 2 ml. of 90% acetic acid and lyophilized. The dry powder is washed twice with 15 ml. of anhydrous diethyl ether and stored *in vacuo* over P_2O_5 at $-15°$. In the author's laboratory, the yields of free amino acyl adenylate have varied from 40 to 81% of theory based on the free amino acids.[26, 32]

Properties and Analysis of Products[26]

The carbobenzoxy amino acyl adenylate preparations contain less than 5% of adenosine 5'-phosphate and give elemental analytical values close to the theoretical. Treatment of these products with hydroxylamine (1 mg. mixed with $0.05 M$ hydroxylamine, pH 7) gives the corresponding hydroxamate derivatives, which may be identified by paper chromatography. The spots may be rendered visible by spraying the dried chromatogram with a solution of 95% ethanol containing $1 N$ hydrochloric acid and $0.185 M$ ferric chloride. Treatment of the free amino acyl adenylates with hydroxylamine in the same manner yields the corresponding amino acyl hydroxamates which may also be chromatographed.

---

even after the first extraction. The extraction procedure may be followed by applying the hydroxamic acid–ferric chloride procedure to the residual solid material and to the extracts.

[32] See also *Biochem. Preparations* **8**, 11 (1961).

Application of the quantitative hydroxylamine–ferric chloride procedure (see above) to free amino acyl adenylate preparations has been found to give 62 to 95% of values obtained with the authentic amino acyl hydroxamic acids. Some hydrolysis of the anhydrides undoubtedly occurs during hydrogenation, lyophilization, and storage at $-15°$. Free amino acyl adenylates may also be characterized by paper ionophoresis.[26] Ionophoresis at pH 4.5 separates adenosine 5′-phosphate, free amino acids, and amino acyl adenylates. Preparations of amino acyl adenylates may contain variable quantities of the isomeric adenosine 5′-phosphate-2′(3′)-amino acid esters. Paper ionophoresis at pH 5.9 separates these esters from the anhydrides. Ester formation has been observed on storage of amino acyl adenylate preparations at $-15°$. Amino acyl adenylates may also be characterized by analyses for amino acids and for adenosine 5′-phosphate.[26]

The $α$-amino acyl adenylates are rapidly hydrolyzed in neutral aqueous solution to yield adenosine 5′-phosphate and amino acid, as judged by disappearance of the ability to form hydroxamic acids; however, they exhibit much greater stability at much lower values of pH. Under conditions often employed in enzymatic studies, e.g., 0.1 M potassium phosphate buffer (pH 7.2) and 37°, the $α$-amino acyl adenylates are rapidly hydrolyzed. The half-life of the anhydrides under these conditions is in the neighborhood of 4 to 6 minutes. Carbobenzoxy amino acyl adenylates and $β$-alanyl adenylate are hydrolyzed much more slowly than are the $α$-amino acyl adenylates. The high reactivity of amino acyl adenylates is indicated not only by their rapid reaction with hydroxylamine but also by their ability to acylate the free amino groups of amino acids and proteins and by their reaction with nucleotides and ribonucleic acid preparations. Incubation of amino acyl adenylates with concentrated ammonium hydroxide at 26° for 3-minutes gives the corresponding amino acid amides.

[103] The Preparation of C^{14}-Amino Acyl soluble-RNA

By KIVIE MOLDAVE

Principle. The incorporation of amino acids into s-RNA occurs by a two-step reaction (see, for example, reviews by Hoagland[1] and Berg[2]).

[1] M. B. Hoagland, *in* "The Nucleic Acids." (E. Chargaff and J. N. Davidson, eds.), Vol. III, p. 349. Academic Press, New York, 1960.
[2] P. Berg, *Ann. Revs. Biochem.* **30**, 293 (1961).

$$\text{Amino acid} + \text{ATP} + \text{enzyme} \overset{Mg^{++}}{\rightleftarrows}$$
$$\text{(Enzyme–amino acyl adenylate)} + PP_i \quad (1)$$

$$\text{(Enzyme–amino acyl adenylate)} + s\text{-RNA} \rightleftarrows$$
$$\text{Amino acyl } s\text{-RNA} + \text{AMP} + \text{enzyme} \quad (2)$$

$$\text{Sum: Amino acid} + \text{ATP} + s\text{-RNA} \overset{\text{enzyme}}{\underset{Mg^{++}}{\rightleftarrows}}$$
$$\text{Amino acyl } s\text{-RNA} + PP_i + \text{AMP} \quad (3)$$

The first step [reaction (1)] involves the carboxyl activation of the amino acid, the elimination of pyrophosphate from ATP, and the formation of an amino acyl adenylate anhydride. The second step [reaction (2)] involves the transfer of the amino acyl group from the enzyme-bound complex to s-RNA where it occurs esterified to the $2'(3')$-hydroxyl group of the terminal (adenylate) nucleotide.

The "pH 5 enzymes" fraction of rat liver contains both amino acid-activating enzymes and s-RNA; in the presence of ATP, it catalyzes the activation and incorporation of labeled amino acids into amino acid-specific sites on s-RNA [reaction (3)]. The s-RNA extracted from such incubations contains, in addition to the isotopic amino acids incorporated, a spectrum of endogenous, unlabeled, bound amino acids.[3] The procedures described below are based essentially on those of Hoagland et al.[4] for the labeling of rat liver s-RNA and of Kirby[5] and Gierer and Schramm[6] for the isolation of RNA.

Reagents

Medium A: 0.35 M sucrose, 0.035 M KHCO$_3$, 0.025 M KCl, 0.004 M MgCl$_2$, 0.02 M K$_2$HPO$_4$; pH 7.4.[7]

Medium B: 0.35 M sucrose, 0.025 M KCl, 0.004 M MgCl$_2$.[7]

1 N acetic acid.

1 M 2-amino-2-hydroxymethylpropane-1,3-diol (Tris), buffer, pH 7.4.

0.2 M ATP·MgCl$_2$. Solution containing ATP is adjusted to pH 7.4 with solid KHCO$_3$, made up to volume, and an equivalent concentration of MgCl$_2$ is added.

[3] G. Acs, G. Hartmann, H. G. Bowman, and F. Lipmann, *Federation Proc.* 18, 178 (1959).

[4] M. B. Hoagland, M. L. Stephenson, J. F. Scott, L. I. Hecht, and P. C. Zamecnik, *J. Biol. Chem.* 231, 241 (1958).

[5] K. S. Kirby, *Biochem. J.* 64, 405 (1956).

[6] A. Gierer and G. Schramm, *Nature* 177, 702 (1956).

[7] E. B. Keller and P. C. Zamecnik, *J. Biol. Chem.* 221, 45 (1956).

Water-saturated phenol (at 4°).
20% potassium acetate, adjusted to pH 5 with acetic acid.
2 to 3 M salt-free hydroxylamine.[8]
0.05 M Tris buffer, pH 10.
0.01 M KOH.

Procedure

Preparation of "pH 5 Enzymes." All steps are carried out at 4°. Approximately 50 g. of finely minced rat liver tissue are homogenized in 110 ml. of medium A; it is advisable to carry out the homogenization in small portions. The homogenate is centrifuged at 12,000 × g for 15 minutes; the supernatant is diluted with 3 vol. of media B and centrifuged for 2 hours in a preparative ultracentrifuge at 144,000 × g. The nonparticulate supernatant is carefully aspirated off and further diluted with an equal volume of medium B. Cold 1 N acetic acid is then added dropwise to the supernatant, with continuous stirring, to a pH value of 5.2, and the resulting turbid solution is centrifuged at 10,000 × g for 30 minutes. The residue ("pH 5 enzymes" fraction) is resuspended in 20 ml. of medium A, homogenized briefly to ensure thorough resuspension, and used directly or stored at −14° for several days prior to use.

Labeling of "pH 5 Enzymes" s-RNA. Twenty milliliters of solution containing the "pH 5 enzymes" fraction obtained from 50 g. of rat liver are incubated with 1000 micromoles of Tris buffer (pH 7.4), 200 micromoles of ATP·MgCl₂, and 25 μc. of C^{14}-L-amino acid. The incubation is carried out for 20 minutes at 37° in a total volume of 23 ml. The flask contents are then chilled and diluted with 2 vol. of cold distilled water, and the pH is adjusted to 5.2 with 1 N acetic acid as described above. The precipitate is obtained by centrifugation at 15,000 × g for 20 minutes and resuspended in 20 ml. of medium A. The suspension is diluted with 2 vol. of cold distilled water, and the isoelectric precipitation at pH 5.2 is repeated. The precipitate, obtained by centrifugation, is resuspended in 20 ml. of medium A.

Isolation of C^{14}-Labeled s-RNA. All steps are carried out at 4°. To the resuspended fraction, obtained from the incubation of "pH 5 enzymes" and C^{14}-amino acid, an equal amount of cold water-saturated phenol is added, and the suspension is vigorously shaken for 1 hour. The emulsion is then centrifuged at 20,000 × g for 25 minutes, and the top aqueous layer is removed by aspiration. If adequate separation of the two phases is not obtained, 2 to 3 ml. of water are added, the mixture is shaken briefly, and the centrifugation is repeated for 20 minutes. After

[8] H. Beinert, D. E. Green, P. Hele, H. Hift, R. W. Von Kroff, and C. V. Ramakrishnan, *J. Biol. Chem.* 203, 35 (1953).

removal of the aqueous phase, an equal amount of water is added to the bottom phenol layer; it is shaken for 20 minutes and centrifuged at 20,000 × g for 25 minutes. The aqueous phase is obtained by aspiration and combined with the first aqueous layer; 0.1 vol. of cold 20% potassium acetate, pH 5.0, and 2 vol. of ethanol are added to the combined fractions and allowed to stand overnight at −14°. The precipitate is collected by centrifugation at low speed, resuspended in 20 ml. of water, and reprecipitated with potassium acetate and ethanol. The precipitate thus obtained (C^{14}-amino acyl s-RNA) is resuspended in 1 ml. of water and dialyzed for 4 hours against running water at 4° or several changes of distilled water.

Assay of C^{14}-Amino acyl s-RNA.

The routine analyses and characterization of s-RNA-bound amino acids by several criteria, some of which are described below, are recommended. Approximately 15 mg. of s-RNA containing 2 to 7% of the initial labeled amino acid are recovered from the incubated "pH 5 enzymes" fraction as described above. Most of the RNA-bound radioactivity is alkali-labile and hydroxylamine-reacting. Approximately 25% of the C^{14}-amino acid is incorporated into protein when C^{14}-amino acyl s-RNA is incubated with cytoplasmic ribonucleoprotein in the presence of GTP and an amino acyl-transferring preparation.

1. *Specific Activity.* A 0.01-ml. aliquot of the dialyzed C^{14}-s-RNA

AMINO ACID INCORPORATION INTO *soluble*-RNA

C^{14}-substrate (25 μc.)		C^{14}-Amino acyl s-RNA recovered	
Amino acid	Specific activity, μc./micromole	Total activity, c.p.m.	Specific activity, c.p.m./mg. s-RNA
L-Leucine	8	98,000	9,000
L-Leucine	30	566,000	51,000
L-Alanine	69	154,000	13,000
L-Aspartate	92	368,000	37,000
L-Lysine	138	453,000	38,000
L-Threonine	92	356,000	31,000
L-Phenylalanine	165	264,000	21,000
L-Amino acid mixture[a]		260,000	26,000

[a] C^{14}-*Chlorella* protein hydrolyzate; specific activity, 88 μc./mg.

solution is assayed for radioactivity, and a similar aliquot is diluted to 1 ml. and analyzed spectrophotometrically (at 260 mμ) for RNA content. The table lists the results obtained with several typical preparations.

2. *Amino Acyl Transfer.* A small aliquot of C^{14}-amino acyl s-RNA

(containing approximately 1000 c.p.m.) is incubated with 2 to 4 mg. of rat liver microsomes or ribosomes, 0.5 micromole of $GTP \cdot MgCl_2$, 10 micromoles of $ATP \cdot MgCl_2$, and 5 to 7 mg. of "pH 5 supernatant" protein in a total volume of 2 ml. as described.[9, 10] Analysis of the ribonucleoprotein particles after 20 minutes of incubation at 37° reveals that 20 to 30% of the C^{14}-amino acid is incorporated into the residual protein obtained after rigorous extractions and removal of RNA.[11]

3. *Release of C^{14}-Amino Acid with Hydroxylamine.* An aliquot of C^{14}-amino acyl s-RNA is incubated in salt-free neutral 2 M hydroxylamine for 20 minutes at 37°. Two milligrams of carrier C^{12}-s-RNA are then added, followed by 0.1 vol. of 20% potassium acetate, pH 5.0, and 2 vol. of cold ethanol. The preparation is allowed to stand at $-14°$ for at least 4 hours; the resulting precipitate and the supernatant are obtained by centrifugation. Radioactivity determinations reveal that the C^{14} is almost quantitatively released into the ethanol-soluble fraction. When this radioactive fraction is concentrated and chromatographed on paper, most of the radioactivity migrates in a single area with a mobility corresponding to that of standard amino acid hydroxamates.

4. *Release of C^{14}-Amino Acid with Mild Alkali.* An aliquot of C^{14}-amino acyl s-RNA is incubated at pH 10 in 0.05 M Tris buffer or in 0.01 M KOH for 20 minutes at 37°. Carrier RNA, potassium acetate, and ethanol are then added as described above, and the ethanol-insoluble and ethanol-soluble fractions are obtained. Most of the radioactivity is released from s-RNA under these conditions, whereas in control incubations, at pH values of 5.6 to 6.0, most of the C^{14} is recovered with the RNA in the ethanol-insoluble fraction. Paper chromatography of the radioactive fraction indicates that free amino acid is the principal labeled product released from s-RNA.

ADDENDUM

The preparation of C^{14}-amino acyl $soluble$-RNA from rat liver is described in detail in this article. The preparation from $E.\ coli$ has been described by von Ehrenstein and Lipmann.[12]

[9] L. G. Grossi and K. Moldave, *J. Biol. Chem.* **235**, 2370 (1960).
[10] J. M. Fessenden and K. Moldave, *Biochemistry* **1**, 485 (1962).
[11] K. Moldave, *J. Biol. Chem.* **235**, 2365 (1960).
[12] K. von Ehrenstein and F. Lipmann, *Proc. Natl. Acad. Sci. U. S.* **47**, 491 (1961).

[104] Preparation and Properties of Acyl Adenylates

By WILLIAM P. JENCKS

Preparation

Principle. Acyl adenylates (mixed anhydrides of 5'-adenylic acid and a carboxylic acid) have been prepared from acyl chlorides and the silver salt of AMP (5'-adenylic acid),[1,2] from the reaction of carboxylic acid, AMP, and dicyclohexylcarbodiimide,[3-5] and from the pyridine-catalyzed reaction of acid anhydrides with AMP, based on the method described by Avison for the acylation of phosphates.[6] Compounds which have been prepared in high yield by the acid anhydride method include acetyl-,[1] propionyl-,[7] butyryl-,[8] hexanoyl-,[9,10] octanoyl-,[10] benzoyl-,[2,11] lipoyl-,[12] and phenylacetyl adenylates.[11] The procedure is simple and generally useful, if the acid anhydride is readily available or can be easily prepared, and is the method described here. For rare or sensitive acids the carbodiimide method, which has been used for oxyluciferyl-[3] and oleyladenylates,[5] may be preferable. The preparation and properties of amino acyl adenylates are described elsewhere in this volume (Vol. VI [102]) and will not be discussed here.

Reagents

5'-AMP.
Anhydride of the desired acid.
Pyridine.
1.0 M lithium hydroxide.

Procedure. The method for acetyl adenylate will be described: 700 mg. (2 millimoles) of 5'-AMP is dissolved in 10 ml. of 50% pyridine and 2 ml. of 1.0 M lithium hydroxide in a small beaker standing in an

[1] P. Berg, *J. Biol. Chem.* **222**, 1015 (1956).
[2] G. M. Kellerman, *J. Biol. Chem.* **231**, 427 (1958).
[3] W. C. Rhodes and W. D. McElroy, *J. Biol. Chem.* **233**, 1528 (1958).
[4] P. T. Talbert and F. M. Huennekens, *J. Am. Chem. Soc.* **78**, 4671 (1956).
[5] B. Borgström, *Acta Chem. Scand.* **12**, 1533 (1958).
[6] A. W. D. Avison, *J. Chem. Soc.* 732 (1955).
[7] H. S. Moyed and F. Lipmann, *J. Bacteriol.* **73**, 117 (1957).
[8] C. H. Lee Peng, *Biochim. et Biophys. Acta* **22**, 42 (1956).
[9] W. P. Jencks and F. Lipmann, *J. Biol. Chem.* **225**, 207 (1957).
[10] M. Whitehouse, H. Moeksi, and S. Gurin, *J. Biol. Chem.* **226**, 813 (1957).
[11] K. Moldave and A. Meister, *J. Biol. Chem.* **229**, 463 (1957).
[12] L. J. Reed, F. R. Leach, and M. Koike, *J. Biol. Chem.* **232**, 123 (1958).

ice bath on a magnetic stirrer. When the mixture is cold, 2.0 ml. of acetic anhydride is added over 1 to 2 minutes with good stirring, and the mixture is immediately extracted three times with cold ether to remove the pyridine, which will cause decomposition of the product. The aqueous layer is poured into 50 ml. of cold acetone, and 100 ml. of cold acetone is added slowly with stirring. After standing for an hour in the cold, the precipitate is collected on a Büchner funnel, rapidly washed several times with cold acetone and ether, and dried overnight under vacuum in a desiccator over P_2O_5 in the cold room. It is stored over a desiccant at −15°.

The procedure for other acyl adenylates is the same, except that the concentration of pyridine may be increased up to 80% to increase the solubility of the more hydrophobic anhydrides in the reaction mixture, and the reaction time may be increased for insoluble or very unreactive anhydrides, such as benzoic anhydride.[2] Prolonging the reaction time after the anhydride has gone into solution with acetic and hexanoic anhydrides can lead only to degradation of the product, since the anhydride disappears completely within a few seconds after solution in the presence of high concentrations of pyridine.[13] The reported yields are 60 to 80%, of products of approximately 90% purity, based on hydroxamic acid assay for activated acyl groups[14] and ultraviolet absorption at 260 mμ for adenine; the products may contain an additional 10 to 20% of nonabsorbing material, presumably water of hydration. For rare anhydrides or nucleotides other than AMP, the whole procedure may conveniently be carried out on a microscale in a glass-stoppered test tube; the ether is removed by suction, and the precipitate is collected by centrifugation.[13] The products obtained are generally sufficiently pure for use as substrates in enzymatic reactions; some further purification may be obtained by chromatography on anion-exchange resins or cellulose columns or by paper electrophoresis.[1, 2, 4, 5, 8]

Properties

Stability. The solid is stable for at least several weeks at −15° in the absence of moisture. Prolonged storage of the solid or of solutions results in decomposition; this may not be evident from the hydroxamic acid assay, since one of the decomposition products, 2′(3′)-acyl-5′-adenylate (ribose-acylated AMP), will react with hydroxylamine to give a hydroxamic acid, although at a slower rate than acyl adenylate.[15]

[13] W. Allison and W. P. Jencks, unpublished experiments (1959).
[14] F. Lipmann and L. C. Tuttle, *Biochim. et Biophys. Acta* 4, 301 (1950); cf. Vol. V [63] and Vol. III [39].
[15] W. P. Jencks, *Biochim. et Biophys. Acta* 24, 227 (1957).

In solution the material is stable at neutrality or in dilute acid at room temperature for several hours, but it is very rapidly destroyed by dilute base (1 minute at pH 10) or hot dilute acid (2 minutes at pH 1 at 100°).[1] It reacts rapidly with most buffers at neutral pH to acylate the buffer, and measured rates of hydrolysis generally represent largely the rate of reaction with the buffer, unless extrapolated to zero buffer concentration.[15, 16] The reactions with phosphate, carboxylate, and tris(hydroxymethyl)aminomethane are slow enough to permit use of these buffers in reaction mixtures. Benzoyl adenylate, like other benzoyl compounds, is considerably more stable.[2]

Chemical Properties. The pure material reacts with periodate and shows complexing with borate on paper electrophoresis in borate buffer, indicating that the *cis*-hydroxyl groups of the ribose are free. The ultraviolet spectrum is identical to that of AMP (with the exception of the benzoyl compounds[2]), the compound is deaminated by nitrous acid, and the mobility on paper electrophoresis at pH 3.7 is the same as that of AMP, indicating the absence of acylation of the adenine portion of the molecule. The secondary acidic dissociation of the phosphate group of AMP is absent, as shown by titration and by paper electrophoresis in 0.01 M citrate buffer, pH 6.6, or 0.05 M phosphate buffer, pH 6.8, at 0°, in which acyl adenylates (charge -1) migrate behind AMP (charge -2).

Enzymatic Reactions. Acyl adenylates behave like intermediates in reactions of acyl-activating enzymes in that they will act as acyl or adenylate donors to coenzyme A or pyrophosphate, respectively, according to the general equation[2, 4, 5, 7-11, 17]

$$
\text{RCOO}^- + \text{ATP} + \text{enzyme} \underset{\pm \text{PP}}{\rightleftarrows} \begin{array}{c} \overset{\text{O}}{\underset{\parallel}{}} \\ \text{RC} \sim \text{AMP} \\ \downarrow \\ \text{Enzyme-X} \end{array} \underset{\pm \text{HSCoA}}{\rightleftarrows} \text{RCO} \sim \text{SCoA} + \text{AMP} + \text{enzyme}
$$

There is some lack of specificity in respect to their action; acetyl-AMP, for example, reacts readily at low concentration with the fatty acid-activating enzyme, whereas acetate itself is relatively unreactive with this enzyme.[9] Most enzyme preparations contain as a contaminant an enzyme which catalyzes the hydrolysis of acyl adenylates with P—O bond breaking.[2, 18] Acyl adenylates are not attacked by adenylic acid deaminase.[1]

[16] W. P. Jencks and J. Carriuolo, *J. Biol. Chem.* **234**, 1272 (1959).
[17] P. Berg, *J. Biol. Chem.* **222**, 991 (1956).
[18] G. M. Kellerman, *Biochim. et Biophys. Acta* **33**, 101 (1959).

Reactions with Nucleophilic Reagents. Acyl adenylates react rapidly with compounds which are good nucleophilic reagents toward the carbonyl group; these include most amines, amino acids, hydroxylamine, and imidazole, as well as hydroxide ion and thiol anions.[2,15] The reactions with amino compounds are with the free base form. The reaction with imidazole is particularly rapid, and, since the product is itself a highly reactive "energy-rich" compound, imidazole acts as a catalyst for acyl transfer to a number of acceptor molecules, such as phosphate and thiols, which themselves react more slowly; imidazole can thus catalyze the same reactions as transacetylase and the acyl-activating enzymes.[15,16] Attack by hydroxide ion is on the carbonyl group and results in C—O bond breaking.[2]

Free Energy of Hydrolysis. At neutral pH the hydrolysis of acyl adenylates results in the release of two protons according to the equation

$$\underset{\overset{\|}{R C}}{\overset{O}{}} \sim AMP^- + H_2O \rightarrow RCOO^- + AMP^{--} + 2H^+$$

so that the apparent free energy of hydrolysis is highly pH-dependent. Between pH 6 and 7 acetyl adenylate is in reversible equilibrium with acetylimidazole, which puts the free energy of hydrolysis of the former compound some 5000 calories above ATP at this pH and 7000 to 8000 calories above ATP at pH 8.[15] The apparent free energy of hydrolysis of oxyluciferyl adenylate at pH 7.1 has been estimated by Rhodes and McElroy to be 13,100 calories.[3] This unfavorable equilibrium is one reason why accumulation of free acyl adenylates in enzymatic reactions is exceedingly small or absent.

2'(3')-Acyl-5'-adenylates

Old preparations of acyl adenylates or preparations which have been allowed to stand in imidazole buffers contain 2'(3')-acyl-5'-adenylates (ribose-acylated AMP) as the principal hydroxylamine-reactive component.[15] 2'(3')-Acetyl-5'-adenylate reacts with neutral 1.1 M hydroxylamine at 37° with a half-time of 5 minutes, compared to 0.6 minute for acetyl adenylate, but the reaction is base-catalyzed, and at pH 5.5 the rate difference is greater (cf. Zachau et al.[19]). In contrast to acetyl adenylate, this compound does not react with periodate nor complex with borate on paper electrophoresis, and it moves with the same mobility as adenylic acid on paper electrophoresis at pH 6.6. The spectrum appears to be identical to that of AMP, although the extinction coefficient is not known.

[19] H. G. Zachau, G. Acs and F. Lipmann, *Proc. Natl. Acad. Sci. U.S.* **44**, 885 (1958).

2'(3')-Acetyl-5'-adenylate may conveniently be prepared by taking advantage of the fact that the equilibrium between acetyl adenylate and acetylimidazole is readily reversible and favors acetylimidazole, whereas acylation of the ribose hydroxyl group by acetylimidazole is essentially irreversible. AMP, (0.41 g.) is dissolved in 4 ml. of water and brought to approximately pH 10 with NaOH. Imidazole (0.34 g.) is added, the mixture is placed in a 37° water bath, and 0.24 ml. of acetic anhydride is added, in portions, over 30 minutes. The product, after precipitation with barium acetate and alcohol, contains one mole of alkaline hydroxylamine-reactive acetyl groups, measured by the Hestrin procedure for esters,[20] for each mole of ultraviolet-absorbing adenine. The yield, based on adenine absorption, is 90%, but the product is contaminated with nonabsorbing material.

[20] S. Hestrin, *J. Biol. Chem.* **180**, 249 (1949).

[105] Preparation and Properties of Sulfuryl Adenylates

By PHILLIPS W. ROBBINS

The sulfate-activating enzymes and sulfokinases are described in Vol. V [129] and [130]. The preparation and properties of adenosine-5'-phosphosulfate and of 3'-phosphoadenosine-5'-phosphosulfate (PAPS) are described below.

I. Adenosine-5'-phosphosulfate (APS)

Small amounts of APS may be prepared enzymatically from ATP and inorganic sulfate by using purified ATP-sulfurylase and inorganic pyrophosphatase.[1,2] However, for the preparation of substrate amounts of material, the chemical methods are preferable. Two synthetic methods for APS have been described. The procedure of Baddiley *et al.*,[3] which is described with slight modification below, has the advantage of complete freedom from side reactions, so that the only nucleotides to be separated after the reaction are unreacted AMP and the product, APS. A method involving the condensation of inorganic sulfate with AMP in the presence of dicyclohexylcarbodiimide has been reported by Reichard and Ringertz.[4]

[1] P. W. Robbins and F. Lipmann, *J. Biol. Chem.* **233**, 686 (1958).
[2] L. G. Wilson and R. S. Bandurski, *J. Biol. Chem.* **233**, 975 (1958).
[3] J. Baddiley, J. G. Buchanan, and R. Letters, *J. Chem. Soc.* 1067 (1957).
[4] P. Reichard and N. R. Ringertz, *J. Am. Chem. Soc.* **79**, 2025 (1957).

Chemical Synthesis of APS

Principle. APS is formed by the reaction of the pyridine–SO_3 complex with AMP in aqueous bicarbonate solution. There is probably a direct reaction between the anhydride and the substituted phosphate anion with the displacement of pyridine. The method should be a general one for the preparation of phosphate–sulfate anhydrides. The reaction is similar to that described by Avison for the preparation of mixed anhydrides between carboxylic and phosphoric acids.[5]

Procedure.[3] The pyridine–SO_3 complex[6] (800 mg., m.p. 172°) is added with stirring to a solution of adenosine-5′-phosphate[7] (400 mg.) and sodium bicarbonate (1.0 g.) in 10 ml. of water at 40° to 50°. After 30 minutes at this temperature, stirring is discontinued, and the solution is cooled in ice water and diluted to 100 ml. with cold water.

The solution is passed through a column of washed Dowex 1–chloride or formate (10×3 cm.).[8] The effluent is discarded, and the column is washed with 300 ml. of water, which is also discarded. APS is bound much more tightly to Dowex 1 than is AMP. Since these are the only two nucleotides present, almost any simple gradient or stepwise elution system will achieve complete separation. The eluting solution should have a pH of 5.0 or higher, however, since APS is acid-labile but stable in neutral and alkaline solution in the cold. The elution is followed by determining the absorbancy of the effluent fractions at 259 mμ. According to the method of Baddiley *et al.*,[3] elution is carried out with ammonium formate solution (pH 5) in an apparatus designed to give an approximately linear concentration/volume gradient during the delivery of about 3 l. of eluate. The final concentration of ammonium formate is

---

[5] A. W. D. Avison, *J. Chem. Soc.* **732** (1955).

[6] P. Baumgarten, *Ber.* **59**, 1166 (1926). It has also been found satisfactory to substitute the mixture of anhydride and pyridine hydrochloride obtained by a 2:1 molar combination of pyridine and chlorosulfonic acid. In the present case the procedure would be as follows: anhydrous pyridine (10 millimoles, 0.81 ml.) is placed in a 50-ml. beaker in an ice bath, and chlorosulfonic acid (5 millimoles, 0.32 ml.) is added immediately. As soon as the mixture has solidified, the aqueous solution of AMP and bicarbonate is added. The mixture is then stirred at 40° to 50° for 30 minutes and worked up by the usual procedure. The reaction has been scaled up by as much as tenfold over the amounts given here with satisfactory results.

[7] 5′-AMP from muscle (Sigma Chemical Co.) is most satisfactory if material completely free of PAP is needed. 5′-AMP prepared from yeast has been found to contain traces of PAP which can be detected by the very sensitive method of Gregory (Vol. V [130]).

[8] If the preparation is scaled up tenfold over the amounts given here, the volume of resin need only be doubled. Even this amount is probably an excess, since APS is bound strongly by Dowex 1.

$2\,M$. Elution of AMP is complete after about 1200 ml. of eluate. APS is eluted in the range 2400 to 3000 ml.

The fractions containing APS are combined and stirred for 1 hour in the cold with 3 g. of acid-washed Norit A charcoal. The charcoal is recovered by filtration and washed with 300 ml. of water. The charcoal is eluted three times with 20-ml. portions of 50% ethanol containing $0.01\,N$ ammonia. The charcoal eluates are combined and, if necessary, are centrifuged at room temperature to remove traces of charcoal. The solution is evaporated to dryness under reduced pressure, taken up in about 2 ml. of water, and adjusted to pH 7 to 8. This APS solution is stable when kept frozen. For preparation of the solid lithium salt the solution is passed through a column $(2 \times 1$ cm.) of Dowex 50 ($\mathrm{Li^+}$). The column is washed with water until free from material absorbing at 259 mμ, and the combined eluate and washings are lyophilized. The resulting solid is dissolved in 2 ml. of water, and 20 ml. of ethanol are added. The precipitated lithium salt of APS is collected by centrifugation and dried *in vacuo*. This preparation gives the expected analyses for the di lithium salt of APS, $\mathrm{C_{10}H_{12}O_{10}N_5SPLi_2}$.

Determination of APS

Qualitative Detection. In common with other nucleotides, APS may be detected on paper chromatograms and paper electrophoresis patterns by its quenching of ultraviolet fluorescence. Paper electrophoresis is especially valuable when solutions containing buffers and salts are being examined. In $0.025\,M$ sodium citrate, pH 5.8, or $0.05\,M$ ammonium acetate, pH 5.5, APS has a mobility slightly greater than ADP. Detection of APS in the presence of ADP is difficult under these conditions. In $0.025\,M$ $\mathrm{Na_3}$ citrate (unbuffered), or in $0.05\,M$ ammonium acetate, pH 8.5, APS has a mobility between that of AMP and ADP and is readily separated from these two nucleotides. Baddiley *et al.*[3] give the following paper chromatography data:

Solvent A: *n*-Propanol–ammonia (d 0.88)–water (6:3:1)
Solvent B: Isobutyric acid–0.5 N ammonia (5:3)

	R_f values in solvent	
	A	B
AMP	0.40	0.50
ADP	0.30	0.40
APS	0.63	0.32

The fact that APS is completely hydrolyzed in 0.1 N HCl in 60 minutes at 37°, whereas most other nucleotides are stable under these conditions, may be of help in deciding whether or not APS is present in a nucleotide mixture; i.e., electrophoresis or chromatography is carried out before and after exposure to acid conditions.

Quantitative Determination of APS. Since there are no chemical methods available for the determination of phosphate-sulfate anhydrides, it is necessary to rely on enzymatic methods. Either ATP-sulfurylase or APS-kinase may be used for the measurement of APS.[1] However, sulfurylase is better for this purpose because the enzyme is more stable and the method is less tedious. The reaction and the apparent equilibrium constant are as follows:

$$APS^{2-} + PP^{3-} \rightleftarrows ATP^{4-} + SO_4^{2-} + H^+$$

$$K = \frac{[ATP][SO_4]}{[APS][PP]} = 10^8$$

The favorable equilibrium and high affinity of sulfurylase for APS lead to a quantitative and rapid conversion of APS to ATP in the presence of excess PP.

Reagents

1 M Tris hydrochloride, pH 8.5.
0.012 M PP.
Enzyme. Yeast ATP-sulfurylase solution containing 100 to 500 units/ml. The description of the enzyme preparation and definition of the enzyme unit are given in Vol. V [129]. Any of the fractions II to VII are suitable. If fraction II is used, it should be dialyzed overnight before use.
0.1 M $MgCl_2$.
Purified pyrophosphatase, 50 μg./ml. (see Vol. II [91]).
Reagents for the determination of inorganic phosphate.[9]

Procedure. Tris (20 μl.), PP (50 μl.), enzyme solution (30 μl.), and the APS solution to be tested (100 μl. of a 0.5 to 5.0 mM solution) are placed in a test tube calibrated at 10 ml. and incubated for 20 minutes at 37°. The reaction is stopped by heating the tube in a boiling-water bath for 90 seconds. After the tube has been chilled in ice, 20 μl. of $MgCl_2$ and 100 μl. of pyrophosphatase solution are added, and the tube is incubated at 37° for 5 minutes to convert pyrophosphate to inorganic phosphate. Phosphate analysis is carried out directly in the calibrated tube. Appropriate controls for each determination are prepared by boiling the solution of Tris, PP, and enzyme before the addition of the APS solution, or by omitting APS from the incubation. The APS sample must

[9] L. F. Leloir and C. E. Cardini, Vol. III [115].

be free of organic solvents, heavy metal ions, or other materials that might inactivate the enzyme. The disappearance of 1 micromole of PP ($\backsimeq$ 2 micromoles P_i) is equivalent to 1 micromole of APS. The omission of Mg^{++} from the first incubation decreases the interference from ATPase and pyrophosphatase activities that may be present in the sulfurylase preparation.

A number of other final analytical methods may be used for the determination of the amount of PP that has disappeared or the amount of ATP that has formed as a result of the first incubation. For example, the ATP may be measured with glucose, hexokinase, and glucose-6-phosphate dehydrogenase.[10] The agreement between this method and pyrophosphate disappearance has been shown.[1] Alternatively, the ATP could be absorbed to charcoal and measured as inorganic phosphate after the charcoal has been heated in $1\,N$ HCl at $100°$ for 12 minutes.[11] The use of radioactive pyrophosphate and measurement of the radioactivity absorbable to charcoal after the incubation with APS would undoubtedly increase the sensitivity of the method.

The degree of purity of an APS preparation may be judged from the ratio of the amount of APS, as determined above, to the total amount of adenine nucleotide in the sample ($a_M = 15,400$ [259 mμ, pH 7]).

II. 3'-Phosphoadenosine-5'-phosphosulfate (PAPS)

Since the chemical synthesis of PAPS has been described,[12] only the enzymatic preparation from ATP is presented here. The procedure for S^{35}-labeled PAPS is described below, since more often than not the labeled substrate is required for metabolic studies.

Enzymatic Synthesis and Isolation of PAPS[35]

Principle. Crude enzyme systems from both liver and yeast contain the activating system for the synthesis of PAPS from ATP and inorganic sulfate. The reaction and preparation of the enzymes are described in detail in Vol. V [129]. The yeast enzyme is used for making labeled PAPS because it can be freed of ammonium sulfate by dialysis without loss of activity.

Reagents

1 M Tris hydrochloride, pH 8.
0.05 M K_2SO_4.

---

[10] A. Kornberg and W. E. Pricer, Jr., *J. Biol. Chem.* **193**, 365 (1951).
[11] R. K. Crane and F. Lipmann, *J. Biol. Chem.* **201**, 235 (1953); W. P. Jencks and F. Lipmann, *ibid.* **225**, 207 (1957).
[12] J. Baddiley, J. G. Buchanan, and R. Letters, *Proc. Chem. Soc.* p. 147 (1957).

0.2 M MgCl$_2$.

0.1 M ATP, neutralized.

5 mc. of carrier-free S^{35}-SO$_4$$^{2-}$ from Oak Ridge National Laboratory, neutralized to pH 7 to 8 with NaOH. Although this material contains traces of other radioactive compounds, these are eliminated in the purification of PAPS.

Enzyme. Yeast sulfate-activating system (Vol. V [129], Section III.B). The enzyme is dialyzed for 18 to 24 hours against one or two changes of 20 mM Tris–HCl, pH 8, to remove the ammonium sulfate. The protein concentration should be 15 to 30 mg./ml. after dialysis.

Procedure. One milliliter each of Tris, K$_2$SO$_4$, MgCl$_2$, and ATP are placed in a centrifuge tube. The enzyme (3 ml.), neutralized S^{35}-SO$_4$$^{2-}$ solution, and water are then added to give a final volume of 10 ml. The inorganic sulfate concentration used here (5 mM) is equal to the concentration giving a half-maximal initial rate of PAPS formation (Vol. V [129], Section III.B). This represents a compromise between the need for a significant molar yield and the need for PAPS of high specific activity. The specific activity of the final product may be lower than the theoretical 100 μc/micromole, probably because of the presence of inorganic sulfate in the enzyme solution. If PAPS of higher or lower specific activity is required, more or less K$_2$SO$_4$ may be added to the incubation. If nonradioactive material is being prepared, a K$_2$SO$_4$ concentration of 100 mM may be used.

After 90 minutes of incubation at 37°, the solution is heated in boiling water for 3 minutes, cooled, and centrifuged. The supernatant solution is diluted to 100 ml. with water and passed through a 1 × 15-cm. column of washed Dowex 1–chloride. The column is washed with 70 ml. of water which is discarded. The original eluate and wash are completely free of nucleotide and radioactivity.

Separation of the PAPS from the other anions is carried out by essentially the method of Brunngraber.[13] The column is eluted with 400 ml. of 0.5 M NaCl which is discarded. During this elution with 0.5 M NaCl, inorganic sulfate emerges as a sharp peak between 80 and 120 ml. The nucleotide peaks appear between 120 and 250 ml. The final 150 ml. of NaCl should bring the optical density at 259 mμ to 0.1 or less. PAPS is eluted with 300 ml. of 1 M NaCl. The nucleotide and radioactivity peaks occur at about 150 ml. of 1 M NaCl. The PAPS fraction is passed through a 1-g. column of acid-washed Norit A–Celite 535 (1:2). The column is washed with 40 ml. of water which is discarded. The PAPS is

[13] E. G. Brunngraber, *J. Biol. Chem.* **233**, 472 (1958).

eluted with 50 ml. of 50% ethanol containing 0.01 N ammonia. The solvent is removed by flash evaporation at 30° or similar means, and the residue is dissolved in 3 ml. of water and adjusted to pH 7 to 8. If necessary, the solution is centrifuged and then is stored frozen. In a typical preparation, the final PAPS concentration was 0.5 mM, and the specific activity was 10^8 c.p.m./micromole in a windowless gas-flow counter.

Properties. The PAPS preparation contains less than 5% inorganic sulfate. The only significant nucleotide and radioactive impurity is APS. The APS varies from less than 1% to 10% of the PAPS. PAPS has a half-life of 6 minutes in 0.1 N HCl at 37°. There is no detectable hydrolysis in 0.1 N NaOH at 37° in 2 hours. The alkali stability of the phosphosulfate bond is shown by the demonstration of Baddiley *et al.*[14] that hydrolysis of PAPS is incomplete even after 2 hours in 0.1 N NaOH at 100°. APS and PAPS are stable indefinitely when kept at pH 8 in frozen solution. The phosphosulfate anhydride is not attacked at an appreciable rate by hydroxylamine. PAPS is hydrolyzed to APS by rye grass 3'-nucleotidase,[15] but the phosphosulfate bonds of PAPS and APS are quite resistant to enzymatic hydrolysis. Only snake venom[3] and bull semen 5'-nucleotidase[15] have been found to hydrolyze APS. In both cases the products are adenosine, inorganic sulfate, and inorganic phosphate. Neither enzyme hydrolyzes PAPS.

Qualitative Detection. As with APS, paper electrophoresis and paper chromatography are the most convenient methods for qualitative detection. On paper, PAPS may be seen as an ultraviolet fluorescence-quenching area, or, if radioactive sulfate has been used for the preparation, radioautography may be used. There is no specific test for the phosphosulfate anhydride on paper, but exposure of an aliquot of the material to be tested to 0.1 N HCl at 37° for 60 minutes should lead to nearly complete hydrolysis. Electrophoresis or chromatography before and after this treatment may be of help in deciding whether or not a nucleotide is PAPS.

Under the usual conditions of paper electrophoresis (pH 4 to 9) PAPS has greater mobility than ATP. Inorganic sulfate has a much greater mobility than the adenine nucleotides and is easily distinguished from PAPS. Electrophoresis with 0.025 M citrate, pH 5.8, has the advantage of distinguishing PAP, the hydrolysis product of PAPS, from ATP, ADP, and PAPS. At pH below 5.8, PAP has a mobility similar to ADP, whereas at higher pH separation of PAP from ATP is difficult.

---

[14] J. Baddiley, J. G. Buchanan, R. Letters, and A. R. Sanderson, *J. Chem. Soc.* 1731 (1959).

[15] P. W. Robbins and F. Lipmann, *J. Biol. Chem.* 229, 837 (1957).

Baddiley *et al.*[14] give the following data for electrophoresis carried out for 6 hours on Whatman No. 1 paper soaked in 0.025 M sodium citrate buffer, pH 5.8. The voltage gradient was 10 volts/cm.

	Distance moved toward anode, cm.
Adenosine-5'-phosphate	18.7
Adenosine-3'-phosphate	20.2
Adenosine-2'-phosphate	20.2
Adenosine-2',5'-diphosphate	34.0
Adenosine-3',5'-diphosphate	34.1
Adenosine-5'-pyrophosphate	29.6
Adenosine-5'-triphosphate	37.4
Adenosine-3'-phosphate-5'-phosphosulfate	44.1

The same authors give the following paper chromatography data:

Solvent A: *n*-Propanol–ammonia (*d* 0.88)–water (6:3:1)
Solvent B: Saturated ammonium sulfate–0.1 M ammonium
    acetate–isopropanol (79:19:2)

	R_f in solvent A	B
Adenosine-5'-phosphate	0.28	0.36
Adenosine-2'-phosphate	0.33	0.31
Adenosine-3'-phosphate	0.33	0.21
Adenosine-2',5'-diphosphate	0.10	0.48
Adenosine-3',5'-diphosphate	0.11	0.40
Adenosine-5'-pyrophosphate	0.10	0.50
Adenosine-5'-triphosphate	0.17	—
Adenosine-2',3'-phosphate-5'-phosphate	0.23	—
Adenosine-3'-phosphate-5'-phosphosulfate	0.18	0.29

Quantitative Determination of PAPS. PAPS has the typical adenosine absorption spectrum, and concentration of pure solutions may be determined from the molar absorbancy index of 15,400 at 259 mμ (pH 7). This should be used only after the purity of the preparation has been checked by paper electrophoresis or paper chromatography.

The problems involved in the quantitative measurement of PAPS are discussed in Vol. V [129] and [130]. For preparations free of PAP, a simple and accurate method is that of Gregory.[16] In this procedure, PAPS or PAP acts as a catalyst for the transfer of sulfate from nitrophenyl sulfate to phenol in the presence of phenol sulfokinase. Gregory's

[16] J. D. Gregory, Vol. V [130].

article should be consulted for a discussion of the principle and details of the method. Although the fact that PAP responds to this assay as well as PAPS is a disadvantage, it has been found that the assay gives accurate results in enzymatic reaction mixtures. The probable reason for this is that no enzymatic hydrolysis of the phosphosulfate bond takes place in the yeast and liver extracts that have been used for sulfate activation. In fact, enzymatic hydrolysis of this bond has not yet been observed with any tissue extract or protein preparation. The conversion of PAPS to PAP takes place only if a sulfokinase and the appropriate acceptor are present, or if the preparation is exposed to acid. Brief boiling at pH 8, the method usually used to stop the activation reaction, does not give detectable hydrolysis. Although the hydrolysis of PAPS to APS and inorganic phosphate does take place in tissue extracts, APS is not detected and does not interfere in the catalytic PAP–PAPS assay. When the assay is applied to material that has been purified by Dowex 1 chromatography, the absence of PAP may be demonstrated by paper electrophoresis or paper chromatography.

In principle, PAPS can be determined by transferring the sulfate group to a phenol or steroid in the presence of the appropriate sulfokinase, followed by the determination of the amount of phenyl or steroid sulfate formed. This method is generally satisfactory, but its accuracy is limited by the presence of sulfatase and 3'-phosphatase activity in the crude sulfokinase preparations that are used for this purpose. The method given below uses p-nitrophenol as the sulfate acceptor.[15] A more suitable acceptor from the point of view of the final equilibrium is m-aminophenol, and Brunngraber has described a transfer method with this phenol as sulfate acceptor.[13]

Reagents

1 M imidazole hydrochloride, pH 7.0.
0.1 M cysteine hydrochloride freshly neutralized to pH 7 with NaOH.
0.01 M p-nitrophenol.
95% ethanol.
0.1 N KOH.
Enzyme. Phenol sulfokinase is prepared according to Robbins and Lipmann[15] or Gregory,[16] or by another method.[13, 17]

Procedure. Imidazole, cysteine, and p-nitrophenol (0.05 ml. each) are placed in a small test tube. The PAPS sample, water, and enzyme are then added to give a final volume of 0.5 ml. The PAPS sample should

[17] Y. Nose and F. Lipmann, *J. Biol. Chem.* **233**, 1348 (1958).

be free of organic solvents, heavy metal ions, or other materials that might inactivate the enzyme. The amount of enzyme required must be determined in preliminary experiments that measure the rate of transfer when excess PAPS is present. After incubation for 30 minutes at 37°, 2 ml. of ethanol is added, and the mixture is freed of the precipitated protein by centrifugation. To 5.0 ml. of 0.1 N KOH is added 1.0 ml. of the supernatant fluid, and, after mixing, the extinction is measured in the Klett-Summerson photometer (filter 42) or in a spectrophotometer at 400 mμ. The sample is compared with an unincubated control, or to a blank without PAPS added. Nitrophenol disappearance is equivalent to nitrophenyl sulfate formation. No metabolism of nitrophenol occurs other than sulfurylation under these conditions.

The initial concentration of PAPS must be 0.5 mM or less. Otherwise, the transfer is incomplete because the reverse reaction, between nitrophenyl sulfate and PAP, becomes appreciable. Gregory[16] has shown that the apparent equilibrium constant for the transfer of sulfate from PAPS to p-nitrophenol is 26:

$$K = \frac{[p\text{-NPS}][\text{PAP}]}{[p\text{-NP}][\text{PAPS}]} = 26$$

Therefore, if 50% of the nitrophenol has reacted with PAPS at equilibrium, the analytical error in the determination of PAPS will be 4%. When more than 50% of the nitrophenol is converted to the sulfate, the necessary correction becomes larger. The amount of enzyme added should be enough to establish equilibrium in the 30-minute incubation period. The value for PAPS concentration by this method will be too low if transfer is not complete, or if PAPS or nitrophenyl sulfate is hydrolyzed by interfering activities in the sulfokinase preparation.

[106] Preparation and Properties of Firefly Luciferyl and Oxyluciferyl Adenylic Acid

By W. D. McElroy

Preparation

The preparation of LH$_2$-AMP and L-AMP described is based on the original report of Rhodes and McElroy.[1,2] The evidence that these two

[1] W. C. Rhodes and W. D. McElroy, *Science* **128**, 253 (1958).
[2] W. C. Rhodes and W. D. McElroy, *J. Biol. Chem.* **233**, 1528 (1958).

acyl adenylates are products of the firefly luciferase-catalyzed ATP reaction has been reported by McElroy and Green[3] and by Airth *et al.*[4]

The procedure used in preparing LH_2-AMP is as follows: 50 μg. of solid crystalline luciferin[5] are mixed with approximately 100 μg. of solid crystalline sodium adenylate in a small test tube. To this mixture is added 0.1 ml. of pyridine containing 10 mg. of dicyclohexylcarbodiimide.[6] The mixture is incubated at room temperature for about 1 hour with occasional shaking, after which 0.5 ml. of 0.05 M bicarbonate buffer at pH 7.8 is added. The result is a white cloudy mixture which is then clarified by three successive extractions with a total of 1.5 ml. of ether. The solution is adjusted to a pH between 2 and 3 with 1 N HCl and extracted three times with a total volume of 1.5 ml. of redistilled ethyl acetate. This procedure removes luciferin and luciferyl anhydride.

The extraction of luciferin may be perceived by observing the blue fluorescence of the ethyl acetate layer. The fluorescence of the LH_2-AMP remaining in the water phase is yellow. Traces of ethyl acetate are removed by passing a stream of nitrogen into the mixture. The pH of the solution is adjusted to 5.0 with NaOH. The oxyluciferyl adenylate was prepared in the same manner.

Properties

LH_2-AMP hydrolyzes very rapidly at alkaline pH. The half-life at pH 7.5 (0.1 M Tris–maleate buffer) is approximately 10 minutes. The residual LH_2-AMP was determined by measuring light emission in the presence of luciferase. The assay mixture consisted of 0.1 ml. of LH_2-AMP, 0.01 ml. of luciferase (5 mg. of protein per milliliter), and Tris–maleate buffer at pH 7.0 to give a final volume of 2.5 ml. The over-all reaction may be described as follows:

$$LH_2\text{-AMP} + E \overset{K_m}{\rightleftharpoons} E\text{–}LH_2\text{-AMP} \overset{k_3}{\rightarrow} E\text{–}L\text{-AMP} + \text{light}$$

The K_m value was calculated to be $2.3 \times 10^{-7}\ M$, and k_3 equals 1.9×10^{-1} sec.$^{-1}$. The dissociation constant of the luciferase–L-AMP complex is 5×10^{-10}, and the free energy of hydrolysis of L-AMP is -13.1 kcal.

The L-AMP when tightly bound to the enzyme readily reacts reversibly with CoA to form oxyluciferyl CoA and adenylic acid. When

[3] W. D. McElroy and A. A. Green, *Arch. Biochem. Biophys.* **64**, 257 (1956). See Vol. VI [63].
[4] R. L. Airth, W. C. Rhodes, and W. D. McElroy, *Biochim. et Biophys. Acta* **27**, 519 (1958).
[5] B. Bitler and W. D. McElroy, *Arch. Biochem. Biophys.* **72**, 358 (1957).
[6] H. G. Khorana, *J. Am. Chem. Soc.* **76**, 3517 (1954).

pyrophosphate reacts with E–L-AMP, free oxyluciferin and ATP are formed.

In acid pH, LH_2-AMP gives a fluorescence emission at 565 mμ which corresponds closely to the bioluminescence emission at neutral pH. Excitation wavelength is at 327 mμ (peak absorption). L-AMP under similar conditions (excitation at 349 mμ) gives a red fluorescence with a peak at 620 mμ which corresponds to the bioluminescence emission at pH 5 to 5.5. Oxyluciferin has a brilliant fluorescence emission at 540 mμ, but this is greatly depressed when the adenylate is formed. Consequently one can readily follow the enzymatic formation of L-AMP from L and ATP by observing the decrease in fluorescence at this wavelength. The equilibrium constant for this activation process is 2.5×10^5, which corresponds to a free energy of formation of E–L-AMP of -7.2 kcal.

An enzyme has been found in the firefly which catalyzes the hydrolysis of oxyluciferyl adenylate. However, luciferase-bound oxyluciferyl adenylate will not act as a substrate. The use of this enzyme and the known increase in fluorescence when L-AMP is hydrolyzed makes it possible to measure the rate of dissociation of the luciferase–oxyluciferyl adenylate complex. The constant (k) was found to be 1.3×10^{-3} sec.$^{-1}$.

[107] Isolation of Uridine Diphosphate Glucose, Uridine Diphosphate Acetylglucosamine, and Guanosine Diphosphate Mannose

By LUIS F. LELOIR and ENRICO CABIB

Principle

An alcoholic extract from toluene-autolyzed yeast is fractionated on anion-exchange columns. The nucleotides are recovered by adsorption on charcoal and subsequent elution with aqueous ethanol.[1]

Yeast Extract

Ten kilograms of baker's yeast[2] are brought to 36° and intimately mixed with 1000 ml. of warm toluene. After incubation at 35° to 37° for

---

[1] H. G. Pontis, E. Cabib, and L. F. Leloir, *Biochim. et Biophys. Acta* **26**, 146 (1957); see also Vol. III [143].

[2] Ordinary starch-free baker's yeast may be used. Higher and more uniform yields were obtained with "starter" yeast, that is, the yeast used as inoculum in large scale industrial culture. The amount indicated is usually enough for two or three chromatographic runs.

40 minutes, 10 l. of 95% ethanol are added, and the mixture is heated with stirring until it boils. After standing overnight in the refrigerator, it is filtered through a 32-cm. Büchner funnel with a filter aid. The extract is brought to pH 7 with bromothymol blue as indicator just before chromatography. Sometimes a turbidity develops on standing or after neutralization. In such cases the extract is again filtered.

Column Chromatography

The columns are 120 cm. long. The diameter is 0.9 cm. for the analytical columns and 4.5 cm. for the preparative ones.

Automatic fraction collectors are employed; an automatic recorder of the ultraviolet absorption, like the one attached to the Gilson fraction collector (Gilson Medical Electronics, Madison, Wisconsin), is very convenient.

The resin used is Dowex 1-X4, 200 to 400 mesh, in the chloride form. The resin is freed from fines by repeated decantation from water, and the fraction that sediments in 20 minutes in a 4-l. beaker is separated for use. This fraction is washed a few times alternately with 1 N sodium hydroxide and 1 N hydrochloric acid before use, and this treatment is repeated after several runs. If analytical reagent resin is available (California Corporation for Biochemical Research, Los Angeles, Cal.), it can be used without previous treatment. The resin is poured as a slurry into a column and allowed to settle until a height of 100 cm. is attained.

The nucleotides are eluted with a chloride gradient at constant pH. A solution 0.01 N in hydrochloric acid and 0.1 N in sodium chloride is fed into a mixing chamber containing a solution of the same acid concentration and 0.02 N in sodium chloride. The flow rate is regulated by air pressure, with a control system such as the Moore Nullmatic Pressure Regulator (Moore Products Co., Philadelphia, Pennsylvania). After each run the darkened top layer of the column is removed and replaced with fresh resin. The column is then washed with 1 N hydrochloric acid until the absorbancy at 260 mμ drops under 0.05. The resin is removed and washed batchwise two or three times with water. Then the resin is replaced in the column and is washed with water until the pH of the effluent is above 5.

Analytical Columns. Since the content of the nucleotides to be isolated in the yeast available at the authors' laboratory is rather variable, it has been found advantageous to run an analytical column for each new batch, in order to ascertain the amount of extract to be used in the large-scale preparation. The mixing chamber is a 2000-ml. round-bottomed flask. An amount of extract containing about 600 micromoles (calculated as uridine from the absorbancy at 260 mμ) is passed through the column at a rate of 0.4 ml./min. The chloride gradient is immediately started, and 15-ml. fractions are collected at the same rate. A large amount of ultraviolet material is eluted in the beginning, followed by many fractions of low absorbancy, and, after about 2000 ml., the GDPM peak appears, followed by the UDP-acetylglucosamine and UDPG peaks as shown in Fig. 1. The nucleotides are identified from the ultraviolet

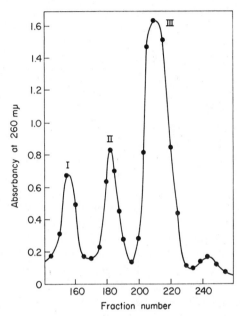

Fig. 1. Separation of nucleotides on an analytical column. Peak I, GDPM; peak II, UDP-acetylglucosamine; peak III, UDPG. The small peak after UDPG is uridine diphosphate mixed with an unidentified uridine compound.

spectrum. The fractions belonging to each individual peak are pooled, and the absorbancy at 260 mμ is determined in order to ascertain the total amount of each nucleotide. It should be noted that it is also possible to run an analytical column only 30 cm. long, with a considerable saving of time. Although the nucleotides are only partly separated in this case, it is possible to calculate the approximate amount by estimating the area under the curves.

Preparative Columns. In this case the mixing chamber is a 12-gallon Pyrex or polyethylene bottle. An amount of extract containing up to 1200 micromoles of UDPG is passed through the column at a rate not greater than 8 ml./min. The gradient is started at the same rate. The first 40 l. are discarded, and then 500-ml. fractions are collected. The absorbancy of the fractions is read at 260 mμ. The individual peaks are pooled and adjusted to about pH 6.5.

Concentration

Charcoal columns are prepared as follows: a suspension of 25 g. of Celite (Hyflo Super-Cel) in water is poured into a Büchner funnel of 9.5-cm. diameter, fitted with a filter-paper disk (a medium-porosity fritted-glass funnel of the same size can be used with advantage). The water is sucked off, and the layer of Celite is covered with another filter-paper disk. Then a slurry made up with 30 g. of Norit A and 40 g. of Celite in water is poured into the funnel, and the water is withdrawn with gentle vacuum, care being taken that air does not enter into the column. Finally, the charcoal layer is covered with a filter-paper disk, held in place with a glass ring. Some water is added, and the funnel is closed with a rubber stopper provided with an inlet tube. The stopper is tied onto the bottom of the funnel with string, to resist the pressure. In order to prevent the column from drying out a syphon tube is attached to the stem of the funnel by means of a rubber connection. All the Celite used for these columns is previously weighed and then suspended in 1 l. of water and allowed to settle for 15 minutes. The supernatant fluid is discarded.

A column of this size can be used for at least 1200 micromoles of nucleotide. The pooled and neutralized peak is passed through the column in the refrigerator at a rate of 5 ml./min. Then 400 ml. of 0.01 M EDTA at pH 7 are percolated at 2 ml./min., followed by 400 ml. of water. The absorbancy of these liquids at 260 mμ is measured to check the procedure. Each time the remaining liquid on top of the column is withdrawn with gentle suction. Finally aqueous ethanol (50 ml. of 95% ethanol brought to 100 ml. with water) is passed through the column at room temperature, at a rate of 1 ml./min., and 125-ml. fractions are collected. Usually 80 to 90% of the nucleotide is recovered in the first three to four fractions. These are brought to pH 6.7 to 6.9 and evaporated to dryness at 35° under reduced pressure, with frequent checks of the pH. The residue is dissolved in 2.5 ml. of water and centrifuged if necessary to sediment a small amount of Celite and charcoal that may have seeped through. The supernatant liquid is brought to 3.5 ml. with water, and 2.5 M barium bromide solution is added (about 1.2 ml./ millimole of nucleotide). A small precipitate usually appears at this point. The suspension is acidified to about pH 4.7 with hydrobromic acid, and the precipitate is centrifuged off. The supernatant fluid is transferred to a 50-ml. centrifuge tube and neutralized with 0.3 N barium hydroxide, and 20 to 25 ml. of 95% ethanol are added. The UDPG precipitate is usually sticky and has a tendency to conglomerate, so that it has to be broken up by pressing it against the wall of the tube with a spatula. The tube is stoppered and left in the refrigerator overnight. After centrifugation, the precipitate is washed twice with 95% ethanol and twice with ether.

Then the tube is placed in a desiccator, and the remaining ether is pumped out. After several hours the desiccator is opened and the substance weighed.

Properties and Purity

The properties of UDPG have been described in Vol. III [143].

The liberation of acetylglucosamine from UDP-acetylglucosamine[3] in 0.01 N acid at 100° is practically complete in 15 minutes. The lability in acid of the phosphate linked to the sugar is the same as for UDPG.

Mannose is very easily hydrolyzed from GDPM[4] by 0.01 N acid at 100°. Ten minutes are sufficient for quantitative splitting. Previous reports[4] on the lack of reducing power of intact GDPM were found to be erroneous.[5] The reducing power seems to be due to splitting of mannose by the alkaline reagents used for these determinations.

About 50% of the total phosphate of GDPM is liberated by 1 N acid in 20 minutes or by 0.1 N acid in 120 minutes.

On the basis of the phosphate content the nucleotides prepared by the above procedure are about 65% pure. The ratio of ultraviolet absorbancy to total phosphate is usually 0.6 to 0.65 (theoretical 0.5). Sometimes faint foreign spots appear under ultraviolet light on paper chromatograms of the nucleotides in ethanol–ammonium acetate solvents.[6] GDPM is frequently contaminated with adenylosuccinic acid.[7] This compound can be eliminated by preparative paper chromatography on Whatman 3 MM paper by using ethanol–ammonium acetate of pH 7.5[6] as solvent. Adenylosuccinic acid runs behind GDPM under these conditions. Before the GDPM is eluted with water, most of the ammonium acetate is washed out of the paper with 95% ethanol.

The barium salts do not seem to be hygroscopic.

Analysis

The nucleotides are usually analyzed for purity by determination of the ultraviolet spectrum (uridine or guanosine as standard) of the acid-labile and total phosphate and of the acid-labile sugar.

[3] E. Cabib, L. F. Leloir, and C. E. Cardini, *J. Biol. Chem.* **203**, 1055 (1953).
[4] E. Cabib and L. F. Leloir, *J. Biol. Chem.* **206**, 779 (1954).
[5] H. G. Pontis and E. Cabib, unpublished observations, 1959.
[6] A. C. Paladini and L. F. Leloir, *Biochem. J.* **51**, 426 (1952).
[7] C. E. Carter and L. H. Cohen, *J. Am. Chem. Soc.* **77**, 499 (1955).

The acetylglucosamine of UDP-acetylglucosamine may be assayed after acid hydrolysis by the method of Reissig et al.[8]

For the mannose of GDPM, after being heated for 15 minutes at 100° in 0.05 N sulfuric acid and neutralized with 0.06 N barium hydroxide, the nucleotides are precipitated by adding equal volumes of 5% zinc sulfate and 0.3 N barium hydroxide. If an unheated sample is used as blank, it must be submitted to the same treatment; otherwise GDPM will interfere in the succeeding determination. After centrifugation, mannose is assayed in the supernatant liquid by the reducing power.

[8] J. L. Reissig, J. L. Strominger, and L. F. Leloir, J. Biol. Chem. **217**, 959 (1955).

[108] Preparation of UDP-D-Xylose and UDP-L-Arabinose

By D. S. FEINGOLD, E. F. NEUFELD, and W. Z. HASSID

Principle

The UDP-pentoses (UDP-D-xylose and UDP-L-arabinose) can be prepared by two types of reaction catalyzed by plant enzymes. One of these types is catalyzed by UDP-D-glucuronic acid decarboxylase,[1] reaction (1), and the other by UDP-pentose pyrophosphorylase,[2] reactions

$$UDP\text{-}D\text{-glucuronic acid} \rightarrow UDP\text{-}D\text{-xylose} + CO_2 \qquad (1)$$

(2) and (3).

$$UTP + \alpha\text{-}D\text{-xylose 1-phosphate} \rightleftarrows UDP\text{-}D\text{-xylose} + \text{pyrophosphate} \quad (2)$$

$$UTP + \beta\text{-}L\text{-arabinose 1-phosphate} \rightleftarrows UDP\text{-}L\text{-arabinose} + \text{pyrophosphate} \qquad (3)$$

Since many crude plant preparations also contain a UDP-L-arabinose 4-epimerase,[2] which catalyzes the interconversion of UDP-D-xylose and UDP-L-arabinose, a mixture of the two UDP-pentoses is usually obtained.

Assay

The two UDP-pentoses can be separated from all other components present in the reaction mixtures by paper electrophoresis in 0.2 M ammonium formate buffer, pH 3.6 (6.3 g. of ammonium formate and 3.75 ml. of 98% HCOOH per liter of solution), at a minimal voltage of 25

[1] D. S. Feingold, E. F. Neufeld, and W. Z. Hassid, J. Biol. Chem. **235**, 910 (1960).
[2] E. F. Neufeld, V. Ginsburg, E. W. Putman, D. Fanshier, and W. Z. Hassid, Arch. Biochem. Biophys. **63**, 602 (1957).

volts/cm. on Whatman No. 1 paper, washed with either oxalic or acetic acid. The electrophoretic mobility (Table I) of the UDP-pentoses in this

TABLE I

ELECTROPHORETIC MOBILITIES OF REACTION COMPONENTS RELATIVE TO THE MOBILITY OF PICRIC ACID ($M_{\text{P.A.}}$) IN 0.2 M AMMONIUM FORMATE BUFFER, pH 3.6

Substance	$M_{\text{P.A.}}$
Inorganic pyrophosphate	2.1
UTP	1.6
α-D-glucuronic acid 1-phosphate	1.6
UDP-D-glucuronic acid	1.5
Inorganic phosphate	1.5
UDP	1.4
UDP-D-xylose or UDP-L-arabinose	1.2
α-D-xylose 1-phosphate or β-L-arabinose 1-phosphate	1.0
D-glucuronic acid	0.9
UMP	0.9

buffer is 1.2 times that of picric acid, which is used as a visible mobility indicator. A quantity of 0.1 to 0.4 micromole of nucleotide material in a maximal volume of 25 μl. per spot is applied to the paper. An adequate separation of the components is attained when the picric acid has migrated about 15 cm. from the point of application. The sugar nucleotide zones are located by visual inspection or by contact printing under an ultraviolet lamp emitting light in the 260-mμ region. The sugar nucleotides are eluted from the paper, diluted to an appropriate volume, and the concentration of the eluate is determined by the optical density at 262 mμ (a_M for uridine nucleotides $= 10.0 \times 10^{-3}$ at pH 7).[3] A correction must be made for ultraviolet-absorbing material eluted from the paper.

Preparation of UDP-Pentose by the UDP-D-Glucuronic Acid Decarboxylase Reaction

Preparation of Enzyme. Although UDP-D-glucuronic acid decarboxylase occurs in particulate preparations from many plants, it is most conveniently prepared from commercial wheat germ. Wheat germ (50 g.) is stirred at 2° for 2 hours with 100 ml. of 0.05 M Tris buffer, pH 7.5; the slurry is filtered through cheesecloth and centrifuged in the cold at 18,000 $\times$ g for 1 hour. The slightly turbid supernatant liquid containing the enzyme is pipetted off and used for the preparation of UDP-pentose. The crude preparation is active at −10° for at least a year.

[3] Pabst Laboratories, Circular OR-15 (1959).

Preparation of UDP-Pentose

Reagents

UDP-D-glucuronic acid, 0.1 M, neutralized with $NaHCO_3$ to pH 7.5.
Wheat germ enzyme preparation.

Ten volumes of enzyme preparation are mixed with 1 vol. of UDP-D-glucuronic acid and incubated at 25°. Since the sugar nucleotides are slowly degraded by nucleotide pyrophosphatase present in wheat germ extracts, the time of incubation which will result in a maximal yield of UDP-pentose must be determined by a preliminary experiment. Usually an incubation period of 2 to 4 hours is required.

Preparation of UDP-Pentoses by the UDP-Pentoses Pyrophosphorylase Reaction

Preparation of Enzyme. All operations are carried out at 0° to 4°. Four- to five-day-old mung bean seedlings (100 g.) are homogenized for 1 minute in a Waring blendor with 70 ml. of 0.01 M phosphate buffer at pH 7. The slurry is filtered through cheesecloth and then centrifuged at 18,000 $\times$ g for 30 minutes. The supernatant solution is fractionated with solid ammonium sulfate, and the protein precipitating between 55% and 65% saturation is taken up in 1 to 2 ml. of 0.1 M Tris buffer, pH 7.5. The resulting suspension is dialyzed overnight against 1 l. of the same buffer, and any residual precipitate is removed by centrifugation at 18,000 $\times$ g for 30 minutes and discarded. The supernatant liquid contains UDP-pentose pyrophosphorylase activity, which is stable at −10° for many months.

A freshly made preparation of UDP-pentose pyrophosphorylase possesses UDP-L-arabinose 4-epimerase activity. The 4-epimerase activity is gradually lost during storage; however, no completely reproducible method for removing the 4-epimerase from the pyrophosphorylase has yet been devised.

An active preparation of UDP-pentose pyrophosphorylase may also be obtained from wheat germ. A wheat germ extract is prepared in the manner described for the preparation of UDP-D-glucuronic acid decarboxylase and fractionated with solid ammonium sulfate. The protein precipitating between 50% and 70% saturation is suspended in 1 to 2 ml. of 0.1 M Tris buffer, pH 7.5, and dialyzed overnight against 1 l. of the same buffer. Insoluble material is removed by centrifugation at 18,000 $\times$ g for 30 minutes and discarded. The supernatant liquid is then used for the preparation of UDP-pentose.

Preparation of α-D-Xylose 1-Phosphate and β-L-Arabinose 1-Phos-

phate. The barium salts of α-D-xylose 1-phosphate and β-L-arabinose 1-phosphate are prepared by the method of Cori *et al.*,[4] as described by Posternak (see Vol. III [16A]) for the preparation of α-D-glucose 1-phosphate, except that smaller quantities of reagents (multiplied by a factor of 0.83) are used. The sugar phosphates are then purified by conversion to the dicyclohexylammonium salts and recrystallization.[5]

1,2,3,4-Tetra-O-acetyl-β-D-xylopyranose and 1,2,3,4-tetra-O-acetyl-α-L-arabinopyranose are first prepared by treatment of the pentoses with acetic anhydride and anhydrous sodium acetate. The crystalline acetylated D-xylose and the sirupy acetylated L-arabinose (the latter crystallizes with difficulty) are converted to the respective 2,3,4-tri-O-acetyl-α-D-xylopyranosyl bromide and 2,3,4-tri-O-β-L-arabinosyl bromide derivatives by treatment with hydrogen bromide.

The triacetyl pentosyl bromides are then treated with trisilver phosphate, and the resulting products, mainly triglycosyl phosphates, are converted to the pentose 1-phosphates by partial hydrolysis and deacetylation; the crude esters are isolated as the barium salts are converted to the dicyclohexylammonium salts for further purification as follows[5]:

A 1.5-g. sample of each of the barium salt preparations is dissolved in 15 ml. of water and passed through a small column of Dowex 50 (H⁺) to adsorb the barium. The effluent is neutralized with a 0.5 M alcoholic solution of cyclohexylamine and evaporated to dryness under reduced pressure. Both compounds crystallize as the cyclohexylammonium salts during concentration of the solutions.

The crude cyclohexylammonium salts of the ester preparations are collected by filtration, washed with ethanol, and air-dried. The compounds are purified as follows: The dry salts are dissolved in water, treated with Darco G-60, filtered, and concentrated in a vacuum desiccator. The dicyclohexylammonium salts crystallize during concentration of their aqueous solutions and are recrystallized from 80% ethanol.

Preparation of UDP-Pentoses

Reagents

0.1 M dicyclohexylammonium α-D-xylose 1-phosphate or 0.1 M dicyclohexylammonium β-L-arabinose 1-phosphate.
0.05 M MgCl₂.
Yeast inorganic pyrophosphatase (see Vol. II [91]), 3 units/ml. in 0.1 M Tris, pH 7.5.

[4] C. F. Cori, S. P. Colowick, and G. T. Cori, *J. Biol. Chem.* **121**, 465 (1937).
[5] E. W. Putman and W. Z. Hassid, *J. Am. Chem. Soc.* **79**, 5057 (1957).

0.1 M uridine triphosphate (UTP), neutralized to pH 7.5 with NaHCO$_3$.

Mung bean enzyme preparation.

The reagents are mixed in the following proportions (by volume): 0.1 M UTP, 1.0; 0.1 M sugar phosphate, 1.0; 0.05 M MgCl$_2$, 0.5; inorganic pyrophosphatase, 0.5, mung bean preparation, 2.0. The reaction is run at 37°. The action of inorganic pyrophosphatase displaces the equilibrium reaction [cf. Eqs. (2) and (3)] in favor of UDP-pentose formation, thus increasing the yield of the sugar nucleotide. The optimal reaction time may be determined by a preliminary test. A maximal yield is usually obtained in 30 to 60 minutes.

Purification of UDP-Pentoses

For the preparation of small quantities of UDP-pentoses, electrophoretic separation, as described in the assay section, is the most direct and convenient method. The reaction mixture, containing up to 10 micromoles of nucleotides, may be directly streaked on a strip of paper 15 cm. wide. At the completion of the electrophoresis, the paper is air-dried. The ammonium formate buffer can be removed by placing the strip of paper in a vacuum desiccator over NaOH and H$_2$SO$_4$ for 1 to 2 days, and the pure sugar nucleotide can then be eluted with water.

For the preparation of large quantities of the UDP-pentoses, an anion-exchange column is required. The procedure described by Leloir and Palladini (see Vol. III [143]) for the preparation of UDP-D-glucose is followed. The nucleotides are converted to the ammonium salts, adsorbed on an anion-exchange resin and eluted with solutions of increasing acidity and salt concentration. The operations are carried out at 4° to minimize decomposition of the sugar nucleotides. UDP-pentoses are eluted under the same conditions as UDP-D-glucose (0.03 N NaCl–0.01 N HCl).[6] After elution from the column and concentration by charcoal treatment, the UDP-pentoses may be subjected to a final purification by paper electrophoresis as described above.

The two UDP-pentoses can be separated from each other by paper chromatography. A solvent consisting of 15 vol. of ammonium acetate buffer, pH 5.1 (77 g. of ammonium acetate per liter, to which enough glacial acetic acid—about 21 ml.—is added to reach the specified pH), 85 vol. of 95% ethanol, and 0.1 vol. of 1 M ammonium ethylenediaminetetraacetate, pH 5, is used. Acid-washed Whatman No. 1 paper should be used, and the spots should contain no more than 0.04 micromole of each nucleotide. In 96 hours, the UDP-D-xylose spot moves about 7.3 cm.,

[6] V. Ginsburg, P. K. Stumpf, and W. Z. Hassid, *J. Biol. Chem.* **223**, 977 (1956).

and the UDP-L-arabinose spot moves 5.7 cm. For complete separation it may be necessary to rechromatograph each of the UDP-pentoses.

Properties of UDP-Pentoses

The behavior of the UDP-pentoses in acid and alkali is similar to that of UDP-D-glucose (see Vol. III [143]).

Biological Activities. The two UDP-pentoses undergo pyrophosphorolysis in the presence of inorganic pyrophosphate [Eqs. (2) and (3)]. They are interconverted by the action of UDP-L-arabinose 4-epimerase, reaction (4). The equilibrium constant for this reaction is 1.0.[1]

$$\text{UDP-D-xylose} \rightleftharpoons \text{UDP-L-arabinose} \tag{4}$$

UDP-D-xylose serves as a xylosyl donor to β-1,4-linked xyloöligosaccharides:

$$\text{UDP-D-xylose} + (\text{xylose})_n \rightarrow \text{UDP} + (\text{xylose})_{n+1}$$

where $n = 2$ to 5.[7]

[7] D. S. Feingold, E. F. Neufeld, and W. Z. Hassid, *J. Biol. Chem.* **234**, 488 (1959).

[109] Synthesis of Cytidine Diphosphate Diglyceride

By EUGENE P. KENNEDY

Principle. Cytidine diphosphate diglycerides (CDP-diglycerides) are a group of closely related compounds of the general structure shown in (I). The length and degree of saturation of the fatty acids attached to

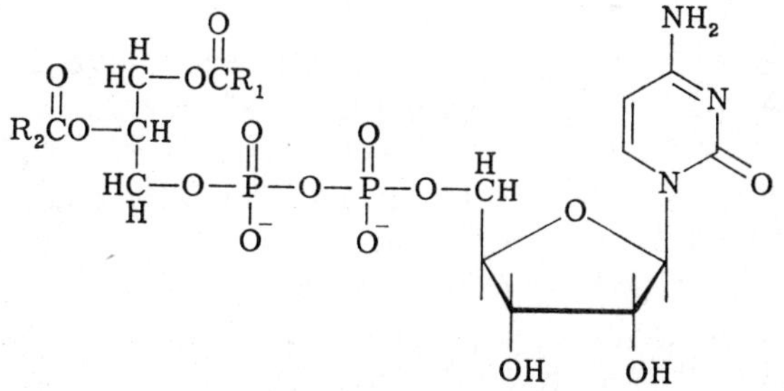

(I)

the glycerol residue may be varied, thus giving rise to a number of CDP-diglycerides. Compounds of this type are now known to be essential intermediates in the biosynthesis of inositol monophosphatide[1,2] and of polyglycerol phosphatides.[3] The synthesis of CDP-diglycerides by the carbodiimide method has been described by Paulus and Kennedy.[2]

Reagents. Dipalmitoyl-L-α-glycerophosphoric acid is synthesized by using the general procedures of Howe and Malkin[4] and of Baer.[5] Phosphatidic acids are unstable when stored as the free acids, unless rigid precautions are taken to exclude moisture. The dipalmitoyl-L-α-glycerophosphoric acid should either be freshly prepared or stored as the more stable disodium salt and converted to the free acid by suspension in excess aqueous hydrochloric acid and extraction of the free acid with ether. The ether solution is then washed several times with water and evaporated to dryness under vacuum in a rotary evaporator.

Cytidine-5'-monophosphate (CMP) may be synthesized by the method of Baddiley *et al.*[6] or obtained commercially (Pabst Co., Milwaukee, Wisconsin). *N,N'*-Dicyclohexylcarbodiimide may be synthesized by the method of Schmidt *et al.*[7] or obtained commercially (Aldrich Chemical Co., Milwaukee, Wisconsin).

The silicic acid used in chromatography is a product of the Bio-Rad Corporation, Berkeley, California, and is activated at 110° for 12 hours before use. Solvents are CP grade used without further purification.

Procedure. Dipalmitoyl-L-glycerophosphoric acid (600 mg.) and 350 mg. of CMP (free acid) in 44 ml. of pyridine plus 6 ml. of water are stirred with 10 g. of *N,N'*-dicyclohexylcarbodiimide for 2 days at 40° to 45°. At the end of that period, 100 ml. of water are added, and the precipitated dicyclohexylurea is removed by centrifugation and washed with 100 ml. of water. To the combined turbid supernatant solutions, 2.5 g. of barium acetate are added slowly with stirring, and the suspension is allowed to stand in the icebox overnight. The precipitate is then collected by centrifugation (ca. 0.5 g.) and dissolved in a mixture of 1 ml. of 4 N HCl, 10 ml. of methanol, and 10 ml. of chloroform. Water (about 50 ml.) is then added, and the chloroform phase which separates out is washed several times with water and finally dried over anhydrous sodium sulfate.

The dry chloroform solution is then passed over a column of 10 g. of silicic acid suspended in chloroform and eluted by gradient elution, in an apparatus similar to that of Busch *et al.*[8] The mixing chamber initially contains 250 ml. of chloroform, and the upper reservoir contains

[1] B. W. Agranoff, R. M. Bradley, and R. O. Brady, *J. Biol. Chem.* **233,** 1077 (1958).
[2] H. Paulus and E. P. Kennedy, *J. Biol. Chem.* in press.
[3] J. Kiyasu, H. Paulus, and E. P. Kennedy, *Federation Proc.* in press.
[4] R. J. Howe and T. Malkin, *J. Chem. Soc.* 2663 (1951).
[5] E. Baer, *J. Biol. Chem.* **189,** 235 (1951).
[6] J. Baddiley, J. G. Buchanan, and A. R. Sanderson, *J. Chem. Soc.* 3107 (1958).
[7] D. Schmidt, F. Hitzler, and E. Lahde, *Ber. chem. Ges.* **71,** 1933 (1938).
[8] H. Busch, R. B. Hurlbert, and V. R. Potter, *J. Biol. Chem.* **196,** 717 (1952).

methanol. Fractions are collected with an automatic fraction collector and examined for compounds with absorption in the ultraviolet at 280 mμ. CDP-dipalmitin emerges as a discrete peak at about 20% methanol in chloroform with the absorption at 280 mμ characteristic of cytidine compounds.

Fractions containing the product are pooled and taken to dryness *in vacuo*. The yield is 61 mg., or 6%. The purity at this point is about 93%. Further purification is achieved by dissolving the CDP-diglyceride in water as the potassium salt by the cautious addition of two equivalents of KOH, followed by slight excess of barium acetate. The insoluble barium salt is collected by centrifugation, thoroughly washed with water, and dried under high vacuum.

Properties of CDP-Dipalmitin. The barium salt prepared in the manner described shows the following analysis:

$C_{44}H_{79}N_3O_{15}P_2Ba$ Calculated: N, 3.86;   P, 5.69;   H, 7.31;   C, 48.51

(1089.4)   Found: N, 3.85;   P, 5.70;   H, 7.31;   C, 47.30

CDP-dipalmitin is characterized by its solubility in chloroform as the free acid; this is a remarkable property in a derivative of a nucleotide. The dipotassium salt forms a clear aqueous dispersion, in which it can be stored at 0° for several days without significant breakdown, whereas the free acid decomposes to a considerable extent in chloroform at room temperature.

When aliquots of the potassium salt are suitably diluted with 0.01 N HCl, the absorption spectrum closely resembles that of CMP, with a molar absorbance of 13×10^6 mole^{-1} cm.^{2} at 280 mμ.

Ester determinations by the method of Stern and Shapiro,[9] with dipalmitoyl-glycerol-DL-α-benzyl ether as a standard, give values very close to the theoretical.

[9] I. Stern and B. Shapiro, *J. Clin. Pathol.* **6**, 158 (1953).

[110] Adenylosuccinase and Adenylosuccinic Acid

By C. E. Carter

The product of the reversible enzymatic reaction between fumaric acid and adenosine 5′-phosphate, adenylosuccinic acid, is prepared in millimole quantities by use of purified yeast adenylosuccinase.[1]

Fumaric acid + adenylic acid ⇌ Adenylosuccinic acid

[1] C. E. Carter and L. H. Cohen, *J. Biol. Chem.* **222**, 17 (1956).

Preparation of Adenylosuccinase

One kilogram of Fleischmann's active dry baker's yeast was suspended in 3 l. of $0.1 N$ $NaHCO_3$ and vigorously stirred at $37°$ for 5 to 7 hours. The mixture was frozen, thawed, and stirred at $37°$ for 3 more hours. It was again frozen, thawed, and centrifuged at $3500 \times g$ for 1 hour at $4°$. The supernatant solution was collected, and, for each 100 ml., 33 g. of ammonium sulfate were added slowly with stirring at room temperature. The solution was stirred for an additional 30 minutes and then centrifuged at $3500 \times g$ for 20 minutes at $25°$. The supernatant solution was discarded, and the precipitate was dissolved in 400 ml. of distilled water. This solution was chilled to $4°$, 80 g. of ammonium sulfate were added slowly, and the solution was kept at this temperature for an additional 30 minutes. After centrifugation at $3500 \times g$ at $4°$ for 20 minutes, the precipitate was discarded, and 30 g. of ammonium sulfate were added slowly to the supernatant solution, while the temperature of the solution was brought to $25°$. After standing at this temperature for 20 minutes, the mixture was centrifuged ($25°$) for 20 minutes at $3500 \times g$, the supernatant solution discarded, and the precipitate dissolved in 100 ml. of distilled water.

Fumarase was removed from the preparation at this stage by heat inactivation. Ten minutes of heating at $60°$ resulted in complete inactivation of fumarase with about 90% retention of adenylosuccinase activity. After chilling of the preparation to $4°$, denatured protein was removed by centrifugation. The supernatant solution was dialyzed against $0.1 M$ phosphate buffer, pH 6.5, and this fraction, which was stable for several months at $-10°$, was employed for synthesis of adenylosuccinate. The pH optimum for synthesis was 5.9, for the reverse reaction 7.0. The enzyme does not require metal ions for optimal activity.

Assay of Enzyme

Synthesis. A solution containing 3 micromoles of fumaric acid and 0.5 micromole of adenylic acid in 3 ml. of $0.01 M$ phosphate buffer, pH 5.9, was incubated with 0.01 to 0.05 ml. of purified enzyme, and increment absorption at 280 mμ was determined.

Degradation. Adenylosuccinate (0.05 micromole/ml. in $0.01 M$ phosphate buffer, pH 7.0) was incubated with 0.01 to 0.05 ml. of purified enzyme, and the decrease in absorption at 280 mμ was determined. Under these conditions the conversion of adenylosuccinic to adenylic acid results in a change in absorbancy at 280 mμ of 10.7 per micromole per milliliter.

Physical Constants of Adenylosuccinase Reaction

The equilibrium constant at 35° is 6.8×10^{-3}. The K_m values are as follows: adenylosuccinate, $1.2 \times 10^{-5} M$; adenylate, $4.8 \times 10^{-5} M$; fumarate, $5.2 \times 10^{-4} M$.

Preparation of Adenylosuccinic Acid

One millimole of adenosine 5'-phosphate and 9 millimoles of fumaric acid were dissolved in 50 ml. of distilled water by adjusting the pH of the solution to 6.0. Then 50 ml. of enzyme solution were added, and the mixture was incubated at 37°. The progress of adenylosuccinic acid synthesis was followed at intervals by transferring 0.02 ml. of incubation mixture to 5 ml. of $0.01 M$ NH$_4$OH and determining the absorption at 280 mμ. During a 2-hour incubation period the absorption at 280 mμ increased approximately threefold as adenylosuccinate was formed. When the maximum was reached, the solution was heat-treated in a boiling-water bath, denatured protein was removed by centrifugation, and the precipitate was washed with distilled water. To the combined supernatant solutions were added 1 ml. of concentrated NH$_4$OH and sufficient water to bring the final volume to 200 ml. This solution was then percolated through a column of Dowex 1–acetate, 2% crosslinked, 200 to 400 mesh, which was 15 cm. long and had a diameter of 1.5 cm. The original effluent, containing no ultraviolet-absorbing material, was discarded, and the column was treated with a solution containing $2 M$ acetic acid and $0.25 M$ ammonium acetate until the absorption at 260 mμ of the effluent solution fell below 0.20. Usually 1000 to 1200 ml. of solution were required. This fraction contained residual adenylic and fumaric acids. A solution of $4.5 M$ ammonium acetate and $1.4 M$ acetic acid was then run through the column at the rate of 1 ml./min. and collected in 10-ml. fractions. Absorption at 267 mμ was determined, and fractions with optical densities lower than 10 were discarded. As determined by spectrophotometry, 0.725 millimole of adenylosuccinic acid was recovered in 50 to 60 ml. of effluent. The pooled fractions were concentrated to a sirup by vacuum distillation at 50°, and 400 ml. of absolute ethanol were added with stirring. After standing for 10 minutes at room temperature, the precipitate was collected by centrifugation at 4°, and the supernatant solution was discarded. The precipitate was suspended with stirring in 400 ml. of absolute ethanol at 40° for 5 minutes and again collected by centrifugation. The precipitate in the centrifuge bottle was dried in a vacuum desiccator. The yield of adenylosuccinate, containing 1.4 moles of ammonium ion per mole of phosphorus, in several preparations averaged 300 to 350 mg., or approximately 0.6 millimole. The purity varied between 88 and 95%, as calculated from the phosphorus content. The compound sintered at about 130° and melted with decomposition at about 155°; the melting point was not sharp.

Dissociating groups which influence spectra have pK values at 2.3, 4.1, and 5.1. Maximum absorption is at 266 to 268 mμ and shows a hyperchromic shift with decreasing concentration of hydrogen ion ($\alpha_M \times 10^{-3}$, pH 2, 16.9; pH 3 to 3.5, 18.5; pH 4.5 to 13, 19.2).

Adenylosuccinic acid is hydrolyzed to 6-succinoaminopurine and ribose 5'-phosphate by heating in 1 N HCl at 90° for 15 minutes. Longer periods of heating result in degradation of the aglycone to aminoimidazolecarboxamide.[2]

[2] C. E. Carter, *J. Biol. Chem.* **223**, 139 (1956).

[111] Measurement of Pyridine Nucleotides by Enzymatic Cycling

By OLIVER H. LOWRY and JANET V. PASSONNEAU

Principle. The coenzyme to be determined is made to catalyze an enzymatic dismutation between two substrates. After several thousand cycles, one of the products is measured. The nucleotides during cycling are used at concentrations well below their Michaelis constants; consequently, reaction rates are proportional to nucleotide concentrations. The final product is again a pyridine nucleotide, so the cycling process can be repeated if necessary.

The system used for TPN measurement utilizes glucose-6-P dehydrogenase and glutamic dehydrogenase, reactions (1) and (2). The 6-P-

$$\text{TPNH} + \alpha\text{-ketoglutarate} + \text{NH}_4^+ \rightarrow \text{TPN}^+ + \text{glutamate} \qquad (1)$$

$$\text{TPN}^+ + \text{glucose-6-P} \rightarrow \text{TPNH} + \text{6-P-gluconate} \qquad (2)$$

gluconate formed is oxidized in a second step with 6-P-gluconate dehydrogenase and extra TPN+. The TPNH produced is measured by its fluorescence.

DPN is measured with lactic dehydrogenase and glutamic dehydrogenase, reactions (3) and (4). The pyruvate formed is reduced in a

$$\text{DPNH} + \alpha\text{-ketoglutarate} + \text{NH}_4^+ \rightarrow \text{DPN}^+ + \text{glutamate} \qquad (3)$$

$$\text{DPN}^+ + \text{lactate} \rightarrow \text{DPNH} + \text{pyruvate} \qquad (4)$$

second step with added DPNH and lactic dehydrogenase. The DPN⁺ produced is measured fluorometrically.

Either the reduced or oxidized forms of the pyridine nucleotides, or the sum of the two, can be measured. To measure TPN⁺ (DPN⁺), destruction of TPNH (DPNH) and interfering enzymes in tissues can be accomplished with brief acid treatment. The reduced forms can be measured after destruction of TPN⁺ (DPN⁺) and interfering enzymes with mild alkaline treatment.[1]

Determination of TPN⁺ or TPNH: Sample Procedure

Stock Cycling Reagent

0.1 M Tris–HCl buffer, pH 8.0.
5 mM α-ketoglutarate.
1 mM glucose-6-P.
0.1 mM ADP.
0.025 M ammonium acetate.
0.2 mg. of bovine plasma albumin per milliliter.

To the stock reagent are added within a few hours of use 0.1 mg. of crystalline beef liver glutamic dehydrogenase per milliliter and sufficient yeast glucose-6-P dehydrogenase to give a calculated activity in the reagent at 25° of 0.3 mole/l./hr. with optimal TPN⁺ and glucose-6-P concentrations. The required activity can be provided by a concentration (per milliliter) of 0.025 mg. of the glucose-6-P dehydrogenase supplied (1962) by Boehringer and Sons, Mannheim, Germany. (This preparation appears to have about 30% of the activity of the crystalline enzyme.[2]) The glutamic dehydrogenase concentration is sufficient to give a calculated velocity with optimal substrate and TPNH concentrations of approximately 0.2 mole/l./hr. at 25°.

If the enzymes are suspended in ammonium sulfate, they are centrifuged and resuspended in sufficient 2 M ammonium acetate to provide the necessary NH_4^+ concentration indicated above. This reduces the sulfate concentration to 5 mM or less, which is desirable, since sulfate inhibits glucose-6-P dehydrogenase in proportion to the square of its concentration.

Volumes of 100 μl. of the complete cycling reagent kept at 0° are pipetted into 3-ml. fluorometer tubes in a rack in ice. TPN⁺ or TPNH, in a volume of 1 to 20 μl., is added to give a concentration in the range

[1] O. H. Lowry, J. V. Passonneau, and M. K. Rock, *J. Biol. Chem.* **236**, 2756 (1961).
[2] E. A. Noltmann, C. J. Gubler, and S. A. Kuby, *J. Biol. Chem.* **236**, 1225 (1961).

of 1×10^{-9} to $1 \times 10^{-8} M$. Water, or better a medium identical to that containing the TPN, is added to bring all the samples to the same volume $\pm 2\%$. The rack of tubes is transferred to a 38° bath for 60 minutes and then to a 100° bath for 2 minutes. To each tube is then added 1 ml. of 6-P-gluconate reagent: 0.02 M Tris–HCl buffer (pH 8.0), 0.03 M ammonium acetate, 0.03 mM TPN⁺, 0.1 mM EDTA, 0.02% bovine plasma albumin, and 6-P-gluconate dehydrogenase to give a calculated maximal activity of 0.16 millimole/l./hr.

After 30 minutes at room temperature, the fluorescence of each sample is measured together with control samples containing 6-P-gluconate in the range anticipated. Standards which increase in steps of two or three to cover the chosen concentration range are provided. Standards and blanks are carried through the entire process, including any procedure before cycling. (If the 6-P-gluconate dehydrogenase is contaminated with glutamic dehydrogenase, the TPNH formed will decrease with time, owing to the presence of α-ketoglutarate from the cycling reagent. See preparation of 6-P-gluconate dehydrogenase below.)

The yield of 6-P-gluconate is 7000- to 10,000-fold, under the conditions described. The 6-P-gluconate concentration should not exceed 10^{-5} M in the fluorometer, as this is the limit of proportionality of TPNH fluorescence.

The stock reagents without enzymes may be stored for 2 weeks at $-20°$ or for 2 months at $-85°$. Longer storage results in loss of α-ketoglutarate.

Permissible Variations. The time of incubation may be varied between 15 and 60 minutes. The enzyme concentrations may be reduced to 0.2 of those given, with a reduction of cycling yield to 2000- to 3000-fold. This will permit increasing the concentration of TPN during cycling to $5 \times 10^{-8} M$, which may at times be a convenience. (Some samples of glucose-6-P dehydrogenase have been contaminated with TPN. Reducing the enzyme concentrations, and increasing TPN concentrations accordingly, minimizes this difficulty.) The volume may be decreased to 1 μl. in an appropriately smaller tube with corresponding increase in sensitivity.

If the amount of 6-P-gluconate formed is less than 5×10^{-10} mole, the TPNH may be measured indirectly, with consequent 10-fold increase in sensitivity. In this case, after cycling and heating to 100°, the sample is treated for 30 minutes with 3 to 10 vol. of 6-P-gluconate dehydrogenase reagent. A solution of 0.3 M Na$_3$PO$_4$–0.3 M K$_2$HPO$_4$ is added in an amount equal to twice the cycling volume and heated for 10 minutes at 60° to destroy excess TPN⁺. An aliquot is then added to 0.2 ml. of 6 N NaOH containing 0.03% H$_2$O$_2$ in a fluorometer tube. After 10 minutes of heating at 60°, 1 ml. of water is added, and the fluorescence is measured.

The temperature coefficient is 6.4% per degree between 0° and 25° and 7.7% per degree between 25° and 38°. The rate at 0° is 8% of that at 38°. Therefore, the time between addition of the first and last samples to the cycling mixture should not exceed half the subsequent incubation time if the limit of tolerance is 4%. The optimal pH is 8.0, but the rate is only 7% lower at pH 7.7 and 11% lower at pH 8.3.

The glucose-6-P concentration may be increased to 5 mM to permit TPN concentrations during cycling up to $3 \times 10^{-7} M$ with full cycling rate. In this case, if the cycling procedure given as an example is followed, an aliquot of the cycling mixture must be transferred after heating at 100° to another fluorometer tube in order not to exceed a final concentration of $10^{-5} M$ TPNH. There may be a slight addition to the cycling blank from the extra glucose-6-P.

The α-ketoglutarate concentration cannot be substantially altered without decreased cycling. The ADP level is not critical but is used to protect glutamic dehydrogenase. The NH$_4^+$ level is not critical, but, at a concentration of 0.1 M, cycling is decreased owing to an increase in the apparent K_m for TPN$^+$ with glucose-6-P dehydrogenase.

Source of Blanks. At concentrations of TPN as low as 1 or $2 \times 10^{-9} M$ during cycling, blank values became critical. The over-all blank need not exceed the equivalent of $3 \times 10^{-9} M$ TPN calculated as concentration during cycling. There are three sources of blank. One is from the cycling reagent itself, which contains materials that fluoresce slightly at pH 8.0. This may account for a third of the blank. The 6-P-gluconate dehydrogenase reagent may have a fluorescent blank equivalent to $3 \times 10^{-7} M$ TPNH measured directly, or equivalent to $10^{-9} M$ TPN during cycling. (A third of this is due to water, a third to the Tris buffer, and a third to the TPN$^+$ and enzyme. Higher readings indicate dirty tubes or contaminated solutions. It is recommended that all tubes be cleaned by heating them for 15 minutes at 100° first in half-concentrated HNO$_3$, then in distilled water, followed by a rinse with redistilled water.) There may also be a blank resulting from incubation of the complete cycling reagent. This presumably indicates the presence of minute amounts of TPN. This need not exceed the equivalent of $5 \times 10^{-10} M$ TPN. The first two contributions to the blank may be reduced 10-fold by measuring the TPNH indirectly as indicated.

Contamination of the materials used with the coenzymes themselves must be avoided. Pipets used for cycling should not be used to pipet strong coenzyme solutions, or should be specially cleaned by soaking in 0.1 N NaOH followed by rinsing with 0.1 N HCl.

Determination of DPN$^+$ or DPNH: Sample Procedure

Stock Reagent

0.2 M Tris–HCl buffer, pH 8.4.
100 mM sodium lactate.
0.3 mM ADP.
5 mM α-ketoglutarate.
0.05 M ammonium acetate.

To the stock reagent are added within an hour of use 0.2 mg. of crystalline beef liver glutamic dehydrogenase per milliliter and 0.025 mg. of crystalline beef heart lactic dehydrogenase per milliliter (charcoal-treated, see below). If the enzymes are suspended in $(NH_4)_2SO_4$, this can furnish the required NH_4^+. The amounts of enzymes used are such as to give calculated rates in the cycling mixture at 25° (measured with optimal substrate and coenzyme levels) of 0.5 mole/l./hr. for glutamic dehydrogenase and 0.06 mole/l./hr. for lactic dehydrogenase. Volumes of 100 μl. of complete cycling mixture kept at 0° are placed in 3-ml. fluorometer tubes in a rack in ice. DPN$^+$ or DPNH (1 to 20 μl.) is added to give a concentration in the range of $3 \times 10^{-9} M$ to $4 \times 10^{-8} M$. Water, or the medium used for the nucleotide, is added to bring all the samples to the same volume ±2%. The rack is transferred to a bath at 25° for 1 hour and then to 100° for 2 minutes. The cycling rate at 0° is 15 to 20% of that at 25°; therefore the difference in time between addition of the first and last samples should not exceed 10 minutes.

To each tube in ice is added 100 μl. of a reagent containing 0.65 M NaH_2PO_4, 0.15 M K_2HPO_4, 1.5 μg. of crystalline rabbit skeletal muscle lactic dehydrogenase per milliliter, and DPNH at a concentration three to ten times the expected pyruvate level (2500 moles of pyruvate per mole of DPN$^+$). The lactic dehydrogenase is added to ice-cold phosphate reagent within an hour of use, and DPNH within 15 minutes of use. The rack of samples is incubated at 20° to 30° for 15 minutes and returned to the ice bath. To each sample is added immediately 25 μl. of 5 N HCl with *thorough mixing*, after which the tubes are brought to room temperature. This step is to destroy the excess DPNH. Finally, 1 ml. of 6 N NaOH is added with *immediate* mixing. After heating for 10 minutes at 60°, the tubes are brought exactly to room temperature, and the fluorescence is read. Light must be kept subdued after the addition of strong NaOH, since the fluorescent form of DPN$^+$ is sensitive to destruction by light,[3] although the fluorescence is stable for hours in darkness. The fluorometer is set to use the minimum amount of light necessary to read

[3] O. H. Lowry, N. R. Roberts, and J. Kapphahn, *J. Biol. Chem.* **224**, 1047 (1957).

the samples. If fading should occur during reading, the tube may be remixed without much total loss of fluorescence.

Since pyruvate formation during cycling is not strictly linear with DPN concentration, standards of DPN+ are included which increase in steps of two to cover the range of assay. Blanks and standards are carried through the entire procedure and are treated as nearly as possible like the samples to be analyzed.

Permissible Variations. The cycling procedure is not so flexible as that for TPN, owing primarily to a fall-off in rate as pyruvate accumulates. Arbitrarily, a pyruvate concentration of 0.1 mM at the end of cycling has been set as a practical upper limit. Incubation for less than 30 minutes is not recommended, owing to the substantial cycling rate at 0°, which makes it difficult to handle large numbers of samples. The temperature coefficient is 7% per degree between 0° and 25°, but only 3% per degree between 25° and 32°, and 2.2% per degree between 32° and 38°. Therefore, there is only a modest gain in raising the temperature above 25°. However, the complete reagent is quite stable, and cycling can be carried out satisfactorily at 38°.

If higher levels of DPN+ are used, the lactic dehydrogenase may be cut in half with consequent lower yield (1200-fold). To decrease the cycling rate, the lactic dehydrogenase should be lowered more than the glutamic dehydrogenase to keep the DPN+-to-DPNH ratio high and thus minimize the fall-off in pyruvate formation. A reagent containing 50 mM lactate, 0.1 mg. of glutamic dehydrogenase per milliliter, and 0.005 mg. of lactic dehydrogenase per milliliter gives a yield of 600 moles of pyruvate per mole of DPN in 60 minutes. This permits use of high DPN levels without excessive departure from linearity and reduces the danger of variability from contamination with DPN.

Cycling volumes may be changed. No difficulty has been encountered with volumes of 5 μl. in appropriately smaller tubes and 10^{-8} M DPN.

The optimal pH is 8.4, but the rate is diminished only 2% and 7% by changing the pH to 8.8 and 8.0, respectively.

The lactate concentration is kept high to minimize the inhibitory effects of pyruvate accumulation. Since the best commercial lactate contains 1 part of pyruvate in 40,000, there is some advantage in regard to blank in reducing lactate when measuring very low DPN levels (1 to 5 $\times$ 10^{-9} M). The initial rate is as high with 50 mM lactate as with 100 mM lactate. Substantial change in α-ketoglutarate concentration will decrease glutamic dehydrogenase activity but will not have much effect on the cycling rate, since lactate oxidation is the rate-limiting step. The concentration of NH$_4$+ is also less critical than for the TPN method.

Source of Blanks. With 100 mM lactate, 1 part of pyruvate in 40,000 is equal to $2.5 \times 10^{-6} M$ pyruvate or the equivalent of $10^{-9} M$ DPN with 2500-fold cycling. The heart lactic dehydrogenase used, after charcoal treatment, contributed the equivalent of $2 \times 10^{-10} M$ DPN. The remainder of the cycling components do not contribute appreciably to the blank. The DPNH used for pyruvate measurement adds a blank equal to 1% or 2% of the amount used. However, if the DPNH has become oxidized, it may contribute a much greater blank. The fluorescence from the alkali used in the last step need not exceed 1 or $2 \times 10^{-8} M$ DPN+ (direct reading) or the equivalent of 1 or $2 \times 10^{-10} M$ DPN as cycled in sample procedure. (Fluorescence in the strong NaOH can be removed by exposure to strong daylight for a few hours.) The over-all blanks tend to be more erratic for DPN than for TPN. This is attributed, on the basis of considerable evidence, to contamination by DPN itself. Suggestions for avoidance of contamination are given in the TPN section.

Lactate Pyruvate Equilibrium and Pyruvate Measurement. The equilibrium constant for (pyruvate) (DPN)/(lactate) (DPN+) was determined to be 8.8×10^{-4} at pH 8.4 in $0.1 M$ Tris buffer. Lactate is used at a very high concentration during cycling to compensate for the unfavorable equilibrium. When it comes to the pyruvate measurement, the high level of lactate is a disadvantage, and the pH value is lowered as much as possible. At the pH chosen, 6.5, (pyruvate) (DPNH)/(lactate) (DPN+) is 1×10^{-5}. Any pyruvate remaining represents a negative error. This error, as a fraction of the total initial pyruvate, is equal to the final equilibrium value for (pyruvate)/(DPN+). With a lactate concentration of $0.05 M$ after dilution in the second step, the equilibrium constant for (pyruvate)/(DPN+) is equal to $(5 \times 10^{-7} M)/(\text{DPNH})$. This means that, to prevent a negative error of more than 5%, the excess DPNH cannot be less than $10^{-5} M$, or 0.02% of the lactate concentration. To provide sufficient DPNH levels becomes important only when measuring lowest DPN levels. For example, a DPN concentration of $10^{-9} M$ with 2000-fold cycling results in $10^{-6} M$ pyruvate at the pH 6.5 step, and would require a 10-fold excess of DPNH.

The phosphate solution used to shift the pH and thus the equilibrium has a pH value of 6.2, and this solution destroys DPNH at a rate of 40% per hour at 25°, but only 5% per hour at 0°. It is for this reason that DPNH is added to cold reagent just before use.

Reproducibility and Proportionality. The cycling procedures are reproducible almost to within the limits imposed by the fluorometer. The standard deviation for the nucleotides determined in rat tissues ranged from 2 to 3%.[4]

[4] O. H. Lowry, J. V. Passonneau, D. W. Schulz, and M. K. Rock, *J. Biol. Chem.* **236**, 2746 (1961).

The TPN cycle is proportional within a 100-fold range of coenzymes with the same cycling reagent. The DPN cycle is proportional without serious departure from linearity over a 20-fold range, i.e., a conversion of 0.005 to 0.1% lactate to pyruvate. With a sufficient number of standards, the range can be extended.

Specificity. DPN added to the TPN cycle at ten times the TPN level affects neither the blank nor the cycling. TPN does not affect the DPN cycle when added at one hundred times the DPN level.

If TPN$^+$ is heated in 0.02 N H$_2$SO$_4$ for 30 minutes at 60°, less than 1 part in 2000 is converted to DPN$^+$. The alkaline degradation products of TPN$^+$ affect the cycling of TPN. TPN at 5×10^{-8} M was inhibited 16% by 2×10^{-6} M TPN$^+$ and 33% by 10^{-5} M TPN$^+$ that had been destroyed by alkali. On the other hand, 2×10^{-7} M DPN was inhibited less than 10% by 4×10^{-5} M DPN$^+$ that had been destroyed by alkali.

Tissue blanks have been found to be largely eliminated by the cycling procedure. The blank contributions of the liver, brain, and blood of rat were found to be negligible in relation to the native values of pyridine nucleotides.[4]

Special Preparations

6-P-Gluconate Dehydrogenase.[4] (Available commercial preparations of 6-P-gluconate dehydrogenase from yeast are contaminated with glucose-6-P dehydrogenase and are therefore unsuitable for the present purpose. It seems probable that crystalline yeast enzyme[5] would be highly satisfactory.) Fresh rat livers, 50 to 200 g., were homogenized in 9 vol. of 0.025 M phosphate buffer at pH 7.5 containing 0.2 mM EDTA. (Throughout, the preparation was kept at 0° to 4°. All centrifugations were made at approximately 10,000 $\times$ g for 20 to 30 minutes.) The initial activity was 21 millimoles/l./hr. measured in the fluorometer at 25° with 0.5 mM 6-P-gluconate, 0.05 mM TPN$^+$, 0.01% bovine plasma albumin, and 1 mM EDTA in 0.05 M Tris at pH 8. The precipitates were discarded after centrifugation of the original homogenate and after addition of solid ammonium sulfate to concentrations of first 1 M and then 2 M. The precipitate obtained at an ammonium sulfate concentration of 3 M contained 70% of the original activity. This measured approximately 0.3 ml./g. of liver. It was dissolved in 3 vol. of phosphate buffer (0.025 M, pH 7.5) and dialyzed for 5 hours at 4° against this same buffer containing 0.2 mM EDTA. The solution was diluted to a protein concentration of 1% (measured with Folin phenol reagent), and nucleic acid was removed with 0.04 vol. of 1% protamine sulfate. After cen-

[5] S. Pontremoli, A. de Flora, E. Grazi, G. Mangiarotti, A. Bonsignore, and B. L. Horecker, *J. Biol. Chem.* **236**, 2973 (1961).

trifugation, solid ammonium sulfate was added to the supernatant fluid to a concentration of $2 M$ with enough $1 N$ NH_4OH to bring the pH value to 7. The precipitate was discarded, and the activity (55% of the original) was precipitated at $2.8 M$ ammonium sulfate concentration (neutralized). After dialysis as before, the sample was brought to a protein concentration of 1% and adsorbed on $Ca_3(PO_4)_2$ gel. The gel was added in three steps (1.5 ml. of 0.4% gel per milliliter of sample at each step). The gel was removed by centrifuging after each addition. The third gel treatment adsorbed half the activity remaining at that stage, and 60% of this activity was recovered by elution with $0.2 M$ phosphate buffer, pH 7.4 (40 ml./g. of gel), and precipitation with 3 vol. of $4 M$ ammonium sulfate (neutralized). The activity was 30 moles/kg. of protein per hour, assayed as above. The yield was approximately 12% with a 25-fold purification. Further ammonium sulfate fractionation resulted in loss of activity without gain in specific activity. Glucose-6-P dehydrogenase and hexokinase activity were 1 part in 10,000 and 1 part in 900, respectively, of the 6-P-gluconate dehydrogenase activity. The preparation also contained isocitrate dehydrogenase and a trace of malic enzyme. Some preparations also contained traces of glutamic dehydrogenase, which could disturb the final assay (see above). With fractional absorption by $Ca_3(PO_4)_2$ gel, the later gel fractions contained less glutamic dehydrogenase. In $0.02 M$ Tris buffer at pH 8 the Michaelis constant for 6-P-gluconate is approximately $10^{-5} M$, and that for TPN$^+$ is about $3 \times 10^{-7} M$.

Purification of Heart Lactic Dehydrogenase. Beef heart lactic dehydrogenase (8 ml. of a 2% suspension in $2.5 M$ ammonium sulfate; Worthington Biochemical Corporation) was diluted to 40 ml. with 2% Norit. The enzyme was then recrystallized from the supernatant fluid according to Neilands.[6] This treatment reduced DPN$^+$ concentration from 8×10^{-4} to 1×10^{-5} mole/kg. of protein. Recrystallization without charcoal treatment did not remove appreciable DPN$^+$.

Preparation of Sodium Lactate. Five grams of calcium [L(+)-lactate]$_2 \cdot 4H_2O$ (California Corporation for Biochemical Research) was suspended in 32 ml. of H_2O. After addition of 11 ml. of $2 M$ Na_2CO_3, the suspension was shaken vigorously and filtered. The alkaline filtrate was brought to pH 7 with approximately 1 ml. of $5 N$ HCl. The solution assayed $0.78 M$ with acetyl-DPN and lactic dehydrogenase.

⁶ J. B. Neilands, Vol. I [69].

[112] Preparation and Properties of Dihydrofolic Acid

By Sidney Futterman

Preparation

Principle. Dihydrofolic acid has generally been prepared by catalytic hydrogenation of folic acid in 0.1 N NaOH over platinum oxide.[1,2] Folic acid is also readily reduced to dihydrofolic acid by sodium hydrosulfite.[3] The products are identical, and the latter procedure is described here.

Reagents

1 N KOH.
Potassium ascorbate solution, pH 6.0, 100 mg. of ascorbic acid per milliliter.
2 N HCl.
0.005 N HCl.
Sodium hydrosulfite.
Folic acid.

Procedure. Folic acid (20 mg.) is suspended in 2 ml. of water and dissolved by the addition of 1 N KOH dropwise with stirring. Then 5 ml. of potassium ascorbate solution is added, and 200 mg. of sodium hydrosulfite is dissolved in the reaction mixture. After 5 minutes the solution is cooled to 0°, and 2 N HCl is added dropwise until the pH is lowered to approximately 2.8 (thymol blue). After several minutes at 0°, the precipitated dihydrofolic acid is recovered by centrifugation. The precipitate is dissolved in 5 ml. of potassium ascorbate solution and reprecipitated from the chilled solution by the dropwise addition of 2 N HCl as before. The precipitate is then washed four times with 10-ml. portions of cold 0.005 N HCl. Freshly prepared dihydrofolic acid is a flocculent white precipitate which can be stored for several days at refrigerator temperature as a suspension in 0.005 N HCl. After the precipitate has been washed once with cold water, it can be lyophilized to yield about 15 mg. of dry material that gradually turns yellow on storage.

[1] B. L. O'Dell, J. M. Vandenbelt, E. S. Bloom, and J. J. Pfiffner, *J. Am. Chem. Soc.* **69**, 250 (1947).
[2] M. J. Osborn and F. M. Huennekens, *J. Biol. Chem.* **233**, 969 (1958).
[3] S. Futterman, *J. Biol. Chem.* **228**, 1031 (1957).

Properties

Freshly prepared dihydrofolic acid contains no diazotizable amine,[3] (Bratton and Marshall procedure[4]), indicating that the preparation is not significantly contaminated with tetrahydrofolic acid. However, solutions of dihydrofolic acid at pH 6 decompose on standing at room temperature and accumulate overnight about 0.5 micromole of diazotizable amine for each micromole of dihydrofolic acid.

In alkaline solutions dihydrofolic acid is slowly oxidized, the oxygen consumption being in excess of the amount calculated for oxidation to folic acid.

The absorption spectrum of dihydrofolic acid has been determined at various pH values.[1,2,5] At pH 7.5[2] there is an absorption minimum at 253 mμ, and the molecular extinction at the absorption maximum (283 mμ) is 19,000, whereas at pH 11[5] the molecular extinction at the absorption maximum (284 mμ) is 22,600.

In the presence of folic acid, dihydrofolic acid can be determined by a fluorometric method[5] or by a spectrophotometric method[5] based on the absorption at 420 mμ of a degradation product which forms in strong acid.

Paper chromatography can be carried out with hydrogen gas in the jars and 0.2% ascorbic acid present in each solvent system to retard decomposition of the dihydrofolic acid[5]: (1) 0.1 M sodium acetate buffer (pH 4.5)–ethanol (1:1), R_f 0.26; (2) isopropanol–pyridine–water (1:1:1), R_f 0.17; (3) 0.1 M K$_2$HPO$_4$, R_f 0.15. Each of the following solvent systems contains 0.2% mercaptoethanol to retard air oxidation[2]: (1) 0.1 M phosphate buffer, pH 8.0, R_f 0.18. (2) 1.0 M sodium formate–2% formic acid, R_f 0.07. (3) 0.1 M glycine buffer, pH 9.5, containing 0.2% sodium ethylenediaminetetraacetate, R_f 0.24. Dihydrofolic acid is detected by its bluish white fluorescence in ultraviolet light.

[4] A. C. Bratton and E. K. Marshall, Jr., J. Biol. Chem. 128, 537 (1939).
[5] B. E. Wright, M. L. Anderson, and E. C. Herman, J. Biol. Chem. 230, 271 (1958).

[113] Preparation and Properties of Tetrahydrofolic Acid

By F. M. HUENNEKENS, C. K. MATHEWS, and K. G. SCRIMGEOUR

Principle. 5,6,7,8-Tetrahydrofolic acid (tetrahydro-L-pteroylglutamic acid) can be prepared chemically by the catalytic hydrogenation of folic acid (L-pteroylglutamic acid) over platinum oxide in glacial acetic acid.

This procedure results in the preparation of dl,L-tetrahydrofolic acid[1] (I), which is only 50% reactive in enzymatic assay systems.[2-5] Alterna-

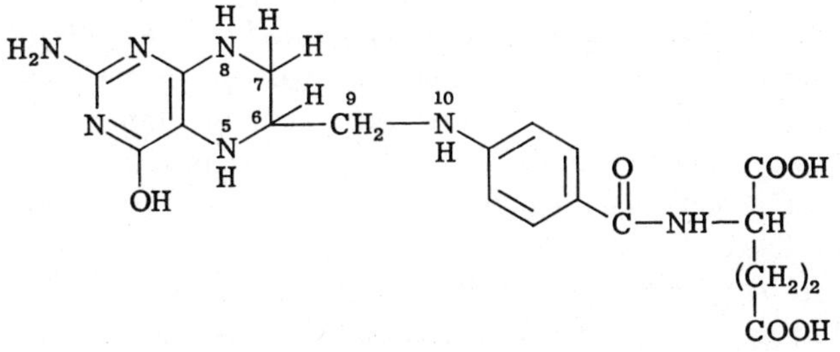

Tetrahydrofolic acid

(I)

tively, the l,L-diastereoisomer of tetrahydrofolic acid is prepared by the TPNH-linked reduction of dihydrofolic acid catalyzed by dihydrofolic reductase.[6]

Chemical Synthesis of Tetrahydrofolic Acid[7,8]

Reagents

Platinum oxide, obtained from the American Platinum Works, Newark, New Jersey.

Glacial acetic acid.

---

[1] Commercial folic acid (pteroyl-L-glutamic acid), and all related compounds, contain an optically active center (L-configuration) at the α-carbon atom of the glutamic acid residue. There is a second potential asymmetric center when reduction occurs at the 6-position of these compounds. However, until the absolute configuration at the 6-position has been determined, compounds with the reduced pyrazine ring can only be characterized as dl, d, or l, with reference to the measured optical rotation. Naturally occurring folinic acid (N^5-formyl-5,6,7,8-tetrahydrofolic acid) is levorotatory [D. B. Cosulich, J. M. Smith, Jr., and H. P. Broquist, J. Am. Chem. Soc. 74, 4215 (1952)] and has been designated, therefore, as the l,L-diastereoisomer.

[2] Y. Hatefi, M. J. Osborn, L. D. Kay, and F. M. Huennekens, J. Biol. Chem. 227, 637 (1957).

[3] J. M. Peters and D. M. Greenberg, J. Am. Chem. Soc. 80, 6679 (1958).

[4] G. K. Humphreys and D. M. Greenberg, Arch. Biochem. Biophys. 78, 275 (1958).

[5] L. D. Kay, M. J. Osborn, Y. Hatefi, and F. M. Huennekens, J. Biol. Chem. 235, 195 (1960).

[6] See Vol. VI [48].

[7] B. L. O'Dell, J. M. Vandenbelt, E. S. Bloom, and J. J. Pfiffner, J. Am. Chem. Soc. 69, 250 (1947).

[8] Y. Hatefi, P. T. Talbert, M. J. Osborn, and F. M. Huennekens, Biochem. Preparations 7, 89 (1959).

Folic acid,[9] obtained from California Corporation for Biochemical Research.

Procedure. Two grams of platinum oxide is suspended in 30 ml. of dry glacial acetic acid in the hydrogenation vessel.[10] The catalyst is reduced at 25° and atmospheric pressure until no more hydrogen is adsorbed. Two grams (4.54 millimoles) of folic acid, suspended in 20 ml. of dry glacial acetic acid, is introduced through the side arm into the hydrogenation flask. In approximately 3 to 4 hours the hydrogen uptake (corrected to STP) has reached a limiting value of 2 moles/mole of folic acid. During the course of the reduction, the appearance of the solution changes as the yellow suspension of folic acid is replaced by a nearly colorless solution of tetrahydrofolic acid.

At the conclusion of hydrogenation, the apparatus is flushed with nitrogen. The flask is removed from the apparatus and stoppered immediately to prevent contact of the solution with air.[11] The flask is opened, preferably in a "dry box" which is flushed continuously with nitrogen, and the contents are filtered through coarse paper directly into a round-bottomed flask which is immersed in a dry ice–acetone bath. The flask is then transferred to a lyophilization apparatus, and the contents are taken to dryness. The fluffy white product (1.95 g., 95% of theory) is dried in a vacuum desiccator over P_2O_5 and stored in evacuated sealed ampules where it is stable for many months at room temperature.

Characteristics. Prepared in this manner, tetrahydrofolic acid contains 2 moles[12] of bound acetic acid per mole of product. The molecular weight, therefore, is 565.4, if the absence of bound water is assumed. When freshly dissolved in 0.01 M phosphate buffer, pH 7.0, tetrahydrofolate has a single absorption maximum at 298 mμ and an estimated extinction coefficient of $28 \times 10^3 M^{-1}$ cm.$^{-1}$. Unless the solution is stabilized by the addition of $10^{-2} M$ mercaptoethanol, tetrahydrofolate is oxidized within 40 minutes to dihydrofolate (absorption maximum at 282 mμ). It is difficult to characterize tetrahydrofolate by paper chromatography, owing to its oxidation to dihydrofolate during the process even

---

[9] Commercial folic acid should be examined by paper chromatography[5] and, if impure, should be recrystallized from a dilute acid solution. Other purification procedures for folic acid have been described by R. L. Blakley (*Biochem. J.* **65**, 331 (1957)] and by W. Sakami and R. Knowles [*Science* **129**, 274 (1959)].

[10] The hydrogenation apparatus was patterned after that described by Dunlop [*Ann. N. Y. Acad. Sci.* **53**, 1087 (1951)]. The hydrogenation procedure may also be carried out in a Parr hydrogenator at room temperature (see Vol. VI [114]).

[11] Air oxidation of tetrahydrofolic acid results in the formation of a reddish-brown degradation product.

[12] R. H. Himes and J. C. Rabinowitz, *J. Biol. Chem.* **237**, 2903 (1962).

when mercaptoethanol is included in the solvent system. Column chromatography, as described below, can be used to quantitate tetrahydrofolate. Attempts have also been made to quantitate tetrahydrofolate[13] by converting it to N^{10}-formyltetrahydrofolate via an excess of the other components of the formate-activating enzyme system:

$$\text{Formate} + \text{ATP} + \text{tetrahydrofolate} \overset{\text{Mg}^{++},\ \text{K}^{+}}{\rightleftarrows} N^{10}\text{-Formyltetrahydrofolate} + \text{ADP} + \text{P} \quad (1)$$

Enzymatic Synthesis of Tetrahydrofolic Acid[13]

Reagents

Dihydrofolic acid, prepared according to the method of Futterman, Vol. VI [112].
TPN.
Glucose-6-P.
2-Mercaptoethanol.
1 M Tris buffer, pH 7.5.
Glucose-6-P dehydrogenase (about 1 unit/mg.), obtained from Sigma Chemical Company.
Dihydrofolic reductase (10 to 12 units/mg. of protein), prepared according to Mathews *et al.*, Vol. VI [48].
DEAE-cellulose.
1.0 M Na$_2$HPO$_4$.
5 $\times$ 10^{-3} M Tris buffer, pH 7.0.
0.2 M Tris buffer, pH 7.0.

Procedure. The following components are mixed in a 50-ml. Erlenmeyer flask: 25 mg. (56 micromoles) of dihydrofolate, 8 mg. (10 micromoles) of TPN, 200 micromoles of glucose-6-P, 100 micromoles of 2-mercaptoethanol, 500 micromoles of Tris buffer (pH 7.5), 3 mg. of glucose-6-P dehydrogenase, 12 mg. of dihydrofolic reductase, and water to make 20 ml. After the clear yellow solution is incubated at 37° for 150 minutes without shaking, sufficient mercaptoethanol is added to raise the concentration to 0.2 M, and the solution is either stored in the frozen state or chromatographed immediately. It is not necessary to remove the protein before chromatography.

DEAE-cellulose is washed with 1.0 M Na$_2$HPO$_4$ until the washings are colorless, washed with water until the pH of the suspension is approximately 7, and stored under water. Before use in chromatography, the cellulose is washed by suction filtration with a large volume of 5 $\times$

[13] C. K. Mathews and F. M. Huennekens, *J. Biol. Chem.* **235**, 3304 (1960).

$10^{-3} M$ Tris buffer, pH 7.0. A slurry of the adsorbent in this buffer is poured into a 2.5 × 30-cm. glass column equipped with a 19/22 standard taper joint at the bottom. The column bed is formed without suction or pressure to a height of about 20 cm. The column then is washed with approximately 1 l. of 5 × $10^{-3} M$ Tris buffer, pH 7.0, which also contains 2-mercaptoethanol at a concentration of 0.2 M, and allowed to drain to incipient dryness, whereupon the above reaction mixture (20 ml.) is layered carefully at the top of the adsorbent. After the solution has percolated into the column, and approximately the upper one-third of the bed has become yellow, the column is eluted by a gradient method in which the mixing chamber contains 800 ml. of 5 × $10^{-3} M$ Tris buffer, pH 7.0, and the reservoir contains 1000 ml. of 0.2 M Tris buffer, pH 7.0. Both buffer solutions also contain mercaptoethanol at a concentration of 0.2 M. The flow rate of the column under gravity is about 1 ml./min. The effluent is collected in 22.5-ml. portions (about fifty tubes) over a period of 10 to 12 hours with an automatic fraction collector. Tetrahydrofolate is found in the region of tubes 20 to 29. The contents of these tubes are pooled and frozen, or else lyophilized, and the residue is redissolved in a minimal volume (approximately 5 ml.) of 0.2 M mercaptoethanol (total yield, 47 micromoles, or about 83% of theory).

[114] Preparation and Properties of "Active Formaldehyde" and "Active Formate"

By F. M. HUENNEKENS, P. P. K. HO, and K. G. SCRIMGEOUR

Principle. "Active formaldehyde," an adduct between formaldehyde and tetrahydrofolic acid, has been identified[1,2] as N^5,N^{10}-methylenetetra-

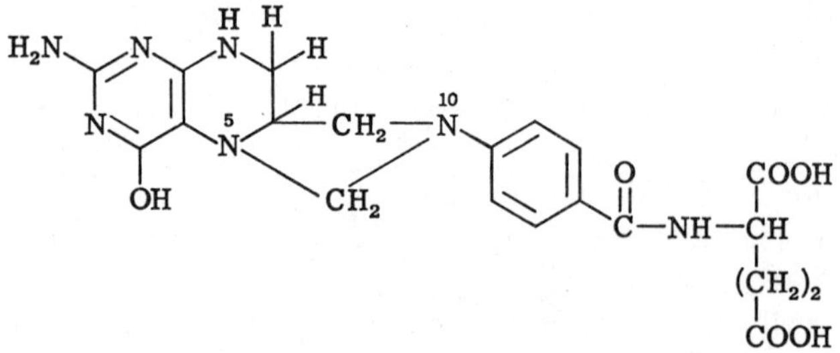

N^5,N^{10}-Methylenetetrahydrofolic Acid

(I)

[1] M. J. Osborn, P. T. Talbert, and F. M. Huennekens, *J. Am. Chem. Soc.* **82,** 4921 (1960).

[2] R. L. Blakley, *Biochem. J.* **74,** 71 (1960).

hydrofolic acid (I). The compound is synthesized most easily by allowing tetrahydrofolic acid to interact with an excess of formaldehyde at a slightly acidic pH and purifying the product by column chromatography.

There are three known forms of "active formate," namely, N^5-formyltetrahydrofolic acid (folinic acid), N^{10}-formyltetrahydrofolic acid, and N^5,N^{10}-methenyltetrahydrofolic acid, which are encountered in enzymatic systems.[3,4] Folinic acid is commercially available. N^5,N^{10}-Methenyltetrahydrofolic acid (II) is readily prepared as a stable, crystalline

N^5,N^{10}-Methenyl tetrahydrofolic acid
(II)

material by the procedure given below, and N^{10}-formyltetrahydrofolate can be generated *in situ* by adjusting solutions of the methenyl derivative to pH 7 or higher.

Preparation of "Active Formaldehyde"

Reagents

dl,L-Tetrahydrofolate,[5] prepared according to F. M. Huennekens *et al.*, Vol. VI [113].

0.025 M formaldehyde.

5 N NaOH.

1 N NaOH.

Solka-floc,[6] obtained from the Brown Co.

Ethanol–sodium bicarbonate ($5 \times 10^{-2} M$, pH 9.3) mixture (40:60), containing $10^{-2} M$ mercaptoethanol.

0.1 M sodium phosphate buffer, pH 8, containing 0.2% mercaptoethanol.

[3] F. M. Huennekens and M. J. Osborn, *Advances in Enzymol.* **21**, 369 (1959).
[4] J. C. Rabinowitz, *in* "The Enzymes" (P. D. Boyer, H. Lardy, and K. Myrbäck, eds.), 2nd ed., Vol. 2, p. 185. Academic Press, New York, 1960.
[5] For an explanation and discussion of these symbols, see footnote 1 in Vol. VI [113].
[6] Solka-floc is washed according to the directions of P. N. Campbell, T. S. Work and E. Mellanby *Biochem. J.* **48**, 106 (1951).

1 M sodium formate–2% formic acid, containing 0.2% mercapto-ethanol.

Ethanol–water mixture (70:30), containing 0.2% mercaptoethanol.

10^{-3} M TPN.

N^5,N^{10}-methylenetetrahydrofolic dehydrogenase.[7]

Procedure.[1] Twenty milligrams of *dl*,L-tetrahydrofolate is suspended in 2.0 ml. of 0.025 M formaldehyde and brought into solution by the dropwise addition of 5 N NaOH to pH 5. After standing at room temperature for 10 to 15 minutes, the mixture is cooled to 2°, and the pH is adjusted to 9.5 by the dropwise addition of 1 N NaOH. The solution is adsorbed onto a 1 × 10-cm. column of washed Solka-floc, and the column is eluted with a mixture of 40% ethanol–60% bicarbonate buffer containing 10^{-2} M mercaptoethanol. One-milliliter samples of the effluent are collected with an automatic fraction collector. When the fractions are examined for total pteridine by absorbancy at 295 mμ, a single asymmetrical peak (tubes 10 to 20) is obtained in the elution profile. The first fractions of the peak contain unreacted tetrahydrofolate plus some "active formaldehyde"; the small amount of material in the trailing fractions is identified by absorption spectrum as dihydrofolate (λ_{max} at 283 mμ). The center fractions represent 30 to 50% of the total pteridine material and are largely "active formaldehyde" (85 to 100% pure) as judged by enzymatic assay.

Properties. As prepared above, the material contains one mole of bound formaldehyde (as estimated by the acetyl acetone method[1]) per mole of tetrahydrofolate. At pH 7, "active formaldehyde" has a single absorption maximum at 290 to 295 mμ, with an extinction coefficient[8] of 32×10^3 M^{-1} cm.$^{-1}$. Relative to tetrahydrofolate, the compound at neutral or basic pH is quite stable to air oxidation. The association constant[1] for the formation of "active formaldehyde" [Eq. (1)] at pH

$$HCHO + \text{tetrahydrofolate} \rightleftarrows N^5,N^{10}\text{-Methylenetetrahydrofolate} \quad (1)$$

4.3 and 22° is 1.3×10^4 M^{-1}. A value of 2.1×10^4 at pH 7.2 has been reported previously.[8]

Paper chromatography may be used to distinguish "active formaldehyde" from tetrahydrofolate, although considerable degradation of both compounds is encountered. Ascending chromatography is carried out on sheets of Whatman No. 1 paper with the following solvent systems: (1) 0.1 M phosphate buffer, pH 8; (2) 1 M sodium formate–2% formic acid; and (3) 70% ethanol–30% H_2O. R_f values of 0.53, 0.62, and 0.32 are

[7] K. G. Scrimgeour and F. M. Huennekens, Vol. VI [49].

[8] R. L. Blakley, *Nature* 182, 1719 (1958).

observed for tetrahydrofolate in solvent systems 1, 2, and 3; the corresponding values for "active formaldehyde" are 0.27, 0.20, and 0.25.

"Active formaldehyde" may be quantitated by its activity in the N^5,N^{10}-methylenetetrahydrofolic dehydrogenase system[7] [Eq. (2)]. The

$$N^5,N^{10}\text{-Methylenetetrahydrofolate} + TPN^+ \rightleftarrows$$
$$(N^5,N^{10}\text{-Methenyltetrahydrofolate})^+ + TPNH \quad (2)$$

following components are added to a Corex cuvette of 1-cm. optical path: 0.1 micromole of "active formaldehyde," 0.3 micromole of TPN, 10 micromoles of 2-mercaptoethanol, 50 micromoles of bicarbonate buffer, pH 9.5, and water to make 1.2 ml. The optical blank contains all components except TPN. Enzyme (0.03 ml.) is added to start the reaction, and the changes in optical density at 340 mμ (ΔE_{340}) are determined as a function of time. The extinction coefficient (ϵ) for TPNH is 6.2×10^3 M^{-1} cm.$^{-1}$. The rate is further corrected for a control omitting "active formaldehyde."

Preparation of "Active Formate"

Reagents

Folic acid.
98% formic acid.
Platinum oxide, obtained from the American Platinum Works, Newark, New Jersey.
0.1 M formic acid–0.01 M mercaptoethanol.
Whatman cellulose powder.
0.1 N HCl–0.1 M mercaptoethanol.

Procedure. Two hundred and fifty milligrams (565 micromoles) of folic acid is dissolved in 50 ml. of 98% formic acid contained in a 250-ml. round-bottomed flask equipped with a reflux condenser, and heated to 50° to 60° for 3 hours.[9] As the reaction proceeds, the color of the solution changes from yellow to orange-red (N^{10}-formylfolic acid, λ_{max} at 250 mμ and 316 mμ in formic acid). The solution is cooled to room temperature and transferred to a Parr apparatus[10] containing 250 mg. of platinum oxide where it is hydrogenated for 20 minutes under 35 pounds of pressure. At the conclusion of the hydrogenation step, the solution is separated from the catalyst by suction filtration through a Büchner funnel

---

[9] The flask, covered with aluminum foil to minimize photodecomposition of both folic acid and the N^{10}-formyl derivative, is kept at the desired temperature in a stream of hot tap water.

[10] Pressure Reaction Apparatus, Parr Instrument Co., Moline, Illinois.

into a 250-ml. round-bottomed flask. The yellow[11] filtered solution is lyophilized to dryness, and the granular residue (240 mg.) is dissolved in 30 ml. of $0.1\,M$ formic acid–$0.01\,M$ 2-mercaptoethanol. By measurement of the optical density of this solution at 355 mμ, the amount of product, N^5,N^{10}-methenyltetrahydrofolic acid, is calculated[12] to be 250 micromoles (44% yield).

A glass chromatographic column (4 × 30 cm.) is packed under moderate pressure to a height of 20 cm. with Whatman cellulose powder.[13] The column is washed with 500 ml. of $0.1\,M$ formic acid–$0.01\,M$ mercaptoethanol, drained to incipient dryness, and the above solution (30 ml.) containing the N^5,N^{10}-methenyltetrahydrofolic acid is adsorbed onto the column. With a flow rate of about 1 ml./min., elution is carried out with $0.1\,M$ formic acid–$0.01\,M$ mercaptoethanol, and the effluent is collected in 5-ml. portions with an automatic fraction collector. The elution profile, obtained by plotting the optical density at 355 mμ of each fraction, reveals a single peak centered at about tube 34 and skewed slightly on the trailing edge. In addition to the desired product, unknown degradation products characterized by absorption maxima near 280 mμ occur in tubes 20 to 30. All tubes in the elution profile (about thirty-four to fifty) which show a value greater than 1.6 for the ratio of light absorption at 355 and 280 mμ (E_{355}/E_{280}) contain adequate amounts of product relative to contaminants. The contents of these tubes are pooled and lyophilized to dryness (72 mg.; 136 micromoles of product; dry-weight purity, 0.88).

The above amorphous material is dissolved with warming in a minimum volume (about 3 ml.) of $0.1\,M$ HCl–$0.1\,M$ mercaptoethanol, and the solution is cooled to room temperature and then placed in a refrigerator at about 5°. Yellow rod-shaped crystals of N^5,N^{10}-methenyltetrahydrofolic acid appear almost immediately and, after standing for 36 hours, are removed by centrifugation, dried by washing successively with abso-

---

[11] Catalytic hydrogenation of N^{10}-formylfolic acid in acid solution yields N^{10}-formyltetrahydrofolic acid, but the latter compound, in the presence of strong acid, cyclizes immediately to N^5,N^{10}-methenyltetrahydrofolic acid (λ_{max} at 355 mμ in formic acid). Above pH 7, N^5,N^{10}-methenyltetrahydrofolic acid opens selectively to yield N^{10}-formyltetrahydrofolic acid [L. D. Kay, M. J. Osborn, Y. Hatefi, and F. M. Huennekens, *J. Biol. Chem.* **235**, 195 (1960)], which is unstable to air oxidation.

[12] In 0.01 N HCl the extinction coefficient of N^5,N^{10}-methenyltetrahydrofolic acid at 335 mμ is reported[4] to be $25 \times 10^{-3}\,M^{-1}$ cm.$^{-1}$.

[13] Whatman No. 50 filter paper was placed over the sintered-glass plate at the bottom of the column in order to prevent fine particles of cellulose powder from appearing in the effluent.

lute ethanol and ether, and stored in a vacuum desiccator. Yield, 50.5 mg.; 18.2% of theory.

Properties. In $1.0 N$ HCl the crystalline material exhibits a principal absorption maximum at 348 mμ and a smaller peak at 290 mμ.[14] The ratio of absorbancy at 348 mμ (maximum) and 305 mμ (minimum) may be taken as an index of purity; for the pure material, a value of 2.45 is obtained. A sample of the crystalline material is dried over P_2O_5 in a vacuum pistol at 50°, and a weighed amount is dissolved in $1.0 N$ HCl; on the basis of the molecular weight of 491.9 for the chloride salt, an extinction coefficient of $26.5 \times 10^3 M^{-1}$ cm.$^{-1}$ at 348 mμ is calculated from the optical density measurement. A value of $26.2 \times 10^3 M^{-1}$ cm.$^{-1}$ has been reported previously[4] for N^5,N^{10}-methenyltetrahydrofolic acid prepared by an alternative procedure.[15]

When the product is examined by descending paper chromatography on Whatman No. 1 paper with $1.0 M$ formic acid–$0.01 M$ mercapto-ethanol as the solvent system, a single white fluorescent spot ($R_f = 0.50$) is observed under ultraviolet light. The failure of this spot to respond to the spray reagent,[17] which reductively cleaves many folic acid compounds at the C^9—N^{10} linkage, is referable to the stabilizing effect of the methenyl bridge between the N^5- and N^{10}-positions.

[14] Several isomeric forms of N^5,N^{10}-methenyltetrahydrofolic acid have been reported. When crystallized from HCl, anhydroleucovorin (the most commonly used isomer) contains 1 mole of chloride ion. Anhydroleucovorin A[15, 16] is formed by recrystallizing anhydroleucovorin chloride from hot water; this compound is believed to have a betaine-type structure in which a glutamate carboxyl group is the counterion to the quaternary nitrogen. By treatment of either anhydroleucovorin or anhydroleucovorin A with pH 4 buffer at 100°, anhydroleucovorin B is obtained; the structure of this isomer is unknown. Anhydroleucovorin and the A isomer have nearly equal absorption maxima at 350 to 355 mμ, whereas the A isomer has a weaker band at 350 mμ and a new band at 275 mμ.

[15] D. B. Cosulich, B. Roth, J. M. Smith, Jr., M. E. Hultquist, and R. P. Parker, *J. Am. Chem. Soc.* **73,** 5006 (1951).

[16] D. B. Cosulich, B. Roth, J. M. Smith, Jr., M. E. Hultquist, and R. P. Parker, *J. Am. Chem. Soc.* **74,** 3252 (1952).

[17] Y. Hatefi, P. T. Talbert, M. J. Osborn, and F. M. Huennekens, *Biochem. Preparations* **6,** 89 (1960).

[115] Preparation and Properties of 5-Formimino-tetrahydrofolic Acid

By Jesse C. Rabinowitz

Preparation

Principle. *l*-5-Formiminotetrahydrofolic acid is formed from *dl*-tetrahydrofolic acid and formiminoglycine by the action of formiminoglycine formimino-transferase.[1] The product is isolated from the reaction mixture by chromatography on Dowex 1–acetate.

Reagents

dl-Tetrahydrofolic acid diacetate.[2-4]
0.05 M 2-mercaptoethanol.
1.0 M potassium maleate buffer, pH 7.0.
0.5 M ethylenediaminetetraacetic acid.
0.54 M formiminoglycine.[5]
Formiminoglycine formimino-transferase.[6]
2.0 N KOH.

Procedure. Two grams of *dl*-tetrahydrofolic acid are dissolved in 50 ml. of 2-mercaptoethanol, and the solution is adjusted to pH 7.0 with KOH. To the solution are added 6.25 ml. of maleate buffer, 0.25 ml. of ethylenediaminetetraacetic acid, and 100 ml. of formiminoglycine. The pH of the solution is adjusted to 7.0 with KOH, and the solution is placed in a flask protected from light with a black cloth. The reaction is initiated by the addition of about 1000 units (80 ml., $Q = 200$) of formiminoglycine formimino-transferase. The mixture is held at room temperature. The progress of the reaction is followed by removing 10-μl. aliquots of the reaction mixture at 3-minute intervals and determining the amount of 5-formiminotetrahydrofolic acid present according to the procedure described below. When the reaction is complete, or after 10 to 15 minutes, the reaction mixture is chilled in an ice bath.

[1] J. C. Rabinowitz and W. E. Pricer, Jr., *Federation Proc.* 16, 236 (1957).
[2] J. C. Rabinowitz and W. E. Pricer, Jr., *J. Biol. Chem.* 229, 321 (1957).
[3] R. H. Himes and J. C. Rabinowitz, *J. Biol. Chem.* 237, 2903 (1962).
[4] See Vol. VI [113].
[5] H. Tabor and J. C. Rabinowitz, *Biochem. Preparations* 6, 100 (1957); see also Vol. VI [97].
[6] J. C. Rabinowitz and W. E. Pricer, Jr., *J. Am. Chem. Soc.* 78, 5702 (1956); see also Vol. VI [10].

Purification Procedure

The chilled mixture, without deproteinization or further treatment, is placed on a column of Dowex 1–acetate (2 × 30 cm.) under air pressure. The entire operation must be carried out at 2° to 4°, and the material should be protected from exposure to light. The column is then washed with 1 l. of water. This procedure serves to remove the excess formimino-glycine, which is not retained by the column, and also serves to oxidize the unreacted d-isomer of tetrahydrofolic acid to products which remain on the column throughout the elution procedure. The column is then developed with 0.2 N acetic acid, and 15-ml. fractions are collected with an automatic fraction collector. The fractions are analyzed for absorption at 285 mμ and for 5-formiminotetrahydrofolic acid according to the following procedure. A 0.01- to 0.10-ml. aliquot of each fraction is diluted in 3.0 ml. of water, and the absorbancy is determined at 285 mμ. Those fractions which show absorbancy are treated with 0.13 ml. of 6 N HCl, and the absorbancy at 350 mμ is determined immediately. The acidified solution is then heated for 50 seconds at 100° and cooled, and the absorbancy is redetermined. The increase in absorbancy at 350 mμ is due to the conversion of 5-formiminotetrahydrofolic acid to 5,10-methenyl-tetrahydrofolic acid.

Formiminotetrahydrofolic acid is essentially the only component with absorption at 285 mμ obtained under these conditions. It appears after approximately 400 ml. of the 0.2 N acetic acid have been passed over the column and is completely eluted by 250 ml. more of the eluting fluid. The fractions containing the 5-formiminotetrahydrofolic acid are pooled and lyophilized. The product is obtained as a light-yellow powder in a yield of 43 to 55% based on the amount of l-tetrahydrofolic acid added to the incubation mixture. Roughly 370 mg. of material are obtained from 2 g. of dl-tetrahydrofolic acid diacetate. The product is stored at −20°.

Properties

5-Formiminotetrahydrofolic acid has an absorption maximum at 285 mμ with a molar extinction coefficient at pH 7.0 of 35,400.[7] The ultraviolet as well as the infrared spectrum[7] is very similar to that of 5-formyltetrahydrofolic acid. Preparations obtained in the manner described show variable absorption at 356 mμ at neutral pH. This is probably due to contamination of the material with varying amounts of 5,10-methenyltetrahydrofolic acid formed during the isolation procedure.

The half-life of the material in solution over the pH range of 5 to 9

[7] K. Uyeda, Ph.D. Thesis, University of California, Berkeley, 1962.

is only 1 hour at 37° and is less than 1 minute at 100°. Solutions may be stored frozen, but considerable destruction occurs at 0° over a 12-hour period. 5-Formiminotetrahydrofolic acid, like 5-formyltetrahydrofolic acid, is not affected by exposure to atmospheric oxygen, in contrast to the behavior of tetrahydrofolic acid.

Analysis

5-Formiminotetrahydrofolic acid may be converted to 5,10-methenyltetrahydrofolic acid and ammonia by treatment with acid as described above and determined from the increase in absorption at 350 mμ under these conditions. However, both 5-formyltetrahydrofolic acid and 10-formyltetrahydrofolic acid are also converted to 5,10-methenyltetrahydrofolic acid by this treatment, although they do not liberate ammonia. The most specific method for the analysis of this compound is by using the enzyme formiminotetrahydrofolic acid cyclodeaminase.[8]

[8] See Vol. VI [52].

[116] Preparation and Properties of 5,10-Methenyltetrahydrofolic Acid and 10-Formyltetrahydrofolic Acid

By JESSE C. RABINOWITZ

Preparation

Principle. The three formyl derivatives of tetrahydrofolic acid— 5-formyl-,[1] 10-formyl-, and 5,10-methenyltetrahydrofolic acid[2]—are in equilibrium.[3,4] The 10-formyl- and 5,10-methenyl- derivatives are most conveniently prepared from 5-formyltetrahydrofolic acid. 5-Formyltetrahydrofolic acid is converted to the insoluble 5,10-methenyltetrahydrofolic acid in dilute acid. The latter compound is converted to 10-formyltetrahydrofolic acid at neutral pH in the presence of mercaptoethanol.

[1] *dl*-5-Formyltetrahydrofolic acid has also been referred to as folinic acid-SF and leucovorin. The product isolated from natural sources is the *l*-form and has been called *citrovorum factor.*
[2] 5,10-Methenyltetrahydrofolic acid has also been referred to as N^5,N^{10}-anhydroformyltetrahydrofolic acid, anhydroleucovorin, and isoleucovorin.
[3] M. May, T. J. Bardos, F. L. Barger, M. Lansford, J. M. Ravel, G. L. Sutherland, and W. Shive, *J. Am. Chem. Soc.* **73**, 3067 (1951).
[4] D. B. Cosulich, B. Roth, J. M. Smith, Jr., M. E. Hultquist, and R. P. Parker, *J. Am. Chem. Soc.* **74**, 3252 (1952).

Reagents

5-Formyltetrahydrofolic acid, calcium salt.[3, 5, 6]
1 *M* 2-mercaptoethanol.
1.0 *M* HCl.
1.0 *M* KOH.

Procedure. *dl*-5-Formyltetrahydrofolic acid (50 mg.) is dissolved in 4 ml. of 2-mercaptoethanol, and the solution is adjusted to pH 1.5 with HCl. The *dl*-5,10-methenyltetrahydrofolic acid formed precipitates in the cold and may be removed by filtration after several hours. For the preparation of 10-formyltetrahydrofolic acid, the suspension of 5,10-methenyltetrahydrofolic acid obtained as described above is neutralized with KOH and stored in an evacuated vessel at 4°. The conversion of the 5,10-methenyltetrahydrofolic acid to 10-formyltetrahydrofolic acid is determined by the disappearance of absorption at 356 mμ and is usually complete after 4 hours. The solution of 10-formyltetrahydrofolic acid may be stored at −20° for periods up to 3 days. 10-Formyltetrahydrofolic acid has not been isolated in solid form.

Properties

5,10-Methenyltetrahydrofolic acid has an absorption maximum at 345, 352, and 360 mμ in 1.0, 0.01, and 0.001 *M* HCl, respectively. The molar extinction coefficients under these conditions are approximately 26,000, 25,000, and 25,100, respectively. 10-Formyltetrahydrofolic acid has an absorption maximum at 258 mμ in neutral solution. 10-Formyltetrahydrofolic acid is readily degraded by exposure to oxygen in the absence of 2-mercaptoethanol. 5,10-Methenyltetrahydrofolic acid is relatively stable to oxygen. The rate of hydrolysis of 5,10-methenyltetrahydrofolic acid to 10-formyltetrahydrofolic acid at neutral pH is markedly affected by anions. The compound is most stable in maleate buffer and is most readily hydrolyzed in phosphate, arsenate, or pyrophosphate buffers.[7]

[5] B. Roth, M. E. Hultquist, M. J. Fahrenbach, D. B. Cosulich, H. P. Broquist, J. A. Brockman, Jr., J. M. Smith, Jr., R. P. Parker, E. L. R. Stokstad, and T. H. Jukes, *J. Am. Chem. Soc.* **74**, 3247 (1952).
[6] The most common source of this material has been the generous gifts from the Lederle Laboratories. The material is now available commercially.
[7] J. C. Rabinowitz, *in* "The Enzymes," (P. D. Boyer, H. Lardy, and K. Myrbäck, eds.), 2nd ed., Vol. 2, p. 185. Academic Press, New York, 1960.

SPECIAL TECHNIQUES

[117] Chromatographic Determination of Amino Acids by the Use of Automatic Recording Equipment

By Stanford Moore and William H. Stein

Introduction

Quantitative determination of amino acids bears a relationship to the chemistry of proteins similar to that which elementary analysis bears to the chemistry of simpler organic molecules. Since the amino acid composition is a primary characteristic of a given protein, accuracy in the analysis is of greater importance than the speed or convenience of the determination. The ease of the analysis increases in significance in the study of small peptides when less accurate data may suffice to define the composition.

Coupled with the results of other physical and chemical procedures, amino acid analysis can provide evidence as to the purity of a protein preparation. A single molecular species should yield analytical values which can be expressed in terms of integral numbers of amino acid residues per molecule. If the protein is large, the number of amino acid residues may be so great that analytical data cannot be conclusive in determining whether or not integral molar ratios are present. In this latter instance, analyses may be useful in evaluating the purity of a protein preparation if each step in the isolation procedure is monitored by amino acid analysis. Once a pure preparation has been achieved, further attempts at fractionation will produce no change in amino acid composition.

An analysis for amino acids is one of the first steps in the elucidation of the chemical structure of a protein molecule. Knowledge of the composition is necessary as a guide before structural analysis is begun. In particular, the nature of the approach to the structural study will be decisively influenced by the presence or absence of certain amino acids, notably cystine, cysteine, and tryptophan. It has been found that quantitative amino acid analyses are usually essential if unambiguous results are to be obtained in the elucidation of the sequence of amino acid residues in peptides isolated by fragmentation of the parent protein.

Studies on the nature of the alterations brought about by chemical treatment of a protein can be followed in many instances by amino acid analysis. Examples are found in attempts to modify the so-called "active site" of an enzyme by the use of selective reagents.

It is to facilitate such studies that attention has been directed in recent years to the development of accurate automatic recording apparatus for the chromatographic determination of amino acids. The present section is concerned primarily with the use of this equipment. References to the many other procedures that may be considered are included in Vol. III of this series and in recent reviews.[1,2]

Hydrolysis

As the precision of analytical methods for the determination of amino acids has increased, the limiting factor in deriving the amino acid composition of a protein has become the extent to which the composition of the hydrolyzate is a true reflection of the composition of the parent protein. Discrepancies arise owing to the decomposition of certain labile amino acids during hydrolysis and as a result of the fact that some amino acids are liberated from peptide linkage with difficulty. Conditions for hydrolysis that currently are giving fairly satisfactory and reproducible results are the following:

Procedure. A 5-mg. sample of the air-dried or lyophilized protein is placed into a 16×125-mm. heavy-walled Pyrex test tube (Corning Cat. No. 9860). It is convenient to wash several dozen of these tubes at one time with chromic acid, and to rinse them with distilled water and finally with $1 N$ HCl. The residual HCl is removed in an air oven at $100°$, and the tubes are stored inverted in a covered container to prevent deposition of NH_4Cl from the laboratory air.

Samples are also weighed out at the same time for the determination of moisture and ash (on ca. 5 mg.), if the results are to be expressed on a moisture and ash-free basis, and for the determination of total nitrogen (on ca. 3 mg.), when the percentage recovery of the total nitrogen as amino acid nitrogen is desired. Amino acid analysis of a 5-mg. sample permits the peaks of the chromatographic pattern to be integrated with maximum accuracy. Reasonably good accuracy can be obtained with a 1-mg. sample.

The protein in the tube is suspended in 1 ml. of $6 N$ HCl (a $1:1$ dilution of reagent concentrated HCl). With a small glass rod sealed inside the rim of the tube, as a handle, a section of the tube about 3 cm. from the top is constricted (oxygen flame) to about 1-mm. bore. The handle is removed and the lower half of the tube is inserted in a bath of solid carbon dioxide and ethanol. When the sample is frozen, the tube is connected to a vacuum line through a short sleeve of Tygon tubing of ½-inch inner diameter. The system is evacuated with an oil pump

[1] P. Edman, *Ann. Rev. Biochem.* **28**, 69 (1959).
[2] R. L. Hill, J. R. Kimmel, and E. L. Smith, *Ann. Rev. Biochem.* **28**, 97 (1959).

to below 50 microns (McCleod gage, or more conveniently a Pirani gage[3]) and the tube is sealed under vacuum.

For the highest recovery of *S*-carboxymethylcysteine and tyrosine, it is desirable to remove the traces of dissolved air in the HCl before the tube is sealed. The following additional step is included in the procedure in this laboratory as a part of the standard technique for sealing off the tubes: When the oil pump has brought the pressure down to 60 microns, the tube is withdrawn from the dry ice bath and the frozen solution is allowed to thaw slowly with the pump still on. As bubbles form and rise up in the tube from the viscous solution, momentary immersion of the tube in the dry ice mixture will break the bubbles and allow the liquid to drain back. The pressure increases to about 80 microns during this degassing process. When the pressure is back down to 60 microns (in 10 to 20 minutes), the tube is shaken to make certain that gas removal is complete, and then sealed off. Refreezing of the solution is usually unnecessary.

The hydrolysis is conducted at 110° ± 1° for 20 hours or 70 hours. The temperature should be accurately controlled. If an air oven is used, a model providing forced air circulation from a fan is needed.

After the tube has cooled to room temperature, any liquid on the walls is spun down by gentle centrifugation. (If the analysis is not to be performed immediately, the hydrolysis tube can be stored at 4° or in the deepfreeze.) The tube is scored with a sharp file at a point below the tapered end and cracked by the use of a hot glass rod. The cut end of the tube is fire-polished and the HCl is removed under reduced pressure (water pump) by attaching the tube with a short section of Tygon (½-inch inner diameter) to the condenser of a rotary evaporator[4] which can be operated with the condenser axis at a downward tilt of about 30°. Alternatively, the hydrolyzate can be transferred to a 10-ml round-bottom flask prior to concentration on an evaporator operated horizontally. Most of the HCl is removed in about 20 minutes; the tube or flask is left on the evaporator at 40° for about 20 minutes more to ensure thorough removal of the excess acid. Before analysis, the residue is dissolved in 0.5 ml. of water, and the solution is brought to about pH 6.5 by the addition of 0.5 ml. of 0.2 M sodium phosphate buffer of pH 6.5. The nearly neutral solution (there is no loss of NH_3 at this pH) is allowed to stand for 4 hours to permit air oxidation of any cysteine to cystine[5,6] and is then brought to about pH 2 by the addition

[3] Pirani gage, Type 516, from National Research Corporation, Newton 61, Mass.
[4] L. C. Craig, J. D. Gregory, and W. Hausmann, *Anal. Chem.* **22**, 1462 (1950).
[5] W. H. Stein and S. Moore, *J. Biol. Chem.* **211**, 915 (1954).
[6] C. H. W. Hirs, W. H. Stein, and S. Moore, *J. Biol. Chem.* **211**, 941 (1954).

of 0.06 ml. of freshly prepared (to minimize contamination with ammonia) N HCl. The resulting solution is transferred quantitatively to a 5-ml. volumetric flask by the use of 1-ml. aliquots of pH 2.2 buffer[7] containing 5 ml./l. of thiodiglycol. A 2-ml. aliquot of this solution is used for each chromatogram.

Comments. The rate of decomposition of serine and threonine during acid hydrolysis varies with, among other factors, the purity of the HCl.[6,8] In our experience, however, redistillation of reagent-grade HCl from glass has not diminished the amount of decomposition. It is important to conduct the hydrolysis at a temperature of $110° \pm 1°$. At 105°, for example, hydrolysis is usually incomplete in 20 hours, and there is chromatographic evidence for the presence of peptides in the hydrolyzate. The maintenance of constant temperature is essential if results after 20 hours and 70 hours of hydrolysis are to be extrapolated to zero time in order to correct for the decomposition of the labile amino acids. These include cystine and tyrosine,[6] as well as serine and threonine.

If the tubes are not well evacuated (below 100 microns), oxidation may lead to the formation of appreciable amounts of cysteic acid, methionine sulfoxide, and chlorotyrosine. The removal of HCl by placing the tube in a desiccator overnight[6] has been found (A. M. Crestfield, unpublished observations) to be accompanied by additional decomposition of serine and threonine and the appearance of several small new peaks near the positions of tryptophan, serine, cystine, and methionine. The size of these artifact peaks increases as the time of storage in the desiccator is lengthened. If the HCl is removed rapidly on the rotary evaporator, this decomposition is avoided.

The decomposition of *S*-carboxymethylcysteine is particularly sensitive to oxygen (A. M. Crestfield, W. H. Konigsberg, and G. Guidotti, unpublished observations). If a water pump is used to evacuate the tubes, the recovery of *S*-carboxymethylcysteine may be less than 70%. The recovery runs about 94% if an oil pump is used but the tubes are sealed without removal of dissolved air; the recovery is 99% when the solution is thawed and degassed, as described previously.

Formic acid should not be added when hydrolysis is carried out in sealed tubes; excessive pressure may be developed by the carbon monoxide formed. The presence of sulfate ion ($0.1 M$ or more) during hydrolysis may lead to the formation of sulfoserine which emerges from the ion exchange column at the same position as cysteic acid.

[7] S. Moore and W. H. Stein, *J. Biol. Chem.* **211**, 893 (1954).
[8] M. W. Rees, *Biochem. J.* **40**, 632 (1946).

Chromatographic Analysis

Procedure. The automatic recording equipment used in conjunction with the ion exchange columns is illustrated in Fig. 1. The tubes are packed with Amberlite IR-120, a sulfonated polystyrene resin. Two columns are used;[9] one, 150 cm. long, for the determination of the neutral and acidic amino acids, and one, 15 cm. long, for the basic amino acids. The automatic recording equipment is designed and operated as described by Spackman *et al.*[10, 11] Influent buffers are drawn from reservoir bottles and pumped through the ion-exchange columns at a constant speed of 30 ml. per hour. Ninhydrin reagent is pumped into the effluent stream at 15 ml. per hour, and the effluent-ninhydrin mixture is carried in a coil of capillary Teflon tubing through a boiling-water bath to develop the colors formed when ninhydrin reacts with amino acids. The absorbancy of the colored solution is measured by a three-unit photometer and registered on a multipoint recorder. The top unit of the photometer measures the absorbancy at 570 mμ; the second has a filter that transmits at 440 mμ to measure the intensity of the yellow color formed with proline and hydroxyproline; and the third unit measures the absorbancy at 570 mμ, except that in this instance the cell is one-third the diameter of the top one, so that, if the reading in the top cell exceeds an absorbancy of 1.4, the curve from the third cell can be integrated.

A typical recorded curve is shown in Fig. 2. The recovery of each amino acid is quantitative.[9] The area under a peak is proportional to the product of the height of the peak and the width of the peak at half the height. The width is measured accurately by counting the dots on the curve above the half-height line. The instrument is calibrated by analyzing a synthetic mixture of the pure amino acids,[10, 12] thus establishing the height-width constants for each amino acid. When the instrument is first placed in operation, it is well to run a chromatogram with the calibration mixture each time the reservoir of ninhydrin reagent is filled and again when the reservoir is nearly empty, about ten analyses later. If the ninhydrin reagent (stored under nitrogen) is adequately

[9] S. Moore, D. H. Spackman, and W. H. Stein, *Anal. Chem.* **30**, 1185 (1958).

[10] D. H. Spackman, W. H. Stein, and S. Moore, *Anal. Chem.* **30**, 1190 (1958).

[11] Among the current suppliers of commercial models of instruments of this type are Spinco Division, Beckman Instruments, Palo Alto, California and Phoenix Precision Instrument Company, Philadelphia 40, Pennsylvania. The instruments should have a vapor-tight glass covered vessel for the 100° water bath and two pairs of 8-liter aspirator bottles to protect the ninhydrin reagent fully from air. The deaerators on the buffer lines can be omitted.

[12] Ampoules of a synthetic mixture of pure amino acids for calibration purposes are available through the Spinco organization.

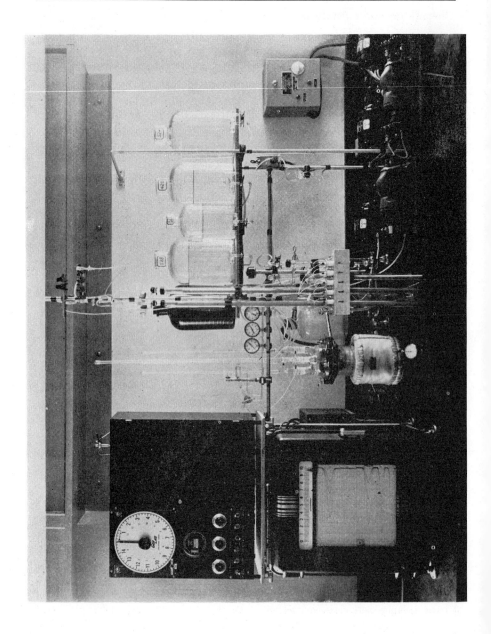

protected from air, the results at the beginning and end of a reservoir should be reproducible to $100 \pm 3\%$ for loads ranging from 0.25 to 2 micromoles of each amino acid. If an analysis runs low, the recovery of proline is an indication of whether the difficulty comes from access of oxygen to the ninhydrin reagent; under such a condition the color yield from proline remains relatively constant while the color production from the α-NH$_2$ acids is decreased.

An instrument of the design shown in Fig. 1 is capable of providing one analysis of the type illustrated in Fig. 2 per day.

Calculations. The results from a typical chromatographic analysis of a hydrolyzate of bovine pancreatic ribonuclease A are given in the table. Control experiments on the "hydrolysis" of synthetic mixtures of amino acids have shown that, under the conditions used for these experiments, the decomposition of threonine, cystine, and tyrosine is about 5% in 20 hours, and that of serine is about 10%. These approximate corrections can be applied to obtain an estimate of the composition of a protein from a single analysis. To obtain more accurate figures, the results of a companion hydrolyzate heated for 70 hours are required. The values obtained for the labile amino acids at the two times of hydrolysis are extrapolated to zero time, with first-order kinetics assumed.[6] In the calculation, A_1, A_2, and A_0 are the quantities of amino

$$\log A_0 = \left(\frac{t_2}{t_2 - t_1}\right) \log A_1 - \left(\frac{t_1}{t_2 - t_1}\right) \log A_2$$

acids present after t_1, t_2, and zero hours of hydrolysis, respectively. When applied to proteins, this method of calculation does not take into account the possibility that the rate of decomposition of free amino acids may differ from that prevailing during the initial stages of protein hydrolysis. It does, however, provide some compensation for the variations that have been observed in the extent of the decomposition of labile amino acids from one protein sample to another, and even from one laboratory to another. Therefore, this method of calculation probably comes as close to giving accurate results as is possible at present.

The use of a longer time of hydrolysis is also required in order to obtain accurate values for amino acids that are liberated unusually slowly by acid hydrolysis. For example, the isoleucine in ribonuclease

FIG. 1. Automatic recording apparatus for use in the chromatography of amino acids.[11] The ion-exchange columns are in the center, the pumps on the right, and the boiling-water bath and the photometer on the left. A set of five Teflon microvalves (center, in front of columns) replaces the stopcock manifold of the initial instruments.[9] From S. Moore, D. H. Spackman, and W. H. Stein, *Federation Proc.* **17**, 1107 (1958).

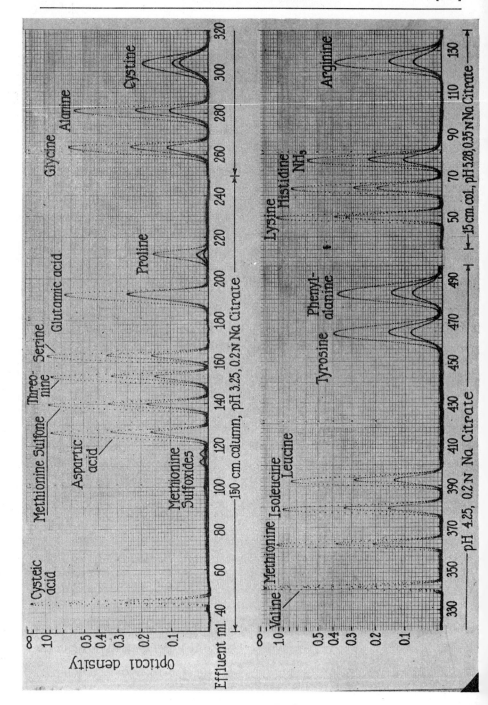

is liberated very slowly[13] because of the presence of an isoleucyl-isoleucyl sequence.[14] Alloisoleucine, always formed in very small amounts (1 to 5% of the isoleucine) by the racemization of isoleucine during hydrolysis, gives rise to a peak between methionine and isoleucine. The small amount of alloisoleucine found should be added to the isoleucine value in calculating the molar ratios. Similarly, small amounts of the sulfoxides of methionine, if present, are included with methionine.

Since the number of amino acid residues per molecule of ribonuclease is known,[14, 15] it is possible to compare (see the table) the values calculated from the single analysis with the number of residues to be expected in a chromatographically purified sample of the enzyme. The results of this representative analysis agree within about 3% with the expected integral values even though three possible sources of variation are involved—the purity of the sample, the hydrolytic procedure, and the accuracy of the chromatographic analysis per se.

Cysteine and Cystine. The cystine peak obtained when a hydrolyzate is analyzed is not symmetrical; racemization of cystine during hydrolysis gives rise to *meso*-cystine as well as to the D,L-racemic mixture,[6] and two overlapping zones are obtained in the cystine range, with the *meso* form occurring in the most rapidly moving zone. The combined peak is integrated by addition of the absorbancy values[10] instead of by the height-width method.

Air-oxidation of cysteine in the hydrolyzate to cystine, before chromatography, is desirable in order to obtain a more accurate value for the total cystine + cysteine content of the protein and also because cysteine may interfere with the determination of proline. Cysteine is not fully stable during passage through the ion-exchange column. It is slowly oxidized, but a detectable quantity of the —SH compound may emerge at the proline position and cause a slightly high result for proline. The presence of cysteine can be recognized by comparing the ratio of the readings obtained at 440 mμ and 570 mμ in the proline peak with the ratio found when a control mixture containing no cysteine is analyzed. A lower-than-normal ratio occurs when cysteine is present. Even though the protein may initially contain only cystine, some cysteine may be formed as a result of reactions occurring during hydrolysis.

Fig. 2. Chromatographic analysis of a synthetic mixture of amino acids automatically recorded in 22 hours by the equipment shown in Fig. 1. From D. H. Spackman, W. H. Stein, and S. Moore, *Anal. Chem.* **30**, 1190 (1958).

[13] E. J. Harfenist, *J. Am. Chem. Soc.* **75**, 5528 (1953).
[14] C. H. W. Hirs, S. Moore, and W. H. Stein, *J. Biol. Chem.* **235**, 633 (1960).
[15] C. H. W. Hirs, W. H. Stein, and S. Moore, *J. Biol. Chem.* **211**, 151 (1956).

AN AMINO ACID ANALYSIS OF RIBONUCLEASE A

(Data from a single chromatographic analysis of a 20-hour hydrolyzate of a sample of chromatographically purified bovine pancreatic ribonuclease A. The calculations are given to illustrate the information derivable from a single analysis on the ion-exchange columns. The results refer to the micromoles of each constituent in a 2-ml. aliquot of the hydrolyzate. More accurate calculations would require data on a companion hydrolyzate heated for 70 hours.)

Amino acid	H, net height	W, width (dots)	$H \times W$	Micro-moles[a] $\frac{H \times W}{C^b}$	Found[c]	Corrected for destruction during hydrolysis[d]	Number of residues per molecule (m.w. 13,683)[e,f]
Lysine	0.900	26.7	24.0	1.02	9.8	9.8	10
Histidine	0.259	33.2	8.60	0.408	3.92	3.92	4
Ammonia	0.820	49.7	40.8	1.96	18.8	16.8[d]	17
Arginine	0.128	66.0	8.45	0.408	3.92	3.92	4
Aspartic acid	0.978	30.8	30.1	1.56	15.0	15.0	15
Threonine	0.652	30.0	19.6	1.01	9.7	10.2[d]	10
Serine	0.979	29.3	28.7	1.42	13.6	15.1[d]	15
Glutamic acid	0.658	37.9	24.9	1.27	12.2	12.2	12
Proline	0.053	43.0	2.28	0.407	3.91	3.91	4
Glycine	0.147	43.0	6.32	0.324	3.11	3.11	3
Alanine	0.546	46.5	25.4	1.28	12.3	12.3	12
Half-cystine + Cysteine	Opt. dens. units = 1.094[g]			0.781	7.50	7.90[d]	8
Valine	0.279	65.1	1.82	0.973	9.3	9.3	9
Methionine	0.316	24.6	7.77	0.409	3.93	3.93	4
Isoleucine	0.168	28.7	4.82	0.230	2.20	2.20[h]	3
Leucine	0.136	32.4	4.41	0.210	2.02	2.02	2
Tyrosine	0.205	57.5	11.8	0.548	5.53	5.82[d]	6
Phenylalanine	0.109	61.5	6.58	0.321	3.08	3.08	3
Tryptophan	Absent						

[a] When moisture, ash, and total nitrogen have been determined on the sample, the recovery can be calculated from the micromoles in terms of nitrogen as per cent of total nitrogen and in terms of weight of amino acid *residues* as per cent of total weight of the sample.

[b] C is the constant for integration of a given peak, as determined in the calibration of the amino acid analyzer.[10]

[c] The relative molar quantities of the amino acid residues were calculated in this instance by assuming that the micromoles of aspartic acid correspond to 15.0 residues per molecule.

[d] The approximate corrections applied for decomposition of amino acids in 20 hours of hydrolysis (see text) were 5% for threonine, cystine, and tyrosine, and 10% for serine. These calculations indicated that a correction of 2.0 should be applied to the value for ammonia to compensate for the ammonia liberated by decomposition of serine and threonine.

The analysis described above does not distinguish between cysteine and half-cystine residues in the protein. An independent and more accurate determination of the sum of cystine and cysteine can be made either by determining the cysteic acid content of the performic acid-oxidized protein[16] or the carboxymethylcysteine content of the reduced protein that has been treated with iodoacetate.[17,18] Both cysteic acid (Fig. 2) and carboxymethylcysteine[17] can be determined with the automatic recording equipment.

Differentiation between cysteine and cystine may sometimes be accomplished chromatographically by determining the yield of carboxymethylcysteine obtained when the protein (without oxidation or reduction) is alkylated with iodoacetate in the presence of a denaturing agent.[19]

Tryptophan. A special problem is presented by tryptophan, which is markedly labile during acid hydrolysis. The chromatogram often affords a convenient qualitative means of determining whether tryptophan is present. When hydrolysis is conducted in a sealed tube, some tryptophan and an acid decomposition product from it usually are present after 20 hours, and give rise to two peaks on the chromatogram (15-cm. column) just ahead of lysine.[17] If tryptophan is present it can be determined chromatographically in alkaline hydrolyzates by a modification of the procedure of Drèze,[20] but the details of the technique are still receiving study. When the amount of tyrosine is known independently, for example, as a result of the chromatographic analysis, the spectrophotometric determination of tryptophan in the intact protein[21] becomes

---

e C. H. W. Hirs, S. Moore, and W. H. Stein, *J. Biol. Chem.* **235**, 633 (1960).

f C. H. W. Hirs, W. H. Stein, and S. Moore, *J. Biol. Chem.* **211**, 151 (1956).

g The asymmetrical peak from DL-cystine and *meso*-cystine is integrated by addition of the optical density units, and the result is divided by 0.106 × C (instrument of Spackman *et al.*,[10] 4-second printing time on recorder) or 0.133 × C (Spinco or Phoenix instruments, 5-second printing time).

h Only two-thirds of the total isoleucine in ribonuclease is liberated in 20 hours (see text).

[16] E. Schram, S. Moore, and E. J. Bigwood, *Biochem. J.* **57**, 33 (1954).

[17] S. Moore, R. D. Cole, H. G. Gundlach, and W. H. Stein, Symposium on Proteins, 4th International Congress of Biochemistry, Vienna, 1958, Pergamon Press, p. 52 (1960).

[18] A. M. Crestfield, J. Skupin, S. Moore, and W. H. Stein, *Federation Proc.* **19**, 341 (1960).

[19] R. D. Cole, W. H. Stein, and S. Moore, *J. Biol. Chem.* **233**, 1359 (1958).

[20] A. Drèze, *Biochem. J.* **62**, 3P (1956).

[21] G. H. Beavan and E. R. Holiday, *Advances in Protein Chem.* **7**, 320 (1952).

more accurate than when both tyrosine and tryptophan are estimated by this means.

Methionine Sulfoxide. Methionine sulfoxide reverts to methionine on acid hydrolysis (Ray and Koshland[22]). It may be detected after alkaline hydrolysis, or by alkylation of the thio-ether sulfur of methionine residues with iodoacetate and determination of the sulfoxide as the sulfone after oxidation with performic acid (Neumann[23]). The alkylation protects the methionine sulfur from oxidation; the sulfone is fully stable under the conditions of acid hydrolysis.

Amide Ammonia. The ammonia content of the hydrolyzate, even when corrected for the amount of ammonia contributed by the decomposition of serine and threonine, is only an approximate measure of the amide ammonia content of the protein. The precise calibration of the recorder for ammonia presents a problem, since a synthetic mixture of amino acids may contain small amounts of ammonia in addition to that contributed by the NH_4Cl that has been added as a standard. If ammonia is of special concern, a solution containing only NH_4Cl dissolved in pH 2.2 buffer should be used for the standardization, and the ammonia value obtained from a blank chromatographic analysis of 2 ml. of the pH 2.2 buffer should be subtracted in order to obtain the true constant for ammonia. An analysis of a "blank" hydrolyzate can also be useful in determining the ammonia content of the HCl used for hydrolysis. Independent determination of the amide ammonia of the protein can be made by a modification[6] of the procedure described by Laki *et al.*[24]; the Conway microdiffusion technique is employed, and ammonia is determined by the photometric ninhydrin method.

Amino Sugars. If the hydrolyzate of the protein preparation contains an amino sugar, the first indication will be the appearance of a peak ahead of the position of emergence of tryptophan from the short column. More definitive qualitative identification of glucosamine and galactosamine can then come from the presence of peaks which emerge at 510 ml. and 580 ml., respectively, from the 150-cm. column. The severe conditions used for the hydrolysis of proteins result in low yields of the amino sugars, and milder conditions of hydrolysis have to be used if they are to be determined accurately.[25]

[22] W. J. Ray, Jr., and D. E. Koshland, Jr., *Brookhaven Symp. in Biol.* No. 13, 135 (1960).

[23] N. P. Neumann, *Brookhaven Symp. in Biol.*, No. 13, 149 (1960).

[24] K. Laki, D. R. Kominz, P. Symonds, L. Lorand, and W. H. Seegers, *Arch. Biochem. Biophys.* 49, 276 (1954).

[25] S. Gardell, *Acta chem. Scand.* 7, 207 (1953).

Other Applications

The manner in which the recording equipment has been used to study the composition of the peptides obtained during the elucidation of the complete structure of ribonuclease is described in the papers of Hirs *et al.*[14] and Spackman *et al.*[26] An example of the use of the method to study the results of chemical alteration of an enzyme may be found in the experiments of Gundlach *et al.*[27] on the reaction of iodoacetate with ribonuclease. The peaks on the effluent curve that could be attributed to the carboxymethyl derivatives of histidine, lysine, and methionine permitted the reaction of iodoacetate with these three different types of residues to be followed. Stark *et al.*[28] studied the reaction of lysine residues in proteins with cyanate as evidenced by homocitrulline in the hydrolyzates.

Other uses of amino acid analysis in enzymology may include the determination of the free amino acids in tissues and physiological fluids. The recording equipment is applicable to the determination of the fifty principal ninhydrin-positive constituents occurring in mammalian tissues, urine, and blood plasma.[10, 29]

[26] D. H. Spackman, W. H. Stein and S. Moore, *J. Biol. Chem.* **235**, 648 (1960).
[27] H. G. Gundlach, W. H. Stein, and S. Moore, *J. Biol. Chem.* **234**, 1754 (1959).
[28] G. R. Stark, W. H. Stein, and S. Moore, *J. Biol. Chem.* **235**, 3177 (1960).
[29] S. Moore, D. H. Spackman, and W. H. Stein, *Federation Proc.* **17**, 1107 (1958).

[118] Sequence Methods*

By VERNON M. INGRAM

There have been many developments in the methods available for the determination of amino acid sequences in peptides. In the short space available, only five of these will be mentioned: (1) Partial hydrolysis with dilute acetic acid. (2) Digestion with elastase. (3) Digestion with leucine amino peptidase. (4) Sjöquist's modification of the Edman stepwise degradation method. (5) Fingerprinting of mixtures of peptides. Details of actual examples will be given, but it should be realized that no two sequence problems are alike. If the reader is a newcomer to this subject, he is strongly urged to consult the reading list before attempting to apply the methods to his material.

* Submitted May 1960; partially revised July 1963.

Partial Hydrolysis with Dilute Acetic Acid[1]

Principle. Partridge and Davis[1] reported that certain proteins when boiled in dilute oxalic or acetic acids release aspartic acid preferentially. This method has been applied by Hunt[2,3] to the determination of the amino acid sequence of the tridecapeptide "A-T 26" isolated from tryptic digests[4] of human hemoglobin A. This peptide has the sequence

Val·Asn·Val·Asp·Glu·Val·Gly·Gly·Glu·Ala·Leu·Gly·Arg

The second aspartic acid and both glutamic acid residues have free carboxyl groups.

Reagents (all analytical-reagent grade)

0.25 M acetic acid in water.
"pH 6.4 buffer"[5]—pyridine–acetic acid–water, 25:1:225 by volume.
"pH 3.6 buffer"[5]—pyridine–acetic acid–water, 1:10:90 by volume.
n-Butanol–acetic acid–water chromatographic solvent, 3:1:1 by volume.
"Redfield-2" chromatographic solvent[6]—t-butyl alcohol–methyl ethyl ketone–water–diethylamine, 10:10:5:1 by volume.
Formic acid, pH 2.1, a 2.2% aqueous solution.

Procedure. About 0.1 to 0.2 micromole of the peptide A-T 26 was dissolved in about 0.2 ml. of 0.25 M acetic acid, sealed in a glass capillary tube, and heated at 120° for 11 hours. Some precipitate was formed, and the whole hydrolyzate was evaporated and re-evaporated with water in a vacuum desiccator. The hydrolyzate was fractionated by paper iono-phoresis at pH 6.4 and subsequently at pH 3.6 (Fig. 1). The peptides obtained were estimated quantitatively (see later in this section) in the butanol–acetic acid–Redfield-2 system.[2] The results are summarized in Table I. Peptide 2 gave only a very faint ninhydrin color and was most easily located by the use of the specific arginine stain.[7] An attempt was made to determine the N-terminal amino acid of peptide

[1] M. Partridge and H. F. Davis, *Nature* 165, 67 (1950).
[2] J. A. Hunt and V. M. Ingram, *Biochim. et Biophys. Acta* 42, 409 (1960).
[3] J. A. Hunt and V. M. Ingram, *Biochim. et Biophys. Acta* 49, 520 (1961).
[4] V. M. Ingram, *Biochim. et Biophys. Acta* 28, 539 (1958).
[5] H. Michl, *Monatsh. Chem.* 82, 489 (1951).
[6] R. R. Redfield, *Biochim. et Biophys. Acta* 10, 344 (1953).
[7] R. J. Block, E. L. Durum, and G. Zweig, "Paper Chromatography and Paper Electrophoresis," p. 128. Academic Press, New York, 1958.

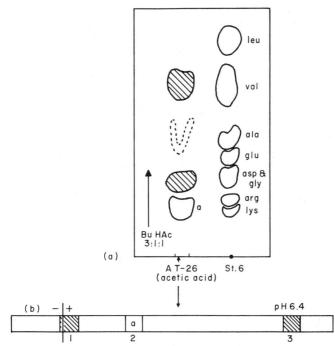

FIG. 1. Separation of the products formed by the incubation of peptide A-T 26 with dilute acetic acid: (a) by chromatography, (b) by paper electrophoresis at pH 6.4 (a marks peptides which stain for arginine).

fragment 2 by the FDNB procedure.[8] No DNP-amino acid could be detected after hydrolysis of the DNP-peptide, and the residue had the same composition as the original peptide. This apparent lack of an N-terminal amino group is presumably due to the fact that this peptide begins with a glutamic acid residue which is in the pyrrolidone form.

TABLE I

Peptide fragment	Arg	Asp	Glu	Gly	Ala	Val	Leu	Yield of fragment, micromole
1-1 (unhydrolyzed)	—	—	—	—	—	1.6	—	0.110
2	1.0	—	1.9	2.7	1.0	0.9	1.0	0.067
3	—	1.3	—	—	—	—	—	0.090

It seems that the first four amino acids have been split from the N-terminal end of peptide A-T 26 and that this tetrapeptide has been

[8] F. Sanger and H. Tuppy, *Biochem. J.* **49**, 463 (1951).

hydrolyzed to free aspartic acid and valine. Clearly, peptide fragment 2 is the peptide A-T 26 minus its two aspartic acids and minus two of the three valine residues. Hence A-T 26 has the structure

$$(Val_2, Asp_2)(Glu_2, Gly_3, Val_1, Ala_1, Leu_1, Arg_1)$$

Compare the complete sequence:

Val—Asn—Val—Asp—Glu—Val—Gly—Gly—Glu—Ala—Leu—Gly—Arg

The important information contributed by the dilute acetic acid degradation is the split between aspartic and glutamic acids in positions 4 and 5.

Quantitative Amino Acid Analysis on Paper[2]

Quantitative analysis of most amino acids could be achieved by one-dimensional chromatography on Whatman No. 1 paper in n-butanol–acetic acid–water, 4:1:5 by volume. However, aspartic acid and glycine did not separate. The ninhydrin color was developed and determined as described by Kay et al.[9]

Although quantitative amino acid analysis on paper does not have the accuracy which can be achieved by ion exchange chromatography, the use of paper should be considered for the analysis of peptides produced by the degradation of proteins. Here the ultimate is accuracy is not required, and the small quantities which are often available can be handled more easily on paper than on ion exchange columns. Many systems have been proposed, and some are reviewed in Leggett Bailey's monograph (see Reading List). The four methods described below are chosen because the author has more direct experience of them, not because they are necessarily better.

Method 1. A complete separation of the amino acids alanine, aspartic acid, arginine, glutamic acid, glycine, leucine, lysine, and valine, as well as a mixture containing glutamic acid, histidine, threonine, leucine, lysine, proline, and valine, in a manner suitable for quantitative analysis, was achieved by using a modification of the Redfield two-dimensional chromatography system.[6] Amino acid mixtures were obtained by hydrolysis of the peptide in 6 N HCl (0.2 ml. for up to 1 micromole of peptide) in sealed glass capillaries heated at 110° in an oven for some 16 hours. The acid was removed in a vacuum desiccator. The hydrolyzates were spotted onto 20 × 20-cm. squares of Whatman No. 1 paper and neutralized by exposure to ammonia, as described for the Redfield system. The papers were developed by ascending chromatography in the n-butanol–acetic acid–water (3:1:1) solvent for 7 to 10 hours. They

[9] R. E. Kay, D. E. Harris, and C. Entenman, *Arch. Biochem. Biophys.* **63**, 14 (1956).

were dried overnight at room temperature in a hood and then developed by ascending chromatography in the Redfield-2 solvent. The papers were dried in the Kodak film-drying cabinet for at least 45 minutes and then steamed in an open autoclave or similar apparatus for 10 minutes to remove diethylamine, which interferes with the ninhydrin coloration. The ninhydrin color of the amino acids was developed by dipping the papers into a 0.5% solution of ninhydrin in acetone containing 5% by volume of a pH 7.2 0.05 M phosphate solution[10] and heating the papers for 22 minutes at 65°. The spots were cut out with a razor blade and eluted into 4 ml. of 71% ethanol.[9] The optical density of each eluate was read at 575 mμ against a blank cut from one of the papers (usual value for the blank was less than 0.01). Aliquots of 0.010 ml. of standard solutions of the appropriate amino acids of 2, 5, 8, and 10 mM strength were run together with the hydrolyzates. Standard curves were drawn for each amino acid so that the amount of each amino acid in the hydrolyzates could be determined. This method is accurate for as little as 0.02 micromole of a small peptide. The reproducibility of the ninhydrin colors was not of prime importance, since standards were run with each determination. It was, however, quite good.

By replacing the first Redfield solvent[6] by butanol–acetic acid, two advantages have been gained over the original system. Firstly, the system is much more tolerant of small amounts of salt in the hydrolyzates, especially after dinitrophenylation and Edman degradation. Secondly, the separation of arginine, lysine, aspartic acid, and glutamic acid is better because the spots are more compact. On the other hand, the size of the spots varies more than in the Redfield system, and the whole operation takes longer to perform. It is believed that this system will separate most of the common amino acids, except for leucine and isoleucine.

Method 2. A variant of this system of analysis has been developed by Stretton.[11] Chromatography of the amino acids is replaced by ionophoresis on Whatman No. 1 paper, held horizontally between polyethylene sheets, cooled underneath by a water-cooled brass plate. A rubber cushion, a glass plate, and a weight complete the assembly. Extra paper at the ends dip into solvent vessels. Samples of 0.01 to 0.1 micromole of amino acids applied to 1 inch of a starting line are run at about 100 volts/inch for 3 hours in 2.2 per cent formic acid, pH 2.0. Obviously, the dimensions and running conditions can be modified easily. The papers are air-dried, and the color is developed and estimated as above. Well-separated bands of the following amino acids are obtained: lysine (the

[10] V. M. Ingram and M. R. J. Salton, *Biochim. et Biophys. Acta* **24**, 9 (1956).
[11] V. M. Ingram and A. O. W. Stretton, *Biochim. et Biophys. Acta* **63**, 20 (1962).

fastest band), arginine + histidine, glycine, alanine, serine + valine, threonine + leucine + isoleucine, glutamic acid + methionine + proline, aspartic acid + phenylalanine + cysteine, and tyrosine. Although the resolving power of this system is limited, it is often useful for the analysis of peptides with a limited range of amino acids. It has the advantage of being rapid and relatively tolerant of contamination by salt.

Method 3. Resolution of a more complete mixture could be achieved by a two-dimensional procedure.[11] The amino acid mixture was separated in ionophoresis as in Method 2. A strip of paper containing all fractions was cut in the direction of ionophoresis and was sewn onto the bottom edge of a sheet of Whatman No. 1 paper so that, by ascending chromatography in butanol–acetic acid–water (3:1:1, v/v), the amino acids moved out of the strip and separated on the new sheet of paper. All the amino acids found in acid hydrolyzates of proteins were separated by this procedure with the exception of histidine and arginine; leucine and isoleucine; aspartic acid and methionine sulfone.

Amino acids could be estimated with an accuracy of about 10% by developing the ninhydrin color under carefully controlled conditions, similar to those described by Kay, Harris, and Entenman.[9] The papers were dipped in 0.5% ninhydrin in acetone containing 5% by volume of a 0.05 M phosphate buffer (pH 7.0) and the dry papers heated at 65–70° for 20 min. The spots were cut out and eluted in 4 ml. of 75% ethanol in test tubes, and the absorbancy of the eluates was read at 575 mμ.

It was necessary to use standards in each experiment; usually these contained 0.02, 0.04, and 0.08 μmole of each amino acid. Standard curves were drawn for each amino acid and the unknowns estimated from these.

Method 4. A convenient method of two-dimensional paper ionophoresis of peptides and amino acids exists in which the peptide material from the first dimension is transferred onto a fresh sheet of paper by sewing the paper with a domestic sewing machine.[12] This method is capable of application beyond the preparation of analytical two-dimensional ionograms; for example:

(1) A region of paper containing a group of peptides which do not separate at one pH can be cut out, sewn onto a fresh sheet, and rerun in the same direction at a different pH. This allows a number of samples to be fractionated in parallel at the same time.

(2) It has also been found possible to digest peptides *in situ* on the paper with proteolytic enzymes.

All ionograms were prepared on Whatman No. 3MM paper in the

[12] M. A. Naughton and H. Hagopian, *Anal. Biochem.* 3, 276 (1962).

Varsol-cooled ionophoresis tanks described by Katz *et al.*,[13] with the modification that the level of buffer in both electrode compartments was 2 in. from the bottom of the tank, thus allowing a 3-ft. length of paper to be run. Two tanks were used, one containing buffer at pH 6.4 (pyridine, acetic acid, water 10:0.4:90)[5] and the other at pH 1.9 (2.5% formic acid, 8.7% acetic acid).

The amount of amino acid in a given band was determined by the method of Dreyer.[14] The spots were developed by dipping the ionogram in 0.5% ninhydrin dissolved in acetone, drying at room temperature, and heating at 60°C for 20 min. The strip containing the spots was then scanned with a Beckman Analytrol fitted with 550 mμ filter.

The strip of paper containing peptides to be run in the second dimension was sewn onto a fresh sheet of paper with a domestic sewing machine. A machine equipped to give a zigzag stitch was preferred as this stitch held the overlapping edges of paper together and prevented loss of peptide material. It was found easier to sew the strip onto the sheet and then to cut away the paper from underneath the strip with scissors. When a very narrow strip (1–2 cm. in width) was used, one edge of the strip was sewn onto the sheet, and an incision extending almost to the sides of the paper just in front of the stitch line was made with a razor blade. The other edge of the strip was then sewn down on the paper and the cut edge folded back and cut away.

The paper was placed on a glass plate with a glass rod under each side of the sewn strip and the sheet wetted with buffer to the edge of the strip. The buffer was allowed to flow into the strip from both sides, thus concentrating the peptide bands into a thin line at the center of the strip where there is little danger of a warped electric field. Excess buffer was removed from the areas on either side of the strip containing peptides, with blotting paper. The sheet was then subjected to ionophoresis.

The samples of amino acids were applied as ½-in. bands across the center of a 3-ft. length of paper (12–18 in. wide). A standard mixture of amino acids containing ϵ-DNP-lysine (a visible marker) was applied at the same time. The paper was wetted with pH 6.4 buffer and subjected to ionophoresis at 50 volts/cm. for 45 min. The region containing the neutral amino acids was located by the yellow ϵ-DNP-lysine band. The neutral band had moved slightly from the origin due presumably to electroendosmosis. After the sheet was dried, a 1½-in. wide strip containing the neutral bands was cut out and sewn into a new 3-ft. length of paper at a distance of 10 cm. from one end. This paper was then subjected to ionophoresis in pH 1.9 buffer for 1.5 hr. at 5 kv. The two ends of the pH 6.4 ionogram and the pH 1.9 ionogram were developed

[13] A. M. Katz, W. J. Dreyer, and C. B. Anfinsen, *J. Biol. Chem.* **234**, 2897 (1959).
[14] W. J. Dreyer, *Brookhaven Symposia in Biol.* **13**, 243 (1960).

with ninhydrin. As suggested by Dreyer,[14] the amount of amino acid in each band can be calculated by scanning the strip with a Beckman Analytrol. The relationship between concentration and density is linear, if amounts greater than 0.01 μmole are measured.

Because of the difficulty of reproducing the exact conditions for the development of the amino acids on different runs, standard amino acids were run with the unknown and the amounts calculated with reference to these standards. This standard contained 0.02 μmole of each amino acid. It should be noted that methionine is converted to the sulfone during drying on the paper at pH 6.4, and this results in the appearance of a new spot just behind tryptophan. Proline shows up as a yellow spot which is not scanned by the Analytrol at the wavelength used.

The group "serine, isoleucine, and leucine" run in that order but they are not resolved and need a longer run for positive identification. If serine and leucine only are present, they are resolved. Asparagine and glutamine run close together in that order; asparagine when first developed is yellow but gradually becomes brown. Tryptophan and methionine sulfone are also close, but the presence of methionine sulfone is characterized by its gray color. The ϵ-DNP-lysine marker runs behind all the other amino acids. This method can be used in the refractionation of peptide bands as well as amino acids. It has the advantage that there is not so much loss of material as there is by eluting the material and reapplying it to fresh paper.

Digestion with Elastase[15]

Principle. Enzymatic digestion of a peptide with elastase produces fragments which help in elucidating the amino acid sequence. The enzyme splits preferentially between two adjacent nonpolar aliphatic amino acids. The example is the same peptide A-T 26 from human hemoglobin A[3] as above.

Reagents

    Elastase. See Vol. V [90].
    Ionophoresis buffers and chromatographic solvents, as in previous
      section.

Procedure. The peptide A-T 26 was purified by paper electrophoresis at pH 6.4, and a quantity equivalent to 1 to 2 micromoles was dissolved in 1.5 ml. of water. The pH was adjusted to 9 in an autotitrator,[16] and

---

[15] M. A. Naughton and F. Sanger, *Biochem. J.* **70**, 4P (1958).
[16] C. F. Jacobsen, J. Leonis, K. Linderstrøm-Lang, and M. Ottesen, *Methods of Biochem. Anal.* **4**, 171 (1957).

0.025 mg. of pure elastase in a small volume of water was added. The course of the reaction was followed by the amount of alkali consumed at pH 9 and 38°. About 5 micromoles of alkali were used after 2½ hours of reaction.

The reaction mixture was divided into two equal portions and evaporated. The resulting peptide fragments were separated by paper ionophoresis at pH 6.4 and, where necessary, at pH 3.6 for further purification (Fig. 2). The amino acid composition of each peptide was

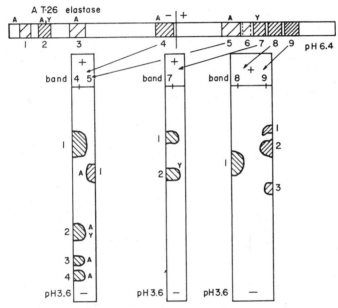

Fig. 2. Separation of the peptides obtained by elastase digestion of peptide A-T 26 by paper electrophoresis at pH 6.4 and at pH 3.6.

determined quantitatively by using the chromatographic system outlined above in Method 1. The results are summarized in Table II. End groups of the peptides 5, 8, 9-1, and 9-2, were determined by using the fluoro-dinitrobenzene method.[8] Peptides 5, 8, 9-1, and 9-2 all gave DNP-valine after hydrolysis of their DNP derivatives. No DNP-amino acid could be found for peptide 7-2, probably because of the known lability of DNP-glycine to the hydrolysis conditions.

Thus the action of elastase is to produce the peptides shown in Table III. Peptide 5 is undigested A-T 26.

The recovery of the peptides containing N-terminal valine is rather lower than that of the peptides containing C-terminal arginine (0.15 against 0.23 micromole).

TABLE II

Peptide fragment[a]	Arg	Asp	Glu	Gly	Ala	Val	Leu	Yield of fragment, micromole
2	0.9	—	—	1.1	—	—	—	0.124
3	1.1	—	—	1.2	—	—	0.7	0.027
4-1	—	—	—	—	—	—	1	0.145
4-2	1.0	—	1.0	2.6	1.0	—	1.0	0.084
5	0.9	1.7	2.0	2.8	1.0	2.8	1.1	0.074
7-2	—	—	1.1	2.4	0.9	—	—	0.014
8	—	1.7	1	—	—	2.9	—	0.080
9-1	—	1.7	2.0	2.0	1	2.8	—	0.073
9-2								
9-3	(small amount of Glu and Ala)							

[a] Fragments 4-3 and 4-4 contain less than 0.01 micromole of Arg, Glu, Leu, and Arg, Glu, respectively.

The bonds split are in agreement with the known specificity of elastase[17] (see Fig. 3). In peptide A-T 26 they are the valyl–glycyl,

TABLE III

Val(Asp$_2$,Glu$_2$,Val$_2$,Gly$_2$,Ala$_1$)	(9-1,9-2)
Val(Asp$_2$,Glu$_1$,Val$_2$)	(8)
(Gly$_2$,Glu$_1$,Ala$_1$)	(7-2)
(Gly$_3$,Glu$_1$,Ala$_1$,Leu$_1$,Arg$_1$)	(4-2)
(Gly$_1$,Leu$_1$,Arg$_1$)	(3)
(Gly$_1$,Arg$_1$)	(2)

alanyl–leucyl, and leucyl–glycyl peptide bonds. The structure of A-T 26 is as follows:

Val—Asp—Val—Asp—Glu—Val—Gly—Gly—Glu—Ala—Leu—Gly—Arg

Digestion with Leucine Amino Peptidase

Principle. Digestion of a peptide with leucine amino peptidase (LAP)[18] liberates free amino acids sequentially starting from the N-terminus. However, it is not usually possible to deduce a sequence from the amino acids which appear. An advantage of the method is the fact that glutamine and asparagine are liberated in these forms; hence, the presence of side-chain amide groups can be detected.

In the example, LAP has been used to locate the proline residue in

[17] M. A. Naughton and F. Sanger, *Biochem. J.* **78**, 156 (1961).
[18] R. L. Hill and E. L. Smith, *J. Biol. Chem.* **228**, 577 (1957).

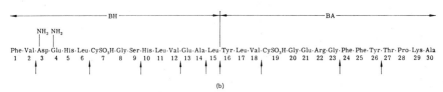

FIG. 3. Structure of the A and B chains of beef insulin, showing the peptide linkages which are attacked by elastase.[π] (a) A chain, (b) B chain.

the given peptide, since this amino acid stops enzyme action at the particular enzyme concentration used. Free aspartic or glutamic acid residues in a peptide also virtually stop the enzyme.

Procedure. In this example, Hill and Schwartz[19] examined peptide A-T 4, obtained by tryptic digestion of human hemoglobin A.[4] This peptide has eight amino acids–valine, histidine, leucine, threonine, proline, (glutamic acid)$_2$, and lysine. It was prepared and purified by paper ionophoresis at pH 6.4 (see buffer in first section of this article) and subsequent paper chromatography in *n*-butanol–acetic acid–water (200:30:75).

Peptide (0.2 to 0.4 micromole), dissolved in 0.1 ml. of water, was incubated with 0.01 ml. of 0.025 M MgCl$_2$, 0.01 ml. of 0.5 M Tris buffer, pH 8.5, and 0.0025 ml. of LAP ($C_1 = 55$, 10 mg./ml.)[18] at 40° for 4 hours. The products of the hydrolysis were examined by two methods:

1. By a fingerprint (see later), which clearly showed the presence of free amino acids valine, histidine, and leucine in apparently equal amounts, and threonine in trace quantities, along with a single peptide. This peptide could be shown to contain one threonine, one proline, two glutamic acid, and one lysine residue after it was removed from the paper, hydrolyzed in 6 N HCl, and analyzed on the Spinco automatic amino acid analyzer or by paper chromatography in butanol–acetic acid–water (200:30:75).

2. By direct analysis of the entire reaction mixture on the Spinco automatic amino acid analyzer.

Either method yielded essentially the same results. The choice of method of course depends on the level of precision desired.

With this method of digestion, only small quantities of LAP are

[19] R. L. Hill and H. C. Schwartz, *Nature* 184, 641 (1959); also unpublished observations.

required, and the entire reaction can be carried out in the bottom of a conical 15-ml. centrifuge tube. The LAP does not interfere with either kind of analysis; consequently, deproteinization is unnecessary. Larger amounts of LAP are, of course, necessary for complete degradation (5 to 10 mg., 0.2 to 0.4 micromole).

These results locate the proline residue at position 5 in this peptide, with threonine in position 4. Lysine is assumed to be C-terminal, because this is a tryptic peptide. Hence, the amino acid composition and the LAP experiments alone give the partial formula

$$(\text{Val,His,Leu})\text{Thr·Pro·Glu·Glu·Lys}$$

which compares with the full structure

$$\text{Val·His·Leu·Thr·Pro·Glu·Glu·Lys}$$

Stepwise Degradation (Sjöquist's modification[20] of Edman's procedure)

Note (July 1963): Since this section was written, many modifications in this very important method were made; a good detailed description may be found in Leggett Bailey's monograph (see Reading List), pp. 169–186. The technique is still undergoing major changes (Sjöquist, personal communication) so that a description will not be attempted here. The reader who wishes to use this stepwise degradation will find the necessary details and the bibliography in Leggett Bailey's book.

Fingerprinting[4, 21]

Principle.[4] The denatured protein is digested with trypsin, and the resulting mixture of peptides is characterized by a combination of paper electrophoresis and paper chromatography. Each peptide occupies a position on the final map or "fingerprint" which is characteristic of its amino acid composition and, to a much smaller degree, of its sequence. The method is equally applicable to mixtures of peptides obtained from the original enzymatic digestion of a protein, to the products of further degradation of a previously prepared peptide, or to mixtures of amino acids.

Reagents

"Buffer pH 6.4." See section on dilute acetic acid.

n-Butanol–acetic acid–water solvent. See section on dilute acetic acid.

Pyridine–isoamyl alcohol–water, 35:35:30 by volume.[22]

[20] J. Sjöquist, *Arkiv Kemi* **14**, 291, 323 (1959).
[21] C. Baglioni, *Biochim. et Biophys. Acta* **48**, 392 (1961).
[22] H. G. Wittmann and G. Braunitzer, *Virology* **9**, 726 (1959).

Procedure. In order to render hemoglobin susceptible to trypsin attack, it first has to be denatured. Four milliliters of a 2.5% hemoglobin solution are brought to pH 7.75 with 0.1 N NaOH in a polyethylene vial and heated in a water bath at 90° for 6 minutes. The sample is then immediately cooled. A fine light-brown suspension is obtained; this shows little tendency to settle, and trypsin attacks it readily.

The trypsin used for these digestions is the TCA-purified protein obtained from Worthington Biochemicals, Freehold, New Jersey. It is used without further purification as a 0.5% solution in 0.001 N HCl, stored frozen. Samples of the denatured hemoglobin solution at pH 8 are stirred in an autotitrator continuously at 38° in an atmosphere of nitrogen; 0.10 ml. of the above trypsin solution is added at zero time. The vessel contains a combined calomel glass electrode (Radiometer GK 2021B). The splitting of peptide bonds liberates hydrogen ions; 0.5 N NaOH is added automatically by the autotitrator. The digestion is equally successful if the NaOH is added manually by means of an Agla micrometer buret to keep the pH between 7.95 and 8.00; a direct-reading pH meter is used. It is well known that, when alkali consumed is plotted against time, the slope of this curve never flattens completely. In preliminary experiments, the end point was determined by adding further aliquots of trypsin solution and observing that after a certain time—about 90 minutes—there was no additional alkali uptake (Fig. 4).

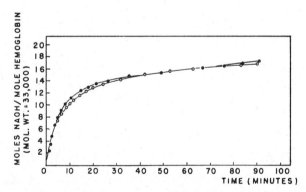

Fig. 4. Course of the trypsin digestion of hemoglobins A and S. ○——○ hemoglobin A; ●——● hemoglobin S.

After the digestion, the pH is adjusted to 6.5 with 1 N HCl, the solution is frozen and thawed, and the precipitate of the insoluble trypsin-resistant "core" of hemoglobin is removed by centrifugation at high speed in the cold. The supernatant may be stored frozen until used for fingerprinting and for the preparation of various peptides.

The separation of the peptides obtained by tryptic digestion may be achieved by a two-dimensional combination of paper electrophoresis and chromatography—fingerprinting. Paper electrophoresis is carried out on Whatman No. 3MM paper which has been dipped into Michl's volatile buffer[5]: pyridine–glacial acetic acid–water, pH 6.4. Excess liquid must be removed carefully by blotting firmly between two sheets of fresh blotting paper. The moist No. 3MM is immediately placed onto a horizontal piece of ¼-inch polished plate glass which rests on the two buffer vessels, each 3-inches high.

The samples of peptides—0.10 ml. of the original solution which has been evaporated and dissolved in 0.020 ml. of water—should be applied with a constriction pipet at the point indicated in Fig. 4. Similarly, a second peptide mixture could be applied to the other paper. At once a second sheet of plate glass is placed on top of the first one, thus enclosing the moist filter paper as a sandwich. Figure 5 shows a plan of the ar-

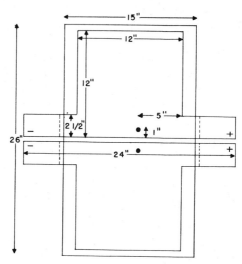

Fig. 5. Dimensions of filter paper used for fingerprinting.

rangement and in particular the way the paper is cut. The tags dip into buffer vessels (not shown). This arrangement allows two separations to be carried out under identical conditions; it is now generally used to compare peptide mixtures from normal hemoglobins with those from the abnormal proteins. Great care is taken to handle pairs of paper in the same way. Preferably they are cut from a single sheet of paper, 24 × 24 inches, with the machine direction of the paper always pointing in the same chromatographic direction. Other dimensions are, of course, also useful, e.g., footnote 23.

After the sample has been applied, 15 minutes are allowed to elapse to give the buffer time to diffuse into the sample, even though the latter was already at pH 6.5. A potential of 19 volts/cm. is applied, and, as the current rises, the voltage falls to about 14 volts/cm. after 150 minutes. The glass sandwich gets comfortably warm during the electrophoresis. The tags are trimmed off the papers, and the buffer is removed by drying at room temperature in a current of air for not less than 2 hours, and preferably longer. The dry papers, now 12×12 inches, are hung for 2 hours in the air space above the butanol–acetic acid chromatographic solvent, which is freshly prepared each Friday or Saturday for use in the next week beginning on Monday. Alternatively, the pyridine solvent should be used at once.

Ascending chromatography is carried out overnight, usually for 15 hours. The papers are again dried in a current of air at room temperature. The peptide spots can be revealed by dipping the papers into 0.2% ninhydrin in acetone and allowing development to take place at room temperature or in a warm place (Fig. 6). This usually takes 24 hours to reach maximum intensity. Since it is difficult to preserve these "fingerprints," reflex prints or photographs are made.

One-dimensional paper ionophoresis on 3MM paper in the above volatile pH 6.4 buffer is frequently used for examining peptide or amino acid mixtures and for isolating peptides. Another volatile buffer[9] at pH 3.6 (pyridine–glacial acetic acid–water, 1:10:90 by volume) is also useful for one-dimensional separations. The arrangement[23] is a sandwich similar to the fingerprinting apparatus but measuring only 12×16 inches or 17×24 inches. The bottom member of the sandwich is a flat brass plate $\frac{1}{4}$ inch thick and heavily chromium-plated, cooled by copper coils soldered underneath, carrying tap water. A thin layer of silicone grease and another of thick polythene film separate the brass plate from the paper. Over the paper is placed another polythene sheet, then a $\frac{1}{2}$-inch rubber sponge pad, and a weighted glass plate. Up to 35 volts/cm. can be used without noticeable rise in temperature. When the isolation of a peptide is required, the papers are usually first washed by descending irrigation with the pH 6.4 buffer, followed by soaking in water, and drying. Such papers are developed by spraying with 0.025% ninhydrin in butanol and gentle heating. If isolation of peptides is unimportant, then 0.2% ninhydrin is used to develop all the spots. The colors are allowed to develop at room temperature. Alternatively, guide strips may be used. The peptides are eluted with water or 20% aqueous acetic acid.

[23] V. M. Ingram and A. O. W. Stretton, *Biochim. et Biophys. Acta* **62,** 456 (1962).

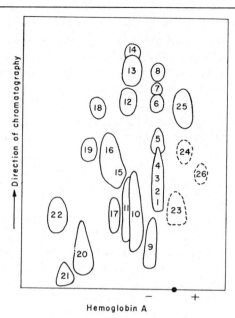

Hemoglobin A

FIG. 6. Hemoglobin A fingerprint. Note: One or two of these numbered spots contain an additional peptide.

Stage I   Paper Electrolysis at pH 6.4

Stage 2   Paper   Chromatography

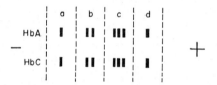

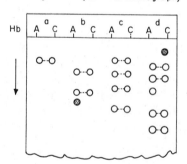

FIG. 7. "Dissected fingerprint" (see text).

Dissected Fingerprints

For the close examination of the hemoglobin C fingerprints, Hunt and Ingram[2] found it useful to prepare what may be called "dissected" fingerprints of tryptic digests hemoglobins A and C. This refinement of the original method[4] involves the separation of the tryptic peptides from both hemoglobins into two sets of groups of peptides (see Fig. 7). One-dimensional paper ionophoresis at pH 6.4 may be used for this. The paper is then cut into strips of convenient width so that each contains one or more peptide bands. This is done for peptides from both hemoglobin A and C, care being taken that the electrophoresis and the subse-

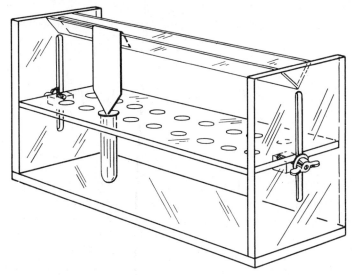

Fig. 8. Plastic rack for the elution of peptides from paper (after B. S. Hartley, unpublished). Over-all length 10 to 12 inches.

quent cutting into strips are done in a comparable manner. The strips are next eluted, preferably in the apparatus shown in Fig. 8, covered by a large plastic box. It is now possible to compare these eluates by placing the corresponding peptides from the normal and the abnormal hemoglobin side by side on filter paper and developing them chromatographically. Although similar to the original fingerprinting method, the modification adds considerably to the sensitivity of the method, because peptides from the two proteins can be compared side by side on the same paper and because the transfer of eluate allows the use of Whatman No. 1 paper and a different chromatographic solvent (n-butanol–acetic acid–water, 4:1:5 by volume) both of which have greater resolving power.

It has been possible to detect by this technique the replacement of threonine by serine as the only difference between a pair of octadeca-peptides derived from hemoglobins A and A_2.[23]

READING LIST

J. I. HARRIS AND V. M. INGRAM, Methods of Sequence Analysis in Proteins. *In* "Analytical Methods of Protein Chemistry" (P. Alexander and R. J. Block, eds.), Vol. II, p. 421. Pergamon, New York, 1960.

H. FRAENKEL-CONRAT, J. I. HARRIS, AND A. L. LEVY, *Methods of Biochem. Anal.* **2**, 359 (1955).

F. SANGER, *Advances in Protein Chem.* **7**, 1 (1952).

C. H. W. HIRS, W. H. STEIN, AND S. MOORE, *in* "Symposium on Protein Structure" (A. Neuberger, ed.), p. 211. Methuen, London, 1958.

J. LEGGETT BAILEY, "Techniques in Protein Chemistry." Elsevier, Amsterdam, 1962.

[119] Two-Dimensional Immunodiffusion

By DAVID STOLLAR and LAWRENCE LEVINE

Many proteins, including enzymes, are potentially antigenic, being able to elicit antibody production when parenterally injected into animals. The antibody formed as a result of this injection reacts specifically with the protein administered. Although the chemical basis of this serological specificity of globular proteins is not known, immunochemical methods offer the biochemist a powerful tool for the study of antigenic proteins.

The reaction between an antigen and its antibody, in liquid medium or in a gel, may give rise to a visible precipitate. The technique of double diffusion in agar in two dimensions, as developed by Ouchterlony,[1] is a simple and inexpensive method of studying precipitation in gels. Some uses of double diffusion in agar and the method used in our laboratory are presented here.

Principles. If a protein and its specific antiserum are allowed to diffuse toward each other from two separate depots in agar, a precipitate will form at the point where antigen and antibody meet in optimal proportions. A precipitating band forms first at a spot along the shortest line between the depots and then spreads laterally. The band may become curved, concave toward either depot. Factors affecting the position and the curvature of the band are (1) concentrations of the antigen and antibody and (2) diffusion coefficients of the antigen and antibody.

[1] O. Ouchterlony, *Acta Pathol. Microbiol. Scand.* **26**, 507 (1949).

Antigen–antibody combination at the molecular level occurs over a wide range of concentrations of reactants. Visible precipitation, however depends on aggregation into a complex three-dimensional structure, and maximal aggregation requires optimal proportions of antigen and antibody. In double diffusion in agar, the position between the reactant depots where this optimal proportion occurs varies with the initial concentrations of the reactants and their diffusion constants. The optimal molecular ratio varies with different immune systems. For example, a ratio of 70 molecules of ribonuclease to 200 molecules[2] of antiribonuclease is optimal for this system. With thyroglobulin, however, the optimal ratio is 1 molecule of antigen to 200 molecules of antithyroglobulin.[3]

Technique. There are many variations of the double-diffusion technique of Ouchterlony. Most of these variations are concerned with the procedure by which holes for deposit of antigens and antisera are made. The following is the method used in our laboratory.

To prepare the agar, 10 g. of Ionagar No. 2 (purchased from Consolidated Laboratories, Inc., Chicago Heights, Illinois), 8.5 g. of NaCl, and 200 mg. of merthiolate are used for each liter of solution and are dissolved in water at 100°. While this solution is cooling, but still liquid, 1 N NaOH is added to give a pH of 8.0. At this pH, the solution and the resulting gel are sufficiently clear for observation of precipitating bands as well as for photographic recording. The agar solution is distributed into tubes in 10-ml. portions and stored in the refrigerator.

For use, about 10 ml. of the agar is melted and poured into a glass or plastic petri dish and allowed to harden at room temperature. Hollow stainless-steel tubes, 8 mm. in diameter and 8 mm. in height, are placed on top of the solidified layer of agar in the desired pattern, about 0.5 to 1 cm. apart. A second 10-ml. portion of agar is then melted and poured, forming a second layer in which the steel tubes serve as molds for the reactant depots. After this second layer has hardened, the tubes are removed. The use of the stainless-steel tubing prevents sticking of the agar to the mold surface and subsequent tearing of the agar on removal of the molds, as sometimes happens when glass tubing or penicillin assay cups are used as templates.

In general, we have observed sharper and faster precipitation with thinner agar layers. However, care must be taken not to have the lower agar layer so thin that the bottom of the depot is in contact with the surface of the petri dish. If this happens, the reactants may spread in a liquid phase underneath the agar instead of diffusing into it. The dimen-

---

[2] R. K. Brown, R. Delaney, L. Levine, and H. Van Vunakis, *J. Biol. Chem.* **234,** 2043 (1959).

[3] H. E. Stokinger and M. Heidelberger, *J. Exptl. Med.* **66,** 251 (1937).

sions of the depots may be set as convenient. The size described will accept 0.1 ml. of solution with no spilling over the edge. The depot capacity and quantity of reactant added should be great enough so that the solution does not dry up before effective diffusion takes place.

Applications

Studies of Immunochemical Homogeneity. If an immune system consists of a single antigen and its antibody, one band of precipitation will form in a double-diffusion experiment. When several different antigens and their antibodies are present, it is probable that, owing to differences in concentrations and differences in diffusion characteristics of the various antigens, several bands will form. Two precautions in interpretation should be noted. If only one experiment is performed, some antigen–antibody systems which are present may not occur in adequate concentration and may not be visible. To reduce this possibility, several plates with varying amounts of antigen and serum should be prepared. Second, it is possible that two different systems may form superimposed precipitates. The two systems may sometimes be separated by varying the antiserum and antigen concentrations. Thus, the number of visible bands represents the *minimum* number of antigen–antibody systems present. In general, however, double diffusion is a sensitive test for immunochemical heterogeneity. Nonantigenic impurities cannot, of course, be detected or ruled out.

Two approaches which favor the demonstration of heterogeneity may be used. Under these conditions the formation of a *single* line indicates a high probability of immunochemical homogeneity.

In one approach, one may immunize with a crude preparation from the same source as a purified antigen being tested for immunochemical purity. The antiserum obtained contains antibodies to the test antigen and to several other crude preparation antigens, traces of which may be detected in the purified material.

For example, in the study of a purified calf serum protein, fetuin,[4] an antiserum to crude fetuin-containing material was obtained which gave three bands of precipitation when tested with the crude sample (Fig. 1a). The purified fetuin, however, gave a single line of precipitation (Fig. 1b).

In another study,[5] in which the purity of a sample of λ bacteriophage prepared from *Escherichia coli* was tested, an antiserum to crude *E. coli* extract was obtained. The zone between this antiserum and the purified

[4] F. Bergmann, L. Levine, and R. Spiro, *Biochim. et Biophys. Acta* **58**, 41 (1962).
[5] A. Soller, L. Levine, and H. Epstein, unpublished data.

bacteriophage contained several bands of precipitation, indicating the presence of impurities from *E. coli* material in the virus preparation.

The second approach to this problem involves immunization with purified material. Any antigenic impurities will also stimulate antibody

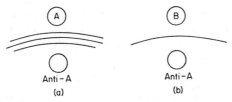

Anti–A   Anti–A
(a)   (b)

FIG. 1. Precipitin bands formed in agar on two-dimensional double diffusion with antiserum to crude material. A, crude fetuin-containing material (calf serum); anti-A, antiserum to calf serum; B, purified fetuin.

production. The antiserum obtained is then tested against a crude preparation from the same source as the purified antigen. This crude sample contains the antigens potentially present as impurities in the purified preparation. It may detect, in the antiserum against purified antigen, antibodies also directed against such impurities. For example, rabbits were immunized with crystalline pancreatic deoxyribonuclease, and the antiserum obtained was tested in double diffusion with crude extract as well as with purified enzyme. A single line was found in both cases, suggesting that there was no antigenic component of crude pancreatic extract present as an impurity in the crystalline enzyme.[6] In contrast, antiserum to a purified triosephosphate dehydrogenase from lobster gave several bands with the crude extract.[7]

Comparisons of Protein Structure. The reactivity of two antigen preparations with a given antiserum may be compared if three reactant depots are arranged in a triangular pattern. Three major patterns of comparative reactivity may be observed. If two antigens A and B are identical, the "reaction of identity" pattern of Fig. 2a will be obtained when antibody to either A or B is placed in the third well. If two antigens such as C and D are immunologically unrelated, only one band will form if an antiserum to one of the antigens is placed in the third well. If antibodies to both antigens are present in the third well, two bands will form, cross each other, and continue to grow laterally (Fig. 2b). If the antigens are related, sharing some common antigenic determinants but differing in others, as E and F, the reaction of partial identity (Fig. 2c) will occur, one band forming a spur beyond the junction with the other. These comparisons may be useful in enzyme studies such as the following:

[6] M. Reichlin, L. Levine, and L. Grossman, unpublished data.
[7] W. Allison and N. O. Kaplan, unpublished data.

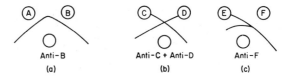

Fig. 2. Comparisons of protein structure. (*a*) Pattern of identity: A, crude pancreatic extract; B, purified pancreatic deoxyribonuclease. (*b*) Pattern of nonidentity: C, purified pancreatic deoxyribonuclease; D, bovine serum albumin. (*c*) Pattern of partial identity: E, human serum albumin; F, bovine serum albumin.

1. Comparison of proteins with similar activity, isolated from different organs of the same animal. The reaction pattern of lactic dehydrogenases of heart and skeletal muscle with the antibody against skeletal muscle LDH is shown in Fig. 3, revealing nonidentity.[8]

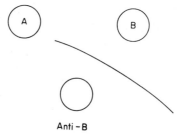

Fig. 3. Comparison of enzymes with similar function from different organs of the same species. A, chicken heart muscle lactic dehydrogenase; B, chicken skeletal muscle lactic dehydrogenase.

2. Comparison of enzymes with the same function, from the same organ, of different species. For example, antiserum to porcine gastric pepsinogen[9] revealed partial identity of structure in porcine, human, and horse gastric pepsinogens (Fig. 4).

3. Comparison of enzymes with the same function, from the same organ, of embryonic and adult animals of the same species. For example, crude breast muscle extract of a 6-day-old chick embryo showed a pattern of identity with adult heart muscle LDH in reaction with antibodies to the adult enzyme. The embryonic extract did not react with antibodies to adult breast muscle LDH.[10]

Diffusion Coefficients. As indicated above, the diffusion characteristics of the antigen and antibody are important in determining the position and curvature of precipitin bands forming in double-diffusion experiments. In general, the band will be convex toward the depot contain-

[8] R. D. Cahn, Ph.D. Dissertation, Brandeis University, 1962.
[9] H. Van Vunakis and L. Levine, *Ann. N. Y. Acad. Sci.* **103**(2), 735 (1963).
[10] R. D. Cahn, N. O. Kaplan, L. Levine, and E. Zwilling, *Science* **136**, 962 (1962).

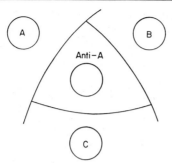

FIG. 4. Comparison of enzymes with the same function, from the same organ, of different species. A, porcine gastric pepsinogen; B, horse gastric pepsinogen; C, human gastric pepsinogen.

ing the reactant with a higher diffusion coefficient (Fig. 5) and will be straight if the antigen and antibody have equal diffusion coefficients. However, the correlation between the degree of curvature and these coefficients is not exact.

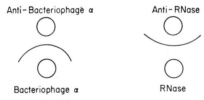

FIG. 5. Relationship of precipitin band curvature to relative diffusion coefficients of antigen and antibody.

More precise relationships exist if the antigen and antibody are placed in the agar in troughs at right angles to each other, as described by Allison and Humphrey.[11] Precipitation then occurs along a straight

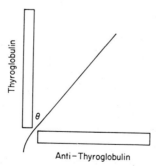

FIG. 6. Two-dimensional double diffusion from troughs at right angles, for estimation of the diffusion coefficient of an antigen.

[11] A. C. Allison and J. H. Humphrey, *Nature* **183**, 1590 (1959).

line extending outward between the troughs from the point where the reactants are closest (Fig. 6). The angle (θ) between the straight precipitin band and the antigen trough is such that $\tan \theta = (D_g/D_b)^{1/2}$, where D_g and D_b are the diffusion coefficients of antigen and antibody, respectively. The diffusion coefficient for all rabbit antibodies may be considered the same and taken as 3.8×10^{-7} cm.^{2}/sec.

The best straight line will be obtained if the amounts of antigen and antibody in the troughs are near optimal proportions. Gross deviations from these amounts will result in the band's curving and spreading toward the trough with the weaker reactant.

With this technique, the diffusion coefficients of unrelated antigens may be measured in a mixture, as the precipitin bands form independently. The values obtained for several proteins, as measured in our laboratory, are listed in the table and compared with values obtained by free diffusion techniques.

DIFFUSION COEFFICIENTS OF PROTEINS AS MEASURED BY
TWO-DIMENSIONAL IMMUNODIFFUSION

Antigen	Angle θ	D$_2$OH$_2$O Immuno	Free
Human serum albumin	52.7°	6.54×10^{-7}	6.1×10^{-7}
T$_2$ bacteriophage	9.7°	0.11×10^{-7}	0.3×10^{-7}
Thyroglobulin	40.5°	2.77×10^{-7}	2.65×10^{-7}
Rhodospirillum heme protein	58.0°	9.73×10^{-7}	8.65×10^{-7}
Cytochrome (Rhodospirillum rubrum)	58.4°	10.03×10^{-7}	
Pepsinogen	56.6°	8.7×10^{-7}	9.69×10^{-7} (pepsin)
P22 bacteriophage	18.1°	0.4×10^{-7}	

[120] Sedimentation Equilibrium in a Buoyant Density Gradient

By JEROME VINOGRAD

Introduction

Equilibrium ultracentrifugation in a density gradient is a recently developed method[1] for the study of macromolecules, viruses, and particulate materials. A homogeneous mixture of a concentrated binary solvent and a dilute macrospecies are brought to sedimentation-diffusion equi-

[1] M. Meselson, F. W. Stahl, and J. Vinograd, *Proc. Natl. Acad. Sci. U.S.* **43**, 581 (1957).

librium in the ultracentrifuge. The new distribution of the concentrated binary solvent presents a stable density gradient. At the same time the macrospecies concentrates to form a band in the redistributed binary solvent. After equilibrium is established, the macrospecies at band center is neutrally buoyant in the redistributed binary solvent. The *buoyant density* of the macrospecies and the density of the binary solvent at band center are identical.

With a homogeneous macrospecies a Gaussian concentration is formed about the center of the band. At this position centrifugal forces on the macrospecies vanish. Only diffusional forces cause the macrospecies to spread away from band center. The spreading is opposed by both centrifugal and centripetal forces, and an equilibrium concentration distribution results. Larger macromolecules diffuse less readily and form narrower bands than smaller macromolecules. The variance, σ^2, of the Gaussian distribution has been shown to be given by Eq. (1).[2] The gas

$$\sigma^2 = \frac{RT}{M_{s,0}(d\rho/dr)_{\text{eff},0}\bar{v}_{s,0}\omega^2 r_0} \tag{1}$$

constant, temperature, angular velocity, and radial distance to band center are R, T, ω, and r_0, respectively. The quantities $M_{s,0}$, $(d\rho/dr)_{\text{eff},0}$ and $\bar{v}_{s,0}$ are the solvated molecular weight, the effective density gradient, and the partial specific volume of the solvated species, respectively; all the foregoing quantities are given at band center as indicated by subscript zero. The quantity $\bar{v}_{s,0}$ is identical with the reciprocal of the buoyant density, $\rho_{s,0}$.

Mixtures of materials of identical buoyant densities and differing molecular weights form symmetrical but non-Gaussian bands. From the shape of the bands, weight and number average molecular weights, both solvated and nonsolvated, may be evaluated.[1,2]

A skewed band indicates buoyant density heterogeneity. Materials of sufficiently different buoyant densities in a mixture form bimodal or polymodal bands. We have, therefore, a general method for recognizing heterogeneity among macromolecules. In a special but important exception, density heterogeneity may be contained within Gaussian bands and, at first sight, may not be recognized.[3,4]

Since this procedure was first described, it has come into extensive use. Most of the applications derive from the fact that the macrospecies are separable and identifiable on the basis of buoyant density, a newly accessible physical property of a dissolved macrospecies. The procedures

[2] J. E. Hearst and J. Vinograd, *Proc. Natl. Acad. U.S.* **47**, 999 (1961).
[3] R. L. Baldwin, *Proc. Natl. Acad. Sci. U.S.* **45**, 939 (1959).
[4] N. Sueoka, *Proc. Natl. Acad. Sci. U.S.* **45**, 1480 (1959).

for carrying out these experiments are simple, involve relatively little personal skill, and are easy to reproduce.

Density gradient experiments are performed in either preparative or analytical ultracentrifuges. The concentration distributions at equilibrium are independent of the shapes of the containers, and Eq. (1) is valid in both types of equipment.

At the present time aqueous CsCl is the most widely used buoyant solvent. Deoxyribonucleic acids, polypeptides, proteins, viruses, and ribosomes have been studied in CsCl density gradients. Ribonucleic acids and some polynucleotides are too dense for even saturated aqueous CsCl at 25°. These macromolecules have been studied in aqueous cesium formate[5, 6] and cesium sulfate[7] solutions. The choice of the medium depends on experimental and theoretical considerations which are discussed below. In the experimental section aqueous CsCl will be regarded as the buoyant solvent unless otherwise stated.

Experimental Procedure

1. *Preparation of Solutions.* The solutions are prepared by volumetric combination of a concentrated CsCl stock solution, water, buffer, and macromolecule solutions. Graduated serological pipets and micropipets are used. Occasionally when the macromolecular material is available only in dilute solution, weighed amounts of solid dried salt are added. For volumetric work, additive mixing relations are satisfactory [Eqs. (2a) and (2b)]. The subscripts w and c refer to water and the concen-

$$v_{1,w} = (\rho_c - \rho°)/(\rho_c - 0.997) \qquad (2a)$$

$$v_w = v_c(\rho_c - \rho°)/(\rho° - 0.997) \qquad (2b)$$

trated salt solution. The quantity $\rho°$ is the desired density. The subscript 1 in Eq. (2a) indicates the volume required to prepare 1 ml. of solution. Stock solutions are stored in plastic bottles, and solutions are made up in well-stoppered glass vials. Densities are checked refractometrically before the experiments are set up. The relations between refractive index and density and between weight composition and density are useful aids in the preparation of solutions [Eqs. (3a) and (3b)]. These rela-

$$\rho°^{,25°} = 10.8601 n_D{}^{25°} - 13.4974 \qquad (3a)$$

$$\text{Wt.}\% = 137.48 - 138.11(1/\rho°^{,25°}) \qquad (3b)$$

[5] H. Dintzis, H. Borsook, and J. Vinograd, *in* "Microsomal Particles and Protein Synthesis" (R. B. Roberts, ed.). Pergamon, New York, 1958.
[6] C. I. Davern and M. Meselson, *J. Mol. Biol.* 2, 153 (1960).
[7] R. G. Wake and R. L. Baldwin, *J. Mol. Biol.* submitted.

tions are valid in the density range 1.2 to 1.9 g./ml. Several such relations for other salts in water have been collected.[8] Densities accurate to ± 0.001 g./ml. are easily measured with a pair of calibrated 0.3-ml. micropipets used as pycnometers.

The pH of the solutions is adjusted with buffer solution or with acid or base and may be measured with standard glass and reference electrodes. Cesium chloride, like KCl, suppresses the liquid junction potential and does not interfere with the measurement of pH. The high salt concentrations affect the ionization constants of the buffer acids and bases. Shifts in pH of 0.5 unit have been observed[9] on dilution of some buffers into 7 molal CsCl instead of water.

The concentration of the macrospecies, C_0, at band center in a Gaussian concentration distribution is related to the initial concentration, C_u, of the uniform solution, the standard deviation of the band, and the length of the liquid column, L [Eq. (4a)]. The extreme concentration

$$C_0 = 0.40(L/\sigma)C_u \tag{4a}$$

gradients at the inflections are given by Eq. (4b). These equations were

$$\left(\frac{dc}{dr}\right)_{\pm\sigma} = \mp 0.24(L/\sigma^2)C_u \tag{4b}$$

derived for cells with parallel walls but are sufficiently accurate for the preparation of solutions. The maximum concentration at band center is limited by experimental considerations and by the virial coefficients if molecular weights are desired. The minimum concentration detectable by absorption optics in the analytical ultracentrifuge corresponds in the case of narrow bands in a 12-mm. centerpiece to an optical density of approximately 0.02. A suitable initial concentration of DNA is approximately 1 μg./ml.

2. *Time of Approach to Equilibrium.* The three-component solution is initially homogeneous. With the application of the field, redistribution of the binary medium to form the density gradient is followed by two-directional accumulation of the macrospecies to form the band. The relative rates of the two processes depend on the sedimentation properties of the components. If the macrospecies has a large sedimentation coefficient, e.g., viruses, the species will within an hour or less find its buoyant position. The rate-limiting feature is then the time required for the medium to reach equilibrium. At 44,770 r.p.m., a 1.2-cm. liquid column of CsCl at density 1.70 g./ml. is substantially at equilibrium at

[8] J. Vinograd and J. Hearst, *Progr. in Chem. Org. Nat. Prod.* **20**, 372 (1962).
[9] J. Vinograd, J. Morris, N. Davidson, and W. F. Dove, *Proc. Natl. Acad. Sci. U.S.* **49**, 12 (1963).

12 hours. Procedures for calculating rate of attainment of equilibrium in two-component systems are given by Van Holde and Baldwin.[10] With macrospecies of lower sedimentation coefficients the rate of band formation is slower. A good estimate of the time required to attain equilibrium within 1% everywhere between band center and $\pm 2\sigma$ is given by the Eq. (5).[1,11] The diffusion coefficient is D, and the length of

$$t^* = \frac{\sigma^2}{D} \ln\left(\frac{L}{D} + 1.26\right) \qquad L \gg \sigma \qquad (5)$$

the liquid column is L. Equation (5) was derived with the assumption that the equilibrium density distribution pre-exists before band formation begins. For globular proteins and most DNA samples, 24 hours is adequate at high angular velocities. At lower velocities the time required increases rapidly because the lead term, σ^2, is inversely proportional to the fourth power of the angular velocity.

3. *Equipment.* Substantially standard commercial ultracentrifuges are used for density gradient sedimentation. The machines should be well maintained so that they may be left unattended for the 12-hour to 4-day periods required for the runs. The schlieren light source with its water supply in the analytical ultracentrifuge is normally turned off during overnight runs. Metal centerpieces are avoided because of corrosion problems. Centerpieces fabricated from Kel-F and Epon resins are satisfactory for most runs. Cell leaks are avoided in experiments at high speeds and high densities by tightening with a torque of 115 inch-pounds. Cell parts are carefully cleaned and lubricated before assembly. In the author's laboratory it has been found that wear on the rotating parts of the ultracentrifuges is small in long runs, and that drives need replacement less frequently than in short-run operation.

4. *Setting Up Runs.* Standard procedures for sedimentation velocity experiments are used in setting up density gradient experiments in the analytical ultracentrifuge. The density gradient generates a prism in the liquid column, causing the light rays to be bent away from the axis of rotation. If the effective prism is large enough, the light is obstructed at the camera lens holder. A $-1°$ quartz window at the top of the cell alleviates this problem but does not correct for errors in apparent concentration which arise from light rays passing through the cell at a small angle to the axis.[12] Window breakage traceable to high pressures, 56,100 r.p.m., density 2.05 g./ml., has not been encountered. Monochromaticity of the ultraviolet light is improved with a Corning No. 9683 filter. High-

[10] K. E. Van Holde and R. L. Baldwin, *J. Phys. Chem.* **62**, 734 (1958).
[11] M. Meselson, Doctoral Dissertation, California Institute of Technology, 1957.
[12] J. E. Hearst and J. Vinograd, *J. Phys. Chem.* **65**, 1069 (1961).

quality double-cell runs are now performed with small light-source slits, side-wedge bottom quartz windows, a long exposure clock, and an alternator.[13] Double-cell schlieren optical runs are routinely performed with upper windows differing in wedge angle by 1°. Double-sector cells of filled Epon or Kel-F have been routinely used at 56,100 r.p.m. and density 1.3 g./ml.

Preparative ultracentrifuge runs are performed in swinging bucket rotors. The SW-39 rotor is normally filled with 2 ml. of liquid and overlayered with 2.5 ml. of light mineral oil to prevent tube collapse. The rotors are run at 35,000 r.p.m. to compensate for the increased load and to provide an extra safety factor. Three experiments can be run at once.

5. *Recording of Results.* Three methods are available for recording results in the analytical ultracentrifuge. These are absorption optical photographs, schlieren optical photographs, and autoradiograms.[14] Autoradiographic experiments may also be performed in the preparative ultracentrifuge with a modified analytical rotor.

In the various kinds of experiment, different approaches are taken to the problem of recording results. When buoyant positions are sought, experiments may be performed at high speeds and at relatively high concentrations. When accurate band shapes are of interest, approximately optimum quantities of macrospecies must be introduced so as to obtain concentration distributions with accuracy. In the absorption optical system a maximum optical density change through the band should not exceed a value of 1.5, so that the image remains in the linear range of the characteristic curve of the film. According to Eq. (4a), the concentration at band center is inversely proportional to the square of the angular velocity. Similarly with the schlieren optical system the initial concentration is limited so that refractive index gradients do not exceed the limits of the optical system. Combining Eqs. (1), (4b), and (13), we obtain an expression for the magnitude of the extreme concentration gradients in the band [Eq. (6)]. An initial concentration of 0.1% bovine

$$\left(\frac{dc}{dr}\right)_{\pm\sigma} = \mp 0.24L\,\frac{M\beta_{eff}\omega^4 r_0^2}{\rho_0 RT} \tag{6}$$

mercaptalbumin is satisfactory for runs in CsCl, $\rho° = 1.3$ g./ml., at 56,100 r.p.m.[15] A change in the molecular weight or in the buoyant medium will require the indicated change in the angular velocity. Changes in refractive index increment also have to be considered with

[13] R. Inman and R. L. Baldwin, *J. Mol. Biol.* **5**, 185 (1962).
[14] J. Vinograd and R. Kent, unpublished observations.
[15] J. B. Ifft and J. Vinograd, *J. Phys. Chem.* **66**, 1990 (1962).

Eq. (6) when the buoyant solvent is changed. Wales[16] has shown that the second viral coefficient causes the band to widen with concentration. The band width derived from the separation between the maximum and minimum gradients for polystyrene in an organic binary buoyant solvent could be extrapolated to zero concentration in a linear plot of the standard deviation versus square root of the concentration.

The ultraviolet optical system is usually chosen to record results in the analytical ultracentrifuge with nucleic acids and virus. Careful alignment of the optical elements so as to obtain light parallel with the optic axis and uniform illumination of the cell are required for accurate work. Optical elements must be clean. At the end of the run a series of exposures of increasing time is made; the film is developed and traced with a densitometer linear in optical density. Linearity is verified with the trace of an exposure made by an exponential aperture[17] in the rotor or the counterbalance. An alternative test is the superimposability of tracings from successive exposures. Pedersen[18] has presented an extensive discussion of the ultraviolet optical system, and the reader is referred to this article for details. The problem of obtaining accurate records in density gradient experiments for molecular weight calculations is comparable with that encountered in two-component sedimentation equilibrium experiments. It is simpler in that base lines may be interpolated but more difficult in that smaller linear distances are involved. A density gradient experiment is essentially a short liquid-column experiment, with the length of liquid column corresponding to 4 to 6 σ. It is free of difficulties normally encountered near the top and bottom bounds of the liquid.

Photographic records from the schlieren optical system are of value in several types of experiment. DNA and viral bands may be located with precision.[19] Banded precipitates and dissolved macrospecies may be easily differentiated.[20] A main use is encountered in the study of small proteins[15] and nonabsorbing macrospecies. The requirements that the buoyant solvent not absorb at 265 mμ, that the cells be uniformly illuminated, and that the films be linearly developed may be relaxed.

For quantitative work, interference boundary-forming double-sector cells are used.[15] The reference sector is filled with a slight excess of salt

[16] M. Wales, *J. Appl. Polymer Sci.* submitted.
[17] E. Robkin, M. Meselson, and J. Vinograd, *J. Am. Chem. Soc.* **81**, 1305 (1959).
[18] K. O. Pedersen, *in* "The Ultracentrifuge" (T. Svedberg and K. O. Pedersen, eds.). Clarendon, Oxford, 1940.
[19] J. E. Hearst, J. B. Ifft, and J. Vinograd, *Proc. Natl. Acad. Sci. U.S.* **47**, 1015 (1961).
[20] J. Vinograd, J. Morris, and R. Greenwald, unpublished observations.

solution of the same density as the solution containing the macrospecies. Liquid levels become identical on acceleration of the rotor. An accurate base line is then superimposed in photographs taken at equilibrium. The photographic plate is traced under an enlarger, the base line is subtracted from the diphasic schlieren curve, and the concentration distribution is calculated by numerical integration procedures. The final photographs are exposed at phase plate angles of 85° to 55°.

Disks of photograph film protected from the CsCl solution by 0.5-mil Mylar can be centrifuged without distortion in standard analytical ultracentrifuge cell assemblies.[14] The cell assemblies are provided with Dural windows and are filled under red light. Because the radioactive macrospecies is initially dilute relative to the final distribution in the band [cf. Eq. (4a)], it is a simple matter to adjust the time of ultracentrifugation and the dose of radioactive species so that clean bands appear on the photographic film. The experiments have been performed with standard 12-mm. Kel-F, double-sector 12-mm. Epon, double-sector 3-mm. Epon, and special triple-sector 1.5-mm. Kel-F centerpieces. No-screen X-ray film is used to form autoradiograms of P^{32}-labeled viruses and DNA. A direct image of the liquid column appears as a gray background when C^{14}-labeled leucine is present in the CsCl solution. Because of the long range of P^{32} electrons, the resolution of bands is poorer than in photographic procedures.

These experiments require 24 to 48 hours and need very small amounts of radioactive macrospecies—40 to 1800 c.p.m. or 1.7 to 2.5 c.p.m. per 0.001 ml. of solution. Approximately 250 to 500 c.p.m. of C^{14}-leucine per microliter are used. The entire experiment is easier to perform than the preparative ultracentrifuge experiments. In the analytical ultracentrifuge the maximum angular velocity is one and one-half times as great as in the preparative ultracentrifuge with swinging bucket rotors. Obvious advantages result in the shorter time of approach to equilibrium, narrow band width, higher concentrations at band center, and greater range of densities. The preparative ultracentrifuge with a modified analytical rotor may be used at 50,000 r.p.m. for autoradiogram experiments.

Experiments in swinging bucket rotors are indicated whenever isolation of the materials is required for either preparative or analytical purposes. The plastic test tubes are removed from centrifuge containers and mounted in a rubber stopper sleeve on a ring stand. The plastic tubes are connected with a smaller rubber stopper to a container with variable air pressure so that the rate of drop formation is controllable. The plastic test tube is pierced with a sharp needle, and drops are collected individually or in groups in small test tubes previously lined up in racks. Three simple pressure-control systems used at the California Institute of Technology are shown in Fig. 1. Szybalski[21] has described a somewhat more complicated device.

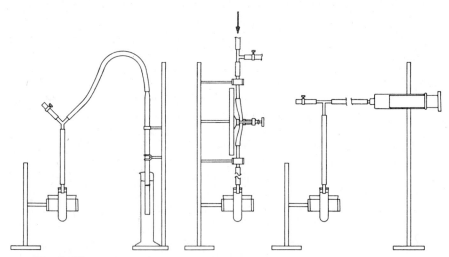

FIG. 1. Three types of apparatus for pressure control in dropwise analysis of density gradient experiments.

The fractions obtained with relatively little mixing are analyzed for macrospecies by optical density, radioactivity, and biological activity, and for density by refractive index. The combination of the former three measurements provides an elegant procedure[22] for specifying density homogeneity of biologically active materials such as viruses and infective nucleic acids.

Calculations and Theory

1. *Buoyant Density Determinations.* The buoyant density of a macrospecies was defined for Eq. (1) as the density of the medium at band center. This solution is normally at a pressure of 100 to 200 atm. and is several thousandths of a unit more dense than a solution of the same composition at atmospheric pressure. The error introduced by ignoring pressure effects, $1/\bar{v}_{s,0} = \rho_{0,s} \cong \rho_0{}^\circ$, is less than 1% and at present is of no significance in applying Eq. (1). The quantity $\rho_0{}^\circ$ is the density at atmospheric pressure of the solution at band center.

Buoyant densities have come to be of substantial importance in desig-

[21] S. Szybalski, *Experientia* 16, 164 (1960).
[22] L. Levintow and J. E. Darnell, Jr., *J. Biol. Chem.* 235, 70 (1960).

nating the composition and structure of DNA from various sources. Buoyant positions and buoyant densities can be measured with an accuracy of ±0.001 g./ml. Differences in buoyant density can be measured with still greater accuracy. It is important, therefore, that the nature of the pressure effects be understood.

It has been shown[19] that increasing the hydrostatic pressure over a CsCl density gradient causes the liquid column to compress. Otherwise the salt distribution does not change. Thus pressure merely adds a *compression gradient* to the *composition gradient*. The relative position in the liquid column for any salt composition is independent of pressure. A buoyant density may therefore be expressed as the composition or density at atmospheric pressure with no ambiguity arising from the effects of pressure on the gradient column. However, the buoyant species compresses differently than does the CsCl. Bands of DNA and TMV move toward the top of the liquid column when hydrostatic pressure is applied.[19] There is no way of avoiding the specification of pressure in specifying a buoyant density. The buoyant density of reference DNA should be determined with cells that are about 90% full (0.70 ml. in a 4° sector) and run at a speed of 44,770 r.p.m. The pressure at the root-mean-square position in the liquid column is 150 atm. Alternatively, experiments at various pressures may be performed so that results can be expressed as the density of the salt solution at band center for a band which is itself at atmospheric pressure. This quantity is the reciprocal of the partial specific volume of the solvated species at atmospheric pressure. Pressure correction terms arising from variable speeds can now be made only for native T-4 DNA and for TMV.

The next problem in expressing buoyant densities of DNA is to establish the isoconcentration position, r_e, in the gradient column.[23] It is at this position that the original salt composition will be found. For DNA in CsCl the isoconcentration distance is readily calculated with Eq. (7),

$$r_e = \sqrt{(r_t^2 + r_b^2)/2} \qquad (7)$$

where the subscripts t and b refer to the radial distances at top and bottom of the liquid column. For other densities and for other salts a general procedure for calculating r_e has been described.[23]

In a 1.2-cm. liquid column in the analytical cell the isoconcentration distance is 0.03 cm. below the center of the liquid column. The choice of the center of the cell as the isoconcentration distance leads to buoyant densities which are 0.004 g./ml. too high with a density gradient of 0.12 g./ml.

[23] J. B. Ifft, D. H. Voet, and J. Vinograd, *J. Phys. Chem.* **65**, 1138 (1961).

For a new material, buoyant densities are best determined by using two solutions which form bands on either side of the isoconcentration distance and then interpolating. If the distances are not large, the assumption that the density gradient is constant over the interpolation is satisfactory.

2. *The Density Gradient and the Use of Density Markers.* The sum of the composition density gradient and the compression density gradient is called the *physical density gradient.* In CsCl solutions the compression gradient accounts for 8 to 10% of the total gradient, and this relative contribution is independent of the angular velocity. Whereas the physical gradient is used in calculating densities in the liquid column, and the composition gradient is used in calculating compositions in the column, the *buoyancy density gradient*[15] [Eq. (8)] is used to calculate the

$$\left(\frac{d\rho}{dr}\right)_{b,0} = \left[\frac{1}{\beta_0{}^\circ} + \psi\rho_0{}^{\circ 2}\right]\omega^2\bar{r}_0 = \frac{\omega^2\bar{r}_0}{\beta_{b,0}{}^\circ} \tag{8}$$

composition of the solution that bands the macrospecies at the isoconcentration distance from data relating the band position and solution density. The quantity $\beta_0{}^\circ$ is the coefficient in the differential equation for sedimentation equilibrium in a two-component incompressible system [Eq. (9)]. The quantity $\rho_0{}^\circ$ is the density at band center, and $\bar{r}_0$ is the

$$\left(\frac{d\rho}{dr}\right)_{\text{comp}} = \frac{\omega^2 r}{\beta_0{}^\circ} \tag{9}$$

average of the radial distances to band center and the isoconcentration distance. The buoyancy density gradient is also obtained in a two-cell experiment from the distance between bands and the densities of the two solutions [Eq. (10)], where $\rho_{e,2}{}^\circ$ and $\rho_{e,1}{}^\circ$ are the densities of the initial

$$\left(\frac{d\rho}{dr}\right)_{b,0} = \frac{\rho_{e,2}{}^\circ - \rho_{e,1}{}^\circ}{(r_{0,2} - r_{e,2}) - (r_{0,1} - r_{e,1})} \tag{10}$$

solutions, $r_{0,2}$ and $r_{0,1}$ are the distances at band center, and $r_{e,2}$ and $r_{e,1}$ are the isoconcentration distances.

The term ψ arises from the fact[2] that bands seek out slightly different salt compositions in buoyant media of different density, ρ°, because of the variation of band position with pressure. The pressure dependence of band position[2] results from the difference in the compressibility of the buoyant solvent and the apparent compressibility of buoyant macrospecies, and also from the variation of solvation with salt concentration.

A convenient and accurate method of determining buoyant densities of DNA is to band a DNA together with a well-studied reference DNA in the same solution.[24] The difference in buoyant density is then given by Eq. (11), where $\bar{\beta}_{b,0}°$ is the average value of $\beta_{b,0}°$. The quantity Δr

$$\Delta\rho = \frac{1}{\bar{\beta}_{b,0}°} \omega^2 \bar{r}_0 \, \Delta r \tag{11}$$

is the distance between the bands. The assumption is made in the foregoing procedure that ψ for all DNA samples will be the same. The alternative procedure is to ignore the pressure contribution. This is equivalent to assuming that $\psi = 0$ in all cases. The equation then used is Eq. (12), where $\bar{\beta}_0°$ is the term in brackets in Eq. (8).

$$\Delta\rho = \frac{1}{\bar{\beta}_0°} \omega^2 \bar{r}_0 \, \Delta r \tag{12}$$

3. *Molecular Weight Determinations.* Inspection of the derivation of Eq. (1) reveals that the net result of any phenomena affecting the buoyant density, $1/v_{0,s}$, or the solvated molecular weight, $M_{0,s}$, linearly about band center is to change the density gradient, $(d\rho/dr)_{\mathrm{eff}}$. The effective density gradient with two such variables taken into account—the dependence of solvation on salt composition and the effects of pressure—has been shown[2, 25] to be

$$\left(\frac{d\rho}{dr}\right)_{\mathrm{eff},0} = \left(\frac{1}{\beta_0°} + \psi\rho^{02}\right)(1 - \alpha)\omega^2 r \tag{13}$$

where

$$\alpha = \left(\frac{\partial\rho_0°}{\partial a_1}\right)_p \left(\frac{da_1°}{d\rho°}\right) \quad \text{and} \quad \psi = \frac{(\kappa - \kappa_s)}{(1 - \alpha)}$$

where $(\partial\rho_0°/\partial a_1)_p$ is the slope of the plot of buoyant density versus water activity, and $(da_1°/d\rho°)$ is the reciprocal of the slope of the plot of solution density versus water activity. Both slopes are taken at the intersection of the plots. The quantity ψ is evaluated from the pressure dependence of band position. Curves for DNA in a variety of cesium salts[26] are shown in Fig. 2. In CsCl, $\alpha = 0.24$, and $\psi\rho^{02} = 0.66 \times 10^{-10}$ c.g.s. The effective density gradient, Eq. (13), is decreased 7% by ignoring the $\psi\rho^{02}$ in the first term and increased 31% by ignoring the α in the second term. The solvated molecular weight will then be 23% too low. The solvation for T-4 bacteriophage DNA is, however, 28%. Thus by a coincidental cancellation of errors, Eq. (14) gives within 5% the

[24] R. Rolfe and M. Meselson, *Proc. Natl. Acad. Sci. U.S.* **45**, 1039 (1959).

[25] J. W. Williams, K. E. Van Holde, R. L. Baldwin, and H. Fujita, *Chem. Revs.* **58**, 715 (1958).

[26] J. E. Hearst and J. Vinograd, *Proc. Natl. Acad. Sci. U.S.* **47**, 1005 (1961).

$$\sigma^2 = \frac{RT}{M_{\mathrm{anh}}(d\rho/dr)_{0,\mathrm{comp}}\bar{v}_{s,0}\omega^2 r} \tag{14}$$

anhydrous molecular weight of cesium DNA in CsCl. It is not to be expected that similar good fortune will attend the study of other systems.

4. *Quantitative Analysis of Macrospecies.* The photographic record of absorbance may, after densitometry, be used to calculate the amount of macrospecies in a band. The calculation procedures require data from the simultaneous exposure of an exponential aperture in the rotor. Directions for carrying out these calculations have been published.[17] The accuracy of the procedure is limited by small effects due to the failure of the reciprocity law for photographic film.

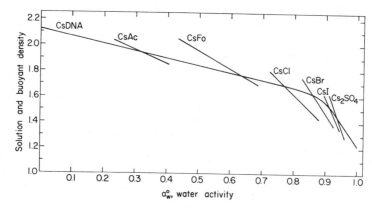

Fig. 2. Buoyant density of T-4 bacteriophage DNA and solution density of various cesium salts versus water activity.[26]

Concentrations may be calculated from schlieren photographs, providing the refractive index increment of the macrospecies in the buoyant solvent and the instrument constants are known. The evaluation of total infectivity, radioactivity, or optical density in a band in the preparative ultracentrifuge requires only simple summation of analytical results on the single drops. The hazardous assumption of constant drop volume is involved in the summation procedure. The assumption may be avoided by weighing the individual fractions.

5. *Resolution of Macrospecies.* The resolution of two homogeneous macrospecies in a buoyant density experiment is independent of angular velocity. Increasing the velocity narrows the bands but causes them to approach each other. Resolution, however, may be improved by using salts which give lower density gradients at a given speed.[8, 23]

Applications

Several applications of equilibrium sedimentation in a density gradient have been developed in the last five years. Studies have been made of biological materials, of biological processes, of synthetic high polymers, and also of some physical-chemical phenomena uniquely revealed in the procedure. In this section only selected examples of the various applications are given to illustrate the range of the method.

1. *Solvation.* Unambiguous thermodynamic relations have been derived which permit the evaluation of the net solvation of the buoyant species.[25] The solvation parameter, Γ, sometimes called the selective or preferential solvation, is the number of molecules of one of the two components of the binary solvent that acts in establishing the density of the macrospecies, as though it is attached. Cesium DNA has an anhydrous density of 2.12 g./ml.[26, 27] The buoyant density is approximately 1.70 g./ml. From numbers such as these, together with an assumption regarding the magnitudes of the specific or the partial specific volumes of water in the solvate layer, the quantity Γ' is calculated from Eq. (15).

$$\frac{1}{\rho_0} = \frac{\bar{v}_3 + \Gamma'\bar{v}_1}{1 + \Gamma'} \equiv \frac{v_3 + \Gamma'v_1}{1 + \Gamma'} \tag{15}$$

The quantity Γ' is the net solvation expressed on a weight basis, ρ_0 is the buoyant density, and $\bar{v}$ and v are the partial specific and specific volumes. The subscripts 3 and 1 refer to the anhydrous macrospecies and to the component in excess in the solvate layer. From the results in Fig. 2 the net hydration of buoyant cesium DNA from T-4 bacteriophage is about 10% by weight in cesium acetate solution, 28% in cesium chloride solution, and 68% in cesium sulfate solution.[26]

Solvations of these magnitudes have been observed with bovine mercaptalbumin[28] and cesium poly-L-glutamate.[20]

2. *Study of Structural Changes.* Denaturation of DNA causes the buoyant density to increase.[29] On renaturation the original buoyant density is substantially restored.[30] Although the reasons for these changes are not understood at present, the density shifts provide a basis for recognizing the extent of denaturation of a DNA sample and for separating denatured from undenatured DNA. A similar density shift has been observed[31] after heat denaturation of bovine serum albumin. In

[27] J. E. Hearst and J. Vinograd, *Proc. Natl. Acad. Sci. U.S.* **47**, 825 (1961).
[28] J. B. Ifft and J. Vinograd, unpublished observations.
[29] M. Meselson and F. W. Stahl, *Proc. Natl. Acad. Sci. U.S.* **44**, 671 (1958).
[30] P. Doty, J. Marmur, J. Eigner, and C. Schildkraut, *Proc. Natl. Acad. Sci. U.S.* **46**, 461 (1960).
[31] D. J. Cox and V. N. Schumaker, *J. Am. Chem. Soc.* **83**, 2439 (1961).

the latter case the density shift was observed in a three-component solvent—CsCl, $(NH_4)_2SO_4$, and water.

3. *Effects of Composition Changes.* The two main factors governing buoyant density are the anhydrous density and the net solvation. The anhydrous density may be changed by an increment in mass without volume change by isotopic substitution; or, as in the substitution of adenine-thymine for guanine–cytosine pairs in DNA, the anhydrous density may be changed by an increment in volume with substantially no mass change. In both instances the density shifts are enhanced by shifts in net solvation. Buoyant density shifts also occur when simultaneous changes in mass and volume are introduced. Examples of these are (1) shifts occurring on acid-base titration of the macrospecies in CsCl (in this event hydrogen and cesium atoms are exchanged); (2) specific binding of anions to proteins; and (3) the biological substitution of bromines for methyl groups in thymine in DNA. Buoyant density is a property which may be readily changed in a variety of ways. Some of these variations have formed the basis for new areas of investigation.

A. STABLE ISOTOPE SUBSTITUTIONS. The combined use of high concentrations of stable isotopes and density gradient sedimentation provides a procedure[29] for separating new from old biological macromolecules and cell particulates. In a series of *transfer* experiments, Meselson and Stahl,[29] Sueoka,[32] and Simon[33] demonstrated that DNA replicates semi-conservatively and that, after one and two divisions, "hybrid" DNA exists. These experiments were performed with *Escherichia coli* B, *Chlamydomonas reinhardi*, and HeLa (human tissue culture) cells, respectively. The hybrid DNA most likely consists of a double-stranded DNA. The dense strand derives from the parent molecule, and the light strand is formed on replication after dilution with light isotope. The results of Meselson and Stahl and of Sueoka were obtained with the N^{15} isotope, which changes the buoyant density by 0.015 g./ml. Simon's result was obtained with 5-bromouridine, which changes the buoyant density by approximately 0.10 g./ml. Davern and Meselson,[6] with C^{13} and N^{15} isotopes, showed that ribosomal RNA is conserved after two divisions of *E. coli*. The mixture of isotopes gave a density shift between new and old ribosomal RNA of 0.05 g./ml. in cesium formate.

Brenner *et al.*[34] also used both the isotopes C^{13} and N^{15} to show that messenger RNA and new protein could be found in old ribosomes. These experiments were performed with CsCl in the preparative ultracentrifuge, and the messenger RNA and new protein were recognized with P^{32} and

[32] N. Sueoka, *Proc. Natl. Acad. Sci. U.S.* **46**, 83 (1960).
[33] E. H. Simon, *J. Mol. Biol.* **3**, 101 (1961).
[34] S. Brenner, F. Jacob, and M. Meselson, *Nature* **190, 576** (1961).

S^{35} radioactive isotopes. A twenty-drop interval occurred between the modes of the corresponding dense and light ribosomes.

Hybrid DNA was first observed in the *in vivo* growth experiments of Meselson and Stahl. This hybrid has a buoyant density midway between N^{14}-DNA and N^{15}-DNA and therefore contained equal amounts of each isotope. When the hybrid was heated to 100° for 30 minutes in 7 molal CsCl, two bands having buoyant densities of denatured N^{15}-DNA and N^{14}-DNA were obtained. This result showed that the hybrid consisted of two subunits of DNA which were not definitely identified as DNA strands. Doty et al.[30] investigated this problem by the sedimentation velocity-viscosity method and concluded that the *in vitro* hybrid did indeed separate into single strands. These authors showed that the separate strands would recombine on slow cooling in dilute salt solution. Hybrid bands were observed to form between the closely related organisms *E. coli* and *Shigella*, but not between *E. coli* and *Diplococcus pneumoniae* or *Serratia marcescens*. Marmur and Lane,[35] in studies with transforming DNA, also concluded that heating and slow cooling lead to hybrid molecules. Schildkraut et al.[36] used deuterium and the N^{15} isotope to enhance greatly the buoyant density difference between labeled and unlabeled DNA. This difference was 0.040 g./ml.

B. THE GUANINE–CYTOSINE DISTRIBUTION IN DNA. Sueoka *et al.*[37] and Rolfe and Meselson[24] have shown that DNA from a variety of organisms, known from earlier work to contain widely different average guanine–cytosine contents, formed narrow bands of different buoyant density. Among the organisms studied, several of the DNA samples showed no detectable overlap in the concentration distribution in CsCl. This significant biological result has not yet been explained. The chemical result has been proposed as a method for base analyses.[24, 38] Only 1 μg. of DNA is required for the analysis. Between 25 and 75% guanine–cytosine content, the buoyant density varies linearly with guanine–cytosine content. The slope corresponds closely to 0.00100 g./ml. per % guanine-cytosine. DNA molecules from microbial sources are by no means necessarily homogeneous in base composition. Evidence in support of this view has been reported by Sueoka.[4]

C. BUOYANT DENSITY TITRATIONS. When acidic hydrogens in a polyanion or polyampholyte are titrated with base and converted to salts, it is to be expected that the macrospecies will become more dense in

[35] J. Marmur and D. Lane, *Proc. Natl. Acad. Sci.* **46**, 453 (1960).
[36] C. L. Schildkraut, J. Marmur, and P. Doty, *J. Mol. Biol.* **3**, 595 (1961).
[37] N. Sueoka, J. Marmur, and P. Doty, *Nature* **183**, 1429 (1959).
[38] P. Doty, J. Marmur, and N. Sueoka, *Brookhaven Symposia in Biol.* **No. 12**, 1 (1959).

buoyant cesium chloride.[20] Such density shifts have been observed in alkaline CsCl for DNA[9] and in acid CsCl for poly-L-glutamic acid.[20] In the former case complete titration of guanine and thymine residues at pH 11.6 increases the buoyant density 0.063 g./ml. A density shift of 0.20 g./ml. was observed for poly-L-glutamic acid on titration of the γ-carboxylic acid with base. These procedures appear to be general and include, for example, buoyant density changes on hydrolysis of ester side chains such as exist in pectins.

4. *Isolation, Purification, and Characterization of Macrospecies.* Macrospecies can sometimes be readily separated from accompanying impurities by banding procedures. Viruses usually band rapidly and form narrow bands at characteristic buoyant densities different from those of host materials. Several reports of banding viruses in concentrated salt solutions with retention of biological activity have appeared. These include tobacco mosaic virus,[39] turnip yellow virus,[40] the bacterial viruses φX-174,[41] T-4, and T-7,[42] λ and several transducing λ viruses,[43] and the animal viruses polio[23] and rous sarcoma.[44] For biological assay it is usually sufficient to dilute the recovered drops from preparative runs by a factor of 50 or 100, to avoid interfering effects of the salt.

Purification from small molecules is less efficient than from large molecules. For nonsedimenting impurities the fraction remaining in the sample is $4\sigma/L$ with 95% recovery or $6\sigma/L$ with 99% recovery of the macrospecies. In experiments with viruses, globular proteins, or globular particulates, the time required for an experiment may be considerably shortened by preforming the density distribution by layering of solutions or with a gradient-making machine. This is especially the case when long liquid columns are used. The desired density distributions may be calculated or interpolated from published data.[23]

The numerical values of buoyant density at a given temperature and pressure are data that characterize materials. Such data may be expected to accumulate in the future. Investigators will do well to specify the purity of the buoyant solvent, the angular velocity, the length of the liquid column, the temperature, and the method of calculation.

[39] A. Siegel and W. Hudson, *Biochim. et Biophys. Acta* **34**, 254 (1959).
[40] R. E. F. Matthews, *Nature* **184**, 530 (1959).
[41] R. L. Sinsheimer, *J. Mol. Biol.* **1**, 37 (1959).
[42] M. Meselson, *in* "The Cell Nucleus" (Proc. informal meeting by Faraday Soc., Cambridge, 1959). Butterworths, London, 1960.
[43] J. Weigle, M. Meselson, and K. Paigen, *J. Mol. Biol.* **1**, 379 (1959).
[44] L. V. Crawford, *Virology* **12**, 143 (1960).

[121] Preparation of Tritium-Labeled Substrates

By JOHN M. LOWENSTEIN

The ready availability of sodium borohydride-2-H^3 makes possible simple syntheses of many substances that are labeled with tritium in specific positions. Sodium borohydride-2-H^3 itself is readily prepared and, provided that it is kept dry, is a very stable substance.

Preparation of Sodium Borohydride-H^3

Principle. Unlabeled sodium borohydride is heated in a sealed tube with tritium gas. This results in an exchange of borohydride hydrogen with the tritium gas.[1] On cooling and breaking of the tube, the material is ready for use without further purification.

Reagents

Sodium borohydride. A pure grade that has not been exposed to moisture is essential.

Tritium gas. This is obtainable in glass ampules with breakoff seal from Oak Ridge National Laboratory, Oak Ridge, Tennessee. The specific activity is 2.59 curies/cc. at S.T.P., and the material is >99% pure.

Procedure. Sodium borohydride (100 mg.) is introduced through *A* into the special tube shown in Fig. 1, and the tube is sealed off at A. The volume of the sealed-off vessel should be about 6 ml. The tube is attached to the vacuum line at *B* and evacuated. About 2 ml. or less of tritium gas is introduced into the tube from the storage bulb by means of a Toepler pump. The tube is pulled off at capillary constriction *C*. It is heated at 350° for 48 hours. The tube is cooled to room temperature and attached to the vacuum line at *D*. After the line has been evacuated and closed, the breakoff seal, *E*, is broken, and the gas is transferred back to the storage bulb by means of the Toepler pump. Dry air is then admitted to the tube. The labeled borohydride is divided into suitable portions and stored in evacuated, sealed ampules. On storage, sodium borohydride-H^3 turns pink and eventually a deep purple color. This is due to the excitation of electrons out of their normal valence bands and the trapping of

[1] N. H. Smith, K. E. Wilzbach, and W. G. Brown, *J. Am. Chem. Soc.* **77**, 1033 (1955).

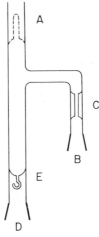

Fig. 1.

these electrons in the new bands. It does not mean that the material has become unusable.

Preparation of D-Glucose-1-H^3

Principle. Glucono-δ-lactone is reduced with sodium borohydride-H^3 to yield glucose-1-H^3.

Reagents

> Glucono-δ-lactone (0.75 g.) and a crystal of bromothymol blue are dissolved in 5 ml. of water shortly before use. The solution is kept on ice.
>
> Sodium borohydride-H^3 (0.20 g.) is dissolved in 5 ml. of ice-cold water immediately before use. The solution is kept on ice.
>
> Dowex 1 ion-exchange resin, acetate salt, 200 to 400 mesh, 10% crosslinked.
>
> Amberlite IRC-50 ion-exchange resin, hydrogen form, <200 mesh, washed by decantation until free of slowly sedimenting material.

Procedure.[2] The solution of glucono-δ-lactone and bromothymol blue is placed in a small vessel which is stirred mechanically and kept on ice. The solution of borohydride is added to the solution of the lactone, 2 drops at a time, over a period of 50 minutes. The reaction mixture is kept at a greenish-yellow color by the careful addition of 1 N H_2SO_4. This is best done with a micrometer-operated microburet, but the cautious addi-

[2] J. M. Lowenstein, *J. Biol. Chem.* **236**, 1213 (1961).

tion of very small drops from a drawn-out dropper is adequate. The solution should not be allowed to become alkaline (blue color) for any appreciable length of time; otherwise a portion of the lactone will hydrolyze. After the addition of the borohydride solution has been completed, the mixture is stirred for 10 minutes, and a further 0.5 ml. of 1 N H_2SO_4 is then added. The resulting solution contains the labeled glucose in over 95% yield. It is passed through a column of Dowex 1–acetate resin with a resin bed 1.5 cm. in diameter and 15 cm. high. The column is washed with water until over 99% of the radioactivity has been eluted. The combined effluents are placed on a second column of Amberlite IRC-50, hydrogen form, with a resin-bed size as above. The column is again washed with water until 99% of the radioactivity has been eluted. The combined effluents are evaporated on a rotary drier. Absolute ethanol (13 ml.) is added, and the mixture is heated at 80°. All or virtually all the material dissolves; any insoluble residue is removed by centrifugation and discarded. The solution is again evaporated to dryness on a rotary drier. The sirup is dissolved in 5 ml. of water and evaporated as before three times; this removes traces of labile tritium. At this point the sirup contains almost pure glucose, which may contain very small amounts of nonradioactive gluconic acid. If desired, this can be removed by electrophoresis.[2] Alternatively, the compound may be crystallized as follows.[3] After the last evaporation, the sirup is dissolved in 0.5 ml. of water. This is followed by 0.9 ml. of methanol, and then dropwise isopropanol (about 0.9 ml.). Crystallization takes place on seeding with anhydrous α-D-glucose. The yield of crystals is no greater than 60%. The mother liquors should be saved, since they contain radioactive glucose of apparently the same, or nearly the same, purity as the crystalline material. The resulting radioactive glucose contains tritium exclusively in the 1-position. To minimize chemical and radiation decomposition, the material, whether crystalline or sirupy, is stored as a dilute solution in 95% ethanol.

Preparation of D,L-Malate-2-H^3

Principle. Oxalacetate is reduced to D,L-malate-2-H^3 with sodium borohydride-2-H^3.

Reagents

Oxalacetic acid (1 millimole, 132 mg.) and a crystal of bromothymol blue are suspended in about 1.5 ml. of water. Then 2 N NaOH is added dropwise until the oxalacetic acid has dissolved completely, and the solution is a green color. The volume is now

[3] R. Bentley and D. S. Bhate, *J. Biol. Chem.* **235**, 1225 (1960).

about 2.5 ml. This solution is prepared shortly before use and is kept on ice.

Sodium borohydride-H^3 (10 mg.) is dissolved in 0.5 ml. of ice-cold water. This solution is prepared immediately before use and is kept on ice.

Dowex 1 ion-exchange resin, formate salt, 200 to 400 mesh, 10% crosslinked.

Approximately 0.25 M formic acid.

Procedure. The solution of borohydride is added 20 to 30 μl. at a time, with stirring, to the solution of oxalacetate. After each addition of borohydride the pH is adjusted back to a green color by the addition of 1 N H_2SO_4 from a micrometer-operated microburet. When the addition of borohydride has been completed, enough acid is added to bring the color of the solution to yellow. The solution is allowed to stand at room temperature for 10 minutes. It is then diluted to 10 ml. with water and adjusted to approximately pH 9 with NaOH. The solution is placed on a column of Dowex 1–formate ion-exchange resin with a resin bed 1 cm. in diameter and 10 cm. in height. After the solution has soaked in, the column is washed with 40 ml. of water. It is then eluted with two fractions of 50 ml. of 0.25 M formic acid. This is followed by four fractions of 50 ml. of 0.50 M formic acid. Malate is eluted chiefly in the second and to some extent in the third 0.50 M formic acid fraction. The second 0.50 M formic acid fraction is freeze-dried, and the residue is dissolved in water. The solution so obtained contains malate-2-H^3 that is pure radiochemically.

The purity of the product may be checked by subjecting an aliquot of the solution to two-dimensional paper chromatography followed by autoradiography.[4] Both short and prolonged exposure to X-ray film reveals a single radioactive spot, which corresponds exactly with the position of authentic DL-malate. The chromatographic system is capable of separating malate from lactate, fumarate, succinate, β-hydroxybutyrate, and isocitrate, besides many other compounds. On prolonged exposure, these compounds can be detected at radioactivity levels below 0.1% that of malate-2-H^3. Hence, these substances are absent as contaminants in the malate-2-H^3.

Preparation of Isocitrate-2-H^3

Principle. Triethyloxalosuccinate is reduced with sodium borohydride-H^3 to yield triethylisocitrate-2-H^3. The latter is hydrolyzed and

---

[4] A. A. Benson, J. A. Bassham, M. Calvin, T. C. Goodale, V. A. Haas, and W. Stepka, *J. Am. Chem. Soc.* **72**, 1710 (1950).

purified by ion-exchange chromatography. Reduction of triethyloxalosuccinate is preferable to reduction of oxalosuccinate, since oxalosuccinate is not readily obtained in pure form.

Reagents

Triethyloxalosuccinate (174 mg.) and a crystal of bromothymol
blue are dissolved in 0.87 ml. of ethanol. Then 1.57 ml. of water
are added, and the whole is mixed until a homogeneous solution
is obtained. This is facilitated by slight warming. The solution
is kept at room temperature.

Sodium borohydride-H^3 (6 mg.) is dissolved in 0.5 ml. of ice-cold
water, and the solution is stored on ice. This solution is prepared
just before use.

Dowex 1 ion-exchange resin, chloride salt, 200 to 400 mesh, 10%
crosslinked.

Procedure for Triethyloxalosuccinate.[5] Potassium metal (39.5 g.) is
cut to pieces under xylene in a wide evaporating dish. The xylene is
poured off, and the metal is washed three times with 50 ml. of anhydrous
ether. The metal is transferred quickly into a three-necked flask containing 650 ml. of anhydrous ether. The vessel is fitted with a reflux
condenser, a mercury-sealed stirrer, and a dropping funnel containing
150 ml. of anhydrous ethanol. The ethanol is added dropwise over a
period of 1½ hours. Stirring is not necessary. After most of the ethanol
has been added, the flask is warmed on a water bath to ensure complete
solution of the potassium metal. This takes up to 3 or 4 hours. The solution is cooled to room temperature, and 146 g. of diethyloxalate are
added rapidly with stirring. A yellow color develops. Stirring is continued
for 10 minutes. Then 174 g. of diethylsuccinate are added rapidly with
vigorous stirring. After a few minutes, the potassium salt of the enol
form of triethyloxalosuccinate crystallizes, and stirring becomes impossible. The salt is collected on a filter funnel and washed with ether until
colorless. It is then dissolved in 270 ml. of water, and 100 ml. of concentrated HCl are added. The resulting oil is removed with 100-ml. portions
of ether until the aqueous solution is almost colorless. The ether extracts
are dried over anhydrous Na_2SO_4, the solution is filtered, and the ether
is distilled off under reduced pressure. This yields 225 g. of the product,
which is distilled under reduced pressure (less than 1 mm.).

Procedure for Triethylisocitrate. The solution of borohydride is
added, 20 to 30 μl. at a time, with stirring, to the solution of triethyloxalosuccinate (see Reagents). After each addition, the pH is adjusted

[5] L. Friedman and E. Kosower, *Organic Syntheses Coll. Vol.* 3, 510 (1955).

back to a green color by the addition of $1 N$ H_2SO_4 from a micrometer-operated microburet. When the addition of borohydride has been completed, the mixture is extracted two times with 5 ml. of ether. The combined ether extracts are evaporated cautiously under a gentle stream of nitrogen. When the evaporation of the ether appears to be complete, 2 ml. of $1 N$ NaOH are added, and the mixture is heated at 80° for 2 hours. The solution is diluted to about 10 ml. and is then neutralized to a pH between 8 and 9 with $1 N$ HCl. The resulting solution is poured on a column of Dowex 1–chloride with a resin bed 1 cm. in diameter and 10 cm. in height. After the liquid has soaked in, the column is washed two times with 25 ml. of water. It is then eluted with 25-ml. fractions of $0.01 N$ HCl. Isocitrate is eluted in fractions 6 and 7. The most active fractions are combined and freeze-dried. The residue is taken up in several milliliters of water and is carefully neutralized with NaOH. It is then heated at 90° for 30 minutes to hydrolyze any isocitric lactone which may have formed during the freeze-drying. The resulting solution may be used as such.

Enzymatic analysis with isocitrate dehydrogenase (Vol. I [116]) shows a quantitative yield in terms of the triethyloxalosuccinate, if it is assumed that the four possible stereoisomers are formed in equal proportions during the reduction.

Preparation of DL-β-Hydroxybutyrate-3-H^3

Principle. Acetoacetate is reduced with sodium borohydride-H^3 to yield DL-β-hydroxybutyrate-3-H^3.

Reagents

> Approximately $1 M$ acetoacetate[6]: 2.6 ml. of recently distilled ethyl acetoacetate plus 10.2 ml. of $2 N$ NaOH are diluted to 20 ml. with water and kept at 40° for 1 hour. The solution is then neutralized to pH 7 approximately.
> Sodium borohydride-H^3 (10 mg.) is dissolved in 1 ml. of ice-cold water, and the solution is stored on ice. It is prepared just before use.

Procedure. The borohydride solution is added a small drop at a time, with stirring, to 0.25 ml. of ice cold M acetoacetate solution containing a small crystal of bromothymol blue. When the addition is complete, the pH of the mixture is carefully lowered to about 4 by the addition of very small amounts of $1 N$ H_2SO_4. It is then allowed to stand at room temperature for 10 minutes. The resulting solution is applied in a line to

[6] H. A. Krebs and L. V. Eggleston, *Biochem. J.* **39**, 408 (1945).

Whatman No. 3 paper which has previously been washed with water and is subjected to ascending chromatography with butanol–propionic acid–water (19:9:12 by volume). A spot of bromocresol purple is used as reference compound. After the paper has dried, it is exposed to Kodak No-Screen Medical X-ray film for several days. When the film is developed, a single radioactive band is observed. It coincides with the position of authentic β-hydroxybutyrate. The distance traveled by the β-hydroxybutyrate divided by the distance traveled by the reference compound (R_{BCP}) is 1.14. A major impurity found in some commercial preparations of β-hydroxybutyrate-C^{14} gives an R_{BCP} of 1.32. This may be the dimer of β-hydroxybutyrate. It has not been observed in our preparations but is mentioned in case it is encountered. Alternatively, the radioactive area may be detected by cutting out a narrow strip of the paper and passing it through a strip counter, or by counting the strip in sections. The area containing β-hydroxybutyrate-H^3 is cut out and eluted with water. The resulting solution, which may be concentrated by freeze-drying, contains essentially pure DL-β-hydroxybutyrate-H^3. In order to minimize chemical changes, it is best stored in frozen solution at a pH of about 7.

Preparation of Glucose-6-phosphate-1-H^3

Principle. Glucose-1-H^3 is phosphorylated enzymatically to yield glucose-6-phosphate-1-H^3.

Reagents

0.024 M glucose-1-H^3.
0.1 M creatine phosphate.
0.1 M ATP.
0.5 M MgCl$_2$.
1.0 M Tris–HCl buffer, pH 7.4.
Dowex 1 ion-exchange resin, chloride salt, 200 to 400 mesh, 10% crosslinked.

Procedure. One milliliter of 0.024 M glucose-1-H^3, 0.5 ml. of 0.1 M creatine phosphate, 0.1 ml. of 0.1 M ATP, 0.02 ml. of 0.5 M MgCl$_2$, and 0.1 ml. of 1.0 M Tris–HCl buffer, pH 7.4, are mixed together. The pH is checked and adjusted, if necessary, to 7.4. Then 0.1 ml. of creatine phosphokinase (4 mg./ml.) (Vol. II [100]) and 0.1 ml. of crystalline hexokinase (10 mg./ml.) (Vol. V [25]) are added to start the reaction. The mixture is incubated at 38° for 45 minutes. It is then diluted to 10 ml. with water and adjusted to pH 8 by the addition of small drops of 2 M Tris base. This solution is allowed to soak into a column of Dowex 1–

chloride with a resin bed 1 cm. in diameter, and 10 cm. in height. After the solution has soaked in, the column is washed with 50 ml. of water. Fractions of 25 ml. are then collected as follows: fractions 2 to 6, water; fractions 7 to 14, 0.006 N HCl. Most of the glucose-6-phosphate is eluted in fractions 8 and 9. The radioactivity elution pattern shows that more than 99% of the glucose has been converted to glucose-6-phosphate. The fractions containing glucose-6-phosphate-1-H^3 are free of glucose-1-H^3 and adenine nucleotides but may contain small amounts of orthophosphate. Most or all of the orthophosphate is, however, eluted after glucose-6-phosphate. The glucose-6-phosphate-1-H^3 fractions can be neutralized to pH 7 with a suitable base, concentrated on a rotary drier, and used as such. Alternatively, the fractions containing glucose-6-phosphate-1-H^3 may be neutralized to pH 7.5 with barium hydroxide solution and reduced in volume to about 4 ml. by freeze-drying or on a rotary drier. The pH of the concentrated suspension is checked and adjusted if necessary by the addition of barium hydroxide. An equal volume of ethanol is added, and, after standing overnight in the cold, the suspension is centrifuged. The residue is washed by resuspension and centrifugation in 8 ml. of 95% ethanol, 8 ml. of ethanol–ether (1:1 by volume), and 8 ml. of ether.

Preparation of Glucose-6-phosphate-2-H^3

Principle. Glucose-6-phosphate is incubated in tritium-labeled water with phosphoglucose isomerase (see Vol. I [37]). An equilibrium mixture of glucose-6-phosphate-2-H^3 and fructose-6-phosphate-1-H^3 results. The specific activity of the sugar phosphates (in counts per minute per micromole) is about the same as that of the water (in counts per minute per microatom of hydrogen). Only one hydrogen becomes affixed to the substrates, in agreement with the complete stereospecificity of attack on only one of the C-1 hydrogens of fructose-6-phosphate.[7,8]

Reagents

> 0.1 M glucose-6-phosphate, free of barium, adjusted to pH 8.
> Muscle phosphoglucose isomerase (see Vol. I [37]).
> Tritium-labeled water, specific activity 1 curie/ml., or as desired.

Procedure. A reaction mixture consisting of 0.2 ml. of 0.1 M glucose-6-phosphate and 0.2 ml. of water-H^3 (1 curie/ml.) is incubated with 0.1 ml. (16 units) of muscle phosphoglucose isomerase in a test tube (tube A) about 1.5 cm. in diameter, ending in a standard taper, at 25° for 10 hours. The solution is evaporated in vacuum. Because of the large

[7] I. A. Rose and E. L. O'Connell, *Biochim. et Biophys. Acta* **42**, 159 (1960).
[8] I. A. Rose and E. L. O'Connell, *J. Biol. Chem.* **236**, 3086 (1961).

amounts of radioactivity to be distilled off, this is best done as follows. Tube A is attached to a Y-shaped tube supplied with standard-taper endings on two ends and a vacuum stopcock on the third end. A second but empty standard-taper test tube (tube B) is attached to the second standard-taper ending of the Y-shaped tube. Tube A is immersed about half-way in liquid nitrogen. When the radioactive mixture in tube A is at the temperature of liquid nitrogen, the Y-tube is evacuated. The stopcock is closed, tube A is removed from the liquid nitrogen, and test tube B is immersed in the liquid nitrogen. As tube A begins to warm up, the radioactive water distills from tube A to tube B. The material in tube A stays frozen until all the water has been distilled off, or it may melt toward the end of the distillation. When the distillation has been completed, air is admitted through the stopcock, and both tubes are removed. The material in tube A is dissolved in 1 ml. of 0.2 N HCl, and the solution is evaporated again as above. This removes most of the remaining labile hydrogen. The residue is dissolved in water and neutralized in the cold with dilute triethanolamine. A slight excess of barium acetate is added. After several hours have been allowed for crystallization, the precipitate is washed with cold water, dissolved in dilute HCl, carefully neutralized in the cold, and allowed to recrystallize.

The supernatant from the first crystallization may be used for the preparation of fructose-6-phosphate-2-H^3.

Other Compounds

Many other compounds of biological interest have been labeled with tritium in specific positions. For a summary the reader is referred to Murray and Williams.[9] The labeling of steroids with deuterium and tritium has been described by Bloch (Vol. IV [29]).

Tritium Gas Exposure Labeling

Principle. The exposure of an organic compound to tritium gas results in the labeling of the compound with tritium.[10, 11]

Reagents

Tritium gas, see Reagents for sodium borohydride-H^3.

Procedure. The procedure is similar to that described for the labeling of sodium borohydride, except that the sealed tube containing the com-

[9] A. Murray and D. Lloyd Williams, "Organic Syntheses with Isotopes," Part II, p. 1660. Interscience, New York, 1958.
[10] K. E. Wilzbach, *J. Am. Chem. Soc.* **79**, 1013 (1957).
[11] K. E. Wilzbach, "Tritium in the Physical and Biological Sciences," Vol. II, p. 3. International Atomic Energy Agency, Vienna, 1962.

pound and carrier-free tritium gas is not heated. The compound to be labeled may be used in amounts anywhere between a milligram and several grams. If it is desired to introduce larger amounts of tritium gas than those indicated for sodium borohydride, a bigger reaction vessel can be employed. The compound is exposed to the tritium gas from 2 to 21 days. The gas is then removed as described for sodium borohydride. The chief difficulty of this method lies in the removal of radioactive impurities from the desired compound. The radioactive by-products often occur in only small amounts of very high specific activity. Those most difficult to remove are products of hydrogenation and isomerization, including racemization. In the words of Wilzbach, the number of tritium-labeled impurities that are likely to be present is so great that it is remarkable that radiochemical purity can be achieved for any but the simplest compounds. Great care must therefore be taken to ensure and confirm the radiochemical purity of complex compounds labeled by this procedure. Although this may appear to condemn the method, this is far from being the case. It permits the labeling of compounds that cannot be labeled synthetically, including compounds of unknown structure. The principle disadvantages are that the activities attainable are not so great as those obtained by synthetic methods, and the purification of a product labeled by this method might be a more difficult task than its synthesis.

A number of attempts have been made to accelerate the labeling by the use of electric discharges, ultraviolet light, and irradiation by other isotopes. It has not so far been demonstrated that these modifications provide higher activities or fewer impurities.[11] A comprehensive list of examples of compounds that have been labeled by this method is given by Whisman and Eccleton.[12] The reader is also referred to the book by Wenzel and Schulze, which appeared recently.[13]

Catalytic Exchange Labeling

Principle. The exposure of an organic compound to solvents such as labeled water (H_2^3O) or glacial acetic acid (CH_3—$COOH^3$) in the presence of a catalyst in a sealed tube at an elevated temperature leads to a random labeling of the compound.

Procedure. Between 1 and 50 mg. of the compound to be labeled are introduced into a tube such as that shown in Fig. 1 together with 50 mg. of a catalyst such as platinum black. The tube is attached to a vacuum line, and 100 to 1000 mg. of radioactive solvent are distilled into it *in vacuo.* The tube is sealed and then heated at 100° to 140° for several days. Many variations in quantity, solvent, temperature, and duration are possible, and the success of the method depends initially on the

[12] M. L. Whisman and B. H. Eccleton, *Nucleonics* **20**, 98 (1962).
[13] M. Wenzel and P. E. Schulze, "Tritium Markierung." De Gruyter, Berlin, 1962.

stability of the compound under the conditions employed for the exchange. After the heating has been stopped, the tube is again attached to the vacuum line, the breakoff seal is broken, and the solvent is removed by distillation. The compound is then purified by suitable methods. The same considerations apply to the purification of compounds labeled by this method as were mentioned in connection with the gas exposure method of labeling. The application of this method to steroids has been described by Bloch (Vol. IV [29]).

[122] Neutron Activation Chromatography of Phosphorus Compounds[1]

By A. A. Benson

Effectiveness of the chromatographic method depends upon sensitive and specific detection of the separated compounds. The Hanes-Isherwood phosphate spray test[2,3] is useful where the phosphorus density is 0.5 to 1.0 μg. of phosphorus per square centimeter of paper. Neutron activation of P^{31} compounds allows detection and assay of 0.01 to 0.05 μg. of P^{31} per square centimeter. Although the method requires some time, it is simple and useful.

In neutron activation chromatography,[4-6] the components of a mixture are separated on paper prior to activation. Those having elements with higher neutron-capture cross sections leading to radioisotopes of useful half-lives are identified by their chromatographic coordinates and estimated by their induced radioactivity.

Procedure

Purification. Compounds to be analyzed are freed as far as possible from elements yielding radioisotopes of half-lives over 15 hours and

[1] Work described in this paper was supported by the U.S. Atomic Energy Commission, the National Science Foundation, Research Grant A-2567 from the Institute for Arthritis and Metabolic Diseases of the Public Health Service, and the Pennsylvania Agricultural Experiment Station.
[2] C. S. Hanes and F. A. Isherwood, *Nature* **164**, 1107 (1949).
[3] R. S. Bandurski and B. Axelrod, *J. Biol. Chem.* **193**, 405 (1951).
[4] F. P. W. Winteringham, A. Harrison, and R. G. Bridges, *Nucleonics* **10**, No. 3, 52 (1952).
[5] K. Schmeiser and D. Jerchel, *Angew. Chem.* **65**, 366, 490 (1953).
[6] A. A. Benson, B. Maruo, R. J. Flipse, H. W. Yurow, and W. W. Miller, *Proc. 2nd U.N. Conf. Peaceful Uses Atomic Energy, Geneva, 1958* **24**, 289 (1959); F. Nakayama and R. Blomstrand, **8**, 230 (1961).

having high cross sections for slow neutron capture. Treatment with Dowex 50 (H⁺) is generally sufficient, since anions containing elements with high neutron-capture cross sections and leading to long-lived products are not abundant. Elements having high cross sections may be determined from the cross-section graphs compiled by Meinke and Maddock.[7]

Activation Chromatographic Analysis of Phosphatides

Purification. Phosphatides, being organic-soluble, are readily freed of offending ions and lend themselves to neutron activation analysis. They may be chromatographed on silicic acid-impregnated paper.[8] In preparation of this paper for neutron activation, the use of nitric acid rather than hydrochloric acid leads to lower induced backgrounds. Thorough washing with distilled or deionized water is advisable.

A chloroform solution of phosphatides is washed thoroughly in a tall cylinder by a slow stream of distilled water. The interfacial solid material is redissolved in the chloroform by addition of a small amount of ethanol to the surface. The chloroform and residual water are removed at reduced pressure in the evaporator (Fig. 1). If water remains, toluene and ethanol are added and evaporated until the concentrate does not appear turbid. Any solid residue is taken up in toluene followed by ethanol. In this way lipids of 10 g. of fatty tissue, such as mammary gland, may be prepared in 1.0 ml. of standard solution.

Deacylation of Glycerolphosphatides.[9,10] An aliquot portion containing washed lipids of 25 to 200 mg. of wet plant tissue or of 0.1 to 1.0 g. of fresh mammalian tissue in 100 μl. of ethanol (containing about 10% toluene) is diluted with 100 μl. of 0.2 N ethanolic potassium hydroxide and maintained at 37° for 15 minutes. If acidic components or a large triglyceride content are anticipated, a microliter of solution may be removed with a capillary and its pH measured on test paper wet with water. A pH of over 11 should be maintained by addition of further base as required.

The alcoholysis is stopped by addition of a small drop of water and 0.5 ml. of petroleum ether. Dry Dowex 50 (H⁺) (4 mg. per 100 μl. of potassium hydroxide solution) is added and mixed by shaking or with a dropper. The pipet tip and test paper may be used to ascertain acidification (pH 4) of the aqueous suspension. The 2-ml. conical tube is centrifuged to separate resin and liquid phases. The petroleum is discarded if alkali-stable products are of no concern. The aqueous phase is applied to

[7] W. W. Meinke and R. S. Maddock, *Anal. Chem.* **29**, 1171 (1957).
[8] G. V. Marinetti, J. Erbland, and J. Kochen, *Federation Proc.* **16**, 837 (1957).
[9] R. M. C. Dawson, *Biochim. et Biophys. Acta* **14**, 374 (1954).
[10] B. Maruo and A. A. Benson, *J. Biol. Chem.* **234**, 254 (1959).

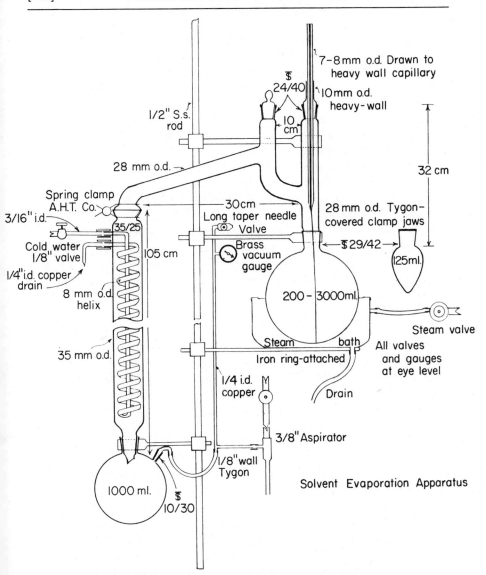

FIG. 1. Apparatus for rapid concentration of solutions.

the origin of a two-dimensional chromatogram on Whatman No. 4 paper (Schleicher and Schuell 589 White Ribbon acquires less induced background activity but, being acid-washed, cannot tolerate as much γ-radiation). The Dowex precipitate is washed with 50% ethanol to remove the absorbed products. Care must be taken to minimize contact time with the resin which causes cleavage of the diesters and production of cyclic 1,2-glycerophosphate. In some cases it is advisable to take up the residual Dowex in phenol–water chromatographic solvent and apply it to the chromatogram origin as a thick slurry. Development in phenol–water (100:40) completely elutes all phosphorus compounds from the Dowex.

Chromatography. The chromatogram is developed 17 inches in phenol–water (100:40) and dried at room temperature in an air stream. Care must be taken to avoid or purify phenol containing hypophosphorus acid stabilizer. Distillation in glass is effective. For effective separation of diglycerophosphorylglycerol (GPGPG),[11] 1% formic acid is added to the phenol solvent. The second solvent is butanol–propionic acid–water (142:71:100). Maximal separation of deacylated glycerolphosphatides is attained when a spot of "Tropaeolin 0000" orange dye applied at the origin is moved 20 to 24 inches. Sphingolipids are determinable when this dye is developed for 18 inches. After development in the second solvent, care must be taken to avoid fingerprints and contact with certain metals. Chromatographic troughs of stainless steel, glass, or polyethylene are recommended.

Neutron Activation[6, 11]

Standardization of P^{31} Activation. The chromatogram is well dried. P^{31} control samples containing 1.0 μg. of P are applied to the paper along a vacant edge parallel to the y-axis. A solution of ammonium phosphate in 50% ethanol containing 1.0 μg. of P^{31} per microliter is used for this purpose. Activity induced in these spots is a measure of the neutron flux applied to the paper.

Activation Containers. The chromatogram is rolled parallel to its y-direction and inserted in a straight, clean polyethylene tube, 1 inch in inside diameter and about 4 feet long. The lower end, especially, is neatly sealed, since radiation renders it nonsealable. A warm burner flame is used to heat the polyethylene to transparency. It is then lightly compressed in the jaws of a clean vise until cool. The upper end is outside the intense gamma flux and can be resealed many times after opening.

Neutron Irradiation. The tube containing the chromatogram at its lower end is tied with 3S aluminum wires to a lead weight and to a cotton or nylon cord. After being suspended at the face of the swimming pool type of reactor for more than 15 to 20 hours at a flux of 5×10^{11} neutrons/cm.²/sec., the paper becomes fragile and difficult to unroll. Irradiation times of 3 hours suffice for larger phosphorus samples. The short-lived paper impurities (chlorine, sodium, potassium, silicon) decay during 5 days, after which the tube is removed from the pool and the

[11] E. H. Strickland and A. A. Benson, *Arch. Biochem. Biophys.* **88,** 344 (1960).

upper end is cut off with a saw. Care should be taken to avoid ignition of the hydrogen (1 to 2 atm.) expelled as the tube is opened. The paper is unrolled and carefully flattened with a steam iron if necessary. Although the major part of the induced activity is short-lived, contamination of the laboratory by dust from the active paper should be minimized. Activated phosphorus compounds and the 1.0-μg. P^{31} spots are usually seen as spots fluorescing in 3600-A. light (two F40T12BLB lamps in a standard fixture).

Analysis

Measurement of Induced Radioactivities. P^{32} radioactivity may be measured with large-diameter G-M tubes (Anton type 1001-T, Lionel Electronic Laboratories, Inc., Brooklyn, New York). Correction for background radioactive impurities in the paper requires special consideration. Counting shields of $\frac{1}{32}$-inch stainless-steel (3 $\times$ 4 inches) having round or rectangular holes of various sizes are convenient for counting induced radioactivity within defined areas. A circular hole 1 cm. in diameter is used for counting activity of the 1.0-μg. P standard spots. This activity is corrected for background of an immediately adjacent area. Since radioautograms often reveal inhomogeneities in background impurities, it is advisable to examine several similar areas for variations. An appreciable fraction of the background radiation is very soft and can be absorbed by a thin (0.003-inch) sheet of clear plastic with little diminution of P^{32} radiation. Induced P^{32} activity in chromatographic spots is determined in the same manner by using larger rectangular apertures. It is corrected for background of adjacent identical areas. Since G-M tubes may not have radially linear sensitivities, it is advisable to count as small areas as practical. Large-diameter gas-flow proportional counters are well adapted for this purpose.

Variations in neutron flux over the length of the rolled paper may be as high as 20%. Comparison of activities of the 1.0-μg. P standards are used to correct for these variations in calculating the P^{31} content of the chromatographic spots. The 1.0-μg. P sample receiving the same neutron flux, i.e., with identical y-coordinate, is chosen as the standard for that spot.

Slow neutron fluxes attenuate in water with a relaxation length (distance to diminish to $1/e$) of 4 to 20 cm. Fortunately, the flux within 5 cm. of the core is approximately constant because of reflection and scattering effects, and flux variation is limited to that in the y-direction as determined by position and lengths of the chromatogram and reactor.

Scintillation Counting of Induced P^{32} Radioactivity. Large-diameter plastic phosphor scintillators are useful for measuring P^{32} β-radiation.

Since multiplier phototubes up to 5 inches in diameter are available, it is possible to determine activities in large areas not readily measurable with the larger G-M tubes. The scintillator is uniformly sensitive over its area, and, with a pulse-height discriminator, the system may be adjusted to count only radiations of energies characteristic of P^{32}. The background induced in paper impurities appears to be due largely to P^{32}, with significant amounts of weaker and more energetic radiations.

A ⅛-inch-thick phosphor (NE-102 Plastic Phosphor, Nuclear Enterprises, Ltd., Winnipeg, Man.) is attached to a multiplier phototube with silicone fluid (Dow Corning 200 Fluid). The tube and phosphor are enclosed in a compact light-tight cylinder, and the phosphor is covered with a 0.001-inch aluminum foil as described by Fig. 2. Scintillation

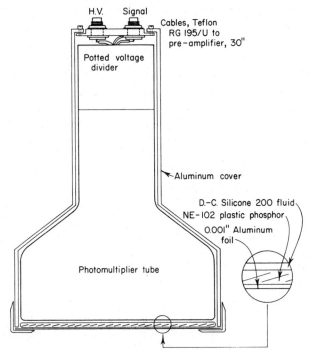

H.V.   Signal

Cables, Teflon
RG 195/U to
pre-amplifier, 30"

Potted voltage
divider

Aluminum cover

D.-C. Silicone 200 fluid
NE-102 plastic phosphor
0.001" Aluminum
foil

Photomultiplier tube

FIG. 2. Plastic phosphor scintillation counter for determination of induced P^{32} radioactivity in neutron-activated paper chromatograms.

counting of soft β-emitters may be extended to include C^{14} and S^{35} by use of aluminized thin (0.003-inch) plastic phosphors.[12] An adherent and abrasion-resistant coating on polystyrene phosphors is achieved by suc-

[12] K. Steenberg and A. A. Benson, *Nucleonics* 14, No. 12, 40 (1956).

cessive vacuum deposition of silicon monoxide, stainless steel (1 A.), silicon monoxide, aluminum, silicon monoxide, and aluminum.[13]

The output from the preamplifier is amplified by a linear amplifier, and the energy spectrum is examined by using an integral or differential pulse-height analyzer. The instrument is calibrated for pure P^{32} radiation, and an optimum energy range is chosen for counting the chromatogram with minimum background interference. In this way the background compared to that for a G-M tube may be reduced several-fold.

Liquid scintillation counting of excised areas of the activated chromatogram is the most sensitive and reproducible method for measurement of induced activities. Although the paper is no longer available as a record, the automatic operation and pulse-height selectivity of such counting systems provide simple and effective analyses.

Radioautography. Best radiograms are prepared 7 to 10 days after activation. Single-coated X-ray film (Eastman Kodak Co., Blue

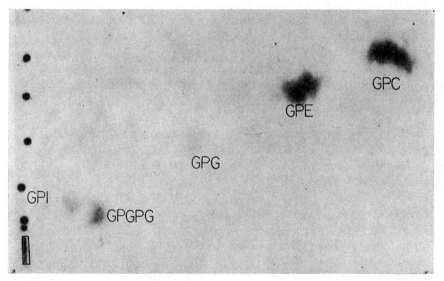

Fig. 3. "Neutron radiogram" of deacylated lipids of sheep heart mitochondria.[11] Chromatogram was developed in the x-direction with phenol–water and in the y-direction with butanol–propionic acid–water. Spots on left margin record P^{32} activity induced in 1-μg. control samples of P^{31}.

Sensitive) is placed with its *back* to the chromatogram. The image formed on No-screen (double-coated) film generally has better contrast on the far side than on the side nearest the paper. Exposure times of up to 1 half-life are recommended except where the induced activities

[13] Leland Eminhizer, H. R. B.-Singer, Inc., personal communication, 1961.

are too high. Activities as high as 30 mr./hr. may be induced in good chromatographic spots of glycerophosphoryl choline from lecithin in animal tissues. A typical neutron radiogram is shown in Fig. 3.

Other Applications

Activation of Short-Lived Isotopes on Paper. Potassium, chlorine, bromine, sodium, cobalt, vanadium, and many other elements are readily activated by slow neutron capture on paper. The radioisotopes produced have half-lives of 4 minutes to 15 hours. In these cases the chlorine impurity of filter paper reduces sensitivity of the method. Of commercially available papers, Schleicher and Schuell 589 Black Ribbon paper was found to have the lowest short-lived induced background. The other types of 589 papers are comparable and possess somewhat better chromatographic properties.

Bromine-containing derivatives of biochemically important compounds offer excellent opportunities for analysis.[14] High cross section ($\sigma = 8.5$ barns) and short life (4.4 hours) of Br^{80} allow sensitive assays of chromatographically separated derivatives. Bromoacetyl and *m*-bromobenzoyl derivatives of hydroxy compounds, bromophenylazo derivatives of proteins, *p*-bromophenylhydrazones of sugars and keto acids, *p*-bromophenylsulfonamides of amino acids and *p*-bromophenacyl esters of carboxylic acids may be analyzed by neutron activation.

Cobalt compounds are among the most sensitively analyzed.[14] The 100% natural abundance of C^{59} and its high cross section lead to good yields of Co^{60m} ($t_{1/2} = 10.7$ minutes) which transforms to Co^{60} emitting a 0.06-Mev. electron. This is readily detected on chromatograms with thin mica end-window G-M tubes, by aluminized plastic phosphor scintillation counters, or, most sensitively, by liquid scintillation counting. One microgram of cobalt yields 10^5 c.p.m. with a background of 500 c.p.m. after 10 minutes of activation at 5×10^{11} neutrons/cm.^{2}/sec. and 20 minutes of decay. The cut-out paper area is immersed in phosphor solution and measured in the Packard Tri-Carb liquid scintillation counter.[15]

[14] H. W. Yurow, Neutron Activation Analysis of Chromatographically Separated Elements on Paper. Thesis, The Pennsylvania State University, 1960; J. M. Steim, Neutron Activation Chromatographic Analysis, Thesis, The Pennsylvania State University, 1962; A. A. Benson, W. W. Miller, and J. M. Steim, Biochemical Application of Neutron Activation Chromatographic Analysis. *Proc. 5th Japan Conf. on Radioisotopes, Tokyo, May, 1963;* J. M. Steim and A. A. Benson, *Arch. Biochem. Biophys.,* in press (1963).
[15] Packard Instrument Co., Inc., La Grange, Illinois.

[123] Histochemical Methods for Dehydrogenases

By Arnold M. Seligman

Selection of Tetrazolium Salt

Although a histochemical method for the succinic dehydrogenase system was proposed with 2,2′,5,5′-tetraphenyl-3,3′-(3,3′-dimethoxy-4,4′-biphenylene) ditetrazolium chloride (BT)[1] and later modified by the use of other tetrazolium salts,[2,3] all the requirements for a satisfactory histochemical method were not fulfilled until the agent 2,2′-di-*p*-nitrophenyl-5,5′-diphenyl-3,3′-(3,3′-dimethoxy-4,4′-biphenylene) ditetrazolium chloride (Nitro-BT) had been synthesized[4,5] and made available.[6] This tetrazole has a more favorable redox potential[7] and competes successfully with oxygen in very thin frozen sections because it accepts electrons earlier in the chain of electron transport than the other tetrazoles.[8] Furthermore, on reduction it yields a diformazan that is exceptionally insoluble in lipid and exhibits substantive properties for tissue protein, thus markedly decreasing any tendency to crystal formation in the tissue sections.[9] Artifacts of diffusion into lipid are eliminated, and the possibility of dehydrating and clearing the sections in order to permit mounting in balsam is provided.[9] With this agent or one of its iodine-containing isomers,[5] it is possible to demonstrate dehydrogenase activity in intracellular organelles, such as mitochondria, not only by light microscopy[9-11] but by electron microscopy as well.[5] Recently developed tetrazoles which

[1] A. M. Seligman and A. M. Rutenburg, *Science* **113**, 317 (1951).

[2] H. A. Padykula, *Am. J. Anat.* **91**, 107 (1952).

[3] B. Pearson and V. Defendi, *J. Histochem. and Cytochem.* **2**, 248 (1954).

[4] K. C. Tsou, C. S. Cheng, M. M. Nachlas, and A. M. Seligman, *J. Am. Chem. Soc.* **78**, 6139 (1956).

[5] S. S. Karmarkar, R. J. Barrnett, M. M. Nachlas, and A. M. Seligman, *J. Am. Chem. Soc.* **81**, 377 (1959).

[6] May be obtained from Dajac Laboratories, Chemical Division, The Borden Company, 5000 Langdon Street, Philadelphia 24, Pennsylvania.

[7] S. S. Karmarkar, A. G. E. Pearse, and A. M. Seligman, *J. Org. Chem.* **25**, 575 (1960).

[8] M. M. Nachlas, S. Margulies, and A. M. Seligman, *J. Biol. Chem.* **235**, 2739 (1960).

[9] M. M. Nachlas, K. C. Tsou, E. DeSouza, C. S. Cheng, and A. M. Seligman, *J. Histochem. and Cytochem.* **5**, 420 (1957).

[10] M. M. Nachlas, D. G. Walker, and A. M. Seligman, *J. Biophys. Biochem. Cytol.* **4**, 29 (1958).

[11] M. M. Nachlas, D. G. Walker, and A. M. Seligman, *J. Biophys. Biochem. Cytol.* **4**, 467 (1958).

yield formazans that chelate with cobalt and other metals have also been used for this purpose with some success.[7, 12-14]

Preparation of Tissue

Small blocks or slices of tissue 3 to 5 mm. thick were either placed in 10 to 20% neutral formalin[15] at 5° for 15 minutes or rapidly frozen by immersion in isopentane brought to −70° with a mixture of acetone and dry ice.[9] The blocks were then placed in the cryostat (−20°) for sec-

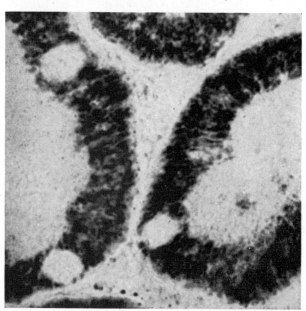

FIG. 1. Proximal convoluted tubule of rat kidney, fixed for 15 minutes in 10% formalin at 5°. Frozen sections were incubated in the medium for succinic dehydrogenase for 25 minutes. Linearly arranged accumulations in rodlike mitochondria may be seen extending the full length of the cell on either side of the reactionless nuclei. × 1500.

tioning. Sections 3 to 6 μ thick were cut with the rotary microtome and mounted on untreated clean glass slides. Care was taken to stain the sections promptly to avoid the damaging effects of drying. Although rapidly frozen tissue gave good tissue localization, superior preservation of mito-

[12] A. G. E. Pearse, *J. Histochem. and Cytochem.* **5**, 515 (1957).
[13] D. G. Scarpelli, R. Hess, and A. G. E. Pearse, *J. Biophys. Biochem. Cytol.* **4**, 747 (1958).
[14] R. Hess, D. G. Scarpelli, and A. G. E. Pearse, *J. Biophys. Biochem. Cytol.* **4**, 753 (1958).
[15] D. G. Walker and A. M. Seligman, *J. Biophys. Biochem. Cytol.* **9**, 415 (1961).

chondria was obtained by treating tissue slices 1 mm. thick with cold formalin for a brief period prior to preparing frozen sections.[15] The sectioning and staining should be done within 24 hours. No obvious inhibi-

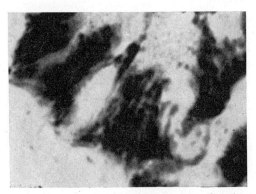

Fig. 2. Same as Fig. 1. Photographed through an orange filter. ×2000.

tion of the dehydrogenases was observed with this procedure.[15] Figures 1 and 2 illustrate mitochondria in the proximal convoluted tubules of the rat kidney stained for succinic dehydrogenase. The kidney was treated with cold formalin before frozen sections were made.

Succinic Dehydrogenase System[9]

Reagents

Phosphate buffer (0.20 M), pH 7.6, 1.0 ml.
Sodium succinate (0.20 M), 1.0 ml.
Nitro-BT (1 mg./ml.), 2.0 ml.

Procedure. A stock solution of buffered succinate could be stored in the refrigerator for months. This solution was prepared by combining equal volumes of phosphate buffer and sodium succinate. Into a Coplin jar containing 10 ml. of the buffered succinate solution and 10 ml. of the Nitro-BT solution were placed four to six mounted sections. The incubation was conducted at 37° for 5 minutes to an hour, depending on the activity of the tissue and the density of pigment production desired. For study of mitochondria it was important not to incubate too long (15 to 30 minutes). The sections were rinsed in 0.85% saline for 1 minute and mounted in glycerol–gelatin or dehydrated in graded concentrations of alcohol and mounted in Canada balsam or Permount. Counter-staining cell nuclei of some sections with 1% aqueous saffranin O solution was found to be helpful in studying the sections. Complete inhibition of enzymatic activity was produced by heating to 80° for 1 hour or by prior immersion in iodoacetate ($10^{-2}\,M$) for 30 minutes.

Procedure for Electron Microscopy.[5] Small blocks of tissue were either washed in $0.44\,M$ sucrose, frozen, and thawed, or treated with cold 10% formalin for 15 minutes. The blocks were then incubated in the phosphate-buffered succinate–Nitro-BT medium for 30 to 60 minutes and were then fixed briefly in buffered osmium tetroxide solution, embedded in *n*-butyl methacrylate catalyzed with α,α-azodimethyliso-butyronitrile. Sections $0.2\,\mu$ thick were cut with a glass knife in a Porter-Blum microtome, and the sections were viewed with an electron microscope.

Diphosphopyridine Nucleotide Diaphorase[10]

Reagents

Phosphate buffer ($0.2\,M$), pH 7.4, 1.0 ml.
Nitro-BT (1.5 mg./ml.), 1.0 ml.
$\begin{cases} \text{DPN (5 mg./ml.), 0.3 ml.} \\ \text{Sodium lactate } (0.5\,M),\ 0.5\ \text{ml.} \\ \text{Lactic dehydrogenase } (1.5\%),\ 0.2\ \text{ml.} \end{cases}$
    or
DPNH (6 mg./ml.), 1.0 ml.

DPN-Linked Dehydrogenases[10]

Reagents

Sodium lactate ($0.5\,M$), 0.6 ml.
Sodium malate ($2.5\,M$), 0.3 ml.
Sodium β-hydroxybutyrate ($2.0\,M$), 0.6 ml.
Ethanol (20%), 0.15 ml.
Sodium glutamate ($0.5\,M$), 0.6 ml.

One of these substrates, adjusted to pH 7.4, was added to the following medium: phosphate buffer ($0.2\,M$), pH 7.4, 1.0 ml.; Nitro-BT (1.5 mg./ml.), 1.0 ml.; DPN (5 mg./ml.), 0.3 ml.; and substrate and water to make 3.0 ml.

Although each dehydrogenase is capable of reducing the tetrazolium salt only via DPN diaphorase, some of the dehydrogenases have a more limited distribution than that of DPN diaphorase, allowing for a measure of selective demonstration.

Procedure. The stock solution of Nitro-BT was adjusted to pH 7.4 and could be stored at 4° for 2 to 3 months. The solutions of DPN and the dehydrogenases were made up in small quantities without pH adjustment and stored for as long as a month at —20°. If DPNH is preferred to the lactic dehydrogenase–lactate and DPN medium, it is prepared fresh. Sections were mounted on half cover slips so that the incubations could be carried out in a small volume of reagents. To ensure adherence of the sections to the cover slip, a jet of air was directed over the mounted section for about 5 seconds just before it was inserted into the incubation solution. The sections were incubated aerobically at room temperature for 5 to 30 minutes. They were then rinsed briefly in 0.85% sodium chloride solution, fixed in formalin for 10 minutes, and mounted in glycerol–gelatin or dehydrated through graded alcoholic solutions, xylol, and mounted in Canada balsam or Permount. Control sections were prepared by immersion in boiling water for 5 minutes, or the fresh sections were incubated in medium devoid of substrate.

Steroid 3β-ol Dehydrogenase[16, 17]

Reagents

Dehydroepiandrosterone (0.2 mg./ml. of acetone).
Propylene glycol, 1.0 ml.
Nitro-BT (1.0 mg./ml.), 0.7 ml.
Nicotinamide (1.6 mg./ml.), 0.7 ml.
DPN (3 mg./ml.), 0.8 ml.
Phosphate buffer (0.1 M), pH 7.4, 4.0 ml.

Procedure. The acetone solution of dehydroepiandrosterone was evaporated to dryness at 37°, and the other constituents were added. Fresh-frozen sections were first placed for 5 minutes in 0.1 M phosphate buffer (pH 7.4) at room temperature to remove endogenous substrates. Incubation in the above medium was carried out at 37° for 5 to 60 minutes. Control sections were incubated in the same mixture lacking the substrate. After incubation, the sections were fixed for 30 minutes in a mixture containing 50% ethanol and 10% formalin, then mounted with glycerol–gelatin.

Triphosphopyridine Nucleotide Diaphorase[11]

Reagents

Veronal buffer (0.05 M), pH 7.4, 1.1 ml.
Nitro-BT (5 mg./ml.), 0.3 ml.

[16] L. W. Wattenberg, *J. Histochem. and Cytochem.* **6,** **225** (1958).
[17] H. Levy, H. W. Deane, and B. L. Rubin, *Endocrinology* **65,** 932 (1959).

Manganese chloride (0.005 M), 0.3 ml.

$\begin{cases} \text{TPN (5 mg./ml.), 0.2 ml.} \\ \text{Sodium DL-isocitrate (0.1 } M\text{), 0.6 ml.} \\ \text{Sodium L-malate (2.5 } M\text{), 0.5 ml.} \end{cases}$

or

TPNH (5 mg./ml.), 1.3 ml.

Procedure. Stock solutions were stored at 4°, and enzyme preparations were stored at —20°. In order to conserve reagents, two sections at a time were mounted on a half cover slip and incubated in a test tube containing enough medium to cover them. A reagent mixture prepared in the morning was used throughout the day for repeated incubations with as many as ten sections.

Both substrates, DL-isocitrate and L-malate, were used in the incubation medium to ensure reduction of TPN to TPNH in all possible sites of TPN diaphorase activity. The alternative of using a soluble TPN-linked dehydrogenase and its substrate was not possible because of inavailability of pure enzyme. The availability of TPNH makes either alternative unnecessary. With TPNH the procedure is convenient, although expensive.

Sites of dehydrogenase activity are better accredited to a specific TPN-linked dehydrogenase rather than to TPN diaphorase, when they are sharply limited to special areas of localization with certain substrates. This has been shown to be the case with 6-phosphogluconic dehydrogenase which is located only in the macula densa of the cortex of rat kidney.[11] TPN diaphorase has a widespread localization in the cortex. Optimal activity was obtained when the barium salt of 6-phosphogluconic acid, at 5.0 mg./ml., was added to the medium instead of both isocitrate and malate.

Note added in proof: Since this section was submitted, further improvement in methodology has been published by Walker and Seligman.[18]

[18] D. G. Walker and A. M. Seligman, *J. Cell Biol.* **16**, 455 (1963).

[124] Applications of Nuclear Magnetic Resonance and Electron Spin Resonance to Enzymology

By LUTHER E. ERICKSON and ROBERT A. ALBERTY

Introduction

The discovery of electron spin resonance (ESR) in 1945[1] and nuclear magnetic resonance (NMR) in 1946[2] opened up the possibility of solving a number of old problems and of answering new types of questions. A broad range of phenomena has been investigated, and it is impossible to treat all the applications of ESR and NMR adequately here. Therefore, this discussion will be limited to high-resolution NMR[3] and the ESR[4] of free radicals, these being the areas of chief interest in biochemical applications.

I. Nuclear Magnetic Resonance

A. Basis of Method

Resonance Condition. In common with other spectroscopic techniques, NMR spectroscopy depends on the absorption of certain frequencies of radiation by molecules resulting in transitions from one energy level to another. Certain nuclei have a spin angular momentum as a result of rotation about their own axes. Associated with this spin angular momentum is a magnetic moment. Consequently, in a magnetic field these nuclei behave like magnets. In contrast to macroscopic magnets, which always tend to point in a particular direction when placed in a magnetic field, these nuclear magnets have more than one allowed orientation with respect to the direction of the magnetic field. Since the energy of the nuclear magnet in a magnetic field is given by the product of the field strength, H_0, and the component of the magnetic moment in the direction

[1] E. K. Zavoisky, *J. Phys. (U.S.S.R.)*, **9**, 447 (1945).

[2] E. M. Purcell, H. C. Torrey, and R. V. Pound, *Phys. Rev.* **69**, 37 (1946); F. Block, W. W. Hansen, and M. E. Packard, *ibid.* **70**, 474 (1946).

[3] For a more complete discussion, see J. A. Pople, W. G. Schneider, and H. J. Bernstein, "High Resolution Nuclear Magnetic Resonance." McGraw-Hill, New York, 1959; J. D. Roberts, "Nuclear Magnetic Resonance." McGraw-Hill, New York, 1959; J. E. Wertz, *Chem. Rev.* **55**, 904 (1955).

[4] Also referred to as electron paramagnetic resonance (EPR). For a more thorough treatment, see D. J. E. Ingram, "Free Radicals as Studied by Electron Spin Resonance." Academic Press, New York, 1958; J. E. Wertz, *Chem. Rev.* **55**, 904 (1955).

of the field, the energies will differ for the different orientations. Transitions between these energy levels (Zeeman levels) give rise to the NMR absorption.

The nuclear spin, I (which is characteristic of the nuclear species like atomic number and atomic weight), can always be represented by some integer multiple of $\frac{1}{2}$. The maximum component of spin angular momentum in the direction of the applied field is $I\hbar$, where $\hbar$ is Planck's constant divided by 2π. For a nucleus of spin $\frac{1}{2}$ (for example, H^1, F^{19}, and P^{31}), there are only two orientations for the magnetic moment with respect to an applied magnetic field. Therefore, there are only two energy levels, and the NMR absorption results from transitions between these two levels.

In general, for a nucleus of spin I, there are $(2I+1)$ values for the component of spin angular momentum along any fixed axis, and these are given by $I\hbar$, $(I-1)\hbar$, . . . $(-I+1)\hbar$, $-I\hbar$. Corresponding to these values for the spin angular momentum are $(2I+1)$ values for the magnetic moment given by $m\mu/I$, where m, the magnetic quantum number, can take on values I, $I-1$, . . . , $-I+1$, $-I$, and μ is the maximum measurable component of the magnetic moment in the direction of the applied field. In a magnetic field of strength H_0, the energy of a magnetic moment, whose component in the direction of the field is μ, is given by $-\mu H_0$. For a nucleus of spin I, there result $2I+1$ energy levels given by

$$-\mu H_0, \ (-I+1)/I\,\mu H_0 \ldots,$$

$$(I-1)/I\,\mu H_0, \mu H_0$$

In the NMR experiment, transitions occur only between states for which $\Delta m = \pm 1$. Applying the Bohr frequency condition, $\Delta E = h\nu$, we have

$$\nu_0 = \mu H_0/Ih \tag{1}$$

where ν_0 is the resonance frequency in cycles per second. Since μ and I have characteristic values for each nuclear species, it is evident that the resonance frequency will be different for each nuclear species and will be proportional to the strength of the applied magnetic field. For example, in a magnetic field of 10 kilogauss, the proton resonance occurs at about 42.577 Mc. (Mc. = megacycles per second) and the N^{14} resonance occurs at about 3.076 Mc.; in a 5-kilogauss field the resonance frequencies for H^1 and N^{14} are about 21.288 and 1.538 Mc., respectively.

Properties of Representative Nuclei. Nuclei whose spins are zero (C^{12}, O^{16}, and all other nuclei having an even number of protons and neutrons) have no magnetic moment and hence cannot be studied by nuclear magnetic resonance. The nuclear properties of several commonly encountered isotopes whose spins are not zero are summarized in Table I. Proton and

TABLE I[a]
PROPERTIES OF SOME COMMONLY ENCOUNTERED NUCLEI

Isotope	NMR frequency for a 10-kilogauss field	Spin	Magnetic moment, μ, in nuclear magnetons	Relative sensitivity for equal numbers of nuclei at constant frequency
H^1	42.577	1/2	2.79270	1.000
H^2	6.536	1	0.85738	0.409
B^{11}	13.660	3/2	2.6880	1.60
C^{13}	10.707	1/2	0.70216	0.251
N^{14}	3.076	1	0.40357	0.193
O^{17}	5.772	5/2	-1.8930	1.58
F^{19}	40.055	1/2	2.6273	0.941
P^{31}	17.235	1/2	1.1305	0.405
Cl^{35}	3.266	3/2	0.82089	0.490

[a] Taken from complete table published by Varian Associates.

fluorine NMR spectra have been most frequently investigated. Although the detection of oxygen requires the rare isotope O^{17} and the detection of carbon requires C^{13}, P^{31}, the common isotope of phosphorus, is readily detected. The illustrations in the following sections involve proton spectra. With appropriate generalizations and the data in Table I, the conclusions reached can readily be extended to other nuclei.

General Features of NMR Spectra. NMR spectra are usually obtained by linearly varying the magnetic field strength of an external magnetic field through a very small range in the neighborhood of H_0, while the oscillator frequency is maintained at $\mu H_0/Ih$. A plot of intensity of resonance signal versus field strength is obtained.

The resonance frequency is determined by the magnetic field strength *at the nucleus* which depends on (1) the field strength of the externally applied magnetic field, (2) the shielding by orbital electrons, and (3) the contributions from other magnetic nuclei in the molecule. As a result of factors 2 and 3, the same nuclear species in different molecular environments exhibit resonance at *slightly* different applied magnetic field strengths. Since frequencies in the megacycle range can be measured with such high accuracy, and since means have been developed for producing very homogeneous and stable magnetic fields, a difference in resonance fields of 0.01 p.p.m. can be determined with sharp resonances.

The effect of diamagnetic shielding by orbital electrons which results in slightly different resonance fields for nuclei of the same species in different environments is called the *chemical shift*.[5] For example, the methyl protons of acetaldehyde exhibit resonance at a higher field strength than the aldehyde at a given frequency. The difference in the strength of the applied field required to produce resonance for the two types of proton is proportional to the oscillator frequency.

The effect of other magnetic nuclei in the molecule in determining the magnetic field strength at the nucleus being observed is referred to as *spin-spin splitting* or *spin coupling*. The magnetic effect of other nuclei is not transmitted directly when molecules are in rapid Brownian motion but is transmitted by the orbital electrons and ordinarily cannot be detected between nuclei more than three or four bonds apart.[6,7] In contrast to the chemical shift, the magnitude of spin-spin splitting is *independent* of oscillator frequency.

The effects of the chemical shift and spin-spin splitting are clearly evident in the proton NMR spectrum of acetaldehyde shown in Fig. 1.

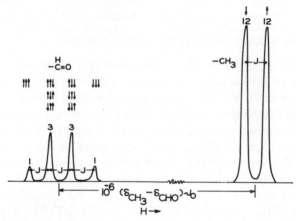

Fig. 1. Proton magnetic resonance spectrum of acetaldehyde. The numbers over the peaks represent the relative areas.

The less shielded proton of the aldehyde group is responsible for the quartet at lower fields, whereas the more shielded methyl protons give rise to the doublet at higher fields. The resonance pattern in each region is a multiplet because of the interaction of methyl and aldehyde protons (spin-spin splitting). The quantitative treatment of these effects will be considered under Determination of Spectra. It should be emphasized that the range in magnetic field strength covered in the whole spectrum is only 10^{-5} of the average absolute field strength.

[5] W. D. Knight, *Phys. Rev.* **76**, 1259 (1949).
[6] H. S. Gutowsky, D. W. McCall, and C. P. Slichter, *Phys. Rev.* **84**, 589 (1951).
[7] E. L. Hahn and D. E. Maxwell, *Phys. Rev.* **84**, 1286 (1951).

B. Experimental Procedure

Equipment. Although it is not the only way to observe nuclear magnetic resonance, the crossed-coil induction technique originally used by Block[2] is most commonly used, since it is the basis of the high-resolution spectrometers available from Varian Associates, Palo Alto, California. A block diagram of the spectrometer is shown in Fig. 2. The sample

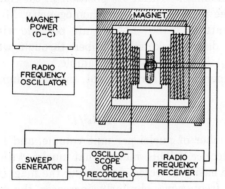

FIG. 2. Block diagram of NMR spectrometer.

solution, contained in a sealed glass tube, is placed in the gap between the pole pieces of a large electromagnet. The sample tube fits into a probe which contains transmitter and receiver coils wound around the sample so that their axes and the direction of the applied field are mutually perpendicular. A constant radiofrequency signal is supplied to the transmitter coil by a crystal-controlled oscillator. A linearly varying output from a "sweep" generator is applied to coils wound around the magnet pole pieces in order to vary the magnetic field. When the magnetic field is such that the resonance condition, Eq. (1), is satisfied for a certain type of nucleus in the sample, a voltage change is induced in the receiver coil. This signal is amplified and presented on an oscilloscope or a recorder.

Since the differences in resonance fields for the different nuclei of the same species in a sample are very small, extremely high resolution is necessary to separate the individual lines in the spectrum. The resolution obtainable is usually limited by the homogeneity of the magnetic field through the sample. The resolution can be considerably improved by rotating the sample tube at several hundred revolutions per minute with an air turbine so that a given nucleus experiences an average field.[8] Small samples (normally 0.5 ml. in a 5-mm. o.d. tube) are used, since these require a smaller region of constant magnetic field strength.

[8] W. A. Anderson and J. T. Arnold, *Phys. Rev.* **94**, 497 (1954).

Various "cycling" procedures, in which the magnetic field is taken to an appropriate higher level and returned to the resonance region, are used to obtain the desired homogeneity in the magnetic field within the probe containing the sample. Additional control can be effected by mechanical adjustment of the pole pieces and by careful control of the temperature of the magnet assembly.

The probe is mounted so that its position in the magnet gap can be varied to locate the region of optimum homogeneity. Measurements over a range of temperatures ($-60°$ to $20°$) are possible with special probes equipped with a dewar-jacketed chamber through which heated (or cooled) air (or N_2) is passed over the sample.[9]

For a given experimental arrangement, the maximum signal intensity is proportional to

$$N(I + 1)\mu^3 H_0^2/I^2 \qquad (2)$$

where N is the number of nuclei per unit volume, and I, μ, and H_0 have their usual significance.[10] This means that nuclei with large magnetic moments are more easily detected. Increasing the magnetic field strength (and transmitter frequency) similarly increases the sensitivity. Operating at higher frequency has the added advantage of increasing chemical shifts, making interpretation of the resulting spectra simpler in many cases. However, the difficulty of maintaining the necessary homogeneity in large magnetic fields limits the practical working range to magnetic fields up to about 14,000 gauss where proton resonances occur at about 60 Mc.

Sample Preparation. From Formula (2) it is evident that the signal intensity depends on the number of nuclei present. The size of the sample is limited to about 1 ml. by the difficulty of maintaining an adequately homogeneous field over larger samples. Therefore, it is desirable to use a pure liquid sample or to use as high a concentration as possible for solutions of solids.

Several restrictions limit the choice of solvent considerably. The compound must be soluble to the extent of about $0.05 M$ for detection of a

[9] J. N. Shoolery and J. D. Roberts, *Rev. Sci. Instr.* **28**, 61 (1957).
[10] Formula (2) is applicable only under conditions where $T_1 = T_2$ (usually true for liquids) and for low transmitter power where saturation is not significant; T_1, the transverse or spin-lattice relaxation time, is related to the rate at which the spin system attains its equilibrium distribution (with a slight excess in lower energy states) when placed in a magnetic field; T_2, the longitudinal or spin-spin relaxation time, gives a measure of the line width. In solids the line width is determined largely by magnetic interaction among neighboring nuclear spins, but the line widths in liquid samples are usually determined by magnetic field inhomogeneities.

single line due to one proton per molecule at 40 Mc. For other nuclei and different frequencies, the necessary concentration can be estimated from Formula (2) and Table I. When several nuclei give rise to a single line (same chemical shifts), the necessary molar concentration is decreased proportionally. Conversely, if the signal from a single nucleus is split into several components, a higher concentration is required. For the highest resolution a concentration of the order of 1 to 2 M is desirable for the study of proton resonances at 40 Mc. At very high concentrations high viscosities often result in line broadening.

Since the solvent is always present in larger amounts than the solute, it should have no nuclei of the species being investigated or should be chosen so that its resonance signals occur at quite different magnetic field strengths. In proton resonance investigations, deuterated solvents are commonly employed, since the deuterium resonance frequency is about one-sixth the proton resonance frequency.

Determination of Spectra. In order to interpret NMR spectra quantitatively, measurements of the line separations and their relative intensities are needed. In comparing chemical shifts of different compounds, it is also necessary to measure the positions of the lines relative to some standard.

The most satisfactory method of measuring line separations is the side-band technique of Packard and Arnold.[11] By modulating the applied magnetic field with an audiofrequency low-amplitude signal, side bands are obtained on either side of the resonance signal. The frequency separation of the side bands from the original signal is just the modulating frequency which can be measured accurately with a frequency counter. By assuming a linear sweep of the magnetic field, this separation can be used to calibrate the chart recording. For greater accuracy, it is desirable to adjust the audiofrequency so as to overlap successively several lines in the NMR spectrum. With this technique, line separations can be measured to better than 1 c.p.s.

When resonance fields for different compounds (at constant frequency) are compared, the fields must be measured relative to some standard compound which gives a single easily recognized peak. Either an internal standard in the solution itself or an external standard contained in a capillary within the sample tube or in the annular space between coaxial tubes can be used. Internal reference standards have the advantage that they experience the same magnetic field as nuclei of the sample. External reference standards, on the other hand, will experience

---

[11] M. E. Packard and J. T. Arnold, *Phys. Rev.* **83**, 210A (1951).

a somewhat different field from the sample because of differences in the bulk susceptibility of the sample and reference compound.[12]

Relative intensities are normally measured by the areas under peaks. Measurements can seldom be made to better than 5 to 10% in routine work.

C. Interpretation of Spectra

Intensities as Related to Structure. In contrast to optical frequency spectroscopy (ultraviolet, visible, or infrared) in which the relative intensities of the different absorption bands differ widely and are not simply related, in NMR spectroscopy the relative intensities of different lines (or groups of lines) are directly proportional to the number of nuclei responsible for these lines. For example, in Fig. 1 the total area of the groups of lines due to CH_3 is three times as great as that due to the proton in CHO. When D_2O is added to fumarate in the fumarase-catalyzed hydration to L-malate, the proton NMR spectrum of the resulting L-malate shows two absorption regions (each consisting of a doublet) of equal intensity. This indicates that the hydration is stereospecific, since the methine proton and one proton at the methylene position of the L-malate are unaffected by repeated dehydration and rehydration.[13]

Chemical Shift Scale. The chemical shift is measured quantitatively by the difference in resonance fields (at constant frequency) for nuclei of type i and the nuclei of some reference group, $H_i - H_{ref}$. However, since the resonance fields are proportional to the oscillator frequency and will have different numerical values for different frequencies, chemical shifts are usually expressed (in parts per million) in terms of the dimensionless parameter, δ_i, defined by Eq. (3), where $\nu_i - \nu_{ref}$ is the fre-

$$\delta_i = 10^6(H_i - H_{ref})/H_0 = 10^6(\nu_i - \nu_{ref})/\nu_0 \qquad (3)$$

quency difference (measured by the side-band technique), and $\nu_0 \approx \nu_i$ or ν_{ref} is the oscillator frequency. The factor 10^6 makes proton chemical shifts of the order of unity. With this definition, the chemical shift is independent of frequency. Large chemical shifts indicate a high degree

---

[12] In accurate work the following correction should be made for chemical shifts measured with respect to an external reference compound.

$$\delta_{corr} = \delta_{obs} + 2\pi/3 \ (\chi_{V_{ref}} - \chi_V)$$

where δ_{corr} and δ_{obs} are the corrected and observed chemical shifts in parts per million and $\chi_{V_{ref}}$ and χ_V are the volume susceptibilities of the reference compound and sample. By extrapolating δ_{obs} to infinite dilution, χ_V of the solvent can be used. See W. C. Dickinson, *Phys. Rev.* **81,** 717 (1951).

[13] R. A. Alberty and P. Bender, *J. Am. Chem. Soc.* **80,** 542 (1958); R. A. Alberty, W. G. Miller, and H. F. Fisher, *ibid.* **79,** 3973 (1957).

of shielding of the nucleus by orbital electrons, and small chemical shifts indicate less shielding.

Several workers have attempted to correlate proton chemical shifts with the functional groups in which the protons are found so that these groups may be identified in the NMR spectrum. Table II gives typical

TABLE II[a]

TYPICAL PROTON CHEMICAL SHIFTS FOR PURE LIQUID COMPOUNDS

Compound	Parts per million
$-SO_3H$	-6.7 ± 0.3[b]
$-CO_2H$	-6.4 ± 0.8
RCHO	-4.7 ± 0.3
RCONH$_2$	-2.9
ArOH	-2.3 ± 0.3
ArH	-1.9 ± 1.0
$=CH_2$	-0.6 ± 0.7
ROH	-0.1 ± 0.7
H_2O	0
$-OCH_3$	$+1.6 \pm 0.3$
$-CH_2X$	$+1.7 \pm 1.2$
$\equiv C-H$	$+2.4 \pm 0.4$
$=C-CH_3$	$+3.3 \pm 0.5$
$-CH_2-$	$+3.5 \pm 0.5$
RNH$_2$	$+3.6 \pm 0.7$
$\overset{\mid}{\underset{\mid}{-C-CH_3}}$	$+4.1 \pm 0.6$

[a] Taken from data of L. H. Meyer, A. Saika, and H. S. Gutowsky, *J. Am. Chem. Soc.* **75**, 4567 (1953).

[b] The indicated ranges include the several compounds containing that group which were studied.

proton chemical shifts (measured with respect to H_2O) obtained for a series of pure liquid compounds.[14] In general, the more acidic (and less shielded) protons have smaller chemical shifts, but there are notable exceptions such as the aldehyde and acetylenic protons. It should be emphasized that caution is necessary in associating particular signals with a given functional group on the basis of chemical shifts alone, especially if the compounds are not very similar. For example, in AlCl$(CH_2CH_3)_2$, we have $\delta_{CH_3} < \delta_{CH_2}$, whereas in the compounds summarized in Table II, we have $\delta_{CH_2} < \delta_{CH_2}$.[15]

[14] G. V. D. Tiers has compiled a table of proton chemical shifts measured with respect to $(CH_3)_4Si$ as an internal standard. This eliminates any necessity of making any susceptibility correction. See G. V. D. Tiers, *J. Phys. Chem.* **62**, 1151 (1958).

[15] E. B. Baker, *J. Chem. Phys.* **26**, 960 (1957).

Much larger chemical shifts (up to 100 to 1000 p.p.m.) are often observed for C^{13}, N^{14}, F^{19}, and P^{31}.

Spin-Spin Coupling. For cases in which the different groups of nuclei have widely different chemical shifts (compared with spin-spin coupling constants), the effects of spin-spin coupling of neighboring nuclei are readily interpreted. In the case of acetaldehyde, the spacing between the two components of the CH_3 doublet is equal to the separations of the four lines of the CHO quartet and is equal to 2.85 c.p.s., which is referred to as the spin-spin coupling constant, J.

In general, to a first approximation the number of components arising from spin-spin interaction with n nuclei of spin I is given by the number of unique ways of permuting the n nuclei among their $(2I + 1)$ spin states. Relative intensities of the different components depend on the number of identical permutations. Thus, the CHO proton signals of acetaldehyde consist of a quartet with relative intensities of $1:3:3:1$ because of the eight possible permutations of the two spin states of the three protons; two groups of three are identical, since the three CH_3 protons are identical in their coupling with the CHO proton. This is shown in Fig. 2, which also emphasizes that the total intensity of the two CH_3 peaks is three times that of the four CHO peaks.

This simple analysis holds as long as the relative chemical shifts are much greater than the spin-spin coupling constants (both expressed in cycles per second) and complications due to time averaging (next section) are not present. When the spin-spin coupling constant and chemical shift are more nearly equal, as is often true, the observed spin-spin multiplets deviate significantly from those predicted by these simple rules. However, the expected spectra may be calculated provided that the number of nuclei with spin is not too large.[16]

Figure 3 illustrates the calculated spectra for a system of two non-equivalent nuclei (A and B) of spin ½ for different values of the ratio $J_{AB}/10^{-6}(\delta_A - \delta_B)\nu_0$, where J_{AB} is the spin-spin coupling constant in cycles per second, and $(\delta_B - \delta_A)$ is the chemical shift difference in parts per million. When $J_{AB} \ll 10^{-6}(\delta_B - \delta_A)\nu_0$, the chemical shift for each nucleus corresponds to the center of the doublet with which it is associated, and the separation of the outer pairs is just J_{AB}. However, when J_{AB} is of the same order of magnitude as $10^{-6}(\delta_B - \delta_A)\nu_0$, the chemical shifts no longer correspond to the centers of each doublet, although the separation of the outer lines still equals J_{AB}. The intensities also deviate widely from the simple prediction. In any case the four-line spectrum is symmetrical (with respect to separations and intensities) about the

[16] For a complete discussion see Chapter 6 of J. A. Pople *et al.*[3]

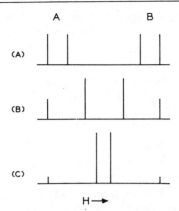

FIG. 3. Variation in the NMR spectrum of two nonequivalent nuclei, A and B, of spin ½ as a function of $J_{AB}/10^{-6}(\delta_B - \delta_A)\nu_0$. (A) $J_{AB} \ll 10^{-6}(\delta_B - \delta_A)\nu_0$. (B) $J_{AB} = 10^{-6}(\delta_B - \delta_A)\nu_0$. (C) $J_{AB} \gg 10^{-6}(\delta_B - \delta_A)\nu_0$.

average of the chemical shifts, and the central pair of peaks are separated by[7]

$$\{[10^{-6}(\delta_B - \delta_A)\nu_0]^2 + J_{AB}{}^2\}^{\frac{1}{2}} - J_{AB}$$

Effect of Rapid Chemical Interconversion. When two forms of a substance in solution are interconverted with a frequency much greater than the chemical shifts and/or coupling constants, the observed spectrum will be intermediate between that of the two forms. The transition between discrete spectra for the two components and a single average spectrum occurs when

$$\tau \approx (\tfrac{1}{2\pi})\Delta\nu_0 \tag{4}$$

where τ is the mean lifetime of the nuclei in one of the forms, and $\Delta\nu_0$ is the frequency separation (in cycles per second) of the lines for the two components in the absence of the interconversion.

For example, if the absorption lines of two forms are separated by 100 c.p.s., a chemical reaction interconverting them with a half-life of much less than 0.01 second will cause the observed spectrum to be an average spectrum. Thus for acids in H_2O only a single proton line is observed. Its chemical shift is a weighted average of the chemical shift of the H_2O protons and those of the acid.[17]

Between the extremes illustrated by acetaldehyde (discrete lines for CHO and CH_3 protons) and water–acid mixtures (a single average line) lie many systems where the exchange rate can be varied by varying the

[17] H. S. Gutowsky and A. Saika, *J. Chem. Phys.* **21**, 1688 (1953).

temperature or catalyst concentrations so as to produce a transition from discrete lines for each form to an average spectrum. With $\Delta\nu_0$, the separation of the components in the absence of exchange, known, the rate of the process can be estimated. This principle was used by Grunwald and co-workers[18] in their investigation of the rates and mechanism of proton transfer in aqueous amine solutions and in the following example from Ogg's work.[19]

The proton spectrum of liquid NH_3 consists of a triplet (due to spin-spin splitting by N^{14}, with $I = 1$). Trace amounts of H_2O or NH_2^- are sufficient to cause a rapid exchange of the protons resulting in a single line whose position is the average of the original NH_3 triplet. The transition from a triplet to a singlet occurs when the NH_2^- concentration is approximately equal to $10^{-7} M$ and $2\pi\Delta\nu = 46$ radians sec.$^{-1}$. Therefore $\tau = 1/46$ second. With a bimolecular mechanism assumed, the second-order rate constant for the proton transfer reaction is about 4.6×10^8 liter mole^{-1} sec.$^{-1}$.

Rates of internal rotation as well as rates of chemical exchange can be investigated by this technique. For example, the two methyl peaks in the *N,N*-dimethyl formamide proton spectrum (different chemical shifts) were observed to coalesce as the temperature was raised, indicating an increase in the rate of rotation about the C—N bond.[20]

When splitting in a spectrum is due to a type of nucleus other than that being studied, the spectrum may be simplified by use of a spin decoupler.[21] A spin decoupler is an additional radiofrequency source which is adjusted to the resonance frequency of the nuclei which are causing complications in the spectrum but which are not being observed directly in the experiment. The effect of this irradiation is to uncouple the spin coupling between the two types of nucleus and simplify the spectrum.

Isotopic Substitution. In order to identify chemical shifts with nuclei of particular molecular sites and spin-spin coupling constants with particular pairs of nuclei, it is often helpful to substitute another isotope which, having a different magnetic moment and spin, will not exhibit resonance in the same region of the spectrum. However, since the electronic configuration is essentially unaltered, the chemical shifts and spin-spin coupling constants for the remaining nuclei will remain virtually unchanged. Analysis of the pyridine spectrum was simplified by such a substitution of deuterium at the 4-position.[22]

[18] E. Grunwald, A. Lowenstein, and I. Meiboom, *J. Chem. Phys.* **27,** 630 (1957).
[19] R. A. Ogg, *Discussions Faraday Soc.* **17,** 215 (1954).
[20] W. D. Phillips, *J. Chem. Phys.* **23,** 1363 (1955).
[21] F. Block, *Phys. Rev.* **93,** 944 (1954).
[22] W. G. Schneider, H. J. Bernstein, and J. A. Pople, *Can. J. Chem.* **35,** 1487 (1959).

D. Range of Applications

Structure and Conformation. Since NMR spectra reveal the different types of nucleus (protons, for example) in a molecule, the numbers of these different types of nucleus, and their spin-spin coupling constants, very important structural deductions can be made. Some examples have already been cited.

It appears that, in special cases, subtle differences in spectra can be used to assign conformational isomers. In several substituted cyclohexanes and pyranose sugars, protons on adjacent carbons which are both axial (*trans*) give much larger coupling constants (two to three times as great) than proton pairs in other relative conformations.[23] Similarly, coupling constants between nuclei which are *trans* are greater than those between nuclei which are *cis* in substituted ethylenes. Theoretical calculations support these generalizations.[24]

These generalizations also apply where there is the possibility for many conformations but where a large fraction of the molecules are in one preferred conformation. The dipotassium monodeutero-L-malate (K_2D_cM) prepared by the fumarase-catalyzed addition of D_2O to fumarate provides a simple example. The spin-spin coupling constant for the two protons on adjacent carbons is 10.2 c.p.s. (at infinite dilution). For the monodeutero-L-malate which has a deuterium atom at the other methylene position (K_2D_bM), the spin-spin coupling constant is 2.9 c.p.s.[13] If the largest fraction of the malate ions are in the conformation with the negatively charged carboxyl group *trans*, this result can be taken as evidence that the two protons in K_2D_cM are *trans* rather than *gauche* (when the carboxyl groups are *trans*). Therefore, the addition of water to fumarate must occur by a *trans* mechanism.[25]

Association in Solution. Because the chemical shift is very sensitive to the electronic environment, the NMR spectrum will be strongly affected by any association in solution. For example, chemical shifts of hydrogen bonded protons are concentration- and temperature-dependent. Dilution with an inert solvent reduces the fraction of hydrogen-bonded species and shifts the resonance signal to higher fields. The amount by which the chemical shift differs from that in an infinitely dilute solution gives some measure of the extent of hydrogen bonding.

Similarly, investigations of the Na^{23} resonance in aqueous solutions

[23] R. U. Lemieux, R. K. Kullnig, H. J. Bernstein, and W. G. Schneider, *J. Am. Chem. Soc.* **80**, 6098 (1958).

[24] M. Karplus, *J. Chem. Phys.* **30**, 11 (1959).

[25] This conclusion, in contrast to the earlier deduction based on the proton NMR line width of solid monodeutero-L-malic acid, has been confirmed by the stereospecific synthesis of monodeutero-L-malic acid by O. Gawron and T. P. Fondy. *J. Am. Chem. Soc.* **81**, 6333 (1959).

of lactate, pyruvate, and citrate indicate that specific complexes are formed with Na^+ ion.[26] The concentration dependence of the proton spectrum of alkali metal malates also suggests ion pair formation.[27] Increasing the concentration of malate or addition of alkali metal chlorides has the same effect on chemical shifts and spin-spin coupling constants as decreasing the pH (which increases the fraction of the carboxyl groups which are protonated).

Extent of Reaction or Isotopic Substitution. In addition to measurement of rates of exchange reactions or of internal rotation by observation of the coalescence of spectral lines, NMR can be used as a conventional analytical method to determine the extent of isotopic substitution and to thereby measure reaction rates. Although not so accurate as mass spectrometric analysis, the NMR analysis has the advantage of indicating the location of the substitution without the necessity of selective degradation of the molecule. Rates of deuterium exchange at particular sites can be measured by observing the decrease in intensity of the proton lines associated with those sites. The following example illustrates the advantage of NMR in determining both the extent and position of deuterium exchange and the effect of the spin decoupler in simplifying spectra.

When completely deuterated dipotassium malate is heated in aqueous KOH, the methylene deuterons are exchanged for protons in the medium.[28] Figure 4 shows the proton spectra obtained for (a) malic acid (H_2M) recovered from the reaction mixture after partial exchange and dissolving in D_2O, and (c) the corresponding dipotassium malate (K_2M) dissolved in D_2O. The large peak at the extreme left (low field) is due to the protons in the medium which also spend part of their time at the carboxyl and/or hydroxyl position. Since the chemical shifts of the two methylene protons are nearly the same, only a single line is obtained for the methylene protons. In the salt form, where the chemical shifts of the two methylene protons differ significantly, four lines arising from the two-proton species K_2D_aM (the subscripts a, b, and c refer to the methine and two methylene positions, respectively) are evident in the methylene region. When complications due to unresolved, weak H-D spin-spin coupling are removed by the spin decoupler, the single lines arising from the one proton species $K_2D_aD_bM$ and $K_2D_aD_cM$ become clearly evident (d). The fact that the areas under the $K_2D_aD_bM$ and $K_2D_aD_cM$ lines are nearly equal indicates that the two methylene deuterons of malate exchange in aqueous KOH solutions at essentially equal rates.

[26] O. Jardetzky and J. B. Wertz, *Arch. Biochem. Biophys.* **65**, 569 (1956).
[27] L. E. Erickson, Ph.D. Thesis, University of Wisconsin, 1959.
[28] L. E. Erickson and R. A. Alberty, *J. Phys. Chem.* **63**, 705 (1959).

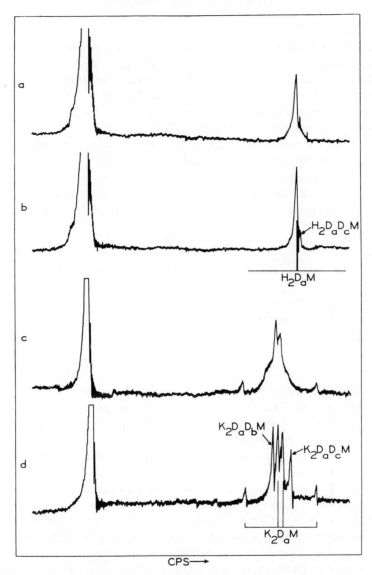

FIG. 4. Proton NMR spectra of D₂O solutions of completely deuterated malic acid and its dipotassium salt after partial exchange of methylene deuterons in aqueous KOH: (a) acid form, no spin decoupler; (b) acid form, with deuterium spin decoupler; (c) dipotassium salt, no spin decoupler; (d) dipotassium salt with deuterium spin decoupler.

II. Electron Spin Resonance

A. Basis of Method[4]

Resonance Condition. Just as nuclei which have magnetic moments exhibit a resonance absorption in a magnetic field, so also the magnetic moments associated with the unpaired electrons in paramagnetic sub- stances give rise to a similar absorption. In the magnetic field each unpaired electron precesses like a gyroscope at a frequency which de- pends on the magnetic field strength. The radiofrequency oscillation (which is in the microwave region for electron spin resonance for the magnetic field strengths usually used) is equivalent to a rotating mag- netic field, and when the frequency of the rotation coincides with that of the precession, energy is absorbed by the electrons. Formally, the resonance condition is the same [i.e., Eq. (5)], where S denotes the

$$\nu_0 = \frac{\mu}{Sh} H_0 \tag{5}$$

electron spin. However, this equation is usually written in terms of the gyromagnetic ratio, $g = \mu/S$. When μ, the magnetic moment, is expressed in units of the Bohr magneton, $eh/4\pi mc = \beta$, Eq. (5) can be written as Eq. (6). For a free electron for which $\mu = 1$ Bohr magneton and $S = \frac{1}{2}$, $g = 2.0023$.

$$\nu_0 = \frac{g\beta}{h} H_0 \tag{6}$$

Since the magnetic moment of an unpaired electron is much larger than that of a nucleus (658 times that of the proton), electron spin resonance (ESR)[4] is observed at a considerably higher frequency than NMR. In a magnetic field of 10 kilogauss, where the proton resonance is observed at 42.6 Mc., a free electron would give resonance at 28,000 Mc. Fields of this magnitude are required in order to get the desired resolu- tion. Thus, whereas NMR is observed in the radiofrequency region, ESR is most commonly observed at microwave frequencies.

Substances Which Can Be Studied by Electron Spin Resonance. Any substance containing an unpaired electron gives rise to ESR absorption. Included in this group are free radicals, odd molecules, transition ele- ments and rare-earth ions, color centers in alkali metal halides, and triplet states. Whereas for free radicals and electrons in metals g values are found to be very near the free electron value (2.0023), these other substances yield g values from 0.2 to 8, indicating large contributions of orbital motion to the magnetic moment. Measurements can be made with crystals, amorphous solids, or solutions. In the case of crystals the results will, in general, depend on the orientation of the crystal.

B. Experimental Procedure

Apparatus. It is possible to observe ESR at radiofrequencies with a magnetic field of the order of 10 gauss. Since ESR line widths are of the order of 10 gauss, measurement at higher field strength (of the order of 10 kilogauss), where the resonance field is many times as great as the line width, is necessary to achieve resolution of fine structure, and microwave frequencies must be used.

The experimental equipment resembles the NMR apparatus in Fig. 1 and is shown in Fig. 5. In both cases an electromagnet provides a ho-

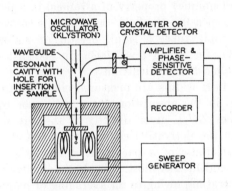

Fig. 5. Block diagram of ESR spectrometer.

mogeneous magnetic field which can be varied from zero to 10,000 gauss. The magnetic field strength is varied slowly over a narrow range by use of the sweep coils. The source of microwave frequencies is a klystron tube whose output frequency is about 10,000 Mc. A sample cavity resonating at the klystron frequency serves the same function as the transmitter coil in the NMR experiment. A bolometer of crystal detector is used. Audio modulation or higher frequencies (100 kc.) may be used; considerably increased sensitivity is obtained at higher modulation frequencies. Plots of the derivative of the absorption versus field strength are obtained directly and can be displayed on an oscilloscope or recorded on a chart.

Sample Preparation and Size. Samples are normally examined in quartz tubes 1 to 5 mm. in diameter. The concentration required for detection depends on the line width. For most radicals, presently available spectrometers operating at 10 kilogauss are able to detect about 10^{-10} mole of unpaired electrons. Depending on the diameter of the tube, the volume needed is in the range 0.10 to 0.30 cm.^{3}. The sensitivity of detecting unpaired electrons in aqueous solutions is increased by freezing. In frozen aqueous solutions radicals may be detected at about $2 \times$

$10^{-5} M$, in aqueous solutions at about $5 \times 10^{-4} M$, and in nonaqueous solutions at about $10^{-7} M$.[29] In some cases the lines are so broad that much higher concentrations are required. The following metal ions of biological interest may be detected, but concentrations of about $10^{-4} M$ are required: Mn^{++}, Mo^{III}, Mo^{V}, Fe^{III}, Fe^{II}, Cu^{++}.

C. Interpretation of ESR Spectra

The g Value. A measurement of both the resonance frequency and field strength makes it possible to calculate the g value which is a characteristic (but not unique) property of a radical in a particular environment. In order to measure the field strength it is convenient to include, affixed to the sample tube, a small crystal of α,α'-diphenyl-β-picryl hydrazil (DPPH), for which $g = 2.0036$. In contrast to paramagnetic ions in crystals, almost all radicals have g values near 2.

Line Widths. Line widths are measured either by the width in gauss at half-maximum in the absorption curve, $\Delta H_{1/2}$, or by the separation of maximum slopes, ΔH_{ms}, given directly by the separation of the peak and the valley in the derivative curve usually obtained. Although line widths are characteristic of a particular system, they are, in general, concentration-dependent. For most radicals the line widths decrease with increasing concentration, owing to an averaging of magnetic interactions to zero ("exchange narrowing").

Hyperfine Splitting. Just as lines in an NMR spectrum are split when there are other nuclei with spin in the molecule, similarly ESR spectra show hyperfine splitting due to the various orientations of nuclear spins.

Such hyperfine splitting was first observed for the manganous ion which gave a six-line pattern resulting from spin splitting by the six $(2I + 1)$ possible orientations of Mn^{56} $(I = 5/2)$.[30] The expected splittings and relative line intensities can be predicted in the same way as outlined earlier for nuclear spin-spin coupling. For example, in irradiated polymethylmethacrylate and polystyrene, three resonance peaks with relative intensities 1:2:1 are observed.[31] These are attributed to interaction of an unpaired electron with the two protons of the CH_2 group with which it is associated.

Just as in the case of NMR, the hyperfine splitting is eliminated by exchange of the unpaired electron with a frequency greater than the separation of the multiplet components in the absence of exchange. Thus the multiplet structure of naphthalene radicals, formed by reaction with

[29] H. Beinert, personal communication.
[30] E. K. Zavoisky, *Doklady Akad. Nauk. U.S.S.R.* **57**, 887 (1947).
[31] E. E. Schneider, *Discussions Faraday Soc.* **19**, 158 (1959).

Na metal, disappears when the concentration of radicals is increased.[32] Rates of electron transfer in which the lifetime of the radicals is about 10^{-6} second can be measured by observing the collapse of multiplet structure in this way.

D. Range of Applications

Free Radical Detection and Identification. In many cases just the demonstration of the existence of free radicals is significant. Identification of radicals is usually made on the basis of the hyperfine splitting pattern. In cases where the radicals are present only at very low concentrations and react rapidly, as in studies of radiation damage, operation at liquid air or even lower temperatures is desirable in order to trap the reactive species and increase the sensitivity. In order to get a measure of the concentration of radicals, known amounts of some relatively stable radical such as DPPH can be used as a reference in nonaqueous solutions. Because of the instability of aqueous solutions of radicals, it is very difficult to provide such reference solutions for studies in water. As in NMR, the area under the absorption peak is proportional to the concentration of unpaired electrons.

Biochemical Applications. The concentration of paramagnetic ions may be decreased by complexing, and the absorption eliminated or shifted to another frequency. This makes it possible to determine complexing constants.[33] Electron spin resonance studies have shown the production of unpaired electrons in plant tissues on irradiation.[34, 35] The orientation of heme groups in hemoglobin has been revealed by electron spin resonance studies.[36]

Free radicals have been detected in a flavoprotein enzyme after light irradiation.[37] Experiments with flavoprotein enzymes suggest that flavin free radicals may be intermediates in the catalytic action.[38, 39] Electron spin resonance studies have been used to study the complexing of manganese and enolase[40] and have shown the state of copper in lactase.[41] From

[32] R. L. Ward and S. I. Weissman, *J. Am. Chem. Soc.* **76,** 3612 (1954).
[33] M. Cohn and J. Townsend, *Nature* **173,** 1090 (1954).
[34] P. B. Sogo, N. G. Pon, and M. Calvin, *Proc. Natl. Acad. Sci. U.S.* **43,** 387 (1957).
[35] G. Tollin and M. Calvin, *Proc. Natl. Acad. Sci. U.S.* **43,** 895 (1957).
[36] M. J. Lyons, J. F. Gibson, and D. J. E. Ingram, *Nature* **181,** 1003 (1958).
[37] B. Commoner and B. B. Lippincott, *Proc. Natl. Acad. Sci. U.S.* **44,** 1110 (1958).
[38] B. Commoner, B. B. Lippincott, and J. V. Passonneau, *Proc. Natl. Acad. Sci. U.S.* **44,** 1099 (1958).
[39] A. Ehrenberg and G. Ludwig, *Science* **127,** 1177 (1958).
[40] B. G. Malmström, T. Vänngård, and M. Larsson, *Biochim. et Biophys. Acta* **30,** 1 (1958).
[41] B. G. Malmström, R. Mosbach, and T. Vänngård, *Nature* **183,** 321 (1959).

electron spin resonance studies of xanthine oxidase solutions it has been concluded that the iron of the enzyme is always in the ferrous state and that the molybdenum is reduced and free radicals are formed when the substrate is added.[42]

As an illustration of an electron spin resonance study, the signals for a sample of cytochrome reductase are shown in Fig. 6.[43] The signal shown

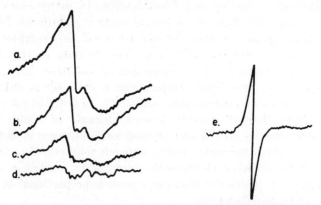

FIG. 6. Electron spin resonance signals of cytochrome reductase in aqueous solution at pH 8 and −100°. (From Beinert and Sands,[43] with permission of Academic Press.)

in *a* is due to iron. When DPNH was added, the signal progressively disappeared as shown in *b* to *d*. When after disappearance of the iron signal more DPNH was added, a free radical signal (shown in *e*) appeared which is believed to be a flavin semiquinone. An electron spin resonance signal has also been obtained from copper in cytochrome oxidase.[44]

[42] R. C. Bray, B. G. Malmström, and T. Vänngård, *Biochem. J.* **73**, 193 (1959).
[43] H. Beinert and R. H. Sands, *Biochem. Biophys. Research Communs.* **1**, 171 (1959).
[44] R. H. Sands and H. Beinert, *Biochem. Biophys. Research Communs.* **1**, 175 (1959).

[125] Infrared Measurements in Aqueous Media

By WILLIAM P. JENCKS

Introduction

Infrared spectroscopy, since it depends on energy absorption from vibrations of individual bonds and chemical groups rather than from electronic excitations, has the advantage over visible and ultraviolet spectroscopy that it can be used for the identification, characterization, and quantitative analysis of specific chemical groups. Its principal disadvantages from a biochemical point of view are the relatively high concentration (but not amount) of material required for analysis, the considerable expense (of the order of $15,000) for a high-quality spectrophotometer, and the instability of the commonly used cells with sodium chloride windows to many substances of biochemical interest, particularly water. These disadvantages have been partly overcome in recent years. Infrared spectrophotometers are now standard equipment in most chemistry departments, and instruments which are satisfactory for many purposes are available in the $5000 range. A considerable number of materials for constructing cells which are transparent to a considerable range of infrared radiation, but are resistant to water and to commonly used biochemical solutions, have been developed.

Excellent reviews on the theory and general applications of infrared spectrophotometry have been published by Jones and Sandorfy[1] and by Bellamy,[2] and these should be consulted for the detailed assignment and interpretation of group frequencies. A review of the application of infrared spectroscopy to enzymology has appeared in an earlier volume of *Methods in Enzymology*[3] and Sutherland has reviewed earlier work on amino acids and proteins, principally in the solid state, up to 1952.[4]

The most obvious advantage of the use of aqueous solutions as solvents for infrared spectroscopy is the high solubility of most biochemical materials in water, compared to the organic solvents ordinarily used. Spectroscopic examination in aqueous solution avoids the sometimes large and unpredictable shifts and splitting of absorption bands and the consequent difficulties in interpretation which are encountered in spectra of solids by replacing the forces of mutual interaction of crystals by a large, but relatively constant, interaction with the aqueous solvent. In biochemical studies it is particularly advantageous to examine compounds and reactions in their normal aqueous environment; this is especially important in the study of proteins and nucleic acids. Chemical and enzymatic reactions may be followed by directly measuring the appearance or disappearance of the absorption of specific chemical groups. As in ordinary infrared spectroscopy, infrared spectroscopy in aqueous solution may be used to determine the structure and properties of known and unknown compounds of biochemical importance.

[1] R. N. Jones and C. Sandorfy, *in* "Chemical Applications of Spectroscopy" (W. West, ed.); *in* "Technique of Organic Chemistry" (A. Weissberger, ed.), Vol. IX. Interscience, New York, 1956.
[2] L. J. Bellamy, "The Infrared Spectra of Complex Molecules," 2nd ed. Methuen, London, 1958.
[3] D. L. Wood, Vol. IV [3].
[4] G. B. B. M. Sutherland, *Advances in Protein Chem.* 7, 291 (1952).

Aside from the problem of water-resistant materials for cell windows, the principal disadvantages of aqueous solvents are the broadening and decreased resolution observed with some peaks in water, which may be due to varying amounts of interaction of the absorbing group with the solvent, and the infrared absorption of water itself, which occurs in several regions of great importance. The latter problem was largely solved by Gore et al.[5] in 1949 by the use of deuterium oxide as a solvent. Deuterium oxide is similar to water in most of its properties, but because of the increased mass of deuterium compared to the hydrogen atom, the O—D stretching and bending frequencies are considerably lower than the corresponding O—H frequencies of water so that deuterium oxide is transparent to infrared radiation in the regions of greatest interest. By use of both water and deuterium oxide as solvents it is possible to cover essentially the whole range of infrared absorption in common use for spectroscopy.

Gore et al.[5] measured the absorption spectra and state of ionization of several acids and amino acids as a function of pH and were able to follow hydrogen–deuterium exchange spectrophotometrically in deuterium oxide solution. The most extensive studies of materials of biological importance have been carried out by Blout and Lenormant. They have shown that aqueous infrared spectrophotometry can be used to study the structure and state of ionization of amino acids, peptides, and proteins[6-11]; the rates and conditions for hydrogen–deuterium exchange on the peptide bond and spectral shifts associated with helix-coil transitions and denaturation of proteins and polypeptides[9, 11, 12, 12a]; infrared flow dichroism as an indicator of polypeptide structure[13, 14]; the structure of DNA as a function of pH, temperature, and molecular size[15]; and even whole bacteria and yeast.[10, 16] Much of this work has been reviewed by Blout.[17] The technique has been particularly useful in determining the

[5] R. C. Gore, R. B. Barnes, and E. Petersen, *Anal. Chem.* **21**, 382 (1949).

[6] H. Lenormant, *Compt. rend. acad. sci.* **234**, 1959 (1952).

[7] H. Lenormant and J. Chouteau, *Compt. rend. acad. sci.* **234**, 2057 (1952).

[8] H. Lenormant, *J. chim. phys.* **49**, 635 (1952).

[9] H. Lenormant and E. R. Blout, *Nature* **172**, 770 (1953).

[10] E. R. Blout and H. Lenormant, *J. Opt. Soc. Am.* **43**, 1093 (1953).

[11] P. Doty, A. Wada, J. T. Yang, and E. R. Blout, *J. Polymer Sci.* **23**, 851 (1957).

[12] H. Lenormant and E. R. Blout, *Bull. soc. chim. France* 859 (1954).

[12a] G. D. Fasman and E. R. Blout, *J. Am. Chem. Soc.* **82**, 2262 (1960).

[13] G. R. Bird, M. Parrish, Jr., and E. R. Blout, *Rev. Sci. Instr.* **29**, 305 (1958).

[14] G. R. Bird and E. R. Blout, *J. Am. Chem. Soc.* **81**, 2499 (1959).

[15] E. R. Blout and H. Lenormant, *Biochim. et Biophys. Acta* **17**, 325 (1955).

[16] H. Lenormant, *Compt. rend. soc. biol.* **147**, 406 (1953).

[17] E. R. Blout, *Ann. N. Y. Acad. Sci.* **69**, 84 (1957).

structure of certain molecules in aqueous solution. Information as to the predominant tautomeric structure of amino and carbonyl groups of nucleotides has been obtained by comparison with the appropriate O-methyl and N-methyl compounds,[18-21a] and it has been shown that the mutual interaction of polyribonucleotide chains leads to characteristic alterations in their infrared spectra.[22-23b] Ehrlich and Sutherland have studied the state of ionization of polyelectrolytes and appropriate model compounds in aqueous solution,[24,25] and Susi et al.[26] have been able to confirm the previously suggested presence of undissociated "buried" carboxyl groups in β-lactoglobulin near neutrality by double-beam spectrophotometry of aqueous protein solutions at different pH values. Infrared spectroscopy has been used to show that carbon dioxide exists predominantly as CO_2, rather than as carbonic acid, in water[27]; that pyruvate is present principally in the unhydrated form in aqueous solutions[28]; that phthalaldehydic acid, but not the corresponding anion, exists in a cyclic form in water[28a]; and that the reaction of a number of carbonyl compounds with nitrogen bases proceeds with an initial rapid addition of the nitrogen base to the carbonyl group, followed by a slow dehydration step.[29] Within compounds of a given chemical class there is a rough correlation between carbonyl frequency in deuterium oxide solution and the "energy-rich" nature or reactivity toward nucleophilic reagents of activated acyl groups, because these parameters are affected in a similar manner by varying inductive and resonance effects; among compounds of different chemical nature such correlations generally do not hold because of the varying relative importance of these and other factors in

[18] H. Lenormant and E. R. Blout, Compt. rend. acad. sci. 239, 1281 (1954).
[19] R. L. Sinsheimer, R. L. Nutter, and G. R. Hopkins, Biochim. et Biophys. Acta 18, 13 (1955).
[20] H. T. Miles, Biochim. et Biophys. Acta 22, 247 (1956).
[21] H. T. Miles, Biochim. et Biophys. Acta 27, 46 (1958).
[21a] C. L. Angell, J. Chem. Soc. 504 (1961).
[22] H. T. Miles, Biochim. et Biophys. Acta 30, 324 (1958).
[23] H. T. Miles, Biochim. et Biophys. Acta 35, 274 (1959).
[23a] H. T. Miles, Biochim. et Biophys. Acta 45, 196 (1960).
[23b] T. Shimanouchi, M. Tsuboi, Y. Kyogoku, and I. Watanabe, Biochim. et Biophys. Acta 45, 195 (1960).
[24] G. Ehrlich, J. Am. Chem. Soc. 76, 5263 (1954).
[25] G. Ehrlich and G. B. B. M. Sutherland, J. Am. Chem. Soc. 76, 5268 (1954).
[26] H. Susi, T. Zell, and S. N. Timasheff, Arch. Biochem. Biophys. 85, 437 (1959).
[27] L. H. Jones and E. McLaren, J. Chem. Phys. 28, 995 (1958).
[28] W. P. Jencks and J. Carriuolo, Nature 182, 598 (1958).
[28a] E. Bernatek, Acta Chem. Scand. 14, 785 (1960).
[29] W. P. Jencks, J. Am. Chem. Soc. 81, 475 (1959).

determining carbonyl frequency.[30] The rate of mutarotation of carbohydrates[30a] and the binding of lactate to metal ions in water[30b] have been studied by infrared spectroscopy.

The use of infrared spectrophotometry for following enzymatic reactions, either by analysis of aliquots or by direct observation of the reaction in the infrared cell, is just beginning, but it promises to be a very

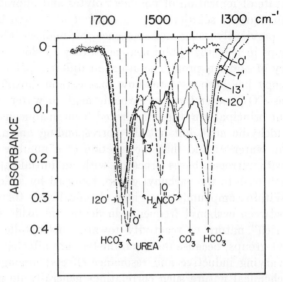

FIG. 1. Infrared spectra taken during the course of the urease-catalyzed hydrolysis of urea in D₂O. The reaction mixture contained 0.2 M urea, 2 mg. of urease (Sigma, Type V), and 0.4 mg. of Diamox (a carbonic anhydrase inhibitor) in 1.0 ml.; the reaction was allowed to proceed in a 0.1-ml. aliquot in a calcium fluoride cell with a path length of 0.05 mm. The final pH, after dilution with water, was 9.3.

useful tool, since the nature as well as the rate of formation of intermediates and reaction products may be followed directly. The rates of disappearance and formation of urea, carbamate, and bicarbonate in the urease-catalyzed hydrolysis of urea are shown in Figs. 1 and 2 as an example.[31] Urea hydrolysis occurs according to Eq. (1). Spectra taken

$$\underset{\overset{\displaystyle \parallel}{O}}{H_2NCNH_2} \rightarrow NH_4^+ + H_2NCO_2^- \rightarrow HCO_3^- + NH_3 \tag{1}$$

[30] W. P. Jencks, C. Moore, F. Perini, and J. Roberts, Arch. Biochem. Biophys. **88**, 193 (1960).

[30a] F. S. Parker, Biochim. et Biophys. Acta **42**, 513 (1960).

[30b] J. D. S. Goulden, Spectrochem. Acta **16**, 715 (1960).

[31] W. P. Jencks, M. Montemezzi, and J. Carriuolo, unpublished experiments (1960).

at intervals after the addition of urease to an unbuffered solution of urea show first the disappearance of urea peaks at 1604 and 1490 cm.$^{-1}$ and the appearance of carbamate peaks at 1441 and 1540 cm.$^{-1}$, along with some bicarbonate. At the end of the reaction, the urea and carbamate bands have been replaced by strong bicarbonate absorption at 1363 and 1630 cm.$^{-1}$. Figure 2 shows the time course of the appearance and

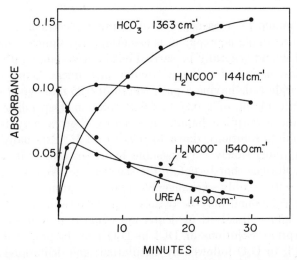

FIG. 2. The time course of the appearance and disappearance of urea, carbamate, and bicarbonate absorption during the urease-catalyzed hydrolysis of urea. Experimental conditions are similar to those for the experiment described in Fig. 1, except for a urea concentration of 0.1 M.

disappearance of the individual peaks in a similar experiment; carbamate is formed rapidly and then slowly disappears, while bicarbonate is formed more slowly and approaches a maximum value. The absorption in the region of the 1441 peak of carbamate does not disappear completely because of the appearance of an absorption band of CO_3^{--}, which is in equilibrium with HCO_3^-, at 1416 cm.$^{-1}$. Cyanate is a possible intermediate in urea hydrolysis, but no change in the region of cyanate absorption, near 2150 cm.$^{-1}$, could be detected in the course of the enzyme-catalyzed reaction.

Infrared and Raman spectra are similar in that they measure frequencies associated with vibrations of chemical bonds, but they are complementary in that they depend on different properties of the bond to give the observed spectra. Infrared absorption requires a change in dipole moment, and the Raman effect depends on a change in polarizability of the bond during vibration; consequently, although most absorption frequencies are the same or closely similar by the two methods, some bands are observed with only one technique. Raman spectra of many compounds of biochemical importance in aqueous solution have been available for some time,[32-35] but the technical demands of the method are such that it has not come into wide use.

Procedure

The technique for measurement of infrared spectra in aqueous solution, once the sample is prepared, is essentially the same as for any other spectra and should generally be carried out in accordance with the manufacturer's instructions for the particular instrument being used. With calcium fluoride cells having a path length of 0.05 mm., good spectra are obtained with 0.1 ml. of a 0.25 to 2.0 M solution. Depending on the intensity of the absorption bands, satisfactory spectra may be obtained with more-dilute solutions, down to 0.05 to 0.1 M. Especially with substances containing slowly exchangeable hydrogen atoms, it is advisable to prepare solutions shortly before the spectra are taken. For nonvolatile compounds with readily exchangeable hydrogen atoms or water of hydration, the compound may be dissolved in deuterium oxide, evaporated to dryness, and redissolved in deuterium oxide if it is desired to avoid HDO absorption. Solutions of DCl in D_2O may be prepared by slowly adding $POCl_3$ to D_2O followed by distillation; and deuterated carboxylic acids, RCOOD, may be prepared by dissolving the corresponding acid anhydride in D_2O. It is desirable to have a thoroughly cleaned and dried cell and to rinse it first with a little of the solution to be measured, particularly for quantitative work. The cells are filled with a small syringe or by placing the sample in one of the cell openings and applying a gentle steady suction with an eyedropper bulb or syringe to the other opening. If the cells are clean, it is not difficult to fill them without bubbles. After use the cells are washed out immediately with three to five changes of solvent, and they may be dried with a stream of dry air. Thorough cleaning is important and may be difficult; it is particularly difficult to remove the last traces of one solvent, e.g., water, so that its spectrum does not contaminate a subsequent run with another solvent, e.g., deuterium oxide. It is advisable to rinse the cells thoroughly with the solvent to be used in a particular run if they have previously been used with

[32] J. T. Edsall, *J. Chem. Phys.* **4**, 1 (1936).
[33] J. T. Edsall, *J. Chem. Phys.* **5**, 508 (1937).
[34] J. T. Edsall, D. Garfinkel, and co-workers, *J. Am. Chem. Soc.* **80**, 3807, 4833 (1958), and references therein.
[35] J. H. Hibben, "The Raman Effect and Its Chemical Applications." Reinhold, New York, 1939.

another solvent. It is our practice to run a solvent spectrum on the sample cell after use, to be certain that it has been properly cleaned. Wavelength calibration should be carried out on each spectrum; this may be conveniently done by utilizing the sharp peaks of atmospheric water vapor at 5.884 and 6.107 μ or the readily available polystyrene film standards.

Cells. The sodium chloride cells ordinarily used for infrared spectroscopy with organic solvents are obviously not suitable for work with aqueous solutions; they may, however, be used with care for spectra in dry alcohols or saturated aqueous sodium chloride. The commonly used, commercially available materials for aqueous infrared spectrophotometry are calcium fluoride, which is transparent up to 9 μ, and barium fluoride, which is transparent to 13 μ. The chief disadvantages of these materials are their instability to acid and concentrated ammonium salts and their considerable expense: a pair of calcium fluoride cells costs about $500, and barium fluoride is nearly twice this amount. Thallium bromide–iodide is also satisfactory, but it is toxic and is very expensive. Recently, a number of materials have become available for cell windows which show considerable promise, although they have not yet been widely used. Eastman Kodak makes a water-resistant, arsenic-modified selenium glass which is relatively inexpensive and is transparent from 2 to 18 μ, except for a sharp peak at 12.7 μ. The same firm makes Irtran 1 and 2, which have reasonably good transmission in the 2- to 8-μ (1) and 1 to 13-μ (2) regions, are resistant to dilute acids, moderately concentrated bases, and organic solvents, and are relatively inexpensive. Cells with Irtran 2 windows are now commercially available for about $160. An arsenic trisulfide glass which is stable to dilute acids, but not to fairly concentrated base, and is transparent in the 2- to 12-μ region, is marketed under the trade name Servofrax by the Servo Corporation of America, Hicksville, New York, and may be obtained at moderate expense already cut and drilled to fit most commercial spectrophotometer cells. A method for the construction of silver chloride cells at very little expense has been described by Nachod and Martini.[36] Silver chloride is transparent from 2 to 15 μ and was used in the early investigation of Gore *et al.*[5] as well as in many later studies, but it is subject to darkening or attack by light and amines and is too soft to give cells of reproducible thickness. Robinson has described polyethylene cells which can be prepared at negligible expense and which are transparent except for bands at 3.3 μ, 6.8 μ, and 13 to 14 μ.[37] J. Gordon, in this laboratory, has found it convenient to prepare such cells by fusing together two small

[36] F. C. Nachod and C. M. Martini, *Appl. Spectroscopy* **13**, 45 (1959).
[37] T. Robinson, *Nature* **184**, 448 (1959).

pieces of polyethylene with a Parafilm liner. A piece of Parafilm, cut out in the center to the size of the desired window, is placed between two pieces of polyethylene. These are in turn placed between two brass plates with a hole in the center the size of the desired window and heated with a hot iron. The sample and a solvent blank, 0.05 to 0.1 ml., are placed in two such cells, which are then taped onto the brass plates and placed in the light path of the spectrophotometer. Although light scattering decreases the transmission through such cells, they give satisfactory spectra and may be used for materials, such as acids, which cannot be used in CaF_2 cells. Similar cells may also be prepared from Teflon tape,[36] but this material is more difficult to seal satisfactorily. A heat-sensitive fluorocarbon film (DuPont, Teflon FEP) has recently become available, which can be readily made into cells by sealing with a soldering iron and is transparent, except for absorption at 7.5 to 9.0 μ and at 10.2 μ. The principal difficulty with such flexible cells is the difficulty in obtaining even, matched, and reproducible distribution of the liquid sample in the light beam.

Interpretation

A summary of some of the commonly observed absorption frequencies in deuterium oxide solution and their assignments is given in the table. The spectrum of a 0.5 M solution of acetylcholine chloride is given in Fig. 3 as an example.

Vibrations Involving Hydrogen (X—H). Spectra of liquid H_2O and D_2O (which are reproduced in Blout[17]) show that D_2O is generally the more transparent solvent for infrared spectroscopy, particularly in the important carbonyl region. With a double-beam instrument and a compensating solvent cell, the weak D_2O absorption near 1550 cm.$^{-1}$ does not interfere appreciably. In regions of intense solvent absorption a double-beam instrument is unresponsive, and at either side of such regions either positive or negative absorption may be observed due to changes in the O—D or O—H bond length and frequency in the sample compared to the blank cell; such changes result from changes in the structure of the solvent which are caused by interaction with the solute.[38] Hydrogen atoms on nitrogen, oxygen, sulfur, and acidic carbon atoms may be expected to exchange rapidly with the solvent when dissolved in D_2O, so that their X—H absorption bands are not observed. A relatively small amount of H_2O dissolved in D_2O, from contamination or from exchange of labile hydrogen atoms of the dissolved material, is converted to HDO, which gives a characteristic peak at 3388 cm.$^{-1}$ and a broader, less intense

---

[38] E. Ellenbogen, *Abstr. 135th Natl. Meeting, Am. Chem. Soc.* 24c (1959); *Biochim. et Biophys. Acta* **39**, 174 (1960).

SOME INFRARED FREQUENCY ASSIGNMENTS IN AQUEOUS SOLUTION

Bond	Frequency, cm⁻¹
X—H	
O—H of H_2O (solvent)	2800–3800, 2100, 1600–1800
O—H of HDO	3380, 1455(b)[a]
O—D of D_2O (solvent)	2200–2850, 1550(w), 1150–1250
CH_3	2950–2980(m), 1420–1470(m), 1365–1390(m)
CH_2	2915–2950(m), 1430–1480
CH_2, α to C=O	2915–2950(m), 1405–1435
C—H, aromatic and olefinic	3010–3130(m)
C=O (all strong or very strong)	
Esters	
Normal	1710
Ethyl carbamate	1681
Acetylcholine chloride	1735
Alanine ethyl ester hydrochloride	1743
Ethyl trifluoroacetate	1786
Acids	
Normal	1710
α-Keto and amino acids	1720–1730
Acyl phosphates	
Acetyl phosphate dianion	1713
Acetyl phosphate monoanions	1735–1750
Carbamyl phosphate	1670
Thiol esters	1670–1680
Amides	
Normal	1625–1680
Urea	1604
Acetylimidazole	1740
Aldehydes and ketones	1665–1730
Carboxylate ions	
Normal	1560–1590, 1405–1420
α-Keto and amino acids	1610–1630, 1400–1410
Bicarbonate	1630, 1363
Carbamate	1540, 1441
Carbonate	1416
C—O	
Esters and acids	1100–1350(s)
Alcohols and ethers	1050–1200(s)
C=C	1600–1680(w–m)
Aromatic ∼1600, ∼1500, ∼1450 (variable intensity)	
Nucleotides (principal ring absorptions at neutrality; all strong intensity)	
AMP	1626
UMP	1645–1658, 1678–1692
IMP	1674
CMP	1493–1505, 1610, 1649
TMP	1620, 1650
Phosphates: Dianions	975–985(s), 1070–1090(s). 1105–1120 (variable)
Monoanions	1050–1090(s), 1205–1240(s), 915–940(?)

[a] b = broad, w = weak, m = medium, s = strong.

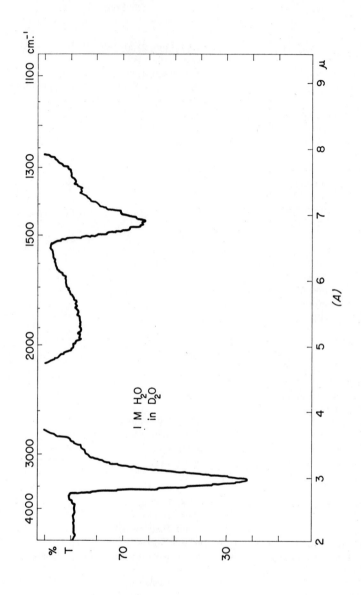

(A)

1 M H_2O in D_2O

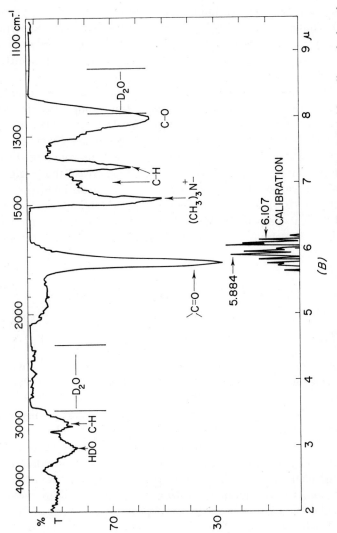

Fig. 3. Infrared spectra in deuterium oxide solution using a calcium fluoride cell, path length 0.05 mm. A, 1.0 M H₂O (i.e, 2.0 M HDO). B, 0.5 M acetylcholine chloride.

peak at 1455 cm.$^{-1}$ (Fig. 3.4). If the HDO peak is not too intense, it is often possible to identify C—H absorption bands in the 2900- to 3100-cm.$^{-1}$ region. The disappearance of these bands or the appearance of the HDO band may be used to follow the rate and extent of hydrogen–deuterium exchange with the solvent. Sharp C—H bending absorptions may generally be observed in the 1370- to 1480-cm.$^{-1}$ region. In the α-position to a carbonyl group these bands are intensified and shifted to lower frequencies.

$C{=}O$. The intense carbonyl peaks are usually the most prominent feature of spectra of compounds containing this group. The frequency of carbonyl absorption is a function of the C=O bond length and, like the bond length, is influenced by inductive and resonance effects of neighboring atoms. Adjacent groups which are electron-donating through either inductive or resonance effects will tend to increase the single-bond character of the carbonyl group by favoring the single-bonded contributing structures (b) and (c), whereas groups which are electron-withdrawing or which compete for the electrons involved in carbonyl resonance will have the opposite effect. Hydrogen bonding by the solvent will

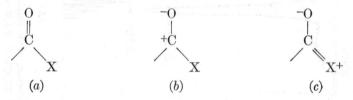

also tend to increase the C=O bond length and decrease the absorption frequency, and ionic resonance forms (b) and (c) will be favored by polar solvents; consequently, carbonyl absorption bands in D_2O tend to be at considerably lower frequencies than in nonpolar, nonhydrogen-bonding solvents.[30] Infrared absorption frequencies may also be affected by mass or coupling effects of neighboring atoms or groups, such as C—H or C—N in aldehydes and amides. On solution in D_2O, the labile hydrogen atoms of compounds such as amides and acids will exchange with the solvent, and the carbonyl absorption may be shifted to lower frequencies to the extent that it is affected by the substitution of N—D or O—D for N—H or O—H. The presence of mass and coupling effects requires that comparisons between compounds of different structural classes must be made with caution.

Normal esters, amides, and carboxylate ions show a progressive decrease in carbonyl absorption frequency due to the increasing relative importance of resonance form (c) in this series of compounds. Monomeric acids are structurally similar to esters and absorb at approximately the same frequency; carboxylic acids exist primarily as hydrogen-bonded dimers in nonpolar solvents but are monomeric in dilute aqueous solution. The absorption of thiol esters is at a considerably lower frequency than oxygen esters, probably because of a rather complex interaction of opposing influences on the carbonyl group.[30, 38a] The acetyl phosphate dianion absorbs at 1713 cm.[-1], but substituted acetyl phosphates and the acetyl phosphate monoanion, which are more electron-withdrawing on the phosphate group, absorb at 1735 to 1750 cm.[-1]. Aldehydes and ketones absorb in the range 1665 to 1730 cm.[-1]. Carboxylate ions show a strong symmetrical vibration frequency at 1400 to 1420 cm.[-1] which tends to shift in the opposite direction from the higher frequency asymmetrical absorption, with varying substituents. The examples given in the table show that, within a particular group of compounds, electron-withdrawing substituents on either side of a carbonyl group cause the expected increase in carbonyl absorption frequency, unless the electron-withdrawing effect is overcome by a larger resonance effect (c), as in compounds containing the amide group.

$C = X$. The C=C and C=N bonds give considerably weaker absorption bands than the carbonyl group and may be difficult to identify. Aromatic compounds show characteristic bands near 1600, 1500, and 1450 cm.[-1], which are usually intense and easily identified, but may be of variable intensity. Nucleic acids and nucleotides show absorption bands which may be assigned to the component parts of the molecule; the principal absorption bands of the aromatic base component of a series of nucleotides at neutral pH are given in the table.[15, 18-23] On changing the ionic species by changing the pH or by incorporating the nucleotides into an ordered structure, as in DNA or RNA, characteristic changes in these absorption bands appear. Conversely, changes in the structure of DNA, as in denaturation, may be followed by the associated changes in infrared absorption frequency and intensity. DNA may be differentiated from RNA by the presence of a band at 1060 cm.[-1] in the former compound.

$C - O$. The C—O band of esters and acids is strong and broad but usually falls at least partly in the region of D_2O absorption. This band and the similar band of alcohols and ethers may be obscured for this reason. The absorption frequency of this type of single bond is particularly sensitive to mass and coupling effects, so that correlations must be restricted to compounds of similar chemical structure.

Ionic phosphates show characteristic absorption bands, which may be

[38a] A. W. Baker and G. H. Harris, *J. Am. Chem. Soc.* 82, 1923 (1960).

used to differentiate mono- and dianions, in the 915- to 1120-cm.$^{-1}$ region.[39] Unfortunately, some of these bands fall in regions of D_2O and CaF_2 absorption.

Proteins and polypeptides show the expected absorption bands of their component parts. The amide I band of the carbonyl group of normal proteins and polypeptides appears at 1640 to 1660 cm.$^{-1}$ in deuterium oxide solution. In acid solution, undissociated carboxylic acid groups absorb at 1710 cm.$^{-1}$. On the addition of alkali this band disappears and carboxylate ion absorption frequencies appear at 1575 and 1410 cm.$^{-1}$. The intensity of these bands may be roughly correlated with the carboxylate content of the protein.[12, 17] The amide II band, which appears to result from vibrations associated with both N—H and amide groups, is found at 1550 cm.$^{-1}$ in native protein but is shifted to 1450 cm.$^{-1}$ on deuteration because of the substitution of N—D for N—H. At neutral or slightly acid pH this change occurs slowly in native proteins, presumably because some of the amide N—H groups are tied up in a helical, hydrogen-bonded structure and are not free to exchange with the solvent D_2O. On denaturation or in slightly alkaline solution, exchange occurs rapidly and may be followed directly by spectrophotometry.[9, 12, 25]

[39] M. Tsuboi, *J. Am. Chem. Soc.* **79**, 1351 (1957).

[126] Optical Rotatory Dispersion

By GERALD D. FASMAN*

Optical rotation has been used for many years[1-5] as a sensitive probe to follow conformational transitions in proteins, e.g., denaturation. The specific rotation, $[\alpha]_D$, measured at the sodium D-line has long been utilized as a criterion of purity for many proteins. The basis of this rotatory power of polypeptides and proteins lies firstly in the presence of

* Contribution No. 234 of the Graduate Department of Biochemistry, Brandeis University, Waltham 54, Massachusetts. This article was written during the tenure of an Established Investigatorship of the American Heart Association. This work was supported in part by grants from the National Institutes of Health (AM-05852-02) and the National Science Foundation (GB-428).

[1] D. M. Greenberg, *in* "The Chemistry of the Amino Acids and Proteins" (C. A. Schmidt, ed.). C. C Thomas, Springfield, Illinois, 1938.
[2] H. J. Almquist and D. M. Greenberg, *J. Biol. Chem.* **105**, 419 (1934).
[3] E. Straus and W. A. Collier, *Tabulae Biol.* **3**, 271 (1926).
[4] J. H. Clark, *J. Gen. Physiol.* **27**, 101 (1943).
[5] P. Doty and P. Geiduschek, *in* "The Proteins" (H. Neurath and K. Bailey, eds.), Vol. I, Part A, Chapter 5. Academic Press, New York, 1953.

an asymmetric carbon in the majority of the amino acids (except glycine and sarcosine), i.e., the configuration L or D, and secondly in the asymmetric spatial arrangement of the peptide backbone, or, as it is commonly termed, the conformation of the protein.[6] Until recently the majority of measurements were usually made at one wavelength, but in the period 1950–1960 instruments became available capable of making optical rotatory dispersion measurements, that is, measuring the variation of optical rotatory power with wavelength, which has yielded much greater information on conformation. A great stimulus was given to these studies when synthetic polypeptides,[7,8] capable of assuming the helical conformation predicted by Pauling,[9] became available, and theoreticians tackled the interpretation of these data. Since that time, the optical rotatory properties of many proteins, fibrous and globular, enzymes, etc., have been related to their helical contents. Improved instrumentation within the 1961–1963 period has enabled investigators[10] to extend the region of measurement from the near ultraviolet to the far ultraviolet (195 mμ) and thus enter the region of the main optically active absorption bands—the peptide unit. In this region, the rotation about the Cotton effects (see below) have extremely large values (80,000°), and thus the sensitivity of the method is greatly enhanced.

Numerous excellent articles and reviews[5,6,11–23] on the optical rota-

[6] E. R. Blout, *in* "Optical Rotatory Dispersion" (C. Djerassi, ed.), Chapter 17, McGraw-Hill, New York, 1960.

[7] C. H. Bamford, A. Elliott, and W. E. Hanby, "Synthetic Polypeptides." Academic Press, New York, 1956.

[8] E. Katchalski and M. Sela, *Advances in Protein Chem.* 13, 243 (1958).

[9] L. Pauling, R. B. Corey, and H. R. Branson, *Proc. Natl. Acad. Sci. U.S.* 37, 205 (1951); L. Pauling and R. B. Corey, *ibid.* 37, 251, 729 (1951).

[10] E. R. Blout, I. Schmier, and N. S. Simmons, *J. Am. Chem. Soc.* 84, 3193 (1962); E. R. Blout, J. P. Carver, and J. Gross, *ibid.* 85, 644 (1963).

[11] P. Urnes and P. Doty, *Advances in Protein Chem.* 16, 401 (1961)

[12] B. Jirgensons, *Makromol. Chem.* 44/46, 123 (1961).

[13] B. Jirgensons, *Tetrahedron* 13, 106 (1961).

[14] S. J. Leach, *Rev. Pure Appl. Chem.* 9, 33 (1959).

[15] E. M. Shooter, *Progr. in Biophys. and Biophys. Chem.* 10, 195 (1960).

[16] J. A. Schellman and C. G. Schellman, *Arch. Biochem. Biophys.* 65, 58 (1956).

[17] J. A. Schellman and C. G. Schellman, *J. Polymer Sci.* 49, 129 (1961).

[18] E. Katchalski and I. Z. Steinberg, *Ann. Rev. Phys. Chem.* 12, 433 (1961).

[19] W. Kauzmann, *Ann. Rev. Phys. Chem.* 8, 413 (1957).

[20] A. R. Todd, *in* "A Laboratory Manual of Analytical Methods in Protein Chemistry" (P. Alexander and R. J. Block, eds.), Vol. II, Chapter 8. Pergamon, New York, 1960.

[21] J. T. Yang, *Tetrahedron* 13, 143 (1961).

[22] W. F. Harrington and P. von Hippel, *Advances in Protein Chem.* 16, 1 (1961).

[23] P. Doty and P. Geiduschek, *in* "The Proteins" (H. Neurath and K. Bailey, eds.), Vol. I, Part A, Chapter 5. Academic Press, New York, 1953.

tions of proteins and synthetic polypeptides are available, so only a summary of the salient points will be discussed.

The Measurement of Optical Rotatory Power and Analysis of Results

The optical activity, or rotatory power, at a fixed wavelength is defined in terms of the specific rotation, $[\alpha]_\lambda$,

$$[\alpha]_\lambda = \frac{100}{lc} \alpha_\lambda \tag{1}$$

where α_λ is the observed rotation in degrees at wavelength λ, l is the path length in decimeters, and c is the concentration of the optically active solute in grams per 100 ml. As $[\alpha]_\lambda$ is temperature-dependent, this variable should be carefully controlled and recorded.

For low-molecular-weight substances the molar rotation, $[M]_\lambda$, is defined as

$$[M]_\lambda = \frac{M}{100} [\alpha]_\lambda \tag{2}$$

where M is the molecular weight of the solute. However, for the comparison of rotations of proteins and polypeptides which differ greatly in molecular weights, a more meaningful unit, the mean residue rotation, is used, $[m]$:

$$[m] = \frac{MRW}{100} [\alpha]_\lambda \tag{3}$$

in degrees centimeters per decimole, where MRW is the mean residue weight. The residue weight is the sum of the atomic weights in the unit

$$-[\overset{\displaystyle C}{\underset{\displaystyle O}{\|}}-CHR-NH]-$$

The MRW for a large body of proteins has been found to be approximately 115, and this value is used for comparative purposes, when the exact amino acid composition is not known.

The final variable on which the optical rotatory power is dependent is the refractive index of the solvent. Therefore, to compare observed rotations in a variety of media, they are reduced to the value they would have in a vacuum by means of the Lorentz correction factor. The reduced mean residue rotation, $[m']$, at wavelength λ, incorporates the refractive index correction in the following manner:

$$[m'] = \frac{3}{(n^2 + 2)} \frac{MRW}{100} [\alpha]_\lambda \tag{4}$$

where n is the refractive index of the solvent at wavelength λ.

The dispersion of the refractive index—i.e., its variation with the wavelength of the light used—of the solvent should be taken into account for the most accurate measurements, and when these values are not available, from tables such as the International Critical Tables, they can be approximated by use of the Sellmeier equation,

$$n^2 = 1 + \frac{a\lambda^2}{\lambda^2 - \lambda_v^2} \tag{5}$$

which can be solved for a and λ_v by measurement of the refractive index at two wavelengths. The value of n varies slightly for water between 578 and 365 mμ (see Table III), so usually the n_D value is used. However, in solutions of high salt concentrations, such as $8\,M$ urea, this correction can be of a significant magnitude (see Tables V to IX for refractive indices in various solvents). There is only a very small change of refractive index of water with temperature (Table IV), so this can usually be neglected.

Typical dispersion plots are seen in Fig. 1 for three substances, each in different conformations; poly-γ-benzyl-L-glutamate (PBG) and poly-L-glutamic acid (PGA) in the helical and random coil conformation, and bovine serum albumin (BSA) in the native and denatured states.

The analyses of these optical rotatory dispersion curves are generally handled by the use of two equations, the Drude equation and the Moffitt equation.

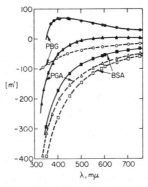

FIG. 1. Optical rotatory dispersions for ordered and disordered forms of two synthetic polypeptides and a globular protein. PBG, poly-γ-benzyl-L-glutamate; helical structure in m-cresol, ●——●; random coil in dichloroacetic acid, ○——○. PGA, poly-L-glutamic acid; helical structure at pH 4.72 in 1:2 water–dioxane mixture, 0.2 M NaCl, ▲——▲; random coil at pH 6.56 in the same solvent, △——△. BSA, bovine serum albumin: native protein in water at pH 5.47, ■——■; denatured protein in 8 M urea at pH 5.5, □——□. From P. Urnes and P. Doty, *Advances in Protein Chem.* **16**, 404 (1961).

The Drude Equation

The Drude equation[24] often describes the optical rotatory dispersion in spectral regions far from the optically active absorption bands (the peptide chromophore absorption bands lie at 145 mμ ($N \rightarrow V_2$), 185 mμ ($N \rightarrow V_1$) and 225 mμ ($n \rightarrow \pi^*$).[25-27] The theories of optical activity will not be discussed here, as there are many excellent reviews on the subject.[28-30]

The Drude equation takes the form

$$[\alpha]_\lambda = \frac{A}{\lambda^2 - \lambda_c^2} \tag{6}$$

where $[\alpha]_\lambda$ is the specific rotation, the A term includes the refractive index term, and λ_c represents the mean wavelength of the optically active electronic transitions (absorption bands) or the position of the strongest rotatory band which contributes to the visible rotations. However, it must be remembered that not all absorption bands are optically active and that strong rotations may be associated with very weak absorption bands[31] (see, for example, below, the cause of the 225-mμ Cotton effect).

To analyze whether a dispersion curve fits this simple equation, a variety of plots can be made.[32] The procedure of Yang and Doty[33] gives accurate results, when $[\alpha]_\lambda \lambda^2$ versus $[\alpha]_\lambda$ is plotted; when Eq. (6) is valid, λ_c is obtained from the slope of the resulting straight line. In Fig. 2 we see such a plot for poly-γ-benzyl-L-glutamate and poly-L-glutamic acid in the random coil and helical conformation, and for bovine serum albumin in the native and denatured forms.

The disordered or random conformations of synthetic polypeptides usually fit this Drude equation, and are thus termed simple, whereas helical conformations generally do not, and their dispersions are termed complex.[34] Both the native and denatured forms of bovine serum albumin

[24] P. Drude, "Lehrbuch der Optik," 2nd ed. S. Hirzel Verlag, Leipzig, 1906.

[25] J. S. Ham and J. R. Platt, *J. Chem. Phys.* **20**, 335 (1952).

[26] D. L. Peterson and W. T. Simpson, *J. Am. Chem. Soc.* **79**, 2375 (1957).

[27] W. B. Gratzer, G. M. Holzwarth, and P. Doty, *Proc. Natl. Acad. Sci. U.S.* **47**, 1785 (1961).

[28] E. V. Condon, *Revs. Modern Phys.* **9**, 432 (1937).

[29] A. Moscowitz, *in* "Optical Rotatory Dispersion" (C. Djerassi, ed.), Chapter 12. McGraw-Hill, New York, 1960.

[30] W. Heller and D. D. Fitts, *in* "Techniques of Organic Chemistry—Physical Methods" (A. Weissberger, ed.), Vol. 1, Part 3, Chapter 33. Interscience, New York, 1960.

[31] W. Kuhn, *Ann. Rev. Phys. Chem.* **9**, 417 (1958).

[32] T. M. Lowry, "Optical Rotatory Power." Longmans Green, London, 1935.

[33] J. T. Yang and P. Doty, *J. Am. Chem. Soc.* **79**, 761 (1957).

[34] T. M. Lowry and T. W. Dickson, *Trans. Faraday Soc.* **10**, 96 (1914).

obey the Drude equation, yielding different λ_c values. The random conformation of synthetic polypeptides display simple dispersion with $\lambda_c = 212$. Many globular proteins also obey the Drude equation in both native and denatured states, the latter giving λ_c values often approach-

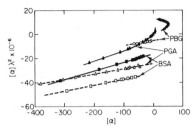

FIG. 2. Graphical treatment of dispersion data in Fig. 1 according to the simple Drude equation, Eq. (6). The disordered forms of the synthetic polypeptides, poly-γ-benzyl-L-glutamate (PBG, ○——○) and poly-L-glutamic acid (PGA, △——△), give linear plots, which specify their dispersions as simple. The dispersion of bovine serum albumin (BSA) in both native (■——■) and denatured (□——□) forms are likewise simple. In contrast, the helical structures of poly-γ-benzyl-L-glutamate (●——●) and poly-L-glutamic acid (▲——▲) display obvious curvature when plotted in this manner, so that their dispersions are defined as complex. From P. Urnes and P. Doty, *Advances in Protein Chem.* **16**, 411 (1961).

ing that found in the random conformation of synthetic polypeptides. Both polypeptides and proteins with low (but not zero) helical content may show simple dispersion and yield λ_c values higher than 212 mμ (see below).

The Moffitt Equation

Analysis of helical conformations can be made by use of the Moffitt equation.[35,36] Cohen[37] pointed out, in 1955, that the changes in optical rotation accompanying denaturation of proteins may be due to changes from helical to more random conformations. Moffitt considered the α-helix as a rigid array of identical chromophores which interact to form a cooperative unit acting as a single exciton system. He developed the phenomenological equation,

$$[m']_\lambda = [m] \frac{3}{n^2 + 2} = \frac{a_0\lambda_0^2}{\lambda^2 - \lambda_0^2} + \frac{b_0\lambda_0^4}{(\lambda^2 - \lambda_0^2)^2} \qquad (7)$$

where b_0 and λ_0 are principally functions of the helical backbone, independent of side chains and environment-independent, and a_0 represents

[35] W. Moffitt, *J. Chem. Phys.* **25**, 467 (1956); *Proc. Natl. Acad. Sci. U.S.* **42**, 736 (1956).

[36] W. Moffitt and J. T. Yang, *Proc. Natl. Acad. Sci. U.S.* **42**, 596 (1956).

[37] C. Cohen, *Nature* **175**, 129 (1955).

both intrinsic residue rotations, present irrespective of the helix, and rotations due to interactions within the helix.[35,36] Thus, a_0 should be solvent-dependent. Although this treatment is oversimplified,[38] there is an excellent empirical correlation between helical content and the b_0 value.[11] When a plot is made of $[m'](\lambda^2 - \lambda_0^2)$ versus $(\lambda^2 - \lambda_0^2)^{-1}$, b_0 is obtained from the slope and a_0 from the intercept.

This equation was first tested on poly-γ-benzyl-L-glutamate and poly-L-glutamic acid[36] (Fig. 3) and since has been tested on many other

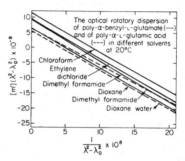

FIG. 3. Moffitt-Yang plot of poly-γ-benzyl-L-glutamate and poly-L-glutamic acid in the helical conformation. From W. Moffitt and J. T. Yang, *Proc. Natl. Acad. Sci. U. S.* **42**, 596 (1956).

synthetic polypeptides.[6,11] It was found that $\lambda_0 = 212$ mμ gave the best linear plots, and b_0 averaged about -630 for a right-handed helix. This choice of 212 mμ for λ_0 was satisfactory in the wavelength region 600 to 350 mμ; however, for lower regions, 280 to 240 mμ, it was found that $\lambda_0 = 216$ was required for a 5% L-tyrosine:95% L-glutamic acid copolymer,[11] and this has been verified for poly-L-glutamic acid.[39] The constant a_0 changes with solvent and polypeptide, e.g., poly-γ-benzyl-L-glutamate: in ethylene dichloride $a_0 = 205$, $b_0 = -635$; in dioxane $a_0 = 135$, $b_0 = -630$[36] (also see Table I). The b_0 value for a random chain usually is zero; however, many instances have been reported where small positive values ($+50$) have been recorded. The value of $b_0 = -630$ for the helical conformation seems valid in aqueous solution; however, in aqueous–organic solvent mixtures, higher values have been observed,[39] as illustrated in Table I. Consequently, caution should be used in interpreting an increase in the magnitude of the $-b_0$ value as an increase in helical content on the addition of organic solvents.

The sign of the b_0 value is indicative of the sense of helix.[35] A value

[38] W. Moffitt, D. D. Fitts, and J. G. Kirkwood, *Proc. Natl. Acad. Sci. U.S.* **43**, 723 (1957).

[39] G. D. Fasman, unpublished data, 1963.

of —630 is associated with the right-handed helix of the L-amino acids, whereas conversely +630 is associated with the left-handed helix of the D-amino acids.[40] Previously, work on poly-L-alanine[41,42] had led to the same conclusion. Studies on myoglobin verified the association of the negative b_0 value with the right-handed helix. The X-ray determination of the structure,[43,44] coupled with the optical rotatory dispersion of myo-

TABLE I

OPTICAL ROTATORY PARAMETERS b_0 AND a_0 IN VARIOUS SOLVENT SYSTEMS[a]

Solvent	b_0	a_0
Poly-L-glutamic Acid (pH 4.88, 20°)		
H_2O	−625	+ 69
0.2 M NaCl	−460	− 25
0.2 M NaCl–dioxane, 2:1	−685	+150
0.2 M NaCl–ethylene glycol, 2:1	−700	+137
H_2O–chloroethanol, 2:1	−703	+112
H_2O–dioxane, 2:1	−694	+120
Copoly-L-glutamic Acid-L-leucine 75:25 (pH 4.88, 20°)		
0.2 M NaCl	−555	− 10
0.2 M NaCl–dioxane, 2:1	−642	− 4
0.2 M NaCl–ethylene glycol, 2:1	−755	+ 2

[a] G. D. Fasman, unpublished data.

globin,[45,46] confirmed this conclusion. A helical content of approximately 75% was found by both methods. The estimation of partial helical content by measuring the b_0 values of synthetic polypeptides, e.g., for copolymers of L-lysine and L-glutamic acid[47,48] and poly-L-lysine,[49] with various parameters, e.g., pH, has justified the use of b_0 in estimating the helical content of materials which are partially helical and partially random, as illustrated in the case of myoglobin.

[40] E. R. Blout, *Tetrahedron* **13**, 123 (1961).
[41] L. Brown and I. F. Trotter, *Trans. Faraday Soc.* **52**, 537 (1956).
[42] A. Elliott and B. R. Malcolm, *Nature* **178**, 912 (1956); *Proc. Roy. Soc.* **A249**, 30 (1958).
[43] J. C. Kendrew, R. E. Dickerson, B. E. Strandberg, R. G. Hart, D. R. Davies, D. C. Phillips, and V. C. Shore, *Nature* **185**, 422 (1960).
[44] J. C. Kendrew, H. C. Watson, B. E. Strandberg, R. E. Dickerson, D. C. Phillips, and V. C. Shore, *Nature* **190**, 666 (1961).
[45] P. J. Urnes, K. Imahori, and P. Doty, *Proc. Natl. Acad. Sci. U.S.* **47**, 1635 (1961).
[46] S. Beychok and E. R. Blout, *J. Mol. Biol.* **3**, 769 (1961).
[47] E. R. Blout and M. Idelson, *J. Am. Chem. Soc.* **80**, 4909 (1958).
[48] P. Doty, K. Imahori, and E. Klemperer, *Proc. Natl. Acad. Sci. U.S.* **44**, 424 (1958).
[49] J. Applequist and P. Doty *in* "Polyamino Acids, Polypeptides and Proteins" (M. Stahmann, ed.), p. 161. Univ. of Wisconsin Press, Madison, Wisconsin, 1962.

Studies on synthetic polypeptides have revealed some anomalies. Poly-β-benzyl-L-aspartate has a b_0 value of $+611$ and the helix was shown to be left-handed instead of right-handed,[50,51] (as are other L-amino acids). There are several exceptions to the sign of b_0 and the helical sense associated with it. Poly-L-tyrosine[52,53] has a positive b_0 value; however, it was demonstrated to have a right-handed helical structure.[54] The probable cause of this anomaly is the interaction of the chromophores of the aromatic side chains with the polypeptide backbone. Poly-L-tryptophan[55] also has a b_0 value of $+410$. Poly-1-benzyl-L-histidine dichloroacetate is thought to be a left-handed helix,[56] based on infrared and optical rotatory studies. Poly-L-proline exists in two helical modifications; however, this material cannot exist as an α-helix owing to the presence of a five-membered imino group[22,57,58] incapable of hydrogen bonding.

One of the greatest gaps in our knowledge of polypeptide conformation is the optical rotatory properties of the β-conformation (or pleated sheet structure). The β-conformation in low-molecular-weight synthetic polypeptides is known to be concentration-dependent.[59-62] Poly-O-acetyl-L-serine,[63] poly-L-serine,[64] and poly-O-benzyl-L-serine[65] exist in the β-conformation in the solid state and in solution. Poly-O-benzyl-L-serine[65] has an a_0 of $+600$ and a b_0 of $+190$ for the β-conformation. Although these values differ from other values reported,[62] it should be pointed out that it has not yet been possible to obtain a true solution of a β-polymer without the presence of a strong hydrogen-bonding agent such as dichloroacetic acid. Therefore, the absolute values should be taken with

[50] E. R. Blout and R. H. Karlson, *J. Am. Chem. Soc.* **80**, 1259 (1958); R. H. Karlson, K. S. Norland, G. D. Fasman, and E. R. Blout, *ibid.* **82**, 2268 (1960).
[51] E. M. Bradbury, A. R. Downie, A. Elliott, and W. E. Hanby, *Proc. Roy. Soc.* **A259**, 110 (1960).
[52] A. Elliott, W. E. Hanby, and B. R. Malcolm, *Nature* **180**, 1340 (1957).
[53] J. D. Coombes, E. Katchalski, and P. Doty, *Nature* **185**, 534 (1960).
[54] G. D. Fasman, *Nature* **193**, 681 (1962).
[55] M. Sela, I. Z. Steinberg, and E. Daniel, *Biochim. et Biophys. Acta* **46**, 433 (1961).
[56] K. S. Norland, G. D. Fasman, E. Katchalski, and E. R. Blout, "Biopolymers" (in press); see also ref. 6.
[57] J. Kurtz, A. Berger, and E. Katchalski, *Nature* **178**, 1066 (1956).
[58] E. R. Blout and G. D. Fasman, *in* "Recent Advances in Gelatin and Glue Research" (G. Stainsby, ed.), p. 122. Pergamon, London, 1958.
[59] E. R. Blout and A. Asadourian, *J. Am. Chem. Soc.* **78**, 955 (1956).
[60] J. T. Yang and P. Doty, *J. Am. Chem. Soc.* **79**, 761 (1957).
[61] P. Doty, A. M. Holtzer, J. H. Bradbury, and E. R. Blout, *J. Am. Chem. Soc.* **76**, 4493 (1954).
[62] A. Wada, M. Tsuboi, and E. Konishi, *J. Phys. Chem.* **65**, 1119 (1961).
[63] G. D. Fasman and E. R. Blout, *J. Am. Chem. Soc.* **82**, 2262 (1960).
[64] Z. Bohak and E. Katchalski, *Biochemistry* **2**, 228 (1963).
[65] E. M. Bradbury, A. Elliott, and W. E. Hanby, *J. Mol. Biol.* **5**, 487 (1962).

reservation. The important problem of the interpretation of the optical rotatory dispersion of polypeptides which contain helical, β, and random regions has not been solved. Schellman and Schellman[17] have discussed various possible combinations of the parameters for the α-helical, β-, and random conformations.

One further observation on synthetic polypeptides which, in all likelihood, will bear significantly on the interpretation of optical rotatory dispersion is the following: On the basis of infrared measurements, the amino acids can be classified into helix and nonhelix formers when incorporated into poly-α-amino acids.[66] Those amino acids which have disubstitution on the β-carbon (e.g., valine) or have a hetero-atom on the β-carbon (e.g., serine) are all β-formers. Thus, by these criteria, valine, isoleucine, serine, cysteine, and threonine are non-α-helix formers, and also, for other reasons, glycine, proline, and hydroxyproline. Protein sequences of these amino acids probably cannot form α-helical structures.

If polypeptides and proteins were constituted of only disordered regions and helical regions, the mean residue rotations could be expressed as a sum of $[m']^D$, disordered, and $[m']^H$, helical. Thus, $[m'] = f_D [m']^D + f_H [m']^H$, where f_D is the fraction in the disordered chain, and f_H is the fraction in helical chain.[11] However, it is known that other structures probably exist (see above), and also there is probably a contribution of the end effects, which are not likely to be hydrogen-bonded into a definitive structure. The contribution of the end effects will be dependent on the solvent. Studies have shown[67] that the residue rotation changes for increasing lengths of lysyl low-molecular-weight polymers. Further work has demonstrated that in nonaqueous media twenty-two residues were required to form a helix for γ-methyl-L-glutamate.[68] However, the stability of any structure will depend on the solvent in which it is dissolved and also on whether there is any stabilization in addition to hydrogen bonding such as hydrophobic forces.[69]

Optical Rotatory Dispersion of Proteins

Whereas the analysis of synthetic polypeptide conformation can be handled with a fair degree of competence (excluding the β-structure

[66] E. R. Blout, C. de Loze, S. M. Bloom, and G. D. Fasman, *J. Am. Chem. Soc.* **82**, 3787 (1960); S. M. Bloom, G. D. Fasman, C. de Loze, and E. R. Blout, *ibid.* **84**, 458 (1962).

[67] B. Erlanger and E. Brand, *J. Am. Chem. Soc.* **73**, 4025 (1951).

[68] M. Goodman, E. E. Schmitt, and D. Yphantis, *J. Am. Chem. Soc.* **82**, 3483 (1960); M. Goodman and E. E. Schmitt, *ibid.* **81**, 5307 (1959).

[69] W. Kauzmann, *Advances in Protein Chem.* **14**, 1 (1959); G. D. Fasman, C. Lindblow, and E. Bodenheimer, *J. Am. Chem. Soc.* **84**, 4977 (1962); and references cited therein.

and perhaps other forms), the attempted analyses of proteins into helical and disordered forms have had varying success.

The optical rotatory dispersion of all native globular proteins measured obey the simple Drude equation in the spectral range 350 to 600 $m\mu$. The λ_c values fall between 280 $m\mu$ and less than 210 $m\mu$. On denaturation this dispersion remains simple, and the λ_c values approach that found for randomly coiled synthetic polypeptides, $\lambda_c = 212$.

Schellman and Schellman[17] have divided the globular proteins into three groups: (1) λ_c greater than 230 $m\mu$; e.g., bovine serum albumin, $\lambda_c = 264$; (2) λ_c 210 to 230, the same region as denatured proteins, e.g., pepsin, $\lambda_c = 216$; (3) λ_c less than 210, e.g., γ-lactoglobulins, $\lambda_c = 205$ to 215 (see also Jirgensons[13]).

Excellent collections of data are found in reviews by Schellman and Schellman,[17] Jirgensons,[13] Blout,[6] Todd,[20] and Urnes and Doty.[11] Yang and Doty[33] have demonstrated with poly-L-glutamic acid that, up to a 40% helical content, a simple Drude relationship is obeyed (in the 700- to 350-$m\mu$ range), and λ_c changes from 212 for the random chain to 268 at the 40% helical composition. Above a 40% helical conformation, a curved Drude plot is obtained. Thus, one can roughly estimate helical content by comparing the λ_c value in the native state to that found in the denatured state. Below 300 $m\mu$, protein optical rotatory dispersion curves become more complex.

The spectrum of the optical rotatory dispersion measurements on proteins may be divided up as follows: (1) In the visible range the simple dispersion is obeyed, the native form having λ_c and $[\alpha]_{589}$ values greater than those found in the disordered chain. (2) In the ultraviolet region down to 300 $m\mu$, the dispersion becomes more complex for globular proteins, with λ_c higher than those derived from visible measurements. (3) In the far ultraviolet the globular proteins, e.g., serum albumin, have shown complex dispersion; $[\alpha]_\lambda$ becomes extremely levorotatory $(-15,000°)$, with a sharp minimum at 233 $m\mu$ and then shows a maximum at 196 $m\mu$ ($[\alpha] = 75,000°$).[10] This is due to the Cotton effect and will be treated below. The data in the visible and near ultraviolet have been fitted to the Moffitt equation, with $\lambda_0 = 212$, and estimates of helical content have been made on the basis of the b_0 values obtained. Foster et al.[70] have demonstrated that the most suitable λ_0 for bovine serum albumin is 218 ± 2 $m\mu$.

The fibrous proteins, however, are more analogous to the α-helical synthetic polypeptides. *Penna nobiles* tropomyosin has a b_0 value of

[70] M. Sogami, W. L. Leonard, Jr., and J. F. Foster, *Arch. Biochem. Biophys.* **100**, 206 (1963).

—650,[71,72] and for these muscle proteins a λ_c of 210 ± 10 mμ has been found for the denatured form. The Cotton effect at 233 mμ also has the same magnitude as a completely helical synthetic polypeptide.[73]

Denatured serum albumin and ovalbumin display positive b_0 and a_0 values, and this has been interpreted as evidence of β-structures (see above).[74]

There are numerous examples of the effect of various solvents on protein conformation. When a nonaqueous solvent is added to an aqueous solution of a protein, an increase in the magnitude of $-b_0$ is often observed, which has been interpreted as an increase in helical content, e.g., chloroethanol,[75-79] dioxane,[78-83] and ethanol.[79,81,83] However, caution should be exercised in such interpretations, since increased $-b_0$ values have been found with copolymers of L-glutamic acid and L-leucine in mixed solvents (aqueous–organic), and values as high as -890 have been observed[39] (see also Table I). The meaning of such values is not at present understood but clearly cannot be interpreted if -630 is taken to represent a 100% helical polymer.

The Cotton Effect[84]

When the measurement of rotation is carried out through an optically active absorption band, the phenomenon termed the Cotton effect is observed (Fig. 4). As the absorption maximum of the chromophore is reached, the rotation rises with longer wavelengths, falls sharply, passing through zero at the center of the band with a point of inflection, then shows a minimum, and rises again. This is termed a positive Cotton

[71] C. M. Kay and K. Bailey, *Biochim. et Biophys. Acta* 31, 20 (1959).
[72] C. Cohen and A. Szent-Gyorgi, *J. Am. Chem. Soc.* 79, 248 (1957).
[73] N. S. Simmons, C. Cohen, A. G. Szent-Gyorgi, D. B. Wetlaufer, and E. R. Blout, *J. Am. Chem. Soc.* 83, 4766 (1961).
[74] K. Imahori, *Biochim. et Biophys. Acta* 37, 336 (1960).
[75] P. Doty, *Proc. 4th Intern. Congr. Biochem. Vienna, 1958* 8, 8 (1960); P. Doty, K. Imahori, and E. Klemperer, *Proc. Natl. Acad. Sci. U.S.* 44, 424 (1958).
[76] R. Weber and C. Tanford, *J. Am. Chem. Soc.* 81, 3255 (1959).
[77] M. Marsh, *J. Am. Chem. Soc.* 84, 1896 (1962).
[78] J. Brahms and C. M. Kay, *J. Biol. Chem.* 237, 3449 (1962).
[79] C. Tanford, P. K. De, and V. G. Taggart, *J. Am. Chem. Soc.* 82, 6028 (1960).
[80] C. Tanford, *J. Am. Chem. Soc.* 84, 1747 (1962).
[81] C. Tanford and P. K .De, *J. Biol. Chem.* 236, 1711 (1961).
[82] S. E. Bressler, *Discussions Faraday Soc.* 25, 158 (1958).
[83] C. Tanford, C. E. Buckley, P. K. De, and E. P. Lively, *J. Biol. Chem.* 237, 1168 (1962).
[84] A. Cotton, *Ann. chim. et phys.* 8, 347 (1896).

effect. The reverse, showing a minimum first at longer wavelength, is termed a negative Cotton effect.

Cotton effects have been observed due to several phenomena.

1. *The peptide chromophore absorption bands.* Simmons and Blout[85] measured a minimum in the optical rotatory dispersion of tobacco mosaic virus protein at 232 mμ (a negative Cotton effect) and extended this observation to polypeptides and proteins.[73] This Cotton effect is due to

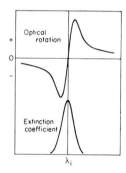

FIG. 4. Idealized Cotton effect at an isolated, optically active absorption band with its maximum at λ_i. If the positive limb of the rotatory dispersion is to the high wavelength side of the band, the Cotton effect is termed positive; if the negative limb is at higher wavelength, the effect is negative. In regions distant from the absorption band, the rotatory dispersion approaches simple Drude behavior. From P. Urnes and P. Doty, *Advances in Protein Chem.* **16**, 419 (1961).

an $n \rightarrow \pi^*$ transition[86] of the amide chromophore, which shows extremely weak absorption.[25–27] The value of this minimum is about $[m'] = 12,000°$. Another Cotton effect[10] (positive) at 190 mμ has been found in poly-L-glutamic acid ($[m'] = 80,000°$) and serum albumin, and this is thought to be due to the main $N \rightarrow V_1$ transition of the amides and polypeptides. Both these Cotton effects are conformation-dependent and vanish in the random coil, where a new negative weaker Cotton effect with trough at 204 mμ appears. This is probably due to the peptide bond, from the asymmetric α-amino acids, without superimposed conformational effects.

2. *Binding of dyes or other chromophores to a helical structure.* Blout and Stryer[87] have demonstrated that one or more 'Cotton effects are observed when the cationic dyes acriflavene, Rhodamine 6, and pseudocsacyanine are asymmetrically bound to the helical conformation of poly-α-L-glutamic acid. The polymer of the L- and D-amino acids gave the opposite Cotton effect, and it was suggested that this method may be

[85] N. S. Simmons and E. R. Blout, *Biophys. J.* **1**, 55 (1960).
[86] J. A. Schellman and P. Oriel, *J. Chem. Phys.* **37**, 2114 (1962).
[87] E. R. Blout and L. Stryer, *Proc. Natl. Acad. Sci. U.S.* **45**, 1591 (1959).

useful to determine the sense of helix. The random coil did not display the Cotton effect, although the dye remained bound.

3. *The Cotton effect due to binding of prosthetic groups.* Proteins containing prosthetic groups which are presumably asymmetrically bound display Cotton effects at the wavelengths of the most prominent absorption bands of these groups. Cotton effects have been observed with the following proteins containing heme as the prosthetic groups; cytochrome c,[88] myoglobin and hemoglobin,[46,89] catalase,[89] and sickle cell hemoglobin.[90]

4. *The binding of coenzymes containing chromophores to enzymes.* Ulmer et al.[91,92] have demonstrated that Cotton effects are observed due to the binding of reduced diphosphopyridine nucleotide to yeast alcohol dehydrogenase. They used this procedure to demonstrate the stoichiometry of the binding and, also, the inhibition by chelation of the cofactor, zinc, with 1,10-phenanthroline.

Summary of Methods for Measuring the Helical Content of Proteins

1. The best established procedure is the measurement of the b_0 value by use of the Moffitt equation; $b_0{}^H$ (H = helix) is -630 for the helical conformation of many synthetic polypeptides, and $b_0 = 0$ for the disordered chain. This implies that λ_0 and λ_c are equal, with the value of 212 mμ. Here λ_0 represents the value for the disordered chain; however, λ_c might not equal this for several reasons. Denatured proteins may not necessarily be random coils, e.g., due to disulfide groups. X-Ray analysis of myoglobin has shown that the disordered regions have definite orientation with respect to the rest of the molecule and thus may contribute to λ_c in a manner other than that of the random coil. The λ_0 value of 212 mμ is valid above 300 mμ, but below 300 mμ down to 240 mμ $\lambda_0 = 216$ mμ is more suitable. For example, in the wavelength range 240 to 280 mμ for tropomysin, a b_0 value of -900 is obtained[11] with $\lambda_0 = 212$; however, with $\lambda_0 = 216$, a b_0 value of -560 is obtained.

2. The standard value of $\Delta[m']_{589}^{H-D} = 88°$. This is the change in mean residue rotation on changing from the helical to the random conformation; the specific rotation in water with a mean residue weight of 115, $[\alpha]_{589}^{H-D}$, is $97°$. $\Delta[m']^{H-D}$ may not be a good indication of helical

[88] G. Eichhorn and J. F. Cairns, *Nature* 181, 994 (1958).
[89] G. Eichhorn, *Tetrahedron* 13, 208 (1961).
[90] M. Murayama, *Federation Proc.* 20(1), 384 (1961).
[91] D. D. Ulmer and B. L. Vallee, *J. Biol. Chem.* 236, 730 (1961).
[92] D. D. Ulmer, T. K. Li, and B. L. Vallee, *Proc. Natl. Acad. Sci. U.S.* 47, 1155 (1961).

content, as this measure may represent changes in species of residue, change in ionic strength, solvent, etc.

3. The measurement of the depth of the Cotton effect at 233 mμ may be taken as representative of the helical content. The $[m']_{233}$ value for a 100% helical polypeptide is $-12,700°$, whereas the random coil is $-1800°$. Good agreement between estimates based on $[m']_{233}$ and b_0 values of helical content have been reported. Future work on the Cotton effect at 190 mμ where $[m']_{190} = 80,000°$ may establish this as an additional criterion.

Instrumentation and Methods

The theory and principles involved in polarimetry have been discussed by several authors and will not be reviewed here.[29,30,93,94] There are at present several photoelectric spectropolarimeters available, and two more are to be marketed soon. A brief description of each follows:

1. The Rudolph photoelectric spectropolarimeter[95] (Model 200 AS/80Q/650) was the first photoelectric instrument commercially available, and much of the original work reported was performed on this machine. The principle employed in this instrument, which differentiates it from other instruments, is the mechanical oscillation of the polarizer between two positions (intervals of 2 to 3 seconds between swings). The analyzer is adjusted by hand until the photomultiplier registers the same light intensity on either side of the extinction point. This is known as the method of symmetrical angles. The accuracy of this instrument is between 0.001° and 0.004° of arc. The Rudolph Instruments Engineering Company, Little Falls, New Jersey, is producing an automatic recording spectropolarimeter using the same principle. This instrument is capable of making measurements from the visible down to 220 to 240 mμ. Details of these instruments can be found elsewhere.[93,94]

A later model (O. C. Rudolph and Sons, Model 220/200/1012/658-313/100), which employs a double-prism monochromator, has been successfully used down to 185 mμ with nitrogen flushing.[10]

2. The Keston unit[96] is an attachment for the Beckmann D.U. spectrophotometer. This instrument can be used only to 400 mμ and is not recording. It has found only limited use in work with proteins.[93] A detailed description can be found in Klyne and Parker.[94]

[93] C. Djerassi, "Optical Rotatory Dispersion." McGraw-Hill, New York, 1960.
[94] W. Klyne and A. C. Parker, in "Techniques of Organic Chemistry, Physical Methods" (A. Weissberger, ed.), 3rd ed., Vol. 1, Part 3, p. 2335. Interscience, New York, 1960.
[95] O. C. Rudolph & Sons, Caldwell, New Jersey.
[96] Manufactured by the Standard Polarimeter Co., New York.

3. The Perkin-Elmer recording spectropolarimeter attachment.[97] This attachment is for use with existing double-beam spectrophotometers (Spectrachord Model 4000A). This instrument carries a larger error when compared with the Rudolph spectropolarimeter and also has found little use in protein work. A detailed description of this attachment is found in Djerassi.[93]

New Instruments

At the time of writing, two new recording instruments were about to be marketed, the Cary recording spectropolarimeter[98] and the Bendix-Ericcson-Polarimatic 62.[99] These instruments employ the magneto-optic effect, or Faraday effect, in which the direction of polarization of light in a transparent medium is rotated by means of a magnetic field applied parallel to the beam. This is used to compensate for the rotation caused by the sample, thus allowing the photomultiplier tube to be subjected to a constant beam intensity. The Bendix-Ericsson instrument is a new design in which the functions of monochromator and polarimeter are combined by using two crystalline quartz prisms both to disperse the radiation and to polarize it. The two prisms are fixed in position, and wavelength scanning is achieved by the rotation of two plane mirrors in unison about a common axis. The optical rotation of the specimen is compensated by means of a Faraday cell and is thus measured in terms of current. This instrument is estimated to read to ±0.0002° in the wavelength range 600 to 185 mμ.

The Cary instrument also has a double monochromator. A specially designed Rochon polarizer and analyzer with an automatic compensating system is used to maintain the polarizer in a null or "crossed" orientation. This system includes a motor energized by the amplified current from the ultraviolet-sensitive photomultiplier. The motor moves the polarizer via a compound mechanical linkage substantially free from backlash and effects of friction. This instrument is sensitive to 1⅓ millidegrees and operates from 600 mμ to 185 mμ. The prospects of a recording instrument with the sensitivities claimed are eagerly awaited. For details of the principles involved in these instruments the literature references should be consulted.[100-102]

[97] Perkin-Elmer Corp., Norwalk, Connecticut.
[98] Manufactured by the Applied Physics Corp., 2724 South Peck Road, Monrovia, California.
[99] Manufactured by the Bendix-Ericsson U.K. Ltd. and distributed in the United States by the Bendix Corp., Cincinnati Division, Cincinnati 8, Ohio.
[100] M. B. Trageser, *J. Opt. Soc. Am.* **43**, 866 (1952).
[101] E. J. Gillham, *Nature* **178**, 1412 (1956).
[102] E. J. Gillham and R. J. King, *J. Sci. Instr.* **38**, 21 (1961).

General Methods

1. *Solvents.* A solvent must be chosen which transmits throughout the wavelength region of interest. Water is the solvent of choice for protein work. This should be degassed if measurements in the far-ultraviolet (200 mμ) region are required, to remove dissolved oxygen which absorbs in this region.

2. *Solutions.* The solutions to be measured must be optically pure. Therefore the solutions should be clarified by either filtration or centrifugation. Solutions containing suspended material or gel-like particles may be birefringent and cause optical anomalies. Frequently it will be necessary to determine concentrations of such clarified solutions. This can be conveniently done by Kjedahl nitrogen or Nessler analysis, or by using the optical density for protein solutions whose molar extinction coefficients are accurately known.

The optical density of the solutions must be within certain limits. With the Rudolph spectropolarimeter one must use solutions of low absorbance, below the limit where stray light gives rise to artifacts.[11] Also, the lack of spectral purity at high absorbance can give rise to errors in the amplitude of the Cotton effect. Solutions of high absorbance eliminate light for which the monochromator is set, while permitting stray light outside the pass band to reach the analyzer and give rise to rotations characteristic for those wavelengths. Stray light can be polarized by reflection in the optical path.[103] Thus stray light can give rise to artifacts at absorption bands that simulate Cotton effects. In instruments such as the Rudolph spectropolarimeter employing a Beckmann D.U. spectrophotometer, the absorbance must be less than 2. To investigate whether an observed Cotton effect is genuine, one employs the Beer's law test. One measures the observed effect at two different concentrations of material in the light path. When a plot is made of the specific rotation, there should be no change in amplitude, shape, or wavelength of the Cotton effect, at both concentrations. Another source of error may be due to the lamp, if there are wavelengths at which it fails to emit light, e.g., resonance absorption by the mercury itself between 255 and 260 mμ—a situation similar to absorption by a chromophore in the light path.

Light Sources

One of the greatest problems in all photoelectric polarimeters is the light source. Stable, high-intensity arcs are usually difficult to obtain and maintain in working order.

---

[103] G. Markus and F. Karush, *J. Am. Chem. Soc.* **80**, 89 (1958).

The most satisfactory light source to date is the Xenon arc. In the author's laboratory a 150-watt Hanovia DC Xenon Compact Arc #901Cl[104] placed in the water-jacketed Beckmann D.U. spectrophotometer light housing has given excellent performance. Other light sources of higher intensity have been used, e.g., the 450-watt d-c Xenon, Osram Berlin.[105]

Polarimeter Tubes

A great variety of polarimeter tubes can be obtained from several manufacturers Rudolph & Sons[95] supply various path lengths in plain or jacket tubes for temperature studies. For work in the visible range, tubes of 1 or 2 decimeters are usually used. It is essential to maintain temperature equilibrium during the period measurements are taken. Thus, a 10- to 15-minute period should be allowed for equilibrium at low temperatures (20° to 50°) and even longer periods at higher temperatures (60° to 70°). A convenient polarimeter tube for observing the temperature of the solution is the center-filling type.[95] A thermometer placed in a stopper can be inserted in the well of the tube, and the temperature of the solution under study can be carefully followed. For work in the ultraviolet and far-ultraviolet, shorter path lengths are desired for two reasons. Firstly, one wishes to cut down on the absorbance of the solution, especially if absorption bands in the wavelength range to be studied are present; and secondly, the rotations usually become much larger with decreasing wavelength, and long path lengths are not desired. As the absorption increases logarithmetrically with concentration and path length, while the rotation increases linearly with these variables, then by reduction of the path length a gain is achieved in cutting the absorption versus the loss in rotation. Cells of 1 cm and 1 mm (fused quartz) are obtainable[106] for this spectral range. These should be checked for birefringence by running cell blanks throughout the spectrum range of interest.

Measurements should be made with the solutions under study and then a cell blank with the identical solvent run immediately afterward, with the measurement repeated at each wavelength. If a polarimeter tube with removable end plates is used, great care should be taken not to cause any strain on these end plates by excessive tightening. Consequently the center-filling type is desirable, as both solvent and solution can be run without the dangers involved in removing end plates.

[104] Hanovia Lamp Division, Engelhard Hanovia, Inc., 100 Chestnut St., Newark, New Jersey.
[105] S. B. Zimmerman and J. A. Schellman, *J. Am. Chem. Soc.* **84**, 2259 (1962).
[106] Optical Cell Co. Inc., 4204 37th St., Brentwood, Maryland.

Calculation of the Constants of the Drude ánd Moffitt Equations

1. *The Specific Rotation:*

$$[\alpha]_\lambda^T = \frac{\alpha_{obs} \times 10^2}{l \times c} \tag{1}$$

where T = temperature.

λ = wavelength.

α_{obs} = polarimeter reading of solution minus polarimeter tube solvent blank.

l = length of tube in decimeters.

c = concentration in grams per 100 ml.

2. *Mean Residue Rotations:*

$$[m]_\lambda = \frac{MRW}{100} [\alpha]_\lambda \tag{3}$$

where MRW = mean residue weight . For most proteins, 115 is used.

3. *Reduced Mean Residue Rotation:*

$$[m'] = [m] \frac{3}{(n^2 + 2)} = \frac{MRW}{100} \times \frac{3}{(n^2 + 2)} [\alpha]_\lambda \tag{4}$$

where n is the refractive index at wavelength λ.

4. *Drude Equation:*

$$[\alpha]_\lambda = \frac{A}{\lambda^2 - \lambda_c^2} \tag{6}$$

where $[\alpha]_\lambda$ = specific rotation as defined above.

5. *Moffitt Equation:*

$$[\alpha]_\lambda = \frac{100}{MRW} \frac{(n^2 + 2)}{3} \left[\frac{a_0 \lambda_0^2}{\lambda^2 - \lambda_0^2} + \frac{b_0 \lambda_0^4}{(\lambda^2 - \lambda_0^2)^2} \right] \tag{8}$$

where $[\alpha]_\lambda$ = specific rotation as defined above.

MRW = mean residue weight.

n = refractive index of solvent at wavelength λ.

λ_0 = 2120 A or 2160 A (see text above).

a_0 = constant (see below)

b_0 = constant (see below)

Calculations

1. *Drude Plot:*

$[\alpha]_\lambda^T$ is calculated as indicated above at various λ's.

$[\alpha]\lambda^2$ is calculated from Table II.

Plot $[\alpha]\lambda^2$ versus $[\alpha]$; the square root of the slope is λ_c.

2. Moffitt Plot:

Rearranging Eq. (8), we have

$$[\alpha](\lambda^2 - \lambda_0^2) = \frac{100}{MRW}\left(\frac{n^2 + 2}{3}\right)\left[a_0\lambda_0^2 + \frac{b_0\lambda_0^4}{\lambda^2 - \lambda_0^2}\right] \qquad (9)$$

and from Eq. (4):

$$[m'](\lambda^2 - \lambda_0^2) = a_0\lambda_0^2 + \frac{b_0\lambda_0^4}{(\lambda^2 - \lambda_0^2)}$$

$[\alpha] (\lambda^2 - \lambda_0^2)$ is calculated from Table II.

Alternatively, $[m'] (\lambda^2 - \lambda_0^2)$ may be calculated in a similar fashion. Plot either of these against $1/(\lambda^2 - \lambda_0^2)$ (Table II).

3. b_0 Calculation:

If $[\alpha] (\lambda^2 - \lambda_0^2)$ was plotted:

$$b_0 = \frac{\text{slope}}{K\lambda_0^4}$$

where $K = \dfrac{100}{M}\left(\dfrac{n^2 + 2}{3}\right)$ (K obtained from Tables I to IX)

$\lambda_0^4 = 20.2 \times 10^{12}$ ($\lambda_0 = 212$).

If $[m'] (\lambda^2 - \lambda_0^2)$ was plotted:

$$b_0 = \frac{\text{slope}}{\lambda_0^4}$$

For example, see a Moffitt plot of the rotatory dispersion of paramyosin, as shown in Fig. 5.

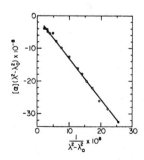

FIG. 5. A Moffitt plot of the rotatory dispersion of paramyosin in 0.3 M KCl, 0.01 M, pH 7.4, phosphate buffer at 20°, $b_0 = -570$. From L. M. Riddeford and H. A. Scheraga, *Biochemistry* 1, 108 (1962).

Sample Calculation for b_0

Suppose we have a solution of polybenzyl-L-glutamate (PBG) in $CHCl_3$:

$$MRW = 218$$

$$100 \left(\frac{n^2 + 2}{3} \right) = 136.3 \text{ from Table IX, CHCl}_3 \text{ solution}$$

$$K_{PBG}^{CHCl_3} = \frac{100}{M} \left(\frac{n^2 + 2}{3} \right) = 0.621$$

$$\lambda_0^4 = (2.12 \times 10^3)^4 = 20.2 \times 10^{12}$$

$$K_{PBG}^{CHCl_3} \lambda_0^4 = 12.55 \times 10^{12}$$

$$b_0 = \frac{\text{slope}}{12.55 \times 10^{12}} = \frac{\text{slope}}{K\lambda_0^4}$$

Values for the refractive index corrections of water, urea solutions, LiBr solutions, and various organic solvents are found in Tables III to IX.

Sample Calculation for a_0

If $[\alpha] (\lambda^2 - \lambda_0^2)$ is plotted:

$$a_0 = \frac{\text{Zero intercept}}{K\lambda_0^2}$$

where K is obtained as above.

If $[m'] (\lambda^2 - \lambda_0^2)$ is plotted:

$$a_0 = \frac{\text{Zero intercept}}{\lambda_0^2}$$

TABLE II
For b_0 Calculations[a]
$(\lambda_c = 212)$

λ, mμ	$\lambda^2 \times 10^8$ (A^2)	$(\lambda^2 - \lambda^2_{212}) \times 10^8$ (A^2)	$(\lambda^2 - \lambda^2_{212})^{-1} \times 10^{-8}$ (A^2)$^{-1}$
224	0.050176	0.005232	192.0
225	0.050625	0.005681	176.2
226	0.051076	0.006132	161.5
227	0.051529	0.006575	152.2
228	0.051984	0.007040	142.2
229	0.052441	0.007497	133.9
230	0.0529	0.007956	126.0
232	0.053824	0.00888	112.0
234	0.054756	0.009812	102.0
236	0.055696	0.010752	93.2
237	0.0560	0.0111	90.0
238	0.056644	0.0117	85.25
240	0.05760	0.0126	78.99
242	0.0585	0.0136	73.5
244	0.0595	0.0146	68.5
245	0.06003	0.01509	66.27
246	0.0605	0.0156	64.0
248	0.0615	0.0166	60.3
250	0.0625	0.01757	56.92
252	0.0635	0.0185	54.0
253.6	0.06431	0.01937	51.63
254	0.064516	0.019572	51.0
255	0.06500	0.02056	49.60
256	0.065536	0.020592	48.7
257	0.066049	0.021105	47.4
258	0.066564	0.021620	46.4
260	0.06760	0.02266	44.13
265	0.07023	0.02529	39.54
270	0.07290	0.02796	35.77
275	0.07563	0.03069	32.58
278	0.077284	0.032340	31.00
280	0.07840	0.03346	29.89
281	0.078961	0.034017	29.4
285	0.08122	0.03628	27.56
289.4	0.08375	0.03881	25.77
296.7	0.08803	0.04309	23.21
302.1	0.09126	0.04632	21.59
310	0.09610	0.05116	19.55
313.2	0.09809	0.05315	18.81
320	0.1024	0.0575	17.39
325	0.1056	0.0607	16.47

[a] Underlined wavelength denotes Hg lines.

TABLE II (*Continued*)

λ, mμ	$\lambda^2 \times 10^8$ (A^2)	$(\lambda^2 - \lambda^2_{212}) \times 10^8$ (A^2)	$(\lambda^2 - \lambda^2_{212})^{-1} \times 10^{-8}$ (A^2)$^{-1}$
330	0.1089	0.0640	15.63
334.1	0.1116	0.0667	14.99
335	0.1122	0.0674	14.88
340	0.1156	0.0707	14.14
345	0.1190	0.0741	13.50
350	0.1225	0.0776	12.89
355	0.1260	0.0811	12.33
360	0.1296	0.0847	11.81
365.2	0.1334	0.0885	11.30
366.5	0.1343	0.0894	11.17
370	0.1369	0.0920	10.87
375	0.1406	0.0957	10.45
380	0.1444	0.0995	10.05
385	0.1482	0.1033	9.68
391.5	0.1533	0.1084	9.23
400	0.1600	0.1151	8.69
404.8	0.1639	0.1190	8.40
405	0.1640	0.1191	8.39
410	0.1681	0.1232	8.12
415	0.1722	0.1273	7.86
420	0.1764	0.1315	7.60
425	0.1806	0.1357	7.37
430	0.1849	0.1400	7.14
435.8	0.1899	0.1450	6.90
436	0.1900	0.1451	6.89
440	0.1936	0.1487	6.72
445	0.1980	0.1531	6.53
450	0.2025	0.1576	6.35
460	0.2116	0.1667	6.00
470	0.2209	0.1760	5.68
475	0.2256	0.1807	5.53
480	0.2304	0.1855	5.39
486	0.2360	0.1913	5.23
492	0.2421	0.1962	5.07
493	0.2431	0.1981	5.05
500	0.2500	0.2051	4.88
510	0.2601	0.2152	4.65
520	0.2704	0.2255	4.43
525	0.2756	0.2307	4.33
530	0.2809	0.2360	4.24
540	0.2916	0.2467	4.05
546	0.2981	0.2532	3.95
548	0.3003	0.2554	3.92

TABLE II (*Continued*)

λ, mμ	$\lambda^2 \times 10^8$ (A²)	$(\lambda^2 - \lambda^2_{212}) \times 10^8$ (A²)	$(\lambda^2 - \lambda^2_{212})^{-1} \times 10^{-8}$ (A²)⁻¹
550	0.3025	0.2576	3.88
560	0.3136	0.2687	3.72
570	0.3249	0.2800	3.57
575	0.3306	0.2857	3.50
578	0.3341	0.2892	3.46
589	0.3469	0.3020	3.31
600	0.3600	0.3151	3.17
625	0.3906	0.3457	2.89
650	0.4225	0.3776	2.65
675	0.4556	0.4107	2.43
700	0.4900	0.4451	2.25
725	0.525		
750	0.562		

TABLE IIA
For b_0 CALCULATIONS
($\lambda_c = 216$)

λ, mμ	$\lambda^2 \times 10^8$ (A²)	$(\lambda^2 - \lambda^2_{216}) \times 10^8$ (A²)	$(\lambda^2 - \lambda^2_{216})^{-1} \times 10^{-8}$ (A²)⁻¹
294	0.0864	0.03974	25.16
286	0.08179	0.03513	28.47
278	0.07728	0.03062	32.66
270	0.07290	0.02624	38.11
263	0.06917	0.02251	44.42
257	0.06605	0.01939	51.57
250	0.0625	0.01584	63.13
244	0.0595	0.01284	77.88
238	0.05664	0.00998	100.20
233	0.05429	0.00763	131.06
227	0.05153	0.00487	205.34
222	0.04928	0.00262	381.68
217	0.04709	0.00043	
213	0.04537		
208.5	0.04347		
204	0.04162		
200	0.04000		
196	0.03842		
192	0.03686		

TABLE III
DISPERSION OF THE REFRACTIVE INDEX OF WATER $(n^{20})^a$

λ, mµ	n^{20}	$\dfrac{n^2 + 2}{3}$	$\dfrac{3}{n^2 + 2}$
182.9	1.4640	1.3811	0.7241
199.0	1.4257	1.3445	0.7438
231.3	1.3888	1.3098	0.7635
242.8	1.3810	1.3024	0.7678
257.3	1.3745	1.2964	0.7714
267.6	1.3690	1.2914	0.7744
274.9	1.3664	1.2890	0.7758
308.2	1.3567	1.2802	0.7811
340.4	1.3504	1.2745	0.7846
361.1	1.3474	1.2718	0.7863
394.4	1.3437	1.2685	0.7883
396.8	1.3435	1.2683	0.7885
434.1	1.3404	1.2656	0.7901
436.0	1.3403	1.2655	0.7902
441.6	1.3398	1.2650	0.7905
467.8	1.3382	1.2636	0.7914
480	1.3375	1.2630	0.7918
486.1	1.3371	1.2626	0.7920
533.8	1.3350	1.2607	0.7932
535.0	1.3349	1.2607	0.7932
546.0	1.3345	1.2603	0.7935
577.0	1.3334	1.2593	0.7941
579.0	1.3333	1.2592	0.7942
589.3	1.3330	1.2589	0.7943
656.3	1.3312	1.2574	0.7953
670.8	1.3308	1.2570	0.7955
768.2	1.3289	1.2553	0.7966
871	1.3270	1.2536	0.7977
943	1.3258	1.2526	0.7983

[a] From "International Critical Tables," Vol. VII, p. 13, 1930.

TABLE IV
VARIATION OF REFRACTIVE INDEX OF H_2O, $(n)^T$, WITH TEMPERATURE[a]

Temperature	λ_{434}	λ_{486}	λ_{589}
0°			1.3340
5°			1.3339
10°	1.3411	1.3378	1.3337
20°	1.3404	1.3371	1.3330
60°	1.3346	1.3315	1.3272
70°	1.3325	1.3294	1.3252

[a] From "International Critical Tables," Vol. VII, p. 13, 1930.

TABLE V

REFRACTIVE INDEX CALCULATIONS OF VARIOUS SOLVENTS[a]

Solvent	n_D^{20}	$\dfrac{n^2 + 2}{3}$	$\dfrac{3}{n^2 + 2}$
Acetic acid	1.3718	1.2939	0.7729
Chloroethanol	1.4419	1.3597	0.7355
Chloroform	1.446	1.3636	0.7334
Dichloroacetic acid[b]	1.4659	1.3830	0.7231
Dimethyl formamide[c]	1.4280	1.3464	0.7427
Dimethyl sulfoxide[d]	1.4787	1.3955	0.7166
Dioxane	1.4232	1.3418	0.7453
Ethylene dichloride	1.4443	1.3620	0.7342
Ethylene glycol[c]	1.4306	1.3489	0.7413
Trifluoroacetic acid	1.285	1.2171	0.8216
Water	1.3330	1.2590	0.7943

[a] From "Merck Index."
[b] 22°.
[c] 25°.
[d] 21°.

TABLE VI

REFRACTIVE INDEX OF UREA (n_D^{20}) VERSUS CONCENTRATION[a]

M^b	n_D^{20}	$\dfrac{n^2 + 2}{3}$	$\dfrac{3}{n^2 + 2}$
1.2534	1.3442	1.2690	0.7880
2.5763	1.3560	1.2796	0.7815
5.7296	1.3855	1.3065	0.7654
7.2611	1.4000	1.3200	0.7576

[a] From Landolt-Börnstein's Physikalische-Chemische Tabellen, EII, 5th ed. Springer, Berlin, 1923.
[b] M = molarity.

TABLE VII
REFRACTIVE INDEX OF SODIUM CHLORIDE ($n_D{}^{25}$) VERSUS CONCENTRATION[a]

Grams per 100 g. of mixture	$n_D{}^{25}$	$\dfrac{n^2 + 2}{3}$	$\dfrac{3}{n^2 + 2}$
0.5280	1.3334	1.2593	0.7941
0.5493	1.3334	1.2593	0.7941
0.9980	1.3342	1.2600	0.7937
1.0618	1.3343	1.2601	0.7936
1.1068	1.3344	1.2602	0.7935
5.3562	1.3417	1.2667	0.7895
5.4131	1.3418	1.2668	0.7894
14.344	1.3575	1.2809	0.7807

[a] From "International Critical Tables," Vol. VII, p. 73, 1930.

TABLE VIII
REFRACTIVE INDEX OF LiBr ($n_D{}^{25}$) VERSUS CONCENTRATION[a]

Grams per 100 g. of mixture	$n_D{}^{25}$	$\dfrac{n^2 + 2}{3}$	$\dfrac{3}{n^2 + 2}$
0.1980	1.3327	1.2587	0.7945
0.4313	1.3331	1.2591	0.7942
1.0244	1.3340	1.2599	0.7937
1.8718	1.3353	1.2610	0.7930
3.7527	1.3383	1.2637	0.7913
4.2994	1.3391	1.2644	0.7909
14.966	1.3566	1.2801	0.7812
32.55	1.3919	1.3125	0.7619

[a] From Landolt-Börnstein's Physikalische-Chemische Tabellen, EII, 5th ed. Springer. Berlin.

TABLE IX
DISPERSION OF REFRACTIVE INDEX OF VARIOUS SOLVENTS[a]

λ, mμ	n	$\dfrac{n^2 + 2}{3}$	$\dfrac{3}{n^2 + 2}$
	CCl_4 (Carbon Tetrachloride)		
265.5			
289.4			
313.1	1.4985	1.4152	0.7066
365.0	1.4831	1.3999	0.7143
435.8	1.4706	1.3876	0.7207
546.1	1.4603	1.3775	0.7260
	2-Chloroethanol		
265.5	1.4940	1.4107	0.7089
289.4	1.4823	1.3991	0.7148
313.1	1.4731	1.3900	0.7194
365.0	1.4608	1.3780	0.7257
435.8	1.4511	1.3686	0.7307
546.1	1.4423	1.3601	0.7352
	$CHCl_3$ (Chloroform)		
265.5	1.5051	1.4218	0.7033
289.4	1.4911	1.4078	0.7103
313.1	1.4806	1.3974	0.7156
365.0	1.4661	1.3832	0.7230
435.8	1.4546	1.3720	0.7289
546.1	1.4454	1.3631	0.7336
	C_6H_{12} (Cyclohexane)		
265.5	1.4741	1.3910	0.7189
289.4	1.4631	1.3802	0.7245
313.1	1.4549	1.3723	0.7287
365.0	1.4432	1.3609	0.7348
435.8	1.4335	1.3516	0.7399
546.1	1.4249	1.3435	0.7443
	Dichloroacetic Acid		
265.5			
289.4			
313.1	1.5045	1.4212	0.7036
365.0	1.4893	1.4060	0.7112
435.8	1.4776	1.3944	0.7172
546.1	1.4673	1.3843	0.7224
	Dimethylformamide		
265.5			
289.4	1.4913	1.4080	0.7102

[a] We are indebted to Drs. J. Foss, Y. Kang, and J. Schellman for permission to publish these tables.

TABLE IX (*Continued*)

λ, mμ	n	$\dfrac{n^2 + 2}{3}$	$\dfrac{3}{n^2 + 2}$
	Dimethylformamide (cont.)		
313.1	1.4761	1.3930	0.7179
365.0	1.4564	1.3737	0.7280
435.8	1.4419	1.3597	0.7355
546.1	1.4313	1.3495	0.7410
	p-Dioxane		
265.5	1.4699	1.3869	0.7210
289.4	1.4583	1.3755	0.7270
313.1	1.4500	1.3675	0.7313
365.0	1.4384	1.3563	0.7373
435.8	1.4293	1.3476	0.7421
546.1	1.4207	1.3395	0.7466
	Ethylene Dichloride		
265.5	1.5002	1.4169	0.7058
289.4	1.4878	1.4045	0.7120
313.1	1.4778	1.3946	0.7171
365.0	1.4648	1.3819	0.7236
435.8	1.4539	1.3713	0.7292
546.1	1.4447	1.3694	0.7340
	Formamide		
265.5	1.5379	1.4551	0.6872
289.4	1.5139	1.4306	0.6990
313.1	1.4980	1.4147	0.7069
365.0	1.4772	1.3940	0.7174
435.8	1.4619	1.3791	0.7251
546.1	1.4495	1.3670	0.7315
	Formic Acid		
265.5	1.4178	1.3367	0.7481
289.4	1.4063	1.3259	0.7542
313.1	1.3982	1.3183	0.7586
365.0	1.3874	1.3083	0.7644
435.8	1.3785	1.3001	0.7692
546.1	1.3709	1.2931	0.7733
	Furan		
265.5			
289.4			
313.1	1.4766	1.3935	0.7176
365.0	1.4537	1.3711	0.7293
435.8	1.4369	1.3549	0.7381
546.1	1.4241	1.3427	0.7448

TABLE IX (*Continued*)

λ, mμ	n	$\dfrac{n^2 + 2}{3}$	$\dfrac{3}{n^2 + 2}$
Hydrazine			
265.5			
289.4			
313.1			
365.0	1.4980	1.4147	0.7069
435.8	1.4837	1.4005	0.7140
546.1	1.4714	1.3883	0.7203
Methylene Chloride			
265.5	1.4786	1.3954	0.7167
289.4	1.4661	1.3832	0.7230
313.1	1.4561	1.3734	0.7281
365.0	1.4431	1.3609	0.7348
435.8	1.4323	1.3505	0.7405
546.1	1.4237	1.3423	0.7450
8 M Urea (Aqueous Solution)			
265.5	1.4572	1.3745	0.7276
289.4	1.4433	1.3610	0.7347
313.1	1.4340	1.3521	0.7396
365.0	1.4208	1.3396	0.7465
435.8	1.4105	1.3298	0.7520
546.1	1.4022	1.3221	0.7564

[127] The Use of Starch Electrophoresis in Dehydrogenase Studies

By I. H. FINE and L. A. COSTELLO

Introduction

The use of starch as a supporting medium for electrophoretic separation is of growing importance in enzyme chemistry. Certain advantages over paper, agar, silica, asbestos fiber, or glass powder make this the preferred tool for analytical studies as well as for preparative isolation. In general, starch electrophoresis has proved to be a gentle-enough procedure for proteins so that denaturation effects are minimal; gels or homogeneous pastes are easily prepared with buffers. Adsorption of most proteins on starch is slight; electroösmosis is lower than in most other supporting media, and resolving power is great.

A simplified theoretical explanation for phenomena involved in electrophoresis is presented by Kunkel.[1] For more rigorous treatment, Bier may be consulted.[2] Specific applications and experimental conditions for a variety of substances are available in the literature.[3-5]

In what follows we shall discuss the most important procedural aspects, using illustrations from our experience with dehydrogenases, for which starch electrophoresis has been employed as a routine technique in our laboratory.

Starch Block

Preparation of the Starch

The starch can be prepared in small quantities for individual experiments or in large batches for routine use and storage.

Raw potato starch is suspended in approximately 3 vol. of distilled H_2O. The starch is allowed to settle, and the supernatant water is decanted, removing insoluble impurities and small starch particles. After it is washed with H_2O three times, the starch is partially dried by suction on a large Büchner funnel. It is then washed three times with the buffer in which electrophoresis will be conducted. If immediate use is desired,

[1] H. G. Kunkel, *Methods Biochem. Anal.* 1, 141 (1954).
[2] M. Bier (ed.), *Electrophoresis.* Academic Press, New York, 1959.
[3] H. Bloemendal, *J. Chromatog.* 3, 509 (1960).
[4] H. Bloemendal, *J. Chromatog.* 2, 121 (1959).
[5] I. Smith, *Chromatographic and Electrophoretic Techniques*, Vol. II. Interscience, New York; William Heinemann Medical Books, 1960.

an adequate quantity is mixed once more with just enough cold buffer to yield a very thick homogeneous paste. To the remaining quantity of starch, enough buffer is added to cover the surface, and this is stored in the cold room. To avoid bacterial contamination of stored starch, 0.05 ml. of toluene is added per liter of buffer. For dehydrogenases, barbital buffer, pH 8.6, with an ionic strength of 0.05, is employed.

Preparation of the Block

Blocks are prepared in Lucite molds $(53 \times 7 \times 0.5$ cm.$)$ that have a cellophane sheet pressed to the inner surfaces and extended over the edges. For ease of handling, the molds are stored in the cold room. When they are brought to room temperature for construction of the block, water vapor from the ambient air condenses in a thin film on the mold and facilitates adherence of the cellophane sheet. A sheet of cellophane approximately 10 cm. longer than the mold at each end and three times as wide (this is the standard width for commercially available packages of cellophane like Saran wrap) is lightly placed over the mold. Heavy cardboard, cut to fit the inner dimensions of the mold, is placed over the cellophane and then pushed into the mold, thus forcing the cellophane to adhere firmly to the inner surfaces. The cardboard is removed, and the barely liquid mixture of prepared starch is poured into the cellophane-lined mold. The mold is overfilled to that level above the height of the mold at which surface tension will just maintain the starch mixture without allowing it to flow out. This will result in a block whose geometry favors minimum flow of condensed buffer over the surface of the block during the separation experiment. Excess buffer is removed by placing heavy blotting paper over the block and applying a hard rubber roller over the length of the blotting paper. The blotting paper is replaced enough times to result in a surface which is just damp enough to be stable, but not so dry as to be powdery. The block is now ready for use; if desired, the cellophane widths can be folded over the top of the block, the ends folded under the mold, and the block stored in the cold room. Best results are obtained when the blocks are not stored for periods longer than 24 hours.

Application of the Sample

Sample volumes of up to 4 ml. of sucrose tissue extracts, varying in concentration from 200 mg. to 2 g. of tissue per milliliter of extract, can be used. However, best results are obtained with sample volumes of 1 to 2 ml.

A ruler can be placed on the block to guide a spatula which is used to cut a thin slit across the width of the block. The slit should not extend to the edges of the block. A hypodermic syringe with a No. 25 needle is used to apply the sample to the block. Maximum control is obtained if the cylinder is held with the thumb and middle finger while the piston is operated by the index finger. The needle is placed deep within the slit at one end and guided along the slit to the other end, while gentle, even pressure is applied to the piston. The sample should not be delivered in one step but in a series of steps, with time allowed between each fractional delivery of sample to the slit for the sample to be evenly absorbed without flooding the slit. A piece of filter paper is placed on the block a few centimeters from the slit to absorb buffer displaced by the sample.

After the total sample is applied to the block and is evenly absorbed, the slit is filled in with starch. A spatula is used to lift thin strips of starch, equal in length to the width of the block, from one end of the block. These strips are packed into the slit and manipulated until a flat even surface, level with the rest of the block, is achieved.

For lactic dehydrogenase the origin is situated in the middle of the block or near the middle on the cathodal side.

The Separation Experiment

Wicks made of eight thicknesses of Whatman No. 1 filter paper cut to fit the width of the block are dampened with buffer and pressed to the top of the block at both ends. The cellophane sides are folded over the top of the block and cut at the outer ends of the protruding wicks, thus completely encasing block and wicks.

Simple electrode vessels can be constructed with Lucite. Each vessel is divided into two compartments by a central Lucite plate which has holes drilled along its lower length to allow communication between the two compartments. Platinum wire electrodes are sinosoidally mounted on small Lucite pegs which protrude from one side of the central plates.

The Lucite molds containing the blocks are supported by the ends of the electrode vessels. The wicks are immersed in the vessel compartments on the side of the central plate away from the platinum electrodes. Arranged in this manner, the central plates should adequately baffle pH changes without need for plugging the communicating holes with cotton wicks. The vessels should be filled with buffer as completely as ease of handling will allow, since there is a large drop in electrical potential along the wicks between the surface of the buffer and the surface of the block. Buffer levels in the vessels are balanced by connecting cathode and anode with a rubber tube which must be closed during the separation experiment.

Separation is conducted in the cold room at 4° for the desired time (4 to 20 hours) with a voltage of 410 volts and resultant current of 10 to 15 ma. per block.

Localization of Dehydrogenase Activity on the Block

At the conclusion of the separation experiment, the cellophane is unfolded, the wicks are removed, the cellophane is wiped clean of condensed moisture and adhering starch, and the block is blotted lightly to withdraw excess moisture. A spatula is used to cut a narrow slit, about 5 mm. from one side, along the length of the block. The slit should penetrate no deeper than just below the surface of the block. Approximately 1.5 ml. of a solution consisting of 0.005 M acetylpyridine adenine dinucleotide (AcPyAD), free of fluorescing impurities, 0.08 M lithium lactate, and 0.1 M Tris–HCl, pH 9, is applied along the slit, by traversing the slit just once with a hypodermic syringe. A few minutes are allowed for the solution to become evenly absorbed, and then the block is rewrapped with the cellophane and allowed to develop at 4° for 1 to 24 hours. After development the block is viewed under long-wave ultraviolet light. The bands of lactic dehydrogenase are indicated by fluorescent spots of reduced AcPyAD along the slit. Figure 1 shows the resolution of rat tissue lactic dehydrogenase on starch block.

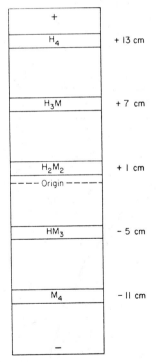

FIG. 1. Migration of rat tissue lactic dehydrogenase on starch block,[6] 20 hours at 420 volts.

[6] I. H Fine, N. O. Kaplan, and D. Kuftinec, *Biochemistry* **2**, 118 (1963).

This method of localizing enzyme activity can be adapted to other dehydrogenases by substitution in the streaking solution of the appropriate substrate for lactate. Enzymes which do not react with AcPyAD may be localized by adding substrate, nicotinamide adenine dinucleotide (NAD), and a pyridine nucleotide transhydrogenase to the AcPyAD–Tris solution.

Mr. Barry Kitto of our laboratory has used the following scheme for localization of α-glycerophosphate dehydrogenase:

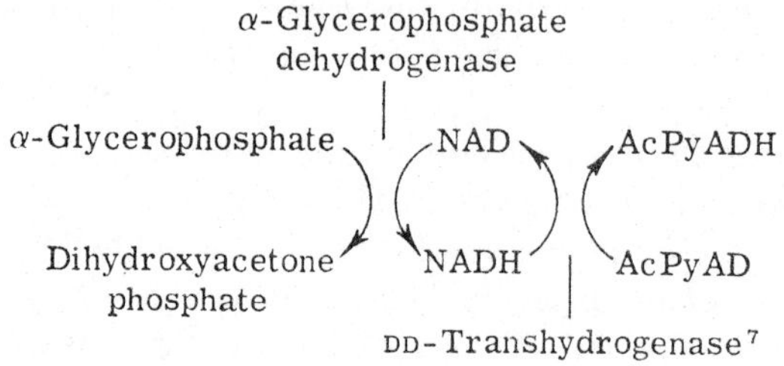

The streaking solution for this scheme consists of 1 millimole of L-α-glycero-phosphate, 1 millimole of Tris buffer, pH 8.5, 20 mg. of AcPyAD, 2 mg. of DPNH, and distilled H_2O, resulting in a final volume of 10 ml.

Elution from Starch Grain

The block is sliced transversely on both sides of the fluorescent spots, and the segments medial to the fluorescent spots are taken. Each segment is packed in a medium-grain sintered-glass filter. The enzyme is eluted by the addition, above the starch, of aliquots of 0.1% bovine serum albumin in 0.05 M potassium phosphate buffer, pH 7.5, followed by the application of gentle suction. Each segment is eluted three times with 2-ml. aliquots of the buffer (small starch segments may be eluted with smaller aliquots of buffer), resulting in total displacement of enzyme from the starch.

Starch Gel

Preparation of the Gel

Hydrolyzed starch ready for use in gels is commercially available (Connaught Medical Research Laboratories, University of Toronto, Canada), or, alternatively, raw potato starch may be treated so as to obtain characteristics required for gel application.[3,8]

[7] Purified fraction from *Pseudomonas fluorescens,* prepared by Dr. Takashi Kawasaki of this laboratory. See also Vol. II [152], p. 868.
[8] O. Smithies, *Biochem. J.* **61**, 629 (1955).

Twenty-one grams of hydrolyzed starch are added to a 500-ml. Pyrex vacuum filtration flask containing 150 ml. of pH 7.0 buffer which consists of 7.0 ml. of $0.2\,M$ citric acid plus 43.0 ml. of $0.2\,M$ Na_2HPO_4 plus 950 ml. of H_2O. The flask is connected to an aspirator, and the contents are thoroughly mixed by constant vigorous swirling as the bottom of the flask is evenly heated over the naked flame of a Bunsen burner. The swirling must be maintained to ensure a homogeneous suspension and even heating as the mixture changes in consistency, first becoming very viscous and opaque and finally becoming more fluid and translucent. As soon as the first small bubbles indicating boiling appear, the flask is removed from the flame, a rubber stopper is applied, and the suspension is allowed to boil under reduced pressure for 1 minute. The vacuum line is then disconnected, the rubber stopper removed, and the hot fluid gently decanted into a Lucite tray, $19.5 \times 7 \times 0.6$ cm. While the gel is being poured, the mouth of the flask is held as near to the inside of the tray as possible, close to one end of the tray. This end of the tray is filled almost to the point of overflow. The flask is then advanced toward the other end of the tray slowly enough so that all the tray area away from the direction of advance is evenly overfilled. This procedure, resulting in a flat even surface, will minimize the possibility of trapping air bubbles, both when the gel is being poured and when a cover is being applied to the gel. Small bubbles that may be present can be teased out of the hot starch with a thin spatula.

The gel is allowed to cool to that point at which the surface begins to resist being distorted by a lightly applied finger. A small amount of the original hot solution, poured into a petrie dish after the gel is poured, can be used for such testing. If no time is allowed for cooling before application of the cover, subsequent cooling may result in shrinkage of the gel away from the sides of the tray. The cover, a flat glass plate large enough to cover the tray completely, is gently lowered so that one end makes contact first with one end of the tray. In this manner air bubbles are excluded, and on pressing the cover flush to the top of the tray, excess starch is expressed, and a gel of uniform thickness is obtained.

A weight is placed on the cover, and the gel is allowed to cool at room temperature for at least 30 minutes before being moved into the cold room for at least 2 hours at $4°$. Gels may be stored for periods up to 24 hours.

Insertion of the Sample

The undersurfaces of the protruding edges of the cover are scraped clean of any adhering gel, and the cover is slid across the width of the gel until free of the gel surface.

Levels of dehydrogenase activity in tissue extracts are determined by spectrophotometrically measuring the oxidation of NADH in quartz cuvettes of 1-cm. light path. The reaction mixture consists of enzyme solution plus $0.1\,M$ potassium phosphate buffer, pH 7.5, $1.5 \times 10^{-4}\,M$ NADH, and $3 \times 10^{-4}\,M$ pyruvate in final concentrations. Sample solutions containing enzyme activity optimal for gel analysis catalyze the oxidation of 3.10 micromoles of NADH per minute per milliliter of solution containing enzyme. Overloading the gel with higher levels of enzyme activity may result in incomplete resolution, trailing of bands, and/or the appearance of spurious bands.

Four thicknesses of Whatman No. 1 filter paper, cut into 5×5-mm. squares, are held by fine forceps, dipped into the sample solution, and introduced into the gel in the following manner. A small spatula, oriented so that its flat surface is parallel to the width of the gel, is inserted perpendicularly into the gel to the point of contact with the inner surface of the tray. The spatula is inclined slightly away from the perpendicular so as to broaden slightly the width of the resulting slit, but care is taken not to allow splitting of the gel, which would increase the length of the slit. The sample-soaked filter paper square is guided along the flat surface of the spatula into the gel and released by the forceps, and the spatula is withdrawn. In this manner up to eight samples can be inserted in a line across the width of the gel. A greater number of samples may result in splitting the gel along a line joining the samples. If solutions of sufficient activity cannot be prepared, six thicknesses of filter paper may be inserted into the gel.

For most lactic dehydrogenase assays, the samples are inserted in a line approximately one-third the length of the gel away from the cathodal end.

The Separation Experiment

A Lucite box, partitioned into four compartments by three parallel Lucite plates, not so tall as the sides of the box, serves as an electrode apparatus. Platinum wire electrodes are mounted along the bottom of the two central compartments. The electrode buffer consists of 30 ml. of $0.2\,M$ citric acid plus 240 ml. of $0.2\,M$ Na_2HPO_4 and 1230 ml. of distilled H_2O; the resulting pH is 7.0. The buffer is balanced between the four compartments by momentarily tilting the box up on one end so that the buffer can flow freely over the partitions at the lowered end.

Wicks made of eight thicknesses of Whatman No. 1 filter paper cut to fit the width of the gel are dampened with buffer and pressed to the top of the gel at each end. The mold and protruding wicks are wrapped in cellophane so as to be completely and firmly encased, thus protecting the gel from loss of water by evaporation. The gel is placed in the electrode apparatus by situating it across the tops of the partitions. The wicks are immersed in the outer compartments, and filter paper bridges, made of eight thicknesses of Whatman No. 1 filter paper and wrapped in cellophane, are used for electrical connections between the outer compartments and the electrode compartments. A cover is placed over the electrode apparatus to achieve a completely closed system.

Separation is carried out for 10 to 20 hours at 4° with a voltage gradient of 7.5 to 12.5 volts/cm. along the gel and a constant current of 24 to 27 ma.

Localization of Activity on the Gel

After the separation experiment, the gel is unwrapped and the wicks are removed. A spatula is then run along the inner surfaces of the sides of the tray to disrupt adhesion of the gel. The gel can be lifted at one end, and a 2-mm.-thick Lucite plate cut to fit the inner dimensions of the tray is carefully slid under the gel, thus elevating it within the tray. A second Lucite plate is placed over the gel to prevent distortion while the gel is being sliced. To section the gel, a wire cheese slicer, with a wishbone handle, is used in the following manner. The handle of the slicer straddles the top plate while the wire rests on top of the sides of the tray at one end. Gentle, even pressure is applied to the top plate as the cutter, always maintained in contact with the tray, is advanced through the elevated gel toward and then beyond the other end of the tray. The top plate is carefully removed after it has been freed with a spatula from the surface of the gel at one corner. One end of the top section of the gel is teased free, and then the entire top section is lifted from the remainder of the gel. Another Lucite plate is inserted under the bottom one, and the gel is sliced again, so that three equal sections are obtained. The middle section is usually stained, as surface aberrations may be present in the other sections, which are wrapped in cellophane and refrigerated.

Although little practice is required to master manipulation of the gel, certain alterations in the gel tray can be introduced to facilitate sectioning. A small hole can be drilled through the bottom of the tray near one end, and a thin Lucite plate covering the bottom of the tray can be left in this position throughout the entire experiment. The Lucite plates used to elevate the gel for sectioning can easily be inserted under this plate, which is elevated through the hole in the bottom of the tray.

Lactic dehydrogenase is localized by a modification of the method of Dewey and Conklin.[9] The center slice is incubated in the following

[9] M. M. Dewey and J. L. Conklin, *Proc. Soc. Exptl. Biol. Med.* **105**, 492 (1960).

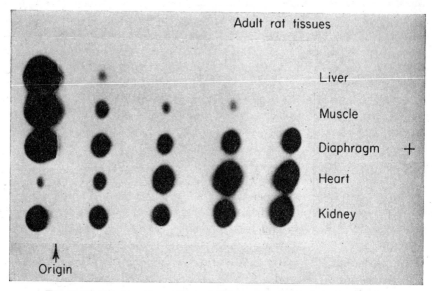

FIG. 2. Electrophoretic pattern of different rat tissues on starch gel.[6]

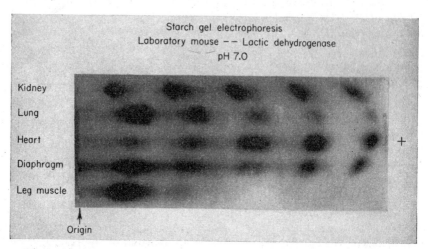

FIG. 3. Migration of laboratory mouse lactic dehydrogenase on starch gel. The appearance of multiple sub-bands has been described.[10] Resolution and migration can be increased by decreasing the ionic strength of the buffer in which the gel is prepared and/or decreasing the per cent of starch in the gel. However, gels of less than 11% starch are unstable.

[10] L. A. Costello and N. O. Kaplan, *Biochim. et Biophys. Acta* **73** (1963).

solution for 15 to 120 minutes at 37° in the dark: Tris buffer, pH 8.0, 0.1 M, 26.5 ml.; lithium lactate, 2.0 M, 1.5 ml.; phenazine methosulfate (PMS), 5 mg./ml., 0.12 ml.; nitro blue tetrazolium (NBT), 10 mg./ml., 1.0 ml.; NAD, 30 mg./ml., 0.6 ml. ˙ actic dehydrogenase is indicated in the gel by discrete purple spots, whose size and color intensity are proportional to enzyme activity (Figs. 2 and 3). Resolution is great enough to show the presence of sub-bands (Fig. 3).

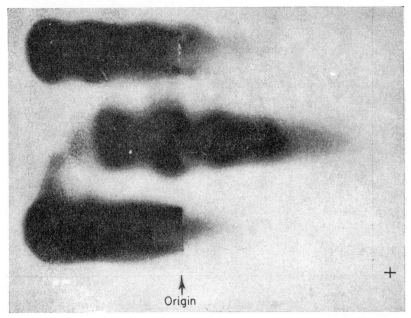

Origin

FIG. 4. Starch gel electrophoresis of malic dehydrogenase. Top: crude preparation from pig heart mitochondria; middle: purified malic dehydrogenase from horse heart mitochondria; bottom: purified malic dehydrogenase from pig heart mitochondria.[11]

If preservation of the stained gel is desired, it must be washed free of the unbound staining solution in order to avoid light-catalyzed background discoloration. This can be accomplished by submerging the gel for 15 to 30 minutes in a dish which is being overflowed by a gentle stream of tap water.

Table I indicates substitutions in the staining solution for localization of other dehydrogenases (Fig. 4).

[11] C. J. R. Thorne, L. Grossman, and N. O. Kaplan, *Biochim. et Biophys. Acta* **73**, 193 (1963).

TABLE I
STAINING SOLUTIONS FOR LOCALIZATION OF DEHYDROGENASES ON STARCH-GEL[a]

Staining solutions	Milliliters
1. Lactic dehydrogenase	
0.1 M Tris, pH 8.5	23.3
2 M Lithium lactate	1.5
30 mg./ml. NAD	0.6
5 mg./ml. PMS	0.12
10 mg./ml. Nitro BT	1.0
2. 6-Phosphogluconic dehydrogenase	
0.1 M Tris, pH 7.0	12.1
2 M 6-Phosphogluconic acid (neutralized)	3.0
10 mg./ml. NADP[b]	1.0
5 mg./ml. PMS	0.06
10 mg./ml. Nitro BT	0.5
3. Glutamic dehydrogenase	
0.1 M Tris, pH 8.5	23.3
2 M Glutamic acid (neutralized)	1.5
30 mg./ml. NAD	0.6
5 mg./ml. PMS	0.12
10 mg./ml. Nitro BT	1.0
4. Alcohol dehydrogenase	
0.1 M Tris, pH 8.5	23.3
Alcohol-Tris:	
3.0 ml. 100% ethanol ⎱ Dilute to 100 ml. with H_2O	
3.3 ml. 3 M Tris ⎰	1.5
30 mg./ml. NAD	0.6
5 mg./ml. PMS	0.12
10 mg./ml. Nitro BT	1.0
5. Malic dehydrogenase	
0.1 M Tris, pH 8.5	23.3
2 M Malic acid (neutralized)	1.5
30 mg./ml. NAD	0.6
5 mg./ml. PMS	0.12
10 mg./ml. Nitro BT	1.0

[a] Gel slices containing enzymes 1–6 are incubated in the staining solution for 15 to 120 minutes in the dark (see text). Gels containing triosephosphate dehydrogenase are incubated for $\frac{1}{2}$ hour at room temperature, then viewed under long-wave ultraviolet light. Activity is indicated by fluorescent spots.

[b] NADP = nicotinamide adenine dinucleotide phosphate.

TABLE I (*Continued*)

Staining solutions	Milliliters
6. Isocitric dehydrogenase	
0.1 M Tris, pH 8.5	26.6
0.25 M MnCl$_2$	1.2
0.10 M DL-Sodium isocitrate	0.8
10 mg./ml. NADP	0.4
5 mg./ml. PMS	0.12
10 mg./ml. Nitro BT	1.0
7. Triosephosphate dehydrogenase	
0.05 M Pyrophosphate, pH 8.5	15.
0.2 M Hexose phosphate	0.5
0.3 M Dibasic arsenate	0.3
1 mg./ml. Aldolase	0.5
30 mg./ml. NAD	0.2

Elution from the Gel

A variety of techniques for recovering samples from starch gel are available[12]; perhaps the easiest is homogenization of the gel.

The unstained sections of the gel can be compared with the stained section, and appropriate segments taken. These segments are added, along with buffer, to a small ground-glass homogenizer, which is gently hand-operated. Centrifugation of the resultant homogenate will yield a supernatant fluid containing the recovered enzyme. In general, the per cent recovery from gel is considerably smaller than that from grain.

Vertical Starch Gel Electrophoresis

Vertical arrangement of the gel as introduced by Smithies[13] is particularly valuable for separation of serum lactic dehydrogenase where the low levels of enzyme activity necessitate the use of larger sample volumes. This technique permits introduction of a liquid sample directly into the gel without the use of a supporting medium, eliminates the problem of electrodecantation, and improves resolution.

Apparatus

A modification of the apparatus designed by Smithies[13] is commercially available (Buchler Instruments Inc., Fort Lee, New Jersey). This apparatus (Fig. 5) consists of a Lucite mold with removable end

---

[12] V. Bocci, *J. Chromatog.* **6**, 367 (1961).
[13] O. Smithies, *Biochem. J.* **71**, 585 (1959).

plates and a removable combed sample slot-former, two baffled electrode chambers with platinum electrodes, and a supporting stand with shelves

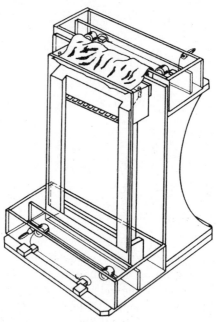

Fig. 5. Vertical starch gel apparatus. Reproduced by permission from Buchler Instruments, Inc., Fort Lee, New Jersey.

for the electrode vessel and attachments for positioning the mold. This apparatus differs from that of Smithies in that samples are introduced through a window in the underside of the mold.

Preparation of the Gel

Seventy grams of hydrolyzed starch are added to standard 1000-ml. round-bottomed Pyrex flask containing 500 ml. of pH 8.6 buffer. The buffer is prepared by the addition of 121 g. of Tris, 12 g. of EDTA, and 9.2 g. of boric acid to 6 l. of distilled H_2O. A round-bottomed flask is preferable, as it facilitates even heating. The gel mixture is heated, degassed, and poured as described above. To avoid air bubbles in the area of the sample slots, pouring should proceed from the end distant from the slot-former. Overflow along both sides and both ends must be allowed. Immediately after the gel is poured, a glass plate, large enough to cover the entire surface of the mold, is applied to the surface of the gel. The glass plate is tilted up at an angle, and one end first makes contact in the taller end plate (distant from the slot-former); then the other end is gently lowered to a horizontal position. Pressure applied to the plate, if at all, should be minimal and just sufficient to provide a gel of uniform thickness. The gel is allowed to cool for 45 minutes at room temperature and then for 2 hours at 4°.

Insertion of the Sample

The mold containing the cooled gel is carefully inverted, and the slot-former is removed by first freeing one end and then gently lifting it from the mold. It is important that the glass plate (now covering the under surface of the gel) not be disturbed. Pasteur pipets are used to

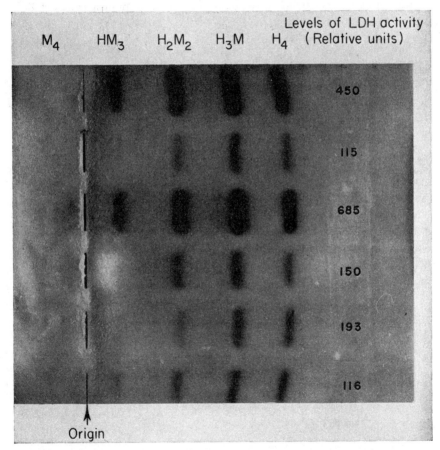

Fig. 6. Resolution by vertical starch gel electrophoresis of human serum from individuals with different levels of lactic dehydrogenase.[14] Relative units are per milliliter of serum.

[14] L. A. Costello, unpublished data.

introduce ten samples of 0.03 to 0.05 ml. into the sample slots. The pipet must not be inserted too deeply into the slot, for it may puncture the gel and allow leakage of the sample between the glass plate and the under-surface of the gel. The slots should be filled to approximately 4 mm. from the surface of the gel. Liquefied petroleum jelly (40°) is poured over the exposed samples gently enough so as not to displace them and then allowed to solidify.

The Separation Experiment

The end plates of the mold are removed, and the mold is stood upright in the lower (anodal) buffer vessel on cut strips of Whatman No. 1 filter paper. Electrical connection with the upper (cathodal) vessel is made by a wick of eight thicknesses of Whatman No. 1 filter paper cut to fit the width of the mold held in place by refastening the end plate over it. A spirit level is used to ensure that the gel is perpendicular in both planes.

Separation is conducted at 4° for 14 to 16 hours with a voltage gradient, 7.5 to 12.5 volts/cm., along the gel at a constant current of 24 to 27 ma.

Localization of Activity

After the separation experiment, the wicks with petroleum jelly are removed, the end plates are replaced, and the glass cover is freed at one end and then lifted from the surface of the gel. The gel is sectioned and stained as described above, in this case with three times the volume of staining solution required for the smaller horizontal gel. Figure 6 shows the resolution of human serum lactic dehydrogenase by vertical starch gel.

ACKNOWLEDGMENT

This work was supported in part by the National Institutes of Health Contract No. SA-43-ph-2440.

Author Index

Numbers in parentheses are reference numbers and indicate that an author's work is referred to although his name is not cited in the text.

A

Abrams, R., 28, 107
Abramson, C., 156
Acs, G., 758, 765
Adams, M. H., 133
Adler, J., 39, 242, 720, 721(9), 723, 724, 725, 746, 747, 748(17)
Agranoff, B. W., 508, 509, 788
Ahrens, E. H., 522
Airth, R. L., 776
Ajl, S. J., 289
Alberts, A. W., 546
Alberty, R. A., 902, 907(13), 908
Aleman, V., 202
Alexander, B., 673, 700
Alexander, M., 27
Allen, M. B., 308, 309, 310, 311(7), 318, 440
Allison, W., 763, 851, 853
Allsebrook, W. E., 708
Almquist, H. J., 928
Ames, B. N., 256, 478, 577, 578, 707
Aminoff, D., 463, 464
Amos, H., 157
Anders, M., 23, 24(8)
Anderson, A. D., 557
Anderson, D. G., 575
Anderson, E. H., 259
Anderson, E. I., 567, 569, 570(4)
Anderson, E. P., 252, 253, 254(23)
Anderson, I. C., 313, 314(6), 315(6), 317(6)
Anderson, M. L., 802
Anderson, W. A., 899
Anderson, W. W., 295, 306(3)
Anfinsen, C. B., 63, 139, 146(9), 837
Angell, C. L., 917, 927(21a)
Angulo, J., 625, 626(9)
Applequist, J., 935
Archer, A. A. P. G., 594
Archibald, R. M., 184
Arison, B. H., 305, 306(37b)
Arnold, J. I., 899, 901
Arnon, D. I., 308, 309, 310, 311, 312, 318, 407, 440

Arsove, S., 459
Asadourian, A., 936
Asano, A., 292
Ashton, D. M., 57, 60, 61, 679
Ashwell, G., 340
Atkin, L., 204
Auerbach, F., 482
Avison, A. W. D., 762, 767
Avron, M., 318, 431
Axelrod, B., 881
Ayala, W., 459
Ayengar, P., 165, 235, 612

B

Bachhawat, B. K., 553
Baddiley, J., 574, 575, 577, 766, 767, 768, 770, 772, 773, 788
Baer, E., 788
Baglioni, C., 842
Bailey, K., 939
Baker, A. W., 927
Baker, E. B., 903
Bakerman, H. A., 584(9), 585, 587
Baldwin, R. L., 855, 856, 858, 859
Ballantine, J., 294, 306
Ballio, A., 345
Ballou, C. E., 479, 481, 483(1), 484
Baltscheffsky, M., 317
Bamberger, E., 606, 607(8)
Bamford, C. H., 929
Bandurski, R. S., 766, 881
Bardawill, C. J., 224
Bardawill, C. S., 5
Bardos, T. J., 814
Barger, F. L., 814
Barker, C. C., 562, 563(6)
Barker, H. A., 199, 381, 563, 703, 705
Barkulis, S. S., 565
Barner, H. D., 131, 135
Barnes, R. B., 916, 921(5)
Barrnett, R. J., 889, 892(5)
Bartsch, R. G., 391, 397, 402
Basford, R. E., 307, 308(45)

T

Tabor, C. W., 615, 616, 619(1, 2), 621(2)
Tabor, H., 96, 377, 386, 582, 584, 585, 587(7, 8), 589, 615, 616, 619(2), 621 (2), 707, 711, 812
Tafel, J., 611
Tagawa, O., 312
Taggart, V. G., 939
Takagaki, G., 564
Takki-Luukkainen, I.-T., 460
Talbert, P. T., 762, 763(4), 764(4), 803, 806, 808(1), 811
Tallan, H. H., 562, 563(2), 564, 566, 615
Tanford, C., 939
Taniguchi, S., 397
Tannenbaum, S., 50
Tapley, D. F., 551
Tarr, H. L. A., 161, 162(9)
Taylor, Z., 272
Tchen, T. T., 505, 506(2), 510, 511, 512
Tener, G. M., 24, 242, 477, 653
Teranishi, R., 518
Thayer, S. A., 303
Thomas, Y., 565
Thorne, C. J. R., 967
Tice, S. V., 612
Tiegs, O. W., 713
Tiers, G. V. D., 903
Tietz, A., 4
Timasheff, S. N., 917
Tipson, R. S., 577
Tisdale, H. D., 307, 308(45)
Tiselius, A., 9, 14, 78, 366
Tishler, M., 295
Tissières, A., 285, 288, 289, 290(15)
Titus, E., 598, 605
Todd, A. R., 647, 655(5), 716, 929, 938
Toennies, G., 545, 548(3), 575
Tollin, G., 913
Tolmach, L. J., 727
Toribara, T. Y., 255
Torrey, H. C., 895
Towne, J. C., 484, 485(1), 487
Townsend, J., 913
Trageser, M. B., 943
Trams, E. G., 539
Trautner, T., 739, 741(7), 748(7)
Trenner, N. R., 305, 306(37b)
Troll, W., 586

Trotter, I. F., 935
Tsou, K. C., 889, 890(9), 891(9)
Tsuboi, M., 917, 928, 936
Tsugita, A., 17
Tsukada, Y., 564
Tuppy, H., 833
Turner, A. F., 646, 664
Turner, J. F., 165, 166(7)
Turner, J. M., 334
Turnquest, B. W., 609
Tustanovski, A. A., 635
Tuttle, L. C., 324, 325, 326(1), 327(1), 543, 546, 548, 610, 763

U

Udenfriend, S., 469, 598, 599, 600, 603, 604(12), 605
Uhlenbruck, G., 455
Ulmer, D. D., 941
Urech, J., 511
Urnes, P., 929, 931, 933, 934(11), 935, 937(11), 938, 940, 944(11)
Utter, M. F., 166, 288, 289, 293(16)
Uyeda, K., 97(10), 99, 813

V

Vagelos, P. R., 539, 545, 546, 549
Vallee, B. L., 941
Vandenbelt, J. M., 126, 301, 801, 802(1), 803
van der Vlugt, M. J., 609
van Heyningen, W. E., 464
Van Holde, K. E., 858, 865
Vänngård, T., 913, 914
Van Vunakis, H., 50, 849, 852, 864(2), 865(2)
Vasington, F. D., 411, 414
Veldstra, L., 308
Vennesland, B., 181, 197, 202, 203, 405, 406, 409(2, 5)
Vernon, L. P., 391, 397(3, 5)
Villar-Palasi, C., 356
Vinograd, J., 735, 854, 855, 856, 857, 858, 859, 860, 861(14), 863, 864(15), 865, 866(23, 26), 867, 870(9, 20, 23)
Vizsolyi, J. P., 664
Voet, D. H., 863, 866(23), 870(23)
Vogel, H. J., 557, 592
Vollmayer, E., 157
von Ehrenstein, K., 761

Subject Index

A

Absorption spectra, 5-hydroxyindoles and, 598

Acetaldehyde,
  diacetylmethylcarbinol assay and, 491
  nuclear magnetic resonance spectrum of, 898
  spin-spin coupling and, 904

Acetate,
  acetylimidazole and, 608
  phosphodiesterase II and, 247, 248

Acetic acid
  diacetylmethylcarbinol and, 489–491
  dilute, hydrolysis with 832–838
  refractive index calculations, 953

Acetoacetate,
  assay of, 267–268
  oxidative phosphorylation and, 266
  tritium-labeled β-hydroxybutyrate from, 876–877

Acetoin, diacetylmethylcarbinol and, 489, 490, 492

Acetokinase, metaphosphate synthesis and, 262–263

Acetothiokinase, acyl phosphatase and, 327

p-Acetoxybenzoic acid, hydrolysis, imidazoles and, 609

Acetoxymethylacetylacetone,
  formation of, 487, 488
  hydrolysis of, 489

Acetylacetone, methylation of, 488

1-Acetyl-N-acetylhistamine, preparation of, 606

Acetyl adenylate,
  acyl phosphatase and, 326
  fatty acid activating enzyme and, 764
  hydrolysis, free energy of, 765
  preparation of, 762–763

N-Acetylaspartic acid,
  analysis of, 563–566
  biosynthesis of, 752
  ion-exchange chromatography of, 564–566
  paper chromatography of, 564, 565

preparation of, 562–563
properties of, 563

N-Acetylbenzimidazole,
  p-acetoxybenzoic acid hydrolysis and, 609–610
  formation of, 607
  hydrolysis of, 609

L-Acetyl-N-benzoylhistidine, amino acids and, 608

Acetylcholine chloride, infrared absorption of, 923, 924

Acetyl coenzyme A, acyl phosphatase and, 326

Acetyl coenzyme A carboxylase, preparation of, 540–541

N-Acetyl-O-diacetylneuraminic acid,
  preparation of, 455–457
  properties of, 458
  resorcinol and, 462
  thiobarbituric acid and, 464

Acetylenic compounds, proton chemical shift for, 903

Acetylglutamic acid,
  electrophoresis of, 560
  enzymatic assay of, 559–562
  functions of, 557
  isolation of, 557–559
  physical properties of, 559
  synthesis of, 557

N-Acetylimidazole,
  hydrolysis, 608
    free-energy of, 765
  infrared absorption of, 923
  p-nitrophenyl acetate hydrolysis and, 609
  synthesis of, 606

N-Acetylimidazole hydrochloride, preparation of, 607–608

N-Acetylmannosamine, formation of, 458, 465

N-Acetylmethionine, methionine-activating enzyme and, 572–573

N-Acetylnaphth-(1,2)-imidazole, formation of, 607

N-Acetylneuraminic acid,
  cleavage of, 472–473

G

Methylene blue,
  anthranilic deoxyribulotide and, 593
  bacterial photophosphorylation and, 317
  microbial phosphorylation and, 293
  triphosphopyridine nucleotide diaphorase and, 431, 433
Methylene chloride, refractive index, dispersion of, 957
Methylene group,
  infrared absorption of, 923
  proton chemical shift for, 903
N^5,N^{10}-Methylene tetrahydrofolic acid,
  chemical synthesis of, 806–809
  thymidylate synthetase and, 124, 126, 128–129
N^5,N^{10}-Methylenetetrahydrofolic dehydrogenase,
  "active formaldehyde" assay and, 809
  assay of, 369–370
  dihydrofolic reductase and, 368
  N^{10}-formyltetrahydrofolic deacylase and, 373, 375
  properties of, 372
  purification of, 370–372
  pyridine nucleotides and, 368–370
Methyl group, infrared absorption of, 923
γ-Methyl-L-glutamate, helix formation and, 937
5-Methylimidazole, 4-amino-5-imidazolecarboxylic acid and, 708
Methyl iodide, S-adenosylmethionine synthesis and, 576
Methyl linoleates, retention times of, 533
Methylmalonyl coenzyme A,
  preparation of, 538–539, 549
  properties of, 539
Methylmalonyl pantetheine, preparation of, 539
Methyl p-nitrophenyl phosphate, phosphodiesterase I and, 241
N-Methyl-N-nitroso-p-toluenesulfonamide, diazomethane and, 515
Methyl oleate, retention times of, 533, 534
5-Methylorotate, dihydroörotic dehydrogenase and, 202–203
  orotidine 5'-phosphate pyrophosphorylase and, 151
Methyl stearate, retention times of, 533
10-Methyltetrahydrofolate, formimino-

tetrahydrofolic cyclodeaminase and, 383
5'-Methylthioadenosine, 5-adenosylmethionine and, 576
Methylthioribose, preparation of, 577
5-Methyluridine, uridine kinase and, 197
N-Methyluridine-5'-diphosphate, polynucleotide phosphorylase and, 716
Methyl viologen, photosynthetic pyridine nucleotide reductase and, 444
Mevalonic acid,
  assay of, 505–506
  isopentenylpyrophosphate formation from, 508
  lanosterol preparation from, 511–512
  paper chromatography of, 510
  stability of, 509
Mevalonic kinase,
  mevalonate assay and, 505–506
  phosphomevalonate preparation and, 506
Micrococcus aerogenes, cyclodehydrase in, 383
Micrococcus lysodeikticus,
  deoxyribonucleic acid of, 736
  oxidative phosphorylating system of, 292
Microorganisms,
  deoxyribonucleic acid, isolation of, 726–738
  oxidative phosphorylating systems, 284–295
    particulate fraction, 288–290
    soluble fraction, 290–292
  purine biosynthetic enzymes from, 56, 66
Microsomes,
  L-gulono-γ-lactone oxidase in, 339–340
  phosphodiesterase I in, 242–244
Mitochondria,
  adenosine triphosphate-phosphate exchange and, 319, 323
  beef heart, isolation of, 418–419
  citrulline synthesis and, 561
  disintegration,
    digitonin and, 265–266
    mechanical, 280–281
    sonication and, 275–276
  malic dehydrogenase of, 967
  Nitro-BT and, 899, 892